43.	$O(g(n))$	big-Oh of $g(n)$	
44.	δ_{ij}	Kronecker δ	

■ CIRCUIT DIAGRAMS

45.		and gate	page 39
46.		or gate	page 39
47.		not gate	page 39
48.		nand gate	page 40
49.		nor gate	page 40
50.	$p \cdot q$	Boolean multiplication (p and q)	page 38
51.	$p + q$	Boolean addition (p or q)	page 38
52.	p'	Boolean complement (not p)	page 38

■ GRAPHS AND TREES

53.	$G(V, E)$	graph with vertices V and edges E	pages 75 and 227
54.	$\{a, b\}$	edge connecting vertices a and b	pages 75 and 227
55.	(a, b)	directed edge from vertex a to vertex b	page 233
56.	$deg(v)$	degree of a vertex v	page 229
57.	K_n	complete graph with n vertices	page 231
58.	$K_{m,n}$	complete bipartite graph with m and n vertices	page 231
59.	$v_0 v_1 v_2 \cdots v_n$	path of length n from v_0 to v_n	page 230
60.	$G^c(V, E')$	complement of graph $G(V, E)$	page 519
61.	$C_G(\lambda)$	number of ways of coloring graph $G(V, E)$ using λ colors	page 536
62.	$G_{\hat{e}}$	graph G with edge e removed	page 536
63.	$G_{/e}$	graph G with edge $e = \{a, b\}$ removed and vertices a and b identified	page 536
64.	$cl(G)$	closure of graph $G(V, E)$	page 547
65.	$c(e)$	capacity of an edge e	page 620
66.	$f(e)$	flow in an edge e	page 621
67.	$val(f)$	value of a flow f	page 623
68.	(S, T)	cut of a network into S and T	page 623
69.	$C(S, T)$	capacity of a cut (S, T)	page 623

■ NUMBER THEORY

70.	Z	set of integers	page 105
71.	N or Z^+	set of positive integers	page 107
72.	$a \mid b$	a divides b	page 119
73.	$\gcd(a, b)$	greatest common divisor of a and b	page 121
74.	$\text{lcm}(a, b)$	least common multiple	page 122
75.	$a \equiv b \pmod{n}$	a congruent to b modulo n	page 129
76.	$[a]$	equivalence class containing a	page 84
77.	$[a] \oplus [b]$	addition of congruence classes	page 131
78.	$[a] \odot [b]$	multiplication of congruence classes	page 131
79.	$[[m]]_n$	smallest positve integer congruent to m modulo n	page 131
80.	ϕ	Euler ϕ function	page 401
81.	$\text{ord}_n a$	order of a modulo n	page 406

■ MATRICES

82.	$[A_{ij}]$	matrix with entry A_{ij}	page 146
83.	$A + B$	matrix sum	page 148
84.	aA	matrix scalar product	page 148

Discrete Mathematics with Combinatorics

SECOND EDITION

Discrete Mathematics with Combinatorics

SECOND EDITION

James A. Anderson
University of South Carolina, Spartanburg

With the assistance of
Jerome Lewis
and
O. Dale Saylor
University of South Carolina, Spartanburg

Upper Saddle River, New Jersey 07458

Library of Congress Cataloging-in-Publication Data

Anderson, James A. (James Andrew)
    Discrete mathematics with combinatorics.—2nd ed. / James A. Anderson; with the assistance of Jerome Lewis and O. Dale Saylor.
        p. cm.
    Includes bibliographical references and index.
    ISBN: 0-13-045791-4
    1. Mathematics.   2. Combinatorial analysis.   I. Lewis, Jerome (Jerome L.) and Saylor, O. Dale.   II. Title.
QA39.2.A534 2004
510—dc21   2003054831

Executive Acquisitions Editor: *George Lobell*
Editor-in-Chief: *Sally Yagan*
Production Editor: *Bob Walters, Prepress Management, Inc.*
Senior Managing Editor: *Linda Mihatov Behrens*
Executive Managing Editor: *Kathleen Schiaparelli*
Vice President/Director of Production and Manufacturing: *David W. Riccardi*
Production Assistant: *Nancy Bauer*
Assistant Manufacturing Manager/Buyer: *Michael Bell*
Manufacturing Manager: *Trudy Pisciotti*
Marketing Manager: *Halee Dinsey*
Marketing Assistant: *Rachel Beckman*
Editorial Assistant/Print Supplements Editor: *Jennifer Brady*
Art Director: *Jonathan Boylan*
Interior Designers: *David Ellman, Mary Siener*
Cover Designer: *Alamini Design*
Art Editor: *Thomas Benfatti*
Creative Director: *Carole Anson*
Cover Photo: *Adolph Gottlieb (1903–1974), Composition, 1955. The Museum of Modern Art/Licensed by Scala-Art Resource, NY.* © *VAGA, NY*
Text Art: *Kristin and Philip Muzik*

© 2004, 2001 Pearson Education, Inc.
Pearson Prentice Hall
Pearson Education, Inc.
Upper Saddle River, New Jersey 07458

All rights reserved, No part of this book may be reproduced, in any form or by any means, without permission in writing from the publisher.

Pearson Prentice Hall® is a trademark of Pearson Education, Inc.

Printed in the United States of America
10 9 8 7 6 5 4 3 2 1

ISBN 0-13-045791-4

Pearson Education LTD., *London*
Pearson Education Australia PTY, Limited, *Sydney*
Pearson Education Singapore, Pte. Ltd
Pearson Education North Asia Ltd. *Hong Kong*
Pearson Education Canada, Ltd., *Toronto*
Pearson Educación de Mexico, S.A. de C.V.
Pearson Education—Japan, *Tokyo*
Pearson Education Malaysia, Pte. Ltd.

In Memory of
Elwood W. Stone and Orville H. Wiebe

To My Family
Marilyn, Andy, Kristin, and Phil
and to
Tom Head, Naoki Kimura,
Leonard S. Laws, and Edward Lee Dubowsky,
Teachers, mentors, and friends

CONTENTS

Preface xi

1 Truth Tables, Logic, and Proofs 1
1.1 Statements and Connectives 1
1.2 Conditional Statements 9
1.3 Equivalent Statements 13
1.4 Axiomatic Systems: Arguments and Proofs 19
1.5 Completeness in Propositional Logic 29
1.6 Karnaugh Maps 33
1.7 Circuit Diagrams 38

2 Set Theory 51
2.1 Introduction to Sets 51
2.2 Set Operations 54
2.3 Venn Diagrams 61
2.4 Boolean Algebras 67
2.5 Relations 72
2.6 Partially Ordered Sets 80
2.7 Equivalence Relations 83
2.8 Functions 87

3 Logic, Integers, and Proofs 97
3.1 Predicate Calculus 97
3.2 Basic Concepts of Proofs and the Structure of Integers 105
3.3 Mathematical Induction 111
3.4 Divisibility 119
3.5 Prime Integers 124
3.6 Congruence Relations 129

4 Functions and Matrices 141
4.1 Special Functions 141
4.2 Matrices 146
4.3 Cardinality 156
4.4 Cardinals Continued 158

5 Algorithms and Recursion 168
- 5.1 The ``for'' Procedure and Algorithms for Matrices **168**
- 5.2 Recursive Functions and Algorithms **171**
- 5.3 Complexity of Algorithms **181**
- 5.4 Sorting Algorithms **187**
- 5.5 Prefix and Suffix Notation **196**
- 5.6 Binary and Hexadecimal Numbers **199**
- 5.7 Signed Numbers **210**
- 5.8 Matrices Continued **213**

6 Graphs, Directed Graphs, and Trees 227
- 6.1 Graphs **227**
- 6.2 Directed Graphs **232**
- 6.3 Trees **237**
- 6.4 Instant Insanity **242**
- 6.5 Euler Paths and Cycles **244**
- 6.6 Incidence and Adjacency Matrices **250**
- 6.7 Hypercubes and Gray Code **258**

7 Number Theory 271
- 7.1 Sieve of Eratosthenes **271**
- 7.2 Fermat's Factorization Method **272**
- 7.3 The Division and Euclidean Algorithms **273**
- 7.4 Continued Fractions **278**
- 7.5 Convergents **283**

8 Counting and Probability 290
- 8.1 Basic Counting Principles **290**
- 8.2 Inclusion-Exclusion Introduced **296**
- 8.3 Permutations and Combinations **302**
- 8.4 Generating Permutations and Combinations **313**
- 8.5 Probability Introduced **316**
- 8.6 Generalized Permutations and Combinations **322**
- 8.7 Permutations and Combinations with Repetition **326**
- 8.8 Pigeonhole Principle **330**
- 8.9 Probability Revisited **334**
- 8.10 Bayes' Theorem **346**
- 8.11 Markov Chains **348**

9 Algebraic Structures 357
- 9.1 Partially Ordered Sets Revisited **357**
- 9.2 Semigroups and Semilattices **360**
- 9.3 Lattices **365**
- 9.4 Groups **371**
- 9.5 Groups and Homomorphisms **377**
- 9.6 Linear Algebra **383**

10 Number Theory Revisited 391
- 10.1 Integral Solutions of Linear Equations **391**
- 10.2 Solutions of Congruence Equations **393**
- 10.3 Chinese Remainder Theorem **396**
- 10.4 Properties of the Function ϕ **401**
- 10.5 Order of an Integer **405**

11 Recursion Revisited 416
- 11.1 Homogeneous Linear Recurrence Relations **416**
- 11.2 Nonhomogeneous Linear Recurrence Relations **425**
- 11.3 Finite Differences **434**
- 11.4 Factorial Polynomials **437**
- 11.5 Sums of Differences **447**

12 Counting Continued 455
- 12.1 Occupancy Problems **455**
- 12.2 Catalan Numbers **459**
- 12.3 General Inclusion-Exclusion and Derangements **464**
- 12.4 Rook Polynomials and Forbidden Positions **470**

13 Generating Functions 485
- 13.1 Defining the Generating Function (Optional) **485**
- 13.2 Generating Functions and Recurrence Relations **487**
- 13.3 Generating Functions and Counting **496**
- 13.4 Partitions **501**
- 13.5 Exponential Generating Functions **507**

14 Graphs Revisited 516
- 14.1 Algebraic Properties of Graphs **516**
- 14.2 Planar Graphs **530**
- 14.3 Coloring Graphs **534**
- 14.4 Hamiltonian Paths and Cycles **543**
- 14.5 Weighted Graphs and Shortest Path Algorithms **550**

15 Trees 564
- 15.1 Properties of Trees **564**
- 15.2 Binary Search Trees **569**
- 15.3 Weighted Trees **575**
- 15.4 Traversing Binary Trees **582**
- 15.5 Spanning Trees **589**
- 15.6 Minimal Spanning Trees **607**

16 Networks 620
- 16.1 Networks and Flows **620**
- 16.2 Matching **632**
- 16.3 Petri Nets **639**

17 Theory of Computation ... 648
- 17.1 Regular Languages ... 648
- 17.2 Automata ... 652
- 17.3 Finite State Machines with Output ... 662
- 17.4 Grammars ... 672
- 17.5 Turing Machines ... 682

18 Theory of Codes ... 689
- 18.1 Introduction ... 689
- 18.2 Generator Matrices ... 692
- 18.3 Hamming Codes ... 700

19 Enumeration of Colors ... 708
- 19.1 Burnside's Theorem ... 708
- 19.2 Polya's Theorem ... 712

20 Rings, Integral Domains, and Fields ... 720
- 20.1 Rings and Integral Domains ... 720
- 20.2 Integral Domains ... 727
- 20.3 Polynomials ... 730
- 20.4 Algebra and Polynomials ... 736

21 Group and Semigroup Characters ... 750
- 21.1 Complex Numbers ... 750
- 21.2 Group Characters ... 751
- 21.3 Semigroup Characters ... 755

22 Applications of Number Theory ... 760
- 22.1 Application: Pattern Matching ... 760
- 22.2 Application: Hashing Functions ... 767
- 22.3 Application: Cryptography ... 772

Bibliography ... 779

Answers ... A-1

Index ... I-1

PREFACE

As in the first edition, the purpose of this book is to present an extensive range and depth of topics in discrete mathematics and also work in a theme on how to do proofs. Proofs are introduced in the first chapter and continue throughout the book. Most students taking discrete mathematics are mathematics and computer science majors. Although the necessity of learning to do proofs is obvious for mathematics majors, it is also critical for computer science students to think logically. Essentially, a logical bug-free computer program is equivalent to a logical proof. Also, it is assumed in this book that it is easier to use (or at least not misuse) an application if one understands why it works. With few exceptions, the book is self-contained. Concepts are developed mathematically before they are seen in an applied context.

Additions and alterations in the second edition:

- More coverage of proofs, especially in Chapter 1.

- Added computer science applications, such as a greedy algorithm for coloring the nodes of a graph, a recursive algorithm for counting the number of nodes on a binary search tree, or an efficient algorithm for computing a^b mod n for very large values of a, b, and n.

- An extensive increase in the number of problems in the first eight chapters.

- More problems are included that involve proofs.

- Additional material is included on matrices

- Inclusion of finite states with output and Turing machines.

- True-False questions at the end of each chapter.

- Summary questions at the end of each chapter.

- A glossary at the end of each chapter.

- Functions and sequences are introduced earlier (in Chapter 2).

Calculus is not required for any of the material in this book. College algebra is adequate for the basic chapters. However, although this book is self-contained, some of the remaining chapters require more mathematical maturity than do the basic chapters, so calculus is recommended more for giving maturity, than for any direct uses.

This book is intended for either a one- or two-term course in discrete mathematics. The first eight chapters of this book provide a foundation in discrete mathematics and would be appropriate for a first-level course for freshmen or sophomores. These chapters are essentially independent, so that the instructor can pick the material he/she wishes to cover. The remainder of the book contains appropriate material for a second course in discrete mathematics. These chapters expand concepts introduced earlier and introduce numerous advanced topics. Topics are explored from different points of view to show how they may be used in different settings. The range of topics include:

Logic—Including truth tables, propositional logic, predicate calculus, circuits, induction, and proofs.

Set Theory—Including cardinality of sets, relations, partially ordered sets, congruence relations, graphs, directed graphs, and functions.

Algorithms—Including complexity of algorithms, search and sort algorithms, the Euclidean algorithm, Huffman's algorithm, Prim's algorithms, Warshall's algorithm, the Ford-Fulkerson algorithm, the Floyd-Warshall algorithm, and Dijkstra's algorithms.

Graph Theory—Including directed graphs, Euler cycles and paths, Hamiltonian cycles and paths, planar graphs, and weighted graphs.

Trees—including binary search trees, weighted trees, tree transversal, Huffman's codes, and spanning trees.

Combinatorics—including permutations, combinations, inclusion-exclusion, partitions, generating functions, Catalan numbers, Sterling numbers, Rook Polynomials, derangements, and enumeration of colors.

Algebra—Including semigroups, groups, lattices, semilattices, Boolean algebras, rings, fields, integral domains, polynomials, and matrices.

There is extensive number theory and algebra in this book. I feel that this is a strength of this book, but realize that others may not want to cover these subjects. The chapters in these areas are completely independent of the remainder of the book and can be covered, or not, as the instructor desires. This book also contains probability, finite differences, and other topics not usually found in a discrete mathematics text.

■ Organization

The first three chapters cover logic and set theory. It is assumed in this book that an understanding of proofs is necessary for the logical construction of advanced computer programs.

The basic concepts of a proof are given and illustrated with numerous examples. In Chapter 2, the student is given the opportunity to prove some elementary concepts of set theory. In Chapter 3, the concept of an axiom system for number theory is introduced. The student is given the opportunity to prove theorems in a familiar environment. Proofs using induction are also introduced in this chapter. Throughout the remainder of the book, many proofs are presented and many of the problems are devoted to proofs. Problems, including proofs, begin at the elementary level and advance in level of difficulty throughout the book.

Relations, functions, and graphs are introduced in Chapter 2. Functions are then continued in Chapter 4. However, the development of functions in Chapter 4 is independent of the material in Chapter 2. Similarly, the development of graphs in Chapter 6 does not depend on their development as relations in Chapter 2.

Matrices, permutations, and sequences are introduced in Chapter 4 as special types of functions. Further properties of functions and matrices follow in Chapter 5. Algorithms for matrices are introduced and further properties of matrices are developed, which will be used in later chapters on algebra, counting, and theory of codes.

Permutations are used for counting in Chapter 8 and also for applications in algebra and combinatorics in later chapters. Again, the material in Chapter 8, while related to Chapter 4, can be studied independently.

Chapter 5 is independent of the previous chapters except for the matrices in the previous chapter. Algorithms are developed, including sorting algorithms. The complexity of algorithms is also developed in this chapter. Prefix and suffix notation are introduced here. They are, again, discussed in Chapter 15 with regard to traversing binary trees. Binary and hexadecimal numbers are also introduced in Chapter 5.

Many elementary concepts of graphs, directed graphs, and trees are covered in Chapter 6. These concepts are covered in more depth in Chapters 14–16. Chapter 6 is independent of the previous chapters.

In Chapters 7 and 10, the basics of number theory are developed. These chapters are necessary for applications of number theory in Chapter 22, but are otherwise completely independent of the other chapters and may be omitted if desired.

Chapter 8 is the beginning of extensive coverage of combinatorics. This is continued in many of the chapters including Chapters 12, 13, and 17. Chapter 8 also introduces basic ideas in probability which is not common in most other discrete mathematics books.

Chapters 9 and 20 cover the basic concepts of algebra, including semigroups, groups, semilattices, lattices, rings, integral domains, and fields. These chapters use Sections 3.6, and 4.3 for examples of groups and rings. Chapter 9 is necessary for the applications in Chapters 17–21.

In many ways Chapters 11, 12, and 13 form a cluster. Recursion is continued in Chapter 11. In addition to the standard linear recurrence relations normally covered in a discrete mathematics text, the theory of finite difference is also covered. Chapter 6 should be covered before this chapter unless the student already has some knowledge of recursion. Chapter 12 continues the counting introduced in Chapter 8. It covers topics introduced in Chapter 8, such as occupancy problems and inclusion-exclusion. It also introduces derangements and rook polynomials. It is closely related to Chapter 11. Many of the same topics are covered from different points of view. One example of this is Stirling numbers. However neither chapter is dependent on the other.

Chapters 11 and 12 are tied together in Chapter 13, where generating functions are used to continue the material in both chapters. In particular, generating functions provide a powerful tool for the solution of occupancy problems.

Chapters 14–16 continue the study of trees and graphs begun in Chapter 6. They obviously depend on the material in Chapter 6, but are virtually independent of most of the preceding chapters. One exception is the use of matrices in some of the algorithms. Many of the standard topics of graphs and trees are covered, including planar graphs, Hamiltonian cycles, binary trees, spanning trees, minimal spanning trees, weighted trees, shortest path algorithms, and network flows.

Chapters 17–22 form another cluster consisting of number theory, algebra, combinatorics, and their application. The theory of computation is introduced in Chapter 17. This includes codes, regular languages, automata, grammars, Turing machines, and their relationship. This chapter uses semigroups from Section 9.2. Chapter 18 introduces special codes, such as error detecting codes and error correcting codes. This chapter requires knowledge of group theory, found in Section 9.4, and some knowledge of matrices, found in Chapters 4 and 5. Codes are explored from yet another direction in Chapter 22 where cryptography is introduced. This chapter is dependent on the previous chapters on number theory.

In Chapter 19, algebra and combinatorices are combined for the development of Burnside's Theorem and Polya's Theorem for the enumeration of colors. It primarily depends on a knowledge of permutations found in Section 9.4.

Chapter 21 is a simple application of groups and semigroups and their mapping onto the complex plane. The prerequisites for this chapter are Sections 9.2 and 9.4.

Chapter 22 gives three important applications of number theory. The study of Hashing functions and cryptography are particularly relevant to computer science.

When teaching a beginning course, I normally cover Chapters 1–5 in their entirety, Sections 8.1–8.5, and the first three sections Chapter 6. As mentioned previously, the material in the first eight chapters is arranged for maximal flexibility. The following chart shows the required prerequisites for each chapter.

Chapter	Prequisite Chapters or Sections
Chapter 1	None
Chapter 2	None
Chapter 3	Sections 1.1–1.4 and 2.1
Chapter 4	None
Chapter 5	Sections 4.1–4.3
Chapter 6	None
Chapter 7	Chapter 3
Chapter 8	None
Chapter 9	Sections 2.6, 2.7, and 3.6
Chapter 10	Chapter 7
Chapter 11	Sections 5.1–5.3
Chapter 12	Chapter 8
Chapter 13	Chapters 11 and 12
Chapter 14	Chapter 6
Chapter 15	Chapter 6
Chapter 17	Chapter 9
Chapter 18	Chapters 5 and 9
Chapter 19	Chapter 9
Chapter 20	Chapter 9
Chapter 21	Chapter 9
Chapter 22	Chapter 10

■ Supplements

A solutions manual is available from the publisher with complete solutions to all problems. A website is available at www.prenhall.com/janderson. This website includes links to other interesting sites in discrete mathematics, quizzes, and supplementary problems. In addition, there are two problems oriented paperbacks that can be used with the textbook: Practice Problems in Discrete Mathematics (407 pp.) by B. Obrenic and Discrete Mathematics Workbook (316 pp.) by J. Bush. The first consists entirely of problems with answers/solutions. The second contains an outline of subject, sample worked out problems, and problem sets (with answers). Each of these two supplements is free when shrinkwrapped with the text. As stand-alone items, they have prices. So the order ISBN for the textbook plus the free Obrenic supplement is 013-117279-4. The order ISBN for the textbook plus the free Bush supplement is 013-117278-6.

■ Acknowledgments

First, I would like to thank Dale Saylor for the tremendous amount of work he has contributed to the book. Without him, this book would probably not have been revised. I also want to especially thank Jerome Lewis for the computer science applications which he has contributed to the took. I am very grateful to Douglas Shier for the use of his collection of methods for counting spanning trees. I would like to thank Kristin and Philip Muzik for their excellent artwork. I am especially grateful to James Bell for the tremendous amount of work that he contributed to the first edition. I would also like to thank my colleagues Rick Chow, Dale Saylor, Debabrata Mukherjee, and Jerome Lewis for their help in checking answers to the exercises. I would like to thank Alex Osipovk, Ole-Kristian S. Losvik, Timothy Haven, and others who have sent me corrections to the first edition of the book. I would also like to thank George Lobell and Jennifer Brady at Prentice Hall and Bob Walters of Prepress Management, Inc., for their help with the book.

With apologies for the delay, I would like to thank the following reviewers of the first edition:

Anthony B. Evans, Wright State University
Madeleine Schep, Columbia College
Akihiro Kanamori, Boston University
Krishnaiya Thulasiraman, University of Oklahoma
Gabor Sarkozy, Worcester Polytechnic Institute
Alvin Swimmer, Arizona State University

I would also like to thank the following reviews of this edition:

Beth Novick, Clemson University
Thomas Hughes, Vanier College/Concordia University
John Konvalina, University of Nebraska-Omaha
George M. Butler, Louisiana Tech University
Michael Neubauer, California State University, Northridge
Myron Hlynka, University of Windsor
William G. Brown, McGill University
Ted Wilcox, Rochester Institute of Technology

James A. Anderson
janderson@gw.uscs.edu

Discrete Mathematics with Combinatorics

SECOND EDITION

CHAPTER 1

Truth Tables, Logic, and Proofs

1.1 Statements and Connectives

In this section we develop truth tables and use them to begin the first step in logic. We will find as we continue in this chapter that truth tables are also a basic tool for other important concepts in discrete mathematics. Logic, developed by Aristotle (384–322 BCE), has been used throughout the centuries in the development of many areas of learning including theology, philosophy, and mathematics. It is the foundation on which the whole structure of mathematics is built. Basically it is the science of reasoning, which may allow us to determine statements about mathematics that are true and statements that are false based on a set of basic assumptions called axioms. Some statements about mathematics cannot be proven either true or false using axioms. Logic is also used in computer science to construct computer programs and to show that the programs do what they are designed to do. One of the primary goals of this book is to develop logic and show how to use it in computer science and to develop techniques for analyzing and proving theorems in mathematics.

A **proposition** is a statement or declarative sentence that may be assigned a true or false value. It must make sense to consider the statement being true or false. The true or false value assigned to a statement is called its **truth value**.

Sentences that are not propositions include

Who are you?

(a question).

Read this chapter before the next class.

(a command or exclamation).

This sentence is not true.

(self-contradictory).

We will use $p, q, r, \ldots$ to represent propositions. For example, p could represent the statement *It is going to rain tomorrow* and q could represent the statement *The square of an integer is positive*.

In English, sentences are combined using connectives and clauses to form more complex compound sentences. Common connectives are *and, or, not, if ... then, only if,* and *if and only if*. The logical meaning that we will give to these connectives will be completely determined. The truth of a compound proposition is determined completely by the truth or falsity of the component parts. A statement that contains no connectives is called a **simple** statement. A statement that contains connectives is called a **compound** statement.

Let p and q refer to the propositions

p: Jane drives a Ford.

q: Bob has gray hair.

The compound statement

Jane drives a Ford **and** *Bob has gray hair.*

has two parts joined with the connective *and*. This statement may be expressed symbolically as

p **and** q

or, more simply, as

$p \wedge q$

where the symbol $\wedge$ represents the word *and* in the translation from English to symbolic expressions. The expression $p \wedge q$ is called the **conjunction** of p and q.

Similarly, the statement

Jane drives a Ford **or** *Bob has gray hair.*

is expressed symbolically by

p **or** q

or

$p \vee q$

where $\vee$ represents the word *or* in the translation from English to symbols. The expression $p \vee q$ is called the **disjunction** of p and q.

The negation or denial of p is indicated by

$\sim p$

using the tilde to indicate negation. Thus, if p is the statement *Jane drives a Ford*, then $\sim p$ is the statement *Jane does not drive a Ford*. Although the negation is considered to be a connective, it differs from the other connective in the sense that it does not connect two statements together. It simply changes a statement and does not necessarily make it either more or less complex. For example, if p is the statement *Joe does not like spinach*, then $\sim p$ is the statement *Joe does like spinach*, which is actually a simpler statement.

If r is the statement *Joe likes computer science*, then *Jane does not drive a Ford and Bob has gray hair, or Joe likes computer science* would be symbolically indicated by $((\sim p) \wedge q) \vee r$. Conversely, the expression $p \wedge (\sim q) \wedge r$ represents the statement *Jane drives a Ford and Bob does not have gray hair and Joe likes computer science*.

Consider the expression $p \wedge q$. Clearly, when someone says, "Jane drives a Ford and Bob has gray hair," we expect both a gray-headed Bob and a Ford that is driven by Jane. If any other situation occurs, we would declare that speaker to be incorrect.

There are four possible cases to consider. The proposition p could be true (T) or false (F) and, regardless of which value is assigned to p, the proposition q could also be true (T) or false (F). A **truth table** lists all possible combinations of the truth and falsity of the component propositions.

Case	p	q	$p \wedge q$
1	T	T	T
2	T	F	F
3	F	T	F
4	F	F	F

It was decided earlier that only for case 1 would $p \wedge q$ be true. The other cases yield a false truth value for $p \wedge q$. For example, case 3 describes how to obtain the truth value of $p \wedge q$ when Jane does not drive a Ford and Bob has gray hair. If p is the statement *John is rich* and q is the statement *John is handsome*, then if John's blind date has been assured that the statement "John is rich and John is handsome" or "John is rich and handsome" is true, then she should expect John to be both rich and handsome.

Similarly, consider the statement

Jane drives a Ford or Bob has gray hair.

expressed symbolically by $p \vee q$. A person saying "Jane drives a Ford or Bob has gray hair" would be incorrect only when Jane fails to drive a Ford and also Bob is not gray headed. Only one of the two component propositions needs to be true to make this statement true. Therefore, $p \vee q$ has the truth table

Case	p	q	$p \vee q$
1	T	T	T
2	T	F	T
3	F	T	T
4	F	F	F

and, only in case 4, when both p and q are false, is $p \vee q$ false.

If p is the statement *John is rich* and q is the statement *John is handsome*, then if John's blind date has been assured the statement "John is rich or John is handsome" or "John is rich or handsome" is true, then she should expect one of the statements to be true but not necessarily both. She should feel misled only if she finds that John is poor and ugly. It is sometimes helpful to form tables for $\wedge$ and $\vee$ that are similar to the addition and multiplication tables that we were forced to memorize at some time in our lives. Given the multiplication table

×	0	1	2	3	4	5	6	7	8	9
0	0	0	0	0	0	0	0	0	0	0
1	0	1	2	3	4	5	6	7	8	9
2	0	2	4	6	8	10	12	14	16	18
3	0	3	6	9	12	15	18	21	24	27
4	0	4	8	12	16	20	24	28	32	36
5	0	5	10	15	20	25	30	35	40	45
6	0	6	12	18	24	30	36	42	48	54
7	0	7	14	21	28	35	42	49	56	63
8	0	8	16	24	32	40	48	56	64	72
9	0	9	18	27	36	45	54	63	72	81

we know that if $x = 3$ and $y = 7$, we find $x \times y$ by using the multiplication table to find the product of 3 and 7.

We may form the following tables for $\wedge$ and $\vee$:

$\wedge$	T	F
T	T	F
F	F	F

$\vee$	T	F
T	T	T
F	T	F

Thus, if $r = T$ and $s = F$, we may use the first table to find that $r \wedge s = T \wedge F = F$. Similarly, we may use the second table to find that $r \vee s = T \vee F = T$. We may combine tables so that if $r = F$, $s = T$, and $t = F$, and we wish to find $(r \vee s) \wedge t$, we use the $\vee$ table to determine that $r \vee s = F \vee T = T$. We then use the $\wedge$ table to determine that $(r \vee s) \wedge t = T \wedge F = F$.

The truth table for the negation or denial of p that has been indicated by

$$\sim p$$

is

Case	p	$\sim p$
1	T	F
2	F	T

The truth value of p is always the opposite of the truth value of $\sim p$. The negation is always evaluated first in a truth table unless it is followed by a statement in parentheses. Therefore, $\sim p \vee q$ is interpreted as $(\sim p) \vee q$ so that the negation applies only to p. If we wish to negate the entire statement, we write it as $\sim(p \vee q)$.

The symbols $\wedge$ and $\vee$ are **binary** connectives since they connect two propositions as in $p \wedge q$ and in $p \vee q$. The symbol $\sim$ is a **unary** connective since it is applies to only one proposition.

Another binary connective is the **exclusive or**, denoted $\underline{\vee}$. The statement $p \underline{\vee} q$ has the meaning p or q **but not both**. It has the truth table

Case	p	q	$p \underline{\vee} q$
1	T	T	F
2	T	F	T
3	F	T	T
4	F	F	F

When we use the word *or*, we may mean *exclusive or*. For example, when we say "He will pass the course or he will fail the course," we certainly assume he will not do both. Similarly, when we say p is true or false, we certainly assume it is not both. The *exclusive or* table is given by

$\underline{\vee}$	T	F
T	F	T
F	T	F

We shall not use the *exclusive or* often. In the remainder of this discussion, we shall not use *exclusive or*.

Next consider the statement

Either Sam makes his car payment or else Sam will both lose his car and walk to work.

If we let p represent the statement *Sam makes his car payments*, q represent the statement *Sam will keep his car*, and r represent the statement *Sam walks to work*, then the symbolic representation of the compound statement is

$$p \vee ((\sim q) \wedge r)$$

where we have used parentheses to indicate clearly which statements are the components of each connective.

A truth table will enable us to know exactly those situations when the statement $p \vee ((\sim q) \wedge r)$ is true; however, we must be sure to list all possibilities. Since there are three fundamental propositions p, q, and r in the compound proposition, there are eight cases.

1.1 Statements and Connectives

Case	p	q	r	$\sim q$	$(\sim q) \wedge r$	$p \vee ((\sim q) \wedge r)$
1	T	T	T	F	F	T
2	T	T	F	F	F	T
3	T	F	T	T	T	T
4	T	F	F	T	F	T
5	F	T	T	F	F	F
6	F	T	F	F	F	F
7	F	F	T	T	T	T
8	F	F	F	T	F	F

In constructing the truth table for the $(\sim q) \wedge r$ column, we refer to the columns for $(\sim q)$ and r and the truth table definition for $\wedge$. The truth table for $\wedge$ indicates that only when both $(\sim q)$ and r are true will the statement $(\sim q) \wedge r$ be true. This happens in cases 3 and 7 only.

To determine truth values for the $p \vee ((\sim q) \wedge r)$ column, we note that only the truth of p and of $(\sim q) \wedge r$ matters. The truth table defining $\vee$ indicates that the only case in which an "or" statement is false is when both parts are false. Such a combination occurs only for cases 5, 6, and 8.

If Sam does not make his car payments (i.e., p is false or has the truth value F), Sam does lose his car (q is F), and Sam walks to work (r is T), then we have case 7 and a person saying "Either Sam makes his car payment or else Sam will both lose his car and walk to work" would be making a true statement.

An alternative, but equivalent, way to construct a truth table is to express the truth value of an expression under the connective. Again consider the expression $p \vee ((\sim q) \wedge r)$. We first put the truth value under the variables p, q, and r. The 1's at the bottom of the columns indicate that these columns are given truth values first. In general, the number at the bottom of the column will denote the level or order in which the operations are performed.

Case	p	q	r	p	$\vee$	$((\sim$	$q)$	$\wedge$	$r)$
1	T	T	T	T			T		T
2	T	T	F	T			T		F
3	T	F	T	T			F		T
4	T	F	F	T			F		F
5	F	T	T	F			T		T
6	F	T	F	F			T		F
7	F	F	T	F			F		T
8	F	F	F	F			F		F
				1			1		1

We then put in the values for $\sim q$ under the $\sim$ symbol:

Case	p	q	r	p	$\vee$	$((\sim$	$q)$	$\wedge$	$r)$
1	T	T	T	T		F	T		T
2	T	T	F	T		F	T		F
3	T	F	T	T		T	F		T
4	T	F	F	T		T	F		F
5	F	T	T	F		F	T		T
6	F	T	F	F		F	T		F
7	F	F	T	F		T	F		T
8	F	F	F	F		T	F		F
				1		2	1		1

The value for $(\sim q) \wedge r$ is now placed under the $\wedge$ symbol.

Case	p	q	r	p	∨	((∼	q)	∧	r)
1	T	T	T	T		F	T	**F**	T
2	T	T	F	T		F	T	**F**	F
3	T	F	T	T		T	F	**T**	T
4	T	F	F	T		T	F	**F**	F
5	F	T	T	F		F	T	**F**	T
6	F	T	F	F		F	T	**F**	F
7	F	F	T	F		T	F	**T**	T
8	F	F	F	F		T	F	**F**	F
				1		2	1	3	1

Finally, the true values for $p \vee ((\sim q) \wedge r)$ are now placed under the $\vee$ symbol.

Case	p	q	r	p	∨	((∼	q)	∧	r)
1	T	T	T	T	**T**	F	T	F	T
2	T	T	F	T	**T**	F	T	F	F
3	T	F	T	T	**T**	T	F	T	T
4	T	F	F	T	**T**	T	F	F	F
5	F	T	T	F	**F**	F	T	F	T
6	F	T	F	F	**F**	F	T	F	F
7	F	F	T	F	**T**	T	F	T	T
8	F	F	F	F	**F**	T	F	F	F
				1	*	2	1	3	1

EXAMPLE 1.1

Let p represent the proposition *Fred likes football*, q represent the proposition *Fred likes golf*, and r represent the proposition *Fred likes tennis*. Translate the proposition *Fred likes football and it is not true that he likes golf or he likes tennis* to symbolic form and find the truth table.

First change the proposition to *Fred likes football and it is not true that Fred likes golf or Fred likes tennis*. *Fred likes golf or Fred likes tennis* is expressed symbolically as $q \vee r$. *It is not true that Fred likes golf or Fred likes tennis* is expressed symbolically as $\sim(q \vee r)$ because the negation applies to the entire "that" clause. Thus, the proposition is expressed as $p \wedge (\sim(q \vee r))$. Its truth table is

Case	p	q	r	p	∧	(∼	(q ∨ r))		
1	T	T	T	T	F	F	T	T	T
2	T	T	F	T	F	F	T	T	F
3	T	F	T	T	F	F	F	T	T
4	T	F	F	T	T	T	F	F	F
5	F	T	T	F	F	F	T	T	T
6	F	T	F	F	F	F	T	T	F
7	F	F	T	F	F	F	F	T	T
8	F	F	F	F	F	T	F	F	F
				1	*	3	1	2	1

Exercises

State which of the following are propositions. Give the truth value of the propositions.

1. What time is it?
2. The integer 1 is the smallest positive integer.
3. If $x = 3$, then $x^2 = 6$.
4. Look out for that car!

5. South Dakota is a southern state.
6. All even numbers are divisible by 2.
7. Load the packages in the car.
8. This statement cannot possibly be true.
9. Jupiter is the closest planet to the sun.
10. Compact disks should never be stored in the microwave.

Let p, q, and r be the following statements:

p: Traveling to Mars is expensive.
q: I will travel to Mars.
r: I have money.

Express the following English sentences as symbolic expressions:

11. I have no money and I will not travel to Mars.
12. I have no money and traveling to Mars is expensive, or I would travel to Mars.
13. It is not true that I have money and will travel to Mars.
14. Traveling to Mars is not expensive and I will go to Mars, or traveling to Mars is expensive and I will not go to Mars.

Let p, q, and r be the following statements:

p: My computer is very fast.
q: I will finish my project on time.
r: I will pass the course.

Express the following English sentences as symbolic expressions:

15. My computer is not very fast or I would finish my project on time.
16. I will not finish my project on time and I will not pass the course.
17. It is not true that I will finish my project on time and pass the course.
18. My computer is very fast or I will not finish my project on time and pass the course.
19. Find the truth table for the sentence in problem 11.
20. Find the truth table for the sentence in problem 12.
21. Find the truth table for the sentence in problem 13.
22. Find the truth table for the sentence in problem 14.
23. Find the truth table for the sentence in problem 15.
24. Find the truth table for the sentence in problem 16.
25. Find the truth table for the sentence in problem 17.
26. Find the truth table for the sentence in problem 18.

Let p, q, r, and s be the following statements:

p: He is brave.
q: He is afraid of dogs.
r: He is running.
s: It is a small dog.

Express the following in symbolic form:

27. He is brave but he is afraid of dogs and he is running, or he is not brave and it is a small dog.
28. He is brave or he is not afraid of dogs, and he is not running or it is a small dog.
29. He is afraid of dog but he is brave and it is a small dog.

Let p, q, r, and s be the following statements:

p: She likes to travel by train.
q: Her husband is afraid to fly.
r: They will travel by train.
s: They will fly.

Express the following in symbolic form:

30. She likes to travel by train and they will travel by train or they will fly and her husband is afraid to fly.
31. She likes to travel by train or her husband is afraid to fly, and they will not fly.
32. It is not true that she likes to travel by train and her husband is afraid to fly, but they will travel by train and they will not fly.
33. Let p and q be the following statement:

p: A student can play basketball.
q: A student can play in the band.

Find two symbolic statements for the sentence:

A student can play basketball or in the band but not both.

Let p, q, and r be the following statements:

p: This game is very difficult.
q: I play chess.
r: It takes time to play chess.

Express the following symbolic expressions as English sentences:

34. $q \wedge r$
35. $\sim p \vee \sim q$
36. $(p \vee r) \wedge q$
37. $p \wedge q \wedge r$

Let p, q, and r be the following statements:

p: Great Danes are large dogs.
q: I have a small house.
r: I have a Great Dane.

Express the following symbolic expressions as English sentences:

38. $p \wedge q \wedge \sim r$
39. $p \wedge (\sim q \vee \sim r)$
40. $(p \vee \sim q) \wedge r$
41. $(p \wedge r) \vee (q \wedge \sim r)$
42. Construct truth tables for the statement in problem 34.

8 Chapter 1 Truth Tables, Logic, and Proofs

43. Construct truth tables for the statement in problem 35.
44. Construct truth tables for the statement in problem 36.
45. Construct truth tables for the statement in problem 37.
46. Construct truth tables for the statement in problem 38.
47. Construct truth tables for the statement in problem 39.
48. Construct truth tables for the statement in problem 40.
49. Construct truth tables for the statement in problem 41.

Let p, q, r, and s be the following statements:

   p: Cabbages have one head.
   q: Two heads are better than one.
   r: A cabbage cannot use a computer.
   s: I like cabbage.

Express the following symbolic expressions as English sentences.

50. $\sim(p \wedge q) \vee (r \wedge s)$
51. $(p \wedge q) \wedge (\sim r \wedge s)$
52. $\sim((p \vee \sim s) \wedge (\sim q \vee r))$

Let p, q, r, and s be the following statement:

   p: She likes to quote Milton.
   q: He likes Bogart movies.
   r: Bogart doesn't quote Milton.
   s: They go to movies.

Express the following symbolic expressions as English sentences:

53. $\sim(((p \wedge q) \wedge \sim r) \vee \sim s)$
54. $((p \wedge \sim q) \wedge \sim r) \vee (((p \wedge q) \wedge r) \wedge \sim s)$

Construct truth tables for the following propositions:

55. $p \wedge (q \vee \sim r)$
56. $(q \wedge \sim r) \vee (\sim p \wedge r)$
57. $\sim(p \wedge r) \vee (\sim q \wedge r)$
58. $\sim(\sim p \vee (q \wedge \sim r))$
59. $(p \wedge r) \vee (p \wedge \sim q)$
60. $(p \vee q) \wedge (r \vee q)$
61. $(\sim q \wedge r) \vee \sim(p \wedge r)$
62. $\sim((p \wedge r) \vee \sim q)$
63. $\sim(\sim p \wedge (q \vee \sim r))$
64. $(p \vee \sim r) \wedge \sim(p \vee \sim q)$
65. $(p \veebar q) \wedge (p \veebar r)$
66. $\sim((p \wedge \sim q) \vee (p \wedge \sim r)) \veebar (q \wedge \sim r)$
67. $\sim(p \veebar q) \vee \sim(\sim(p \veebar r) \vee \sim(q \veebar r))$
68. $(p \wedge (q \vee \sim r)) \veebar \sim((p \wedge \sim q) \vee r)$

69. (a) Negate the statement
      On the ranch they raise cattle and sheep.
   (b) Let p and q be the following statements:
      p: On the ranch they raise cattle.
      q: On the ranch they raise sheep.
      Express the original statement in symbolic form.
      Express the negation of the original statement in symbolic form.
   (c) Find the truth table of both statements and determine whether or not you have properly negated the original statement.

70. (a) Negate the statement
      This city is dirty or noisy.
   (b) Let p and q be the following statement:
      p: This city is dirty.
      q: This city is noisy.
      Express the original statement in symbolic form.
      Express the negation of the original statement in symbolic form.
   (c) Find the truth table of both statements and determine whether or not you have properly negated the original statement.

71. For a truth table with one variable, we have

p
T
F

with two cases.

For a truth table with two variables, we have

p	q
T	T
T	F
F	T
F	F

with four cases;

p	q	r
T	T	T
T	T	F
T	F	T
T	F	F
F	T	T
F	T	F
F	F	T
F	F	F

with eight cases.

(a) How many cases are there with four variables? How many with k variables?

(b) Examine the foregoing patterns for one, two, and three variables and determine a pattern for increasing the number of variables from one to two and two to three.

(c) Use this pattern to show all cases when increasing the number of variables to four.

(d) Describe how to increase the number of variables from k to $k+1$.

72. For one variable and two variable, respectively, we have

p
T
F

and

p	q
T	T
T	F
F	F
F	T

Find a way to rearrange the cases for three and four variables so that each case can be obtained from the previous case by changing only one T to an F or one F to a T, and also the first case can be obtained from the last case in the same manner. Then try to describe how to increase the number of variables from k to $k+1$ and retain this property.

73. How many different truth tables are possible with two variables? With three variables? With four variables? With n variables?

1.2 Conditional Statements

Assume someone makes the statement that if one event happens, then another event will happen. This statement often takes the form of a threat, but to be more positive, suppose a father makes the statement that "if you make straight As this semester, then I will buy you a new car." Note that this statement is in the form if p then q, where p is the statement *You make straight As this semester* and q is the statement *I will buy you a new car*. Symbolically we will denote this compound statement as $p \to q$. Under what conditions is the father telling the truth? Suppose both statements p and q are true. In this case the happy student gets straight As and the shocked father buys a new car. Surely no one would question the fact that the statement made by the father was true. There are still, however, three other cases to consider. Suppose that the student did make straight As and the father did not buy the new car. Obviously the kindest thing to be said about the father is that he has lied. Hence, if p is true and q is false, then $p \to q$ is false. Suppose the student did not make all As but the father still bought the student a new car. The father might be considered extremely generous, perhaps, but surely he would not be considered a liar. Hence, if p is false and q is true then *if p, then q* or $p \to q$ is true. In the final case, suppose that the student does not get straight As and the relieved father does not buy the new car. Since the student did not fulfill his part of the agreement, the father is released from his obligation. Thus, if p is false and q is false, then $p \to q$ is assumed to be true. In summary, the only time the father has lied is if he made a promise and did not keep it.

The truth table for $p \to q$ is, therefore, defined as follows:

Case	p	q	$p \to q$
1	T	T	T
2	T	F	F
3	F	T	T
4	F	F	T

The symbol $\to$ is called the **conditional** connective.

Another example that may motivate our truth table for the conditional connective is the statement

If an integer is 3, then its square is 9.

We would certainly want this statement to be always true. Let p be the statement *An integer is 3* and q be the statement *The square of the integer is 9*. If p is true and q is true, then the integer is 3, and its square is 9, so case 1 is true. If the integer is -3, then its square is still 9. In this case p is false and q is true, which is case 3. Therefore, we would want case 3 to be true. If the integer is 4, then both p and q are false, which is case 4. We, therefore, want case 4 to be true. Notice that case 2 does not occur, and that in each of the other cases the preceding truth table gives us the desired result.

It may appear that $p \to q$ is a cause-effect relationship, but this is not necessarily the case. To see the absence of cause and effect, let's return to our example where p is the statement *Jane drives a Ford* and q is the statement *Bob has gray hair*. Then the statement *If Jane drives a Ford, then Bob has gray hair* is translated as

if p then q

or

$$p \to q$$

We hope that Jane driving a Ford has no causal relationship with whether Bob's hair turns gray; however, we need to remember that the truth or falsity of a binary compound proposition depends only upon the truth of its component proposition parts and not upon any other relationship or nonrelationship between the parts.

EXAMPLE 1.2 Find the truth table for the expression

$$(p \to q) \land (q \to r)$$

Using the preceding truth table definition for $\to$, we first find the truth tables for $(p \to q)$ and $(q \to r)$, remembering that the only time either is false is when $T \to F$:

Case	p	q	r	(p	→	q)	∧	(q	→	r)
1	T	T	T	T	**T**	T		T	**T**	T
2	T	T	F	T	**T**	T		T	**F**	F
3	T	F	T	T	**F**	F		F	**T**	T
4	T	F	F	T	**F**	F		F	**T**	F
5	F	T	T	F	**T**	T		T	**T**	T
6	F	T	F	F	**T**	T		T	**F**	F
7	F	F	T	F	**T**	F		F	**T**	T
8	F	F	F	F	**T**	F		F	**T**	F
				1	2	1		1	2	1

Finally, we use the $\land$ table to connect the truth tables for the statement

$$(p \to q) \land (q \to r)$$

Case	p	q	r	(p	→	q)	∧	(q	→	r)
1	T	T	T	T	T	T	**T**	T	T	T
2	T	T	F	T	T	T	**F**	T	F	F
3	T	F	T	T	F	F	**F**	F	T	T
4	T	F	F	T	F	F	**F**	F	T	F
5	F	T	T	F	T	T	**T**	T	T	T
6	F	T	F	F	T	T	**F**	T	F	F
7	F	F	T	F	T	F	**T**	F	T	T
8	F	F	F	F	T	F	**T**	F	T	F
				1	2	1	*	1	2	1

The statement $(p \to q) \land (q \to p)$ is denoted $p \leftrightarrow q$. The symbol $\leftrightarrow$ is called the **biconditional**. The truth table of $(p \to q) \land (q \to p)$ will determine the truth table definition of $p \leftrightarrow q$.

Case	p	q	$(p \to q)$	$\land$	$(q \to p)$	$p \leftrightarrow q$
1	T	T	T	T	T	T
2	T	F	F	F	T	F
3	F	T	T	F	F	F
4	F	F	T	T	T	T
				*		*

From this definition we see that the biconditional, $p \leftrightarrow q$, is true only when p and q agree in truth value.

One might wonder how to interpret the statements $\sim p \lor q$, $p \land q \lor r$, $p \land q \to r$, and $p \land q \leftrightarrow q \lor r$, where there are no parentheses. It is usually better to use parentheses to avoid misunderstanding. However, as in algebra, there is an order of precedence. The operations are performed in the following order: $\sim$, $\land$, $\lor$, $\to$, and $\leftrightarrow$. Therefore, the preceding statements would be interpreted to mean $(\sim p) \lor q$, $(p \land q) \lor r$, $(p \land q) \to r$, and $(p \land q) \leftrightarrow (q \lor r)$.

Exercises

Let p, q, and r be the following statements:

p: He will buy a new computer.
q: He will celebrate all night.
r: He will win the lottery.

Express the following English sentences as symbolic expressions:

1. If he wins the lottery, then he will buy a new computer and he will celebrate all night.

2. If he doesn't buy a new computer, then he will not celebrate all night.

3. If he wins the lottery, then he will celebrate all night; and if he doesn't win the lottery, then he will not get a new computer.

4. If he doesn't win the lottery or buy a new computer, then he will not celebrate all night.

Let p, q, and r be the following statements:

p: He reads comic books.
q: He loves science fiction.
r: He is a computer scientist.

Express the following English sentences as symbolic expressions:

5. If he reads comic books and loves science fiction, then he is a computer scientist.

6. If he doesn't read comic books or love science fiction, then he is not a computer scientist.

7. If he reads comic books, then he loves science fiction; and if he does not read comic books, then he is a computer scientist.

8. If he is a computer scientist, then he reads comic books or does not love science fiction.

Change the following statements to symbolic form:

9. If it is not true that dentists are tidy or sharks do not wear shoes, then tidy people shine their shoes and shoes should be worn at night.

10. If dentists are not tidy and sharks wear shoes, and if shoes should not be worn at night, then either tidy people do not shine their shoes or sharks do not wear shoes.

Let p, q, r, and s be the following statements:

p: Carrots do not brush their teeth.
q: Everyone should brush their teeth.
r: People without teeth do not chew gum.
s: Carrots never have gum on their toothbrush.

Change the following statements to symbolic form:

11. If people without teeth chew gum, then it is not true that carrots do not brush their teeth if and only if carrots never have gum on their toothbrush, or everyone should brush their teeth.

12. If carrots do not brush their teeth, then people without teeth do not chew gum, and if carrots brush their teeth then either carrots have gum on their toothbrush or everyone should brush their teeth but not both.

Let p, q, and r be the following statements:

p: He loves purple ties.
q: He is popular.
r: He has strange friends.

Express the following symbolic expressions as English sentences:

13. $(p \land q) \to r$

14. $q \to \sim r$

15. $p \to (q \lor r)$

16. $(p \to \sim q) \land (q \to r)$

Let p, q, and r be the following statements:

p: He is successful.
q: He is popular.
r: He is rich.

Express the following symbolic expressions as English sentences:

17. $\sim(p \rightarrow q)$
18. $(p \vee r) \rightarrow q$
19. $q \leftrightarrow (p \wedge r)$
20. $(p \rightarrow q) \wedge (\sim r \rightarrow (\sim p \vee \sim q))$

Let p, q, r, and s be the following statements:

p: Castles are drafty.
q: Barns are drafty.
r: Kings live in castles.
s: Cattle live in barns.

Express the following symbolic expressions as English sentences.

21. $(p \wedge r) \leftrightarrow (\sim q \rightarrow (r \veebar s))$
22. $((p \wedge r) \vee s) \rightarrow (\sim q \vee r)$
23. $(r \wedge s) \rightarrow (\sim p \vee \sim q)$
24. $(((p \vee q) \wedge (p \rightarrow r)) \wedge (q \rightarrow s)) \rightarrow (r \vee s)$

Let p, q, r, and s be the following statements:

p: x is less than 5.
q: x is greater than 1.
r: $x = 2$.
s: $x = 6$.

Express the following symbolic expressions as English sentences.

25. $(p \wedge q) \rightarrow (r \wedge \sim s)$
26. $(r \rightarrow p) \wedge (s \rightarrow q)$
27. $\sim((p \rightarrow r) \wedge (q \rightarrow s))$
28. $(r \veebar s) \leftrightarrow (((p \rightarrow r) \vee (q \rightarrow s)) \wedge (p \vee q))$

If p is true, q is false, and r is true, which of the following are true and which are false?

29. $p \rightarrow (q \rightarrow r)$
30. $(p \rightarrow q) \rightarrow r$
31. $(p \wedge q) \rightarrow r$
32. $p \wedge (q \rightarrow \sim r)$
33. $(p \rightarrow \sim q) \veebar (q \rightarrow r)$

If p is false, q is false, and r is true, which of the following are true and which are false?

34. $p \vee (q \rightarrow r)$
35. $(p \rightarrow \sim q) \rightarrow r$
36. $(p \wedge r) \rightarrow \sim q$
37. $p \vee (q \rightarrow r)$
38. $(p \rightarrow q) \veebar (q \rightarrow r)$

Find truth tables for the following expressions:

39. $(p \rightarrow q) \rightarrow r$
40. $p \rightarrow (q \rightarrow r)$
41. $q \rightarrow (p \wedge r) \leftrightarrow ((q \rightarrow p) \wedge (q \rightarrow r))$
42. $((p \rightarrow q) \vee r) \rightarrow (\sim p \vee \sim q)$
43. $(p \rightarrow q) \rightarrow (q \rightarrow r)$
44. $(p \rightarrow q) \vee \sim(r \wedge q)$
45. $(p \vee r) \rightarrow (p \wedge q)$
46. $\sim((p \rightarrow q) \wedge \sim r) \rightarrow (p \vee \sim r)$
47. $((p \rightarrow q) \wedge \sim(r \vee p)) \rightarrow (\sim p \vee \sim q)$
48. $(p \rightarrow q) \leftrightarrow (\sim q \rightarrow \sim p)$
49. $(p \wedge \sim(q \vee \sim r)) \leftrightarrow (p \rightarrow q)$
50. $(p \vee r) \rightarrow q$
51. $((p \rightarrow q) \wedge (q \rightarrow \sim r)) \rightarrow (r \rightarrow p)$
52. $(p \wedge (q \vee r)) \rightarrow ((p \wedge q) \vee (p \wedge r))$
53. $(\sim(p \wedge \sim r) \vee q) \rightarrow (q \vee r)$
54. $\sim((p \wedge \sim q) \vee r) \leftrightarrow (r \rightarrow q)$
55. $(\sim(p \rightarrow q) \rightarrow (q \rightarrow r))$
56. $(p \vee q) \wedge (p \rightarrow r)$
57. $((p \vee r) \rightarrow q) \rightarrow ((p \rightarrow q) \vee (p \rightarrow r))$

Tell whether each of the following statements is true or false

58. If $2^2 = 4$, then $3^2 = 9$.
59. If $2^2 = 5$, then $3^2 = 9$.
60. If $2^2 = 5$, then $3^2 = 10$.
61. If $2^2 = 4$, then $3^2 = 10$.
62. Find values for p, q, r, and s for which the following statement is true: $((p \vee q) \vee r) \rightarrow s$.
63. Find values for p, q, r, and s for which the following statement is true: $(p \vee q) \rightarrow (r \wedge s)$.
64. Find values for p, q, r, and s for which the following statement is false: $(p \rightarrow q) \wedge (r \rightarrow s.)$.
65. Find values for p, q, r, and s for which the following statement is false: $(p \rightarrow s) \rightarrow (r \veebar s.)$.

Assume the connective | has the truth table

Case	p	q	$p \mid q$
1	T	T	F
2	T	F	T
3	F	T	T
4	F	F	T

Find truth tables for the following:

66. $(p \rightarrow q) \mid (q \rightarrow r)$
67. $((p \mid q) \mid r) \leftrightarrow (p \mid (q \mid r))$
68. $(q \vee r) \mid (p \wedge q)$

Assume the connective $\downarrow$ has the truth table

Case	p	q	$p \downarrow q$
1	T	T	F
2	T	F	F
3	F	T	F
4	F	F	T

Find truth tables for the following:

69. $(q \downarrow r) \downarrow (p \wedge r)$

70. $((p \downarrow q) \downarrow r) \leftrightarrow (p \downarrow (q \downarrow r))$

71. $((p \vee q) \downarrow (q \rightarrow r))$

1.3 Equivalent Statements

We are particularly interested in compound statements that are expressed differently but are, in fact, true in exactly the same cases. Such propositions are said to be **logically equivalent**. Equivalence of two propositions can easily be established by constructing truth tables for both propositions and then comparing the two.

As an example, let p and q be the propositions

p: It rained today.

q: It snowed today.

and consider the compound propositions

It is not true that it rained or snowed today.

or

$$\sim(p \vee q)$$

and

It didn't rain today and it didn't snow today.

or

$$\sim p \wedge \sim q$$

We can construct truth tables for both:

Case	p	q	$\sim$	$(p \vee q)$	$\sim p$	$\wedge$	$\sim q$
1	T	T	F	T	F	F	F
2	T	F	F	T	F	F	T
3	F	T	F	T	T	F	F
4	F	F	T	F	T	T	T
			*			#	

Thus, we see that in all four cases, the truth values for $\sim(p \vee q)$ (indicated by *) are, respectively, the same as the truth values for $\sim p \wedge \sim q$ (indicated by #). This result means that the two propositions in question are logically equivalent. We shall denote this by

$$\sim(p \vee q) \equiv \sim p \wedge \sim q$$

This equivalence is a very useful fact. Namely, to negate an "or" statement, just negate each part and change the "or" to "and."

Associated with the conditional statement $p \rightarrow q$ are three other statements: the converse, inverse, and contrapositive of $p \rightarrow q$. They are defined as follows:

$p \rightarrow q$ **conditional**
$q \rightarrow p$ **converse** of $p \rightarrow q$
$\sim p \rightarrow \sim q$ **inverse** of $p \rightarrow q$
$\sim q \rightarrow \sim p$ **contrapositive** of $p \rightarrow q$

Given the statement "If he plays football, then he is popular," we have the following:

> converse: "If he is popular, then he plays football."
> inverse: "If he does not play football, then he is not popular."
> contrapositive: "If he is not popular, then he does not play football."

It is important to understand that the statements *If he lives in Detroit, then Bob will visit him* and *Bob will visit him if he lives in Detroit* are really the same statement. However, *If Bob visits him, then he lives in Detroit* is not the same statement. The order in which p and q occur in the sentence is not important but it is important which part is the "if" part and which is the "then" part. It may seem that one can find the converse of a statement by changing *if p, then q* to *q, if p*, but logically it is still the original statement.

EXAMPLE 1.3 Given the statement "If he tries, he will not fail," we have the following:

> converse: "If he will not fail, then he tries."
> inverse: "If he does not try, then he will fail."
> contrapositive: "If he fails, then he did not try." ∎

Part (g) of the following theorem states that a conditional statement and its contrapositive statement are logically equivalent. The equivalence of a conditional statement and its contrapositive statement is an important concept in mathematics. Often it is much easier to prove the contrapositive of a theorem rather than to prove the theorem directly. Using the equivalence of a statement and its contrapositive, the student should show that the converse and inverse of a statement have the same truth table. However, a conditional statement and its converse (or inverse) do not have the same truth table. This mistaken assumption is a common source of error in "logical" arguments.

THEOREM 1.4 The following list of logically equivalent propositions can be established using truth tables.

(a) *Idempotent Laws*

$$p \wedge p \equiv p$$
$$p \vee p \equiv p$$

(b) *Double Negation*

$$\sim(\sim p) \equiv p$$

(c) *De Morgan's Laws*

$$\sim(p \vee q) \equiv \sim p \wedge \sim q$$
$$\sim(p \wedge q) \equiv \sim p \vee \sim q$$

(d) *Commutative Properties*

$$p \wedge q \equiv q \wedge p$$
$$p \vee q \equiv q \vee p$$

(e) *Associative Properties*

$$p \wedge (q \wedge r) \equiv (p \wedge q) \wedge r$$
$$p \vee (q \vee r) \equiv (p \vee q) \vee r$$

(f) *Distributive Properties*

$$p \wedge (q \vee r) \equiv (p \wedge q) \vee (p \wedge r)$$
$$p \vee (q \wedge r) \equiv (p \vee q) \wedge (p \vee r)$$

(g) *Equivalence of Contrapositive*

$$p \to q \equiv \sim q \to \sim p$$

(h) *Other Useful Properties*

$$p \to q \equiv \sim p \vee q$$
$$p \leftrightarrow q \equiv (p \to q) \wedge (q \to p)$$

■

Note that because of the associative property, either $(p \wedge q) \wedge r$ or $p \wedge (q \wedge r)$ may simply be written as $p \wedge q \wedge r$. Similarly, either $(p \vee q) \vee r$ or $p \vee (q \vee r)$ may simply be written as $p \vee q \vee r$.

A proposition that is true in every case is said to be **logically true** or to be a **tautology**; a proposition that is constructed to be false in every case is said to be **logically false** or to be a **contradiction**. Theorems in mathematics are examples of tautologies. It would be rather disturbing if a theorem in mathematics were true only 80 percent of the time. The statement *He will pass or he will not pass* is an example of a tautology, since one event or the other must occur. The logical statement $p \vee \sim p$ is then a tautology. Shakespeare should have said "To be or not to be, that is a tautology." The statement *If he passes, then he passes* is also a tautology, although not a very profound one. The statement *If he is rich and successful, then he is successful* is also an example of a tautology. The statement *She is moving to Connecticut and she is not moving to Connecticut* is never true since she cannot do both. Therefore, it is a contradiction.

Consider a proposition of the form

$$(p \wedge (p \to q)) \to q$$

The truth table for this proposition is

Case	p	q	$(p$	$\wedge$	$(p \to q))$	$\to$	q
1	T	T	T	T	T	T	T
2	T	F	T	F	F	T	F
3	F	T	F	F	T	T	T
4	F	F	F	F	T	T	F
						*	

The column marked with an asterisk gives the truth value of the entire compound proposition. This proposition is true in all four possible cases and, hence, is a tautology.

Once we have a proposition that is always true, it is easy to construct one that is always false—just negate the logically true proposition. The statement

$$\sim((p \wedge (p \to q)) \to q)$$

is logically false.

We now give several properties dealing with logically true and logically false statements. The symbol **T** will represent a statement that is a tautology and so has a truth table that consists of all Ts. The symbol **F** will represent a statement that is a contradiction and so has a truth table that consists of all Fs. The following can be shown using truth tables:

$$p \wedge T \equiv p$$
$$p \wedge F \equiv F$$
$$p \vee T \equiv T$$
$$p \vee F \equiv p$$
$$p \wedge \sim p \equiv F$$
$$p \vee \sim p \equiv T$$
$$p \to p \equiv T$$

Any component of a compound statement can be replaced by any statement logically equivalent to that statement without changing the truth value of the statement since the truth value of the statement depends only on the truth value of its component parts (and not upon their form or complexity). For example,

$$
\begin{aligned}
(q \vee r) \vee (p \wedge \sim r) &\equiv q \vee (r \vee (p \wedge \sim r)) && \text{associative property} \\
&\equiv q \vee ((r \vee p) \wedge (r \vee \sim r)) && \text{distributive property} \\
&\equiv q \vee ((r \vee p) \wedge T) && \text{equivalence} \\
&\equiv q \vee (r \vee p) && \text{equivalence} \\
&\equiv q \vee (p \vee r) && \text{commutative property} \\
&\equiv (q \vee p) \vee r && \text{associative property} \\
&\equiv (p \vee q) \vee r && \text{commutative property}
\end{aligned}
$$

At this point we have taken our first step in using logic and a fixed set of "true" statements to derive new "true" statements.

The conditional is expressed in English in several ways, all of which are written symbolically as $p \to q$. Some of these follow:

If p, then q.

p is sufficient for q.
p is a sufficient condition for q.

q is necessary for p.
q is a necessary condition for p.

p only if q (or only if q then p).

From the table for $p \to q$, we know that if $p \to q$ is true and p is true, then q must be true so that p being true is sufficient for q being true. Therefore, *p is sufficient for q* has the same meaning as $p \to q$. Similarly, if q is false and *q is necessary for p*, then p must be false. Therefore, if $\sim q$ is true, then $\sim p$ must be true and $\sim q \to \sim p$. But this last statement is the contrapositive of $p \to q$; therefore, *q is necessary for p* has the same meaning as $p \to q$.

The meaning of *p only if q* is similarly analyzed. It means that p is true only if q is true. If q is not true, then p cannot be true. But this is equivalent to saying that if $\sim q$ is true, then $\sim p$ must be true and $\sim q \to \sim p$. This is the contrapositive of $p \to q$ and, therefore, *p only if q* has the same meaning as $p \to q$.

As stated before, the order that the clauses occur in the sentence is not important, but only which clause follows the *if* (or in this case *only if*) and which clause follows the *then*. Therefore, *Only if Fred saves money then he can afford to go to college* and *Fred can afford to go to college only if he saves money* are considered to be the same statement.

EXAMPLE 1.5

Change each of the following to the form $p \to q$ or $q \to p$:

(a) *He will succeed only if he works hard.*

(b) *He will be happy only if he drives a Ford.*
(c) *Having money is sufficient for being happy.*
(d) *Having money is necessary for being happy.*
(e) *To win an election, it is necessary to get enough votes.*

The solutions are

(a) If p represents *He will succeed* and q represents *He works hard*, then we have p only if q or if p then q or $p \to q$, which we would state in the form *If he succeeds, then he works hard.*

(b) If p represents *He drives a Ford* and q represents *He will be happy*, then we have q only if p or $q \to p$, which we would state in the form *If he is happy, then he drives a Ford.*

(c) If p represents *He has money* and q represents *He is happy*, then p is sufficient for q or $p \to q$, which we would state as *If he has money, then he is happy.*

(d) If p represents *He has money* and q represents *He is happy*, then p is necessary for q or $q \to p$, which we would state as *If he is happy, then he has money.*

(e) If p represents *He will win an election* and q represents *He will get enough votes*, then q is necessary for p or if p, then q or $p \to q$, which we would state in the form *If he wins the election, then he gets enough votes.* ∎

We again consider the biconditional $\leftrightarrow$. Since $p \leftrightarrow q$ is equivalent to $(p \to q) \wedge (q \to p)$, $p \leftrightarrow q$ is the same as p if and only if q. Therefore, if p is the statement *Jim will play football* and q is the statement *Jane is a cheerleader*, then $p \leftrightarrow q$ may be expressed as

Jim will play football if and only if Jane is a cheerleader.

We have the following equivalent ways of writing the biconditional $p \leftrightarrow q$:

p if and only if q.
p is necessary and sufficient for q.
p is a necessary and sufficient condition for q.

If someone makes the statement

Sam will play golf if and only if it is warm

then if it is warm we should expect to see Sam play golf and if Sam plays golf, then we should expect it to be warm.

Similarly, if someone makes the statement

Being lucky is a necessary and sufficient condition for being successful.

then they are saying if a person is lucky, then they will be successful and if a person is successful, then they have been lucky.

Exercises

Use the truth table method to establish the following equivalences:

1. De Morgan's Law

$$\sim(p \wedge q) \equiv \sim p \vee \sim q$$

2. The Associative property for $\vee$

$$p \vee (q \vee r) \equiv (p \vee q) \vee r$$

3. The Distributive property for "or" over "and"

$$p \vee (q \wedge r) \equiv (p \vee q) \wedge (p \vee r)$$

4. The conversion of conditional to an "or" statement

$$p \to q \equiv \sim p \vee q$$

5. Without using truth tables, show that the negation of $p \to q$ is $p \wedge \sim q$.

6. As previously stated, there are several related conditional statements involving two component propositions p and q. They are

$p \to q$ the conditional
$q \to p$ the converse (of $p \to q$)
$\sim q \to \sim p$ the contrapositive (of $p \to q$)
$\sim p \to \sim q$ the inverse (of $p \to q$)

Using truth tables, prove that

$$p \to q \not\equiv q \to p$$

The conditional is equivalent to its contrapositive; however, the conditional is not equivalent to its converse. Often we use the phrasing "if p, then q; and conversely." This statement really means "if p, then q; and, if q, then p" or

$$(p \to q) \wedge (q \to p)$$

which is equivalent to $p \leftrightarrow q$ or "p if and only q." Prove, without direct use of truth tables, that $p \to q \equiv \sim q \to \sim p$. Use this result to prove that the inverse of a conditional statement is equivalent to its converse.

Using logically equivalent statements without the direct use of truth tables, show that

7. $p \equiv \sim(p \wedge s) \to (\sim s \wedge p)$
8. $\sim(p \leftrightarrow q) \equiv (p \wedge \sim q) \vee (q \wedge \sim p)$

Convert the following statements to if-then form:

9. He is a Martian only if he has six legs.
10. To be a successful politician, it is necessary to be elected.
11. Having money is sufficient for being popular.

Convert the following statements to if-then form:

12. It is necessary to have a helmet to play football.
13. Only if I read Shakespeare, then I am literary.
14. Passing this course is sufficient for me to graduate.
15. To enjoy downhill skiing, it is necessary to have snow.
16. Being a spider is sufficient for being disliked.
17. I am able to wake up in the morning only if I drink three cups of coffee.
18. To really mess things up, it is necessary to have a computer.
19. One is truly educated only if one knows Latin.
20. To get to sleep quickly, it is sufficient to read this book.

Given the statement If I vote, then I am a good citizen,

21. State the converse of this statement.
22. State the inverse of this statement.
23. State the contrapositive of this statement.

Given the statement If I don't pay my loan, then I will have to leave town,

24. State the converse of this statement.
25. State the inverse of this statement.
26. State the contrapositive of this statement.
27. If u, v, and w are (possibly compound) propositions, explain why the following are true:
    (a) $u \equiv u$.
    (b) If $u \equiv v$, then $v \equiv u$.
    (c) If $u \equiv v$ and $v \equiv w$, then $u \equiv w$.
28. Determine whether or not the following statements are equivalent: $p \to (q \vee r)$ $(p \to q) \vee (p \to r)$.
29. Determine whether or not the following statements are equivalent: $p \to (q \wedge r)$ $(p \to q) \wedge (p \to r)$.
30. Determine whether or not the following statements are equivalent: $(p \vee q) \to r$ $(p \to r) \vee (q \to r)$.
31. Determine whether or not the following statements are equivalent: $(p \wedge q) \to r$ $(p \to r) \wedge (q \to r)$.
32. Determine whether or not the following statements are equivalent: $(p \vee q) \to r$ $(p \to r) \wedge (q \to r)$.
33. Determine whether or not the following statements are equivalent: $(p \wedge q) \to r$ $(p \to r) \vee (q \to r)$.
34. Determine whether or not the following statements are equivalent: $(p \to q) \to r$ $p \to (q \to r)$.
35. Determine whether or not the following statements are equivalent: $\sim((p \wedge q) \to r)$ $\sim p \vee \sim q \vee r$.
36. Determine whether or not the following statements are equivalent: $p \veebar (q \veebar r)$ $(p \veebar q) \veebar r$.
37. Determine whether or not the following statements are equivalent: $p \wedge (q \veebar r)$ $(p \wedge q) \veebar (p \wedge r)$.
38. Determine whether or not the following statements are equivalent: $p \wedge (q \to r)$ $\sim(p \vee (\sim q \to \sim r))$.

Determine which of the following are logically true and which are logically false:

39. $(p \wedge q) \to p$ 40. $(p \vee q) \to p$
41. $((p \vee q) \wedge \sim p) \to q$ 42. $(p \to q) \wedge (p \wedge \sim q)$
43. $(((p \to q) \wedge (q \to r)) \wedge r) \to p$
44. $p \to T$ 45. $T \to p$
46. $p \to F$ 47. $F \to p$
48. $(((p \to q) \wedge (q \to r)) \wedge \sim r) \to p$
49. $((p \leftrightarrow q) \wedge (p \veebar q)$ 50. $((p \leftrightarrow q) \vee (p \veebar q)$
51. $(\sim p \to F) \to p$

Using properties of equivalence, show the following pairs of propositions are equivalent without using truth tables:

52. $p \to (q \wedge r) \quad (p \to q) \wedge (p \to r)$
53. $p \to (q \vee r) \quad (p \to q) \vee (p \to r)$
54. $(p \vee q) \to r \quad (p \to r) \wedge (q \to r)$
55. $(p \wedge q) \to r \quad (p \to r) \vee (q \to r)$
56. $(p \wedge q) \to r \quad p \to (q \to r)$

1.4 Axiomatic Systems: Arguments and Proofs

Much of mathematics deals with theorems and proofs of theorems. Theorems are "true" statements about the mathematical system being considered. For example, the statement

> The hypotenuse of a right triangle is longer than either of the other two sides.

is a theorem in Euclidean geometry. The statement is considered to be true because it is "derivable" or "deducible" from previously accepted, or derived, truths in Euclidean geometry.

A mathematical system begins with undefined terms and statements precisely describing the fundamental characteristics or truths about these terms which the mathematician uses to create the system. We call these fundamental characteristics axioms and postulates. The statements derived (proved) using only these fundamental properties (axioms and postulates), previously proved statements, and the rules of logic are called **theorems**.

In a mathematical system, therefore, all of the information necessary to prove a theorem must be contained in axioms and previously proven theorems. When working in a particular area of a mathematical system, all the original axioms and previously proven theorems may not be included. Instead, we may accept some of the previously proven theorems as axioms and begin at that point. For example, the integer axioms and the Peano axioms for the positive integers implicitly assume the axioms of set theory, but since the emphasis in number theory is on the properties of integers, it would be distracting to attempt a full development of set theory as well. So we may start in "midstream" with the properties of either the natural numbers or the integers.

It is essential that the rules of logic we use to derive new theorems from axioms, postulates, and previously proven theorems in the system do not produce false statements as "theorems." The rules of logic we use to derive new theorems from axioms, postulates, and previously proven theorems are called the **rules of inference**. An **argument** consists of a collection of statements called hypotheses and a statement called the conclusion. A **valid argument** is an argument whose conclusion is true whenever all the hypotheses are true. Rules of inference are selected so that they are valid arguments.

The argument is often displayed as

$$\begin{array}{ll} H_1 & \\ H_2 & \text{hypotheses} \\ \underline{H_3} & \\ \therefore\ C & \text{conclusion} \end{array}$$

in order to highlight the parts. The symbol $\therefore$ means "therefore." The hypotheses are a list of one or more propositions or **premises**. The argument is valid if

whenever $H_1, H_2,$ and H_3 are true, then C is true

or, equivalently,

whenever $H_1 \wedge H_2 \wedge H_3$ is true, then C is true

We shall have two methods of showing that an argument is valid. The first is to construct a truth table and show that whenever all of the hypotheses are true, then the conclusion is true. The second is to use the truth table in the method just mentioned to validate rules of inference and then use the rules of inference to prove the validity of the conclusion. For long arguments it is often easier to use the rules of inference. For example, we shall see that it is easy to test the validity of the argument

$$\begin{array}{c} p \to q \\ q \to r \\ r \to s \\ \underline{s \to t} \\ \therefore\ p \to t \end{array}$$

using rules of inference but extremely tedious using truth tables. We cannot, however, show that an argument is invalid using rules of inference as we can using truth tables.

Consider the argument

$$\begin{array}{c} p \\ p \to q \\ \underline{q \to r} \\ \therefore\ p \wedge q \wedge r \end{array}$$

We construct the truth table for the hypothesis and conclusion as follows:

Case	p	q	r	p	$p \to q$	$q \to r$	$p \wedge q \wedge r$
1	T	T	T	T	T	T	T
2	T	T	F	T	T	F	F
3	T	F	T	T	F	T	F
4	T	F	F	T	F	T	F
5	F	T	T	F	T	T	F
6	F	T	F	F	T	F	F
7	F	F	T	F	T	T	F
8	F	F	F	F	T	T	F
				1	2	3	*

Note that each time all of the hypotheses are true (which occurs in case 1), the conclusion is also true and the argument is valid.

Consider the argument

$$\begin{array}{c} p \vee q \\ p \to r \\ \underline{q \to r} \\ \therefore\ r \end{array}$$

We again construct the truth table for the argument.

Case	p	q	r	$p \vee q$	$p \to r$	$q \to r$	r
1	T	T	T	T	T	T	T
2	T	T	F	T	F	F	F
3	T	F	T	T	T	T	T
4	T	F	F	T	F	T	F
5	F	T	T	T	T	T	T
6	F	T	F	T	T	F	F
7	F	F	T	F	T	T	T
8	F	F	F	F	T	T	F
				1	2	3	*

We again find that when all of the hypotheses are true (which occurs in cases 1, 3, and 5), then the conclusion is true, and the argument is valid.

If we consider the following argument, however,

$$p \to q$$
$$q \to r$$
$$r$$
$$\therefore p$$

and look at the truth table,

Case	p	q	r	$p \to q$	$q \to r$	r	p
1	T	T	T	T	T	T	T
2	T	T	F	T	F	F	T
3	T	F	T	F	T	T	T
4	T	F	F	F	T	F	T
5	F	T	T	T	T	T	F
6	F	T	F	T	F	F	F
7	F	F	T	T	T	T	F
8	F	F	F	T	T	F	F
				1	2	3	*

we see that, although all of the hypotheses and the conclusion are true in case 1, in cases 5 and 7 the hypotheses are all true but the conclusion is false. Hence, the argument is not valid.

We also introduce an alternative to the truth table method for testing the validity of arguments. We shall not try to justify the method at this point but simply accept it on faith and later explain why it is a valid method. This method is indirect since we try to show that an argument is invalid. If we succeed, then the argument is invalid. If we are unable to show the argument is invalid, then we assume it is valid. For example, consider the argument

$$p \lor q$$
$$p \to r$$
$$q \to r$$
$$\therefore r$$

If the argument is invalid, then there are truth values for p, q, and r that make the hypotheses true and the conclusion false. If the conclusion is false, then r is false. If $q \to r$ is true and r is false, then q must be false. Similarly if $p \to r$ is true, then p must be false. But then $p \lor q$ must be false so that it is impossible to have the conclusion false and all of the hypotheses true and we conclude that the argument is valid.

Consider the argument

$$p \to q$$
$$q \to r$$
$$r \to s$$
$$s \to t$$
$$\therefore p \to t$$

We again try to show a case where the conclusion is false and the hypotheses are true. If $p \to t$ is false, then p must be true and t must be false. Since t is false, then for $s \to t$ to be true, s must be false. If s is false and $r \to s$ is true, then r must be false. If r is false and $q \to r$ is true, then q must be false. But since p

must be true and q is false, then $p \to q$ is false. Again it is impossible to have the conclusion false and all of the hypotheses true and we conclude that the argument is valid.

Consider the argument

$$\begin{array}{c} p \to q \\ q \to r \\ \hline \therefore p \vee r \end{array}$$

If the conclusion is false, then p and r must both be false. However, if q is false, then $p \to q$ and $q \to r$ are both true. Since the hypotheses are both true and the conclusion is false, then the argument is invalid. If we look at the truth table for this argument

Case	p	q	r	$p \to q$	$q \to r$	$p \vee r$
1	T	T	T	T	T	T
2	T	T	F	T	F	T
3	T	F	T	F	T	T
4	T	F	F	F	T	T
5	F	T	T	T	T	T
6	F	T	F	T	F	F
7	F	F	T	T	T	T
8	F	F	F	T	T	F
				1	2	*

we see that we have described case 8 in the truth table, which shows that the argument is invalid.

Consider the argument

$$\begin{array}{c} p \to q \\ q \to r \\ r \\ \hline \therefore p \end{array}$$

If the conclusion is false, then p is false. For the hypotheses to be true, r must be true. But $p \to q$ and $q \to r$ are true whether q is true or false. Since the hypotheses are true and the conclusion is false either when p is false, q is true, and r is true or when p is false, q is false, and r is true, there are two cases where we can show the argument is invalid. If we look at the truth table

Case	p	q	r	$p \to q$	$q \to r$	r	p
1	T	T	T	T	T	T	T
2	T	T	F	T	F	F	T
3	T	F	T	F	T	T	T
4	T	F	F	F	T	F	T
5	F	T	T	T	T	T	F
6	F	T	F	T	F	F	F
7	F	F	T	T	T	T	F
8	F	F	F	T	T	F	F
				1	2	3	*

we see that we have described cases 5 and 7.

Note that in any argument, with hypotheses $H_1, H_2, H_3, \ldots, H_n$ and conclusion C, the argument is valid if and only if the proposition

$$(H_1 \wedge H_2 \wedge H_3 \wedge \ldots \wedge H_n) \to C$$

is a tautology. The order of listing the premises is unimportant since

$$H_1 \wedge H_2 \equiv H_2 \wedge H_1$$

It is easily shown that the following is a valid argument, which we will adopt as a rule of inference.

$$\begin{array}{c} p \\ p \to q \\ \hline \therefore\ q \end{array}$$

This particular valid argument is called the **law of detachment** or **modus ponens**.

As an example of the law of detachment, suppose that b is a particular integer. Let p and q be given by

$$\begin{array}{ll} p: & b \text{ is even} \\ q: & b \text{ is divisible by } 2 \end{array}$$

so that

$$p \to q: \quad \text{if } b \text{ is even, then } b \text{ is divisible by } 2$$

The law of detachment gives

$$\begin{array}{ll} p \to q & \text{if } b \text{ is even, then } b \text{ is divisible by } 2 \\ p & b \text{ is even} \\ \hline \therefore\ q & \therefore\ b \text{ is divisible by } 2 \end{array}$$

Assume the statement *if b is even, then b is divisible by* 2 is derived as a property of the integers and $b = 12$, then both premises are true; so we know for sure that 12 is divisible by 2. On the other hand, if $b = 13$, then p is false, and even though the argument is valid, we cannot be sure whether $b = 13$ is divisible by 2. Valid arguments say nothing about what happens to the conclusion if one of the premises is false.

For future reference, we list several rules of inference:

(a) *Law of Detachment (Modus Ponens)*

$$\begin{array}{c} p \to q \\ p \\ \hline \therefore\ q \end{array}$$

(b) *Syllogism*

$$\begin{array}{c} p \to q \\ q \to r \\ \hline \therefore\ p \to r \end{array}$$

(c) *Modus Tollens*

$$\begin{array}{c} p \to q \\ \sim q \\ \hline \therefore\ \sim p \end{array}$$

(d) *Addition*

$$\begin{array}{c} p \\ \hline \therefore\ p \vee q \end{array}$$

(e) *Specialization*

$$\frac{p \wedge q}{\therefore p}$$

(f) *Conjunction*

$$\begin{array}{c} p \\ q \\ \hline \therefore p \wedge q \end{array}$$

(g) *Cases*

$$\begin{array}{c} p \\ p \to (r \vee s) \\ r \to q \\ s \to q \\ \hline \therefore q \end{array}$$

(h) *Case Elimination*

$$\begin{array}{c} p \vee q \\ p \to (r \wedge \sim r) \\ \hline \therefore q \end{array}$$

(i) *Reductio ad Absurdum*

$$\frac{\sim p \to (r \wedge \sim r)}{\therefore p}$$

All of the preceding arguments can be shown to be valid by using truth tables.

The proof of a statement is a process that uses statements that have already been accepted as true and a series of valid arguments to derive more statements as conclusions until the statement to be proven is reached as a conclusion. When all of the statements are propositions, proofs are easy since truth tables can be used to directly show that a proposition can be derived from other propositions. However, we shall latter find that proving true statements can be very complex, if not impossible. We shall find that the types of valid arguments we have introduced are very helpful when more complex techniques for proof are needed.

The reductio ad absurdum argument is used for the method of proof called **proof by contradiction**. We assume the negation of the statement we are trying to prove. We then try to reach a contradiction. If we can do so, then we have proved the original statement. Note that this is the method that we were using in the alternate method for determining validity previously. Thus, the alternative method is a valid way of reasoning.

Consider the following argument:

$$\begin{array}{c} \text{If the apple is red, then it is ripe.} \\ \text{The apple is ripe.} \\ \hline \therefore \text{The apple is red.} \end{array}$$

Using the notation

$p:$ the apple is red
$q:$ the apple is ripe

the form of the argument is

$$\begin{array}{c} p \to q \\ q \\ \hline \therefore p \end{array}$$

From the truth table

Case	p	q	$((p \rightarrow q)$	$\wedge$	$q)$	$\rightarrow$	p
1	T	T	T	T	T	T	T
2	T	F	F	F	F	T	T
3	F	T	T	T	T	F	F
4	F	F	T	F	F	T	F
						*	

it is evident that the argument is not valid. In case 3, both the premise $p \rightarrow q$ and the premise q are true; however, the conclusion p is not. Thus, the argument is not valid. This particular invalid argument is called the **converse fallacy**. Similarly, the argument

$$\sim p \rightarrow \sim q$$
$$\underline{p}$$
$$\therefore q$$

is not valid and is called the **inverse fallacy**.

In most axiomatic systems, the theorems and axioms can be very complex and the weaving together of theorems, axioms, and rules of inference can be complicated. The objective of the following discussion is to make clear what is happening in proofs and how proofs are constructed. In a complex proof, it is usually necessary to use many of these valid argumentative forms over and over again. This application to proofs can be summarized by describing a proof to be a sequence of statements each of which is

(a) True by hypothesis

(b) An axiom or definition

(c) A previously proven theorem or proposition

(d) A statement implied by previous statements as a conclusion of a valid argument

(e) Logically equivalent to a previous statement

As an example using logic symbolism, we will show that

$$p \rightarrow q$$
$$\sim r \rightarrow \sim q$$
$$\underline{\sim r}$$
$$\therefore \sim p$$

is a valid argument.

Number	Statements	Reason
1	$p \rightarrow q$	given
2	$\sim r \rightarrow \sim q$	given
3	$\sim r$	given
4	$\sim q$	2, 3, and law of detachment
5	$\sim q \rightarrow \sim p$	1 and equivalence $p \rightarrow q \equiv \sim q \rightarrow \sim p$
6	$\sim p$	4, 5, and law of detachment

Therefore, the three premises imply the conclusion $\sim p$ and we have proved or derived $\sim p$.

In most mathematical proofs the logic is "hidden" in that it is not specifically mentioned. It is understood that the reader can follow the logic without help and that including discussion of the logic would unnecessarily complicate the proof. In some of the proofs to follow later, we will look behind the scenes to illustrate how the logic is used.

At this point we demonstrate some of the techniques of proofs. Since we are interested only in the techniques, we are using material from other sections so than some of the definitions and results will be presented more fully later. Many of the statements are not propositions. Do not worry about what is known and what is not known. Just try to follow the pattern of the proof.

EXAMPLE 1.6 As an example of a direct proof, we show that if n is an odd integer, then n^2 is an odd integer.

n is an odd integer	given
$n = 2k + 1$ for some integer k	definition of odd integers
$n^2 = (2k + 1)^2 = 4k^2 + 4k + 1$	properties of multiplication in algebra
$n^2 = 2(2k^2 + 2k) + 1$	distributive law
n^2 is odd	n^2 has the form $2m+1$ and, by definition, is an odd number

EXAMPLE 1.7 This is an example of a proof using the contrapositive. To prove that $p \to q$, we prove that $\sim q \to \sim p$. We prove that if n^2 is an odd integer, then n is an odd integer. To do this we assume that n is not odd and then show that n^2 is not odd.

n is an even integer	given that n is not odd
$n = 2k$ for some integer k	definition of even integers
$n^2 = 4k^2$	properties of multiplication in algebra
$n^2 = 2(2k^2)$	associative law
n^2 is even	n^2 has the form $2m$ and, by definition, is an even number.
n^2 is not odd	definition of odd number

EXAMPLE 1.8 This is an example of a proof by cases. If we can show that in every case a statement is true, then the statement must be true. We shall show that for every integer n, $n^2 + 1$ is positive. There are three possible cases. The integer n can be positive, negative, or zero.

n is a positive integer	case 1
n^2 is a positive integer	The product of two positive integers is positive.
$n^2 + 1$ is a positive integer	The sum of two positive integers is positive.
n is a negative integer	case 2
$-n$ is a positive integer	definition of negative number
$(-n)^2$ is a positive number	The product of two positive integers is positive.
n^2 is a positive integer	$(-n)^2 = n^2$
$n^2 + 1$ is a positive integer	The sum of two positive integers is positive.
$n = 0$	case 3
$n^2 = 0$	property of 0
$n^2 + 1 = 1$	property of 0
$n^2 + 1$ is a positive number	1 is a positive number.
Therefore, $n^2 + 1$ is a positive number	All cases shown

EXAMPLE 1.9 Finally we consider proof by contradiction. We first show that there is no largest positive integer. We assume that there is a largest positive integer and then show that we reach a contradiction. If there is a largest positive integer, let c be that integer.

But $c+1$ is larger than c so that c is not the largest positive integer. Hence, we have a contradiction so that we have shown there is no largest positive integer. ∎

EXAMPLE 1.10

$\sqrt{2}$ is not a rational number. (This is a classical proof going back to Euclid or Pythagorus.)

Again we use proof by contradiction and assume $\sqrt{2}$ is a rational number. Therefore, $\sqrt{2} = \dfrac{a}{b}$, which has been reduced to lowest terms.

Therefore, $a = b\sqrt{2}$ and $a^2 = 2b^2$.
Therefore, a^2 is even by definition of an even integer.
Therefore, a is even as shown previously.
Therefore, $a = 2c$ for some integer c, by definition of an even integer.
Therefore, $a^2 = 4c^2$ and $4c^2 = 2b^2$. Since $a^2 = 2b^2$, dividing both sides of the equation by 2, we have $b^2 = 2c^2$.
Therefore, b^2 is even.
Therefore, b is even.

But this contradicts the assumption that $\dfrac{a}{b}$ has been reduced to lowest terms.

Therefore, $\sqrt{2}$ is not a rational number. ∎

Exercises

Using the truth table criterion, prove that the following arguments are valid:

1. The syllogism

$$\begin{array}{c} p \to q \\ q \to r \\ \hline \therefore p \to r \end{array}$$

2. The cases argument

$$\begin{array}{c} p \\ p \to (r \vee s) \\ r \to q \\ s \to q \\ \hline \therefore q \end{array}$$

3. The reductio ad absurdum argument

$$\begin{array}{c} \sim w \to (r \wedge \sim r) \\ \hline \therefore w \end{array}$$

4. Show that the following argument is not valid.

$$\begin{array}{c} p \to q \\ q \to r \\ \hline \therefore r \end{array}$$

Determine which of the following arguments are valid:

5. $\begin{array}{c} p \to q \\ p \to r \\ q \vee r \\ \hline \therefore p \end{array}$

6. $\begin{array}{c} \sim p \vee q \\ \sim q \vee r \\ \sim r \\ \hline \therefore \sim p \end{array}$

7. $\begin{array}{c} p \to q \\ p \to r \\ \sim(p \wedge q) \\ \hline \therefore \sim p \end{array}$

8. $\begin{array}{c} \sim p \vee \sim q \\ r \vee \sim q \\ \sim p \\ \hline \therefore r \vee \sim p \end{array}$

Determine which of the following arguments are valid:

9. $\begin{array}{c} \sim r \\ p \to r \\ q \to r \\ \hline \therefore \sim(p \wedge q) \end{array}$

10. $\begin{array}{c} p \vee q \\ \sim q \vee r \\ \sim r \\ \hline \therefore p \end{array}$

11. $\begin{array}{c} p \to q \\ q \to r \\ q \vee r \\ \hline \therefore p \end{array}$

12. $\begin{array}{c} p \vee \sim q \\ r \vee \sim q \\ \sim p \\ \hline \therefore \sim r \end{array}$

Use the alternative to the truth table method to determine the validity of the following:

13. $\begin{array}{c} s \vee t \\ t \to r \\ s \to w \\ \hline \therefore r \vee w \end{array}$

14. $\begin{array}{c} p \to q \\ q \to r \\ r \\ \hline \therefore p \end{array}$

15. $\begin{array}{c} p \to q \\ \sim q \to \sim s \\ s \to t \\ t \vee q \\ \hline \therefore p \vee s \end{array}$

16. $\begin{array}{c} \sim p \vee q \\ r \vee \sim q \\ p \\ s \\ \hline \therefore r \vee s \end{array}$

Use the alternative to the truth table method to determine the validity of the following:

17. $s \vee t$
    $\sim (s \wedge q)$
    q
    $\therefore t$

18. $s \vee \sim t$
    $\sim (t \wedge q)$
    t
    $\therefore s \wedge q$

19. $p \rightarrow \sim q$
    $s \rightarrow q$
    $q \rightarrow r$
    $r \rightarrow s$
    $\therefore s \vee \sim p$

20. $\sim p \vee q$
    $q \rightarrow r$
    $\sim s \rightarrow \sim r$
    $s \rightarrow t$
    $\therefore \sim p \vee t$

Use the rules of inference and equivalent statements to show that the following arguments are valid:

21. $\sim (s \wedge t)$
    $\sim w \rightarrow t$
    $\therefore s \rightarrow w$

22. $p \rightarrow q$
    $\sim r \rightarrow \sim q$
    $\sim r$
    $\therefore \sim p$

23. $\sim (\sim p \vee q)$
    $\sim z \rightarrow \sim s$
    $(p \wedge \sim q) \rightarrow s$
    $\sim z \vee r$
    $\therefore r$

24. $\sim x \rightarrow \sim w$
    $(x \vee \sim w) \rightarrow z$
    $\sim p \rightarrow \sim z$
    $p \rightarrow (\sim r \vee \sim s)$
    $\therefore \sim r \vee \sim s$

25. Prove that stating that two propositions are logically equivalent means the same as stating that the biconditional of the two propositions is a tautology. That is, for propositions p and q, prove that

    $$p \equiv q$$

    means the same as

    $p \leftrightarrow q$ is a tautology

    Furthermore, prove that this last statement is equivalent to the statement

    both $p \rightarrow q$ and $q \rightarrow p$ are tautologies

    When the conditional $p \rightarrow q$ is a tautology, we say that p implies q and often write $p \Rightarrow q$.

Determine which of the following are tautologies:

26. $p \rightarrow (p \wedge q)$
27. $p \rightarrow (p \vee q)$
28. $(p \rightarrow q) \rightarrow (q \rightarrow p)$
29. $((p \rightarrow q) \wedge \sim q) \rightarrow \sim p$
30. $((p \rightarrow q) \wedge q) \rightarrow p$

31. Give rules of inference to justify the statements in the following proof.

$p \vee q$	given
$q \rightarrow r$	given
$\sim s \rightarrow \sim r$	given
$s \rightarrow t$	given
$\sim t$	given
$\sim s$	
$\sim r$	
$\sim q$	
p	

32. Give rules of inference to justify the statements in the following proof.

$\sim p$	assumed
$p \vee q$	given
$q \rightarrow t$	given
$\sim r \vee \sim s$	given
$(s \wedge t) \rightarrow r$	given
$q \rightarrow s$	given
q	
s	
t	
$s \wedge t$	
r	
$\sim r$	
$r \wedge \sim r$	
p	

Using the rules of inference, find conclusions for the following:

33. If you do not brush your teeth, then the dentist is angry.
    Angry dentists drive too fast.
    My dentist drives slowly.

34. Anteaters do not like to fish.
    Anyone who is jovial likes to fish.
    All policemen are jovial.

35. All Martians are vegetarians.
    Vegetarians do not eat rodents.
    All squirrels are rodents.

36. People who believe in the tooth fairy are either foolish or under 10 years of age.
    Charlie believes in the tooth fairy.
    Charlie is 30 years old.

37. Sharks cannot play golf.
    Gary Player is an excellent golfer.
    Only sharks are dangerous fish.

38. All mermaids are strange.
    They read fiction only if they are literate.
    Those who don't read fiction are not strange.
    All literate people have been to college.

39. Bears cannot count.
    Those who are not honest can become politicians.
    Those who are worthy can count.
    Only if they are worthy then they are honest.

40. Smoking is sufficient for being unhealthy.
    If one swims, then one is healthy.
    Only if one is a fish then one is in a school.
    To be a fish, it is necessary to swim.

41. *To wear a bow tie, it is necessary to attend a party.
    Those who do not live on the coast do not act silly.
    To look intelligent, one must wear a bow tie.

No bassoon players chew gum.
People who attend parties act silly.
People who do not chew gum look intelligent.

42. *Artists only attend concerts on Saturday night.
Jenkins never drives to a concert unless Brahms is played.
No sober person drives in heavy traffic.
If the traffic is not heavy, it is not Saturday night.
People who listen to Brahms are sober drivers.

43. Use logically equivalent propositions and other rules of inference to derive the rule of modus tollens.

44. Prove that there is no largest even positive integer.

45. *Prove that there is an infinite number of prime numbers.

46. Prove that $\sqrt{3}$ is not a rational number.

1.5 Completeness in Propositional Logic

In addition to logic, one of the important uses of truth tables is in the design of circuit diagrams. Before beginning the study of circuit diagrams, let us consider the minimal number of logical connectives actually needed to express any proposition formed by the connectives we have now defined. We know that $p \leftrightarrow q$ can be expressed as $(p \rightarrow q) \wedge (q \rightarrow p)$ so that $\leftrightarrow$ is convenient but not necessary. In addition $p \veebar q$ is equivalent to

$$(p \wedge \sim q) \vee (\sim p \wedge q)$$

Also $p \rightarrow q$ is equivalent to $\sim p \vee q$ so we do not need $\rightarrow$ if we have $\sim$ and $\vee$. Also $p \wedge q$ is equivalent to $\sim(\sim p \vee \sim q)$ and $p \vee q$ is equivalent to $\sim(\sim p \wedge \sim q)$. Hence, only the connectives $\sim$ and $\wedge$ or $\sim$ and $\vee$ are needed to express any proposition, and in either case both connectives are needed. There are, however, two connectives each of which has the property that any proposition can be written using only the single connective. These are $|$, called the **Sheffer stroke** and $\downarrow$, called **Peirce's arrow**. They were named after mathematicians H. M. Sheffer and C. S. Peirce. These connectives have the following truth tables:

Case	p	q	$p \mid q$	Case	p	q	$p \downarrow q$
1	T	T	F	1	T	T	F
2	T	F	T	2	T	F	F
3	F	T	T	3	F	T	F
4	F	F	T	4	F	F	T

To show that the connective $|$ can replace any connective, it is only necessary to show that $\sim$ and $\wedge$ or $\sim$ and $\vee$ can be expressed using only $|$ since we have seen that any connective can be expressed using either of these pairs of connectives. From the truth tables

Case	p	q	p	$\mid$	p
1	T	T	T	F	T
2	T	F	T	F	T
3	F	T	F	T	F
4	F	F	F	T	F
				*	

and

Case	p	q	$(p$	$\mid$	$p)$	$\mid$	$(q$	$\mid$	$q)$
1	T	T	T	F	T	T	T	F	T
2	T	F	T	F	T	T	T	F	T
3	F	T	F	T	F	T	T	F	T
4	F	F	F	T	F	F	F	T	F
						*			

we see that $p \mid p$ is equivalent to $\sim p$ and $(p \mid p) \mid (q \mid q)$ is equivalent to $p \vee q$. It can also be shown that $(p \mid q) \mid (p \mid q)$ is equivalent to $p \wedge q$.

Similarly, if we show that $\sim$ and $\wedge$ or $\sim$ and $\vee$ can be expressed using only $\downarrow$, then any connective can be expressed using only $\downarrow$. It is left to the reader to show that $p \downarrow p$ is equivalent to $\sim p$, $(p \downarrow p) \downarrow (q \downarrow q)$ is equivalent to $p \wedge q$, and $(p \downarrow q) \downarrow (p \downarrow q)$ is equivalent to $p \vee q$. Notice that $p \mid q$ is equivalent to $\sim(p \wedge q)$ and $p \downarrow q$ is equivalent to $\sim(p \vee q)$. Hence in the future we will refer to the | connective as **nand** and the $\downarrow$ connective as **nor**.

Assume we are given an arbitrary truth table. There is a straightforward way to find a proposition that has this truth table. For example, suppose we have the truth table

Case	p	q	
1	T	T	T
2	T	F	T
3	F	T	F
4	F	F	T

We know that $p \wedge q$ is true in case 1 and false in all the other cases. Similarly, $p \wedge \sim q$ is true only in case 2, $\sim p \wedge q$ is true only in case 3, and $\sim p \wedge \sim q$ is true only in case 4. If, for each case that we want to be true, we select the proposition that is true only in that case and connect these propositions with $\vee$, then we will have a statement that is true only for the desired cases. In the preceding example the proposition

$$(p \wedge q) \vee (p \wedge \sim q) \vee (\sim p \wedge \sim q)$$

has the desired truth table.

For a truth table with three variables we have an analogous situation. For each case in the following table, the proposition that is true only in that case is given.

Case	p	q	r	
1	T	T	T	$p \wedge q \wedge r$
2	T	T	F	$p \wedge q \wedge \sim r$
3	T	F	T	$p \wedge \sim q \wedge r$
4	T	F	F	$p \wedge \sim q \wedge \sim r$
5	F	T	T	$\sim p \wedge q \wedge r$
6	F	T	F	$\sim p \wedge q \wedge \sim r$
7	F	F	T	$\sim p \wedge \sim q \wedge r$
8	F	F	F	$\sim p \wedge \sim q \wedge \sim r$

Notice that in each case a variable is negated in the corresponding proposition only if it is listed as false for that particular case. If we wish to have a proposition with a particular truth table, we select the expressions corresponding to the cases that are true and connect them with $\vee$. For example, to construct a proposition with the truth table

Case	p	q	r	
1	T	T	T	T
2	T	T	F	T
3	T	F	T	F
4	T	F	F	F
5	F	T	T	T
6	F	T	F	F
7	F	F	T	F
8	F	F	F	T

we would form the proposition

$$(p \wedge q \wedge r) \vee (p \wedge q \wedge \sim r) \vee (\sim p \wedge q \wedge r) \vee (\sim p \wedge \sim q \wedge \sim r)$$

To construct a proposition with the truth table

Case	p	q	r	
1	T	T	T	F
2	T	T	F	T
3	T	F	T	T
4	T	F	F	F
5	F	T	T	F
6	F	T	F	T
7	F	F	T	F
8	F	F	F	F

we form the proposition

$$(p \wedge q \wedge \sim r) \vee (p \wedge \sim q \wedge r) \vee (\sim p \wedge q \wedge \sim r)$$

This form of expression for a proposition is called **disjunctive normal form**. The expressions $p \wedge q \wedge \sim r$, $p \wedge \sim q \wedge r$, and $\sim p \wedge q \wedge \sim r$ are called **minterms**. More formally, we have the following definitions.

DEFINITION 1.11

> If simple statements $p_1, p_2, p_3, \ldots, p_n$ are considered, then $x_1 \wedge x_2 \wedge x_3 \wedge \cdots \wedge x_n$, where $x_i = p_i$ or $\sim p_i$, is called a **minterm**. An expression that is the disjunction of minterms is in **disjunctive normal form** so that if $m_1, m_2, m_3, \ldots, m_n$ are minterms, then $m_1 \vee m_2 \vee m_3 \vee \cdots \vee m_n$ is in disjunctive normal form.

Although every proposition can be expressed in disjunctive normal form, it is usually not the simplest form of the expression. In the next section, we will study Karnaugh maps, which allow us to simplify expressions of propositions in disjunctive normal form.

In similar fashion, we note that $p \vee q \vee r$ is false only where p, q, and r are all false. In general, in the table

Case	p	q	r	
1	T	T	T	$\sim p \vee \sim q \vee \sim r$
2	T	T	F	$\sim p \vee \sim q \vee r$
3	T	F	T	$\sim p \vee q \vee \sim r$
4	T	F	F	$\sim p \vee q \vee r$
5	F	T	T	$p \vee \sim q \vee \sim r$
6	F	T	F	$p \vee \sim q \vee r$
7	F	F	T	$p \vee q \vee \sim r$
8	F	F	F	$p \vee q \vee r$

each expression is false on the line where it is located and true elsewhere. If we then want to find a statement having a given truth table, for each case where the truth table is false, we take the statement using that line on the preceding table corresponding to that case and connect these statements with $\wedge$. For example, to construct a proposition with the truth table

Case	p	q	r	
1	T	T	T	T
2	T	T	F	T
3	T	F	T	F
4	T	F	F	F
5	F	T	T	T
6	F	T	F	F
7	F	F	T	F
8	F	F	F	T

we would form the proposition

$$(\sim p \vee q \vee \sim r) \wedge (\sim p \vee q \vee r) \wedge (p \vee \sim q \vee r) \wedge (p \vee q \vee \sim r)$$

This form of expression for a proposition is called **conjunctive normal form**. The expressions $\sim p \vee q \vee \sim r$, $\sim p \vee q \vee r$, $p \vee \sim q \vee r$, and $p \vee q \vee \sim r$ are called **maxterms**. More formally, we have the following definitions.

DEFINITION 1.12 If simple statements $p_1, p_2, p_3, \ldots, p_n$ are considered, then $x_1 \vee x_2 \vee x_3 \vee \ldots \vee x_n$, where $x_i = p_i$ or $\sim p_i$, is called a **maxterm**. An expression that is the conjunction of miniterms is in **conjunctive normal form** so that if $m_1, m_2, m_3, \ldots, m_n$ are maxterms, then $m_1 \wedge m_2 \wedge m_3 \wedge \ldots \wedge m_n$ is in conjunctive normal form.

Exercises

Use the truth table method to establish the following equivalences:

1. $p \downarrow p$ is equivalent to $\sim p$.
2. $(p \downarrow p) \downarrow (q \downarrow q)$ is equivalent to $p \wedge q$.
3. $(p \downarrow q) \downarrow (p \downarrow q)$ is equivalent to $(p \vee q)$.

Find propositions in disjunctive normal form which have the following truth tables:

4.
Case	p	q	r	
1	T	T	T	T
2	T	T	F	T
3	T	F	T	F
4	T	F	F	F
5	F	T	T	F
6	F	T	F	T
7	F	F	T	F
8	F	F	F	T

5.
Case	p	q	r	
1	T	T	T	T
2	T	T	F	F
3	T	F	T	T
4	T	F	F	F
5	F	T	T	F
6	F	T	F	T
7	F	F	T	F
8	F	F	F	T

6.
Case	p	q	r	
1	T	T	T	T
2	T	T	F	F
3	T	F	T	T
4	T	F	F	T
5	F	T	T	F
6	F	T	F	F
7	F	F	T	F
8	F	F	F	T

Find propositions which have the following truth tables:

7.
Case	p	q	r	
1	T	T	T	F
2	T	T	F	T
3	T	F	T	T
4	T	F	F	T
5	F	T	T	F
6	F	T	F	F
7	F	F	T	F
8	F	F	F	T

8.
Case	p	q	r	
1	T	T	T	F
2	T	T	F	F
3	T	F	T	F
4	T	F	F	F
5	F	T	T	T
6	F	T	F	T
7	F	F	T	F
8	F	F	F	T

1.6 Karnaugh Maps

9.

Case	p	q	r	
1	T	T	T	T
2	T	T	F	F
3	T	F	T	T
4	T	F	F	T
5	F	T	T	F
6	F	T	F	F
7	F	F	T	T
8	F	F	F	F

Find propositions in conjunctive normal form that have the following truth tables:

10.

Case	p	q	r	
1	T	T	T	T
2	T	T	F	F
3	T	F	T	F
4	T	F	F	T
5	F	T	T	F
6	F	T	F	T
7	F	F	T	T
8	F	F	F	F

11.

Case	p	q	r	
1	T	T	T	T
2	T	T	F	T
3	T	F	T	F
4	T	F	F	T
5	F	T	T	F
6	F	T	F	T
7	F	F	T	T
8	F	F	F	F

12.

Case	p	q	r	
1	T	T	T	T
2	T	T	F	F
3	T	F	T	F
4	T	F	F	T
5	F	T	T	F
6	F	T	F	T
7	F	F	T	F
8	F	F	F	T

13.

Case	p	q	r	
1	T	T	T	F
2	T	T	F	T
3	T	F	T	F
4	T	F	F	T
5	F	T	T	F
6	F	T	F	T
7	F	F	T	T
8	F	F	F	T

14.

Case	p	q	r	
1	T	T	T	T
2	T	T	F	T
3	T	F	T	F
4	T	F	F	F
5	F	T	T	T
6	F	T	F	T
7	F	F	T	F
8	F	F	F	T

15.

Case	p	q	r	
1	T	T	T	T
2	T	T	F	F
3	T	F	T	T
4	T	F	F	T
5	F	T	T	T
6	F	T	F	F
7	F	F	T	T
8	F	F	F	F

16. Find an equivalent statement to $p \rightarrow q$ using only Peirce's arrows.

17. Find an equivalent statement to $p \leftrightarrow q$ using only Peirce's arrows.

18. Find an equivalent statement to $p \veebar q$ using only Peirce's arrows.

19. Find an equivalent statement to $p \rightarrow q$ using only Sheffer strokes.

20. Find an equivalent statement to $p \leftrightarrow q$ using only Sheffer strokes.

21. Find an equivalent statement to $p \veebar q$ using only Sheffer strokes.

1.6 Karnaugh Maps

If simple statements $p_1, p_2, p_3, \ldots$ and p_n are considered, then there are 2^n distinct minterms. (We shall show this in Chapter 8.) For example, for statements p and q, there are minterms $p \wedge q$, $p \wedge \sim q$, $\sim p \wedge q$, and $\sim p \wedge \sim q$. A **Karnaugh map** is a table in which each entry represents a minterm. For example, for statements p and q, a Karnaugh map would have the form in Figure 1.1 where the boxes inside represent the minterms in Figure 1.2.

	q	~q
p		
~p		

Figure 1.1

34 Chapter 1 Truth Tables, Logic, and Proofs

	q	$\sim q$
p	$p \wedge q$	$p \wedge \sim q$
$\sim p$	$\sim p \wedge q$	$\sim p \wedge \sim q$

Figure 1.2

	q	$\sim q$
p	×	
$\sim p$		×

Figure 1.3

	q	$\sim q$
p	×	×
$\sim p$		

Figure 1.4

To represent a statement in disjunctive normal form in a Karnaugh map, we place an × in the boxes representing minterms in the statement. For example, the statement

$$(p \wedge q) \vee (\sim p \wedge \sim q)$$

has the Karnaugh map in Figure 1.3.

Notice that if the Karnaugh map for a statement has two ×'s adjacent in the same row or column, then the expression can be simplified so that two minterms are reduced to one term that contains one less statement (i.e., either p or q does not occur in the expression). For example, the statement $(p \wedge q) \vee (p \wedge \sim q)$, which has the Karnaugh map in Figure 1.4, is equivalent to the statement p because $(p \wedge q) \vee (p \wedge \sim q) \equiv p \wedge (q \vee \sim q) \equiv p \wedge T \equiv p$.

A Karnaugh map for p, q, and r may have the form in Figure 1.5 where the boxes inside represent the minterms in Figure 1.6.

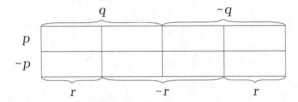

Figure 1.5

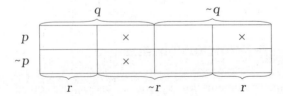

Figure 1.6

Hence, the statement

$$(p \wedge q \wedge \sim r) \vee (p \wedge \sim q \wedge r) \vee (\sim p \wedge q \wedge \sim r)$$

would have the Karnaugh map in Figure 1.7.

Figure 1.7

Again, since two checks are adjacent, the two minterms represented by the checks may be reduced to one minterm with one less of the statements p, q, and r. In this case, $(p \wedge q \wedge \sim r) \vee (\sim p \wedge q \wedge \sim r)$ is reduced to $(q \wedge \sim r)$, so the expression becomes $(q \wedge \sim r) \vee (p \wedge \sim q \wedge r)$.

1.6 Karnaugh Maps

If four checks occur together in a rectangle such as shown in Figure 1.8

Figure 1.8

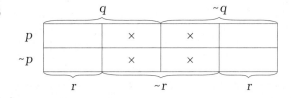

or in Figure 1.9,

Figure 1.9

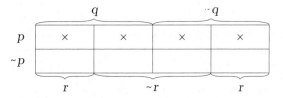

then the four minterms represented by the checks can be reduced to one term containing only one of the statements p, q, and r. For example, the first Karnaugh map represents

$$(p \wedge q \wedge \sim r) \vee (p \wedge \sim q \wedge \sim r) \vee (\sim p \wedge q \wedge \sim r) \vee (\sim p \wedge \sim q \wedge \sim r)$$

which reduces to $\sim r$.

The second Karnaugh map represents

$$(p \wedge q \wedge r) \vee (p \wedge q \wedge \sim r) \vee (p \wedge \sim q \wedge \sim r) \vee (p \wedge \sim q \wedge r)$$

which reduces to p.

Since r appears at both ends of the Karnaugh map, the map "wraps around" so that the checks in the Karnaugh map in Figure 1.10 are considered to form a rectangle of four checks so that the expression

$$(p \wedge q \wedge r) \vee (\sim p \wedge q \wedge r) \vee (p \wedge \sim q \wedge r) \vee (\sim p \wedge \sim q \wedge r)$$

reduces simply to r.

Figure 1.10

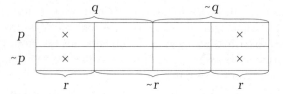

The expression

$$(p \wedge q \wedge r) \vee (\sim p \wedge q \wedge r) \vee (p \wedge q \wedge \sim r) \vee (\sim p \wedge q \wedge \sim r) \vee (p \wedge \sim q \wedge \sim r)$$

represented by Figure 1.11 can be reduced to $q \vee (p \wedge \sim q \wedge \sim r)$, using the right four-block of checks. But it can be further reduced to $q \vee (p \wedge \sim r)$ by using the two-block of checks in the middle of row 1.

We now list four steps for using a Karnaugh map:

1. For each minterm, check the appropriate box in the map.

2. Cover the checks using as few rectangular boxes as possible.

3. Make the boxes as large as possible without changing the number of boxes.

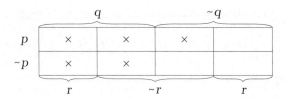

Figure 1.11

4. Evaluate the boxes with the corresponding proposition that describes it and connect these propositions with the ∨ symbol.

We will illustrate this process as we construct a Karnaugh map for four statements p, q, r, and s in Figure 1.12 where the boxes inside represent minterms composed of the margin statements that pertain to the row and column of the boxes. For example, in the second row and third column, the corresponding margin statements are $p, \sim q, \sim r$, and $\sim s$, which give rise to the minterm $p \wedge \sim q \wedge \sim r \wedge \sim s$.

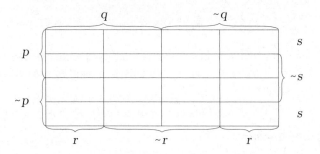

Figure 1.12

Consider the expression

$$(p \wedge q \wedge r \wedge s) \vee (p \wedge q \wedge \sim r \wedge s) \vee (p \wedge q \wedge r \wedge \sim s)$$
$$\vee (p \wedge q \wedge \sim r \wedge \sim s) \vee (\sim p \wedge q \wedge r \wedge \sim s) \vee (\sim p \wedge q \wedge \sim r \wedge \sim s)$$
$$\vee (\sim p \wedge q \wedge r \wedge s) \vee (\sim p \wedge q \wedge \sim r \wedge s) \vee (p \wedge \sim q \wedge \sim r \wedge s)$$
$$\vee (p \wedge \sim q \wedge r \wedge s) \vee (p \wedge \sim q \wedge r \wedge \sim s)$$

Placing the corresponding checks in the Karnaugh map, we have the map in Figure 1.13, which can be covered by an eight-block, a four-block, and a two-block. An eight-block can be described by using only one of the statements p, q, r, or s. In this case, the eight-block is described by q. A four-block can be described using only two statements. In this case, the four-block is described by $p \wedge s$. A two-block can be described using three statements. In this case the two-block is described by $p \wedge \sim q \wedge r$. However, it is simpler to use the four-block $p \wedge r$. Hence, the original statement can be simplified to $q \vee (p \wedge s) \vee (p \wedge r)$. Since r occurs at both ends of the row and s occurs at both ends of the column, the Karnaugh map "wraps around" both ways so the top is considered adjacent to the bottom and the left side is considered adjacent to the right side.

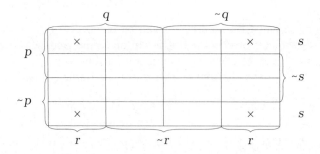

Figure 1.13

1.6 Karnaugh Maps

Figure 1.14

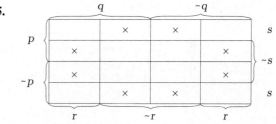

For example, in the Karnaugh map in Figure 1.14, the four-block is described by $r \wedge s$.

Exercises

Simplify the propositions expressed by the following Karnaugh maps:

1.

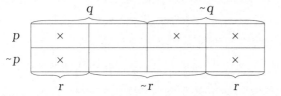

2.

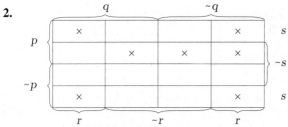

3.

4.

5.

6.

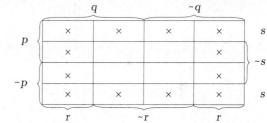

Use Karnaugh maps to simplify the following expressions:

7. $(p \wedge q \wedge r) \vee (\sim p \wedge q \wedge r) \vee (\sim p \wedge q \wedge \sim r) \vee$
$(\sim p \wedge \sim q \wedge \sim r) \vee (p \wedge q \wedge \sim r)$

8. $(p \wedge q \wedge \sim r \wedge s) \vee (p \wedge \sim q \wedge \sim r \wedge s)$
$\vee (p \wedge q \wedge \sim r \wedge \sim s) \vee (p \wedge \sim q \wedge \sim r \wedge \sim s)$
$\vee (\sim p \wedge q \wedge \sim r \wedge \sim s) \vee (\sim p \wedge q \wedge \sim r \wedge s)$
$\vee (\sim p \wedge \sim q \wedge \sim r \wedge s) \vee (\sim p \wedge q \wedge r \wedge \sim s)$
$\vee (\sim p \wedge \sim q \wedge r \wedge \sim s)$

9. $(p \wedge q \wedge r \wedge s) \vee (\sim p \wedge q \wedge r \wedge s)$
$\vee (p \wedge \sim q \wedge \sim r \wedge \sim s) \vee (p \wedge \sim q \wedge r \wedge \sim s)$
$\vee (\sim p \wedge \sim q \wedge \sim r \wedge \sim s) \vee (\sim p \wedge \sim q \wedge r \wedge \sim s)$

10. $(p \wedge q \wedge r) \vee (\sim p \wedge q \wedge \sim r) \vee (\sim p \wedge q \wedge r)$
$\vee (p \wedge \sim q \wedge \sim r) \vee (\sim p \wedge \sim q \wedge r)$

11. $(p \wedge q \wedge {\sim}r \wedge s) \vee (p \wedge {\sim}q \wedge {\sim}r \wedge s)$
$\vee ({\sim}p \wedge q \wedge r \wedge s) \vee ({\sim}p \wedge q \wedge {\sim}r \wedge s)$
$\vee ({\sim}p \wedge {\sim}q \wedge {\sim}r \wedge s) \vee ({\sim}p \wedge {\sim}q \wedge r \wedge s)$

12. $(p \wedge q \wedge r \wedge s) \vee (p \wedge q \wedge r \wedge {\sim}s)$
$\vee ({\sim}p \wedge q \wedge r \wedge {\sim}s) \vee (p \wedge {\sim}q \wedge r \wedge s)$
$\vee (p \wedge {\sim}q \wedge r \wedge {\sim}s) \vee ({\sim}p \wedge {\sim}q \wedge r \wedge {\sim}s)$

1.7 Circuit Diagrams

It has become common to express the corresponding propositions for circuit diagrams using the notation for Boolean algebra, which will be described in Section 2.4. Hence, before beginning the study of circuit diagrams, we will convert from logical notation to Boolean notation. The symbols $\wedge$, $\vee$, and $\sim$ are replaced respectively by $\cdot$, $+$, and $'$. Hence, $(p \wedge q) \vee {\sim}r$ becomes $(p \cdot q) + r'$ and

$$(p \wedge q \wedge {\sim}r) \vee (p \wedge {\sim}q \wedge r) \vee ({\sim}p \wedge q \wedge {\sim}r)$$

becomes

$$(p \cdot q \cdot r') + (p \cdot q' \cdot r) + (p' \cdot q \cdot r')$$

As in regular algebra, the product symbol is often suppressed and the product is assumed to be performed before addition so that the preceding expression can be written as

$$pqr' + pq'r + p'qr'$$

In truth tables T is changed to 1 and F to 0 so that the truth table

Case	p	q	r	p	$\vee$	((~	q)	$\wedge$	r)
1	T	T	T	T	T	F	T	F	T
2	T	T	F	T	T	F	T	F	F
3	T	F	T	T	T	T	F	T	T
4	T	F	F	T	T	T	F	F	F
5	F	T	T	F	F	F	T	F	T
6	F	T	F	F	F	F	T	F	F
7	F	F	T	F	T	T	F	T	T
8	F	F	F	F	F	T	F	F	F

becomes

Case	p	q	r	p	+	(q'	$\cdot$	r)
1	1	1	1	1	1	0	0	1
2	1	1	0	1	1	0	0	0
3	1	0	1	1	1	1	1	1
4	1	0	0	1	1	1	0	0
5	0	1	1	0	0	0	0	1
6	0	1	0	0	0	0	0	0
7	0	0	1	0	1	1	1	1
8	0	0	0	0	0	1	0	0

Battery

Bulb

In 1938, C. E. Shannon discovered the relationship between truth tables and electrical circuits. Consider the switching circuit in Figure 1.15, where there are circuit components representing a battery and a lightbulb. Assign the value 1 to the switches p and q if they are closed (i.e., electricity can pass through them). Otherwise assign the value 0. Assign the value 1 to the circuit if the lightbulb lights up (i.e., electricity passes through the lightbulb). Notice that when the circuits p and

Figure 1.15

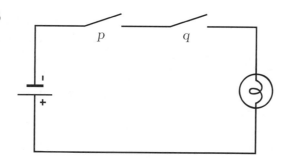

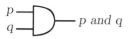

Figure 1.16

q are in series, as they are in the foregoing circuit, the lightbulb can light up so that the value of the circuit equals 1 only if both the p and q switches are closed so that both p and q have the value 1. Thus, this circuit corresponds to the statement $p \cdot q$. We call this arrangement of the switches the *p and q* gate and denote this gate by the symbol in Figure 1.16.

Now consider the switching circuit in Figure 1.17 where the switches p and q are in parallel.

Figure 1.17

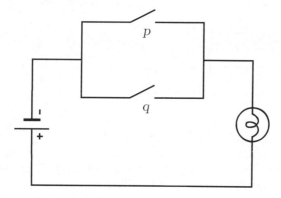

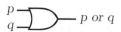

Figure 1.18

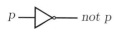

Figure 1.19

Note that the lightbulb lights up so that the value of the circuit equals 1 if either of the switches p or q is closed so that either the value of $p = 1$ or $q = 1$ (or both). This circuit corresponds to the statement $p + q$. We call this arrangement of the switches the *p or q* gate. This gate is denoted by the symbol in Figure 1.18.

Suppose there is a circuit (which we will not attempt to draw) with one switch p that has the property that the lightbulb lights up if and only if p is open. Hence, the circuit has the value 1 when p has the value 0 and has the value 0 when p has the value 1. This circuit corresponds to p' and the gate is called *not p* or the *negation gate*. It is denoted by the symbol in Figure 1.19.

EXAMPLE 1.13

The circuit in Figure 1.20 consists of the gate *p and q* followed by the negation gate so that it corresponds to $(p \cdot q)'$. Notice that the negation gate negates the entire circuit preceding it. ∎

EXAMPLE 1.14

The circuit in Figure 1.21 consists of the *p or q* gate connected to *not r* with the and gate. Hence, it corresponds to $(p + q) \cdot r'$. ∎

EXAMPLE 1.15

The Boolean expression for the circuit in Figure 1.22 is $(p' \cdot q) + (p \cdot r')'$. ∎

40 Chapter 1 Truth Tables, Logic, and Proofs

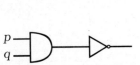

Figure 1.20

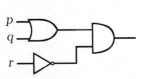

Figure 1.21

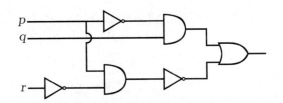

Figure 1.22

> **EXAMPLE 1.16** The circuit diagram corresponding to $(p' \cdot q) + r$ is shown in Figure 1.23.

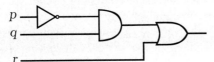

Figure 1.23

> **EXAMPLE 1.17** The circuit diagram corresponding to $((p+q)' \cdot (p+r)) + r'$ is shown in Figure 1.24.

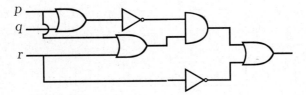

Figure 1.24

> **EXAMPLE 1.18** To design a three-way switch for a light, that is, a light that can be turned on by three different on-off switches, we first consider the Boolean expression for the circuit. We want the light to come on when all three switches are on, so we want pqr. If one of the switches is turned off, then we want the circuit to be off. However, if another switch is turned off, we want the switch to come on. Hence, the expression we are seeking is $pqr + pq'r' + p'q'r + p'qr'$. To simplify the circuit, instead of using the circuit in Figure 1.25, for pqr, we shall use the circuit in Figure 1.26.
>
> Instead of using the circuit in Figure 1.27, for $p + q + r$, we shall use the circuit in Figure 1.28.
>
> The circuit for this expression is then shown in Figure 1.29.

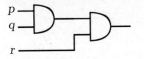

Figure 1.25

Figure 1.26

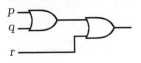

Figure 1.27

It was noted earlier that the Sheffer stroke denoted by | had the same truth table as $(pq)'$ (using Boolean notation) and, hence, was referred to as **nand** and that Peirce's arrow, denoted by ↓, had the same truth table as $(p+q)'$ and, hence, was referred to as **nor**. For this reason, nand and nor have the symbols shown in Figures 1.30 and 1.31.

Therefore, the expression $(p|q) \downarrow (p|r)$ would have the circuit in Figure 1.32.

> **EXAMPLE 1.19** The **half-adder** adds two binary numbers 1 and 0, which have the following addition table:

Figure 1.28

+	0	1
0	0	1
1	1	10

Figure 1.29
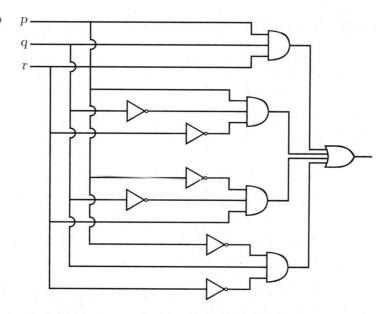

Figure 1.30

Figure 1.31

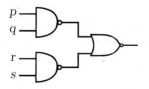

Figure 1.32

It is called the half-adder because if we wish to add two binary numbers having more than one digit (which will be discussed in a later chapter), we can only add the rightmost digits since we could not add a number that is "carried." It is convenient to have the sum of two single-digit binary numbers so that our foregoing addition table is changed to

+	0	1
0	00	01
1	01	10

If we let p and q represent the numbers to be added and d_1, d_0 be the first and second digits of the sum, then we have the following truth tables:

Case	p	q	d_0
1	1	1	0
2	1	0	1
3	0	1	1
4	0	0	0

Case	p	q	d_1
1	1	1	1
2	1	0	0
3	0	1	0
4	0	0	0

Hence, $d_0 \equiv pq' + p'q$, which is equivalent to $(p+q) \cdot (pq)'$. This can be shown using truth tables or using the equivalence laws (see Exercises). Also $d_1 \equiv pq$. The circuit diagram for the half-adder is shown in Figure 1.33.

Since the half-adder is the sum of two numbers, we shall denote it by the symbol in Figure 1.34. ∎

Figure 1.33

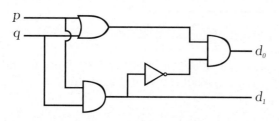

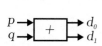

Figure 1.34

EXAMPLE 1.20

The **full-adder** adds three single digit binary numbers. Hence, it can add two binary numbers and a number that has been "carried." At this point we need only consider it as adding three single-digit binary numbers. If we let p, q, and r represent the single-digit binary numbers to be added and $d_1^\#, d_0^\#$ be the first and second digits of the sum, then we have the following truth tables:

Case	p	q	r	$d_1^\#$	$d_0^\#$
1	1	1	1	1	1
2	1	1	0	1	0
3	1	0	1	1	0
4	1	0	0	0	1
5	0	1	1	1	0
6	0	1	0	0	1
7	0	0	1	0	1
8	0	0	0	0	0

$d_0^\#$ is really the result of adding the d_0 in the sum of p and q to r. Hence, its circuit is easily described. The value of $d_1^\#$ is the result of taking d_0 of the d_1 obtained from adding r to the d_0 in the sum of p and q, and d_1 obtained from adding p and q. Much more simply, using the preceding truth table,

$$d_1^\# \equiv pqr + pqr' + pq'r + p'qr$$

and using the Karnaugh map on this expression for $d_1^\#$, we have the map shown here:

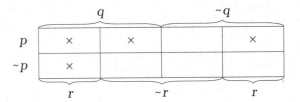

Thus, $d_1^\# \equiv pq + pr + qr \equiv pq + (p+q)r$ so that the circuit may be expressed as shown in Figure 1.35,

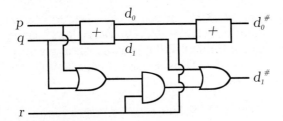

Figure 1.35

or in more detail as shown in Figure 1.36. Since the full-adder adds three numbers, it will have the symbol shown in Figure 1.37. ∎

Figure 1.36

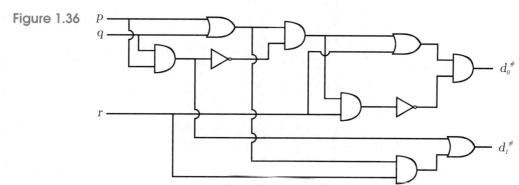

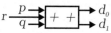

Figure 1.37

Exercises

Convert the following propositions to Boolean algebra notation:

1. $(p \wedge q) \wedge (q \vee \sim r)$
2. $(q \vee \sim r) \wedge (\sim p \vee r) \wedge (q \vee \sim r \vee s)$
3. $\sim(p \wedge \sim q \wedge r) \vee (\sim q \wedge r) \vee (p \wedge \sim q \wedge \sim r)$
4. $\sim(\sim p \vee (q \wedge \sim r))$
5. $(p \wedge r) \vee (p \wedge \sim q) \wedge (p \vee r \vee \sim s)$

Convert the following propositions to Boolean algebra notation:

6. $(p \vee q \vee \sim r) \wedge \sim(r \vee \sim q)$
7. $(\sim q \wedge r) \vee \sim(p \wedge r)$ 8. $\sim((p \wedge r) \vee \sim q)$
9. $(p \wedge q \wedge r) \vee (\sim p \wedge q \wedge \sim r) \vee (\sim p \wedge q \wedge r)$
   $\vee (\sim p \wedge q \wedge \sim r)$
10. $(p \wedge \sim q \wedge \sim r \wedge s) \vee (\sim p \wedge q \wedge r \wedge s)$
    $\vee (\sim p \wedge q \wedge \sim r \wedge s)$

Give the Boolean expressions corresponding to the circuit diagrams:

11.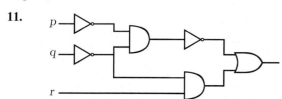

12.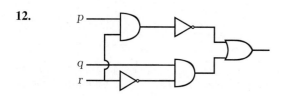

13.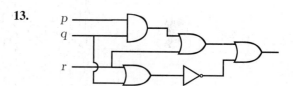

14.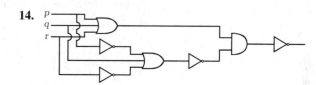

15.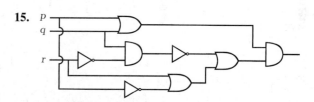

Give the Boolean expressions corresponding to the circuit diagrams:

16.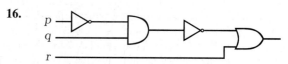

17.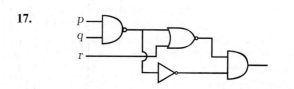

18.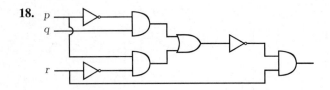

19.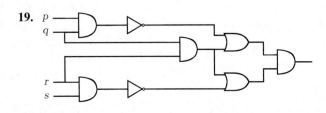

20.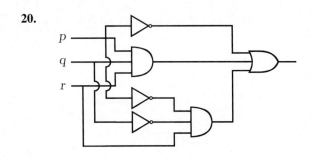

Construct the circuit diagrams corresponding to the following Boolean expressions:

21. $(p' + q)(p + (qr))$ **22.** $(pq') + ((qr') + (p'r))$

23. $(q'r') + ((pq)'(q'r))$

24. $(q(r + s))((p'q') + (qr'))$

25. $(pq) + ((pq'r') + r)'$ **26.** $(pq')' + (qr)'$

27. $((p + q)z)'$

28. $((p'(q + p'r')) + (pq' + r)$

29. $(p'q + (q'r))' + ps'$ **30.** $((pq')' + (r's))(p + r')$

31. The city council contains five members. Each member has a yes button and a no button for voting. An issue passes if the majority vote for it. Construct a circuit diagram which would determine whether or not an issue passes by having a light come on if the issue passes. *Hint*: Construct circuits using switches in series and parallel as shown at the beginning of the section.

32. The city council contains five members including a chairperson. Each member has a yes button and a no button for voting. An issue passes if the majority vote for it. However, the chairperson votes only if there is a tie vote. Construct a circuit diagram that would determine whether or not an issue passes by having a light come on if the issue passes. *Hint*: Construct circuits using switches in series and parallel as shown at the beginning of the section.

33. The city council contains five members including a chairperson. Each member has a yes button and a no button for voting. An issue passes if the majority votes for it except that the chairperson has veto power. Construct a circuit diagram that would determine whether or not an issue passes by having a light come on if the issue passes.

34. A light circuit has four off-on switches. Construct a circuit that enables the light to be turned off or on at each switch.

For the following switching circuits, find the corresponding circuit output table and find the Boolean expression in disjunctive normal form describing the circuit. Simplify the Boolean expression using Karnaugh maps and give the simplified circuit.

35.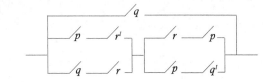

36.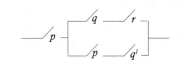

37.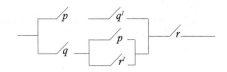

38.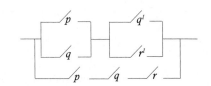

39.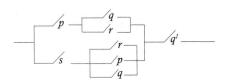

40.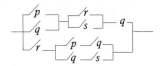

Find switching circuits for each of the following Boolean expressions:

41. $(p + q)(q' + pr)$ **42.** $(ps + qt)(rq + ps)$

43. $(p(q+p'r'))+(pr+q)$ **44.** $(pq+rs)(q+p)$

45. $((pq)(p'q+r))+(r+s)$

For each of the following circuit output tables, use Karnaugh maps to find a simplified Boolean expression to describe the table. Then draw a circuit diagram and a switching circuit of the expression.

46.

p	q	r	
1	1	1	1
1	1	0	0
1	0	1	0
1	0	0	1
0	1	1	1
0	1	0	1
0	0	1	0
0	0	0	1

47.

p	q	r	
1	1	1	0
1	1	0	1
1	0	1	1
1	0	0	0
0	1	1	1
0	1	0	0
0	0	1	1
0	0	0	1

48.

p	q	r	s	
1	1	1	1	0
1	1	1	0	1
1	1	0	1	1
1	1	0	0	0
1	0	1	1	1
1	0	1	0	0
1	0	0	1	1
1	0	0	0	1
0	1	1	1	1
0	1	1	0	1
0	1	0	1	1
0	1	0	0	1
0	0	1	1	1
0	0	1	0	1
0	0	0	1	1
0	0	0	0	1

49.

p	q	r	s	
1	1	1	1	1
1	1	1	0	0
1	1	0	1	1
1	1	0	0	1
1	0	1	1	1
1	0	1	0	0
1	0	0	1	1
1	0	0	0	0
0	1	1	1	1
0	1	1	0	1
0	1	0	1	1
0	1	0	0	1
0	0	1	1	1
0	0	1	0	0
0	0	0	1	1
0	0	0	0	0

GLOSSARY

Chapter 1: Truth Tables, Logic, and Proofs

1.1 Statements and Connectives

Axioms	A set of basic assumptions.
Binary connective	A connective such as "and" or "or" that connects two propositions.
Compound statement	A statement that contains connectives.
Conjunction	A compound statement that contains two simple statements connected by the word "and."
Connectives	Unary or binary operations that change or combine a statement or compound statement.
Disjunction	A compound statement that contains two simple statements connected by the word "or."
Exclusive or	A binary connective meaning p or q but not both.
Negation	The statement that is the denial of a statement.
Proposition	A statement or declarative sentence that may be assigned a true or false value.
Simple statement	A statement that contains no connectives.
Truth table	A table containing all the possible combinations of truth and falsity of the component propositions.
Truth value	The true or false value assigned to a statement.
Unary connective	A connective such as the negation that is applied to only one proposition.

1.2 Conditional Statements

Biconditional statement	A statement, indicated symbolically by $(p \leftrightarrow q)$, which is true only when the statements p and q agree in truth value.
Conditional statement	A statement saying that if one event happens, then another event will happen. This is indicated symbolically by $(p \rightarrow q)$.

1.3 Equivalent Statements

Conditional (basic)	$(p \to q)$, p implies q.
Contrapositive of a conditional	$(\sim q \to \sim p)$. Not q implies not p.
Converse of a conditional	$(q \to p)$. q implies p.
Inverse of a conditional	$(\sim p \to \sim q)$. Not p implies not q.
Logically equivalent	Compound statements that may be expressed differently but are, in fact, true in exactly the same cases (i.e., they have the same truth value).

1.4 Axiomatic Systems: Arguments and Proofs

Addition	$$\frac{p}{\therefore p \vee q}$$
Argument	A collection of statements called hypotheses and a statement called a conclusion.
Case elimination	$$\begin{array}{c} p \vee q \\ p \to (r \wedge \sim r) \\ \hline \therefore q \end{array}$$
Cases	$$\begin{array}{c} p \\ p \to (r \vee s) \\ r \to q \\ s \to q \\ \hline \therefore q \end{array}$$
Conjunction	$$\begin{array}{c} p \\ q \\ \hline \therefore p \wedge q \end{array}$$
Converse fallacy	$((p \to q) \wedge q) \to p$
Inverse fallacy	$((\sim p \to \sim q) \wedge p) \to q$
Law of detachment	A particular valid argument (also called **modus ponens**): $$\begin{array}{c} p \\ p \to q \\ \underline{q} \\ \therefore I \end{array}$$
Modus tollens	$$\begin{array}{c} p \to q \\ \sim q \\ \hline \therefore \sim p \end{array}$$
Peano axioms	A set of axioms about the positive integers.
Proof	Sequence of statements each of which is: a. True by hypothesis b. An axiom or definition c. A previously proven theorem or proposition d. A statement implied by previous statements as a conclusion of a valid argument e. Logically equivalent to a previous statement

Reductio ad absurdum	Also called the **proof by contradiction**. $$\frac{\sim p \to (r \land \sim r)}{\therefore p}$$
Rules of inference	The rules of logic we use to derive new theorems from axioms, postulates, and previously proved theorems.
Specialization	$$\frac{p \land q}{\therefore p}$$
Syllogism	$$\frac{\begin{array}{c} p \to q \\ q \to r \end{array}}{\therefore p \to r}$$
Theorem	Statements derived (proved) using only fundamental properties (axioms and postulates), previously proved statements, and the rules of logic. True statements about the mathematical system being considered. The theorem is considered to be true because it is "derivable" or "deducible" from previously accepted, or derived, truths.
Valid argument	An argument whose conclusion is true whenever all the hypotheses are true.

1.5 Completeness in Propositional Logic

Conjunctive normal form	An expression that is the conjunction of maxterms is in **conjunctive normal form**, so that if $m_1, m_2, m_3, \ldots m_n$ are maxterms, then $m_1 \land m_2 \land m_3 \land \cdots \land m_n$ is in conjunctive normal form.
Disjunctive normal form	An expression that is the disjunction of minterms is in **disjunctive normal form**, so that if $m_1, m_2, m_3, \ldots m_n$ are minterms, then $m_1 \lor m_2 \lor m_3 \lor \cdots \lor m_n$ is in disjunctive normal form.
Maxterm	If simple statements $p_1, p_2, p_3, \ldots p_n$ are considered, then $x_1 \lor x_2 \lor x_3 \ldots \lor x_n$, where $x_i = p_i$ or $\sim p_i$, is called a **maxterm**.
Minterm	If simple statements $p_1, p_2, p_3, \ldots p_n$ are considered, then $x_1 \land x_2 \land x_3 \ldots \land x_n$, where $x_i = p_i$ or $\sim p_i$, is called a **minterm**.
Pierce's arrow (nor)	Case p q $p \downarrow q$ 1 T T F 2 T F F 3 F T F 4 F F T
Sheffer stroke (nand)	Case p q $p\|q$ 1 T T F 2 T F T 3 F T T 4 F F T

1.6 Karnaugh Maps

Karnaugh map	A Karnaugh map is a table in which each entry represents a minterm. For example, $(p \land q) \lor (\sim p \land \sim q)$ has the Karnaugh map

	q	$\sim q$
p	x	
$\sim p$		x

Karnaugh map usage		1. For each minterm, check the appropriate box in the map. 2. Cover the checks using as few rectangular boxes as possible. 3. Make the boxes as large as possible without changing the number of boxes. 4. Evaluate the boxes with the corresponding proposition that describes it and connect these propositions with the $\vee$ symbol.

1.7 Circuit Diagrams

Full-adder	A circuit that adds three single-digit binary numbers as in the following: | Case | p | q | r | $d_1^\#$ | $d_2^\#$ | |---|---|---|---|---|---| | 1 | 1 | 1 | 1 | 1 | 1 | | 2 | 1 | 1 | 0 | 1 | 0 | | 3 | 1 | 0 | 1 | 1 | 0 | | 4 | 1 | 0 | 0 | 0 | 1 | | 5 | 0 | 1 | 1 | 1 | 0 | | 6 | 0 | 1 | 0 | 0 | 1 | | 7 | 0 | 0 | 1 | 0 | 1 | | 8 | 0 | 0 | 0 | 0 | 0 |
Half-adder	A circuit that adds two binary numbers that have the following addition table: | + | 0 | 1 | |---|---|---| | 0 | 00 | 01 | | 1 | 01 | 10 |
Nand	$(pq)'$
Negation gate	This circuit corresponds to p'. Not p.
Nor	$(p+q)'$

TRUE-FALSE QUESTIONS

1. The conjunction of p and q is true if p is true.
2. The disjunction of p and q is true if either is true.
3. If $p \wedge q$ is true, then $p \vee q$ is true.
4. If p and q are true, then $p \veebar q$ is true.
5. The statement $p \rightarrow q$ is true if p is false and q is true.
6. The statement $\sim (p \rightarrow q)$ has the same truth table as $\sim p \rightarrow \sim q$.
7. If the statement $p \rightarrow q$ is true, then q is true.
8. The converse of the statement "If he drives too fast, then he will get a ticket" is "He will get a ticket if he drives too fast."
9. The inverse of a conditional statement is logically equivalent to the converse of the statement.
10. The contrapositive of the statement "If he does not finish early, then he will miss lunch" is "He will finish early if he does not miss lunch."
11. The inverse of the contrapositive of a conditional statement is the converse of the statement.
12. $\sim (p \leftrightarrow q)$ is equivalent to $p \veebar q$.
13. $p \rightarrow q$ is logically equivalent to $q \vee \sim p$.
14. $(p \wedge (q \vee r)) \leftrightarrow ((p \wedge q) \vee (p \wedge r))$ is a tautology.
15. $p \rightarrow \sim p$ is a contradiction.
16. $\sim (p \wedge q)$ is logically equivalent to $\sim p \wedge \sim q$.
17. "p is necessary for q" is equivalent to "q only if p."
18. If p is necessary for q and p is false, then q must be false.
19. The statement only if p then q is true if p is true and q is false.
20. If p is sufficient for q and p is false, then q must be false.
21. $(p \wedge (q \vee r)) \equiv ((p \vee q) \wedge (p \vee r))$ is one of the distributive laws.

22. The argument

$$\begin{array}{c} p \to q \\ p \\ \hline \therefore q \end{array}$$

is a true argument.

23. If the conclusion of a valid argument is false, then one of the hypotheses is false.

24. The rule of inference

$$\begin{array}{c} p \to q \\ \sim q \\ \hline \therefore \sim p \end{array}$$

is called modus tollens.

25.
$$\begin{array}{c} q \land \sim q \\ \hline \therefore p \end{array}$$

is a valid argument.

26. $(p \downarrow q) \downarrow (p \downarrow q)$ is equivalent to $p \land q$.

27. Every possible truth table in p, q, and r is the truth table of some proposition using only p, q, r, and the connectives $\sim$ and $\land$.

28. $(p \mid q)$ is equivalent to a proposition using p, q, and the connective $\downarrow$.

29. $(p \land q \land \sim r) \lor (p \land \sim q \land r) \lor (\sim p \land q \land r)$ has the truth table $FTTFTFFF$.

30. $p \to (q \to r)$ is logically equivalent to $(p \to q) \to r$.

31. Every proposition is equivalent to a statement that consists of minterms connected by $\lor$.

32. If a truth table has propositions p, q, r, and s, then there are 12 distinct minterms.

33. $(p \land q) \lor (p \land \sim r)$ is expressed in disjunctive normal form.

34. There is a proposition that cannot be expressed in proper disjoint normal form.

35. There are 16 nonequivalent propositions containing p, q, and r.

36. $(p \mid p)$ is logically equivalent to $(p \downarrow p)$.

37. The argument

$$\begin{array}{c} \sim p \to \sim q \\ p \\ \hline \therefore q \end{array}$$

is called the converse fallacy.

38. The argument

$$\begin{array}{c} p \\ p \to r \lor q \\ \sim r \\ \hline \therefore q \end{array}$$

is a valid argument.

39. In a Karnaugh map for p, q, r, and s, a four-block can be described using exactly three of the four variables.

40. In a Karnaugh map for p, q, r, and s, the top row and the bottom row form an eight-block.

41. The simplified expression described by the Karnaugh map

	q	q	$\sim q$	$\sim q$	
p	×	×			s
p	×	×			$\sim s$
$\sim p$		×	×	×	$\sim s$
$\sim p$			×	×	s
	r	$\sim r$	$\sim r$	r	

is $(p \land q) \lor (\sim p \land \sim q) \lor (q \land \sim r \land \sim s) \lor (\sim p \land \sim r \land \sim s)$.

42. The simplified expression described by the Karnaugh map

	q	q	$\sim q$	$\sim q$	
p	×		×	×	s
p	×				$\sim s$
$\sim p$	×				$\sim s$
$\sim p$	×		×	×	s
	r	$\sim r$	$\sim r$	r	

is $(q \land r) \lor (\sim p \land \sim q \land s) \lor (p \land \sim q \land s)$.

43. The simplest expression described by a Karnaugh map is always unique.

44. Using a Karnaugh map, one can always simplify an expression in disjoint normal form.

45. A "blank" Karnaugh map would correspond to a contradiction.

46. It is possible to construct a logical circuit using only nor gates.

47. Karnaugh maps can be used to simplify logical circuits.

48. Switching to Boolean notation, the expression $\sim (p \land \sim q) \lor r$ becomes $(pq')' + r$.

49. The Boolean expression corresponding to

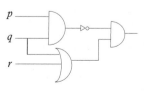

is $(pq')(p+q)$.

50. In a circuit, two switches in sequence correspond to an "or" gate.

SUMMARY QUESTIONS

1. Determine if $\sim (p \leftrightarrow q)$ is logically equivalent to $p \veebar q$.

2. Determine the validity of the argument
$$p \vee q$$
$$q \rightarrow r$$
$$\underline{\sim r}$$
$$\therefore q$$

3. If p is true, q is false, and r is true, determine the truth value of $p \rightarrow \sim ((q \wedge (r \vee \sim q)))$.

4. Explain why we know that the only connectives necessary for finding the truth table for a proposition are and, or, and not.

5. Explain why we know that the only connectives necessary for finding the truth table for a proposition are $\wedge$ and $\sim$.

6. Let p, q, and r be the statements:
   p: I like to fish.
   q: I don't have a boat.
   r: It will probably rain.
   Express the following symbolic expression as an English sentence.
   $$(p \wedge \sim q) \vee r$$

7. State De Morgan's laws.

8. State the contrapositive of the sentence "If he is honest, then he will not be elected."

9. Find the circuit diagram for the Boolean expression $(pq' + qr')(r + s')$.

10. Simplify the expression
$$(p \wedge q \wedge r) \vee (\sim p \wedge q \wedge r) \vee (p \wedge q \wedge \sim r)$$
$$\vee (\sim p \wedge q \wedge \sim r) \vee (p \wedge \sim q \wedge \sim r)$$
$$\vee (p \wedge \sim q \wedge r)$$

11. What invalid argument is called the converse fallacy?

12. Convert the sentence "To practice medicine, it is necessary to be a doctor" to a conditional statement.

13. Give an expression for $p \rightarrow q$ using the connectives $\sim$ and $\vee$.

14. Convert the statement "I am a good citizen only if I vote" to a conditional statement.

15. Explain how, when using truth tables, we can determine that an argument is invalid.

ANSWERS TO TRUE-FALSE

1. F 2. T 3. T 4. F 5. T 6. F 7. F 8. F 9. T 10. T
11. T 12. T 13. T 14. T 15. F 16. F 17. T 18. T
19. T 20. F 21. F 22. F 23. T 24. T 25. T 26. F
27. T 28. T 29. T 30. F 31. T 32. F 33. F 34. T
35. F 36. T 37. F 38. T 39. F 40. T 41. F 42. F
43. F 44. F 45. T 46. T 47. T 48. T 49. F 50. F

CHAPTER 2

Set Theory

2.1 Introduction to Sets

A **set** is a well-defined collection of objects or elements. In reality this statement is not a definition but the avoidance of a definition. This "definition" is similar to the one given by Georg Cantor, who was instrumental in the early development of set theory. The inadequacy of this definition became apparent when paradoxes or contradictions were developed by the Italian logician Burali-Forti in 1879 and later by Bertrand Russell. Axiomatic systems have been developed for set theory in hope of avoiding further contradictions and paradoxes. These systems include the Zermelo-Fraenkel-von Neumann system, the Gödel-Hilbert-Bernays system, and the Russell-Whitehead system. These systems are beyond the scope of this book, so we will essentially leave sets as undefined objects and assume that sets have been selected so that we can determine their elements and so that no contradictions are produced.

Finite sets may be described by listing their elements. This is done by listing the objects in the set between two braces and placing commas between them. For example, $\{1, 2, 3, 4\}$ is the set containing 1, 2, 3, and 4. The set of vowels is represented by $\{a, e, i, o, u\}$. We will normally assign capital letters to represent sets. $A = \{$Bob, Jane, Nancy$\}$ is the set consisting of Bob, Jane, and Nancy. The set of positive integers may be denoted by $\{1, 2, 3, 4, \ldots\}$ and the first n positive integers by $\{1, 2, 3, \ldots, n\}$, where three dots are used to indicate the continuation of a pattern. Often when the sets are listed, an arbitrary element is used to describe the pattern. For example, $C = \{1, 8, 27, \ldots, k^3, \ldots\}$ describes the set of all cubes of positive integers and $S = \{1, 4, 9, \ldots, n^2\}$ describes the set of squares of all positive integers less than or equal to n. Obviously the listing of elements is convenient only if the set is small or has an easily described pattern. One could not easily describe the citizens of the United States using this method and could not possibly describe the set of real numbers.

Set builder notation is also used to describe sets. Braces are again used to denote the set, and within these braces, the characteristic describing the set is given. Thus, the set $\{x : x \text{ has property } P\}$ is intended to contain exactly those objects having the property P. For example, $\{x : x \text{ is a football player for Southwestern College}\}$ is the set consisting of all the football players at Southwestern College. The notation

$\{x : x$ is a citizen of England$\}$ describes the set of all citizens of England. The definition of the set must be adequate to completely determine the set. This is no problem if the objects in the set are listed, but consider $A = \{x : x$ is a tall student in this class$\}$ or $B = \{x : x$ is a good student in this class$\}$. If different members of the class were to select the members of A and B, they might not select the same people. In the case of $C = \{x : x$ is an attractive (or handsome) member of the class$\}$, it is not only difficult to select the members of C but also we do not even recommend trying. However, when $A = \{x : x$ is a student of this class who is over 6 feet tall$\}$ and $B = \{x : x$ is in this class and has a grade point of 3.0 or better$\}$, then it is possible to tell definitely whether or not an object is in A or in B so that A and B are sets.

We now formalize some of the ideas discussed previously in the following definition.

DEFINITION 2.1 If a is one of the objects of the set A, we say that a is an **element** of A or a **belongs** to A. The statement that a belongs to A is denoted by $a \in A$. If it is not true that a is an element of A, we write $a \notin A$.

Thus, $3 \in \{1, 2, 3, 4\}$ but $5 \notin \{1, 2, 3, 4\}$. If P is the set $\{x : x$ was a president of the United States$\}$, then Abraham Lincoln $\in P$ but Patrick Henry $\notin P$.

DEFINITION 2.2 A set A is a **subset** of a set B (denoted by $A \subseteq B$) if every element of A is an element of B; that is, if $x \in A$, then $x \in B$. In particular, every set is a subset of itself. If it is not true that A is a subset of B, we write $A \nsubseteq B$. Thus, $A \nsubseteq B$ if there is an element of A that is not in B.

Hence, $\{1, 2, 3\} \subseteq \{1, 2, 3, 4\}$, but $\{1, 2, 5\} \nsubseteq \{1, 2, 3, 4\}$. If $A = \{x : x$ is a football player in college$\}$, $B = \{x : x$ is an athlete in college$\}$, and $C = \{x : x$ is a mathematics major in college$\}$, then $A \subseteq B$ but $C \nsubseteq B$.

Sets are equal if they contain exactly the same elements. If A is the set $\{2, 4, 6\}$ and B is the set $\{x : x$ is an even positive integer less than 7$\}$, then A and B are equal sets. This discussion is formalized in the following definition.

DEFINITION 2.3 If A and B are sets, then we say that A **equals** B, written $A = B$, whenever, for any x, $x \in A$ if and only if $x \in B$. An alternative definition is that $A = B$ if and only if $A \subseteq B$ and $B \subseteq A$. If $A \subseteq B$ and $A \neq B$, we write $A \subset B$ and say that A is a **proper subset** of B.

Thus, proving that two sets A and B are equal is a two-step process:

Prove that A is a subset of B.
Prove that B is a subset of A.

Since a set is defined only by the elements that it contains, there is no order implied in the listing of the elements. Thus, $\{1, 2, 4, 6\} = \{2, 1, 6, 4\}$. Also an element is either contained in a set or it is not. There is no concept of an element being in a set more than once.

At this point we introduce two new sets, the universal set and the empty set. In a sense they are opposites since the empty set contains no elements and the universal set contains "all" elements.

DEFINITION 2.4 The **empty set**, denoted by $\varnothing$ or by $\{\}$, is the set that contains no elements. The **universal set** U is a set that has the property that all sets under consideration are subsets of it.

In discussions about number theory, the universal set is usually the set of all integers or the positive integers. In calculus, the universal set may be the set of all real numbers or the set of all points in n-dimensional space. Note that although called the universal set, U is not uniquely defined unless precisely stated. Certainly any set containing U could be used as the universal set.

We shall also be concerned by the number of elements in a set or the size of a set. In the case of infinite sets, this is rather a complex subject and will be discussed in later chapters. For the present we shall limit ourselves to finite sets.

DEFINITION 2.5 The **size** or **cardinality** of a finite set, A, denoted by $|A|$ is the number of elements in the set A.

Thus, $|\varnothing| = 0$, and $|\{a, b, c, d\}| = 4$.

By definition, every set is a subset of the universal set. The empty set is a subset of any given set A since every element of the empty set is contained in A. Perhaps it is better to say that there is no element of the empty set that is not also in A.

Exercises

1. List the elements in the set $\{x : x \text{ is an integer and } x^2 < 100\}$.
2. List the elements in the set $\{x : x \text{ is a president of the United States since 1940}\}$.
3. List the elements in the set $\{x : x \text{ is a vowel}\}$.
4. List the elements in the set $\{x : x \text{ is a positive even integer less than 21}\}$.
5. Express in set builder notation the set $\{3, 6, 9, 12, 15, 18, 21, 24\}$.
6. Express in set builder notation the set $\{a, b, c, d, e, f, g, h, i, j\}$.
7. Express in set builder notation the set $\{$Ohio, Oklahoma, Oregon$\}$.
8. Express in set builder notation the set $\{1, 4, 9, 16, 25, 36, \ldots\}$.
9. List the subsets of $\{a\}$.
10. List the subsets of $\{a, b\}$.
11. List the subsets of $\{a, b, c\}$.
12. List the subsets of $\{a, b, c, d\}$.
13. List the subsets of $\varnothing$.
14. Based on the five previous exercises, determine the number of subsets of a set with n elements.

State whether each of the following is true or false:

15. $\varnothing \subseteq \varnothing$
16. $\varnothing \subset \varnothing$
17. $\varnothing \in \varnothing$
18. $\varnothing \subseteq A$, where A is an arbitrary set.
19. $\varnothing \in A$, where A is an arbitrary set.
20. $\varnothing \in \{\{\varnothing\}, 1, 2, 3, 4, 5\}$.
21. $\varnothing \in \{\{\varnothing\}, \varnothing, 1, 2, 3, 4, 5\}$.
22. $\{2\} \in \{1, 2, 3, 4, 5\}$
23. $\{2\} \subseteq \{1, 2, 3, 4, 5\}$
24. $\varnothing = \{\varnothing\}$
25. $\{1, 2, 3\} \in \{1, 2, 3, \{1, 2, 3\}\}$
26. $\{1, 2, 3\} \subseteq \{1, 2, 3, \{1, 2, 3\}\}$
27. $\{1, 2\}, \in \{1, 2, 3, \{1, 2, 3\}\}$.
28. $\{1, 2\} \subseteq \{1, 2, 3, \{1, 2, 3\}\}$.
29. $\{\varnothing, \{\varnothing\}\} \in \{\varnothing, \{\varnothing\}, \{\varnothing, \{\varnothing\}\}\}$.
30. $\{\varnothing, \{\varnothing\}\} \subseteq \{\varnothing, \{\varnothing\}, \{\varnothing, \{\varnothing\}\}\}$.
31. If $|A| = |B|$, then $A = B$.
32. $\{a, b, c\} = \{c, b, a\}$.
33. $\{a, \{b, c\}\} = \{\{c, b\}, a\}$.
34. $\{a, \{b, c\}\} = \{\{a, b\}, c\}$.

Find the cardinality of the following:

35. $\{\varnothing, \{\varnothing\}\}$
36. $\{\{\varnothing, \{\varnothing\}\}\}$
37. $\{1, 2, 3, \{1, 2, 3\}\}$
38. $\{\varnothing, \{\varnothing\}, a, b, \{a, b\}, \{a, b, \{a, b\}\}\}$
39. $\{\varnothing, \{\varnothing\}, \{\varnothing, \{\varnothing\}\}\}$
40. Find a set that has a subset but no proper subset.
41. *If there is a set S that is the set of all sets, then $S \in S$. It is a little disturbing for a set to belong to itself. For example, if $A = \{a, b, A\}$, then $A = \{a, b, \{a, b, A\}\} = \{a, b, \{a, b, \{a, b, \{a, b, A\}\}\}\} \ldots$. We know most sets do not belong to themselves. For example, $\varnothing \notin \varnothing$. Let $\omega = \{x : x \notin x\}$, so that ω is the set of all sets that do not belong to themselves. Does $\omega \in \omega$?

2.2 Set Operations

The following operations on sets allow us to use existing sets to create new sets.

DEFINITION 2.6

The **intersection** of sets A and B is the set of all elements that are in both A and B. The intersection of sets A and B is denoted $A \cap B$. Equivalently, $A \cap B = \{x : x \in A \text{ and } x \in B\}$.

Computer Science Application

Let f1 and f2 be two input files of unknown length whose records are integers. If the records in both files are in ascending order, the files can be merged to produce a third file f3 whose records are likewise in ascending order. The following algorithm shows how this can be done without using arrays or other costly internal data structures.

function merge (file f1, file f2, file f3)

    get next record from f1 into x
    get next record from f2 into y

    while (not eof(f1) or not eof(f2))
      if $x < y$ then
        write x to f3
        get next record from f1 into x
      else
        write y to f3
        get next record from f2 into y
      end-if
    end-while

end-function

The preceding algorithm makes no attempt to deal with duplicates. If f1 contains three records equal to 15 and f2 contains five records equal to 15, file f3 will contain eight records equal to 15. If files f1 and f2 are each free of duplicates, the following algorithm writes their mathematical union to file f3.

function union (file f1, file f2, file f3)

    get next record from f1 into x
    get next record from f2 into y

    while (not eof(f1) or not eof(f2))
      if $x < y$ then
        write x to f3
        get next record from f1 into x
      else if $y < x$ then
        write y to f3
        get next record from f2 into y
      else

```
            write x to f3
            get next record from f1 into x
            get next record from f2 into y
        end-if
    end-while

end-function
```

For example, if $A = \{1, 2, 3, 4, 5\}$ and $B = \{1, 3, 5, 7, 9\}$, then $A \cap B = \{1, 3, 5\}$. If $C = \{x : x \text{ is over 6 feet tall}\}$ and $D = \{x : x \text{ likes to play chess}\}$, then $C \cap D = \{x : x \text{ is over 6 feet tall and likes to play chess}\}$. Note that in describing the intersection $B \cap C$, we need to use the word "and." The symbols $\cap$, introduced previously, and $\wedge$, introduced in Chapter 1, are similar symbols and we shall see that they are related and have many similar properties.

If A_1, A_2, and A_3 are subsets, then we indicate their intersection by

$$B = A_1 \cap (A_2 \cap A_3)$$

It will be shown later that $A_1 \cap (A_2 \cap A_3) = (A_1 \cap A_2) \cap A_3$ so that we can write

$$B = A_1 \cap A_2 \cap A_3$$

Evidently, $x \in B$ if and only if $x \in A_1$, $x \in A_2$, and $x \in A_3$; that is, $x \in B$ if and only if x is in all of the three sets A_1, A_2, or A_3. Let $J = \{1, 2, 3\}$. Then, $x \in B$ if and only if $x \in A_j$ for all $j \in J$, or equivalently,

$$B = \{x : x \in A_j \text{ for all } j \in J\}$$

We define a more general intersection as follows.

DEFINITION 2.7

If $I = \{1, 2, 3, \ldots, k\}$, then

$$\bigcap_{i \in I} A_i = A_1 \cap A_2 \cap A_3 \cap \cdots \cap A_k$$

$$= \{x : x \in A_i \text{ for all } i \in I\}$$

EXAMPLE 2.8

The **union** of sets A and B is the set of all elements that are in either A or B. The union of sets A and B is denoted $A \cup B$. Equivalently, $A \cup B = \{x : x \in A \text{ or } x \in B\}$. ∎

For example, if $A = \{1, 2, 6, 7\}$ and $B = \{2, 3, 5, 6\}$, then $A \cup B = \{1, 2, 3, 5, 6, 7\}$. The union $A \cup B$ is formed from A and B by combining the elements of A and B.

If $A = \{x : x \text{ is a politician}\}$ and $B = \{x : x \text{ is a college graduate}\}$, then $A \cup B = \{x : x \text{ is a politician or is a college graduate}\}$. Note that in describing the union $A \cup B$, we use the word "or" just as for intersection we used "and."

If A_1, A_2, and A_3 are subsets, then we indicate their union by

$$B = A_1 \cup (A_2 \cup A_3)$$

It will be shown later that $A_1 \cup (A_2 \cup A_3) = (A_1 \cup A_2) \cup A_3$ so that we can write

$$B = A_1 \cup A_2 \cup A_3$$

56 Chapter 2 Set Theory

Evidently, $x \in B$ if and only if $x \in A_1$, $x \in A_2$, or $x \in A_3$; that is, $x \in B$ if and only if x is in at least one of the three sets A_1, A_2, or A_3. Thus, $x \in B$ if and only if for some $j \in \{1, 2, 3\}$, $x \in A_j$, or equivalently,

$$B = \{x : x \in A_j \text{ for some } j \in \{1, 2, 3\}\}$$

We define a more general union as follows:

DEFINITION 2.9

If $I = \{1, 2, 3, \ldots, k\}$, then

$$\bigcup_{i \in I} A_i = A_1 \cup A_2 \cup A_3 \cup \cdots \cup A_k$$
$$= \{x : \text{there is an } i \in I \text{ such that } x \in A_i\}$$

DEFINITION 2.10

Let A and B be sets. The **set difference** $A - B$ is the set of all elements that are in A but are not in B. Equivalently, $A - B = \{x : x \in A \text{ and } x \notin B\}$. The **symmetric difference** of A and B, denoted by $A \triangle B$, is the set $(A - B) \cup (B - A)$.

For example, if $A = \{1, 2, 4, 6, 7\}$ and $B = \{2, 3, 4, 5, 6\}$, then $A - B = \{1, 7\}$ and $A \triangle B$ is the set $\{1, 3, 5, 7\}$. The symmetric difference of A and B is comprised of those elements that are in one and only one of the two sets A and B. If $A = \{x : x \text{ plays tennis}\}$ and $B = \{x : x \text{ plays golf}\}$, then $A - B = \{x : x \text{ plays tennis but not golf}\}$. The set $A \triangle B = \{x : x \text{ plays tennis or golf but not both}\}$. Notice the similarity to the "exclusive or" of Chapter 1.

DEFINITION 2.11

The **complement** of a set A, denoted by A', is the set of all elements of the universe which are not in A. Hence

$$A' = U - A = \{x : x \in U \text{ and } x \notin A\}$$

If U is the set of positive integers and $A = \{2, 4, 6, 8, \ldots\}$ is the set of all even positive integers, then $A' = \{1, 3, 5, 7, \ldots\}$ is the set of all odd positive integers. If U is the set of the letters in the English alphabet and $V = \{a, e, i, o, u\}$, then V' is the set of all consonants. If $A = \{x : x \text{ likes to read science fiction}\}$, then $A' = \{x : x \text{ does not like to read science fiction}\}$. Notice that the complement of a set is related to the $\sim$ symbol in logic. The complement of a set is the set of all elements *not* in the set.

Following the definition of equality of sets, the following steps are used to show two sets are equal:

Prove that the first set is a subset of the second.

Prove that the second set is a subset of the first.

In the proofs of the following theorems, we demonstrate the use of these steps.

THEOREM 2.12 For arbitrary sets A and B, $A - B = A \cap B'$.

Proof To show the two sets are equal, we need to show that each is a subset of the other. This is done by picking an arbitrary element of each set and showing that it is in the other. This is done more easily in the following argument since each step is reversible.

$$a \in A - B \Leftrightarrow (a \in A) \land (a \notin B) \quad \text{definition of } A - B$$
$$\Leftrightarrow (a \in A) \land (a \in B') \quad \text{definition of complement}$$
$$\Leftrightarrow a \in (A \cap B') \quad \text{definition of intersection} \quad \blacksquare$$

Observe that in the proof of the following theorem we are actually using one of De Morgan's laws of logic to prove the corresponding De Morgan's law for set theory.

THEOREM 2.13 For arbitrary sets A and B,

(a) $(A \cap B)' = A' \cup B'$.

(b) $(A \cup B)' = A' \cap B'$.

Proof Part (a) is proved here. Part (b) is left to the reader. Again we show that each set in the equality is a subset of the other.

$$a \in (A \cap B)' \Leftrightarrow a \notin (A \cap B) \quad \text{definition of complement}$$
$$\Leftrightarrow \sim(a \in (A \cap B)) \quad \text{definition of } \notin$$
$$\Leftrightarrow \sim((a \in A) \land (a \in B)) \quad \text{definition of intersection}$$
$$\Leftrightarrow \sim(a \in A) \lor \sim(a \in B) \quad \text{De Morgan's law for logic}$$
$$\Leftrightarrow (a \notin A) \lor (a \notin B) \quad \text{definition of } \notin$$
$$\Leftrightarrow (a \in A') \lor (a \in B') \quad \text{definition of complement}$$
$$\Leftrightarrow a \in (A' \cup B') \quad \text{definition of union} \quad \blacksquare$$

Observe that in the proof of the following theorem we are actually using a distributive law of logic to prove the corresponding distributive law of set theory.

THEOREM 2.14 For arbitrary sets A, B, and C,

(a) $A \cap (B \cup C) = (A \cap B) \cup (A \cap C)$

(b) $A \cup (B \cap C) = (A \cup B) \cap (A \cup C)$

Proof Part (a) is proved here. Part (b) is left to the reader. Again we show that each set in the equality is a subset of the other.

$$a \in A \cap (B \cup C) \Leftrightarrow (a \in A) \land (a \in (B \cup C)) \quad \text{definition of intersection}$$
$$\Leftrightarrow (a \in A) \land ((a \in B) \lor (a \in C)) \quad \text{definition of union}$$
$$\Leftrightarrow ((a \in A) \land (a \in B)) \quad \text{definition of intersection.}$$
$$\lor ((a \in A) \land (a \in C)) \quad \text{De Morgan's law for logic}$$
$$\Leftrightarrow (a \in (A \cap B)) \lor (a \in (A \cap C)) \quad \text{definition of intersection}$$
$$\Leftrightarrow a \in ((A \cap B) \cup (A \cap C)) \quad \text{definition of union} \quad \blacksquare$$

We have seen that the properties just shown for set theory have their counterparts in logic. In Section 2.4 we generalize the operations of both set theory and logic when we discuss Boolean algebras.

DEFINITION 2.15 The **power set** of a set A, denoted by $\mathcal{P}(A)$, is the set consisting of all subsets of A.

Computer Science Application

A set S with n elements has 2^n subsets, counting the empty set and the set S itself. Is there a systematic way to list all the subsets? If, for example, the 5 elements of $S = \{a, b, c, d, e\}$ are placed into a 5-component array A, the $2^5 = 32$ subsets correspond to the 32 5-digit binary strings from 0 (00000) to 31 (11111). The following algorithm displays the subsets of S, one per line.

function power-set (character array A)

```
    for i = 0 to 31
        k = i
        j = 5
        while k > 0 do
            if k mod 2 = 1 then
                Print A(j)
            end-if
            j = j - 1
            k = k div 2
        end-while
        Print end-of-line
    end-for

    return

end-function
```

When $i = 25 = (11001)_2$, the function generates the subset $\{e, b, a\}$.

We seldom need to actually list subsets; we might instead need to examine each subset to see which ones possess some specified property. The foregoing function shows how easily the subsets can be generated without repetition or omission.

Hence, the power set of the set $A = \{1, 2, 3\}$ is the set

$$\mathcal{P}(A) = \{\varnothing, \{1\}, \{2\}, \{3\}, \{1, 2\}, \{2, 3\}, \{1, 3\}, \{1, 2, 3\}\}$$

When A has three elements, $\mathcal{P}(A)$ has $2^3 = 8$ elements; or equivalently, A has $2^3 = 8$ subsets. This will be shown in Chapter 8. In general, if A has n elements, the set $\mathcal{P}(A)$ has 2^n elements and A has 2^n subsets. This is one reason that $\mathcal{P}(A)$ is often denoted by 2^A.

Another operation between sets that is commonly used is the Cartesian product, which is defined as follows.

DEFINITION 2.16 The **Cartesian product** of the sets A and B, denoted by $A \times B$, is the set $\{(a, b) : a \in A \text{ and } b \in B\}$. The object (a, b) is called an **ordered pair** with first component a and second component b.

The set $A \times B$ consists of all ordered pairs having the first component an element of A and the second component an element of B. This is essentially the same ordered pair that you have used in algebra. When plotting graphs, we know that plotting the point (1, 2) gives a different result then plotting the point (2, 1).

Let $A = \{1, 2, 3\}$ and $B = \{r, s\}$; then

$$A \times B = \{(1, r), (1, s), (2, r), (2, s), (3, r), (3, s)\}$$

If A and B are both the set of real numbers, then $A \times B$ is the Cartesian plane on which we use ordered pairs of real numbers to graph functions in algebra and calculus.

If A contains n elements and B contains m elements, then $A \times B$ contains $n \cdot m$ elements. Also, by the definition, if either A or B is empty, then $A \times B$ is empty.

Computer Science Application

The following algorithm produces in file f3 the mathematical intersection of files f1 and f2. Notice the use of "and" rather than "or" in the condition controlling the loop.

function intersection (file f1, file f2, file f3)

```
    get next record from f1 into x
    get next record from f2 into y

    while (not eof(f1) and not eof(f2))
        if x < y then
            get next record from f1 into x
        else if y < x then
            get next record from f2 into y
        else
            write x to f3
            get next record from f1 into x
            get next record from f2 into y
        end-if
    end-while

end-function
```

When implementing these algorithms, the obvious adjustments can easily be made to accommodate records other than simple integers. Also, an appropriate sentinel value should be placed into x if the end-of-f1 is encountered before end-of-f2, and vice- versa. This keeps the while-loop executing until all remaining records from the other file have been read.

Exercises

Let $A = \{1, 2, 3, 4, 5, 6, 7\}$, $B = \{4, 5, 6, 7, 8, 9, 10\}$, $C = \{2, 4, 6, 8, 10\}$, and $U = \{1, 2, 3, 4, 5, 6, 7, 8, 9, 10\}$. Find the following sets:

1. $A \cup C$
2. $A \cap B$
3. $A \cap (B \cup C)$
4. $(A \cap B) \cup C$
5. $(A \cap B)'$
6. $A' \cap B'$
7. $A \triangle B$
8. $A - B$
9. $A - C$
10. $(A - B) \cup (B - A)$
11. $A \cap B \cap C'$
12. $(A \cup C) - B'$
13. $(A - \emptyset) \cup (A - A)$

14. $B \triangle C$
15. $C - A$

Let $A = \{1, 2, 3\}$ and $B = \{a, b\}$.

16. Find $A \times B$.
17. Find $B \times B$.
18. Find $A \times \varnothing$.
19. Find $A \times A$.
20. Find $B \times A$.
21. Find $A \triangle B$.
22. Find $\mathcal{P}(A)$ when $A = \varnothing$.
23. Find $\mathcal{P}(A)$ when $A = \{\varnothing, \{\varnothing\}\}$.
24. Find $\mathcal{P}(\mathcal{P}(A))$ when $A = \varnothing$.

Determine which of the following are true and which are false:

25. $A \cap \varnothing = A$.
26. $A \triangle A = \varnothing$.
27. If $A \subseteq B$, then $A \cap B = A$.
28. If $A \cap B = A$, then $B \subseteq A$.
29. $A - A = A$.
30. $(A \times B)' = A' \times B'$.
31. $A \cup \varnothing = A$.
32. $A \triangle \varnothing = A$.
33. If $A \subseteq B$, then $A \cup B = A$.
34. If $A \cup B = A$, then $B \subseteq A$.
35. $A - \varnothing = A$.
36. Prove that $A \cup (B \cap C) = (A \cup B) \cap (A \cup C)$.
37. Prove that $(A \cup B)' = A' \cap B'$.
38. Define (a, b) to be the set $\{a, \{a, b\}\}$. Show that we are justified in referring to "first" and "second" components because $(a, b) = (c, d)$ if and only if $a = c$ and $b = d$. Therefore in the above definition, we are producing something that has order using the concept of set, which has no order.

The successor of a set A is the set $A \cup \{A\}$. Find the successor of the following sets:

39. $\varnothing$
40. $\{\varnothing\}$
41. $\{\varnothing, \{\varnothing\}\}$
42. $\{\varnothing, \{\varnothing\}, \{\varnothing, \{\varnothing\}\}\}$.
43. $\{\varnothing, \{\varnothing\}, \{\varnothing, \{\varnothing\}\}, \{\varnothing, \{\varnothing\}, \{\varnothing, \{\varnothing\}\}\}\}$.

Determine which of the following are power sets of a set:

44. $\{\varnothing\}$.
45. $\{\varnothing, \{a\}, \{a, b\}\}$.
46. $\{\varnothing, \{\varnothing\}, \{\varnothing, \{\varnothing\}\}\}$.
47. $\{\varnothing\}, \{\varnothing\}, \{\varnothing, \{\varnothing\}\}, \{\{\varnothing\}, \{\varnothing, \{\varnothing\}\}\}$.
48. Prove that if $A \subseteq B$, then $\mathcal{P}(A) \subseteq \mathcal{P}(B)$.
49. Prove or disprove that $\mathcal{P}(A \cup B) = \mathcal{P}(A) \cup \mathcal{P}(B)$.
50. Prove or disprove that $\mathcal{P}(A \cap B) = \mathcal{P}(A) \cap \mathcal{P}(B)$.
51. What can be said about the sets A and B if $A - B = B - A$?
52. What can be said about the sets A and B if $A \cap B = B$?
53. What can be said about the sets A and B if $A \cup B = B$?
54. What can be said about the sets A and B if $A \times B = B \times A$?
55. If I is a set and A_i is a set for each $i \in I$, then
$$\bigcap_{i \in I} A_i = \{x : x \in A_i \text{ for all } i \in I\}$$
The set $\bigcap_{i \in I} A_i$ is called the **generalized intersection** of the collection of sets $\{A_i : i \in I\}$, which is sometimes abbreviated by $\{A_i\}$. The set I is called the **indexing set** and the collection $\{A_i\}$ is said to be **indexed** by I.

   (a) Let $N = \{1, 2, 3, \ldots\}$ and $I = \{\square, \triangle, \bigcirc\}$. Let
   $$A_\square = \{x : x = 2k \text{ for some positive integer } k\}$$
   $$A_\triangle = \{x : x = 3k \text{ for some positive integer } k\}$$
   $$A_\bigcirc = \{x : x = 5k \text{ for some positive integer } k\}$$
   Find C where
   $$C = \bigcap_{i \in I} A_i = \{x : x \in A_i \text{ for all } i \in \{\square, \triangle, \bigcirc\}\}$$

   (b) Let $I = N = \{1, 2, 3, \ldots\}$ and
   $$A_i = \{x : x \in N \text{ and } x \geq i\} = \{i, i+1, i+2, \ldots\}$$
   Find $\bigcap_{i \in I} A_i$.

56. If I is a set and A_i is a set for each $i \in I$, then
$$\bigcup_{i \in I} A_i = \{x : \text{ there is an } i \in I \text{ such that } x \in A_i\}$$
The set $\bigcup_{i \in I} A_i$ is called the **generalized union** of the collection of sets $\{A_i : i \in I\}$. Let $I = N = \{1, 2, 3, \ldots\}$ and
$$A_i = \{x : x \in N \text{ and } x \geq i\} = \{i, i+1, i+2, \ldots\}$$
Find $\bigcup_{i \in I} A_i$

2.3 Venn Diagrams

A useful tool for representing sets and observing how the operations work is the Venn diagram. In a Venn diagram, sets are represented as interiors of circles, intersections of circles, unions of circles, and so on. The universe is rectangular. In Figure 2.1, a Venn diagram for only one set, say A, is represented as the interior of a circle. The area outside the circle, but inside the rectangle, represents A'.

As shown in Figure 2.2, for a Venn diagram for two sets, say A and B, each set is represented as a circle and the two circles intersect.

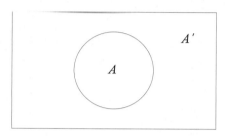

Figure 2.1   Figure 2.2

The diagram is divided into four parts as shown. The set $A \cap B$ is shaded in the diagram shown in Figure 2.3. The shaded area shown in Figure 2.4 represents $A \cup B$.

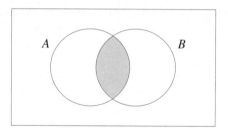 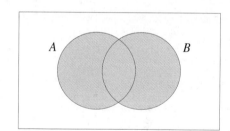

Figure 2.3   Figure 2.4

The set $A - B$ is represented by the shaded area shown in Figure 2.5. The Venn diagram for three sets, say A, B, and C, is shown in Figure 2.6.

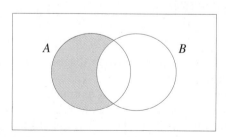

Figure 2.5   Figure 2.6

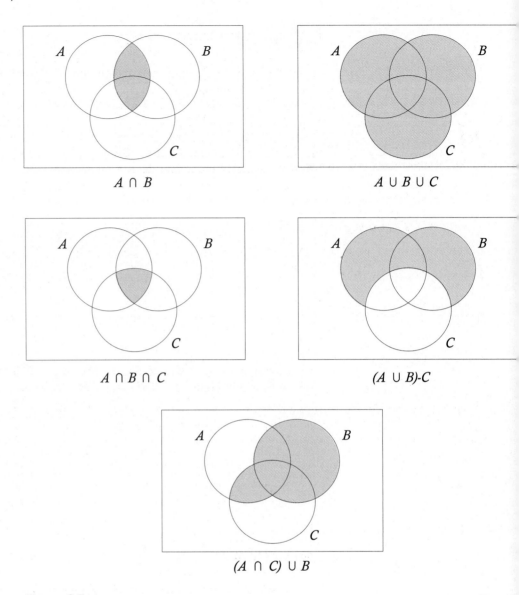

Figure 2.7

It is divided into eight parts. The shaded areas in Figure 2.7 represent $A \cap B$, $(A \cup B \cup C)$, $(A \cap B \cap C)$, $(A \cup B) - C$, and $(A \cap C) \cup B$, respectively.

EXAMPLE 2.17 Note that the eight parts of the Venn diagram for three sets can be represented as shown in Figure 2.8.

In a manner similar to expressing a proposition with a given truth table in disjoint normal form, the shaded area of a Venn diagram can be similarly expressed. Simply take the expressions in the sections that are shaded and connect them with union. Thus, Figure 2.9 can be expressed as $(A' \cap B' \cap C') \cup (A' \cap B' \cap C) \cup (A' \cap B \cap C) \cup (A' \cap B \cap C')$ and Figure 2.10 can be expressed as $(A \cap B' \cap C') \cup (A \cap B \cap C') \cup (A' \cap B \cap C') \cup (A \cap B \cap C)$. Once these expressions are formed, they can be simplified using Karnaugh maps. ∎

One of the uses of Venn diagrams is to show that two sets are equal.

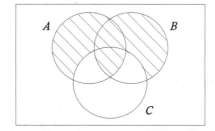

Figure 2.8

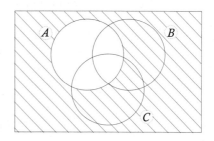

Figure 2.9

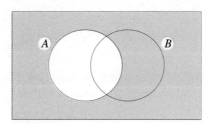

Figure 2.10

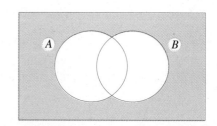

Figure 2.11

EXAMPLE 2.18

Show that $(A \cup B)' = A' \cap B'$. The set $(A \cup B)'$ is the complement of $A \cup B$, whose Venn diagram is given in Figure 2.4 and, hence, is represented by the shaded area in Figure 2.11.

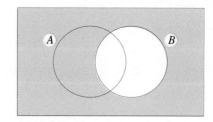

Figure 2.12   Figure 2.13

The set A' is represented by the shaded area in Figure 2.12. The set B' is represented by the shaded area in Figure 2.13.

The set $A' \cap B'$ in Figure 2.14 is the part that is shaded in both diagrams and is the darker colored shaded area.

Hence, $(A \cup B)'$ and $A' \cap B'$ both have the same representation in the Venn diagram and $(A \cup B)' = A' \cap B'$.

Figure 2.14

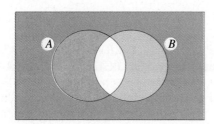

EXAMPLE 2.19 Show that $A \cap (B \cup C) = (A \cap B) \cup (A \cap C)$. The set A is represented by the shaded area in Figure 2.15. The set $B \cup C$ is represented by the shaded area in Figure 2.16.

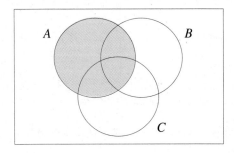

Figure 2.15

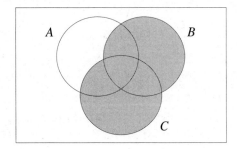

Figure 2.16

Hence, the set $A \cap (B \cup C)$ is the part in Figure 2.17 that is shaded in both diagrams and is represented by the darker colored shaded area. The set $A \cap B$ is represented by the shaded area in Figure 2.18.

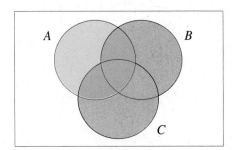

Figure 2.17

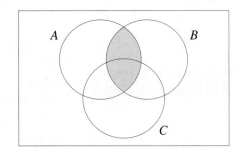

Figure 2.18

The set $A \cap C$ is represented by the shaded area in Figure 2.19. The set $(A \cap B) \cup (A \cap C)$ is the entire shaded area in Figure 2.20.

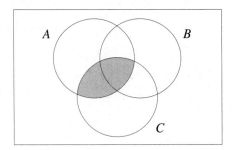

Figure 2.19

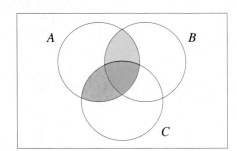

Figure 2.20

Hence, $A \cap (B \cup C)$ and $(A \cap B) \cup (A \cap C)$ both have the same representation in the Venn diagram and $A \cap (B \cup C) = (A \cap B) \cup (A \cap C)$. ∎

The properties given in the following theorem can be proved using formal proofs or Venn diagrams. Notice that they are parallel to similar properties of propositional calculus (see Theorem 1.4).

THEOREM 2.20 Let A, B, and C be subsets of the universal set U.

(a) *Idempotent Laws*
$$A \cap A = A$$
$$A \cup A = A$$

(b) *Double Complement* $(A')' = A$

(c) *De Morgan's Laws*
$$(A \cup B)' = A' \cap B'$$
$$(A \cap B)' = A' \cup B'$$

(d) *Commutative Properties*
$$A \cap B = B \cap A$$
$$A \cup B = B \cup A$$

(e) *Associative Properties*
$$A \cap (B \cap C) = (A \cap B) \cap C$$
$$A \cup (B \cup C) = (A \cup B) \cup C$$

(f) *Distributive Properties*
$$A \cap (B \cup C) = (A \cap B) \cup (A \cap C)$$
$$A \cup (B \cap C) = (A \cup B) \cap (A \cup C)$$

(g) *Identity Properties*
$$A \cup \varnothing = A$$
$$A \cap U = A$$

(h) *Complement Properties*
$$A \cup A' = U$$
$$A \cap A' = \varnothing$$

■

Exercises

For each of the following sets, use a Venn diagram for two sets A and B and shade in the portion that describes the given set:

1. A'
2. $(A \cap B)'$
3. $(A \cup B) - (A \cap B)$
4. $A - B'$
5. B'
6. $(A \cup B)'$
7. $A - A \cap B$
8. $A \triangle B$

For each of the following sets use a Venn diagram for three sets A, B, and C and shade in the portion that describes the given set:

9. $A - B$
10. $(A \cap B)'$
11. $(A \cup B) - A \cap B$
12. $A \cup (B \cap C)$
13. $(B \cap C) - A$
14. $B - (A \cup C)$
15. $(A \cap B \cap C')'$
16. $A \triangle B$
17. $(A \cup B)'$
18. $A - A \cap B$
19. $(A \cap B) \triangle C$
20. $(A \cup B \cup C) - (A \cap B \cap C)$
21. $(A \cap B) \cup (B \cap C) \cup (A \cap C)$
22. $(A - B) \cup (B - C)$

Describe the set shaded in each of the Venn diagrams:

23.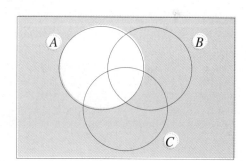

24.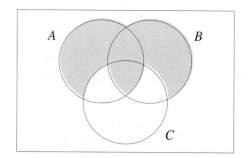

25.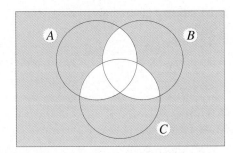

26.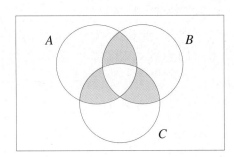

27.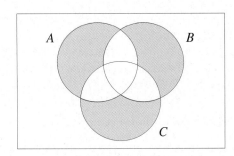

28.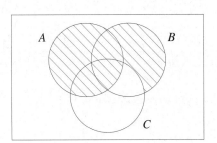

29.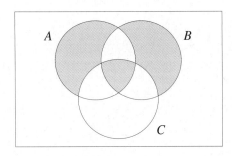

30.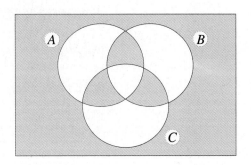

31.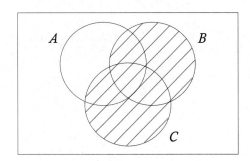

32. Use Venn diagrams to show that $(A \cap B)' = A' \cup B'$

33. Use Venn diagrams to show that $A \cup (B \cap C) = (A \cup B) \cap (A \cup C)$

Use Venn diagrams to determine whether or not the following pairs of sets are equal:

34. $(A \triangle B) \triangle C$, $A \triangle (B \triangle C)$

35. $(A \triangle B) \cap C$, $(A \triangle C) \cap (B \triangle C)$

36. $(A \cap B) \triangle C$, $(A \triangle B) \cap (A \triangle C)$

37. $A - (B \cup C)$, $(A - B) - C$

38. $A - (B \cup C)$, $(A - B) \cap (A - C)$

39. $A - (B \cup C)$, $(A - B) \cup (A - C)$

40. $A - (B \cap C)$, $(A - B) \cup (A - C)$

41. $A - (B \cap C)$, $(A - B) \cap (A - C)$

Use the properties in Theorem 2.20 along with definitions of the binary operations for sets and other properties of sets, to show the following pairs of sets are equal:

42. $(A' \triangle B')$, $(A \cap B') \cup (A' \cap B)$

43. $(A \triangle B) \cup B$, $(A \cup B)$

44. $((B - A') \cup (A - B')) \cap C$, $A \cap B \cap C$

45. $(A \cap B \cap C) \cup (A \cap B' \cap C) \cup (B - (A \cap C))$, $(A \cap C) \cup B$

*Use the process in Example 2.17 and Karnaugh maps to find simple expressions for the following Venn diagrams.

46.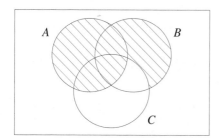

48.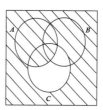

47.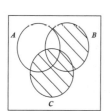

49.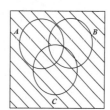

50.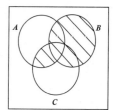

2.4 Boolean Algebras

If the basic properties of sets and propositional logic are compared, they are seen to have many common properties. We generalize these properties in a general theory known as **Boolean algebra**. Boolean algebra is named after George Boole, who is well known as a pioneer in logic. The basic purpose of this chapter is to define the basic axioms of a Boolean algebra and use them to prove theorems that are true for any system that has these properties. In many of the examples and problems we will assume that elementary properties of the integers are known even though they are not formally developed until the next chapter.

DEFINITION 2.21 An operator on a set is **binary** if it combines or operates on two elements of a set to produce another element of the set.

DEFINITION 2.22 An operator on a set is **unary** if it operates on one element of a set to produce another element of the set.

The following notation was introduced in Chapter 1.

DEFINITION 2.23 A **Boolean algebra** is a set B containing special elements 1 and 0 together with binary operators, $+$ and $\cdot$ and a unary operator $'$ on B, which satisfy the following axioms for all x, y, and z in B:

(a) **Commutative Laws**
$$x \cdot y = y \cdot x$$
$$x + y = y + x$$

(b) **Associative Laws**
$$x \cdot (y \cdot z) = (x \cdot y) \cdot z$$
$$x + (y + z) = (x + y) + z$$

(c) **Distributive Laws**

$$x \cdot (y + z) = (x \cdot y) + (x \cdot z)$$
$$x + (y \cdot z) = (x + y) \cdot (x + z)$$

(d) **Identity Laws**

$$x + 0 = x$$
$$x \cdot 1 = x$$

(e) **Complement Laws**

$$x + x' = 1$$
$$x \cdot x' = 0$$

The element 1 is called **unity**, the element 0 is called **zero**, and x' is called the **complement** of x. The binary operation $\cdot$ is often omitted so $x \cdot y$ is written simply as xy.

Using the preceding axioms, we can immediately prove a number of theorems about Boolean algebras including the following.

THEOREM 2.24 For all elements x and y of a Boolean algebra:

(a) *Idempotent Laws*

$$x + x = x$$
$$x \cdot x = x$$

(b) *Null Laws*

$$x + 1 = 1$$
$$x \cdot 0 = 0$$

(c) *Absorption Laws*

$$x + (x \cdot y) = x$$
$$x \cdot (x + y) = x$$

Proof In each case only half of the theorem is proved. The other half is left to the reader.

(a) $\quad x + x = (x + x) \cdot 1 \qquad$ identity law
$\qquad\qquad = (x + x) \cdot (x + x') \qquad$ complement law
$\qquad\qquad = x + (x \cdot x') \qquad$ distributive law
$\qquad\qquad = x + 0 \qquad$ complement law
$\qquad\qquad = x \qquad$ identity law

(b) $\quad x + 1 = (x + 1) \cdot 1 \qquad$ identity law
$\qquad\qquad = (x + 1) \cdot (x + x') \qquad$ complement law
$\qquad\qquad = x + (1 \cdot x') \qquad$ distributive law
$\qquad\qquad = x + (x' \cdot 1) \qquad$ commutative law
$\qquad\qquad = x + x' \qquad$ identity law
$\qquad\qquad = 1 \qquad$ complement law

(c) $x + (x \cdot y) = (x \cdot 1) + (x \cdot y)$ identity law
$= x \cdot (1 + y)$ distributive law
$= x \cdot (y + 1)$ commutative law
$= x \cdot 1$ null law
$= x$ identity law

THEOREM 2.25   **(Uniqueness of Complement Law)**   The complement of an element x of a Boolean algebra is uniquely defined by its properties; that is, if $x + x' = 1$, $x \cdot x' = 0$, $x + x^* = 1$, and $x \cdot x^* = 0$, then $x' = x^*$.

Proof   If $x + x' = 1$ and $x + x^* = 1$, then

$x' = x' \cdot 1$ identity law
$= x' \cdot (x + x^*)$ given
$= x' \cdot x + x' \cdot x^*$ distributive law
$= x \cdot x' + x' \cdot x^*$ commutative law
$= 0 + x' \cdot x^*$ given
$= x' \cdot x^* + 0$ commutative law
$= x' \cdot x^*$ identity law

and

$x^* = x^* \cdot 1$ identity law
$= x^* \cdot (x + x')$ given
$= x^* \cdot x + x^* \cdot x'$ distributive law
$= x \cdot x^* + x' \cdot x^*$ commutative law
$= 0 + x' \cdot x^*$ given
$= x' \cdot x^* + 0$ commutative law
$= x' \cdot x^*$ identity law

so that $x^* = x' x^* = x'$.   ■

THEOREM 2.26   For all elements x and y of a Boolean algebra:

(a) *Involution Law*

$$(x')' = x$$

(b) *Complement of Identities Laws*

$$0' = 1$$
$$1' = 0$$

(c) *De Morgan's Laws*

$$(x + y)' = x' \cdot y'$$
$$(x \cdot y)' = x' + y'$$

Proof For the proof of part (a) we note that

$$x' + x = x + x' \quad \text{commutative law}$$
$$= 1 \quad \text{complement law}$$
$$x' \cdot x = x \cdot x' \quad \text{commutative law}$$
$$= 0 \quad \text{complement law}$$

Hence, x is the complement of x' and by the uniqueness of complement law,

$$(x')' = x$$

Part (b) is left to the reader. Part (c) is shown in the discussion following. ∎

Observe that each axiom for a Boolean algebra consists of a pair of equations that are dual in the sense that if we replace $+$ by $\cdot$, $\cdot$ by $+$, 0 by 1, and 1 by 0 in one equation, we get the other equation. As a result, every theorem is dual in the sense that if we replace $+$ by $\cdot$, $\cdot$ by $+$, 0 by 1, and 1 by 0 in any theorem in Boolean algebra, we get another theorem (although it may not be distinct). This correspondence occurs because every step in the proof of the dual theorem is the dual of the corresponding step in the original theorem. For example, consider the following proofs of the first of De Morgan's laws

$$(x + y)' = x' \cdot y'$$

and its dual

$$(x \cdot y)' = x' + y'$$

First, we prove $(x + y)' = x' \cdot y'$. We first show that $(x + y) + x' \cdot y' = 1$ and $(x + y) \cdot (x' \cdot y') = 0$. When we have done this, by the uniqueness of complement law, we have shown that $(x + y)' = x' \cdot y'$.

$$
\begin{aligned}
(x + y) + x' \cdot y' &= ((x + y) + x') \cdot ((x + y) + y') & \text{distributive law} \\
&= ((y + x) + x') \cdot ((x + y) + y') & \text{commutative law} \\
&= (y + (x + x')) \cdot (x + (y + y')) & \text{associative law} \\
&= (y + 1) \cdot (x + 1) & \text{complement law} \\
&= 1 \cdot 1 & \text{null law} \\
&= 1 & \text{identity law}
\end{aligned}
$$

$$
\begin{aligned}
(x + y) \cdot (x' \cdot y') &= (x' \cdot y') \cdot (x + y) & \text{commutative law} \\
&= ((x' \cdot y') \cdot x) + ((x' \cdot y') \cdot y) & \text{distributive law} \\
&= (x \cdot (x' \cdot y')) + ((x' \cdot y') \cdot y) & \text{commutative law} \\
&= ((x \cdot x') \cdot y') + (x' \cdot (y' \cdot y)) & \text{associative law} \\
&= ((x \cdot x') \cdot y') + (x' \cdot (y \cdot y')) & \text{commutative law} \\
&= (0 \cdot y') + (x' \cdot 0) & \text{complement law} \\
&= (y' \cdot 0) + (x' \cdot 0) & \text{commutative law} \\
&= 0 + 0 & \text{null law} \\
&= 0 & \text{identity law}
\end{aligned}
$$

Next we prove $(x \cdot y)' = x' + y'$. We first show that $(x \cdot y) \cdot (x' + y') = 0$ and $(x \cdot y) + (x' + y') = 1$. When we have done this, by the uniqueness of complement law, we have shown $(x \cdot y)' = x' + y'$.

$$
\begin{aligned}
(x \cdot y) \cdot (x' + y') &= ((x \cdot y) \cdot x') + \cdot((x \cdot y) \cdot y') & \text{distributive law} \\
&= ((y \cdot x) \cdot x') + ((x \cdot y) \cdot y') & \text{commutative law} \\
&= (y \cdot (x \cdot x')) + (x \cdot (y \cdot y')) & \text{associative law} \\
&= (y \cdot 0) \cdot (x \cdot 0) & \text{complement law} \\
&= 0 \cdot 0 & \text{null law} \\
&= 0 & \text{identity law}
\end{aligned}
$$

$$\begin{aligned}
(x \cdot y) + (x' + y') &= (x' + y') + (x \cdot y) && \text{commutative law} \\
&= ((x' + y') + x) \cdot ((x' + y') + y) && \text{distributive law} \\
&= (x + (x' + y')) \cdot ((x' + y') + y) && \text{commutative law} \\
&= ((x + x') + y') \cdot (x' + (y' + y)) && \text{associative law} \\
&= ((x + x') + y') \cdot (x' + (y + y')) && \text{commutative law} \\
&= (1 + y') \cdot (x' + 1) && \text{complement law} \\
&= (y' + 1) \cdot (x' + 1) && \text{commutative law} \\
&= 1 \cdot 1 && \text{null law} \\
&= 1 && \text{identity law}
\end{aligned}$$

Note that every line in the second proof is the dual of the corresponding line in the first proof.

Hence, when we prove a theorem about a Boolean algebra, we really prove two theorems, both the theorem and its dual (if they are distinct). Thus, we prove a theorem and get one free.

From the rules of set theory (Theorem 2.20) it is easily shown that the subsets of a set A form a Boolean algebra where $\cap$ and $\cup$ correspond to the binary operators $\cdot$ and $+$, respectively, and $A - B$ corresponds to B'. The set A is the 1 element and the empty set is the 0 element of this Boolean algebra.

THEOREM 2.27 In a Boolean algebra, $x + y = y$ if and only if $xy = x$.

Proof Assume $x + y = y$. $x = (x + y)x$ by the absorption law. Therefore, $xy = (x + y)x = x$. The converse is the dual of the first part of the proof. ∎

Exercises

Let x and y be elements of a Boolean Algebra.

1. Prove directly that $x \cdot x = x$.
2. Prove directly that $x \cdot 0 = 0$.
3. Prove directly that $x \cdot (x + y) = x$.
4. Prove that $0' = 1$.
5. Prove that $1' = 0$.

Find the dual of the following expressions:

6. $x \cdot y' + x \cdot z' + y \cdot x'$
7. $x \cdot y \cdot z' + x \cdot y' \cdot z$
8. $x \cdot y \cdot (x + 0 + (z \cdot 1))$
9. $(x + y') \cdot (z' + y)'$
10. $(1 + x) \cdot y + x \cdot y' \cdot z$
11. $(x \cdot y + 1) \cdot (0 + x) \cdot z$
12. Describe a Boolean algebra with two elements, 0 and 1.
13. *Let B_n denote the set of all strings of length n consisting of 0's and 1's. For example, 10010101 is an element of B_8. For fixed n, describe operations $\cdot$ and $+$ on B_n so that B_n is a Boolean algebra.
14. *Let B consist of all factors of 30. For $a, b \in B$, let $a \cdot b = \gcd(a, b)$, and $a + b = \text{lcm}(a, b)$. Prove that B with these operations is a Boolean algebra. What are the zero and unit elements? What is the complement of an element a?
15. Prove that if $xy = xz$ and $x'y = x'z$, then $y = z$.
16. State the dual of the previous exercise.
17. Prove $xy' = 0$ if and only if $xy = x$.
18. State the dual of the previous exercise.
19. Prove that the 0 element and the 1 element are uniquely defined by their properties.
20. A set is **cofinite** if its complement is finite. Let the universal set U be the set of all finite and cofinite subsets of the positive integers. Prove that the subsets of U, together with the operations union, intersection, and complement, form a Boolean algebra.

2.5 Relations

One of the set operations that was described earlier is the Cartesian product of two sets A and B, which was denoted by $A \times B$. This is the set $\{(a, b) : a \in A \text{ and } b \in B\}$. The set $A \times B$, thus, consists of all ordered pairs having the first component an element of A and the second component an element of B.

DEFINITION 2.28   A **relation** R **between** A **and** B is a subset of $A \times B$. If $(a, b) \in \mathbf{R}$, we write $a\mathbf{R}b$ and say that a is related to b by means of $\mathbf{R}$ or simply a is **related** to b. If $A = B$, so that the relation is a subset of $A \times A$, then the relation is said to be a **relation on** A.

If $A = \{1, 2, 3\}$ and $B = \{r, s\}$ so that

$$A \times B = \{(1,r), (1,s), (2,r), (2,s), (3,r), (3,s)\}$$

then $R = \{(1, r), (1, s), (3, s)\}$ is a relation between A and B. We may also write $3Rs$ because $(3, s) \in R$. The set $A \times B$ has six elements so there are $2^6 = 64$ subsets of $A \times B$. Therefore, there are 64 different relations on $A \times B$.

In the examples and exercises we will assume that elementary properties of integers, real numbers, and functions are known.

The following are examples of relations:

1. The entire set $A \times B$ is a relation between A and B.
2. If A is the set of real numbers, then $\{(x, y) \in A \times A : x^2 + y^2 = 4\}$ is a relation on A.
3. If A is the set of merchandise in a store and B is the set of rational numbers, then $\{(x, y) \in A \times B : y \text{ is the price of } x\}$ is a relation between A and B.
4. If A is the set of women and B is the set of men, then $\{(x, y) : y \text{ is the husband of } x\}$ is a relation between A and B.
5. If A is the set of people, then $\{(x, y) \in A \times A : y \text{ is the cousin of } x\}$ is a relation on A.

DEFINITION 2.29   The **domain** of a relation R between A and B is the set of all $x \in A$ such $(x, y) \in R$ for some $y \in B$; that is, the domain of R is the set of all first coordinates of ordered pairs in R. The **range** of a relation R between A and B is the set of all $y \in B$ such $(x, y) \in R$ for some $x \in A$; that is, the range of R is the set of all second coordinates of ordered pairs in R.

In the preceding examples of relations, in (1), the domain is the entire set A and the range is the entire set B. In (2), the domain and range are both the set $\{t : -2 \leq t \leq 2\}$. In (3), the domain is the set A and the range is the set of all rational numbers that are the price of some piece of merchandise in the store. In (4), the domain is the set of all married women and the range is the set of all married men. In (5), the domain and range are both the set of all people who have cousins.

Associated with each relation $R \subseteq A \times B$ is a relation $R^{-1} \subseteq B \times A$.

2.5 Relations

DEFINITION 2.30 Let $R \subseteq A \times B$. Then the relation $R^{-1} \subseteq B \times A$ is defined by
$$R^{-1} = \{(b, a) : (a, b) \in R\}$$
That is, $(b, a) \in R^{-1}$ if and only if $(a, b) \in R$; or, equivalently, $bR^{-1}a$ if and only if aRb. The relation R^{-1} is called the **inverse relation** of R.

If $R = \{(1, r), (1, s), (3, s)\}$, then $R^{-1} = \{(r, 1), (s, 1), (s, 3)\}$. If R is the relation $\{(x, y) : y \text{ is the husband of } x\}$, then R^{-1} is the relation $\{(x, y) : y \text{ is the wife of } x\}$. If R is the relation $\{(x, y) : y \text{ is the cousin of } x\}$ or R is the relation $\{(x, y) : x^2 + y^2 = 4\}$, then $R^{-1} = R$.

Two given relations may be combined to generate a new relation as follows.

DEFINITION 2.31 Let $R \subseteq A \times B$ be a relation from A to B and $S \subseteq B \times C$ be a relation from B to C. The **composition** of S and R is the relation $T \subseteq A \times C$ defined by
$$T = \{(a, c) : \text{there is an element } b \text{ of } B \text{ such that } (a, b) \in R \text{ and } (b, c) \in S\}$$
This set is denoted by $T = S \circ R$.

EXAMPLE 2.32 Let $A = \{1, 2, 3\}$, $B = \{x, y\}$, and $C = \{\square, \triangle, \bigcirc, \bigstar\}$, and let the relations $R \subseteq A \times B$ and $S \subseteq B \times C$ be given by
$$R = \{(1, x), (1, y), (3, x)\}$$
$$S = \{(x, \square), (x, \triangle), (y, \bigcirc), (y, \bigstar)\}$$
Then
$$S \circ R = \{(1, \square), (1, \triangle), (1, \bigcirc), (1, \bigstar), (3, \square), (3, \triangle)\}$$
since
$$(1, x) \in R \text{ and } (x, \square) \in S \text{ imply that } (1, \square) \in S \circ R$$
$$(1, x) \in R \text{ and } (x, \triangle) \in S \text{ imply that } (1, \triangle) \in S \circ R$$
$$(1, y) \in R \text{ and } (y, \bigcirc) \in S \text{ imply that } (1, \bigcirc) \in S \circ R$$
$$\vdots$$
$$(3, x) \in R \text{ and } (x, \triangle) \in S \text{ imply that } (3, \triangle) \in S \circ R$$

■

EXAMPLE 2.33 Let R and S be relations on the set of positive integers defined by $S = \{(x, x+2) : x \text{ is a positive integer}\}$ and let $R = \{(x, x^2) : x \text{ is a positive integer}\}$. Then $S \circ R = \{(x, x^2 + 2) : x \text{ is a positive integer}\}$ and $R \circ S = \{(x, (x + 2)^2) : x \text{ is a positive integer}\}$.

■

THEOREM 2.34 Composition of relations is associative; that is, if A, B, and C are sets and if $R \subseteq A \times B$, $S \subseteq B \times C$, and $T \subseteq C \times D$, then $T \circ (S \circ R) = (T \circ S) \circ R$.

Proof First show that $T \circ (S \circ R) \subseteq (T \circ S) \circ R$. Let $(a, d) \in T \circ (S \circ R)$, then there exists $c \in C$ such that $(a, c) \in S \circ R$ and $(c, d) \in T$. Since $(a, c) \in S \circ R$, there exists $b \in B$ so that $(a, b) \in R$ and $(b, c) \in S$. Since $(b, c) \in S$ and $(c, d) \in T$, $(b, d) \in T \circ S$. Since $(b, d) \in T \circ S$ and $(a, b) \in R$, $(a, d) \in (T \circ S) \circ R$. Thus, $T \circ (S \circ R) \subseteq (T \circ S) \circ R$. The second part of the proof showing that $(T \circ S) \circ R \subseteq T \circ (S \circ R)$ is similar and is left to the reader.

■

74 Chapter 2 Set Theory

We now consider special properties of relations on A.

DEFINITION 2.35

A relation R on A is **reflexive** if for all a in A, (a, a) is in R. A relation R is **antireflexive** if $(a, b) \in R$ implies $a \neq b$. R is **symmetric** if for all a and b in A, (a, b) in R implies that (b, a) is in R. R is **transitive** if for all a, b, and c in A, whenever (a, b) and (b, c) are in R, then (a, c) is in R. R is **antisymmetric** if for all a and b in A, whenever (a, b) and (b, a) are in R, then $a = b$.

EXAMPLE 2.36

Let $A = \{1, 2, 3, 4, 5, 6\}$ and let the relation $R_1 \subseteq A \times A$ be the set $R_1 = \{(1, 1), (2, 2), (3, 3), (4, 4), (5, 5), (6, 6), (1, 2), (1, 4), (2, 1), (2, 4), (3, 5), (5, 3), (4, 1), (4, 2)\}$.

The relation R_1 is reflexive since for each $a \in A$, we have $(a, a) \in R_1$.

The relation R_1 may be shown to be symmetric by considering all possible cases

Case	$(a, b) \in R_1$	(b, a)	$(b, a) \in R_1$?
1	(1,2)	(2,1)	Yes
2	(1,4)	(4,1)	Yes
3	(2,1)	(1,2)	Yes
⋮	⋮	⋮	⋮

and showing that in every case if $(a, b) \in R_1$ then $(b, a) \in R_1$.

Also R_1 may be shown to be transitive by "inspection" or by exhaustion, as illustrated in the following table:

Case	$(a, b) \in R_1$	$(b, c) \in R_1$	(b, c)	$(a, c) \in R_1$?
1	(1,2)	(2,1)	(1,1)	Yes
2	(1,2)	(2,2)	(1,2)	Yes
3	(1,2)	(2,4)	(1,4)	Yes
4	(1,4)	(4,1)	(1,1)	Yes
5	(1,4)	(4,2)	(1,2)	Yes
⋮	⋮	⋮	⋮	⋮

Every possible case of $(a, b) \in R_1$ and $(b, c) \in R_1$ is examined, and it is found that $(a, c) \in R_1$.

R_1 is not antisymmetric because $(1, 2) \in R_1$ and $(2, 1) \in R_1$ but $1 \neq 2$. ∎

EXAMPLE 2.37

Let $A = \{\square, \triangle, \bigcirc, \bigstar\}$ and let $R_2 \subseteq A \times A$ be defined by

$R_2 = \{(\square,\square), (\square,\triangle), (\square,\bigstar), (\triangle,\square), (\bigstar,\square), (\bigstar,\bigstar), (\bigcirc,\bigstar), (\bigcirc,\bigcirc)\}$

R_2 is not reflexive because $\triangle \in A$, but $(\triangle, \triangle) \notin R_2$. R_2 is not symmetric because $(\bigcirc, \bigstar) \in R_2$, but $(\bigstar, \bigcirc) \notin R_2$. R_2 is not antisymmetric because $(\triangle, \square) \in R_2$ and $(\square, \triangle) \in R_2$, but $\triangle \neq \square$. R_2 is not transitive because $(\triangle, \square) \in R_2$ and $(\square, \bigstar) \in R_2$ but $(\triangle, \bigstar) \notin R_2$. ∎

EXAMPLE 2.38

Let A be the set of positive integers and define the relation R by $(x, y) \in R$ if y is a multiple of x. R is reflexive since for every positive integer n, $n = 1 \cdot n$ and $(n, n) \in R$. R is not symmetric since $(2, 4) \in R$, but $(4, 2) \notin R$; however, R is antisymmetric since if $(m, n) \in R$ and $(n, m) \in R$, then n is a multiple of n and m is a multiple of m, so that $m = m$. R is transitive for if $(m, n) \in R$ and $(n, p) \in R$ then n is a multiple of m and p is a multiple of n so that p is a multiple of m and $(m, p) \in R$. ∎

2.5 Relations

DEFINITION 2.39 Let R be a relation on a set A. The **reflexive closure** of R is the smallest reflexive relation on A containing R as a subset. The **symmetric closure** of R is the smallest symmetric relation on A containing R as a subset. The **transitive closure** of R is the smallest transitive relation on A containing R as a subset.

THEOREM 2.40 Let R be a relation on a set A and $I = \{x : x = (a, a) \text{ for some } a \in A\}$.

(a) The reflexive closure of R is $R \cup I$.

(b) The symmetric closure of R is $R \cup R^{-1}$.

(c) If A is a finite set containing n elements, then the transitive closure of R is the relation $R \cup R^2 \cup R^3 \cup \cdots \cup R^n$ where $R^2 = R \circ R, \ldots, R^{k+1} = R \circ R^k$.

Proof The proofs of parts (a) and (b) are left to the reader. Let the transitive closure of R be denoted by $\overline{R}$. To prove part (c), first we show that $R \cup R^2 \cup R^3 \cup \cdots \cup R^n \subseteq \overline{R}$. We do this using induction on n. For $n = 1$, we have $R \subseteq \overline{R}$, which is certainly true. Assume $R \cup R^2 \cup R^3 \cup \cdots \cup R^k \subseteq \overline{R}$. We need to show that $R \cup R^2 \cup R^3 \cup \cdots \cup R^{k+1} \subseteq \overline{R}$ or equivalently that $R^{k+1} \subseteq \overline{R}$. Let $(a, c) \in R^{k+1}$. Then there exists b so that $(a, b) \in R^k$ and $(b, c) \in R$. But, by the induction hypothesis, (a, b) and $(b, c) \in \overline{R}$. Since $\overline{R}$ is transitive, $(a, c) \in \overline{R}$. Therefore $R \cup R^2 \cup R^3 \cup \cdots \cup R^{k+1} \subseteq \overline{R}$. To show that $\overline{R} \subseteq R \cup R^2 \cup R^3 \cup \cdots \cup R^{k+1}$, we simply show that $R \cup R^2 \cup R^3 \cup \cdots \cup R^{k+1}$ is transitive. Let $(a, b) \in R^j$ and $(b, c) \in R^k$. Then $(a, c) \in R^{j+k}$. If $a = c$, we are done. Otherwise there exists $b_2, b_3, b_4, \ldots, b_{j+k-1} \in A$ such that $(a, b_2), (b_2, b_3), (b_3, b_4), \ldots, (b_{j+k-2}, b_{b_{j+k-1}}), (b_{j+k-1}, c) \in R$. Denote a by b_1 and c by b_{j+k}. If any of the b_i are equal, say for example $b_p = b_q$, we can remove $(b_p, b_{p+1}), (b_{p+1}, b_{p+3}), \ldots, (b_{q-1}, b_q)$ from the sequence of related ordered pairs above and still have $a, b_2, b_3, b_{p-1}, b_q, \ldots, b_{j+k-1}, c$ where each element in this sequence is R-related to the next one. Thus we can continue until all of the elements are distinct, but each is R-related to the next one. Since there are only n distinct elements in A, we have $(a, c) \in R^n$, and $R \cup R^2 \cup R^3 \cup \cdots \cup R^{k+1}$ is transitive. ∎

One of the ways to represent a finite antireflexive symmetric relation is to use a graph. We shall see that it is easy to go back and forth between finite antireflexive symmetric relations and graphs. However, graphs take on a life of their own and graph theory is an important field in mathematics.

DEFINITION 2.41 A **graph** is a finite set V called the **vertex set**, a symmetric antireflexive relation R on V and a collection E of two-element subsets of V defined by $\{a, b\} \in E$ if and only if $(a, b) \in R$. The set E is called the **edge set**. An element of E is called an **edge**. A graph is denoted by $G(V, E)$. The elements a and b of V are **joined** or **connected** by the edge $\{a, b\}$ if $\{a, b\} \in E$.

Figure 2.21

A finite graph is usually represented by a diagram in which the vertices are represented by dots and the edges connecting two vertices are represented by lines between their dots.

EXAMPLE 2.42 The graph with $V = \{a, b, c\}$ and $E = \{\{a, b\}, \{b, c\}\}$ may be shown as Figure 2.21 or Figure 2.22.

$$R = \{(a, b), (b, a), (b, c), (c, b)\}.$$ ∎

76 Chapter 2 Set Theory

EXAMPLE 2.43

The graph with $V = \{a, b, c, d, e\}$ and
$$E = \{\{a, b\}, \{a, e\}, \{b, e\}, \{b, d\}, \{b, c\}, \{c, d\}\}$$
has the diagram shown in Figure 2.23. $R = \{(a, b), (b, a), (e, a), (a, e), (e, b), (b, e), (b, d), (d, b), (b, c), (c, b), (d, c), (c, d)\}$.

Figure 2.22

We saw that a graph described a symmetric relation R on a set V in which no element was related to itself. For a more general relation R, we need to represent an element $(a, b) \in R$ where possibly $(b, a) \notin R$. We also need to represent an element $(a, a) \in R$. We do this with a directed graph.

DEFINITION 2.44

A **directed graph** or **digraph** G, denoted $G(V, E)$, consists of a set V of vertices, together with a relation E on V called the set of **directed edges** or simply **edges** if it is understood that the graph is directed. An element of E is called a **directed edge**. If $(a, b) \in E$, then a is called the **initial vertex** of (a, b) and b is called the **terminal vertex**.

Figure 2.23

Note that we include loops for the directed graph, which we did not do for graphs. We basically do this for two reasons: First, we can include directed edges naturally in our definition since a loop at a vertex a is simply the edge (a, a). We could not do this for undirected graphs since an edge of an undirected graph has the form $\{a, b\}$, so that a loop would have the form $\{a, a\}$. Second, generally in a relation, elements may be related to themselves.

The edge (a, b) of a digraph is denoted on the diagram by an arrow from a to b. Note that in a simple graph, an edge is represented by a two element subset to emphasize that the relation is symmetric, while in a directed graph, an edge is represented by an ordered pair to emphasize that order is important and that (a, b) may be an edge in a digraph while (b, a) is not.

Figure 2.24

EXAMPLE 2.45

The digraph with
$$V = \{a, b, c\}$$
and
$$E = \{(a, a), (a, b), (b, c), (c, b), (c, a)\}$$
is shown in Figure 2.24.

EXAMPLE 2.46

The digraph with
$$V = \{a, b, c, d\}$$
and
$$E = \{(a, b), (b, c), (c, c), (b, d), (d, b), (c, d), (d, a)\}$$
is shown in Figure 2.25.

Figure 2.25

Exercises

Find the range and domain of the following relations:

1. $\{(a, 1), (a, 2), (c, 1), (c, 2), (c, 4), (d, 5)\}$

2. $\{(1, 2), (2, 4), (3, 6), (4, 8), \dots\}$

3. $\{(x, y) : x, y \in R \text{ and } x = y^2\}$

4. $\{(a, 1), (a, 2), (a, 3), (a, 4), (a, 5), (a, 6)\}$

5. $\{(x, y) : x, y \in I \text{ and } x^2 + y^2 \leq 16\}$.

6. $\{(x, y) : 0 \leq x, y \leq 10 \text{ and } x > 2y\}$

Let $A = \{1, 2, 3, 4, 5\}$
$B = \{6, 7, 8, 9\}$
$C = \{10, 11, 12, 13\}$
$D = \{\square, \triangle, \bigcirc, *\}$

Let $R \subseteq A \times B$, $S \subseteq B \times C$, and $T \subseteq C \times D$ be defined by
$R = \{(1, 7), (4, 6), (5, 6), (2, 8)\}$
$S = \{(6, 10), (6, 11), (7, 10), (8, 13)\}$
$T = \{(11, \triangle), (10, \triangle), (13, *), (12, \square), (13, \bigcirc)\}$

Compute the relations:

7. R^{-1} and S^{-1}
8. $S \circ R$
9. $S \circ S^{-1}$ and $S^{-1} \circ S$
10. $R^{-1} \circ S^{-1}$
11. $T \circ (S \circ R)$
12. $T \circ S$
13. $(T \circ S) \circ R$

Let $A = \{(b, a), (c, e), (d, i), (f, o)(g, u)\}$ and $B = \{(v, a), (w, e), (x, i), (y, o)(z, u)\}$.

14. Describe the relation A^{-1}.
15. Describe the relation B^{-1}.
16. Describe the relation $A^{-1} \circ B$.
17. Describe the relation $B^{-1} \circ A$.

Let relations $U, V \subseteq R \times R$ be defined by $U = \{(x, y) : y = x^2 + 5)\}$ and $V = \{(x, y) : y = 3x\}$.

18. Describe the relation $U \circ V$.
19. Describe the relation $V \circ U$.
20. Describe the relation U^{-1}.
21. Describe the relation V^{-1}.
22. Find the range of U.
23. Let $A = \{a, b, c, d, e\}$. Describe the smallest reflexive relation on A.

Let $A = \{a, b, c, d, e\}$ and $S = \{(a, a), (a, b), (b, c), (b, d), (c, e), (e, d), (c, b)\}$.

24. Describe the smallest symmetric relation on A containing S.
25. Describe the smallest reflexive, symmetric relation on A containing S.
26. Describe the largest symmetric relation contained in S.
27. Describe the smallest transitive relation on A containing S.

Let $A = \{a, b, c, d, e\}$ and S, T, U, and V be relations on A where

$S = \{(a, a), (a, b), (b, c), (b, d), (c, e), (e, d), (c, a)\}$
$T = \{(a, b), (b, a), (b, c), (b, d), (e, e), (d, e), (c, b)\}$
$U = \{(a, b), (a, a), (b, c), (b, b), (e, e), (b, a), (c, b),$
$(c, c), (d, d), (a, c), (c, a)\}$
$V = \{(a, b), (b, c), (b, b), (e, e), (b, a), (c, b),$
$(d, d), (a, c), (c, a)\}$

28. Which of the relations are symmetric?
29. Which of the relations are reflexive?
30. Which of the relations are transitive?
31. Which of the relations are antisymmetric?
32. Describe $U \cap V$.
33. Describe $S \cup T$.
34. Describe $U - T$.
35. Describe $U \triangle S$.
36. Prove that the intersection of reflexive relations is reflexive.
37. Prove that the intersection of symmetric relations is symmetric.

Let $A = \{a, b, c, d, e\}$.

38. Describe a relation on A that is reflexive but not symmetric nor transitive.
39. Describe a relation on A that is symmetric but not reflexive nor transitive.
40. Describe a relation on A that is transitive but not reflexive nor symmetric.
41. Describe a relation on A that is reflexive and symmetric but not transitive.
42. Describe a relation on A that is symmetric and transitive but not reflexive.
43. Describe a relation on A that is reflexive and transitive but not symmetric.

State whether the following are true or false. For each statement that is false, give a counterexample.

44. If relations R and S are reflexive, then $R \cap S$ is reflexive.
45. If relations R and S are reflexive, then $R \cup S$ is reflexive.
46. If relations R and S are reflexive, then $R \circ S$ is reflexive.
47. If relations R and S are reflexive, then $R - S$ is reflexive.
48. If relations R and S are reflexive, then $R \triangle S$ is reflexive.

78   Chapter 2   Set Theory

State whether the following are true or false. For each statement that is false, give a counterexample.

49. If relations R and S are symmetric, then $R \cap S$ is symmetric.
50. If relations R and S are symmetric, then $R \cup S$ is symmetric.
51. If relations R and S are symmetric, then $R \circ S$ is symmetric.
52. If relations R and S are symmetric, then $R - S$ is symmetric.
53. If relations R and S are symmetric, then $R \triangle S$ is symmetric.
54. If relations R and S are antisymmetric, then $R \cap S$ is antisymmetric.
55. If relations R and S are antisymmetric, then $R \cup S$ is antisymmetric.
56. If relations R and S are antisymmetric, then $R \circ S$ is antisymmetric.
57. If relations R and S are antisymmetric, then $R - S$ is antisymmetric.
58. If relations R and S are antisymmetric, then $R \triangle S$ is antisymmetric.
59. If relations R and S are transitive, then $R \cap S$ is transitive.
60. If relations R and S are transitive, then $R \cup S$ is transitive.
61. If relations R and S are transitive, then $R \circ S$ is transitive.
62. If relations R and S are transitive, then $R - S$ is transitive.
63. If relations R and S are transitive, then $R \triangle S$ is transitive.

Construct the graph for each of the following relations on A:

64. $A = \{a, b, c, d, e\}$
    $R = \{(a, b), (b, a), (b, c), (c, b), (c, a), (a, c),$
    $(d, e), (e, d)\}$
65. $A = \{a, b, c, d, e\}$
    $R = \{(a, b), (b, a), (b, c), (c, b), (c, d), (d, c),$
    $(c, a), (a, c)\}$
66. $A = \{a, b, c, d, e\}$
    $R = \{(a, b), (b, a), (b, c), (c, b), (c, d), (d, c), (d, e),$
    $(e, d), (b, e), (e, b), (b, d), (d, b)\}$
67. $A = \{a, b, c, d\}$
    $R = \{(a, b), (b, a), (b, c), (c, b), (c, d), (d, c), (d, a),$
    $(a, d), (b, d), (d, b), (a, c), (c, a)\}$
68. $A = \{a, b, c, d, e\}$
    $R = \{(a, b), (b, a), (b, c), (c, b), (c, d), (d, c),$
    $(d, a), (a, d)\}$
69. $A = \{a, b, c, d, e, f\}$
    $R = \{(a, b), (b, a), (b, c), (c, b), (c, a), (a, c),$
    $(d, e), (e, d)\}$
70. $A = \{a, b, c, d, e\}$
    $R = \{(a, b), (b, a), (b, c), (c, b), (c, d), (d, c), (d, e),$
    $(e, d), (a, d), (d, a)\}$
71. $A = \{a, b, c, d, e, f\}$
    $R = \{(a, b), (b, a), (b, c), (c, b), (c, a), (a, c), (a, e),$
    $(e, a), (b, e), (e, b), (b, f), (f, b), (c, d), (d, c),$
    $(c, f), (f, c), (d, f), (f, d), (e, f), (f, e)\}$

Find the vertex set, edges, and the symmetric relation for the graphs in

72.   73.

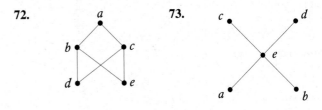

74.   75.

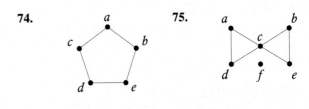

76.   77.

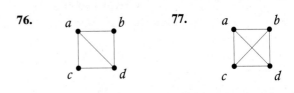

78.   79.

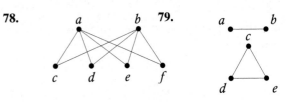

Construct digraphs with the following properties:

80. Vertex set $\{a, b, c, d, e, f\}$ and relation R of edges where $R = \{(a, b), (b, c), (a, c), (b, e), (c, f), (c, d), (d, f), (f, e)\}$

81. Vertex set $\{a, b, c, d\}$ and relation R of edges where $R = \{(b, c), (a, d), (b, a), (d, c), (b, d), (c, a)\}$

82. Vertex set $\{a, b, c, d\}$ and relation R of edges where $R = \{(b, a), (a, a), (b, c), (c, d), (d, c), (d, b), (d, a)\}$

83. Vertex set $\{a, b, c, d, e\}$ and relation R of edges where $R = \{(a, b), (a, c), (a, d), (c, a), (d, e), (e, d)\}$

84. Vertex set $\{a, b, c, d, e, f\}$ and relation R of edges where $R = \{(a, b), (b, c), (d, c), (d, e), (f, e), (f, a), (b, e)\}$

85. Vertex set $\{a, b, c, d\}$ and relation R of edges where $R = \{(a, c), (a, b), (d, c), (d, b), (a, d), (b, c), (a, a), (c, c)\}$

86. Vertex set $\{a, b, c\}$ and relation R of edges where $R = \{(a, b), (a, a), (b, c), (b, b), (c, c), (c, a)\}$

87. Vertex set $\{a, b, c, d, e\}$ and relation R of edges where $R = \{(a, b), (a, c), (a, d), (c, d), (d, e), (b, e)\}$

Find the vertices and directed edges for the following digraphs.

88.

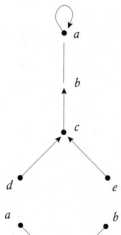

89.

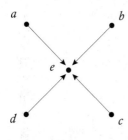

90.

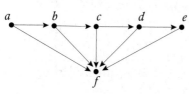

91.

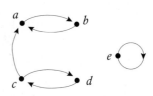

92.

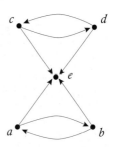

93.

94.

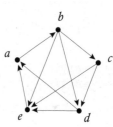

95.

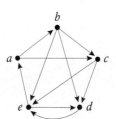

96. The following argument supposedly shows that if a relation R is symmetric and transitive, then it is reflexive. Is the proof correct? Why or why not?

Let R be a symmetric and transitive relation on a set A. Let $a \in A$ and $(a, b) \in R$. Since $(a, b) \in R$ and R is symmetric, $(b, a) \in R$. Therefore, since R is transitive, $(a, a) \in R$ and R is reflexive.

97. Complete the proof of Theorem 2.34.

98. Complete the proof of Theorem 2.40.

2.6 Partially Ordered Sets

For the remainder of this chapter, we shall assume that elementary properties of the integers, real numbers, and functions are known even though they have not been formally discussed at this time.

DEFINITION 2.47 A relation R on A is a **partial ordering** if it is reflexive, antisymmetric, and transitive. If the relation R on A is a partial ordering, then (A, R) is a **partially ordered set** or **poset** with ordering R. If the ordering R is understood, then (A, R) may simply be denoted by A.

It has already been shown in Example 2.38 that if A is the set of positive integers and the relation R is defined by $(x, y) \in R$ if x divides y evenly, then R is a partial ordering and (A, R) is a partially ordered set.

EXAMPLE 2.48 Let $C = \{1, 2, 3\}$ and X be the power set of C:

$$X = P(C) = \{\varnothing, \{1\}, \{2\}, \{3\}, \{1, 2\}, \{1, 3\}, \{2, 3\}, \{1, 2, 3\}\}$$

Define the relation R on X by $(T, V) \in R$ if $T \subseteq V$. Thus, $(\{2\}, \{1, 2\}) \in R$ because $\{2\} \subseteq \{1, 2\}$ and $(\{2, 3\}, \{3\}) \notin R$ because $\{2, 3\} \not\subseteq \{3\}$. One can easily verify that R is a partial ordering and (A, R) is a partially ordered set. ∎

EXAMPLE 2.49 Let S be the set of real numbers and R_1 be the relation defined by $(x, y) \in R_1$ if $x \leq y$. It is easily shown that R_1 is a partial ordering and (S, R_1) is a partially ordered set. ∎

It is customary to denote a partial ordering by $\leq$ and a partially ordered set by $(S, \leq)$ where $\leq$ is the partial ordering on the set S. If $(a, b) \in \leq$, then by our previous notation for relations, $a \leq b$.

DEFINITION 2.50 Two elements a and b of the partially ordered set (poset) $(S, \leq)$ are **comparable** if either $a \leq b$ or $b \leq a$. If every two elements of a partially ordered set $(S, \leq)$ are comparable, then $(S, \leq)$ is a **total ordering** or a **chain.**

EXAMPLE 2.51 Let T be the set of all positive divisors of 30 and $\leq_1$ be the relation $m \leq_1 n$ if m divides n evenly. The integers 5 and 15 are comparable because 5 divides 15 evenly, but 5 and 6 are not comparable. ∎

EXAMPLE 2.52 Let A be the set of integers and $R = \leq_2$ be the relation $x \leq_2 y$ if x is less than or equal to y. The ordered set $(A, \leq_2)$ is a chain. ∎

EXAMPLE 2.53 Let S be the set of all subsets of $\{a, b, c\}$, and $\leq_3$ be the partial ordering defined in Example 2.48. The sets $\{a, b\}$, $\varnothing$, and $\{a, b, c\}$ are all comparable, but $\{a, b\}$ and $\{b, c\}$ are not comparable. Thus, $(S, \leq_3)$ is a partially ordered set that is not a chain. ∎

The usual device for illustrating a poset is the **Hasse diagram.** For a given poset $(A, \leq)$, a Hasse diagram consists of a collection of points and lines in which the points represent the elements of A and if $a \leq c$ for elements a and c of A, then a is placed below c and they are connected by a line if there is no $b \neq a, c$ such that $a \leq b \leq c$. When relations are restricted to posets, a Hasse diagram is simply a directed graph for which loops are not shown; and if $a \leq b \leq c$, then the line from a to c is not shown. For example, the Hasse diagram representing $(T, \leq_1)$ above is given in Figure 2.26.

Figure 2.26

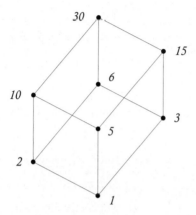

The Hasse diagram representing $(S, \leq_3)$ is shown in Figure 2.27.

Figure 2.27

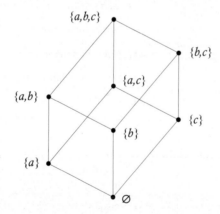

Figure 2.28

The Hasse diagram for a chain with four elements is shown in Figure 2.28.

Exercises

For which of the following is the relation R a partial ordering on $A = \{a, b, c, d\}$?

1. $R = \{(a, a), (b, b), (c, c), (d, d), (a, c), (b, c), (c, d), (a, d), (b, d)\}$

2. $R = \{(a, a), (b, b), (c, c), (d, d), (a, b), (b, c), (c, d), (d, a)\}$

3. $R = \{(b, b), (c, c), (d, d), (a, c), (b, c), (c, d), (a, d), (b, d)\}$

4. $R = \{(a, a), (b, b), (c, c), (d, d), (a, b), (b, c), (a, c), (a, d), (b, d), (c, d)\}$

For which of the following is the relation R a partial ordering on A

5. A is the set of all people and R is defined by xRy if x is older than y.

6. A is the set of all citizens of the United States and R is defined by xRy if x has a larger social security number than y.

82 Chapter 2 Set Theory

7. A is the set of all integers and R is defined by xRy if $x \geq 2y$.

8. A is the set of all people and R is defined by xRy if x and y are siblings.

9. A is the set of all ordered pairs of positive integers and $(a,b)R(c,d)$ if $a \leq c$ and if $a = c$, then $b \leq d$.

Construct the Hasse diagram for the following partially ordered sets $(A, \leq)$ where

10. $A = \{a, b, c, d\}$ and $\leq = \{(a,a), (b,b), (c,c), (d,d), (a,c), (b,c), (c,d), (a,d), (b,d)\}$

11. $A = \{a, b, c, d\}$ and $\leq = \{(a,a), (b,b), (c,c), (d,d), (a,c), (b,c)\}$

12. $A = \{a, b, c, d\}$ and $\leq = \{(a,a), (b,b), (c,c), (d,d)\}$

13. $A = \{a, b, c, d\}$ and $\leq = \{(a,a), (b,b), (c,c), (d,d), (a,b), (b,c), (a,c), (c,d), (a,d), (b,d)\}$

14. $A = \{1, 2, 3, 4, 5, 6, 7, 8, 9, 10, 11\}$ and $x \leq y$ if x divides y evenly.

15. $A = B \times B$ where $B = \{1, 2, 3, 6, \}$. Define $a \leq_1 b$ if a divides b evenly and $(a, b) \leq (c, d)$ if $a \leq_1 c$ and if $a = c$, then $b \leq_1 d$.

16. A is the set of all positive integers that divide 27 evenly and $x \leq y$ if x divides y evenly.

17. A is the set of all positive integers that divide 54 evenly and $x \leq y$ if x divides y evenly.

List A and express $\leq$ as a set of ordered pairs for each of the following Hasse diagrams:

18.

19.

20.

21.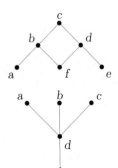

22.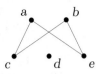

For each partial order in problem 18, list the elements:

23. Not comparable to a. 24. Not comparable to b.

25. Not comparable to c. 26. Not comparable to d.

For each partial order in problem 19, list the elements:

27. Not comparable to a. 28. Not comparable to b.

29. Not comparable to c. 30. Not comparable to d.

For each partial order in problem 20, list the elements:

31. Not comparable to a. 32. Not comparable to b.

33. Not comparable to c. 34. Not comparable to d.

For each partial order in problem 21, list the elements:

35. Not comparable to a. 36. Not comparable to b.

37. Not comparable to c. 38. Not comparable to d.

For each partial order in problem 22, list the elements:

39. Not comparable to a. 40. Not comparable to b.

41. Not comparable to c. 42. Not comparable to d.

43. Describe relations that are symmetric partial orderings.

44. *Let B be a Boolean algebra. For $x, y \in B$, define $x \leq y$ if $xy = x$. Show that $(B, \leq)$ is a partially ordered set.

Let B be a Boolean algebra with the ordering defined in problem 44. An element $x \in B$ is called an **atom** of B if for all $y \in B$, $y \leq x$ implies $y = 0$ or $y = x$.

45. *Let B be a Boolean algebra. Prove that if x is an atom of B, and $y \in B$, then either $xy = 0$ or $xy = x$.

46. *Let B be a Boolean algebra. Prove that if x and y are atoms of B, then $xy = 0$.

47. Given a linearly ordered set S a **lexicographical ordering** $\preceq$ on the strings of elements of S of is defined by $a \prec\!\!\prec b$ if $a = a_1a_2a_3\ldots a_m$, $b = b_1b_2b_3\ldots b_n$ and any of the following occur

    (a) $a_1 < b_1$.

    (b) $a_i = b_i$ for all $1 \leq i \leq k$ and $a_{k+1} < b_{k+1}$.

    (c) $a_i = b_i$ for all $1 \leq i \leq m$ and $m < n$.

    Prove that a lexicographical ordering is a partial ordering. Is it a linear ordering?

2.7 Equivalence Relations

In Section 2.5 we defined the following properties of relations. A relation R on A is **reflexive** if for all a in A, (a, a) is in R. R is **symmetric** if for all a and b in A, (a, b) in R implies that (b, a) is in R. R is **transitive** if for all a, b, and c in A, whenever (a, b) and (b, c) are in R, (a, c) is in R. These properties are combined in the following definition.

DEFINITION 2.54 A relation R on A is an **equivalence relation** if it is reflexive, symmetric, and transitive.

Let $A = \{1, 2, 3, 4, 5, 6\}$. The relation R_1 on A defined by
$$R_1 = \{(1, 1), (2, 2), (3, 3), (4, 4), (5, 5), (6, 6), (1, 2),$$
$$(1, 4), (2, 1), (2, 4), (3, 5), (5, 3), (4, 1), (4, 2)\}$$
was previously found in Example 2.36 to be reflexive, transitive, and symmetric; therefore, R_1 is an equivalence relation on the set A.

EXAMPLE 2.55 Let A be the set of all integers. Define the relation $R_3 \subseteq A \times A$ by $R_3 = \{(a, b) : a - b = 5 \cdot k \text{ for some integer } k\}$. For example, $(7, 2) \in R_3$ because $7 - 2 = 5 = 5 \cdot 1$, and $(-11, 4) \in R_3$ because $(-11) - 4 = -15 = 5(-3)$.

The relation R_3 is reflexive. If a is an integer (i.e., $a \in A$), then $a - a = 0 = 5 \cdot 0 = 5 \cdot k$ for $k = 0$ so that $(a, a) \in R_3$.

The relation R_3 is symmetric. Assume $(a, b) \in R_3$. Then there is an integer m so that $a - b = 5 \cdot m$ and
$$b - a = -(a - b)$$
$$= -(5 \cdot m)$$
$$= 5 \cdot (-m)$$
for the integer $-m$. Thus, $(b, a) \in R_3$.

The relation R_3 is transitive. Assume that a, b, and c are integers, $(a, b) \in R_3$, and $(b, c) \in R_3$. By definition,

$(a, b) \in R_3$ implies that $a - b = 5 \cdot k$ for some integer k

and

$(b, c) \in R_3$ implies that $b - c = 5 \cdot m$ for some integer m

Adding these two equalities gives
$$(a - b) + (b - c) = 5 \cdot k + 5 \cdot m$$
or
$$a - c = 5 \cdot (k + m)$$
for the integer $k + m$. By definition of R_3, $(a, c) \in R_3$, and R_3 is transitive.

Since R_3 is reflexive, symmetric, and transitive, it is an equivalence relation. ∎

An equivalence relation R on A separates the elements of A into subsets where the elements in a given subset are all related to each other through the relation R, but the elements in different sets are not related. In the context of equivalence relations, these subsets are called the equivalence classes of the relation R.

A physical analogy would be for A to be a collection of different colored balls and R to be the relation defined by $(a, b) \in R$ if and only if ball a has the same color as ball b. Since R is an equivalence relation, each equivalence class will consist of balls having the same color. If R is the relation defined by $(a, b) \in R$ if and only if a and b have the same diameter, then each equivalence class of R will consist of balls having the same size.

DEFINITION 2.56 Let $a \in A$ and R be an equivalence relation on A. Let $[a]$ denote the set $\{x : xRa\} = \{x : (x, a) \in R\}$, called the **equivalence class** containing a. The symbol $[A]_R$ denotes the set of all equivalence classes of A for the equivalence relation R.

EXAMPLE 2.57 Previously R_1, defined previously, was shown to be an equivalence relation on the set $A = \{1, 2, 3, 4, 5, 6\}$. We obtain the equivalence classes of R_1 by determining the equivalence class of each element of A:

$$[1] = \{x : (x, 1) \in R_1\} = \{x : xR_1 1\}$$
$$= \{1, 2, 4\}$$

where $1 \in [1]$ because $(1, 1) \in R_1$, $2 \in [1]$ because $(2, 1) \in R_1$, and $4 \in [1]$ because $(4, 1) \in R_1$, and there is no other x in A such that $(x, 1) \in R_1$. Similarly, we obtain

$$[2] = \{x : (x, 2) \in R_1\}$$
$$= \{2, 1, 4\}$$

$$[3] = \{x : (x, 3) \in R_1\}$$
$$= \{3, 5\}$$

$$[4] = \{x : (x, 4) \in R_1\}$$
$$= \{4, 1, 2\}$$

$$[5] = \{x : (x, 5) \in R_1\}$$
$$= \{5, 3\}$$

$$[6] = \{x : (x, 6) \in R_1\}$$
$$= \{6\}$$

There are only three distinct equivalence classes:

$$[1] = [2] = [4] = \{1, 2, 4\}$$
$$[3] = [5] = \{3, 5\}$$

and

$$[6] = \{6\}$$

so that

$$[A]_{R_1} = \{[1], [3], [6]\} = \{\{1, 2, 4\}, \{3, 5\}, \{6\}\}$$

In this example we see that any member of an equivalence class generates the equivalence class; that is, if $b \in [a]$, then $[a] = [b]$. Because of this property, we say that any member of an equivalence class represents the class. Every equivalence class contains at least one element because the relation is reflexive so the set of all elements related to an element a must contain a. We also see that no element is in two distinct equivalence classes. ∎

2.7 Equivalence Relations

EXAMPLE 2.58 Consider the equivalence relation R_3 of Example 2.55. For the set A of all integers, $R_3 \subseteq A \times A$ was defined by $R_3 = \{(a, b) : a - b = 5 \cdot k \text{ for some integer } k\}$. Since

$$[a] = \{x : (x, a) \in R_3\} = \{x : xR_3a\}$$
$$= \{x : x - a = 5 \cdot k \text{ for some integer } k\}$$
$$= \{x : x = a + 5 \cdot k \text{ for some integer } k\}$$

we see that

$$[0] = \{\ldots, -15, -10, -5, 0, 5, 10, 15, 20, 25, \ldots\}$$
$$= \cdots = [-5] = [0] = [5] = [10] = [15] = \cdots$$

$$[1] = \{\ldots, -14, -9, -4, 1, 6, 11, \ldots\}$$
$$= \cdots = [-9] = [-4] = [1] = [6] = \cdots$$

$$[2] = \{\ldots, -13, -8, -3, 2, 7, 12, \ldots\}$$
$$= \cdots = [-3] = [-2] = [7] = [12] = \cdots$$

$$[3] = \{\ldots, -12, -7, -2, 3, 8, 13, \ldots\}$$
$$= \cdots = [-2] = [3] = [8] = [13] = \cdots$$

$$[4] = \{\ldots, -11, -6, -1, 4, 9, 14, \ldots\}$$
$$\cdots = [-6] = [-1] = [4] = [9] = \cdots$$

are the distinct equivalence classes for the equivalence relation R_3. Thus,

$$[A]_{R_3} = \{[0], [1], [2], [3], [4]\}$$

The elements of [0] are "alike" in the sense that each is a multiple of 5. The elements of any equivalence class are "alike" in the sense that when any element of the set is divided by 5, the remainder is the same. ∎

As mentioned previously, the collection of equivalence classes divides up all of the elements of the set A into nonempty sets that are **mutually exclusive** or **disjoint**, meaning that no two of them have an element in common. Such a division of a set is called a partition of the set.

DEFINITION 2.59 Let A and I be sets and let $\langle A \rangle = \{A_1 : i \in I\}$, with I nonempty, be a set of nonempty subsets of A. The set $\langle A \rangle$ is called a **partition** of A if both of the following are satisfied:

(a) $A_i \cap A_j = \varnothing$ for all $i \neq j$.
(b) $A = \bigcup_{i \in I} A_i$; that is, $a \in A$ if and only if $a \in A_i$ for some $i \in I$.

The next theorem shows that the concept of partition and the concept of equivalence relation are the same.

THEOREM 2.60 A nonempty set of subsets $\langle A \rangle$ of a set A is a partition of A if and only if $\langle A \rangle = [A]_R$ for some equivalence relation R.

Proof Let $\langle A \rangle = \{A_1 : i \in I\}$ be a partition of A. Define a relation R on A by aRb if and only if a and b are in the same subset A_i for some i. Certainly for all a in A, aRa and R is reflexive. If a and b are in the same subset A_i, then b and a

are in the subset A_i and R is symmetric. Since the sets $A_i \cap A_j = \varnothing$ for $i \neq j$, if a and b are in the same subset and b and c are in the same subset, then a and c are in the same subset. Hence, R is transitive and R is an equivalence relation.

Conversely, assume that R is an equivalence relation. We need to show that $[A]_R = \{[a] : a \in A\}$ is a partition of A. Certainly, for all a, $[a]$ is nonempty since $a \in [a]$. Obviously, A is the union of the $[a]$, such that $a \in A$. Assume that $[a] \cap [b]$ is nonempty and let $x \in [a] \cap [b]$. Then xRa and xRb, and by symmetry, aRx. But since aRx and xRb, by transitivity, aRb. Therefore, $a \in [b]$. If $y \in [a]$, then yRa and since aRb, by transitivity, yRb. Therefore, $[a] \subseteq [b]$. Similarly, $[b] \subseteq [a]$ so that $[a] = [b]$, and we have a partition of A. ∎

EXAMPLE 2.61 Let $A = \{\square, \triangle, \bigcirc, \bigstar\}$ and consider the partition

$$A_1 = \{\square\}, A_2 = \{\triangle, \bigcirc, \bigstar\}$$

According to the proof of the preceding theorem, we should define R by $R = \{(a, b) : a \in A_i \text{ and } b \in A_i \text{ for some } i\}$. Thus,

$$R = \{(\square, \square), (\bigstar, \triangle), (\bigstar, \bigcirc), (\bigstar, \bigstar), (\bigcirc, \triangle),$$
$$(\bigcirc, \bigcirc), (\bigcirc, \bigstar), (\triangle, \triangle), (\triangle, \bigcirc), (\triangle, \bigstar)\}$$

is the relation generated by the given partition. ∎

Exercises

Determine whether each of the following relations on A is an equivalence relation. For each equivalence relation, give the equivalence classes.

1. A is the set of integers and R is the relation, $(a, b) \in R$ if $a + b = 0$.

2. A is the set of integers and R is the relation, $(a, b) \in R$ if $a + b = 5$.

3. A is the set of propositions and R is the relation, $(p, q) \in R$ if p is logically equivalent to q.

4. A is the set of ordered pairs of integers and $(a, b)R(c, d)$ if $ad = bc$.

5. $A = \{-10, -9, -8, -7, \ldots 0, 1, \ldots, 9, 10\}$ and $(a, b) \in R$ if $a^2 = b^2$.

6. $A = \{-10, -9, -8, -7, \ldots 0, 1, \ldots, 9, 10\}$ and $(a, b) \in R$ if $a^3 = b^3$.

Let $A = \{1, 2, 3\}$. Determine whether each of the following relations on A is an equivalence relation. For each equivalence relation, give the equivalence classes.

7. $R_1 = \{(2, 2), (1, 1)\}$

8. $R_2 = \{(1, 1), (2, 2), (3, 3)\}$

9. $R_3 = \{(1, 1), (2, 2), (3, 3), (1, 2), (2, 1), (3, 1), (1, 3)\}$

10. $R_4 = \{(1, 1), (2, 2), (3, 3), (1, 2), (3, 2), (2, 1)\}$

11. $R_5 = \{(1, 1), (2, 2), (3, 3), (1, 2), (2, 1), (2, 3), (3, 2), (1, 3), (3, 1)\}$

Determine whether each of the following relations on A is an equivalence relation. For each equivalence relation, give the equivalence classes.

12. Let A be the power set of the set $\{a, b, c, d\}$. Define the relation R by sRt if s and t contain the same number of elements.

13. Let $A = \{1, 2, 3, 4, 5, 6, 7, 8, 9, 10\}$. Define the relation R by aRb if $a + b$ is positive.

14. Let $A = \{1, 2, 3, 4, 5, 6, 7, 8, 9, 10\}$. Define the relation R by aRb if $a + b$ is even.

15. Let A be the set of lines in a plane. Define the relation R by nRm if lines n and m intersect.

16. Let A be the set of lines in a plane. Define the relation R by nRm if lines n and m are parallel.

17. Let A be the set of computer programs written for NASA. Define the relation R by pRq if p and q are written in the same computer language.

Let $A = \{\square, \triangle, \bigcirc, w, t, z, h, \bigstar\}$. The relation R on A is defined by $R = \{(\square, \square), (\triangle, \triangle), (\bigcirc, \bigcirc), (w, w), (t, t), (z, z), (h, h), (\bigstar, \bigstar), (\square, \bigcirc), (\bigcirc, z), (w, h), (t, \bigstar), (\square, z), (z, \square), (z, \bigcirc), (h, w), (\bigstar, t), (\bigcirc, \square)\}$.

18. Show that R is an equivalence relation on A.

19. Calculate all the equivalence classes of R.

Let $f(x) = x^2 + 1$, where $x \in [-2, 4] = \{x : x \text{ is a real number and } -2 \leq x \leq 4\} = A$. Define the relation R on A as follows:

$$(a, b) \in R \text{ if and only if } f(a) = f(b)$$

20. Show that R is an equivalence relation on $A = [-2, 4]$.
21. Determine the equivalence classes:

$$[1], [2], [0], [3], [-1/2], \text{ and } [4]$$

For $X \neq \emptyset$, define the relation I to be $I = \{x : x = (a, a) \text{ for some } a \in X\} = \{(a, a) \in X \times X : a \in X\}$. I is an equivalence relation on X and is called the identity relation on X. If R is a relation on X, prove that

22. R is reflexive if and only if $I \subseteq R$.
23. R is symmetric if and only if $R = R^{-1}$.
24. R is transitive if and only if $R \circ R \subseteq R$.

State which of the following collections are partitions of the set $A = \{1, 2, 3, 4, 5, 6, 7\}$. For any that are not partitions, state why not. For those which are partitions, list the elements of the corresponding relation R or if the list in R is long describe the set of ordered pairs rather than listing them.

25. $\{\{1, 2\}, \emptyset, \{3, 4, 5\}, \{6, 7\}\}$
26. $\{\{1, 2\}, \{3, 4, 5\}, \{6, 7\}\}$ 27. $\{\{1, 7\}, \{3, 4, 6\}\}$
28. $\{\{1, 5\}, \{3, 4, 5\}, \{2, 6, 7\}\}$ 29. $\{\{1, 2, 3, 4, 5, 6, 7\}\}$

State which of the following collections are partitions of the set $A = \{1, 2, 3, 4, 5, 6, 7\}$. For any which are not partitions, state why not. For those which are partitions, list the elements of the corresponding relation R or if the list in R is long describe the set of ordered pairs rather than listing them.

30. $\{\{1\}, \{2\}, \{3\}, \{4\}, \{5\}, \{6\}, \{7\}\}$
31. $\{\{1, 4\}, \{3, 5, 8\}, \{6, 2, 7\}\}$
32. $\{\{1, 6\}, \{3, 4\}, \{5, 2\}\}$
33. $\{\{1, 7\}, \{3, 5\}, \{2, 4\}, \{6\}\}$
34. $\{\{1, 2, 3, 4\}, \{4, 5, 6, 7\}\}$

35. Let A be the set of nonnegative integers and B be the set of positive integers. Define a relation R on $A \times B$ by $(a, b)R(c, d)$ if $ad = bc$. Is R an equivalence relation? If so prove it. If not, state why not. What are the elements of $A \times B$ related to $(1, 2)$?

36. Let A be the set of nonnegative integers and define R on $A \times A$ by $(a, b)R(c, d)$ if $a + d = b + c$. Is R an equivalence relation? If so prove it. If not state why not. What are the elements related to $(1, 2)$?, to $(6, 3)$?

37. Let A be the set of strings of the alphabet and define R on $A \times A$ by uRv if the first four letters in their strings are the same. Is R an equivalence relation?

38. Let A be the set of strings of the alphabet and define R on $A \times A$ by uRv if their strings begin and end with the same letter. Is R an equivalence relation?

39. Let A be the set of strings of 1's and 0's. Define R on $A \times A$ by uRv if u and v contain the same number of zeroes. Is R an equivalence relation?

2.8 Functions

For sets A and B, a function is a special relation on $A \times B$ that has the following properties:

1. The domain of the relation is the entire set A. Hence, for every element a of A, there is an element b of B so that a is related to b.
2. If a is related to b and a is related to b', then $b = b'$. In terms of ordered pairs this statement means that if (a, b) and (a, b') are in the relation. then $b = b'$. We define a function more formally as follows.

DEFINITION 2.62 A relation f on $A \times B$ is a **function** from A to B, denoted by $f : A \to B$, if for every $a \in A$, there is one and only one $b \in B$ so that $(a, b) \in f$. If $f : A \to B$ is a function and $(a, b) \in f$, we say that $b = f(a)$. A is called the **domain** of the function f and B is called the **codomain**. If $E \subseteq A$, then $f(E) = \{b : f(a) = b \text{ for some } a \text{ in } E\}$ is called the **image** of E. The image of A itself is called the **range** of f. If $F \subseteq B$, then $f^{-1}(F) = \{a : f(a) \in F\}$ is called the **preimage** of F. A function $f : A \to B$ is also called a **mapping** and we speak of the domain A being mapped into B by the mapping f. If $(a, b) \in f$ so that $b = f(a)$, then we say that the element a is mapped to the element b.

EXAMPLE 2.63 Let $A = \{-2, -1, 0, 1, 2\}$ and $B = \{0, 1, 2, 3, 4, 5\}$. Define the relation $f \subseteq A \times B$ as $f = \{(-2, 5), (-1, 2), (0, 1), (1, 2), (2, 5)\}$. The relation f is a function from A

to B because $f \subseteq A \times B$ and each of the five elements of A appears as the first component of an ordered pair in f exactly once. ∎

EXAMPLE 2.64   Let $A = \{-2, -1, 0, 1, 2\}$ and $B = \{0, 1, 2, 3, 4, 5\}$. Let $f : A \to B$ be defined by $f(x) = x^2 + 1$. If $E = \{1, 2\}$, then

$$f(E) = \{b : (a, b) \in f \text{ for some } a \text{ in } E\}$$
$$= \{b : b = f(a) \text{ for some } a \text{ in } E\}$$
$$= \{2, 5\}$$

is the image of E under f.

If $F = \{0, 2, 3, 4, 5\} \subseteq B$, then

$$f^{-1}(F) = \{b : \text{ there is an } a \in A \text{ such that } f(a) = b\}$$
$$= \{-1, 1, -2, 2\}$$

is the preimage of F where $-1 \in f^{-1}(F)$ because $f(-1) = 2$, $1 \in f^{-1}(F)$ because $f(1) = 2$, $-2 \in f^{-1}(F)$ because $f(-2) = 5$, and $2 \in f^{-1}(F)$ because $f(2) = 5$. Note that the elements 0, 3, and 4 do not contribute any elements to $f^{-1}(F)$ because they are elements of B that are not in the range of the function f. The preimage can be empty as it would be for $W = \{0, 3\}$, since there is no $a \in A$ for which $f(a) = 0$ or $f(a) = 3$.

The range of f is

$$f(A) = \{b : f(a) = b \text{ for some } a \in A\}$$
$$= \{1, 2, 5\}$$

The elements of $f(A)$ are just those elements of B that are "used" by the function f. ∎

We saw previously that if R is a relation on $A \times B$ and S is a relation on $B \times C$, then we could define a relation $S \circ R$ on $A \times C$, called the composition of S and R. If R and S are functions, then so is $S \circ R$. The next theorem gives some useful properties of composite functions.

THEOREM 2.65   Let $g : A \to B$ and $f : B \to C$; then

(a) The composition $f \circ g$ is a function from A into C, denoted by

$$f \circ g : A \to C$$

(b) If $a \in A$, then $(f \circ g)(a) = f(g(a))$.

Proof   (a) If $g : A \to B$ and $f : B \to C$, then $g \subseteq A \times B$ and $f \subseteq B \times C$. By definition of the composition of relations, $f \circ g \subseteq A \times C$. Let $(a, c) \in f \circ g$ and $(a, c') \in f \circ g$. We need to show that $c = c'$.

Since $(a, c) \in f \circ g$, there is a member $b \in B$ with $(a, b) \in g$ and $(b, c) \in f$. Similarly, since $(a, c') \in f \circ g$, there is a member $b' \in B$ with $(a, b') \in g$ and $(b', c') \in f$. But g is a function so that $(a, b) \in g$ and $(a, b') \in g$ imply that $b = b'$. Now we have $(b', c') = (b, c') \in f$ and $(b, c) \in f$. Since f is also a function, $c = c'$. Therefore, $f \circ g$ is a function from A into C.

(b) If $g : A \to B$ and $a \in A$, then there is a b in B with $b = g(a)$. Since $b \in B$ and $f : B \to C$, there is a $c \in C$ with $(b, c) \in f$, or equivalently, with $c = f(b)$. But $b = g(a)$ so that $f(g(a)) = f(b) = c = (f \circ g)(a)$ because $(a, b) \in g$ and $(b, c) \in f$ imply that $(a, c) \in f \circ g$. ∎

Since composition of relations has been shown to be associative (Theorem 2.34), we have as a special case the following theorem.

THEOREM 2.66 Let $f : A \to B, g : B \to C$, and $h : C \to D$. Then $h \circ (g \circ f) = (h \circ g) \circ f$ so that composition of functions is associative. ∎

EXAMPLE 2.67 Let $f(x) = \sqrt{x}$ and $g(x) = x + 3$ be real valued functions. The function $f(g(x)) = f(x+3) = \sqrt{x+3}$. The function $g(f(x)) = g(\sqrt{x}) = \sqrt{x} + 3$. ∎

A function f from A to B can be further classified by whether there is more than one element of A associated with any of the elements of B and whether all of the elements of the codomain B are associated with corresponding members of the domain A.

DEFINITION 2.68 A function $f : A \to B$ is **one-to-one** or **injective** if $f(a) = f(a')$ implies that $a = a'$. A function f is **onto** or **surjective** if for every $b \in B$ there exists an $a \in A$ so that $f(a) = b$. A function that is both one-to-one and onto is a **one-to-one correspondence** or a **bijection**. If $A = B$ and $f : A \to B$ is a one-to-one correspondence, then f is said to be a **permutation** of A.

EXAMPLE 2.69 Let A and B be the set of real numbers and $f : A \to B$ be defined by $f(x) = 3x + 5$. The function f is one-to-one, for if $f(a) = f(a')$, then $3a + 5 = 3a' + 5$ and, therefore, $a = a'$. The function f is also onto. For any real number b we want a so that $f(a) = b = 3a + 5$. Solving this equation for a, we see that if $a = (1/3)(b-5)$, then $f(a) = b$. Therefore, f is a one-to-one correspondence and, since $A = B$, f is also a permutation. ∎

EXAMPLE 2.70 Let A and B be the set of real numbers and $f : A \to B$ be defined by $f(x) = x^2$. The function f is not one-to-one since $f(2) = f(-2)$ but $2 \neq -2$. Also f is not onto since there is no real number a so that $f(a) = -1$. Note that if A and B are the set of nonnegative real numbers, then f is both one-to-one and onto. ∎

Let f be a function from A into the set B or $f : A \to B$. Certainly, $f \subseteq A \times B$ so that f is a relation on $A \times B$. The **inverse relation** $f^{-1} \subseteq B \times A$ is defined by $f^{-1} = \{(b, a) : (a, b) \in f\}$. The relation f^{-1} may not be a function from B to A even though f is a function from A to B. The following theorems show when f^{-1} is a function.

THEOREM 2.71 If a function $f : A \to B$ is one-to-one and onto, then the inverse relation f^{-1} is a function from B into A that is also one-to-one and onto. Conversely, for $f : A \to B$, if f^{-1} is a function from B into A, then f is one-to-one and onto.

Proof To show that f^{-1} is a function from B into A, we need to show that the domain of f^{-1} is B and that if (b, a) and $(b, a') \in f^{-1}$, then $a = a'$. To show the domain of f^{-1} is B, let $b \in B$. Since f is onto, there exists $a \in A$ such that $f(a) = b$ or $(a, b) \in f$. Therefore, $(b, a) \in f^{-1}$ and B is the domain of f^{-1}. Let (b, a) and $(b, a') \in f^{-1}$. Then (a, b) and $(a', b) \in f$. Since f is one-to-one, $a = a'$. Therefore, f^{-1} is a function. It is also onto, for if $a \in A$, then, since f is a function, there is a b so that $f(a) = b$ or $(a, b) \in f$. Therefore, $(b, a) \in f^{-1}$ and a belongs to the range of f^{-1}. Hence, A is the range of f^{-1} and is f^{-1} is onto. To show that f^{-1} is one-to-one, assume (b, a) and $(b', a) \in f^{-1}$ or $f^{-1}(b) = a$ and $f^{-1}(b') = a$. Then $(a, b) \in f$ and $(a, b') \in f$ (or, equivalently, $f(a) = b$ and $f(a) = b'$). Since f is a function, $b = b'$. To show the converse, simply exchange f and f^{-1} in the last argument. ∎

THEOREM 2.72 If $f : A \to B$ is one-to-one and onto, then

$$f(f^{-1}(b)) = b \text{ for each } b \text{ in } B$$

and

$$f^{-1}(f(a)) = a \text{ for each } a \text{ in } A$$

Proof Let $b \in B$ and $a = f^{-1}(b)$; then $f(a) = b$. But since $a = f^{-1}(b)$, we obtain $f(f^{-1}(b)) = f(a) = b$. The proof is similar to show $f^{-1}(f(a)) = a$ for each a in A. ∎

EXAMPLE 2.73 Find the inverse of $y = 3x + 6$. The inverse of the function is

$$\{(y, x) : y = 3x + 6\}$$

But this is equal to

$$\{(x, y) : x = 3y + 6\}$$

Solving for y, we have

$$\left\{(x, y) : y = \frac{x - 6}{3}\right\}$$

DEFINITION 2.74 Let $I : A \to A$ be defined by $I(a) = a$ for all $a \in A$. The function I is called the **identity** function on A.

The proofs of the following theorems are left to the reader.

THEOREM 2.75 If $f : A \to A$ and I is the identity function on A, then $I \circ f = f \circ I = f$. If f has an inverse function, then $f \circ f^{-1} = f^{-1} \circ f = I$. ∎

THEOREM 2.76 Let $g : A \to B$ and $f : B \to C$; then

(a) If g and f are onto B and C, respectively, then $f \circ g$ is onto C.
(b) If g and f are both one-to-one, then $f \circ g$ is one-to-one.
(c) If g and f are both one-to-one and onto, then $f \circ g$ is one-to-one and onto.
(d) $(f \circ g)^{-1} = g^{-1} \circ f^{-1}$.

A special notation is often used when the domain of a function is the set of positive integers or the set of all integers between two integers. For example, assume $A : N \to N$ is defined by $A(n) = n^2 + 1$. Then we denote $A(n)$ by A_n so that we have $A_1 = 2, A_2 = 5, \ldots A_k = k^2 + 1$. Suppose we want the sum of the function values for $5 \leq n \leq 8$. We denote this by $A_5 + A_6 + A_7 + A_8$. If we want the sum of the function values for $5 \leq n \leq k$, we denote this by $A_5 + A_6 + A_7 + \cdots + A_k$. ∎

Exercises

Let $f \subseteq R \times R$ where R is the set of real numbers. Find the range and domain of the following functions:

1. $f(x) = x^2 + 4$
2. $f(x) = \sqrt{x - 2}$
3. $f(x) = \dfrac{1}{\sqrt{x - 2}}$
4. $f(x) = \dfrac{1}{x^2 + 4}$
5. $f(x) = \dfrac{1}{x^2 - 4}$
6. $f(x) = |x|$

Which of the following relations are functions if x and y are real numbers and x is in the domain and y is in the range:

7. $y^2 = x^2 + 4$
8. $y^3 = x^3 + 4$

9. $y = 5$
10. $x = 7$
11. $y = \sqrt{x^2 - 2}$

For real valued functions f and g, find f(g(x)) and g(f(x)) when

12. $f(x) = x^2 + 1$ and $g(x) = x + 3$
13. $f(x) = \sqrt{x^2 + 2}$ and $g(x) = x^2 + 3$
14. $f(x) = \dfrac{1}{x}$ and $g(x) = 2x + 3$
15. Find functions f and g so that $f(g(x)) = \dfrac{7}{x^2 + 3}$.
16. Find functions f and g so that $f(g(x)) = \sqrt{x^2 - 2} + 5$.
17. Find functions f, g and h so that $f(g(h(x))) = \sqrt{\dfrac{1}{x^2 - 2}}$.

Find the inverse of each of the following functions:

18. $y = \dfrac{x+4}{2}$
19. $y = x^3$
20. $y = \dfrac{x-2}{x+3}$

Which of the following functions, whose domain and codomain are the real line, are one-to-one, which are onto, and which have inverses:

21. $f(x) = |x|$
22. $f(x) = x^2 + 4$
23. $f(x) = x^3 + 6$
24. $f(x) = x + |x|$
25. $f(x) = x(x-2)(x+2)$

Let $f : A \to B$, A_1 and A_2 be subsets of A, and B_1 and B_2 be subsets of B. Prove that

26. If $A_1 \subseteq A_2$, then $f(A_1) \subseteq f(A_2)$.
27. If $B_1 \subseteq B_2$, then $f^{-1}(B_1) \subseteq f^{-1}(B_2)$.
28. $f(A_1 \cup A_2) = f(A_1) \cup f(A_2)$.
29. $f^{-1}(B_1 \cup B_2) = f^{-1}(B_1) \cup f^{-1}(B_2)$.
30. $f(A_1 \cap A_2) \subseteq f(A_1) \cap f(A_2)$.
31. $f^{-1}(B_1 \cap B_2) = f^{-1}(B_1) \cap f^{-1}(B_2)$.
32. $f^{-1}(B_1') = (f^{-1}(B_1))'$.

33. Give an example of a function f and sets A_1 and A_2 such that $f(A_1 \cap A_2) \neq f(A_1) \cap f(A_2)$.

34. Let $f : A \to B$ be one-to-one and onto and $g : B \to A$ be any function having the properties

$$(g \circ f)(a) = a \text{ for all } a \in A$$
$$(f \circ g)(b) = b \text{ for all } b \in B$$

Prove that $g = f^{-1}$. Note that this statement says that f^{-1} is the only function from B into A such that $f \circ f^{-1} = I_B$ and $f^{-1} \circ f = I_A$, where I_A and I_B are, respectively, the identity mappings (relations) on A and B. (*Hint*: Use associativity of functions.)

35. If $f : A \to B$, then prove that f is one-to-one if and only if for all subsets X and Y of A, $f(X \cap Y) = f(X) \cap f(Y)$.

For $f : A \to B$, prove that

36. The function f is one-to-one if and only if $(f^{-1} \circ f)(W) = W$ for every $W \subseteq A$, where $f^{-1}(K)$ indicates the preimage of the set K.

37. The function f is onto if and only if $(f \circ f^{-1})(V) = V$ for every $V \subseteq B$.

Let $g : A \to B$ and $f : B \to C$. Then prove

38. If g and f are onto B and C, respectively, then $f \circ g$ is onto C.

39. If g and f are both one-to-one, then $f \circ g$ is one-to-one.

40. If g and f are both one-to-one and onto, then $f \circ g$ is one-to-one and onto.

41. $f \circ g = g^{-1} \circ f^{-1}$.

42. Prove that if $f : A \to A$ and I is the identity function on A, then $I \circ f = f \circ I = f$. If f has an inverse function, then $f \circ f^{-1} = f^{-1} \circ f = 1$.

Let $g : A \to B$ and $f : B \to C$

43. Prove that if $f \circ g$ is one-to-one then g is one-to-one.

44. Prove that if $f \circ g$ is onto, then f is onto.

GLOSSARY

Chapter 2: Set Theory

2.1 Introduction to Sets

Empty set ($\emptyset$)	The set that contains no elements.
Equal sets	Equal sets have the same elements.
Proper subsets	Set A is a proper subset of set B if every element of set A is an element of set B and there is at least one element in set B that is not in set A.

Set	A **set** is a well-defined collection of objects or elements.
Set builder notation	Braces are used to describe the set and within the braces the characteristic describing the set is given.
Subset	Set A is a subset of set B if every element in set A is also in set B.
Universal set (U)	A set that contains all the elements of the sets under consideration.

2.2 Set Operations

Cartesian product of sets	The Cartesian product of sets A and B, denoted by $A \times B$, is the set $\{(a, b) : a \in A, b \in B\}$.
Complement of a set $A(A')$	The complement of set A, denoted by A', is the set of all elements of the universe that are not in A.
Intersection of sets $(A \cap B)$	The intersection of sets A and B is the set containing all elements that are in both sets A and B.
Ordered pair	(a, b) is called an **ordered pair** with first component a and second component b.
Power set	The power set of a set A, denoted by $\mathcal{P}(A)$, is the set consisting of all subsets of A. If A has n elements, the set $\mathcal{P}(A)$ has 2^n elements.
Set difference $(A - B)$	The set of all elements in set A but not in set B.
Symmetric difference of sets A and B ($A \triangle B$)	$A \triangle B = (A - B) \cup (B - A) = (A \cup B) - (A \cap B)$
Union of sets $(A \cup B)$	The union of sets A and B is the set containing all elements that are in either set A or set B.

2.3 Venn Diagrams

Associative properties	$A \cap (B \cap C) = (A \cap B) \cap C$ $A \cup (B \cup C) = (A \cup B) \cup C$
Commutative properties	$A \cap B = B \cap A$ $A \cup B = B \cup A$
Complement properties	$A \cup A' = U$ $A \cap A' = \varnothing$
De Morgan's laws	$(A \cup B)' = A' \cap B'$ $(A \cap B)' = A' \cup B'$
Distributive properties	$A \cap (B \cup C) = (A \cap B) \cup (A \cap C)$ $A \cup (B \cap C) = (A \cup B) \cap (A \cup C)$
Double complement	$(A')' = A$
Idempotent laws	$A \cap A = A$ $A \cup A = A$
Identity properties	$A \cup \varnothing = A$ $A \cap U = A$

2.4 Boolean Algebras

Absorption laws	$x + (x \cdot y) = x$ $x \cdot (x + y) = x$

Associative laws	$x \cdot (y \cdot z) = (x \cdot y) \cdot z$ $x + (y + z) = (x + y) + z$
Binary operator	An operator on a set is binary if it combines or operates on two elements of a set to produce another element of the set.
Boolean algebra	Boolean algebra, named after George Boole, is a set B containing special elements 1 and 0 together with binary operators, $+$ and $\cdot$ and a unary operator $'$ on B, which satisfy the Boolean algebra axioms for all x, y, and z in B.
Commutative laws	$x \cdot y = y \cdot x$ $x + y = y + x$
Complement laws	$x + x' = 1$ $x \cdot x' = 0$
Complement of identities law	$0' = 1$ $1' = 0$
Complement of x	x'
De Morgan's laws	$(x + y)' = x' \cdot y'$ $(x \cdot y)' = x' + y'$
Distributive laws	$x \cdot (y + z) = (x \cdot y) + (x \cdot z)$ $x + (y \cdot z) = (x + y) \cdot (x + z)$
Idempotent laws	$x + x = x$ $x \cdot x = x$
Identity laws	$x + 0 = x$ $x \cdot 1 = x$
Involution law	$(x')' = x$
Null laws	$x + 1 = 1$ $x \cdot 0 = 0$
Unary operator	An operator on a set is unary if it operates on one element of a set to produce another element of the set.
Unity	The element 1.
Zero	The element 0.

2.5 Relations

Antireflexive	A relation R is **antireflexive** if $(a, b) \in R$ implies $a \neq b$.
Antisymmetric	R is **antisymmetric** if for all a and b in A, whenever (a, b) and (b, a) are in R, then $a = b$.
Composition relation	Let $R \subseteq A \times B$ is a relation on $A \times B$ and $S \subseteq B \times C$ is a relation on $B \times C$. The **composition** of S and R is the relation $T \subseteq A \times C$ defined by $T = \{(a, c) :$ there is an element b of B such that $(a, b) \in R$ and $(b, c) \in S\}$. This set is denoted by $T = S \circ R$.
Directed graph	A **directed graph** or **digraph** G, denoted $G(V, E)$, consists of a set V of vertices, together with a relation $E \subseteq V \times V$ called the set of **directed edges** or simply **edges** if it is understood that the graph is directed. An element of E is called a **directed edge**. If $(a, b) \in E$, then a is called the **initial vertex** of (a, b) and b is called the **terminal vertex**.

Domain	The domain of a relation R is the set of all first coordinates of ordered pairs in R.
Graph	A **graph** is a finite set V called a **vertex set**, a symmetric antireflexive relation R on V and a collection E of two-element subsets of V defined by $\{a, b\} \in E$ if and only if $(a, b) \in R$. The set E is called the **edge set**. An element of E is called an **edge**. A graph is denoted by $G(V, E)$. The elements a and b of V are **joined** or **connected** by the edge $\{a, b\}$ if $\{a, b\} \in E$.
Inverse relation	If $R \subseteq A \times B$, be a relation on $A \times B$. Then the relation R^{-1} on $B \times A$ is defined by $R^{-1} = \{(b, a) : (a, b) \in R\}$. The relation R^{-1} is called the **inverse relation** of R.
Range	The range of a relation R is the set of all second coordinates of ordered pairs in R.
Reflexive	A relation R on $A \times A$ is **reflexive** if for all a in A, (a, a) is in R.
Reflexive closure	If R is a relation on set A, the **reflexive closure** of R is the smallest reflexive relation on A containing R as a subset.
Relation (aRb)	A relation R between sets A and B is a subset of $A \times B$. And if $A = B$, so that the relation is a subset of $A \times B$, then the relation R is said to be a **relation on A**.
Symmetric	Relation R is **symmetric** if for all a and b in A, (a, b) in R implies that (b, a) is in R.
Symmetric closure	If R is a relation on set A, the **symmetric closure** of R is the smallest symmetric relation on A containing R as a subset.
Transitive	R is **transitive** if for all a, b, and c in A, whenever (a, b) and (b, c) are in R, then (a, c) is in R.
Transitive closure	If R is a relation on set A, the **transitive closure** of R is the smallest transitive relation on A containing R as a subset.

2.6 Partially Ordered Sets

Comparable	Two elements a and b of the partially ordered set (poset) $(S, \leq)$ are **comparable** if either $a \leq b$ or $b \leq a$. If every two elements of a partially ordered set $(S, \leq)$ are comparable, then $(S, \leq)$ is a **total ordering** or a **chain**.
Hasse diagram	For a finite poset, $(A, \leq_2)$, a Hasse diagram consists of a collection of points and lines used to display the poset.
Partial ordering	A relation R on A is a **partial ordering** if it is reflexive, antisymmetric, and transitive.
Partially ordered set	If the relation R on A is a partial ordering, then (A, R) is a **partially ordered set** or **poset** with ordering R. If the ordering R is understood, then (A, R) may simply be denoted by A.

2.7 Equivalence Relations

Equivalence class	Let $a \in A$ and R be an equivalence relation on $A \times A$. $[a]$ denotes the set $\{x : xRa\} = \{x : (x, a) \in R\}$, called the **equivalence class** containing a. The symbol $[A]_R$ denotes the set of all equivalence classes of A for the equivalence relation R.
Equivalence relation	A relation R on A is an **equivalence relation** if it is reflexive, symmetric, and transitive.

Partition	Let A and I be sets and let $\langle A \rangle = \{A_1 : i \in I\}$, with I nonempty, be a set of nonempty subsets of A. The set $\langle A \rangle$ is called a **partition** of A if both of the following are satisfied: a. $A_i \cap A_j = \emptyset$ for all $i \neq j$. b. $A = \bigcup_{i \in I} A_i$; that is, $a \in A$ if and only if for some $i \in I$.
Reflexive	A relation R on A is **reflexive** if for all a in A, (a, a) is in R.
Symmetric	R is **symmetric** if for all a and b in A, (a, b) in R implies that (b, a) is in R.
Transitive	R is **transitive** if for all a, b, and c in A, whenever (a, b) and (b, c) are in R, (a, c) is in R.

TRUE-FALSE QUESTIONS

1. For an arbitrary set A, $\{\emptyset\} \subseteq A$.
2. The set $\{\{a, b, c\}, \{d, e\}, \emptyset, \{a, b, \{c, d, e,\}\}\}$ contains five elements.
3. For an arbitrary set A, $A \subseteq A$.
4. If $A \subseteq B$, then $A \cap B = B$.
5. For an arbitrary set A, $A - \emptyset = A$.
6. If $A = \{x : x \text{ plays tennis}\}$ and $B = \{x : x \text{ plays golf}\}$, then $A \cap B = \{x : x \text{ plays tennis and golf}\}$.
7. For an arbitrary set A, $A \triangle \emptyset = \emptyset$.
8. For an arbitrary set A, $A \times \emptyset \subseteq A$.
9. If $A \subseteq B$ and $C \subseteq D$, then $A \times C \subseteq B \times D$.
10. For an arbitrary set A, $A \in \mathcal{P}(A)$.
11. $\mathcal{P}(\emptyset) = \emptyset$.
12. For sets A and B, $(A \cup B)' = A' \cup B'$.
13. The set of even integers and the set of odd integers are disjoint.
14. For sets A and B, $A \triangle B = B \triangle A$.
15. If the set A has n elements, then the power set of A has n^2 elements.
16. If the set A has n elements, then $A \times A$ has n^2 elements.
17. The set $\{a, b, c, d, e\}$ is equal to the set $\{a, e, c, d, b\}$.
18. If the set A has fewer elements than the set B, then $A \subseteq B$.
19. If $A \cap B = \emptyset$, then either $A = \emptyset$ or $B = \emptyset$.
20. In a Boolean algebra, one of the absorption laws is $1 + x = x$.
21. The dual of $(x + y)' \cdot 1$ is $(x \cdot y)' + 1$.
22. In a Boolean algebra, $x + x' = 1$.
23. The collection of all sets in a universe U, together with the operations intersection, union, and complement, form a Boolean algebra.
24. In a Boolean algebra, the complement x' of an element x is unique.
25. In a Boolean algebra A, $x \cdot x = x$ for all $x \in A$.
26. A relation R is symmetric if for all $a, b \in A$. If $(a, b) \in R$ and $(b, a) \in R$, then $a = b$.
27. Let A be the set of all people. Define the relation R by $(a, b) \in R$ if a and b have the same parents. The relation R is transitive.
28. Let A be the set of all people. Define the relation R by $(a, b) \in R$ if a and b are the same age. The relation R is a partial ordering.
29. Let A be the set of all integers. Define R by $(a, b) \in R$ if ab is positive. The relation R is an equivalence relation.
30. Let A be the set of all people. Define the relation R by $(a, b) \in R$ if a and b occasionally have lunch together. The relation R is an equivalence relation.
31. Let A be the power set of a set S. Define the relation R on A by $(B, C) \in R$ if $B \subseteq C$. The relation R is antisymmetric.
32. The relation E of edges of a directed graph is antireflexive.
33. The relation R determining the edges of a graph is symmetric.
34. If $R = \{(a, 1), (b, 1), (c, 2), (d, 4), (e, 2)\}$ and $S = \{(1, c), (3, e), (4, f), (2, b)\}$, then $R \circ S = \{(a, c), (b, c), (c, b), (d, f), (e, b)\}$.
35. If $R = \{(a, 1), (b, 1), (c, 2), (d, 4), (e, 2)\}$, then $R^{-1} = \{(1, a), (1, b), (2, c), (4, d), (2, e)\}$.

36. If $(a, b) \in E$, the edges of a directed graph, then a is the initial vertex of (a, b).

37. If a and b belong to a partially ordered set with relation $\leq$, then either $a \leq b$ or $b \leq a$.

38. The Hasse diagram and the directed graph representing the relation for a partially ordered set are the same.

39. A relation R is transitive if and only if $R = R^{-1}$.

40. $\{\{a, b, c\}, \{d, e\}, \{e, f\}\}$ is a partition of $\{a, b, c, d, e, f\}$.

41. $\{\{a, b, c\}, \{d, e\}, \{f, g\}\}$ is the set of equivalence classes of an equivalence relation on $\{a, b, c, d, e, f, g\}$.

42. The set of equivalence classes of a given equivalence relation on a set A is unique.

43. A partition of a set A forms the set of equivalence classes for a unique equivalence relation on the set A.

44. An equivalence relation can be represented by a graph.

45. For a fixed graph G, the relation defined by aRb if there is an edge from a to b is an equivalence relation.

46. The relation represented by a given graph is unique.

47. If a relation is symmetric and transitive, then it is reflexive.

48. Every relation on a finite set A can be represented as a directed graph.

49. The Hasse diagram

corresponds to the relation $R = \{(a, c), (b, c), (c, d)\}$.

50. The directed graph

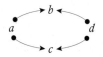

corresponds to the relation $R = \{(a, a), (b, b), (c, c), (d, d), (d, c), (a, c), (a, b), (d, b)\}$.

SUMMARY QUESTIONS

1. State the number of elements in the set $\{a, \{a, b\}, \{a, b, c, d\}, \{c, d\}, \{a, c, d\}\}$.

2. Let $A = \{x : x \text{ reads books}\}$ and $B = \{x : x \text{ listens to music}\}$. Find the set $A \cap B'$.

3. Find the power set of the set $\{a, b, c\}$.

4. If a set A has n elements, how many elements are in the power set of A?

5. Find the Venn diagram for the set $(A \cup B) - C$.

6. State the distributive properties for sets.

7. Give an example of a relation that is transitive and symmetric but not reflexive.

8. Determine whether or not the relation xRy if x and y graduated from the same college is an equivalence relation.

9. Determine whether or not the relation xRy if x is older than y is a partial ordering.

10. Find a partition of the set $\{a, b, c, d, e\}$.

11. In a Boolean algebra, prove $x \cdot x = x$.

12. In a Boolean algebra, find the dual of the expression $(x + y' + 1)(xy + 0)$.

13. What is the difference between the set of edges E in a graph and the set of edges E' in a digraph?

14. Given the relation $R = \{(a, x), (b, x), (c, y), (b, b), (d, e)\}$, find R^{-1}.

15. Find the equivalence relation whose equivalence classes are $\{\{a\}, \{b, c\}, \{d\}, \{e\}\}$.

ANSWERS TO TRUE-FALSE

1. F 2. F 3. T 4. F 5. T 6. T 7. F 8. T 9. T 10. T
11. F 12. F 13. T 14. T 15. F 16. T 17. T 18. F 19. F
20. F 21. F 22. T 23. T 24. T 25. T 26. F 27. T 28. F
29. T 30. F 31. T 32. F 33. T 34. F 35. T 36. T 37. F
38. F 39. F 40. F 41. T 42. T 43. T 44. F 45. F 46. T
47. F 48. T 49. F 50. F

3 Logic, Integers, and Proofs

3.1 Predicate Calculus

Many statements that appear to be propositions are not because they contain variables whose values are not specified. Since such a statement may be true for some values that may be assigned to the variables and not for others, the statement cannot be assigned a truth value. Statements such as

$$P(x): \quad 3 + x = 5$$
$$Q(x, y, z): \quad x^2 + y^2 \geq z^2$$
$$R(x, y): \quad x^2 + y^2 \geq 0$$
$$S(x): \quad -1 \leq \sin(x) \leq 1$$

are called **predicates**. A predicate that has one variable is called a **one-place predicate**. A predicate that has two variables is called a **two-place predicate**, and a predicate that has n variables is called an **n-place predicate**. The predicate $P(x)$ is a one-place predicate since it has one variable. Predicate $Q(x, y, z)$ is a three-place predicate.

Predicates become propositions when their variables are specified. For instance,

$$P(2): \quad 3 + 2 = 5$$

is a proposition and is true, but

$$P(7): \quad 3 + 7 = 5$$

is a proposition and is false. Combinations of variable values causing the predicate to be true are said to **satisfy the predicate**. Thus, 2 satisfies $P(x)$ because $P(2)$ is true but 7 does not satisfy $P(x)$ because $P(7)$ is false.

Some predicates are true for every possible choice of values of the variables that are chosen from a given set. As before, $S(x)$ is the predicate $-1 \leq \sin(x) \leq 1$, where x is allowed to be from the set of real numbers. Since

$$-1 \leq \sin(x) \leq 1$$

for any choice of x, then we can say that

for every x, $S(x)$ (is true)

expressed symbolically as

$$\forall x\, S(x)$$

or

$$\forall x(-1 \leq \sin(x) \leq 1)$$

The symbol $\forall x$ is called a **universal quantifier** and is read "for every x" or "for all x." The set of values from which x may be taken is called the **universe** or **domain of discourse** or simply **domain**. Frequently, the domain is the set of real numbers, the set of integers, or other familiar sets. Generally the truth of a statement with a universal quantifier is dependent on the universe of discourse. For example, the statement

$$\forall x(x^2 > 5)$$

is not true if the universe of discourse is the set of integers; however, it would be true if the universe of discourse is the set of integers greater than 20. Similarly, the statement

$$\forall x(x^2 \geq 0)$$

is true if the universe of discourse is the real numbers; however, it would not be true if the universe of discourse contained the complex number i, since $i^2 = -1$.

The quantified predicate

$$\forall x\, P(x)$$

where $(P(x) : 3 + x$ is $5)$ is not true if the domain is the set of integers, because there is an instance of x in the domain (for example, $x = 4$) such that $P(x)$ is false. For $\forall x\, P(x)$ to be true, $P(x)$ must be true for each instance of x in the domain of discourse. Therefore, $\forall x\, P(x)$ has a truth value and so it is a proposition.

The predicate $\forall x\, \forall y\, R(x, y)$ is read as "for every x for every y, $R(x, y)$" or "for every x and for every y, $R(x, y)$." The predicate $\forall x\, \forall y\, R(x, y)$ is true only if, for every choice of x and of y from their respective domains of discourse, $x^2 + y^2 \geq 0$ is true. If the domain of discourse for both x and y is the set of all real numbers, then

$$\forall x\, \forall y\, R(x, y)$$

is true. Considering the predicate $Q(x, y, z)$ given previously, $(Q(x, y, z) : x^2 + y^2 \geq z^2)$, we see that

$$(\forall x)\,(\forall y)\,(\forall z)\, Q(x, y, z)$$

is not true, where the domain of discourse for x, y, and z is the set of integers, because

$$Q(1, 2, 3)$$

is false. For clarity, we will sometimes surround the quantifier with parentheses, writing $(\forall x)P(x)$ instead of $\forall x\, P(x)$ and $(\forall x)(\forall y)R(x, y)$ instead of $\forall x\forall y R(x, y)$.

Even though $\forall x\, P(x)$ is not true, $P(x)$ is true for at least one choice of x, namely, $x = 2$. Therefore, there exists x such that $P(x)$ is true. We express this symbolically by writing

$$\exists x\, P(x)$$

The symbol $\exists x$ is called the **existential quantifier.** Here, as with $\forall x$, the expression $\exists x$ refers to a value of x within the **domain of discourse**.

The statement

$$\exists x(x^2 = 4)$$

is true if the domain of discourse is the set of real numbers since $x^2 = 4$ is true for $x = 2$. The statement

$$\exists x(x^2 = 5)$$

is also true if the domain of discourse is the set of real numbers; however, it is not true if the domain of discourse is the set of integers, since there is no integer whose square is 5.

Since for the predicate $Q(x, y, z)$ the statements $Q(1, 2, 0)$ and $Q(-3, 4, 5)$ are true, the statement

$$\exists x\, \exists y\, \exists z\, Q(x, y, z)$$

is true.

To be a proposition, all variables appearing in the predicates must be given a value or associated with some quantifier. For example,

$$(\exists x)(\forall z)Q(x, y, z)$$

is not a proposition because the variable y is not associated with any quantifier.

If $D(x)$ is a predicate, we saw that $\forall x\, D(x)$ is true only if, for every instance of x, $D(x)$ is true. The question now is: What does it mean to say that $\forall x\, D(x)$ is not true? We can express the denial or negation of $\forall x\, D(x)$ by

$$\sim \forall x\, D(x)$$

For $\forall x\, D(x)$ to fail to be true, we need to find only one instance of x so that $D(x)$ is false (or, equivalently, so that $\sim D(x)$ is true). Thus, $\forall x\, D(x)$ is false if and only if

$$\exists x(\sim D(x))$$

is true or there exists an x (within the domain of discourse) so that $\sim D(x)$ is true (or so that $D(x)$ is false). For example, the statement

$$\forall x(x^2 > 0)$$

is not true if the domain of discourse is the set of integers, since x^2 is not greater than 0 if $x = 0$. Thus, $\forall x(x^2 > 0)$ is false because $\exists x(x^2 \not> 0)$ is true. But this means that $\forall x(x^2 > 0)$ is false because $\exists x(\sim(x^2 > 0))$ is true.

Similarly, if $G(x)$ is a predicate, we would deny that there exists an x so that $G(x)$ is true by writing

$$\sim(\exists x\, G(x))$$

Clearly, if there is no instance of x making $G(x)$ true, then all values of x must make $G(x)$ false. In this case, $\sim G(x)$ will be true for all values of x or, equivalently,

$$\forall x(\sim G(x))$$

The results of the discussion are expressed as equivalences:

$$\sim \forall x(D(x)) \equiv \exists x(\sim D(x))$$
$$\sim \exists x(G(x)) \equiv \forall x(\sim G(x))$$

To form the negation or denial of a universally quantified predicate, change $\forall x$ to $\exists x$ and negate the predicate that follows. To form the negation of an existentially quantified predicate, change $\exists x$ to $\forall x$ and negate the predicate that follows.

The negation of a statement with more than one quantifier is determined by considering one quantifier at a time beginning with the first quantifier. For example, to negate $(\forall x)(\forall y)R(x, y)$, we have

$$\sim(\forall x)(\forall y)R(x, y) \equiv \exists x \sim(\forall y)R(x, y)$$
$$\equiv (\exists x)(\exists y)\sim R(x, y)$$

Similarly,

$$\sim(\exists x)(\forall y)(\exists z)Q(x, y, z) \equiv (\forall x)\sim(\forall y)(\exists z)Q(x, y, z)$$
$$\equiv (\forall x)(\exists y)\sim(\exists z)Q(x, y, z)$$
$$\equiv (\forall x)(\exists y)(\forall z)\sim Q(x, y, z)$$

Hence, to negate a statement containing quantifiers, $\exists$ is changed to $\forall$ and conversely; also the predicate associated with the quantifier is negated.

There is a relationship between quantified predicates and predicates using particular instances of the variables. For example, consider

$$\forall x\, S(x)$$

or

for every x it is true that $-1 \leq \sin(x) \leq 1$

Assume that this proposition is true. We wish to apply it to the particular value 8 of x and conclude

$$S(8)$$

If $\forall x\, S(x)$ is true and a is any particular instance of x (in the domain of discourse), then

$$S(a)$$

is true. So for the instance 8 for x, we have as true

$$-1 \leq \sin(8) \leq 1$$

Conversely, if b is an arbitrary instance of the universe (that is, the only property of b known or assumed is that b is in the domain of discourse) and we show that $S(b)$ is true, then

for every x, $S(x)$

or

$$\forall x\, S(x)$$

These two rules allow us to go back and forth between universally quantified predicates and predicates as applied to constants in the universe.

The statement $\forall x\, S(x)$ implies that $S(a)$ is true for arbitrary a in the universe (i.e., any one selected).

The statement $S(b)$ for an arbitrary constant b in the universe implies that $\forall x\, S(x)$.

Given the statement $\exists x(x^2 + 2x - 3 = 0)$, one can conclude there is a value a (namely, $a = 1$) such that $a^2 + 2a - 3 = 0$. Conversely, if for a particular value of a, $a^2 + 2a - 3 = 0$, then we can conclude $\exists x(x^2 + 2x - 3 = 0)$.

These two rules allow us to pass back and forth between predicates with particular values to existentially quantified predicates.

The statement $\exists x\, S(x)$ implies that $S(a)$ is true for some a in the universe.

The statement $S(b)$ for some constant b in the universe implies that $\exists x\, S(x)$.

We summarize these four relationships.

1. **Universal Instantiation**
   From $\forall x\, P(x)$ we can infer $P(a)$ for arbitrary a in the universe.
2. **Universal Generalization**
   From having arbitrary a in the universe make $P(a)$ true we can infer that $\forall x(P(x))$ is true.
3. **Existential Instantiation**
   From $\exists x\, P(x)$ we can infer that there is an instance b with $P(b)$ true.
4. **Existential Generalization**
   From an instance c in the universe with $P(c)$ true we can infer that $\exists x\, P(x)$.

The following inferences can be proved using the preceding four inference relationships.

THEOREM 3.1 For arbitrary statements $P(x)$ and $Q(x)$ having the same domain of discourse,

(a) $\forall x(P(x) \wedge Q(x)) \equiv \forall x\, P(x) \wedge \forall x\, Q(x)$.

(b) $\exists x(P(x) \vee Q(x)) \equiv \exists x\, P(x) \vee \exists x\, Q(x)$.

(c) From $\forall x\, P(x) \vee \forall x\, Q(x)$ we can infer that $\forall x(P(x) \vee Q(x))$.

(d) From $\exists x(P(x) \wedge Q(x))$ we can infer that $\exists x\, P(x) \wedge \exists x\, Q(x)$.

Proof To prove
$$\forall x(P(x) \wedge Q(x)) \equiv \forall x\, P(x) \wedge \forall x\, Q(x)$$
we first assume $\forall x(P(x) \wedge Q(x))$ and conclude $\forall x\, P(x) \wedge \forall x\, Q(x)$. From $\forall x(P(x) \wedge Q(x))$, using universal instantiation, we can conclude that for arbitrary a, $P(a) \wedge Q(a)$. Hence, we can conclude $P(a)$ for arbitrary a and $Q(a)$ for arbitrary a, by specialization (propositional logic). By universal generalization, we conclude $\forall x\, P(x)$ and $\forall x\, Q(x)$. Thus, we can conclude $\forall x\, P(x) \wedge \forall x\, Q(x)$ by conjunction (propositional logic).

Conversely, assume $\forall x\, P(x) \wedge \forall x\, Q(x)$. By specialization, we can conclude $\forall x\, P(x)$ and $\forall x\, Q(x)$. By universal instantiation, we can conclude that for arbitrary a, $P(a)$ and $Q(a)$. Hence, by conjunction, we can conclude $P(a) \wedge Q(a)$ for arbitrary a. By universal generalization, we conclude $\forall x(P(x) \wedge Q(x))$.

The remainder of the proof is left to the reader. ∎

If $p(x)$ is the statement "x is a student" and $q(x)$ is the statement "x studies hard," then $\forall x(p(x) \rightarrow q(x))$ could be expressed in English as "All students study hard." Similarly, if $p(x)$ is the statement "x is an elephant" and $q(x)$ is the statement "x prefers chocolate-coated peanuts," then the expression $\exists x(p(x) \wedge q(x))$ could be expressed as "Some elephants prefer chocolate-coated peanuts." The statement "All men are mortal" could be expressed logically as $\forall x(p(x) \rightarrow q(x))$ where $p(x)$ is the statement "x is a man" and $q(x)$ is the statement "x is mortal." Also, "Some integers are divisible by 5" could be converted to $\exists x(p(x) \wedge q(x))$ where $p(x)$ is the statement "x is an integer" and $q(x)$ is the statement "x is divisible by 5." Usually, $\forall x$ statements in logic can be converted to "all" statements in English and $\exists x$ statements in logic can be converted to "some" statements in English.

Because of this translation from logic to English and conversely, the negation of an "all" statement is a "some" statement and the negation of a "some" statement is an "all" statement. For example, the negation of the statement

All integers are prime.

is the statement

Some integers are not prime.

and the negation of the statement

Some people like to eat turnips.

is the statement

All people do not like to eat turnips.

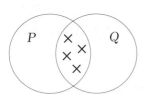

Figure 3.1

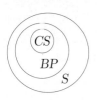

Figure 3.2

If we have an argument consisting of "all" and "some" statements, an informal method for testing the validity of the argument is with the Euler diagram. The Euler diagram consists of circles for sets. The statement "All p are q" would be represented as shown in Figure 3.1 where the circle for p is contained in the circle for q. The statement "Some p are q" would be represented as shown in Figure 3.2 where the intersection of circles for p and q is shown to have nonempty intersection. To test the validity of an argument, one tries to construct an Euler diagram where the hypotheses are true and the conclusion is false. If this construction can be done, then the argument is invalid, and if not, then the argument is valid.

EXAMPLE 3.2   Consider the argument

> All college students are brilliant.
> All brilliant people are scientists.
> ∴ All college students are scientists.

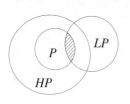

To satisfy the hypotheses, the circle for college students (CS) must be inside the circle for brilliant people (BP) and the circle for brilliant people must be inside the circle for scientists (S). Hence, the circle for college students must be inside the circle for scientists and the argument is valid. ∎

EXAMPLE 3.3   Consider the argument

> All poets are happy.
> Some poets are lazy.
> ∴ Some lazy people are happy.

To satisfy the hypotheses, the circle for poets (P) must be inside the circle for happy people (HP) and the intersection of poets and lazy people (LP) must be nonempty. But this intersection is contained in the circle for poets so the intersection of lazy people and happy people is nonempty and the argument is valid. ∎

EXAMPLE 3.4   Consider the argument

> Some poets are unsuccessful.
> Some athletes are unsuccessful.
> ∴ Some poets are athletes.

The Euler diagram is shown in Figure 3.3.

It can be seen in the figure that it is possible to create a Euler diagram such that the intersection of the circles for poets (P) and unsuccessful people (UP) is

Figure 3.3

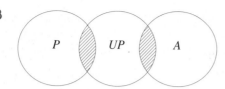

nonempty and the intersection of the circles for athletes (A) and unsuccessful people is nonempty so that the hypotheses are true but for which the circles for poets and athletes do not meet so that the conclusion is not true. Hence, the argument is not valid. ∎

EXAMPLE 3.5 Consider the argument

> All geniuses are illogical.
> Some politicians are illogical.
> ∴ Some politicians are geniuses.

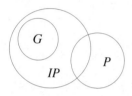

In the figure in the margin, the circle for geniuses (G) is contained in the set of illogical people (IP) and the intersection of the circles for politicians (P) and illogical people is nonempty, but the circles for politicians and geniuses do not intersect. Hence, the hypotheses are true but the conclusion is false and the argument is not valid. ∎

Exercises

Given the predicates

$P(x, y, z): \quad x^2 + y^2 \geq z^2$

$Q(x, y): \quad y = \dfrac{x-4}{x+4}$

$R(x, r): \quad |x - 1| \leq r$

$S(n, y): \quad y = n!$

$T(a, b, c): \quad a$ likes b better than c.

write out the following statements:

1. $P(3, 4, 5)$
2. $Q(8, 2)$
3. $R(3, 7)$
4. $S(4, 24)$
5. T(John, Sue, Mary)

Given the predicates

$P(x, y, z): \quad x^2 + y^2 = z^2$

$Q(x, y): \quad$ If $y^2 = x^2$ then $x = y$.

$R(x): \quad \lfloor \lfloor x \rfloor \rfloor = \lfloor x \rfloor$

$S(a, b, x): \quad a \leq x^2 \leq b$

$T(a, b): \quad a$ is a better tennis player than b.

write out the following statements:

6. $P(3, 4, 5)$
7. $Q(-2, 2)$
8. $R(3.1416)$
9. $S(0, 4, -3)$
10. T(John, Fred)

Using P, Q, R, S, and T from Exercises 1–5, write out the following statements:

11. $(\exists x)(\exists y) P(x, y, 25)$
12. $(\exists x) Q(x, 7)$
13. $(\forall r)(\exists x) R(x, r)$
14. $(\exists n)(\forall y) S(n, y)$
15. $(\forall c) T$(John, Sue, c)

Using P, Q, R, S, and T from Exercises 6–10, write out the following statements:

16. $(\forall x)(\forall y)(\exists z) P(x, y, z)$
17. $(\forall x)(\forall y) Q(x, y)$
18. $(\forall x) R(x)$
19. $(\forall x)(\exists a)(\exists b) S(a, b, x)$
20. $(\forall a) T(a, \text{Ted})$

Assign a symbol to each of the predicates and write the following in symbolic form. If helpful, supply the domain of discourse.

21. Every dog has his day.
22. Some machines are smarter than all men.
23. Everyone plays tennis better than Fred.
24. For every action there is an equal and opposite reaction.

25. Every golfer will eventually be defeated by a better golfer.

Assign symbols to represent the predicates and write the following in symbolic form. If helpful, supply the domain of discourse.

26. Some composers write better symphonies than some other composers.
27. For every n, there exist three integers such that the nth power of one is the sum of the nth powers of the others.
28. There are horses that can be outrun by some men.
29. He is the best athlete in the world.
30. There are no perfect heroes.

For $Q(x, y, z) : x^2 + y^2 \geq z^2$ we showed that

$$\forall x \forall y \forall z \, Q(x, y, z)$$

is false. If the universe is the positive integers, then what can be said about the truth of

31. $\forall x \forall y \exists z \, Q(x, y, z)$?
32. $\forall x \exists y \exists z \, Q(x, y, z)$?
33. $\exists x \forall y \exists z \, Q(x, y, z)$?
34. $\exists x \exists y \exists z \, Q(x, y, z)$?
35. $\exists x \exists y \forall z \, Q(x, y, z)$?

If we know that on the set of integers

$$\forall n (n \text{ is odd} \rightarrow (\exists k)(n = 2k + 1))$$

36. Which of these statements is true on the set of integers?

$$7 \text{ is odd} \rightarrow (\exists k)(7 = 2k + 1)$$
$$6 \text{ is odd} \rightarrow (\exists k)(6 = 2k + 1)$$

37. Which of these statements is true, on the set of integers and why?

$$(\exists k)(7 = 2k + 1)$$
$$(\exists k)(6 = 2k + 1)$$

38. Prove $\exists x (P(x) \vee Q(x)) \equiv \exists x P(x) \vee \exists x Q(x)$.
39. Prove that from $\forall x P(x) \vee \forall x Q(x)$ we can infer that $\forall x (P(x) \vee Q(x))$.
40. Prove that from $\exists x (P(x) \wedge Q(x))$ we can infer that $\exists x P(x) \wedge \exists x Q(x)$.

Use Euler diagrams to test the validity of each of the following arguments.

41. All lawyers are wealthy.
    All wealthy people eat lobster.
    ∴ All lawyers eat lobster.

42. Some lawyers are wealthy.
    Some doctors are wealthy.
    ∴ Some doctors are lawyers.

43. All doctors like music.
    All poets like music.
    ∴ All doctors are poets.

44. Some doctors are intelligent.
    All intelligent people are poets.
    ∴ Some doctors are poets.

45. All men like red meat.
    Some teachers are men.
    ∴ Some teachers like red meat.

46. Some martians are green.
    All crabgrass is green.
    ∴ Some Martians are crabgrass.

47. Some geese are male.
    Some males play golf.
    ∴ Some geese play golf.

48. All men are mortal.
    Geese are not men.
    ∴ Geese are not mortal.

49. All cars are expensive.
    Bicycles are not expensive.
    ∴ Bicycles are not cars.

50. All men watch television.
    Some plumbers are men.
    ∴ Some plumbers watch television.

Negate the following statements.

51. Everyone likes Charlie.
52. Some basketball players are not tall.
53. All men are mortal.
54. Some people like spinach.
55. $\exists x$ such that $x^2 = 9$.
56. $\forall x \exists y$ such that $x + y = 5$.
57. $\exists y \exists y \forall z \forall n, x^n + y^n = z$.
58. $\forall x \; x \geq 0$.

3.2 Basic Concepts of Proofs and the Structure of Integers

In our discussion we want to develop some properties of number theory that may prove to be useful. More importantly, perhaps, we want to use the properties of the integers as a familiar area in which to develop and practice techniques for proving theorems. The first problem is where to begin. What are you to assume known and, hence, usable in the proof of theorems? In a formal course in number theory, we would begin with basic axioms of number theory and develop everything else. This would not be practical in our case. Another problem is to decide which details to include. Do we want to explain why in one line of a proof we have $4 + x + 5 = y$ and in the next line we have $y = x + 9$? If we justify every small detail, then the main points of the proof will be lost in a collection of clutter. At some point we have to assume that everyone understands the details and could fill them in if they really needed to do so. There are generally no exact rules that tell us how much to put in or leave out. Developing an elegant proof for a theorem is to some extent an art form. However, we will assume that for our purposes most of the axioms listed in this section will be assumed known and used but not explicitly stated.

First, we shall assume the following axioms of equality. These axioms apply to the elements of any set, including the set Z of integers.

(E1) For all a, $a = a$.
(E2) For all a and b, if $a = b$, then $b = a$.
(E3) For all a, b, and c, if $a = b$ and $b = c$, then $a = c$.

Axioms E1, E2, and E3 state that equality has the properties of an equivalence relation on any set. These axioms apply to all forms of mathematics and not just to number theory.

We also include the following properties of addition, subtraction, and multiplication of the set Z of integers.

(I1) If a and b are integers, then $a + b$ and $a \cdot b$ are integers. The integers are **closed** under the operations of addition and multiplication.
(I2) If $a = b$ and $c = d$, then $a + c = b + d$ and $a \cdot c = b \cdot d$.
(I3) For all integers a and b, $a + b = b + a$ and $a \cdot b = b \cdot a$. The integers are **commutative** under the operations of addition and multiplication.
(I4) For all integers a, b, and c, $(a + b) + c = a + (b + c)$ and $a \cdot (b \cdot c) = (a \cdot b) \cdot c$. The integers are **associative** under the operations of addition and multiplication.
(I5) For all integers a, b, and c, $a \cdot (b + c) = (a \cdot b) + (a \cdot c)$. Multiplication of integers is **distributive** over addition.
(I6) There exist unique integers 0 and 1 such that for all integers a, $a + 0 = 0 + a = a$ and $a \cdot 1 = 1 \cdot a = a$. The integer 0 is called the **additive identity**, or the **zero**, of the integers and 1 is called the **multiplicative identity**.
(I7) For every integer a there is a unique integer, $-a$, called the **additive inverse** of a, such that $a + (-a) = (-a) + a = 0$.
(I8) If b and c are integers and $a \cdot b = a \cdot c$ for some nonzero integer a, then $b = c$. This statement is called the **multiplicative cancellation property**.

As previously mentioned, it is common in mathematics to prove theorems informally since the reader will be familiar with many of the details; therefore, they do not need to be explicitly stated. To illustrate this concept, we will first give a formal proof that the square of an even number is even. Since the domain of discourse is the set of integers, a formal statement of the theorem would be

$$\forall n(n \text{ is even} \to n^2 \text{ is even})$$

The proof will formally proceed as follows, where we will ultimately use the principle of universal generalization to show that the universally quantified statement is true.

1	a is an integer	given instance in universe
2	a is even	given (assumption)
3	$\forall n(n \text{ is even} \to (\exists k)(n = 2k))$	definition of even integer
4	$a \text{ is even} \to (\exists k)(a = 2k)$	1, 3, and universal instantiation
5	$(\exists k)(a = 2k)$	2, 4, and law of detachment
6	$a = 2L$ for some L	5 and existential instantiation
7	$\forall r \forall s(r = s \to r^2 = s^2)$	follows from axiom I2
8	$(a = 2L) \to a^2 = (2L)^2$	7 and universal instantiation
9	$a^2 = (2L)^2$	6, 8, and law of detachment
10	$(2L)^2 = 2(2L^2)$	9 and associative property of integers (steps to here omitted for brevity)
11	$\forall x \forall y \forall z((x = y \text{ and } y = z) \to x = z)$	transitive property of equality assumed
12	$a^2 = (2L)^2 \text{ and } (2L)^2 = 2(2L^2) \to a^2 = 2(2L^2)$	9, 11, and universal instantiation
13	$a^2 = (2L)^2 \text{ and } (2L)^2 = 2(2L^2)$	9, 10, and conjunction valid argumentative form
14	$a^2 = 2(2L^2)$	12, 13, and law of detachment
15	$(\exists k)(a^2 = 2k)$	14 and existential generalization for the instance $(2L^2)$
16	$\forall x((\exists k)(x = 2k) \to x \text{ is even})$	definition of even integer
17	$(\exists k)(a^2 = 2k) \to a \text{ is even}$	16 and universal instantiation with the instance a^2
18	a^2 is even	15, 17, and law of detachment
19	$a \text{ is even} \to a^2 \text{ is even}$	2, 18, and $((p \land q) \to (p \to q))$ is a tautology)
20	$\forall n(n \text{ is even} \to n^2 \text{ is even})$	1, 19, and universal generalization

The foregoing argument has been written out more formally than is customary except in courses in mathematical logic. This example illustrates how logic, quantifiers, valid arguments, and inference interact in a rather specific example. The more typical mathematical proof that you will see in this book will appear as follows.

THEOREM 3.6 If n is even, then n^2 is even.

Proof Assume that n is an (arbitrary) even integer. By definition of even integer, there is an integer L so that

$$n = 2L$$

If two integers are equal, then their squares are equal so that

$$n^2 = (2L)^2$$

But
$$(2L)^2 = (2L)(2L) = 2(2L^2)$$
so $n^2 = 2(2L^2)$ and $n^2 = 2J$ for some integer J (namely, $J = 2L^2$). By the definition of even integer, n^2 is even. ∎

As another example, we consider the converse of the previous theorem.

THEOREM 3.7 If n^2 is even, then n is even.

Proof The proof of this theorem provides an example of **proving the contrapositive** of the theorem rather than directly proving the theorem. This method of proof is possible since
$$p \to q \equiv \sim q \to \sim p$$
That is, proving $\sim q \to \sim p$ is equivalent to proving $p \to q$. The statement of
$$\sim q \to \sim p$$
is

If n is not even, then n^2 is not even.

Let n be an integer that is not even (we note that even integers have the form $2K$ and odd integers have the form $2L + 1$ and an integer is either even or odd, but not both). We need to show that n^2 is odd. Since n is odd, there exists an integer L so that
$$n = 2L + 1$$
Squaring both sides, we have
$$n^2 = (2L + 1)^2$$
$$= 4L^2 + 4L + 1$$
$$= 2(2L^2 + 2L) + 1$$
So if $J = 2L^2 + 2L$, we have
$$n^2 = 2J + 1$$
By definition of odd integers, n^2 is not even, and the theorem (the contrapositive version) is proved. ∎

The proof of the following theorem is left to the reader.

THEOREM 3.8 Let a, b, and c be integers.

(a) If $b + a = c + a$, then $b = c$.
(b) For each integer a, $a \cdot 0 = 0$.
(c) For each integer a, $-(-a) = a$.
(d) $a \cdot (-b) = -(ab)$.
(e) $(-a) \cdot (-b) = a \cdot b$.
(f) $-(a + b) = (-a) + (-b)$. ∎

The set of integers Z contains a subset N, called the positive integers. We will accept as axioms the following statements about the set N.

(N1) The integer 1 is a positive integer.

(N2) The positive integers are closed under addition and multiplication; that is, if a and b are positive integers, then $a + b$ and $a \cdot b$ are positive integers.

(N3) (Trichotomy Axiom) For every integer a, one and only one of the following is true:

(a) a is a positive integer.
(b) $a = 0$.
(c) $-a$ is a positive integer.

The following is a partial ordering of the integers in which the integers are arranged to illustrate our idea of "preceding" or "greater than." We usually think of the integers as being "ordered" in the following way:

$$\ldots, -4, -3, -2, -1, 0, 1, 2, 3, 4, 5, \ldots$$

where all positive integers "follow" zero and zero "follows" or is "greater than" any negative integer. Since trichotomy allows a determination of whether an integer is positive, we can define the relation $>$ contained in $Z \times Z$ as $a > b$ if and only if $a - b$ is positive. Because of trichotomy, $a - b$ is either positive, zero, or negative so that we can collect two of the possibilities into one category and say that $a \geq b$ if and only if either $a > b$ or $a = b$. This is formalized in the following definition.

DEFINITION 3.9 For integers a and b, $a > b$ if and only if $a - b$ is positive; and $a \geq b$ if and only if $a > b$ or $a = b$. Also, $b < a$ if $a > b$ and $b \leq a$ if $a \geq b$.

Obviously, $a > 0$ is equivalent to stating that a is positive because $a > 0$ if and only if $a - 0 = a$ is positive. Similarly, $a < 0$ is equivalent to stating that a is negative.

Although the following theorems are well known and could easily be included as assumptions, we include them for practice in proving theorems.

THEOREM 3.10 For integers a and b:

(a) $(a \geq b) \wedge (b \geq a) \rightarrow a = b$.
(b) $(a > b) \wedge (b > c) \rightarrow a > c$.

Proof (a) This is an example of the use of reductio ad absurdum. We begin by assuming that the statement in part (a) is not true. If we can then reach a contradiction, we have proved that the statement in part (a) is true. Assume that the statement in part (a) is not true. But if $a \geq b$, $b \geq a$ and $a \neq b$, then $a > b$ and $b > a$. But then $a - b$ is positive and $b - a = -(a - b)$ is positive, contradicting N3, the trichotomy axiom. Since we have reached a contradiction, it must be the case that $a = b$.

(b) If $a > b$ and $b > c$, then $a - b$ and $b - c$ are positive. Hence,

$$(a - b) + (b - c) = a - c$$

is positive by N2, and $a > c$. ■

THEOREM 3.11 Let a, b, c, and d be integers.

(a) $(a > b) \wedge (c > d) \rightarrow a + c > b + d$.
(b) $(a > 0) \wedge (c > d) \rightarrow ac > ad$.
(c) $(a > b > 0) \wedge (c > d > 0) \rightarrow ac > bd$.
(d) $(a \geq b \geq 0) \wedge (c \geq d \geq 0) \rightarrow ac \geq bd$.

Proof (a) If $a > b$ and $c > d$, then $a - b$ and $c - d$ are positive. Hence, $(a - b) + (c - d)$ is positive. But $(a - b) + (c - d) = (a + b) - (c + d)$, and so $a + b > c + d$.

(b) If $c > d$, then $c - d$ is positive. Since a is positive, $a(c - d) = ac - ad$ is positive and $ac > ad$.

(c) If $a > b > 0$ and $c > d > 0$, then $ac > bc$ and $bc > bd$. Therefore, $ac > bd$.

(d) The proof is left to reader. ∎

At this point we illustrate the **case proof**. For three cases, it basically has the form

$$\begin{array}{l} p \\ p \to (r \vee s \vee t) \\ r \to q \\ s \to q \\ t \to q \\ \hline \therefore q \end{array}$$

THEOREM 3.12 Let n be an integer. Then $n^2 \geq 0$.

Proof Since n is an integer, then it is positive, negative, or 0 by the trichotomy axiom. If n is positive, then $n \cdot n$ is positive by axiom N2 and $n^2 \geq 0$. If n is negative, then $n = -m$ for some positive integer m. Therefore, $n^2 = (-m)(-m) = m^2$, which again is positive and $n^2 \geq 0$. If $n = 0$, then $n^2 = 0$, so that $n^2 \geq 0$. Hence, $n^2 \geq 0$. ∎

It is important to note that when we create axioms for concepts such as the integers and the positive integers and then consider all of the theorems that we can prove using these axioms, we are only creating a model of the concept. For example, it is quite possible that a statement about the positive integers is true but cannot be proven using the axioms for the positive integers. Such a statement is said to be not reachable. A mathematical model created by a set of axioms is **consistent** if no theorem can be proven both true and false. In other words, no contradictions can be created. A mathematical model is **complete** if every true statement about the concept being modeled can be proven. Obviously, the more axioms we have, the more theorems we can prove. If a theorem is not reachable, we can always add it as an axiom, so it is now in the model.

It would seem reasonable then to keep adding axioms until every true statement can be proven. However, a mathematician and logician named Kurt Gödel proved a theorem called the **Gödel Incompleteness Theorem**, which destroyed this idea. Simply stated, the theorem says that any consistent formal finite axiomatic system in which the ordinary arithmetic of integers can be performed is incomplete. Thus, we have the unfortunate choice of having an axiomatic system that has contradictions, or we have statements that are true but cannot be proven. The proof of this theorem is much beyond the scope of this book.

We are also due for another disappointment. It would be nice to know when input is put in a computer whether or not we will get an answer. Some computer programs can take a tremendously long time to run. If we have waited a couple of hours for a program to finish, we are faced with two possibilities. One is that the program will finish if we just wait long enough. The other is that the program is in an infinite loop and will never produce a solution but just keep humming along.

The **halting problem** asks if there is a computer program (procedure or algorithm) that accepts a program P and its input I and decides whether or not the program P with input I will eventually halt. Unfortunately, no such program exists. The proof of this problem is somewhat similar to the proof that we shall see in Chapter 4 when we look at Russell's paradox. We shall assume such a program exists and then find a contradiction.

There is a theorem that we shall not prove that states that there is a one-to-one correspondence between computer programs and strings of symbols in the input language. Given a program P we shall let $\hat{P}$ denote the corresponding string of symbols. This helps us in the following Theorem.

THEOREM 3.13 No program exists that accepts a program P and its input I and decides whether or not the program will eventually halt.

Proof Assume that there is a computer program $h(I, \hat{P})$, which for a given input I and program P prints out 1 if the program P with input I halts and prints out 0 if the program doesn't halt. Thus, $h(\hat{P}, \hat{P}) = 1$ if a program halts when accepting itself as input and $h(\hat{P}, \hat{P}) = 0$ if the program doesn't halt when accepting itself as input. We now define a function $k(\hat{P})$, which equals 1 if $h(\hat{P}, \hat{P}) = 0$ and goes into an infinite loop if $h(\hat{P}, \hat{P}) = 1$.

Now consider $k(\hat{k})$. If k with input $\hat{k}$ halts, then $k(\hat{k}) = 1$ and $h(\hat{k}, \hat{k}) = 0$. But this means that k with input $\hat{k}$ doesn't halt. If k with input $\hat{k}$ doesn't halt, then $k(\hat{k})$ goes into an infinite loop so that $h(\hat{k}, \hat{k}) = 1$. But this means that k with input $\hat{k}$ does halt. Either way we have a contradiction so we must conclude that the function h does not exist. ∎

Exercises

Let a, b, and c be integers. Prove the following:

1. If $b + a = c + a$, then $b = c$.
2. For each integer a, $a \cdot 0 = 0$.
3. For each integer a, $-(-a) = a$.
4. $a \cdot (-b) = -(ab)$.
5. $(-a) \cdot (-b) = a \cdot b$.
6. $-(a + b) = (-a) + (-b)$.
7. Prove that if $a + b = a$, then $b = 0$.
8. Prove that if $a \geq b \geq 0$ and $c \geq d \geq 0$, then $ac \geq bd$.
9. Prove that for every integer a, $a^2 \geq a$.
10. Prove that for integers a and b, $a^2 + b^2 \geq 2ab$.

Prove or show a counterexample for the following:

11. If a and b are odd integers, then ab is an odd integer.
12. If a and b are even integers, then ab is an even integer.

Prove or give a counterexample for the following:

13. If $a, b, c,$ and d are integers such that $a \geq b$ and $c \geq d$, then $ac \geq bd$.
14. The additive inverse of an even number is even.
15. The additive inverse of an odd number is odd.
16. There is a smallest positive integer.
17. There is a largest positive integer.
18. If a and ab are positive, then b is positive.
19. If a and ab are negative, then b is negative.
20. If a is positive and ab is negative, then b is negative.
21. For every integer a, $a^2 > a$.
22. For every integer a, $a^2 \geq a$.
23. For all integers m and n, if $mn = 1$, then either $m = n = 1$ or $m = n = -1$.
24. If a, b, c and d are integers, then $(a + b)(c + d) = ac + ad + bc + bd$.
25. If a and b are integers, then $a^3 + b^3 = (a + b)(a^2 - ab + b^2)$.
26. If a and b are integers and $(a + b)$ is even, then $a^3 + b^3$ is even.
27. If a and b are integers and $(a^3 + b^3)$ is even, then $a + b$ is even.

28. If a and b are integers and $(a+b)$ is odd, then $a^3 + b^3$ is odd.

29. If a and b are integers and $(a^3 + b^3)$ is odd, then $a+b$ is odd.

30. If an integer a has a factor that is even, then a is even.

31. If an integer a has a factor that is odd, then a is odd.

3.3 Mathematical Induction

One of the most difficult problems in number theory is to prove that a statement is true for all positive integers. It cannot be assumed that because a statement is true for the first 10 positive integers or even the first 10 billion positive integers, the statement is true for all positive integers. It is usually easy to prove that a statement is not true for all positive integers. Using the fact that the negation of $(\forall x) P(x)$ is $(\exists x) \sim P(x)$, we need find only one positive integer where the statement $P(x)$ is not true. This procedure is known as finding a **counterexample**. The tool we need to prove that a statement is true for all positive integers is the principle of induction, which is our last axiom for the positive integers N. This does not mean that the proof of every statement about the positive integers requires induction. Many theorems are proven using other theorems and axioms. However, the underlying structure for the proof of many theorems is the principle of induction. The principle of induction will later be stated in equivalent forms.

(N4) (Principle of Mathematical Induction) Let $P(n)$ be a statement with the property that if

(a) $P(1)$ is true and

(b) whenever $P(k)$ is true, then $P(k+1)$ is true,

then $P(n)$ is true for every positive integer n.

In symbolic form the principle of mathematical induction has the form

$$(P(1) \land ((\forall k)\ P(k) \to P(k+1))) \to (\forall n)\ P(n)$$

The principle of mathematical induction has been compared to an infinite row of dominoes. Properly aligned dominoes have the property that if any one is pushed over, then the next one falls over. Let $P(n)$ be the statement that the nth domino falls over. Pushing over the first domino makes $P(1)$, the statement that the first domino falls over, true. Since each domino knocks over the next one, if $P(k)$ is true, then $P(k+1)$ is true. Since all the dominoes fall, $P(n)$ is true for every positive integer n.

Since the term *proof by mathematical induction* is lengthy, we shall often simply say proof by induction rather that proof by mathematical induction.

EXAMPLE 3.14 Assume that we want to find a formula for calculating the sum of the first n positive integers, $1 + 2 + 3 + \cdots + n$. Such a formula would be useful if we needed to calculate the sum of the first 100,000 positive integers but did not want to perform 99,999 additions to obtain the answer. Assume we have reason to suspect that the sum of the first n positive integers is $\frac{n(n+1)}{2}$. This assertion is proved using mathematical induction.

Let S_n be the statement

$$1 + 2 + 3 + \cdots + n = \frac{n(n+1)}{2} \tag{3.1}$$

We first show that S_1 is true. The left side of the equality, $1+2+3+\cdots+n$, gives 1 for $n = 1$. On the right side we get

$$\frac{n(n+1)}{2} = \frac{1(1+1)}{2} = 1$$

so S_1 is true. Next we assume that for the positive integer k, S_k is true or, equivalently, that Equation (3.1) is true for $n = k$. That is, we assume that

$$1+2+3+\cdots+k = \frac{k(k+1)}{2}$$

We now must prove that S_{k+1} is true or, equivalently, that the equality S_n is true for $n = k+1$. In other words, we must prove that

$$1+2+3+\cdots+k+(k+1) = \frac{(k+1)((k+1)+1)}{2}$$

Using our assumption that

$$1+2+3+\cdots+k = \frac{k(k+1)}{2}$$

and comparing this equality to the one we need, we see that adding $k+1$ to both sides of the assumed equality will produce the left side of the equality we are seeking:

$$1+2+3+\cdots+k = \frac{k(k+1)}{2}$$

$$1+2+3+\cdots+k+(k+1) = \frac{k(k+1)}{2} + (k+1)$$

$$= (k+1)\left(\frac{k}{2}+1\right)$$

$$= \frac{(k+1)((k+1)+1)}{2}$$

We have proved that

$$1+2+3+\cdots+k+(k+1) = \frac{(k+1)((k+1)+1)}{2}$$

which means that Equation (3.1) holds for $n = k+1$. This result ensures that S_{k+1} is true. Hence, by Axiom N4 (that is, by induction), S_n is true for every positive integer n. ■

EXAMPLE 3.15 Prove that

$$1^2 + 2^2 + 3^2 + \cdots + n^2 = \frac{n(n+1)(2n+1)}{6} \quad (3.2)$$

Let S_n be the statement that the equality in Equation (3.2) holds. We first show that the statement S_n is true for $n = 1$. Substituting 1 in both sides of the equation, we have

$$1^2 = \frac{1(1+1)(2(1)+1)}{6}$$

giving $1 = 1$ and the statement S_n is true for $n = 1$.

Next we assume that the equation S_n is true for $n = k$. That is, we assume that

$$1^2 + 2^2 + 3^2 + \cdots + k^2 = \frac{k(k+1)(2k+1)}{6}$$

We now must prove that the equation S_n is true for $n = k + 1$. In other words, we must prove that

$$1^2 + 2^2 + 3^2 + \cdots + k^2 + (k+1)^2 = \frac{(k+1)((k+1)+1)(2(k+1)+1)}{6}$$
$$= \frac{(k+1)(k+2)(2k+3)}{6}$$

Using our assumption that

$$1^2 + 2^2 + 3^2 + \cdots + k^2 = \frac{k(k+1)(2k+1)}{6}$$

we see that adding $(k+1)^2$ to both sides of the assumed equality will produce the left side of the equality we are seeking. Thus,

$$1^2 + 2^2 + 3^2 + \cdots + k^2 = \frac{k(k+1)(2k+1)}{6}$$

$$1^2 + 2^2 + 3^2 + \cdots + k^2 + (k+1)^2 = \frac{k(k+1)(2k+1)}{6} + (k+1)^2$$

$$= \frac{k(k+1)(2k+1)}{6} + \frac{6(k+1)^2}{6}$$

$$= \frac{k(k+1)(2k+1) + 6(k+1)^2}{6}$$

$$= \frac{(k+1)[k(2k+1) + 6(k+1)]}{6}$$

$$= \frac{(k+1)(2k^2 + 7k + 6)}{6}$$

$$= \frac{(k+1)(2k+3)(k+2)}{6}$$

We have proved that

$$1^2 + 2^2 + 3^2 + \cdots + k^2 + (k+1)^2 = \frac{(k+1)(k+2)(2k+3)}{6}$$

which means that the equation S_n holds for $n = k + 1$. Hence, the equation S_n is true for every positive integer n. ■

EXAMPLE 3.16

Show that $n^3 - n$ is divisible by 3 for every positive integer n. (Note that a positive integer t is divisible by 3 provided that there is a positive integer m so that $t = 3m$.) The proof will proceed by induction. Let S_n be the statement that $n^3 - n$ is divisible by 3.

Show that S_1 is true. For $n = 1$ we calculate $n^3 - n = 1^3 - 1 = 1 - 1 = 0$, which is divisible by 3 because $0 = 3 \cdot 0$. Thus, S_1 is true.

Show that for any positive integer k, whenever S_k is true, then S_{k+1} is true. Assume that S_k is true. Thus, $k^3 - k$ is divisible by 3 so that $k^3 - k = 3m$ for some positive integer m. We must show that S_{k+1} is true or, equivalently, that

$(k+1)^3 - (k+1)$ is divisible by 3 so there exists a positive integer w such that $(k+1)^3 - (k+1) = 3w$. We now focus on the positive integer $(k+1)^3 - (k+1)$. The strategy here is to manipulate $(k+1)^3 - (k+1)$ so that we will be able to use our knowledge about k (namely, that $k^3 - k = 3m$). Thus,

$$\begin{aligned}(k+1)^3 - (k+1) &= (k^3 + 3k^2 + 3k + 1) - (k+1) \\ &= (k^3 - k) + (3k^2 + 3k) \\ &= 3 \cdot m + 3(k^2 + k) \\ &= 3 \cdot w\end{aligned}$$

where $w = m + (k^2 + k)$. Thus, $(k+1)^3 - (k+1)$ is divisible by 3 and, consequently, S_{k+1} is true. Therefore, by induction, S_n is true for all n. ∎

EXAMPLE 3.17 Show that a plane divided up into regions by any number of distinct straight lines can be painted with black and white paint in such a way that any two regions having a common boundary will be painted in different colors as depicted in Figure 3.4.

Since the statement of the problem asks about all possible numbers of regions and lines, we have infinitely many possibilities. This type of problem is one that the principle of induction is designed to handle. We have a choice of whether to show that any number m of such regions can be properly painted or that any number of regions generated by m lines can be properly painted. In either case, if we can show the property for all occurrences of m regions or all occurrences or m lines, we will have shown that any such drawing could be painted. In the first case, we would say that we are proceeding by induction upon the number m of regions. In the second case, we would say that we are proceeding by induction upon the number m of lines. We will use the number of lines involved in a drawing.

Let $P(m)$ be the statement that a drawing made with m lines can be properly painted (i.e., with two colors black and white with no adjacent regions that share a common boundary segment having the same color).

We choose

$$\begin{aligned}M &= \{m : P(m) \text{ is true}\} \\ &= \{m : \text{a drawing made with } m \text{ lines can be colored}\}\end{aligned}$$

Thus, if we show that $M = N$, the positive integers, then a drawing created with any number of lines can be properly painted. $M \subseteq N$ by definition.

(a) Show that $1 \in M$. $1 \in M$ provided that $P(1)$ is true or provided that a drawing made with only one line can be painted. It is known from geometry that a line drawn in the plane divides the plane into two regions. Choose one to paint black and paint the other one white as shown in Figure 3.5. Then $1 \in M$.

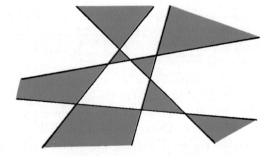

Figure 3.4
Painted regions

Figure 3.5
Painted with one line

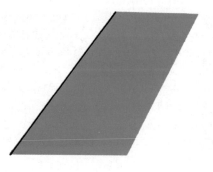

Figure 3.6
$m + 1$ lines

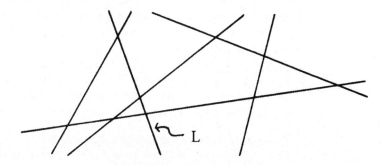

Figure 3.7
Painted with m lines

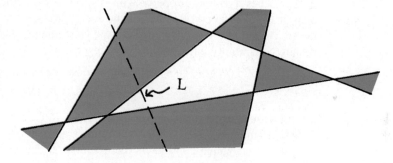

(b) Show that whenever $m \in M$, then $m + 1 \in M$. Assume that m is a positive integer in M, which means that $P(m)$ is true and any drawing made with m lines can be properly painted. Consider the positive integer $m + 1$. We need to show that $P(m+1)$ is true, or, equivalently, that any drawing made with $m + 1$ lines can be properly painted. In doing so, we are allowed to use "the induction hypothesis," namely, the knowledge that any drawing made with m lines can be properly painted. So consider a drawing made with $m + 1$ lines as depicted in Figure 3.6.

To be able to proceed, we must conceive of, or create, a situation for which we can use the knowledge that $m \in M$ or that any drawing with m lines can be properly painted. Toward this end, select one of the lines, say line L, and remove it.

The drawing that remains is made of only m lines, so since we assumed that any drawing made with m lines can be properly painted, we can properly paint the drawing in Figure 3.7.

Now replace the line L in its original position and change the color of each region on one side of the line L to the opposite color (change whites to black and change blacks to white) as shown in Figure 3.8. This drawing is properly colored because each boundary that occurred on the chosen side of L where the changes were made originally had different colors on either side so that if the colors were

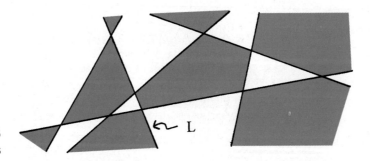

Figure 3.8
Painted with $m+1$ lines

switched, the colors on either side of any such boundary would still be different. The only other boundaries affected were those along L, but if L divided a region, then the color was changed on one side of L, making different colors on either side of such a boundary. Thus, this drawing with $m+1$ lines can be properly painted. This result means that $P(m+1)$ is true and that $m+1 \in M$.

In summary, we have shown parts (a) and (b) of the induction principle. Thus, $M = N$. We often say at this point that "by induction" we have shown that $M = N$. Thus, a drawing with any (finite) number of lines can be properly painted. ∎

There are two other equivalent definitions for the principle of mathematical induction. They are equivalent to the other definition in the sense that any of the three forms can be used to prove the other two. Consider the set $A = \{3, 7, 8\}$. The integer 3 has a unique property with respect to A; namely, 3 is in A and 3 is less than or equal to every member of the set A. We say that 3 is the "least" integer in the set A. Not every set of integers has a least integer. The set $\varnothing$ has no least integer because it is empty and contains no integer at all. The set

$$C = \{n : n \in \mathbb{Z} \text{ and } n = 5k \text{ for some integer } k\}$$
$$= \{\ldots, -15, -10, -5, 0, 5, 10, 15, 20, \ldots\}$$

has no least integer, for if m is a least integer in C, then $m = 5 \cdot k$ for some integer k. But the integer $s = 5(k-1)$ is also in C and, clearly, $s = 5(k-1) < 5 \cdot k = m$ so that $s < m$, which contradicts the assumption that m is the least integer in C.

DEFINITION 3.18 Let A be a set of integers. The integer a is the **least integer** of A if and only if
(a) a is in A.
(b) If b is in A, then $a \leq b$.

THEOREM 3.19 The least integer of a set of integers is unique.

Proof Assume that a and b are both least elements of a set A. Then by definition of least element, $a \leq b$ and $b \leq a$. Hence, $a = b$. ∎

We now state the two equivalent alternatives to the induction principle.

(N5) (The Well-Ordering Principle) Every nonempty set of positive integers contains a least element.

(N6) (Second Principle of Induction) Let $P(n)$ be a statement. If

(a) $P(1)$ is true and
(b) whenever $P(m)$ is true for all $m < k$, then $P(k)$ is true,

then $P(n)$ is true for every positive integer n.

In symbolic form the principle of induction has the form

$$(P(1) \wedge ((\forall k)(\forall m < k) P(m)) \rightarrow P(k)) \rightarrow (\forall n)\ P(n)$$

An example of the use of the second principle of induction N6 will be given later when it is shown that every integer is the product of prime integers. The following theorem is an example of a statement that seems too obvious to bother to prove, but since our model of the integers is defined by a set of rules, the statement may not be obvious or even true for our "integers." We can only find out by proving the theorem using the rules that we have. This proof is a good example of an application of the well-ordering principle, N5. It is also an example of reductio ad absurdum.

THEOREM 3.20 There is no positive integer a such that $0 < a < 1$.

Proof If there is such an integer, let S be the set of all positive integers between 0 and 1. Since S is nonempty, it has, by Axiom N5, a least positive integer, say a. Multiplying the inequality $0 < a < 1$ by a, we have $0 < a^2 < a$. Also since $a < 1$, $a^2 < 1$. But this contradicts the fact that a is the least positive integer less than 1. Hence, S is empty, and there is no positive integer between 0 and 1. ∎

Corollary 3.21 If b is an integer, then there is no integer a having the property $b < a < b + 1$. ∎

Axioms N4 and N6 address the integers starting with 1 and specifying conditions on integers greater than 1. A similar induction theorem applies to the integers greater than or equal to an integer j that is not necessarily the integer 1. This theorem provides induction upon a subset of the integers instead of just upon the positive integers.

THEOREM 3.22 (**Induction Principle for Integers**) Let $P(n)$ be a proposition with the property that

(a) $P(j)$ is true and

(b) whenever $P(k)$ is true for $k \geq j$, then $P(k+1)$ is true

then $P(n)$ is true for every $n \geq j$.

Proof The proof follows from Axiom N4 by letting $n = t - j + 1$. ∎

In this example we are proving a statement containing an inequality. When this occurs, we often have to use the property from Theorem 3.11:

$$(a > b > 0) \wedge (c > d > 0) \rightarrow ac > bd$$

EXAMPLE 3.23 Prove that $n! > 2^n$ for every integer $n \geq 4$ using the induction principle for integers.

If we want to use induction on n, we cannot start with $n = 1$ since the statement is not true for $n = 1$. Our starting place must be $n = 4$ since the statement is also not true for $n = 2$ and $n = 3$. In the notation of the induction principle for integers, $j = 4$, $P(n)$ is the statement that $n! > 2^n$.

First prove the statement for $n = 4$. When $n = 4$ we have $4! = 24$ and $2^4 = 16$, so $4! > 2^4$.

We have $k! > 2^k$ by the induction hypotheses and we want to prove $(k+1)! > 2^{k+1}$. We can see that if we multiply the left side of the equality by $(k+1)$ and the right side by 2, we would get the desired result. Therefore, if we can show that $(k+1) > 2$, we will have $k! > 2^k$ and $(k+1) > 2$, so that we can conclude that $(k+1)! > 2^{k+1}$. Since $k \geq 4$, we know that $k > 2$. Hence, $(k+1) \cdot k! > 2 \cdot 2^k$ and $(k+1)! > 2^{k+1}$. Thus, $n! > 2^n$ for every $n \geq 4$. ∎

Exercises

1. Prove that $1 + 4 + 7 + 10 + \cdots + (3n - 2) = \dfrac{n(3n - 1)}{2}$.

2. Prove that $\dfrac{1}{1 \cdot 3} + \dfrac{1}{3 \cdot 5} + \dfrac{1}{5 \cdot 7} + \dfrac{1}{7 \cdot 9} + \cdots + \dfrac{1}{(2n - 1)(2n + 1)} = \dfrac{n}{2n + 1}$.

3. Prove that $1 + 2 + 2^2 + 2^3 + \cdots + 2^{n-1} = 2^n - 1$.

4. Prove that $1 + r + r^2 + r^3 + \cdots + r^{n-1} = \dfrac{1 - r^n}{1 - r}$.

5. Prove that $a + ar + ar^2 + ar^3 + \cdots + ar^{n-1} = \dfrac{a - ar^n}{1 - r}$.

6. Prove that $1 + 3 + 5 + 7 + \cdots + (2n - 1) = n^2$.

7. Prove that $1^3 + 2^3 + 3^3 + 4^3 + \cdots + n^3 = \dfrac{n^2(n + 1)^2}{4}$.

8. Prove that $1^4 + 2^4 + 3^4 + \cdots + n^4 = \dfrac{n(n + 1)(2n + 1)(3n^2 + 3n - 1)}{30}$.

9. Prove that $1 \cdot 2 + 2 \cdot 3 + 3 \cdot 4 + \cdots + n(n + 1) = \dfrac{n(n + 1)(n + 2)}{3}$.

10. Prove that $1 + 5 + 9 + \cdots + (4n - 3) = n(2n - 1)$.

11. Prove that $2 + 6 + 10 + \cdots + (4n - 2) = 2n^2$.

12. Prove that $a + (a + d) + (a + 2d) + \cdots + (a + (n - 1)d) = \dfrac{n(2a + (n - 1)d)}{2}$.

13. Prove, using mathematical induction, that $n^2 > 2n + 1$ for $n \geq 3$.

14. Prove, using mathematical induction, that $2^n > n^2$ for $n \geq 5$.

15. Find the largest set of positive integers for which it is true that $2^n > n!$. Prove that this is true.

16. Prove, using mathematical induction, that $13^n - 6^n$ is divisible by 7 for all $n \geq 1$.

17. Prove, using mathematical induction, that $7^n - 2^n$ is divisible by 5 for all $n \geq 1$.

18. Prove, using mathematical induction, that $2^{3n} - 1$ is divisible by 7 for all $n \geq 1$.

19. Prove, using mathematical induction, that $n^3 + 2n$ is divisible by 3 for all $n \geq 1$.

20. Prove, using mathematical induction, that $a^n - 1$ is divisible by $a - 1$ for all $n \geq 1, a > 1$.

21. Prove that every integer $n \geq 8$ can be written as $n = 3k + 5m$ for some nonnegative integers k and m.

22. Prove that for each positive integer $n \geq 2$,
$$\dfrac{4^n}{n + 1} < \dfrac{(2n)!}{(n!)^2}.$$

23. Using mathematical induction, prove that $n! > n^3$ for all $n \geq 6$.

24. Use mathematical induction to prove that
$$\left(1 - \dfrac{1}{4}\right)\left(1 - \dfrac{1}{9}\right)\left(1 - \dfrac{1}{16}\right) \cdots \left(1 - \dfrac{1}{n^2}\right)$$
$$= \dfrac{n + 1}{2n} \quad \text{for } n \geq 2.$$

25. Prove that if W is a nonempty subset of Z and there is an integer x such that $x \leq w$ for all w in W, then W has a least element.

26. What is wrong with the following proof that all horses are the same color?

    Proof We prove this by showing that, given any n horses, they are all the same color. For $n = 1$, we have only one horse, which is certainly the same color as itself. For $n = k$, assume that given any k horses they are all the same color. Suppose we place $k + 1$ horses in a corral. We remove one horse. There are now k horses in the corral, so they are the same color. Now place the horse that was removed back into the corral and remove another horse. Again there are k horses in the corral so they are the same color. They are also the same color as the ones that were removed, so all $k + 1$ horses are the same color. Therefore, all horses are the same color. ∎

27. Define $\bigcup_{i=1}^{n} A_i$ by $\bigcup_{i=1}^{1} A_i = A_1$ and $\bigcup_{i=1}^{k+1} A_i = \left(\bigcup_{i=1}^{k} A_i\right) \cup A_{k+1}$. Define $\bigcap_{i=1}^{n} A_i$ by $\bigcap_{i=1}^{1} A_i = A_1$ and $\bigcap_{i=1}^{k+1} A_i = \left(\bigcap_{i=1}^{k} A_i\right) \cap A_{k+1}$.

    (a) Prove that
    $$\left(\bigcup_{i=1}^{n} A_i\right) \cap A = \bigcup_{i=1}^{n} (A_i \cap A)$$

    (b) Prove that
    $$\left(\bigcup_{i=1}^{n} A_i\right)' = \bigcap_{i=1}^{n} A_i'$$

28. Using the above definitions,

    (a) Prove that $\left(\bigcap_{i=1}^{n} A_i\right) \cap A = \bigcap_{i=1}^{n} (A_i \cap A)$.

    (b) Prove that $\left(\bigcap_{i=1}^{n} A_i\right)' = \bigcup_{i=1}^{n} A_i'$.

For a positive integer n, define a^n by $a^1 = a$ and $a^{k+1} = a^k \cdot a$.

29. Prove that $a^{m+n} = a^m a^n$.

30. Prove, using induction, that $(ab)^n = a^n b^n$ for ll $n \geq 1$.

31. Prove that for any integer a and positive integers m and n, $a^{mn} = (a^m)^n$.

32. Prove that the first principle of induction implies the well-ordering principle.

33. Prove that the well-ordering principle implies the second principle of induction.

34. Prove that the second principle of induction implies the first principle of induction.

35. The method of "infinite descent" is sometimes used to prove that no positive integer can have a particular property. The method works as follows. Suppose that there is a positive integer, say, a_1, having the property. Then generate a second positive integer, say, a_2, also having the property but such that $a_2 < a_1$. Similarly, generate a positive integer a_3 having the property and such that $a_3 < a_2$, and so on. Thus, we have an "infinitely descending" sequence $a_1 > a_2 > a_3 > \cdots$ of positive integers. Evidently, no such sequence can exist. This sequence was generated by assuming that there was an integer having the given property; therefore, the assumption is false. Prove the following statement, which justifies the argument above: Assume that

    (i) T is a subset of Z^+.

    (ii) If $n \in T$, then there is an $m \in Z^+$ such that $m < n$ and $m \in T$.

    Then $T = \varnothing$.

36. Use the method of infinite descent to prove that there are no positive integers p and q such that $p^2 = 3q^2$ by finding a positive integer $p' < p$ such that $(p')^2 = 3(q')^2$ for some positive integer q'. You may assume it is known that if a and b are positive integers such that $b^2 = 3 \cdot a$, then there is a positive integer k with $b = 3 \cdot k$.

37. Using the addition formulas from trigonometry
    $$\cos(\alpha + \beta) = \cos(\alpha)\cos(\beta) - \sin(\alpha)\sin(\beta)$$
    $$\sin(\alpha + \beta) = \sin(\alpha)\cos(\beta) + \cos(\alpha)\sin(\beta)$$
    prove De Moivre's Theorem, which states that for all $n \geq 1$,
    $$(\cos(\theta) + i\sin(\theta))^n = \cos(n\theta) + i\sin(n\theta)$$

38. Show that for $n \geq 2$, a set containing n elements has $\dfrac{n(n-1)}{2}$ two-element subsets.

39. Show that if there are n distinct numbers to be multiplied together, then $n - 1$ multiplications are needed to complete the multiplications regardless of the location of the parentheses.

40. Show that there are 2^n distinct strings of length n, consisting of 1's and 0's (binary strings).

3.4 Divisibility

Many integers can be expressed as the product of smaller integers, and important characteristics and relationships of integers can often be obtained by examining the structure of integers in terms of their factors.

DEFINITION 3.24 An integer a is a **multiple** of the integer b if $a = bm$ for some integer m. A nonzero integer b divides an integer a, denoted by $b \mid a$, if a is a multiple of b. An integer b that divides an integer a is also said to be a **factor** of a or to be a **divisor** of a.

From the definition, $9 \mid 27$ because $27 = 9 \cdot 3$, but 5 does not divide 12 since there is no integer m such that $12 = 5 \cdot m$. The integers 1, 2, 3, 4, 6, and 12 are all divisors of 12 and are the only positive divisors of 12. The integers $-1, -2, -3, -4, -6$, and -12 are also divisors of 12.

THEOREM 3.25 Let $a, b,$ and c be integers. Then

(a) For all a, $a \mid a$.

(b) For all $a, b,$ and c, if $a \mid b$ and $b \mid c$, then $a \mid c$.

(c) For all a, b, and c, $b \mid a$ and $b \mid c$ if and only if $b \mid (m \cdot a + n \cdot c)$ for all integers m and n.

Proof **(a)** $a = a \cdot 1$.

(b) If $a \mid b$ and $b \mid c$, then $b = a \cdot m$ and $c = b \cdot n$. Hence, $c = b \cdot n = (a \cdot m) \cdot n = a \cdot (m \cdot n)$ and $a \mid c$.

(c) Assume that $b \mid a$ and $b \mid c$. Then $a = b \cdot p$ and $c = b \cdot q$ for some integers p and q. Hence, $m \cdot a + n \cdot c = m \cdot (b \cdot p) + n \cdot (b \cdot q) = b \cdot (m \cdot p + n \cdot q)$ and $b \mid (m \cdot a + n \cdot c)$. Conversely, assume that $b \mid (m \cdot a + n \cdot c)$ for all integers m and n. Then if $m = 1$ and $n = 0$, we have $b \mid a$. If $m = 0$ and $n = 1$, we have $b \mid c$. ∎

THEOREM 3.26 If a and b are positive integers and $a \mid b$, then $a \leq b$.

Proof The proof follows from Theorem 3.11(d). ∎

Corollary 3.27 If, for positive integers a and b, $a \mid b$ and $b \mid a$, then $a = b$. ∎

THEOREM 3.28 For integers a and b, if $a \mid b$ and $b \mid a$, then $a = b$ or $a = -b$.

Proof If $a > 0$ and $b > 0$, by Corollary 3.27 we have $a = b$. If $a < 0$ and $b > 0$, then $-a > 0$ and $(-a) \mid b$ and $b \mid (-a)$ so that $b = -a$ or $a = -b$. The case $a > 0$ and $b < 0$ follows by symmetry. The case $a < 0$ and $b < 0$ is left to the reader. ∎

The preceding theorems imply that, for positive integers, the relation "divides" is reflexive, antisymmetric, and transitive and, hence, is a partial ordering of the positive integers. Notice that the integers are not partially ordered by division because "divides" is not an antisymmetric relation in the set of integers.

In the following theorem, we justify a process that occurs in elementary arithmetic. We can divide a positive integer a into a positive integer b and get a remainder r that is less than a. For example, if we divide 31 by 9, we get a quotient of 3 and a remainder of 4. Hence,

$$31 = 3 \cdot 9 + 4$$

where $4 < 9$.

The proof of the following theorem again illustrates of the use of the well-ordering theorem.

THEOREM 3.29 **(Division Algorithm)** For positive integers a and b there exist unique nonnegative integers q and r with $0 \leq r < b$ such that $a = bq + r$. The integers r and q are, respectively, called the **remainder** and **quotient** when a is divided by b.

Proof Let a and b be positive integers. Consider the set S of all nonnegative integers of the form $a - bq'$ for q' a nonnegative integer; that is,

$$S = \{x : x = a - bq', x \geq 0, \text{ and } q' \geq 0\}$$

S is nonempty because for $q' = 0$ we have $a - bq' = a - b \cdot 0 = a > 0$ so that $a \in S$. The set S has a least element, say r', since if $0 \in S$, it is the least element; and if not, then S contains only positive integers and, hence, has a least element by the well-ordering principle, N5. Since r' is in S, let q' be the nonnegative integer such that $r' = a - bq'$.

If $r' \geq b$, then $r'' = r' - b \geq 0$ and
$$a - bq' = r'$$
$$a - bq' - b = r' - b$$
$$a - b(q' + 1) = r''$$
so that r'' is in S. Also, $r'' < r'$ because
$$b > 0$$
$$-b < 0$$
$$r' - b < r' + 0 = r'$$
But $r'' < r'$ and $r'' \in S$ contradict the assumption that r' is the least element of S. Thus, $0 \leq r' < b$.

To show uniqueness, assume that $a = bq + r = bp + s$ where $0 \leq r \leq s < b$. Since $r \geq 0$, $-r \leq 0$. Hence, $s - r < b + 0 = b$. But $s - r = b(q - p)$, so by Theorem 3.26, $s - r \geq b$ unless $q - p = 0$. Hence, $q - p = 0$, $q = p$, and $r = s$. ∎

If $a < b$, then q will have to be 0 in order for $a = bq + r$ with $0 \leq r < b$ and $q \geq 0$. For example, for $a = 4$ and $b = 7$, the division algorithm gives $q = 0$ and $r = a = 4$ so $4 = 7 \cdot 0 + 4$.

By the uniqueness of q and r, if we can obtain q and r by any means whatsoever with $a = bq + r$, $0 \leq r < b$ and $q \geq 0$, then this q and r must be those guaranteed by the theorem.

Since any positive divisor of a positive integer can be no greater than the integer, we can always find every positive divisor of an integer n simply by testing every integer k, such that $1 \leq k \leq n$, to see whether it divides n. Thus, the positive divisors of 12 were found to be 1, 2, 3, 4, 6, and 12. Similarly, the positive divisors of 90 may be shown to be 1, 2, 3, 5, 6, 9, 10, 15, 18, 30, 45, and 90.

By inspection, 1, 2, 3, and 6 are divisors of both 12 and 90. 1, 2, 3, and 6 are called common divisors of 12 and 90. Further, 6 is the largest or greatest of these common divisors. Notice also that all of the common divisors, 1, 2, 3, and 6 happen to divide the greatest common divisor 6. In the context of greatest common divisors we consider only positive divisors.

DEFINITION 3.30 A positive integer d is a **common divisor** of a and b if $d \mid a$ and $d \mid b$.

DEFINITION 3.31 A positive integer d is the **greatest common divisor** of integers a and b provided that (i) $d \mid a$ and $d \mid b$ and (ii) if $c \mid a$ and $c \mid b$, then $c \mid d$. The greatest common divisor of a and b will be denoted by $\gcd(a, b)$.

An algorithm for finding the greatest common divisor is given in Chapter 7.

THEOREM 3.32 If d and c are greatest common divisors of integers a and b, then $c = d$; that is, there is at most one greatest common divisor.

Proof By definition of greatest common divisor, $d \mid c$ and $c \mid d$. Hence, $c = d$ or $c = -d$. Since c and d are both positive, $c = d$. ∎

THEOREM 3.33 The greatest common divisor of positive integers a and b exists. Further, the greatest common divisor can be written in the form
$$u \cdot a + v \cdot b$$
for some integers u and v. Further, it is the least positive integer of this form.

Proof Let S be the set of all positive integers of the form $na + mb$. Let $d = ua + vb$ be the least element of S. Then $d \leq a$ since $a = 1 \cdot a + 0 \cdot b$ is in S. Also, $a = qd + r$ for some q and r where $q > 0$ and $0 \leq r < d$. So $a = q(ua + vb) + r$. Solving for r, we have $r = (1 - qu)a + (-v)qb$, so r is in S or $r = 0$. But r is less than d, which is the least element of S so that $r = 0$. Therefore, $d \mid a$. Similarly, $d \mid b$. If c is any divisor of both a and b, then since $d = ua + vb$, by Theorem 3.25 (c), $c \mid d$. Hence, d is the greatest common divisor of a and b. ■

An algorithm for finding u and v is given in the Chapter 7.

The properties of the greatest common divisor, given in the following theorem, will be used later.

THEOREM 3.34 If $a = bq + c$, then $\gcd(a, b) = \gcd(b, c)$; or every divisor of a and b is a divisor of b and c, and conversely.

Proof Let $e = \gcd(a, b)$ and $f = \gcd(b, c)$. Since $e \mid a$ and $e \mid b$, then $e \mid c$ by Theorem 3.25(c), because $c = a - bq$. By definition of greatest common divisor, $e \mid f$.

Conversely, since $f \mid b$ and $f \mid c$, then $f \mid a$ by Theorem 3.25(c). By definition of greatest common divisor, $f \mid e$. Hence, $e = f$. ■

THEOREM 3.35 If a, b, and c are integers, $\gcd(a, b) = 1$, and $a \mid bc$, then $a \mid c$.

Proof Since $\gcd(a, b) = 1$, there exist integers u and v such that $au + bv = 1$. Multiply each term by c giving $cau + cbv = c$. Then $a \mid cau$ and $a \mid cbv$ since $a \mid bc$. Hence, $a \mid c$. ■

DEFINITION 3.36   If the greatest common divisor of a and b is 1, then a and b are said to be **relatively prime**.

It follows immediately that if a and b are relatively prime, then there exist integers u and v so that $au + bv = 1$.

Note that if a and b are positive integers then ab is a multiple of both a and b. If we consider the set of all multiples of a and b then by the well-ordering theorem, there is a least multiple of a and b. If c is the least multiple of a and b and d is another multiple of a and b, then $c \mid d$. Otherwise there exist q and r so that $d = qc + r$ where $r < c$ and, since a and b both divide both d and c, they also divide r. But then r would be a multiple of a and b that is less than c, which is a contradiction. Hence, we have the following definitions.

DEFINITION 3.37   A positive integer m is a **common multiple** of integers a and b if $a \mid m$ and $b \mid m$.

DEFINITION 3.38   A positive integer m is the **least common multiple** of integers a and b provided that

(a) $a \mid m$ and $b \mid m$, and

(b) if $a \mid n$ and $b \mid n$, then $m \mid n$.

The least common multiple of a and b will be denoted by $\text{lcm}(a, b)$.

The greatest common divisor is also helpful in finding solutions to equations of the form $ax + by = c$ or showing that they do not exist, as is seen in the next theorem.

An algorithm for finding x and y is given in Chapter 7.

THEOREM 3.39 The equation $ax + by = c$, where a, b, and c are integers, has an integral solution (i.e., there exist integers x and y such that $ax + by = c$) if and only if c is divisible by $\gcd(a, b)$. When c is divisible by $\gcd(a, b)$, a solution of $ax + by = c$ is

$$x_0 = \frac{u \cdot c}{\gcd(a, b)} \qquad y_0 = \frac{v \cdot c}{\gcd(a, b)}$$

where u and v comprise any solution of $\gcd(a, b) = au + bv$.

Proof We know that there exist integers u and v so that $au + bv = \gcd(a, b)$. If c is divisible by $\gcd(a, b)$, then $c = e \cdot \gcd(a, b)$ for some integer e. Hence, $aue + bve = e \cdot \gcd(a, b) = c$ so that $x = u \cdot e$ and $y = v \cdot e$ are solutions to the equation. Conversely, if there exist x and y so that $ax + by = c$, then since $\gcd(a, b)$ divides both a and b, it divides $ax + by$ and, hence, divides c. ∎

Computer Science Application

A *perfect number* is a positive integer that is equal to the sum of its proper divisors. For example, $6 = 1 + 2 + 3$ and $28 = 1 + 2 + 4 + 7 + 14$ are perfect numbers. *Amicable numbers* are pairs (m, n) of distinct integers for which each is equal to the sum of the proper divisors of the other. The pairs (220, 284) and (1184, 1210) are both amicable. The following algorithm returns the sum of the proper divisors of a positive integer n.

```
integer function sum-of-divisors (integer n)

    integer sum = 1

    for i = 2 to n/2
        if n mod i = 0 then
            sum = sum + i
        end-if
    end-for

    return sum

end-function
```

The condition

$$j = \text{sum-of-divisors}\ (j)$$

can be used to determine if j is a perfect number while the compound condition

$$i = \text{sum-of-divisors}\ (j) \text{ AND } j = \text{sum-of-divisors}\ (i)$$

will determine if (i, j) is an amicable pair.

Sociable groups of order n are harder to find. A sequence of n distinct integers $a_1, a_2, \ldots, a_n$ are sociable if the sum of the proper divisors of a_i is equal to a_{i+1}, for $i = 1, 2, \ldots, n - 1$ and the sum of the proper divisors of a_n is equal to a_1. These are scarce and hard to find using ordinary integer variables since most of the time at least one member of the group is too large to represent without special data structures. Using the sum-of-divisors function, the following code produces the sociable group

$$12496, \ 14288, \ 15472, \ 14536, \ 14264$$

function sociable-five ()
   integer a = 2, b, c, d, e
   repeat
      a = a + 1
      b = sum-of-divisors (a)
      c = sum-of-divisors (b)
      d = sum-of-divisors (c)
      e = sum-of-divisors (d)
   until sum-of-divisors (e) = a AND e not equal to a
   print a, b, c, d, e
   return
end-function

Exercises

Find all of the positive divisors of each for the following integers:

1. 54   **2.** 63   **3.** 72   **4.** 73   **5.** 74

For positive integers a and b, find nonnegative integers q and r with $0 \leq r < b$ such that $a = bq + r$.

6. $a = 54, b = 27$ **7.** $a = 47, b = 47$

8. $a = 93, b = 17$ **9.** $a = 43, b = 8$

10. $a = 33, b = 1$

For integers a and b, find $\gcd(a, b)$, $\text{lcm}(a, b)$, and $\gcd(a, b) \cdot \text{lcm}(a, b)$ where they are defined.

11. $a = 54, b = 27$ **12.** $a = 17, b = 0$

13. $a = 6, b = 15$ **14.** $a = 12, b = 16$

15. $a = 33, b = 1$

16. Prove or disprove: For positive integers a and b, $\gcd(a, b) \leq \text{lcm}(a, b)$.

17. If a or b were 0, what would we want $\text{lcm}(a, b)$ to be? Justify your answer.

18. If a and b are relatively prime, prove that for a given integer n, there exist integers x and y so that $ax + by = n$.

19. For positive integers a and b with $a > b$, prove that $\gcd(a, b) = \gcd(a - b, b)$.

20. Consider the theorem: For integers a and b, if $a \mid b$ and $b \mid a$, then $a = b$ or $a = -b$. Prove that the theorem holds in the case that $a < 0$ and $b < 0$.

21. Consider the theorem: For integers a and b, if $a|b$ and $b|a$, then $a = b$ or $a = -b$. Prove that the theorem holds in the case that $a < 0$ and $b < 0$.

22. Prove that for integers a, b, c, and d, if $a|c$ and $b|d$, then $ab|cd$.

23. Prove that for positive integers a, b, and c, if $ac|bc$, then $a|b$.

24. Prove that the product of three consecutive integers is always divisible by 3.

25. Prove that if a^2 is even, then it is divisible by 4.

26. Prove that the product of three even integers is divisible by 8.

27. Prove that if $a|b$, then $a^2|b^2$.

3.5 Prime Integers

Clearly, every integer a is divisible by itself and 1 since $a = 1 \cdot a$. Every integer divides 0, and 0 divides no integer. For some problems in number theory one needs to know whether a particular integer has any divisors besides itself and 1. Some integers cannot be factored into a product of integers except in this trivial way. Such integers are called primes.

DEFINITION 3.40 An integer greater than 1 is called **prime** if its only positive factors are itself and 1. A positive integer greater than 1 is **composite** if it is not a prime integer.

Of the first 10 positive integers, 2, 3, 5, and 7 are prime integers. On the other hand, the integers $4 = 2 \cdot 2$, $6 = 2 \cdot 3$, $8 = 2 \cdot 4$, $9 = 3 \cdot 3$, and $10 = 2 \cdot 5$ are composite. Thus, if $n = r \cdot s$ with $1 < r < n$ and $1 < s < n$, then n is composite. By definition, the integer 1 is neither prime nor composite. The integer 2 is the only even prime. It is easy to determine which of the small integers are prime by attempting a division by all smaller integers because the number of such possibilities is relatively small; however, deciding whether a large integer is prime can be a difficult task.

The following theorem shows us that there is an infinite number of primes. The proof of the theorem is another classic case of the use of reductio ad absurdum.

THEOREM 3.41 (**Euclid**) There are infinitely many prime integers.

Proof Assume that there are only finitely many primes, say, $p_1, p_2, \ldots, p_k$. Consider the integer $(p_1 p_2 \cdots p_k) + 1$. Let p_r be a prime and suppose that $p_r \mid ((p_1 p_2 \cdots p_k) + 1)$. But $p_r \mid (p_1 p_2 \cdots p_k)$, which implies that $p_r \mid 1$, a contradiction since $p_r > 1$. Hence, $(p_1 p_2 \cdots p_k) + 1$ is prime, also a contradiction since it is not one of the finitely many primes. Therefore, our assumption that there are only finitely many primes is false and there must be infinitely many prime integers. ∎

Because the prime factorization of integers is important, we need quick and easy ways to decide whether a given positive integer is prime or composite. The next theorem shows that only some of the possible factors need to be considered to test an integer for primeness.

THEOREM 3.42 If the positive integer n is a composite integer, then n has a prime factor p such that $p^2 \leq n$.

Proof Let p be the smallest prime factor of n. If n factors into r and s, then $p \leq r$ and $p \leq s$. Hence, $p^2 \leq rs = n$. ∎

For example, to determine whether $n = 521$ is prime, we only need to consider primes p that are less than or equal to 22 because $22^2 = 484$ and $23^2 = 529$. The primes less than or equal to 22 are 2, 3, 5, 7, 11, 13, 17, and 19. Trying each of these integers, we find that none of them divides 521. Therefore, 521 itself is prime by the preceding theorem.

The primes form a set of building blocks for the integers as shown in the following theorem.

THEOREM 3.43 Every positive integer is equal to 1, is a prime, or may be written as a product of primes.

Proof We prove this theorem using the second principle of induction. The theorem is certainly true for $n = 1$. Assume that it is true for all positive integers m less than k. If k is a prime, then the theorem is also true for k. If k is not a prime, then it is divisible by some integer p and $k = pq$ where neither p nor q is equal to either k or 1. Since, by Theorem 3.26, p and q are less than k, by the induction hypothesis they are primes or may be written as a product of primes. Hence, $k = pq$ may be written as a product of primes. ∎

The integer 37 is prime. The integer $1554985071 = 3 \cdot 3 \cdot 4463 \cdot 38713$ is a product of four prime factors, two of which are the same prime.

In the next theorem we will see that if an integer n is divisible by a prime p, then it is not possible to factor n in such a way that p does not divide at least one of the factors of n.

THEOREM 3.44 If p is a prime and $p \mid ab$, where a and b are positive integers, then $p \mid a$ or $p \mid b$.

Proof If $p \mid a$, then the conclusion holds. On the other hand, assume that p does not divide a. Since p and a are relatively prime, there exist u, v so that $pu + av = 1$. Therefore, $pub + avb = b$. Since $p \mid pub$ and $p \mid avb$ (because $p \mid ab$), then $p \mid b$. ∎

Lemma 3.45 If a prime number p divides a product of positive integers $q_1 q_2 \cdots q_n$, then p divides q_i for some i, $1 \leq i \leq n$.

Proof We prove this lemma using induction on n, the number of factors in the product. If $n = 1$, the lemma is obviously true. Assume that the lemma is true for $n = k$; that is, if p divides any product of k integers, then p divides one of the k factors. Assume that p divides a product of $k + 1$ integers, say, $p \mid q_1 q_2 \cdots q_k q_{k+1}$ so that $p \mid (q_1 q_2 \cdots q_k) q_{k+1}$. If p divides q_{k+1}, then we are finished. If p does not divide q_{k+1}, then, by Theorem 3.44, $p \mid (q_1 \cdots q_k)$. But since $q_1 \cdots q_k$ is the product of k integers, by the induction hypothesis, $p \mid q_i$ for some $1 \leq i \leq k$. Hence, $p \mid q_i$ for some i, $1 \leq i \leq k + 1$. ∎

Lemma 3.46 If a prime number p divides a product of primes $q_1, q_2, \ldots, q_n$, then $p = q_i$ for some i, $1 \leq i \leq n$.

Proof By Lemma 3.45, p divides q_i for some $1 \leq i \leq n$. Since p and q_i are both prime, $p = q_i$. ∎

The foregoing series of theorems leads to the main result of this chapter, which is called the fundamental theorem of arithmetic or the unique prime factorization theorem.

THEOREM 3.47 (**Unique Prime Factorization**) Any positive integer that is greater than 1 is a prime or can be written as a product of primes where this product is unique except for the arrangement of the primes.

Proof Assume that $q_1 q_2 \cdots q_n$ and $p_1 p_2 \cdots p_s$ are two ways of writing the positive integer m as a product of primes. We shall prove the theorem using induction on n, the number of prime factors in the first product. If $n = 1$, the theorem is trivially true. Assume that the theorem is true when $q_1 q_2 \cdots q_k = p_1 p_2 \cdots p_s$; that is, if $m = q_1 q_2 \cdots q_k = p_1 p_2 \cdots p_s$, then $k = s$ and the product is unique up to the order of primes. Assume that $m = q_1 q_2 \cdots q_{k+1} = p_1 p_2 \cdots p_{s'}$. Since q_{k+1} divides $p_1 p_2 \cdots p_{s'}$, it follows that $q_{k+1} = p_i$ for some $1 \leq i \leq s'$. Divide both products by q_{k+1} or use the cancellation property (Axiom I8). Then $q_1 q_2 \cdots q_k = p_1 p_2 \cdots p_{i-1} p_{i+1} \cdots p_{s'}$. But by induction, $k = s' - 1$ and the product is unique up to the order of primes. Hence, $k + 1 = s'$ and the factorization of m is unique up to order of primes. ∎

For example,
$$n = 39616304$$
$$= 2 \cdot 13 \cdot 7 \cdot 2 \cdot 23 \cdot 13 \cdot 2 \cdot 13 \cdot 2 \cdot 7$$
$$= 2 \cdot 2 \cdot 2 \cdot 2 \cdot 7 \cdot 7 \cdot 13 \cdot 13 \cdot 13 \cdot 23$$

are two factorizations of n; however, the same primes are used the same number of times in both products. Only the order in which the factors are written differs. In fact, there are 12,600 distinct factorizations of n using the 10 prime factors, but there are no factorizations without exactly four 2's, two 7's, three 13's, and one 23. Usually, the prime factors are grouped and combined using exponential notation as in

$$n = 2^4 7^2 13^3 23^1$$

Corollary 3.48 Every positive integer greater than 1 can be written uniquely, except for order, in the form $q_1^{k(1)} q_2^{k(2)} \cdots q_n^{k(n)}$, where $k(1), k(2), \ldots,$ and $k(n)$ are positive integers. ∎

At this point we can see why 1 is not allowed to be a prime since we would not have the unique prime factorization theorem. When discussing the factorization of several integers using the representation given by the preceding corollary, it is often a notational convenience to allow a prime to have zero for an exponent. This practice normally causes no confusion since $q_i^0 = 1$ if $q_i \geq 1$. If the prime factorization of an integer is known, then the primes forming the factorization of any divisor of that integer are a subset of those of the dividend.

The proofs of the following two theorems are left to the reader.

THEOREM 3.49 If $a = p_1^{a(1)} \cdots p_k^{a(k)}$ and $b \mid a$, then $b = p_1^{b(1)} \cdots p_k^{b(k)}$ where $0 \leq b(i) \leq a(i)$ for all i. ∎

THEOREM 3.50 Let $a = p_1^{a(1)} p_2^{a(2)} p_3^{a(3)} \cdots p_k^{a(k)}$ and $b = p_1^{b(1)} p_2^{b(2)} p_3^{b(3)} \cdots p_k^{b(k)}$ where the p_i are primes dividing either a or b and some of the exponents may be 0. Let $m(i) = \min(a(i), b(i))$ and $M(i) = \max(a(i), b(i))$ for $1 \leq i \leq k$. Then

$$\gcd(a, b) = p_1^{m(1)} p_2^{m(2)} p_3^{m(3)} \cdots p_k^{m(k)}$$

and

$$\operatorname{lcm}(a, b) = p_1^{M(1)} p_2^{M(2)} p_3^{M(3)} \cdots p_k^{M(k)}$$

∎

As an application of Theorem 3.50, let $a = 195000$ and $b = 10435750$. The factorizations of a and b are

$$a = 2^3 3^1 5^4 13^1 \quad \text{and} \quad b = 2^1 5^3 13^3 19^1$$

Thus,

$$\gcd(195000, 10435750) = 2^{\min(3,1)} 3^{\min(1,0)} 5^{\min(4,3)} 13^{\min(1,3)} 19^{\min(0,1)}$$
$$= 2^1 3^0 5^3 13^1 19^0 = 2^1 5^3 13^1 = 3250$$

and

$$\operatorname{lcm}(195000, 10435750) = 2^{\max(3,1)} 3^{\max(1,0)} 5^{\max(4,3)} 13^{\max(1,3)} 19^{\max(0,1)}$$
$$= 2^3 3^1 5^4 13^3 19^1 = 626145000$$

The following theorem follows from Theorem 3.50. The proof is left to the reader.

THEOREM 3.51 Let a and b be positive integers. Then $\gcd(a, b) \cdot \operatorname{lcm}(a, b) = ab$. ∎

Computer Science Application

The following algorithm tests a positive integer n for primality by using the fact that when a natural number has factors other than itself and 1, at least one of these factors must be less than or equal to the square root of the number.

boolean function prime(integer n)

    if $n = 1$ or $n = 2$ or $n = 3$ then
        return true
    end-if

    if n mod 2 = 0 then
        return false
    end-if

    for $i = 3$ to sqrt(n) by 2
        if n mod $i = 0$ then
            return false
        end-if
    end-for

    return false

end-function

This is a fairly efficient way to classify numbers that are not too large. The problem with this trial division approach is that it takes too long to return false when the composite number it is testing is very large and has no small factors and too long to return true when the prime it is testing is large. As we shall see in a later chapter, large primes are essential ingredients in modern cryptosystems, and generating them with certainty is challenging. On the fastest computer today, determining that a 100-digit number is prime using this algorithm would take longer than the age of the universe.

Exercises

Factor each of the following integers into the product of primes:

1. 728
2. 1599 (use the fact that $1599 = 1600 - 1$)
3. 4899 (use the fact that $4899 = 4960 - 1$)
4. 131
5. 523

Use the theorems of this chapter to find the gcd and lcm of the following pairs of numbers:

6. 162 and 12
7. 71 and 23
8. 72 and 30
9. $n!$ and $(n+2)!$
10. 75 and 99

11. Two primes a and b are called twin primes if $a + 2 = b$. For example, 3 and 5 are twin primes. Find three other pairs of twin primes.

12. Is the average of twin primes a prime?

13. If a and b are primes, is it necessary that $a^2 + b^2$ be a prime?

14. Explain why, for any positive integer $n \geq 2$, $n!+2$, $n!+3$, $n!+4$, $\ldots$, $n!+n$ are all composite. What does this show about the distance between primes?

15. Use the previous exercise and a calculator to show that the numbers 479001603 and 479001607 are not primes.

16. If $p_1, p_2, p_3, \ldots, p_n$ are the first n consecutive primes, does this necessarily mean that $p_1 \cdot p_2 \cdot p_3 \cdot \ldots \cdot p_n + 1$

(the product of the first n primes plus one) is a prime? Why or why not?

17. Show that there are no prime triplets, that is three consecutive odd numbers each of which is prime, except 3, 5, and 7.

18. If p and q are primes greater than or equal to 5, prove that either $p + q$ or $p - q$ is divisible by 3 and hence $p^2 - q^2$ is divisible by 24. (*Hint* : Either $p + 2$ or $p - 2$ is divisible by 3 and either $q - 2$ or $q + 2$ is divisible by 3.)

19. Prove that if $a = p_1^{a(1)} \cdots p_k^{a(k)}$ and $b \mid a$, then $b = p_1^{b(1)} \cdots p_k^{b(k)}$ where $0 \leq b(i) \leq a(i)$ for all i; and conversely.

20. Let $a = p_1^{a(1)} p_2^{a(2)} p_3^{a(3)} \cdots p_k^{a(k)}$ and $b = p_1^{b(1)} p_2^{b(2)} p_3^{b(3)} \cdots p_k^{b(k)}$ where the p_i are primes dividing either a or b and some of the exponents may be 0. Let $m(i) = \min(a(i), b(i))$ and $M(i) = \max(a(i), b(i))$ for $1 \leq i \leq k$. Then prove that

$$\gcd(a, b) = p_1^{m(1)} p_2^{m(2)} p_3^{m(3)} \cdots p_k^{m(k)}$$

and

$$\text{lcm}(a, b) = p_1^{M(1)} p_2^{M(2)} p_3^{M(3)} \cdots p_k^{M(k)}$$

A set of integers is pairwise relatively prime if any pair of integers in the set is relatively prime. Determine which of the following sets are pairwise relatively prime.

21. $\{8, 9, 26, 31\}$

22. $\{111, 27, 23, 137\}$

23. $\{119, 327, 329, 347, 443\}$

24. $\{247, 319, 413, 509, 634\}$

25. Prove that if $a^2 \mid b^2$, then $a \mid b$ or give a counterexample.

26. If $17 \cdot 19 \cdot 23 \cdot 29 = 43 \cdot 47 \cdot n$, does 29 divide n? Why or why not?

27. If $23 \cdot 47 \cdot 67 = 667 \cdot 571 \cdot n$, does 23 divide n? Why or why not?

28. Is $n! + 1$ always a prime? Why or why not?

29. Prove or disprove that for all integers $n > 2$, $n^2 - n - 1$ is prime.

30. If 14 divides n^2, does 14 divide n?

3.6 Congruence Relations

In Example 2.55, we discussed the relation R_3 on the integers defined by $R_3 = \{(a, b) : a - b = 5 \cdot k\}$. The equivalence relation generated by this partition is called "congruence modulo 5." In fact, we can form such partitions for any positive integer n as we did for $n = 5$.

DEFINITION 3.52 Let n be a positive integer. The integer a is **congruent** to the integer b modulo n, denoted $a \equiv b \pmod{n}$, if n divides $(a - b)$.

The proof for the following theorem is similar the one in Example 2.55.

THEOREM 3.53 The relation $\equiv$ for fixed n is an equivalence relation on the set of integers; that is,

(a) $a \equiv a \pmod{n}$ for every integer a.
(b) If $a \equiv b \pmod{n}$, then $b \equiv a \pmod{n}$ for integers a and b.
(c) If $a \equiv b \pmod{n}$ and $b \equiv c \pmod{n}$, then $a \equiv c \pmod{n}$. ∎

DEFINITION 3.54 Let n be a positive integer. The set of all equivalence classes modulo n is denoted by Z_n and is called the **set of integers modulo n**.

The integers modulo n are new objects. They are equivalence classes. The elements in each equivalence class have the property that they are congruent modulo n.

For example, let $n = 3$. There are three equivalence classes for congruence modulo 3 so that the set

$$Z_3 = \{[0], [1], [2]\}$$

has three members. The members of Z_3 are equivalence classes and, hence, sets. The three contain 0, 1, and 2, respectively, as their names suggest. In each of these equivalence classes, all of the elements are congruent to one another modulo 3 so that $a \equiv b \pmod{3}$ if and only if a and b are in the same equivalence class. Thus,

$$[0] = \{\ldots -6, -3, 0, 3, 6, 9 \ldots\}$$
$$[1] = \{\ldots -8, -5, -2, 1, 4, 7 \ldots\}$$
$$[2] = \{\ldots -7, -4, -1, 2, 5, 8 \ldots\}$$

We will eventually want to define operations between the equivalence classes in the set of integers modulo n. We will find that the congruence relation $\equiv$, as its notational similarity may suggest, has many of the same properties as the equality relation. With congruence modulo n there will often be more restrictions than apply with the analogous properties of equality.

THEOREM 3.55

(a) If $a \equiv b \pmod{n}$ and $c \equiv d \pmod{n}$, then $a + c \equiv b + d \pmod{n}$ and $ac \equiv bd \pmod{n}$.

(b) If $a \equiv b \pmod{mn}$, then $a \equiv b \pmod{m}$ and $a \equiv b \pmod{n}$.

Proof Part (a) is proved here. Part (b) is left to the reader. By definition, if $a \equiv b \pmod{n}$ and $c \equiv d \pmod{n}$, then $a - b = un$ and $c - d = vn$ for some integers u and v. Thus,

$$(a + c) - (b + d) = (a - b) + (c - d) = u \cdot n + v \cdot n = (u + v) \cdot n$$

for some integers u and v and $a + c \equiv b + d \pmod{n}$. Also, $a = b + un$ and $c = d + vn$, so

$$\begin{aligned} ac &= (b + (un))(d + (vn)) \\ &= bd + b(vn) + d(un) + uvn^2 \\ &= bd + (bv + ud + uvn)n \end{aligned}$$

Thus, $ac - bd = (bv + ud + vun)n$ and $ac \equiv bd \pmod{n}$. ∎

EXAMPLE 3.56

Let $n = 3$.

$$1 \equiv 7 \pmod{3}$$
$$2 \equiv -4 \pmod{3}$$

Theorem 3.55 (a) implies that

$$1 + 2 \equiv 7 + (-4) \pmod{3}$$

It also implies that

$$(1)(2) \equiv (7)(-4) \pmod{3}$$

which is true since $2 - (-28) = 30 = 3 \cdot 10 = 3 \cdot k$. ∎

THEOREM 3.57 For a given positive integer n,

(a) If $r \equiv r' \pmod{n}$, $0 \leq r < n$ and $0 \leq r' < n$, then $r = r'$.

(b) If a is any integer and n is a positive integer, then there is an integer r with $0 \leq r < n$ such that $a \equiv r \pmod{n}$. The integer r is the remainder when a is divided by n so that $a = nq + r$.

Proof This theorem follows directly from the division algorithm. ∎

Part (a) guarantees that the equivalence classes modulo n have the property that those equivalence classes generated by the nonnegative integers less than n are distinct. Thus, $[0], [1], [2], \ldots, [n-2], [n-1]$ are different sets relative to the congruence modulo n.

Part (b) shows that every integer a is congruent modulo n to one of the integers $0, 1, 2, \ldots, n-1$. So any integer a is in one of the n distinct equivalence classes $[0], [1], \ldots, [n-1]$. This discussion justifies the next theorem.

THEOREM 3.58 Let n be a given positive integer. Then Z_n, the set of integers modulo n, consists of exactly the n distinct equivalence classes

$$[0], [1], [2], \ldots, [n-1]$$

represented by the distinct remainders that can be obtained with division by n. Further, for $0 \leq r < n$, the equivalence class $[r]$ consists of exactly those integers a such that $a \equiv r \pmod{n}$. ∎

The next theorem is simply a restatement of the preceding theorems, which is often useful.

THEOREM 3.59 If $a = nq + r$ and $b = nq' + r'$ for $0 \leq r < n$ and $0 \leq r' < n$, then $r = r'$ if and only if $a \equiv b \pmod{n}$. ∎

In general when describing an equivalence class modulo n, we want to select the smallest nonnegative integer in the class to represent it. This is particularly important for the addition and multiplication of equivalence classes in Z_n, since we want the sum and product to still be represented by integers between 0 and $n-1$. Hence, we have the following definition.

DEFINITION 3.60 Let Z_n be the set of equivalence classes modulo n. Then given any integer m, there is an integer r such that $0 \leq r \leq n-1$ and $[m] = [r]$ or $m \equiv r \pmod{n}$. We say that $[[m]]_n = r$.

EXAMPLE 3.61 For $n = 5$, we obtain

$$Z_5 = \{[0], [1], [2], [3], [4]\}$$
$$= \{[r] : 0 \leq r < 5\}$$

∎

For the set Z_n we can define the operations of addition and multiplication. If $[a]$ is the equivalence class containing a in the congruence modulo n and $[b]$ is the modulo n equivalence class contain b, we define addition and multiplication by

$$[a] \oplus [b] = [a + b] = [[a + b]]_n$$
$$[a] \odot [b] = [a \cdot b] = [[a \cdot b]]_n$$

where the addition and multiplication in the center and on the right-hand side are between integers and the addition and multiplication on the left-hand side are between equivalence classes.

Using the preceding theorems we see that addition and multiplication of equivalence classes are well defined; that is, the definitions are independent of the representatives of the equivalence classes. For example, if $[a] = [c]$, then the result of addition must be the same whether we use $[a]$ or $[c]$. That is, if $[a] = [c]$ and $[b] = [d]$, then $[a] \oplus [b] = [c] \oplus [d]$ and $[a] \odot [b] = [c] \odot [d]$.

EXAMPLE 3.62

For $n = 5$ and $Z_5 = \{[0], [1], [2], [3], [4]\}$, we see that

$$[2] \oplus [4] = [2 + 4] = [6] = [1] \quad \text{since} \quad 6 \equiv 1 \pmod 5$$

$$[2] \odot [4] = [2 \cdot 4] = [8] = [3] \quad \text{since} \quad 8 \equiv 3 \pmod 5$$

By calculating all possible sums and products, we can produce "addition" and "multiplication" tables for the integers modulo 5.

$[a] \oplus [b]$	[0]	[1]	[2]	[3]	[4]
[0]	[0]	[1]	[2]	[3]	[4]
[1]	[1]	[2]	[3]	[4]	[0]
[2]	[2]	[3]	[4]	[0]	[1]
[3]	[3]	[4]	[0]	[1]	[2]
[4]	[4]	[0]	[1]	[2]	[3]

$[a] \odot [b]$	[0]	[1]	[2]	[3]	[4]
[0]	[0]	[0]	[0]	[0]	[0]
[1]	[0]	[1]	[2]	[3]	[4]
[2]	[0]	[2]	[4]	[1]	[3]
[3]	[0]	[3]	[1]	[4]	[2]
[4]	[0]	[4]	[3]	[2]	[1]

■

Theorem 3.58 shows that the integers may be partitioned into n disjoint sets or equivalence classes for any positive integer n. We will often focus on these sets as the objects of our attention, but sometimes we will be interested in representative integers taken from the n sets.

DEFINITION 3.63

For a positive integer n, if $b \equiv r \pmod n$, then we say that r is a **residue** of b modulo n. A **complete residue system** modulo n is a set $S = \{r_1, r_2, \ldots, r_n\}$ where the intersection of S with each equivalence class modulo n contains exactly one integer; that is, S contains one and only one representative from each such equivalence class. The complete residue system $\{0, 1, 2, \ldots, (n-1)\}$ is called the **primary residue system**. If b is an integer, $b \equiv r \pmod n$, and $0 \leq r \leq n - 1$, then we denote this unique primary residue modulo n by $r = [[b]]_n$. A **reduced residue system** modulo n is the subset of a complete residue system consisting of only those integers that are relatively prime to n; that is, $\{r : r \in S \text{ and } \gcd(r, n) = 1\}$.

A complete residue system is obtained by choosing one integer from each equivalence class $[0], [1], \ldots, [n-1]$ of Z_n. So for $n = 6$, $\{24, 7, -58, 40, 113\}$ is a complete residue system modulo 6 because

$$24 \equiv 0 \pmod{6} \quad \text{giving} \quad 24 \in [0]$$
$$7 \equiv 1 \pmod{6} \quad \text{giving} \quad 7 \in [1]$$
$$-58 \equiv 2 \pmod{6} \quad \text{giving} \quad -58 \in [2]$$
$$15 \equiv 3 \pmod{6} \quad \text{giving} \quad 15 \in [3]$$
$$40 \equiv 4 \pmod{6} \quad \text{giving} \quad 40 \in [4]$$
$$113 \equiv 5 \pmod{6} \quad \text{giving} \quad 113 \in [5]$$

Trivially, $\{0, 1, 2, 3, 4, 5\}$ is also a complete residue system modulo 6 and is the primary residue system. By definition $[[24]]_6 = 0$ and $[[-58]]_6 = 2$.

Since all members of an equivalence class modulo n are congruent to one another but to no member of another equivalence class, every integer is congruent to exactly one member of a complete residue system. This argument gives the following theorem.

THEOREM 3.64 If n is a positive integer, $\{r_1, r_2, \ldots, r_n\}$ is a complete residue system modulo n, and a is an integer, then $a \equiv r_k \pmod{n}$ for one and only one k, $1 \leq k \leq n$. ∎

The members of the complete residue system $\{24, 7, -58, 15, 40, 113\}$ and the primary (complete) residue system $\{0, 1, 2, 3, 4, 5\}$ modulo 6 that are relatively prime to $n = 6$ comprise the sets $\{7, 113\}$ and $\{1, 5\}$, respectively. Therefore, these are both reduced residue systems modulo 6. We say that $\{1, 5\}$ is the primary reduced residue system modulo 6.

The following theorem shows that a reduced residue system captures much of the essence of integers in Z that are relatively prime to n. The proof is left to the reader.

THEOREM 3.65 Let n be a positive integer and $\{r_1, r_2, \ldots, r_k\}$ be a reduced residue system modulo n. If a is an integer relatively prime to n, then $a \equiv r_j \pmod{n}$ for one and only one j, $1 \leq j \leq k$. ∎

Given one residue system, there are several ways to generate other residue systems. One way is to add the same integer to each residue in a complete residue system. Another is shown in the following theorem. The proof is left to the reader.

THEOREM 3.66 Let n be a positive integer and $\{r_1, r_2, \ldots, r_k\}$ be a complete [reduced] residue system modulo n. If a is an integer relatively prime to n, then $\{ar_1, ar_2, \ldots, ar_k\}$ is also a complete [reduced] residue system. ∎

Since $\gcd(6, 35) = 1$, given the complete residue system $\{0, 1, 2, 3, 4, 5\}$ and the reduced residue system $\{1, 5\}$ modulo 6, we have that $\{35 \cdot 0, 35 \cdot 1, 35 \cdot 2, 35 \cdot 3, 35 \cdot 4, 35 \cdot 5\}$ and $\{35 \cdot 1, 35 \cdot 5\}$ are complete and reduced residue systems, respectively. But

$$[[0]]_6 = 0 \quad \text{and} \quad [[35 \cdot 0]]_6 = [[0]]_6 = 0$$
$$[[1]]_6 = 1 \quad\quad\quad\quad [[35 \cdot 1]]_6 = [[35]]_6 = 5$$
$$[[2]]_6 = 2 \quad\quad\quad\quad [[35 \cdot 2]]_6 = [[70]]_6 = 4$$
$$[[3]]_6 = 3 \quad\quad\quad\quad [[35 \cdot 3]]_6 = [[107]]_6 = 3$$
$$[[4]]_6 = 4 \quad\quad\quad\quad [[35 \cdot 4]]_6 = [[140]]_6 = 2$$
$$[[5]]_6 = 5 \quad\quad\quad\quad [[35 \cdot 5]]_6 = [[175]]_6 = 1$$

Thus, the result of multiplying every member r_i of a complete or reduced residue system modulo n by an integer a relatively prime to n produces representatives of the same equivalence classes but possibly in a different order.

In computer applications, frequently we have n sets of information $I_1, I_2, I_3, \ldots I_n$, where each set is "named" by a key, $k_1, k_2, k_3, \ldots k_n$. The keys may be considered to be integers although they may be a finite sequence of alphabetic characters. It is desirable to find the information I_j quickly when the key k_j is found. One solution is the **hash function** $h : K \to \{0, 1, 2, 3, \ldots m\}$ where $m + 1 \geq n$. We then provide $m + 1$ computer storage units where n sets of information are kept. Thus, to find I_j we calculate $h(k_j)$ and go to $h(k_j)$ where information I_j is kept. It is generally difficult to find a hash function that is one-to-one so that $h(k_j) \neq h(k_m)$ when $k_j \neq k_m$. Hence we may have $h(k_j) = h(k_m)$ with $k_j \neq k_m$. When this occurs, it is called a collision. For methods of dealing with collisions and hash functions that are one-to-one see Section 22.2. The hash function that we have considered in this section is $h(k_j) = [[k_j]]_m$.

For example, suppose the units of information concern individual people and the keys are the Social Security numbers of the individuals. We might then select the approximate number of people, say, 2389, and let $h(k_j) = [[k_j]]_{2389}$. Thus, for the Social Security number $213 - 10 - 0408$, $h(213 - 10 - 0408) = [[213100408]]_{2389} = 1608$.

Exercises

Evaluate

1. $[[37]]_4$
2. $[[93]]_5$
3. $[[48]]_{23}$
4. $[[149]]_{27}$
5. $[[33]]_6$
6. $[[47]]_9$
7. $[[49]]_{17}$
8. $[[146]]_{21}$
9. $[[8!]]_6$
10. $[[1 + 2 + 3 + 4 + 5 + 6 + 7 + 8 + 9 + 10]]_3$
11. $[[1^2 + 2^2 + 3^2 + \cdots 25^2]]_2$
12. $[[(146)^2]]_{21}$
13. Show that $a^2 \equiv b^2 \pmod{n}$ need not imply $a \equiv b \pmod{n}$.
14. Show that if a is an odd integer, then $a^2 \equiv 1 \pmod{8}$.
15. Construct an addition table for the integers modulo 4.
16. Construct a multiplication table for the integers modulo 4.
17. Construct an addition table for the integers modulo 7.
18. Construct a multiplication table for the integers modulo 7.

Using the addition and multiplication tables for the integers modulo 5 given earlier, find solutions in Z_5 for the following:

19. $x + [3] = [0]$
20. $[3] \cdot x = [2]$
21. $[4] \cdot x = [3]$
22. $[2] \cdot x = [1]$

Using the addition and multiplication tables for the integers modulo 4 in the above exercises, find solutions in Z_4 for the following:

23. $x + [3] = [2]$
24. $[3] \cdot x = [2]$
25. $[2] \cdot x = [2]$
26. $[2] \cdot x = [0]$

Using the addition and multiplication tables for the integers modulo 7 in the preceding exercises, find solutions in Z_7 for the following:

27. $x + [5] = [2]$
28. $[3] \cdot x = [4]$
29. $[2] \cdot x = [5]$
30. $[4] \cdot x = [6]$
31. Prove that if $a \equiv b \pmod{mn}$, then $a \equiv b \pmod{m}$ and $a \equiv b \pmod{n}$.
32. Prove that if $ac \equiv bc \pmod{n}$ and the integers c and n are relatively prime, then $a \equiv b \pmod{n}$.
33. Prove Theorem 3.65. Let n be a positive integer and $\{r_1, r_2, \ldots, r_k\}$ be a reduced residue system modulo n. If a is an integer relatively prime to n, then $a \equiv r_j \pmod{n}$ for one and only one j, $1 \leq j \leq k$.
34. Prove Theorem 3.66. Let n be a positive integer and $\{r_1, r_2, \ldots, r_k\}$ be a complete [reduced] residue system modulo n. If a is an integer relatively prime to n, then $\{ar_1, ar_2, \ldots, ar_k\}$ is also a complete [reduced] residue system.

Let h be the hash function defined by $h(k_i) = [[k_i]]_{93}$. Find:

35. $h(776)$
36. $h(9076)$
37. $h(78)$
38. For the foregoing hashing function h, find two numbers for which there is a collision.

Let h be the hash function defined by $h(k_i) = [[k_i]]_{876}$. Find:

39. $h(776)$
40. $h(81042)$
41. $h(77623)$
42. For the foregoing hashing function h, find two numbers for which there is a collision.

GLOSSARY

Chapter 3: Logic, Integers, and Proofs

3.1 Predicate Calculus

Existential generalization	From an instance b in the universe with $P(b)$ true, we can infer that $\exists x P(x)$.
Existential instantiation	From $\exists x P(x)$ we can infer that there is an instance b with $P(b)$ true.
Existential quantifier	The symbol $\exists x$ is called an **existential quantifier**. Here the $\exists x$ refers to a value of x within the **domain of discourse**.
Inference relationships	a. $\forall x(P(x) \wedge Q(x)) \equiv \forall x P(x) \wedge \forall Q(x)$ b. $\exists x(P(x) \vee Q(x)) \equiv \exists x P(x) \vee \exists Q(x)$ c. From $\forall x P(x) \vee \forall Q(x)$ we can infer that $\forall x(P(x) \vee Q(x))$ d. From $\exists x(P(x) \wedge Q(x))$ we can infer that $\exists x P(x) \wedge \exists x Q(x)$.
n-place predicate	A predicate that has n variables is called an ***n*-place predicate**.
One-place predicate	A predicate that has one variable is called a **one-place predicate**.
Predicate	A **predicate** is a statement that cannot be assigned a truth value because it contains a variable.
Two-place predicate	A predicate that has two variables is called a **two-place predicate**.
Universal generalization	If there is an arbitrary a in the universe making $P(a)$ true, we can infer that $\forall x P(x)$ is true.
Universal instantiation	From $\forall x P(x)$ we can infer $P(a)$ for arbitrary a in the universe.
Universal quantifier	The symbol $\forall x$ is called a **universal quantifier** and is read "for every x" or "for all x." The set of values from which x may be taken is called the **universe** or **domain of discourse** or simply **domain**.

3.2 Basic Concepts of Proofs and the Structure of Integers

Axioms for the positive integers	**(N1)** The integer 1 is a positive integer.
	 (N2) The positive integers are closed under addition and multiplication; that is, if a and b are positive integers, then $a+b$ and $a \cdot b$ are positive integers.
	 (N3) (Trichotomy Axiom) For every integer a, one and only one of the following is true:
	     (a) a is a positive integer.
	     (b) $a = 0$.
	     (c) $-a$ is a positive integer.

Axioms of equality	**(E1)** For all a, $a = a$. **(E2)** For all a and b, if $a = b$, then $b = a$. **(E3)** For all a, b, and c, if $a = b$ and $b = c$, then $a = c$.
Partial ordering of the integers	For integers a and b, $a > b$ if and only if $a - b$ is positive; and $a \geq b$ if and only if $a > b$ or $a = b$. Also, $b < a$ if and only if $a > b$ and $b \leq a$ if $a \geq b$.

Properties of the set Z of integers	**(I1)** If a and b are integers, then $a+b$ and $a \cdot b$ are integers. The integers are **closed** under the operations of addition and multiplication.
	(I2) If $a = b$ and $c = d$, then $a+c = b+d$ and $a \cdot c = b \cdot d$.
	(I3) For all integers a and b, $a+b = b+a$ and $a \cdot b = b \cdot a$. The integers are **commutative** under the operations of addition and multiplication.
	(I4) For all integers a, b, and c, $(a+b)+c = a+(b+c)$ and $a(bc) = (ab)c$. The integers are **associative** under the operations of addition and multiplication.
	(I5) For all integers a, b, and c, $a \cdot (b+c) = (a \cdot b) + (a \cdot c)$. Multiplication of integers is **distributive** over addition.
	(I6) There exist unique integers 0 and 1 such that for every integer a, $a+0 = 0+a = a$ and $a \cdot 1 = 1 \cdot a = a$. The integer 0 is called the **additive identity**, or the **zero**, of the integers and 1 is called the **multiplicative identity**.
	(I7) For every integer a, there is a unique integer $-a$, called the **additive inverse** of a, such that $a+(-a) = (-a)+a = 0$.
	(I8) If b and c are integers and $a \cdot b = a \cdot c$ for some nonzero integer a, then $b = c$. This statement is called the **multiplication cancellation property**.

3.3 Mathematical Induction

Counterexample	Using the fact that the negation of $(\forall x)P(x)$ is $(\exists x) \sim P(x)$, we need only find one positive integer where the statement $P(x)$ is not true.
Induction principle for integers	Let $P(n)$ be a proposition with the property that a. $P(j)$ is true and b. whenever $P(k)$ is true for $k \geq j$, then $P(k+1)$ is true, then $P(n)$ is true for every $n \geq j$.
Least integer	Let A be a set of integers. The integer a is the **least integer** of A if and only if a. a is in A. b. If b is in A, then $a \leqslant b$.
Principle of mathematical induction	Let $P(n)$ be a statement with the property that if a. $P(1)$ is true, and b. whenever $P(k)$ is true, then $P(k+1)$ is true, then $P(n)$ is true for every positive integer n.
Second principle of induction	Let $P(n)$ be a statement. If a. $P(1)$ is true and b. whenever $P(m)$ is true for all $m < k$, then $P(k)$ is true, then $P(n)$ is true for every positive integer n.
Well-ordering principle	Every nonempty set of positive integers contains a least element.

3.4 Divisibility

Common divisor	A positive integer d is a **common divisor** of a and b if $d|a$ and $d|b$.
Common multiple	A positive integer m is a **common multiple** of integers a and b if $a|m$ and $b|m$.

Division algorithm	For positive integers a and b, there exist unique nonnegative integers q and r with $0 \leq r < b$ such that $a = bq + r$. The integers r and q are, respectively, called the **remainder** and **quotient** when a is divided by b.
Divisor/factor	An integer b that divides an integer a is also said to be a **factor** of a or to be a **divisor** of a.
Greatest common divisor	A positive integer d is the **greatest common divisor** of integers a and b provided that i. $d\|a$ and $d\|b$ and ii. if $c\|a$ and $c\|b$, then $c\|d$. The greatest common divisor of a and b will be denoted by $\gcd(a, b)$.
Least common multiple	A positive integer m is the **least common multiple** of integers a and b provided that a. $a\|m$ and $b\|m$, and b. if $a\|n$ and $b\|n$, then $m\|n$. The least common multiple of a and b will be denoted by $\mathrm{lcm}(a, b)$.
Multiple	An integer a is a multiple of the integer b if $b = bm$ for some integer m. A nonzero integer b divides an integer a, denoted by $b\|a$, if a is a multiple of b.
Relatively prime	If the greatest common divisor of a and b is 1, then a and b are said to be **relatively prime**.

3.5 Prime Integers

Composite	A positive integer greater than 1 is **composite** if it is not a prime integer.
Prime	An integer greater than 1 is called **prime** if its only positive factors are itself and 1.
Unique prime factorization	Any positive integer that is greater than 1 is a prime or can be written as a product of primes where this product is unique except for the arrangement of the primes.

3.6 Congruence Relations

Congruent	Let n be a positive integer. The integer a is **congruent** to the integer b modulo n, denoted by $a \equiv b \pmod{n}$, if n divides $(a - b)$.
Set of integers modulo n	Let n be a positive integer. The set of all equivalence classes modulo n is denoted by Z_n and is called the **set of integers modulo n**.
Equivalence classes modulo n	Let Z_n be the set of equivalence classes modulo n. Then given any integer m, there is an integer r such that $0 \leq r \leq n - 1$ and $[m] = [r]$ or $m \equiv r \pmod{n}$. We say that $[[m]]_n = r$.
Residue	For a positive integer n, if $b \equiv r \pmod{n}$, then we say that r is a **residue** of b modulo n.
Complete residue system	A **complete residue system** modulo n is a set $S = \{r_1, r_2, ..., r_n\}$ where the intersection of S with each equivalence class modulo n contains exactly one integer; that is, S contains one and only one representative from each such equivalence class.

Primary equivalence system	The complete residue system $\{0, 1, 2, ..., (n-1)\}$ is called the **primary residue system**. If b is an integer, $b \equiv r \pmod{n}$, and $0 \leq r \leq n-1$, then we denote this unique primary residue modulo n by $r = [[b]]_n$.
Reduced residue system	A **reduced residue system** modulo n is the subset of a complete residue system consisting of only those integers that are relatively prime to n; that is, $\{r : r \in S \text{ and } \gcd(r, n) = 1\}$

3.7 Functions

Function	A relation f on $A \times B$ is a **function** from A to B, denoted by $f : A \to B$, if for every $a \in A$, there is only one $b \in B$ so that $(a, b) \in f$. If $f : A \to B$ and $(a, b) \in f$, we say that $b = f(a)$.
Domain/codomain	A is called the **domain** of the function f and B is called the **codomain** for $f : A \to B$.
Image/range	If $E \subseteq A$, then $f(E) = \{b : f(a) = b \text{ for some } a \text{ in } E\}$ is called the **image** of E. The image of A itself is called the **range** of f.
Preimage	If $F \subseteq B$, then $f^{-1}(F) = \{a : f(a) \in F\}$ is called the **preimage** of F.
Mapping	A function $f : A \to B$ is also called a **mapping** and we speak of the domain A being mapped into B by the mapping f. If $(a, b) \in f$ so that $b = f(a)$, then we say that the element a is mapped to the element b.
One-to-one (injective)	A function $f : A \to B$ is **one-to-one** or **injective** if $f(a) = f(a')$ implies that $a = a'$.
Onto (surjective)	A function $f : A \to B$ is **onto** or **surjective** if for every $b \in B$ there exists an $a \in A$ so that $f(a) = b$.
One-to-one correspondence (bijection)	A function that is both one-to-one and onto is a **one-to-one correspondence** or a **bijection**.
Permutation	If $A = B$ and $f : A \to B$ is a one-to-one correspondence, then f is said to be a **permutation** of A.
Inverse relation	The **inverse relation** $f^{-1} \subseteq B \times A$ is defined by $f^{-1} = \{(b, a) : (a, b) \in f\}$.
Identity function	Let $I : A \to A$ be defined by $I(a) = a$ for all $a \in A$. The function I is called the **identity function on** A.

TRUE-FALSE QUESTIONS

1. $\exists x \forall y (x + y = 4)$ is true.

2. The statement $\exists x P(x)$ does not depend on the domain of discourse.

3. If the domain of discourse is increased, the statement $\forall P(x)$ is less likely to be true.

4. $\sim \forall x P(x) \equiv \forall x \sim P(x)$

5. From $\exists x P(x)$, we can infer $P(a)$ for arbitrary a in the universe.

6. $\exists x (P(x) \wedge Q(x)) \equiv \exists x P(x) \wedge \exists x Q(x)$

7. The argument

   All crocodiles are illogical.
   No carrots are illogical.
   ∴ No crocodiles are carrots.

   is valid.

8. The principle of mathematical induction is one of the axioms of the positive integers.

9. Using mathematical induction, one does not always begin with $n = 1$.

10. The least common multiple of positive integers a and b is always divisible by their greatest common divisor.

11. If a and b are relatively prime and b and c are relatively prime, then a and c are relatively prime.

12. If a and b are relatively prime, then there exist integers u and v so that $au + bv = 2$.

13. If a and b are relatively prime, then $\text{lcm}(a, b) = ab$.

14. The integers u and v such that $au + bv = \gcd(a, b)$ are unique.

15. If $a \equiv b \pmod{mn}$, then $a \equiv b \pmod{m}$ and $a \equiv b \pmod{n}$.

16. If $a \equiv b \pmod{m}$ and $a \equiv b \pmod{n}$, then $a \equiv b \pmod{mn}$.

17. If $ac \equiv bc \pmod{n}$, then $a \equiv b \pmod{n}$.

18. Every integer in a complete residue system modulo n is less than n.

19. If a reduced residue system modulo n contains $n - 1$ integers, then n is prime.

20. If $[r_1, r_2, r_3, \cdots, r_k]$ is a complete residue system modulo n and a is an integer relatively prime to n, then $[ar_1, ar_2, ar_3, \cdots, ar_k]$ is a complete residue system modulo n.

21. In Z_7, the equation $[2] \cdot x = 5$ has a unique solution.

22. In Z_8, the equation $[3] \cdot x = 5$ has a unique solution.

23. In Z_8, the equation $[2] \cdot x = 5$ has a solution.

24. If a and b are primes, then $a^2 + b^2$ is a prime.

25. 1599 is a prime.

26. Every integer greater than 2 is a product of primes.

27. To determine whether or not an integer is prime, one need only check for prime factors less than or equal to the square root of n.

28. The number 1 is a prime.

29. In Z_8, if $[a] \cdot [a] = 1$, then $[a] = 1$.

30. In an axiomatic development of the integers, the fact that there is no integer between 0 and 1 is obvious and does not need to be proved.

31. $n! \leq 5^n$ for every positive integer n.

32. The well-ordering principle is equivalent to the second principle of induction.

33. Since the first and second principles of induction are equivalent, when using the second principle of induction, one might as easily use the first principle.

34. To show that a theorem is not true for every n, one would usually use induction.

35. $1 + r + r^2 + \cdots + r^{n-1} = \dfrac{1 - r^n}{1 - r}$ for every positive integer n.

36. In Z_n, if $[a] = [b]$, then $[a]^2 = [b]^2$.

37. $1 + 3 + 5 + \cdots + (2n - 1) = n^2$ for every positive integer n.

38. There exists an integer k so that every integer greater than k can be written as $3u + 5v$ for nonnegative integers u and v.

39. The integers are associative under the operation of subtraction.

40. The even integers satisfy the Trichotomy Axiom.

41. $2^n + 3$ is prime for every integer n.

42. $n^3 \equiv n \pmod{6}$ for every positive integer n.

43. If m and n are prime, then $m^2 - n^2$ is never prime.

44. 4861 is a prime.

45. In the set of positive integers if $\forall n(P(n))$ is false, then there is a least positive integer k for which $P(k)$ is false.

46. In the set of positive integers if $\forall n(P(n))$ is false, then there is a largest integer for which it is true.

47. If m and n are nonzero integers, then $m^2 + n^2$ is never prime.

48. Every set of integers has a least element.

49. Every integer can be written in the form $3m + 5n$ for integers m and n.

50. If p is a prime integer, then $p! + 1$ is prime.

SUMMARY QUESTIONS

1. Negate the statement "All tennis players love to eat steak."

2. Test the validity of the argument:
   All college students are brilliant.
   <u>Some politicians are not brilliant.</u>
   ∴ Some politicians are not college students.

3. Given the statement $P(x, y, z) : x^2 \geq yz$, write out the statement $P(3, 5, 2)$.

4. Using induction, prove that $2 + 4 + 6 + \cdots + 2n = n(n + 1)$.

5. State the well-ordering principle.

6. Use the well-ordering principle to prove that every integer greater than 2 is either a prime or a product of primes.

7. Find the positive divisors of 476.

8. Find the greatest common divisor of 936 and 1242.
9. Factor 2499 into primes.
10. Evaluate $[[153]]_{31}$.
11. Find five numbers congruent to 7 modulo 9.
12. Find a reduced residue system modulo 21.
13. In Z_6, find all x so that $[4] \cdot x = [2]$.
14. Prove that if the least common multiple of two numbers exists, then it is unique.
15. Prove that $n^3 - n$ is divisible by 6.

ANSWERS TO TRUE-FALSE

1. F 2. F 3. T 4. F 5. F 6. F 7. T 8. T 9. T 10. T 11. F
12. T 13. T 14. F 15. T 16. F 17. F 18. F 19. T 20. T
21. T 22. T 23. F 24. F 25. F 26. F 27. T 28. F 29. F
30. F 31. F 32. T 33. F 34. F 35. T 36. T 37. T 38. T
39. F 40. T 41. F 42. T 43. T 44. T 45. T 46. F 47. F
48. F 49. T 50. F

CHAPTER 4
Functions and Matrices

4.1 Special Functions

Previously a permutation was defined to be a function on a set that is one-to-one and onto. If f is a permutation on the set $\{1, 2, 3, \ldots, n\}$, then it can be represented in the form

$$\begin{pmatrix} 1 & 2 & \ldots & n \\ f(1) & f(2) & \ldots & f(n) \end{pmatrix}$$

Recall that there is one special function, the identity function I, defined by $I(a) = a$ for all $a \in A$. It is represented as

$$\begin{pmatrix} 1 & 2 & \ldots & n \\ 1 & 2 & \ldots & n \end{pmatrix}$$

EXAMPLE 4.1 If $A = \{1, 2, 3\}$ and $f : A \to A$ is defined by $f(1) = 3$, $f(2) = 2$, and $f(3) = 1$, then f may be represented as

$$\begin{pmatrix} 1 & 2 & 3 \\ 3 & 2 & 1 \end{pmatrix}$$ ■

If $g : A \to A$ is defined by $g(1) = 2$, $g(2) = 3$, and $g(3) = 1$, then g may represented as

$$\begin{pmatrix} 1 & 2 & 3 \\ 2 & 3 & 1 \end{pmatrix}$$

The composition function

$$f \circ g = \begin{pmatrix} 1 & 2 & 3 \\ 3 & 2 & 1 \end{pmatrix} \circ \begin{pmatrix} 1 & 2 & 3 \\ 2 & 3 & 1 \end{pmatrix} = \begin{pmatrix} 1 & 2 & 3 \\ 2 & 1 & 3 \end{pmatrix}$$

and

$$g \circ f = \begin{pmatrix} 1 & 2 & 3 \\ 2 & 3 & 1 \end{pmatrix} \circ \begin{pmatrix} 1 & 2 & 3 \\ 3 & 2 & 1 \end{pmatrix} = \begin{pmatrix} 1 & 2 & 3 \\ 1 & 3 & 2 \end{pmatrix}$$

It was shown in the previous problems that $I = \begin{pmatrix} 1 & 2 & 3 \\ 1 & 2 & 3 \end{pmatrix}$ is also the identity in the sense that $I \circ f = f \circ I = f$ for every permutation f on the set A. To find the inverse of a permutation, find the number that is above 1 and place it below 1, find the number that is above 2 and place it below 2, and find the number that is above 3 and place it below 3. Thus, the inverse of the permutation

$$\begin{pmatrix} 1 & 2 & 3 \\ 2 & 3 & 1 \end{pmatrix} \quad \text{is} \quad \begin{pmatrix} 1 & 2 & 3 \\ 3 & 1 & 2 \end{pmatrix}$$

At this point we introduce three special functions.

DEFINITION 4.2

The function $f: A \to B$ where A is the set of real numbers and B is the set of integers is called the **floor function**, denoted by $f(x) = \lfloor x \rfloor$, if it assigns to each $a \in A$ the largest integer less than or equal to A. It is called the **ceiling function**, denoted by $f(x) = \lceil x \rceil$, if it assigns to each $a \in A$ the smallest integer greater than or equal to A.

Thus, $\lfloor 2.99 \rfloor = 2$, $\lfloor 4 \rfloor = 4$, $\lfloor -4 \rfloor = -4$, and $\lfloor -4.1 \rfloor = -5$. Also $\lceil 2.99 \rceil = 3$, $\lceil 4 \rceil = 4$, $\lceil -4 \rceil = -4$, and $\lceil -4.1 \rceil = -4$.

DEFINITION 4.3

Let A and B both be the set of nonnegative integers. The **factorial function** $f: A \to B$, denoted by $f(n) = n!$, is defined by

$$\begin{aligned} 0! &= 1 \\ 1! &= 1 &&= 1 \cdot 0! \\ 2! &= 1 \cdot 2 &&= 2 &&= 2 \cdot 1! \\ 3! &= 1 \cdot 2 \cdot 3 &&= 6 &&= 3 \cdot 2! \\ 4! &= 1 \cdot 2 \cdot 3 \cdot 4 &&= 24 &&= 4 \cdot 3! \\ &\vdots &&\vdots &&\vdots \\ k! &= 1 \cdot 2 \cdot 3 \cdot 4 \cdots k &&&&= k \cdot (k-1)! \end{aligned}$$

EXAMPLE 4.4

$$\frac{12!}{10!} = \frac{12 \cdot 11 \cdot 10!}{10!} = 12 \cdot 11 = 132 \qquad \blacksquare$$

EXAMPLE 4.5

$$\frac{7!}{4! \cdot 3!} = \frac{7 \cdot 6 \cdot 5 \cdot 4!}{4! \cdot 3 \cdot 2 \cdot 1} = 35 \qquad \blacksquare$$

Operations such as addition, subtraction, multiplication, and division are also examples of functions. In the case of subtraction of integers, $5 - 3$ means that an integer, namely 2, is assigned to 5 and 3 by the subtraction operation. We could write $-((5, 3)) = 2$ to emphasize the functional nature of the subtraction operation. Subtraction assigns a unique integer to each pair (r, s) of integers. This special application of a function of two variables is called a binary operation and is defined next.

DEFINITION 4.6

A **binary operation** on the set A is a function $b: A \times A \to A$. The image of (r, s) under b is written as $b((r, s))$ or as $r\, b\, s$.

Because the range of a binary operation on A is a subset of A, by definition, a binary operation exhibits the property of closure wherein the operation on two members r and s of A is also a member of A.

Sequences are special functions whose domain is an initial subset of the positive integers or nonnegative integers. It is assumed that the reader is familiar with sequences such as 1, 2, 4, 9, 16, ... or 2, 4, 6, 8, 10, ... although they have probably never thought of them as functions.

DEFINITION 4.7 A **finite sequence** is a function from $\{1, 2, 3, 4, \ldots, n\}$ to some set S. An **infinite sequence** is a function from $\{1, 2, 3, 4, \ldots\}$ to some set S. Either a finite sequence or an infinite sequence is called a **sequence**.

In many cases the set S is the set of positive integers, integers, rationals, or real numbers. A sequence is usually displayed as a list. If A is the function, it is assumed the first element listed is $A(1)$ or simply denoted as A_1, the second element listed is $A(2)$ or A_2, and so forth. If A is a finite sequence, it may be listed as

$$A(1), A(2), A(3), \ldots, A(n)$$

or

$$A_1, A_2, A_3, \ldots, A_n$$

If A is an infinite sequence, its range may be listed as

$$A(1), A(2), A(3), \ldots$$

or

$$A_1, A_2, A_3, \ldots$$

Hence, 1, 4, 9, 16, ... corresponds to the sequence $A_i = i^2$. Thus, in this example, $A(1) = 1$, $A(2) = 4$, $A(3) = 9$, $A(4) = 16$, and so forth. Note that 1, 4, 9, 16 and 1, 9, 4, 16 are not the same sequence.

EXAMPLE 4.8 Let $A(n) = n + 5$. The first five terms in the sequence are $A(1) = 1 + 5 = 6$, $A(2) = 2 + 5 = 7$, $A(3) = 3 + 5 = 8$, $A(4) = 4 + 5 = 9$, and $A(5) = 5 + 5 = 10$. ∎

EXAMPLE 4.9 Let $A(n) = n^2 - 3$. The first five terms in the sequence are $A(1) = 1^2 - 3 = -2$, $A(2) = 2^2 - 3 = 1$, $A(3) = 3^2 - 3 = 6$, $A(4) = 4^2 - 3 = 13$, and $A(5) = 5^2 - 3 = 22$. ∎

EXAMPLE 4.10 Given the first five terms of a sequence are 1, 6, 11, 16, 21, we wish to describe the function. Since each term is 5 more than the last term, $A(n) = 1 + 5(n - 1)$ for $1 \leq n \leq 5$. ∎

EXAMPLE 4.11 Given the first five terms of a sequence are 0, 3, 8, 15, 24, we wish to describe the function. Since the first five terms of a sequence $B(n) = n^2$ are 1, 4, 9, 16, 25, this sequence may be $A(n) = n^2 - 1$. ∎

EXAMPLE 4.12 Given the first five terms of a sequence are 1, −2, 3, −4, 5, the sequence may be described by $A(n) = (-1)^{n+1} n$. If a sequence alternates in sign and begins with a negative number, we would use $(-1)^n$ in place of $(-1)^{n+1}$. ∎

Two sequences that are of particular interest are the **arithmetic sequence** and the **geometric sequence**. In an arithmetic sequence, each element in the sequence may be obtained from the previous element by adding a constant c. Thus, an arithmetic sequence has the form $a + (n-1)c$ where a is the first element in the sequence. The sequence 3, 5, 7, 9, 11, ..., is an arithmetic sequence where $a = 3$ and $c = 2$.

In a geometric sequence, each element in the sequence may be obtained from the previous element by multiplying by a constant r. Thus, a geometric sequence has the form $A(n) = ar^{(n-1)}$ where a is the first element in the sequence. The sequence 4, 12, 36, 108, 324, ... is a geometric sequence with $a = 4$ and $r = 3$. The sequence 32, 16, 8, 4, 2, ... is a geometric sequence with $a = 32$ and $r = \frac{1}{2}$.

DEFINITION 4.13

> The sum $A_r + A_{r+1} + A_{r+2} + A_{r+3} + \cdots + A_{r+k}$ may be written in **summation notation** as $\sum_{i=r}^{r+k} A_i$.

For example,

$$\sum_{i=3}^{7} i^2 = 3^2 + 4^2 + 5^2 + 6^2 + 7^2$$

$$\sum_{i=4}^{4} i^3 = 4^3$$

$$\sum_{i=1}^{4} i = 1 + 2 + 3 + 4$$

$$\sum_{i=1}^{3} (i+3) = (1+3) + (2+3) + (3+3)$$

It is often desirable to add the first n terms of a sequence A as in $\sum_{i=1}^{n} A_i$. To find a formula for the sum of the first n terms of an arithmetic series

$$S = a + (a+c) + (a+2c) + \cdots + (a+(n-2)c) + (a+(n-1)c)$$

reverse the terms of S and add S to itself so that we have

$$S = a + (a+c) + (a+2c) + \cdots + [a+(n-1)c]$$

$$S = [a+(n-1)c] + [a+(n-2)c] + [a+(n-2)c] + \cdots + a$$

$$2S = [a + (a+(n-1)c)] + [a + (a+(n-1)c)] + [a + (a+(n-1)c)]$$
$$+ \cdots + [a + (a+(n-1)c]$$

$$= [2a + (n-1)c] + [2a + (n-1)c] + [2a + (n-1)c] + \cdots + [2a + (n-1)c]$$

and

$$2S = n(2a + (n-1)c)$$

so

$$S = \frac{n}{2}(2a + (n-1)c))$$

Since $A_1 = a$ and $A_n = a + (n-1)c$,
$$S = \frac{n}{2}(A_1 + A_n)$$

EXAMPLE 4.14
$$S = 3 + 5 + 7 + 9 + 11 + 13 = \frac{6}{2}(3 + 13) = 48$$ ∎

To sum a geometric series $S = a + ar + ar^2 + ar^3 + ar^4 + \cdots + ar^{n-2} + ar^{n-1}$, note that $rS = ar + ar^2 + ar^3 + ar^4 + ar^{54} + \cdots + ar^{n-1} + ar^n$. If we subtract S from rS, we have

$$rS = ar + ar^2 + ar^3 + ar^4 + \cdots + ar^{n-1} + ar^n$$
$$S = a + ar + ar^2 + ar^3 + \cdots + ar^{n-2} + ar^{n-1}$$

so that $rS - S = ar^n - a$ and

$$S = \frac{a(r^n - 1)}{r - 1}$$

EXAMPLE 4.15
$2 + 6 + 18 + 54 + 162 = \dfrac{2(3^5 - 1)}{3 - 1} = \dfrac{2(486 - 1)}{3 - 1} = 243$, since $a = 2$, $r = 3$ and $n = 5$. ∎

> **Computer Science Application**
>
> The floor function is commonly available in most high-level programming languages. The following simple algorithm uses the floor function to return the integer nearest the real number x.
>
> integer function nearest-integer(real x)
>
> > integer i
> >
> > $i = $ floor$(x + 0.5)$
> > return i
>
> end-function
>
> This trick preserves the convention that real numbers of the form $a.5$ round up to $a + 1$ rather than down to a. The function above can easily be modified to round x to the nearest specified digit past the decimal. The ^ is used to denote exponentiation.
>
> real function nearest-fixed(real x, integer n)
>
> > integer i
> > real f
> >
> > $i = $ floor$(x*(10\wedge n) + 0.5)$
> > $f = i/(10\wedge n)$
> > return f
>
> end-function
>
> When n is 2, this function can be used to perform standard nearest-penny arithmetic on real variables representing monetary amounts.

Exercises

For permutations f and g in the following exercises, find the permutations $f \circ g$, $g \circ f$, f^{-1}, and g^{-1}.

1. $f = \begin{pmatrix} 1 & 2 & 3 \\ 2 & 3 & 1 \end{pmatrix} \quad g = \begin{pmatrix} 1 & 2 & 3 \\ 3 & 1 & 2 \end{pmatrix}$

2. $f = \begin{pmatrix} 1 & 2 & 3 & 4 \\ 2 & 1 & 4 & 3 \end{pmatrix} \quad g = \begin{pmatrix} 1 & 2 & 3 & 4 \\ 4 & 3 & 2 & 1 \end{pmatrix}$

3. $f = \begin{pmatrix} 1 & 2 & 3 & 4 \\ 4 & 2 & 1 & 3 \end{pmatrix} \quad g = \begin{pmatrix} 1 & 2 & 3 & 4 \\ 3 & 2 & 4 & 1 \end{pmatrix}$

4. $f = \begin{pmatrix} 1 & 2 & 3 & 4 \\ 3 & 2 & 1 & 4 \end{pmatrix} \quad g = \begin{pmatrix} 1 & 2 & 3 & 4 \\ 1 & 2 & 3 & 4 \end{pmatrix}$

Evaluate the following:

5. $\dfrac{10!}{8!}$

6. $\dfrac{12!}{8! \cdot 4!}$

7. $\dfrac{n!}{(n-2)!}$

8. $-\lfloor -2.999 \rfloor$

9. $\lfloor 1.001 \rfloor$

10. $\lceil 3.5 \rceil$

11. $\lceil -4.01 \rceil$

Evaluate the following:

12. $\dfrac{14!}{13! \cdot 0!}$

13. $\dfrac{15!}{5! \cdot 4! \cdot 6!}$

14. $\lfloor -9 \rfloor$

15. $-\lfloor 2.999 \rfloor$

16. $\lfloor -1.001 \rfloor$

17. $\lceil -3.7 \rceil$

18. $\lceil 4.01 \rceil$

Find the first five terms of the following sequences:

19. $A_n = n^3 + 3$

20. $A_n = (-1)^{n+1} \dfrac{n^2}{n!}$

21. $A_n = \left\lceil \dfrac{n}{3} \right\rceil$

22. $A_n = \dfrac{(n+3)!}{(n!)(3!)}$

Find the first five terms of the following sequences:

23. $A_n = n^2 + 2n + 3$

24. $A_n = \dfrac{n(n+1)}{n!}$

25. $A_n = \left\lfloor \dfrac{n-5}{n} \right\rfloor$

26. $A_n = (-1)^{n+1} \left(\dfrac{1}{n} \right)$

Given the first five terms of A, find an expression for A_n:

27. 3, 8, 13, 18, 23

28. $\dfrac{1}{2}, -\dfrac{2}{3}, \dfrac{3}{4}, -\dfrac{4}{5}, \dfrac{5}{6}$

29. 4, 12, 36, 78, 144

30. 2, 2, 4, 12, 48

31. 2, 8, 18, 32, 50

32. $\dfrac{1}{2}, -\dfrac{1}{6}, \dfrac{1}{18}, -\dfrac{1}{54}, \dfrac{1}{162}$

33. 1, −1, 2, −2, 3

34. 3, 6, 11, 18, 27

4.2 Matrices

The functions in this chapter have been "functions of one variable." They are often denoted using the symbolism $f(x)$, where x is called the "variable." For example, $f(x) = x^2$, with domain R is a function of one variable. If the domain of f is the Cartesian product of two sets, say $C \times D$, then $f : C \times D \to B$ is said to be a "function of two variables" and is often denoted using the symbolism $f(x, y)$, where x and y are called the variables with x selected from C and y selected from D. Thus, if $(c, d) \in C \times D$, then we write $f((c, d))$ or simply $f(c, d)$. For example, $f(x, y) = x^2 + y^2$, with domain $R \times R$, is a function of two variables. Binary operations are also examples of functions of two variables. Similarly, if the domain of f is $C \times D \times E$ and $f : C \times D \times E \to B$, then f is called a function of three variables. For example, $f(x, y, z) = x^2 + y^2 + 2yz$, with domain $R \times R \times R$, is a function of three variables. We now give a description of a special function of two variables called a matrix or array. Data in computers are often stored as a matrix or array.

DEFINITION 4.16

For positive integers m and n, an $m \times n$ **matrix** or **array** is a function $A : \{1, 2, \ldots, m\} \times \{1, 2, \ldots, n\} \to D$, where D is usually the real numbers, complex numbers, rational numbers, or integers. The elements of D are called **scalars**. Therefore, for each i, $1 \le i \le m$, and each j, $1 \le j \le n$, there is an element of D, denoted by $A(i, j)$, which, we think of as being in the ith row and the jth column of a rectangular array. The image $A(i, j)$ of the domain element (i, j), is usually shortened to A_{ij}. Thus, the $n \times m$ matrix A is represented by a rectangular array where the images of $(i, j) \in \{1, 2, \ldots, m\} \times \{1, 2, \ldots, n\}$ under A are listed as follows:

$$A = \begin{bmatrix} A_{11} & A_{12} & A_{13} & \cdots & A_{1n} \\ A_{21} & A_{22} & A_{23} & \cdots & A_{2n} \\ \vdots & & & & \vdots \\ A_{m1} & A_{m2} & A_{m3} & \cdots & A_{mn} \end{bmatrix}$$

where the first row of the representation consists of A_{1j} for $j = 1$ to n, the second row consists of A_{2j} for $j = 1$ to n, and so on. We say that A has m rows and n columns and has **dimension** $m \times n$. Sometimes we write $A = [A_{ij}]$ for short or even $A = [a_{ij}]$. The value A_{ij} is called a **component**, **entry**, or **element** of the matrix A. A matrix of dimension $1 \times m$ is called a **row matrix**; and a matrix of dimension $n \times 1$ is called a **column matrix**. If the number of rows of a matrix is the same as the number of columns, then the matrix is called a **square matrix**. If A is a row matrix, then the subscripts for the row are often suppressed and we write

$$A = \begin{bmatrix} A_{11} & A_{12} & \cdots & A_{1m} \end{bmatrix} = \begin{bmatrix} A_1 & A_2 & \cdots & A_m \end{bmatrix}$$

Similarly, we write

$$A = \begin{bmatrix} A_{11} \\ A_{21} \\ \vdots \\ A_{n1} \end{bmatrix} = \begin{bmatrix} A_1 \\ A_2 \\ \vdots \\ A_n \end{bmatrix}$$

suppressing the column indices if A is a column matrix.

For example, $A = \begin{bmatrix} 2 & 1 & 7 \\ 4 & 0 & 6 \end{bmatrix}$ is a 2×3 matrix and $B = \begin{bmatrix} -2 & 5 & 7 \\ -3 & 9 & 0 \\ 25 & 2 & 9 \end{bmatrix}$ is a 3×3 square matrix. By definition, $A_{13} = 7$, $A_{21} = 4$, $B_{12} = 5$, and $B_{31} = 25$. The matrix $C = \begin{bmatrix} 7 & 1 \\ 3 & 9 \end{bmatrix}$ is a 2×2 square matrix.

DEFINITION 4.17   Two $m \times n$ matrices $A = [A_{ij}]$ and $B = [B_{ij}]$ are **equal** if the corresponding elements are equal; that is, $A = B$ if and only if $A_{ij} = B_{ij}$ for all i with $1 \leq i \leq m$ and j with $1 \leq j \leq n$.

For example, $\begin{bmatrix} A_{11} & A_{12} & A_{13} \\ A_{21} & A_{22} & A_{23} \\ A_{31} & A_{32} & A_{33} \end{bmatrix} = \begin{bmatrix} B_{11} & B_{12} & B_{13} \\ B_{21} & B_{22} & B_{23} \\ B_{31} & B_{32} & B_{33} \end{bmatrix}$ if and only if $A_{ij} = B_{ij}$ for $1 \leq i, j \leq 3$. We note that this definition of matrix equality is just a restatement of when two functions A and B are equal: $A((i, j)) = B((i, j))$ for all (i, j).

Definitions and theorems will be stated in full generality in some cases and in other cases they will be stated only for 2×2 and 3×3 matrices in this section and in full generality in later chapters; however, in order to reduce the complexity of the presentation and for convenience and definiteness, many theorems will be proved only for 2×2 and 3×3 matrices.

148 Chapter 4 Functions and Matrices

DEFINITION 4.18 If d is a scalar and $A = [A_{ij}]$ is an $m \times n$ matrix, then dA is the $m \times n$ matrix $D = [D_{ij}]$, where $D_{ij} = dA_{ij}$, so that every component of D is obtained by multiplying the corresponding component of A by d. The product of a number d and a matrix A is called a **scalar product**.

EXAMPLE 4.19 Let $A = [A_{ij}] = \begin{bmatrix} 1 & -3 & 5 \\ 6 & 0 & -2 \end{bmatrix}$. Then

$$7A = 7 \begin{bmatrix} 1 & -3 & 5 \\ 6 & 0 & -2 \end{bmatrix} = \begin{bmatrix} (7)(1) & (7)(-3) & (7)(5) \\ (7)(6) & (7)(0) & (7)(-2) \end{bmatrix}$$

$$= \begin{bmatrix} 7 & -21 & 35 \\ 42 & 0 & -14 \end{bmatrix}$$

DEFINITION 4.20 If $A = [A_{ij}]$ and $B = [B_{ij}]$ are $m \times n$ matrices, then $A + B$ is the $m \times n$ matrix $C = [C_{ij}]$ where $C_{ij} = A_{ij} + B_{ij}$, that is, addition is componentwise. C is called the **matrix sum** of A and B. The notation $A - B$ means $A + (-1)B$.

EXAMPLE 4.21 Let

$$A = \begin{bmatrix} -1 & 3 \\ 2 & 7 \\ 4 & -5 \end{bmatrix} \quad \text{and} \quad B = \begin{bmatrix} 3 & 11 \\ -5 & 4 \\ 8 & 2 \end{bmatrix}$$

Then

$$A + B = \begin{bmatrix} -1 & 3 \\ 2 & 7 \\ 4 & -5 \end{bmatrix} + \begin{bmatrix} 3 & 11 \\ -5 & 4 \\ 8 & 2 \end{bmatrix} = \begin{bmatrix} (-1)+3 & 3+11 \\ 2+(-5) & 7+4 \\ 4+8 & (-5)+2 \end{bmatrix}$$

$$= \begin{bmatrix} 2 & 14 \\ -3 & 11 \\ 12 & -3 \end{bmatrix}$$

DEFINITION 4.22 We define the **product** of two matrices as follows:

(a) If V is either a row or column matrix with n entries and W is either a row or column matrix with n entries so that

$$V = \begin{bmatrix} V_1 \\ V_2 \\ \vdots \\ V_n \end{bmatrix} \quad \text{or} \quad V = \begin{bmatrix} V_1 & V_2 & \cdots & V_n \end{bmatrix}$$

and

$$W = \begin{bmatrix} W_1 \\ W_2 \\ \vdots \\ W_n \end{bmatrix} \quad \text{or} \quad W = \begin{bmatrix} W_1 & W_2 & \cdots & W_n \end{bmatrix}$$

then the **dot product** or **inner product** of V and W, denoted by $V \bullet W$, is the number $V_1 W_1 + V_2 W_2 + \cdots + V_n W_n$.

(b) If
$$A = \begin{bmatrix} A_{11} & A_{12} & \cdots & A_{1p} \\ A_{21} & A_{22} & \cdots & A_{2p} \\ \vdots & & & \vdots \\ A_{m1} & A_{m2} & \cdots & A_{mp} \end{bmatrix}$$

is an $m \times p$ matrix and

$$B = \begin{bmatrix} B_{11} & B_{12} & \cdots & B_{1n} \\ B_{21} & B_{22} & \cdots & B_{2n} \\ \vdots & & & \vdots \\ B_{p1} & B_{p2} & \cdots & B_{pn} \end{bmatrix}$$

is an $p \times n$ matrix, then the (matrix) **product** of A and B, denoted by AB, is the $m \times n$ matrix $C = [C_{ij}]$, where C_{ij} is the dot product of the ith row of A and the jth column of B; that is,

$$C_{ij} = \begin{bmatrix} A_{i1} & A_{i2} & A_{i3} & \cdots & A_{ip} \end{bmatrix} \bullet \begin{bmatrix} B_{1j} \\ B_{2j} \\ B_{3j} \\ \vdots \\ B_{pj} \end{bmatrix} = \sum_{k=1}^{p} A_{ik} B_{kj}$$

EXAMPLE 4.23 Let

$$A = [A_{ij}] = \begin{bmatrix} 1 & -3 & 5 \\ 6 & 0 & -2 \end{bmatrix} \quad \text{and} \quad B = [B_{ij}] = \begin{bmatrix} -2 & 4 & 0 & 8 \\ 3 & -1 & -2 & 1 \\ 0 & 5 & 7 & 0 \end{bmatrix}$$

Calculate the matrix product $C = [C_{ij}] = AB$ as follows. The matrix A is 2×3 and B is 3×4 so that the product $AB = C$ is defined and will be a 2×4 matrix. C_{11} is the dot product of the first row of A and the first column of B and C_{23} is the dot product of the second row of A and the third column of B. Thus,

$$C_{11} = \begin{bmatrix} 1 & -3 & 5 \end{bmatrix} \bullet \begin{bmatrix} -2 \\ 3 \\ 0 \end{bmatrix} = (1)(-2) + (-3)(3) + (5)(0) = -11$$

$$C_{23} = \begin{bmatrix} 6 & 0 & -2 \end{bmatrix} \bullet \begin{bmatrix} 0 \\ -2 \\ 7 \end{bmatrix} = (6)(0) + (0)(-2) + (-2)(7) = -14$$

Continuing in this manner, we obtain

$$AB = \begin{bmatrix} 1 & -3 & 5 \\ 6 & 0 & -2 \end{bmatrix} \begin{bmatrix} -2 & 4 & 0 & 8 \\ 3 & -1 & -2 & 1 \\ 0 & 5 & 7 & 0 \end{bmatrix} = \begin{bmatrix} -11 & 32 & 41 & 5 \\ -12 & 14 & -14 & 48 \end{bmatrix} \blacksquare$$

We shall state the following theorem without proof. For a proof, see any linear algebra book.

The proof of the following theorem is left to the reader.

THEOREM 4.24 The following are true for all $m \times n$ matrices A, B, and C:

(a) $A + B = B + A$ (commutative property of addition)

(b) $A + (B + C) = (A + B) + C$ (associative property of addition) ∎

THEOREM 4.25 The following are true for all matrices A, B, and C and real numbers r and s, where A is an $m \times p$ matrix and B, C are $p \times n$ matrices:

(a) $A \cdot (B + C) = (A \cdot B) + (A \cdot C)$ (distributive property of matrices)

(b) $A \cdot (rB + sC) = r(A \cdot B) + s(A \cdot C)$ (linear property of matrices)

Proof Let $(A \cdot (B + C))_{ij}$ be the ijth element of $A \cdot (B + C)$. Thus, it is the element in the ith row and jth column of $A \cdot (B + C)$. Then

$$(A \cdot (B+C))_{ij} = \sum_{k=1}^{p} A_{ik}(B+C)_{kj}$$

$$= \sum_{k=1}^{p} A_{ik}(B_{kj} + C_{kj})$$

$$= \sum_{k=1}^{p} A_{ik}B_{kj} + A_{ik}C_{kj}$$

$$= (A \cdot B)_{ij} + (A \cdot C)_{ij}$$

$$= ((A \cdot B) + (A \cdot C))_{ij}$$

which is the ijth element of $(A \cdot B) + (A \cdot C)$, so $(A \cdot (B+C)) = ((A \cdot B) + (A \cdot C))$. Part (b) is left to the reader. ∎

THEOREM 4.26 Let A, B, and C be matrices such that BC and $A(BC)$ are defined. Then AB and $(AB)C$ are defined and $A(BC) = (AB)C$.

Proof If $A(BC)$ is defined and is an $m \times n$ matrix, then A is an $m \times p$ matrix and BC is a $p \times n$ matrix for some p. Since BC is defined, there exists q so that B is a $p \times q$ matrix and C is a $q \times n$ matrix. Since A is an $m \times p$ matrix and B is a $p \times q$ matrix, AB is defined and is an $m \times q$ matrix. Since AB is an $m \times q$ matrix and C is a $q \times n$ matrix, $(AB)C$ is defined and is an $m \times n$ matrix.

Let $(A(BC))_{ij}$ be the ijth element of $(A(BC))$, then

$$(A(BC))_{ij} = \sum_{r=1}^{p} A_{ir}(BC)_{rj}$$

$$= \sum_{r=1}^{p} A_{ir} \left(\sum_{s=1}^{q} B_{rs}C_{sj} \right)$$

$$= \sum_{r=1}^{p} \sum_{s=1}^{q} A_{ir}B_{rs}C_{sj}$$

$$= \sum_{r=1}^{p} (A_{ir}B_{rs}) \sum_{s=1}^{q} C_{sj}$$

$$= \sum_{s=1}^{q} (AB)_{is}C_{sj}$$

$$= ((AB)C)_{ij}$$

Since the ijth element of $(A(BC))$ is equal to the ijth element of $(AB)C$, $(A(BC)) = (AB)C$. ∎

DEFINITION 4.27 If A is an $n \times m$ matrix where the entry in the ith row and jth column is A_{ij}, then A^t, the **transpose** of A, is the $m \times n$ matrix with $A^t{}_{ij} = A_{ji}$. If A is an $n \times n$ matrix and $A_{ij} = A_{ji}$ for all $1 \leq i, j \leq n$, then A is **symmetric** or is said to be a **symmetric matrix**. An equivalent statement is that A is symmetric if and only if $A = A^t$.

EXAMPLE 4.28 Let $A = \begin{bmatrix} 1 & -3 & 5 \\ 6 & 0 & -2 \end{bmatrix}$. Then A^t is the 3×2 matrix obtained from A by writing the rows of A as columns.

$$A^t = \begin{bmatrix} 1 & -3 & 5 \\ 6 & 0 & -2 \end{bmatrix}^t = \begin{bmatrix} 1 & 6 \\ -3 & 0 \\ 5 & -2 \end{bmatrix}$$

■

THEOREM 4.29 If A and B are matrices whose product is defined, then $(AB)^t = B^t A^t$.

Proof The proof is left to reader. ■

DEFINITION 4.30 Let $A = \{a_1, a_2, \ldots, a_n\}$ and $B = \{b_1, b_2, \ldots, b_m\}$. Let R be a relation $\subseteq A \times B$. The **representation matrix** for R is the matrix $M_{nm} = [M_{ij}]$ defined by

$$M_{ij} = \begin{cases} 1 & \text{if } (a_i, b_j) \in R \\ 0 & \text{if } (a_i, b_j) \notin R \end{cases}$$

EXAMPLE 4.31 Let $A = \{a_1, a_2, a_3\}$ and $B = \{b_1, b_2\}$. Let R be the relation $\{(a_1, b_1), (a_2, b_1), (a_2, b_2), (a_3, b_2)\}$. Then the representation matrix is

$$\begin{bmatrix} 1 & 0 \\ 1 & 1 \\ 0 & 1 \end{bmatrix}$$

■

EXAMPLE 4.32 The smallest reflexive relation on $A = \{a_1, a_2, a_3\}$ is the relation $\{(a_1, a_1), (a_2, a_2), (a_3, a_3)\}$. Its representation matrix is

$$\begin{bmatrix} 1 & 0 & 0 \\ 0 & 1 & 0 \\ 0 & 0 & 1 \end{bmatrix}$$

The largest relation on A is $A \times A$ and its representation matrix is

$$\begin{bmatrix} 1 & 1 & 1 \\ 1 & 1 & 1 \\ 1 & 1 & 1 \end{bmatrix}$$

■

DEFINITION 4.33 Define the Boolean operations $\vee$ and $\wedge$ on the set $\{0, 1\}$ by

$\vee$	0	1
0	0	1
1	1	1

and

$\wedge$	0	1
0	0	0
1	0	1

A **Boolean matrix** is a matrix whose entries are all either 0 or 1. Let A and B be $m \times n$ Boolean matrices and C an $n \times p$ Boolean matrix:

(i) $U = A \vee B$ is defined by
$$U_{ij} = A_{ij} \vee B_{ij} \text{ for } 1 \leq i \leq m, 1 \leq j \leq n$$

(ii) $I = A \wedge B$ is defined by
$$I_{ij} = A_{ij} \wedge B_{ij} \text{ for } 1 \leq i \leq m, 1 \leq j \leq n$$

(iii) $D = A \odot C$ is defined by
$$D_{ij} = (A_{i1} \wedge C_{1j}) \vee (A_{i2} \wedge C_{2j}) \vee \cdots \vee (A_{in} \wedge C_{nj})$$
for $1 \leq i \leq m, \; 1 \leq j \leq p$.

EXAMPLE 4.34 Let $A = \{a_1, a_2, a_3\}$, $R = \{(a_1, a_1), (a_1, a_2), (a_2, a_1), (a_2, a_2), (a_3, a_1)\}$, and $S = \{(a_1, a_1), (a_1, a_3), (a_3, a_1), (a_2, a_2), (a_2, a_3), (a_3, a_2)\}$. The representation matrices for R and S, respectively, are

$$\begin{bmatrix} 1 & 1 & 0 \\ 1 & 1 & 0 \\ 1 & 0 & 0 \end{bmatrix} \text{ and } \begin{bmatrix} 1 & 0 & 1 \\ 0 & 1 & 1 \\ 1 & 1 & 0 \end{bmatrix}$$

then

$$\begin{bmatrix} 1 & 1 & 1 \\ 1 & 1 & 0 \\ 0 & 0 & 0 \end{bmatrix} \wedge \begin{bmatrix} 1 & 0 & 1 \\ 0 & 1 & 1 \\ 1 & 1 & 0 \end{bmatrix} = \begin{bmatrix} 1 & 0 & 1 \\ 0 & 1 & 0 \\ 0 & 0 & 0 \end{bmatrix}$$

$$\begin{bmatrix} 1 & 1 & 1 \\ 1 & 1 & 0 \\ 0 & 0 & 0 \end{bmatrix} \vee \begin{bmatrix} 1 & 0 & 1 \\ 0 & 1 & 1 \\ 1 & 1 & 0 \end{bmatrix} = \begin{bmatrix} 1 & 1 & 1 \\ 1 & 1 & 1 \\ 1 & 1 & 0 \end{bmatrix}$$

$$\begin{bmatrix} 1 & 1 & 1 \\ 1 & 1 & 0 \\ 0 & 0 & 0 \end{bmatrix} \odot \begin{bmatrix} 1 & 0 & 1 \\ 0 & 1 & 1 \\ 1 & 1 & 0 \end{bmatrix} = \begin{bmatrix} 1 & 1 & 1 \\ 1 & 1 & 1 \\ 0 & 0 & 0 \end{bmatrix}$$ ∎

THEOREM 4.35 Let R and S be relations on a finite set $A = \{a_1, a_2, \ldots, a_n\}$ with respective representation matrices M and N.

(a) If R is reflexive, then $M_{ii} = 1$ for $1 \leq i \leq n$.

(b) If R is symmetric, then $M_{ij} = M_{ji}$ for $1 \leq i, j \leq n$. Therefore, $M = M^t$ and M is a symmetric matrix.

(c) Let $M^{\odot 2} = M \odot M$. If R is transitive, then whenever $M^{\odot 2}_{ik} = 1$, $M_{ik} = 1$ for $1 \leq i, k \leq n$.

(d) $M \vee N$ is the representation matrix for $R \cup S$.

(e) $M \wedge N$ is the representation matrix for $R \cap S$.

Proof (a) Since R is reflexive, $(a_i, a_i) \in R$ for all $1 \leq i \leq n$. Therefore, $M_{ii} = 1$ for $1 \leq i \leq n$.

(b) If $M_{ij} = 1$, then $(a_i, a_j) \in R$; and since R is symmetric, $(a_j, a_i) \in R$ so that $M_{ji} = 1$. Conversely, if $M_{ji} = 1$, then $(a_j, a_i) \in R$; and since R is symmetric, $(a_i, a_j) \in R$ so that $M_{ij} = 1$. Since $M_{ij} = 1$ if and only if $M_{ji} = 1$, $M = M^t$ and M is a symmetric matrix.

(c) Assume $M^{\odot 2}_{ik} = 1$. Since $M^{\odot 2}_{ik} = (A_{i1} \wedge A_{1k}) \vee (A_{i2} \wedge A_{2k}) \vee \cdots \vee (A_{in} \wedge A_{nk})$, there exists some m so that $(A_{im} \wedge A_{mk}) = 1$. Therefore, $A_{im} = 1$ and $A_{mk} = 1$.

By definition of representation matrix, $(a_i, a_m) \in R$ and $(a_m, a_k) \in R$. Since R is transitive, $(a_i, a_k) \in R$ and $M_{ik} = 1$.

The remainder of the proof is left to the reader. ∎

The proof of the following theorem is left to the reader.

THEOREM 4.36 Let R be a relation from A to B and S be a relation from B to C. If M and N are the representation matrices for R and S, respectively, then $M \odot N$ is the representation matrix for $S \circ R$. ∎

EXAMPLE 4.37 Let $A = \{a_1, a_2\}$, $B = \{b_1, b_2, b_3\}$, and $C = \{c_1, c_2, c_3, c_4\}$. Let
$$R = \{(a_1, b_1), (a_1, b_2), (a_2, b_3)\}$$
and
$$S = \{(b_1, c_1), (b_1, c_2), (b_2, c_3), (b_3, c_1), (b_3, c_2)\}$$

Then the representation matrices for R and S, respectively, are

$$\begin{bmatrix} 1 & 1 & 0 \\ 0 & 0 & 1 \end{bmatrix} \text{ and } \begin{bmatrix} 1 & 1 & 0 & 0 \\ 0 & 0 & 1 & 0 \\ 1 & 1 & 0 & 0 \end{bmatrix}$$

$$\begin{bmatrix} 1 & 1 & 0 \\ 0 & 0 & 1 \end{bmatrix} \odot \begin{bmatrix} 1 & 1 & 0 & 0 \\ 0 & 0 & 1 & 0 \\ 1 & 1 & 0 & 0 \end{bmatrix} = \begin{bmatrix} 1 & 1 & 1 & 0 \\ 1 & 1 & 0 & 0 \end{bmatrix}$$

and $S \circ R = \{(a_1, c_1), (a_1, c_2), (a_1, c_3), (a_2, c_1), (a_2, c_2)\}$. ∎

The following theorem follows from Theorems 4.35, 4.36, and 2.40.

THEOREM 4.38 Let A be the representation matrix for the relation R.
(a) The reflexive closure of R has representation matrix $A \vee I$.
(b) The symmetric closure of R has representation matrix $A \vee A^t$.
(c) If A is a finite set containing n elements, then the transitive closure of R has representation matrix $A \vee A^{\odot 2} \vee A^{\odot 3} \vee \cdots \vee A^{\odot n}$. ∎

DEFINITION 4.39 Let M be an $n \times n$ matrix such that in each row and each column, one entry is 1 and the others are 0. M is called a **permutation matrix**.

THEOREM 4.40 Let M be a permutation matrix and I be a $n \times 1$ matrix with distinct entries. Then MI is a $n \times 1$ matrix whose entries are a permutation of the entries in I.

Proof Let $I = \begin{bmatrix} a_1 \\ a_2 \\ a_3 \\ \vdots \\ a_n \end{bmatrix}$ and $I' = MI$. Consider a_j in the jth row. Since

$$C_i = \begin{bmatrix} M_{i1} & M_{i2} & M_{i3} & \cdots & M_{in} \end{bmatrix} \bullet \begin{bmatrix} a_1 \\ a_2 \\ a_3 \\ \vdots \\ a_n \end{bmatrix} = \sum_{k=1}^{n} M_{ik} a_k$$

and there is one row p where $M_{pj} = 1$ and $M_{pk} = 0$ for $k \neq j$, the pth row of I' is a_j. Further, since $M_{kj} = 0$ for all $k \neq p$, a_j does not appear in any other row of I'. ∎

EXAMPLE 4.41 Let $M = \begin{bmatrix} 0 & 0 & 1 \\ 1 & 0 & 0 \\ 0 & 1 & 0 \end{bmatrix}$ and $I = \begin{bmatrix} a_1 \\ a_2 \\ a_3 \end{bmatrix}$, then $MI = \begin{bmatrix} a_3 \\ a_1 \\ a_2 \end{bmatrix}$. ∎

Computer Science Application

When a matrix A of dimension $m \times n$ is multiplied (on the right) by a matrix B of dimension $n \times p$, the product is a matrix C of dimension $m \times p$ that will require mnp scalar multiplications. The matrix product $ABCD$, where A is 13×5, B is 5×89, C is 89×3 and D is 3×34 produces a matrix E of dimension 13×34. The number of scalar multiplications required to produce E will depend on the order in which the four matrices are multiplied. Since matrix multiplication is associative, E can be computed in the following five ways:

$((AB)C)D$	10,582 scalar multiplications
$(AB)(CD)$	54,201 scalar multiplications
$(A(BC))D$	2,856 scalar multiplications
$A((BC)D)$	4,055 scalar multiplications
$A(B(CD))$	26,418 scalar multiplications

For long chains of matrices of varying dimensions the difference between the extremes can be profound. In *Algorithmics: Theory and Practice*, G. Brassard and P. Bratley present a dynamic programming algorithm that determines the minimal number of scalar multiplications required to produce such a product.

When computing powers of a square matrix A, the order obviously does not matter. If A is of dimension 10×10, A^5 will require 4000 multiplications. If instead we view A^5 as $(A^2)^2 A$, then there are only 3000. For larger powers such as $A^{25} = ((A^2 A)^2)^2)^2 A$, the reduction from 24,000 multiplications to 6000 is even more impressive.

Exercises

Calculate the following:

1. $\begin{bmatrix} 3 & -1 \\ 2 & 7 \\ -5 & 0 \end{bmatrix} \begin{bmatrix} 2 & -1 & 0 & 4 \\ 5 & 1 & 3 & 0 \end{bmatrix}$

2. $\begin{bmatrix} 1 & -1 & 2 \\ 3 & 1 & 4 \\ 6 & -1 & 5 \end{bmatrix} \begin{bmatrix} 2 & -2 & 8 \\ 5 & 5 & 1 \\ 0 & 6 & 4 \end{bmatrix}$

3. $\begin{bmatrix} 1 \\ 3 \\ 4 \\ 5 \end{bmatrix} \begin{bmatrix} -2 & 3 & -5 & 0 \end{bmatrix}$

4. $\begin{bmatrix} -2 & 3 & -5 & 0 \end{bmatrix} \begin{bmatrix} 1 \\ 3 \\ 4 \\ 5 \end{bmatrix}$

5. $(-2) \begin{bmatrix} -5 & 1 \\ 6 & -7 \end{bmatrix}$

6. $\begin{bmatrix} -1 & 3 \\ 2 & 7 \end{bmatrix} - 8 \begin{bmatrix} 3 & 4 \\ -8 & 1 \end{bmatrix}$

Let $A = \begin{bmatrix} -2 & 4 & 7 \\ 6 & 1 & 5 \\ 2 & 3 & -4 \end{bmatrix}$. *Calculate the following:*

7. A^t 8. AA^t 9. $A^t A$

Let $A = \begin{bmatrix} -3 & 2 \\ 7 & 5 \end{bmatrix}$ and $B = \begin{bmatrix} 4 & 1 \\ -2 & 9 \end{bmatrix}$. Calculate the following:

10. AB
11. BA
12. AB^t
13. $A^t B$
14. $(AB)^t$
15. $B^t A^t$
16. $A - B$
17. $B - A$
18. $5A$
19. $2B - 3A$

Let $A = \begin{bmatrix} 1 & 0 & 5 \\ 2 & -4 & 6 \end{bmatrix}$ and $B = \begin{bmatrix} -1 & 2 \\ 3 & 9 \\ 10 & 0 \end{bmatrix}$. Calculate the following:

20. AB
21. BA
22. A^t
23. B^t
24. $A^t B^t$
25. $(BA)^t$

26. Prove or show a counterexample for the statement *For matrices A, B, and C, if $AB = AC$, then $B = C$.*

27. Prove or show a counterexample for the statement *For matrices A and B, if AB is the zero matrix, then either A or B is the zero matrix.*

28. Prove that if A and B are matrices whose product is defined, then $(AB)^t = B^t A^t$.

29. Prove Theorem 4.24 that the following are true for all $m \times n$ matrices A, B, and C:
    (a) $A + B = B + A$ (commutative property of addition)
    (b) $A + (B + C) = (A + B) + C$ (associative property of addition)

30. Prove that for real number r and $m \times n$ matrices A, B, $r(A + B) = rA + rB$.

31. Prove that for real numbers r and s and $m \times n$ matrix A, $(r+s)A = r \cdot A + s \cdot A$.

32. Prove that for real numbers r and s and $m \times n$ matrix A, $r(sA) = (rs)A$.

33. Prove that for $m \times n$ matrices A and B, there exists a matrix C so that $A + C = B$.

34. Prove Theorem 4.25 (b) for all matrices A, B, and C and real numbers r and s, where A is an $m \times p$ matrix and B, C are $p \times n$ matrices, $A \cdot (rB + sC) = r(A \cdot B) + s(A \cdot C)$ (linear property of matrices).

35. Prove that if R and S are relations on a finite set $A = \{a_1, a_2, \ldots, a_n\}$ with respective representation matrices M and N, then $M \vee N$ is the representation matrix for $R \cup S$.

36. Prove that if R and S are relations on a finite set $A = \{a_1, a_2, \ldots, a_n\}$ with respective representation matrices M and N, then $M \wedge N$ is the representation matrix for $R \cap S$.

37. Prove that if $R \subseteq A \times B$ and $S \subseteq B \times C$ and if M and N are the representation matrices for R and S, respectively, then $M \odot N$ is the representation matrix for $S \circ R$.

Let $A = \{1, 2, 3, 4, 5\}$
$B = \{6, 7, 8, 9\}$
$C = \{10, 11, 12, 13\}$
$D = \{\square, \triangle, \bigcirc, *\}$
and

let $R \subseteq A \times B$, $S \subseteq B \times C$, and $T \subseteq C \times D$ be defined by
$R = \{(1, 7), (4, 6), (5, 6), (2, 8)\}$
$S = \{(6, 10), (6, 11), (7, 10), (8, 13)\}$
$T = \{(11, \triangle), (10, \triangle), (13, *), (12, \square), (13, \bigcirc)\}$

Find the representation matrices for R, S, and T, then find the representation matrices for

38. R^{-1} and S^{-1}
39. $S \circ R$
40. $S \circ S^{-1}$ and $S^{-1} \circ S$
41. $R^{-1} \circ S^{-1}$
42. $T \circ (S \circ R)$
43. $T \circ S$
44. $(T \circ S) \circ R$

Let $A = \{a, b, c, d, e\}$ and $B = \{f, g, h, i\}$, $C = \{u, v, w, x, y\}$. Let $R \subseteq B \times A$ and $S \subseteq B \times C$ be defined by
$R = \{(f, a), (g, b), (h, b), (h, a), (i, d), (h, e), (g, c)\}$
$S = \{(f, u), (f, v), (g, x), (h, y), (h, w), (h, u)\}$

Find the representation matrices for R and S. Then find the representation matrices for

45. R^{-1}
46. S^{-1}
47. $S \circ R^{-1}$
48. $R \circ S^{-1}$

Let $A = \{a, b, c, d, e\}$ and $S, T, U,$ and V relations on A where
$S = \{(a, a), (a, b), (b, c), (b, d), (c, e), (e, d), (c, a)\}$
$T = \{(a, b), (b, a), (b, c), (b, d), (e, e), (d, e), (c, b)\}$
$U = \{(a, b), (a, a), (b, c), (b, b), (e, e), (b, a),$
    $(c, b), (c, c), (d, d), (a, c), (c, a)\}$
$V = \{(a, b), (b, c), (b, b), (e, e), (b, a), (c, b),$
    $(d, d), (a, c), (c, a)\}$

Find the representation matrices for S, T, U, and V. Then use these matrices to determine

49. Which of the relations are symmetric.
50. Which of the relations are reflexive.
51. Which of the relations are transitive.
52. Which of the relations are antisymmetric.

Let $A = \{a, b, c, d, e\}$ and $S, T, U,$ and V relations on A where
$S = \{(a, a), (a, b), (b, c), (b, d), (c, e), (e, d), (c, a)\}$
$T = \{(a, b), (b, a), (b, c), (b, d), (e, e), (d, e), (c, b)\}$
$U = \{(a, b), (a, a), (b, c), (b, b), (e, e), (b, a),$
    $(c, b), (c, c), (d, d), (a, c), (c, a)\}$
$V = \{(a, b), (b, c), (b, b), (e, e), (b, a), (c, b),$
    $(d, d), (a, c), (c, a)\}$

53. Find the representation matrices for S, T, U, and V.

54. Find the representation matrices for $S \circ T$ and $V \circ U$.

55. Find the representation matrices for $S \cap T$ and $U \cap V$.

56. Find the representation matrix for $S \triangle U$.

Let $A = \{1, 2, 3, 4,\}$
$B = \{a, b, c\}$

Find the relations on $A \times B$ with relation matrices

57. $\begin{bmatrix} 1 & 0 & 1 \\ 0 & 1 & 0 \\ 0 & 0 & 1 \\ 1 & 1 & 0 \end{bmatrix}$

58. $\begin{bmatrix} 0 & 1 & 1 \\ 1 & 1 & 0 \\ 1 & 0 & 1 \\ 1 & 0 & 1 \end{bmatrix}$

59. $\begin{bmatrix} 1 & 0 & 1 \\ 0 & 1 & 0 \\ 1 & 0 & 1 \\ 0 & 1 & 0 \end{bmatrix}$

Let $A = \{a, b, c, d, e\}$. Find the relations on A with relationship matrices

60. $\begin{bmatrix} 1 & 0 & 0 & 1 & 0 \\ 0 & 1 & 0 & 0 & 0 \\ 0 & 0 & 1 & 0 & 0 \\ 1 & 0 & 0 & 1 & 1 \\ 0 & 0 & 1 & 0 & 1 \end{bmatrix}$

61. $\begin{bmatrix} 0 & 0 & 1 & 0 & 1 \\ 0 & 1 & 0 & 1 & 0 \\ 1 & 0 & 1 & 0 & 1 \\ 0 & 1 & 0 & 1 & 0 \\ 1 & 0 & 1 & 0 & 0 \end{bmatrix}$

62. $\begin{bmatrix} 0 & 1 & 0 & 0 & 0 \\ 1 & 0 & 0 & 0 & 0 \\ 0 & 1 & 0 & 1 & 0 \\ 0 & 1 & 0 & 0 & 1 \\ 0 & 0 & 1 & 0 & 0 \end{bmatrix}$

63. $\begin{bmatrix} 1 & 0 & 0 & 0 & 1 \\ 0 & 1 & 0 & 1 & 0 \\ 1 & 0 & 1 & 0 & 0 \\ 0 & 1 & 0 & 1 & 0 \\ 1 & 0 & 1 & 0 & 1 \end{bmatrix}$

Let $A = \{a, b, c, d, e\}$. Construct graphs with the following representation matrices:

64. $\begin{bmatrix} 1 & 0 & 1 & 1 & 0 \\ 0 & 0 & 0 & 0 & 0 \\ 1 & 0 & 1 & 0 & 1 \\ 1 & 0 & 0 & 0 & 1 \\ 0 & 0 & 1 & 1 & 1 \end{bmatrix}$

65. $\begin{bmatrix} 0 & 0 & 1 & 0 & 1 \\ 0 & 1 & 0 & 1 & 0 \\ 1 & 0 & 1 & 0 & 1 \\ 0 & 1 & 0 & 1 & 0 \\ 1 & 0 & 1 & 0 & 0 \end{bmatrix}$

66. $\begin{bmatrix} 0 & 1 & 0 & 0 & 1 \\ 1 & 0 & 1 & 1 & 0 \\ 0 & 1 & 0 & 1 & 1 \\ 0 & 1 & 1 & 0 & 1 \\ 1 & 0 & 1 & 1 & 0 \end{bmatrix}$

Let $A = \{a, b, c, d, e\}$. Construct digraphs with the following representation matrices:

67. $\begin{bmatrix} 1 & 0 & 1 & 1 & 0 \\ 0 & 1 & 0 & 1 & 0 \\ 1 & 0 & 0 & 0 & 0 \\ 1 & 1 & 0 & 0 & 1 \\ 0 & 1 & 1 & 0 & 1 \end{bmatrix}$

68. $\begin{bmatrix} 0 & 0 & 1 & 0 & 1 \\ 1 & 0 & 0 & 1 & 0 \\ 1 & 1 & 1 & 0 & 1 \\ 0 & 1 & 0 & 0 & 0 \\ 1 & 0 & 1 & 0 & 0 \end{bmatrix}$

69. $\begin{bmatrix} 0 & 0 & 0 & 0 & 0 \\ 1 & 0 & 0 & 1 & 0 \\ 0 & 1 & 0 & 1 & 0 \\ 0 & 1 & 0 & 0 & 1 \\ 1 & 0 & 0 & 0 & 0 \end{bmatrix}$

For each relation R described by a representation matrix in the problem, find $R \circ R$ using representation matrices

70. $\begin{bmatrix} 1 & 0 & 1 & 1 & 0 \\ 0 & 1 & 0 & 1 & 0 \\ 1 & 0 & 0 & 0 & 0 \\ 1 & 1 & 0 & 0 & 1 \\ 0 & 1 & 1 & 0 & 1 \end{bmatrix}$

71. $\begin{bmatrix} 0 & 0 & 1 & 0 & 1 \\ 1 & 0 & 0 & 1 & 0 \\ 1 & 1 & 1 & 0 & 1 \\ 0 & 1 & 0 & 0 & 0 \\ 1 & 0 & 1 & 0 & 0 \end{bmatrix}$

72. $\begin{bmatrix} 0 & 0 & 0 & 0 & 0 \\ 1 & 0 & 0 & 1 & 0 \\ 0 & 1 & 0 & 1 & 0 \\ 0 & 1 & 0 & 0 & 1 \\ 1 & 0 & 0 & 0 & 0 \end{bmatrix}$

4.3 Cardinality

In this section we discuss the concept of the size of a set. If a set is finite the size of the set is simply the number of the elements in the set. To determine the size of the set, we simply count the number of elements in the set. We state this more formally in the following definition.

DEFINITION 4.42 The empty set is a **finite set of cardinality 0**. If there is a one-to-one correspondence between a set A and the set $\{1, 2, 3, \ldots n\}$, then A is a **finite set of cardinality** n.

EXAMPLE 4.43 The set $\{a, p, r, x, z\}$ has cardinality 5 since there exists a one-to-one correspondence $\phi : \{1, 2, 3, 4, 5\} \to \{a, p, r, x, z\}$ defined by $\phi(1) = a$, $\phi(2) = p$, $\phi(3) = r$, $\phi(4) = x$, $\phi(5) = z$. ∎

DEFINITION 4.44 A set A is **countably infinite** if there is a one-to-one correspondence between a set A and the set of positive integers $\{1, 2, 3, \ldots, n, \ldots\}$. A set is **countable** if it is finite or countably infinite.

THEOREM 4.45

(a) Let A and B be disjoint finite sets. Then $A \cup B$ is finite. If A has cardinality n, and B has cardinality m, then $A \cup B$ has cardinality $m + n$.

(b) Let A and B be disjoint countably infinite sets. Then $A \cup B$ is countably infinite.

(c) Let A and B be disjoint countable sets. Then $A \cup B$ is countable.

Proof

(a) Since A and B are finite, there exist one-to-one correspondences $\phi_1 : \{1, 2, 3, \ldots, n\} \to A$ and $\phi_2 : \{1, 2, 3, \ldots, m\} \to B$ for integers n and m. Define $\phi : \{1, 2, 3, \ldots, m+n\} \to A \cup B$ by $\phi(k) = \phi_1(k)$ for $1 \leq k \leq n$ and $\phi(n+i) = \phi_2(i)$ for $1 \leq i \leq m$. The function ϕ is a one-to-one correspondence between $\{1, 2, 3, \ldots, m+n\}$ and $A \cup B$.

(b) Since A and B are countably infinite, there exist one-to-one correspondences $\phi_1 : \{1, 2, 3, \ldots n, \ldots\} \to A$ and $\phi_2 : \{1, 2, 3, \ldots m, \ldots\} \to B$. Define $\phi : \{1, 2, 3, \ldots\} \to A \cup B$ by $\phi(n) = \phi_1\left(\frac{n+1}{2}\right)$ if n is odd and $\phi(n) = \phi_2\left(\frac{n}{2}\right)$ if n is even. The function ϕ is a one-to-one correspondence between $\{1, 2, 3, \ldots\}$ and $A \cup B$.

(c) Let A and B be disjoint countable sets. Then the only case not covered by (a) and (b) is the case where either A or B is finite and the other is countably infinite. Assume A is finite. Then there exists a one-to-one correspondence $\phi_1 : \{1, 2, 3, \ldots, n\} \to A$; and since B is countably infinite, there exists $\phi_2 : \{1, 2, 3, \ldots, m, \ldots\} \to B$. Define $\phi : \{1, 2, 3, \ldots\} \to A \cup B$ by $\phi(k) = \phi_1(k)$ for $1 \leq k \leq n$ and $\phi(k) = \phi_2(k - n)$ for $k > n$. The function ϕ is a one-to-one correspondence between $\{1, 2, 3, \ldots\}$ and $A \cup B$. ∎

The proof of the following theorem is left to the reader.

THEOREM 4.46 The set of nonnegative integers and the set of integers are both countably infinite. ∎

Other properties of cardinality will be discussed in the next section. It will be shown that there are sets that are not countable and that there are an infinite number of unique infinite cardinal numbers.

Exercises

1. Using the definition, find the cardinality of the following sets:

    (a) $\{a, b, c, d, e, f, g\}$

    (b) $\{3, 6, 9, 12, 15, 18, 21, 24, 27, 30, 33\}$

    (c) $\{5, 10, 15, 20, 25, 30, \ldots\}$

2. Find the cardinality of the set $\{\emptyset, a, b, \{a, b\}\}$.

3. Find the cardinality of the set $\{\emptyset\}$.

4. Find the cardinality of the set $\{\emptyset, \{\emptyset\}\}$.

5. Find the cardinality of the set $\{\emptyset, \{\emptyset\}, \{\emptyset, \{\emptyset\}\}\}$.

6. Find the cardinality of the set $\{\emptyset, \{\emptyset\}, \{\emptyset, \{\emptyset\}\}, \{\emptyset, \{\emptyset\}, \{\emptyset, \{\emptyset\}\}\}\}$.

7. Show that the nonnegative integers are countably infinite.

8. Show that the negative integers are countably infinite.
9. Show that the integers are countably infinite.
10. Show that the set $\{-10, -9, -8, -7, -6, -5, -4, -3, -2, -1, 0, 1, 2, 3, \ldots\}$ is countably infinite.
11. Show that if A is countably infinite and there is a one-to-one correspondence between A and B, then B is countably infinite.
12. Show that the set $\{\ldots, -10, -9, -8, -7, -6, -5\}$ is countably infinite.
13. Show that the set $\{\ldots, -6, -3, 0, 3, 6, 9, 12\}$ is countably infinite.
14. Show that the set $\{0, 5, 10, 15, 20, \ldots\}$ is countably infinite.

4.4 Cardinals Continued

We now explore the cardinality of sets more thoroughly. We begin by showing that a subset of a countable set is countable. From this it follows that any set that is in one-to-one correspondence with a subset of a countable set is countable.

THEOREM 4.47 A subset of a countable set is countable.

Proof We show that if B is a countable set and $A \subseteq B$, then A is countable. Assume B is finite. Then we show that A is finite. Since B is finite, there exists a one-to-one correspondence ϕ between $\{1, 2, 3, \ldots, n\}$ and B for some positive integer n. Hence, B can be expressed in the form $\{b_1, b_2, b_3, \ldots, b_n\}$, where $b_i = \phi(i)$. We now form a one-to-one correspondence $\overline{\phi}$ from $\{1, 2, 3, \ldots, m\}$ to A, where $m \leq n$. Begin with b_1 and find the first b_i, say b_{a_1}, in the sequence $b_1, b_2, b_3, \ldots, b_n$ such that $b_i \in A$. Let $\overline{\phi}(1) = b_{a_1}$. Continue along the sequence until the next b_i, say b_{a_2} is found such that $b_i \in A$. Let $\overline{\phi}(2) = b_{a_2}$. Continue this process until b_n is reached. Let m be the last integer that $\overline{\phi}$ maps onto an element of A. Since $i \leq a_i$ for all i, $m \leq n$ and certainly $\overline{\phi}$ maps $\{1, 2, 3, \ldots, m\}$ to A.

Now assume B is countably infinite. If A is finite, then it is countable, so assume it is not finite. Since B is countably infinite, there exists a one-to-one correspondence ϕ between $\{1, 2, 3, \ldots\}$ and B. Hence, B can be expressed in the form $\{b_1, b_2, b_3, \ldots\}$, where $b_i = \phi(i)$. Using the well-ordering principle, we select the smallest i, such that $b_i \in A$. Let $\overline{\phi}(1) = b_i$ and $B_1 = B - \{b_i\}$. Obviously $1 \leq i$. Assume we have unique $\overline{\phi}(1), \overline{\phi}(2), \overline{\phi}(3), \ldots, \overline{\phi}(k) \in A$, $B_k = B - \{\overline{\phi}(1), \overline{\phi}(2), \overline{\phi}(3), \ldots, \overline{\phi}(k)\}$, $\overline{\phi}(i) \leq b$ for all $1 \leq i \leq k$ and all $b \in B_k$, and if $\overline{\phi}(k) = b_r$ then $k \leq r$. Select the smallest j such that $b_j \in A \cap B_k$. Let $\overline{\phi}(k) = b_j$ and $B_{k+1} = B_k - \{b_j\}$. Obviously we have unique $\overline{\phi}(1), \overline{\phi}(2), \overline{\phi}(3), \ldots, \overline{\phi}(k), \overline{\phi}(k+1) \in A$, $B_{k+1} = B - \{\overline{\phi}(1), \overline{\phi}(2), \overline{\phi}(3), \ldots, \overline{\phi}(k), \overline{\phi}(k+1)\}$, $\overline{\phi}(i) \leq b$ for all $1 \leq i \leq k+1$ and all $b \in B_{k+1}$, and if $\overline{\phi}(k+1) = b_r$, then $k+1 \leq r$. Inductively we have defined $\overline{\phi}$ and obviously $\overline{\phi}(i) \in A$ for all i, and $\overline{\phi}(i) \neq \overline{\phi}(j)$ when $i \neq j$. We need only show that $\overline{\phi}$ is onto. Let $a \in A$, then $a = b_n$ for some n. Hence, for some $m \leq n$, $\overline{\phi}(m) = b_n$. ∎

THEOREM 4.48 If a set S is countably infinite, then $S \times S$ is countably infinite.

Proof We first show that if N is the set of positive integers, then $N \times N$ is countable.

Using Figure 4.1, beginning with the first diagonal and following the arrows, we define ϕ by $\phi(1) = (1, 1)$, $\phi(2) = (1, 2)$, $\phi(3) = (2, 1)$, $\phi(4) = (1, 3)$, $\phi(5) = (2, 3)$, and so on. The function ϕ is the desired one-to-one correspondence. The ordered pair (m, n) occurs along diagonal $m + n - 1$ and is the mth entry along that diagonal line or ordered pairs show. Hence, the set in Figure 4.1 is countable. Since

Figure 4.1

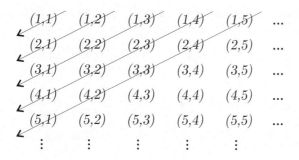

$N \times N$ is a subset of this set, it is countable. Since S is countably infinite, there exists a one-to-one correspondence $\theta : N \to S$. Define $\theta \times \theta : N \times N \to S \times S$ by $\theta \times \theta(a,b) = (\theta(a), \theta(b))$. The function $\theta \times \theta$ is easily shown to be a one-to-one correspondence between $N \times N$ and $S \times S$. ∎

THEOREM 4.49 The set Q^+ of positive rationals is countably infinite.

Proof Consider the subset M of $N \times N$ consisting of $\{(a,b) : (a,b) \in N \times N$ and the integers a and b are relatively prime$\}$. The function $\phi : Q^+ \to M$ defined by $\phi\left(\dfrac{a}{b}\right) = (a,b)$ is obviously a one-to-one correspondence. Since $N \times N$ is countably infinite, M is countably infinite and, hence, Q^+ is countably infinite. ∎

THEOREM 4.50 If A and B are countable, then $A \cup B$ is countable.

Proof Since $A - B$ is a subset of the countable set A, $A - B$ is countable. But $A - B$ and B are disjoint, so by Theorem 4.45 (c), $(A - B) \cup B = A \cup B$ is countable. ∎

At this point it might seem that all sets are countable, however, in the following theorem it is shown that there is an infinite but not countable set.

THEOREM 4.51 Let R be the set of real numbers. The set $I = \{x : x \in R$ and $0 < x < 1\}$ is not countable.

Proof This is a classic use of the reductio ad absurdum method of proof. We assume that the set I is countable. Then I can be expressed in the form $\{a_1, a_2, a_3, a_4, a_5, \ldots\}$ where a_i is the image of i in the one-to-one correspondence between N and I. Let $.a_{i1}a_{i2}a_{i3}a_{i4}a_{i5}\ldots$ be the decimal expansion of a_i as demonstrated in the following table:

$$a_1 = .a_{11}a_{12}a_{13}a_{14}a_{15}\ldots$$
$$a_2 = .a_{21}a_{22}a_{23}a_{24}a_{25}\ldots$$
$$a_3 = .a_{31}a_{32}a_{33}a_{34}a_{35}\ldots$$
$$a_4 = .a_{41}a_{42}a_{43}a_{44}a_{45}\ldots$$
$$a_5 = .a_{51}a_{52}a_{53}a_{54}a_{55}\ldots$$

Define $b \in I$ where $b = .b_1b_2b_3b_4b_5\ldots$ by $b_i = 5$ if $a_{ii} \neq 5$ and $b_i = 8$ if $a_{ii} = 5$. Consider the following demonstration:

$$a_1 = .5135897623\ldots$$
$$a_2 = .1425958273\ldots$$
$$a_3 = .2837462982\ldots$$
$$a_4 = .1985723986\ldots$$
$$a_5 = .2309847621\ldots$$
$$\vdots \quad \vdots \quad \vdots$$

Since $a_{11} = 5$, $b_1 = 8$. Since $a_{22} = 4$, $b_2 = 5$. Since $a_{33} = 3$, $b_3 = 5$. Since $a_{44} = 5$, $b_4 = 8$. Since $a_{55} = 8$, $b_5 = 5$, and so forth. Hence,

$$b = .85585\ldots$$

The number b is not in the set $\{a_1, a_2, a_3, a_4, a_5, \ldots\}$ for b cannot equal a_m for any m since $b_m \neq a_{mm}$ and, hence, b and a_m disagree in the mth digit. But this contradicts the assumption that $I = \{a_1, a_2, a_3, a_4, a_5, \ldots\}$ since $b \in I$ but $b \notin \{a_1, a_2, a_3, a_4, a_5, \ldots\}$. Hence, we must assume that the statement that I is countably infinite is false and I is not countably infinite. ∎

THEOREM 4.52 The set R of real numbers is not countably infinite.

Proof If R were countably infinite then, since I is a subset of R, I is countably infinite. However, I is not countably infinite. Hence R is not countably infinite. ∎

In the following we develop a method to generate an infinite number of infinite sets, no two of which are in one-to-one correspondence.

THEOREM 4.53 There is no one-to-one correspondence between a set S and its power set, $\mathcal{P}(S)$.

Proof There is certainly no one-to-one correspondence between the empty set $\varnothing$ and its power set $\{\varnothing\}$. Hence, we will assume the set S is nonempty. Again we use reductio ad absurdum. Assume there is a one-to-one correspondence ϕ between S and its power set. Hence, ϕ maps elements of S to subsets of S. If S is the set of positive integers then we might have

$$\phi(1) = \{2, 5, 7, 9, 10\}$$
$$\phi(2) = \{2, 4, 6, 8, \ldots\}$$
$$\phi(3) = \varnothing$$
$$\phi(4) = \{1, 2, 3, 4, 5, 6, \ldots\}$$
$$\phi(5) = \{1, 2\}$$
$$\vdots$$

In some cases $i \in \phi(i)$. In our example $2 \in \phi(2)$ and, $4 \in \phi(4)$. However, $1 \notin \phi(1)$, $3 \notin \phi(3)$, and $5 \notin \phi(5)$. In general, whatever element i is mapped by ϕ to the empty set will have the property $i \notin \phi(i)$ and whatever element j is mapped by ϕ to S will have the property that $j \in \phi(j)$. Let $W = \{x : x \in S$ and $x \notin \phi(x)\}$. Since ϕ is onto, let $\phi(a) = W$. Is $a \in W$? If $a \in W$, then a belongs to the set of elements of S that ϕ does not map to sets that contain them. Hence, $a \notin \phi(a) = W$, a contradiction. If $a \notin W$, then $a \notin \phi(a) = W$ and, hence, meets the requirements for belonging to W. Hence, $a \in W$, a contradiction. Either way we have a contradiction and must assume that the statement that there is a one-to-one correspondence ϕ between S and its power set is false. ∎

EXAMPLE 4.54   Let $S = \{1, 2, 3, 4\}$ and

$$\phi(1) = \{2, 4\}$$
$$\phi(2) = \{1, 2, 3, 4\}$$
$$\phi(3) = \{1, 3\}$$
$$\phi(4) = \varnothing$$

Then $W = \{1, 4\}$ and nothing maps to it. ∎

An important example whose argument is very similar to the proof of the previous theorem is called Russell's paradox. It is important because it demonstrates that we must be careful how we define a set or we will have a contradiction in our mathematical system. Especially, we must not let sets get too large.

EXAMPLE 4.55   (*Russell's Paradox*) Let S be the set of all sets. Let $W = \{x : x \notin x\}$. The empty set belongs to W since it does not belong to itself. In fact most sets that we are familiar with belong to W. However, S does not belong to W since $S \in S$. The question is, does $W \in W$? If $W \in W$, then it belongs to the set of all sets that do not belong to themselves and, hence, $W \notin W$. However, if we assume that $W \notin W$, then W belongs to the set of all sets that do not belong to themselves. But this set is W so that $W \in W$. Either way we reach a contradiction. Therefore, W is not a well-defined set. ∎

DEFINITION 4.56   Two sets S and T have the same **cardinality** if there is a one-to-one correspondence between the two sets.

Obviously all countably infinite sets have the same cardinality. We know that the real numbers do not have the same cardinality as countably infinite sets.

Notation 4.57   If there is a one-to-one function $f : S \to T$, we denote this by $S \preceq T$. If $S \preceq T$ but there is no one-to-one correspondence between S and T, then $S \prec T$.

In Chapter 3, we mentioned that the Gödel Incompleteness Theorem gives us a choice between a mathematical system that has contradictions. Russell's paradox is an example of a possible contradiction. Assuming that our present mathematical system has no contradictions, there is a statement that we cannot prove true or false. We now know that the power set of the integers has greater cardinality than the integers in the sense that there is a one-to-one function from the integers to its power set but not conversely. We also know that the set of real numbers has greater cardinality than the integers. Actually the set of real numbers and the power set of the integers have the same cardinality. The proof of this is beyond the scope of this book. We might ask if there is a set S that has greater cardinality than the integers and smaller cardinality than the reals so that $Z \prec S \prec \mathcal{P}(Z)$. We cannot prove that this statement is either true of false.

The remainder of this section is more advanced and might not be suitable for a freshman class.

The following theorem shows us that if there is a one-to-one function $f : S \to T$ and a one-to-one function $g : T \to S$, then S and T have the same cardinality.

THEOREM 4.58 If $S \preceq T$ and $T \preceq S$, then there is a one-to-one correspondence between S and T.

Proof Assume $f : S \to T$ and $g : T \to S$. For each $s \in S$, we find $g^{-1}(s)$ if it exists. We then find $f^{-1}g^{-1}(s)$ if it exists. Then find $g^{-1}f^{-1}g^{-1}(s)$ if it exists. We continue this process. There are three possible results: (1) The process continues indefinitely. (2) The process ends because for some s_i in the process, there is no $g^{-1}(s_i)$. (3) The process ends because for some t_i in the process, there is no $f^{-1}(t_i)$. Let S_1 be the elements of S for which the first result occurs. Let S_2 be the elements of S for which the second result occurs. Let S_3 be the elements of S for which the third result occurs. Obviously these sets are disjoint. Similarly, form T_1, T_2, and T_3. f is a one-to-one correspondence from S_1 to T_1. f is also a one-to-one correspondence from S_2 to T_2. g^{-1} is a one-to-one correspondence from S_3 to T_{31}. Let $\theta : S \to T$ be defined by

$$\theta(s) = f(s) \text{ if } s \in S_1$$
$$= f(s) \text{ if } s \in S_2$$
$$= g^{-1}(s) \text{ if } s \in S_3$$

θ is a one-to-one correspondence from S to T. ∎

The following theorem is intuitively obvious but not particularly easy to prove. It does allow us in the problems to show that a set is infinite if and only if it can be placed in one-to-one correspondence with itself.

THEOREM 4.59 *If m and n are positive integers, then there exists a one-to-one correspondence between the integers less than or equal to m and a proper subset of the integers less than or equal to n if and only if $m < n$.

Proof If $m < n$, then the function $f(i) = i$ is a one-to-one correspondence between the integers less than or equal to m and a proper subset of the integers less than or equal to n.

To show the inverse, we use induction on m. If $m = 1$, and $f(i) = i$ is a one-to-one correspondence between the integers less than or equal to 1 (hence, just the integer 1) and a proper subset of the integers less than or equal to n, then obviously $n > 1$.

Assume there is a one-to-one correspondence f between the integers less than or equal to $m = k$ and a proper subset of the integers less than or equal to n. We want to find a one-to-one correspondence g between the integers less than or equal to $k - 1$ and a proper subset of the integers less than or equal to $n - 1$. To do this we define g on $\{1, 2, 3, \ldots m - 1\}$ as follows.

$$g(i) = f(i) \text{ if } f(i) \neq n$$
$$= f(k) \text{ if } f(i) = n$$

g is obviously a one-to-one correspondence between $\{1, 2, 3, \ldots k - 1\}$ and a proper subset of integers less than $n - 1$. This follows from the fact that there is an integer $1 \leq j \leq n$ not mapped onto by f, so it is not mapped onto by g. Therefore, $k - 1$ is less than $n - 1$ and k is less than n. ∎

Exercises

1. Let $S = \{1, 2, 3, 4, 5, 6\}$ and ϕ be a one-to-one function from S to its power set $\mathcal{P}(S)$ defined by

   $\phi(1) = \{1, 2, 3, 4\}$

   $\phi(2) = \{1, 4\}$

   $\phi(3) = \{2, 3, 4\}$

   $\phi(4) = \varnothing$

   $\phi(5) = \{1, 2, 3, 4, 5, 6\}$

   $\phi(6) = \{1, 3, 6\}$

What is the set which Theorem 4.53 guarantees us will not be mapped onto?

2. Let $S = \{1, 2, 3, 4, 5, 6\}$ and ϕ be a one-to-one function from S to its power set $\mathcal{P}(S)$ defined by

$$\phi(1) = \{2, 3, 4\}$$
$$\phi(2) = \{1, 4\}$$
$$\phi(3) = \{2, 4\}$$
$$\phi(4) = \varnothing$$
$$\phi(5) = \{1, 2, 3, 4, 6\}$$
$$\phi(6) = \{1, 3\}$$

What is the set which Theorem 4.53 guarantees us will not be mapped onto?

3. Let $S = \{1, 2, 3, 4, 5, 6\}$ and ϕ be a one-to-one function from S to its power set $\mathcal{P}(S)$ defined by

$$\phi(1) = \{1, 2, 3, 4\}$$
$$\phi(2) = \{1, 2, 4\}$$
$$\phi(3) = \{2, 3, 4, 5\}$$
$$\phi(4) = \{4\}$$
$$\phi(5) = \{1, 2, 3, 4, 5, 6\}$$
$$\phi(6) = \{1, 3, 6\}$$

What is the set which Theorem 4.53 guarantees us will not be mapped onto?

4. Prove that the set of nonnegative rational numbers is countable.

5. Prove that the set of rational numbers is countable.

6. Let $S = \{a + bi : a, b \in Z\}$ where $i = \sqrt{-1}$. Prove that S is countable.

7. Let $Z_i = Z$. Show that $Z_1 \times Z_2 \times Z_3 \times \cdots \times Z_n$ is countable.

8. Prove that the set of all polynomials with integral coefficients of degree 5 is countable.

9. Prove that the set of all polynomials with integral coefficients of degree n or less is countable.

10. Prove that the set of all $n \times n$ matrices with integral coefficients is countable.

11. Prove that any countably infinite set can be put into one-to-one correspondence with a proper subset of itself.

12. Prove that any infinite set contains a countably infinite subset.

13. Prove that $A \prec P(A)$ for any set A.

14. *Using Theorem 4.58, prove that there is no one-to-one correspondence between $\{1, 2, 3, 4, \ldots n\}$ and one of its proper subsets.

15. *Using the results of the previous problems, prove that a set S is infinite if and only if it there is a one-to-one correspondence between S and a proper subset of S.

GLOSSARY

Chapter 4: Functions and Matrices

4.1 Special Functions

Arithmetic sequence	In an **arithmetic sequence**, each element in the sequence may be obtained from the previous element by adding a constant d (difference).
Binary operation	A **binary operation** on the set A is a function $b : A \times A \to A$. The image of (r, s) under b is written as $b((r, s))$ or as rbs.
Ceiling function	The function $f : A \to B$ where A is the set of real numbers and B is the set of integers is called the **ceiling function**, denoted by $f(x) = \lceil x \rceil$, if it assigns to each $a \in A$ the smallest integer greater than or equal to a.
Factorial function	Let A and B both be the set of nonnegative integers. The **factorial function** $f : A \to B$, denoted by $f(n) = n!$, is defined by $$0! = 1$$ $$1! = 1$$ $$2! = 1 \cdot 2$$ $$3! = 1 \cdot 2 \cdot 3$$ $$\vdots$$ $$k! = 1 \cdot 2 \cdot 3 \cdot 4 \cdot \cdots \cdot k$$

Finite sequence	A **finite sequence** is a function from $\{1, 2, 3, \ldots n\}$ to some set S.
Floor function	The function $f : A \to B$ where A is the set of real numbers and B is the set of integers is called the **floor function**, denoted by $f(x) = \lfloor x \rfloor$, if it assigns to each $a \in A$ the largest integer less than or equal to a.
Geometric sequence	In a **geometric sequence**, each element in the sequence may be obtained from the previous element by multiplying by a constant r.
Infinite sequence	An **infinite sequence** is a function from $\{1, 2, 3, \ldots\}$ to some set S.
Summation notation	The sum $A_r + A_{r+1} + A_{r+2} + A_{r+3} + \cdots + A_{r+k}$ may be written in **summation notation** as $\sum_{i=r}^{r+k} A_i$.

4.2 Matrices

Boolean matrix	A **Boolean matrix** is a matrix whose entries are all either 0 or 1. All operations on the matrices are defined by Boolean operations on their respective components.
Column matrix	A matrix of dimension $n \times 1$ is called a **column matrix**.
Component	The value A_{ij} is called a **component**, **entry**, or **element** of the matrix A.
Dimension	If matrix A has m rows and n columns, then it is said to have **dimension** $m \times n$.
Equal matrices	Two matrices are equal if their corresponding elements are equal.
Matrix or array	For positive integers m and n, an $m \times n$ **matrix** or **array** is a function $A : \{1, 2, \ldots, m\} \times \{1, 2, \ldots, n\} \to D$, where D is usually the set of real numbers, complex numbers, rational numbers, or integers. For each i, $1 \leq i \leq m$, and each j, $1 \leq j \leq n$, there is an element of D, denoted by $A(i, j)$, which we think of as being in the ith row and the jth column of a rectangular array. The image of $A(i, j)$ of the domain element (i, j), is usually shortened to A_{ij}. Thus, the $m \times n$ matrix A is represented by a rectangular array where the images of $(i, j) \in \{1, 2, \ldots, m\} \times \{1, 2, \ldots, n\}$ under A are listed as follows: $$A = \begin{bmatrix} A_{11} & A_{12} & A_{13} & \ldots & A_{1n} \\ A_{21} & A_{22} & A_{23} & \ldots & A_{2n} \\ \vdots & & & & \vdots \\ A_{m1} & A_{m2} & A_{m3} & \ldots & A_{mn} \end{bmatrix}$$ Sometimes we write $A = [A_{ij}]$ or $A = [a_{ij}]$.
Matrix product	The **matrix product** of matrices A and B, denoted by AB, is the $m \times n$ matrix $C = [C_{ij}]$, where C_{ij} is the dot product of the ith row and the jth column of B.
Matrix sum	If $A = [A_{ij}]$ and $B = [B_{ij}]$ are $m \times n$ matrices, then $A + B$ is the $m \times n$ matrix $C = [C_{ij}]$, where $C_{ij} = A_{ij} + B_{ij}$, that is, addition is componentwise. C is called the **matrix sum** of A and B.
Permutation matrix	Let M be an $m \times n$ matrix such that in each row and each column, one entry is 1 and the others are 0. M is called a **permutation matrix**.

Representation matrix	Let $A = \{a_1, a_2, \ldots, a_n\}$ and $B = \{b_1, b_2, \ldots, b_m\}$. Let R be a relation on $A \times B$. The **representation matrix** for R is the matrix $M_{nm} = [M_{ij}]$ defined by $$M_{ij} = \left\{ \begin{array}{ll} 1 & \text{if } (a_i, b_j) \in R \\ 0 & \text{if } (a_i, b_j) \notin R \end{array} \right\}$$
Row matrix	A matrix of dimension $1 \times m$ is called a **row matrix**.
Scalar product	The product of a scalar d and a matrix A is called a **scalar product**.
Scalars	The elements of a matrix are called **scalars**.
Square matrix	If the number of rows of a matrix is the same as the number of columns, then the matrix is called a **square matrix**.
Symmetric matrix	If A is an $n \times m$ matrix and $A_{ij}^t = A_{ji}$ for all $1 \leq i, j \leq n$ then A is a **symmetric matrix**. A is symmetric if and only if $A = A^t$.
Transpose of A	If A is an $n \times m$ matrix where the entry in the ith row and the jth column is A_{ij}, then A^t, the **transpose** of A, is the $m \times n$ matrix with $A_{ij}^t = A_{ji}$.

4.3 Cardinality

Cardinality	Two sets have the same cardinality if there is a 1-1 correspondence between them.
Cardinality 0	The empty set is a finite set of **cardinality 0**.
Cardinality n	If there is a one-to-one correspondence between a set A and the set $\{1, 2, 3, \ldots n\}$, then A is a finite set of **cardinality n**.
Countably infinite	A set A is **countably infinite** if there is a one-to-one correspondence between a set A and the set of positive integers $\{1, 2, 3, \ldots\}$.
Countable	A set is **countable** if it is finite or countably infinite.

4.4 Cardinals Continued

Russell's paradox	Let S be the set of all sets. Let $W = \{x : x \notin x\}$. The empty set belongs to W since it does not belong to itself. In fact most sets that we are familiar with belong to W. However, S does not belong to W since $S \in S$. The question is, does $W \in W$? If $W \in W$, then it belongs to the set of all sets that do not belong to themselves and, hence, $W \notin W$. However, if we assume that $W \notin W$, then W belongs to the set of all sets that do not belong to themselves. But this set is W so that $W \in W$. Either way we reach a contradiction.

TRUE-FALSE QUESTIONS

1. Let $f : A \to B$ be a function. The range of f is contained in the codomain of f.

2. Let $f : A \to B$ be a function and E be the image of f. Then $f^{-1}(E) = A$.

3. Let $f : A \to B$ be a function and B_1, B_2 be subsets of B. Then $f^{-1}(B_1 \cup B_2) = f^{-1}(B_1) \cup f^{-1}(B_2)$.

4. Let $f : A \to B$ be a function and A_1, A_2 be subsets of A. Then $f(A_1 \cap A_2) = f(A_1) \cap f(A_2)$.

5. Let $f : A \to B$ be a function and A_1 be a subset of A. Then $f^{-1}(f(A_1)) = A_1$.

6. Let $f = x + |x|$ be defined on the real numbers. The range of f is R, the set of real numbers.

7. A permutation is a bijection on a finite set.

8. Given permutations $f = \begin{pmatrix} 1 & 2 & 3 & 4 & 5 \\ 2 & 3 & 4 & 5 & 1 \end{pmatrix}$ and $g = \begin{pmatrix} 1 & 2 & 3 & 4 & 5 \\ 2 & 3 & 1 & 5 & 4 \end{pmatrix}$, $f \circ g = \begin{pmatrix} 1 & 2 & 3 & 4 & 5 \\ 3 & 1 & 5 & 4 & 2 \end{pmatrix}$.

9. $\lfloor -4.1 \rfloor = -4$.

10. $\lceil \lfloor 5.7 \rfloor \rceil = \lfloor \lceil 5.7 \rceil \rfloor$.

11. $\lfloor a \rfloor = a$ if and only if a is an integer.

12. $\dfrac{15!}{14!0!} = 15$.

13. $\sum\limits_{i=2}^{10} a_n = \sum\limits_{i=2}^{5} a_n + \sum\limits_{i=5}^{10} a_n$.

14. $\sum\limits_{i=0}^{10}(i+1)^2 = \sum\limits_{i=1}^{11} i^2$.

15. A matrix A is symmetric if and only if $A^t = A$.

16. A permutation matrix has exactly one 1 in each row and exactly one 1 in each column.

17. The product of two permutation matrices is a permutation matrix.

18. If A is a 3×4 matrix, then A^2 is a 3×4 matrix.

19. $\begin{bmatrix} 1 & 1 & 0 \\ 0 & 0 & 1 \end{bmatrix} \odot \begin{bmatrix} 1 & 1 & 0 & 0 \\ 0 & 0 & 1 & 0 \\ 1 & 1 & 0 & 0 \end{bmatrix} = \begin{bmatrix} 1 & 1 & 1 & 0 \\ 1 & 1 & 0 & 0 \end{bmatrix}$.

20. $\begin{bmatrix} 1 & 1 & 0 \\ 0 & 0 & 1 \end{bmatrix} \begin{bmatrix} 1 & 1 & 0 & 0 \\ 0 & 0 & 1 & 0 \\ 1 & 1 & 0 & 0 \end{bmatrix} = \begin{bmatrix} 1 & 1 & 1 & 0 \\ 1 & 1 & 0 & 0 \end{bmatrix}$.

21. Let $A = \{a_1, a_2, a_3\}$, $B = \{b_1, b_2, b_3\}$, $C = \{c_1, c_2, c_3\}$, $R = \{(a_3, b_1), (a_2, b_1), (a_2, b_2), (a_1, b_3)\}$ and $S = \{(b_2, c_1), (b_2, c_2), (b_1, c_2), (b_3, c_1), (b_3, c_2)\}$. The representation matrix for $S \circ R$ is $\begin{bmatrix} 1 & 1 & 0 \\ 1 & 1 & 1 \\ 1 & 1 & 1 \end{bmatrix}$.

22. A matrix A represents a transitive relation if and only if $A \circ A = A$.

23. A set A is countable if and only if there is a one-to-one correspondence between A and the positive integers.

24. A set A is countably infinite if and only if there is a one-to-one correspondence between A and the positive rational numbers.

25. If a set A is countably infinite, then its power set is countably infinite.

26. If A is an arbitrary set and $f : A \to A$ is one-to-one, then it is onto.

27. The set of all polynomials with rational coefficients and degree less than or equal to a fixed number n is countable.

28. The set of irrational numbers is countable.

29. $A \cup B$ is countable if and only if A and B are both countable.

30. Let R be a relation on $A \times B$ and S be a relation on $B \times C$. If M and N are representation matrices for R and S, respectively, then $N \odot M$ is the representation matrix for $S \circ R$.

31. Let a relation R on a set have representation matrix M. R is reflexive if and only if $M \vee I = M$.

32. For $n \times n$ matrices A and B, $(AB)^t = A^t B^t$.

33. For matrices A and B, if AB is the zero matrix, then either A or B is the zero matrix.

34. For matrices A and B, if $AB = A$, then B is the identity matrix.

35. Let R be a relation on A and M be its representation matrix. The representation matrix for $R \circ R^{-1}$ is the identity matrix.

36. For a Boolean matrix M, $M \vee M = M$.

37. For a Boolean matrix M, $M \odot M$ is always defined.

38. For matrices A and B, $(A + B)^t = A^t + B^t$.

39. For $n \times n$ matrices A and B, $AB = BA$.

40. $\left\lfloor \dfrac{m^2}{m^2 + 1} \right\rfloor = 0$ for every real number m.

41. $\left\lfloor \dfrac{m^2 + 1}{m^2} \right\rfloor = 1$ for every integer m.

42. Let R be a relation on A and M be its representation matrix. R is symmetric if and only if $M \vee M^t = M$.

43. Every relation has an inverse.

44. A function $f : A \to A$ has an inverse if and only if it is one-to-one.

45. If A is a finite set, then $f : A \to A$ has an inverse if and only if it is onto.

46. The function $f = 2x + |x|$, defined on the real numbers, is one-to-one.

47. The function $f = 2x + |x|$, defined on the real numbers, is onto.

48. Let $R = A \times A$ and S be a relation on A. $R \circ S = S$.

49. Let $R = A \times A$ and S be a relation on A. The range of $S \circ R$ is equal to the range of S.

50. If R is a relation with representation matrix M, then R^{-1} has representation matrix M^t.

SUMMARY QUESTIONS

1. Find the product of matrices A and B where

$$A = \begin{bmatrix} 1 & 2 & 5 & 1 \\ 3 & 2 & 4 & 4 \end{bmatrix} \text{ and } B = \begin{bmatrix} 3 & 2 & 2 \\ 0 & 1 & 3 \\ 1 & 0 & 1 \\ 3 & 5 & 0 \end{bmatrix}.$$

2. Evaluate $\lceil 3.0001 \rceil$.

3. Prove that the set of integers greater than -5 is countable.

4. Prove that the set of all numbers of the form $a + b\sqrt{3}$ where a and b are integers is countable.

5. Let $A = \begin{bmatrix} 1 & 2 & 5 & 1 \\ 3 & 2 & 4 & 4 \end{bmatrix}$. Find A^t.

6. Find $\begin{bmatrix} 1 & 0 & 0 \\ 0 & 0 & 1 \\ 1 & 0 & 1 \end{bmatrix} \odot \begin{bmatrix} 0 & 1 & 0 \\ 1 & 1 & 1 \\ 0 & 1 & 0 \end{bmatrix}$.

7. For permutations
$$f = \begin{pmatrix} 1 & 2 & 3 & 4 \\ 4 & 3 & 2 & 1 \end{pmatrix} \text{ and } g = \begin{pmatrix} 1 & 2 & 3 & 4 \\ 3 & 4 & 1 & 2 \end{pmatrix}$$
find $f \circ g$.

8. Evaluate $\lfloor -1.314 \rfloor$.

9. Find the inverse of the permutation $f = \begin{pmatrix} 1 & 2 & 3 & 4 \\ 3 & 2 & 4 & 1 \end{pmatrix}$.

10. Find $\begin{bmatrix} 1 & 1 & 0 \\ 0 & 1 & 1 \\ 1 & 0 & 1 \end{bmatrix} \vee \begin{bmatrix} 0 & 0 & 0 \\ 1 & 0 & 1 \\ 0 & 1 & 0 \end{bmatrix}$.

11. Find $\dfrac{14!}{8!4!2!}$.

12. Evaluate $\sum_{i=4}^{8} 3i$.

13. Find the sum of the series $1.02 + (1.02)^2 + (1.02)^3 + (1.02)^4 + (1.02)^5 + (1.02)^6$.

14. Find $3 \cdot \begin{bmatrix} 2 & 1 & 3 \\ 4 & 1 & -5 \\ 16 & -3 & 1 \end{bmatrix} + 2 \cdot \begin{bmatrix} 4 & 3 & 1 \\ 1 & 2 & 1 \\ 3 & 1 & 3 \end{bmatrix}$.

15. Find the first seven terms of the sequence $A_n = \lfloor \frac{n}{2} \rfloor$.

ANSWERS TO TRUE-FALSE

1. T 2. T 3. T 4. F 5. F 6. F 7. F 8. F 9. F 10. F 11. T
12. T 13. F 14. T 15. T 16. T 17. T 18. F 19. T 20. T
21. F 22. F 23. F 24. T 25. F 26. F 27. T 28. F 29. T
30. F 31. T 32. F 33. F 34. F 35. F 36. T 37. F 38. T
39. F 40. T 41. F 42. T 43. T 44. F 45. T 46. T 47. T
48. F 49. T 50. T

CHAPTER 5
Algorithms and Recursion

5.1 The "for" Procedure and Algorithms for Matrices

An algorithm is a set of instructions about how to perform a task. The sets of instructions that we get with appliances telling us how to use or install them, though usually unintelligible, are algorithms. A computer needs a set of algorithms to tell it how to perform tasks. A computer program is an algorithm. It is desirable for an algorithm to be as general as possible. For example, if an algorithm tells us how to multiply two matrices together, it is important that it work for any two matrices that can be multiplied. As we shall soon see, many algorithms contain a repetitive process. Again this is true of matrices, where the multiplication of two $n \times n$ involves repeating a process n^2 times to find all of the elements in the product. The computer is especially valuable in performing tasks whose algorithm has this repetitive process.

The summation notation introduced earlier is actually a form of algorithm. The sum $\sum_{i=1}^{n} p(i)$ tells us to calculate $p(i)$ when $i = 1$, then let $i = 2$, calculate $p(i)$, and add it to the previous sum. Continue adding 1 to i until $i = n$ and repeat the previous process. In a similar manner we introduce the "for" procedure. In the following set of instructions,

> For $r = 1$ to n
>   Step (1)
>   Step (2)
> $\vdots$
>   Step (m)
> Endfor

let $r = 1$ and perform step (1) through step (m). We then let $r = 2$ and repeat the process. After completing the process for $r = k$, we let $r = k + 1$ and repeat the process until we have completed the process when $r = n$.

Thus, we could compute $S = \sum_{i=1}^{n} p(i)$ as follows:

Let $S = 0$.
    For $i = 1$ to n
        Replace the value of S with that of $S + p(i)$.
    Endfor

At the end, S has the value $\sum_{i=1}^{n} p(i)$.

EXAMPLE 5.1

Calculate $M = r^n$.

Let $M = 1$.
For $i = 1$ to n
    Replace the value of M with that of $M * r$.
Endfor ∎

EXAMPLE 5.2

Calculate $p(a)$, where $p(x)$ is a polynomial using Horner's polynomial evaluation algorithm. We first observe that

$$a_4x^4 + a_3x^3 + a_2x^2 + a_1x + a_0 = x(x(x(a_4x + a_3) + a_2) + a_1) + a_0$$

Using this example, we develop the following algorithm that evaluates $S = p(a)$, where $p(x) = a_nx^n + a_{n-1}x^{n-1} + \cdots + a_2x^2 + a_1x + a_0$:

Let $S = a_n$.
For $i = 1$ to n
    Replace S with $aS + a_{n-i}$.
Endfor ∎

DEFINITION 5.3

Given partial orderings, $\leq_1$ and $\leq_2$, the ordering $\leq_2$ is a **topological sorting** of the ordering $\leq_1$ if $\leq_2$ is a total ordering and if whenever $a \leq_1 b$, then $a \leq_2 b$.

At this point we will introduce a rather simple algorithm for creating a topological ordering of an ordering $\leq_1$ on a set S. The algorithm basically tells us to pick any minimal element of our partial ordering and consider it as the least element on our total ordering. Then pick the another minimal element and place it above the first element. Continue doing this until all the elements have been selected. Since we are always picking a minimal element, if $a \leq b$ in the original ordering, this will still be true in the new ordering, because a must be selected first. We now state our algorithm formally:

Procedure Topological Sort $(S, \leq_1, \leq_2)$
    Pick a minimal element s of $(S, \leq_1)$.
    Remove s from S.
    Perform the following steps until S is empty.
        Pick a minimal element t of $(S, \leq_1)$ and remove it.
        Let $s \leq_2 t$.
        Let $s = t$.
    end.
End.

If $A = [A_{ij}]$ is an $m \times n$ matrix, then the scalar product $[B_{ij}] = a[A_{ij}]$ was defined in Chapter 2 to be equal to $[aA_{ij}]$. We can describe this algorithm for this product as follows:

Procedure Scalar(a, A, m, n)
    For $i = 1$ to m
        For $j = 1$ to n
            $B_{ij} = aA_{ij}$
        Endfor
    Endfor
End

We have given the procedure the name scalar and indicated that the input is the scalar a and the $m \times n$ matrix A.

If $A = [A_{ij}]$ and $B = [B_{ij}]$ are $m \times n$ matrices, then $A + B$ was defined in Chapter 2 to be the $m \times n$ matrix $C = [C_{ij}]$ where $C_{ij} = A_{ij} + B_{ij}$. We can describe this algorithm for adding two $m \times n$ matrices A and B as follows:

Procedure Matrixadd(A, B, m, n)
    For $i = 1$ to m
        For $j = 1$ to n
            $C_{ij} = A_{ij} + B_{ij}$
        Endfor
    Endfor
End

Recall that if

$$A = \begin{bmatrix} A_{11} & A_{12} & \cdots & A_{1p} \\ A_{21} & A_{22} & \cdots & A_{2p} \\ \vdots & & & \vdots \\ A_{n1} & A_{n2} & \cdots & A_{np} \end{bmatrix}$$

is an $n \times p$ matrix and

$$B = \begin{bmatrix} B_{11} & B_{12} & \cdots & B_{1m} \\ B_{21} & B_{22} & \cdots & B_{2m} \\ \vdots & & & \vdots \\ B_{p1} & B_{p2} & \cdots & B_{pm} \end{bmatrix}$$

is an $p \times m$ matrix, then the matrix product of A and B, denoted by AB, is the $n \times m$ matrix $C = [C_{ij}]$, where C_{ij} is the dot product of the ith row of A and the jth column of B; that is,

$$C_{ij} = \begin{bmatrix} A_{i1} & A_{i2} & A_{i3} & \cdots & A_{ip} \end{bmatrix} \bullet \begin{bmatrix} B_{1j} \\ B_{2j} \\ B_{3j} \\ \vdots \\ B_{pj} \end{bmatrix} = \sum_{k=1}^{p} A_{ik} B_{kj}$$

We can describe this algorithm for multiplying an $n \times p$ matrix A and $p \times m$ matrix B as follows:

Procedure MatrixMult (A, B, m, p, n)
    For $i = 1$ to m
        For $j = 1$ to n
            $C_{ij} = 0$
            For $k = 1$ to p
                The value of C_{ij} is replaced by the value of $C_{ij} + A_{ik}B_{kj}$
            Endfor
        Endfor
    Endfor
End

Exercises

1. Write an algorithm to find the transpose of a matrix.

Calculate:

2. $\begin{bmatrix} 1 & -1 \\ 3 & 4 \\ 0 & -5 \end{bmatrix} \begin{bmatrix} 3 & 0 & -1 & 2 \\ 3 & 1 & 5 & 0 \end{bmatrix}$

3. $\begin{bmatrix} 1 & -1 & 2 \\ 3 & 1 & 4 \\ 4 & -1 & 5 \end{bmatrix} \begin{bmatrix} 2 & -2 & 3 \\ 5 & 3 & 1 \\ 0 & 0 & 4 \end{bmatrix}$

4. $\begin{bmatrix} 1 \\ -2 \\ 3 \\ -5 \end{bmatrix} \begin{bmatrix} 4 & 3 & -5 & 0 \end{bmatrix}$

5. $\begin{bmatrix} -4 & 3 & -5 & 0 \end{bmatrix} \begin{bmatrix} 1 \\ 2 \\ 4 \\ 3 \end{bmatrix}$

6. $(-4) \begin{bmatrix} -3 & 1 \\ 2 & -4 \end{bmatrix}$

7. $\begin{bmatrix} -1 & 3 \\ 3 & 2 \end{bmatrix} - 3 \begin{bmatrix} 1 & 4 \\ -2 & 1 \end{bmatrix}$

Let $A = \begin{bmatrix} -2 & 4 & 3 \\ 1 & 1 & 7 \\ 2 & -4 & 1 \end{bmatrix}$. Calculate:

8. A^t 9. AA^t 10. A^tA

Let $A = \begin{bmatrix} 1 & 2 \\ 4 & 5 \end{bmatrix}$ and $B = \begin{bmatrix} 4 & 1 \\ -2 & -3 \end{bmatrix}$. Calculate:

11. AB 12. BA 13. AB^t 14. A^tB

15. $(AB)^t$ 16. B^tA^t 17. $A - B$ 18. $B - A$

19. $5A$ 20. $2B - 3A$

Let $A = \begin{bmatrix} 1 & 0 & -5 \\ 2 & -4 & 0 \end{bmatrix}$ and $B = \begin{bmatrix} -1 & 2 \\ 3 & 4 \\ 6 & 0 \end{bmatrix}$. Calculate the following:

21. AB 22. BA 23. A^t

24. B^t 25. A^tB^t 26. $(BA)^t$

27. Construct an algorithm for finding the largest integer in a sequence.

28. Construct an algorithm for finding the smallest integer in a sequence.

29. Construct an algorithm for finding the smallest and largest integers in a sequence.

30. Construct an algorithm that determines if a matrix A is a symmetric matrix.

31. Using the representation matrix of a relation, construct an algorithm to determine if the relation is reflexive.

32. Construct an algorithm for finding the average of n numbers.

33. Write an algorithm for adding two integers.

34. Write an algorithm for subtracting two integers

35. Write an algorithm for multiplying two integers.

36. Write an algorithm for dividing an integer by another integer.

37. Write an algorithm for adding two integers in binary notation.

38. Write an algorithm for multiplying two integers in binary notation.

39. Given two propositions p and q and their truth tables, construct an algorithm to find the truth table of $p \wedge q$.

40. Given two propositions p and q and their truth tables, construct an algorithm to find the truth table of $p \vee q$.

41. Given two propositions p and q and their truth tables, construct an algorithm to find the truth table of $p \rightarrow q$.

42. Given two propositions p and q and their truth tables, construct an algorithm to find the truth table of $p \leftrightarrow q$.

5.2 Recursive Functions and Algorithms

Although it is usually preferable to define a function directly in terms of the variable, it is sometime convenient or even necessary to use a method called recursion. Because of the principle of induction, we know that if

(a) A function is defined for a given starting value a (usually 0 or 1) and

(b) If, when it is defined for a value k greater than a, it can then be defined for the value $k + 1$,

then the function can be defined for all integers greater than a.

It is probably more accurate to state that the function can be defined recursively rather than to state that the function is a recursive function since many functions can be defined either recursively or directly.

Note that this form of recursive definition is used for functions defined on the positive integers. Later a more general form of recursive definition is discussed. Also note that it is not always obvious that a recursive function has been defined. For example, consider the function defined by

$$\begin{cases} f(0) = 2 \\ f(k+1) = \dfrac{f(k)!}{(k+1)!} \end{cases}$$

It may appear that we have satisfied parts (a) and (b) previously so that we have defined a recursive function for all positive integers. We see, however, that $f(1) = 2$, $f(2) = 1$, $f(3) = \frac{1}{6}$, but $f(4)$ is undefined, since $\left(\frac{1}{6}\right)!$ has no meaning.

Similarly if an algorithm can be performed for a fixed value a and when the algorithm can be performed for the value k greater than a, it can then be performed for the value $k + 1$. The algorithm can then be performed for all values $n \geq a$.

We begin with examples of recursive functions.

Consider the function $f(n) = a^n$ defined on the set of nonnegative integers. It may be defined as the product of n values of a. It may also be defined recursively as follows:

$$f(0) = 1$$
$$f(k+1) = a \cdot f(k)$$

EXAMPLE 5.4 Consider the factorial function, $f(n) = n!$. It is easy to express the function by

$$n! = 1 \cdot 2 \cdot 3 \cdots n$$

but it is hard to explain to a computer that may be performing the calculation what the ellipsis or three dots means. It is, of course, possible to write a procedure or algorithm that directly computes $n!$. A description of such an algorithm could be

    Procedure Factorial(n)
        If $n = 0$ then Factorial(n) = 1.
        Let Factorial(n) = 1
        For $k = 1$ to n
            The value of Factorial(n) is replaced by the value
            of $k \cdot$ Factorial(n).
        Endfor
    End

If we write the factorial function recursively, it would take the form

$$\text{Factorial}(0) = 1$$
$$\text{Factorial}(k+1) = (k+1) \cdot \text{Factorial}(k)$$

A description of such a computer program could be

    Procedure Factorial(n)
        If $n = 0$ then Factorial(n) = 1.
        If $n > 0$ then Factorial(n) = $n \cdot$ Factorial($n - 1$).
    End

Notice that in this program the procedure Factorial(n) calls itself, which is allowed in many computer languages. When the procedure Factorial(n) is called for $n \geq 1$, it proceeds until it gets to Factorial($n - 1$). It then stops the program (storing all information in stacks until the program can be resumed) and calls Factorial($n - 1$).

If $n - 1 \geq 0$, then Factorial$(n - 1)$ proceeds until it gets to Factorial$(n - 2)$ in the program. It then stops the program (storing all information in stacks until the program can be resumed) and calls Factorial$(n - 2)$. This continues until Factorial(0) is reached and evaluated. This enables the continuation of Factorial(1), which is evaluated, and in turn enables the continuation of Factorial(2), which is evaluated. This process continues until Factorial$(n - 1)$ is evaluated and finally Factorial(n) is evaluated. ∎

EXAMPLE 5.5

Another function usually described recursively is the Fibonacci sequence. The first element in the sequence is 1, the second element is 1, and each succeeding element of the sequence is the sum of the two previous terms. Hence, the first 10 terms of the sequence are 1, 1, 2, 3, 5, 8, 13, 21, 34, and 55.

The following procedure describes the program for calculating Fib(n), the value of the nth term in the sequence:

> Procedure Fib(n)
> If $n = 1$, then Fib$(n) = 1$.
> If $n = 2$, then Fib$(n) = 1$.
> If $n > 2$, then Fib$(n) =$ Fib$(n - 1) +$ Fib$(n - 2)$.
> End

Although easily stated, this procedure is very inefficient since at each stage k it calculates Fib$(k - 1)$ and starts over to calculate Fib$(k - 2)$, ignoring the fact that it already calculated Fib$(k - 2)$ in calculating Fib$(k - 1)$. It is instructive to go through the process and hand calculate Fib(n) following the preceding procedure for some small value of n, such as 5 or 6. ∎

EXAMPLE 5.6

The sequence of **Catalan numbers** is given by the formula

$$\text{Cat}(n) = \frac{(2n)!}{(n+1)!(n!)}$$

If we begin our sequence with the 0th term, the first 11 elements in the sequence are 1, 1, 2, 5, 14, 42, 132, 429, 1430, 4862, and 16796. The Cat function can be defined recursively by

$$\text{Cat}(0) = 1$$

$$\text{Cat}(n+1) = \frac{2(2n+1)}{(n+2)} \text{Cat}(n)$$

To prove this we first evaluate $\text{Cat}(0) = \dfrac{0!}{1!0!} = 1$. We next show that

$$\frac{2(2n+1)}{(n+2)} \text{Cat}(n) = \text{Cat}(n+1)$$

Simplifying the left side of the equation, we get

$$\frac{2(2n+1)}{(n+2)} \text{Cat}(n) = \frac{2(2n+1)}{(n+2)} \cdot \frac{(2n)!}{(n+1)!(n!)}$$

$$= \frac{(n+1)2(2n+1)(2n)!}{(n+1)(n+2)!(n!)}$$

$$= \frac{(2n+2)(2n+1)(2n)!}{(n+2)!(n+1)(n!)}$$

$$= \frac{(2n+2)!}{(n+2)!(n+1)!}$$

$$= \text{Cat}(n+1)$$

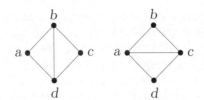

Figure 5.1

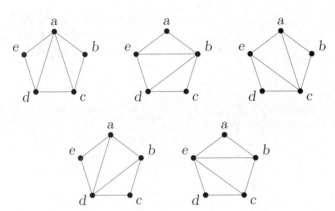

Figure 5.2

At this point we mention two cases where the Catalan numbers appear.

1. Given an $(n + 2)$-sided convex polygon, the number of ways that the polygon can be cut into triangles by connecting the corners of the polygon with $n - 1$ intersecting straight lines is equal to Cat(n). Given a three-sided figure ($n = 1$), there is only one way to form a triangle using 0 lines and Cat$(1) = 1$. Given a four-sided figure ($n = 2$), we see by Figure 5.1 that there are only two ways this can be done using one line.

   We note that Cat$(2) = 2$. Given a five-sided figure ($n = 3$), there are five ways that the triangle can be divided using three lines as shown in Figure 5.2.
   In this case we have Cat$(3) = 5$.

2. The number of ways that n pairs of parentheses can be inserted into the expression $a_1 a_2 a_3 \cdots a_n a_{n+1}$ to form n products of pairs of numbers from the factors is equal to Cat(n). When $n = 1$, we have only $(a_1 a_2)$ or one way. When $n = 2$, we have $((a_1 a_2) a_3)$ and $(a_1 (a_2 a_3))$ so there are two ways. When $n = 3$, we have $(((a_1 a_2) a_3) a_4)$, $(a_1 (a_2 (a_3 a_4)))$, $((a_1 a_2)(a_3 a_4))$, $((a_1 (a_2 a_3)) a_4)$, and $(a_1 ((a_2 a_3) a_4))$ so there are five ways. ∎

EXAMPLE 5.7

Another important function is Ackermann's function denoted by Ack $: N \times N \rightarrow N$. It is defined recursively on $N \times N$, where N is the set of nonnegative integers, by

$$\text{Ack}(0, n) = n + 1$$
$$\text{Ack}(m, 0) = \text{Ack}(m - 1, 1) \text{ if } m > 0$$
$$\text{Ack}(m, n) = \text{Ack}(m - 1, \text{Ack}(m, n - 1)) \text{ if } n, m > 0$$

The procedure could be written as follows:

    Procedure Ack(m, n)
        If $m = 0$, then Ack$(m, n) = n + 1$.
        If $m > 0$ and $n = 0$ then Ack$(m, n) = $ Ack$(m - 1, 1)$.
        If $m > 0$ and $n > 0$ then Ack$(m, n) = $ Ack$(m - 1, \text{Ack}(m, n - 1))$.
    End

Try computing by hand such values as Ack$(3, 2)$. ∎

EXAMPLE 5.8 Another example of a function that may be defined recursively is the function that determines the value of an annuity where an amount P has been invested each period for n periods at a rate of interest i for the length of the period and compounded each period. We begin by noting that if an amount P is invested for a period at a rate of interest i per period expressed as a decimal, then at the end of the period the value of the investment is the principal plus the interest earned. The interest is equal to iP, so that the value of the investment is $P + iP = (1+i)P$. For example if \$1000 is invested for one year at 5% interest so that $i = .05$, then at the end of a year the value of the investment would the \$1000 invested plus $.05 \times \$1000 = \$50$. If no further money is invested, then at the end of the second period, the investment is equal to $(1+i)P + i(1+i)P = (1+i)^2 P$, where $i(1+i)P$ is the interest on $(1+i)P$. At the end of two periods, the initial investment is worth $(1+i)^2 P$. In our example, at the end of two years, the investment of \$1000 is equal to \$1050 plus interest $.05(1050)$. At the end of n years, the initial investment is worth $(1+i)^2 P$.

Now suppose amount P has been invested at the end of each period for n periods at a rate of interest i for the length of the period and compounded each period. The amount P just invested has accumulated no interest and is still worth P. The amount P invested the previous period has accumulated interest for one period and is worth $(1+i)P$. The amount invested a period earlier has accumulated interest for two periods and is worth $(1+i)^2 P$. If we continue we find that the value of the total amount is equal to

$$P + (1+i)P + (1+i)^2 P + \cdots + (1+i)^{n-1} P = P \cdot S(n, i)$$

where $S(n, i) = 1 + (1+i) + (1+i)^2 + \cdots + (1+i)^{n-1}$. But this is a geometric series of the form

$$1 + r + r^2 + r^3 + \cdots + r^{n-1} = \frac{r^n - 1}{r - 1}$$

Hence,

$$S(n, i) = \frac{(1+i)^n - 1}{1 + i - 1} = \frac{(1+i)^n - 1}{i}$$

It is obviously more efficient to compute $S(n, i)$ directly; however, we can also compute it recursively. Observe that

$$S(n+1, i) = 1 + (1+i) + (1+i)^2 + \cdots + (1+i)^{n-1} + (1+i)^n$$
$$= 1 + (1+i)(1 + (1+i) + (1+i)^2 + \cdots + (1+i)^{n-1})$$
$$= 1 + (1+i)S(n, i)$$

Hence, $S(n, i)$ can be defined recursively by

$$S(1, i) = 1$$
$$S(n, i) = 1 + (1+i)S(n-1, i) \text{ for } n > 1$$ ∎

At this point we consider a recursive algorithm that is not a recursive function. The Tower of Hanoi consists of three pegs, a collection of disks, and a myth. In the game or puzzle, there are three pegs. There are disks, each a different size, placed on one of the pegs in descending order, with the largest disk on the bottom and the smallest on top. The object of the puzzle is to place all the disks on another peg in the same order by moving them one at a time on the various pegs and not placing a larger disk on top of a smaller disk. This puzzle is obviously recursive since if, for example, there are six disks on the peg and I know how to move five of them, I can move the top five to the middle peg, move the last disk to the unoccupied peg, and

176 Chapter 5 Algorithms and Recursion

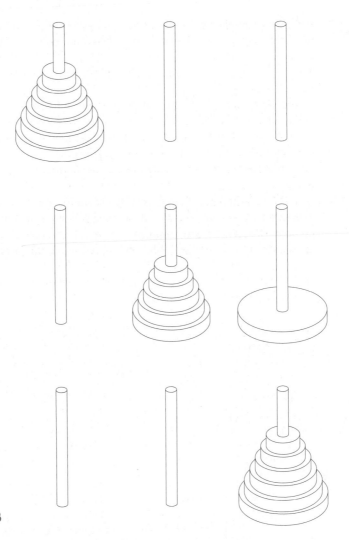

Figure 5.3

then move the five from the middle peg on top of the largest disk. The situation is depicted in Figure 5.3.

The following is an algorithm for moving n disks from the first peg to the third peg. The peg will be numbered from 1 to 3. They will also be labeled A, B, and C where these variables will assume the values from 1 to 3.

    Procedure Hanoi(A, C, n)
        If $n = 1$, Move the single disk from peg A to peg C.
        Let $B = 6 - A - C$.
        If $n > 1$
            Use Hanoi($A, B, n - 1$) to move the top $n - 1$ disks to peg B.
            Use Hanoi($A, C, 1$) to move the last disk from peg A to peg C
            Use Hanoi($B, C, n - 1$) to move the top $n - 1$ disks to peg C
        Endif
    End

For example, if $n = 3$, begin with Hanoi(1, 3, 3) so $A = 1$, $B = 2$, and $C = 3$. Within Hanoi(1, 3, 3), first call Hanoi(1, 2, 2) to move the top two disks to peg 2. In Hanoi(1, 2, 2), $A = 1$, $B = 3$, and $C = 2$. Within Hanoi(1, 2, 2), first call Hanoi(1, 3, 1) to move the top disk from peg 1 to peg 3. Then call Hanoi(1, 2, 1) to move the second disk to peg 2. Then call Hanoi(1, 3, 1) to move the second disk from peg 1 to peg 3. Now call Hanoi(3, 2, 1) to move the top disk from peg 2 onto the second disk on peg 3. Return to Hanoi(1, 3, 3) and call Hanoi(1, 3, 1) to remove the bottom disk from peg 3 to peg 1. Next call Hanoi(2, 3, 2) to move the two disks from peg 2 to peg 3. At this point, $A = 2$, $B = 1$, and $C = 3$. Within Hanoi(2, 3, 2), call Hanoi(2, 1, 1) to move the top disk from peg 2 to peg 1. Next call Hanoi(2, 3, 1) to move the second disk from peg 2 to peg 3. Finally call Hanoi(1, 3, 1) to move the top disk from peg 1 onto peg 3. This completes Hanoi(2, 3, 2) and Hanoi(1, 3, 3) so we are finished.

The ancient legend (begun in the late nineteenth century) states that there were 64 golden disks on one of three diamond pins and that they were being moved using the foregoing rules to a second pin by monks at the rate of one move per second. According to the legend, when the mission is completed, the world will end. The question is, how long will it take? We now develop a function that gives the number of moves (and, hence, the number of seconds) needed to transfer n disks from one peg to another. Call the function H. We know that if $n = 1$, it takes only one move to transfer it from one peg to another. Hence, $H(1) = 1$. Assume $H(k-1)$ is known; then $H(k)$, the number of moves to transfer k disks from peg 1 to peg 3, involves transferring $k-1$ disks from peg 1 to peg 2, which takes $H(k-1)$ moves, then moving the bottom disk from peg 1 to peg 3, using one move, and then transferring $k-1$ disks from peg 2 to peg 3, which takes $H(k-1)$ moves. Thus, $H(k) = 2H(k-1) + 1$ and we have our recursive function. We would like to change this to a direct function if possible. To do this we determine the first few values of H and try to find a pattern:

$$H(1) = 1 \qquad\qquad = 1$$
$$H(2) = 2 \cdot 1 + 1 \ \ = 3$$
$$H(3) = 2 \cdot 3 + 1 \ \ = 7$$
$$H(4) = 2 \cdot 4 + 1 \ \ = 15$$
$$H(5) = 2 \cdot 15 + 1 = 31$$

If we add 1 to each of the sequence 1, 3, 7, 15, 31, we get 2, 4, 8, 16, 32. Hence, it appears that $H(n) = 2^n - 1$. We prove this using induction. Obviously the statement is true for $n = 1$. Assume it is true for $n = k$; that is, $H(k) = 2^k - 1$. We wish to prove that it is true for $n = k + 1$; that is, $H(k+1) = 2^{k+1} - 1$,

$$H(k + 1) = 2 \cdot H(k) + 1$$
$$= 2 \cdot (2^k - 1) + 1$$
$$= 2^{k+1} - 1$$

so our guess is true. The reader is asked to determine when the world will end according to the legend.

In the previous discussion, we changed the description of a function from recursive to direct. This procedure is also referred to as **solving** the recursive function. This was done by evaluating the function for the first few values in its domain and then seeking a pattern. We shall discuss more of these examples. We shall also change the description of a function from direct to recursive. We shall also solve the function by evaluating the function and seeking a pattern.

EXAMPLE 5.9   Solve the recursive function

$$f(1) = 1$$
$$f(k) = f(k-1) + k$$

Since

$$f(1) = 1 \qquad = \frac{1 \cdot 2}{2}$$

$$f(2) = 1 + 2 \qquad = \frac{2 \cdot 3}{2}$$

$$f(3) = (1+2) + 3 \qquad = \frac{3 \cdot 4}{2}$$

$$f(4) = (1+2+3) + 4 \qquad = \frac{4 \cdot 5}{2}$$

$$f(5) = (1+2+3+4) + 5 = \frac{5 \cdot 6}{2}$$

it appears that the function may be

$$f(n) = 1 + 2 + 3 + \cdots + n = \frac{n(n+1)}{2}$$

To prove this formula holds we must show that this function satisfies the definition of the recursive function; that is, we show that the function $f(n) = \frac{n(n+1)}{2}$ satisfies

$$f(1) = 1$$
$$f(k) = f(k-1) + k$$

We first show that $f(1) = 1$:

$$f(1) = \frac{1 \cdot 2}{2} = 1$$

Next we show that $f(k) = f(k-1) + k$:

$$f(k-1) + k = \frac{k(k-1)}{2} + k$$

$$= \frac{k^2 - k}{2} + \frac{2k}{2}$$

$$= \frac{k^2 + k}{2}$$

$$= \frac{k(k+1)}{2}$$

$$= f(k)$$

Thus, $f(n) = \frac{n(n+1)}{2}$ satisfies the given recursive function. ∎

EXAMPLE 5.10

Solve the recursive function

$$f(1) = 2$$
$$f(k) = 2 \cdot k \cdot f(k-1)$$

5.2 Recursive Functions and Algorithms

Since

$$f(1) = 2 \qquad\qquad = 2^1 \cdot 1!$$
$$f(2) = 2 \cdot 2 \cdot 2 \qquad\qquad = 2^2 \cdot 2!$$
$$f(3) = 2 \cdot 2 \cdot 2 \cdot 2 \cdot 3 \qquad\qquad = 2^3 \cdot 3!$$
$$f(4) = 2 \cdot 2 \cdot 2 \cdot 2 \cdot 3 \cdot 2 \cdot 4 \qquad\qquad = 2^4 \cdot 4!$$
$$f(5) = 2 \cdot 2 \cdot 2 \cdot 2 \cdot 3 \cdot 2 \cdot 4 \cdot 2 \cdot 5 = 2^5 \cdot 5!$$

the function appears to be $f(n) = 2^n \cdot n!$. To prove this we must show that this function satisfies the definition of the recursive function. That is, we show that the function $f(n) = 2^n \cdot n!$ satisfies

$$f(1) = 2$$
$$f(k) = 2kf(k-1)$$

We first show that $f(1) = 2$:

$$f(1) = 2^1 \cdot 1! = 2$$

Next we show that $f(k) = 2kf(k-1)$. But

$$2kf(k-1) = 2k(2^{k-1} \cdot (k-1)!)$$
$$= 2(2^{k-1}) \cdot k(k-1)!$$
$$= 2^k \cdot k!.$$

and $f(n) = 2^n \cdot n!$ satisfies the given recursive function. ∎

EXAMPLE 5.11 Given the function $f(n) = \dfrac{n(n+1)(2n+1)}{6}$, express it as a recursive function.

$$f(1) = 1 \ = 1^2$$
$$f(2) = 5 \ = 1^2 + 2^2$$
$$f(3) = 14 = 1^2 + 2^2 + 3^2$$
$$f(4) = 30 = 1^2 + 2^2 + 3^2 + 4^2$$
$$f(5) = 55 = 1^2 + 2^2 + 3^2 + 4^2 + 5^2$$

and the function appears to be

$$f(1) = 1$$
$$f(k) = f(k-1) + k^2$$

To show this we use induction. Certainly $f(1) = 1$. Assume $f(k-1) = \dfrac{(k-1)(k)(2k-1)}{6}$; then we need to show that $f(k) = \dfrac{k(k+1)(2k+1)}{6}$:

$$f(k-1) + k^2 = \dfrac{(k-1)(k)(2k-1)}{6} + k^2$$
$$= \dfrac{(k-1)(k)(2k-1) + 6k^2}{6}$$

$$= k\frac{(k-1)(2k-1)+6k}{6}$$
$$= k\frac{2k^2-4k+1+6k}{6}$$
$$= k\frac{(k+1)(2k+1)}{6}$$
$$= f(k)$$ ∎

Exercises

Find $f(1)$, $f(2)$, $f(3)$, and $f(4)$ for the following recursive functions:

1. $\begin{cases} f(0) = 3 \\ f(k) = 3f(k-1) \quad k > 0 \end{cases}$

2. $\begin{cases} f(0) = 1 \\ f(k) = 2f(k-1) + 3k \quad k > 0 \end{cases}$

3. $\begin{cases} f(0) = 2 \\ f(k) = (f(k-1))^2 \quad k > 0 \end{cases}$

4. $\begin{cases} f(0) = 1 \\ f(k) = k^2 f(k-1) \quad k > 0 \end{cases}$

5. $\begin{cases} f(0) = 1 \\ f(k) = 2^{f(k-1)} \quad k > 0 \end{cases}$

6. $\begin{cases} f(0) = 0 \\ f(k) = k + f(k-1) \quad k > 0 \end{cases}$

7. $\begin{cases} f(0) = 1 \\ f(k) = \left\lfloor \dfrac{f(k-1)}{2} \right\rfloor + 3k \quad k > 0 \end{cases}$

8. $\begin{cases} f(0) = 2 \\ f(k) = 8k^2 - (f(k-1))^2 \quad k > 0 \end{cases}$

9. $\begin{cases} f(0) = 4 \\ f(k) = \dfrac{f(k-1)}{k^2} \quad k > 0 \end{cases}$

10. $\begin{cases} f(0) = 2 \\ f(k) = \dfrac{f(k-1)}{k!} \quad k > 0 \end{cases}$

Find $f(2)$, $f(3)$, $f(4)$, and $f(5)$ for the following recursive functions:

11. $\begin{cases} f(0) = 1 \\ f(1) = 3 \\ f(k) = 2f(k-1) - f(k-2) \quad k > 1 \end{cases}$

12. $\begin{cases} f(0) = 0 \\ f(1) = 1 \\ f(k) = (f(k-1))^2 - f(k-2)^2 \quad k > 1 \end{cases}$

13. $\begin{cases} f(0) = 1 \\ f(1) = 2 \\ f(k) = (f(k-1))^2 - f(k-2) + k^2 \quad k > 1 \end{cases}$

14. $\begin{cases} f(0) = 1 \\ f(1) = 2 \\ f(k) = (f(k-1) - f(k-2))k! \quad k > 1 \end{cases}$

15. $\begin{cases} f(0) = -1 \\ f(1) = 1 \\ f(k) = f(k-1) \div f(k-2) \quad k > 1 \end{cases}$

16. $\begin{cases} f(0) = 2 \\ f(1) = 4 \\ f(k) = 3f(k-1) - 2f(k-2) \quad k > 1 \end{cases}$

17. $\begin{cases} f(0) = 1 \\ f(1) = 2 \\ f(k) = (f(k-1))! - (f(k-2))! \quad k > 1 \end{cases}$

18. $\begin{cases} f(0) = 0 \\ f(1) = 2 \\ f(k) = (f(k-1))! \div (f(k-2))! \quad k > 1 \end{cases}$

19. $\begin{cases} f(0) = 10 \\ f(1) = 20 \\ f(k) = \left\lfloor \dfrac{f(k-1) + f(k-2)}{k!} \right\rfloor \quad k > 1 \end{cases}$

20. $\begin{cases} f(0) = -1 \\ f(1) = 1 \\ f(k) = f(k-1) \div (f(k-2))^2 \quad k > 1 \end{cases}$

Using pattern recognition, solve the recursive functions by finding explicit nonrecursive expressions for $f(n)$:

21. $\begin{cases} f(0) = 1 \\ f(k) = 2f(k-1) \quad k > 0 \end{cases}$

22. $\begin{cases} f(0) = 1 \\ f(k) = 2 + f(k-1) \quad k > 0 \end{cases}$

23. $\begin{cases} f(0) = 1 \\ f(k) = f(k-1) \div k \quad k > 0 \end{cases}$

24. $\begin{cases} f(0) = 1 \\ f(k) = f(k-1) + 2 \quad k > 0 \end{cases}$

25. $\begin{cases} f(0) = 1 \\ f(k) = 5f(k-1) \quad k > 0 \end{cases}$

26. $\begin{cases} f(0) = 2 \\ f(k) = \dfrac{f(k-1)}{k} \quad k > 0 \end{cases}$

27. $\begin{cases} f(0) = 1 \\ f(k) = 2 + f(k-1) \quad k > 0 \end{cases}$

28. $\begin{cases} f(0) = -1 \\ f(k) = \dfrac{-1}{f(k-1)} \quad k > 0 \end{cases}$

29. $\begin{cases} f(0) = 1 \\ f(k) = -3f(k-1) \quad k > 0 \end{cases}$

30. $\begin{cases} f(0) = 1 \\ f(k) = 5 + 2f(k-1) \quad k > 0 \end{cases}$

31. $\begin{cases} f(0) = -1 \\ f(1) = 1 \\ f(k) = \dfrac{f(k-1)}{f(k-2)} \quad k > 1 \end{cases}$

32. Prove that $a_n = 7 - 2^{n+1}$ satisfies the recursive function
$$\begin{cases} a_0 = 5 \\ a_k = 2a_{k-1} - 7 \text{ for } k > 0 \end{cases}$$

33. Prove that $a_n = 3^n - n3^{n+1}$ satisfies the recursive function
$$\begin{cases} a_0 = 1 \\ a_1 = -6 \\ a_k = 6a_{k-1} - 9a_{k-2} \text{ for } k > 1 \end{cases}$$

34. Prove that $a_n = -1 - 2^{n+1}$ satisfies the recursive function
$$\begin{cases} a_0 = -3 \\ a_1 = -5 \\ a_k = 6a_{k-1} - 8a_{k-2} - 3 \text{ for } k > 1 \end{cases}$$

35. Prove that
$$a_n = \frac{1 - r^{n+1}}{1 - r} \quad r \neq 1$$
satisfies the recursive function
$$\begin{cases} a_0 = 1 \\ a_k = a_{k-1} + r^k \text{ for } k > 0 \end{cases}$$

36. Prove that
$$a_n = (n^2 + n + 1)$$
satisfies the recursive function
$$\begin{cases} a_1 = 3 \\ a_k = a_{k-1} + 2k \text{ for } k > 1 \end{cases}$$

37. Prove that
$$a_n = n^2(n+1)^2$$
satisfies the recursive function
$$\begin{cases} a_1 = 4 \\ a_k = a_{k-1} + 4k^3 \text{ for } k > 1 \end{cases}$$

38. Prove that
$$a_n = -2(-1)^n + 2 \cdot 3^n$$
satisfies the recursive function
$$\begin{cases} a_0 = 0 \\ a_1 = 8 \\ a_k = 2a_{k-1} + 3a_{k-2} \text{ for } k > 1 \end{cases}$$

39. Prove that $a_n = 2(-3)^n + 5 \cdot 2^n$ satisfies the recursive function
$$\begin{cases} a_0 = 7 \\ a_1 = 4 \\ a_k = -a_{k-1} + 6a_{k-2} \text{ for } k > 1 \end{cases}$$

40. Prove that $a_n = 3(-2)^n + 2 \cdot 3^n - 3 \cdot (2)^n$ satisfies the recursive function
$$\begin{cases} a_0 = 2 \\ a_1 = -6 \\ a_k = a_{k-1} + 6a_{k-2} + 3 \cdot (2)^k \text{ for } k > 1 \end{cases}$$

41. Prove that
$$a_n = \frac{1}{\sqrt{5}}\left(\left(\frac{1+\sqrt{5}}{2}\right)^n - \left(\frac{1-\sqrt{5}}{2}\right)^n\right)$$
satisfies the recursive function
$$\begin{cases} a_1 = 1 \\ a_2 = 1 \\ a_k = a_{k-1} + a_{k-2} \text{ for } k > 2 \end{cases}$$

5.3 Complexity of Algorithms

In the discussion of the Tower of Hanoi it was shown that it requires $2^n - 1$ individual moves of disks to transfer n disks from one peg to another. It is often important to be able to measure the number of elementary operations a computer requires or the time a computer requires to run an algorithm. These are not necessarily the same thing since some operations may take longer than others to perform. Also other factors must often be considered such as space required on the computer to run the algorithm, accuracy, and ease of programming. Obviously the real concern is with

182 Chapter 5 Algorithms and Recursion

programs that require a significant amount of time to run. Often the amount of time required is based upon the size of some factor such as the amount of data, or as before, the number of disks, where the running time or number of operations required as a function of this factor increases as the size n of the factor increases. We begin by comparing the relative rate at which some functions increase as n increases.

DEFINITION 5.12 Let f and g be functions whose domain is the positive integers and range is the real numbers. The function g **dominates** the function f if there is a real number k and a positive integer m such that $|f(n)| \leq k|g(n)|$ for all $n \geq m$. If g dominates f, this is denoted by $f(n) = O(g(n))$. The symbol $O(g(n))$ is read "**big-Oh**" of $g(n)$ and $f(n)$ is said to have order big-Oh of $g(n)$.

DEFINITION 5.13 Let f and g be functions whose domain is the positive integers and range is the real numbers. The function g is **dominated** by the function f if there is a real number k and a positive integer m such that $|f(n)| \leq k|g(n)|$ for all $n \geq m$. If g is dominated by f, this is denoted by $f(n) = \Omega(g(n))$. The symbol $\Omega(g(n))$ is read "**big-Omega**" of $g(n)$ and $f(n)$ is said to have order **big-Omega** of $g(n)$. If $f(n) = O(g(n))$ and $f(n) = \Omega(g(n))$, then $f(n) = \Theta(g(n))$ and $f(n)$ is said to have order **big-Theta** of $g(n)$.

If r and s are positive integers, then the following theorem can be shown using induction. It can then be shown that the theorem is true when r and s are positive rational numbers without using logarithms (see exercises).

THEOREM 5.14 If r and s are real numbers, $r \leq s$ and $n \geq 1$, then $n^r \leq n^s$. Hence, $n^r = O(n^s)$.

Proof Since the natural logarithm function is an increasing function, $a \leq b$ if and only if $\ln(a) \leq \ln(b)$. Hence, $n^r \leq n^s$ if and only if $\ln(n^r) \leq \ln(n^s)$ if and only if $r \ln(n) \leq s \ln(n)$ if and only if $r \leq s$ since $\ln(n)$ is positive when $n > 1$. ∎

The following theorems show that the property of being equal to $O(g(n))$ is closed under addition and under multiplication by a scalar.

THEOREM 5.15 If $f(n) = O(g(n))$, then $cf(n) = O(g(n))$.

Proof By definition, $|f(n)| \leq k|g(n)|$ for some constant real number k and all n greater than or equal to some integer m. Therefore,

$$|cf(n)| \leq k|c||g(n)|$$

and $cf(n) = O(g(n))$. ∎

THEOREM 5.16 If $f(n) = O(g(n))$ and $h(n) = O(g(n))$, then $(f+h)(n) = O(g(n))$.

Proof By definition, for some constant k and some integer m_1, $|f(n)| \leq k|g(n)|$ for all $n > m_1$. Also by definition, for some constant l and some integer m_2, $|h(n)| \leq l|e(n)|$ for all $n > m_2$, Let $m = \max(m_1 m_2)$. Hence, for all $n > m$,

$$|f(n) + h(n)| \leq |f(n)| + |h(n)|$$
$$\leq k|g(n)| + l|g(n)|$$
$$= (k+l)|g(n)|$$

∎

and $(f+h)(n) = O(g(n))$.

THEOREM 5.17 If $f(n) = O(g(n))$ and $h(n) = O(e(n))$, then $(f \cdot h)(n) = O((g \cdot e)(n))$.

Proof By definition, for some constant k and some integer m_1, $|f(n)| \leq k|g(n)|$ for all $n > m_1$. Also by definition, for some constant l and some integer m_2, $|h(n)| \leq l|e(n)|$ for all $n > m_2$. Let $m = \max(m_1 m_2)$. Therefore, for all $n > m$,

$$|f(n) \cdot h(n)| = |f(n)| \cdot |h(n)|$$
$$\leq k|g(n)| l|e(n)|$$
$$= kl|g(n) \cdot e(n)|$$

and $(f \cdot h)(n) = O((g \cdot e)(n))$. ∎

In the following theorem we see that the degree of a polynomial uniquely determines its dominance.

THEOREM 5.18 If $p(n) = a_k n^k + a_{k-1} n^{k-1} + \cdots + a_2 n^2 + a_1 n + a_0$, then $p(n) = O(n^k)$.

Proof

$$|p(n)| \leq \left| a_k n^k + a_{k-1} n^{k-1} + \cdots + a_2 n^2 + a_1 n + a_0 \right|$$
$$\leq \left| a_k n^k \right| + \left| a_{k-1} n^{k-1} \right| + \cdots + \left| a_2 n^2 \right| + |a_1 n| + |a_0|$$

by the triangle inequality: $|A + B| \leq |A| + |B|$

$$= |a_k| n^k + |a_{k-1}| n^{k-1} + \cdots + |a_2| n^2 + |a_1| n + |a_0|$$
$$\leq |a_k| n^k + |a_{k-1}| n^k + \cdots + |a_2| n^k + |a_1| n^k + |a_0| n^k$$

by Theorem 5.14

$$= (|a_k| + |a_{k-1}| + \cdots + |a_2| + |a_1| + |a_0|) n^k$$

and $p(n) = O(n^k)$. ∎

Corollary 5.19 If $p(n) = a_k n^k + a_{k-1} n^{k-1} + \cdots + a_2 n^2 + a_1 n + a_0$ and $q(n) = b_k n^k + b_{k-1} n^{k-1} + \cdots + b_2 n^2 + b_1 n + b_0$, then $p(n) = \Theta(q(n))$. ∎

THEOREM 5.20 For integers a, b, and n greater than 1, $\log_a(n) = O(\log_b(n))$.

Proof Follows immediately since $\log_a(n) = \dfrac{\log_b(n)}{\log_b(a)}$. ∎

Corollary 5.21 For integers a, b, and n greater than 1, $\log_a(n) = \Theta(\log_b(n))$. ∎

THEOREM 5.22 If n is a nonnegative integer, then $n < 2^n$; therefore, $n = O(2^n)$.

Proof Using induction, for $n = 0$, $0 < 2^0 = 1$. Assume $k < 2^k$, then

$$k + 1 \leq k + k \leq 2^k + 2^k = 2 \cdot 2^k = 2^{k+1}$$

and by induction $n < 2^n$. ∎

Let R be a relation on the set of positive integer valued functions defined by fRg if $f(n) = O(g(n))$. Obviously R is reflexive. In the following theorem, we see that it is also transitive. The proof of the theorem is left to the reader.

THEOREM 5.23 If $f(n) = O(g(n))$ and $g(n) = O(h(n))$, then $f(n) = O(h(n))$. ∎

In the following theorems we see which functions dominate others.

THEOREM 5.24 For integers a and n greater than 1, $\log_a(n) = O(n)$.

Proof By Theorem 5.22, $n < 2^n$ and so $\log_2(n) < \log_2(2^n) = n$ and $\log_2(n) = O(n)$. Since $\log_a(n) = O(\log_2(n))$, by Theorem 5.20, $\log_a(n) = O(n)$. ∎

The proofs of the following theorems are left to the reader.

THEOREM 5.25 If n is a positive integer, then $n! \leq n^n$ and, therefore, $n! = O(n^n)$. ∎

THEOREM 5.26 If $a > 1$ and n is a positive integer, then $\log_a(n!) \leq n\log_a(n)$ and so $\log_a(n!) = O(n\log_a(n))$. ∎

Previously we found the number of steps required to move the disks for the Tower of Hanoi. In the following examples we determine the number of arithmetic steps required to add two matrices and the number of steps required to multiply two matrices.

EXAMPLE 5.27 If we examine the algorithm for matrix addition

    Procedure MatrixAdd(A, B, m, n)
        For $i = 1$ to m
            For $j = 1$ to n
                $C_{ij} = A_{ij} + B_{ij}$
            Endfor
        Endfor
    End

we see that for each i and for each j, an addition is performed. Since there are m values for i and n values for j, there are mn additions performed. Let $N = \max(m, n)$; then the number of arithmetic operations has order $O(N^2)$. ∎

EXAMPLE 5.28 We can describe an algorithm for multiplying an $m \times p$ matrix A and $p \times n$ matrix B as follows:

    Procedure MatrixMult(A, B, m, p, n)
        For $i = 1$ to m
            For $j = 1$ to n
                $C_{ij} = 0$
                For $k = 1$ to p
                      The value of C_{ij} is replaced by the value
                      of $C_{ij} + A_{ik}B_{kj}$.
                Endfor
            Endfor
        Endfor
    End

As k ranges from 1 to p, there are p additions and p multiplications. Since k ranges from 1 to p for each i and each j, there are mn times when k ranges from 1 to p. Hence, there are mnp additions and mnp multiplications. Therefore, there are $2mnp$ operations. Let $N = \max(m, n, p)$; then the number of arithmetic operations has order $O(N^3)$. ∎

EXAMPLE 5.29 Compare the number of computations performed in calculating $p(c)$, where $p(x) = a_n x^n + a_{n-1} x^{n-1} + \cdots + a_2 x^2 + a_1 x + a_0$ by direct computation and using Horner's polynomial evaluation algorithm.

If $p(c)$ is calculated directly, computing c^k requires $k-1$ multiplications. Multiplication by a_k requires an additional multiplication so computation of $a_k c^k$ requires k multiplications. Hence, there are $1 + 2 + 3 + \cdots + n = \dfrac{n(n+1)}{2}$ multiplications. Also addition of the $n+1$ terms requires n additions so there is a total of $\dfrac{n(n+1)}{2} + n$ arithmetic operations, which is $O(n^2)$.

If we calculate $p(a)$, where $p(x)$ is a polynomial using Horner's polynomial evaluation algorithm, we first observe that $a_4 x^4 + a_3 x^3 + a_2 x^2 + a_1 x + a_0 = x(x(x(a_4 x + a_3) + a_2) + a_1) + a_0$ has four additions and four multiplications. We see that $p(x) = x(x(\cdots x((a_n x + a_{n-1}) + a_{n-2}) + \cdots + a_2) + a_1) + a_0$ has n additions and n multiplications for a total of $2n$ operations, which is $O(n)$. ∎

The following is a summary in order of the complexity of an algorithm where n is the size of the input.

$O(c)$	Constant complexity
$O(\log(n))$	Logarithmic complexity
$O(\log(n!))$	
$O(n \log(n))$	$n \log(n)$ complexity
$O(n^k)$	Polynomial complexity
$O(c^n)$	Exponential complexity (for $c > 1$)
$O(n!)$	Factorial complexity
$O(n^n)$	

Obviously it is desirable to have an algorithm as efficient as possible with all other factors considered. It is important that, for a given problem, an algorithm be able to solve the problem in polynomial time (i.e., with polynomial complexity). The set of all algorithms that can be solved by an algorithm in polynomial time is denoted by **P**. The algorithms that we have discussed so far and can be used in computers are called **deterministic algorithms**. In these algorithms, the computer has only one choice in what it can do next. In some machines, all of which are theoretical, including some we will discuss in later chapters, the machine at a given step may have several choices about what it does next. When this happens, the algorithm is called a **nondeterministic algorithm**. The set of all problems that can be solved in polynomial time by a nondeterministic algorithm is denoted by **NP**. Obviously $\mathbf{P} \subseteq \mathbf{NP}$. It seems likely that there are problems in **NP** that are not in **P**; however, no one has been able to prove this.

There are some problems called **NP** complete that, if could be shown that they belong to **P**, would enable us to prove that $\mathbf{P} = \mathbf{NP}$. This is done by polynomial reducibility, which means that, given the solution of an **NP**-complete problem, other problems in **NP** can be solved in polynomial time. We list two such problems:

1. **Traveling Salesman Problem**. Given a set of cities and the distances between them, a tour of all the cities may be found with distance less than a constant k.

2. **Hamiltonian Cycle** (see Section 14.4). Given a graph, find a simple cycle that includes all of the vertices.

Exercises

1. How many moves are necessary to move eight rings on the Tower of Hanoi?

2. As previously noted, the legend states that there were 64 golden disks on one of three diamond pins and they were being moved to a second pin by monks at the rate of one move per second. According to the legend, when the mission is completed, the world will end. How long will it take?

Estimate the number of arithmetic operations necessary to evaluate the following polynomials (i) using direct substitution; (ii) using Horner's method:

3. $f(3)$ when $f(x) = 3x^2 + 4x + 5$
4. $f(2)$ when $f(x) = 2x^4 + 4x^3 + 3x^2 + 2x + 3$
5. $f(4)$ when $f(x) = x^5 + 4x^4 + 2x^3 + 6x^2 + x + 3$

Estimate the number of arithmetic operations necessary to evaluate the following product of matrices:

6. $\begin{bmatrix} 3 & -1 \\ 2 & 7 \\ -5 & 0 \end{bmatrix} \begin{bmatrix} 2 & -1 & 0 & 4 \\ 5 & 1 & 3 & 0 \end{bmatrix}$

7. $\begin{bmatrix} 1 & -1 & 2 \\ 3 & 1 & 4 \\ 6 & -1 & 5 \end{bmatrix} \begin{bmatrix} 2 & -2 & 8 \\ 5 & 5 & 1 \\ 0 & 6 & 4 \end{bmatrix}$

8. $\begin{bmatrix} 1 \\ 3 \\ 4 \\ 5 \end{bmatrix} \begin{bmatrix} -2 & 3 & -5 & 0 \end{bmatrix}$

Which of the following functions equals $O(n^3)$?

9. $n^4 - 3n + 5$
10. $n^2 - 6n + 5$
11. $(\ln(n))^3$
12. $n(\ln(n))^2$
13. $n^2 \ln(n)$

Which of the following functions equals $O(n^3)$?

14. $6n^3 + 3n^2 + 2n + 5$
15. $n^3 \ln(n)$
16. $\ln(n^{n \ln(n)})$
17. $\log_a(n!)$
18. $n!$

Which of the following functions equals $\Theta(n^2)$?

19. $6n^2 - 100n + 5$
20. $\lfloor (n - \frac{1}{2})^2 \rfloor$
21. $\lceil 3n^2 - 6n + 5 \rceil$
22. $\lfloor n - 3.6 \rfloor \lceil 3n + 5.1 \rceil$
23. $n \ln n$

Which of the following functions equals $\Theta(n)$?

24. $\dfrac{6n^2 - 100n + 5}{4n - 5}$
25. $\lfloor n + \frac{1}{n} \rfloor$
26. $\ln \lceil 3n^2 - 6n + 2.5 \rceil$
27. $\lfloor n - 6.3 \rfloor \lceil \frac{3}{n} + 5 \rceil$
28. $n \ln n$

Find the least n so that $f(m) = O(m^n)$.

29. $f(m) = m^3 \ln m$
30. $f(m) = m^3 \lceil m^2 - 3.5 \rceil$
31. $f(m) = \dfrac{m^5 + 77m + 5}{m^3 - 3m + 1}$
32. $f(m) = \dfrac{\lceil m - .5 \rceil}{\lfloor m + .5 \rfloor}$

33. Prove that for any given positive integer n, $n! \leq n^n$ and therefore $n! = O(n^n)$.

34. Prove or show a counterexample for the statement: If $f(n) = O(g(n))$ and $h(n) = O(k(n))$, then $\dfrac{f(n)}{h(n)} = O\left(\dfrac{g(n)}{k(n)}\right)$.

35. Prove or show a counterexample for the statement: If $f(n) = \Theta(g(n))$ and $h(n) = \Theta(k(n))$ where $h(n)$ and $k(n)$ are always positive, then $\dfrac{f(n)}{h(n)} = \Theta\left(\dfrac{g(n)}{k(n)}\right)$.

36. Prove that for $a > 1$ and $n \in N$, $\log_a(n!) \leq n \log_a(n)$ and so $\log_a(n!) = O(n \log_a(n))$.

37. Prove that if $f(n) = O(g(n))$ and $g(n) = O(h(n))$, then $f(n) = O(h(n))$.

38. Prove that for integer a greater than one, $\log_a(n) = O(n)$.

39. Using induction prove that if r and s are positive integers, $r \leq s$ and $n > 1$, then $n^r \leq n^s$. Hence $n^r = O(n^s)$.

40. Using the previous problem prove that if r and s are positive rational numbers, $r \leq s$ and $n > 1$, then $n^r \leq n^s$. Hence $n^r = O(n^s)$.

41. Prove that in the set of functions defined on the positive integers the relation R defined by $f \, R \, g$ if and only if $f(n) = \Theta g(n)$ is an equivalence relation.

42. Give a big-Oh estimate of the product of the first n odd positive integers.

43. Find functions $f(n)$ and $g(n)$ so that neither $f(n) = O(g(n))$ nor $g(n) = O(f(n))$ if possible.

44. Determine $O(S(n))$ where $S(n)$ is the number of additions needed to add two integers expressed in binary notation where the length of each of the binary notations has length less than or equal to n.

45. Determine $O(M(n))$ where $M(n)$ is the number of multiplications needed to multiply two integers expressed in binary notation where the length of each of the binary notations has length less than or equal to n.

46. Determine $O(s(n))$ where $s(n)$ is the number of shifts needed to multiply two integers expressed in binary notation where the length of each of the binary notations has length less than or equal to n.

5.4 Sorting Algorithms

In this section, we explore some of the methods for sorting data. We make two assumptions. The first is that the data are numbered in some way so that we have a first item of data, a second item of data, and so on, up to an nth item of data. This may be done by placing the data in an array, using pointers, or whatever way is desired. We are not concerned with the method used. The second assumption is there is a type of total ordering and we want to arrange the items of data using this order. This ordering could be alphabetical (lexicographic) or some sort of numerical value attached to the data. Again, we really don't care what the ordering is as long as we have one. We shall use the notation $a < b$ to mean that a should be placed before b in the ordering.

We begin with the **selection sort**. Let $a_1, a_2, a_3, \ldots, a_n$ be the items being sorted. The process is really very simple. We begin with a_1. We now want to find the smallest of the items and exchange it with a_1. We then move to a_2. We now compare a_2 and all the items after it to find the smallest item. When we find the smallest item, we exchange it with a_2. We continue this process until we are finished. More specifically, if we want to replace a_i we shall have a variable min, which we shall use to keep track of the subscript of the smallest value of a_i and the items following it. We begin by letting $min = i$, and start comparing the remainder of the items to a_{min}. When we find one that is smaller, say a_k, we then let $min = k$, and continue looking for a smaller item by comparing it with a_{min}. When we find one, we repeat the process. When we reach n, we exchange a_i and a_{min}, and proceed to a_{i+1}. We now state the algorithm.

Procedure Selection Sort
    For $i = 1$ to $n - 1$,
        Let $min = i$.
        For $j = 1$ to n
            If $a_j < a_{min}$, let $min = j$.
        Endfor
        Exchange a_i and a_{min}.
    Endfor
End

EXAMPLE 5.30 Let c, b, a, v, h, d be a list to be sorted. We begin with c in position 1, and let $min = 1$. We compare c with b, and since $b < c$ and b is in the second position, we let $min = 2$. We then compare b with a, and since $a < b$, we let $min = 3$. We find no letters less than a, so we exchange a and c getting a, b, c, v, h, d. We repeat the process starting with b, so $min = 2$. As we compare b with the remaining letters, none of them is less than b, so min remains 2, and we exchange b with itself. We now proceed to c, and let $min = 3$. Again nothing is less than c, so c remains in the same place. We now proceed to v and let $min = 4$. Since $h < v$, we let $min = 5$. We then compare h and d. Since $d < v$, we let $min = 6$. We then exchange d and v, getting the sequence a, b, c, d, h, v. We continue the process with h; since $h < v$, we are finished. ∎

If we look at the comparisons in this search of n items, we compare the first item to $n - 1$ items. We compare the second item to $n - 2$ items. We continue until we compare the $(n - 1)$th item with one item. Therefore, there are

$$(n - 1) + (n - 2) + \cdots + 2 + 1 = \frac{n(n - 1)}{2} \quad \text{comparisons}$$

so the number of comparisons is $O(n^2)$.

We now consider a popular sorting algorithm that is rather similar to the selection sort. It is called the **bubble sort** because the small numbers rise to the top. In this case the top is to the left. In this case we start at the right and compare the last two items, moving the lesser item to the left. We then compare the smaller item to the one on its left and again move the smaller item to the left. We continue until we reach the first item. The result is that the smallest item is the first item. We then repeat the process until we get back to the second item. The second item now becomes the second smallest item. We continue this until they are all sorted.

Our problem this time is that we are working from both directions. We start at the right and work our way left to the first item. We then start at the right and work our way left to the second item. We continue to work our way left to the jth item, where j is increasing. We shall need to introduce a new *for* statement. When we say "For $i = n$ to k step -1," where $n > k$, we mean begin at n and keep decreasing by 1 until you get to k.

    Procedure Bubble Sort
        For $i = 2$ to n
            For $j = n$ to i step -1
                If $a_j < a_{j-1}$ exchange a_j and a_{j-1}
            Endfor
        Endfor
    End

EXAMPLE 5.31 Again let c, b, a, v, h, d be a list to be sorted. We begin by exchanging h and d, since d is less than h. We then exchange d and v, since d is less than v. At this point, our list is c, b, a, d, v, h. Since d is not less than a, we do not exchange a and d. We now exchange b and a, since a is less than b. Finally, we exchange a and c, since a is less than c. At this point our list is a, c, b, d, v, h.

Repeating the process, we exchange h and v, since h is less than v. Since h is not less than d, we leave d alone. Now our list is a, c, b, d, h, v. We also do not exchange d and b, since d is not less than b. Since b is less c, we exchange b and c, so that we have a, b, c, d, h, v. Although we repeat the process three more times, we make no further changes, and we are finished. ∎

If we look at the comparisons in this search of n items, we make $n - 1$ the first round. We make $n - 2$ comparisons the second round. We continue until we make one comparison the last round. Therefore, there are

$$(n - 1) + (n - 2) + \cdots + 2 + 1 = \frac{n(n - 1)}{2} \quad \text{comparisons}$$

so the number of comparisons is $O(n^2)$. Thus, we get the same value as for the selection process. Although the comparisons are the same, the actual speed of the two sorts may vary.

The next sorting algorithm we investigate is called **quick sort**. It is the first type of sort that we define recursively. The basic process in quick sort is to pick a_l the last element in the list of items; then place every item (excluding a_l) that is less than or equal to a_l to the left of a_l. Call this list S_1. Place every element greater than a_l to the right of a_l. Call this list S_2. Now apply quick sort to S_1, forming S_{11}, S_{12}, and again apply quick sort to the left list, S_{11}. Continue until there is a list, say, S_x, for which, when divided into S_{x1} and S_{x2}, then S_{x1} is sorted or empty. Quick Sort is then to applied to S_{x2}, the right side of S_x. For S_{x2}, the process begins again by dividing it into a left list and a right list and applying quick sort to the left list. Again continue until a segment, say S_y with left and right lists S_{y1} and S_{y2}, has S_{y1} sorted or empty. Quick Sort is then applied to S_{y2}, the right side of S_y, and the process continues. When both the left list and the right list of a list, say, S_y, have been sorted the segment, then the list S_y is sorted and we return to the list from which S_y was a left or right segment and continue the same process on that list. Hence, the process is to continue to break a list into two lists and use the left list until it is sorted; then go to the right list and repeat the process until it is sorted. Then return to the list for which the left and right lists have been sorted and continue the process.

We first give the procedure and then give examples. We assume the numbers to be sorted are in an array A between numbers Left and Right.

```
Quick Sort(A, Left, Right)
   Boolean Done = false
  if(Left ≥ Right)
    return
  endif
    v = A[Right]
    i = Left + 1
    j = Right
    repeat
      repeat
        i = i + 1
      until (A[i] ≥ v)
      repeat
        j = j - 1
      until (j ≤ Left OR A[j] ≤ v
      If (i ≥ j)
        done = true
        else exchange (A[i], A[j]
      endif
    until done
    exchange (A[i], A[Right]
    Quick Sort(A, Left, i - 1)
    Quick Sort(A, i - 1, Right)
    return
    end Quick Sort.
```

EXAMPLE 5.32 Given 4, 7, 9, 3, 8, 12, 10, 2, 1, 11, 6 use Quick Sort on the list.
First exchange 7 and 1 to get 4, 1, 9, 3, 8, 12, 10, 2, 7, 11, 6.
Next exchange 9 and 2 to get 4, 1, 2, 3, 8, 12, 10, 9, 7, 11, 6.
Next exchange 8 and 6 to get 4, 1, 2, 3, 6, 12, 10, 9, 7, 11, 8, and we break into left list 4, 1, 2, 3 and right list 12, 10, 9, 7, 11, 8.

We now call Quick Sort for the list 4, 1, 2, 3. We exchange 4 and 2 to get 2, 1, 4, 3, and exchange 4 and 3, to get 2, 1, 3, 4, which we break into left list 2, 1 and right list 4. We now call Quick Sort for the list 2, 1 to get 1, 2. Since it is sorted, we call Quick Sort for the right list 4, but it is sorted, so that we have 1, 2, 3, 4. Since left list 4, 1, 2, 3 is now sorted, we now return to the right list 12, 10, 9, 7, 11, 8.

We exchange 12 and 7 to get 7, 10, 9, 12, 11, 8. We now exchange 10 and 8 to get 7, 8, 9, 12, 11, 10. We now get left list 7 and right list 9, 12, 11, 10. We call Quick Sort for the list 7, but it is sorted. We then call Quick Sort for the right list 9, 12, 11, 10. We exchange 10 and 12 to get list 9, 10, 11, 12. We now call Quick Sort for the left list 9, but it is sorted. We then call Quick Sort for the right list 11, 12, but it is sorted. Right list 12, 10, 9, 7, 11, 8 is now sorted and our result is 1, 2, 3, 4, 6, 7, 8, 9, 10, 11, 12. ∎

EXAMPLE 5.33

Again let c, b, a, r, v, h, d be a list to be sorted. We exchange r and d to get c, b, a, d, v, h, r getting left list c, b, a and right list v, h, r. We now call Quick Sort for the left list c, b, a and exchange c and a to get a, b, c. We now have an empty left list and the right list is already sorted so calling Quick Sort has no effect. We now have ordered the original left list so we call Quick Sort for v, h, r. We exchange h and v, getting h, v, r. We now exchange v and r getting h, r, v. We now have left list h and right list r, which are already sorted, so they are not affected by calling Quick Sort. We now have ordered the original right list so we have a, b, c, d, h, r, v. ∎

We now consider the number of comparisons used by this process. The best time with Quick Sort occurs when $n = 2^m$ for some m, and each partition divides the list of items exactly in half. Assumes this happens. If Q_n is the number of comparisons for n items, then there are n comparisons of the items to determine to which list they should be partitioned. When there are $Q_{\frac{n}{2}}$ comparisons for each of the two lists of $n/2$ items, therefore, $Q_n = 2Q_{\frac{n}{2}} + n$. But $n = 2^m$, so that

$$Q_n = Q_{2^m} = 2Q_{2^{m-1}} + 2^m$$

so that

$$\frac{Q_{2^m}}{2^m} = \frac{Q_{2^{m-1}}}{2^{m-1}} + 1$$
$$= \frac{Q_{2^{m-2}}}{2^{m-2}} + 1 + 1$$
$$= \frac{Q_{2^{m-2}}}{2^{m-2}} + 2$$
$$= \frac{Q_{2^{m-3}}}{2^{m-3}} + 3$$
$$\vdots$$
$$= \frac{Q_{2^{m-m}}}{2^{m-m}} + m$$
$$= Q_1 + m$$

Since $n = 2^m$, $m = \log_2(n)$, so that $Q_n = n(Q_1 + \log_2(n))$, which is approximately $n \ln(n)$. Thus, the number of matchings is $O(n \log_2(n))$. In the worst possible case, the number of matchings occurs when the item that is used to divide the item set is always on the end. In this case, the number of matchings is the same as the number in the sorts preceding so that in the worst case the number of matchings is $O(n^2)$.

A similar but simpler to understand procedure is the **merge sort**. The procedure is to divide the items to be sorted into halves, or as nearly so as possible. Then take each of these two lists and divide them in half. Continue this process until each set contains only one item. Now reverse the process, putting the lists together exactly as they were divided but merging them as they are put together. We shall first give the procedure and then give an example showing how to use it. In most of our cases using recursion, the procedure calls itself when the name appears in the procedure.

To emphasize the fact that the procedure is calling itself, we shall use the command "call." In this procedure, the variables l, m, and r will represent left, middle, and right items in the current list.

> Procedure Merge Sort (l, r)
>   If $l < r$ then
>   Let $m = \left\lfloor \dfrac{l+r}{2} \right\rfloor$
>   Call Merge Sort (l, m)
>   Call Merge Sort $(m + 1, r)$
>   Call Merge (l, m, r)
>   Endif
> end

Before explaining the procedure Merge Sort, we need to describe the procedure Merge. Merge simply takes two lists, which are already ordered, and compares the first element of each list. The least of these two elements is removed from its list and placed as the first element in the new sorted list. The procedure is then repeated and each time, the element removed is added to the end of newly sorted list. This is continued until one of the lists is empty. Then the remainder of the other list is added to the new sorted list. We shall not use a formal procedure since the actual method used for implementing Merge is dependent on computer techniques used.

> Procedure Merge (l, m, r)
>   Let $A = (a_l, a_{l+1}, a_{l+2}, \ldots, a_m)$, $B = (a_{m+1}, a_{m+2}, a_{m+i}, \ldots, a_r)$,
>   and C be the new merged list.
>   If both A and B both contain items in their list
>   Compare the first items of A and B and remove the lesser of
>   these items from their set.
>   Add the removed element to the end of the list C.
>   Endif
>   If either A or B contain no items, add the rest of the other list to the
>   end of list C. Relabel this list as $(a_l, a_{l+1}, a_{l+2}, \ldots, a_r)$.
> End

Consider the list c, b, a, v, h, d, x, e to be sorted. Letting the symbol | denote the partitioning, we first divide the list into $c, b, a, v | h, d, x, e$. We then call Merge Sort again to partition the left side, so that we have $c, b | a, v | h, d, x, e$, since this command comes first in the call statements. Again we call Merge Sort to divide the left side, which gives us the partition $c | b | a, v | h, d, x, e$. Finally, we can no longer divide the left side, so we try to divide the right side, which we also cannot do. We then merge c and b, so that we have $b, c | a, v | h, d, x, e$. We now return to the call of Merge Sort, which divided b, c and a, v. We have completed the call for the left side with "Call Merge Sort (l, m)," so we drop down to "Call Merge Sort (m, r)," which partitions the right side, so we have $b, c | a | v | h, d, x, e$. We can no longer call Merge Sort, so we now merge a and v to get a, v so we now have $bc | a, v | h, d, x, e$. We again returned to the level of Merge Sort that divided b, c and a, v. We have finished "Call Merge Sort (l, m)" and "Call Merge Sort (m, r)," so we now merge b, c and a, v to get $a, b, c, v, | h, d, x, e$. We now return to the level of Merge Sort that divided c, b, a, v and h, d, x, e. We have completed "Call Merge Sort (l, m)," so we now use "Call Merge Sort (m, r)" to produce $a, b, c, v, | h, d | x, e$. We now call Merge Sort again to partition h and d giving $a, b, c, v, | h | d | x, e$. We can no longer use either of the Merge Sort instructions, so we merge h and d, giving $a, b, c, v, | d, h | x, e$. We now return to the level of Merge Sort that partitioned h, d and x, e. We can no longer use "Call Merge Sort (l, m)," so we now drop down to "Call Merge Sort (m, r)" to produce $a, b, c, v, | d, h | x | e$. Now we drop down to the command Merge to merge x

and e, so we have $a, b, c, v, |d, h|e, x$. We again return to the level of Merge Sort that partitioned h, d and x, e. We have completed both of the Merge Sort commands, so we now merge d, h and e, x, getting $a, b, c, v|d, e, h, x$. Finally, we return to the level of Merge Sort that divided c, b, a, v and h, d, x, e. We have completed both of the Merge Sort commands at this level so we merge a, b, c, v and d, e, h, x, getting a, b, c, d, e, h, v, x and we are finished.

Merge Sort is the most efficient sorting algorithm as far as the number of comparisons are concerned. To show this, assume that $n = 2^m$ for some value m. Therefore, $m = \log_2 n$. For a fixed m, this is a worst-case situation. If we consider the merging done by the initial call of the sorting procedure, it merges $2 (= 2^1)$ lists each containing 2^{m-1} items. We consider this to be the first level. In general, at the kth level, we have 2^k lists, each containing 2^{m-k} items. We then have to make 2^{k-1} comparisons of pairs of lists, each containing 2^{m-k} items. Each comparison of pairs of lists requires at most $2^{m-k+1} - 1$ comparisons, since the remaining item does not have to be compared with anything. Therefore, there are at most

$$2^{k-1} \cdot (2^{m-k+1} - 1) = 2^{k-1} \cdot (2^{m-(k-1)} - 1)$$

comparisons at level k. Therefore, the total number of comparisons is at most

$$\sum_{k=1}^{m} 2^{k-1} \cdot (2^{m-(k-1)} - 1) = \sum_{i=0}^{m-1} 2^i \cdot (2^{m-i} - 1)$$
$$= \sum_{i=0}^{m-1} 2^m - 2^i$$
$$= m2^m - (2^m - 1)$$
$$= n \log_2 n - (n - 1)$$

so that the number of comparisons in the worst case is $O(n \ln n)$.

Our last sort is called **insertion sort**. The procedure is really quite simple. For this sort, the items of data are taken one at a time and placed where they belong in the list already created. Thus, a place is found where all elements to the left of the item being inserted are less than that item, and all items to the right are greater. Again consider the list c, b, a, v, h, d, x, e. We first select c to place in our new list. Then since b is less than c, we place it before c, forming b, c. Next we select the a. Since it is less than b, we place it before b, forming a, b, c. We next select v. Since it is greater than c, we place it after c, forming a, b, c, v. Since h is less than v and greater than c, we place it between c and v forming a, b, c, h, v. Now d is selected. Since it is greater than c and less than h, we place it between c and h, giving a, b, c, d, h, v. Since x is greater than v, we place it after v, giving a, b, c, d, h, v, x. Finally, since e is greater than d and less than h, we place it between d and h, giving a, b, c, d, e, h, v, x.

In the worst case, the number of comparisons at each stage is the same as for the bubble sort, so this number is $O(n^2)$.

Both the quick sort and the merge sort involve recursively dividing sets into smaller sets, sorting them, and then combining the sorted sets. This technique of breaking sets into forming sets, performing the process on them, and then putting the sets back together is called **divide and conquer**. It is often a very efficient method for performing tasks.

Often divide and conquer procedures, such as those previously, satisfy a recursive formula of the form $Q_n = CQ_{\frac{n}{2}} + f(n)$, where Q_n may represent number of operations, number of comparisons, amount of time, or other measurements. We shall simply use number of operations in the following discussion. The function

$f(n)$ represents the number of operations involved in dividing the set into two sets and putting them back together again. When we use $\frac{n}{2}$, we are assuming that we are dividing the set on which the task is performed into two equal parts. This is fairly common since we often use binary trees or other concepts involving binary operations. If we were dividing the set into k equal parts, we would use $\frac{n}{k}$. We are also assuming that n is a power of 2.

We are primarily interested in the cases where $f(n)$ is constant and where $f(n) = Dn$, for some constant D. When we use $f(n) = Dn$, we are assuming that the number of operations in dividing n objects and putting them back together is directly proportional to the number of elements in the set. In our discussion of the efficiency of quick sort, we showed that the solution of the recursive relation $Q_n = 2Q_{\frac{n}{2}} + n$, when n is a power of 2, is given by $n(Q_1 + \log_2(n))$. This is the solution of $Q_n = CQ_{\frac{n}{2}} + Dn$ when $C = 2$ and $D = 1$. For $C > 2$, the solution of the recursive relation is $Q_n = An^{\log_2 C} + \left(\frac{2D}{2-C}\right)n$, where A is selected so that $Q_1 = A + \frac{2D}{2-C}$.

When $f(n) = D$, for some constant D, the solution to the recursive relation $Q_n = CQ_{\frac{n}{2}} + D$ for $C \neq 1$ is given by $Q_n = \frac{D(Cn^{\log_2 C} - 1)}{C - 1}$. In particular, $Q_n = D(2n - 1)$ when $C = 2$. These equations are discussed further in the problems.

Unfortunately, the integer n is not generally a power of 2. However, if we assume that Q_n is an increasing function, and n is less than a, where a is a power of 2, then $Q_n < Q_a$ and we have bounded Q_n.

Computer Science Application

Sorting arrays involves a lot of comparing and swapping. Often the array components are large, and swapping two of them can be a time-consuming operation. It is possible in these cases to create an index that will provide sorted access to the array components without actually rearranging them. The following algorithm performs a traditional bubble sort on an array x of integers:

function bubble-sort(integer array x, integer n)

```
  for i = n to 1 by -1
    for j = 2 to i
      if x[j - 1] > x[j] then
        swap x[j - 1] with x[j]
      end-if
    end-for
  end-for
```

end-function

This can easily be modified to create an index array that can be subsequently used to access the array components in ascending or descending order. By changing $x[j]$ to $x[index[j]]$ in the comparisons and $x[j]$ to $index[j]$ in the swaps, the components of the array are not rearranged. Since index is always an integer array, this can save considerable time when x's components are large objects.

function indirect-sort(integer array x, integer n, integer array index)

```
  for i = 1 to n
    index[i] = i
  end-for
```

```
for i = n to 1 by -1
    for j = 2 to i
        if x[index[j - 1]] > x[index[j]] then
            swap index[j - 1] with index[j]
        end-if
    end-for
end-for
```

end-function

Often it is not necessary to actually rearrange the array for the purpose at hand. When it is, the index can still be used to do the rearranging afterwards with far fewer swaps. Other sorting algorithms can be modified in a similar way to perform indirect sorting.

In the following theorem, we summarize our result and add one final result when the elements in the set are divided into m parts.

THEOREM 5.34

1. The solution for $Q_n = CQ_{\frac{n}{2}} + Dn$ for $C > 2$ and $n = 2^m$ for some m is given by $Q_n = An^{\log_2 C} + \left(\frac{2D}{2-C}\right)n$ where A is selected so that $Q_1 = A + \left(\frac{2D}{2-C}\right)$.

2. The solution for $Q_n = CQ_{\frac{n}{2}} + Dn$ for $C = 2$ and $n = 2^m$ for some m is given by $Q_n = n(Q_1 + D\log_2(n))$.

3. The solution for $Q_n = CQ_{\frac{n}{b}} + Q_1$ for $C > 2$ and $n = 2^m$ for some m is given by

    (a) $Q_n = Q_1(\log_b n + 1)$ when $C = 1$

    (b) $Q_n = \dfrac{Q_1(Cn^{\log_b C} - 1)}{C - 1}$

Proof

1. See problems.
2. See discussion of efficiency of quick sort.

    Prelude to 3a and 3b

$$\begin{aligned}
Q_n &= CQ_{\frac{n}{b}} + Q_1 \\
&= C^2 Q_{\frac{n}{b^2}} + CQ_1 + Q_1 \\
&= C^3 Q_{\frac{n}{b^3}} + C^2 Q_1 + CQ_1 + Q_1 \\
&\vdots \\
&= C^m Q_{\frac{n}{b^m}} + C^{m-1}Q_1 + \cdots + C^2 Q_1 + CQ_1 + Q_1 \\
&= C^m Q_1 + C^{m-1} Q_1 + \cdots + C^2 Q_1 + CQ_1 + Q_1 \\
&= Q_1(C^m + C^{m-1} + \cdots + C^2 + C + 1) \\
&= Q_1 \frac{C^{m+1} - 1}{C - 1} \\
&\quad \frac{Q_1(Cn^{\log_b C} - 1)}{C - 1}
\end{aligned}$$

3a. If $C = 1$, from the prelude, $= Q_1(C^m + C^{m-1} + \cdots + C^2 + C + 1) = Q_n = mQ_1 + Q_1$, but since $b^m = n$, $m = \log_b n$ and $Q_n = Q_1(\log_b n + 1)$.

3b. If $C > 1$, the result is given by the prelude. ∎

Computer Science Application

Assume that a single-dimensional array A has 1000 student records, each with a field for the final exam score, an integer between 0 and 100. A special sorting algorithm called distribution counting can be used in this context, which is superior to even the best general-purpose sorts like quick sort and merge sort. The following five steps will sort array A in ascending order of final exam score.

(1) for $j = 0$ to 100
    counter(j) = 0
    end-for

(2) for $i = 1$ to 1000
    add 1 to counter($A(i)$.finalexam)
    end-for

(3) for $j = 1$ to 100
    counter(j) = counter(j-1) + counter(j)
    end-for

(4) for $i = 1000$ to 1 by -1
    B(counter($A(i)$.finalexam)) = $A(i)$
    subtract 1 from counter($A(i)$.finalexam)
    end-for

(5) for $i = 1$ to 1000
    $A(i) = B(i)$
    end-for

Step 1 initializes an array of 101 counters corresponding to the 101 scores that are possible on the final exam. Step 2 counts the number of times each score occurred. At the conclusion of step 3 counter(i) contains the number of students who made a score of i or less. Step 4 places the ith record of A strategically in B in a way that guarantees that the record will never have to be moved, while leaving room for all other records in A that have the same final exam score. Step 5 simply overwrites A with B.

It may seem curious that the pass through A in step 4 was done in reverse. In this way the original order of A is preserved as a secondary sort.

Exercises

Sort the sequence 7, 11, 4, 0, 3, 1, 9, 4, 2, 8 *using*

1. selection sort
2. bubble sort
3. quick sort
4. merge sort
5. insertion sort

Sort the sequence 12, 50, −1, −10, 10, 11, 52, 30, 2, 8, −12 *using*

6. selection sort
7. bubble sort
8. quick sort
9. merge sort
10. insertion sort

Sort the sequence $x, a, c, y, p, z, f, t, m, y, u, b, d, n, s, r$ *using*

11. selection sort
12. bubble sort
13. quick sort
14. merge sort
15. insertion sort

Sort the sequence Johnson, Brown, Black, Jackson, Murphy, Smith, Jones using

16. selection sort
17. bubble sort
18. quick sort
19. merge sort
20. insertion sort

Construct a worst case sequence for each of the following:

21. selection sort
22. bubble sort
23. quick sort
24. merge sort
25. insertion sort

26. Show that at most $2n - 1$ comparisons are needed to merge two already sorted lists of length n.

27. Write a computer implementation for selection sort.

28. Write a computer implementation for bubble sort.

29. Write a computer implementation for quick sort.

30. Write a computer implementation for merge sort.

31. Write a computer implementation for insertion sort.

32. Show that $Q_n = An + D$ satisfies the recursive relation $Q_n = 2Q_{\frac{n}{2}} - D$.

33. Show that $Q_n = Dn(\log_2 n + A)$ satisfies the recursive relation $Q_n = 2Q_{\frac{n}{2}} + Dn$.

34. Show that $Q_n = An^{\log_2 C} + \left(\frac{2D}{2-C}\right)n$ satisfies the recurrence relation $Q_n = CQ_{\frac{n}{2}} + Dn$ for $C > 2$.

35. Give reasons, fill in the gaps, and complete the proof that

$$Q_n = An^{\log_2 C} + \left(\frac{2D}{2-C}\right)n$$

is the solution of $Q_n = CQ_{\frac{n}{2}} + Dn$ for $C > 2$, and A is selected so that $Q_1 = A + \frac{2D}{2-C}$. Assume that $n = 2^m$.

$$\frac{Q_{2^m}}{2^m} = \frac{C}{2}\left(\frac{Q_{2^{m-1}}}{2^{m-1}}\right) + D$$

$$= \left(\frac{C}{2}\right)^2 \frac{Q_{2^{m-2}}}{2^{m-2}} + \left(\frac{C}{2}\right)D + D$$

$$= \left(\frac{C}{2}\right)^3 \frac{Q_{2^{m-3}}}{2^{m-3}} + \left(\left(\frac{C}{2}\right)^2 + \frac{C}{2} + 1\right)D$$

$$\vdots$$

$$= \left(\frac{C}{2}\right)^m Q_1 + \left(1 + \frac{C}{2} + \left(\frac{C}{2}\right)^2 + \cdots + \left(\frac{C}{2}\right)^{m-1}\right)D$$

$$= \frac{C^{\log_2 n}}{n}Q_1 + \frac{1 - \left(\frac{C}{2}\right)^m}{1 - \left(\frac{C}{2}\right)}D$$

$$= \frac{n^{\log_2 C}}{n}Q_1 + \frac{1 - \left(\frac{C}{2}\right)^m}{1 - \left(\frac{C}{2}\right)}D$$

$$\vdots$$

36. Derive the solution $Q_n = \frac{D(Cn^{\log_2 C} - 1)}{C - 1}$ for the recursive relation $Q_n = CQ_{\frac{n}{2}} + D$ when $C \neq 1$.

5.5 Prefix and Suffix Notation

In arithmetic, as in logic, there are certain rules regarding precedence of the operations on the integers that enable us to reduce the number of parentheses used in an expression without causing ambiguity. Some properties of the integers also reduce the need for parentheses. For example, from the associative law for addition we know that the expression $3 + 5 + 7$ needs no parentheses since $(3 + 5) + 7 = 3 + (5 + 7)$. Also it is understood that exponent operations are to be performed first, followed by multiplication and division, and finally by addition and subtraction. Even with this convention it is necessary to use parentheses. For example, we cannot express $(a + b)^2(4 + 5b)$ without using parentheses. Parentheses can be eliminated by the use of either prefix or postfix notation. In contrast to these we define the usual notation for arithmetic, where the binary operations are placed between the values on which they operate, as infix notation. Thus, $3 + 4$ and $6 \div 3$ are described using infix notation.

DEFINITION 5.35 An expression is in **prefix** (or **Polish**) **notation** if the operation immediately proceeds the values on which it operates. It is in **suffix** (or **postfix** or **reverse Polish notation**) if the operation immediately follows the values on which it operates. An expression is in **infix notation** if the operation is between the values on which it operates.

For example, the infix expression $a + b$ would be expressed as $+ab$ in prefix notation and as $ab+$ in postfix notation. The infix expression $a \times (b + c)$ would be expressed as $\times a + bc$ in prefix notation and as $abc + \times$ in postfix notation. Note that one should "read" postfix notation from left to right and "read" prefix notation from right to left. Each time a binary operator is reached, the two proceeding results are "processed." For convenience we shall use the symbol ^ to denote explicitly the exponent operation. Thus, a^5 is expressed as a^5. The expression $(a+b) \times (c+d)$^3 is expressed as $\times + ab$^$ + cd3$ using prefix notation and as $ab + cd + 3$^$\times$ using postfix notation.

In this section, algorithms for changing from infix to postfix notation and from postfix to infix notation are developed. Similar algorithms may be developed symmetrically for changing from infix to prefix notation and conversely.

DEFINITION 5.36   An arithmetic expression is **fully parenthesized** if, for every operation in the expression, there corresponds a pair of parentheses that encloses the operator and the values on which it operates.

Put simply, an arithmetic expression is fully parenthesized if there is a parenthesis everywhere it makes sense to put one. For example, $(a \times (b + c))$ and $((a + b) \times ((c + d)$^$x))$ are fully parenthesized but $a + (b \times c)$ and $(a \div b)$^c are not fully parenthesized.

An infix expression may be changed to a postfix expression by the following procedure:

1. Add parentheses to the expression until it is fully parenthesized.
2. Beginning with the innermost parentheses, remove the pair of parentheses and move the corresponding operator to replace the right parenthesis corresponding to the operation. If there is more than one innermost parenthesis, begin with the left one.
3. Moving out to the next pair of parentheses, again remove the parentheses and move the corresponding operator to replace the right parenthesis corresponding to the operation.
4. Continue step (3) until all parentheses are removed.

EXAMPLE 5.37   Given the infix expression $(a + b) \times (c + d)$^x, we add parentheses until it is fully parenthesized so we have the expression $((a + b) \times ((c + d)$^$x))$. Using step (2), we obtain $(ab + \times (cd +$^$x))$. Using step (3), we obtain $(ab + \times cd + x$^$)$. Repeating step (3), we obtain $ab + cd + x$^$\times$. ∎

EXAMPLE 5.38   Given the infix expression $a \times b + c \times d$, we add parentheses until it is fully parenthesized so we have the expression $((a \times b) + (c \times d))$. Using step (2), we obtain $(ab \times + cd \times)$. Repeating step (3), we obtain $ab \times cd \times +$. ∎

A **stack** will be informally defined as an area where data are stored such that the last data item stored is the first retrieved. This is often referred to as LIFO or last-in, first-out. Adding and removing data from the stack are often referred to as the **push** and **pop** operations, respectively.

A postfix expression may be changed to an infix expression by the following procedure:

1. Begin reading the expression from left to right.
2. If the item read is not an operation, push (or place) it in the stack and continue reading.

3. If the item read is an operation,
    (i) Pop (or remove) the two top items from the stack.
    (ii) Place the operation between the second and first item in the stack and place parentheses around this expression.
    (iii) Place the expression from (ii) back in the stack.
4. Continue reading from left to right and performing steps (2) and (3) until finished.

EXAMPLE 5.39

Given the expression $abc \times +$, first read a and, since it is not an operation, place it in the stack as shown in Figure 5.4. Next read b and, since it is not an operation, place it in the stack as shown in Figure 5.5.

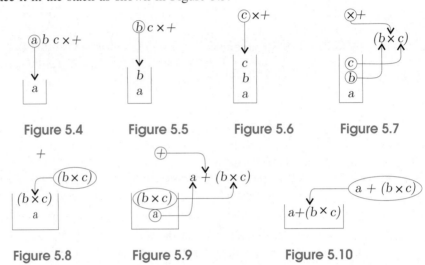

Figure 5.4   Figure 5.5   Figure 5.6   Figure 5.7

Figure 5.8   Figure 5.9   Figure 5.10

Next read c and, since it is not an operation, place it in the stack as shown in Figure 5.6. Next read $\times$ and, since it is an operation, pop (remove) c (the top item) and b (the next item) from the stack and place the $\times$ between b and c as shown in Figure 5.7.

Place $(b \times c)$ in the stack as shown in Figure 5.8. Next read $+$ and, since it is an operation, pop (remove) $(b \times c)$ and a (the next item) from the stack. Put $+$ between a and $(b \times c)$ as shown in Figure 5.9.

Place $(a + (b \times c))$ in the stack as shown in Figure 5.10. This is the desired expression. Obviously, unneeded parentheses can then be removed. ∎

EXAMPLE 5.40

Given the expression $ab + cd + 3\hat{\,}\times$, first read a and, since it is not an operation, place it in the stack. Next read b and, since it is not an operation, place it in the stack. Next read $+$ and, since it is an operation, pop (remove) b (the top item) and a (the next item) from the stack and place the $+$ between a and b. Place $(a + b)$ in the stack. Next read c. Since c is not an operation, place it in the stack. Next read d and, since it is not an operation, place it in the stack. Next read $+$ and, since it is an operation, pop (remove) d (the top item) and c (the next item) from the stack and place the $+$ between c and d. Place $(c + d)$ in the stack. Next read 3 and, since it is not an operation, place it in the stack. Next read $\hat{\,}$ and, since it is an operation, pop 3 the top item and $(c + d)$ (the next item) from the stack and place the $\hat{\,}$ between $(c + d)$ and 3. Place $((c + d)\hat{\,}3)$ in the stack. Read $\times$ and, since it is an operation, pop $((c + d)\hat{\,}3)$ and $(a + b)$ from the stack. Place $\times$ between $(a + b)$ and $((c + d)\hat{\,}3)$ to form $(a + b) \times ((c + d)\hat{\,}3)$. Place $((a + b) \times ((c + d)\hat{\,}3))$ in the stack. This is the desired result. Obviously unneeded parentheses can then be removed. ∎

Exercises

Change the following from infix to postfix notation:

1. $a^2 + b^2$
2. $a^3 + 3a^2 + 7$
3. $(a^2 + b)(c + d^2)$
4. $(a^2 + b^3)^4$
5. $a^2(b + c)$
6. $(c + d) - (a + b)^2$
7. $((a + b) - c)^3$
8. $(a + b)^2 - (c + d)^2$
9. $(a - b)(a + b)$
10. $a - b^{(c + d)}$

Change the following from postfix to infix notation:

11. $ab + cd - \div$
12. $ab2\hat{\ } + cd2\hat{\ } + \times$
13. $3ab \times \times 4cd2\hat{\ } \times \times +$
14. $ab + \hat{\ }2cd\hat{\ } + \div$
15. $a2\hat{\ }b + cd2\hat{\ } + \times$
16. $abc + + 3\hat{\ }$
17. $ab + 2\hat{\ }c + cd2\hat{\ } + -$
18. $3ab \times \times 4cd2\hat{\ } \times \times +$
19. $ab + 2\hat{\ }cd + \div$
20. $a2\hat{\ }b + c + cd2\hat{\ } + \div$

Change the following from infix to prefix notation:

21. $a^2 + b^2$
22. $a^3 + 3a^2 + 7$
23. $(a^2 + b)(c + d^2)$
24. $(a^2 + b^3)^4$
25. $a^2(b + c)$
26. $(c + d) - (a + b)^2$
27. $((a + b) - c)^3$
28. $(a + b)^2 - (c + d)^2$
29. $(a - b)(a + b)$
30. $a - b^{(c + d)}$

Change the following from prefix to infix notation:

31. $\div + ab - cd$
32. $\div + a2\hat{\ }d2 + \hat{\ }c2\hat{\ }b2$
33. $+ \times \times 3ab + \times \times 4cd \times \times 5ad$
34. $\hat{\ } + \hat{\ }a2 + \hat{\ }b2\hat{\ }c23$
35. $\times + a\hat{\ }2b + cd\hat{\ }2$
36. $\hat{\ } + \times abc3$
37. $\times + + abc + + cde$
38. $\hat{\ } + \times \times 3ab \times \times 4cd2$
39. $\hat{\ } \div + a\hat{\ }b2 + c\hat{\ }d22$
40. $\times + - abc - - abc$

41. Write a procedure to change an expression from infix notation to prefix notation.

42. Write a procedure to change an expression from prefix notation to infix notation.

5.6 Binary and Hexadecimal Numbers

One of the major advances in mathematics was the use of the number 0 as a placeholder in a number. Because of this advance, it was possible to distinguish between the numbers 47, 4700, 4007, and 4070. In this example, the value of the 4 and the 7 are determined by their position. In the number 47, the 4 represents 4 tens and the 7 represents 7 ones because of the position of the 4 and 7. In the number 4700 the 4 represents 4 thousands and the 7 represents 7 hundreds. We can, therefore, write 47 as $4 \cdot 10 + 7$ and the number 4700 as $4 \cdot 1000 + 7 \cdot 100$ or $4 \cdot 10^3 + 7 \cdot 10^2$. Similarly, we can write 4007 as $4 \cdot 10^3 + 7 \cdot 1$ and 4070 as $4 \cdot 10^3 + 7 \cdot 10$. In general, if a number n is represented in base 10 or decimal form as $a_m a_{m-1} a_{m-2} \cdots a_2 a_1 a_0$, then

$$n = a_m(10)^m + a_{m-1}(10)^{m-1} + \cdots + a_2(10)^2 + a_1(10) + a_0$$

where a_i is an integer between 0 and 9 for all $0 \leq i \leq m$. For example, $3124 = 3(10)^3 + 1(10)^2 + 2(10) + 4$. Similarly, if a number n is represented in base 2 or **binary** form, then every digit must be either 1 or 0 and again the position of the digit determines its value. Thus, $1101 = 1(2)^3 + 1(2)^2 + 0(2) + 1 = 13$ where 13 is in base 10. The first 20 positive integers written in binary are 1, 10, 11, 100, 101, 110, 111, 1000, 1001, 1010, 1011, 1100, 1101, 1110, 1111, 10000, 10001, 10010, 10011, and 10100. In general when $n = a_m a_{m-1} a_{m-2} \cdots a_2 a_1 a_0$, expressed in binary, then if we express n in our usual decimal form,

$$n = a_m(2)^m + a_{m-1}(2)^{m-1} + \cdots + a_2(2)^2 + a_1(2) + a_0$$

where a_i is either 1 or 0 for all $0 \leq i \leq m$. When there is a possibility of confusion, a number such as 1101101 in binary will be denoted 1101101_2 and a decimal number such as 312 will be denoted 312_{ten}.

On the odometer of a car, when we reach 9999 and add one more mile, the nines all roll over and we get 10 000. Similarly when we have 11111_2 and add 1, the ones all roll over and we get 100000_2. Thus, when we add 1 to 10111_2, we get 11000_2. To convert $a_m a_{m-1} a_{m-2} \cdots a_2 a_1 a_0$ from binary to decimal form, since $a_m a_{m-1} a_{m-2} \cdots a_2 a_1 a_0 = a_m(2)^m + a_{m-1}(2)^{m-1} + \cdots + a_2(2)^2 + a_1(2) + a_0$, we simply evaluate this number in our usual decimal number system. Thus, 1101011_2 is written as

$$\begin{array}{ccccccc} 1 & 1 & 0 & 1 & 0 & 1 & 1 \\ 2^6 & 2^5 & 2^4 & 2^3 & 2^2 & 2^1 & 1 \end{array} = 1(2)^6 + 1(2)^5 + 0(2)^4 + 1(2)^3$$
$$+ 0(2)^2 + 1(2)^1 + 1$$
$$= 64 + 32 + 8 + 2 + 1$$
$$= 107$$

and

$$10101 = 1(2)^4 + 0(2)^3 + 1(2)^2 + 0(2)^1 + 1$$
$$= 16 + 4 + 1$$
$$= 21$$

This method of finding the value of a binary number is often called "counting change" since it is similar to determining how much money we have if we have a quarter, a dime, and a penny. The first method to be described for converting from decimal to binary is often called "making change." If we want to determine which coins to give in change for 42 cents, we first select a quarter since it is the largest coin less than 42 cents. We subtract 25 cents and still need 17 cents. We now select a dime since it is the largest coin less than 17 cents. We subtract 10 cents from 17 cents and still need 7 cents. We next select a nickel and subtract 5 cents, leaving 2 cents to add. Similarly, if we have a number such as 83_{ten} that we wish to convert to binary, we first select the highest power of 2 that is less than 83_{ten}. In this case it is $2^6 = 64_{ten}$. Place a 1 in the 2^6 column. Now subtract 64_{ten} from 83_{ten} to get 19_{ten}:

$$\begin{array}{ccccccc} 1 & & & & & & \\ 2^6 & 2^5 & 2^4 & 2^3 & 2^2 & 2^1 & 1 \end{array}$$

We do not need $2^5 = 32_{ten}$, since it is too large, so we place a 0 in the 2^5 column,

$$\begin{array}{ccccccc} 1 & 0 & & & & & \\ 2^6 & 2^5 & 2^4 & 2^3 & 2^2 & 2^1 & 1 \end{array}$$

but we do need $2^4 = 16_{ten}$ so we place a 1 in the 2^4 column:

$$\begin{array}{ccccccc} 1 & 0 & 1 & & & & \\ 2^6 & 2^5 & 2^4 & 2^3 & 2^2 & 2^1 & 1 \end{array}$$

We subtract 16_{ten} from 19_{ten} to get 3_{ten}. We do not need $2^3 = 8_{ten}$ or $2^2 = 4_{ten}$, since they are too large, so 0's are placed in these columns, but we do need a 2 and a 1 and so 1's are placed in these columns:

$$\begin{array}{ccccccc} 1 & 0 & 1 & 0 & 0 & 1 & 1 \\ 2^6 & 2^5 & 2^4 & 2^3 & 2^2 & 2^1 & 1 \end{array}$$

Hence, our number is 1010011_2.

Similarly, $199_{ten} = 11000111$, since we need one $2^7 = 128_{ten}$, one $2^6 = 64_{ten}$, one $2^2 = 4_{ten}$, one 2_{ten}, and one 1_{ten}.

An alternate method to convert from decimal to binary may be seen if we observe that when we divide 234 by 10 in ordinary arithmetic we get quotient 23 and remainder 4. If we then divide 23 by 10, we get quotient 2 with remainder 3. Finally, if we divide 2 by 10, we get quotient 0 and remainder 2. Note that the remainders in reverse order form the original number. Using the following algorithm, an integer n should produce the same number n.

1. Divide n by 10, getting quotient q_1 and remainder r_1.

2. Divide q_1 by 10, getting quotient q_2 and remainder r_2.

3. As long as q_k is not 0, divide q_k by 10, getting quotient q_{k+1} and remainder r_{k+1}.

4. If $q_n = 0$, then $r_n r_{n-1} \cdots r_3 r_2 r_1 = n$.

Using the preceding algorithm in binary, where 10 is the binary number 2, should produce the same results. But since 10_2 is the decimal number 2, dividing by 10 in binary should give the same remainders as dividing by 2 in decimal arithmetic. Consider the following corresponding operations converting 57 to binary, where all the numbers are the same, except in the first column they are expressed in binary and in the second they are expressed in decimal:

Divide 111001 by 10 so $q_1 = 11100$, $r_1 = 1$ Divide 57 by 2 so $q_1 = 28$, $r_1 = 1$
Divide 11100 by 10 so $q_2 = 1110$, $r_2 = 0$ Divide 28 by 2 so $q_2 = 14$, $r_2 = 0$
Divide 1110 by 10 so $q_3 = 111$, $r_3 = 0$ Divide 14 by 2 so $q_3 = 7$, $r_3 = 0$
Divide 111 by 10 so $q_4 = 11$, $r_4 = 1$ Divide 7 by 2 so $q_4 = 3$, $r_4 = 1$
Divide 11 by 10 so $q_5 = 1$, $r_5 = 1$ Divide 3 by 2 so $q_5 = 1$, $r_5 = 1$
Divide 1 by 10 so $q_6 = 0$, $r_6 = 1$ Divide 1 by 2 so $q_6 = 0$, $r_6 = 1$
111001 = 111001   57 = 111001

The following algorithm should then convert a number n from decimal to binary:

1. Divide n by 2, getting quotient q_1 and remainder r_1.

2. Divide q_1 by 2, getting quotient q_2 and remainder r_2.

3. As long as q_k is not 0, divide q_k by 2, getting quotient q_{k+1} and remainder r_{k+1}.

4. If $q_n = 0$, then $(r_n r_{n-1} \cdots r_3 r_2 r_1)_2 = n_{ten}$.

When a positive integer is expressed in **hexadecimal** or base 16 form, we need new symbols for integers because we do not "start over" until we reach 16. In other words, 10_{hex} is equal to 16 in decimal. We include the "numbers" A, B, C, D, E, and F, so that the first 20 positive integers in hexadecimal are 1, 2, 3, 4, 5, 6, 7, 8, 9, A, B, C, D, E, F, 10, 11, 12, 13, and 14. If a number n is represented in hexadecimal form as $a_m a_{m-1} a_{m-2} \cdots a_2 a_1 a_0$, then, expressed in decimal form,

$$n = a_m(16)^m + a_{m-1}(16)^{m-1} + \cdots + a_2(16)^2 + a_1(16) + a_0$$

where for all $0 \leq i \leq m$, $0 \leq a_i < 16$. Therefore, the number $n = 2A3B4_{hex}$ expressed in decimal form would be $n = 2(16)^4 + A(16)^3 + 3(16)^2 + B(16) + 4$.

Use the following table,

$$2(16)^4 = 2 \times 65536 = 130072$$
$$A(16)^3 = 10 \times 4096 = 40960$$
$$3(16)^2 = 3 \times 256 = 768$$
$$B(16) = 11 \times 16 = 176$$
$$4 = 4 \times 1 = 4$$
$$\text{Total} = 171980$$

and $2A3B4_{hex} = 171980_{10}$.

Similarly, $3C9_{hex} = 3(16)^2 + 12(16) + 9 = 3 \times 256 + 12 \times 16 + 9 = 969_{10}$. In a manner analogous to "rollover" for binary numbers, if we add 1 to $FFFFF$, we get 100000. Hence, if we add 1 to $23CFFFF$, we get $23D0000$.

The first method of converting from decimal to hexadecimal is similar to converting from decimal to binary. If we have a decimal number such as 530328, the largest power of 16 that divides 530328 is $16^4 = 65536$. Since 530328 divided by 66536 has quotient 8, put 8 in the first column:

$$\begin{array}{ccccc} 8 & & & & \\ 65536 & 4096 & 256 & 16 & 1 \\ 16^4 & 16^3 & 16^2 & 16^1 & 16^0 \end{array}$$

Since $530328 - 8(65536) = 6040$, divide 6040 by 4096. Since the quotient is 1 and the remainder is 1944, place a 1 in the second column:

$$\begin{array}{ccccc} 8 & 1 & & & \\ 66536 & 4096 & 256 & 16 & 1 \\ 16^4 & 16^3 & 16^2 & 16^1 & 16^0 \end{array}$$

and divide 1944 by 256. Since the quotient is 7 and the remainder is 152, place a 7 in the third column:

$$\begin{array}{ccccc} 8 & 1 & 7 & & \\ 66536 & 4096 & 256 & 16 & 1 \\ 16^4 & 16^3 & 16^2 & 16^1 & 16^0 \end{array}$$

and divide 152 by 16.

Since the quotient is 9 and the remainder is 8, place a 9 in the fourth slot and an 8 in the last column:

$$\begin{array}{ccccc} 8 & 1 & 7 & 9 & 8 \\ 66536 & 4096 & 256 & 16 & 1 \\ 16^4 & 16^3 & 16^2 & 16^1 & 16^0 \end{array}$$

Hence, $530328_{ten} = 81798_{hex}$.

To use the alternative method for converting from hexadecimal to decimal, we simply divide by $16_{ten} = 10_{hex}$ in the previous algorithm for converting decimal to binary instead of $2_{ten} = 10_2$ and we have the following algorithm:

1. Divide n by 16, getting quotient q_1 and remainder r_1.

2. Divide q_1 by 16, getting quotient q_2 and remainder r_2.

3. As long as q_k is not 0, divide q_k by 16, getting quotient q_{k+1} and remainder r_{k+1}.

4. If $q_n = 0$, then $(r_n r_{n-1} \cdots r_3 r_2 r_1)_2 = n_{ten}$.

5.6 Binary and Hexadecimal Numbers

EXAMPLE 5.41

Again consider converting 530328 to hexadecimal. Since $530328 = 16(33145) + 8$, $q_1 = 33145$ and $r_1 = 8$. Since $33145 = 16(2071) + 9$, $q_2 = 2071$ and $r_2 = 9$. Since $2071 = 16(129) + 7$, $q_3 = 129$ and $r_3 = 7$. Since $129 = 16(8) + 1$, $q_4 = 8$ and $r_4 = 1$. Hence, $r_5 = 8$ and $530328_{ten} = 81798_{hex}$. ∎

A rational number using decimal notation, such as 326.149, has the form

$$3(10)^2 + 2(10)^1 + 6(10)^0 + 1(10)^{-1} + 4(10)^{-2} + 3(10)^{-3}$$

or

$$\begin{array}{cccccc} 3 & 2 & 6 & 1 & 4 & 3 \\ (10)^2 & (10)^1 & (10)^0 & (10)^{-1} & (10)^{-2} & (10)^{-3} \end{array}$$

Similarly, a rational number in binary notation, such as 1011.0111, has the form

$$1(2)^3 + 0(2)^2 + 1(2)^1 + 1(2)^0 + 0(2)^{-1} + 1(2)^{-2} + 1(2)^{-3} + 1(2)^{-4}$$

or

$$\begin{array}{cccccccc} 1 & 0 & 1 & 1 & 0 & 1 & 1 & 1 \\ (2)^3 & (2)^2 & (2)^1 & (2)^0 & (2)^{-1} & (2)^{-2} & (2)^{-3} & (2)^{-4} \end{array}$$

Therefore, to change a rational number from binary to decimal notation convert each digit to decimal notation and add.

EXAMPLE 5.42

Convert 1011.1111_2 to decimal notation.

$$\begin{aligned} 1011.1111_2 &= 1(2)^3 + 0(2)^2 + 1(2)^1 + 1(2)^0 + 1(2)^{-1} \\ &\quad + 1(2)^{-2} + 1(2)^{-3} + 1(2)^{-4} \\ &= 8 + 0 + 2 + 1 + \frac{1}{2} + \frac{1}{4} + \frac{1}{8} + \frac{1}{16} \\ &= 11 + .5 + .25 + .125 + .0625 \\ &= 11.9375 \end{aligned}$$
∎

EXAMPLE 5.43

Convert 1010.0101_2 to decimal notation.

$$\begin{aligned} 1010.0101_2 &= 1(2)^3 + 0(2)^2 + 1(2)^1 + 0(2)^0 + 0(2)^{-1} \\ &\quad + 1(2)^{-2} + 0(2)^{-3} + 1(2)^{-4} \\ &= 8 + 0 + 2 + 0 + 0 + \frac{1}{4} + 0 + \frac{1}{16} \\ &= 10 + .25 + .0625 \\ &= 10.3125 \end{aligned}$$
∎

Notice that .3729 multiplied by 10 is 3.729 where the first digit to the right of the decimal point is moved to the left of the decimal point. If we then take .729 and multiply by 10, the 7 is moved to the left of the decimal point. Again taking .29 and multiplying by 10, the 2 is moved to the left of the decimal point. Finally taking .9 and multiplying by 10, the 9 is moved to the left of the decimal point giving 0. The numbers removed from the left of the decimal point form the original digits of .3729. Therefore, given a decimal less than 1, the process of multiplying by 10, selecting the number to the left of the decimal point, removing it, and then repeating the process until no digits are left produces the original digits of the decimal.

Given a binary number less than 1, the same process occurs. For example, given the binary number .1101, multiplying this number by 10_2 or two, we get 1.101. Remove the 1 and multiply .101 by 10 to get 1.01. Remove the 1 and multiply

.01 by 10 to get 0.1. Finally, remove the 0 and multiply .1 by 10 to get 1. Again the numbers removed form the original digits. If we have a decimal number less than 1, and we multiply by 2, then we should get the same number to the left of the decimal point that we get by multiplying the same number in binary notation by 10_2, which is the first digit of the number in binary notation. Thus, by multiplying the number in decimal notation by 2, removing the digit to the left of the decimal point and continuing the process, we should get the digits that form the number in binary notation. Consider the following corresponding operations converting .125 to binary, where all the numbers are the same, except in the first column they are expressed in binary and in the second they are expressed in decimal:

Binary	Decimal
Multiply .001 by 10 to get 0.01	Multiply .125 by 2 to get 0.250
Remove the 0 and save it as the 1st digit.	Remove the 0 and save it as the 1st digit.
Multiply .01 by 10 to get 0.1	Multiply .250 by 2 to get 0.500
Remove the 0 and save it as the 2nd digit.	Remove the 0 and save it as the 2nd digit.
Multiply .1 by 10 to get 1.0	Multiply .500 by 2 to get 1.000
Remove the 1 and save it as the 3rd digit.	Remove the 1 and save it as the 3rd digit.
$.001_2 = .001_2$	$.125_{ten} = .001_2$

EXAMPLE 5.44 Change .875 from decimal to binary. First multiply .875 by 2 to get 1.750. Remove the 1 and keep it as the first binary digit. Multiply .750 by 2 to get 1.500. Remove the 1 and keep it as the second binary digit. Multiply .500 by 2 to get 1.000. Remove the 1 and keep it as the third binary digit. Therefore, the number in binary is .111.

In the following table, the first column shows the multiplication by 2 at each step; the second column shows the result of the multiplication; and the third column shows the integer that is to be removed to form the decimal in base 2:

$$.875 \times 2 \quad 1.750 \quad 1$$
$$.750 \times 2 \quad 1.500 \quad 1$$
$$.500 \times 2 \quad 1.000 \quad 1$$

Thus, $.875_{ten} = .111_2$. ∎

EXAMPLE 5.45 Change .4 from decimal representation to binary representation. Using the preceding procedure, we have

$$.4 \times 2 \quad 0.8 \quad 0$$
$$.8 \times 2 \quad 1.6 \quad 1$$
$$.6 \times 2 \quad 1.2 \quad 1$$
$$.2 \times 2 \quad 0.4 \quad 0$$
$$.4 \times 2 \quad 0.8 \quad 0$$
$$.8 \times 2 \quad 1.6 \quad 1$$
$$.6 \times 2 \quad 1.2 \quad 1$$
$$.2 \times 2 \quad 0.4 \quad 0$$

Notice that when 0.8 is repeated in the second column, the entire process repeats so that the binary representation consists of the digits 0110 repeated. Therefore, $.4_{ten} = .0110011001100110\cdots = .\overline{0110}$. ∎

In a fashion similar to that for converting from binary to decimal, we note that in hexadecimal notation 32F.12 has the form

$$3(16)^2 + 2(16)^1 + F(16)^0 + 1(16)^{-1} + 2(16)^{-2}$$

or

$$3(16)^2 + 2(16)^1 + 15(16)^0 + 1(16)^{-1} + 2(16)^{-2}$$

which is equal to

$$3 \times 256 + 2 \times 16 + 15 + \frac{1}{16} + \frac{2}{256} = 815 + .0625 + .0078125$$
$$= 815.0703125_{ten}$$

Therefore, $32F.12_{hex} = 815.0703125_{ten}$. In converting from decimal to hexadecimal for numbers less than 1, we observe that multiplying by 16 is really multiplying by 10 in the hexadecimal system. Therefore, similar to the conversion from decimal to binary where we multiplied by $2 (= 10_2)$, we multiply by $16 (= 10_{hex})$, remove the digit to the left of the decimal point to use as the first digit, and continuing to multiply by 16, use digits left of the decimal points as our digits in hexadecimal. For example, consider $.0703125_{ten}$. Multiply by 16 to get 1.1250000 so the first digit in hexadecimal is 1. Now multiply $.125 \times 16$ to get 2.0. Therefore, $.0703125_{ten} = .12_{hex}$:

$$\begin{array}{lll} .0703125 \times 16 & 1.125 & 1 \\ .125 \times 16 & 2.0 & 2 \end{array}$$

EXAMPLE 5.46 Convert .875 from decimal to hexadecimal. Multiply .875 by 16 to get $14 = E_{hex}$. Therefore, $.875_{ten} = .E_{hex}$. ∎

EXAMPLE 5.47 Convert .0078125 from decimal to hexadecimal. We have

$$\begin{array}{lll} .0078125 \times 16 & 0.125 & 0 \\ .125 \times 16 & 2.0 & 2 \end{array}$$

and $.0078125_{ten} = .02_{hex}$. ∎

EXAMPLE 5.48 Convert .8 from decimal to hexadecimal. We have

$$\begin{array}{lll} .8 \times 16 & 12.8 & C \\ .8 \times 16 & 12.8 & C \\ .8 \times 16 & 12.8 & C \end{array}$$

and $.8_{ten} = .CCCC\cdots = .\overline{C}$. ∎

From the preceding examples it may be seen that whether or not the representation of a rational repeats or terminates is a function of the base in which the rational is represented and not on the rational number alone.

It is often important in computer science to be able to convert from hexadecimal notation to binary notation and conversely. Observe that if we have the binary representation $\cdots a_{11}a_{10}a_9a_8a_7a_6a_5a_4a_3a_2a_1a_0$, then

$$\cdots a_{11}a_{10}a_9a_8a_7a_6a_5a_4a_3a_2a_1a_0 = \cdots a_{11}2^{11} + a_{10}2^{10} + a_9 2^9 a_8 2^8$$
$$+ a_7 2^7 + a_6 2^6 + a_5 2^5 + a_4 2^4$$
$$+ a_3 2^3 + a_2 2^2 + a_1 2^1 + a_0$$
$$= \cdots (a_{11}2^3 + a_{10}2^2 + a_9 2^1 a_8)2^8$$
$$+ (a_7 2^3 + a_6 2^2 + a_5 2^1 + a_4)2^4$$
$$+ a_3 2^3 + a_2 2^2 + a_1 2^1 + a_0$$
$$= \cdots (a_{11}2^3 + a_{10}2^2 + a_9 2^1 a_8)16^2$$
$$+ (a_7 2^3 + a_6 2^2 + a_5 2^1 + a_4)16$$
$$+ a_3 2^3 + a_2 2^2 + a_1 2^1 + a_0$$

Then if $\cdots a_{11}2^3+a_{10}2^2+a_9 2^1+a_8$, $a_7 2^3+a_6 2^2+a_5 a^2+a_4$, and $a_3 2^3+a_2 2^2+a_1 2^1+a_0$ are converted, respectively, to hexadecimal numbers $\cdots h_2$, h_1, and h_0, we have the number $\cdots +h_3(16)^2+h_2(16)+h_1$, which is the hexadecimal number $\cdots h_2 h_1 h_0$. Therefore, by collecting the binary digits into groups of four and converting them to hexadecimal, we have converted the entire number from binary to hexadecimal. For example, consider the binary number 1010011001010111. The first four digits 1010 form the hexadecimal number A, the next four digits, 0110, form the hexadecimal number 6, the next four digits, 0101, form the hexadecimal number 5, and the last four digits, 0111, form the hexadecimal number 7. Therefore, the hexadecimal representation is $A657$:

$$\underbrace{1010}_{A} \quad \underbrace{0110}_{6} \quad \underbrace{0101}_{5} \quad \underbrace{0111}_{7}$$

If the number of digits in the binary representation is not a multiple of 4, add zeros to the front so that the number of digits in the binary representation is a multiple of 4. For example, given the binary number 1010011110, add 00 to the front to form 001010011110 and we have the conversion

$$\underbrace{0010}_{2} \quad \underbrace{0110}_{6} \quad \underbrace{1001}_{9} \quad \underbrace{1110}_{E}$$

Given a binary representation of a rational number such as 101100.0111101, begin at the decimal point, and going each direction form groups of four binary digits adding 0's at each end if necessary. Then convert each four-digit binary number to hexadecimal. The resulting hexadecimal number is the representation of the original number. Hence, we have

$$\underbrace{0010}_{2} \quad \underbrace{1100}_{C} \quad \underbrace{.0111}_{.7} \quad \underbrace{1010}_{A}$$

and $101100.0111101_2 = 2C.7A_{hex}$. Similarly $011100.11_2 = 00011100.1100 = 1C.C_{hex}$.

Converting from hexadecimal to binary, we reverse the process. Given the hexadecimal number $72F5$, since $7_{hex} = 0111_2$, $2_{hex} = 0010_2$, $F_{hex} = 1111_2$, and $5_{hex} = 0101_2$, then $72F5_{hex} = 0111001011110101_2$. Similarly, $2A.3C = 00101010.00111100 = 101010.001111$.

Addition, subtraction, multiplication, and division follow the same rules using binary representation as when using decimal representation except that the following tables are used for addition and multiplication:

+	0	1
0	0	1
1	1	10

×	0	1
0	0	0
1	0	1

It is important to remember that, in subtraction, when 1 is "borrowed" from 10, only 1 is left since 10_2 is really the number 2.

EXAMPLE 5.49 Perform the following operations in binary:

(a)
```
  1011
  0110
 +0101
 10110
```

(b)
```
  1100
 -1001
   011
```

(c)
```
    1101
   ×101
    1101
   0000
  1101
 1000001
```

Similarly addition, subtraction, multiplication, and division are the same for hexadecimal except the addition and multiplication tables in Figures 5.11 and 5.12 must be used.

Figure 5.11

+	1	2	3	4	5	6	7	8	9	A	B	C	D	E	F
1	2	3	4	5	6	7	8	9	A	B	C	D	E	F	10
2	3	4	5	6	7	8	9	A	B	C	D	E	F	10	11
3	4	5	6	7	8	9	A	B	C	D	E	F	10	11	12
4	5	6	7	8	9	A	B	C	D	E	F	10	11	12	13
5	6	7	8	9	A	B	C	D	E	F	10	11	12	13	14
6	7	8	9	A	B	C	D	E	F	10	11	12	13	14	15
7	8	9	A	B	C	D	E	F	10	11	12	13	14	15	16
8	9	A	B	C	D	E	F	10	11	12	13	14	15	16	17
9	A	B	C	D	E	F	10	11	12	13	14	15	16	17	18
A	B	C	D	E	F	10	11	12	13	14	15	16	17	18	19
B	C	D	E	F	10	11	12	13	14	15	16	17	18	19	1A
C	D	E	F	10	11	12	13	14	15	16	17	18	19	1A	1B
D	E	F	10	11	12	13	14	15	16	17	18	19	1A	1B	1C
E	F	10	11	12	13	14	15	16	17	18	19	1A	1B	1C	1D
F	10	11	12	13	14	15	16	17	18	19	1A	1B	1C	1D	1E

Figure 5.12

×	1	2	3	4	5	6	7	8	9	A	B	C	D	E	F
1	1	2	3	4	5	6	7	8	9	A	B	C	D	E	F
2	2	4	6	8	A	C	E	10	12	14	16	18	1A	1C	1E
3	3	6	9	C	F	12	15	18	1B	1E	21	24	27	2A	2D
4	4	8	C	10	14	18	1C	20	24	28	2C	30	34	38	3C
5	5	A	F	14	19	1E	23	28	2D	32	37	3C	41	46	4B
6	6	C	12	18	1E	24	2A	30	36	3C	42	48	4E	54	5A
7	7	E	15	1C	23	2A	31	38	3F	46	4D	54	5B	62	69
8	8	10	18	20	28	30	38	40	48	50	58	60	68	70	78
9	9	12	1B	24	2D	36	3F	48	51	5A	63	6C	75	7E	87
A	A	14	1E	28	32	3C	46	50	5A	64	6E	78	82	8C	96
B	B	16	21	2C	37	42	4D	58	63	6E	79	84	8F	9A	A5
C	C	18	24	30	3C	48	54	60	6C	78	84	90	9C	A8	B4
D	D	1A	27	34	41	4E	5B	68	75	82	8F	9C	A9	B6	C3
E	E	1C	2A	38	46	54	62	70	7E	8C	9A	A8	B6	C4	D2
F	F	2E	2D	3C	4B	5A	69	78	87	96	A5	B4	C3	D2	E1

Computer Science Application

The following algorithm uses recursion to write the binary representation of a positive integer n. Note that each binary digit is computed as a remainder upon division by 2 but in the reverse order of occurrence.

function writebase2(integer n)

    if $n > 1$ then
        writebase2($n/2$)
    end-if

    print n mod 2

end-function

The division $n/2$ here is taken as integer division. The obvious changes can be made to produce an algorithm that prints the octal representation of n.

function writebase8(integer n)

   if $n > 7$ then
      writebase8($n/8$)
   end-if

   print n mod 8

end-function

A more versatile algorithm that prints the base-x representation of n for any x between 2 and 16 is almost as easy but requires the use of a character array in order to handle the nonstandard digits in representations where $x > 10$.

function writebasex(integer n, integer x)

   character array $c[0..15] = \{0, 1, 2, 3, 4, 5, 6, 7, 8, 9, A, B, C, D, E, F\}$

   if $n > x - 1$ then
      writebasex($n/x, x$)
   end-if

   print $c[n \bmod x]$

end-function

EXAMPLE 5.50

The following operations are performed in hexadecimal:

$$
\begin{array}{ccc}
(a)\quad \begin{array}{r} F051 \\ 315A \\ +4163 \\ \hline 1630E \end{array} &
(b)\quad \begin{array}{r} FD32 \\ -475A \\ \hline B5D8 \end{array} &
(c)\quad \begin{array}{r} 28 \\ \times 34 \\ \hline A0 \\ 78 \\ \hline 820 \end{array}
\end{array}
$$

Computer Science Application

The unique binary representation of 25 is $(11001)_2$, meaning that

$$25 = 2^4 + 2^3 + 2^0$$

From this simple observation we express

$$x^{25} = x^{16} x^8 x^1$$

which is the central idea behind an algorithm that more efficiently raises x to a large power. The following function returns x^n, for $n > 0$.

integer function power(integer x, integer n)

   integer $p = n$, $y = x$, result $= 1$

   while $p > 0$ do
      if p mod $2 = 1$ then
         result $= y*$result
      end-if
      $p = p$ div 2

$$y = y * y$$
end-while

return result

end-function

When $n = 25$, the while-loop executes five times. The five binary digits of n are computed one at a time from right to left as p mod 2 and then discarded by the integer division $p = p$ div 2. Of course, special data structures to handle large-integer arithmetic will be necessary when n is large, but as n gets larger the reduction in the number of multiplications is greater.

Exercises

1. Count from 1 to 25 in binary.
2. Count from 1 to 25 in hexadecimal.

Convert the following from decimal to binary:

3. 45 4. 86 5. 243 6. 194
7. 312 8. 53 9. 99 10. 287
11. 186 12. 334

Convert the following from binary to decimal:

13. 1011 14. 11010 15. 100100 16. 111011
17. 110101 18. 1101 19. 10110 20. 101101
21. 110101 22. 1010101

Convert the following from decimal to hexadecimal:

23. 63 24. 116 25. 376 26. 734
27. 5246 28. 183 29. 297 30. 458
31. 861 32. 36304

Convert the following from hexadecimal to decimal:

33. $4E$ 34. $2F3$ 35. $4FA$ 36. $41FD$
37. $D1E3$ 38. AF 39. $3A4$ 40. $AB2$
41. $D13A$ 42. $21F2D$

Convert the following from decimal to binary:

43. 23.875 44. 43.75 45. 86.125 46. 94.4
47. 52.3 48. 3.1875 49. 22.625 50. 87.71875
51. 16.53125 52. 34.28125

Convert the following from binary to decimal:

53. 1110.11 54. 1010.101 55. 100.1001
56. 11.0111 57. 1101.0101 58. 101.01011
59. 11100.1010 60. 1011.01110 61. 1101.1011

62. 10101.01111

Convert the following from decimal to hexadecimal:

63. 36.625 64. 161.78125 65. 592.15625
66. 1876.54625 67. 5246.34765625 68. 273.9375
69. 349.796875 70. 576.640625 71. 1997.35
72. 36304.140625

Convert the following from hexadecimal to decimal:

73. $5C.3A$ 74. $F12.AE$ 75. $4AF.C48$
76. $32AB.C11$ 77. $13CE.D24$ 78. $14.D2$
79. $11F.E33$ 80. $47F.344$ 81. $237F.755$
82. $233B.FDA$

Convert the following from binary to hexadecimal:

83. 101.1010 84. 110111.010101
85. 110110110.011011111
86. 1110101110.0011011011
87. 1111010100.001000101
88. 110.11001 89. 100101.011110
90. 101110101.010101011
91. 11010101010.11001101101
92. 10001010111.1101001010101

Convert the following from hexadecimal to binary:

93. $2C.4B$ 94. $1F2.A3C$ 95. $AF8.48C$
96. $A12E.11C$ 97. $C14A.2D4$ 98. $CE.2D5$
99. $F1A.E3D$ 100. $B7E.3C4$ 101. $FA37.2E5$
102. $C46B.AC4$

5.7 Signed Numbers

In the preceding section only positive rational numbers were represented in binary and hexadecimal notation. In this section, it is shown how to represent both positive and negative integers using the **two's complement method**. To use this method we need to represent integers using a fixed number of bits (or binary digits). Hence, a number expressed in binary will always be expressed using the same number of digits. If 8 bits are used, then the binary representation of the number 3 is 00000011. For any positive integer, enough 0's are placed in front of the number to obtain the required 8 bits. In the remainder of this section we will use both 8 bits, using binary notation, and 16 bits, using hexadecimal notation. We begin with 8 bits.

A positive integer must begin with a 0, and a negative integer will begin with a 1. Therefore, 10110011 will represent a negative integer and 01110101 will represent a positive integer. Since the largest positive integer is 01111111, there are 127 positive integers. Given a positive integer represented using n bits, its negation is found by subtracting it from 2^n expressed in binary. This is done by subtracting the number from $2^n - 1$ (called finding the 1's complement) and then adding 1. Therefore, for 8 bits, the positive integer is subtracted from

$$2^n - 1 = 11111111$$

and then 1 is added. For example, to find the negation of the number 00101011, subtract it from 11111111:

$$\begin{array}{r} 11111111 \\ -\ 00101011 \\ \hline 11010100 \end{array}$$

getting 11010100 and then add 1. Therefore, the negation of the number 00101011 is 11010101.

Notice that subtracting 00101011 from 11111111 simply changes 0's to 1's and 1's to 0's. Is this always true when we subtract an 8-bit binary string from 11111111?

EXAMPLE 5.51 Find the representation of -56 in 8-bit two's complement. First 56 is represented as 00111000. Subtracting from 11111111, we have 11000111. Adding 1 gives 11001000 so that -56 is represented as 11001000. ∎

Given a negative signed number, to find the positive integer for which it is the negative, simply perform the same operations on the negative number. Thus, to find what positive number for which 10110111 represents the negation, simply subtract from 11111111, getting 01001000 and add 1, getting 01001001, which is the binary representation of 72. Therefore, 10110111 represents -72. The negative two's complement numbers range from 11111111 to 10000000 so there are 128 of them. Thus, using 8 bits, all integers between -128 and 127 can be represented. In general if there are n bits, all integers between -2^{n-1} and $2^{n-1} - 1$ can be represented.

When using 16 bits (or more) to represent integers, it is more convenient to use hexadecimal representation. Since each hexadecimal digit requires 4 binary digits, for 16 bits, 4 hexadecimal digits are needed. Thus, $21A5$, $F9AB$, and $004F$ are examples of 16-bit signed integers using hexadecimal representation. Since the binary representation of a positive integer must begin with a 0, the largest that the first digit can be in hexadecimal representation is 7 ($= 0111_2$). Therefore, $691F$, $7FFF$, and 0031 represent positive integers while 8124, $A105$, and $FFF1$ represent negative integers. As before, if a positive integer is represented using n bits, its negation is found by subtracting it from 2^n. This is done for 16 bits by subtracting the positive integer in hexadecimal from $FFFF$ and then adding 1. As before, to convert the negative integer back to its positive counterpart, we again perform the two's complement operation, so it is subtracted from

$$2^{16} - 1 = FFFF$$

and 1 is added.

EXAMPLE 5.52

Find the 16-bit hexadecimal representation of the negation of the hexadecimal number $78F$. First, change to the 16 bit hexadecimal number $078F$. Next subtract $078F$ from $FFFF$ getting $F870$, and then add 1. The representation is, therefore, $F871$. ∎

EXAMPLE 5.53

Find the 16-bit hexadecimal representation of -1270_{ten}. Since $1270_{ten} = 486_{hex} = 0486$ represented in 16-bit hexadecimal, subtract 0486 from $FFFF$ getting $FB79$ and add 1 getting $FB7A$. Hence, the 16-bit hexadecimal representation of -1270_{ten} is $FB7A$. ∎

Addition using both binary and hexadecimal fixed-length representations of signed numbers is the same as the usual addition of binary and hexadecimal numbers except that if the sum produces an extra digit, then the first digit is removed. For example, if 0111 and 1011 are added as binary numbers,

$$\begin{array}{r} 0111 \\ \underline{1011} \\ [1]\,0010 \end{array}$$

the number 1 on the left is removed. If hexadecimal signed numbers $FF14$ and $0F1A$ are added as hexadecimal numbers,

$$\begin{array}{r} FF14 \\ \underline{0F1A} \\ [1]\,0E2E \end{array}$$

the number 1 on the left is again removed.

If we attempt to add $7F14$ and $6E23$ as before, we have

$$\begin{array}{r} 7F14 \\ \underline{6E23} \\ ED37 \end{array}$$

so we have added two positive numbers and obtained a negative number. In this case we say that overflow has occurred. The sum is too large to be expressed with 16 bits. A signed integer expressed with 16 binary bits or 4 digits hexadecimal must be between -32768 and 32767.

In decimal arithmetic, to subtract 25 from 75, we add -25 to 75. In many textbooks it says to "change the sign of the subtrahend [number to be subtracted] and proceed as in addition." When adding signed integers we follow a similar process. To subtract one integer from another, we find the two's complement of the number to be subtracted and then add.

For example in the following problem,

$$\begin{array}{r} 01010110 \\ -\ \underline{01110001} \end{array}$$

first find the two's complement of 01110001, which is 10001111, and then add:

$$\begin{array}{r} 01010110 \\ +\ 10001111 \\ \hline 11100101 \end{array}$$

As expected, the result is negative.

Using the same process in the following problem,

$$\begin{array}{r} 01010001 \\ -\ 11101001 \\ \hline \end{array}$$

first find the two's complement of 11101001, which is 00010111, and then add:

$$\begin{array}{r} 01010001 \\ +\ 00010111 \\ \hline 01101000 \end{array}$$

In this case we have subtracted a negative integer from a positive integer, so we change the negative number to a positive number and add.

EXAMPLE 5.54

Subtract 56 from 23 using 8-bit signed numbers. Since $56 = 00111000$ and $23 = 00010111$, we have

$$\begin{array}{r} 00010111 \\ -\ 00111000 \\ \hline \end{array}$$

First find the two's complement of 00111000, which is 11001000, and then add:

$$\begin{array}{r} 00010111 \\ +\ 11001000 \\ \hline 11011111 \end{array}$$

Since the two's complement of 11011111 is 00100001, which is equal to 33, 11011111 is equal to -33. ■

EXAMPLE 5.55

Subtract 7328 from 3614 using 16-bit hexadecimal signed numbers. Since $7328_{dec} = 1CA0_{hex}$ and $3614 = E1E_{hex}$, we have

$$\begin{array}{r} 0E1E \\ -1CA0 \\ \hline \end{array}$$

First find the two's complement of $1CA0$, which is $E360$, and then add:

$$\begin{array}{r} 0E1E \\ +E360 \\ \hline F17E \end{array}$$

Since the two's complement of $F1E$ is $0E82$ and $0E82 = 3714_{dec}$, $3614 - 7328 = -3714_{dec}$. ■

Exercises

Find the two's complement of the following in 8-bit binary form:

1. 01110110
2. 10110101
3. 00110111
4. 11101011
5. 01100100
6. 11011001
7. 00111011
8. 10010000

Express the following decimals as 8-bit binary signed numbers:

9. 73
10. -101
11. -37
12. -14
13. 48
14. -33
15. -58
16. -114

Perform the following operations using 8-bit binary signed numbers:

17. 10110100
    + 01110010

18. 00110100
    + 01001010

19. 00110101
    − 01110110

20. 00110101
    − 01110000

21. 11001001
    + 01011110

22. 00101001
    + 01010011

23. 01011100
    − 01110111

24. 01101101
    − 01111000

Convert the following to 8-bit binary signed numbers and perform the indicated operations:

25. 43 + 28
26. 12 + 81
27. 43 − 67
28. 86 − 73
29. 29 + 61
30. 34 − 56
31. 13 − 87
32. 22 − 98

Find the two's complement of the following signed numbers expressed in hexadecimal form:

33. $723A$
34. $AB20$
35. $FA12$
36. $23BC$
37. $23CF$
38. $1B5C$
39. $FF13$
40. 1020

Express the following decimals as signed numbers in 16-bit hexadecimal form:

41. 270
42. −893
43. −17
44. −9872
45. 48
46. −237
47. −8858
48. −12499

Perform the following operations using signed numbers in 16-bit hexadecimal form:

49. $3C14$ + $89AC$
50. $C12F$ + $359A$
51. $B127$ − $C1AB$
52. $34C5$ − $FF12$
53. 4713 + $0F23$
54. 5379 + $FA12$
55. $37B4$ − $AA44$
56. $C129$ − $D48A$

Convert the following decimal numbers to 16-bit hexadecimal signed numbers and perform the indicated operations:

57. 936 + 258
58. 1639 + 3926
59. 8836 − 19923
60. 994 − 1713

Convert the following to 16-bit hexadecimal signed numbers and perform the indicated operations:

61. 829 + 1499
62. 778 − 569
63. 13 − 87
64. 19237 − 18

65. Find the range of integers that can be expressed as signed numbers using 32 bits.

66. Find the range of integers that can be expressed as signed numbers using 64 bits.

5.8 Matrices Continued

The determinant is an important function from the set of real valued matrices to the set of real numbers. It is used in the solution of simultaneous equations (Cramer's rule) and in determining the inverse of a matrix (if it has one).

Given an $n \times n$ matrix,

$$A = \begin{bmatrix} A_{11} & A_{12} & \cdots & A_{1n} \\ A_{21} & A_{22} & \cdots & A_{2n} \\ \vdots & & & \vdots \\ A_{n1} & A_{n2} & \cdots & A_{nn} \end{bmatrix},$$

the symbol $\overline{A_{ij}}$ will denote the $(n-1) \times (n-1)$ matrix obtained by removing the ith row and the jth column from A.

EXAMPLE 5.56 Let

$$A = \begin{bmatrix} A_{11} & A_{12} \\ A_{21} & A_{22} \end{bmatrix}$$

Then $\overline{A_{11}} = [A_{22}]$, $\overline{A_{12}} = [A_{21}]$, $\overline{A_{21}} = [A_{12}]$, and $\overline{A_{22}} = [A_{11}]$. ∎

EXAMPLE 5.57 Let
$$A = \begin{bmatrix} A_{11} & A_{12} & A_{13} \\ A_{21} & A_{22} & A_{23} \\ A_{31} & A_{32} & A_{33} \end{bmatrix}$$

Then
$$\overline{A_{11}} = \begin{bmatrix} A_{22} & A_{23} \\ A_{32} & A_{33} \end{bmatrix}, \overline{A_{22}} = \begin{bmatrix} A_{11} & A_{13} \\ A_{31} & A_{33} \end{bmatrix}, \text{ and } \overline{A_{31}} = \begin{bmatrix} A_{12} & A_{13} \\ A_{22} & A_{23} \end{bmatrix} \quad \blacksquare$$

DEFINITION 5.58 The **determinant** of the $n \times n$ matrix
$$A = \begin{bmatrix} A_{11} & A_{12} & \cdots & A_{1n} \\ A_{21} & A_{22} & \cdots & A_{2n} \\ \vdots & & & \vdots \\ A_{n1} & A_{n2} & \cdots & A_{nn} \end{bmatrix}$$

denoted by $\det(A)$, by $|A|$, or by
$$\begin{vmatrix} A_{11} & A_{12} & \cdots & A_{1n} \\ A_{21} & A_{22} & \cdots & A_{2n} \\ \vdots & & & \vdots \\ A_{n1} & A_{n2} & \cdots & A_{nn} \end{vmatrix}$$

is defined by

(a) If $n = 1$, then $\det(A) = |A_{11}| = A_{11}$.

(b) If $n \geq 2$, then $\det(A) = \sum_{j=1}^{n} A_{1j}(-1)^{1+j} \det(\overline{A_{1j}})$.

Let $M_{ij} = \det(\overline{A_{ij}})$. M_{ij} is called the **minor** of A_{ij}. $C_{ij} = (-1)^{i+j} \det(\overline{A_{ij}})$ is called the **cofactor** of A_{ij}. Hence, $\det(A) = \sum_{j=1}^{n} A_{1j}(-1)^{1+j} \det(\overline{A_{1j}}) = \sum_{j=1}^{n} A_{1j} C_{1j}$.

Note that in the preceding definition, $|A_{11}|$ is not the absolute value of A_{11} even though the notation appears to be the same.

At this point we are interested in an algorithmic procedure for the determinant. First we note that an $n \times n$ determinant is defined in terms of $(n-1) \times (n-1)$ determinants. This is an example of a recursive algorithm, which will be discussed more fully in later sections. The procedure for the determinant is given as follows:

Procedure Determinant(A, n)
    For $i = 1$ to n
        If $n = 1$, then $\det(A) = |A_{11}| = A_{11}$.
        If $n \geq 2$, then $\det(A) = 0$
        For $j = 1$ to n
            The value of $\det(A)$ is replaced by
            $\det(A) + A_{1j}(-1)^{1+j}$ Determinant$(\overline{A_{1j}}, n-1)$.
        Endfor
    Endfor
End

Since determinant (A, n) is determined in terms of determinant$(\overline{A_{1j}}, n-1)$, we must apply determinant $(\overline{A_{1j}}, n-1)$ first to evaluate determinant (A, n). Many computer programs have the property that they can "call" themselves. This enables us to write a computer program very similar to the preceding procedure. This will be discussed more fully in later sections.

Hence, if
$$A = \begin{bmatrix} A_{11} & A_{12} \\ A_{21} & A_{22} \end{bmatrix}$$
then $\det(A) = A_{11} \cdot |A_{22}| - A_{12} \cdot |A_{21}| = A_{11} \cdot A_{22} - A_{12} \cdot A_{21}$.

If
$$A = \begin{bmatrix} A_{11} & A_{12} & A_{13} \\ A_{21} & A_{22} & A_{23} \\ A_{31} & A_{32} & A_{33} \end{bmatrix}$$
then
$$\det(A) = A_{11} \cdot \begin{vmatrix} A_{22} & A_{23} \\ A_{32} & A_{33} \end{vmatrix} - A_{12} \cdot \begin{vmatrix} A_{22} & A_{23} \\ A_{31} & A_{33} \end{vmatrix} + A_{13} \cdot \begin{vmatrix} A_{21} & A_{22} \\ A_{31} & A_{32} \end{vmatrix}$$

EXAMPLE 5.59 Let $A = \begin{bmatrix} 2 & 3 \\ 1 & 2 \end{bmatrix}$. Then

$$\det(A) = \begin{vmatrix} 2 & 3 \\ 1 & 2 \end{vmatrix} = 2|2| + 3(-|1|) = 2 \times 2 - 3 \times 1 = 1$$

In general, if $A = \begin{bmatrix} a & b \\ c & d \end{bmatrix}$, then $\det(A) = ad - bc$. ∎

EXAMPLE 5.60 Let $A = \begin{bmatrix} 1 & 2 & 0 \\ 1 & 1 & 2 \\ 3 & 0 & 1 \end{bmatrix}$. Then

$$\det(A) = \begin{vmatrix} 1 & 2 & 0 \\ 1 & 1 & 2 \\ 3 & 0 & 1 \end{vmatrix}$$

$$= 1 \begin{bmatrix} 1 & 2 \\ 0 & 1 \end{bmatrix} - 2 \begin{bmatrix} 1 & 2 \\ 3 & 1 \end{bmatrix} + 0 \begin{bmatrix} 1 & 2 \\ 3 & 1 \end{bmatrix}$$

$$= 1(1 \times 1 - 2 \times 0) - 2(1 \times 1 - 2 \times 3) + 0$$

$$= 1 - 2(-5) = 11$$ ∎

Define δ_{ij} by
$$\delta_{ij} = \begin{cases} 1 \text{ if } i = j \\ 0 \text{ if } i \neq j \end{cases}$$

DEFINITION 5.61 Let I_n be the $n \times n$ matrix $[I_{ij}]$ where $I_{ij} = \delta_{ij}$ so that I_n has 1's down the diagonal and 0's elsewhere. I_n is called the $n \times n$ **identity matrix** or simply the **identity matrix** and has the property that for any $n \times n$ matrix A, $AI_n = I_n A = A$.

EXAMPLE 5.62 The 2×2 and 3×3 identity matrices are $I_2 = \begin{bmatrix} 1 & 0 \\ 0 & 1 \end{bmatrix}$ and $I_3 = \begin{bmatrix} 1 & 0 & 0 \\ 0 & 1 & 0 \\ 0 & 0 & 1 \end{bmatrix}$. ∎

DEFINITION 5.63 Let A be an $n \times n$ matrix. The **multiplicative inverse** A^{-1} of A (if it exists) is an $n \times n$ matrix such that $AA^{-1} = A^{-1}A = I_n$.

EXAMPLE 5.64 The matrix $\begin{bmatrix} 2 & 5 \\ 1 & 3 \end{bmatrix}$ is the inverse of $\begin{bmatrix} 3 & -5 \\ -1 & 2 \end{bmatrix}$ since

$$\begin{bmatrix} 2 & 5 \\ 1 & 3 \end{bmatrix}\begin{bmatrix} 3 & -5 \\ -1 & 2 \end{bmatrix} = \begin{bmatrix} 3 & -5 \\ -1 & 2 \end{bmatrix}\begin{bmatrix} 2 & 5 \\ 1 & 3 \end{bmatrix} = \begin{bmatrix} 1 & 0 \\ 0 & 1 \end{bmatrix}$$

■

EXAMPLE 5.65 The matrix $\begin{bmatrix} 2 & 1 & 0 \\ -1 & 2 & 2 \\ 0 & 1 & 1 \end{bmatrix}$ is the inverse of $\begin{bmatrix} 0 & -1 & 2 \\ 1 & 2 & -4 \\ -1 & -2 & 5 \end{bmatrix}$ since

$$\begin{bmatrix} 2 & 1 & 0 \\ -1 & 2 & 2 \\ 0 & 1 & 1 \end{bmatrix}\begin{bmatrix} 0 & -1 & 2 \\ 1 & 2 & -4 \\ -1 & -2 & 5 \end{bmatrix} = \begin{bmatrix} 0 & -1 & 2 \\ 1 & 2 & -4 \\ -1 & -2 & 5 \end{bmatrix}\begin{bmatrix} 2 & 1 & 0 \\ -1 & 2 & 2 \\ 0 & 1 & 1 \end{bmatrix}$$
$$= \begin{bmatrix} 1 & 0 & 0 \\ 0 & 1 & 0 \\ 0 & 0 & 1 \end{bmatrix}$$

■

The following is an algorithm for finding the inverse of an $n \times n$ matrix A:

1. Find $\det(A)$. If it is zero, there is no inverse.
2. For each A_{ij}, find the cofactor C_{ij} and form the matrix $C = [C_{ij}]$, called the matrix of cofactors or **cofactor matrix**.
3. Find the transpose C^t of C, so that $C^t_{ij} = C_{ji}$. The matrix C^t is called the **adjoint** of A and is denoted adj(A).
4. Divide each element of adj(A) by $\det(A)$.
5. $A^{-1} = \dfrac{1}{\det(A)}\text{adj}(A)$.

Hence, if $B = \begin{bmatrix} a & b \\ c & d \end{bmatrix}$, then

$$B^{-1} = \begin{bmatrix} \dfrac{d}{\det(B)} & \dfrac{-b}{\det(B)} \\ \dfrac{-c}{\det(B)} & \dfrac{a}{\det(B)} \end{bmatrix}$$

where $\det(B) = ad - bc$. It is straightforward to show by direct computation that $BB^{-1} = B^{-1}B = I_2$. Similarly, to find the inverse of A if

$$A = \begin{bmatrix} A_{11} & A_{12} & A_{13} \\ A_{21} & A_{22} & A_{23} \\ A_{31} & A_{32} & A_{33} \end{bmatrix}$$

we define the matrix of cofactors $C = [C_{ij}]$ by letting C_{ij} be formed by taking the determinant of the matrix obtained from $A = [A_{ij}]$ by removing the row and column containing A_{ij}, and multiplying by the correct sign, namely, $(-1)^{i+j}$. Finally, we take the transpose of C, and divide by $\det(A)$ to get A^{-1} and

$$A^{-1} = \frac{1}{\det(A)} \cdot \begin{bmatrix} \begin{vmatrix} A_{22} & A_{23} \\ A_{32} & A_{33} \end{vmatrix} & -\begin{vmatrix} A_{12} & A_{13} \\ A_{32} & A_{33} \end{vmatrix} & \begin{vmatrix} A_{12} & A_{13} \\ A_{22} & A_{23} \end{vmatrix} \\ -\begin{vmatrix} A_{21} & A_{23} \\ A_{31} & A_{33} \end{vmatrix} & \begin{vmatrix} A_{11} & A_{13} \\ A_{31} & A_{33} \end{vmatrix} & -\begin{vmatrix} A_{11} & A_{13} \\ A_{21} & A_{23} \end{vmatrix} \\ \begin{vmatrix} A_{21} & A_{22} \\ A_{31} & A_{32} \end{vmatrix} & -\begin{vmatrix} A_{11} & A_{12} \\ A_{31} & A_{32} \end{vmatrix} & \begin{vmatrix} A_{11} & A_{12} \\ A_{21} & A_{22} \end{vmatrix} \end{bmatrix}$$

EXAMPLE 5.66 Let $A = \begin{bmatrix} 3 & -5 \\ -1 & 2 \end{bmatrix}$. Then $\det(A) = 1$. The cofactor matrix of A is

$$C = \begin{bmatrix} 2 & -(-1) \\ -(-5) & 3 \end{bmatrix} = \begin{bmatrix} 2 & 1 \\ 5 & 3 \end{bmatrix}.$$

The adjoint of A is

$$\text{Adj}(A) = C^t = \begin{bmatrix} 2 & 5 \\ 1 & 3 \end{bmatrix}$$

The inverse of A is

$$A^{-1} = \frac{1}{1} \begin{bmatrix} 2 & 5 \\ 1 & 3 \end{bmatrix} = \begin{bmatrix} 2 & 5 \\ 1 & 3 \end{bmatrix}$$ ∎

EXAMPLE 5.67 Let $A = \begin{bmatrix} 0 & -1 & 2 \\ 1 & 2 & -4 \\ -1 & -2 & 5 \end{bmatrix}$. Then

$$\det(A) = 0 \cdot \begin{vmatrix} 2 & -4 \\ -2 & 5 \end{vmatrix} - (-1) \cdot \begin{vmatrix} 1 & -4 \\ -1 & 5 \end{vmatrix} + 2 \cdot \begin{vmatrix} 1 & 2 \\ -1 & -2 \end{vmatrix}$$
$$= 0 + 1 + 0 = 1$$

The cofactor matrix of A is

$$C = \begin{bmatrix} \begin{vmatrix} 2 & -4 \\ -2 & 5 \end{vmatrix} & -\begin{vmatrix} 1 & -4 \\ -1 & 5 \end{vmatrix} & \begin{vmatrix} 1 & 2 \\ -1 & -2 \end{vmatrix} \\ -\begin{vmatrix} -1 & 2 \\ -2 & 5 \end{vmatrix} & \begin{vmatrix} 0 & 2 \\ -1 & 5 \end{vmatrix} & -\begin{vmatrix} 0 & -1 \\ -1 & -2 \end{vmatrix} \\ \begin{vmatrix} -1 & 2 \\ 2 & -4 \end{vmatrix} & -\begin{vmatrix} 0 & 2 \\ 1 & -4 \end{vmatrix} & \begin{vmatrix} 0 & -1 \\ 1 & 2 \end{vmatrix} \end{bmatrix}$$

$$= \begin{bmatrix} 2 & -1 & 0 \\ 1 & 2 & 1 \\ 0 & 2 & 1 \end{bmatrix}$$

Thus, the adjoint of A

$$\text{Adj}(A) = C^t = \begin{bmatrix} 2 & 1 & 0 \\ -1 & 2 & 2 \\ 0 & 1 & 1 \end{bmatrix}$$

and

$$A^{-1} = \frac{1}{1}\begin{bmatrix} 2 & 1 & 0 \\ -1 & 2 & 2 \\ 0 & 1 & 1 \end{bmatrix} = \begin{bmatrix} 2 & 1 & 0 \\ -1 & 2 & 2 \\ 0 & 1 & 1 \end{bmatrix}$$ ■

If we want to solve the set of simultaneous equations

$$x_1 + 2x_2 = 5$$
$$3x_1 + 4x_2 = 11$$

we see that this is equivalent to

$$\begin{bmatrix} 1 & 2 \\ 3 & 4 \end{bmatrix}\begin{bmatrix} x_1 \\ x_2 \end{bmatrix} = \begin{bmatrix} 5 \\ 11 \end{bmatrix}$$

The inverse of $\begin{bmatrix} 1 & 2 \\ 3 & 4 \end{bmatrix}$ is $\begin{bmatrix} -2 & 1 \\ \frac{3}{2} & -\frac{1}{2} \end{bmatrix}$. Multiplying both sides of the previous equation by $\begin{bmatrix} -2 & 1 \\ \frac{3}{2} & -\frac{1}{2} \end{bmatrix}$, we have

$$\begin{bmatrix} -2 & 1 \\ \frac{3}{2} & -\frac{1}{2} \end{bmatrix}\begin{bmatrix} 1 & 2 \\ 3 & 4 \end{bmatrix}\begin{bmatrix} x_1 \\ x_2 \end{bmatrix} = \begin{bmatrix} -2 & 1 \\ \frac{3}{2} & -\frac{1}{2} \end{bmatrix}\begin{bmatrix} 5 \\ 11 \end{bmatrix}$$

or

$$\begin{bmatrix} 1 & 0 \\ 0 & 1 \end{bmatrix}\begin{bmatrix} x_1 \\ x_2 \end{bmatrix} = \begin{bmatrix} -2 & 1 \\ \frac{3}{2} & -\frac{1}{2} \end{bmatrix}\begin{bmatrix} 5 \\ 11 \end{bmatrix}$$

so that

$$\begin{bmatrix} x_1 \\ x_2 \end{bmatrix} = \begin{bmatrix} -2 & 1 \\ \frac{3}{2} & -\frac{1}{2} \end{bmatrix}\begin{bmatrix} 5 \\ 11 \end{bmatrix} = \begin{bmatrix} 1 \\ 2 \end{bmatrix}$$

and we have $x_1 = 1$ and $x_2 = 2$. Thus, if we know the inverse of the matrix of coefficients of a system of linear equations, it is easy to find the solutions.

Consider the set of simultaneous equations

$$x_1 + 2x_2 + 3x_3 = 8$$
$$2x_1 + 5x_2 + 4x_3 = 16$$
$$x_1 - x_2 + 10x_3 = 9$$

which we describe as

$$\begin{bmatrix} 1 & 2 & 3 \\ 2 & 5 & 4 \\ 1 & -1 & 10 \end{bmatrix}\begin{bmatrix} x_1 \\ x_2 \\ x_3 \end{bmatrix} = \begin{bmatrix} 8 \\ 16 \\ 9 \end{bmatrix}$$

Multiplying both sides of the equation by the inverse of $\begin{bmatrix} 1 & 2 & 3 \\ 2 & 5 & 4 \\ 1 & -1 & 10 \end{bmatrix}$, which is $\begin{bmatrix} 54 & -23 & -7 \\ -16 & 7 & 2 \\ -7 & 3 & 1 \end{bmatrix}$, we have

$$\begin{bmatrix} 54 & -23 & -7 \\ -16 & 7 & 2 \\ -7 & 3 & 1 \end{bmatrix}\begin{bmatrix} 1 & 2 & 3 \\ 2 & 5 & 4 \\ 1 & -1 & 10 \end{bmatrix}\begin{bmatrix} x_1 \\ x_2 \\ x_3 \end{bmatrix} = \begin{bmatrix} 54 & -23 & -7 \\ -16 & 7 & 2 \\ -7 & 3 & 1 \end{bmatrix}\begin{bmatrix} 8 \\ 16 \\ 9 \end{bmatrix}$$

or

$$\begin{bmatrix} 1 & 0 & 0 \\ 0 & 1 & 0 \\ 0 & 0 & 1 \end{bmatrix} \begin{bmatrix} x_1 \\ x_2 \\ x_3 \end{bmatrix} = \begin{bmatrix} 54 & -23 & -7 \\ -16 & 7 & 2 \\ -7 & 3 & 1 \end{bmatrix} \begin{bmatrix} 8 \\ 16 \\ 9 \end{bmatrix} = \begin{bmatrix} 1 \\ 2 \\ 1 \end{bmatrix}$$

so that

$$\begin{bmatrix} x_1 \\ x_2 \\ x_3 \end{bmatrix} = \begin{bmatrix} 54 & -23 & -7 \\ -16 & 7 & 2 \\ -7 & 3 & 1 \end{bmatrix} \begin{bmatrix} 8 \\ 16 \\ 9 \end{bmatrix} = \begin{bmatrix} 1 \\ 2 \\ 1 \end{bmatrix}$$

and we have $x_1 = 1$, $x_2 = 2$, and $x_3 = 1$.

Thus, we see that if we have n linear equations in n variables

$$A_{11}x_1 + A_{12}x_2 + A_{13}x_3 + \cdots A_{1n}x_n = B_1$$
$$A_{21}x_1 + A_{22}x_2 + A_{23}x_3 + \cdots A_{2n}x_n = B_2$$
$$A_{31}x_1 + A_{32}x_2 + A_{33}x_3 + \cdots A_{3n}x_n = B_3$$
$$\vdots$$
$$A_{n1}x_1 + A_{n2}x_2 + A_{n3}x_3 + \cdots A_{nn}x_n = B_n$$

where the determinant of A, the matrix of the coefficients of the variables, has an inverse, then

$$\begin{bmatrix} x_1 \\ x_2 \\ x_3 \\ \vdots \\ x_n \end{bmatrix} = A^{-1} \begin{bmatrix} B_1 \\ B_2 \\ B_3 \\ \vdots \\ B_n \end{bmatrix}$$

We now give an algorithm that may be used to calculate the inverse of a matrix. This method is much easier for hand calculation than the algorithm presented previously. We shall not give a proof of this algorithm. Intuitively we may note that when certain operations are performed on the identity matrix, it passes it on. For example, if row 1 is added to row 2 of the 3×3 identity matrix so we have

$$\begin{bmatrix} 1 & 0 & 0 \\ 1 & 1 & 0 \\ 0 & 0 & 1 \end{bmatrix}$$

and we then multiply

$$\begin{bmatrix} 1 & 0 & 0 \\ 1 & 1 & 0 \\ 0 & 0 & 1 \end{bmatrix} \begin{bmatrix} a \\ b \\ c \end{bmatrix} = \begin{bmatrix} a \\ a+b \\ c \end{bmatrix}$$

we note that the third row of the second matrix has the first row added to the second row. Similarly, if we interchange row 1 and row 3 of the identity matrix so we have

$$\begin{bmatrix} 0 & 0 & 1 \\ 0 & 1 & 0 \\ 1 & 0 & 0 \end{bmatrix}$$

and we then multiply

$$\begin{bmatrix} 0 & 0 & 1 \\ 0 & 1 & 0 \\ 1 & 0 & 0 \end{bmatrix} \begin{bmatrix} a \\ b \\ c \end{bmatrix} = \begin{bmatrix} c \\ b \\ a \end{bmatrix}$$

we note that the first row and the third row of the second matrix are interchanged.

Basically we are going to perform operations on the identity matrix that change a given matrix A to the identity. This will change the identity matrix to the inverse since, when it is multiplied times A, it gives the identity matrix.

We are going to do this using three basic rules.

1. A row of a matrix may be multiplied by a scalar.
2. A multiple of a row may be added to another row.
3. Two rows may be interchanged.

To do this we place a given $n \times n$ matrix A and the identity side by side forming an $n \times 2n$ matrix. We then apply the foregoing operations 1–3 to change the matrix A to the identity.

For example, to find the inverse of the matrix

$$\begin{bmatrix} 1 & 3 \\ 1 & 2 \end{bmatrix}$$

we form the matrix

$$\left[\begin{array}{cc|cc} 1 & 3 & 1 & 0 \\ 1 & 2 & 0 & 1 \end{array} \right]$$

We now perform operations to change A to the identity matrix. To do this we first subtract row 1 from row 2, denoted by $\mathbf{R}_2 - \mathbf{R}_1$, to get

$$\left[\begin{array}{cc|cc} 1 & 3 & 1 & 0 \\ 0 & -1 & -1 & 1 \end{array} \right]$$

We then add 3 times row 2 to row 1, denoted $3\mathbf{R}_2 + \mathbf{R}_1$, to get

$$\left[\begin{array}{cc|cc} 1 & 0 & -2 & 3 \\ 0 & -1 & -1 & 1 \end{array} \right]$$

We then multiply row 2 by -1, denoted $-\mathbf{R}_2$, to get

$$\left[\begin{array}{cc|cc} 1 & 0 & -2 & 3 \\ 0 & 1 & 1 & -1 \end{array} \right]$$

We now claim that $\begin{bmatrix} -2 & 3 \\ 1 & -1 \end{bmatrix}$ is the inverse of $\begin{bmatrix} 1 & 3 \\ 1 & 2 \end{bmatrix}$.

Now consider a 3×3 matrix

$$\begin{bmatrix} 1 & 2 & 3 \\ 2 & 5 & 4 \\ 1 & -1 & 10 \end{bmatrix}$$

We first form

$$\left[\begin{array}{ccc|ccc} 1 & 2 & 3 & 1 & 0 & 0 \\ 2 & 5 & 4 & 0 & 1 & 0 \\ 1 & -1 & 10 & 0 & 0 & 1 \end{array} \right]$$

Then we perform operations $\mathbf{R}_2 - 2\mathbf{R}_1$ and $\mathbf{R}_3 - \mathbf{R}_1$ to get

$$\left[\begin{array}{ccc|ccc} 1 & 2 & 3 & 1 & 0 & 0 \\ 0 & 1 & -2 & -2 & 1 & 0 \\ 0 & -3 & 7 & -1 & 0 & 1 \end{array} \right]$$

We now perform operations $\mathbf{R}_1 - 2\mathbf{R}_2$ and $\mathbf{R}_3 + 3\mathbf{R}_2$ to get

$$\left[\begin{array}{ccc|ccc} 1 & 0 & 7 & 5 & -2 & 0 \\ 0 & 1 & -2 & -2 & 1 & 0 \\ 0 & 0 & 1 & -7 & 3 & 1 \end{array} \right]$$

We now perform operations $R_1 - 7R_3$ and $R_2 + 2R_3$ to get

$$\begin{bmatrix} 1 & 0 & 0 & | & 54 & -23 & -7 \\ 0 & 1 & 0 & | & -16 & 7 & 2 \\ 0 & 0 & 1 & | & -7 & 3 & 1 \end{bmatrix}$$

and the inverse of

$$\begin{bmatrix} 1 & 2 & 3 \\ 2 & 5 & 4 \\ 1 & -1 & 10 \end{bmatrix} \text{ is } \begin{bmatrix} 54 & -23 & -7 \\ -16 & 7 & 2 \\ -7 & 3 & 1 \end{bmatrix}$$

Exercises

Find the determinant of the matrices:

1. $\begin{bmatrix} 1 & 3 \\ 2 & 5 \end{bmatrix}$

2. $\begin{bmatrix} 3 & 1 \\ -4 & 6 \end{bmatrix}$

3. $\begin{bmatrix} 1 & 3 \\ 2 & 5 \end{bmatrix}$

4. $\begin{bmatrix} 3 & 1 \\ -4 & 6 \end{bmatrix}$

5. $\begin{bmatrix} 3 & 2 \\ 9 & 6 \end{bmatrix}$

6. $\begin{bmatrix} 3 & -2 & -3 \\ -6 & 1 & 0 \\ 2 & 0 & -1 \end{bmatrix}$

7. $\begin{bmatrix} 6 & -4 & -5 \\ 1 & -11 & 2 \\ 3 & -1 & 1 \end{bmatrix}$

8. $\begin{bmatrix} 3 & -2 & -1 \\ -6 & 1 & 2 \\ 9 & 0 & -3 \end{bmatrix}$

9. $\begin{bmatrix} 1 & -6 & 3 \\ -6 & -1 & 2 \\ 3 & 2 & 1 \end{bmatrix}$

10. $\begin{bmatrix} 0 & 0 & 1 & 0 \\ 0 & 0 & 0 & 1 \\ 0 & 1 & 0 & 0 \\ 1 & 0 & 0 & 0 \end{bmatrix}$

Find A^{-1} if $A = $

11. $\begin{bmatrix} 2 & 3 \\ 5 & 7 \end{bmatrix}$

12. $\begin{bmatrix} 3 & -5 \\ -4 & 7 \end{bmatrix}$

13. $A = \begin{bmatrix} 2 & -5 \\ -4 & 10 \end{bmatrix}$

14. $A = \begin{bmatrix} 3 & -5 \\ 2 & 6 \end{bmatrix}$

15. $\begin{bmatrix} 4 & -1 & 3 \\ -3 & 1 & -3 \\ 2 & 0 & 1 \end{bmatrix}$

16. $\begin{bmatrix} 7 & -7 & 3 \\ 4 & -3 & 2 \\ 2 & -2 & 1 \end{bmatrix}$

17. $A = \begin{bmatrix} 1 & 4 & -1 \\ 0 & 1 & 1 \\ 1 & 1 & 6 \end{bmatrix}$

18. $A = \begin{bmatrix} 3 & -1 & 3 \\ 4 & 6 & 3 \\ 9 & -3 & 9 \end{bmatrix}$

19. Find the inverse of the matrix of coefficients and use it to solve

$$2x_1 + 5x_3 = 14$$
$$x_1 + 3x_2 = 8$$

20. Find the inverse of the matrix of coefficients and use it to solve

$$-x_2 + 2x_3 = 8$$
$$x_1 + 2x_2 - 4x_3 = -23$$
$$-x_1 - 2x_2 + 5x_3 = -31$$

21. Find the inverse of the matrix of coefficients and use it to solve

$$4x_1 - x_2 + 3x_3 = 4$$
$$-3x_1 + x_2 - 3x_3 = 6$$
$$-x_1 - 2x_2 + 5x_3 = 7$$

22. Find the inverse of the matrix of coefficients and use it to solve

$$2x_1 - 1x_2 = 4$$
$$-x_1 + 2x_2 + 2x_3 = 6$$
$$x_2 + x_3 = 8$$

23. Find the inverse of the matrix of coefficients and use it to solve

$$7x_1 - 7x_2 + 3x_3 = 14$$
$$4x_1 - 3x_2 + 2x_3 = 16$$
$$2x_1 - 2x_2 + x_3 = 24$$

24. A permutation matrix is described in Definition 4.39. Prove that if matrix A is a permutation matrix, then $A^{-1} = A^t$.

25. Prove that if A has an inverse, then it is unique.

26. Prove that if A is an $n \times n$ matrix with inverse, then $((A)^{-1})^{-1} = A$.

Assume that if A and B are matrices, then $\det(AB) = \det(A)\det(B)$.

27. Prove that the determinant of the identity matrix is 1.

28. Prove that if A has an inverse, then $\det(A^{-1}) = \dfrac{1}{\det(A)}$.

29. Prove that a matrix A has an inverse if and only if $\det(A) \neq 0$.

30. A matrix is **invertible** if it has an inverse. Prove that the product of two invertible matrices is invertible.
31. Assume that A and B are $n \times n$ matrices with inverses. Prove that $(AB)^{-1} = B^{-1}A^{-1}$.
32. Assume that A is an $n \times n$ matrix with inverse and $k \neq 0$. Prove that the inverse of kA is $\frac{1}{k}A^{-1}$.
33. Prove that if A^t has an inverse, then $(A^t)^{-1} = (A^{-1})^t$.
34. Prove that if A has an inverse and $A^2 = A$, then A is the identity matrix.
35. Prove that if A has an inverse and B is not the zero matrix, then $AB \neq 0$.
36. Prove that if AB has an inverse, then B has an inverse.
37. Prove that if AB has an inverse, then A has an inverse.

Show that the following matrices do not have inverses:

38. $\begin{bmatrix} 1 & -1 & 2 \\ 0 & 0 & 0 \\ 2 & 0 & 1 \end{bmatrix}$

39. $\begin{bmatrix} 1 & 1 & 7 \\ 2 & 2 & 4 \\ 3 & 3 & 9 \end{bmatrix}$

40. $\begin{bmatrix} 0 & 2 & 4 \\ 0 & 1 & 4 \\ 0 & 3 & 7 \end{bmatrix}$

41. $\begin{bmatrix} 1 & 1 & 1 \\ 2 & 2 & 2 \\ 3 & 1 & 7 \end{bmatrix}$

GLOSSARY

Chapter 5: Algorithms and Recursion

5.1 The "for" Procedure and Algorithms for Matrices

Topological sorting	Given partial orderings, $\leq_1$ *and* $\leq_2$, the ordering $\leq_2$ is a **topological sorting** of the ordering $\leq_1$ *if* $\leq_2$ is a total ordering and if whenever $a \leq_1 b$ then $a \leq_2 b$.

5.2 Recursive Functions and Algorithms

Ackermann's function	**Ackermann's function**, denoted by Ack $: N \to N$, is defined recursively on $N \times N$, where N is the set of nonnegative integers, by $\text{Ack}(0, n) = n + 1$ $\text{Ack}(m, 0) = \text{Ack}(m - 1, 1)$ if $m > 0$ $\text{Ack}(m, n) = \text{Ack}(m - 1, \text{Ack}(m, n - 1))$ if $n, m > 0$.
Catalan numbers	The sequence of **Catalan numbers** is given by the formula $$\text{Cat}(n) = \frac{(2n)!}{(n+1)!(n!)}$$
Solving a recursive function	This involves changing the description of a function from a recursive function to a nonrecursive function.
Tower of Hanoi	The **Tower of Hanoi** puzzle consists of three pegs, a collection of disks, and a myth. Disks are moved from one peg to another without placing a larger disk on a smaller disk.

5.3 Complexity of Algorithms

Big-Oh	If g dominates f, this is denoted by $f(n) = O(g(n))$. The symbol $O(g(n))$ is read **"big-Oh"** of $g(n)$ and $f(n)$ is said to have order **big-Oh** of $g(n)$.
Big-Omega	If g is dominated by f, this is denoted by $f(n) = \Omega(g(n))$. The symbol $\Omega(g(n))$ is read **"big-Omega"** of $g(n)$ and $f(\text{n})$ is said to have order **big-Omega** of $g(n)$.
Big-Theta	If $f(n) = O(g(n))$ and $f(n) = \Omega(g(n))$, then $f(n) = \Theta(g(n))$ and $f(\text{n})$ is said to have *big-Theta* of $g(n)$.
Domination of functions	Let f and g be functions whose domain is the positive integers and range is the real numbers. The function g **dominates** the function f if there is a real number k and a positive integer m such that $|f(n)| \leq k |g(n)|$ for all $n \geq m$.

5.4 Sorting Algorithms

Bubble sort	Start at the right and compare the last two items, moving the lesser item to the left. Continue until the first item is reached. The result is that the smallest item is the first term. The process is repeated until it reaches the second item. The second item now becomes the second smallest item. This is continued until the items are all sorted.
Divide and conquer	The technique of breaking sets into forming sets, performing the process on them, and putting the sets back together.
Insertion sort	**Insertion sort** begins by taking the data items one at a time and placing them where they belong in the list already created. Thus, a place is found where all elements to the left of the item being inserted are less than the item, and all items to the right are greater.
Merge sort	Divide the items to be sorted into halves or as nearly so as possible. Then take each of these two lists and divide them in half. Continue this process until each set contains only one item. Now reverse the process, putting the lists together exactly as they were divided but merging them as they are put together.
Quick sort	Pick a_1 the first element in the list of items, then place every item (excluding a_1) that is less than or equal to a_1 to the left of a_1. Call this list S_1. Place every element greater than a_1 to the right of a_1. Call this list S_2. Now apply the same process to S_1 and S_2. Continue this process until each list has only one element in it. Then put the list back together by concatenating $S_i a_i S_{i+1}$ where a_1 is the item that created S_i and S_{i+1}.
Selection sort	Let $a_1, a_2, a_3, ..., a_n$ be the items being sorted. Begin with a_1. Find the smallest of the items and exchange it with a_1. We then move to a_2. Now compare a_2 and all the items after it to find the smallest item. When the smallest item is found, it is exchanged with a_2. The process is continued until the sort is complete.

5.5 Prefix and Suffix Notation

Fully parenthesized	An arithmetic expression is **fully parenthesized** if, for every operation in the expression, there corresponds a pair of parentheses that enclose the operator and the values on which it operates.
Infix notation	An expression is in **infix notation** if the operation is between the values on which it operates.
LIFO	Last-In, First Out.
Pop	Removing data from the stack.
Prefix notation Polish notation	An expression is in **prefix notation** if the operation immediately precedes the values on which it operates.
Push	Adding data to the stack.
Stack	A **stack** (informally) is defined as an area where data are stored such that the last data stored are the first retrieved.
Suffix notation Reverse Polish notation	An expression is in **suffix notation** if the operation immediately follows the values on which it operates.

5.6 Binary and Hexadecimal Numbers

Binary form	A number n represented in base 2 or binary form where every digit is either 1 or 0.
Hexadecimal	A number expressed in base 16 is in **hexadecimal** form (i.e., 1, 2, 3, 4, 5, 6, 7, 8, 9, A, B, C, D, E, and F).

5.7 Signed Numbers

Two's complement	The negation of a number is found by subtracting the number from $2^n - 1$ (called the 1's complement) and then adding 1.

5.8 Matrices Continued

$\overline{A_{ij}}$	$\overline{A_{ij}}$ represents the $(n-1) \times (n-1)$ matrix obtained by removing the ith row and the jth column.
Cofactor	$C_{ij} = (-1)^{i+j} \det(\overline{A_{ij}})$ is called the **cofactor** of A_{ij}.
Determinant of an $n \times n$ matrix	The **determinant** of the $n \times n$ matrix. $$A = \begin{bmatrix} A_{11} & A_{12} & A_{13} & \ldots & A_{1n} \\ A_{21} & A_{22} & A_{23} & \ldots & A_{2n} \\ \vdots & & & & \vdots \\ A_{m1} & A_{m2} & A_{m3} & \ldots & A_{mn} \end{bmatrix}$$ denoted by $\det(A)$, or by $|A|$. $$A = \begin{vmatrix} A_{11} & A_{12} & A_{13} & \ldots & A_{1n} \\ A_{21} & A_{22} & A_{23} & \ldots & A_{2n} \\ \vdots & & & & \vdots \\ A_{m1} & A_{m2} & A_{m3} & \ldots & A_{mn} \end{vmatrix}$$ is defined by a. If $n = 1$, then $\det(A) = |A_{11}| = A_{11}$. b. If $n \geq 2$, then $\det(A) = \sum_{j=1}^{n} A_{1j}(-1)^{i+j} \det(\overline{A_{1j}})$. $$\det(A) = \sum_{j=1}^{n} A_{1j}(-1)^{1+j} \det(\overline{A_{1j}}) = \sum_{j=1}^{n} A_{ij} C_{1j}$$
Identity matrix	If I_n is the $n \times n$ matrix $[I_{ij}]$ where $I_{ij} = \delta_{ij}$ so that I_n has 1's down the diagonal and 0's elsewhere. I_n is called the $n \times n$ **identity matrix** or simply the **identity matrix** and has the property that for any $n \times n$ matrix A, $AI_n = I_n A = A$.
Minor	$M_{ij} = \det(\overline{A_{ij}})$. M_{ij} is called the **minor** of A_{ij}.
Multiplicative inverse	If A is an $n \times n$ matrix, then the **multiplicative inverse** A^{-1} of A (if it exists) is an $n \times n$ matrix such that $AA^{-1} = A^{-1}A = I_n$.

TRUE-FALSE QUESTIONS

1. A topological sorting on a set A has no elements that are not comparable.
2. Given a partially ordered set, there is a unique topological sorting.
3. $f(k+1) = k \cdot f(k)$ uniquely defines a function on the positive integers.
4. Given $\text{Cat}(0) = 1$ and $\text{Cat}(n+1) = \dfrac{n(2n+1)}{n+2} \text{Cat}(n)$, the 6th Catalan number $\text{Cat}(6) = 8$.
5. $a_n = (2n^2 + 1)$ satisfies the recursive function
$$\begin{cases} a_0 = 1 \\ a_k = a_{k-1} + 2k \text{ for } k > 1 \end{cases}$$

6. $a_n = -1 - 2^{n+1}$ satisfies the recursive function
$$\begin{cases} a_0 = -3 \\ a_1 = -5 \\ a_k = 6a_{k-1} - 8a_{k-2} - 3 \text{ for } k > 1 \end{cases}$$

7. $a_n = -2(-1)^n + 2 \cdot 3^n$ satisfies the recursive function
$$\begin{cases} a_0 = 0 \\ a_1 = 7 \\ a_k = 2a_{k-1} + 3a_{k-2} \text{ for } k > 2 \end{cases}$$

8. $n^2 = O(\ln(n!))$.

9. $\ln(n^{10}) = O(\log_{10}(n))$.

10. If $T(n)$ is the time to move n objects in the Tower of Hanoi puzzle, then $T(n) = T(n-1) + T(n-2)$.

11. Given
$$\begin{aligned} \text{Ack}(0, n) &= n + 1 \\ \text{Ack}(m, 0) &= \text{Ack}(m - 1, 1) && \text{if } m > 0 \\ \text{Ack}(m, n) &= \text{Ack}(m - 1, \text{Ack}(m, n-)) && \text{if } n, m > 0 \end{aligned}$$

Ack(2, 1) = Ack(1, 2).

12. If $f(n) = O(g(n))$ and $f(n) = O(h(n))$, then $f(n) = O(g(n) + h(n))$.

13. In general, Quick Sort is more efficient than Merge Sort.

14. Bubble Sort and Quick Sort use the divide and conquer method.

15. In a stack, the last item entered is the first removed.

16. There is not a sort algorithm that is the most efficient in all cases.

17. In general, Merge Sort is more efficient than Bubble Sort for large amounts of data.

18. $3 + (4 \cdot (5 - 6) + 7)$ is fully parenthesized.

19. The "pop" operation removes an item from the stack.

20. The expression $(a + b) \div (cd + e)$ in infix notation corresponds to $ab + cd \cdot e + \div$ in postfix notation.

21. The expression $\div + ab + \cdot cd2$ in prefix notation corresponds to $(a + b) \div (c + 2d)$ in infix notation.

22. $\div a + b + \cdot cd2$ is a proper prefix statement.

23. $ab + \div cd \cdot e+$ is a proper postfix statement.

24. The postfix expression corresponding to a given infix statement is unique.

25. Prefix and infix notations do not require parentheses.

26. Postfix notation is also known as Polish notation.

27. If a positive integer in decimal notation is divided by 2, the remainder will be the first digit of the integer expressed in binary notation.

28. If a rational number expressed in decimal notation terminates (eventually becomes 0 beyond some decimal place), then it will also terminate in hexadecimal and binary notation.

29. If a rational number expressed in hexadecimal notation terminates, then it also terminates, in binary notation.

30. The number FF in hexadecimal notation corresponds to 11111111 in binary notation.

31. $1101100_{bin} = 108_{ten}$.

32. $111111_{bin} + 11_{bin} = 1111110_{bin}$.

33. $FA64_{hex} = 64100_{ten}$.

34. $2A3E_{hex} + 78BC_{hex} = AF2A_{hex}$.

35. $2FEC.78_{hex} = 10111111101100.01111$.

36. In 8-bit binary form, the two's complement of 11011010 is 00100110.

37. In 16-bit hexadecimal form, the two's complement of $2C3F$ is $D3C0$.

38. $3BC4 - 7C3D = BF87$.

39. A signed integer expressed with 16 binary bits must be between -32767 and 32767.

40. Using the recursive form of a function is preferred because it is more efficient in time and storage space.

41. Solving a recursive function means to find its value for a given integer n.

42.
$$\begin{vmatrix} 1 & 2 & 0 \\ 1 & 1 & 2 \\ 3 & 0 & 1 \end{vmatrix} = 11$$

43. The multiplicative inverse of a permutation matrix is a permutation matrix.

44. In the matrix $A = \begin{bmatrix} 2 & 3 & 2 \\ 1 & 1 & 2 \\ 4 & 1 & 3 \end{bmatrix}$ the cofactor of A_{21} is $\begin{vmatrix} 3 & 2 \\ 1 & 3 \end{vmatrix}$.

45. The multiplicative inverse of $\begin{bmatrix} 1 & 2 \\ 1 & 3 \end{bmatrix}$ is $\begin{bmatrix} 3 & -2 \\ -1 & -1 \end{bmatrix}$.

46. If A has an inverse, then $det(A) = det(A^{-1})$.

47. Using Selection Sort on the data k, n, w, p, y, a, c, after the second pass through the data, we should have a, c, k, n, p, w, y.

48. Using Bubble Sort on the data y, z, k, a, c, v, b, after the first pass through the data, we should have a, y, z, k, b, c, v.

49. Of the sorting algorithms Selection Sort, Bubble Sort, Quick Sort, Merge Sort, and Insertion Sort, only Bubble Sort and Merge Sort use partitioning.

50. Quick Sort is most efficient when the number of items sorted is a power of 2.

SUMMARY QUESTIONS

1. Change $2cd + \div 5ab3^\wedge + -\times$ from postfix to infix notation.
2. Convert 58 from decimal to binary notation.
3. Convert $5D2F$ from hexadecimal to binary notation.
4. Change $\div \times +acd - \times bef$ from prefix to infix notation.
5. Find the two's complement of 10110100.
6. Compute using signed numbers in 16-bit hexadecimal form:
   $$\begin{array}{r} 43CF \\ -5BC6 \\ \hline \end{array}$$
7. Find the determinant of the matrix:
   $$\begin{bmatrix} 2 & 1 & 3 \\ 3 & 1 & -2 \\ 1 & 0 & 2 \end{bmatrix}$$
8. Find the inverse of the matrix:
   $$\begin{bmatrix} 2 & 1 & 0 \\ -1 & 2 & 2 \\ 0 & 1 & 1 \end{bmatrix}$$
9. Find the two's complement of $4C78$.
10. Sort the sequence $a, z, h, x, j, p, y, w, q$ using Quick Sort.
11. Sort the sequence $a, z, h, x, j, p, y, w, q$ using Bubble Sort.
12. Using pattern recognition, solve the recursive function by finding explicit nonrecursive expressions for $f(n)$.
    $$\begin{cases} f(1) = 3 \\ f(k) = 3kf(k-1) \text{ for } k > 1 \end{cases}$$
13. Prove that $a_n = n^2 + n + 1$ satisfies the equation
    $$\begin{cases} a_0 = 1 \\ a_k = a_{k-1} + 2k \text{ for } k > 0 \end{cases}$$
14. Prove that all polynomials of degree 3 equal $O(n^3)$.
15. Estimate the number of arithmetic operations necessary to evaluate the product of the matrices
    $$\begin{bmatrix} 1 & 2 & 5 & 1 \\ 3 & 2 & 4 & 4 \end{bmatrix} \begin{bmatrix} 3 & 2 & 2 \\ 0 & 1 & 3 \\ 1 & 0 & 1 \\ 3 & 5 & 0 \end{bmatrix}$$

ANSWERS TO TRUE-FALSE

1. T 2. F 3. F 4. F 5. F 6. T 7. F 8. F 9. T 10. F
11. F 12. F 13. F 14. F 15. T 16. T 17. T 18. F 19. T
20. T 21. F 22. T 23. F 24. T 25. F 26. F 27. F 28. F
29. T 30. T 31. T 32. F 33. T 34. F 35. T 36. T 37. F
38. T 39. F 40. F 41. F 42. T 43. T 44. F 45. F 46. F
47. F 48. T 49. F 50. T

CHAPTER 6

Graphs, Directed Graphs, and Trees

6.1 Graphs

In Chapter 2, we showed how graphs could be used to represent relations. Graph theory, however, takes on a life of its own and is an important field in mathematics. In this chapter, we will define graphs directly without using relations.

DEFINITION 6.1   A **graph** is a finite set V called the **vertex set** and a collection E of two-element subsets of V. The set E is called the **edge set**. An element of E is called an **edge**. A graph is denoted by $G(V, E)$. The elements a and b of V are **joined** or **connected** by the edge $\{a, b\}$ if $\{a, b\} \in E$.

A finite graph is usually represented by a diagram in which the vertices are represented by dots and the edges connecting two vertices are represented by lines between their dots.

DEFINITION 6.2   If $\{a, b\}$ is an edge, then vertices a and b are called the **endpoints** of the edge $\{a, b\}$. The edge $\{a, b\}$ is also said to be **incident** on the vertices a and b. Conversely, vertices a and b are said to be incident on the edge $\{a, b\}$. Two vertices are **adjacent** if they are endpoints of an edge or equivalently they are incident on the same edge. Two edges are **adjacent** if they are incident on a common vertex.

EXAMPLE 6.3   The graph with $V = \{a, b, c\}$ and $E = \{\{a, b\}, \{b, c\}\}$ may be shown as Figure 6.1 or Figure 6.2.   ∎

EXAMPLE 6.4   The graph with $V = \{a, b, c, d, e\}$ and

$$E = \{\{a, b\}, \{a, e\}, \{b, e\}, \{b, d\}, \{b, c\}, \{c, d\}\}$$

has the diagram shown in Figure 6.3.   ∎

Figure 6.1

Figure 6.2

Figure 6.3

In many texts these graphs are referred to as simple graphs. This type of graph is the one that is most commonly used in graph theory. The restriction to only one edge between two vertices allows us to represent an edge as a set of two elements, namely, the vertices of the edge. We now mention general types of graphs, which will be used in other chapters. Many of the theorems in this chapter, however, will also be true for the more general types of graphs. An edge from a vertex to itself is called a **loop**. If we wish to include loops, the structure will be called a **graph with loops**. If we allow more than one edge between two vertices, we call this a **multigraph**. If each edge is labeled, we call it a **labeled graph**. If we allow both loops and more than one edge between two vertices, we call this a **pseudograph**.

Graph theory began with the solution of the Königsberg bridge problem by the famous mathematician Euler in 1736. On the surface, there seem to be two versions of the problem. The problem originated in the Prussian village of Königsberg on the Pregel River. The citizens of Königsberg liked to walk on paths that included seven bridges over the Pregel River. People wondered if they could start on one of the land masses, cross all of the bridges only once, and return to the point where they started without having to swim. The difference in the two versions is in the geography. In one version the Pregel River contained an island called Kneiphof and a region of land caused by the branching of the Pregel River. The seven bridges and the region are shown in Figure 6.4.

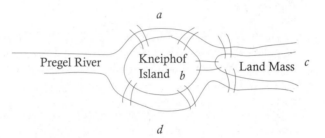

Figure 6.4

In the second version, the Pregel River contained two islands, one of which was called Kneiphof Island and the other was not. This version is in Figure 6.5.

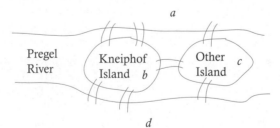

Figure 6.5

The correct version is unknown to the authors. The point is that it really does not matter. Euler produced the multigraph in Figure 6.6, which would work in either case. In this multigraph, Euler let the land masses be vertices and the paths over the bridges be edges. The problem, thus, becomes beginning at any vertex, crossing each edge exactly once, and returning to the original vertex. Using this multigraph we will later show how Euler solved the problem. This illustration points out two of the benefits of mathematical abstraction. First, abstraction eliminates nonessential parts of a problem and focuses on the concepts in the problem that are really necessary. Second, abstraction often shows that problems that appear to be different are actually

Figure 6.6

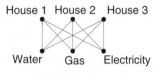

Figure 6.7

essentially the same or at least have similar solutions. This is particularly true in graph theory.

We shall look at the solution for the Königsberg bridge problem when we study Euler graphs in Section 6.2.

Another problem that may be modeled using graph theory is the three house–three utilities problem. Three houses are each to receive three utilities, say water, gas, and electricity, by means of underground lines consisting of cables or pipes. The question is whether the utilities can be supplied to the three houses without any of the lines crossing. The graph that models this problem is shown in Figure 6.7.

A more practical example occurs when designing a chip. At each level we don't want the wires crossing. We will examine this type of problem when we study planar graphs in a later chapter.

DEFINITION 6.5

The **degree** of a vertex v, denoted by $\deg(v)$, is the number of edges that are incident on that vertex. A vertex with degree 0 is called **isolated**.

EXAMPLE 6.6

In the graph shown in Figure 6.8, a and c are adjacent vertices and e_1, e_2, and e_3 are all adjacent edges. However, a and f are not adjacent vertices and e_2, and e_5 are not adjacent edges. Vertices b, c, and d have degree 2, whereas vertices a and f have degree 3. ∎

THEOREM 6.7 The sum of the degrees of the vertices of a graph is always even.

Proof Since each edge in a graph has two endpoints, the degree of each of the two endpoints is increased by 1 due to the given edge. Thus, each edge contributes 2 to the sum of the degrees of the vertices and this sum must be twice the number of edges. Hence, it is an even number. ∎

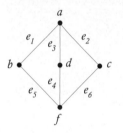

Figure 6.8

We now use this theorem to prove the next theorem.

THEOREM 6.8 In any graph, there is an even number of vertices whose degree is odd.

Proof This is an example of the use the **reductio ad absurdum** or **proof by contradiction** form of proof. We begin by assuming that the theorem is not true and then reach a contradiction. We then conclude that the theorem is true. If this theorem is not true, then there is an odd number of vertices whose degree is odd. Thus, the sum of the degrees of the vertices with odd degrees is odd. But the sum of the degrees of the vertices with even degrees is even. The sum of the degrees of all of the vertices is the sum of the degrees of the vertices with odd degrees added to the sum of the degrees of the vertices with even degrees. Therefore, since the sum of an odd number and an even number is odd, the sum of the degrees of all of the vertices is odd. But this contradicts Theorem 6.7 so we have reached a contradiction. Hence, we conclude that the theorem is true. ∎

DEFINITION 6.9

A graph $G'(V', E')$ is a **subgraph** of a graph $G(V, E)$, denoted by $G'(V', E') \preceq G(V, E)$, if $V' \subseteq V$ and $E' \subseteq E$. Thus, every vertex in G' is a vertex in G and every edge in G' is an edge in G.

EXAMPLE 6.10

The graphs in Figures 6.9, 6.10, and 6.11 are all subgraphs of the graph in Figure 6.12. ∎

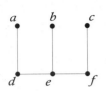
Figure 6.9

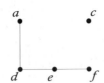
Figure 6.10

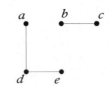
Figure 6.11

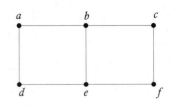
Figure 6.12

A path in a graph is simply a collection of edges that are joined together at the vertices so that one can move along them smoothly. In the following definition, we define a path more formally. For convenience, in the remainder of this section, we shall use $v_0, v_1, v_2, v_3, \ldots, v_k, u,$ and v for vertices.

DEFINITION 6.11   Let $G = G(V, E)$ be a graph with vertices $v_0, v_k \in V$. A **path of length k** from v_0 to v_k (or between v_0 and v_k) is a sequence $v_0 e_1 v_1 e_2 v_2 e_3 v_3 \ldots, v_k e_k v_k$ such that vertices $v_0, v_i, \ldots v_k \in V$; edges $e_1, e_2, \ldots, e_k \in E$ and $e_i = \{v_{i-1}, v_i\}$. Thus a path of length k has k edges. Because of the redundancy in this notation for a graph, this path will generally be denoted by $v_0 v_1 v_2 v_3 \ldots v_k$. Each two consecutive edges in a path share a common vertex and are, therefore, adjacent. A **simple path** from v_0 to v_k is a path in which no vertex is repeated.

EXAMPLE 6.12   In the graph shown in Figure 6.13, paths from v_0 to v_7 include $v_0 v_1 v_2 v_5 v_7$, $v_0 v_1 v_2 v_5 v_4 v_1 v_2 v_5 v_7$, $v_0 v_1 v_4 v_5 v_4 v_5 v_7$, and $v_0 v_3 v_4 v_6 v_7$, which are paths of length 4, 8, 6, and 4, respectively. The paths $v_0 v_1 v_2 v_5 v_7$ and $v_0 v_3 v_4 v_6 v_7$ are simple paths. ∎

DEFINITION 6.13   A graph G is **connected** if there is a path between any two distinct vertices in G.

EXAMPLE 6.14   The graph in Figure 6.14 is not connected. For example, there is no path from v_0 to v_3 or v_2 to v_4. ∎

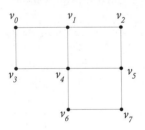
Figure 6.13

Returning to the graph in Figure 6.13, we note that the path $v_0 v_1 v_2 v_5 v_4 v_1 v_2 v_5 v_7$ could be shortened to $v_0 v_1 v_2 v_5 v_7$. Since the vertex v_1 was repeated, we simply removed the part between the two occurrences of v_1 and went from the first occurrence of v_1 directly to v_2. Similarly, the path $v_0 v_1 v_4 v_5 v_4 v_5 v_7$ could be reduced to $v_0 v_1 v_4 v_5 v_7$. Thus, if any vertex v_i occurs more than once on a path, we can remove one of them and the vertices between the two occurrences of the vertex v_i. We can continue doing this until we have removed repetitions of any of the vertices. This gives us the following theorem.

THEOREM 6.15   Let $G = G(V, E)$ be a graph. If there is a path from a vertex v_i to a vertex v_j, then there is a simple path from v_i to v_j. ∎

Combining Definition 6.13 and Theorem 6.15, we immediately have the following corollary (a corollary is a theorem that follows directly from another theorem).

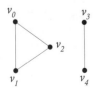
Figure 6.14

Corollary 6.16   A graph G is connected if and only if there is a simple path between any two vertices in G. ∎

6.1 Graphs

DEFINITION 6.17 Let $G = G(V, E)$ be a graph. A subgraph G' of G is a **component** of G if
(1) G' is a nonempty connected graph and
(2) if G'' is a connected subgraph of G and $G' \preceq G''$ then $G' = G''$. Hence, G' is a maximal connected subgraph of G.

EXAMPLE 6.18 In the graph shown in Figure 6.14, the components are shown in figures in Figures 6.15 and 6.16.

DEFINITION 6.19 Let $G = (V, E)$ be a graph. A **cycle** is a path of length greater than zero from a vertex to itself with no repeated edges. A **simple cycle** is a cycle from a vertex v to itself such that v is the only vertex that is repeated. An ***n*-cycle** is a cycle containing n edges and n distinct vertices.

EXAMPLE 6.20 Returning to the graph in Figure 6.13, the paths $v_0v_1v_4v_3v_0$, $v_0v_1v_4v_5v_7v_6v_4v_3v_0$, $v_1v_2v_5v_7v_6v_4v_1$, and $v_0v_1v_2v_5v_7v_6v_4v_3v_0$ are cycles. All are simple cycles except $v_0v_1v_4v_5v_7v_6v_4v_3v_0$.

At this point we want to introduce some specific graphs that will be useful in future sections.

DEFINITION 6.21 A graph is a **complete graph** if there is an edge between every two distinct vertices. A complete graph with n vertices is denoted by K_n.

EXAMPLE 6.22 The graphs K_2, K_3, K_4, and K_5 are shown, respectively, in Figure 6.17.

DEFINITION 6.23 A graph $G = (V, E)$ is called a **bipartite** graph if V can be expressed as the disjoint union of nonempty sets, say $V = A \cup B$, such that every edge has the form $\{a, b\}$ where $a \in A$ and $b \in B$. Thus, every edge connects a vertex in A to a vertex in B and no vertices both in A or both in B are connected. A bipartite graph is called a **complete bipartite** graph $K_{m,n}$ if A contains m vertices, B contains n vertices, and for every $a \in A$, $b \in B$, $\{a, b\} \in E$. Thus, for every $a \in A$ and $b \in B$, there is an edge connecting them.

EXAMPLE 6.24 The graphs $K_{1,2}$, $K_{2,3}$, $K_{2,2}$, and $K_{3,3}$, respectively, are shown in Figure 6.18.

Figure 6.15   Figure 6.16   Figure 6.17

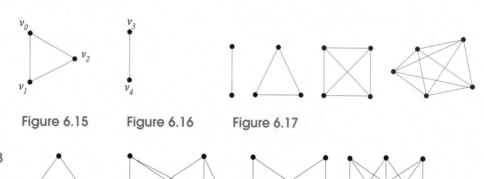

Figure 6.18

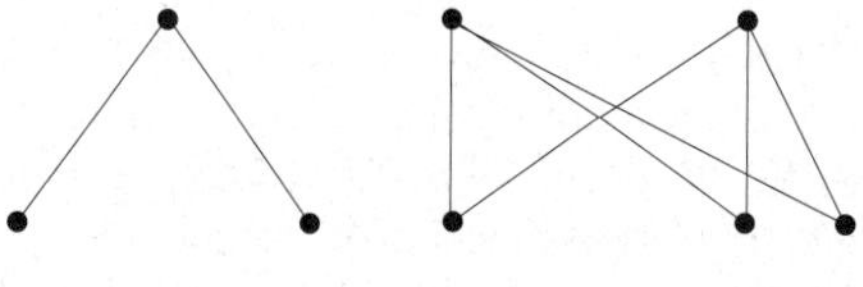

Exercises

Which are paths of the graph in Figure 6.19? Which are simple paths? Give the length of each path.

1. $aebfcd$
2. $aecdaec$
3. $aebecfbd$
4. $aecfbdafc$

Figure 6.19

Which are paths of the graph in Figure 6.20? Which are simple paths? Give the length of each path.

5. $abcabcd$
6. $bcdeca$
7. $debace$
8. $decab$

Figure 6.20

Which are cycles of the graph in Figure 6.21? Which are simple cycles? For each n-cycle, give the value of n.

9. $dabcfbed$
10. $bfcedbfcb$
11. $abcfebfca$
12. $aecfbda$

Figure 6.21

Which are cycles of the graph in Figure 6.22? Which are simple cycles? For each n-cycle, give the value of n.

13. $abcdbaea$
14. $ebcdbcdae$
15. $adcbea$
16. $adbcdea$

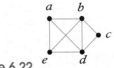

Figure 6.22

Draw the following graphs:

17. K_6
18. $K_{1,3}$
19. $K_{1,4}$
20. $K_{3,4}$

21. Prove that if a graph contains a cycle from a vertex v to itself, then it contains a simple cycle from the vertex v to itself.

22. In the game of pickup sticks, a set of "sticks" is tossed on the table and one tries to pick them up one at a time without disturbing the other "sticks." Let the "sticks" be vertices of a graph. There is an edge between two vertices if one stick is touching the other stick. What would be the components of this graph?

23. Show that every reflexive relation has a corresponding graph with loops and conversely every graph with loops *at each vertex* describes a reflexive relation. Is a graph with loops necessarily reflexive?

24. What is the maximal degree of a vertex in a graph containing n vertices?

25. What is the maximal number of edges in a graph containing n vertices?

26. Is a connected graph with at least two vertices bipartite if it is a subgraph of a bipartite graph?

27. *Prove that a bipartite graph with n vertices has a maximum of $\frac{n^2}{4}$ edges.

28. How many edges does a graph with 10 vertices contain if each vertex has degree 2?

29. *Prove that a graph with at least two vertices has a pair of distinct vertices with the same degree.

30. How many edges are there on the graph K_n?

31. How many edges are there on the graph K_{mn}?

6.2 Directed Graphs

In many cases we need a graph in which the edges are essentially a one-way street. By this we mean that one considers an edge going from vertex a to vertex b but not going from vertex b to vertex a. For example, if a graph is simulating the flow of oil in a pipeline, if the oil flows from point a to point b, we would usually not

6.2 Directed Graphs

want it also flowing from point b to point a. In later chapters, we encounter directed graphs in networks and in automata theory.

DEFINITION 6.25

A **directed graph** or **digraph** G, denoted by $G(V, E)$, consists of a set V of vertices, together with a set E of ordered pairs of V called the set of **directed edges** or simply **edges** if it is understood that the graph is directed. An element of E is called a **directed edge**. If $(a, b) \in E$, then a is called the **initial vertex** of (a, b) and b is called the **terminal vertex**.

Note that we include loops for the directed graph, which we did not do for simple graphs. We basically do this for the following reason: We can include directed edges naturally in our definition since a loop at a vertex a is simply the edge (a, a). We could not do this for undirected graphs since an edge of an undirected graph has the form $\{a, b\}$, so that a loop would have the form $\{a, a\}$.

The edge (a, b) of a digraph is denoted on the diagram by an arrow from a to b. Note that in a simple graph, an edge is represented by a two-element subset to emphasize that the relation is symmetric, whereas in a directed graph, an edge is represented by an ordered pair to emphasize that order is important and that (a, b) may be an edge in a digraph whereas (b, a) is not.

EXAMPLE 6.26

The digraph with $V = \{a, b, c\}$ and $E = \{(a, b), (b, c), (c, b), (c, a)\}$ is shown in Figure 6.23. ∎

EXAMPLE 6.27

The digraph with $V = \{a, b, c, d\}$ and $E = \{(a, b), (b, c), (c, c), (b, d), (d, b), (c, d), (d, a)\}$ is shown in Figure 6.24. ∎

DEFINITION 6.28

If (a, b) is an edge of a directed graph $G(V, E)$, so vertex a is the initial vertex and b is the terminal vertex of (a, b), then vertices a and b are **incident** on the edge (a, b). The vertex a is **adjacent** to the vertex b. Also vertex b is **adjacent** from the vertex a.

DEFINITION 6.29

The **outdegree** of a vertex v, denoted by outdeg(v), is the number of edges for which v is the initial vertex. The **indegree** of a vertex v, denoted by indeg(v), is the number of edges for which v is the terminal vertex. If indeg(v) = 0, then v is called a **source**. If outdeg(v) = 0, then v is called a **sink**.

EXAMPLE 6.30

In the directed graph of Figure 6.25, indeg(v_0) = 0, indeg(v_1) = 1, indeg(v_3) = 2, indeg(v_2) = 2, and indeg(v_4) = 3. Also outdeg(v_0) = 3, outdeg(v_1) = 2, outdeg(v_2) = 2, outdeg(v_3) = 1, and outdeg(v_4) = 0. Thus, v_0 is a source, and v_4 is a sink. ∎

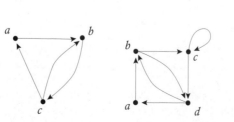

Figure 6.23   Figure 6.24   Figure 6.25

234 Chapter 6 Graphs, Directed Graphs, and Trees

Note that although loops are included in our definition of a directed graph, we still do not allow more than one edge from a vertex a to a vertex b. A directed graph with more than one edge from one vertex to another vertex is called a **multigraph**, or more precisely, a **directed multigraph**. If each edge is labeled, we refer to it as a **labeled directed graph** or simply as a **labeled graph** with the understanding that it is a directed graph. Since it really makes no sense to have two edges from a vertex a to a vertex b if we cannot distinguish them, we normally label them in some way. We formally define a labeled graph as follows.

DEFINITION 6.31 A **labeled graph** $G = G(V, L, E)$ is a set of vertices V, a set of labels L, and a set E of edges, which is a subset of $V \times L \times V$. An edge e of G has the form (a, l, b) where l is a label and a, b are vertices.

Graphically an edge $e = (a, l, b)$ of a labeled graph is denoted by Figure 6.26, or by Figure 6.27 if the edge is a loop.

Figure 6.26

Figure 6.27

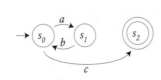

Figure 6.28

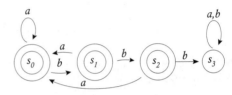

Figure 6.29

Figures 6.28 and 6.29 are examples of a type of labeled graph called an automation.

We will study these in Chapter 17.

DEFINITION 6.32 A directed graph $G'(V', E')$ is a **directed subgraph** of a directed graph $G(V, E)$, denoted by $G'(V', E') \preceq G(V, E)$, if $V' \subseteq V$ and $E' \subseteq E$. Thus, every vertex in G' is a vertex in G and every directed edge in G' is a directed edge in G.

A directed path in a directed graph is similar to a path in a graph except, that, when traveling the directed path, one must travel the same direction as the directed edges in the path.

DEFINITION 6.33 A **directed path** from a to b is described by a sequence of vertices $v_0 v_1 v_2 v_3 \cdots v_n$ where $a = v_0$, $b = v_n$, and (v_{i-1}, v_i) is a directed edge for $1 \leq i \leq n$. The **length** of a directed path is the number of directed edges in the path.

EXAMPLE 6.34 Given the graph G shown in Figure 6.30, the graphs in Figures 6.31 and 6.32 are subgraphs of G. Directed paths of G include $v_0 v_1 v_2 v_4$, $v_1 v_2 v_4$, and $v_0 v_3 v_4$. ∎

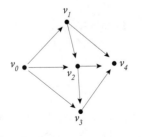

Figure 6.30

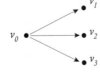

Figure 6.31

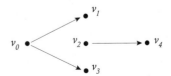

Figure 6.32

6.2 Directed Graphs

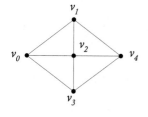

Figure 6.33

At this point, for a given directed graph G, we wish to describe an undirected graph G^s such that each directed edge of G (except loops) becomes an undirected edge of G^s.

For each directed graph $G(V, E)$, let $E' = E - \{(v, v) : v \in V\}$ so that $G'(V, E')$ is the directed subgraph of $G(V, E)$ with the loops removed. Let R be the symmetric closure of E', so that if $(a, b) \in E'$, then $(a, b), (b, a) \in R$, and E^s be the set of edges representing the relation R. Then the graph $G^s(V, E^s)$ is called the **underlying graph** of the directed graph $G(V, E)$. Stated less formally, the set of edges E^s of the underlying graph $G^s(V, E^s)$ is defined by $\{a, b\} \in E^s$ if and only if $(a, b) \in G$ or $(b, a) \in G$ for distinct vertices a and b.

DEFINITION 6.35 A directed graph $G(V, E)$ is **connected** if its underlying graph is connected. A directed graph is **strongly connected** if for every pair of vertices $a, b \in V$, there is a directed path from a to b.

EXAMPLE 6.36 For the directed graph in Figure 6.30, the underlying graph is shown in Figure 6.33. The directed graph is connected since its underlying graph is connected. However, it is not strongly connected since v_3 and v_1 have no directed path between them. ■

Exercises

Find the vertices and directed edges for the following digraphs. Find the indegree and outdegree for each vertex. Are there any sinks or sources?

1.

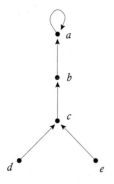

2.

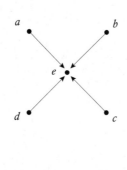

3.

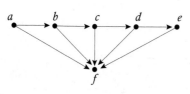

4.

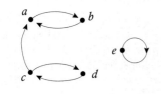

5.

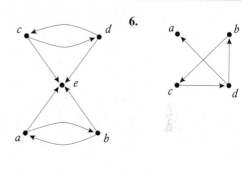

6.

7.

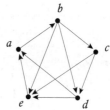

8.

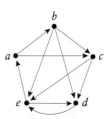

9. Find four subgraphs for each graph in Exercise 5.

10. Find four subgraphs for each graph in Exercise 6.

11. Find four subgraphs for each graph in Exercise 7.

Find the vertices and directed edges for the following digraphs. Find the indegree and outdegree for each vertex. Are there any sinks or sources?

12.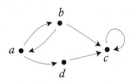

13.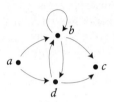

14.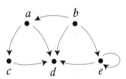

15.

16. Define a simple directed path.

17. Define a directed cycle.

18. Define a simple directed cycle.

Which of the following digraphs are connected? Which are strongly connected?

19.    20.

21.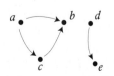

22.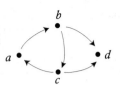

23. For each digraph in Exercise 19, if possible, find a directed path of length 2, a directed path of length 3, a directed path of length 4, and a directed path of length 5. Give a simple path of maximal length. What (if any) is the longest simple cycle that can be drawn?

24. For each digraph in Exercise 20, if possible, find a directed path of length 2, a directed path of length 3, a directed path of length 4, and a directed path of length 5. Give a simple path of maximal length. What (if any) is the longest simple cycle that can be drawn?

25. For each digraph in Exercise 22, if possible, find a directed path of length 2, a directed path of length 3, a directed path of length 4, and a directed path of length 5. Give a simple path of maximal length. What (if any) is the longest simple cycle that can be drawn?

Which of the following digraphs are connected? Which are strongly connected?

26. 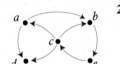   27.

28.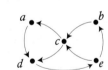

29.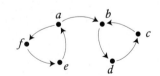

30. For the digraph in Exercise 26, if possible, find a directed path of length 2, a directed path of length 3, a directed path of length 4, and a directed path of length 5. Give a simple path of maximal length. What (if any) is the longest simple cycle that can be drawn?

31. For the digraph in Exercise 27, if possible, find a directed path of length 2, a directed path of length 3, a directed path of length 4, and a directed path of length 5. Give a simple path of maximal length. What (if any) is the longest simple cycle that can be drawn?

32. For the digraph in Exercise 28, if possible, find a directed path of length 2, a directed path of length 3, a directed path of length 4, and a directed path of length 5. Give a simple path of maximal length. What (if any) is the longest simple cycle that can be drawn?

33. Can a subgraph of a directed graph that is not a regular graph be a regular graph?

34. Can there be a directed graph with six vertices in which the outdegrees of the vertices are 1,2,3,3,2,5 and the indegrees are 4,1,2,1,5,2?

35. Can there be a directed graph with six vertices in which the outdegrees of the vertices are 1,2,3,4,0,5 and the indegrees are 5,5,1,1,1,0?

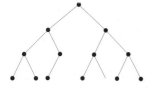

Figure 6.34

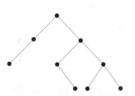

Figure 6.35

6.3 Trees

When Joyce Kilmer wrote, "I think that I shall never see a poem lovely as a tree," it is possible that he was not talking about the tree we are describing. However, we think that he should be given the benefit of the doubt. A **tree** is a connected graph that has no cycles. A **forest** is a graph whose components are trees (why not!). Trees have many uses in numerous areas including mathematics and computer science. For example, they are used as a tool in counting and in storage of data for convenience in sorting the data or searching for data. Some of these uses will be shown in later sections. It is called a tree because, when drawn, it looks like a tree, except that it is upside down. Figure 6.34 is an example of a tree.

We shall see in a later chapter that this tree can be used to demonstrate the results of tossing a coin three times. The graph in Figure 6.35 is not a tree since it has a cycle. The graph in Figure 6.36 is a forest.

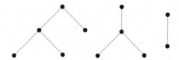

Figure 6.36

A family tree appropriately enough forms a tree. If we begin with a fixed (hopefully famous) person and form edges between a parent and each son or daughter, then a tree is formed. One has to be careful in designing a family tree, however, so that marriages between distant cousins do not create cycles. Another example is an organizational chart. The tree in Figure 6.37 is a typical partial organizational tree for a university. In this section, however, we shall consider trees as graphs.

A **directed tree** T is a loop-free directed graph, whose underlying graph is a tree, such that if there is a path from a vertex a to a vertex b, it is unique. We first note that if there is an edge (a, b) in a directed tree, then there is no edge (b, a), for if there were, then the path aba would be a cycle, and the path from a to b would not be unique. Thus, E, which is both the set of edges for the tree and the relation for the tree, has the property that if $(a, b) \in E$, then $(b, a) \notin E$. Such a relation is called **asymmetric**.

Returning to (undirected) trees, we first observe that a tree with at least one edge has at least two vertices with degree 1. To see this, we consider possible paths for a given tree T. If a tree has n vertices, then the length of a simple path cannot exceed $n - 1$, since each vertex can be used only once. Otherwise we would have a cycle and a tree does not have cycles. There is, therefore, a maximal path that cannot be extended to form a longer path. Assume the path begins at vertex a and ends at vertex b. Both a and b must have degree 1, for otherwise the path could be extended, which is a contradiction since the path is maximal. The vertices of degree 1 are called **leaves**. Other vertices are called **internal vertices**. A maximal path is not necessarily the longest path in the tree. For example, in the tree in Figure 6.38, the path $v_0v_2v_5$ is a maximal path. The vertices v_3, v_4, v_5, v_6, and v_7 are leaves.

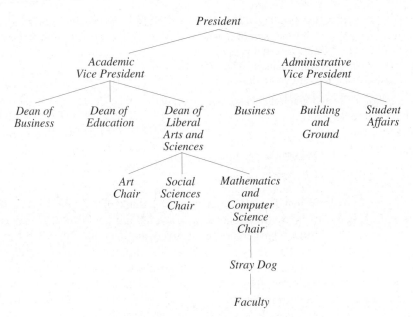

Figure 6.37

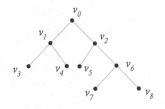

Figure 6.38

A tree has a rather interesting property that we shall prove next.

THEOREM 6.37 For any two vertices a and b of a tree T, there is a unique simple path from a to b.

Proof For this proof, we used the fact that the contrapositive of a statement is logically equivalent to the statement. Thus, we shall assume that the path from a to b is not unique for some vertices a and b of T and show that T is not a tree. Assume that there are two different paths $v_0 v_1 v_2 \cdots v_n$ of length n and $v_0 v_1' v_2' \cdots v_m'$, of length m, where $a = v_0$ and $b = v_n = v_m'$. There must be a first vertex in each path where the corresponding vertices do not agree, say, where $v_i \neq v_i'$, and there must be a point in each where the vertices are the same again, say, where $v_j = v_k'$. Then $v_{i-1} v_i v_{i+1} v_{i+2} \cdots v_j v_k' v_{k-1}' v_{k-2}' v_i' v_{i-1}$ is a cycle and T is not a tree. ∎

Using the following theorem, we see that the converse of the preceding theorem is also true. Hence, we could have defined a tree as a graph for which there is a unique simple path between any two of its vertices.

THEOREM 6.38 If for any two vertices in a graph G there is a unique simple path from a to b, then G is a tree.

Proof We again prove the theorem by proving its contrapositive. Assume that G is not a tree. Then either G is not connected or it contains a cycle. If G is not connected, then there are vertices $a, b \in G$ for which there is no path from a to b. Hence, there is certainly no unique path. If G has a cycle $v_0 v_1 v_2 v_3 v_4 \cdots v_{k-1} v_k$, then $v_2 v_3 \cdots v_{k-1} v_k v_0$ and $v_2 v_1 v_0$ are both paths from v_2 to v_0. Letting $a = v_2$ and $b = v_0$, we have found the vertices a and b, which do not have a unique path between them. ∎

Suppose that we consider a tree as a physical object that is flexible at the vertices and hold the tree by a vertex so that the rest of the tree hangs down below it. For example, given the tree in Figure 6.39, if we hold it by vertex v_3, we have the tree represented in Figure 6.40. If we hold it by vertex v_4, the tree is represented as shown in Figure 6.41.

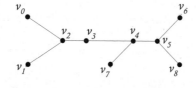

Figure 6.39

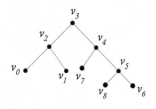

Figure 6.40

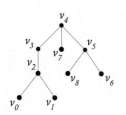

Figure 6.41

The vertex at the top of one of these representations is called the **root** of the tree. Once the root of the tree has been determined, the tree is called a **rooted tree**. We can, if convenient, change a rooted tree T to a directed tree T' where the tree in Figure 6.42 is changed to the tree in Figure 6.43.

This is called the **rooted directed tree T' derived** from the rooted tree T. In doing so, one should remember that this is not the same tree as the undirected tree and that the directed tree depends on the choice of the root.

Once a root is chosen, the **level** of a vertex v is the length of the unique path from the root to v. The **height** of a tree is the length of the longest path from the root of the tree to a leaf. If we consider the rooted tree T' derived from a given rooted tree T, then the vertex u is the **parent** of v and v is the **child** of u, if there is a directed edge from u to v. If u is the parent of v and v', then v and v' are called **siblings**. If there is a directed path from vertex u to vertex v, then u is called an **ancestor** v and v is called a **descendant** of u. If the largest outdegree for any vertex of the tree is m, then the tree is called an ***m*-ary tree**. In particular if $m = 2$, then the tree is called a **binary tree**. In every binary tree, each child of a parent is designated as either a **left child** or **right child** (but not both).

EXAMPLE 6.39

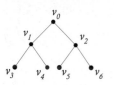

Figure 6.42

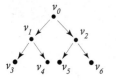

Figure 6.43

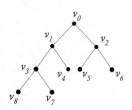

Figure 6.44

The graph in Figure 6.44 is a binary tree. Vertex v_6 has level 2 and vertex v_8 has level 3. The height of the tree is 3, since $v_0 v_1 v_3 v_8$ has length 3 and there is no longer path from a root to a leaf. Vertex v_1 is the parent of v_3 and v_4. Vertices v_3 and v_4 are siblings. So are v_1 and v_2, v_5 and v_6, and v_7 and v_8. Vertex v_1 is the ancestor of v_3, v_7, and v_8 and vertices v_3, v_7, and v_8 are descendants of v_1. Vertex v_7 is a left child of v_3, and v_4 is a right child of v_1. ∎

We now wish to prove that every tree has one more vertex than it has edges. To show this, assume we have a tree T. As we have seen, we change any tree to a rooted tree and this will certainly not change either the number of edges or the number of vertices. We now consider the directed tree T' derived from T. Each edge of T has one and only one terminal vertex. Conversely, every vertex, except the root, is the terminal vertex for one and only one edge. Hence, there is the same number of edges and vertices excluding the root vertex. When we include the root vertex, we have one more vertex than edges. Hence, we have proven the following theorem.

THEOREM 6.40 If a tree T has e edges and v vertices, then $v = e + 1$. ∎

The converse of this theorem is also true.

THEOREM 6.41 If a connected graph G has e edges and v vertices and $v = e + 1$, then G is a tree.

Proof If G has a cycle then, in the proof of Theorem 6.38, we found that if the edge $\{v_i, v_j\}$ is in the cycle, then there are two paths from v_i, to v_j. Thus, the edge $\{v_i, v_j\}$ may be removed from the cycle, and there is still a path from vertex v_i to vertex v_j. Let a and b be any points in G. Since G is connected, there is a path from a to b. If the edge $\{v_i, v_j\}$ is removed, there is still a path from a to b, since if the edge $\{v_i, v_j\}$ occurred in the path, replace this edge with the alternate route from v_i to v_j. Remove the edge $\{v_i, v_j\}$ from G, and if the remaining graph still contains a cycle, remove another edge using the same procedure. Continue until there are no more cycles. We then have a connected graph, say G', which has no cycles. Therefore, G' is a tree and by Theorem 6.40, $v = e' + 1$ where e' is the number of edges of G'. Since no vertices have been removed, the number of vertices is the same as before. If n edges were removed, then $e = e' + n$. But since $v = e + 1$ and $v = e' + 1$, $e = e'$ and $n = 0$. Hence, no edges were removed and G is a tree. ∎

The tree G' that was obtained from G in the previous proof is called a **spanning tree** for G. More formally,

DEFINITION 6.42 A tree T is a **spanning tree** for a graph G, if T is a subgraph of G, and every vertex in G is a vertex in T.

Thus, we have proved the following theorem:

THEOREM 6.43 Every connected graph has a subgraph, which is a spanning tree. ∎

So far we have considered only directed trees produced from rooted (undirected) trees. A directed tree is a **rooted directed tree** if there is a unique vertex v_0 such that $\text{indeg}(v_0) = 0$ and there is a path from v_0 to every other vertex in the tree. Note that the directed rooted trees that we have considered so far do fit the definition. However, consider the directed tree in Figure 6.45. It is a directed tree but not a rooted directed tree. Most of the directed trees we consider, however, will be rooted directed trees.

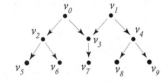

Figure 6.45

Exercises

Which of the following graphs are trees?

1.
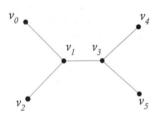

2.
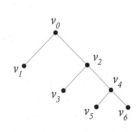

3.
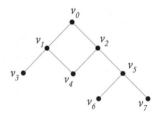

4.
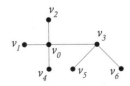

5. For Exercises 1–4 if the graph is a tree, use v_1 as the root and draw the rooted tree.

6. For Exercises 1–4 if the graph is a tree, use v_1 as the root and draw the rooted directed tree.

7. For Exercises 1–4 if the graph is a tree, use v_3 as the root and draw the directed tree.

8. For Exercises 1–4 if the graph is a tree, use v_3 as the root and draw the rooted directed tree.

Which of the following graphs are trees?

9.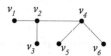

10.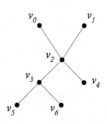

11.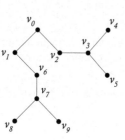

12.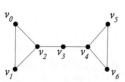

13.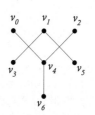

14. For Exercises 9–13 if the graph is a tree, use v_2 as the root and draw the rooted tree.

15. For Exercises 9–13 if the graph is a tree, use v_2 as the root and draw the rooted directed tree.

16. For Exercises 9–13 if the graph is a tree, use v_3 as the root and draw the rooted tree.

17. For Exercises 9–13 if the graph is a tree, use v_3 as the root and draw the rooted directed tree.

18. If a forest contains m components, show that $v = e + m$.

Which of the following graphs are rooted directed trees?

19.

20.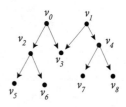

21.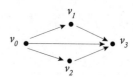

22.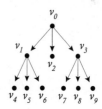

23.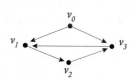

Given the rooted directed tree in Figure 6.46,

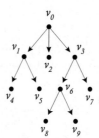

Figure 6.46

24. Find the descendants of v_3.
25. Find the ancestors of v_8.
26. Find the parent of v_5.
27. Find the level of v_6.
28. Find the children of v_3.
29. Find the height of the tree.
30. Find the leaves of the tree.
31. Is this a binary tree (why or why not)?

Given the rooted directed tree in Figure 6.47,

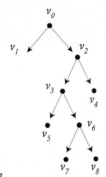

Figure 6.47

32. Find the descendants of v_2.
33. Find the ancestors of v_5.
34. Find the parent of v_1.
35. Find the level of v_5.

36. Find the children of v_2.
37. Find the height of the tree.
38. Find the leaves of the tree.
39. Draw a family tree beginning with one of your great grandfathers.
40. Draw a family tree beginning with one of your great grandmothers (not the wife of the grandfather in the previous problem).
41. Using induction, prove Theorem 6.40, which states that if a tree T has e edges and v vertices, then $v = e + 1$.
42. If $m \geq 2$, for what values of n is $K_{m,n}$ a tree?
43. Prove that a connected graph that has n vertices is a tree if and only if the sum of the degrees of the vertices is $2(n-1)$.
44. Prove that a tree with more than one vertex is a bipartite graph.
45. Prove that a tree with two vertices of degree 3 must have at least four vertices of degree 1 but need not have five vertices of degree 1.
46. Prove that a graph is connected if and only if it has a spanning tree.
47. Prove that a graph is a tree if and only if it is connected but that if any edge is removed, the resulting graph is not connected.
48. Define an order R on a directed tree T, by aRb if $a = b$ or if there is a directed path from a to b. Prove that R is a partial ordering.

6.4 Instant Insanity

The game of **instant insanity** is a game played with four cubes with each of the six edges of each cube painted with one of the colors red, blue, green, or yellow (or any four colors). The object of the game is to stack the cubes so that each of the four colors appears on each of the four sides of the stack. The solutions (if any) depend on the colors chosen for the edges of the cubes. There may be one or more solutions or there may be none. The edges of a cube will be designated by front(f), back(ba), top(t), bottom(bo), left(l), and right(r), which will be shown "spread out" in Figure 6.48.

The parts of the cube that are to be selected to appear will be called Left, Right, Front, and Back (to distinguish them from the edges as they originally appeared in the puzzle). Thus, the Front and Back are opposite each other and Left and Right are opposite each other. Suppose we have the following cubes in a given game shown in Figure 6.49.

Figure 6.48

Figure 6.49

6.4 Instant Insanity

We will show a relatively easy way to find a solution using graphs. Before beginning, however, it is important to realize that selecting one pair of opposites for a cube, say, Front and Back, in no way restricts which of the other two pairs may be Left and Right or which side of the selected pair will be Left and which will be Right. It is perhaps easiest to see this by grasping a cube by opposite sides (Front and Back) and rotating the cube. It is easily seen that the four rotations give all possibilities for the other two sides (Left and Right). Thus, if we select Front and Back for each cube, we have complete freedom in selecting any of the other opposite edges as Left and Right, respectively.

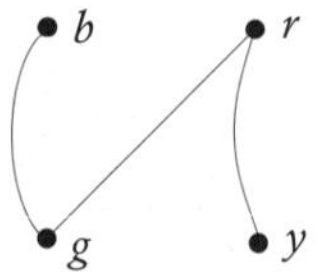

The graph we create has the four colors as vertices. Edges connect opposite colors on a block. Thus, for the first block we want 1-edges from red to blue, green to red, and red to yellow giving Figure 6.50. The following graph appears as shown in Figure 6.51.

When drawing the graph, often the graph will appear less messy if the vertices are rearranged.

We will now construct a subgraph to let us select the front and back for each cube. We want a subgraph such that

1. Each vertex appears in the subgraph.
2. An edge from each cube appears in the subgraph.
3. Each vertex has order 2.

Each vertex must appear so that all four colors appear on the front and back sides. An edge from each cube must appear so that we get a front and back from each cube. Each vertex has order 2 so that each color appears once on the front and once on the back. Some possible subgraphs are given in Figure 6.52.

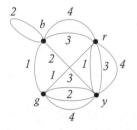

Figure 6.50

Figure 6.51

Figure 6.52

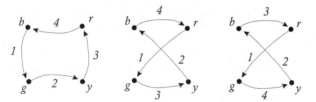

(Try to find others.) We begin by trying the first one shown. By writing this as a directed graph around the circuit, we can consider the initial vertex to be the front and the terminal vertex to be the back, so that we have

Front	Back
b	g
g	y
y	r
r	b

We now need another subgraph with the same properties. The only restriction on the second subgraph is that it must not contain an edge from the first subgraph. These edges have already been used as fronts and backs for the blocks and cannot also appear as sides. To simplify the graph, we remove the used edges so that we have Figure 6.53. From this graph, we find that the subgraph in Figure 6.54 may be used.

244 Chapter 6 Graphs, Directed Graphs, and Trees

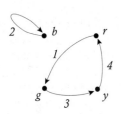

Figure 6.53

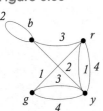

Figure 6.54

This gives the following colors for the Left and Right as shown in the table and we have found a solution. If the graph selected for Front and Back had not worked, or if we wanted to find other solutions, we would try another of the possibilities for Front and Back. In finding the various possibilities, it helps to note that the subgraphs must have one of the basic forms shown in Figure 6.55.

Left	Right
r	g
b	b
g	y
y	r

Figure 6.55

Exercises

Find the solutions (if any) for the following sets of cubes:

1.

2.

3.

4.

5.

6.5 Euler Paths and Cycles

DEFINITION 6.44 Let $G = (V, E)$ be a graph. A cycle that includes all of the edges and the vertices of G is called an **Euler cycle**. When this occurs, we say that the graph G has an Euler cycle.

If we look again at the Königsberg bridge problem, we see that we are really trying to decide whether or not the graph representing this problem has an Euler cycle. To help us we need the following theorem. The following theorem is also true for multigraphs and pseudographs, except that a pseudograph may have only

6.5 Euler Paths and Cycles

one vertex. Furthermore, the proof is the same. For clarity we shall use the term *graph* with the understanding that every statement is also true for multigraphs and pseudographs.

THEOREM 6.45 A graph with more than one vertex has an Euler cycle if and only if it is connected and every vertex has even degree.

Proof Assume that a graph G has an Euler cycle. The graph is connected since every vertex is on the cycle. For any vertex v of G, each time the Euler cycle passes through v, it contributes 2 to the degree of v. Therefore, v has even degree.

Conversely, we wish to show that every connected graph whose vertices have even degree has an Euler cycle. We shall prove this theorem using induction on the number of vertices. Since the theorem is trivially true when $n \leq 3$, we begin our induction with $n = 3$. Assume that every connected graph whose vertices have even degree and with fewer than k vertices has an Euler cycle. Let G be a connected graph containing k vertices each of which has even degree. Assume v_1 and v_2 are vertices in G. Since G is connected, there is a path from v_1 to v_2. Since v_2 has even degree, there is an unused edge that may be used to further continue the path. Since the graph is finite, eventually the path must return to v_1 and an Euler cycle C_1 has been constructed. If C_1 is an Euler cycle for G, then we are done. If not, let G' be the subgraph of G formed by removing all of the edges of C_1. Since C_1 contains an even number of edges incident to each vertex, every vertex in G' also has even degree.

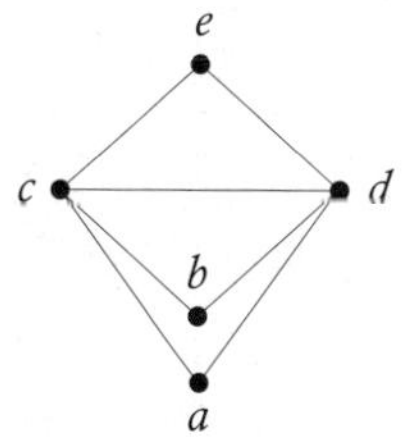

Let e be an edge of G' and G_e the component of G' containing e. Since G_e has fewer than k vertices and each vertex of G_e has even degree, G_e has an Euler cycle, say, C_2. Furthermore, C_1 and C_2 have vertex, say, a, in common. (Why?) We can now extend our Euler cycle by beginning at a, traversing C_1 back to a, then traversing C_2 back to a. If the new Euler cycle is not an Euler cycle for G, we continue to use the same process to extend our Euler cycle until we finally have an Euler cycle for G. ∎

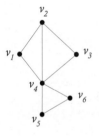

Figure 6.56

Figure 6.57

EXAMPLE 6.46 The graph in Figure 6.56 has an Euler cycle since every vertex has even degree. ∎

EXAMPLE 6.47 The graph in Figure 6.57 has no Euler cycle since vertices v_2 and v_4 have odd degree. ∎

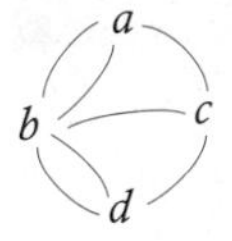

Returning to the Königsberg bridge problem, we see that the multigraph in Figure 6.58, designed by Euler to describe the bridges of Königsberg, has all vertices of odd degree. Hence, it has no Euler cycle so that it is impossible to cross each bridge exactly once and return to the original vertex. We might also note that while this problem was solved using a multigraph, we could also solve it using a simple graph. Where there is more than one edge between vertices, simply include a vertex to represent the center of each bridge in the original multigraph. We then have the simple graph shown in Figure 6.59, which also describes the Königsberg bridge problem.

With regard to the bridges of Königsberg, one might also ask whether one might not cross each bridge exactly once but not necessarily return to the original point of departure. This conjecture leads us to the following definition and theorem.

Figure 6.58

DEFINITION 6.48 Let $G = (V, E)$ be a graph. A path that includes each of the edges of G exactly once is called an **Euler path**. When this occurs, we say that the graph G has an Euler path.

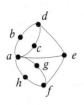

Figure 6.59

At this point we are concerned with Euler paths that are not Euler cycles. We shall call such paths **proper Euler paths**. The proof of the following theorem is left to the reader.

THEOREM 6.49 A graph (multigraph or pseudograph) has a proper Euler path if and only if it is connected and exactly two vertices have odd degree. ∎

Since the graph for the bridges of Königsberg has four vertices with odd degree, we can conclude that it is impossible to cross each bridge exactly once even if we do not have to return to the original point of departure.

EXAMPLE 6.50 The graph in Figure 6.60 has a proper Euler path since exactly two vertices have odd degree. ∎

EXAMPLE 6.51 The graph in Figure 6.61 has no proper Euler path since four vertices have odd degree. ∎

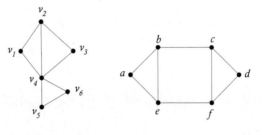

Figure 6.60   Figure 6.61

DEFINITION 6.52 Let $G = (V, E)$ be a directed graph. A **directed cycle** is a directed path of length greater than zero from a vertex to itself with no repeated edges.

DEFINITION 6.53 Let $G = (V, E)$ be a directed graph. A directed cycle that includes all of the edges and the vertices of G is called an **Euler cycle**. When this occurs, we say that the directed graph G has an Euler cycle.

The proof of the following theorem is left to the reader.

THEOREM 6.54 A directed graph has an Euler cycle if and only if it is connected and the indegree of every vertex is equal to its outdegree. ∎

EXAMPLE 6.55 The directed graph in Figure 6.62 has an Euler cycle since the indegree of every vertex is equal to the outdegree. ∎

EXAMPLE 6.56 The directed graph in Figure 6.63 has no Euler cycle since the indegree of v_1 is not equal to its outdegree ∎

Figure 6.62   Figure 6.63

Exercises

1. Which of the following graphs have Euler cycles?

   (a)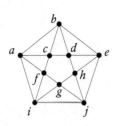

   (b)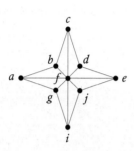

   (c)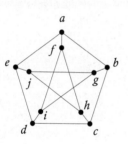

   (d)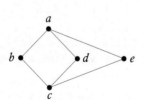

2. Which of the following graphs have Euler cycles?

   (a)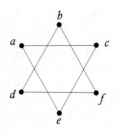

   (b)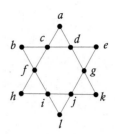

   (c)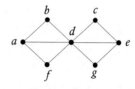

   (d)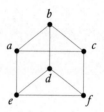

3. Which of the following graphs have proper Euler paths?

   (a)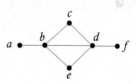

   (b)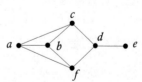

   (c)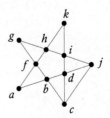

248 Chapter 6 Graphs, Directed Graphs, and Trees

(d)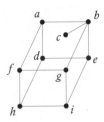

4. Which of the following graphs have proper Euler paths?

(a)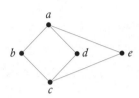

(b)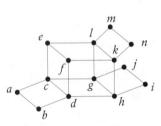

(c)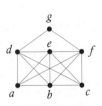

(d)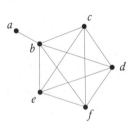

5. Which of the following graphs are strongly connected?

(a)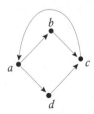

(b)

(c)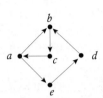

(d)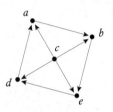

6. Which of the following graphs are strongly connected?

(a)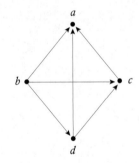

(b)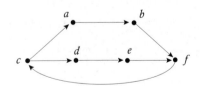

(c)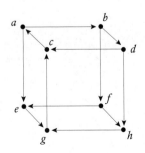

(d)

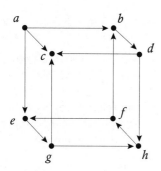

7. Which of the following directed graphs have Euler cycles?

(a)

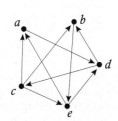

(b)

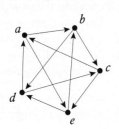

(c)

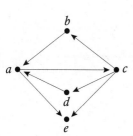

(d)

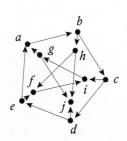

8. Which of the following directed graphs have Euler cycles?

(a)

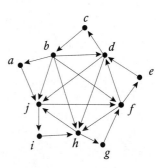

(b)

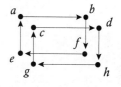

(c)

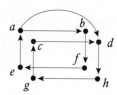

(d)

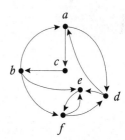

9. Prove Theorem 6.49: A graph (multigraph or pseudograph) has a proper Euler path if and only if it is connected and exactly two vertices have odd degree.

10. Prove that if a graph contains a cycle from a vertex v to itself, then it contains a simple cycle from the vertex v to itself.

11. Prove a directed graph is strongly connected if there is a vertex v in the graph such that every other vertex in the graph is reachable from v and v is reachable from every other vertex in the graph.

12. Prove Theorem 6.54: A directed graph has an Euler cycle if and only if it is strongly connected and the indegree of every vertex is equal to its outdegree.

6.6 Incidence and Adjacency Matrices

In this section we shall define two matrices associated with a graph: the incidence matrix and the adjacency matrix. In each case, we will find that given either matrix, we can again recover the graph. In fact we shall find that the adjacency matrix of a graph is just the representation matrix of the relation represented by that graph. Also, as with the representation matrix for a graph, all of these matrices will have entries of 1 or 0, so the matrices are easily stored in a computer.

DEFINITION 6.57 Let G be a graph. Let B be a matrix whose rows are labeled by the vertices in the graph and whose columns are labeled by the edges in the graph. The entry in the ith row and jth column of B, denoted by B_{ij}, is equal to 1 if the ith vertex is incident to the jth edge and is 0 otherwise. The matrix B is called the **incidence matrix** of the graph G.

EXAMPLE 6.58 Let G be the graph in Figure 6.64, then the incidence matrix is shown in Figure 6.65. ■

It is easily seen that the degree of a vertex is the sum of the entries in the row labeled by that vertex since each 1 in that row represents the incidence of that vertex in an edge. Also each column will have two 1's in it since each edge is incident to two vertices.

We may also include incidence matrices for graphs with loops. It is easy to tell from the incidence matrix if an edge is a loop since an edge is a loop if and only if there is only a single 1 in the column labeled by that edge. Note that in incidence matrices for graphs with loops, the sum of the entries in the row labeled by a given vertex does not represent the degree of that vertex if there is a loop at that vertex.

EXAMPLE 6.59 Let G be the graph in Figure 6.66 with incidence matrix in Figure 6.67. Note that the loops at e_2 and e_5 cause the columns labeled by these edges to have a single 1 in them. ■

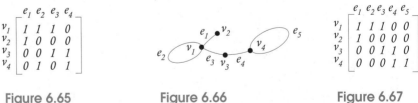

Figure 6.65 Figure 6.66 Figure 6.67

Incidence matrices do not have much value for directed graphs since the incidence matrix gives no clue which way an edge is directed. Thus, an incidence matrix does not allow us to recreate a directed graph. This is not true, however, of the adjacency matrix, which we now define.

DEFINITION 6.60 Let G be a graph (directed graph). Let B be a matrix whose rows are labeled by the vertices in the graph and whose columns are labeled by the same vertices in the same order. The entry in the ith row and jth column of B, denoted by B_{ij}, is equal to 1 if there is an edge (directed edge) from the ith vertex to the jth vertex and is 0 otherwise. The matrix B is called the **adjacency matrix** of the graph G.

Notice that the material for the adjacency matrix is really the same as for the representation matrix in Section 4.3. The only difference is that we use graph terminology and explore the meaning of adjacency matrices for graphs. We also include Warshall's algorithm.

EXAMPLE 6.61 Let G be the graph in Figure 6.68. The adjacency matrix is shown in Figure 6.69. ∎

Since there are no loops, the entries in the diagonal matrix are all 0. Since a graph represents a symmetric relation, the adjacency matrix is symmetric.

EXAMPLE 6.62 Let G be the directed graph in Figure 6.70. The adjacency matrix is shown in Figure 6.71. ∎

In many cases the labels of the vertices are not important. In such a case we will give the matrix without the label. Thus, the matrix

$$\begin{bmatrix} 1 & 0 & 0 & 1 \\ 1 & 0 & 1 & 0 \\ 0 & 0 & 0 & 1 \\ 1 & 1 & 1 & 0 \end{bmatrix}$$

is the adjacency matrix for a directed graph with four vertices and eight edges.

An important use of the adjacency matrix is to find paths of a fixed length k. The following example gives us a clue for doing this.

Let the matrix

$$A = \begin{bmatrix} 1 & 0 & 0 & 1 \\ 1 & 0 & 1 & 0 \\ 0 & 0 & 0 & 1 \\ 1 & 1 & 1 & 0 \end{bmatrix}$$

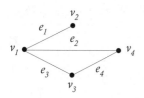

Figure 6.68

$$\begin{array}{c} \, v_1 \, v_2 \, v_3 \, v_4 \\ \begin{array}{c} v_1 \\ v_2 \\ v_3 \\ v_4 \end{array} \begin{bmatrix} 0 & 1 & 1 & 1 \\ 1 & 0 & 0 & 0 \\ 1 & 0 & 0 & 1 \\ 1 & 0 & 1 & 0 \end{bmatrix} \end{array}$$

Figure 6.69

be the adjacency matrix for a directed graph G with vertices v_1, v_2, v_3, and v_4. Consider the matrix

$$A^{\odot 2} = \begin{bmatrix} 1 & 0 & 0 & 1 \\ 1 & 0 & 1 & 0 \\ 0 & 0 & 0 & 1 \\ 1 & 1 & 1 & 0 \end{bmatrix} \odot \begin{bmatrix} 1 & 0 & 0 & 1 \\ 1 & 0 & 1 & 0 \\ 0 & 0 & 0 & 1 \\ 1 & 1 & 1 & 0 \end{bmatrix}$$

Look, for example, at

$$A^{\odot 2}_{12} = (A_{11} \wedge A_{12}) \vee (A_{12} \wedge A_{22}) \vee (A_{13} \wedge A_{32}) \vee (A_{14} \wedge A_{42})$$
$$= (1 \wedge 0) \vee (0 \wedge 0) \vee (0 \wedge 0) \vee (1 \wedge 1)$$
$$= 0 \vee 0 \vee 0 \vee 1$$
$$= 1$$

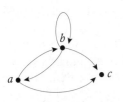

Figure 6.70

$$\begin{array}{c} \, a \, b \, c \\ \begin{array}{c} a \\ b \\ c \end{array} \begin{bmatrix} 0 & 1 & 1 \\ 1 & 1 & 1 \\ 0 & 0 & 0 \end{bmatrix} \end{array}$$

Figure 6.71

Notice the value of $A^{\odot 2}_{12}$ is 1 because A_{14} and A_{42} are both 1, which means there is an edge from vertex v_1 to vertex v_4 and from v_4 to vertex v_2. Therefore, there is a 2-path from vertex v_1 to vertex v_2. In general we can see that $A^{\odot 2}_{ij} = 1$ if and only if there is a k such that $A_{ik} \wedge A_{kj} = 1$, or, in other words, there is an edge from vertex v_i to vertex v_k and from v_k to vertex v_j. Therefore, there is a 2-path from vertex v_i to vertex v_j. But for $A^{\odot 2}_{34}$ we have

$$A^{\odot 2}_{34} = (A_{31} \wedge A_{14}) \vee (A_{32} \wedge A_{24}) \vee (A_{33} \wedge A_{34}) \vee (A_{34} \wedge A_{44})$$
$$= (0 \wedge 1) \vee (0 \wedge 0) \vee (0 \wedge 1) \vee (1 \wedge 0)$$
$$= 0 \vee 0 \vee 0 \vee 0$$
$$= 0$$

so that $A_{34}^{\odot 2} = 0$ because there are no edges from v_3 to v_i and v_i to v_4 for any fixed i. In other words, there are no 2-paths from vertex v_1 to vertex v_2. We conclude then that $A_{ij}^{\odot 2} = 1$ if there is a 2-path from vertex v_i to vertex v_j and $A_{ij}^{\odot 2} = 0$ if there is no 2-path from vertex v_i to vertex v_j. In a similar fashion we see that if we use regular matrix multiplication, then A_{ij}^2 is the number of values k so that A_{ik} and A_{kj} are both 1, so that it is the number of paths from vertex v_i to vertex v_j of length 2.

The reader is asked to prove the following theorems using induction.

THEOREM 6.63 Let G be a (directed) graph with vertices $v_1, v_2, v_3, \ldots, v_n$ and adjacency matrix A. There is a k-path from v_i to v_j for $1 \leq k \leq n$ if and only if $A_{ij}^{\odot k} = 1$. ∎

THEOREM 6.64 Let G be a (directed) graph with vertices $v_1, v_2, v_3, \ldots, v_n$ and adjacency matrix A. There are m k-paths from v_i to v_j for $1 \leq k \leq n$ if and only if $A_{ij}^k = m$. ∎

Using Theorem 6.63, the reader is asked to prove the following theorem.

THEOREM 6.65 Let G be a (directed) graph with vertices $v_1, v_2, v_3, \ldots, v_n$ and adjacency matrix A. Let $\hat{A} = A \vee A^{\odot 2} \vee A^{\odot 3} \vee A^{\odot 4} \vee \cdots \vee A^{\odot n}$. Then $\hat{A}_{ij} = 1$ if and only if there is a path from v_i to v_j. ∎

The following theorem follows directly from the definitions of connected graphs and strongly connected directed graphs. This is really Theorem 6.65 in disguise.

THEOREM 6.66 Let G be a (directed) graph with vertices $v_1, v_2, v_3, \ldots, v_n$ and adjacency matrix A. Let $\hat{A}^I = I \vee A \vee A^{\odot 2} \vee A^{\odot 3} \vee A^{\odot 4} \vee \cdots \vee A^{\odot n}$, where I is the multiplicative identity matrix. Then $\hat{A}_{ij}^I = 1$ for all i and j if and only if G is (strongly) connected. ∎

From Theorem 6.65, we see that if R is the relation described by the (directed) graph G, and A is the adjacency matrix of G, then $\hat{A}$ is the adjacency matrix of the graph, which describes the **transitive closure** of R. For this reason we shall refer to $\hat{A}$ as the transitive closure of A.

In the more general case, where G is a graph that is not necessarily connected, we have the following result for $\hat{A}^I$.

THEOREM 6.67 Let G be a graph with vertices $v_1, v_2, v_3, \ldots, v_n$ and adjacency matrix A. As before, let $\hat{A}^I = I \vee A \vee A^{\odot 2} \vee A^{\odot 3} \vee A^{\odot 4} \vee \cdots \vee A^{\odot n}$. Then the vertices can be arranged (if necessary) so that $\hat{A}^I$ has the form

$$\begin{bmatrix} \hat{A}_1 & & & & \\ & \hat{A}_2 & & 0 & \\ & & \hat{A}_3 & & \\ & 0 & & \ddots & \\ & & & & \hat{A}_m \end{bmatrix}$$

where each $\hat{A}_i$ is a square matrix whose main diagonal is along the main diagonal of $\hat{A}^I$ and has all of its entries equal to 1. As indicated, any entry of $\hat{A}^I$ that is outside all of the $\hat{A}_i$ is equal to 0. Also A may be divided into the exact same size blocks as $\hat{A}^I$ and A has the form

$$\begin{bmatrix} A_1 & & & & \\ & A_2 & & 0 & \\ & & A_3 & & \\ & 0 & & \ddots & \\ & & & & A_m \end{bmatrix}$$

where each A_i has the same shape as $\hat{A}_i$ and is the incidence matrix of a component of G, and any entry of A that is outside all of the $\hat{A}_i$ is equal to 0.

Proof If all of the vertices of G that are in the same component are placed together, then, since there is a path between any two of these vertices, the block of the matrix $\hat{A}^I$ consisting only of these vertices labeling the rows and columns has all 1's as entries. Furthermore, any other entry in the same row or column labeled by one of these vertices must be zero since there is no path from any of the other vertices to a vertex in the component.

Since the blocks $\hat{A}_i$ occur where the vertices labeling the rows and the columns are in the same component, the corresponding block A_i in A represents the graph of a component of G. As before, and for the same reason, all other entries in the same row or column as one of these vertices must be zero. ∎

The matrix $\hat{A}$ can be computed by $\hat{A} = A \vee A^{\odot 2} \vee A^{\odot 3} \vee A^{\odot 4} \vee \cdots \vee A^{\odot n}$, but this is not very efficient method. A much better method is **Warshall's algorithm**, also known as Roy-Warshall's algorithm. To see how it works, consider the adjacency matrix in Figure 6.72.

The matrix A represents the set of all 1-paths. We next want to find all 2-paths where the middle vertex is v_1 to combine with the 1-paths that we already have. We begin with the first column. Ignoring the 1 in the first row, if there is a 1 in the ith row of the first column, then there is an edge or 1-path from v_i to v_1. Since there is a 1 in row 3, there is a 1-path from v_3 to v_1. We now look at the first row. Ignoring the 1 in the first column, if there is a 1 in the jth column, then there is an edge or 1-path from v_1 to v_j. Since there is a 1 in the fourth column, there is an edge from v_1 to v_4. Thus, there is a 2-path from v_3 through v_1 to v_4. We denote this by placing a 1 in the third row of the fourth column, so that we now have the matrix in Figure 6.73.

$$A = \begin{array}{c} \\ v_1 \\ v_2 \\ v_3 \\ v_4 \end{array} \begin{array}{c} v_1 \; v_2 \; v_3 \; v_4 \\ \begin{bmatrix} 1 & 0 & 0 & 1 \\ 0 & 1 & 1 & 0 \\ 1 & 0 & 1 & 0 \\ 0 & 1 & 0 & 0 \end{bmatrix} \end{array}$$

Figure 6.72

Since there are no other 1's in the first column or first row, we have finished this step. If there had been a 1 in any other row of the first column, say, the ith row, or there had been a 1 in any other column of the first row, say, the jth column, then we would have placed a 1 in the ith row of the jth column. In our example using $\vee$ as addition, we have in fact added the first row to the third row. In general if there is a 1 in the ith row of the first column, then we add the first row to ith row.

$$\begin{array}{c} \\ v_1 \\ v_2 \\ v_3 \\ v_4 \end{array} \begin{array}{c} v_1 \; v_2 \; v_3 \; v_4 \\ \begin{bmatrix} 1 & 0 & 0 & 1 \\ 0 & 1 & 1 & 0 \\ 1 & 0 & 1 & 1 \\ 0 & 1 & 0 & 0 \end{bmatrix} \end{array}$$

Figure 6.73

We now want to find all paths of length 3 or less passing through v_1 and/or v_2 (if any). We consider the second column. Ignoring the 1 in the second row, we look for a 1 in any other row of column 2. Since there is a 1 in row 4, there a 1-path from v_4 to v_2 or a 2-path from v_4 through v_1 to v_2, since that is the only way we have produced 1's. Ignoring the 1 in the second column, we look for a 1 any other column of row 2. Since there is a 1 in column 3, there is a 1-path from v_2 to v_3 or a 2-path from v_2 through v_1 to v_3. In any case, there is a path from v_4 to v_3 such that the only vertices that it could have passed through are v_1 and v_2. Again we denote this by placing a 1 in the fourth row of the third column giving the matrix in Figure 6.74.

$$\begin{array}{c} \\ v_1 \\ v_2 \\ v_3 \\ v_4 \end{array} \begin{array}{c} v_1 \; v_2 \; v_3 \; v_4 \\ \begin{bmatrix} 1 & 0 & 0 & 1 \\ 0 & 1 & 1 & 0 \\ 1 & 0 & 1 & 1 \\ 0 & 1 & 1 & 0 \end{bmatrix} \end{array}$$

Figure 6.74

Again we could have accomplished the same thing by adding the second row to the fourth row. If there have been a 1 in the ith row of column 2 and a 1 in the jth column of row 2, we would have added row 2 to row i.

Similarly, we now want all paths of length 4 or less passing through v_1 and/or v_2 and/or v_3 (if any). Consider the third column and third row. If there is a 1 in the ith row of column 3 and a 1 in the jth column 3 of row 3, then there is a path from v_i to v_3 passing only through vertices v_1 and/or v_2 (if any) and a path from v_3 to v_j passing only through vertices v_1 and/or v_2 (if any). Therefore, there is a path from v_i to v_j and a 1 should be placed in the (i, j)th position. This is equivalent to adding row 3 to any row with a 1 in column 3. There is a 1 in the third column of the first and fourth rows, so row 3 is added to each of these rows. This gives us the matrix in Figure 6.75.

$$\begin{array}{c} \begin{array}{cccc} v_1 & v_2 & v_3 & v_4 \end{array} \\ \begin{array}{c} v_1 \\ v_2 \\ v_3 \\ v_4 \end{array} \left[\begin{array}{cccc} 1 & 0 & 0 & 1 \\ 1 & 1 & 1 & 1 \\ 1 & 0 & 1 & 1 \\ 1 & 1 & 1 & 1 \end{array} \right] \end{array}$$

Figure 6.75

Finally, we now want all paths of length 4 or less passing through v_1 and/or v_2 and/or v_3 and/or v_4 (if any). We add the fourth row to every other row in which a 1 appears in that row of column 4. This gives us a 1 in each row and column.

We give two algorithms for computing $\hat{A}$. Both follow from the way we computed our example. The first is the handier if computing by hand. Remember that in the algorithm, addition means Boolean addition.

■ 6.6.1 Warshall Algorithm 1

1. Look at column 1 of A. Where there is a 1 in a row of that column, add row 1 to the row in which that 1 occurred.

2. Look at column 2 of the matrix constructed in (1). Where there is a 1 in a row of that column, add row 2 to the row in which that 1 occurred.

3. Look at column 3 of the matrix constructed in (2). Where there is a 1 in a row of that column, add row 3 to the row in which that 1 occurred.

4. Continue this process of looking at the next column in the previously constructed matrix and where there is a 1 in a row of that column, add the row corresponding to the column being examined to the row in which the 1 occurred.

5. Continue until all columns have been examined.

The second method uses the fact that we began with the first row and column and if there was a 1 in the ith row of the first column and a 1 in the jth column of the first row, we then placed a 1 in the ith row of the jth column of the matrix. In other words if $A_{i1} = 1$ and $A_{1j} = 1$, then we set $A_{ij} = 1$. If it was already 1, then we left it 1. This is equivalent to "for all i and all j, $A_{ij} = A_{ij} \vee (A_{i1} \wedge A_{1j})$, since $A_{i1} \wedge A_{1j}$ is 1 if and only if $A_{i1} = 1$ and $A_{1j} = 1$ and is 0 otherwise." Using the second row and column we would find new values for the A_{ij} by having $A_{ij} = A_{ij} \vee (A_{i2} \wedge A_{2j})$. Continuing this process, we come up with the following algorithm in pseudocode.

■ Warshall Algorithm 2

For $k = 1$ to n
    For $i = 1$ to n
        For $j = 1$ to n
            $A_{ij} = A_{ij} \vee (A_{ik} \wedge A_{kj})$.
        Endfor
    Endfor
Endfor

6.6 Incidence and Adjacency Matrices

Computer Science Application

When G is a directed graph on a large number of nodes, the size of the adjacency matrix may be too large for computational purposes. Consider the following algorithm that computes the product c of the two square $(n \times n)$ matrices a and b:

function matrix-product (integer matrix a, integer matrix b, integer matrix c)

```
    for i = 1 to n
        for j = 1 to n
            c(i, j) = 0
            for k = 1 to n
                c(i, j) = c(i, j) + a(i, k) * b(k, j)
            end-for
        end-for
    end-for

    return

end-function
```

Suppose graph G has 1000 nodes. Then $n = 1000$ in the function above will require 1 billion additions and multiplications to compute the product c. The adjacency matrix A itself has 1 million components as do the various powers of A, which may be computed to determine the number (if any) of paths of a certain length between two specified nodes. Recall that $A^n(i, j) = k$ means that there are exactly k distinct paths of length n from node i to node j. Using powers of A up to 1000 to determine connectivity between nodes in G may involve nearly a trillion additions and multiplications in the course of computing the matrix products that are needed.

A more reasonable data structure for representing a graph with a large number of nodes is something called an adjacency list. A graph G with 1000 nodes would use a 1000-component 1-D array A with $A(i)$ "pointing" to the front of a linked-list consisting of only those integers that node i is adjacent to. For example, if node 15 was adjacent only to nodes 21 and 911,

$$A(15) -> 21 -> 911$$

would suffice to represent the edges of G that initiate from node 15. In this representation no memory is used to denote the absence of edges. Typically, in a graph with a large number of nodes, the adjacency matrix is sparse (i.e., contains mainly zeroes). When this is the case, the adjacency list representation is far more efficient in terms of memory requirements.

Exercises

Find the incidence matrices of the following graphs:

1.

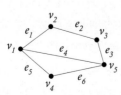

2.

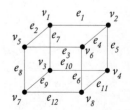

256 Chapter 6 Graphs, Directed Graphs, and Trees

3.
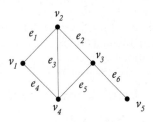

4.
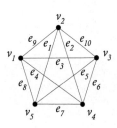

Find the incidence matrices of the following graphs:

5.
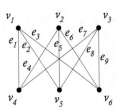

6.
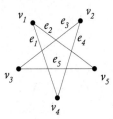

7.
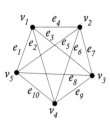

8.
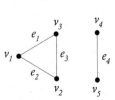

9. Find the adjacency matrix for the graph in Exercise 1.
10. Find the adjacency matrices for the graph in Exercise 2.
11. Find the adjacency matrix for the graph in Exercise 3.
12. Find the adjacency matrices for the graph in Exercise 4.
13. Find the adjacency matrix for the graph in Exercise 5.
14. Find the adjacency matrices for the graph in Exercise 6.
15. Find the adjacency matrix for the graph in Exercise 7.
16. Find the adjacency matrices for the graph in Exercise 8.
17. Given the incidence matrix

$$\begin{bmatrix} 1 & 0 & 1 & 0 & 0 & 1 & 0 & 0 \\ 0 & 1 & 0 & 1 & 0 & 0 & 1 & 0 \\ 0 & 1 & 1 & 0 & 1 & 0 & 0 & 1 \\ 1 & 0 & 0 & 1 & 1 & 0 & 0 & 0 \\ 0 & 0 & 0 & 0 & 0 & 1 & 1 & 1 \end{bmatrix}$$

find the corresponding graph.

18. Given the incidence matrix

$$\begin{bmatrix} 1 & 0 & 1 & 0 & 0 & 0 & 0 & 0 & 0 & 0 & 1 \\ 0 & 1 & 0 & 1 & 1 & 0 & 1 & 0 & 0 & 0 & 0 \\ 0 & 1 & 0 & 0 & 0 & 1 & 0 & 1 & 0 & 0 & 0 \\ 1 & 0 & 0 & 0 & 0 & 0 & 0 & 0 & 1 & 0 & 0 \\ 0 & 0 & 1 & 0 & 0 & 1 & 1 & 0 & 0 & 1 & 0 \\ 0 & 0 & 0 & 1 & 0 & 0 & 0 & 0 & 0 & 0 & 0 \\ 0 & 0 & 0 & 0 & 1 & 0 & 0 & 0 & 1 & 0 \\ 0 & 0 & 0 & 0 & 0 & 0 & 0 & 1 & 1 & 0 & 1 \end{bmatrix}$$

find the corresponding graph.

19. Given the adjacency matrix

$$\begin{bmatrix} 0 & 0 & 1 & 0 & 1 & 0 \\ 0 & 0 & 1 & 0 & 1 & 1 \\ 1 & 1 & 0 & 1 & 0 & 0 \\ 0 & 0 & 1 & 0 & 1 & 1 \\ 1 & 1 & 0 & 1 & 0 & 0 \\ 0 & 1 & 0 & 1 & 0 & 0 \end{bmatrix}$$

find the corresponding graph.

20. Given the adjacency matrix

$$\begin{bmatrix} 0 & 1 & 0 & 0 & 1 & 0 \\ 1 & 0 & 1 & 1 & 0 & 0 \\ 0 & 1 & 0 & 1 & 0 & 0 \\ 0 & 1 & 1 & 0 & 1 & 1 \\ 1 & 0 & 0 & 1 & 0 & 1 \\ 0 & 0 & 0 & 1 & 1 & 0 \end{bmatrix}$$

find the corresponding graph.

Given the graph in Figure 6.76,

21. Find the adjacency matrix.
22. Use the adjacency matrix to find all paths of length 2.
23. Use the adjacency matrix to find all paths of length 3.

6.6 Incidence and Adjacency Matrices   **257**

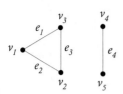

Figure 6.76

Given the graph in Figure 6.77,

24. Find the adjacency matrix.
25. Use the adjacency matrix to find all paths of length 2.
26. Use the adjacency matrix to find all paths of length 3.

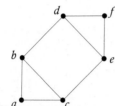

Figure 6.77

Given the graph in Figure 6.78,

27. Find the adjacency matrix.
28. Use the adjacency matrix to find all paths of length 2.
29. Use the adjacency matrix to find all paths of length 3.

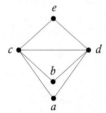

Figure 6.78

Given the graph in Figure 6.79,

30. Find the adjacency matrix.
31. Use the adjacency matrix to find all paths of length 2.
32. Use the adjacency matrix to find all paths of length 3.

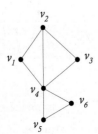

Figure 6.79

33. Use the fact $\hat{A} = A \vee A^{\odot 2} \vee A^{\odot 3} \vee A^{\odot 4} \vee \cdots \vee A^{\odot n}$ to find the transitive closure of the relation represented by the graph in Figure 6.80.

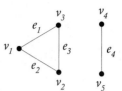

Figure 6.80

34. Use the fact $\hat{A} = A \vee A^{\odot 2} \vee A^{\odot 3} \vee A^{\odot 4} \vee \cdots \vee A^{\odot n}$ to find the transitive closure of the relation represented by the graph in Figure 6.81.

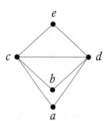

Figure 6.81

35. Use the fact $\hat{A} = A \vee A^{\odot 2} \vee A^{\odot 3} \vee A^{\odot 4} \vee \cdots \vee A^{\odot n}$ to find the transitive closure of the relation represented by the directed graph in Figure 6.82.

Figure 6.82

36. Use the fact $\hat{A} = A \vee A^{\odot 2} \vee A^{\odot 3} \vee A^{\odot 4} \vee \cdots \vee A^{\odot n}$ to find the transitive closure of the relation represented by the directed graph in Figure 6.83.

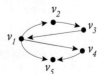

Figure 6.83

37. Use Warshall's algorithm to find the transitive closure of the relation represented by the graph in Figure 6.84.

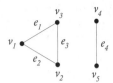

Figure 6.84

38. Use Warshall's algorithm to find the transitive closure of the relation represented by the graph in Figure 6.85.

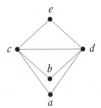

Figure 6.85

39. Use Warshall's algorithm to find the transitive closure of the relation represented by the directed graph in Figure 6.86.

Figure 6.86

40. Use Warshall's algorithm to find the transitive closure of the relation represented by the directed graph in Figure 6.87.

Figure 6.87

41. Use Warshall's algorithm to find the transitive closure of the relation represented by the directed graph in Figure 6.88.

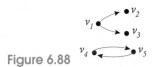

Figure 6.88

42. Use Warshall's algorithm to find the transitive closure of the relation represented by the directed graph in Figure 6.89.

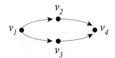

Figure 6.89

43. Incidence matrices for multigraphs may defined in the same way as for regular graphs. How can one tell from the incidence matrix if two edges of a multigraph have the same vertices? (These edges are called parallel edges.)

44. In an incidence matrix for multigraphs, does the sum of the row labeled by a given vertex give the degree of that vertex? What is the significance of the sum of a row?

45. Using induction, prove Theorem 6.63. Let G be a (directed) graph with vertices $v_1, v_2, v_3, \ldots, v_n$ and adjacency matrix A. There is a k-path from v_i to v_j for $1 \leq k \leq n$ if and only if $A_{ij}^{\odot k} = 1$.

46. Using induction, prove Theorem 6.64. Let G be a (directed) graph with vertices $v_1, v_2, v_3, \ldots, v_n$ and adjacency matrix A. There are m k-paths from v_i to v_j for $1 \leq k \leq n$ if and only if $A_{ij}^k = m$.

47. Using Theorem 6.63, prove Theorem 6.65. Let G be a (directed) graph with vertices $v_1, v_2, v_3, \ldots, v_n$ and adjacency matrix A. Let $\hat{A} = A \vee A^{\odot 2} \vee A^{\odot 3} \vee A^{\odot 4} \vee \cdots \vee A^{\odot n}$, then $\hat{A}_{ij} = 1$ if and only if there is a path from v_i to v_j.

48. Describe the adjacent matrix for a tree.

49. Describe the transitive closure of a tree.

6.7 Hypercubes and Gray Code

DEFINITION 6.68 The **distance** between two vertices in a graph is the length of the shortest path between the two vertices.

DEFINITION 6.69 The **diameter** of a graph is the largest distance between any two vertices in the graph.

6.7 Hypercubes and Gray Code

Figure 6.90

Figure 6.91

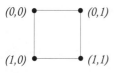

Figure 6.92

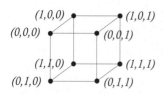

Figure 6.93

Instead of using a single processor that is capable of executing only one program at a time, in many cases a collection of processors can be linked to run in parallel so that different programs can be run at the same time and information exchanged between the processors. One way to connect the computers would be to combine them in series in a ring. This method has the disadvantage that in passing information from one processor to another, it might be necessary to pass through quite a few processors to do so. In the worst case, information might have to pass through half of the processors. A slight improvement would be the use of a grid or rectangular array of processors where a processor is placed at each point on the grid. An example of such a grid is given in Figure 6.90.

This is an improvement, but it is still often necessary to go through a number of processors to pass information from one processor to another. In the previously shown 4×5 grid, it may be necessary to pass through eight processors including the end processors. In general, for an $m \times n$ grid, it might be necessary to pass through $m + n - 2$ processors.

A far better configuration is the hypercube. An n-**hypercube** can be used to connect up to 2^n computers. The graph of an n-hypercube is constructed recursively as follows: For $n = 1$, we represent one vertex by 1 and the other by 0 so that we have the graph in Figure 6.91.

Thus, our vertices consist of all 1-strings of 0 and 1. For $n = 2$, we represent the vertices by 11, 10, 01, and 00 so that we have the graph in Figure 6.92.

Thus, our vertices consist of all 2-strings of 0 and 1. For $n = 3$, we represent the vertices by 111, 110, 101, 100, 011, 010, 001, 000, so that we have the graph in Figure 6.93.

Thus, our vertices consist of all 3-strings of 0 and 1. Note that two vertices are adjacent if one can be changed to the other by changing a single symbol in the string.

In Chapter 1, when constructing truth tables, we found the list all possible combinations of the statements.

If we use Boolean algebra notation and replace T with 1 and F with 0, we find that we produce the vertices of hypercubes of order 1, 2, 3, and 4, which we abbreviate as follows:

order 1	order 2	order 3	order 4
1	11	111	1111
0	10	110	1110
	01	101	1101
	00	100	1100
		011	1011
		010	1010
		001	1001
		000	1000
			0111
			0110
			0101
			0100
			0011
			0010
			0001
			0000

We have even defined what it means for these points to be adjacent for $n = 2, 3,$ and 4. Consider the Karnaugh map:

	q	$\sim q$
p	11	10
$\sim p$	01	00

Our concern is where the boxes are adjacent. If we again use 1 for T and 0 for F and give the value of p followed by the value of q, then the inside values represent where each box is true. Hence, each vertex is adjacent to another vertex if they are adjacent as points on a 2-cube.

Similarly, for three variables where the values of p, q, and r are given in order, we have

	q	q	$\sim q$	$\sim q$
p	111	110	100	101
$\sim p$	011	010	000	001
	r	$\sim r$	$\sim r$	r

and recalling that, for the Karnaugh map, the table is wrapped around so that the two ends are considered adjacent, again two vertices are adjacent in the table only if they are adjacent as points on the 3-cube.

Considering the Karnaugh map for four variables and giving p, q, r, and s in order, we similarly have

	q	q	$\sim q$	$\sim q$	
p	1111	1101	1001	1011	s
p	1110	1100	1000	1010	$\sim s$
$\sim p$	0110	0100	0000	0010	$\sim s$
$\sim p$	0111	0101	0001	0011	s
	r	$\sim r$	$\sim r$	r	

and again recalling that, for the Karnaugh map, the table is wrapped around so that the two ends are considered adjacent and the top and bottom are adjacent, then again two vertices are adjacent in the table only if they are adjacent as points on the 4-cube.

Using the foregoing method, we construct a sequence of vertices for a $k+1$-cube from a sequence for a k-cube as follows:

1. Place a 1 in front of each vertex in the sequence of vertices in the k-cube. Vertices that were adjacent in the k-cube remain adjacent with the 1 in front of them in the $k + 1$-cube.

2. Place a 0 in front of each vertex in the sequence of vertices in the k-cube. Vertices that were adjacent in the k-cube remain adjacent with the 0 in front of them in the $k + 1$-cube.

3. Place the sequence formed in (2) after the sequence formed in (1).

We have now devised a method for constructing hypercubes and, for $n = 1, 2, 3$ and 4, we can use the Karnaugh map to show when vertices are adjacent. For $n > 4$, we can use steps (1)–(3) above to generate vertices.

Suppose that we have a rotating disk, which is divided into sectors, and a series of brushes or laser beams sending back digital information about how far the disk has rotated. If the binary strings recording the numbering of adjacent sectors are substantially different, in the sense that there is a lot of changes of the individual digits in going from one sector to the next, then a reading taken just as the sector was changing could produce a number totally different from the number of either of the sectors. In this case, it is desirable to number the sectors so that the binary string determining the sector has only one digit change between adjacent sectors.

In general, however, the vertices in the foregoing list are not adjacent to each other. We can, however, change this by making a slight change in part (2) above. To form the vertices of the $k + 1$-cube, instead of placing the 0 in front of the list for the k-cube, we *reverse* the list for the vertices of the k-cube before placing the 0 in front of each of these vertices. For example, in forming the vertices of the 2-cube, we place 1 in front of the column for the 1-cube:

$$\begin{matrix} 1 \\ 0 \end{matrix}$$

and then reverse the column to get

$$\begin{matrix} 0 \\ 1 \end{matrix}$$

and place a 0 in front of each so that the final result is

$$\begin{matrix} 1 & 1 \\ 1 & 0 \\ 0 & 0 \\ 0 & 1 \end{matrix}$$

To get the vertices for the 3-cube, we place 1 in front of the foregoing list for the 2-cube and then place a 0 in front of the reverse list for the 2-cube. The final result for the 3-cube is

$$\begin{matrix} 1 & 1 & 1 \\ 1 & 1 & 0 \\ 1 & 0 & 0 \\ 1 & 0 & 1 \\ 0 & 0 & 1 \\ 0 & 0 & 0 \\ 0 & 1 & 0 \\ 0 & 1 & 1 \end{matrix}$$

It is easy to prove that this procedure will always give us a sequence of vertices for the k-cube, which we will call the k-**list**, in which (1) each vertex in the sequence is adjacent to the next one and (2) the first vertex in the sequence is adjacent to the last vertex in the sequence. Using induction we begin by observing that our sequence of vertices for the 1-cube certainly has these sequential properties. Assume that our k-list for the k-cube has these properties. When we place a 1 in front of each vertex in the k-list for the vertices of the k-cube, each vertex in this sequence is certainly adjacent to the next one. Also when we reverse the vertices in the k-list for the k-cube, each vertex is still adjacent to the next one and when a 0 is placed in front of each element in this list of reversed vertices, then each vertex in the list is still adjacent to the next. The first element in the reversed list for the k-cube is the same as the last element of the original k-list for the k-cube so when the first element of the reversed list with a 0 in front of it follows the last element of the original k-list with a 1 in front of it, they differ only in this first digit and hence are adjacent. Similarly, the last element in the reversed k-list is equal to the first element in the k-list, so that when a 0 is placed in front of the last element in the reversed k-list and a 1 is placed in front of the first element of the k-list, these two vertices are adjacent. Thus, our $k + 1$-list has the required properties. This sequence constructed for the n-cube is called a **Gray code** for n.

Thus, our rules for constructing a Gray code for $k+1$ are

1. Place a 1 in front of each vertex in the k-list of the k-cube. Vertices that were adjacent in the k-cube remain adjacent with the 1 in front of them in the $k+1$-cube.
2. Place a 0 in front of each vertex in the reversed k-list of the k-cube. Vertices that were adjacent in the k-cube remain adjacent with the 0 in front of them in the $k+1$-cube.
3. Place the sequence formed in (2) after the sequence formed in (1).
4. Each sequential pair of vertices in $k+1$-list of the $k+1$-cube are adjacent. Also the first vertex in the $k+1$-list is adjacent to the last vertex in the list.

For example, the 3-list and reversed 3-list are, respectively,

$$\begin{array}{ccc} 1 & 1 & 1 \\ 1 & 1 & 0 \\ 1 & 0 & 0 \\ 1 & 0 & 1 \\ 0 & 0 & 1 \\ 0 & 0 & 0 \\ 0 & 1 & 0 \\ 0 & 1 & 1 \end{array} \quad \text{and} \quad \begin{array}{ccc} 0 & 1 & 1 \\ 0 & 1 & 0 \\ 0 & 0 & 0 \\ 0 & 0 & 1 \\ 1 & 0 & 1 \\ 1 & 0 & 0 \\ 1 & 1 & 0 \\ 1 & 1 & 1 \end{array}$$

Note that the first element in each list is equal to the last element in the other list. Adding 1 in front of each vertex in the first list and 0 in front of each vertex in the second list and placing the second list at the end of the first list, we have

$$\begin{array}{cccc} 1 & 1 & 1 & 1 \\ 1 & 1 & 1 & 0 \\ 1 & 1 & 0 & 0 \\ 1 & 1 & 0 & 1 \\ 1 & 0 & 0 & 1 \\ 1 & 0 & 0 & 0 \\ 1 & 0 & 1 & 0 \\ 1 & 0 & 1 & 1 \\ 0 & 0 & 1 & 1 \\ 0 & 0 & 1 & 0 \\ 0 & 0 & 0 & 0 \\ 0 & 0 & 0 & 1 \\ 0 & 1 & 0 & 1 \\ 0 & 1 & 0 & 0 \\ 0 & 1 & 1 & 0 \\ 0 & 1 & 1 & 1 \end{array}$$

and we have constructed the Gray code for 4.

Earlier, we mentioned linking computers in a grid or mesh. By a grid we mean a graph with an $m \times n$ array of vertices such that two adjacent vertices in the same row or column are adjacent as vertices in the graph. Is it possible for $m \leq 2^k$ and $n \leq 2^l$ to construct a subgraph of a $k+l$-cube that is an $m \times n$ grid? We did this when constructing Karnaugh maps. This is easily accomplished by labeling the rows with the first m elements of the Gray code for k and labeling the columns with the first n elements of the Gray code for l. The (i, j)th element of the grid is the ith row label followed by the jth column label. Thus, if we wanted to construct a 3×7 grid, it would have the form

	111	110	100	101	001	000	010
11	11111	11110	11100	11101	11001	11000	11010
10	10111	10110	10100	10101	10001	10000	10010
00	00111	00110	00100	00101	00001	00000	00010

where the (i, j)th element of the grid is the (i, j)th element of the table. We have, thus, demonstrated the following theorem.

THEOREM 6.70 Every $m \times n$ grid is a subgraph of an $i + j$-cube where $m \leq 2^i$ and $n \leq 2^j$. ∎

The proof of the following theorem is left to the reader.

THEOREM 6.71 Each hypercube for $n \geq 1$ is a bipartite graph where the disjoint sets of vertices consist of the set whose vertices are represented by strings containing an even number of 1's and the set whose vertices are represented by strings containing an odd number of 1's. ∎

Exercises

Find the diameter of the following graphs:

1. K_n
2. $K_{m,n}$
3.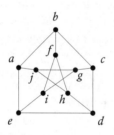
4.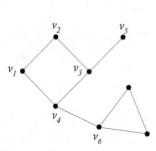
5.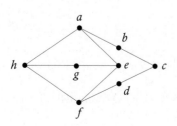
6.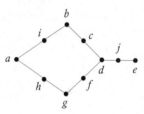
7.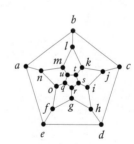
8.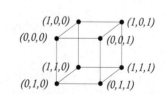
9. a 4-hypercube.
10. Construct the Gray code for 5.
11. Construct the Gray code for 6.
12. Construct a 4×5 grid.

13. Construct a 3 × 8 grid.
14. Construct a 5 × 6 grid.
15. Prove Theorem 6.71: Each hypercube for $n \geq 1$ is a bipartite graph where the disjoint sets of vertices consist of the set whose vertices are represented by strings containing an even number of 1's and the set whose vertices are represented by strings containing an odd number of 1's.
16. Prove that if $m = 2^i$ and $n = 2^j$, then the $m \times n$ grid formed has the property that the corresponding vertices in the first and last rows are adjacent and the corresponding vertices in the first and last columns are adjacent.

GLOSSARY

Chapter 6: Graphs, Directed Graphs, and Trees

Adjacency matrix (6.6)	Let G be a graph (directed graph). Let B be a matrix whose rows are labeled by the vertices in the graph and whose columns are labeled by the same vertices in the same order. The entry in the ith row and jth column of B, denoted by B_{ij}, is equal to 1 if there is an edge (directed edge) from the ith vertex to the jth edge and is 0 otherwise. The matrix B is called the **adjacency matrix** of the graph G.
Adjacent edges (6.1)	Two edges are **adjacent** if they are incident on a common vertex. They have a common endpoint.
Adjacent vertex (6.2)	If (a, b) is an edge of a directed graph $G(V, E)$, so vertex a is the initial vertex and b is the terminal vertex of (a, b), then vertex a is adjacent to vertex b. Also vertex b is **adjacent** to vertex a.
Adjacent vertices (6.1)	Two vertices are **adjacent** if they are the endpoints of an edge or equivalently they are incident on the same edge.
Ancestor vertex/ Descendant vertex (6.3)	If there is a directed path from vertex u to vertex v, then u is called an **ancestor** of v and v is called a **descendant** of u.
Asymmetric relation (6.3)	If E is both the set of edges for the tree and the relation for the tree and has the property that if $(a, b) \in E$, then $(b, a) \notin E$, such a relation is called **asymmetric**.
Bipartite (6.1)	A graph $G = (V, E)$ is called a **bipartite** graph if V can be expressed as the disjoint union of nonempty sets, say, $V = A \cup B$, such that every edge has the form $\{a, b\}$ where $a \in A$ and $b \in B$. Thus, every edge connects a vertex in A to a vertex in B and no vertices both in A or both in B are connected.
Complete bipartite (6.1)	A bipartite graph is called a **complete bipartite** graph $K_{m,n}$ if A contains m vertices, B contains n vertices, and for every $a \in A, b \in B, \{a, b\} \in E$. Thus, for every $a \in A$ and $b \in B$, there is an edge connecting them.
Complete graph (6.1)	A graph is a **complete graph** if there is an edge between every two distinct vertices. A complete graph with n vertices is denoted by K_n.
Component of a graph (6.1)	Let $G = G(V, E)$ be a graph. A subgraph G' of G is a **component** of G if 1. G' is a nonempty connected graph and 2. if G'' is a connected subgraph of G and $G' \preceq G''$, then $G' = G''$. Hence, G' is a maximal connected subgraph of G.
Connected (6.2)	A directed graph $G(V, E)$ is **connected** if its underlying graph is connected.

Connected graph (6.1)	A graph is **connected** if there is a path between any two distinct vertices in G.
Connected/joined (6.1)	The elements a and b of V are **joined** or **connected** by the edge $\{a, b\}$ if $\{a, b\} \in E$.
Cycle (6.1)	A **cycle** is a path of length greater than zero from a vertex to itself with no repeated edges.
Degree of a vertex (6.1)	The **degree** of a vertex v, denoted by $\deg(v)$, is the number of edges that are incident on that vertex.
Diameter of a graph (6.7)	The **diameter** of a graph is the largest distance between any two vertices in the graph.
Directed cycle (6.5)	Let $G = (V, E)$ be a directed graph. A **directed cycle** is a directed path of length greater than zero from a vertex to itself with no repeated edges.
Directed edges/edges (6.2)	An edge of a digraph is called a **directed edge**.
Directed graph/digraph (6.2)	A **directed graph (digraph)** G, denoted by $G(V, E)$, consists of a set V of vertices, together with a set E of ordered pairs of V called the set of **directed edges**.
Directed path (6.2)	A **directed path** from a to b in a directed graph is described by a sequence of vertices $v_0 v_1 v_2 v_3 \ldots v_n$ where $a = v_0$, $b = v_n$, and (v_{i-1}, v_i) is a directed edge for $1 \leq i \leq n$.
Directed subgraph (6.2)	A directed graph $G'(V', E')$ is a **directed subgraph** of a directed graph $G(V, E)$, denoted by $G'(V', E') \preceq G(V, E)$, if $V' \subseteq V$ and $E' \subseteq E$.
Directed tree (6.3)	A **directed tree** T is a loop-free directed graph, whose underlying graph is a tree, such that if there is a path from a vertex a to a vertex b, it is unique.
Distance	The **distance** between two vertices in a graph is the length of the shortest path between the two vertices.
Edge (6.1)	A two-element set of vertices.
Edge set (6.1)	An **edge set** E is a collection of two-element subsets of the vertex set.
Endpoints (6.1)	If $\{a, b\}$ is an edge, then vertices a and b are called the **endpoints** of the edge $\{a, b\}$.
Euler cycle (6.5)	Let $G = (V, E)$ be a graph. A cycle that includes all of the edges and the vertices of G is called an **Euler cycle**. When this occurs, we say the graph G has an Euler cycle.
Euler path (6.5)	Let $G = (V, E)$ be a graph. A path that includes each of the edges of G exactly once is called an **Euler path**. When this occurs, we say the graph G has an Euler path.
Forest (6.3)	A **forest** is a graph whose components are trees.
Graph (6.1)	A **graph** is a finite set V called the **vertex set** and a collection of two-element subsets E, the edge set of V.
Graph with loops (6.1)	If loops are included with a graph, the structure is called a **graph with loops**.
Gray code for n (6.7)	The k-list sequence constructed for the n-cube is called a **Gray code** for n. (See k-list.)
Height of a tree (6.3)	The **height** of a tree is the length of the longest path from the root of the tree to a leaf.

Hypercube/n-hypercube (6.7)	A **hypercube** or **n-hypercube** is an n-tuple consisting of 0's and 1's. Each vertex of the hypercube differs from any adjacent vertex by exactly one coordinate. A hypercube contains 2^n vertices (or nodes), and $n2^{n-1}$ edges. A recursive definition of a hypercube goes as follows: an n-dimensional hypercube is built by connecting like points of two $(n-1)$-dimensional hypercubes. A 0-dimensional hypercube is just a single node.
Incidence matrix (6.6)	Let G be a graph. Let B be a matrix whose rows are labeled by the vertices in the graph and whose columns are labeled by the edges in the graph. The entry in the ith row and jth column of B, denoted by B_{ij}, is equal to 1 if the ith vertex is incident to the jth edge and is 0 otherwise. The matrix B is called the **incidence matrix** of the graph G.
Incident (6.1)	The edge $\{a, b\}$ is said to be **incident** on vertices a and b. Also, vertices a and b are said to be **incident** on the edge $\{a, b\}$.
Incident vertices (6.2)	If (a, b) is an edge of a directed graph $G(V, E)$, so vertex a is the initial vertex and b is the terminal vertex of (a, b), then vertices a and b are **incident** on the edge (a, b).
Indegree (6.2)	The **indegree** of a vertex v, denoted by indeg(v), is the number of edges for which v is the terminal vertex.
Initial vertex (6.2)	If $a, b \in V$, then a is called the **initial vertex** of the edge (a, b).
Instant insanity (6.4)	The game of **instant insanity** is a game played with four cubes with each of the six edges of each cube painted with one of the colors red, blue, green, or yellow (or any four colors). The object of the game is to stack the cubes so that each of the four colors appears on each of the four sides of the stack.
Internal vertices (6.3)	Vertices other than leaves are called **internal vertices**.
Isolated vertices (6.1)	A vertex with degree 0 is called **isolated**.
k-list (6.7)	A **k-list** is a sequence of vertices for the k-cube, in which 1. each vertex in the sequence is adjacent to the next one and 2. the first vertex in the sequence is adjacent to the last vertex in the sequence.
Königsberg bridge problem (6.1)	The citizens of Königsberg on the Pregel River in Prussia liked to walk on the paths that included the seven bridges over the river. People wondered if they could start on one of the land masses, cross all of the bridges only once, and return to the point where they started without having to swim.
Labeled graph (6.1)	A graph that has each edge labeled is called a **labeled graph**.
Labeled graph/labeled direct graph (6.2)	A **labeled graph** $G = G(V, L, E)$ is a set of vertices V, a set of labels L, and a set E, which is a subset of $V \times L \times V$. Thus, an edge e of G has the form (a, l, b) where l is a label and a, b, are vertices. In other words, if each edge is labeled, it is a **labeled directed graph** or simply as a **labeled graph**.
Leaves (6.3)	The vertices of degree 1 are called **leaves**. For a directed tree the outdegree of a leaf is 0.
Left child/right child (6.3)	In every binary tree, each child of a parent is designated as either a **left child** or **right child** (but not both).
Length of a directed path (6.2)	The **length** of a directed path is the number of directed edges in the path.

Level of a vertex (6.3)	Once a root is chosen, the **level** of a vertex v is the length of the unique path from the root to v.
Loop (6.1)	An edge from a vertex to itself is called a **loop**.
m-ary tree/Binary tree (6.3)	If the largest outdegree for any vertex of the tree is m, then the tree is called an ***m*-ary** tree. In particular, if $m = 2$, then the tree is called a **binary tree**.
Multigraph (6.1)	If a graph has more than one edge between two vertices, it is called a **multigraph**.
Multigraph/Directed multigraph (6.2)	A directed graph with more than one edge from one vertex to another vertex is called a **multigraph**, or more precisely a **directed multigraph**.
n-cycle (6.1)	An ***n*-cycle** is a cycle containing n edges and n distinct vertices.
Outdegree (6.2)	The **outdegree** of a vertex v, denoted by outdeg(v), is the number of edges for which v is the initial vertex.
Parent and child vertices (6.3)	If we consider the rooted tree T', then the vertex u is the **parent** of v and v is the **child** of u, if there is a directed edge from u to v.
Path (6.1)	A path in a graph is a collection of edges that are joined together at the vertices so that the sequences of edges have common vertices.
Path of length k (6.1)	A **path of length k** from v_0 to v_k (or between v_0 and v_k) is a sequence $v_0 e_1 v_1 e_2 v_2 e_3 v_3 \ldots v_{k-1} e_k v_k$ such that $e_i = \{v_{i-1}, v_i\}$. A path of length k has k edges.
Proper Euler path (6.5)	Euler paths that are not Euler cycles are called **proper Euler paths**.
Pseudograph (6.1)	A graph with both loops and more than one edge between vertices is a **pseudograph**.
Root (6.3)	The vertex with indegree zero in a directed tree is called the **root** of the tree.
Rooted tree/Rooted directed tree (6.3)	Once the root of the tree has been determined, the tree is called a **rooted tree**. This is called the **rooted directed tree** T' **derived** from the rooted tree T.
Roy-Warshall's algorithm (6.6)	**Roy-Warshall's** algorithm is an algorithm for determining the transitive closure of an adjacency matrix.
Sibling vertices (6.3)	If u is the parent of v and v', then v and v' are called **siblings**.
Simple cycle (6.1)	A **simple cycle** is a cycle from a vertex v to itself such that v is the only vertex that is repeated.
Simple path (6.1)	A **simple path** from v_0 to v_k is a path in which no vertex is repeated.
Sink (6.2)	If outdeg$(v) = 0$, then v is called a **sink**.
Source (6.2)	If indeg$(v) = 0$, then v is called a **source**.
Spanning tree (6.3)	A tree is a **spanning tree** for a graph G, if T is a subgraph of G, and every vertex in G is a vertex in T.
Strongly connected (6.2)	A directed graph is **strongly connected** if for every pair of vertices $a, b \in V$, there is a directed path from a to b.
Subgraph (6.1)	A graph $G'(V', E')$ is a **subgraph** of a graph $G(V, E)$, denoted by $G'(V', E') \preceq G(V, E)$, if $V' \subseteq V$ and $E' \subseteq E$. Every vertex in G' is a vertex in G and every edge in G' is an edge in G.

Terminal vertex (6.2)	If $a, b \in V$, then b is called the **terminal vertex** of the edge (a, b).
Three house–three utilities problem (6.1)	Three houses are to receive three utilities, say, water, gas, and electricity, by means of underground lines consisting of cables or pipes. The question is whether the utilities can be supplied to the three houses without any of the lines crossing.
Tree (6.3)	A **tree** is a graph with no cycles.
Underlying graph (6.2)	Let R be the symmetric closure of E', so that if $(a, b) \in E'$, then $(a, b), (b, a) \in R$, and E^s be the set of edges representing the relation R. Then the graph $G^s(V, E^s)$ is called the **underlying graph** of the directed graph $G(V, E)$.
Vertex set (6.1)	The **vertex** set V is the set of the vertices in a graph G.

TRUE-FALSE QUESTIONS

1. A subgraph of a graph may be formed by removing a single vertex.
2. In a graph with vertices a, b, c, d and e, $abcdcea$ cannot be a cycle.
3. A component of a graph is always connected.
4. If there is a simple path from a vertex a to a vertex b in the graph, then it is always shorter than a path from a to b that is not simple.
5. $K_{2,3}$ is a subgraph of $K_{4,5}$.
6. Every directed graph contains a vertex that is a source.
7. The directed graph

   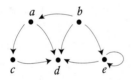

   is connected, but not strongly connected.
8. In the previous graph, the indegree of e is 2.
9. In the previous graph, $acda$ is a directed cycle.
10. In the previous graph, b is a source and d is a sink.
11. A directed graph that is not strongly connected may be divided into components.
12. Every tree has at least two vertices of degree 1.
13. The spanning tree of a graph is unique.
14. Every graph has a spanning tree.
15. $K_{4,3}$ has an Euler cycle.
16. $K_{2,7}$ has a proper Euler path.
17. A directed graph that has a sink cannot have a directed Euler cycle.
18. A graph that has an Euler cycle cannot have a proper Euler path.
19. A directed graph that has a directed Euler cycle is strongly connected.
20. A directed graph that is strongly connected has a directed Euler cycle.
21. A maximal path is the longest path in a tree.
22. A directed tree with more than three vertices cannot have an Euler path.
23. A 3-hypercube has an Euler cycle.
24. The diameter of $K_{7,5}$ is 2.
25. The following is a Gray code for 3.

T	T	T
T	T	F
T	F	T
T	F	F
F	T	T
F	T	F
F	F	T
F	F	F

26. The diameter of an n-hypercube is n.
27. The strings in a 3×7 grid have length 6.
28. An n-hypercube is a bipartite graph.
29. A Gray code for n is an arrangement of the vertices of an n-hypercube.
30. The spanning tree for a graph G is unique if and only if G is a tree.
31. An internal vertex of a tree cannot have degree 2.
32. At least half of the vertices of a tree must be leaves.

33. At most half of the vertices of a tree must be leaves.
34. The spanning tree of a graph G is always a subgraph of G.
35. Warshall's algorithm may be used to find paths of a given length.
36. If no vertex in a graph has degree 0, then $\hat{A} = \hat{A}^I$.
37. If A is the adjacency matrix of a nontrivial graph G, then G is connected if and only if $\hat{A}_{ij} = 1$ for all i, j.
38. I is the adjacency matrix for a directed graph with no edges.
39. If the incidence matrix for a directed graph has exactly one 1 in each column, then the directed graph is a directed tree.
40. If A is the incidence matrix for a directed tree, then A has exactly one 1 in each column.
41. If A is the incidence matrix for a directed graph G and $A_{ii} = 1$ for any i, then the graph contains a loop.
42. If A is the incidence matrix for a graph, then A^n is the incidence matrix for a graph.
43. If row i of an incidence matrix contains k 1's, then vertex v_i has degree k.
44. Every column of an incidence matrix contains exactly two 1's.
45. Let A be an adjacency matrix for a directed tree T. If v_j is a root of T, then $A_{ij} = 0$ for all i.
46. The adjacency matrix for a directed tree has exactly one column that contains only 0's.
47. The adjacency matrix for

is

$$\begin{bmatrix} 1 & 1 & 1 & 0 & 0 \\ 1 & 1 & 1 & 0 & 0 \\ 1 & 1 & 1 & 0 & 0 \\ 0 & 0 & 0 & 1 & 1 \\ 0 & 0 & 0 & 1 & 1 \end{bmatrix}$$

48. If

$$A = \begin{bmatrix} 0 & 1 & 0 & 0 & 0 & 1 & 1 & 0 \\ 1 & 0 & 1 & 0 & 0 & 0 & 0 & 0 \\ 0 & 1 & 0 & 1 & 0 & 0 & 1 & 0 \\ 0 & 0 & 1 & 0 & 1 & 0 & 1 & 1 \\ 0 & 0 & 0 & 1 & 0 & 1 & 0 & 0 \\ 1 & 0 & 0 & 0 & 1 & 0 & 0 & 0 \\ 1 & 0 & 1 & 1 & 0 & 0 & 0 & 0 \\ 0 & 0 & 0 & 1 & 0 & 0 & 0 & 0 \end{bmatrix}$$

Then

$$A^{\odot 2} = \begin{bmatrix} 1 & 1 & 1 & 0 & 0 & 1 & 1 & 0 \\ 1 & 0 & 1 & 0 & 0 & 0 & 0 & 0 \\ 0 & 1 & 1 & 1 & 0 & 0 & 1 & 0 \\ 0 & 0 & 1 & 1 & 1 & 0 & 1 & 1 \\ 0 & 0 & 0 & 1 & 1 & 1 & 0 & 0 \\ 1 & 0 & 0 & 1 & 1 & 0 & 0 & 0 \\ 1 & 1 & 1 & 1 & 0 & 0 & 0 & 0 \\ 0 & 0 & 1 & 1 & 0 & 0 & 0 & 0 \end{bmatrix}$$

49. If

$$A = \begin{bmatrix} 0 & 1 & 0 & 0 & 0 & 1 & 1 & 0 \\ 1 & 0 & 1 & 0 & 0 & 0 & 0 & 0 \\ 0 & 1 & 0 & 1 & 0 & 0 & 1 & 0 \\ 0 & 0 & 1 & 0 & 1 & 0 & 1 & 1 \\ 0 & 0 & 0 & 1 & 0 & 1 & 0 & 0 \\ 1 & 0 & 0 & 0 & 1 & 0 & 0 & 0 \\ 1 & 0 & 1 & 1 & 0 & 0 & 0 & 0 \\ 0 & 0 & 0 & 1 & 0 & 0 & 0 & 0 \end{bmatrix}$$

then

$$A \vee A^{\odot 2} \vee A^{\odot 3} \vee A^{\odot 4}$$

$$= \begin{bmatrix} 1 & 1 & 1 & 1 & 1 & 1 & 1 & 1 \\ 1 & 1 & 1 & 1 & 1 & 1 & 1 & 1 \\ 1 & 1 & 1 & 1 & 1 & 1 & 1 & 1 \\ 1 & 1 & 1 & 1 & 1 & 1 & 1 & 1 \\ 1 & 1 & 1 & 1 & 1 & 1 & 1 & 1 \\ 1 & 1 & 1 & 1 & 1 & 1 & 1 & 1 \\ 1 & 1 & 1 & 1 & 1 & 1 & 1 & 1 \\ 1 & 1 & 1 & 1 & 1 & 1 & 1 & 1 \end{bmatrix}$$

If

$$A = \begin{bmatrix} 0 & 1 & 1 & 0 & 0 & 0 & 0 & 0 \\ 1 & 0 & 1 & 0 & 0 & 0 & 0 & 0 \\ 0 & 1 & 0 & 1 & 0 & 0 & 0 & 0 \\ 0 & 0 & 1 & 0 & 1 & 0 & 0 & 0 \\ 1 & 0 & 0 & 1 & 0 & 0 & 0 & 0 \\ 0 & 0 & 0 & 0 & 0 & 0 & 1 & 1 \\ 0 & 0 & 0 & 0 & 0 & 1 & 0 & 1 \\ 0 & 0 & 0 & 0 & 0 & 1 & 1 & 0 \end{bmatrix}$$

Then

$$\hat{A} = \begin{bmatrix} 1 & 1 & 1 & 1 & 1 & 0 & 0 & 0 \\ 1 & 1 & 1 & 1 & 1 & 0 & 0 & 0 \\ 1 & 1 & 1 & 1 & 1 & 0 & 0 & 0 \\ 1 & 1 & 1 & 1 & 1 & 0 & 0 & 0 \\ 1 & 1 & 1 & 1 & 1 & 0 & 0 & 0 \\ 0 & 0 & 0 & 0 & 0 & 1 & 1 & 1 \\ 0 & 0 & 0 & 0 & 0 & 1 & 1 & 1 \\ 0 & 0 & 0 & 0 & 0 & 1 & 1 & 1 \end{bmatrix}$$

SUMMARY QUESTIONS

1. Draw the graph $K_{2,4}$.

2. What is the sum of the degrees of K_5?

3. Give an example of a directed graph that is connected but not strongly connected.

4. Explain whether or not a directed graph with more than one vertex, one of which is a source, can be connected.

5. What is the spanning tree of a graph?

6. What is the longest possible path in a tree with n vertices?

7. Show that a tree with more than one vertex always has a vertex of degree 1.

8. How many parents may a vertex have?

9. Does $K_{2,5}$ have an Euler path?

10. For what values of m and n does $K_{m,n}$ have an Euler cycle?

11. Is the transitive closure of the relation related to the adjacency matrix

$$\begin{bmatrix} 0 & 1 & 0 & 1 & 1 \\ 0 & 0 & 1 & 0 & 0 \\ 1 & 0 & 0 & 0 & 0 \\ 0 & 0 & 0 & 0 & 1 \\ 0 & 0 & 0 & 1 & 0 \end{bmatrix}$$

represented by a connected graph?

12. What is the diameter of an $m \times n$ grid?

13. Define the root of a rooted directed tree.

14. Give an example of a directed tree in which two vertices have indegree 0.

15. Give an example of a cycle that is not a simple cycle.

ANSWERS TO TRUE-FALSE

1. F 2. T 3. T 4. F 5. T 6. F 7. T 8. T 9. F 10. T
11. F 12. T 13. F 14. F 15. F 16. T 17. T 18. T 19. T
20. F 21. F 22. F 23. F 24. T 25. F 26. T 27. F 28. T
29. T 30. T 31. F 32. F 33. F 34. T 35. F 36. T 37. T
38. F 39. F 40. F 41. T 42. F 43. T 44. F 45. T 46. F
47. F 48. F 49. T 50. T

CHAPTER 7

Number Theory

7.1 Sieve of Eratosthenes

As discussed in Chapter 5, an **algorithm** is a finite set of instructions that can be followed mechanically to solve a problem. Our first example of an algorithm in number theory is the ancient method of determining primes called the sieve of Eratosthenes. It is an algorithm for determining the primes less than a given integer. We illustrate this method to determine the primes between 1 and 100. First list the integers between 1 and 100:

1, 2, 3, 4, 5, 6, 7, 8, 9, 10, 11, 12, 13, 14, 15, 16, 17, 18, 19, 20,
21, 22, 23, 24, 25, 26, 27, 28, 29, 30, 31, 32, 33, 34, 35, 36, 37, 38, 39, 40,
41, 42, 43, 44, 45, 46, 47, 48, 49, 50, 51, 52, 53, 54, 55, 56, 57, 58, 59, 60,
61, 62, 63, 64, 65, 66, 67, 68, 69, 70, 71, 72, 73, 74, 75, 76, 77, 78, 79, 80,
81, 82, 83, 84, 85, 86, 87, 88, 89, 90, 91, 92, 93, 94, 95, 96, 97, 98, 99, 100

Beginning with the first prime, 2, we mark in boldface all multiples of 2 excluding 2 itself. This procedure is straightforward since every second integer larger than 2 is to be marked.

1, 2, 3, *4*, 5, *6*, 7, *8*, 9, *10*, 11, *12*, 13, *14*, 15, *16*, 17, *18*, 19, *20*,
21, *22*, 23, *24*, 25, *26*, 27, *28*, 29, *30*, 31, *32*, 33, *34*, 35, *36*, 37, *38*, 39, *40*,
41, *42*, 43, *44*, 45, *46*, 47, *48*, 49, *50*, 51, *52*, 53, *54*, 55, *56*, 57, *58*, 59, *60*,
61, *62*, 63, *64*, 65, *66*, 67, *68*, 69, *70*, 71, *72*, 73, *74*, 75, *76*, 77, *78*, 79, *80*,
81, *82*, 83, *84*, 85, *86*, 87, *88*, 89, *90*, 91, *92*, 93, *94*, 95, *96*, 97, *98*, 99, *100*

Continuing with the next prime, 3, we mark all multiples of 3. Again this procedure is straightforward since every third integer larger than 3 is to be marked.

1, 2, 3, *4*, 5, *6*, 7, *8*, *9*, *10*, 11, *12*, 13, *14*, *15*, *16*, 17, *18*, 19, *20*,
21, *22*, 23, *24*, 25, *26*, *27*, *28*, 29, *30*, 31, *32*, *33*, *34*, 35, *36*, 37, *38*, *39*, *40*,
41, *42*, 43, *44*, *45*, *46*, 47, *48*, 49, *50*, *51*, *52*, 53, *54*, 55, *56*, *57*, *58*, 59, *60*,
61, *62*, *63*, *64*, 65, *66*, 67, *68*, *69*, *70*, 71, *72*, 73, *74*, *75*, *76*, 77, *78*, 79, *80*,
81, *82*, 83, *84*, 85, *86*, *87*, *88*, 89, *90*, 91, *92*, *93*, *94*, 95, *96*, 97, *98*, *99*, *100*

We now mark all multiples of the next prime, 5.

1, 2, 3, *4*, 5, *6*, 7, *8*, *9*, *10*, 11, *12*, 13, *14*, *15*, *16*, 17, *18*, 19, *20*,
21, *22*, 23, *24*, *25*, *26*, *27*, *28*, 29, *30*, 31, *32*, *33*, *34*, *35*, *36*, 37, *38*, *39*, *40*,
41, *42*, 43, *44*, *45*, *46*, 47, *48*, 49, *50*, *51*, *52*, 53, *54*, *55*, *56*, *57*, *58*, 59, *60*,
61, *62*, *63*, *64*, *65*, *66*, 67, *68*, *69*, *70*, 71, *72*, 73, *74*, *75*, *76*, *77*, *78*, 79, *80*,
81, *82*, 83, *84*, *85*, *86*, *87*, *88*, 89, *90*, 91, *92*, *93*, *94*, *95*, *96*, 97, *98*, *99*, *100*

Next mark all multiples of the next prime, 7.

1, 2, 3, *4*, 5, *6*, 7, *8*, *9*, *10*, 11, *12*, 13, *14*, *15*, *16*, 17, *18*, 19, *20*,
21, *22*, 23, *24*, *25*, *26*, *27*, *28*, 29, *30*, 31, *32*, *33*, *34*, *35*, *36*, 37, *38*, *39*, *40*,
41, *42*, 43, *44*, *45*, *46*, 47, *48*, *49*, *50*, *51*, *52*, 53, *54*, *55*, *56*, *57*, *58*, 59, *60*,
61, *62*, *63*, *64*, *65*, *66*, 67, *68*, *69*, *70*, 71, *72*, 73, *74*, *75*, *76*, *77*, *78*, 79, *80*,
81, *82*, 83, *84*, *85*, *86*, *87*, *88*, 89, *90*, *91*, *92*, *93*, *94*, *95*, *96*, 97, *98*, *99*, *100*

Since 7 is the largest prime whose square is less than or equal to 100, by Theorem 3.42 we need continue no further. The numbers greater than 1 that are not marked are the primes less than 100.

Exercises

1. Create the sieve of Eratosthenes for positive integers less than 200.

2. Use the sieve of Eratosthenes and a calculator to factor 1726.

3. Use the sieve of Eratosthenes and a calculator to factor 481.

4. Use the sieve of Eratosthenes and a calculator to factor 2502.

5. Use the sieve of Eratosthenes and a calculator to factor 1739.

6. Use the sieve of Eratosthenes and a calculator to factor 391.

7. Use the sieve of Eratosthenes and a calculator to factor 3901.

7.2 Fermat's Factorization Method

The next theorem is the basis of an algorithm for factoring primes called Fermat's factorization method. It is used to determine whether an integer is prime.

THEOREM 7.1 An odd integer $n > 1$ is nonprime if and only if there are nonnegative integers p and q such that $n = p^2 - q^2$ with $p - q > 1$.

Proof Obviously, if n can be expressed as the difference of two squares of nonnegative integers, say, $n = p^2 - q^2$, then $n = (p - q)(p + q)$. Since $p - q > 1$, then $p + q > 1$ also; and n is not prime.

Conversely, if $n = rs$ with $r \geq s > 1$, then n can be expressed as $((r+s)/2)^2 - ((r-s)/2)^2$. Since n is odd, r and s are odd and, consequently, $r + s$ and $r - s$ are even. Letting $p = (r+s)/2$ and $q = (r-s)/2$, we see that p and q are nonnegative integers and $p - q = s > 1$. If $n = 1$, let $p = 1$ and $q = 0$. ∎

In using Fermat's factorization method we try to find integers p and q such that $n = p^2 - q^2$ or, equivalently, such that $p^2 = n + q^2$ or $q^2 = p^2 - n$. If the first equation is used, we let $q = 1, 2, \ldots$ until $n + q^2$ is a perfect square. If we have not reached a perfect square before $q = (n-1)/2$, then we will when $q = (n-1)/2$, which gives $n + q^2 = ((n+1)/2)^2$ and factors n into $n \cdot 1$. Obviously, since q has the form $(r-s)/2$ where r and s are factors of n, q cannot exceed $(n-1)/2$. Hence, if we have not reached a perfect square before $q = (n-1)/2$, n is a prime.

If the second equation is used, that is, $q^2 = p^2 - n$, then let m be the smallest integer such that $m^2 \geq n$, and let $p = m, m+1, \ldots$ until $p^2 - n$ is a perfect square. As previously, q cannot exceed $(n-1)/2$, so if we have not reached a perfect square before $p = (n+1)/2$, n is a prime. The advantage of using the second squares method is that we check smaller numbers to see if they are squares.

For example, consider using the form $p^2 = n + q^2$ to test $n = 527$ for being prime. We consider $q = 1, 2, \ldots, (n-1)/2$.

q	$n + q^2$
1	$527 + 1 = 528$
2	$527 + 4 = 531$
3	$527 + 9 = 536$
4	$527 + 16 = 543$
5	$527 + 25 = 552$
6	$527 + 36 = 563$
7	$527 + 49 = 576 = (24)^2$

So $n = 527$ is composite and its factors may be calculated:

$$527 = (24)^2 - 7^2$$
$$= (24-7)(24+7)$$
$$= 17 \cdot 31$$

Not every q from 1 to $(n-1)/2$ needs to be checked unless n is prime.

Exercises

Determine whether the following are primes using Fermat's factorization method:

1. 1001
2. 1349
3. 4851
4. 1079
5. 8051
6. 567
7. 7931

7.3 The Division and Euclidean Algorithms

We earlier proved a theorem that we called the division algorithm. It stated that for positive integers a and b there exist unique nonnegative integers q and r with $0 \leq r < b$ such that $a = bq + r$. It is not an algorithm but there are algorithms for determining q and r. If we have large integers a and b, we may "guess" a number q. If we multiply b by q and it is too large, then replace q by $q-1$ and repeat the process until $bq \leq a$ and let $r = a - bq$. If, however, $bq < a$ but $a - bq \geq b$, replace q by $q+1$ and repeat the process until $a - bq < b$ and let $r = a - bq$. A more systematic method would be the following:

1. Let $q = 1$ and $r = a - bq$.
2. If $r \geq b$, let $q = q + 1$ and $r = a - bq$
3. Continue this process until $r < b$.

There is an algorithm for finding the greatest common divisor of large numbers called the Euclidean algorithm. Finding the greatest common divisor is important for adding fractions and also for solving integral equations.

THEOREM 7.2 **(Euclidean Algorithm)** Assume that a and b are positive integers and the division algorithm is applied repeatedly, giving the following sequence:

$$a = bq_0 + r_0 \qquad 0 \leq r_0 < b$$
$$b = r_0 q_1 + r_1 \qquad 0 \leq r_1 < r_0$$
$$r_0 = r_1 q_2 + r_2 \qquad 0 \leq r_2 < r_1$$
$$r_1 = r_2 q_3 + r_3 \qquad 0 \leq r_3 < r_2$$
$$r_2 = r_3 q_4 + r_4 \qquad 0 \leq r_4 < r_3$$
$$\vdots \qquad\qquad \vdots$$
$$r_k = r_{k+1} q_{k+2} + r_{k+2} \qquad 0 \leq r_{k+2} < r_{k+1}$$
$$\vdots \qquad\qquad \vdots$$

There exists an $r_k = 0$. Let s be the first integer such that $r_s = 0$. Then $r_{s-1} = \gcd(a, b)$ if $s > 0$, and $b = \gcd(a, b)$ if $s = 0$.

Proof Let $S = \{r_0, r_1, \ldots\}$. If $r_0 = 0$, then the result is obvious so assume $r_0 > 0$. We know by the well-ordering principle that S contains a least positive integer, say, r_i. But $0 \leq r_{i+1} < r_i$, and so $r_{i+1} = 0$. Let $s = i + 1$.

By Theorem 3.34,

$$\gcd(a, b) = \gcd(b, r_0) = \gcd(r_0, r_1) = \cdots = \gcd(r_{s-1}, r_s)$$

but since $r_s = 0$, $\gcd(r_{s-1}, r_s) = r_{s-1}$ so that $\gcd(a, b) = r_{s-1}$. ∎

EXAMPLE 7.3 Using the Euclidean algorithm to find $\gcd(203, 91)$, we proceed as follows. First, divide 91 into 203 to obtain

$$203 = 91 \cdot 2 + 21$$
$$a = b \cdot q_0 + r_0$$

We now take the remainder 21 and divide it into the quotient 91:

$$91 = 21 \cdot 4 + 7$$
$$b = r_0 \cdot q_1 + r_1$$

Now dividing 7 into 21, we have

$$21 = 7 \cdot 3 + 0$$
$$r_0 = r_1 \cdot q_2 + r_2$$

So $s = 2$ and $\gcd(203, 91) = \gcd(91, 21) = \gcd(21, 7) = 7 = r_{s-1} = r_1$. ∎

EXAMPLE 7.4 Find $\gcd(99, 205)$. Proceeding as before, we divide 99 into 205, giving

$$205 = 99 \cdot 2 + 7$$

We now divide the remainder 7 into 99, giving

$$99 = 7 \cdot 14 + 1$$

Next divide the remainder 1 into 7 to get

$$7 = 1 \cdot 7 + 0$$

Thus, $\gcd(99, 205) = 1$. ∎

Using the Euclidean algorithm we can find values of u and v so that $au + vb = \gcd(a, b)$. Using the notation of Theorem 7.2, $\gcd(a, b) = r_t$, where $r_{t-2} = r_{t-1} \cdot q_t + r_t$. Hence, $r_t = r_{t-2} - r_{t-1} \cdot q_t$. Similarly, $r_{t-1} = r_{t-3} - r_{t-2} \cdot q_{t-1}$. Substituting in the previous equation, we have

$$r_t = r_{t-2} - (r_{t-3} - r_{t-2} \cdot q_{t-1}) \cdot q_t$$

Similarly, we can solve for r_{t-2} in terms of r_{t-3} and r_{t-4} and substitute it in the preceding equation to eliminate r_{t-2}. Continuing in this manner, we eventually eliminate the r_i until we get back to a and b, and we have $\gcd(a, b)$ in the form $ua + vb$.

EXAMPLE 7.5   Express $\gcd(85, 34)$ in the form $85u + 34v$. First using the Euclidean algorithm, we divide 34 into 85:

$$85 = 34 \cdot 2 + 17$$

Dividing 17 into 34, we get

$$34 = 17 \cdot 2 + 0$$

Thus, $\gcd(85, 34) = 17$ and $\gcd(85, 34) = 17 = (85)(1) + (34)(-2)$. ∎

EXAMPLE 7.6   Express $\gcd(252, 580)$ in the form $252u + 580v$. Divide 252 into 580, obtaining

$$580 = 252 \cdot 2 + 76$$
$$a = b \cdot q_0 + r_0$$

Now divide 76 into 252:

$$252 = 76 \cdot 3 + 24$$
$$b = r_0 \cdot q_1 + r_1$$

Continuing, we get

$$76 = 24 \cdot 3 + 4$$
$$r_0 = r_1 \cdot q_2 + r_2$$

and

$$24 = 4 \cdot 6 + 0$$
$$r_1 = r_2 \cdot q_3 + r_3$$

Back substituting, we find that

$$4 = \underline{76} - \underline{24} \cdot 3$$
$$= \underline{76} - [\underline{252} - \underline{76} \cdot 3] \cdot 3$$
$$= \underline{76} \cdot 10 + \underline{252} \cdot (-3)$$
$$= [\underline{580} - \underline{252} \cdot 2] \cdot 10 + \underline{252} \cdot (-3)$$
$$= \underline{580} \cdot 10 + \underline{252} \cdot (-23)$$

where the r_i are emphasized with underlining. ∎

EXAMPLE 7.7   Express $\gcd(252, 576)$ in the form $252u + 576v$. First, we divide 252 into 576:

$$576 = 252 \cdot 2 + 72$$

Then we divide 72 into 252:

$$252 = 72 \cdot 3 + 36$$

Then, back substituting, we have
$$36 = \underline{252} - \underline{72} \cdot 3$$
$$= \underline{252} - [\underline{576} - \underline{252} \cdot 2] \cdot 3$$
$$= (7)(\underline{252}) + (-3)(\underline{576})$$ ■

In Example 7.3, we showed that $\gcd(91, 203) = 7$. If we divide 91 and 203 by their greatest common divisor 7, obtaining $\frac{91}{7} = 13$ and $\frac{203}{7} = 29$, it is easy to see that the resulting integers 13 and 29 are relatively prime; that is, $\gcd(13, 29) = 1$. So it appears that if two integers are divided by their greatest common divisor, all common factors except 1 are removed. The proof of the following theorem is left to the reader.

THEOREM 7.8 Given nonzero integers a and b, then $\dfrac{a}{\gcd(a, b)}$ and $\dfrac{b}{\gcd(a, b)}$ are relatively prime; that is, $\gcd\left(\dfrac{a}{\gcd(a, b)}, \dfrac{b}{\gcd(a, b)}\right) = 1$. ■

As previously defined, a positive integer m is the **least common multiple** of integers a and b provided that (i) $a \mid m$ and $b \mid m$ and (ii) if n is any common multiple of a and b, then $m \mid n$. The least common multiple of a and b will be denoted by $\text{lcm}(a, b)$.

We need an algorithm to show us how to find the least common multiple of two integers. This goal is accomplished by finding the greatest common divisor of the integers (for which we already have an algorithm) and then using Theorem 3.51. This theorem states that for positive integers a and b, $\gcd(a, b) \cdot \text{lcm}(a, b) = ab$.

To find $\text{lcm}(91, 203)$, first find $\gcd(91, 203)$ using the Euclidean algorithm as was done earlier and then divide it into the product of 91 and 203. Since $\gcd(91, 203) = 7$, we calculate
$$\text{lcm}(91, 203) = \frac{(91)(203)}{7} = 2639$$

At this point, we wish to consider the efficiency of the Euclidean algorithm in terms of the number of divisions that need to be performed. Let us again look at

$$
\begin{array}{rll}
a & = bq_0 + r_0 & 0 \le r_0 < b \\
b & = r_0 q_1 + r_1 & 0 \le r_1 < r_0 \\
r_0 & = r_1 q_2 + r_2 & 0 \le r_2 < r_1 \\
r_1 & = r_2 q_3 + r_3 & 0 \le r_3 < r_2 \\
r_2 & = r_3 q_4 + r_4 & 0 \le r_4 < r_3 \\
& \vdots & \vdots \\
r_{k-2} & = r_{k-1} q_k + r_k & \\
r_{k-1} & = r_k q_{k+1} & 0 \le r_k < r_{k-1}
\end{array}
$$

where r_k is the greatest common divisor of a and b. r_k must be greater than or equal to one by the definition of gcd. Also note than q_i must be greater than or equal to one since $r_i > r_{i+1}$. Therefore, we have

$$
\begin{array}{rll}
r_k & \ge & 1 \\
r_{k-1} & \ge & 2r_k = 2 = 1 + 1 \\
r_{k-2} & \ge & r_{k-1} + r_k \ge 1 + 2 = 3 \\
r_{k-3} & \ge & r_{k-2} + r_{k-1} \ge 2 + 3 \\
\vdots & & \vdots
\end{array}
$$

$$r_{k-i} \geq r_{k-(i+1)} + r_{k-(i+2)}$$
$$\vdots \quad \vdots \quad \vdots$$
$$r_0 \geq r_1 + r_2$$
$$b \geq r_0 + r_1$$

We see that we are forming the Fibonacci sequence where b is greater than or equal to the $(k+3)$ term of the Fibonacci sequence, whose kth term is the nearest integer to $\frac{1}{\sqrt{5}}\left(\frac{1+\sqrt{5}}{2}\right)^k$. Therefore,

$$b \geq \frac{1}{\sqrt{5}}\left(\frac{1+\sqrt{5}}{2}\right)^{k+3} - 1 = \frac{1}{\sqrt{5}}\left(\frac{1+\sqrt{5}}{2}\right)^3 \left(\frac{1+\sqrt{5}}{2}\right)^k - 1$$

so that $\left(\frac{1+\sqrt{5}}{2}\right)^k \leq Ab + B$ where A and B are positive numbers. Let $C = \frac{1+\sqrt{5}}{2}$, so that C is a constant and $C^k \leq Ab + B$. Taking the ln of both sides we have $k \ln C \leq \ln(Ab + B) \leq \ln(b)$, and $k \leq \frac{1}{\ln C} \ln(b)$. Hence, the number of divisions k is $O(\ln(b))$ and is very efficient.

Computer Science Application

The following algorithm returns the greatest common divisor of m and n using Euclid's algorithm, developed over 2000 years ago in Euclid's famous treatise, Elements.

integer function *gcd*(integer m, integer n)

    while $m > 0$
      if $m < n$ then
        temp $= m$
        $m = n$
        $n =$ temp
      end-if
      $m = m - n$
    end-while

    return n

end-function

The algorithm subtracts n from m until m becomes smaller than n. If m is still positive, m and n are swapped and the successive subtraction of n from m repeated. When m is finally reduced to 0, the value of n is returned. This process is time-consuming when m is large and n is small. Though this is the original form of Euclid's algorithm, a more efficient approach is to use the mod operation to compute the remainder when m is divided by n. Accordingly, the statement $m = m - n$ can be replaced by $m = m$ mod n.

Exercises

Given the division algorithm $a = bq + r$, find q and r for the following values of a and b:

1. $a = 75$, $b = 8$
2. $a = 102$, $b = 5$
3. $a = 81$, $b = 9$
4. $a = 16$, $b = 25$

Find the greatest common divisor of the following pairs of numbers:

5. 75, 25
6. 27, 18
7. 621, 437
8. 289, 377
9. 822, 436

10. Find the least common multiple of each of the pairs of numbers in Exercise 5.

11. Find the least common multiple of each of the pairs of numbers in Exercise 6.

12. Find the least common multiple of each of the pairs of numbers in Exercise 7.

13. Find the least common multiple of each of the pairs of numbers in Exercise 8.

14. Find the least common multiple of each of the pairs of numbers in Exercise 9.

For the following pairs of numbers a and b, find u, v, and d so that $au + bv = d$, where d is the greatest common divisor of a and b:

15. 83, 17
16. 361, 418
17. 25, 15
18. 81, 9
19. 216, 324

20. If $\gcd(a, b)$ is expressed as $ax + by$, prove that x and y are relatively prime.

21. Prove that given nonzero integers a and b, then $\frac{a}{\gcd(a,b)}$ and $\frac{b}{\gcd(a,b)}$ are relatively prime; that is, $\gcd\left(\frac{a}{\gcd(a,b)}, \frac{b}{\gcd(a,b)}\right) = 1$ (Theorem 7.8).

22. Write a procedure in pseudocode to find q and r in the division algorithm when an arbitrary positive integer a is divided by a positive integer b.

23. Write a procedure in pseudocode for the Euclidean algorithm.

24. Given integer a and positive integer b, prove that there are integers whose product is a and whose greatest common divisor is b if and only if $b^2 \mid a$.

25. Assume that
$$F(k) = 2^{2^k} + 1 \text{ for each } k \geq 0$$

 (a) Prove that $F(m + 1) = F(0)F(1)F(2) \cdots F(m) + 2$ for each $m \geq 0$.

 (b) Prove that $F(m)$ and $F(n)$ are relatively prime if $m \neq n$.

26. If $\gcd(a, b)$ is expressed as $ax + by$, show that x and y are not unique.

27. If $\gcd(a, b) = 1$, prove that $\gcd(a + nb, b) = 1$.

28. Prove that $n \cdot \gcd(a, b) = \gcd(na, nb)$.

29. Let S be the set all integers of the form $ax + by$. Show that $\gcd(a, b)$ divides all elements of S.

30. Prove that for positive integers a and b, $\gcd(a, b) = a$ if and only if $a \mid b$.

31. Prove that if $a \mid c$, $b \mid c$, and $\gcd(a, b) = 1$, then $ab \mid c$.

7.4 Continued Fractions

The division algorithm and the Euclidean algorithm will be applied frequently in this chapter. We consider first the rational a/b where a and b are integers and $b > 0$. If the division algorithm is applied to the integers a and b with $b > 0$, then there are unique integers t, the quotient, and r, the remainder, such that

$$a = b \cdot t + r \quad \text{and} \quad 0 \leq r < b$$

For $a = -124$ and $b = 35$ we have

$$-124 = (35)(-4) + 16$$

Rewriting these two equations using rational numbers gives

$$\frac{a}{b} = t + \frac{r}{b} \quad \text{and} \quad \frac{-124}{35} = -4 + \frac{16}{35}$$

7.4 Continued Fractions

Using the greatest integer function, the quotient may be written as

$$\left\lfloor \frac{a}{b} \right\rfloor = t \quad \text{and} \quad \left\lfloor \frac{-124}{35} \right\rfloor = -4$$

and the fractional remainder r/b has the property $0 \leq r/b < 1$. If $r \neq 0$, we can rewrite the equations as

$$\frac{a}{b} = t + \frac{1}{b/r} \quad \text{and} \quad \frac{-124}{35} = -4 + \frac{1}{35/16}$$

where $b/r > 1$. We can apply the division algorithm again to get $b = rt' + r'$ and $35 = (16)(2) + 3$. Thus, by inverting the fractional remainder and applying the division algorithm over and over, we obtain these ways of expressing the rational $-124/35$:

$$\frac{-124}{35} = -4 + \frac{16}{35} = -4 + \frac{1}{\left(\frac{35}{16}\right)}$$

$$= -4 + \frac{1}{2 + \frac{3}{16}} = -4 + \frac{1}{2 + \frac{1}{\left(\frac{16}{3}\right)}}$$

$$= -4 + \frac{1}{2 + \frac{1}{\left(5 + \frac{1}{3}\right)}}$$

Since the fraction $1/3$ has 1 as a numerator, we cannot reduce $3/1$ any more by applying the division algorithm. The expression

$$-4 + \frac{1}{2 + \frac{1}{5 + \frac{1}{3}}}$$

is called a continued fraction and is an alternative way of representing the rational number $-124/35$. Because of the regular structure of the continued fraction and because the numerators are always 1, we need to mention only the numbers $-4, 2, 5$, and 3 to specify the continued fraction, and we write $-124/35 = [-4; 2, 5, 3]$ instead. The numbers $-4, 2, 5$, and 3 are called the terms of the continued fraction or the partial quotients since they are produced by the division algorithm. The semicolon distinguishes the special nature of the first term -4. For example, $[3; 7] = 3 + 1/7 = 22/7$, but

$$[0; 3, 7] = 0 + \frac{1}{3 + \frac{1}{7}} = 7/22$$

Returning to the continued fraction $-124/35 = [-4; 2, 5, 3]$, we see that $3 = (3-1) + 1 = (3-1) + 1/1$. Thus, one can write

$$\frac{-124}{35} = -4 + \frac{1}{2 + \frac{1}{5 + \frac{1}{2 + \frac{1}{1}}}}$$

so that $-124/35 = [-4; 2, 5, 2, 1]$ also, but where the division algorithm is not used on the last step. This example suggests that every rational number has two distinct continued fraction representations with integers as terms: one with last term greater than 1 and the other with last term equal to 1.

If x is real but not rational, we cannot apply the division algorithm; however, using the greatest integer function we can obtain a continued fraction representation for x of length n in a way that parallels the rational case. Thus, for x real but not rational, there is a unique integer t_0 with $t_0 \leq x < t_0 + 1$ so that $t_0 = \lfloor x \rfloor$. Then

$$1 > y_1 = x - \lfloor x \rfloor = x - t_0 > 0$$

is the so-called fractional part of x. If $x_1 = 1/y_1$, then

$$x = \lfloor x \rfloor + y_1 = t_0 + y_1 = t_0 + \frac{1}{x_1}$$

and $x_1 > 1$. Let $t_1 = \lfloor x_1 \rfloor$, $y_2 = x_1 - \lfloor x_1 \rfloor$, and $x_2 = 1/y_2$ since $y_2 > 0$. Then $x_1 = t_1 + \frac{1}{x_2}$ with $x_2 > 0$. Combining these equalities gives

$$x = t_0 + \frac{1}{x_1}$$
$$= t_0 + \cfrac{1}{t_1 + \cfrac{1}{x_2}}$$

Continuing in this way, we obtain

$$x = t_0 + \cfrac{1}{t_1 + \cfrac{1}{t_2 + \cfrac{1}{t_3 + \cdots + \cfrac{1}{t_{k-1} + \cfrac{1}{x_k}}}}}$$

where t_i is an integer for each i, $t_i > 0$ for $i \geq 1$, x_k is a nonrational real number, and $x_k > 1$. One could write as well

$$x = [t_0; t_1, t_2, \ldots, t_{k-1}, x_k]$$

For example, let $x = \sqrt{3} = 1.7320508\cdots$. We generate a continued fraction representation for x as follows:

$$t_0 = \lfloor x \rfloor = \lfloor \sqrt{3} \rfloor = 1$$

where

$$y_1 = x - \lfloor x \rfloor = x - t_0 = \sqrt{3} - 1$$
$$x_1 = 1/y_1 = 1/(\sqrt{3} - 1) = (\sqrt{3} + 1)/2$$
$$t_1 = \lfloor x_1 \rfloor = 1$$

where

$$y_2 = x_1 - \lfloor x_1 \rfloor = x_1 - t_1 = (\sqrt{3} + 1)/2 - 1$$
$$= (\sqrt{3} - 1)/2$$
$$x_2 = 1/y_2 = 2/(\sqrt{3} - 1) = \sqrt{3} + 1$$
$$t_2 = \lfloor x_2 \rfloor = 2$$

where

$$y_3 = x_2 - \lfloor x_2 \rfloor = x_2 - t_2 = (\sqrt{3}+1) - 2$$
$$= \sqrt{3} - 1$$
$$x_3 = 1/y_3 = 1/(\sqrt{3}-1) = (\sqrt{3}+1)/2$$

Thus,

$$\sqrt{3} = [t_0; x_1] = [1; (\sqrt{3}+1)/2]$$
$$= [t_0; t_1, x_2] = [t_0; [t_1; x_2]] = [1; 1, (\sqrt{3}+1)]$$
$$= [t_0; t_1, t_2, x_3] = [t_0; t_1, [t_2; x_3]] = [1; 1, 2, (\sqrt{3}+1)/2]$$

But because $x_3 = x_1$, the pattern repeats, giving

$$\sqrt{3} = [1; 1, 2, 1, 2, 1, 2, (\sqrt{3}+1)/2]$$

and so on.

We now give a recursive definition of a continued fraction.

DEFINITION 7.9

For a finite sequence $t_0, t_1, t_2, \ldots, t_n$ of real numbers with $n \geq 0$ and $t_i > 0$ for $i \geq 1$, define the **finite continued fraction** $[t_0; t_1, t_2, \ldots, t_n]$ as follows:

$$[t_0;] = t_0$$

$$[t_0; t_1] = t_0 + \frac{1}{t_1}$$

$$[t_0; t_1, t_2, \ldots, t_k] = [t_0; [t_1; t_2, \ldots, t_k]] \text{ for } 1 < k \leq n$$

The numbers $t_0, t_1, \ldots, t_n$ are called **partial quotients** or **terms** of the continued fraction. The continued fraction $[t_0; t_1, t_2, \ldots, t_n]$ is said to be **simple** if t_i is an integer for each i; that is, every term in the continued fraction is an integer.

The numbers $[t_0;]$, $[t_0; t_1]$, $[t_0; t_1, t_2]$, $\ldots$, $[t_0; t_1, t_2, \ldots, t_k]$, $\ldots$, $[t_0; t_1, t_2, \ldots, t_n]$ are called **convergents** of the continued fraction $[t_0; t_1, t_2, \ldots, t_n]$. $[t_0; t_1, t_2, \ldots, t_k]$ is the kth convergent for $0 \leq k \leq n$. For convenience, if the notation $[t_0; t_1, t_2, \ldots, t_k]$ is used when $k = 0$, we will mean $[t_0;]$. We say that two continued fractions $[t_0; t_1, t_2, \ldots, t_n]$ and $[b_0; b_1, b_2, \ldots, b_m]$ are **equal term by term** provided that $n = m$ and $t_i = b_i$ for $0 \leq i \leq n$. If x is a real number and $x = [t_0; t_1, t_2, \ldots, t_n]$, then we say that $[t_0; t_1, t_2, \ldots, t_n]$ is a **continued fraction representation** of x. Two representations are the **same** or **equal** provided that they are equal term by term.

The next theorem provides another way to decompose a continued fraction into convergents.

THEOREM 7.10 If n is a positive integer and $[t_0; t_1, t_2, \ldots, t_n]$ is a continued fraction, then, for each k, $1 \leq k \leq n$,

$$[t_0; t_1, t_2, \ldots, t_n] = [t_0; t_1, t_2, \ldots, t_{k-1}, [t_k; t_{k+1}, \ldots, t_n]]$$

Proof The equality is proved using mathematical induction. If $n = 1$, the definition gives $[t_0; t_1] = [t_0; [t_1;]]$. Assume the equality of the theorem is true for $n = m$;

that is, for any continued fraction $[b_0; b_1, \ldots, b_m]$,

$$[b_0; b_1, b_2, \ldots, b_m] = [b_0; b_1, b_2, \ldots, b_{j-1}, [b_j; b_{j+1}, \ldots, b_m]]$$

for $1 \leq j \leq m$. Consider $n = m + 1$. If $[t_0; t_1, t_2, \ldots, t_{m+1}]$ is a continued fraction and $1 \leq k \leq m + 1$. Then

$$[t_0; t_1, \ldots, t_{m+1}] = [t_0; [t_1; t_2, \ldots, t_{m+1}]] \text{ by definition}$$
$$= [t_0; [t_1, t_2, \ldots, t_{k-1}, [t_k; t_{k+1}, \ldots, t_{m+1}]]]$$

by the induction hypothesis identifying $b_i = t_{i+1}$ and $j = k - 1$. Thus,

$$[t_0; t_1, \ldots, t_{m+1}] = [t_0; t_1, t_2, \ldots, t_{k-1}, [t_k; t_{k+1}, \ldots, t_{m+1}]]$$

by using the definition of a continued fraction. Hence, by induction, the theorem holds. ∎

Computer Science Application

The following algorithm generates the continued fraction for rational number a/b, $b > 0$. The format that terminates with an integer greater than 1 is used here.

function continued-fraction(integer a, integer b)

   integer t, old_a, new_a, old_b, new_b

   old_a = a
   old_b = b
   t = floor(a/b)
   Print '[', t, ';'

   while old_b is not 1 do
      new_a = old_b
      new_b = old_a − old_b*t
      t = floor(new_a/new_b)
      Print t
      if new_b is not 1 then
         Print ','
      end-if
      old_a = new_a
      old_b = new_b
   end-while

   Print ']'

   return

end-function

Exercises

Find two continued fraction representations, $[t_0; t_1, \ldots, t_n]$, for each of the following rational numbers:

1. 37/11
2. 48/1003
3. −257/2003
4. 11/37
5. −5/44
6. 5

Compute these rational numbers and express them in the form p/q with p and q integers:

7. [3; 5, 2]
8. [0; 3, 5, 2]
9. [−10; 1, 4, 3]
10. [6; 4, 7, 3, 5]
11. [2; 5, 3]
12. [5; 3, 7, 4, 6]

Find the finite continued fraction representation of x of the form $x = [t_0; t_1, t_2, t_3, t_4, t_5, x_6]$ for

13. $x = \sqrt{5}$
14. $x = \sqrt{2}$
15. $x = \pi$
16. $x = (1 + \sqrt{5})/2$

Let $[t_0; t_1, \ldots, t_n]$ be a finite continued fraction with $t_i > 0$ for $1 \leq i \leq n$. If $b > 0$, then prove that

17. (a) $[t_0; t_1, t_2, \ldots, t_n] > [t_0; t_1, t_2, \ldots, t_n + b]$ if n is odd

    (b) $[t_0; t_1, t_2, \ldots, t_n] < [t_0; t_1, t_2, \ldots, t_n + b]$ if n is even

7.5 Convergents

Clearly, $[t_0; t_1, t_2, \ldots, t_k]$ is a real number for every k. We will write explicitly the first four convergents and simplify them by expressing each as a quotient of two real numbers with both the numerator and denominator given in terms of t_i. We will be particularly interested in the form of the numerator and denominator of each convergent.

$$[t_0;] = t_0 = \frac{t_0}{1} = \frac{p_0}{q_0}$$

where we have let $p_0 = t_0$ and $q_0 = 1$.

$$[t_0; t_1] = t_0 + \frac{1}{t_1} = \frac{t_0 t_1 + 1}{t_1} = \frac{p_1}{q_1}$$

where we have let $p_1 = t_0 t_1 + 1$ and $q_1 = t_1$.

$$[t_0; t_1, t_2] = [t_0; [t_1; t_2]]$$
$$= t_0 + \frac{1}{[t_1; t_2]} = t_0 + \frac{1}{t_1 + \frac{1}{t_2}}$$
$$= t_0 + \frac{t_2}{t_1 t_2 + 1} = \frac{t_0 t_1 t_2 + t_0 + t_2}{t_1 t_2 + 1}$$
$$= \frac{(t_0 t_1 + 1) t_2 + t_0}{t_1 t_2 + 1}$$
$$= \frac{p_1 t_2 + p_0}{q_1 t_2 + q_0} = \frac{p_2}{q_2}$$

where we have let $p_2 = p_1 t_2 + p_0$ and $q_2 = q_1 t_2 + q_0$. Similarly, we calculate

$$[t_0; t_1, t_2, t_3] = \frac{t_0 t_1 t_2 t_3 + t_0 t_1 + t_0 t_3 + t_2 t_3 + 1}{t_1 t_2 t_3 + t_1 + t_3}$$
$$= \frac{(t_0 t_1 t_2 + t_0 + t_2) t_3 + (t_0 t_1 + 1)}{(t_1 t_2 + 1) t_3 + t_1}$$
$$= \frac{p_2 t_3 + p_1}{q_2 t_3 + q_1} = \frac{p_3}{q_3}$$

where $p_3 = p_2 t_3 + p_1$ and $q_3 = q_2 t_3 + q_1$. In each case, the number $[t_0; t_1, t_2, \ldots, t_k]$ is "reassembled" from the continued fraction representation without any "cancellation" to form a quotient of two polynomials P and Q in the variables $t_0, t_1, \ldots, t_k$, that is,

$$[t_0; t_1, t_2, \ldots, t_k] = \frac{P(t_0, t_1, \ldots, t_k)}{Q(t_0, t_1, \ldots, t_k)}$$

It is also true that q_0, q_1, q_2, and q_3 are positive because they depend on the multiplication and addition of positive numbers. These examples suggest the following theorem.

THEOREM 7.11 Let n be a nonnegative integer and $[t_0; t_1, t_2, \ldots, t_n]$ be a finite continued fraction with the finite sequences $p_0, p_1, \ldots, p_n$ and $q_0, q_1, \ldots, q_n$ defined recursively as follows:

(a) $p_0 = t_0$,
$q_0 = 1$.

(b) $p_1 = t_0 t_1 + 1$,
$q_1 = t_1$.

(c) $p_k = p_{k-1} t_k + p_{k-2}$,
$q_k = q_{k-1} t_k + q_{k-2}$ for $2 \leq k \leq n$.

Then $q_k > 0$ and $[t_0; t_1, t_2, \ldots, t_k] = \dfrac{p_k}{q_k}$ for $0 \leq k \leq n$.

Proof It is straightforward to show directly by mathematical induction that $q_k > 0$ for $0 \leq k \leq n$. The proof of this part is left to the reader. It was shown in the discussion leading up to this theorem that for $k = 0, 1$, and 2 it is true that $[t_0;] = p_0/q_0$, $[t_0; t_1] = p_1/q_1$, and $[t_0; t_1, t_2] = p_2/q_2$. Assume that $2 \leq k < n$ and, for any for any continued fraction $[b_0; b_1, \ldots, b_k]$ and any j, $0 \leq j \leq k$, it is true that $[b_0; b_1, \ldots, b_j] = p'_j/q'_j$ where the p'_j and q'_j are defined similar to (a) to (c) but for the continued fraction $[b_0; b_1, \ldots, b_k]$. Then

$$[t_0; t_1, \ldots, t_k, t_{k+1}] = [t_0; t_1, \ldots, t_{k-1}, [t_k; t_{k+1}]] \text{ by Theorem 7.10}$$
$$= [b_0; b_1, \ldots, b_{k-1}, b_k]$$

where $b_i = t_i$ for $0 \leq i \leq k-1$ and $b_k = [t_k; t_{k+1}] = t_k + \dfrac{1}{t_{k+1}}$. Thus, by the induction hypothesis,

$$[t_0; t_1, \ldots, t_{k+1}] = \frac{p'_{k-1} b_k + p'_{k-2}}{q'_{k-1} b_k + q'_{k-2}}$$

Since $b_i = t_i$ for $0 \leq i \leq k-1$, we have that $p_i = p'_i$ and $q_i = q'_i$ for such i. Substituting for b_k, p'_i, and q'_i, we obtain

$$[t_0; t_1, \ldots, t_{k+1}] = \frac{p_{k-1}\left(t_k + \dfrac{1}{t_{k+1}}\right) + p_{k-2}}{q_{k-1}\left(t_k + \dfrac{1}{t_{k+1}}\right) + q_{k-2}}$$
$$= \frac{(p_{k-1} t_k + p_{k-2}) + p_{k-1}/t_{k+1}}{(q_{k-1} t_k + q_{k-2}) + q_{k-1}/t_{k+1}}$$
$$= \frac{p_k t_{k+1} + p_{k-1}}{q_k t_{k+1} + q_{k-1}} = \frac{p_{k+1}}{q_{k+1}}$$

Hence, $[t_0; t_1, \ldots, t_k] = p_k/q_k$ for $0 \leq k \leq n$ by mathematical induction. ∎

The numbers p_k and q_k of Theorem 7.11 are defined independently of whether they can be used as the numerator and denominator of an expression equal to the kth convergent. The recurrence relation given by Theorem 7.11 provides a rapid means of calculating the convergents of a given continued fraction since the numerators p_i and denominators q_i may be computed simultaneously and easily. For example, for $x = [-4; 2, 5, 2, 1] = [t_0; t_1, t_2, t_3, t_4]$, the convergents may be calculated in a tableau:

k	t_k	p_k	q_k	p_k/q_k
0	-4	-4	1	$-4/1$
1	2	$(-4)(2) + 1 = -7$	2	$-7/2$
2	5	$(-7)(5) + (-4) = -39$	$(2)(5) + 1 = 11$	$-39/11$
3	2	$(-39)(2) + (-7) = -85$	$(11)(2) + 2 = 24$	$-85/24$
4	1	$(-85)(1) + (-39) = -124$	$(24)(1) + 11 = 35$	$-124/35$

so x is $-124/35$, the number that generated $[-4; 2, 5, 2, 1]$ in Section 5.1.

Because continued fractions are defined recursively, there are many relationships among the convergents in terms of the numerators and denominators p_k and q_k of Theorem 7.11. In order to allow some of these relationships to hold for $k = 0$, it will often be convenient to define p_k and q_k for $k = -1$ as

$$p_{-1} = 1$$
$$q_{-1} = 0$$

Thus, $p_1 = p_0 t_1 + p_{-1} = t_0 t_1 + 1$ and $q_1 = q_0 t_1 + q_{-1} = 1 \cdot t_1 + 0 = t_1$ as in Theorem 7.11.

THEOREM 7.12 If $[t_0; t_1, \ldots, t_n]$ is a finite continued fraction with t_i real and p_k and q_k are given by Theorem 7.11, then

(a) $p_k q_{k-1} - p_{k-1} q_k = (-1)^{k-1}$ for $0 \leq k \leq n$.

(b) $\dfrac{p_k}{q_k} - \dfrac{p_{k-1}}{q_{k-1}} = \dfrac{(-1)^{k-1}}{q_k q_{k-1}}$ for $1 \leq k \leq n$.

(c) $\dfrac{p_k}{q_k} - \dfrac{p_{k-2}}{q_{k-2}} = \dfrac{(-1)^k t_k}{q_k q_{k-2}}$ for $2 \leq k \leq n$.

(d) The even-numbered convergents form an increasing sequence, that is,

$$\frac{p_0}{q_0} < \frac{p_2}{q_2} < \frac{p_4}{q_4} < \cdots$$

and the odd-numbered convergents form a decreasing sequence, that is,

$$\frac{p_1}{q_1} > \frac{p_3}{q_3} > \frac{p_5}{q_5} > \cdots$$

Furthermore,

$$\frac{p_{2j}}{q_{2j}} \leq [t_0; t_1, t_2, \ldots, t_n] \leq \frac{p_{2i+1}}{q_{2i+1}}$$

for $0 \leq j \leq \lfloor n/2 \rfloor$ and $0 \leq i \leq \lfloor (n-1)/2 \rfloor$, with the left equality holding when n is even and the right equality holding when n is odd.

(e) If $[t_0; t_1, \ldots, t_n]$ is simple, then $q_k \geq k$ for $0 \leq k \leq n$.

(f) If $[t_0; t_1, \ldots, t_n]$ is simple, then $q_k < q_{k+1}$ for $1 \leq k \leq n - 1$ and $q_0 \leq q_1$.

Proof (a) For $k = 1$, $p_1 q_0 - p_0 q_1 = (t_0 t_1 + 1)(1) - t_0 t_1 = 1 = (-1)^{1-1}$ by Theorem 7.11. Let $2 \leq m < n$ and assume that the formula in part (a) is true for $k = m$. It is shown that the formula holds for $k = m + 1$. Again using Theorem 7.11,

we have

$$p_{m+1}q_m - p_m q_{m+1} = (p_m t_{m+1} + p_{m-1})q_m - p_m(q_m t_{m+1} + q_{m-1})$$
$$= p_{m-1}q_m - p_m q_{m-1}$$
$$= (-1)(p_m q_{m-1} - p_{m-1}q_m)$$
$$= (-1)(-1)^{m-1} = (-1)^{(m+1)-1}$$

Thus, the formula in part (a) holds for $1 \leq k \leq n$. Also $p_0 q_{-1} - q_0 p_{-1} = t_0 \cdot 0 - 1 \cdot 1 = -1$ and the formula holds for $k = 0$.

(b) This follows from part (a) by dividing by $q_k q_{k-1}$ when $k \geq 1$.

(c) If $k \geq 2$, then, by Theorem 7.11,

$$p_k = p_{k-1}t_k + p_{k-2}$$
$$q_k = q_{k-1}t_k + q_{k-2}$$

Multiplying the first of these equations by q_{k-2} and the second by p_{k-2} gives

$$p_k q_{k-2} = p_{k-1} q_{k-2} t_k + p_{k-2} q_{k-2}$$
$$q_k p_{k-2} = q_{k-1} p_{k-2} t_k + q_{k-2} p_{k-2}$$

Subtracting the last two equations produces

$$p_k q_{k-2} - q_k p_{k-2} = (p_{k-1}q_{k-2} - p_{k-2}q_{k-1})t_k$$

Now, invoking part (a), we obtain

$$p_k q_{k-2} - q_k p_{k-2} = (-1)^{(k-1)-1} t_k = (-1)^k (-1)^{-2} t_k = (-1)^k t_k$$

Dividing by $q_k q_{k-2} > 0$ gives the desired result:

$$\frac{p_k}{q_k} - \frac{p_{k-2}}{q_{k-2}} = \frac{(-1)^k t_k}{q_k q_{k-2}}$$

(d) If $2 \leq k \leq n$ and k is even, then $k-2$ is also even and $k-2 \geq 0$. Thus $(-1)^k = 1$ and part (c) implies that

$$\frac{p_k}{q_k} - \frac{p_{k-2}}{q_{k-2}} = \frac{t_k}{q_k q_{k-2}} > 0$$

because t_k and $q_k q_{k-2}$ are positive for $k \geq 1$. If $3 \leq k \leq n$ and k is odd, then $k-2$ is also odd and $k-2 \geq 1$. In this case $(-1)^k = -1$ so that part (c) implies

$$\frac{p_k}{q_k} - \frac{p_{k-2}}{q_{k-2}} = \frac{(-1)t_k}{q_k q_{k-2}} < 0$$

because t_k and $q_k q_{k-2}$ are positive for $k \geq 3$.

If n is odd, then, by part (b), $\dfrac{p_n}{q_n} > \dfrac{p_{n-1}}{q_{n-1}}$ so that $[t_0; t_1, t_2, \ldots, t_n] = p_n/q_n$ is greater than the largest even convergent, p_{n-1}/q_{n-1}. If n is even, then $\dfrac{p_n}{q_n} < \dfrac{p_{n-1}}{q_{n-1}}$ so that $[t_0; t_1, t_2, \ldots, t_n] = p_n/q_n$ is less than the smallest odd convergent, p_{n-1}/q_{n-1}. The proofs of parts (e) and (f) are left to the reader. ∎

The convergents for $x = [-4; 2, 5, 2, 1]$ were computed in the tableau earlier in this section. Summarizing those results relative to part (d), we see that

$$p_0/q_0 \;<\; p_2/q_2 \;<\; p_4/q_4 \;=\; x \;<\; p_3/q_3 \;<\; p_1/q_1$$
$$-4/1 \;<\; -39/11 \;<\; -124/35 \;=\; x \;<\; -85/24 \;<\; -7/2$$

Also, by direct calculation,
$$p_3q_2 - p_2q_3 = (-85)(11) - (-39)(24) = (-935) - (-936) = 1$$

The import of Theorem 7.12 is that it specifies that consecutive and alternate convergents can differ by no more than a quantity proportional to the reciprocal of the "denominators" of the convergents and that convergents are alternatively greater than and less than $[t_0; t_1, t_2, \ldots, t_n]$.

Several ratios of the numerators and denominators of the convergents p_k/q_k have some utility. They are given in the next theorem whose proof is left to the reader.

THEOREM 7.13 For the finite continued fraction $[t_0; t_1, t_2, \ldots, t_n]$,

(a) $q_k/q_{k-1} = [t_k; t_{k-1}, \ldots, t_2, t_1]$ for $1 \leq k \leq n$.
(b) If $t_0 \neq 0$, then $p_k/p_{k-1} = [t_k; t_{k-1}, \ldots, t_1, t_0]$ for $1 \leq k \leq n$.
(c) If $t_0 = 0$, then $p_k/p_{k-1} = [t_k; t_{k-1}, \ldots, t_2, t_1]$ for $2 \leq k \leq n$. ∎

Exercises

1. Complete the proof of Theorem 7.11 by proving that $q_k > 0$ for $0 \leq k \leq n$.

2. Compute the convergents for
   (a) $[2; 5, 1, 2, 4]$ (b) $[2; 5, 1, 2, 3, 1]$
   (c) $[5; 1, 2, 3]$ (d) $[0; 5, 1, 2, 3]$
   (e) $[-5; 3, 8, 2]$

Compute the convergents p_k/q_k by computing p_k and q_k for $0 \leq k \leq 5$ for $x = [t_0; t_1, t_2, t_3, t_4, t_5, x_6]$ for the following x (see the exercises of Section 7.4):

3. $x = \sqrt{5}$
4. $x = \sqrt{2}$
5. $x = \pi$
6. $x = (1 + \sqrt{5})/2$

7. Prove parts (e) and (f) of Theorem 7.12.

8. Prove Theorem 7.13. *Hint*: Use Theorem 7.11.

9. Let p_k and q_k be given by Theorem 7.11 for the continued fraction $[t_0; t_1, t_2, \ldots, t_n]$. Prove that
$$\begin{vmatrix} p_k & p_{k-1} \\ q_k & q_{k-1} \end{vmatrix} = (-1)^{k+1} \text{ for } 0 \leq k \leq n$$
where the determinant is used on the left side of the equation.

10. In Theorem 7.11, we showed that if $p_{-1} = 1$ and $q_{-1} = 0$, part (c) held for $k = 1$. What value would p_{-2} and q_{-2} have to be in order for part (c) to hold for $k = 0$?

GLOSSARY

Chapter 7: Number Theory

Algorithm (7.1)	An **algorithm** is a finite set of instructions that can be followed mechanically to solve a problem.
Continued fraction representation of x (7.4)	If x is a real number and $x = [t_0; t_1, t_2, \ldots, t_n]$, then we say that $[t_0; t_1, t_2, \ldots, t_n]$ is a **continued fraction representation** of x.
Convergents (7.4)	The numbers $[t_0;], [t_0; t_1], [t_0; t_1, t_2], \ldots [t_0; t_1; t_2, \ldots, t_k], \ldots, [t_0; t_1, t_2, \ldots, t_k]$ are called **convergents** of the continued fraction $[t_0; t_1, t_2, \ldots, t_k]$. $[t_0; t_1, t_2, \ldots, t_k]$ is the kth convergent for $0 \leq k \leq n$.
Division algorithm (7.3)	Every pair of positive integers a, b can be uniquely expressed in the form $a = bq + r$ where $r < b$.

Equal term by term (7.4)	We say that two continued fractions $[t_0; t_1, t_2, \ldots, t_n]$ and $[b_0; b_1, b_2, \ldots, b_m]$ are **equal term by term** provided that $n = m$ and $t_i = b_i$ for $0 \leq i \leq n$.
Euclidean algorithm (7.3)	The **Euclidean algorithm** is an algorithm for determining the gcd of two integers.
Fermat's factorization method (7.2)	**Fermat's factorization method** for determining primes is based on the following theorem: An odd integer $n > 1$ is nonprime if and only if there are nonnegative integers p and q such that $n = p^2 - q^2$ with $p - q > 1$.
Finite continued fractions (7.4)	For a finite sequence $t_0, t_1, t_2, \ldots, t_n$ of real numbers with $i \geq 0$ and $t_i > 0$ for $i \geq 1$, a **finite continued fraction** is defined as follows: $$[t_0;] = t_0$$ $$[t_0; t_1] = t_0 + \frac{1}{t_1}$$ $$[t_0; t_1, t_2, \ldots, t_k] = [t_0; [t_1; t_2, \ldots, t_k]] \text{ for } 1 < k \leq n$$
Least common multiple (7.3)	A positive integer m is the **least common multiple** of integers a and b provided that i. $a \mid m$ and $b \mid m$ and ii. If n is any common multiple of a and b, then $m \mid n$. The least common multiple of a and b is denoted by lcm (a, b).
Partial continued fractions (7.4)	The numbers $t_0, t_1, t_2, \ldots, t_n$ are called **partial quotients** or **terms** of the continued fraction.
Same or equal representations (7.4)	Two continued fraction representations are the **same** or **equal** provided that they are equal term by term.
Sieve of Eratosthenes (7.1)	The **sieve of Eratosthenes** is an algorithm for determining the primes less than a given integer.
Simple continued fractions (7.4)	The continued fraction $[t_0; t_1; t_2, \ldots, t_k]$ is said to be a **simple continued fraction** if t_i is an integer for each I; that is, every term in the continued fraction is an integer.

TRUE-FALSE QUESTIONS

1. $\gcd(a, na + ab) = a$ for every nonnegative integer n.
2. $\gcd(na, ab) = n(\gcd(a, b))$.
3. If c and d are relatively prime, then $\gcd(ca, db) = (\gcd(a, b))$.
4. For all positive integers a and b, $\text{lcm}(a, b) < ab$.
5. 359999 is a prime.
6. $\gcd(3451357, 2254897) = 23$
7. $17459x + 20111y = 18139$ has an integral solution.
8. If $\gcd(a, b) = ax + by$, then x and y are relatively prime.
9. $\gcd(a, b)$ is the least positive integer of the form $ax + by$.
10. $\text{lcm}(8051, 8633) = 716539$
11. $[1; 1, 2, (\sqrt{3} + 1)/2] = [1; 1, 2, 1, 2, 1, 2, (\sqrt{3} + 1)/2)]$
12. $[-4; 2, 5, 3] = [-4; 2, 5, 2, 1]$
13. $[-4; 2, 5, 3] = [-4; 2, 5, 3, 1]$
14. $\dfrac{39}{11} = [3; 2, 1, 3]$
15. $(1 + \sqrt{5})/2 = [1; 1, 1, 1, (1 - \sqrt{5})/2]$
16. If $x = [-4; 2, 5, 2, 1]$, then $\dfrac{p_3}{q_3} = -\dfrac{85}{24}$.
17. If $[t_0; t_1, \ldots, t_n]$, then $p_k q_{k-1} - p_{k-1} q_k = (-1)^k$ for $0 \leq k \leq n$.

18. Given $[2; 4, 1, 2, 3]$, the convergents are given by

k	p_k	q_k
0	2	1
1	9	4
2	11	9
3	29	18
4	104	66

19. Given $x = \sqrt{2}$, the convergents are given by

k	p_k	q_k
0	1	1
1	3	2
2	7	5
3	17	12
4	41	29
$\vdots$	$\vdots$	$\vdots$

20. If x is rational, then $x = [t_0; t_1, \ldots, t_n]$ for some n where t_i is an integer for $0 \leq i \leq n$.

21. If x is irrational, then there is no n such that $x = [t_0; t_1, \ldots, t_n]$ for some n where t_i is an integer for $0 \leq i \leq n$.

22. $[3; 4, 7, 3, 5, 4, 2] = [3; 4, 7, 3, [3; 5, 4, 2]]$

23. $\dfrac{p_k}{q_k} < \dfrac{p_{k+2}}{q_{k+2}}$ if and only if k is even.

24. If $x = [t_0; t_1, \ldots, t_n]$ where t_i is an integer for $0 \leq i \leq n$ and x is negative, then $p_k < 0$ and $q_k > 0$ for $0 \leq k \leq n$.

25. If $x = [t_0; t_1, \ldots, t_n]$ where t_i is an integer for $0 \leq i \leq n$ and x is positive, then $p_k > p_{k-1}$ and $q_k > q_{k-1}$ for $1 \leq k \leq n$.

SUMMARY QUESTIONS

1. Create the sieve of Eratosthenes for the positive integers less than 150.

2. Demonstrate the Euclidean algorithm for integers 426 and 83.

3. Use Fermat's factorization theorem to determine if 551 is a prime.

4. Find the greatest common divisor of 529 and 437.

5. Find u and v so that $u \cdot 435 + v522 = \gcd(435, 522)$.

6. Find $\mathrm{lcm}(85, 34)$.

7. Compute the convergents $\dfrac{p_k}{q_k}$ by computing p_k and q_k for $0 \leq l \leq 5$ for $x = [t_0 : t_1, t_2, t_3, t_4, t_5, x_6]$ for $x = \sqrt{3}$.

8. Compute the convergents for $[5 : 3, 7, 4, 6]$.

9. Find the finite continued fraction representation of x of the form $x = [t_0 : t_1, t_2, t_3, t_4, t_5, x_6]$ for $x = \sqrt{7}$.

10. Find two continued fraction representations, $[t_0 : t_1, t_2, t_3, t_4, t_n,]$ for the rational number $87/15$.

11. Factor 26381.

12. Show u and v so that $ua + bv = \gcd(a, b)$ is not unique.

13. Show that if a and b are relatively prime, then for any c, there exists u and v so that $ua + bv = c$ has a solution.

14. Find the greatest common divisor of 667 and 1147.

15. Solve $247u + 299v = 81$ for integers u and v or show that it has no solution.

ANSWERS TO TRUE-FALSE

1. T **2.** T **3.** F **4.** F **5.** F **6.** F **7.** F **8.** T **9.** T **10.** T
11. T **12.** T **13.** F **14.** F **15.** F **16.** T **17.** F **18.** F **19.** T

20. T **21.** T **22.** F **23.** T **24.** T **25.** T

CHAPTER 8

Counting and Probability

8.1 Basic Counting Principles

Albert Einstein is quoted as saying, "Not everything that can be counted counts, and not everything that counts can be counted." While this statement is not exactly encouraging, we shall still attempt to count things. In fact, an amazing number of problems in this book will begin with the words "how many." How many problems in this book begin with "how many"? There are several reasons why we won't ask this question. First, we don't know the answer. Would the question include itself in the solution? If so, then asking the question changes the answer. Perhaps the main reason for not answering the question is that there is no pattern to use in trying to answer the question. We are really more interested in finding and using patterns in counting than we are in solving an individual problem. Our goal in general is to avoid simply listing the items to be counted and then counting them one by one. This is particularly true if there is a large number of items to be counted. We may, however, choose this method for small numbers of objects in order to help us find a pattern.

Suppose a person, with nothing better to do, tosses a coin and then tosses a die (singular for *dice*). The possible outcomes for the coin are heads (H) and tails (T). The possible outcomes for the die are 1, 2, 3, 4, 5, and 6. The possible outcomes for both events are $(H, 1)$, $(H, 2)$, $(H, 3)$, $(H, 4)$, $(H, 5)$, $(H, 6)$, $(T, 1)$, $(T, 2)$, $(T, 3)$, $(T, 4)$, $(T, 5)$, and $(T, 6)$. One way to represent these outcomes is with a **counting tree** as shown in Figure 8.1.

Outcome $(H, 2)$ is found by taking the left branch to H and then taking the branch to 2. Similarly, we can see that each outcome is represented and that each circled number represents a unique outcome. We can also see that if the coin is H, there are six possible outcomes and if the coin is T, we have six possible outcomes.

Figure 8.1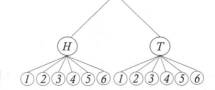

Thus, we can find the total outcomes by multiplying 2×6, so we multiply the number of outcomes for the coin toss times the number of outcomes for the toss of the die. Notice that the outcome of the coin toss does not affect the possible outcomes for the die.

Suppose that a person tosses two dice. There are six possible outcomes for the first die, and for each outcome for the first die, there are six outcomes for the second die. Therefore, there are $6 \times 6 = 36$ possible outcomes when two dice are thrown. If desired, we could again draw a tree. Similarly, suppose a letter is selected from the alphabet and then a second letter is selected, which may be the same as the first letter. To find the total number of pairs of letters, we note that there are 26 choices for the first letter, and for each letter selected, there are 26 choices for the second letter. Hence, there are $26 \times 26 = 676$ possible ways to select pairs of letters from the alphabet. If we do not allow the second letter to be the same as the first, then after we select the first letter, we have only 25 ways to select the second letter. Therefore, the number of possible outcomes is $26 \times 25 = 650$. Another way to approach this problem would be to take the 676 possible ways to select pairs of letters from the alphabet and remove the pairs where both letters are the same. There are obviously 26 of these, so if we subtract 26 from 676, we again get 650. We will often find more than one way to solve a problem.

Suppose that a license plate has three letters of the alphabet followed by three single digit numbers. If we allow repetition, then there are 26 possible choices for each letter and 10 choices for each of the numbers. Thus, the total number of license plates is

$$26 \times 26 \times 26 \times 10 \times 10 \times 10 = 17{,}576{,}000$$

Assume, for absolutely no good reason, that letters and numbers cannot be repeated. Then, after selecting the first letter, we have 25 choices for the second letter and 24 choices for the third letter. Similarly, we have 10 choices for the first number, 9 choices for the second number and 8 choices for the third number. Thus, the total number of license plates is

$$26 \times 25 \times 24 \times 10 \times 9 \times 8 = 11{,}232{,}000$$

We express this method in the following theorem, which we shall accept as an axiom.

THEOREM 8.1 **(Multiplication Counting Principle)** Given a sequence of events

$$E_1, E_2, E_3, \ldots, E_m$$

such that E_1 occurs in n_1 ways and if $E_1, E_2, E_3, \ldots, E_{k-1}$ have occurred, then E_k can occur in n_k ways. There are $n_1 \times n_2 \times n_3 \times \cdots \times n_m$ ways in which the entire sequence of events can occur. ∎

EXAMPLE 8.2 How many one-to-one functions are there from a set S with n elements to a set T with m elements? If $n > m$, then there are no one-to-one functions, so assume $n \leq m$. Assume, without loss of generality, that the elements have been labeled $a_1, a_2, a_3, \ldots, a_n$. There are m choices for mapping a_1. Once the image of a_1 is determined, there are $m - 1$ choices for mapping a_2, since it cannot be mapped to the same element as a_1. There are then $m - 2$ choices for mapping a_3 and $m - 3$ choices for mapping a_4. From the pattern, we can see that there are $m - i + 1$ choices for mapping a_i. Therefore, by the multiplication counting principle, there are

$$(m)(m-1)(m-2)(m-3) \cdots (m-n+1)$$

ways of mapping all of the elements in S to elements of T such that no two elements of S map to the same element. Thus, there are

$$(m)(m-1)(m-2)(m-3)\cdots(m-n+1)$$

one-to-one functions from S to T. ∎

EXAMPLE 8.3 How many functions are there from a set S with n elements to a set T with m elements? This time each of the n elements of S can map to any of the m elements of T. Therefore, there are $(m)(m)(m)(m)\cdots(m) = m^n$ functions from S to T. ∎

EXAMPLE 8.4 A **bit string** is a string of symbols such that each symbol is either 1 or 0. How many bit strings are there of length 5? How many bit strings are there of length k? Since each symbol on the string must be either 1 or 0, there are two choices for each place on the string. Therefore, there are $2 \times 2 \times 2 \times 2 \times 2 = 2^5$ bit strings of length 5. By a similar argument, there are 2^k bit strings of length k. ∎

EXAMPLE 8.5 How many subsets are there of a set S with five elements? How many subsets are there of a set with k elements? There are several ways of counting these subsets. We shall give two. First let a_1, a_2, a_3, a_4, a_5 be the elements of S. Define a one-to-one correspondence from the S to the set of bit strings of length 5, where for each subset s of S, the ith bit of the corresponding string is 1 if $a_i \in S$ and 0 if $a_i \notin S$. Thus, $\{a_1, a_2, a_4\}$ corresponds to the string 11010, $\{a_2, a_5\}$ corresponds to the string 01001, $\varnothing$ corresponds to the string 00000, and S corresponds to the string 11111. Thus, there is a one-to-one correspondence between the subsets of S and the bit strings of length 5. But by the previous example, we know there are 2^5 bit strings of length 5. Therefore, there are 2^5 subsets of a set with five elements. By a similar argument we show there are 2^k subsets of a set with k elements.

The second method of counting is, for each $s \in S$, to define a function f_s from S to the set $\{0, 1\}$. The function f_s is defined by $f_s(a_i) = 1$ if $a_i \in s$ and $f_s(a_i) = 0$ if $a_i \notin s$. Conversely, given a function f from S to the set $\{0, 1\}$, if we let $s = \{a_i : f(a_i) = 1\}$, then f is the function f_s. Therefore, there is a one-to-one correspondence between the subsets of S and the functions from S to $\{0, 1\}$. But by Example 8.3, there are 2^5 functions from S to $\{0, 1\}$. Similarly, if S contains k elements, there are 2^k functions from S to $\{0, 1\}$ and, therefore, 2^k subsets of S. ∎

Assume there are 10 men and 15 women at a meeting and one must be chosen as chairperson. There are $10 + 15 = 25$ ways of selecting the chairperson. Suppose that a person goes to a restaurant where there are 15 beef dishes, 8 pork dishes, and 12 seafood dishes from which to choose. The person can select from $15 + 8 + 12 = 35$ different dishes. Similarly, assume that a student is selecting a book from a shelf on which there are 25 mathematics, 30 computer science, and 15 chemistry books. There are $25 + 30 + 15 = 70$ different ways that the student can select a book. Notice that, in each case, we are assuming that the sets from which the selection is made are disjoint. These examples lead us to the next theorem, which we shall accept without proof.

THEOREM 8.6 (**Addition Counting Principle**) Let $S_1, S_2, S_3, \ldots, S_m$ be pairwise disjoint sets (i.e., $S_i \cap S_j = \varnothing$ for all $i \neq j$), and for each i, let S_i contain n_i elements. The number of elements that can be selected from S_1 or S_2 or S_3 or $\ldots$ or S_m is equal to $n_1 + n_2 + n_3 + \cdots + n_m$. Stated in terms of set theory,

$$|S_1 \cup S_2 \cup S_3 \cup \cdots \cup S_m| = |S_1| + |S_2| + |S_3| + \cdots + |S_m|$$

where $|S|$ is the number of elements in the set S. ∎

EXAMPLE 8.7

How many integers are there between 0 and 1000 that have exactly one digit equal to 6? Let S be the set of integers between 0 and 1000 that have exactly one digit equal to 6. Let S_1 be the subset of S with exactly one digit and exactly one digit equal to 6, S_2 be the subset of S with exactly two digits and exactly one digit equal to 6, and S_3 be the subset of S with exactly three digits and exactly one digit equal to 6. The set S_1 has only one element, the number 6. In S_2 each element having exactly one digit equal to 6, either has a 6 as the first digit or the second digit. If 6 is the second digit, there are 8 choices for the first digit, since the first digit cannot be 0 or 6. If 6 is the first digit, there are 9 choices for the second digit, since it cannot be 6. Therefore, S_2 has $8 + 9 = 17$ elements. An element in S_3 either has a 6 in the first digit, in the second digit, or in the third digit. If the 6 is the first digit, there are 9 choices for the second digit and 9 choices for the third digit. Therefore, by the multiplication counting principle, there are $9 \times 9 = 81$ numbers in S_3 with 6 as the first digit. If the 6 is the second digit, there are 9 choices for the third digit and 8 choices for the first digit, since the first digit cannot be zero. Therefore, there are $9 \times 8 = 72$ numbers in S_3 with 6 as the second digit. Similarly, there are 72 numbers in S_3 with 6 as the third digit. Hence, there are $81 + 72 + 72 = 225$ elements in S_3. Since $S = S_1 \cup S_2 \cup S_3$, there are $1 + 17 + 225 = 243$ elements in S. ∎

EXAMPLE 8.8

How many integers are there between 0 and 1000 that have one or more 6's as digits? To solve this problem we let S be the set of integers between 0 and 1000 that have no 6's as digits. We let S_1 be the subset of S with exactly one digit, S_2 be the subset of S with exactly two digits, S_3 be the subset of S with exactly three digits, and S_4 be the subset of S with exactly four digits. The set S_1 contains nine numbers with no 6, since only the 6 is excluded. The set S_2 contains eight choices for the first digit and nine for the second digit, since the first digit can be neither 6 nor 0 and the second digit cannot be 6. Therefore, S_2 contains $8 \times 9 = 72$ elements with no 6. The set S_3 has eight choices for the first digit, nine for the second digit, and nine for the third digit. Therefore, S_3 contains $8 \times 9 \times 9 = 648$ elements with no 6. The set S_4 contains only the number 1000. Therefore, S contains $9 + 72 + 648 + 1 = 730$ elements. Let T be the set of all numbers between 0 and 1000. Then $T - S$ is the set of all integers between 0 and 1000 that have one or more 6's as digits. Furthermore, S and $T - S$ are disjoint. Therefore, $|S| + |T - S| = |T|$ so that $730 + |T - S| = 1001$. Therefore, $|T - S| = 271$, and there are 271 integers between 1 and 1000 containing one or more 6's as digits. ∎

EXAMPLE 8.9

How many ways can we select two books, each on a different subject, from a shelf that contains 15 computer science, 12 mathematics, and 10 chemistry books? If we select a computer science book and a mathematics book, there are 15 ways to select the computer science book and 12 ways to select the mathematics book, so there are $12 \times 15 = 180$ possibilities. If we select a computer science book and a chemistry book, there are 15 ways to select the computer science book and 10 ways to select the chemistry book, so there are $15 \times 10 = 150$ possibilities. If we select a mathematics book and a chemistry book, there are 12 ways to select the mathematics book and 10 ways to select the chemistry book, so there are $12 \times 10 = 120$ possibilities. Therefore, there are $180 + 150 + 120 = 450$ possible ways of selecting the two books. ∎

EXAMPLE 8.10

How many odd integers are there between 100 and 1000? One way to work this problem is to let S be the set of all odd integers between 100 and 1000. For $i \in \{1, 3, 5, 7, 9\}$, let S_i be the subset of S whose elements end in i. For each i, there are 9 choices for the first digit and 10 for the second digit, so each S_i contains 90 elements. Therefore, since $S = S_1 \cup S_3 \cup S_5 \cup S_7 \cup S_9$,

$$|S| = |S_1| + |S_3| + |S_5| + |S_7| + |S_9|$$
$$= 90 + 90 + 90 + 90 + 90$$
$$= 450$$

so there are 450 odd integers between 100 and 1000.

We can also work this problem using only the multiplication counting principle. There are five choices for the last digit, since it must be an odd number. There are 10 choices for the middle digit and nine choices for the first digit. Therefore, there are $9 \times 10 \times 5 = 450$ odd integers between 100 and 1000. In this example we can see how the multiplication counting principle follows from the addition counting principle. ∎

Finally we finish, as we began, by looking at trees. Suppose teams A and B are playing each other in a baseball tournament. A team must win three out of five games to win the tournament. All possible outcomes are shown by the tree in Figure 8.2, where the letter at each vertex denotes a game won by that team. A circled letter denotes that the team circled has won the tournament. Note that there are 20 possible outcomes with 10 possible ways that team A can win and 10 ways team B can win. Suppose we know that team B won the first game. This gives us the right side of the tree only, beginning with vertex B. This side is enclosed with a line of dashes. Notice that there are four outcomes where A wins and six outcomes where B wins. Suppose that A wins the first game, B the second game, and A the third game. This gives us the part of the tree surrounded by the solid line. Note that there are two outcomes where A wins and only one outcome where B wins.

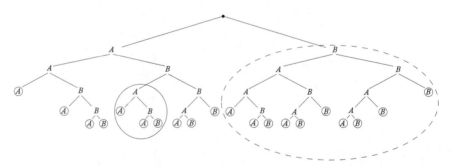

Figure 8.2

EXAMPLE 8.11 Assume teams A and B play in a World Series, where the winning team must win four games out of a possible seven. If team A wins the first two games, in how many ways can A win and in how many ways can B win? The tree in Figure 8.3 shows the possible outcomes. There are ten outcomes where A wins and five possible outcomes where B wins. ∎

EXAMPLE 8.12 Douglas Shier [98] of Clemson University collected a variety of ways of solving the same problem. We shall give the different ways as the appropriate tools are presented in the test. We shall refer to them as the Shier collection.

A telecommunications company has n ground stations and two orbiting satellites. The problem is to determine how many ways the ground stations and satellites can be connected so that all parts of the system can communicate with each other using the smallest number of links. This is shown in Figure 8.4.

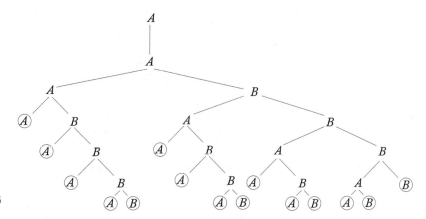

Figure 8.3

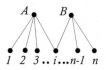

Figure 8.4

For future reference, we note that we can abstract this figure to a subgraph of the complete bipartite graph $K_{2,n}$. Since we want the smallest number of links, we are finding a spanning tree for $K_{2,n}$. The problem can then be reduced to finding the number of spanning trees of $K_{2,n}$.

We begin with the method that Shier calls the direct approach. There is a unique ground station i that is connected to both satellites. There are n choices for i. For $j \neq i$ there are two choices for the ground station, either satellite A or satellite B. By the multiplication principle, there are

$$n \times 2 \times 2 \times \cdots \times 2 = n2^{n-1}$$

choices for selecting the links.

Exercises

1. How many positive integers less than 700 are divisible by 5? How many are divisible by 3? How many are divisible by 5 and 3?

2. If an airline has 15 flights from San Francisco to Chicago and 20 flights from Chicago to New York, how many flights are there from San Francisco to New York passing through Chicago?

3. How many ways can a president, vice president, secretary, and treasurer be chosen from a club with 8 seniors, 10 juniors, 15 sophomores, and 20 freshmen if
   (a) There are no restrictions?
   (b) The president must be a senior?
   (c) The vice president cannot be a senior?
   (d) Freshmen are only eligible to be secretary?

4. A restaurant has a choice of the beverages coffee tea, milk, or cola. There is a choice of soup or salad. There are 10 different steak entrées and 5 different chicken entrées. There is a choice of french fries, baked potato, macaroni and cheese, or rice. For dessert, there is a choice of pie, ice cream, or both.
   (a) How many different meals are possible?
   (b) How many different meals are possible if a steak is selected?
   (c) How many meals are possible if the customer has either steak and potatoes or does not have steak and does not have potatoes?

5. How many distinct four-digit nonnegative numbers are there with possible leading zeroes if at least two of the digits must be the same?

6. If every entry in a 4×5 matrix must be a single-digit nonnegative integer, how many matrices are possible?

7. How many different sets of answers are possible on a test with 30 true-false questions?

8. If a watch displays hours, minutes, seconds, and AM-PM, how many different times can it display?

9. The call letters for radio stations in the United States consist of either three or four letters beginning with either k or w. How many different call letters are possible?

10. How many license plates are there consisting of two letters followed by four digits?

11. If there are four propositions p, q, r, and s,
    (a) How many rows are needed in the truth table?
    (b) In how many rows are p, q, r, and s all true?
    (c) In how many rows are at least one of the four propositions false?

12. How many license plates are there consisting of two letters followed by four digits if the digits cannot be repeated but the letters can?

13. How many different social security numbers are possible?

14. How many license plates are there consisting of two letters followed by four digits if none of the letters can be a vowel? If one or more of the letters must be a vowel?

15. How many functions are there from a set with six elements to a set with three elements?

16. How many one-to-one functions are there from a set with three elements to a set with six elements?

17. How many positive integers less than 700 are not divisible by 8?

18. How many positive integers less than 1000 are not divisible by 5?

19. How many two- or three-letter initials for people are possible?

20. How many two- or three-letter initials for people are possible if no letter can be repeated?

21. How many two or three letter initials for people are available if at least one of the letters must be an A?

22. How many bit strings are there of length 7 or less?

23. How many bit strings are there of length 7 if the first bit and the last bit must be the same?

24. How many bit strings are there of length 7 with two or more 1's in the string?

25. If an area code cannot begin with a 0, 1, or 8, how many telephone numbers are possible?

26. How many four-digit nonnegative integers with possible leading zeroes are there that contain at least one 7?

27. How many four-digit nonnegative integers with possible leading zeroes are there that do not contain a 7?

28. If a computer password must contain seven symbols consisting of either letters or single-digit integers, how many passwords are there which begin with a letter?

29. There are exactly 128 ASCII characters. How many different strings of seven ASCII characters are possible?

30. How many different strings of seven ASCII characters are there if no symbol is repeated?

31. How many different strings of seven ASCII characters are there if K occurs exactly once in the string?

32. How many different strings of seven ASCII characters are there that begin with a digit and end with a capital letter?

33. How many ways are there to arrange the letters a, b, c, d, and e so that a and b are next to each other?

34. How many ways are there to arrange the letters a, b, c, d, and e so that a and b are not next to each other?

8.2 Inclusion-Exclusion Introduced

Up to this point we have only considered disjoint sets when applying the addition counting principle. Suppose that sets S and T are not disjoint and we wish to find $|S \cup T|$. When we add the number of elements in S to the number of elements in T, we count the number of elements in $S \cap T$ twice. Therefore, we must subtract the number of elements in $S \cap T$. This gives us the following theorem.

THEOREM 8.13 Let S and T be sets. The number of elements that can be selected from S or T is equal to $|S|+|T|-|S \cap T|$. In other words, $|S \cup T| = |S|+|T|-|S \cap T|$.

Proof The set $S \cup T = (S-T) \cup (T-S) \cup (S \cap T)$ where $S-T$, $T-S$, and $S \cap T$ are pairwise disjoint. Therefore, $|S \cup T| = |S-T|+|T-S|+|S \cap T|$. Also we have $S = (S-T) \cup (S \cap T)$ and $T = (T-S) \cup (S \cap T)$, so that $|S| = |S-T|+|S \cap T|$ and $|T| = |T-S|+|S \cap T|$. Therefore,

$$|S|+|T|-|S \cap T| = |S-T|+|T-S|+2|S \cap T|-|S \cap T|$$
$$= |S-T|+|T-S|+|S \cap T|$$
$$= |S \cup T| \qquad \blacksquare$$

EXAMPLE 8.14

Suppose that in a group of 100 students, 60 take mathematics, 75 take history, and 45 take both.

(a) How many take mathematics or history?

(b) How many do not take mathematics and do not take history?

(a) Let the universe U be the group of 100 students, M be the set of students who take mathematics, and H be the set of students who take history. The set of students who take mathematics or history is equal to

$$|M \cup H| = |M| + |H| - |M \cap H|$$
$$= 60 + 75 - 45$$
$$= 90$$

(b) The number of students who do not take mathematics and do not take history is equal to $|M' \cap H'|$. But $M' \cap H' = (M \cup H)'$ so

$$|M' \cap H'| = |(M \cup H)'|$$
$$= 100 - 90$$
$$= 10$$

EXAMPLE 8.15

Find the number of positive integers less than 1001 that are divisible by 3 or 5. The number of elements in the set S of positive integers less than 1001 that are divisible by 3 is equal to $\lfloor \frac{1001}{3} \rfloor$ or 333. The number of elements in the set T of positive integers less than 1001 that are divisible by 5 is equal to $\lfloor \frac{1001}{5} \rfloor$ or 200. The elements in $S \cap T$ are the integers less than 1001 that are divisible by both 5 and 3, and so are divisible by 15. Hence, $|S \cap T| = \lfloor \frac{1001}{15} \rfloor$ or 66. Thus,

$$|S \cup T| = 333 + 200 - 66 = 467$$

EXAMPLE 8.16

How many ways are there to arrange the integers

$$0, 1, 2, 3, 4, 5, 6, 7, 8, 9$$

so the first digit is greater than 1 and the last digit is less than 7? Let U be the set of all possible ways to arrange the integers, and S be the set of all arrangements of the integers so the first digit is greater than 1 and the last digit is less than 7. To determine $|U|$ we observe that there are ten ways to select the first digit, nine ways to select the second digit, eight ways to select the third digit,..., two ways to select the ninth digit, and one way to select the tenth digit. Therefore, by the multiplication counting principle, there are $10! = 3,628,800$ ways to select the possible arrangement of the integers, so $|U| = 3,628,800$. As in the first example, we are going to try the "backdoor" approach by finding the number of arrangements of the integers that are not in S and subtracting this number from $|U|$. Let A be set of all arrangements that have the first digit less than or equal to 1. Counting the elements in A, we find that there are two choices for the first digit, namely, 0 and 1. There are then nine choices for the second digit, eight choices for the third digit, ... two choices for the ninth digit, and one choice for the tenth digit. Thus, there are $2 \times 9! = 725,760$ elements in A. Let B be set of all arrangements that have the last digit greater than or equal to 7. Counting the elements in B, we find that there are three choices for the last digit, and, using the same process as for A, we find that there are 9! ways to fill the other digits. Thus, $B = 3 \times 9! = 1,088,640$. The set $A \cap B$ consists of all

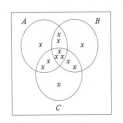

Figure 8.5

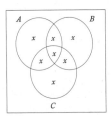

Figure 8.6

arrangements of the integers that have the first digit less than or equal to 1 and the last digit greater than or equal to 7. Counting the elements in $A \cap B$, we find that there are two choices for the first digit and three choices for the last digit. Using the same method as before, we find that there are 8! to fill the other eight digits. Therefore,

$$|A \cap B| = 3 \times 2 \times 8! = 241920$$

and

$$|A \cup B| = 725760 + 1088640 - 241920 = 1572480$$

The arrangement of the integers so the first digit is greater that 1 and the last digit is less than 7 is $A' \cap B' = (A \cup B)'$. Therefore, the number of such arrangements is equal to

$$|U| - |(A \cup B)'| = 3628800 - 1572480 = 2056320 \qquad \blacksquare$$

Having conquered two sets, we now move bravely on to three sets. Given sets A, B, and C, we want to determine $|A \cup B \cup C|$, the number of elements in $A \cup B \cup C$, when the sets are not necessarily disjoint. If we add the number of elements in each of the sets A, B, and C, then we can see in Venn diagram shown in Figure 8.5 that several subsets have been counted twice. If we now subtract $|A \cap B|$, $|A \cap C|$, and $|B \cap C|$, the number of elements in $A \cap B$, $A \cap C$, and $B \cap C$, respectively, then, as seen in the Venn diagram shown in Figure 8.6 the elements in $A \cap B \cap C$ have not been included at all. Therefore, we add $|A \cap B \cap C|$ so that each subset of $A \cup B \cup C$ is counted exactly once. This gives us the following formula:

$$|A \cup B \cup C| = |A| + |B| + |C| - |A \cap B| - |A \cap C| - |B \cap C| + |A \cap B \cap C|$$

Assume that in a survey of 200 people who watch television, it was found that 110 watch sports, 120 watch comedy, 85 watch drama, 50 watch drama and sports, 70 watch comedy and sports, 55 watch comedy and drama, and 30 watch all three. How many people watch sports, comedy, or drama? How many people do not watch any of the above? Let U be the set of 200 people surveyed. Let S be the set of people who watch sports, D be the set of people who watch drama, and C be the set of people who watch comedy. The first question then asks us to find $|S \cup D \cup C|$. We can do this directly using the foregoing formula:

$$|S \cup D \cup C| = |S| + |D| + |C| - |S \cap D| - |D \cap C| - |S \cap C| + |S \cap D \cap C|$$
$$= 110 + 85 + 120 - 50 - 55 - 70 + 30 = 170$$

so that 170 people watch sports, comedy or drama. The second question asks us to find $|(S \cup D \cup C)'| = |U| - |(S \cup D \cup C)| = 200 - 170 = 30$. So 30 people do not watch sports, comedy, or drama on television. $\blacksquare$

There is an alternative method of analyzing the foregoing example that gives us more information. We know that 30 people watch sports, comedy, and drama on television. Since 50 people watch sports and drama, 20 people must watch sports and drama but not comedy. Similarly 55 watch comedy and drama, so 25 must watch comedy and drama but not sports. Also 70 watch sports and comedy, so that 40 must watch sports and comedy but not drama. Thus, we have the following Venn diagram shown in Figure 8.7.

We know that 85 people watch drama. So far, we have accounted for 75 of them, so the other 10 must watch drama only. In similar fashion, we note that of the 110 people that watch sports, we have accounted for 90 of them so the other 20 must watch only sports. Finally, we note that of the 120 people who watch comedy, we

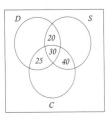

Figure 8.7

have accounted for 95 of them. Thus, 25 people watch comedy only. If we count all the people in all the disjoint subsets, we find there are 170 people. Therefore, 30 people do not watch any of the above. This gives us the Venn diagram shown in Figure 8.8.

With this diagram we can determine, for example, the following facts: 35 people watch drama but not sports; 50 people watch sports or drama but not comedy; 55 people watch exactly one of the three forms of entertainment; 85 watch exactly two of the three forms of entertainment; 85 watch comedy or sports but not drama.

EXAMPLE 8.18

Suppose that in a survey of 100 students, it was found that 50 take chemistry, 53 take mathematics, 42 take physics, 15 take chemistry and physics, 20 take physics and mathematics, 25 take mathematics and chemistry, and 5 take all three.

(a) How many students take at least one of the three areas?
(b) How many students take none of the three areas?
(c) How many students take mathematics only?
(d) How many students take physics or chemistry but not mathematics?
(e) How many do not take either mathematics or chemistry?

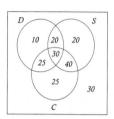

Figure 8.8

Since 5 people take all three, and 15 people take chemistry and physics, this leaves 10 people to take chemistry and physics but not mathematics. Similarly $25 - 5 = 20$ people take mathematics and chemistry but not physics, and $20 - 5 = 15$ people take mathematics and physics but not chemistry. This gives us the Venn diagram shown in Figure 8.9.

Since 50 take chemistry and we have accounted for 35 of them, this leaves 15 who take chemistry only. Similarly, 53 take mathematics and we have accounted for 40 of them, so 13 take mathematics only. Finally, 42 take physics and we have accounted for 30 of them, so 12 take physics only.

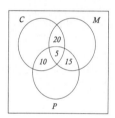

Figure 8.9

(a) Adding the number of people in the seven disjoint sets, we have 90 people who take at least at least one of the three areas.
(b) Since 90 out of 100 people take one or more areas, $100 - 90 = 10$ people take none of the three.
(c) From the Venn diagram, we see that 13 people take mathematics only.
(d) Thirty-seven take chemistry or physics but not mathematics.
(e) From the Venn diagram in Figure 8.10 it may be seen that 75 people take mathematics or physics. Therefore, $100 - 75 = 25$ do not take either mathematics or physics. ∎

EXAMPLE 8.19

How many positive integers less than 1001 are divisible by 2, 3, or 5? How many positive integers less than 1001 are not divisible by 2, 3, or 5? There are

$$\left\lfloor \frac{1001}{2} \right\rfloor = 500$$

integers divisible by 2. There are

$$\left\lfloor \frac{1001}{3} \right\rfloor = 333$$

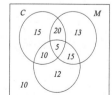

Figure 8.10

integers divisible by 3. There are

$$\left\lfloor \frac{1001}{5} \right\rfloor = 200$$

integers divisible by 5. If an integer is divisible by 2 and 3, then it is divisible by 6, so there are

$$\left\lfloor \frac{1001}{6} \right\rfloor = 166$$

integers divisible by 2 and 3. If an integer is divisible by 2 and 5, then it is divisible by 10, so there are

$$\left\lfloor \frac{1001}{10} \right\rfloor = 100$$

integers divisible by 2 and 5. If an integer is divisible by 3 and 5, then it is divisible by 15, so there are

$$\left\lfloor \frac{1001}{15} \right\rfloor = 66$$

integers divisible by 3 and 5. Finally, if an integer is divisible by 2, 3, and 5, then it is divisible by 30, and there are

$$\left\lfloor \frac{1001}{30} \right\rfloor = 33$$

integers divisible by 2, 3, and 5. Therefore, the number divisible by 2, 3, or 5 is equal to $500 + 333 + 200 - 166 - 100 - 66 + 33 = 734$. The number of positive numbers not divisible by any of the integers is $1001 - 734 = 267$. ∎

EXAMPLE 8.20

How many nonnegative integers less than 100,000 have 3, 6, and 9 among their digits? Let the universe U be the set of all nonnegative integers less than 100,000. Therefore, $|U| = 100{,}000$. Let S_3 be the set of all integers less than 100,000 that do not have 3 as a digit. Let S_6 be the set of all integers less than 100,000 that do not have 6 as a digit. Let S_9 be the set of all integers less than 100,000 that do not have 9 as a digit. We are seeking $|S'_3 \cap S'_6 \cap S'_9|$ but

$$|S'_3 \cap S'_6 \cap S'_9| = |U| - |S_3 \cup S_6 \cup S_9|$$

and

$$|S_3 \cup S_6 \cup S_9| = |S_3| + |S_6| + |S_9| - |S_3 \cap S_6| - |S_3 \cap S_9|$$
$$- |S_6 \cap S_9| + |S_3 \cap S_6 \cap S_9|$$

For S_3, there are nine choices for each of the five digits, since only 3 is excluded. Therefore, $|S_3| = 9^5$. Similarly, $|S_6| = |S_9| = 9^5$. Since each digit of $S_3 \cap S_6$ can be any integer except 3 and 6, there are eight choices for each digit, so $|S_3 \cap S_6| = 8^5$. Similarly, $|S_3 \cap S_9| = |S_6 \cap S_9| = 8^5$. Each digit of $S_3 \cap S_6 \cap S_9$ can be any integer except 3, 6, and 9, so there are seven choices for each digit. Therefore, $|S_3 \cap S_6 \cap S_9| = 7^5$. Hence,

$$|S_3 \cup S_6 \cup S_9| = 9^5 + 9^5 + 9^5 - 8^5 - 8^5 - 8^5 + 7^5$$
$$= 95{,}650$$

and

$$S'_3 \cap S'_6 \cap S'_9 = |U| - (S_3 \cup S_6 \cup S_9)$$
$$= 100{,}000 - 95{,}650$$
$$= 4350$$

∎

Exercises

1. In a class of 200 students, 120 are taking History of Kansas 1995–96, 110 are taking Yogurt Testing II, and 80 are taking both. How many students are taking at least one of the classes? How many are not taking either of the classes?

2. In a class of 50 students, 35 like football and 27 like basketball. If each student likes one of the sports, how many like both football and basketball? How many like football but not basketball?

Fifty people use Pascal and 45 people use C^{++}. How many people use Pascal or C^{++} if

3. Every one who uses Pascal uses C^{++}?

4. No one who uses Pascal uses C^{++}?

5. Twenty-five people use both Pascal and C^{++}?

6. How many positive integers less than 401 are divisible by either 5 or 7?

7. How many positive integers less than 401 are divisible by either 7 or 11?

8. How many positive integers less than 401 are divisible by either 6 or 10?

9. How many positive integers less than 401 are divisible by either 10 or 15?

10. How many positive integers less than 1001 are divisible by 10 but not by 40?

11. How many positive integers less than 1001 are divisible by 10 but not by 14?

12. A year is a leap year if either (a) it is divisible by 4, but not 100, or (b) it is divisible by 400. How many leap years are there between 1001 and 2001?

13. How many arrangements are there of the positive integers less than 10 such that either 4 immediately follows 5 or 5 immediately follows 4? How many arrangements are there in which 4 and 5 are not adjacent?

14. How many arrangements are there of the positive integers less than 10 such that either 4 immediately follows 3 or 7 immediately follows 6?

15. How many positive integers less than 1001 are not cubes or squares of an integer?

16. How many positive integers with five digits or less are there which
    (a) Have their first digit equal to 3?
    (b) Have their last digit equal to 5?
    (c) Have their first digit equal to 3 or their last digit equal to 5?
    (d) Have neither their first digit equal to 3 nor their last digit equal to 5?

17. How many positive integers less than 2003 are divisible by 3, 5, or 7?

18. How many positive integers less than 2003 are divisible by 5, 7, or 11?

19. How many positive integers less than 2003 are divisible by 4, 5, or 6?

20. How many positive integers less than 2003 are divisible by 6, 7, or 8?

21. In a class of 200, 75 take mathematics, 70 take history, 75 take sociology, 35 take mathematics and sociology, 20 take history and sociology, 25 take mathematics and history, and 15 take all three.
    (a) How many take at least one of the three subjects?
    (b) How many take exactly one of the three subjects?
    (c) How many take history or mathematics but not sociology?
    (d) How many do not take exactly two of the three subjects?
    (e) How many do not take either history or mathematics?

22. In a survey of 250 television viewers, 95 like to watch news, 125 like to watch sports, 125 like to watch comedy, 25 like to watch news and comedy, 45 like to watch sports and comedy, 35 like to watch news and sports, and 5 like to watch all three.
    (a) How many like to watch news but not sports?
    (b) How many like to watch news or sports but not comedy?
    (c) How many do not like to watch either news or sports?
    (d) How many do not watch sports only?
    (e) How many watch sports and comedy but not news?

23. How many five-digit integers with no beginning zeros have 3, 5, or 7 as a digit?

24. How many five-digit integers with no beginning zeros begin with 3 and end with 5, or have 7 as a digit?

25. In a class of 100 students, 35 studied French, 42 studied Spanish, 43 studied German, 17 studied French and Spanish, 15 studied Spanish and German, 13 studied French and German, and 20 did not study any of the three languages.
    (a) How many studied French or German but not Spanish?
    (b) How many studied exactly one of the three languages?
    (c) How many studied exactly two of the three languages?
    (d) How many did not study either Spanish or French?
    (e) How many studied Spanish only?

8.3 Permutations and Combinations

Ordinarily when we permute a collection of objects we place them in a different order. A permutation in this sense is a rearrangement of a collection of objects or a function that rearranges a collection of objects. (See Chapter 4, Section 4.2.) We will investigate how many ways we can rearrange a collection of objects. We already have the tools to do this. All that is needed is the multiplication counting rule and, in fact, we have already counted permutations. Consider the number of ways that we can form numbers by rearranging the integers 1, 2, 3, 4, and 5. Some possible arrangements are 12345, 15342, 32415, and 32415. To find the number of possible arrangements or permutations, we know that there are five ways to select the first integer, fours ways to select the second integer, three ways to select the third integer, two ways to select the fourth integer, and one way to select the fifth integer. Therefore, there are 5! ways to form the rearrangements. In similar fashion, if we want to rearrange n objects, it can be done in $n!$ ways. In a permutation, order is important. The numbers 51342 and 32415, formed by permuting the integers 1, 2, 3, 4, and 5, are not the same. Also, since we are considering permutations as rearrangements, each object can be used only once. If we allowed repetition, we would have five choices for each integer selected to form the number, so we would have 5^5 possible numbers.

We also want to consider permutations where there are more objects than places to put them. This is a generalization of our previous definition of permutation since we are no longer considering rearrangements. For example, suppose we have 20 people in an organization and want to select a president, vice president, secretary, and treasurer. We have 20 choices for the president, 19 choices for the vice president, 18 choices for the secretary, and 17 choices for the treasurer. Therefore, there are $20 \times 19 \times 18 \times 17$ ways of selecting the officers. Notice that order is still important. It makes a difference whether Mary Brown is president, vice president, secretary, or treasurer. Each rearrangement of the four people selected is a different set of officers. Each person can still be chosen only once. If Mary Brown is selected as president, she cannot be selected for one of the other three offices. Also notice that

$$20 \times 19 \times 18 \times 17 = \frac{20!}{16!} = \frac{20!}{(20-4)!}$$

since all factors less than or equal to 16 in the numerator are canceled by factors of 16! in the denominator.

Suppose we have n people and want to select r of them and place them in order. Again there are n ways of choosing the first person, $n-1$ ways of choosing the second person, $n-2$ ways of choosing the third person, $n-j+1$ ways of choosing the jth person, and $n-r+1$ ways of choosing the rth person. Therefore, there are

$$(n)(n-1)(n-2)(n-3)\cdots(n-i+1)\cdots(n-r+1)$$

ways of selecting the r people out of n. But

$$(n)(n-1)(n-2)(n-3)\cdots(n-i+1)\cdots(n-r+1) = \frac{n!}{(n-r)!}$$

We state this more formally as follows.

THEOREM 8.21 The number of ways to select r objects in order from n objects is

$$(n)(n-1)(n-2)(n-3)\cdots(n-i+1)\cdots(n-r+1) = \frac{n!}{(n-r)!}$$ ∎

8.3 Permutations and Combinations

DEFINITION 8.22 Let $P(n, r) = \dfrac{n!}{(n-r)!}$. $P(n, r)$ is the number of r **permutations on** n **objects**.

Notice that if we select all n objects and place them in order, then $r = n$ and, since $0! = 1$,

$$P(n, n) = \frac{n!}{(n-n)!} = \frac{n!}{0!} = n!$$

Thus, we get the expected result when permuting n elements.

EXAMPLE 8.23 How many distinct four-digit numbers can be formed from the integers $1, 2, 3, \ldots, 9$ if all of the digits in each four-digit number must be distinct? To form each four-digit number we are simply selecting four integers out of nine, so there are

$$P(9, 4) = \frac{9!}{(9-4)!} = \frac{9!}{5!} = 3024$$

different numbers. ∎

EXAMPLE 8.24 How many ways can five boys and six girls line up for a picture if no two girls are to stand together and no two boys are to stand together? Since the first person must be a girl and the boys and girls must alternate, the line must have the form $GBGBGBGBGBG$. There are $6!$ ways to place girls in the G positions and $5!$ ways to place the boys in the B positions. Therefore, there are $6! \times 5!$ ways to line up for the picture. ∎

EXAMPLE 8.25 How many ways can five boys and five girls line up for a picture if no two girls are to stand together and no two boys are to stand together? Here the first person may be either a boy or a girl. If the first person is a girl and the boys and girls must alternate, the line must have the form $GBGBGBGBGB$. There are $5!$ ways to place girls in the G positions and $5!$ ways to place the boys in the B positions. Therefore, there are $5! \times 5!$ ways to line up for the picture if the girl is first. Similarly, if the first person is a boy, there are $5! \times 5!$ ways to line up for the picture. Thus, there are $2 \times 5! \times 5!$ ways to line up for the picture. ∎

EXAMPLE 8.26 How many ways can 10 people be seated at a round table if we are only concerned with whom sits at each person's right or left? There are several ways to work this problem. First, we observe that if the people are rotated around the table, the seating is not changed since each person will have the same people to their right and left. Assume that *every* seating is unique. Then there are $10!$ ways of seating the people. Now we consider seatings to be the same if the people are rotated, and since there are 10 rotations, we divide $10!$ by 10 to get $9!$ ways of seating the people if we are only concerned with whom sits at each person's right or left.

An alternate way of solving the problem is to simply seat one person. This stops the rotation and we can seat the remaining 9 people in $9!$ ways. ∎

EXAMPLE 8.27 If there are 15 distinct mathematics books, 12 distinct physics books, and 16 distinct computer science books to be placed on a shelf, how many ways can this be done if

(a) There are no restrictions?

(b) All books on the same subject must be placed together?

(c) All books on the same subject must be placed together and the mathematics books and computer science books cannot be next to each other?

   (a) There are 43 books, so there $43!$ different ways of placing them on the shelf.

304 Chapter 8 Counting and Probability

(b) First, assume that we place the mathematics books first, the physics books second, and the computer books last. Denote this formation by *MPC*. There are 15! distinct ways to place the mathematics books, 12! distinct ways to place the physics books, and 16! distinct ways to place the computer science books. Therefore, there are 15! × 12! × 16! ways to place the books. If we place the physics books first, then the computer science books and finally the mathematics books, denoted by *PCM*, we would again get 15!×12!×16! ways to place the books. In fact, each way that we permute *P*, *C*, and *M*, we get 15! × 12! × 16! ways to place the books. Since there are 3! ways to permute *P*, *C*, and *M*, there are 3! × 15! × 12! × 16! different ways to place the books.

(c) Since the physics books must be in the middle, the computer science books must be either first or last. Therefore, the only possible formations are *CPM* and *MPC*. Since for each of these, there are 15! × 12! × 16! different ways to place the books, we have a total of 2 × 15! × 12! × 16! different ways to place the books. ∎

EXAMPLE 8.28 How many permutations of the letters $w, e, d, i, g, m, a, t, h$ are there so that the words "we," "dig," and "math" are not spelled out with consecutive letters? Thus, $d, g, i, w, e, t, h, a, m$ is not allowed since "we" is spelled out with consecutive letters. Let the universe U be the set of all permutations of the letters $w, e, d, i, g, m, a, t, h$. $|U| = 9! = 362,880$. Let S_1 be the set of permutations in which the word "we" is spelled out with consecutive letters, S_2 be the set of permutations in which the word "dig" is spelled out with consecutive letters, and S_3 be the set of permutations in which the word "math" is spelled out with consecutive letters. We are trying to find $|S_1' \cap S_2' \cap S_3'|$ or $|U| - |S_1 \cup S_2 \cup S_3|$. But

$$|S_1 \cup S_2 \cup S_3| = |S_1| + |S_2| + |S_3| - |S_1 \cap S_2| - |S_1 \cap S_3| - |S_2 \cap S_3|$$
$$+ |S_1 \cap S_2 \cap S_3|$$

Since "we" must appear together, the set S_1 consists of all permutations of the eight symbols we, d, i, g, m, a, t, h. Therefore, $|S_1| = 8!$. The set S_2 consists of all permutations of the seven symbols w, e, dig, m, a, t, h. Therefore, $|S_2| = 7!$. The set S_3 consists of all permutations of the six symbols $w, e, d, i, g, math$. Therefore $|S_3| = 6!$. The set $S_1 \cap S_2$ consists of all permutations of the six symbols we, dig, m, a, t, h. Therefore, $|S_1 \cap S_2| = 6!$ The set $S_2 \cap S_3$ consists of all permutations of the four symbols $w, e, dig, math$. Therefore, $|S_2 \cap S_3| = 4!$ The set $S_1 \cap S_3$ consists of all permutations of the four symbols $we, d, i, g, math$. Therefore, $|S_1 \cap S_2| = 5!$ Finally, $S_1 \cap S_2 \cap S_3$ consists of all permutations of the three symbols $we, dig, math$. Therefore, $|S_1 \cap S_2 \cap S_3| = 3!$, and

$$|S_1 \cup S_2 \cup S_3| = 8! + 7! + 6! - 6! - 5! - 4! + 3!$$
$$= 45,222$$

so that

$$|S_1' \cap S_2' \cap S_3'| = |U| - |S_1 \cup S_2 \cup S_3|$$
$$= 362,880 - 45,222$$
$$= 317,658 \quad \blacksquare$$

Let us again consider selecting a president, vice president, secretary, and treasurer from a group of 20 people. We know that there are

$$\frac{20!}{16!}$$

different ways of making the selections. In this case, selecting Mary Brown as president, George Smith as vice president, Jane Jones as secretary, and Joe Jackson as treasurer is different from selecting Jane Jones as president, Joe Jackson as vice president, George Smith as secretary, and Mary Brown as treasurer. Suppose that instead of selecting a president, vice president, secretary, and treasurer, we simply select a committee of four. In this case selecting Mary Brown, George Smith, Jane Jones, and Joe Jackson to the committee is the same as selecting Jane Jones, Joe Jackson, George Smith, and Mary Brown to the committee. Order no longer distinguishes the selections. Each way we could permute Mary Brown, George Smith, Jane Jones, and Joe Jackson was distinct when selecting the four officers, but for the committee, they are all the same. Since there are 4! ways of permuting the four officers, which are all the same committee, we need to divide $\frac{20!}{16!}$ by 4! to find the number of committees. Using the same argument for picking r objects out of n, without order, we get the following theorem.

THEOREM 8.29 The number of ways to select r objects without order from n objects is

$$\frac{n!}{(n-r)!r!}$$

DEFINITION 8.30 For $0 \leq r \leq n$, $C(n,r) = \binom{n}{r} = \frac{n!}{(n-r)!r!}$. $C(n,r)$ is the number of r **combinations on n objects**.

THEOREM 8.31 For $0 \leq r \leq n$, $C(n,r) = C(n, n-r)$.

Proof

$$C(n,r) = \frac{n!}{(n-r)!r!}$$

$$= \frac{n!}{(n-r)!(n-(n-r))!}$$

$$= C(n, n-r)$$

Please observe that with combinations, just as with permutations, each object may be used only once when selecting the r objects.

If we select all n objects without order, then $r = n$. Since $0! = 1$,

$$C(n,n) = \frac{n!}{(n-n)!(n!)} = \frac{n!}{(0!)(n!)} = 1$$

so there is only one way to select n elements: simply take all of them.

EXAMPLE 8.32 If a set has ten elements, how many three-element subsets does it have? Since a set is not ordered, we are choosing three objects out of ten, so there are

$$C(10,3) = \frac{10!}{7!3!} = 120$$

different subsets.

EXAMPLE 8.33 How many strings of length 9 contain exactly five 1's and four 0's? Although the strings themselves have order, this problem can still be worked using combinations. There are nine places in the string to put 1's and 0's. We can select any five of the nine places to put the 1's. There are, therefore,

$$C(9, 5) = \frac{9!}{5!4!} = 126$$

places to put the 1's. Once the 1's are inserted, the remainder of the places are filled with 0's. We see that, although the strings are ordered, there is no order in the way we fill the five places with 1's once the five places have been selected. ∎

EXAMPLE 8.34 How many ways can a committee of 6 men and 8 women be selected from a group of 12 men and 20 women? There are

$$C(12, 6) = \frac{12!}{6!6!}$$

ways of selecting the men and

$$C(20, 8) = \frac{20!}{8!12!}$$

ways of selecting the women. Therefore, by the multiplication counting principle, there are

$$\frac{12!}{6!6!} \times \frac{20!}{8!12!} = 116396280$$

ways of selecting the committee. ∎

EXAMPLE 8.35 How many distinct ways can a hand of 5 cards be dealt from a standard 52-card deck? Since a hand of cards has no order, we are simply selecting 5 objects out of 52, so there are

$$C(52, 5) = \frac{52!}{5!47!} = 2598960$$

possible hands. ∎

EXAMPLE 8.36 How many ways can a hand of 5 cards, which are all clubs, be dealt from a standard 52-card deck? There are 13 clubs from which to select 5, so there are

$$C(13, 5) = \frac{13!}{5!8!} = 1287$$

possible hands that are all clubs. ∎

EXAMPLE 8.37 A hand of five cards has four of a kind if four of the five cards are all aces, all kings, all queens, and so on. The four cards are said to be of the same rank. How many ways can a hand have four of a kind? There are 13 ways to choose the four of a kind and 48 ways to choose the last card. Therefore, there are $13 \times 48 = 624$ different hands that have four of a kind. ∎

EXAMPLE 8.38 A full house contains three of one kind and two of another kind. For example, a hand containing three kings and two sixes would be a full house. How many possible ways are there for a five-card hand to be a full house? Suppose for the moment that we select the three kings and two sixes for our full house. We are selecting three kings out of four, so there are $C(4, 3) = 4$ ways to select the three kings. We are selecting two sixes out of four, so there are $C(4, 2) = 6$ ways to select the two sixes. Therefore, by the multiplication counting principle, there are $4 \times 6 = 24$ ways of selecting three kings and two sixes or any fixed three of a kind and two of a kind.

There are 13 ways to choose the three of a kind, and 12 ways to pick the two of a kind. Therefore, there are $13 \times 12 = 156$ different ways to determine which three of a kind and which two of a kind to fix. Hence, there are $156 \times 24 = 3744$ possible ways for a five-card hand to be a full house. ∎

Consider the expansion of
$$(a+b)^5 = (a+b)(a+b)(a+b)(a+b)(a+b)$$

Each term in the expansion is the result of selecting an a or b from each factor $(a+b)$ and multiplying them together. For example, a^5 is obtained by selecting an a from each factor. If an a is selected from the first factor and a b from each of the other factors, then ab^4 is obtained. Suppose that we want to find the coefficient of a^3b^2. The term a^3b^2 is obtained by selecting three a's and two b's from the five factors. Since there are
$$C(5,3) = \binom{5}{3} = \frac{5!}{3! \times 2!}$$
ways of selecting three a's from the five factors, the coefficient of a^3b^2 is
$$\binom{5}{3}$$

In general we have the following theorem.

THEOREM 8.39 (Binomial Theorem) For a given nonnegative integer n,
$$(a+b)^n = \sum_{r=0}^{n} \binom{n}{r} a^r b^{n-r} = \sum_{r=0}^{n} \binom{n}{r} a^{n-r} b^r$$

Proof Since $a^r b^{n-r}$ is obtained by selecting r of the a's and $n-r$ of the b's from the n factors of $(a+b)^n$, the coefficient of $a^r b^{n-r}$ is equal to the number of ways of selecting r of the a's out of the n factors, or
$$\binom{n}{r}$$
The second equality follows from the fact that
$$\binom{n}{r} = \binom{n}{n-r}$$
∎

EXAMPLE 8.40 Expand $(2x + 3y^2)^4$. Using the binomial theorem,
$$(2x + 3y^2)^4 = \binom{n}{0}(2x)^4 + \binom{4}{1}(2x)^3(3y^2) + \binom{4}{2}(2x)^2(3y^2)^2$$
$$+ \binom{4}{3}(2x)(3y^2)^3 + \binom{4}{4}(3y^2)^4$$
$$= (2x)^4 + 4(2x)^3(3y^2) + 6(2x)^2(3y^2)^2 + 4(2x)(3y^2)^3 + (3y^2)^4$$
$$= 16x^4 + 96x^3y^2 + 216x^2y^4 + 216xy^6 + 81y^8$$
∎

EXAMPLE 8.41 Find the coefficient of x^4y^6 in the expansion of $(3x+4y)^{10}$. Using the binomial theorem, we find that the term containing x^4y^6 is

$$\binom{10}{4}(3x)^4(4y)^6 = 210(3x)^4(4y)^6$$
$$= 210 \cdot 3^4 \cdot 4^6 x^4 y^6$$
$$= 69672960 x^4 y^6$$

so the coefficient of x^4y^6 is 69672960. ∎

EXAMPLE 8.42 Show that for any nonnegative integer n,

$$2^n = \sum_{r=0}^{n} \binom{n}{r}$$

Letting $a = b = 1$ in the binomial theorem, we have

$$(1+1)^n = \sum_{r=0}^{n} \binom{n}{r} 1^r 1^{n-r}$$

and the result is immediate. ∎

The following theorem has many uses. One of these is that it enables us to create Pascal's triangle. The proof that we shall give uses counting techniques. The reader is also asked to prove the theorem using induction.

THEOREM 8.43 For all integers r and n, such that $1 \le r < n$,

$$C(n,r) = C(n-1, r-1) + C(n-1, r)$$

Proof Let m be one of the n objects from which r are to be chosen. Of the $C(n,r)$ ways that r objects can be selected, we consider the number of ways in which m is one of the objects selected and the number of ways in which m is not one of the objects selected. Their sum should equal $C(n,r)$. (Why?) First consider the number of ways in which m is one of the objects selected. In this case, since m has already been selected, there are $r-1$ objects still to be selected out of $n-1$ objects. There are $C(n-1, r-1)$ ways of making these selections. Next consider the number of ways in which m is not one of the objects selected. We must still select r objects but, since m is not one of the objects selected, there are only $n-1$ objects from which to select the r objects. Thus, there are $C(n-1, r)$ ways of making these selections. Adding the number of ways of making the selections in both possible cases, we have

$$C(n,r) = C(n-1, r-1) + C(n-1, r).$$ ∎

The diagram in Figure 8.11 is known as Pascal's triangle. Each of the interior elements of the triangle is the sum of the two elements above it.

This is a direct application of the previous theorem. Thus,

$$\binom{2}{1} = \binom{1}{0} + \binom{1}{1} \quad \text{and} \quad \binom{3}{2} = \binom{2}{1} + \binom{2}{2}$$

In the first case $n = 2$ and $r = 1$, and in the second case, $n = 3$ and $r = 2$. It may be noted that the $n + 1$st row gives the coefficients of $(a+b)^n$. For example,

$$(a+b)^4 = \binom{4}{0}a^4 + \binom{4}{1}a^3b + \binom{4}{2}a^2b^2 + \binom{4}{3}a^3b^3 + \binom{4}{4}b^4$$

Figure 8.11

$$\binom{0}{0}$$

$$\binom{1}{0} \quad \binom{1}{1}$$

$$\binom{2}{0} \quad \binom{2}{1} \quad \binom{2}{2}$$

$$\binom{3}{0} \quad \binom{3}{1} \quad \binom{3}{2} \quad \binom{3}{3}$$

$$\binom{4}{0} \quad \binom{4}{1} \quad \binom{4}{2} \quad \binom{4}{3} \quad \binom{4}{4}$$

$$\binom{5}{0} \quad \binom{5}{1} \quad \binom{5}{2} \quad \binom{5}{3} \quad \binom{5}{4} \quad \binom{5}{5}$$

$$\vdots$$

$$\underbrace{\binom{n\text{-}1}{r\text{-}1} \quad \binom{n\text{-}1}{r}}_{\binom{n}{r}}$$

Figure 8.12

```
              1
           1     1
        1     2     1
     1     3     3     1
  1     4     6     4     1
1    5    10    10    5    1
              ⋮
```

The coefficients are given in the fifth row of the triangle. Figure 8.12 gives the Pascal triangle with the combinations evaluated.

EXAMPLE 8.44 Find the expansion of $(a+b)^5$. Using the sixth row of Pascal's triangle, we find that

$$(a+b)^5 = a^5 + 5a^4b + 10a^3b^2 + 10a^2b^3 + 5ab^4 + b^5 \quad \blacksquare$$

The following theorem, while interesting in itself, produces a rather nice special case.

THEOREM 8.45 (Vandermonde) Let m, n, and r be positive integers such that $r \leq \min(m, n)$. Then

$$\binom{m+n}{r} = \sum_{k=0}^{r} \binom{m}{k}\binom{n}{r-k}$$

Proof The left-hand side of the equation is the number of ways we can pick r objects out of $m+n$ objects. Any way that we pick r objects out of $m+n$ objects we must pick k objects out of m objects and $r - k$ objects out of n objects for some $0 \leq k \leq r$. This can be done in

$$\binom{m}{k}\binom{n}{r-k}$$

ways. Conversely, if for any $0 \leq k \leq r$ we pick k objects out of m objects and $r - k$ objects out of n objects, then we pick r objects out of $m + n$ objects. Therefore,

$$\binom{m+n}{r} = \sum_{k=0}^{r} \binom{m}{k}\binom{n}{r-k}$$

∎

Corollary 8.46 For any positive integer n,

$$\binom{2n}{n} = \sum_{k=0}^{n} \binom{n}{k}^2$$

Proof Letting $n = m = r$ in the previous theorem, we have

$$\binom{n+n}{n} = \sum_{k=0}^{n} \binom{n}{k}\binom{n}{n-k}$$

but

$$\binom{n}{k} = \binom{n}{n-k}$$

so that

$$\binom{2n}{n} = \sum_{k=0}^{n} \binom{n}{k}\binom{n}{n-k} = \sum_{k=0}^{n} \binom{n}{k}^2$$

∎

This enables us to find the sum of the squares of the numbers in a row of a Pascal triangle.

EXAMPLE 8.47 We again provide a solution to Example 8.12 from the Shier collection. This method is called the deletion approach. The problem is to find the number of spanning trees of $K_{2,n}$. Instead of finding the $n+1$ edges to be selected from the $2n$ possible edges, he focuses on the $n-1$ edges to be deleted. The first edge, incident to the ith ground station, can be any of the $2n$ edges. The second edge, incident to the jth ground station, must avoid isolating the ith ground station so there are $2n - 2$ edges from which to select the second deletion. The third deletion must avoid isolating either the ith or jth ground station, so there are $2n - 4$ edges from which to select the third deletion. The $(j + 1)$st edge can be chosen from $2n - 2j$ edges. However, we have overcounted since each order of edges $e_1, e_2, e_3, \ldots e_{n-1}$ gives the same result. Thus, every permutation of these edges gives the same result so that the preceding product needs to be divided by $(n-1)!$. Therefore, the number of spanning trees is

$$\frac{1}{(n-1)!} \prod_{j=0}^{n-2} (2n - 2j) = 2^{n-1} \frac{n(n-1)\cdots 2}{(n-1)!} = n 2^{n-1}.$$

∎

Computer Science Application

The following algorithm returns the value of $C(m, n)$. The recursive identity $C(m, n) = C(m - 1, n - 1) + C(m - 1, n)$ is used to express $C(m, n)$ in simpler terms of itself, thus making the algorithm a candidate for recursion. The base cases of $C(m, 1) = m$ and $C(m, m) = C(m, 0) = 1$ are used to stop the recursive calls.

integer function binomial (integer m, integer n)

    if $m = n$ or $n = 0$ then
      return 1
    else if $n = 1$ then
      return m
    else
      return binomial $(m - 1, n - 1)$ + binomial $(m - 1, n)$
    end-if

end-function

The value of $C(52, 5) = 52!/(47!*5!) = 2{,}598{,}960$ is returned by binomial $(52, 5)$ without having to compute large products like $52!$, which would produce integer overflow.

Exercises

1. Compute
   (a) $P(8, 5)$
   (b) $P(11, 8)$
   (c) $C(12, 7)$
   (d) $C(14, 2)$
   (e) $C(14, 12)$

2. Compute
   (a) $P(8, 3)$
   (b) $P(11, 4)$
   (c) $C(15, 5)$
   (d) $C(12, 7)$
   (e) $C(12, 5)$

3. How many three-digit numbers can be formed using the numbers 2, 3, 4, 5, 6, 8, and 9 without repetition? How many three-digit numbers are less than 450? How many are even? How many are divisible by 4?

4. Find the number of ways that nine people can be lined up for a photograph.

5. Unfortunately, a judge at a flower show knows nothing about orchids. If he picks randomly, and there are 18 entries, how many ways can he award first, second, and third prize?

6. Ten horses are entered in a horse race. In how many ways can the first three horses finish?

7. Five couples go to a movie together. In how many ways can they sit if
   (a) They may sit in any order?
   (b) Each couple sits together?

8. Six boys and six girls attend a concert together.
   (a) How many ways can they sit if no boys sit together?
   (b) How many ways can they sit if no boys sit together and no girls sit together?
   (c) How many ways can they sit if all the boys sit together?
   (d) How many ways can they sit if a boy sits at each end?
   (e) How many ways can they sit if one boy and one girl refuse to sit next to each other?

9. How many six-digit numbers with a possible leading zero, but no repeated numbers are there if
   (a) The last two digits must be 7 or 8?
   (b) The first digit must be a 1 and the last two digits cannot be 7 or 8?
   (c) The digits 7 and 8 must be next to each other?

(d) The number must be divisible by 4?

(e) The number must be divisible by 8?

(f) The digits 5 and 6 must appear in the numbers?

10. How many permutations of the letters $a, c, f, m, p, r, t,$ and x are there if

    (a) There are no restrictions?

    (b) There must be two or three letters between a and c?

    (c) There must not be two or three letters between a and c?

    (d) The first four letters must be chosen from the letters $a, c, f,$ and m?

    (e) The letters $a, c, f,$ and m must be together?

11. In how many ways can five men and five women be seated at a circular table if no two men sit together?

12. How many different ways can a committee of 5 be selected from a club containing 25 members?

13. How many different ways can a committee of 6 men and 7 women be selected from an organization containing 15 men and 20 women?

14. How many eight digit binary strings can be formed containing three zeros and five ones?

15. How many ways are there of dealing a 13-card hand from a standard deck of 52 cards?

16. How many ways are there of dealing a 13-card hand containing 6 cards of one suit?

17. How many ways are there of dealing a 13-card hand containing 7 cards of one suit?

18. How many ways are there of dealing a 13-card hand containing 8 cards of one suit?

19. How many ways are there of dealing a 13-card hand containing 9 cards of one suit?

20. In a five-card hand, how many ways are there of having a exactly two pair (two of one rank and two of another rank, for example, two aces and two kings)?

21. In a five-card hand, how many ways are there of having exactly three of a kind (for example, three 10's)?

22. In a five-card hand, how many ways are there of having a flush (all cards of the same suit)?

23. In how many ways can 10 people be divided into two 5-man teams to play basketball?

24. Let $A = \{a, b, c, d, e, f, g, h\}$.

    (a) How many three-element subsets does A have?

    (b) How many five-element subsets of A are there if b must be in the set?

    (c) How many five-element subsets of A are there if b cannot be in the set?

    (d) How many five-element subsets of A are there if c must belong to the set, and d and e cannot belong to the set?

    (e) How many subsets of A are there that contain at least three elements?

    (f) How many subsets of A are there that contain at most six elements?

25. If a coin is tossed 10 times, how many ways are there of having exactly four heads and six tails? How many ways are there of having at least three heads?

26. At the pet shop, there are 5 turtles, 7 lizards, and 12 mice. In how many ways can Sam select 2 turtles, 3 lizards, and 5 mice?

27. On a true-false test with 30 questions, Fred knows that 20 of the statements are true and 10 are false. Unfortunately, Fred hasn't the slightest idea which is which. How many ways can he answer the questions so that he has the right number of questions answered true?

28. If a regular polynomial has n sides, how many diagonals does it have?

29. On a Little League team with 20 players, every player can play every position equally well. How many ways are there to choose 9 players to start the game?

30. On the Little League team in the previous problem, how many ways are there to select a starting batting lineup?

31. Using Pascal's triangle, expand $(a + b)^8$.

32. In the expansion of $(2x + 3y)^{10}$, find the coefficient of $x^6 y^4$.

33. In the expansion of $(3x - 4y)^{13}$, find the coefficient of $x^8 y^5$.

34. In the expansion of $(x + 2y^2)^{13}$, find the coefficient of $x^5 y^{16}$.

35. In the expansion of $(x^3 - 3y^2)^{10}$, find the coefficient of $x^9 y^{14}$.

36. Try to find the Fibonacci numbers embedded in Pascal's triangle and describe their locations. *Hint*: Look at the second element of the second row and the first element of the third row; then look at the second number of the third row and the first number of the fourth row. Finally, look at the third number of the third row together with the second element of the fourth row and the first element of the fifth row.

8.4 Generating Permutations and Combinations

Suppose that a set S, containing n elements, is linearly ordered, such as the integers or the letters of the alphabet, and that we are interested in finding all permutations of that set. This means that we are looking at all strings containing each element of S exactly once. To help us, we define **lexicographical ordering**. This ordering is called lexicographical because it is the way that words are ordered in the dictionary. For example, *able* comes before *gray* because *a* comes before *g* in the alphabet. The word *goose* comes before *gray* because, although the first letters are the same, the second letter of *goose* comes before the second letter of *gray* in the alphabet. The word *seconded* comes before *seconds* because although, the first six letters are the same, the seventh letter of *seconded* comes before the seventh letter of *seconds* in the alphabet. Similarly, the number 12,346 comes before 12,349.

DEFINITION 8.48 Given a linearly ordered set S, a **lexicographical ordering** $\preceq$ on the strings of elements of S is defined by $a \prec b$ if $a = a_1 a_2 a_3 \ldots a_m$, $b = b_1 b_2 b_3 \ldots b_n$ and any of the following occur:

(a) $a_1 < b_1$.

(b) $a_i = b_i$ for all $1 \leq i \leq k$ and $a_{k+1} < b_{k+1}$.

(c) $a_i = b_i$ for all $1 \leq i \leq m$ and $m < n$.

We now want to generate the set of all permutations of the elements of the linearly ordered set. For simplicity, we shall only consider permutations of the first n integers, but the method would be the same regardless. We let string $a_1 a_2 a_3 \ldots a_n$ denote permutation

$$\begin{pmatrix} 1 & 2 & 3 & \ldots & n \\ a_1 & a_2 & a_3 & \ldots & a_n \end{pmatrix}$$

Thus, 51342 represents the permutation

$$\begin{pmatrix} 1 & 2 & 3 & 4 & 5 \\ 5 & 1 & 3 & 4 & 2 \end{pmatrix}$$

To find all permutations, we want to be able to generate the next permutation in the lexicographical ordering after a given permutation. Since we are interested only in strings of the same length, part (c) of the definition of lexicographical ordering will not occur. Using lexicographical ordering, an increase in the second digit of a string will increase the size of the string lexicographically more than a change in the fifth digit. Similarly, an increase in the size of the first digit increases the size of the string lexicographically more than a change in the seventh digit. For example, if we begin with the string *dare*, a change to *read* puts us much further toward the end of the dictionary than a change to *dear*. Therefore, to keep the next string from getting too big, we want to change the letters as near to the end as possible. Suppose that $a_i > a_{i+1}$ for $m < a_i < n$. If this were true for all nonnegative m, then we would have the largest string lexicographically, and we would be finished. Therefore, assume that there is an m so that $a_i > a_{i+1}$ for $m < i < n$, but $a_m < a_{m+1}$. If we tried to rearrange just the a_i where $i > m$, we would create a smaller string lexicographically. For example, in the string 12348765, any rearrangement of the last four digits will create a smaller number. We, therefore, must increase the digit a_m but not change any a_i where $i < m$, so we must make a_m larger but not select any a_i where $i < m$ to replace it. We, therefore, select the smallest digit, a_i so that $i > m$ and $a_m < a_i$. We exchange a_m and a_i. We then rearrange all of the a_i after the new a_m in ascending order. This will make the new number as small as possible without being less than the original number.

For example, consider the number 12348765. To get the next integer, we exchange 4 and 5, getting 12358764. We then arrange 8, 7, 6, and 4 in ascending order to get 12354678.

Given the 1437652, to get the next number we exchange 3 and 5, getting 1457632. We then rearrange 7, 6, 3, 2 in ascending order getting 1452367. Notice that when we rearrange the remaining digits we simply reverse them, since they are in descending order. We then have the following procedure:

Procedure GenerateNextPermutation$(a_1, \ldots, a_n)$
    Find m so that $a_i > a_{i+1}$ for $m < a_i < n$ but $a_m < a_{m+1}$.
    Select the smallest digit, a_i so that $i > m$ and $a_m < a_i$.
    Exchange a_m and a_i.
    Rearrange all of the a_i after the new a_m in ascending order.
end

We now consider a method for generating all combinations from a set S of n elements. We shall not use lexicographical order of the elements of S. Instead, we shall use lexicographical order on strings of binary numbers. Recall that any combination from a set of n elements corresponds with a binary string of length n. If a set S is listed as $\{a_1, a_2, a_3, \ldots, a_n\}$, then in the binary string, a 1 in the kth bit of the binary string indicates that a_k was selected in the selected subset, while a 0 in the kth bit of the binary string indicates that a_k was not selected in the selected subset. For example, if the set S is $\{a_1, a_2, a_3, a_4, a_5\}$, then 10110 corresponds to the selection of the set $\{a_1, a_3, a_4\}$, and 10001 corresponds to the selection of the set $\{a_1, a_5\}$. Therefore, all that is needed to generate all combinations from a set of n elements is to generate all binary strings of length n. The easiest way is to count from 0 to $2^n - 1$ using binary strings of length n.

An equivalent method is to show how to determine the next binary string of length n from a given binary string of length n. To do this, simply begin at the right of the binary string and when the first 0 to the left is found, replace it with a 1, and change all of the bits to the right of it to 0's. As predicted, the generation of combinations is not lexicographic with respect to the order of the objects in the set. It is with respect to the binary strings.

EXAMPLE 8.49 Let $S = \{a, b, c, d, e\}$. Find the combination after the one that elects $\{a, d, e\}$. Since the binary string corresponding to $\{a, d, e\}$ is 10011, the next string is 10100. Therefore, the next combination selects $\{a, c\}$. ∎

Finally, we consider how to generate all combinations of r elements selected from a set of S elements where the elements are linearly ordered. We shall again determine these lexicographically with respect to the ordering of the elements in S. As before, we shall let $S = \{1, 2, 3, \ldots n\}$ without loss of generality. We shall assume that the combinations are listed in ascending order. As with permutations we shall show how to find the next combination in the lexicographical order. Suppose $n = 5$ and $r = 3$. If we can make the last digit larger, then we shall do so. Therefore, if we have 123, we can change it to 124. If we have 125, we cannot increase the last digit. Therefore, we go to the next digit to see if we can increase it. We can, since we can change the 2 to 3. However, we want to make the number as small as possible and still be greater than 125. Therefore, we make the last digit 1 larger than the previous digit, since that is the smallest we can make it and still have ascending order. Therefore, the next number is 134. Suppose we have the number 145. The last digit cannot be increased. Neither can the next digit to the left. However, 1 can be increased to 2. To make the number as small as possible we make the last digits 3 and 4. Hence, our number is 234.

Beginning at the right, the value of the last digit is as large as possible if its value is equal to $n = n - r + r$. If the last digit is as large as possible, then the next digit is as large as possible if it is equal to $n = n - r + (r - 1)$ or $n - r + i$ where $i = r - 1$ is the next to last digit. In general, the value of each digit i is as large as possible if the digits to the right are as large as possible and the value is equal to $n - r + i$. To make a permutation as large as possible, we therefore begin at the right and, going left, we check to see if each number has the property that it is equal to $n - r + i$. The first one that does not satisfy this equation can be enlarged. Say it is m in the jth place. We increase m by 1 and then increase each digit after the jth digit by one more than the previous digit.

We, therefore, have the following procedure:

> Procedure r-combination (n, r)
> Beginning at the right, find the first digit with value a_i
> such that $a_i \neq n - r + i$.
> Add one to this a_i.
> Beginning at a_{i+1}, let the value equal one more than the value at the previous digit.
> end

Exercises

1. Determine the next permutation generated after 21435.
2. Determine the next permutation generated after 621435.
3. Determine the next permutation generated after 21453.
4. Determine the next permutation generated after 416532.
5. Let $S = \{a, b, c, d, e\}$. Determine the next combination generated after $\{a, c, d\}$.
6. Let $S = \{a, b, c, d, e\}$. Determine the next combination generated after $\{a, c, d, e\}$
7. Let $S = \{1, 2, 3, 4, 5\}$. Determine the next combination generated after $\{1, 4, 5\}$.
8. Let $S = \{1, 2, 3, 4, 5, 6\}$. Determine the next combination generated after $\{1, 3, 5, 6\}$.
9. Let $S = \{1, 2, 3, 4, 5, 6\}$. Determine the next combination generated after $\{1, 6\}$.
10. Let $S = \{1, 2, 3, 4, 5, 6\}$. Determine the next three-combination generated after $\{1, 5, 6\}$.
11. Let $S = \{1, 2, 3, 4, 5, 6\}$. Determine the next three-combination generated after $\{1, 4, 6\}$.
12. Let $S = \{1, 2, 3, 4, 5, 6\}$. Determine the next four-combination generated after $\{1, 4, 5, 6\}$.
13. Let $S = \{1, 2, 3, 4, 5, 6\}$. Determine the next four-combination generated after $\{1, 3, 4, 5\}$.
14. List all permutations of 1, 2, 3, 4 in lexicographical order.
15. List all permutations of x, y, z in lexicographical order.
16. Generate all three-element subsets of $\{1, 2, 3, 4, 5\}$.
17. Generate all four-element subsets of $\{a, b, c, d, e\}$.
18. Give the first and last elements in the listing of the permutations of the integers $\{1, 2, 3, 4, 5, 6, 7\}$.
19. Give the first and last elements in the listing of the permutations of the letters of the alphabet.
20. Give the first and last elements in the listing of the three-element subsets of the integers $\{0, 1, 2, 3, 4, 5, 6, 7, 8, 9\}$.
21. Give the first and last elements in the listing of the five-element subsets of the alphabet.
22. Give the first and last elements in the listing of the subsets of the alphabet.
23. Using the algorithms in this section, devise an algorithm for generating $P(n, r)$ all of the r-permutations of n elements.
24. Using the algorithm in the previous problem generate the three-permutations of $\{a, b, c, d, e\}$.

8.5 Probability Introduced

Many years ago, a student at Stanford University wrote the San Francisco Weather Bureau to ask what it meant by a 60% chance of rain. Did it mean it would rain over 60% of the area? Did it mean that it would rain 60% of the time? The Weather Bureau replied that it meant that there were ten men at the Weather Bureau and six of them thought that it was going to rain. As we shall see later, it all depends on the sample space.

Many of the problems in both counting and probability are concerned with various types of gambling. One reason for this is that the mathematicians who developed much of the theory about probability were hired by gambling firms to determine the probability of winning and losing in various games. Many of the games provide good examples for the use of probability. It also enables readers to determine how badly they are going to lose if they practice their homework in the wrong place.

Our concept of probability will depend on the notion of an experiment being performed. We shall leave the term *experiment* undefined but require that it has the following properties:

(a) It produces more than one outcome.

(b) The outcome is uncertain.

(c) The finite set of all of its outcomes can be determined prior to the performance of the experiment.

(d) The experiment can be repeated under the same conditions.

An example of an experiment could be the tossing of two coins. In that case, the set of outcomes

$$S = \{(H, H), (H, T), (T, H), (T, T)\}$$

where H denotes a coin coming up heads and T denotes the coin coming up tails. Another example of an experiment could be the tossing of a coin and a die. In this case the set of outcomes

$$S = \{(H, 1), (H, 2), (H, 3), (H, 4), (H, 5), (H, 6), (T, 1), (T, 2), (T, 3),$$
$$(T, 4), (T, 5), (T, 6)\}$$

where the result of the tossing of the coin is followed by the result of the tossing of the die.

DEFINITION 8.50 The **sample space** is the set of all outcomes of an experiment. An **event** is a subset of a sample space.

Thus, in our two preceding examples, S is the sample space in each case. In the set theory that follows, the sample space S serves as the universe.

Intuitively, the probability of an event is the likelihood that the event will occur. If the experiment is repeated a number of times, the probability of an event is the frequency that it occurs. In this section and those immediately following, we shall assume that all outcomes of an experiment are equally likely. Note that this does not mean that all events are equally likely. This assumption enables us to use our counting arguments, which is appropriate since probability is one of the major applications of combinatorics. Under this assumption, we have the following definition of probability.

DEFINITION 8.51 Let S be the set of all possible outcomes of an experiment. The **probability** $P(A)$ that an outcome of an event A occurs is given by

$$P(A) = \frac{|A|}{|S|}$$

EXAMPLE 8.52 Let S be the set of all outcomes when two dice are tossed. Let A be the set of all outcomes where the sum appearing on the dice is equal to 6. Thus,

$$A = \{(1, 5), (2, 4), (3, 3), (4, 2), (5, 1)\}$$

so that $|A| = 5$. As we have already seen, since there are six possible outcomes for the first die and six possible outcomes for the second die, there are 6×6 possible outcomes in S. Therefore,

$$P(A) = \frac{|A|}{|S|} = \frac{5}{36}$$ ∎

EXAMPLE 8.53 Let S be the set of all outcomes when two coins are tossed. Let A be the set of all outcomes where both coins come up the same. Thus, $A = \{(H, H), (T, T)\}$, $S = \{(H, H), (H, T), (T, H), (T, T)\}$, and

$$P(A) = \frac{|A|}{|S|} = \frac{2}{4} = \frac{1}{2}$$ ∎

Does this mean that if we toss the two coins four times that the two coins will come up both heads or both tails exactly two times? The answer is no. But if we toss the two coins 400 times, we would expect the coin to come up heads about 200 times.

Again, assuming all outcome are equal likely, we can prove the following theorems using our counting techniques.

THEOREM 8.54 If A and B are disjoint sets of outcomes of an experiment, then $P(A \cup B) = P(A) + P(B)$.

Proof Using the addition counting principle, we have

$$P(A \cup B) = \frac{|A \cup B|}{|S|}$$
$$= \frac{|A| + |B|}{|S|}$$
$$= \frac{|A|}{|S|} + \frac{|B|}{|S|}$$
$$= P(A) + P(B).$$ ∎

The following theorem gives us some of the properties of the probability of an event.

THEOREM 8.55 If S is a sample space, and A' is the complement of A using S as the universe, then

(a) $P(S) = 1$.
(b) $P(A') = 1 - P(A)$.
(c) $P(\emptyset) = 0$.

(d) For any event A, $P(A) \geq 0$.

(e) If $A \subseteq B$ then $P(A) \leq P(B)$.

(f) For any event A, $0 \leq P(A) \leq 1$.

Proof (a) $P(S) = \dfrac{|S|}{|S|} = 1$.

(b) By part (a) and the previous theorem,
$$1 = P(S) = P(A \cup A') = P(A) + P(A').$$

Therefore, $P(A') = 1 - P(A)$.

(c) By part (b), $P(\varnothing) = 1 - P(\varnothing') = 1 - P(S) = 1 - 1 = 0$.

(d) Since there must be at least two outcomes, $|S| > 0$. Also $|A| \geq 0$, since it is the number of elements in the set A. Therefore,
$$P(A) = \dfrac{|A|}{|S|} \geq 0$$

(e) If $A \subseteq B$ then $B = A + (B - A)$. Therefore,
$$P(B) = P(A + (B - A))$$
$$= P(A) + P(B - A)$$

and
$$P(B) - P(A) = P(B - A).$$

But by part (d)
$$P(B - A) \geq 0.$$

Therefore,
$$P(B) - P(A) \geq 0$$

and
$$P(A) \leq P(B)$$

(f) Since $A \subseteq S$, then $P(A) \leq P(S) = 1$. Combining this with part (d), we have
$$0 \leq P(A) \leq 1.$$ ∎

EXAMPLE 8.56 If two coins are tossed, find the probability they are not both heads. Let A be the set of all outcomes where they are not both heads. Then A' is the set of outcomes where they are both heads. Since there are four possible outcomes and only one gives two heads,
$$P(A') = \dfrac{1}{4}, \text{ and } P(A) = 1 - \dfrac{1}{4} = \dfrac{3}{4}$$ ∎

EXAMPLE 8.57 Find the probability that if two dice are tossed, the sum is not seven. Let A be the set of outcomes where the sum is not seven. Then A' is the set of outcomes where the sum is seven. There are 6×6 total outcomes when two dice are tossed. The outcomes that have a sum of seven are $(1, 6)$, $(2, 5)$, $(3, 4)$, $(4, 3)$, $(5, 2)$, and $(6, 1)$. Thus, $|A'| = 6$, so that
$$P(A') = \dfrac{6}{36} = \dfrac{1}{6}, \text{ and } P(A) = 1 - \dfrac{1}{6} = \dfrac{5}{6}$$ ∎

For completeness we include the following theorem. The principles in this theorem are examined in more detail in the sections describing the inclusion-exclusion principle. This is a generalization of Theorem 8.6.

THEOREM 8.58 If A and B are sets of outcomes of an experiment, then

$$P(A \cup B) = P(A) + P(B) - P(A \cap B)$$

Proof Using Figure 8.13 we see that

$$A = (A - B) \cup (A \cap B)$$
$$B = (B - A) \cup (A \cap B)$$
and $A \cup B = (A - B) \cup (A \cap B) \cup (B - A)$

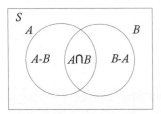

Figure 8.13

where $(A - B)$, $(A \cap B)$, and $(B - A)$ are mutually disjoint sets. Therefore, by Theorem 8.6, we see that

$$P(A) = P((A - B) \cup (A \cap B)) = P(A - B) + P(A \cap B)$$
$$P(B) = P((B - A) \cup (A \cap B)) = P(B - A) + P(A \cap B)$$

and

$$P(A \cup B) = P((A - B) \cup (A \cap B) \cup (B - A))$$
$$= P(A - B) + P(A \cap B) + P(B - A)$$

Therefore,

$$P(A) + P(B) = (P(A - B) + P(A \cap B)) + (P(B - A) + P(A \cap B))$$
$$= (P(A - B) + P(A \cap B) + P(B - A)) + P(A \cap B)$$
$$= P(A \cup B) + P(A \cap B)$$

so that $P(A \cup B) = P(A) + P(B) - P(A \cap B)$. ∎

EXAMPLE 8.59 Find the probability that, if a license plate with three letters and three numbers is selected at random, the three letters will be the same or the three numbers will be the same. Let S be the set of all outcomes where the license plate has three letters and three numbers, A be the set of outcomes where the three letters are the same, and B be the set of outcomes where the three numbers are the same. If the three letters are all the same, then there are 26 ways of choosing the letters and

$$10 \times 10 \times 10 = 1000$$

ways of selecting the numbers. Therefore, $|A| = 26{,}000$. If the numbers are all the same, then there are

$$26 \times 26 \times 26 = 17{,}576$$

ways of selecting the letters and 10 ways of selecting the numbers. Therefore, $|B| = 175{,}760$. If the letters are the same and the letters are the same, then there are 26 ways of selecting the letters and 10 ways of selecting the numbers. Therefore, $|A \cap B| = 260$ as we have previously seen, there are

$$26 \times 26 \times 26 \times 10 \times 10 \times 10 = 17{,}576{,}000$$

different ways of forming license plates with three letters and three numbers. Therefore,

$$P(A) = \frac{26{,}000}{17{,}576{,}000}, \quad P(B) = \frac{175{,}760}{17{,}576{,}000}, \quad \text{and} \quad P(A \cap B) = \frac{260}{17{,}576{,}000}$$

so that

$$P(A \cup B) = \frac{26{,}000}{17{,}576{,}000} + \frac{175{,}760}{17{,}576{,}000} - \frac{260}{17{,}576{,}000}$$

$$= \frac{201{,}500}{17{,}576{,}000}$$

$$= \frac{2{,}015}{175{,}760} \approx .011465 \quad \blacksquare$$

Computer Science Application

A somewhat surprising application of probability is found in the so-called birthday problem. If we ignore leap years and assume that people's birthdays are equally likely to occur among the 365 days of the year, the probability that no two people in a group of n people will have a common birthday is

$$p = (1 - 1/365) * (1 - 2/365) * * * (1 - (n-1)/365)$$

From this we see that in a group of only 23 people the probability that at least two have a common birthday is greater than 1/2. The following simple simulation can be used to demonstrate this. The algorithm assumes a random number generator that produces real numbers in the range [0,1].

real function birthday-problem ()

    integer N, $A[365]$, Sum, GrandSum

    GrandSum = 0
    for i = 1 to 10000
        A = 0
        Counter = 0
        repeat
            N = 365*Random () + 1
            $A[N] = A[N] + 1$
            Counter = Counter + 1
        until $A[N] = 2$
        GrandSum = GrandSum + Counter
    end-for

    return GrandSum/10000

end-function

The simulation returns the average result of 10,000 trials, each generating random integers in the range 1, ... , 365 until the first occurrence of a duplicate. The value returned is always very close to 23.

Exercises

1. Two dice are tossed. Find the probability that the sum is **(a)** 4, **(b)** 7, **(c)** 8, **(d)** 9, **(e)** 10 or greater. Find the probability that both dice are the same. Find the probability that the sum of the dice is even.

2. A card is drawn from a standard 52-card deck.

    (a) Find the probability of drawing a jack, queen, king, or ace (a face card).

    (b) Find the probability of drawing a spade.

    (c) Find the probability of drawing a queen or a spade.

(d) Find the probability of drawing a face card or a heart.

3. An urn contains 12 red, 8 blue, 6 white, and 10 yellow balls. A ball is drawn.
   (a) Find the probability of drawing a red ball.
   (b) Find the probability of drawing a white ball or a yellow ball.
   (c) Find the probability of drawing a ball that is not white.
   (d) Find the probability of drawing a ball that is not blue or white.

4. A letter is randomly picked from the alphabet. What is the probability that it is a vowel?

5. A positive integer less than 302 is picked randomly. Find the probability that
   (a) It is divisible by 3.
   (b) It is divisible by 3 and 7.
   (c) It is divisible by 3 or 7.
   (d) It is not divisible by either 3 or 7.

6. A positive integer less than 101 is picked randomly. Find the probability that
   (a) It is divisible by a prime.
   (b) It is the product of exactly two primes that are not necessarily distinct.
   (c) It is the product of exactly two distinct primes.
   (d) It is the product of exactly three distinct primes.

7. Five positive integers less than 10 are selected at random. What is the probability that they are all distinct?

8. If a license plate contains three letters followed by three numbers, what is the probability that the sum of the three numbers is even? What is the probability that the letters are all vowels? What is the probability that at least one letter is not a vowel?

9. In a survey of 250 television viewers, 86 like to watch news, 100 like to watch sports, 90 like to watch comedy, 23 like to watch news and comedy, 35 like to watch sports and comedy, 33 like to watch news and sports, and 10 like to watch all three. In an interview, a person is picked at random.
   (a) What is the probability that this person watches news but not sports?
   (b) What is the probability that this person watches news or sports but not comedy?
   (c) What is the probability that this person does not watch either news or sports?
   (d) What is the probability that this person watches sports and news but not comedy?

10. If a person is dealt a 13-card bridge hand, what is the probability that
    (a) This person gets exactly 7 cards of one suit?
    (b) This person gets exactly 8 cards of one suit?
    (c) This person gets exactly 9 cards of one suit?

11. If a person is dealt a five-card poker hand, what is the probability that
    (a) This person gets exactly three of a kind (e.g., three aces)?
    (b) This person gets a straight flush (e.g., 2, 3, 4, 5, 6 of same suit)?
    (c) This person gets a royal flush (e.g., $A, K, Q, J, 10$ of same suit)?
    (d) This person gets a flush (e.g., five hearts)?
    (e) This person gets a straight (e.g., 2, 3, 4, 5, 6 of any suit)?

12. In a class of 100 students, 40 studied French, 50 studied Spanish, 40 studied German, 12 studied French and Spanish, 13 studied Spanish and German, 15 studied French and German, and 10 did not study any of the three languages. If a student is selected at random,
    (a) What is the probability that the student studies only French?
    (b) What is the probability that the student studies Spanish or German?
    (c) What is the probability that the student does not study French?
    (d) What is the probability that the student studies German but not Spanish?
    (e) What is the probability that the student does not study any of the three languages?

13. Three urns contain four red balls and five blue balls, six red balls and three blue balls, five red balls and two blue balls, respectively. If a ball is randomly chosen from one of the urns, what is the probability that it is red?

14. What is the probability that a four-digit integer with no beginning zeros has 3, 5, or 7 as a digit?

15. What is the probability that a four-digit integer with no beginning zero begins with 3, ends with 5, or has 7 as a digit?

16. A three-letter word with no letters repeated is drawn randomly. What is the probability that the letters are in alphabetical order?

17. A four-digit number with no leading zeros is drawn randomly. What is the probability that it contains a 7? What is the probability that it contains a 4 or 9?

8.6 Generalized Permutations and Combinations

We now consider a type of counting that is something between a permutation and combination. Really, it is a generalization of both. Suppose we want to determine the distinct arrangements of the letters of the word *mississippi*. Our problem is that not all of the letters are distinct. Our method of counting the number of distinct arrangements is going to be very similar to the method used for counting the number of combinations. First we assume that the letters are all distinct. If this were true then there would be 11! ways to rearrange the 11 letters. Now we look at the four occurrences of i. By assuming that the letters were all distinct, we assumed that each rearrangement of the i's, which left all of the other letters alone, was distinct. If we no longer consider the i's as distinct, then all of these rearrangements are now the same. Thus, the 4! ways that the i's could be rearranged are no longer distinct but count as one arrangement. To count the rearrangements of the 11 letters when the i's are no longer distinct, we must divide 11! by 4!. By a similar argument, to count the arrangements if we no longer consider the four s's as distinct, we must again divide by 4!. Finally there are the two p's to consider. Again, each rearrangement of the two p's, which left all of the other letters alone, was distinct when the p's were distinct but is the same if we no longer consider the p's to be distinct. Since there are 2! ways to rearrange the two p's, we must divide by 2! to count the number of rearrangements where the two p's are no longer distinct. Thus, we get

$$\frac{11!}{4!4!2!}$$

distinct strings.

Notice that in making the letters indistinguishable, we remove the order between them just as we did for combinations.

THEOREM 8.60 Let S contain n objects of k different types such that there are n_1 indistinguishable objects of type 1, n_2 indistinguishable objects of type 2, n_3 indistinguishable objects of type 3, and, in general, n_i indistinguishable elements of type i. Let $P(n; n_1, n_3, n_3, \ldots, n_k)$ be the number of distinct arrangements of S. Then

$$P(n; n_1, n_2, \ldots, n_k) = C(n, n_1)C(n - n_1, n_2)C(n - n_1 - n_2, n_3)$$
$$\cdots C(n - n_1 - n_2 - \cdots - n_{k-1}, n_k)$$
$$= \frac{n!}{n_1! n_2! n_3! \cdots n_k!}$$

Proof Consider the first equation. There are n spaces to fill with elements of S. There are $C(n, n_1)$ ways to select spaces for the n_1 indistinguishable objects of type 1. Once these are selected, there are $n - n_1$ spaces left to fill, so there are $C(n - n_1, n_2)$ ways to select spaces for the n_2 indistinguishable elements of type 2. Once the objects of type 1 and type 2 have been selected, there are $n - n_1 - n_2$ spaces left to fill, so the objects of type 3 can be selected in $C(n - n_1 - n_2, n_3)$ ways. Similarly, the objects of type i for $2 \leq i \leq k$ can be selected in $C(n - n_1 - n_2 - n_3 - \cdots - n_{i-1}, n_i)$ ways. Using the product principle of counting, there are

$$C(n, n_1)C(n - n_1, n_2)C(n - n_1 - n_2, n_3) \cdots C(n - n_1 - n_2 - \cdots - n_{k-1}, n_k)$$

ways of selecting distinct arrangements of S.

To show that

$$P(n; n_1, n_2, \ldots, n_k) = \frac{n!}{n_1! n_2! n_3! \cdots n_k!}$$

we use an argument that we have used several times previously. First, assume that all n of the objects of S are distinct. If this is true, then we have $n!$ ways of arranging these objects. For $1 \leq i \leq k$, the n_i objects are indistinguishable. Therefore, every way that we can rearrange these objects in their spaces without moving objects of any other type is indistinguishable. Since there are $n_i!$ of these rearrangements, we must divide $n!$ by $n_i!$ for each i, to find the number of distinct arrangements possible when all of the n_i objects of type i are indistinguishable. This gives us

$$\frac{n!}{n_1! n_2! n_3! \cdots n_k!}$$

as desired. ∎

EXAMPLE 8.61 Assume that 12 books, consisting of 4 identical mathematics texts, 6 identical computer science texts, and 2 identical chemistry tests are to be placed on a shelf. How many distinct ways can they be placed on the shelf? Using the previous formula, we have

$$P(12; 4, 6, 2) = \frac{12!}{4!6!2!} = 13860$$

∎

EXAMPLE 8.62 How many distinct arrangements can be formed by rearranging the letters in *succeeded*? The nine letters form five types of indistinguishable objects consisting of (1) the letter s, (2) the letter u, (3) the two occurrences of letter c, (4) the three occurrences of the letter e, and (5) the two occurrences of the letter d. Therefore, the number of indistinguishable objects is

$$P(9; 1, 1, 2, 3, 2) = \frac{9!}{1!1!2!3!2!} = 15120$$

Notice that the types consisting of only one object do not affect the answer. Hence, we need only consider the types with more than one object. ∎

We now generalize the binomial theorem to find coefficients of the terms when $(x_1 + x_2 + x_3 + \cdots + x_m)^n$ is expanded. Consider

$$(a + b + c)^4 = (a + b + c)(a + b + c)(a + b + c)(a + b + c)$$

To find the coefficient of ab^2c, we must find all ways of selecting an a, two b's, and one c, where a letter is selected from each of the factors. The possibilities are *abbc, cbba, abcb, dbab, babc, bcba, bbac, bbca, cabb, acbb, bacb*, and *bcab*. But this is just the arrangements of an a, two b's, and a c. Thus, the number of arrangements, which is the coefficient of ab^2c, is

$$P(4; 1, 2, 1) = \frac{4!}{1!2!1!} = 12$$

Similarly, the coefficient of b^2c^2 is

$$P(4; 0, 2, 2) = \frac{4!}{0!2!2!} = 6$$

THEOREM 8.63 For a given positive integer n,

$$(x_1 + x_2 + x_3 + \cdots + x_m)^n = \sum \frac{n!}{n_1! n_2! n_3! \cdots n_m!} x_1^{n_1} x_2^{n_2} x_3^{n_3} \cdots x_m^{n_m}$$

where the sum is taken over all nonnegative integers $n_1, n_2, n_3, \cdots, n_m$ such that

$$n_1 + n_2 + n_3 + \cdots + n_m = n$$

Proof Since $(x_1 + x_2 + \cdots + x_m)^n = (x_1 + x_2 + \cdots + x_m)(x_1 + x_2 + \cdots + x_m) \cdots (x_1 + x_2 + \cdots + x_m)$, each term in the expansion is obtained by selecting an x_i from each factor in the product and multiplying them together. To find the coefficient of $x_1^{n_1} x_2^{n_2} x_3^{n_3} \cdots x_m^{n_m}$ we need to find all possible ways of selecting n_1 of the x_1's, n_2 of the x_2's,... and n_m of the x_m's. But this is the number of all arrangements of n_1 of the x_1's, n_2 of the x_2's, ... and n_m of the x_m's, which is

$$P(n; n_1, n_2, n_3, \cdots, n_m) = \frac{n!}{n_1! n_2! n_3! \cdots n_m!}$$ ∎

EXAMPLE 8.64 Find the coefficient of $ab^2 c^3$ in the expansion of $(a + b + c)^6$. The coefficient is

$$P(6; 1, 2, 3) = \frac{6!}{1! 2! 3!} = 60$$ ∎

EXAMPLE 8.65 Find the coefficient of $a^3 b^2 c^3$ in the expansion of $(2a + 3b + c)^8$. The term is
$$P(8; 3, 2, 3)(2a)^3 (3b)^2 c^3 = 560 \cdot 2^3 \cdot 3^2 a^3 b^2 c^3$$
$$= 4320 a^3 b^2 c^3$$

so the coefficient is 4320. ∎

Suppose that we have seven distinguishable balls and wish to place three balls in one box, two in a second box, and two in a third box. How many ways can we do this? There are $C(7, 3)$ ways to select the three balls for the first box. After these balls are selected, there are four balls left from which two are selected for the second box. There are $C(4, 2)$ ways to do this. Finally, there are two balls left from which two are selected for the last box. This may be done in $C(2, 2)$ or one way. Thus, the number of ways of placing the balls in the boxes is equal to $C(7, 3) \cdot C(4, 2) \cdot C(2, 2)$. But by Theorem 8.60 this is equal to

$$P(7; 3, 2, 2) = \frac{7}{3! 2! 2!}$$

Therefore, there seems to be a relationship between arrangements of indistinguishable objects and partitioning of a set, which is really what we were doing with the balls and boxes. To see this connection, consider the following example. Suppose we have two red, three green, and four blue balls, which are indistinguishable except for color, and we want to find all possible distinct arrangements of these balls. Let $s_1, s_2, s_3, s_4, s_5, s_6, s_7, s_8, s_9$ be the spaces in which the balls are placed. Suppose red balls are placed in spaces s_2 and s_7. We will identify this with placing s_2 and s_7 in box R. Suppose green balls are placed in spaces s_1, s_3, and s_9. We will identify this with placing s_1, s_3, and s_9 in box G. Finally, assume that blue balls are placed in spaces s_4, s_5, s_6, and s_8. We identify this with placing s_4, s_5, s_6, and s_8 in box B. Conversely if s_1 and s_9 are placed in box R, s_2 s_7 and s_8 in box G, and s_3, s_4, s_5, and s_6 in box B, we identify this with placing red balls in spaces s_1 and s_9, green balls in spaces s_2, s_7, and s_8, and blue balls in spaces s_3, s_4, s_5, and s_6. Thus, we have transformed a problem about arrangements of indistinguishable balls into a problem about placing balls in boxes. More generally we have the following theorem.

THEOREM 8.66 Let $C(n; n_1, n_2, n_3, \ldots, n_m)$ be the number of ways of partitioning a set S containing n elements into m sets $S_1, S_2, S_3, \ldots,$ and S_m containing $n_1, n_2, n_3, \ldots,$ and n_m elements, respectively. Then

$$C(n; n_1, n_2, n_3, \ldots, n_m) = P(n; n_1, n_2, n_3, \ldots, n_m) = \frac{n!}{n_1! n_2! n_3! \cdots n_m!}$$

Proof There are $C(n, n_1)$ ways of selecting elements for S_1. Once these are selected, there are $n - n_1$ elements left from which to select the n_2 elements for S_2. This may be done in $C(n - n_1, n_2)$ ways. Once these are selected, there are $n - n_1 - n_2$ elements left from which to select n_3 elements for S_3. Assuming sets $S_1, S_2, S_3, \ldots,$ and S_{i-1} have been selected, there are $n - n_1 - n_2 - \cdots - n_{i-1}$ elements left from which to select the n_i elements for S_i. This may be done in $C(n - n_1 - n_2 - \cdots - n_{i-1}, n_i)$ ways. Using the multiplication principle of counting, we have

$$C(n; n_1, \ldots, n_m) = C(n, n_1) C(n - n_1, n_2) C(n - n_1 - n_2, n_3)$$
$$\cdots C(n - n_1 - \cdots - n_{k-1}, n_k)$$
$$= \frac{n!}{n_1! n_2! n_3! \cdots n_m!}$$
$$= P(n; n_1, n_2, n_3, \ldots, n_m)$$

Notice that the combination $C(n, k) = C(n; k, n - k)$ is just the special case where we partition the set with n elements into a set with k elements (the ones we pick) and a set with $n - k$ elements (the ones left). It is also important to note that we are partitioning into a sequence of subsets having a designated number of elements. We are *not* counting all partitions of the set into subsets of the given sizes. One way to see this is to observe that $C(n, k) = C(n; k, n - k)$ is the number of ways to select k objects out of n objects. We do not include the number of ways of selecting $n - k$ objects, and hence leaving k objects, which is also a partitioning of a set S with n objects into subsets with k and $n - k$ objects. There are also $C(n; k, n - k)$ ways to do this, so there are $2C(n; k, n - k)$ ways of partitioning S into sets containing k objects and $n - k$ objects. Another way to think of the theorem is that of describing the placing of distinguishable balls into distinguishable boxes. ∎

EXAMPLE 8.67

How many ways can four hands, containing five cards each, be dealt from a deck of 52 cards? We are actually partitioning the cards into five sets: the four hands with five cards each and the 32 remaining cards. Therefore, the number of hands is equal to

$$C(52; 5, 5, 5, 5, 32) = \frac{52!}{5! 5! 5! 5! 32!}$$

Exercises

1. How many distinct arrangements can be formed using the letters in *seceded*?

2. How many distinct arrangements can be formed using the letters in *unsuccessful*?

3. How many distinct arrangements can be formed using the letters in *Sussex*?

4. How many distinct arrangements can be formed using the letters in *Tallahassee*?

5. Assume that 26 books, consisting of 8 identical mathematics texts, 6 identical computer science texts, 9 identical physics books, and 3 identical chemistry texts are to be placed on a shelf. How many distinct ways can they be placed on the shelf?

6. An urn contains 6 blue balls, 5 red balls, and 14 yellow balls. If all of the balls of each color are indistinguishable, how many distinct ways can all of the balls be drawn, one at a time, from the urn?

7. In a multiple choice test with 35 questions, in which each question has five choices: *a*, *b*, *c*, *d*, and *e*, how many distinct answer sheets are possible if there are an equal number of answers having each choice?

8. Professor Krank has 20 students in his class. He has decided to give two *A*'s, three *B*'s, ten *C*'s, three *D*'s,

and two F's in a manner known only to himself. In how many ways can he distribute the grades?

9. A baseball team has 24 players. They are staying in six hotel rooms with four players in each room. How many different ways can the players be assigned to the rooms?

10. What is the coefficient of $x^2 y^3 z^4$ in the expansion of $(x + y + z)^9$?

11. What is the coefficient of $x^3 y^6 z^3$ in the expansion of $(x + 2y + 3z)^{12}$?

12. What is the coefficient of $w^3 x^2 y^5 z^7$ in the expansion of $(w + x + y + z)^{17}$?

13. What is the coefficient of $w^4 x^5 y^6 z^3$ in the expansion of $(2w + 4x + y + 5z)^{18}$?

14. What is the coefficient of $x^3 y^6 z^{12}$ in the expansion of $(x + 2y^2 + 4z^3)^{10}$?

15. What is the coefficient of $w^{10} x^{12} y^4 z^3$ in the expansion of

$$(4w^5 + 2x^3 + y^2 + 5z)^{11}?$$

8.7 Permutations and Combinations with Repetition

The idea of a permutation with repetition may seem like an oxymoron, since we have associated permutations with rearrangements of elements that don't seem to allow for repetition. We must indeed reconsider what we mean by a permutation, if we allow repetition. The principal property of a permutation is order. The string cab is different than the string abc. If we return to our problem about license plates consisting of three letters followed by three single-digit numbers and allow no repetition, then we have $26 \times 25 \times 24 \times 10 \times 9 \times 8$ distinct license plates. There are $P(25, 3)$ ways of selecting the letters and $P(10, 3)$ ways of selecting the numbers. Thus, there are $P(25, 3) \times P(10, 3)$ distinct license plates. This is our usual understanding of permutation. Notice that if we allow repetition, we still retain our idea of order. The license tag $FPF199$ is different from the license tag $PFF919$. When we speak of permutations with repetition, we mean that this order is retained. It may be recalled that when we allow repetition of letters and numbers for license plates, then there are 26 possible choices for every choice of letter and 10 choices for each number. Thus, there are $26^3 \times 10^3$ possible license plates if repetition is allowed. In general, if we have k ordered spaces and for each space we can select any of n objects, then there are n^k ways to select the objects. Thus, the number of permutations with repetition when k objects are selected out of n objects is n^k. Perhaps the idea of repetition as used here needs to be clarified. One way to think of repetition is that the object is replaced and can be reused. Another way to consider this type of repetition is that there are enough objects of each type that we do not run out. For example, for the license plate example, we only need three copies of each letter and each number. With this understanding of repetition, we state formally the following theorem.

THEOREM 8.68 Given a set S with n distinguishable objects, the number of distinct permutations formed by selecting k elements with replacement is equal to n^k. ■

EXAMPLE 8.69 In a lottery, a container contains 500 numbers. From this container, a number is selected and recorded. The ball is returned to the container, a second ball is selected, the number is recorded, and the ball is returned to the container. This is continued until five numbers are selected in sequence. How many distinct sequences of numbers may be selected? There are 500 choices for each of the five numbers. Hence, the number of distinct sequences is 500^5. ■

EXAMPLE 8.70 How many possible social security numbers are available? Since numbers must be selected from the nonnegative integers less than ten, we are selecting nine numbers with each selected out of ten numbers with replacement, so there are 10^9 possible social security numbers. ∎

Suppose that a county board of supervisors consists of eight people. When voting on an issue they vote yes, vote no, or abstain. On a given issue how many possible outcomes are there? If we were interested in who voted a particular way, then we would have a permutation problem and there would be three choices for each person, giving 3^8 possible outcomes. Suppose, however, that we are only interested in the outcomes and not who voted a particular way. Thus, four voting yes, three voting no, and one abstaining would be a possible outcome. So would two voting yes, four voting no, and two abstaining. For convenience, we shall list the yes votes first, then the no votes, and finally the abstentions. Thus, we can represent a vote as $YYNNAAAA$ representing two votes yes, two votes no, and four abstentions. We can further partition the votes in the form $YY|NN|AAAA$. Since the order in which the votes are is understood, we could just write $xx|xx|xxxx$ to represent $YY|NN|AAAA$. Thus, $xx||xxxxxx$ would represent two yes votes and six abstentions, and $xxxxxxxx||$ would represent all eight people voting yes. Thus, there is a one-to-one correspondence between possible voting outcomes and the different ways of arranging eight x's and two $|$'s. But this is the number of ways of selecting two places out of ten to place $|$, or equivalently picking eight places out of ten to place an x. Thus, there are

$$C(10, 2) = C(10, 8) = \frac{10!}{8!2!}$$

ways to place the x's and $|$'s. Hence, there are

$$C(10, 2) = C(10, 8) = \frac{10!}{8!2!}$$

possible voting outcomes.

Suppose that we are selecting n objects from k types of objects with unlimited repetition. If we let a_i be the object of type i, then $n_1 a_1 + n_2 a_2 + n_3 a_3 + \cdots + n_k a_k$, where $n_i \geq 0$ for all i and $n_1 + n_2 + n_3 + \cdots + n_k = n$ represents the selection of n_i objects of type i. As before we could write this as

$$a_1 a_1 a_1 \cdots a_1 | a_2 a_2 a_2 \cdots a_2 | \cdots | a_k a_k a_k \cdots a_k$$

where each a_i is repeated n_1 times. Since the spaces where each type is placed is understood, we may write the selection in the form

$$xxx \cdots x | xxx \cdots x | \cdots | xxx \cdots x$$

Note that there is one less partition marker $|$ than number of types. Thus, there are n objects plus $k-1$ partition markers, making $n+k-1$ places to place either an x or a $|$. Each rearrangement of the x's and $|$'s gives a distinct way of selecting n objects from k types of objects with unlimited repetition. Since there are $C(n+k-1, n) = C(n+k-1, k-1)$ ways of selecting places for the x's, or equivalently selecting places for the $|$'s, there are

$$C(n+k-1, n) = C(n+k-1, k-1)$$

distinct ways of selecting n objects from k types of objects with unlimited repetition. We call these selections **combinations of k objects taken n at a time with repetition**. Thus, we have the following theorem.

THEOREM 8.71 The number of distinct combinations of k objects taken n at a time is equal to

$$C(n+k-1, n) = C(n+k-1, k-1) = \frac{(n+k-1)!}{n!(k-1)!}$$
∎

EXAMPLE 8.72 If a bakery sells 10 different types of doughnuts, how many different ways are there to select a dozen doughnuts? Since we are selecting 12 doughnuts from 10 different types with repetition, there are

$$\frac{(10+12-1)!}{12!(10-1)!!} = \frac{21!}{12!9!}$$

different ways to select a dozen doughnuts. ∎

EXAMPLE 8.73 How many solutions are there to the equation

$$n_1 + n_2 + n_3 + n_4 + n_5 = 25$$

such that each n_i is a nonnegative integer? This is equivalent to asking how many solutions there are to

$$n_1 a_1 + n_2 a_2 + n_3 a_3 + n_4 a_4 + n_5 a_5$$

where there are n_i objects of type a_i, and

$$n_1 + n_2 + n_3 + n_4 + n_5 = 25$$

Thus, there are

$$\frac{(25+5-1)!}{25!(5-1)!} = \frac{29!}{25!4!}$$

different solutions to the equation $n_1 + n_2 + n_3 + n_4 + n_5 = 25$. ∎

We already know that the number of distinct combinations of k objects taken n at a time with repetition is equal to

$$C(n+k-1, n) = C(n+k-1, k-1) = \frac{(n+k-1)!}{n!(k-1)!}$$

If at least one of each type of object must be selected, then there are $n - k$ selections to make from k different objects. Therefore, we have $C(n - k + k - 1, k - 1)$ or $C(n - 1, k - 1)$ ways to make the selection.

THEOREM 8.74 The number of distinct combinations of k objects taken n at a time with repetition when one of each object must be selected is equal to

$$C(n-1, n-k) = C(n-1, k-1) = \frac{(n-1)!}{(n-k)!(k-1)!}$$
∎

EXAMPLE 8.75 If a bakery sells 10 different types of doughnuts, how many different ways are there to select two dozen doughnuts if one of each kind must be selected? If there were no restrictions so that we were selecting two dozen doughnuts from 10 different types, there would be

$$C(24+10-1, 24) = C(33, 24)$$

different selections. However, since one of each kind must be selected, we are selecting only $24 - 10 = 14$ doughnuts from 10 different types giving

$$C(14+10-1, 14) = C(33, 14)$$

different selections. ∎

EXAMPLE 8.76

How many different integral solutions to the equation

$$n_1 + n_2 + n_3 + n_4 + n_5 = 25$$

are there if $n_1 \geq 1$, $n_2 \geq 1$, $n_3 \geq 2$, $n_4 \geq 2$, and $n_5 \geq 2$? If we again think of this as having the form

$$n_1 a_1 + n_2 a_2 + n_3 a_3 + n_4 a_4 + n_5 a_5$$

where

$$n_1 + n_2 + n_3 + n_4 + n_5 = 25$$

then at least one a_1, one a_2, two a_3's, two a_4's, and two a_5's, must be selected. This leaves only $25 - 8 = 17$ selections to make out of five types giving

$$\frac{(17 + 5 - 1)!}{17! 4!} = \frac{21!}{17! 4!}$$

solutions to

$$n_1 + n_2 + n_3 + n_4 + n_5 = 25$$ ∎

Exercises

1. How many solutions are there to the equation $n_1 + n_2 + n_3 + n_4 + n_5 = 17$ such that each n_i is a nonnegative integer?

2. How many solutions are there to the equation $n_1 + n_2 + n_3 + n_4 = 23$ such that each n_i is a nonnegative integer?

3. How many solutions are there to the equation $n_1 + n_2 + n_3 + n_4 = 23$ such that $n_1 \geq 2$, $n_2 \geq 3$, $n_3 \geq 4$, $n_4 \geq 5$?

4. How many solutions are there to the equation $n_1 + n_2 + n_3 + n_4 = 21$ such that $n_1 \geq 2$, $n_2 \geq 3$, $n_3 \geq 4$, $n_4 \geq 5$?

5. How many different collections of 10 coins can be made with pennies, nickels, dimes, quarters, and half dollars?

6. Two dice are tossed eight times. How many possible outcomes are there?

7. An element is picked from a set containing eight elements. It is returned and another element is picked. It is replaced and a third element is picked. How many possible outcomes are there?

8. An ice cream store sells 21 different flavors of ice cream in quart containers. Fred wants to take home five quarts of ice cream. If more than one quart can be the same flavor, how many ways may he select the ice cream?

9. If there are 20 red, 20 green, and 20 blue balls in an urn, how many different ways can 10 balls be selected?

10. A florist has 10 different types of flowers in his store. In how many ways can he prepare a bouquet of 12 flowers?

A doughnut shop has 20 types of doughnuts. How many ways are there to choose

11. 4 doughnuts?

12. 6 doughnuts?

13. a dozen doughnuts?

14. Two dozen doughnuts with at least one of each kind?

15. A dozen doughnuts with at most one of each kind?

16. How many positive integers are there less than 10,000 such that the sum of their digits is 12?

17. How many bridge hands (containing 13 cards) are there containing three spades, four hearts, five clubs, and a diamond?

18. If one does not distinguish between the cards of each suit, how many types of bridge hands may be dealt?

19. A man buys 12 different toys to give to his four children. In how many ways may he distribute the toys? How many ways if the toys are distributed evenly?

20. Given the algorithm
    For $i = 1$ to 10
        For $j = 1$ to i
            For $k = 1$ to j
                $F(i,j,k)$
            Endfor
        Endfor
    Endfor

    How many times will $F(i,j,k)$ be executed?

8.8 Pigeonhole Principle

The pigeonhole principle, also known as the Dirichlet drawer principle, is an incredibly simple and yet deceptively powerful idea. It is particularly useful in number theory. This section consists of two fairly simple theorems and some examples that are not so simple. We begin with the best known form of the pigeonhole principle.

THEOREM 8.77 If $n+1$ pigeons are placed in n pigeonholes, then at least one hole contains more than one pigeon.

Proof If each hole contained at most one pigeon, then there cannot be more pigeons than holes. Therefore, there would be a total of at most n pigeons in the holes, which contradicts the hypothesis. ∎

From the theorem, we can conclude that if $m > n$ and there are m pigeons to place in n holes, then one hole contains more than one pigeon.

EXAMPLE 8.78 In a list of $m+1$ positive integers, at least two have the same remainder when divided by m. Let $a_1, a_2, a_3, \ldots, a_m, a_{m+1}$ be a list of positive integers and assume that for $1 \leq i \leq m+1$, $a_i = q_i m + r_i$ where $0 \leq r_i < m$. Therefore, $r_1, r_2, r_3, \ldots, r_m, r_{m+1}$ is a list of $m+1$ nonnegative integers less than m. But there are only m distinct nonnegative integers less than m. Therefore, two of the r_i must be equal. ∎

EXAMPLE 8.79 If $|A| > |B|$, then there cannot exist a one-to-one function $f : A \to B$. Let $|B| = n$. There are n sets $f^{-1}(b) = \{x : f(x) = b\}$ for $b \in B$ and $A = \bigcup_{b \in B} f^{-1}(b)$. But A contains more than n elements. Therefore, at least two elements, say a and a', must be in the same set. But then $f(a) = f(a')$ and f is not one-to-one. ∎

EXAMPLE 8.80 If five points are placed in a rectangle which is 6 inches by 8 inches, then there are two points that are not more than 5 inches apart. Divide the rectangle into four rectangles that are each 3 inches by 4 inches. Since five points must go in or on the edge of four rectangles, two of the points must go in or on the edge of the same 3×4 rectangle. But any two points in or on the edge of the same rectangle are at most 5 inches apart. ∎

EXAMPLE 8.81 In a sequence $a_1, a_2, a_3, \ldots, a_{10}$ of 10 integers, there is a consecutive sequence $a_m, a_{m+1}, a_{m+2}, \ldots, a_{m+n}$ whose sum is divisible by 10. Consider the numbers $s_1 = a_1$, $s_2 = a_1 + a_2$, $s_3 = a_1 + a_2 + a_3$, $s_4 = a_1 + a_2 + a_3 + a_4$, $\ldots$, $s_{10} = a_1 + a_2 + a_3 + a_4 + \cdots + a_{10}$. If one of these sums is divisible by 10, we are finished. If not, then each $s_i = 10q_i + r_i$ where $1 \leq r_i \leq 9$. There are 10 r_i, but only nine values that they may equal. Therefore, there are two r_i that are equal, say $r_j = r_k$ for some $j < k$. Then

$$s_k - s_j = 10q_k + r_k - (10q_j + r_j)$$
$$= 10q_k - 10q_j$$
$$= 10(q_k - q_j)$$

is a multiple of 10. But this is the sum of $a_{j+1} + a_{j+2} + a_{j+3} + \cdots + a_k$, which is the desired sequence. ∎

EXAMPLE 8.82 (Erdös and Szekeres) Let S be a subset of $\{1, 2, 3, \ldots, 2n\}$ containing $n+1$ elements for some positive integer n. There exist two elements in S such that one divides the other. Any positive integer can be expressed in the form $2^i q_i$ where q_i is a positive odd integer. Express each element of S in this form. The set $\{q_1, q_2, \ldots, q_n, q_{n+1}, \}$

contains $n + 1$ positive odd integers, but there are only n odd numbers less than or equal to $2n$. Therefore, $q_i = q_j$ for some $i \neq j$. Thus, there are two distinct integers in S of the form $a = 2^r q_i$ and $b = 2^s q_i$, respectively. If $r < s$, then a divides b, and if $r > s$, then b divides a. ∎

We now state the **strong** forms of the pigeonhole principle.

THEOREM 8.83

(a) If $m > n$ pigeons are placed in n pigeonholes, then some pigeonhole contains at least $\lceil \frac{m}{n} \rceil$ pigeons.

(b) If $m = m_1 + m_2 + m_3 + \cdots + m_n - n + 1$ pigeons are placed in n pigeonholes, where m_i is a positive integer for each $1 \leq i \leq n$, then for some i, the ith pigeonhole contains at least m_i pigeons.

Proof **(a)** If every pigeonhole contains less than $\lceil \frac{m}{n} \rceil$ pigeons, then, since we are not cutting up the pigeons, every pigeonhole contains less than $\frac{m}{n}$ pigeons. Therefore, there are less than $n \times \frac{m}{n} = m$ pigeons, which contradicts the hypothesis.

(b) If the ith pigeonhole contains less than m_i pigeons, then the number of pigeons in the ith pigeonhole is less than or equal to $m_i - 1$. Summing, we have at most

$$(m_1 - 1) + (m_2 - 1) + (m_3 - 1) + \cdots + (m_n - 1)$$
$$= m_1 + m_2 + m_3 + \cdots + m_n - n$$

pigeons, which contradicts the hypothesis. ∎

EXAMPLE 8.84

Suppose that an urn contains 10 red, 10 blue, and 10 green balls. How large a sample must be taken from the urn to be sure of having at least 4 red, or 5 blue, or 6 green balls in the sample? We know that 12 balls is not enough, since we could select 3 red, 4 blue, and 5 green balls. If we identify balls being the same color with pigeons being in the same pigeonhole, then using part (b) of the strong form of the pigeonhole principle, we have $n = 3$ pigeonholes, $m_1 = 4$, $m_2 = 5$, and $m_3 = 5$. Therefore,

$$m = m_1 + m_2 + m_3 + \cdots + m_n - n + 1$$
$$= 4 + 5 + 6 - 3 + 1 = 13$$

is a sufficient number of balls for the sample. ∎

EXAMPLE 8.85

(**Erdös and Szekeres**) A finite sequence $a_1, a_2, a_3, \ldots, a_k$ of integers is increasing if each integer in the sequence is larger that the previous integer in the sequence. More specifically, if $i > j$, then $a_i > a_j$. A finite sequence $a_1, a_2, a_3, \ldots, a_k$ of integers is decreasing if each integer in the sequence is smaller than the previous integer in the sequence. More specifically, if $i > j$, then $a_i < a_j$. Any sequence of $n^2 + 1$ distinct integers has either a decreasing subsequence containing at least $n + 1$ terms or an increasing subsequence containing at least $n + 1$ terms. To show this, let s_i be the number of integers in the longest increasing subsequence of $a_1, a_2, a_3, \ldots, a_{n^2+1}$ beginning with a_i. If $s_i \geq n + 1$ for some i, then we are finished. If not, then for each i, $1 \leq s_i \leq n$. Therefore, there are n values for all $n^2 + 1$ of the s_i where $1 \leq i \leq n^2 + 1$. By part (a) of the strong form of the pigeonhole principle,

$$\left\lceil \frac{n^2 + 1}{n} \right\rceil > n$$

of the s_i are equal. Therefore, at least $n + 1$ of the s_i are equal. Say they are all equal to s. Consider the subsequence of length $n + 1$ consisting of the a_i

such that the s_i are equal to s. This sequence is a decreasing sequence, for if $a_i > a_j$ for $i > j$, then adding a_i onto the front of the sequence of length s, beginning with a_j, would produce an increasing sequence of length $s + 1$, beginning with a_i. But this contradicts the fact that the longest increasing sequence beginning with a_i has length s. Thus, we have a decreasing sequence of length $n + 1$. ∎

EXAMPLE 8.86

Suppose there are six people in a room and that any two of them are either friends or enemies. There are three people who are mutual friends or three people who are mutual enemies. Select one of the people, call him A, and place him in the center of the room. Place all of the enemies of A against the south wall and all of the friends of A against the north wall. By part (a) of the strong pigeonhole principle, there are at least three people against the south wall or at least three people against the north wall. If there are three or more people against the south wall, then there are three people who are mutual friends, in which case we are finished, or two people who are enemies. In that case, these two people, together with A, are three people who are mutual enemies. Similarly, if there are three or more people against the north wall, then there are three people who are mutual enemies, in which case we are finished, or two people who are friends. In that case, these two people, together with A, are three people who are mutual friends. ∎

In the previous example, we have a specific and reasonably easy example from an area of combinatorics called **Ramsey theory**. This is fortunate because there are extremely few easy examples to be found in Ramsey theory. In fact all but a few cases are unsolved. For simplicity we shall explain Ramsey theory in terms of complete graphs with colored edges. Suppose in the previous example we think of the people as vertices of a graph. We create a graph as follows: If two people are friends, there is a red edge between their respective vertices. If two people are enemies, there is a blue edge between their respective vertices. Since any two people are either friends or enemies, there is an edge between any two vertices. Hence, the graph is K_6, the complete graph with six vertices.

The preceding example, thus, states that if all of the edges of K_6 are colored either red or blue, then there is either a red triangle or a blue triangle contained as a subgraph of K_6, where a triangle is simply the graph K_3. The graph K_5 does not have the property that if its edges are colored either red or blue, then it contains a red or blue subgraph K_3. This is shown in by the graph in Figure 8.14, where the dotted lines represent red edges and the blue lines represent blue edges.

We now state the definition of the Ramsey property with two colors in terms of graphs. There is a generalization of the Ramsey property for an arbitrary number of colors.

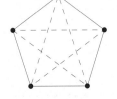

Figure 8.14

DEFINITION 8.87

A complete graph K_n has the **(p,q) Ramsey property** if when it is painted with two colors, say red and blue, it contains either a subgraph K_p with red edges or a subgraph K_q with blue edges.

DEFINITION 8.88

The **Ramsey number**, $R(p,q)$ is the smallest number n so that K_n has the (p, q) Ramsey property.

It is rather obvious that $R(p, q) = R(q, p)$, since we can just change red to blue and blue to red. Also if $m > R(p, q) = n$, then, since K_n is a subgraph of K_m, the graph K_m also has the Ramsey property.

We shall prove only two theorems about Ramsey numbers. These theorems are interesting in themselves, but more importantly, they show the existence of Ramsey numbers.

THEOREM 8.89   For all $p \geq 2$, $R(p, 2) = R(2, p) = p$.

Proof   Consider the graph K_p. If one edge is colored red (or blue), then there is a red (or blue) graph K_2. If not, then all the edges are colored blue (red) and we have a blue (red) graph K_p. ∎

THEOREM 8.90   (Erdös and Szekeres)   For all $p, q \geq 3$,
$$R(p, q) \leq R(p-1, q) + R(p, q-1)$$

Proof   Assume that K is a complete graph with $R(p-1, q) + R(p, q-1)$ vertices whose edges are painted either red or blue. Let v be a vertex selected from K. Let N be the complete subgraph of K whose vertices V_N are connected to v by a red edge. Let S be the complete subgraph of K whose vertices V_S are connected to v by a blue edge. Since $V_N \cup V_S$ contains
$$R(p-1, q) + R(p, q-1) - 1 = R(p-1, q) + R(p, q-1) - 2 + 1$$
vertices divided into $n = 2$ sets, either V_N has $R(p-1, q)$ vertices or V_S has $R(p, q-1)$ vertices. Therefore, either N is a complete graph with $R(p-1, q)$ vertices or S is a complete graph with $R(p, q-1)$ vertices. If N is a complete graph with $R(p-1, q)$ vertices, then it contains either a subgraph K_{p-1} with all red edges or a subgraph K_q with all blue edges. If it contains K_q with all blue edges, we are finished. If it contains K_{p-1} with all red edges, then if we add v, and the red edges from v to all the vertices of the subgraph K_{p-1}. We have a complete graph K_p with all red edges, and we are finished.

Similarly, if S is a complete graph with $R(p, q-1)$ vertices, then it contains either a subgraph K_p with all red edges or a subgraph K_{q-1} with all blue edges. If it contains K_p with all red edges, we are finished. If it contains K_{q-1} with all blue edges, then if we add v and the blue edges from v to all the vertices of the subgraph K_{q-1}. We have a complete graph K_q with all blue edges, and we are finished. ∎

As a result of these two theorems, using induction or the well-ordering property of the positive integers and techniques from the proof of the second theorem, one can prove the following theorem.

THEOREM 8.91   (Ramsey's Theorem)   If p and q are both integers greater than 1, then the Ramsey number $R(p, q)$ exists. ∎

Exercises

1. From a group of 40 women and their husbands, how many people must be chosen to be sure of choosing a married couple?

2. How many numbers must be chosen from the first 210 positive integers to be sure of choosing an even number?

3. Show that in a party of 30 people, there are 2 who that have the same number of mutual friends at the party.

4. In a group of people who come from five different countries, how large must the group be to guarantee that three come from the same country?

5. If a kitchen contains 7 cans of noodle soup, 10 cans of vegetable soup, 14 cans of clam chowder, 18 cans of onion soup, and 20 cans of tomato soup, how many cans must be selected to ensure having 12 cans of the same type of soup?

6. How large a group is needed to ensure that at least two have birthdays in the same month?

7. How large a group is needed to ensure that at least three have birthdays in the same month?

8. How large a group is needed to ensure that at least two have birthdays on the same day of the week?

9. In any list of n positive integers, prove that there always exists a sequence of consecutive numbers whose sum is divisible by n.

10. How many words must be chosen to be sure that at least two begin with the same letter?

11. How many words with two or more letters must be chosen to be sure that at least two begin with the same letter and end with the same letter?

12. A questionnaire is sent to 13 freshmen, 5 sophomores, 15 juniors, and 20 seniors. How many questionnaires must be received to ensure getting 9 from the same class?

13. How many numbers must be chosen from $\{1, 2, 3, 4, \ldots, 19, 20\}$ to ensure that two of these numbers must add up to 21?

14. Show that any subset containing at least n distinct integers between 2 and $2n$ always has two numbers that are relatively prime.

15. If a restaurant has 14 tables and 170 chairs, and n is the largest number of chairs at a table, what is the smallest possible value of n?

16. Prove that the decimal expansion of a rational number is a repeating decimal.

17. Prove that if $n+1$ distinct positive integers are all less than or equal to $2n$, then two of these integers differ by one.

18. Show that if 10 positive numbers have sum 151, then the sum of three of these numbers must be at least 46.

19. Show that if m balls are placed in n boxes and
$$m < \frac{n(n-1)}{2}$$
then at least two boxes must receive the same number of balls.

20. Show that if the sum of $n+1$ positive integers is $2n$, then for any positive integer k less than $2n$, there is a subset of these $n+1$ integers whose sum is k.

8.9 Probability Revisited

In our previous discussion of probability, we assumed that all outcomes of an experiment were equally likely. We now drop this assumption and give a more general development of probability. To do this, our approach must be more axiomatic, and some of our theorems must become axioms and definitions.

DEFINITION 8.92   A probability function on a sample space S is a function P from the subsets of S to the real line such that
(1) $P(A) \geq 0$ for all $A \subseteq S$.
(2) $P(S) = 1$.
(3) If $A, B \subseteq S$ have no outcome in common, then $P(A \cup B) = P(A) + P(B)$.

The proofs of the following theorems are the same as given in Section 8.2. The only difference is that the three requirements of the preceding probability function were proved in Theorem 8.55, using the assumption that all outcomes are equally likely. We must be careful using some of our counting techniques, such as combinations and permutations, so we do not tacitly assume that outcomes are equally likely.

THEOREM 8.93   If S is the set of all outcomes, and A' is the complement of A using S as the universe, then

(a) $P(A') = 1 - P(A)$.
(b) $P(\emptyset) = 0$.
(c) For any event A, $P(A) \geq 0$.

(d) If $A \subseteq B$, then $P(A) \leq P(B)$.

(e) For any event A, $0 \leq P(A) \leq 1$. ∎

The reader is asked to prove the following theorem using induction.

THEOREM 8.94 Let $A_1, A_2, A_3, \ldots, A_m$ be pairwise disjoint events. Then

$$P(A_1 \cup A_2 \cup A_3 \cup \cdots \cup A_m) = P(A_1) + P(A_2) + P(A_3) + \cdots + P(A_m)$$ ∎

The proof of the following theorem is also the same as for the corresponding theorem in Theorem 8.58.

THEOREM 8.95 For events $A, B \subseteq S$, $P(A \cup B) = P(A) + P(B) - P(A \cap B)$. ∎

We have just seen that there is a theorem for probability that is very similar to the addition counting principle. It seems reasonable to ask if there is a theorem for probability that is similar to the multiplication counting theorem. The answer is yes, but we must look closely at what the multiplication counting theorem says. It states that, "Given a sequence of events $E_1, E_2, E_3, \ldots, E_m$ such that E_1 occurs in n_1 ways and if $E_1, E_2, E_3, \ldots, E_{k-1}$ have occurred, then E_k can occur in n_k ways. There are $n_1 \times n_2 \times n_3 \times \cdots \times n_k$ ways in which the entire sequence of events can occur." The number of ways E_k can occur depends on the fact that $E_1, E_2, E_3, \ldots, E_{k-1}$ have occurred. Thus, the number of ways E_k can occur is not isolated but may depend on the previous events.

In a similar way we define conditional probability. Suppose we have two events A and B. We want to determine the probability that B will occur given that A has occurred or is certain to occur. We abbreviate this by saying "the probability of B given A." Intuitively, assuming A has occurred or is certain to occur restricts our sample space to A, as shown in Figure 8.15, so that B becomes restricted to the outcomes in $A \cap B$, and instead of having $P(B) = \dfrac{P(B)}{P(S)}$, we have $\dfrac{P(A \cap B)}{P(A)}$.

Figure 8.15

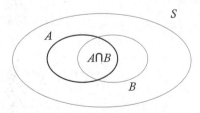

DEFINITION 8.96 The probability of B given A, denoted by $P(B|A)$, is equal to $\dfrac{P(A \cap B)}{P(A)}$.

This can also be stated in the form $P(A \cap B) = P(A)P(B|A)$.

EXAMPLE 8.97 Assume that a card is drawn from a deck of cards, then replaced, and a second card is drawn. What is the probability that they are both spades? What is the probability that at least one of them is a spade? Let A be the set of outcomes in which the first card is a spade and B be the set of outcomes in which the second card is a spade. To answer the first question, we need to find $P(A \cap B)$. We note that

$$P(A) = \frac{13}{52} = \frac{1}{4}$$

since 13 of the 52 cards are spades. Since the card is replaced, we also have

$$P(B|A) = \frac{13}{52} = \frac{1}{4}$$

Therefore, the probability that they are both spades is

$$P(A \cap B) = P(A)P(B|A)$$
$$= \frac{1}{4} \cdot \frac{1}{4}$$
$$= \frac{1}{16}$$

To answer the second question, we need to find $P(A \cup B)$, but

$$P(A) = P(B) = \frac{13}{52} = \frac{1}{4}$$

so that the probability that at least one of them is a spade is given by

$$P(A \cup B) = P(A) + P(B) - P(A \cap B)$$
$$= \frac{1}{4} + \frac{1}{4} - \frac{1}{16} = \frac{7}{16}$$ ∎

EXAMPLE 8.98 Assume that a card is drawn from a deck of cards *but not replaced*, and a second card is drawn. What is the probability that they are both spades? What is the probability that at least one of them is a spade? Let A be the set of outcomes in which the first card is a spade and B be the set of outcomes in which the second card is a spade. To answer the first question, we need to find $P(A \cap B)$. We note that

$$P(A) = \frac{13}{52} = \frac{1}{4}$$

since 13 of the 52 cards are spaces. To determine $P(B|A)$, we know that a spade was drawn and not replaced. Therefore, there are 12 spades that can be chosen from 51 cards and

$$P(B|A) = \frac{12}{51}$$

Therefore, the probability that they are both spades is

$$P(A \cap B) = P(A)P(B|A)$$
$$= \frac{1}{4} \cdot \frac{12}{51}$$
$$= \frac{1}{17}$$

To answer the second question, we need to find $P(A \cup B)$, and we know

$$P(A) = \frac{1}{4}$$

The problem now is to find $P(B)$. If the event B occurred, then either a spade was drawn on the first draw and a spade was drawn on the second draw or a spade was not drawn of the first draw and a spade was drawn on the second draw. Therefore,

$$P(B) = P(A \cap B) + P(A' \cap B)$$

By the previous theorem,
$$P(A' \cap B) = P(A')P(B|A')$$

Since there are 39 cards that are not spades,
$$P(A') = \frac{39}{52} = \frac{3}{4}$$

Once the first card has been selected there are still 13 spades and 51 cards to choose from so that
$$P(B|A') = \frac{13}{51}$$

Therefore,
$$P(A' \cap B) = \frac{3}{4} \cdot \frac{13}{51} = \frac{13}{68}$$

and
$$P(B) = \frac{1}{17} + \frac{13}{68} = \frac{17}{68} = \frac{1}{4}$$

We can now see that $P(B)$, when A is not known, is the same as if B had occurred without A having happened. We can now determine that
$$P(A \cup B) = P(A) + P(B) - P(A \cap B)$$
$$= \frac{1}{4} + \frac{1}{4} - \frac{1}{17} = \frac{15}{34}$$ ∎

EXAMPLE 8.99 Suppose that 6 cards are drawn in succession from a standard deck of 52 cards. What is the probability that the sixth card is the third spade drawn? Let A be the event that exactly two spades are drawn in the first five draws and B be the event that a spade is drawn on the sixth draw. We are then trying to find $P(A \cap B)$. Again we want to use the fact that
$$P(A \cap B) = P(A)P(B|A)$$

The number of ways A can occur is equal to the number of ways that 2 spades are drawn out of 13 possible spades and 3 cards are drawn from the 39 cards that are not spades. Thus,
$$|A| = \binom{13}{2}\binom{39}{3}$$

There are $\binom{52}{5}$ ways to choose 5 cards out of 52. Therefore,
$$P(A) = \frac{\binom{13}{2}\binom{39}{3}}{\binom{52}{5}}$$

To determine $P(B|A)$, since two spades have been selected, there are 11 ways to select a spade. There are 47 cards left to select the sixth card, so
$$P(B|A) = \frac{11}{47}$$

and

$$P(A \cap B) = \frac{\binom{13}{2}\binom{39}{3}}{\binom{52}{5}} \cdot \frac{11}{47} \approx .064$$

We saw earlier that, if a card is drawn and replaced and a second card drawn, then the probability of the second card being a spade was not affected by the selection of the first card. When the outcome of a second event is not affected by the outcome of the first event, then the second event is said to be independent of the first event.

DEFINITION 8.100 If $P(B) = P(B|A)$, then the event B is **independent** of the event A.

Since $P(A \cap B) = P(A)P(B|A)$, then if $P(B|A) = P(B)$, we get the following theorem.

THEOREM 8.101 If the event B is independent of the event A, then $P(A \cap B) = P(A)P(B)$ and conversely. ∎

EXAMPLE 8.102 Suppose a coin is tossed ten times. Let A be the event in which the first coin comes up heads the first nine times and B be the event in which the tenth coin comes up heads. Show that event B is independent of the event A. $P(A \cap B)$ is the probability that a coin is tossed ten times and comes up heads ten times. There are $2^{10} = 1024$ possible equally likely outcomes and only one of them results in ten coins coming up heads. Therefore,

$$P(A \cap B) = \frac{1}{1024}$$

There are two ways in which A can occur. All coins can come up heads or the first nine can come up heads and the last one come up tails. Therefore,

$$P(A) = \frac{2}{1024}$$

and

$$P(B|A) = \frac{\frac{1}{1024}}{\frac{2}{1024}} = \frac{1}{2} = P(B)$$

and event B is independent of event A. ∎

If we were to toss a coin n times, we could assume that the outcome of any given toss would be independent of any other toss. If we were making the third toss, its outcome would in no way be affected by the outcome of the first two tosses. We refer to such an event as a **repetition of independent trials**. Assume that we have an experiment that consists of a repetition of n independent trials and that for each trial there are two possible outcomes. We consider one outcome to be a success and the other to be a failure. Let p be the probability of a success and $q = 1 - p$ be the probability of a failure. We now want to find the probability of k successes in n independent trials. To return to our coin example, if we are tossing a fair coin, and considered a head a success, then $p = q = \frac{1}{2}$. If $n = 5$ and $k = 3$, then we would be trying to find the probability of getting exactly three heads in five tosses of the coin.

Given an experiment of a repetition of n independent trials, if we want to find the probability of k successes, first assume that they occur in some fixed order. For example, assume the k successes occur first and the $n - k$ failures occur second. Since the trials are independent, the probability of k successes followed by $n - k$ failures is $p^k q^{n-k}$. In fact each occurrence of k successes in n independent trials will have probability $p^k q^{n-k}$. Since there are $\binom{n}{k}$ ways of having k successes in n independent trials, the probability of k successes in n trials is $\binom{n}{k} p^k q^{n-k}$. Thus, we have the following theorem.

THEOREM 8.103 In an experiment with n independent trials, called Bernoulli trials, each having two outcomes, success with probability p and failure with probability $q = 1 - p$, the probability of exactly k successes is $\binom{n}{k} p^k q^{n-k}$. ∎

Such an experiment is said to have a **binomial distribution** since

$$\sum_{k=0}^{n} \binom{n}{k} p^k q^{n-k} = (p + q)^n$$

EXAMPLE 8.104 Given a fair coin, what is the probability of getting exactly three heads in five tosses of the coin? Since $p = q = \frac{1}{2}$, the probability of getting exactly three heads in five tosses is

$$\binom{5}{3} \left(\frac{1}{2}\right)^3 \left(\frac{1}{2}\right)^2 = \frac{10}{32}$$

∎

EXAMPLE 8.105 Suppose a fair die is tossed ten times. What is the probability of getting a six exactly three times? The probability p of getting a six is $\frac{1}{6}$. Hence, the probability q of not getting a six is $1 - \frac{1}{6} = \frac{5}{6}$. Therefore, the probability of getting a six exactly three times is

$$\binom{10}{3} \left(\frac{1}{6}\right)^3 \left(\frac{5}{6}\right)^7$$

∎

Suppose that for a sample space, there is a real number associated with each outcome. For example, if two dice are tossed, the number might be the sum of the two dice. If a coin is tossed ten times, the number might be the number of heads that come up out of the ten tosses. We call such a number a random variable and define it as follows.

DEFINITION 8.106 A **random variable** is a function from a sample space to the real numbers.

It is called a random variable because it is neither.

Given a random variable on a sample space, we can then consider the probability that a random variable equals a given value. For example, if we toss a coin seven times, let $R(x)$ be the number of heads that come up. Then the probability that $R(x) = 5$ is equal to

$$\binom{7}{5} \left(\frac{1}{2}\right)^5 \left(\frac{1}{2}\right)^2$$

If we toss a die ten times, and $R(x)$ is the number of fives that come up, the probability that $R(x) = 4$ is

$$\binom{10}{4} \left(\frac{1}{6}\right)^4 \left(\frac{5}{6}\right)^6$$

Suppose that we are playing a game where we are to toss a die and receive the amount of dollars equal to the value of the die. Thus, if a 5 comes up, we receive $5. How much should we expect to pay to play the game? This amount is called the expected value. It is found by taking the sum of the products of values to be received and the probability of receiving that value. In this case, the probability is $\frac{1}{6}$ for each value so the expected value is equal to

$$1 \cdot \frac{1}{6} + 2 \cdot \frac{1}{6} + 3 \cdot \frac{1}{6} + 4 \cdot \frac{1}{6} + 5 \cdot \frac{1}{6} + 6 \cdot \frac{1}{6} = \frac{21}{6} = 3.5$$

so the fair price to play the game is $3.50.

In the preceding example, the random value $R(x)$ is the value received if an x comes up on the die. Thus, $R(1) = \$1$, $R(2) = \$2$, ..., $R(6) = \$6$. In general let R be a random variable defined on outcomes $x_1, x_2, x_3, \ldots, x_n$ that occur with respective probabilities $P_1, P_2, P_3, \ldots, P_n$. The **expected value** or **mathematical expectation** $E(R) = \sum_{i=1}^{n} R(x_i) P_i$. Let

$$a_1, a_2, a_3, \ldots, a_n$$

be the value that the random variable R assumes (i.e., the range of R). The set of all outcomes for which R has the fixed value a_i is denoted by $R = a_i$. The probability that X assumes this value $P(R = a_i)$ is denoted by p_i. A more convenient, but equivalent, form of the expected value $E(R)$ using this notation is $E(R) = \sum_{i=1}^{m} a_i p_i$.

In the following examples, we define two random variables for roulette.

EXAMPLE 8.107   Assume that a roulette wheel has integers 0 through 36 on equally spaced slots on the wheel. In this game, the player pays $1 and picks a number. If he wins, he receives $36. If he loses, he receives nothing. Let the random value R_1 be the amount that the player receives. Let e_1 be the event $R_1 = 36$ and e_2 be the event $R_1 = 0$. Then p_1, the probability the player wins, is $\frac{1}{37}$ and p_2, the probability the player loses, is $\frac{36}{37}$. Therefore,

$$E(R_1) = 36 \cdot \frac{1}{37} + 0 \cdot \frac{36}{37}$$
$$= \frac{36}{37}$$
$$\approx .973$$

Since the player pays $1, on the average, he loses about $0.027 per try.   ■

EXAMPLE 8.108   Now assume that we are playing the same game but let the random value R_2 be the amount that the player wins or loses. Let e_1 be the event $R_2 = 35$ and e_2 be the event $R_2 = -1$. Then p_1, the probability the player wins, is $\frac{1}{37}$ and p_2, the probability the player loses, is $\frac{36}{37}$. Therefore,

$$E(R_2) = 35 \cdot \frac{1}{37} + (-1) \cdot \frac{36}{37}$$
$$= -\frac{1}{37}$$
$$\approx -.027$$

Since R_2 represents the amount that the player wins or loses, on the average, the player loses $0.027 per try, which is consistent with the first example.   ■

8.9 Probability Revisited

EXAMPLE 8.109 Assume we have the same roulette wheel with the odd numbers between 1 and 36 painted black and the even numbers between 1 and 36 painted red. A player pays $1 and picks a color. If he wins, he receives $2. If he loses, he receives nothing. Let the random value R_3 be the amount that the player receives. Let e_1 be the event $R_3 = 2$ and e_2 be the event $R_3 = 0$. Then p_1, the probability the player wins, is $\frac{18}{37}$ and p_2, the probability the player loses, is $\frac{19}{37}$. Therefore,

$$E(R_3) = 2 \cdot \frac{18}{37} + 0 \cdot \frac{19}{37}$$
$$= \frac{36}{37}$$
$$\approx .973$$

Since the player pays $1, on the average, he loses about $0.027 per try. ∎

EXAMPLE 8.110 Now assume that we are playing the same game but let the random value R_4 be the amount that the player wins or loses. Let e_1 be the event $R_4 = 1$ and e_2 be the event $R_4 = -1$. Then p_1, the probability the player wins, is $\frac{18}{37}$ and p_2, the probability the player loses, is $\frac{19}{37}$. Therefore,

$$E(R_4) = 1 \cdot \frac{18}{37} + (-1) \cdot \frac{19}{37}$$
$$= -\frac{1}{37}$$
$$\approx -.027$$

Since R_4 represents the amount that the player wins or loses, on the average, the player loses $0.027 per try, which is consistent with the first example. ∎

We again return to the Bernoulli trials where we let the random variable give the number of successes in n trials. We find the expected value is simply the number of trials multiplied by the probability of a success.

THEOREM 8.111 In an experiment with n independent trials each having two outcomes, success with probability p and failure with probability $q = 1 - p$, let the random variable R represent the number of successes. The expected value $E(R) = np$.

Proof Since the probability of exactly k successes is equal to

$$\binom{n}{k} p^k q^{n-k} = \frac{n!}{k!(n-k)!} p^k q^{n-k}$$

and the first term in the following sum is 0, the expected value is

$$E(R) = \sum_{k=0}^{n} k \frac{n!}{k!(n-k)!} p^k q^{n-k}$$
$$= \sum_{k=1}^{n} k \frac{n!}{k!(n-k)!} p^k q^{n-k}$$
$$= np \sum_{k=1}^{n} \frac{(n-1)!}{(k-1)!(n-k)!} p^{k-1} q^{n-k}$$

Letting $m = k - 1$, we have

$$E(R) = np \sum_{m=0}^{n-1} \frac{(n-1)!}{(m)!(n-(m+1))!} p^m q^{n-(m+1)}$$

$$= np \sum_{m=0}^{n-1} \frac{(n-1)!}{(m)!((n-1)-m)!} p^m q^{(n-1)-m}$$

$$= np(p+q)^{n-1}$$

$$= np$$

since $p + q = 1$. ∎

EXAMPLE 8.112 A coin is tossed 20 times. The player receives \$1 for each head that comes up. A fair price to pay for playing the game is $np = 20 \cdot \frac{1}{2} = 10$ dollars. ∎

We now consider further properties of the expected value, which will be useful shortly.

THEOREM 8.113 Let R and S be random variables defined on outcomes x_1, $x_2, x_3, \ldots, x_n$, which occur with respective probabilities $P_1, P_2, P_3, \ldots, P_n$.

(a) Let c be a constant. Then $E(cR) = c \cdot E(R)$.

(b) Let c be a constant. Then $E(c) = c$.

(c) $E(R + S) = E(R) + E(S)$.

Proof For outcomes $x_1, x_2, x_3, \ldots, x_n$, which occur with respective probabilities, $P_1, P_2, P_3, \ldots, P_n$, $E(R) = \sum_{i=1}^{n} R(x_i) P_i$ and $E(S) = \sum_{i=1}^{n} S(x_i) P_i$. Therefore,

(a) $E(cR) = \sum_{i=1}^{n} cR(x_i) P_i = c \sum_{i=1}^{n} R(x_i) P_i = c \cdot E(R)$.

(b) $E(c) = \sum_{i=1}^{n} cP_i = c \sum_{i=1}^{n} P_i = c$, since $\sum_{i=1}^{n} P_i = 1$.

(c)

$$E(R + S) = \sum_{i=1}^{n} (R(x_i) + S(x_i)) P_i$$

$$= \sum_{i=1}^{n} R(x_i) P_i + S(x_i) P_i$$

$$= \sum_{i=1}^{n} R(x_i) P_i + \sum_{i=1}^{n} S(x_i) P_i$$

$$= E(R) + E(S)$$ ∎

In random sampling we may intuitively think of the mean as the center or most likely value that will occur. In some sense, it is the average. We may note that if each value of a random variable has equal probability, then the expected value would be "average" of the values of the random variable. The expected value is a theoretically computed quantity, however, and is not the average of a set of data. However, the following definition may seem intuitively reasonable.

8.9 Probability Revisited

DEFINITION 8.114 The **mean** μ of a random variable R is the expected value $E(R)$.

The variance of a random variable R indicates how spread out or widely distributed the values of the random value are.

DEFINITION 8.115 The **variance** σ^2 of a random variable is $E((R-\mu)^2) = E((R-E(R))^2)$.

Using Theorem 8.113, we develop an important property of the variance.

THEOREM 8.116 For a random variable R, $\sigma^2 = E(R^2) - \mu^2 = E(R^2) - (E(R))^2$.

Proof

$$\sigma^2 = E((R-\mu)^2)$$
$$= E(R^2 - 2R\mu + \mu^2)$$
$$= E(R^2) - 2\mu E(R) + E(\mu^2) \quad \text{By Theorem 8.113, since } \mu \text{ is a constant.}$$
$$= E(R^2) - 2\mu^2 + \mu^2 \quad \text{By Theorem 8.113, since } \mu \text{ is a constant.}$$
$$= E(R^2) - \mu^2.$$
■

Again we emphasize that both the mean and the variance are theoretical and are not to be confused with the mean and variance of a set of data.

We again select the binomial distribution for our example. We also assume that the random variable R gives the number of successes. We already know that

$$\mu = E(R) = np$$

Since $\sigma^2 = E(R^2) - (E(R))^2$, we next try to find $E(R^2)$. By definition

$$E(R^2) = \sum_{k=0}^{n} k^2 \frac{n!}{k!(n-k)!} p^k q^{n-k}.$$

Writing $k^2 = k(k-1) + k$ and using the fact that the first two terms in the first sum below are 0, we have

$$E(R^2) = \sum_{k=0}^{n} k(k-1) \frac{n!}{k!(n-k)!} p^k q^{n-k} + \sum_{k=0}^{n} k \frac{n!}{k!(n-k)!} p^k q^{n-k}$$

$$= \sum_{k=2}^{n} k(k-1) \frac{n!}{k!(n-k)!} p^k q^{n-k} + \sum_{k=0}^{n} k \frac{n!}{k!(n-k)!} p^k q^{n-k}$$

$$= \sum_{k=2}^{n} \frac{n!}{(k-2)!(n-k)!} p^k q^{n-k} + E(R)$$

$$= n(n-1) \sum_{k=2}^{n} \frac{(n-2)!}{(k-2)!(n-k)!} p^k q^{n-k} + E(R)$$

$$= n(n-1) p^2 \sum_{k=2}^{n-2} \frac{(n-2)!}{(k-2)!(n-k)!} p^{k-2} q^{n-2-(k-2)} + E(R)$$

$$= n(n-1) p^2 \sum_{k=0}^{n-2} \frac{(n-2)!}{k!(n-(k+2))!} p^k q^{n-2-k} + E(R)$$

$$= n(n-1)p^2 \sum_{k=0}^{n-2} \frac{(n-2)!}{k!(n-2-k)!} p^k q^{n-2-k} + E(R)$$
$$= n(n-1)p^2(p+q)^{n-2} + np$$
$$= n(n-1)p^2 + np$$
$$= n^2 p^2 - np^2 + np$$
$$= n^2 p^2 + np(1-p)$$
$$= n^2 p^2 + npq.$$

Therefore,
$$E(R^2) - (E(R))^2 = n^2 p^2 + npq - (np)^2$$
$$= npq$$

and we have the following theorem.

THEOREM 8.117 In an experiment with n independent trials each having two outcomes, success with probability p and failure with probability $q = 1 - p$, let the random variable R represent the number of successes. The variance $\sigma^2 = npq$. ∎

EXAMPLE 8.118 In an experiment a die is to be tossed 72 times. Let the random variable R represent the number of times that 1 appears, so that $p = \frac{1}{6}$ and $q = \frac{5}{6}$. Then

$$E(R) = \mu = np = 72 \cdot \frac{1}{6} = 12$$

and

$$\sigma^2 = npq = 72 \cdot \frac{1}{6} \cdot \frac{5}{6} = 10$$ ∎

We have stated that the variance is a measure of the spread of the values of the random variable. Chebyshev's inequality gives us a more precise measure of the relationship between variance and distribution of the random variable. Before we state and prove Chebyshev's inequality, however, we need the following lemma.

Lemma 8.119 Let R be a random variable defined on a sample space S. Then

$$P(|R| \geq c) \leq \frac{E(R^2)}{c^2} = \frac{\sigma^2}{c^2}.$$

Proof

$$P(|R| \geq c) = \sum_{|R(x_i)| \geq c} P_i$$
$$\leq \frac{1}{c^2} \sum_{|R(x_i)| \geq c} |R(x_i)|^2 P_i$$
$$\leq \frac{1}{c^2} \sum_{x_i \in S} |R(x_i)|^2 P_i$$
$$= \frac{1}{c^2} E(R^2)$$ ∎

Substituting $R - \mu$ for R, we have Chebyshev's inequality.

THEOREM 8.120 (Chebyshev's Inequality) Let R be a random variable defined on a sample space S. Then

$$P(|R - \mu| \geq c) \leq \frac{E((R-\mu)^2)}{c^2} = \frac{\sigma^2}{c^2}.$$

EXAMPLE 8.121 We again provide a solution to Example 8.12 from the Shier collection. The problem is to find the number of spanning trees of $K_{2,n}$. This method is called the conditioning approach. In this method we conditions on the number k of edges that are incident to A. There is only one vertex i that is joined to both A and B. There are $\binom{n}{k}$ ways of selecting k vertices that are joined to A. Also for each selection there are k choices for the special vertex i. Therefore, using the addition principle there are $\sum_{i=1}^{n} k \binom{n}{k}$ possible spanning trees. If a coin with probability p of turning up heads is tossed ten times, then the probability distribution is given by

$$\Pr(X = k) = \binom{n}{k} p^k (1-p)^{n-k}$$

The expected value $E(X) = \sum_{i=1}^{n} k \Pr(X = k) = np$. If $p = \frac{1}{2}$, $\Pr(X = k) = \binom{n}{k}\left(\frac{1}{2}\right)^n$ and

$$E(X) = \frac{1}{2}n = \sum_{i=1}^{n} k \Pr(X = k) = \sum_{i=1}^{n} k \binom{n}{k}\left(\frac{1}{2}\right)^n$$

Multiplying by 2^n, we have $\sum_{i=1}^{n} k \binom{n}{k}\left(\frac{1}{2}\right)^n = 2^{n-1} n$.

Exercises

1. In a shipment of five automobiles, one is defective. Two people come in to purchase automobiles. What is the probability that the second car purchased is not defective given that the first car purchased is not defective?

2. Two dice are tossed. What is the probability that the first die is a 5, given that the sum is 8?

3. Two dice are tossed. What is the probability that the sum is 7 given that the sum is odd?

4. Two dice are tossed. What is the probability that the sum is 8 given that the sum is 7 or larger?

5. An urn contains eight white balls and seven black balls. Four balls are drawn without replacement. What is the probability that the first ball is white given that two white balls were drawn? What is the probability that the first ball is white given that two white balls were drawn if the balls were drawn with replacement?

6. Four aces from a standard deck are shuffled and placed face down on a table. Two are removed without being seen. What is the probability that the third card is the ace of spades?

7. A coin is tossed three times. What is the probability of getting three heads, given that at least one of the tosses is heads?

8. Let A and B be independent events. If $P(A) = .7$ and $P(B) = .6$, find $P(B|A)$.

9. A ball is selected from each of three urns. The first urn contains three white balls and two black balls. The second urn contains two white balls and three black balls. The third urn contains three white balls and three black balls. If two of the balls chosen are white, what is the probability that a white ball was chosen from the first urn?

10. Two cards are drawn from a standard deck of 52 cards. What is the probability that the second card is a king

given that the first card was an ace? What is the probability that the second card is a king given that the first card was the ace of spades?

11. Six balls are numbered. The numbers are 0, 1, 1, 2, 5, and 10, respectively. The balls are placed in an urn and one is randomly selected. A player receives n dollars if the number on the ball selected is n. What is the expected value of the amount won?

12. Six balls are numbered. The numbers are 0, 0, 0, 0, 10, and 20, respectively. The balls are placed in an urn and one is randomly selected. A player receives n dollars if the number on the ball selected is n. What is the expected value of the amount won?

13. Eight balls are numbered. The numbers are 0, 1, 1, 2, 2, 2, 5, and 10, respectively. The balls are placed in an urn and three are randomly selected. The player receives the amount in dollars of the sum of the three balls. What is the expected value of the amount won?

14. Eight balls are numbered. The numbers are 0, 0, 0, 2, 3, 3, 7, and 10, respectively. The balls are placed in an urn and three are randomly selected. The player receives the amount in dollars of the sum of the three balls. What is the expected value of the amount won?

15. Suppose that we are playing a game where we are to toss two dice and be paid the amount of dollars equal to the value of the sum of the dice. Thus, if a 5 comes up, we receive $5. How much should we expect to pay to play the game?

16. Suppose the cards in a deck are given the following values: Ace has value 1, two has value 2, ..., ten has value 10, Jack has value 11, Queen has value 12, and King has value 13. A player selects a card. If it is a heart, the player receives twice the value of the card. If it is a diamond, they receive the value of the card. If the card is black, they pay $10. What is the player's expected value in the game?

17. Ace Cleaning House awards a grand prize of $10 million to one of the contestants entering the contest. If a person pays 33 cents for the stamp (Ace Cleaning House provides the envelope and entry blank) and 50 million people enter the sweepstakes, what is the expected value of the contest?

18. Assume the probability of a girl is 0.51 and the probability of a boy is 0.49. If the Jones family has five children, what is the probability that there are exactly three girls? What is the probability there are at least four girls?

19. If a baseball team wins 60% of the time, what is the probability they will win exactly three of their next five games?

20. If two baseball teams in the world series are equally likely to win any given game and team A loses the first two games, what is the probability that team A will win the world series?

21. If a fair coin is tossed seven times, what is the probability of getting exactly four heads?

22. If a die is tossed seven times, what is the probability that a six will appear exactly two times?

23. If 10% of a product made is defective, what is the probability that if ten samples are selected, exactly two will be defective?

24. Sam Slammer averages a hit one out every three times at bat. If he bats ten times, what is the probability he will get at least three hits?

25. If ten cards are selected sequentially from a deck of cards without replacement, what is the probability that exactly two of the cards are hearts?

26. If a pair of dice is tossed 12 times, what is the probability of getting a 6 exactly 3 times?

27. Assume we are playing the same game of roulette with the same random variable as in Example 8.107 except that we have an additional slot labeled 00. Find the expected value of R_1.

28. Assume we are playing the same game of roulette with the same random variable as in Example 8.108 except that we have an additional slot labeled 00. Find the expected value of R_2.

29. Assume we are playing the same game of roulette with the same random variable as in Example 8.109 except we have an additional slot labeled 00. Find the expected value of R_3.

30. Assume we are playing the same game of roulette with the same random variable as in Example 8.110 except that we have an additional slot labeled 00. Find the expected value of R_4.

31. In an experiment with 10 independent trials each having two outcomes, success with probability .3, let the random variable R represent the number of successes. What is the expected value of R? What is the variance?

8.10 Bayes' Theorem

Bayes' theorem is named after the Reverend Thomas Bayes, a minister and mathematician. His theory was published in 1764, four years after his death, in the paper

Essay towards Solving a Problem in the Doctrine of Chances. It appeared in the *Philosophical Transactions* of the Royal Society of London. Many people have trouble accepting Bayes' theorem. It seems to reverse the process of cause and effect. By learning the effect, one determines the probability of the cause. Bayes' theorem is very useful in determining the causes of diseases such as cancer. It is also useful in determining the effects of a medication upon an illness.

Suppose that we have n disjoint events, $B_1, B_2, B_3, \ldots, B_n$, which partition the sample space, and we know that A has occurred. By definition, $P(B_i|A)$, is equal to $\frac{P(A \cap B_i)}{P(A)}$. But

$$A = (A \cap B_1) \cup (A \cap B_2) \cup (A \cap B_3) \cup \cdots \cup (A \cap B_n)$$

so that

$$P(A) = P(A \cap B_1) + P(A \cap B_2) + P(A \cap B_3) + \cdots + P(A \cap B_n)$$

and

$$P(B_i|A) = \frac{P(A \cap B_i)}{P(A \cap B_1) + P(A \cap B_2) + P(A \cap B_3) + \cdots + P(A \cap B_n)}$$

Since $P(A \cap B_i) = P(A|B_i)P(B_i)$, we have

$$P(B_i|A) = \frac{P(A|B_i)P(B_i)}{P(A|B_1)P(B_1) + P(A|B_2)P(B_2) + \cdots + P(A|B_n)P(B_n)}$$

which tells us that if we know $P(B_i)$ and $P(A|B_i)$ for each i, then we can determine $P(B_i|A)$. Thus, we have "reversed" the conditional probability.

For example, suppose Sam goes out to purchase a garage door closer. The probability that he will go to Andy's Automatic Unopeners is .3; the probability he will go to Ben's Best Buys is .2; and the probability that he will go to Cheap Charley's Cash and Credit is .5. If he goes to Andy's Automatic Unopeners, due to the selection, the probability that he will buy the super automatic garage door closer is .2. If he goes to Ben's Best Buys, the probability that he will buy the super automatic garage door closer is .4. If he goes to Cheap Charley's Cash and Credit, the probability that he will buy the super automatic garage door closer is .3. If he buys the super automatic garage door closer, what is the probability he bought it at Ben's Best Buys?

Let B_1 be the event that the super automatic garage door closer was purchased at Andy's Automatic Unopeners, B_2 be the event that the super automatic garage door closer was purchased at Ben's Best Buys, and B_3 be the event that the super automatic garage door closer was purchased at Cheap Charley's Cash and Credit. Let A be the event that he purchased the super automatic garage door closer. Then $P(B_1) = .3$, $P(A|B_1) = .2$, $P(B_2) = .2$, $P(A|B_2) = .4$, $P(B_3) = .5$, and $P(A|B_3) = .3$. Therefore,

$$P(B_2|A) = \frac{(.2)(.4)}{(.3)(.2) + (.2)(.4) + (.5)(.3)} = \frac{8}{29} \approx .28$$

If events $B_1, B_2, B_3, \ldots, B_n$ are equally likely so that

$$P(B_1) = P(B_2) = P(B_3) = \cdots = P(B_n)$$

Then the equation for $P(B_i|A)$ simplifies to

$$P(B_i|A) = \frac{P(A|B_i)}{P(A|B_1) + P(A|B_2) + P(A|B_3) + \cdots + P(A|B_n)}$$

For example, suppose urn 1 contains three red, four blue, and six white balls, urn 2 contains five red, three blue and two white balls, and urn 3 contains four red, two blue, and four white balls. An urn is selected randomly and a ball is selected. If a red ball is selected, what is the probability that urn 1 was chosen? Let B_1 be the event urn 1 is chosen, B_2 be the event urn 2 is chosen, and B_3 be the event urn 3 is chosen. Let R be the event a red ball is chosen. Since the selection of the urn is random, we assume $P(B_1) = P(B_2) = P(B_3)$. Therefore, we may use the simpler formula. $P(R|B_1) = .3$, $P(R|B_2) = .5$, and $P(R|B_3) = .4$. Therefore,

$$P(B_1|R) = \frac{.3}{.3 + .5 + .4} = .25$$

Exercises

1. A company manufactures tires in three plants. Forty-five percent of the tires are manufactured in plant A, 25 percent are manufactured in plant B, and 30 percent are manufactured in plant C. Suppose 5 percent of the tires manufactured in plant A are defective, 10 percent of the tires manufactured in plant B are defective, and 2 percent of the tires manufactured in plant C are defective. If a tire is purchased and found to be defective, what is the probability that it was manufactured in plant B?

2. Box 1 contains three red, four white, and one blue balls. Box 2 contains two red, four white, and two blue balls. Box 3 contains two red, three white, and three blue balls. If a ball is selected at random and it is blue, what is the probability that it was selected from box 3?

3. A box contains three coins. One has two heads, one is a fair coin, and the other comes up heads 70% of the time. A coin is selected at random and tossed. If it comes up heads, what is the probability that it was the fair coin?

4. A blood test is 90% effective in detecting a disease. It also falsely diagnoses that a healthy person has the disease about 4% of the time. If 10% of the people have the disease, what is the probability that a person who tests positive on the test actually has the disease?

5. Students are equally likely to select one of three liberal arts courses. If they select the History of Kansas 1995–1997, they have a 40% chance of passing. If they select Advanced Finger Counting II, they have a 30% chance of passing. If they select Television III: Channel Selection, they have a 60% chance of passing. If they pass the course, what is the probability that they took History of Kansas 1995–1997?

6. In the manufacturing of a product, 80% of the products made are not defective. Of those inspected, 10% of the good ones are ruled defective and not shipped. Only 5% of the defective products are approved and shipped. If a product is shipped, what is the probability that it is defective?

7. Suppose an insurance company classifies drivers as good, bad, or average. Suppose 25% of the drivers are good, 50% are average, and 25% are bad. Suppose a good driver has a 10% chance of having an accident during the next year, an average driver has a 20% chance of having an accident during the next year, and a bad driver has a 30% chance of having an accident during the next year. If Sam had an accident last year, what is the probability that he is a good driver?

8. A fair coin is tossed. If it comes up heads, a pair of dice is tossed, and the player gets the amount in dollars equal to the number that comes up. If the coin comes up tails, three coins are tossed and the player gets $4 for each head that comes up. If the player wins $8, what is the probability that the original coin came up heads?

8.11 Markov Chains

Suppose that we run a series of one or more experiments with possible outcomes $s_1, s_2, s_3, \ldots, s_n$. We shall call these outcomes **states**. We shall assume that we begin in one of these states and that $p_i^{(0)}$ is the probability that we begin in state s_i. Furthermore, we shall assume that the probability of a given outcome in the current experiment is dependent only on the outcome of the previous experiment. Let p_{ij} be the probability that, as a result of the experiment, the state was changed from state s_i to state s_j. Thus, the experiment began in state s_i and ended in state s_j. Let $p_i^{(1)}$ be the probability that the outcome of the experiment is s_i. Then

$$p_i^{(1)} = p_1^{(0)} p_{1i} + p_2^{(0)} p_{2i} + p_3^{(0)} p_{3i} + \cdots + p_1^{(0)} p_{1i}$$

which means that the probability of ending up in state s_i is the sum of the probabilities of beginning in some state and ending up in state s_i. If the experiment is begun in state s_j, then, as a result of the experiment, the outcome must be some state. Therefore, for each $0 \leq j \leq n$,

$$p_{j1} + p_{j2} + p_{j3} + \cdots + p_{jn} = 1$$

Let $p^{(0)} = (p_1^{(0)}, p_2^{(0)}, p_3^{(0)}, \ldots, p_n^{(0)})$ and $p^{(1)} = (p_1^{(1)}, p_2^{(1)}, p_3^{(1)}, \ldots, p_n^{(1)})$. Suppose, for simplicity, there are three possible states, so $p^{(0)} = (p_1^{(0)}, p_2^{(0)}, p_3^{(0)})$. Let T be the matrix

$$\begin{bmatrix} p_{11} & p_{12} & p_{13} \\ p_{21} & p_{22} & p_{23} \\ p_{31} & p_{32} & p_{33} \end{bmatrix}$$

T is called the **transition matrix**. In general, T is the matrix

$$\begin{bmatrix} p_{11} & p_{12} & p_{13} & \cdots & p_{1n} \\ p_{21} & p_{22} & p_{23} & \cdots & p_{2n} \\ p_{31} & p_{32} & p_{33} & \cdots & p_{3n} \\ \vdots & \vdots & \vdots & \ddots & \vdots \\ p_{n1} & p_{n2} & p_{n3} & \cdots & p_{nn} \end{bmatrix}$$

Let

$$(a, b, c) = (p_1^{(0)}, p_2^{(0)}, p_3^{(0)}) \begin{bmatrix} p_{11} & p_{12} & p_{13} \\ p_{21} & p_{22} & p_{23} \\ p_{31} & p_{32} & p_{33} \end{bmatrix}$$

Then

$$a = p_1^{(0)} p_{11} + p_2^{(0)} p_{21} + p_3^{(0)} p_{31} = p_1^{(1)}$$
$$b = p_1^{(0)} p_{12} + p_2^{(0)} p_{22} + p_3^{(0)} p_{32} = p_2^{(1)}$$
$$c = p_1^{(0)} p_{13} + p_2^{(0)} p_{23} + p_3^{(0)} p_{33} = p_3^{(1)}$$

so that

$$(p_1^{(1)}, p_2^{(1)}, p_3^{(1)}) = (p_1^{(0)}, p_2^{(0)}, p_3^{(0)}) \begin{bmatrix} p_{11} & p_{12} & p_{13} \\ p_{21} & p_{22} & p_{23} \\ p_{31} & p_{32} & p_{33} \end{bmatrix}$$

EXAMPLE 8.122 In the newspaper, one classification of the weather is clear, partly cloudy, or cloudy. If the weather is clear one day, the probability that it will be clear the next is .5; the probability it will be partly cloudy is .4; and the probability it will be cloudy is .1. If the weather is partly cloudy, the probability that it will be clear the next day is .3; the probability that it will stay the same is .5; and the probability that it will become cloudy is .2. If it is cloudy, the probability that it will be cloudy the next day is .4; the probability it will be partly cloudy is .4; and the probability that it will be clear is .2. If on Sunday the probability that it is clear is .6 and the probability that it will be partly cloudy is .4, what is the probability that it will be clear on Monday? What is the probability that it will be partly cloudy on Tuesday?

If the order in which we list the conditions are clear, partly cloudy, and cloudy, then

$$p^{(0)} = (.6, .4, 0)$$

and

$$T = \begin{bmatrix} .5 & .4 & .1 \\ .3 & .5 & .2 \\ .2 & .4 & .4 \end{bmatrix}$$

Therefore,

$$p^{(1)} = (.6, .4, 0) \begin{bmatrix} .5 & .4 & .1 \\ .3 & .5 & .2 \\ .2 & .4 & .4 \end{bmatrix} = (.42, .44, .14)$$

and the probability that it will we clear on Monday is .42.

Let $p_1^{(2)}$ be the probability that it is clear on Tuesday, $p_2^{(2)}$ be the probability that it is partly cloudy on Tuesday, and $p_3^{(2)}$ be the probability that it is cloudy on Tuesday. Let $p^{(2)} = (p_1^{(2)}, p_2^{(2)}, p_3^{(2)})$. Then

$$p^{(2)} = (.42, .44, .14) \begin{bmatrix} .5 & .4 & .1 \\ .3 & .5 & .2 \\ .2 & .4 & .4 \end{bmatrix} = (.37, .444, .186)$$

Therefore, the probability that it is partly cloudy on Tuesday is .444. ∎

Let $p_i^{(m)}$ be the probability that the outcome of the mth performance of the experiment is state s_i, and $p^{(m)} = (p_1^{(m)}, p_2^{(m)}, p_3^{(m)}, \dots, p_n^{(m)})$.

THEOREM 8.123 For any positive integer m, $p^{(m)} = p^{(0)} T^m$.

Proof We prove this using induction. We have already shown that it is true for $m = 1$. Assume that it is true for $n = k$, so that $p^{(k)} = p^{(0)} T^k$. Since

$$p_j^{(k+1)} = p_1^{(k)} p_{1j} + p_2^{(k)} p_{2j} + p_3^{(k)} p_{3j} + p_n^{(k)} p_{n3j}$$

then

$$p^{(k+1)} = p^{(k)} T = p^{(0)} T^k T = p^{(0)} T^{k+1}$$ ∎

EXAMPLE 8.124 In a survey of those who had purchased three brands of American cars, brand A, brand B, and brand C, owners were asked what brand they would choose for their next purchase. Of those who purchased brand A, 20% said they would again purchase brand A, while 50% said they would switch to brand B, and 30% said they would switch to brand C. Of those who purchased brand B, 20% said they would switch to brand A, while 70% said they would again purchase brand B, and 10% said they would switch to brand C. Of those who purchased brand C, 30% said they would switch to brand A, 30% said they would switch to brand B, and 40% said they would remain with brand C.

(a) If someone purchased brand A for his or her first car, what is the probability he or she will purchase brand C for his or her second car?

(b) If a person tossed a coin to decide whether to purchase car B or car C for his or her first car, what is the probability that the person will purchase car A for his or her third car?

(c) If all three choices are equally likely for the purchase of the first car, what is the probability that a person will purchase B for his or her third car?

The transition matrix T for this event is

$$\begin{bmatrix} .2 & .5 & .3 \\ .2 & .7 & .1 \\ .3 & .3 & .4 \end{bmatrix}$$

for (a), $p^{(0)} = (1, 0, 0)$ so

$$p^{(1)} = (1, 0, 0) \begin{bmatrix} .2 & .5 & .3 \\ .2 & .7 & .1 \\ .3 & .3 & .4 \end{bmatrix} = (.2, .5, .3)$$

The probability that the second car will be brand C is .3. For parts (b) and (c), we need to find

$$T^2 = \begin{bmatrix} .23 & .54 & .23 \\ .21 & .62 & .17 \\ .24 & .48 & .28 \end{bmatrix}$$

In part (b), $p^{(0)} = (0, .5, .5)$ and

$$p^{(2)} = (0, .5, .5) \begin{bmatrix} .23 & .54 & .23 \\ .21 & .62 & .17 \\ .24 & .48 & .28 \end{bmatrix} = (.225, .55, .225)$$

so the probability that a person's third car will be brand A is .225. ∎

EXAMPLE 8.125

In part (c), $p^{(0)} = (.333, .333, .333)$ and

$$p^{(2)} = (.333, .333, .333) \begin{bmatrix} .23 & .54 & .23 \\ .21 & .62 & .17 \\ .24 & .48 & .28 \end{bmatrix} = (.227, .547, .227)$$

where all numbers are rounded to three decimal places. Therefore, the probability of selecting brand C for the third car is .227. ∎

Exercises

Let events A, B, and C occurring be represented by states s_1, s_2, and s_3, respectively. If there is a transition every day and the transition matrix is

$$\begin{bmatrix} .2 & .5 & .3 \\ .3 & .6 & .1 \\ .2 & .3 & .5 \end{bmatrix}$$

1. What is the probability that if event A occurred the first day that it will occur again the third day?

2. What is the probability that if event B occurred the first day that event C will occur the fourth day?

3. What is the probability that if event A occurred the first day that event B will occur the fifth day?

Let events A, B, and C occurring be represented by states s_1, s_2, and s_3, respectively. If there is a transition every day and the transition matrix is

$$\begin{bmatrix} .3 & .4 & .3 \\ .4 & .2 & .4 \\ .2 & .5 & .3 \end{bmatrix}$$

4. What is the probability that if event A occurred the first day that B will occur the third day?

5. What is the probability that if event A occurred the first day that event C will occur the fourth day?

6. What is the probability that if event C occurred the first day that event C will occur the fifth day?

A drugstore sells three brands of antacids, A, B, and C. Buyers switch brands according to the transition matrix

$$\begin{bmatrix} .7 & .2 & .1 \\ .2 & .6 & .2 \\ .1 & .1 & .8 \end{bmatrix}$$

If antacids are purchased monthly.

7. What is the probability that if brand A is purchased this month, then brand C will be purchased in three months?

8. What is the probability that if brand B is purchased this month, then brand A will be purchased in two months?

9. What is the probability that if brand C is purchased this month, then brand C will again be purchased in three months?

Suppose that Congress has announced whether or not it will cut taxes. Suppose that the announcement is spread from person to person. If there is a tax cut, the probability that the announcement will be relayed correctly to the next person is .6. If there is not a tax cut, the probability that the announcement will be relayed correctly is .7.

10. If there is a tax cut, what is the probability that the fourth person to hear the rumor will hear it correctly?

11. If there is not a tax cut, what is the probability that the third person to hear the rumor will hear it correctly?

12. If there is not a tax cut, what is the probability that the fifth person to hear the rumor will think that there is a tax cut?

Team A and team B are in the World Series. Each team is equally likely to win the first game. If Team A wins a game, then the probability it will win the next game is 0.6. If Team B wins, the probability it will win the next game is 0.7.

13. What is the probability that team B will win the third game?

14. What is the probability that team A will win the fourth game?

15. An aircraft has four engines. On a combat mission, it must remain in the air for an hour. If the probability of an engine failing in any 15-minute period due to enemy fire is .25 and the plane needs at least two engines to remain in the air, determine the probability that the plane will return. It is assumed that the probability that the plane with three failed engines will remain in the air long enough to have four failed engines is 0.

16. In a college, 80 percent of the freshman class become sophomores their second year. Ten percent drop out or transfer. Ninety percent of the sophomores become juniors the next year. Five percent of the sophomores drop out or transfer. Ninety-five percent of the juniors become seniors the next year. Two percent of the juniors drop out or transfer. One percent of the seniors drop out or transfer and 98 percent graduate. Those who do not drop out, transfer, or advance remain in the same class. What is the probability that an entering freshman will drop out or transfer within four years? If 1000 students enter as freshmen, how many will graduate after four years? How many will graduate after five years?

GLOSSARY

Chapter 8: Counting and Probability

Addition counting principle (8.1)	Let $S_1, S_2, S_3, \ldots, S_m$ be pairwise disjoint sets, and for each I, let S_i contain n_i elements. The number of elements that can be selected from S_1 or S_2 or S_3 or $\ldots$ or S_m is equal to $n_1 + n_2 + n_3 + \cdots + n_m$.
Bayes' theorem (8.10)	$$P(B_i|A) = \frac{P(A|B_i)}{P(A|B_1) + P(A|B_2) + P(A|B_3) + \cdots + P(A|B_n)}$$
Bernoulli trials (8.9)	An experiment with n independent trials.
Binomial distribution (8.9)	$$\sum_{k=0}^{n}\binom{n}{k}p^k q^{n-k} = (p+q)^n$$
Binomial theorem (8.3)	For a given nonnegative integer n, $(a+b)^n = \sum_{r=0}^{n}\binom{n}{r}a^r b^{n-r} = \sum_{r=0}^{n}\binom{n}{r}a^{n-r}b^r$.
Bit string (8.1)	A **bit string** is a string of symbols such that each symbol is either 1 or 0.
Chebyshev's inequality (8.9)	Let R be a random variable defined on a sample space S. Then $$P(|R-\mu| \geq C) \leq \frac{E((R-\mu)^2)}{c^2} = \frac{\sigma^2}{c^2}.$$
Conditional probability (8.9)	The probability of B given A, denoted by $P(A|B)$, is equal to $\frac{P(A \cap B)}{P(A)}$.

Event (8.5)	An **event** is a subset of a sample space.
Expected value/mathematical expectation (8.9)	Let R be a random variable defined on outcomes $x_1, x_2, x_3, \ldots, x_n$, which occur with respective probabilities $P_1, P_2, P_3, \ldots, P_n$. The **expected value** or **mathematical expectation** is $$E(R) = \sum_{i=1}^{n} R(x_i) P_i$$
	If the set of all outcomes for which R has the fixed value a_i is denoted by $R = a_i$, the probability that X assumes this value $P(R = a_i)$ is denoted by p_i. An equivalent form of the expected value $E(R)$ using this notation is $E(R) = \sum_{i=1}^{m} a_i p_i$.
Independent events (8.9)	If $P(B) = P(B\|A)$, then the event B is **independent** of the event A.
Lexicographical ordering (8.4)	The way words are ordered in the dictionary. Given a linearly ordered set, a **lexicographical ordering** $\preceq$ on the strings of elements of S is defined by $a \prec b$ if $a = a_1 a_2 a_3 \ldots a_m$, $b = b_1 b_2 b_3 \ldots b_n$ and any of the following occur: a. $a_1 < b_1$. b. $a_i = b_i$ for all $1 \leq i \leq k$ and $a_{k+1} < b_{k+1}$. c. $a_i = b_i$ for all $1 \leq i \leq m$ and $m < n$.
Markov chains (8.11)	A Markov chain is a sequence of values whose probabilities at a time interval depend on the value of the number at the previous time. An example is the nonreturning walk, where a walker is restricted to not go back to the location previously visited.
Mean (8.9)	The **mean** μ of a random variable R is the expected value $E(R)$.
Multiplication counting principle (8.1)	Given a sequence of events $E_1, E_2, E_3, \ldots, E_m$ such that E_1 occurs in n_1 ways and if $E_1, E_2, E_3, \ldots, E_{k-1}$ have occurred, then E_k can occur in n_k ways. There are $n_1 \times n_2 \times n_3 \times \cdots \times n_m$ ways in which the entire sequence of events can occur.
Pigeonhole principle/ Dirichlet drawer principle (8.8)	If $n + 1$ pigeons are placed in n pigeonholes, then at least one hole contains more than one pigeon.
Probability function (8.9)	A **probability function** on a sample space S is a function P from the subsets of S to the real numbers such that 1. $P(A) \geqslant 0$ for all $A \subseteq S$. 2. $P(S) = 1$. 3. If $A, B \subseteq S$ have no outcome in common, then $P(A \cup B) = P(A) + P(B)$.
Probability of an event (8.5)	If S is the set of all possible outcomes of an experiment, then the **probability** $P(A)$ that an outcome of event A occurs is given by $P(A) = \frac{\|A\|}{\|S\|}$.
r combinations on n objects (8.3)	The number of r combinations (where order does not matter) on n objects is given by $C(n) = \frac{n!}{r!(n-r)!}$.

r permutations on n objects (8.3)	The number of r permutations (where order counts) on n objects is given by $P(n) = \dfrac{n!}{(n-r)!}$.
Ramsey number (8.8)	The **Ramsey number $R(p,q)$** is the smallest number n so that K_n has the (p,q) Ramsey property.
Ramsey property (8.8)	A complete graph K_n has the **(p,q) Ramsey property** if when it is painted with two colors, say, red and blue, it contains either a subgraph K_p with red edges or a subgraph K_q with blue edges.
Ramsey's theorem (8.8)	If p and q are both integers greater than one, then Ramsey's number, $R(p,q)$ exists.
Sample space (8.5)	A sample space is the set of all possible outcomes of an experiment.
Selection of k objects taken n at a time with repetition (8.7)	$C(n+k-1, n) = C(n+k-1, k-1) = \dfrac{(n+k-1)!}{n!(k-1)!}$
States (8.11)	If a series of one or more experiments has possible outcomes $s_1, s_2, s_3, \ldots, s_n$, then those outcomes are called **states**.
Transition matrix (8.11)	The **transition matrix** represents the matrix of probabilities P_{ij} that exists as we go from state i to state j.
Vandermonde theorem (8.3)	Let $m, n,$ and r be positive integers such that $r \leqslant \min(m, n)$, then $\binom{m+n}{r} = \sum_{k=0}^{r} \binom{m}{k}\binom{n}{r-k}$.
Variance (8.9)	The **variance** σ^2 of a random variable is $E((R-\mu)^2) = E((R-E(R))^2)$.

TRUE-FALSE QUESTIONS

1. If four coins are tossed, the number of possible outcomes is 2^4.

2. The number of functions from a set A containing n elements to itself is $n!$.

3. If $|S| = n$ and $|T| = m$, then $|S \cup T| = m + n$.

4. If $|S \cup T| = 50$, $|S - T| = 20$, and $|T - S| = 15$, then $S \cap T = 15$.

5. There are $9!$ different possible social security cards.

6. If a license plate contains three letter and three digits and no repetition of digits is allowed, then there are 1,123,200 possible different license plates.

7. If two dice are tossed, there are 32 ways of getting a sum that is 4 or larger.

8. If eight horses are entered in a race and awards are given for finishing in first, second, and third places, there are $\dfrac{8!}{5!}$ possible ways in which the awards may be given.

9. The number of ways a committee of 6 may be selected from 16 people is equal to the number of ways a committee of 10 may be selected from 16 people.

10. The number of ways to pick a president, vice president, secretary, and treasurer from 16 people is equal to the number of ways to pick a committee of 4 from 16 people.

11. The number of bit strings of length 8 with the first and last digits identical is 2^7.

12. The number integers less than or equal to 501 that are not divisible by either 5 or 7 is 330.

13. The number of ways to arrange the digits $1, 2, 3, \ldots 9$, if 4 and 5 cannot be adjacent is $9! - 8!$.

14. The number of integers between 1 and 1002 not divisible by 5, 7, or 9 is 546.

15. If $|A| = 35, |B| = 33|, C = |32|, |A \cup B \cup C| = 62$, $|A \cap B| = 16, |C \cap B| = 14$, and $|A \cap C| = 17$, then $|A \cap B \cap C| = 9$.

16. In the previous problem, $|A - B| = 17$.

17. $P(n, 0) = n$ for all $n \geq 1$.
18. $\binom{150}{27} = \binom{150}{26} + \binom{149}{27}$.
19. In the expansion of $(a + b)^{20}$, the coefficient of $a^{13}b^7$ is $\binom{20}{7}$.
20. The number of ways a committee of 6 men and 7 women can be picked from 11 men and 14 women is $\binom{25}{13}$.
21. The number of ways of dealing a 13-card hand containing 8 of a suit is $\binom{13}{8}$.
22. The number of ways six boys and six girls can sit together if no two boys sit together is $2 \cdot (5!)^2$.
23. The number of ways six boys and six girls can sit together if no two boys sit together and no girls sit together is $2 \cdot (6!)^2$.
24. The number of 8-bit binary strings containing exactly 3 ones is $2^3 \cdot 2^5$.
25. Let $S = \{a, b, c, d, e\}$. The next element generated after $\{a, b, d\}$ is $\{a, b, e\}$.
26. The next permutation generated after 1347652 is 1357642.
27. For any event A, $P(A') = -P(A)$.
28. $P(A \cup B) = P(A) + P(B)$.
29. If $A \subseteq B$, then $P(A) \leq P(B)$.
30. If four coins are tossed, $\{(H, H, T, T), (H, T, H, T), (H, T, T, H), (T, H, H, T)\}$ is an event.
31. If S is the sample space, then $P(S) = 1$.
32. If a card is selected from a standard 52-card deck, the probability of getting a king, queen, or heart is $\frac{21}{52}$.
33. If a sample space has 12 possible outcomes, then the probability of each event is $\frac{1}{12}$.
34. Let A and B be events in a sample space. Then
$$P(A \cup B) = P(A - B) + P(B - A) + P(A \cap B)$$
35. Five letters are selected from the alphabet. The probability that at least two are the same is
$$1 - \frac{26!}{26^5 \cdot 21!}$$
36. If two dice are tossed, the probability of getting a five or seven is $\frac{11}{36}$.
37. The number of distinct arrangements that can be formed using the letters in *Mississippi* is
$$\frac{11!}{4!4!2!}$$

38. The coefficient of $x^2y^3z^2$ in the expansion of $(x + y + 2z)^7$ is 210.
39. If ten cards are drawn from a standard card deck, there are 12,600 distinct hands containing three hearts, two spades, four diamonds, and a club.
40. Assume an urn contains 7 red, 7 blue, 7 white and 7 green balls. At least 18 balls must be selected to have 3 red, 4 blue, 6 white, or 5 green balls.
41. If an urn contains 13 red, 13 blue, 13 white, and 13 green balls, then there are 12,528 ways of selecting 10 balls.
42. If $P(B|A) = P(B)$, then $P(A \cap B) = P(A) \cdot P(B)$.
43. If A and B are independent events, $P(A) = .4$, and $P(B) = .3$, then $P(A \cup B) = .58$.
44. In an experiment with 8 independent trials, and probability of success .6, the probability of exactly 5 successes is $(.6)^5(.4)^3$.
45. In an experiment with 10 independent trials, with probability of success .7, the expected number of successes is 7.
46. In an experiment with 12 independent trials, with probability of success .7, the variance is 2.52.
47. If a fair coin is tossed 5 times, the probability of getting exactly 3 heads is .5.
48. Markov chains are applicable only when events are independent.
49. Assume we have events A, B, and C, occurring represented by states s_1, s_2, and s_3, respectively, and the transition matrix is $\begin{bmatrix} .2 & .5 & .3 \\ .2 & .7 & .1 \\ .3 & .3 & .4 \end{bmatrix}$. If events A and C each have a 50% chance of occurring the first day and there is a transition every day, the probability that B will occur on the third day is .58.
50. Assume there are 3 urns. The first urn contains 3 red, 5 blue, and 4 green balls. The second contains 4 red, 2 blue, and 3 green balls. The third contains 2 red, 6 blue, and 4 green balls. The probability that a red ball is selected from either urn 1 or urn 3 is $\frac{5}{36}$.
51. Assume there are three urns. The first urn contains 3 red, 5 blue, and 4 green balls. The second contains 6 red, 2 blue, and 4 green balls. The third contains 2 red, 6 blue, and 4 green balls. A blue ball is drawn. The probability that it was chosen from urn 2 is $\frac{1}{12}$.

SUMMARY QUESTIONS

1. How many positive numbers less than 1000 are divisible by 3 or 7 or 11?
2. How many committees with 5 men and 8 women may be selected from a group of 13 men and 19 women?
3. How many positive numbers less than 1000 are relatively prime to 315?
4. How many 3-element subsets are there of the set $\{a, b, c, d, e, f\}$?
5. In how many ways can a president, vice president, secretary, and treasurer be selected from a club with 20 members?
6. How many social security cards are there that have no digit repeated?
7. How many social security cards are there that have at least one digit repeated?
8. In the expansion of $(2x + 3y)^{14}$, find the coefficient of $x^8 y^6$.
9. Determine the next permutation generated after 215436.
10. Let $S = \{1, 2, 3, 4, 5, 6\}$. Find the next combination generated after $\{3, 1, 5, 6\}$.
11. Let two dice be tossed. Find the probability that the sum of the dice is either 5 or 8.
12. How many distinct arrangements can be formed using the letters in *Tennessee*?
13. How many solutions are there to the equation $n_1 + n_2 + n_3 + n_4 = 24$?
14. An urn contains 20 red, 15 blue, and 12 green balls. How large a sample must be taken from the urn to be sure of having 5 red or 6 green or 8 blue balls?
15. If a fair coin is tossed seven times, what is the probability of getting exactly three heads?

ANSWERS TO TRUE-FALSE

1. T 2. F 3. F 4. T 5. F 6. F 7. F 8. T 9. T 10. F 11. T 12. F 13. F 14. F 15. T 16. F 17. F 18. F 19. T 20. F 21. F 22. F 23. T 24. F 25. F 26. T 27. F 28. F 29. T 30. T 31. T 32. F 33. F 34. T 35. T 36. F 37. T 38. F 39. F 40. F 41. F 42. T 43. T 44. F 45. T 46. T 47. F 48. F 49. F 50. T 51. F

CHAPTER 9

Algebraic Structures

9.1 Partially Ordered Sets Revisited

In Definition 2.47, a relation R on A was defined to be a partial ordering if it is reflexive, antisymmetric, and transitive. The set A was said to be a **partially ordered set** or **poset** with ordering R. A relation that is a partial ordering will be denoted by $\leq$ (or $\geq$ where $a \geq b$ if $b \leq a$). The partially ordered set A with ordering $\leq$ is denoted by $(A, \leq)$.

In Definition 2.50, two elements a and b of the partially ordered set (poset) $(S, \leq)$ are **comparable** if either $a \leq b$ or $b \leq a$. If every two elements of a partially ordered set $(S, \leq)$ are comparable, then $(S, \leq)$ is a **total ordering** or a **chain**.

DEFINITION 9.1 For a subset B of a poset A, an element a of A is an **upper bound** of B if $b \leq a$ (or $a \geq b$) for all b in B. The element a is called a **least upper bound** (lub) of B if (i) a is an upper bound of B and (ii) if any other element a' of A is an upper bound of B, then $a \leq a'$. The least upper bound for the entire poset A (if it exists) is called the **greatest element** of A. For a subset B of a poset A, an element a of A is a **lower bound** of B if $a \leq b$ (or $b \geq a$) for all b in B. The element a is called a **greatest lower bound** (glb) of B if (i) a is a lower bound of B and (ii) if any other element a' of A is a lower bound of B, then $a \geq a'$. The greatest lower bound for the entire poset A (if it exists) is called the **least element** of A.

DEFINITION 9.2 An element a of a subset B of a poset A is a **maximal element** of B if for every element b of B, $b \geq a$ implies $b = a$. In other words there is no element of B that is "greater" than a. An element a of a subset B of a poset A is a **minimal element** of B if for every element b of B, $b \leq a$ implies $b = a$. In other words there is no element of B that is "less" than a. Usually the terms minimal and maximal element refer to the entire set, that is, $A = B$.

EXAMPLE 9.3 As in Example 2.48, let $C = \{1, 2, 3\}$ and X be the power set of C:

$$X = P(C) = \{\varnothing, \{1\}, \{2\}, \{3\}, \{1, 2\}, \{1, 3\}, \{2, 3\}, \{1, 2, 3\}\}$$

Define the relation $\leq$ on X by $T \leq V$ if $T \subseteq V$. By definition, $\{1, 2\}$ is the least upper bound of $\{\varnothing, \{1\}, \{2\}\}$ and also of $\{\varnothing, \{1\}, \{2\}, \{1, 2\}\}$. The set $\{1, 2, 3\}$ is the least upper bound of X. The element $\varnothing$ is the greatest lower bound for all three sets. ∎

EXAMPLE 9.4 Consider the set $\hat{X} = \{\varnothing, \{1\}, \{2\}, \{3\}, \{1, 2\}, \{1, 3\}, \{2, 3\}\}$ with the same relation as Example 9.3. In this case $\{1, 2\}$, $\{1, 3\}$, and $\{2, 3\}$ are all maximal elements of $\hat{X}$, but $\hat{X}$ has no greatest element. This situation is possible because none of the three maximal elements are comparable. ∎

EXAMPLE 9.5 Consider the set $\ddot{X} = \{\{1\}, \{2\}, \{3\}, \{1, 2\}, \{1, 3\}, \{2, 3\}, \{1, 2, 3\}\}$ with the same relation as Example 9.3. In this case $\{1\}$, $\{2\}$, and $\{3\}$ are all minimal elements of $\ddot{X}$, but $\ddot{X}$ has no least element. ∎

EXAMPLE 9.6 Let T be the set of all positive divisors of 30 and $\leq_1$ be the relation $m \leq_1 n$ if m divides n evenly. The least upper bound of $\{3, 5\}$ is 15. Obviously the least upper bound 30 and greatest lower bound 1 of T are comparable and $5 \leq_1 15$ because 5 divides 15 evenly, but 5 and 6 are not comparable. ∎

EXAMPLE 9.7 The least upper bound and the greatest lower bound of a set do not have to be members of the set. Let Q be the set of rational numbers and let $\leq$ be the ordinary "less than or equal to" and $<$ be "less than" for rationals. Then the set $A = \{x \in Q : 0 < x < 1\}$ partially ordered by $\leq$ is such that $0 = \text{glb}(A)$ and $1 = \text{lub}(A)$ but $0, 1 \notin A$. ∎

If every two-element subset of a poset A has a least upper bound in A, then the following binary relation can be defined on the set. If a and b belong to A, let $a \vee b = \text{lub}\{a, b\}$. If Boolean algebra notation is used, then $a + b$ is used instead of $a \vee b$.

DEFINITION 9.8 A poset A for which all two-element subsets have a least upper bound in A is called an **upper semilattice** and is denoted by $(A, \vee)$ or $(A, +)$.

If all two-element subsets of a poset A have a greatest lower bound in A, then the following binary relation can be defined on the set. If a and b belong to A, let $a \wedge b = \text{glb}\{a, b\}$. If Boolean algebra notation is used, then $a \cdot b$ is used instead of $a \wedge b$.

DEFINITION 9.9 A poset A for which every two element subset has a greatest lower bound in A is called a **lower semilattice** and is denoted by $(A, \wedge)$ or $(A, \cdot)$.

In Example 9.3, $\{1, 2\} \vee \{1, 3\} = \{1, 2, 3\}$ and $\{1, 2\} \wedge \{1, 3\} = \{1\}$. The least upper bound and greatest lower bound are, respectively, the union and intersection of the sets and the notation is consistent with the notation of set theory.

In Example 9.6, $6 \vee 15 = 30$ and $6 \wedge 15 = 3$. The least upper bound and greatest lower bound are, respectively, the least common multiple and the greatest common divisor.

The proof of the following theorem, which gives algebraic properties of semilattices, is left to the reader.

THEOREM 9.10

(a) Let A be an upper semilattice. Then for all $a, b, c \in A$, $a \vee (b \vee c) = (a \vee b) \vee c$, $a \vee a = a$ and $a \vee b = b \vee a$.

(b) Let A be a lower semilattice. Then for all $a, b, c \in A$, $a \wedge (b \wedge c) = (a \wedge b) \wedge c$, $a \wedge a = a$ and $a \wedge b = b \wedge a$. ∎

Exercises

1. Draw the Hasse diagram (see Chapter 2 see Section 2.6) for the poset in Example 9.3.
2. Draw the Hasse diagram (see Chapter 2 see Section 2.6) for the poset in Example 9.6.
3. Draw the Hasse diagram (see Chapter 2 see Section 2.6) for the poset in Example 9.4.
4. Draw the Hasse diagram (see Chapter 2 see Section 2.6) for the poset in Example 9.5.

Given the Hasse diagram of a poset in Figure 9.1:

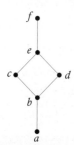

Figure 9.1

5. Find the greatest element of the poset (if it exists).
6. Find the least element of the poset (if it exists).
7. Find the maximal and minimal elements of the poset.
8. Is the poset an upper or lower semilattice (or both)?

Given the Hasse diagram of a poset in Figure 9.2:

Figure 9.2

9. Find the greatest element of the poset (if it exists).
10. Find the least element of the poset (if it exists).
11. Find the maximal and minimal elements of the poset.
12. Is the poset an upper or lower semilattice (or both)?

Given the Hasse diagram of a poset in Figure 9.3:

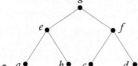

Figure 9.3

13. Find the greatest element of the poset (if it exists).
14. Find the least element of the poset (if it exists).
15. Find the maximal and minimal elements of the poset.
16. Is the poset an upper or lower semilattice (or both)?

Given the Hasse diagram of a poset in Figure 9.4:

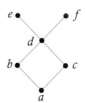

Figure 9.4

17. Find the greatest element of the poset (if it exists).
18. Find the least element of the poset (if it exists).
19. Find the maximal and minimal elements of the poset.
20. Is the poset an upper or lower semilattice (or both)?

Given the Hasse diagram of a poset in Figure 9.5:

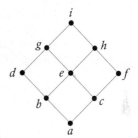

Figure 9.5

21. Find the greatest element of the poset (if it exists).
22. Find the least element of the poset (if it exists).
23. Find the maximal and minimal elements of the poset.
24. Is the poset an upper or lower semilattice (or both)?

Based on the Hasse diagrams of the preceding exercises, describe the feature of a Hasse diagram of a poset that ensures that:

25. A poset has a maximal element.
26. A poset has a least upper bound (or greatest element).
27. A poset has a minimal element.
28. A poset has a greatest lower bound (or least element).

Prove the following:

29. Let A be an upper semilattice. Then $a \vee (b \vee c) = (a \vee b) \vee c$ for all $a, b, c \in A$, $a \vee a = a$ for all $a \in A$ and $a \vee b = b \vee a$ for all $a, b \in A$.

30. Let A be a lower semilattice. Then $a \wedge (b \wedge c) = (a \wedge b) \wedge c$ for all $a, b, c \in A$, $a \wedge a = a$ for all $a \in A$ and $a \wedge b = b \wedge a$ for all $a, b \in A$.

31. Prove that every finite poset has a maximal element and a minimal element.

32. Prove that every finite upper semilattice has a least upper bound (or greatest element).

33. Prove that every finite lower semilattice has a greatest lower bound (or least element).

In Chapter 2 we defined the Boolean operations $\vee$ and $\wedge$ on the set $\{0, 1\}$ by

$\vee$	0	1
0	0	1
1	1	1

and

$\wedge$	0	1
0	0	0
1	0	1

We also defined a Boolean matrix to be matrix whose entries are all either 0 or 1. If $A = [A_{ij}]$ and $B = [B_{ij}]$ are $m \times n$ Boolean matrices

$U = A \vee B$ is defined by $U_{ij} = A_{ij} \vee B_{ij}$

for $1 \leq i \leq m, 1 \leq j \leq n$

$V = A \wedge B$ is defined by $V_{ij} = A_{ij} \wedge B_{ij}$

for $1 \leq i \leq m, 1 \leq j \leq n$

Let S be the set of all $n \times n$ Boolean matrices. Define a relation $\leq$ on S by $A \leq B$ for $A, B \in S$ if $A_{ij} \leq B_{ij}$ for $1 \leq i, j \leq n$.

34. Prove that S with this relation is a poset.

35. Prove that $A \vee B$ is the least upper bound of $\{A, B\}$.

36. Prove that $A \wedge B$ is the greatest lower bound of $\{A, B\}$.

37. Prove that S is both a lower and an upper semilattice.

38. What are the greatest and least elements?

39. Prove that if a partially ordered set has a greatest element then it is unique.

9.2 Semigroups and Semilattices

In Definition 4.6, a binary operation on the set S was defined to be a function $b : S \times S \to S$. Because the range of a binary operation on S is a subset of S, by definition, a binary operation exhibits the property of closure wherein the operation on two members r and s of S is also a member of S. There are many examples of binary operations.

For example, if S is the set of positive integers, then both the product and the sum are binary operations since both the sum and product of positive integers are positive integers. In many cases the operation is associative, for example, in addition of positive integers $(a + b) + c = a + (b + c)$ and in multiplication of positive integers $(a \cdot b) \cdot c = a \cdot (b \cdot c)$. Let S be the set of $n \times n$ matrices whose elements are integers. The sum and the product of S are again elements of S. Further, if A, B, and C are $n \times n$ matrices, then $(A \cdot B) \cdot C = A \cdot (B \cdot C)$ and $(A + B) + C = A + (B + C)$. In the following definition we give a name to binary operators with this property:

DEFINITION 9.11 A set S with binary operation $*$ on S such that for all a, b, and c in S, $(a*b)*c = a*(b*c)$ is called a **semigroup** and is denoted by $(S, *)$ or simply S if the operation is understood. If, in addition, $a*b = b*a$ for all a and b in S, then S with operation $*$ is called an **Abelian** or **commutative semigroup**. If there is an element I in $(S, *)$ such that $I*a = a*I = a$ for all a in A, then I is called the **identity** of $(S, *)$ and $(S, *)$ is called a **semigroup with identity** or a **monoid**.

DEFINITION 9.12 If $(S, *)$ is a semigroup and $\overline{S} \subseteq S$, then $\overline{S}$ is a **subsemigroup** of S if $*$ is a binary operation on $\overline{S}$. Equivalently, $(\overline{S}, *)$ is a subsemigroup of $(S, *)$ if $\overline{S} \subseteq S$ and for every $a, b \in \overline{S}$, $a*b \in \overline{S}$.

We again point out that the fact that $*$ is a binary operation on S indicates that if $s, s' \in S$, then $s * s' \in S$. This property is called **closure**.

9.2 Semigroups and Semilattices

EXAMPLE 9.13 The set S of $n \times n$ matrices, $(S, \cdot)$ is a monoid. The identity I is the matrix whose entries are 1 along the main diagonal and 0 elsewhere so that

$$I_{ij} = \begin{cases} 1 & \text{if } i = j \\ 0 & \text{if } i \neq j \end{cases}$$

Hence, the 3×3 identity matrix is

$$\begin{bmatrix} 1 & 0 & 0 \\ 0 & 1 & 0 \\ 0 & 0 & 1 \end{bmatrix}$$

■

EXAMPLE 9.14 If $(S, \cdot)$ is the semigroup of $n \times n$ matrices, whose elements are rational numbers, whose operation is matrix multiplication and if $(\overline{S}, \cdot)$ is the semigroup of $n \times n$ matrices whose entries are integers, then $(\overline{S}, \cdot)$ is a subsemigroup of $(S, \cdot)$. ■

EXAMPLE 9.15 For a positive integer n, let $S_n^0 = \{x : x \text{ is an integer and } x \geq n\} \cup \{0\}$. S_n^0 is a commutative monoid under integer addition with identity 0. Let $S_n^1 = \{x : x \text{ is an integer and } x \geq n\} \cup \{1\}$. S_n^1 is a commutative monoid under integer multiplication with identity 1. If $m \geq n$, then S_m^0 is a subsemigroup of S_n^0 and S_m^1 is a subsemigroup of S_n^1. ■

EXAMPLE 9.16 Let S be the set of all functions from a nonempty set A to itself with the binary operation composition of functions. Thus, for $f, g \in S$, $f \circ g$ is defined by $(f \circ g)(a) = f(g(a))$ for all $a \in A$. Since for $f, g, h \in S$, $((f \circ g) \circ h)(a) = (f \circ (g \circ h))(a) = f(g(h(a)))$ for all $a \in A$, $(f \circ g) \circ h = f \circ (g \circ h)$ so S is a semigroup. Further, the function $I : A \to A$, defined by $I(a) = a$ for all $a \in A$, is the identity of S so that S is a monoid. ■

EXAMPLE 9.17 Let A be a finite collection of symbols. This set of symbols is called an **alphabet**. For example, A could be a subset of the English alphabet or it might simply consist of the set $\{0, 1\}$. A **string** or **word** of symbols of A has the form $a_1 a_2 a_3 a_4 \ldots a_n$ where $a_i \in A$. Thus, if $A = \{0, 1\}$, then 1, 100, 101, 11011, and 0001 would all be strings of symbols of A. In addition we define an empty string denoted by λ, which has no symbols in it. Let A^* denote the set of all strings of A. Define the binary operation $\circ$ called **concatenation** on A^* as follows: If $a_1 a_2 a_3 a_4 \ldots a_n$ and $b_1 b_2 b_3 b_4 \ldots b_m \in A^*$ then

$$a_1 a_2 a_3 a_4 \ldots a_n \circ b_1 b_2 b_3 b_4 \ldots b_m = a_1 a_2 a_3 a_4 \ldots a_n b_1 b_2 b_3 b_4 \ldots b_m$$

Thus, if $A = \{0, 1\}$, then $11011 \circ 100010 = 11011100010$. In particular, if ω is a string in A^*, then $\lambda \circ \omega = \omega \circ \lambda = \omega$, which simply says that a string preceded or followed by the empty string simply gives the original string.

Let $a_1 a_2 \ldots a_n$, $b_1 b_2 \ldots b_m$ and $c_1 c_2 \ldots c_p \in A^*$. Then

$$(a_1 a_2 \ldots a_n \circ b_1 b_2 \ldots b_m) \circ c_1 c_2 \ldots c_p$$
$$= (a_1 a_2 \ldots a_n \circ b_1 b_2 \ldots b_m) \circ c_1 c_2 \ldots c_p$$
$$= a_1 a_2 \ldots a_n b_1 b_2 \ldots b_m \circ c_1 c_2 \ldots c_p$$
$$= a_1 a_2 \ldots a_n b_1 b_2 \ldots b_m c_1 c_2 \ldots c_p$$
$$= a_1 a_2 \ldots a_n \circ b_1 b_2 \ldots b_m c_1 c_2 \ldots c_p$$
$$= a_1 a_2 \ldots a_n \circ (b_1 b_2 \ldots b_m \circ c_1 c_2 \ldots c_p)$$

and the binary operation concatenation is associative on A^* so that A^* with the concatenation operation is a semigroup. Further, since λ is an identity, A^* is a monoid. Note that this monoid is certainly not commutative. If $\overline{A} \subseteq A$, then $\overline{A}^*$ is a subsemigroup of A^*. The semigroup A^* is called a **free semigroup** on the alphabet A. ∎

EXAMPLE 9.18 Let $Z_n = \{[0], [1], [2], \ldots, [n-1]\}$ be the set of integers modulo n (see Section 3.6). By Theorem 3.55, $(Z_n, +)$ and $(Z_n, \cdot)$ are both well defined. It is left to the reader to prove that they are both semigroups. ∎

Let $(S, \cdot)$ be a semigroup and $a \in S$. Define a^n recursively by $a^1 = 1$ and $a^n = a \cdot a^{n-1}$ for $n \geq 1$. It follows immediately that $a^1 = a$. Using induction, the reader is asked to show that $a^k \cdot a^m = a^{k+m}$ for all integers $k, m \geq 0$. The set

$$\langle a \rangle = \{a^n : n > 0\} = \{a, a^2, a^3, \ldots\}$$

is a subsemigroup of S.

DEFINITION 9.19 The semigroup $\langle a \rangle$ is called a cyclic semigroup. More precisely, it is called the cyclic semigroup generated by a.

The proof of the following theorem is left to the reader.

THEOREM 9.20 Let $(S, \cdot)$ be a semigroup and $a_1, a_2, a_3, \ldots, a_k \in S$. Let $A = \{a_1, a_2, a_3, \ldots, a_k\}$ and $A^* = \langle a_1, a_2, a_3, \ldots, a_k \rangle$ be the set consisting of all finite products of $a_1, a_2, a_3, \ldots, a_k$. Then A^* is a semigroup. Furthermore, A^* is the smallest semigroup of S containing A. ∎

DEFINITION 9.21 The semigroup A^* is called the semigroup generated by A. If for every proper subset B of A, $B^* \neq A^*$, then A is called a **minimal generating set** of A^*.

DEFINITION 9.22 Let $(S, \cdot)$ and $(T, \circ)$ be semigroups and $f : S \to T$ be a function such that $f(s \cdot s') = f(s) \circ f(s')$. The function f is called a **homomorphism** from S to T.

THEOREM 9.23 Let $(S, \cdot)$ and $(T, \circ)$ be semigroups and $f : S \to T$ be a homomorphism from S to T. If $\overline{S}$ is a subsemigroup of S, then $f(\overline{S})$ is a semigroup of T.

Proof Let $t, t' \in f(\overline{S})$. There exist $s, s' \in \overline{S}$ so that $f(s) = t$ and $f(s') = t'$. By definition of semigroup, $s \cdot s' \in \overline{S}$. By definition of homomorphism, $f(s \cdot s') = f(s) \circ f(s') = t \cdot t'$, so $t \cdot t' \in f(\overline{S})$ and $f(\overline{S})$ is a semigroup. ∎

The proof of the following theorem is left to the reader.

THEOREM 9.24 Let $(S, \cdot)$ and $(T, \circ)$ be semigroups and $f : S \to T$ be a homomorphism from S to T. If $\overline{T}$ is a subsemigroup of T, then $f^{-1}(\overline{T})$ is a subsemigroup of S. ∎

DEFINITION 9.25 Let $(S, \cdot)$ be a semigroup and R be an equivalence relation on S. If R has the property that, if $s_1 R s_2$ and $s_3 R s_4$ then $s_1 s_3 R s_2 s_4$ for all $s_1, s_3, s_2, s_4 \in S$, then R is called a **congruence relation**.

9.2 Semigroups and Semilattices

THEOREM 9.26 The equivalence classes of a congruence relation R on a semigroup S form a semigroup defined by $[a] \circ [b] = [ab]$.

Proof Left to the reader. ∎

DEFINITION 9.27 The semigroup formed by a congruence relation R on a semigroup S is called the **quotient semigroup** and is denoted by S/R.

THEOREM 9.28 Let $(S, \cdot)$ and $(T, \circ)$ be semigroups and $f : S \to T$ be a homomorphism from S to T. Define the relation R on S by sRs' if $f(s) = f(s')$. The relation R is a congruence relation.

Proof For all $s \in S$, $f(s) = f(s)$, so R is reflexive. If $f(s) = f(s')$, then $f(s') = f(s)$, and R is symmetric. If $f(s) = f(s')$ and $f(s') = f(s'')$, then $f(s) = f(s'')$, so R is transitive. Assume $s_1 R s_2$ and $s_3 R s_4$. Therefore, $f(s_1) = f(s_2)$ and $f(s_3) = f(s_4)$. Since f is a homomorphism,

$$f(s_1 \cdot s_3) = f(s_1) \circ f(s_3)$$

and

$$f(s_1 \cdot s_3) = f(s_1) \circ f(s_3)$$
$$= f(s_2) \circ f(s_4)$$
$$= f(s_2 \cdot s_4)$$

so

$$(s_1 \cdot s_3) R (s_2 \cdot s_4)$$

∎

DEFINITION 9.29 A commutative semigroup $(S, *)$ is a **semilattice** if $a * a = a$ for all $a \in S$. An element a of a semigroup is called an **idempotent** element if $a * a = a$. A semilattice is, therefore, a commutative semigroup in which every element is an idempotent. If $(S, *)$ is a semilattice and $\overline{S} \subseteq S$, then $\overline{S}$ is a subsemilattice of S if $*$ is a binary operation on $\overline{S}$. Equivalently, $(\overline{S}, *)$ is a subsemilattice of $(S, *)$ if $\overline{S} \subseteq S$ and for every $a, b \in \overline{S}$, $a * b \in \overline{S}$.

It follows from Theorem 9.10 that upper semilattices and lower semilattices are semilattices. Moreover, every semilattice can be considered as either an upper or lower semilattice.

THEOREM 9.30 Let S be a semilattice. Define the relation $\leq$ on S by $a \leq b$ if $a * b = b$ for $a, b \in S$. Then $(S, \leq)$ is a poset and $a * b$ is the least upper bound of a and b. Hence, $(S, *)$ is an upper semilattice. Similarly $(S, *)$ can be considered as a lower semilattice. Define the relation by $a \leq b$ if $a * b = a$.

Proof Since $a * a = a$ for all $a \in S$, $a \leq a$, so $(S, \leq)$ is reflexive. If $a \leq b$ and $b \leq a$, then $a * b = b$ and $a * b = b * a = a$. Therefore, $b = a$ and $(S, \leq)$ is antisymmetric. Assume $a \leq b$ and $b \leq c$. The $a * b = b$ and $b * c = c$. So that

$$a * c = a * (b * c)$$
$$= (a * b) * c$$
$$= b * c$$
$$= c$$

and $a \leq c$. Therefore, $(S, \leq)$ is transitive and so is a poset.

We next show that $a * b$ is the least upper bound of a and b. Since
$$a * (a * b) = (a * a) * b$$
$$= a * b$$
then $a \leq a * b$. Also
$$b * (a * b) = b * (b * a)$$
$$= (b * b) * a$$
$$= b * a$$
$$= a * b$$
so that $b \leq a * b$. Therefore, $a * b$ is an upper bound for a and b. Assume c is an upper bound for a and b, then $a * c = c$ and $b * c = c$. Therefore,
$$(a * b) * c = a * (b * c)$$
$$= a * c$$
$$= c$$
and $a * b \leq c$. Therefore, $a * b$ is the least upper bound of a and b and $(S, \leq)$ is an upper semilattice. ∎

Exercises

1. Let $(S, \cup)$ be the semilattice of all subsets of $\{a, b, c\}$ under the operation union. Find three distinct subsemilattices of $(S, \cup)$.

2. Let $(S, \cdot)$ be the set of all matrices whose entries are real numbers and whose operation is matrix multiplication. Find five subsemigroups of $(S, \cdot)$.

3. Prove that the set of positive integers of the form $4k + 1$ where k is nonnegative integer forms a semigroup under multiplication.

4. Give five subsemigroups of the positive rational numbers under addition.

5. Prove that $(Z_n, +)$ and $(Z_n, \cdot)$ are both semigroups for any integer $n \geq 1$.

6. Prove, using induction, that if S is a semigroup and $a \in S$, then $a^n \cdot a^m = a^{n+m}$ for all integers $m, n \geq 0$.

7. Prove Theorem 9.20. Let $(S, \cdot)$ be a semigroup and $a_1, a_2, a_3, \ldots, a_k \in S$. Let $A = \{a_1, a_2, a_3, \ldots, a_k\}$ and $A^* = \langle a_1, a_2, a_3, \ldots, a_k \rangle$ be the set consisting of all finite products of a_1, a_2, a_3, a_k. Then A^* is a semigroup. Furthermore, A^* is the smallest semigroup of S containing A.

8. Prove Theorem 9.24. Let $(S, \cdot)$ and $(T, \circ)$ be semigroups and $f : S \to T$ be a homomorphism from S to T. If $\overline{T}$ is a subsemigroup of T, then $f^{-1}(\overline{T})$ is a semigroup of T.

   Let $(S, \circ)$ be the semigroup of all functions on a set A with composition as the operation.

9. Show that the set of one-to-one correspondences forms a subsemigroup of S.

10. If $f \in S$ is an idempotent, then $f \circ f = f$. Such a function is called a **retraction map**. The image of f is called a **retract**. Show that a retraction map is a one-to-one map on its image.

11. Show that the real valued functions $f(x) = |x|$, $g(x) = \lfloor x \rfloor$, and $h(x) = \lceil x \rceil$ are all retraction maps. What are the corresponding retracts?

12. Do the functions in Problem 11 commute; that is, does $f \circ g = g \circ f$?

13. Do the functions in Problem 11 form a semilattice under composition?

14. Let S be the set of real numbers and $*$ be the binary operation defined by $a * b = \max(a, b)$. Show that $(S, *)$ is a semigroup. Is it a semilattice?

15. Let $(S, *)$ be the set of $n \times n$ matrices with multiplication as the binary operation. It is known that for $n \times n$ matrices A and B, $\det(A) \cdot (\det B) = \det(AB)$. Accepting this fact, show that the matrices with $\det \neq 0$ form a monoid. What can be said about the matrices whose determinant is equal to 0?

16. Give an example of a binary operation $*$ on the real numbers such that they do not form a semigroup.

17. Show that the identity of a monoid is unique in the sense that there is only one element, say I, such that $a * I = I * a = a$, for all a in the monoid.

18. Let S be a set. Define the operation $*$ on S by $a * b = b$ for all $a, b \in S$. Prove that $(S, *)$ is a semigroup in which every element is an idempotent. Why is it not a semilattice?

19. Prove Theorem 9.26. The equivalence classes of a congruence relation form a semigroup defined by $[a] \circ [b] = [ab]$.

9.3 Lattices

In the previous section, we considered sets with only one binary operations. In this section we consider sets that have two binary operations.

DEFINITION 9.31 A poset $(S, \leq)$ that is both an upper and a lower semilattice is a **lattice**.

In a lattice $(S, \leq)$, denote the least upper bound of $a, b \in S$ by $a \vee b$ (called the **join** of a and b) and the greatest lower bound of $a, b \in S$ by $a \wedge b$ (called the **meet** of a and b). The lattice will be denoted by $(S, \vee, \wedge)$ to emphasize the binary operations involved. Using the properties of least upper bound and greatest lower bound, we prove the following properties of a lattice, which can be used as an alternate definition.

THEOREM 9.32 Let $(S, \vee, \wedge)$ be a lattice; then the following properties are satisfied for all $a, b, c \in S$:

(a) Commutativity

$$a \wedge b = b \wedge a$$
$$a \vee b = b \vee a$$

(b) Associativity

$$(a \wedge b) \wedge c = a \wedge (b \wedge c)$$
$$(a \vee b) \vee c = a \vee (b \vee c)$$

(c) Absorption

$$a \wedge (a \vee b) = a$$
$$a \vee (a \wedge b) = a$$

Proof For parts (a) and (b), see Theorem 9.10 and the exercises at the end of the section. To show $a \wedge (a \vee b) = a$,

$$a \leq a \quad \text{by definition of partial ordering}$$
$$a \leq a \vee b \quad \text{by definition of least upper bound}$$

Hence, a is a lower bound of $\{a, a \vee b\}$. Assume c is a lower bound of $\{a, a \vee b\}$. Then $c \leq a$ so that a is the greatest lower bound of $\{a, a \vee b\}$ and, consequently, $a = a \wedge (a \vee b)$. The proof that $a \vee (a \wedge b) = a$ is left to the reader. ■

EXAMPLE 9.33 Let S be the power set (set of all subsets) of a set A with the partial ordering $U \leq V$ if $U \subseteq V$ for all $U, V \in S$. Then $\text{lub}\{U, V\} = U \cup V$ and $\text{glb}\{U, V\} = U \cap V$, so $(S, \cup, \cap)$ is a lattice. ■

EXAMPLE 9.34 Let S be the set of positive integers with the partial ordering $n \leq m$ if n divides into m evenly. Then $\text{lub}\{m, n\} = \text{lcm}(m, n)$ and $\text{glb}\{m, n\} = \gcd(m, n)$ for all $m, n \in S$. Therefore, S with the binary operators lcm and gcd is a lattice. ∎

EXAMPLE 9.35 The rational numbers, real numbers, and integers with the usual partial ordering all form a lattice where $\text{lub}\{m, n\} = \max(m, n)$ and $\text{glb}\{m, n\} = \min(m, n)$. ∎

DEFINITION 9.36 A nonempty subset $\overline{S}$ of a lattice $(S, \vee, \wedge)$ is a **sublattice** of S if for all $a, b \in \overline{S}$, $a \wedge b$ and $a \vee b$ are in $\overline{S}$.

Therefore, in Example 9.33, if B is a subset of A and $\overline{S}$ is the power set of B, then $(\overline{S}, \cup, \cap)$ is a sublattice of $(S, \cup, \cap)$. In Example 9.34, if $\overline{S}$ is the set of all positive factors of a fixed integer n, then $\overline{S}$ with the binary operators lcm and gcd is a sublattice of S. In Example 9.35, the integers are a sublattice of the rational numbers and the rational numbers are a sublattice of the real numbers.

DEFINITION 9.37 A lattice $(S, \vee, \wedge)$ is **bounded** if the set S, considered as a poset, has a least upper bound and a greatest lower bound. The least upper bound is denoted by 1 and the greatest lower bound is denoted by 0. Equivalently, a lattice is bounded if there exists elements $0, 1 \in S$ such that $0 \wedge a = 0$ and $1 \vee a = 1$ for all $a \in S$.

Since 0 is the least element in the set and 1 is the greatest element in the lattice, we immediately have the following result.

THEOREM 9.38 In a bounded lattice, $1 \wedge a = a$ and $0 \vee a = a$ for all a in the lattice. ∎

EXAMPLE 9.39 The lattice in Example 9.33 is bounded where A is the maximal element for the lattice and $\varnothing$ is the minimal element. ∎

EXAMPLE 9.40 In Example 9.34, if $\overline{S}$ is the lattice of all positive factors of a fixed integer n, then $\overline{S}$ is bounded with maximal element n and minimal element 1. ∎

DEFINITION 9.41 Let $(S, \vee, \wedge)$ and $(S', \vee', \wedge')$ be lattices. A function $f : S \to S'$ is a homomorphism if for all $s, t \in S$, $f(s \wedge t) = f(s) \wedge' f(t)$ and $f(s \vee t) = f(s) \vee' f(t)$. If S and S' are bounded homomorphisms, then $f : S \to S'$ is a homomorphism between bounded lattices S and S' if it is a homomorphism and also $f(0) = 0'$ and $f(1) = 1'$.

DEFINITION 9.42 Let $(S, \vee, \wedge)$ and $(S', \vee', \wedge')$ be lattices. A function $f : S \to S'$ is an isomorphism if f is a homomorphism and f is one-to-one and onto.

EXAMPLE 9.43 Let $(S, \vee, \wedge)$ and $(S', \vee', \wedge')$ be the lattices in Figures 9.6 and 9.7 respectively $f : S \to S'$ defined by $f(0) = f(a) = 0'$, $f(1) = f(d) = 1'$, $f(b) = f(b')$, and $f(c) = f(c')$ is a bounded homomorphism. ∎

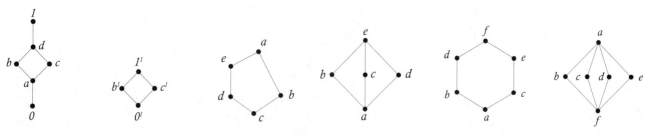

Figure 9.6   Figure 9.7   Figure 9.8   Figure 9.9

EXAMPLE 9.44 Let $A = \{a, b, c\}$ and $B = \{a, b\}$. Let $S = (\mathcal{P}(A), \cup, \cap)$ and $S' = (\mathcal{P}(B), \cup, \cap)$ be the lattices. Then $f : S \to S'$ is an homomorphism where for a given set C in S, $f(C) = f(C')$ where $C' = C - \{c\}$. ∎

In Chapter 2, we noted that for sets, A, B, and C, $A \cap (B \cup C) = (A \cap B) \cup (A \cap C)$ and $A \cup (B \cap C) = (A \cup B) \cap (A \cup C)$, which we called the distributive property of sets. We now define a similar property for lattices.

DEFINITION 9.45 A lattice $(S, \vee, \wedge)$ is a **distributive lattice** if for all $a, b, c \in S$,

$$a \wedge (b \vee c) = (a \wedge b) \vee (a \wedge c)$$
$$a \vee (b \wedge c) = (a \vee b) \wedge (a \vee c)$$

It is left to the reader to show that the lattices in Figure 9.8 are not distributive.

Although it is beyond the scope of this book, it can be shown that a lattice is distributive if and only if it does not contain either of these two lattices as sublattices. Therefore, the two lattices in Figure 9.9 would not be distributive.

A Boolean algebra, which was described in Section 2.3, is a special case of a bounded distributive lattice. At this point we change to the Boolean algebra notation of Section 2.3, replacing the binary operator $\vee$ by $+$ and the binary operator $\wedge$ by $\cdot$. Hence, a lattice is denoted as $(S, +, \cdot)$. We repeat the definition here for convenience. For further details see Section 2.3.

DEFINITION 9.46 A **Boolean algebra** is a bounded distributive lattice $(S, +, \cdot)$, which has a unary operator $'$ on the set S such that $x + x' = 1$ and $x \cdot x' = 0$. For $x \in S$, x' is called the **complement** of x. Hence, a Boolean algebra, which will be denoted by $(S, +, \cdot, ', 1, 0)$, has the following properties for all $x, y, z \in S$:

(1) Associativity

$$x + (y + z) = (x + y) + z$$
$$x \cdot (y \cdot z) = (x \cdot y) \cdot z$$

(2) Commutativity

$$x + y = y + x$$
$$x \cdot y = y \cdot x$$

(3) Distributive Laws

$$x \cdot (y + z) = (x \cdot y) + (x \cdot z)$$
$$x + (y \cdot z) = (x + y) \cdot (x + z)$$

(4) Complements

$$x + x' = 1 \quad x \cdot x' = 0$$

(5) Identities

$$x + 0 = x$$
$$x \cdot 1 = x$$

In the problems in Section 2.6 we gave the following definition of an atom of a Boolean algebra.

DEFINITION 9.47 Let B be a Boolean algebra. An element $x \in B$ is called an **atom** of B if for all $y \in B$, if $y \leq x$, then $y = 0$ or $y = x$.

The student was then asked to prove the following problems, which we list here as theorems.

THEOREM 9.48 Let B be a Boolean algebra. Prove that if x is an atom of B, and $y \in B$, then either $xy = 0$ or $xy = x$. ∎

THEOREM 9.49 Let B be a Boolean algebra. Prove that if x and y are unique atoms of B, then $xy = 0$. ∎

Using the following set of theorems, we prove that if a Boolean algebra has n atoms, then it contains 2^n elements and is isomorphic to a Boolean algebra consisting of the subsets of a set with n elements.

THEOREM 9.50 Let $x_1, x_2, x_3, \ldots x_n$ be the atoms of a Boolean algebra B. If $x \in B$, and $xx_i = 0$ for all $1 \leq i \leq n$, then $x = 0$.

Proof Assume $x \neq 0$ and let $T = \{u : u \in B \text{ and } 0 \leq u \leq x\}$. Since $x \in T$, T is not empty. Since B is finite, there exists a least u such that $0 < u \leq x$. The element u is an atom, since if there were $v \neq u$ $0 < v \leq u \leq x$ this would contradict the definition of u. But $ux = u \neq 0$. This gives a contradiction, so by the principle of Reductio ad Absurdum, we must conclude that $x = 0$. ∎

THEOREM 9.51 Let B be a finite Boolean algebra with atoms $x_1, x_2, x_3, \ldots x_n$. Every nonzero element u of B can be written as a sum of atoms of B. This sum is unique except for the order of the atoms in the sum.

Proof Note that since $+$ is a commutative operation, any rearrangement of the sum of elements in B gives the same result.

For $x \neq 0$, let $T = \{u : u \text{ is an atom of } B \text{ and } 0 \leq u \leq x\} = \{u : u \text{ is an atom of } B \text{ and } ux = x\}$. Let $y = \hat{x}_1 + \hat{x}_2 + \hat{x}_3 + \cdots + \hat{x}_k$ be a sum of the elements of T. $xy = x(\hat{x}_1 + \hat{x}_2 + \hat{x}_3 \cdots + \hat{x}_k) = x\hat{x}_1 + x\hat{x}_2 + x\hat{x}_3 + \cdots + x\hat{x}_k$ by the distributive law.

By Theorem 9.48, $x\hat{x}_i = \hat{x}_i$ for all $1 \leq i \leq k$. Therefore,

$$x\hat{x}_1 + x\hat{x}_2 + x\hat{x}_3 + \cdots + x\hat{x}_k = \hat{x}_1 + \hat{x}_2 + \hat{x}_3 \cdots + \hat{x}_k = y$$

and $xy = y$.

If $x_i \notin T$, then by definition of T and Theorem 9.48, $x_i x = 0$ and $xy'x_i = y'xx_i = 0$. If $x_i \in T$, since $y' = (\hat{x}_1 + \hat{x}_2 + \hat{x}_3 \cdots + \hat{x}_k)' = \hat{x}'_1 \hat{x}'_2 \hat{x}'_3 \cdots \hat{x}'_k$ and

$\hat{x}_i \hat{x}'_i = 0$ for all $1 \leq i \leq k$, $\hat{x}_i \hat{x}'_1 \hat{x}'_2 \hat{x}'_3 \cdots \hat{x}'_k = 0$. Therefore, $\hat{x}_i y' = 0$. Hence, $\hat{x}_i y' x = xy' \hat{x}_i = 0$. Thus, $xy' x_i = 0$ for all $1 \leq i \leq n$. Hence by Theorem 9.50 $xy' = 0$, $x1 = x(y + y') = xy + xy' = xy + 0 = xy = y$, and x can be written as a sum of the elements of T.

If $x_i \notin T$, suppose there are two expressions for u as summands of atoms. Assume x_i is in one summand but not in the other. If x_i appears in the summand, then $x_i u = x_i$. If not, then $x_i u = 0$. This is clearly impossible so the two expressions must contain the same summands of atoms. ∎

The following theorems are left to the reader:

THEOREM 9.52 Let B be a finite Boolean algebra with atoms $x_1, x_2, x_3, \ldots x_n$. Let $B' = (S, \cup, \cap, A, \emptyset)$ be the finite Boolean algebra of all subsets of $A = \{x_1, x_2, x_3, \ldots x_n\}$. The function f, which maps sums of atoms into the subset containing the atoms in the sum, is an isomorphism from B to B'. Thus, every finite Boolean algebra is isomorphic to the Boolean algebra of the subsets of a set. ∎

In the problems in Section 2,4, we showed that β_n denotes the set of all strings $0's$, and $1's$ of length n. For example, 10010101 is an element of β_8. For fixed n, there are operations $\cdot$ and $+$ on β_n so that β_n is a Boolean algebra.

THEOREM 9.53 Let B be a finite Boolean algebra with atoms $x_1, x_2, x_3, \ldots x_n$. B is isomorphic to β_n and, hence, contains 2^n elements. ∎

Exercises

1. Which of the following represent lattices?

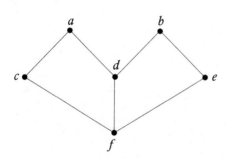

2.

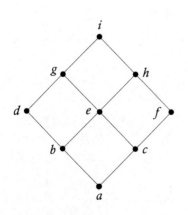

3.

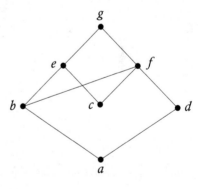

4.

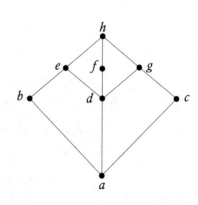

370 Chapter 9 Algebraic Structures

5.

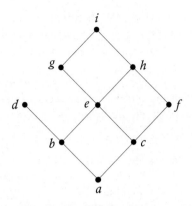

6. Show that the following lattices are not distributive.

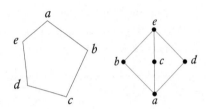

7. Show that the lattice in Figure 9.10 is not distributive.

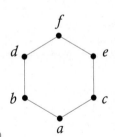

Figure 9.10

8. Show that the lattice in Figure 9.11 is not distributive.

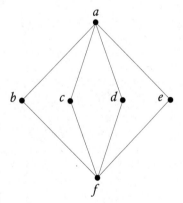

Figure 9.11

9. Which of the following lattices are distributive?

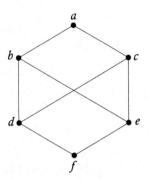

10.

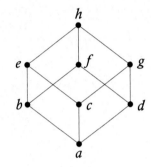

11.

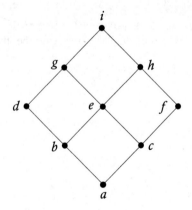

12.

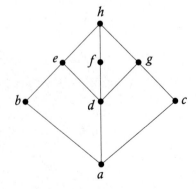

13.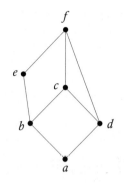

14.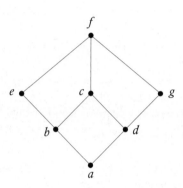

15. Prove that for all a, b in a lattice, $a \vee (a \wedge b) = a$.

16. Let S be the power set (the set of all subsets) of $\{a, b, c\}$. Let $(S, \cup, \cap, ', \varnothing)$ be the Boolean algebra consisting of S and the usual definitions of intersection and union. What are the atoms of the set?

17. Let S be the set of all statements formed from p, q, and r using the usual propositional connectives, $\wedge$, $\vee$, and $\sim$. Show that in the Boolean algebra $(S, \vee, \wedge, \sim, T, F)$, the atoms are simply the miniterms defined in Section 1.6.

18. Find the atoms of the Boolean algebra given in Figure 9.12.

    Figure 9.12

19. Find the atoms of the Boolean algebra given in Figure 9.13.

    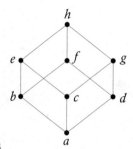

    Figure 9.13

20. Prove Theorem 9.52. Let B be a finite Boolean algebra with atoms $x_1, x_2, x_3, \ldots x_n$. Let $B' = (S, \cup, \cap, ', A, \varnothing)$ be the finite Boolean algebra of all subsets of $A = \{x_1, x_2, x_3, \ldots x_n\}$. The function f, which maps sums of atoms into the subset containing the atoms in the sum, is an isomorphism from B to B'. Thus, every finite Boolean algebra is isomorphic to the Boolean algebra of the subsets of a set.

21. Prove Theorem 9.53. Let B be a finite Boolean algebra with atoms $x_1, x_2, x_3, \ldots x_n$. B is isomorphic to β_n and, hence, contains 2^n elements.

9.4 Groups

We have noted that the nonnegative integers with the operation $+$ forms a monoid. However, if we include the entire set of integers with the operation $+$, we again have a monoid, but we also have a property that we did not have in the first monoid, namely, for each integer a, there exists an integer $-a$ such that $a + (-a) = (-a) + a = 0$, the identity of the monoid. Similarly, if we consider the positive rationals with the multiplication operation, we again have a monoid and for every rational number $\frac{p}{q}$, there exists the rational number $\frac{q}{p}$ such that $\frac{p}{q} \cdot \frac{q}{p} = 1$, the identity of the monoid. Consider the properties involved in solving the equation $2x + 3 = 7$. We first add -3, the additive inverse of 3, to both sides of the equation so that we have $(2x + 3) + (-3) = 7 + (-3)$. We then use associativity to get the equation $2x + (3 - 3) = 4$. We then use the inverse property so that we have $2x + 0 = 4$. We then use the identity property of addition to get $2x = 4$. We then use multiplicative identity, associative and identity properties to finish solving the equation.

Groups are more than just an abstraction of the real numbers under addition or the positive real numbers under multiplication. We shall soon see that the set of permutations on n elements form a group. Groups are important in physics and many other areas. We shall use them in number theory, group codes, and in developing some of the counting methods.

Any monoid $(S, \circ)$, which has the property that for each element s of the monoid there exists an element s^{-1} in the monoid such that $s \circ s^{-1} = s^{-1} \circ s = I$, the identity of the monoid is called a **group**. Formally a group is defined as follows.

DEFINITION 9.54   A **group** is a set G together with a binary operation (or product) $\circ$ on G, which has the following properties:
(1) $a \circ (b \circ c) = (a \circ b) \circ c$ for all a, b, and c in G (i.e., the operation $\circ$ on S is associative).
(2) There exists an element 1 in G, called the **identity**, which has the property that $a \circ 1 = 1 \circ a = a$ for all a in G.
(3) For each element a in G, there exists an element a^{-1}, called the **inverse** of a, in G such that $a \circ a^{-1} = a^{-1} \circ a = 1$.

If a group G has the property that $a \circ b = b \circ a$ for all a, b in G, then G is called a **commutative** or **Abelian** group. A group is a finite group if G is finite.

We again observe that because $\circ$ is a binary operation on G, this implies that if $g, g' \in G$, then $g \circ g' \in G$. This property is called **closure**.

EXAMPLE 9.55   Examples of groups include the following:

1. The even integers with the binary operation addition.
2. The set of $n \times m$ matrices with real entries and the binary operation addition.
3. The integers modulo a positive integer n, forming Z_n, with the binary operation addition. ∎

DEFINITION 9.56   If G is a group with n elements, then n is called the **order** of the group G.

Let $S = \{i, -i, -1, 1\}$ where $i = \sqrt{-1}$. It is easily seen that this set, together with the binary operation multiplication, forms a group of order 4.

Note that every group is a semigroup. The converse, however, is not true.

THEOREM 9.57   The identity of a group G is unique.

Proof   Assume that 1 and e are both identities of G. Then $1 = 1 \circ e = e$ by definition of identity. ∎

THEOREM 9.58   In a group, the inverse of each element is unique.

Proof   Let a be an element of a group G and assume that b and c are both inverses of a. Then
$$b = b \circ 1 = b \circ (a \circ c) = (b \circ a) \circ c = 1 \circ c = c$$
∎

THEOREM 9.59   For each element a of a group G, $(a^{-1})^{-1} = a$.

Proof   Since $a^{-1} \circ a = a \circ a^{-1} = 1$, a satisfies the definition of the inverse of a^{-1}. Since the inverse is unique, a is the inverse of a^{-1}. ∎

THEOREM 9.60 For elements a and b of a group G, $(a \circ b)^{-1} = b^{-1} \circ a^{-1}$.

Proof $(a \circ b) \circ (b^{-1} \circ a^{-1}) = (a \circ (b \circ b^{-1})) \circ a^{-1} = (a \circ 1) \circ a^{-1} = a \circ a^{-1} = 1$. Similarly, $(b^{-1} \circ a^{-1}) \circ (a \circ b) = 1$ and by definition and uniqueness of inverse, $(a \circ b)^{-1} = b^{-1} \circ a^{-1}$. ∎

If a is an element of a group G, denote $(a^{-1})^k$ by a^{-k}. Let a^0 denote the identity 1.

The proof of the following theorems are left to the reader.

THEOREM 9.61 Let G be a group and a be an element of G.

(a) $a^n \circ a^{-n} = 1$ for all positive integers n.
(b) $a^{(m+n)} = a^m \circ a^n$ for all integers n and m.
(c) $(a^m)^n = a^{mn}$ for all integers m and n.
(d) $(a^{-n})^{-1} = a^n$ for all integers n. ∎

THEOREM 9.62 If a is an element of the group $(G, \circ)$ and $a \circ a = a$, then $a = 1$, the identity of G.

Lemma 9.63 If G is a finite group and a is an element of G, then $a^s = 1$ for some positive integer s.

Proof If G is a finite group and a is in G, then we know that $a^j = a^k$ for some positive integers j and k since only a finite number of powers can be distinct. Say $j < k$. Then $a^{-j} a^k = a^{-j} a^j = 1$ and $a^{k-j} = 1$. Let $s = k - j$. ∎

Let p be the least integer greater than 1 such that $a^p = 1$. The elements $1, a, a^2, \ldots, a^{p-1}$ are distinct, for if $a^r = a^s$ for $0 \leq r, s < p$, and $r > s$, then $a^{r-s} = 1$, which contradicts the fact that p is the smallest positive integer such that $a^p = 1$. Also if $n > p$, then $a^n = a^r$ for some $0 \leq r < p$ since if $n = pq + r$ for some $0 \leq r < p$, then

$$a^n = a^{pq+r} = a^{pq} \circ a^r = (a^p)^q \circ a^r = 1^q \circ a^r = 1 \circ a^r = a^r$$

THEOREM 9.64 Let G be a group and a be an element of G such that $a^s = 1$ for some s. If p is the least positive integer such that $a^p = 1$, then $p \mid s$. The integer p is called the **order** of a.

Proof Let p be the least positive integer such that $a^p = 1$. As in the preceding discussion, let $s = pq + r$ for some $0 \leq r < p$; then $a^s = a^{pq+r} = a^{pq} a^r = (a^p)^q a^r = (1)^q a^r = a^r$. Hence, $a^s = 1$ if and only if $r = 0$ because p is the least positive integer with $a^p = 1$. ∎

Again consider the set $S = \{i, -i, -1, 1\}$ under multiplication. What is the order of i? What is the order of -1?

DEFINITION 9.65 A subset H of a group G is a **subgroup** of G if H with the same operation as G is also a group.

EXAMPLE 9.66 Let $(Z, +)$ denote the group of integers under addition. For a positive integer m, let $(mZ, +)$ be the group of all multiples of m. We know this is a group, because if we add two multiples of m, we get a multiple of m. The inverse of mi is $-mi = m(-i)$. ∎

EXAMPLE 9.67 Let $(R, +)$ be the group of real numbers under addition. The group $(Q, +)$ of rationals under addition is a subgroup of $(R, +)$. ∎

EXAMPLE 9.68 Let $(R^+, \cdot)$ be the group of positive real numbers under multiplication. The group $(Q^+, \cdot)$ of positive rationals under multiplication is a subgroup of $(R^+, \cdot)$. ∎

EXAMPLE 9.69 It is easily checked that $(S, +)$ where $S = \{[0], [2], [4]\}$ is a subgroup of $(Z_6, +)$. ∎

To determine if a subset of a group is a subgroup, we do not have to check associativity, since it is inherited from the original group. We have to check that each element has an inverse and that the product of two elements of the set is in the set. We want to be sure that the identity is included, but this follows from the closure and inverse properties as shown in the proof of the next theorem, which gives an equivalent way of checking that a set is a subgroup.

THEOREM 9.70 A nonempty subset H of a group $(G, \cdot)$ is a subgroup if and only if for all $h_1, h_2 \in H$, $h_1 \cdot h_2^{-1} \in H$.

Proof First assume H is a group and $h_1, h_2 \in H$. Then h_2^{-1} is in the group since the inverse of an element of a group is also in the group. Hence, by closure, $h_1 h_2^{-1}$ is in H. Conversely, assume that for all $h_1, h_2 \in H$, $h_1 \cdot h_2^{-1} \in H$. If $h_1 \in H$, then $h_1 h_1^{-1} \in H$, so the identity $I \in H$. Therefore, for all $h_2 \in H$, then $I \cdot h_2^{-1} \in H$, and $h_2^{-1} \in H$. Finally, if $h_1, h_2 \in H$, then $h_1 \cdot (h_2^{-1})^{-1} = h_1 \cdot h_2 \in H$. Therefore, H is a group. ∎

The proof of the following theorem is left to the reader.

THEOREM 9.71 If g is in a group G, $g^n = 1$ for some n, and p is the least positive integer such that $g^p = 1$, then the set $\{g, g^2, \ldots, g^p\}$ is a subgroup of G. ∎

DEFINITION 9.72 If $g^p = 1$ set $\{g, g^2, \ldots, g^p\}$ is called a finite **cyclic group**, or more precisely, the cyclic group generated by g. It is denoted by $\langle g \rangle$.

Once again consider the set $S = \{i, -i, -1, 1\}$ under multiplication. This is a finite cyclic group. What can we use for g? Find two subgroups of S. Are they both cyclic?

The proof of the following theorem is left to the reader.

THEOREM 9.73 Let $(G, \cdot)$ be a group and $a_1, a_2, a_3, \ldots, a_k \in G$. Let $A = \{a_1, a_2, a_3, \ldots, a_k\}$ and $A^* = \langle a_1, a_2, a_3, \ldots, a_k \rangle$ be the set consisting of all finite products of $a_1, a_2, a_3, \ldots, a_k$ and their inverses. Then A^* is a group. Furthermore, A^* is the smallest subgroup of G containing A. ∎

DEFINITION 9.74 If A is a subset of a group G, the subgroup A^* is called the group **generated** by A. If a set A generates a group, H and no proper subset of A also generates H, then A is **minimal generating set** of H. If A contains only one element, then A^* is a cyclic group.

In a cyclic group, a generating set may be generated by one element. It is possible to have a minimal generating set with more than one element and still have a cyclic group.

DEFINITION 9.75 For a subgroup H of a group G and any a in G, $a \circ H = \{x : x = a \circ h$ for some h in $H\}$ is a **left coset** of H in G.

EXAMPLE 9.76 We saw earlier that the set Z of integers with the binary operation of addition, $+$, is a group. Let

$$H = \{n : n \in Z \text{ and } n = 5k \text{ for some integer } k\}$$
$$= \{\ldots, -10, -5, 0, 5, 10, 15, 20, \ldots\}$$

We show that H is a subgroup of Z. Clearly, $H \subseteq Z$. The set H is closed under addition since, if $s \in H$ and $t \in H$, then $s = 5i$ and $t = 5j$ for appropriate integers i and j. Thus, $s + t = 5i + 5j = 5(i+j) = 5k$ for $k = i+j$ so that $s+t$ is in H. The identity element of H is the identity element of G, namely, 0. Every element of H has an inverse with respect to $+$ in H because if $t = 5k \in H$, then $t^{-1} = 5(-k)$.

We now consider the left cosets of H generated by various elements of Z. Let $a \circ H = a + H = \{x : x = a + h$ for some h in $H\}$. So

$$0 + H = \{x : x = 0 + h \text{ for some } h \text{ in } H\}$$
$$= \{\ldots, -10, -5, 0, 5, 10, 15, 20, \ldots\}$$

$$1 + H = \{x : x = 1 + h \text{ for some } h \text{ in } H\}$$
$$= \{\ldots, -9, -4, 1, 6, 11, 16, 21, \ldots\}$$

$$2 + H = \{x : x = 2 + h \text{ for some } h \text{ in } H\}$$
$$= \{\ldots, -8, -3, 2, 7, 12, 17, 22, \ldots\}$$

$$3 + H = \{x : x = 3 + h \text{ for some } h \text{ in } H\}$$
$$= \{\ldots, -7, -2, 3, 8, 13, 18, 23, \ldots\}$$

$$4 + H = \{x : x = 4 + h \text{ for some } h \text{ in } H\}$$
$$= \{\ldots, -6, -1, 4, 9, 14, 19, 24, \ldots\}$$

Thus, we have again created Z_5 since $a + H = [a]$ and the five left cosets of the subgroup H form a partition of the original group Z. ∎

Lemma 9.77 For a fixed subgroup H of G, the left cosets of H in G are a partition of G.

Proof Each left coset is nonempty, since for a left coset $a \circ H$, $a = a \circ 1$ is in $a \circ H$. Assume that the intersection of $a \circ H$ and $b \circ H$ is nonempty; say, c is in the intersection. Hence, $c = a \circ h = b \circ h'$ for some h and h' in H. Multiplying both sides of the equation by h^{-1}, we have $a \circ h \circ h^{-1} = b \circ h' \circ h^{-1}$; so by definition of inverse we have $a = b \circ (h' \circ h^{-1})$. Since $h' \circ h^{-1}$ is in H, a is in $b \circ H$. Hence, $a \circ h$ is in $b \circ H$ for all h in H so that $a \circ H \subseteq b \circ H$. Similarly, $b \circ H \subseteq a \circ H$ and $b \circ H = a \circ H$. Hence, the left cosets form a partition of G. ∎

Note that since the left cosets of H in G are a partition of G, the cosets are equivalence classes of an equivalence relation.

Lemma 9.78 If G is a finite group and H a subgroup of G, then all cosets of H in G contain the same number of elements, namely, the number of elements that are in H.

Proof Let $a \circ H$ be a left coset of H in G. Define $f : H \to a \circ H$ by $f(h) = a \circ h$. It is left to the reader to show that f is one-to-one and onto. ∎

THEOREM 9.79 (**Lagrange**) If G is a finite group and H is a subgroup of G, then the order of H divides the order of G.

Proof If p is the order of H, q is the number of left cosets of H in G, and n is the order of G, then, because of the preceding two lemmas, $n = pq$. ∎

THEOREM 9.80 If G is a group of order n and g is in G, then $g^n = 1$.

Proof Let p be the least positive integer such that $g^p = 1$ and $H = \{g, g^2, \ldots, g^p\}$. Then H is a subgroup with p elements and, hence, p divides n, say $n = pq$. Thus, $g^n = g^{pq} = (g^p)^q = 1^q = 1$. ∎

Exercises

1. Write an addition and a multiplication table for Z_4.
2. Write an addition and a multiplication table for Z_7.
3. Evaluate $[1] \circ [2] \circ [3]$ in Z_4.
4. Evaluate $[1] \circ [2] \circ [3] \circ [4]$ in Z_7.
5. In Z_{10}, find the multiplicative inverse of $[7]$.
6. In Z_{12}, find all a such that $[a] \circ [6] = [0]$.
7. Let $Z_n = \{[0], [1], [2], \ldots, [n-1]\}$ for some integer n. Let $S = \{[k] : k$ is positive and relatively prime to $n\}$. Is S a group under multiplication?
8. Under what conditions is Z_n a group under multiplication?
9. Prove that the multiples of 3 form a subgroup of the integers under addition. Describe the cosets formed with respect to this subgroup.
10. Prove that the multiples of 3 form a semigroup of the integers under multiplication. Why are they not a group?
11. Prove that the integers modulo 6 form a group with respect to addition. Do they also form a group with regard to multiplication?

Prove Theorem 9.61: Let G be a group and a be an element of G.

12. $a^n \circ a^{-n} = 1$ for all positive integers n.
13. $a^{(m+n)} = a^m \circ a^n$ for all integers n and m.
14. $(a^m)^n = a^{mn}$ for all integers m and n.
15. $(a^{-n})^{-1} = a^n$ for all integers n.
16. Prove Theorem 9.71: If g is in a group G, $g^n = 1$ for some n, and p is the least positive integer such that $g^p = 1$, then the set $\{g, g^2, \ldots, g^p\}$ is a subgroup of G.
17. Prove Theorem 9.62: If a is an element of the group $(G, \circ)$ and $a \circ a = a$, then $a = 1$, the identity of G.
18. Prove that if a cyclic group G is generated by an element $a \in G$ of order n, then a^m generates G if and only if $\gcd(m, n) = 1$.
19. Prove that if G is a finite group and $H \subseteq G$, then H is a subgroup of G if and only if $ab \in H$ for every $a, b, \in H$.
20. If G is a subgroup of H and H is a subgroup of K, prove that G is a subgroup of K.
21. Prove that the intersection of two subgroups of a group G is a subgroup of G.
22. Is the union of two groups a group?
23. The **center** of a group is G defined as the set of all $g \in G$ so that $gh = hg$ for all h in G. Prove that the center of a group G is a subgroup of G.

9.5 Groups and Homomorphisms

There is a special function between groups such that certain properties of the domain are preserved in the image of the function. For example, the image of subgroups in the domain are subgroups in the range. Such a function is called a homomorphism. Note that a group is a semigroup and this is the same homomorphism definition as for semigroups.

DEFINITION 9.81 Let $(G, \bullet)$ and $(H, *)$ be groups, where $\bullet$ and $*$ are the operations on G and H, respectively. Let $f : G \to H$ be a function. The function f is a **homomorphism** if $f(g \bullet g') = f(g) * f(g')$ for all g and g' in G. The homomorphism f is said to be a **monomorphism** if f is one-to-one, an **epimorphism** if f is onto, and an **isomorphism** if f is both one-to-one and onto.

EXAMPLE 9.82 Let $(G, \bullet)$ and $(H, *)$ be groups and 1 be the identity of H. If $f : G \to H$ is defined by $f(g) = 1$ for all $g \in G$, then f is a homomorphism. ∎

EXAMPLE 9.83 Let $(G, +)$ and $(H, +)$ each be the group of integers under addition, and $f : G \to H$ be defined by $f(g) = 2g$ for all $g \in G$, then f is a homomorphism. ∎

EXAMPLE 9.84 In Example 9.76, where $H = \{n : n \in Z \text{ and } n = 5k \text{ for some integer } k\}$ is a subgroup of Z, the set of left cosets of H is

$$W = \{0 + H, 1 + H, 2 + H, 3 + H, 4 + H\}$$

We can define addition of left cosets as follows:

$$(a + H) \oplus (b + H) = (a + b) + H$$

where the addition $(a + b)$ is addition of integers a and b. It is straightforward to show that the definition of $\oplus$ is independent of the representation of $a + H$ and $b + H$.

For the group $(Z_5, +)$ of equivalence classes modulo 5, it is easy to show that the function $f : Z_5 \to W$ defined by

$$f([k]) = k + H$$

is a homomorphism between the groups $(Z_5, +)$ and $(W, \oplus)$ and is also an isomorphism.

Consider the two groups $(Z, +)$ and $(Z_5, +)$ and the function $g : Z \to Z_5$ defined by

$$g(a) = [a]$$

g is a homomorphism since $g(a + b) = [a + b] = [a] + [b] = g(a) + g(b)$. The homomorphism g is an epimorphism since, given any element $[c]$ of Z_5, clearly $g(c) = [c]$. ∎

EXAMPLE 9.85 It was stated in the problems in Section 5.8 that an $n \times n$ matrix has an inverse if and only if its determinant is not 0. Although it is beyond the scope of this book, it can be shown that the determinant function is a homomorphism from the set of $n \times n$ real (or rational) valued matrices under multiplication to the set of real (or rational) numbers under multiplication. Hence, if the determinant of two matrices is not zero, then the determinant of their product is not zero. It follows that the product of two matrices with inverses also has an inverse. This set of $n \times n$ matrices, therefore, forms a group under multiplication called the group of nonsingular matrices. If the determinant of a matrix M is a nonzero value r, what is the determinant of M^{-1}? ∎

The proofs of the following theorems are left to the reader.

THEOREM 9.86 Let $f : G \to H$ be a homomorphism from group G to group H and 1 be the identity in G. Then $f(1)$ is the identity in H. ∎

THEOREM 9.87 Let $f : G \to H$ be a homomorphism from group G to group H and g' be the inverse of g in G. Then $f(g')$ is the inverse of $f(g)$ in H. ∎

THEOREM 9.88 If $f : G \to H$ is a homomorphism from group G to group H and K is a subgroup of H, then $f^{-1}(K)$ is a subgroup of G. ∎

THEOREM 9.89 If $f : G \to H$ is a homomorphism from group G to group H and K is a subgroup of G, then $f(K)$ is a subgroup of H. ∎

DEFINITION 9.90 For subsets H and K of a group $(G, \circ)$, let $H \circ K = \{h \circ k : h \in H, k \in K\}$. When $\circ$ is known, $H \circ K$ is often abbreviated by HK. When $H = \{h\}$, then $H \circ K$ is usually denoted by hK.

The proof of the following theorem is left to the reader.

THEOREM 9.91 If H, J, and K are subsets of a group $(G, \circ)$, then $(H \circ J) \circ K = H \circ (J \circ K)$. ∎

We saw in Example 9.84 that a group was divided into cosets and it was possible to "multiply" the cosets (in this case the multiplication of the group was addition of integers). In general, this is not possible. However, we shall see that if the subgroup used to form the cosets has certain properties, then it is possible to multiply the cosets.

DEFINITION 9.92 If H is a subgroup of a group $(G, \circ)$ and has the property that $gHg^{-1} = H$ for all $g \in G$, then H is called a **normal subgroup**.

DEFINITION 9.93 Let $f : G \to H$ be a homomorphism from the group G to the group H. The **kernel**. of G is the set $\{x : x \in G \text{ and } f(x) = 1\} = f^{-1}(\{1\})$, where 1 is the identity of H.

The proof of the following theorem is left to the reader.

THEOREM 9.94 The kernel of $f : G \to H$ is a normal subgroup of G. ∎

THEOREM 9.95 The subgroup H of a group $(G, \circ)$ is a normal subgroup if and only if $gH = Hg$ for all $g \in G$.

Proof Assume H is normal in G so that $gHg^{-1} = H$ for all $g \in G$. Let $gh \in gH$. Since $gHg^{-1} = H$, $ghg^{-1} = h'$ for some $h' \in H$. Therefore, $gh = h'g$, and $gh \in Hg$, so that $gH \subseteq Hg$. Similarly, $Hg \subseteq gH$. The converse is left to the reader. ∎

THEOREM 9.96 If H is a subgroup of a group $(G, \circ)$, then $H \circ H = H$.

Proof Assume H is a group. Then by definition of closure $H \circ H \subseteq H$. Since H is a group, then for $h \in H$, we have $h = h \circ I$, where I is the identity of G and, hence, of H. Therefore, $h \in H \circ H$ and $H \subseteq H \circ H$. Thus, $H = H \circ H$. ∎

The converse is not true, however. If H is the set of positive integers with the usual multiplication, then $H = H \circ H$, but H is not a group.

THEOREM 9.97 If H is a normal subgroup of a group $(G, \circ)$, then $abH = (aH)(bH)$ for all $a, b \in G$.

Proof For $a, b \in G$,

$$\begin{aligned} abH &= a(bH) &&\text{By Theorem 9.91} \\ &= a(b(HH)) &&\text{By Theorem 9.96} \\ &= a(bH)H &&\text{By Theorem 9.91} \\ &= a(Hb)H &&\text{By Theorem 9.95} \\ &= (aH)(bH) &&\text{By Theorem 9.91} \end{aligned}$$

∎

Corollary 9.98 If H is a normal subgroup of a group $(G, \circ)$, then the cosets of H in G form a group under the operation $(aH)(bH) = abH$. This group is called the **factor group** and denoted by G/H. ∎

Corollary 9.99 If $f : G \to G/H$ is defined by $f(a) = aH$, then f is a homomorphism. ∎

THEOREM 9.100 (first isomorphism theorem for groups) Let $f : G \to H$ be an epimorphism with kernel K. Then the quotient group G/K is isomorphic to H.

Proof We define $\theta : G/K \to H$ by $\theta(gK) = f(g)$ for all $g \in G$. We first have to show that θ is well defined, which means that if $\theta(gK) = h$ and $\theta(gK) = h'$, then $h = h'$. Let $gK = g'K$, and $f(g) = h$, $f(g') = h'$. Since $gK = g'K$, $g = g'k$ for some $k \in K$. Therefore,

$$\begin{aligned} h &= f(g) \\ &= f(g'k) \\ &= f(g')f(k) \\ &= f(g') \circ 1 \text{ since } k \text{ is in the kernel of } f \\ &= f(g') \\ &= h' \end{aligned}$$

We now show that θ is a homomorphism.

$$\begin{aligned} \theta(gKg'K) &= \theta(gg'K) \\ &= f(gg') \\ &= f(g)f(g') \\ &= (gK)(g'K) \end{aligned}$$

Obviously θ is an epimorphism. Since f is an epimorphism, for $h \in H$, there exist $g \in G$ so $f(g) = h$ and $\theta f(gK) = h$.

Next we show that θ is a monomorphism. If $\theta(hK) = \theta(h'K)$, then $f(h) = f(h')$ and $f(h)f(h)^{-1} = f(h')f(h)^{-1}$ so $f(hh^{-1}) = f(h'h^{-1})$ or $f(1) = f(h'h^{-1})$. Since $f(1) = 1$, $f(h'h^{-1}) = 1$ and $h'h^{-1} \in K$. Hence $h'K = hK$. ∎

The proofs of the following theorems are left to the reader:

THEOREM 9.101 If $f : G \to G'$ is a homomorphism from group $(G, \circ)$ to group $(G', \circ)$ then $H = \{x : f(x) = 1 \text{ the identity of } G'\}$ is a subgroup of G. ∎

THEOREM 9.102 If $f : G \to H$ is a homomorphism from group G to group H, and K is a subgroup of H, then $f^{-1}(\{1\})$, where 1 is the identity of H, is a normal subgroup of G. ∎

In Section 4.1, it was shown that permutations on a finite set are closed under the binary operation of composition. It was also shown how to find the inverse of a given permutation. As with all other functions, these permutations are associative with respect to composition. Therefore, the set of all permutations for a fixed finite set forms a group under the binary operation of composition. The set of all permutations on n elements is called the symmetric group and denoted by S_n.

EXAMPLE 9.103 Also recall that if f is a permutation on the set $\{1, 2, 3, \ldots, n\}$, then it can be represented in the form

$$\begin{pmatrix} 1 & 2 & \ldots & n \\ f(1) & f(2) & \ldots & f(n) \end{pmatrix}$$

Consider the following subgroup of S_4 based on the square given in Figure 9.14. The permutation

$$\delta_1 = \begin{pmatrix} 1 & 2 & 3 & 4 \\ 3 & 2 & 1 & 4 \end{pmatrix}$$

interchanges 1 and 3. Geometrically it reflects the square around the diagonal between 2 and 4. Similarly, the permutation

$$\delta_2 = \begin{pmatrix} 1 & 2 & 3 & 4 \\ 1 & 4 & 3 & 2 \end{pmatrix}$$

interchanges 2 and 4. Geometrically it reflects the square around the diagonal between 1 and 3. The permutation

$$\phi_1 = \begin{pmatrix} 1 & 2 & 3 & 4 \\ 4 & 3 & 2 & 1 \end{pmatrix}$$

geometrically reflects the square around the horizontal line through the center of the square. The permutation

$$\phi_2 = \begin{pmatrix} 1 & 2 & 3 & 4 \\ 2 & 1 & 4 & 3 \end{pmatrix}$$

geometrically reflects the square around the vertical line through the center of the square. Let ρ_1, ρ_2, and ρ_3 rotate the square clockwise by 90°, 180°, and 270°, respectively. Hence, $\rho_1 = \begin{pmatrix} 1 & 2 & 3 & 4 \\ 2 & 3 & 4 & 1 \end{pmatrix}$, $\rho_2 = \begin{pmatrix} 1 & 2 & 3 & 4 \\ 3 & 4 & 1 & 2 \end{pmatrix}$, and

Figure 9.14

$\rho_3 = \begin{pmatrix} 1 & 2 & 3 & 4 \\ 4 & 1 & 2 & 3 \end{pmatrix}$. I is the identity permutation. The multiplication of these elements gives the following table:

$\circ$	I	ρ_1	ρ_2	ρ_3	δ_1	δ_2	ϕ_1	ϕ_2
I	I	ρ_1	ρ_2	ρ_3	δ_1	δ_2	ϕ_1	ϕ_2
ρ_1	ρ_1	ρ_2	ρ_3	I	ϕ_1	ϕ_2	δ_2	δ_1
ρ_2	ρ_2	ρ_3	I	ρ_1	δ_2	δ_1	ϕ_2	ϕ_1
ρ_3	ρ_3	I	ρ_1	ρ_2	ϕ_2	ϕ_1	δ_1	δ_2
δ_1	δ_1	ϕ_2	δ_2	ϕ_1	I	ρ_2	ρ_3	ρ_1
δ_2	δ_2	ϕ_1	δ_1	ϕ_2	ρ_2	I	ρ_1	ρ_3
ϕ_1	ϕ_1	δ_1	ϕ_2	δ_2	ρ_1	ρ_3	I	ρ_2
ϕ_2	ϕ_2	δ_2	ϕ_1	δ_1	ρ_3	ρ_1	ρ_2	I

This group is called the **octic group** or the **group of symmetries of the square**. The set $H = \{I, \rho_1, \rho_2, \rho_3\}$ is a subgroup of the octic group. Since H has four elements and the number of elements in the coset must divide the order of a group, there is only one other left coset, aH, for any $a \notin H$. It consists of all elements of the octic group that are not in H. But it must also be Ha for any $a \notin H$, since there is only one other right coset. Therefore, H is a normal subgroup of the octic group. It is left to the reader to show that $\{I, \rho_2\}$ is a normal subgroup of the octic group. ∎

Let A be a finite set with n elements and S_n be the group of permutations on A. Since there are $n!$ permutations of the elements of A, S_n contains $n!$ elements. For convenience, let $1, 2, \ldots, n$ be the elements of A. For a fixed permutation σ, define a relation R on A, by aRb if $b = \sigma^k(a)$ for some nonnegative integer k.

THEOREM 9.104 The relation R is an equivalence relation on A.

Proof Since S_n is finite, $\sigma^m = 1$, the identity of S_n for some nonnegative integer m. Hence, $\sigma^m(a) = a$, and aRa. Let aRb, then $b = \sigma^k(a)$ for some nonnegative integer k. Since $\sigma^k \sigma^{m-k} = \sigma^m = 1$, σ^{m-k} is the inverse of σ^k, so that $\sigma^{m-k}(b) = a$ and bRa. Therefore, R is symmetric. Let aRb and bRc. Then $b = \sigma^j(a)$ for some nonnegative integer j and $c = \sigma^k(b)$ for some nonnegative integer k. Therefore, $c = \sigma^{j+k}(a)$ and R is transitive. Hence, R is an equivalence relation. ∎

Since R is an equivalence relation, it divides $\{1, 2, 3, \ldots, n\}$ into equivalence classes. For example, let $A = \{1, 2, 3, \ldots, 8\}$ and

$$\sigma = \begin{pmatrix} 1 & 2 & 3 & 4 & 5 & 6 & 7 & 8 \\ 4 & 3 & 2 & 7 & 8 & 6 & 1 & 5 \end{pmatrix}$$

The set of equivalence classes is easily seen to be $\{\{1, 4, 7\}, \{2, 3\}, \{5, 8\}, \{6\}\}$. The set of equivalence classes created in this manner is called the **orbit** of σ.

At this point, we want to define a particular type of permutation called a **cycle**. If we denote $\sigma_c = (a_1, a_2, a_3, \ldots, a_k)$ as a cycle, then $\sigma_c(a_i) = a_{i+1}$ for $i < k$, and $\sigma_c(a_k) = a_1$. For example, if $\sigma_c = (1, 3, 5, 7)$ is a cycle in S_8, then

$$\sigma_c = \begin{pmatrix} 1 & 2 & 3 & 4 & 5 & 6 & 7 & 8 \\ 3 & 2 & 5 & 4 & 7 & 6 & 1 & 8 \end{pmatrix}$$

Given the permutation σ, we can write it as a product of cycles having no common components by first dividing A into equivalence classes. We pick an element a and form the cycle $(a, \sigma(a), \sigma^2(a), \sigma^3(a), \ldots, \sigma^k(a))$ where $\sigma^{k+1}(a) = a$. The cycles contain no components in common because the equivalence classes are disjoint. We shall say that the cycles are **disjoint**.

For example, in the permutation

$$\sigma = \begin{pmatrix} 1 & 2 & 3 & 4 & 5 & 6 & 7 & 8 \\ 4 & 3 & 2 & 7 & 8 & 6 & 1 & 5 \end{pmatrix}$$

with set of equivalence classes $\{\{1, 4, 7\}, \{2, 3\}, \{5, 8\}, \{6\}\}$, the cycles are $(1, 4, 7)$, $(2, 3)$, $(5, 8)$, and (6). Since each cycle moves only numbers in the cycle and the cycles are disjoint, $\sigma = (1, 4, 7)(2, 3)(5, 8)(6)$. Since the cycles are disjoint, the order of the cycles does not change the permutation. If desired, singleton cycles can be left out, since they do not really move anything. Therefore, we can write $\sigma = (1, 4, 7)(2, 3)(5, 8)$. If singleton cycles are used, then the identity permutation in S_8 may be written as $(1)(2)(3)(4)(5)(6)(7)(8)$.

Exercises

1. Prove Theorem 9.86: Let $f : G \to H$ be a homomorphism from group G to group H, and 1 be the identity in G. Then $f(1)$ is the identity in H.

2. Prove Theorem 9.87: Let $f : G \to H$ is a homomorphism from group G to group H, and g' be the inverse of g in G. Then $f(g')$ is the inverse of $f(g)$ in H.

3. From the proof of Theorem 9.78, let $a \circ H$ be a left coset of H in G. Define $f : H \to a \circ H$ by $f(h) = a \circ h$. Show that f is one-to-one and onto.

4. Prove Theorem 9.94: The kernel of $f : G \to H$ is a normal subgroup of G.

5. Prove that $f : G \to H$ is a monomorphism if and only if the kernel of f is $\{1\}$.

6. Let $Z_5^R = \{[1], [2], [3], [4]\}$ be the set of nonzero elements of Z_5. These elements correspond to integers relatively prime to 5. With the multiplication, $\cdot$, of Z_5, $(Z_5^R, \cdot)$ is a group. Show that $(Z_5^R, \cdot)$ is isomorphic to $(Z_4, +)$ by exhibiting an explicit isomorphism $f : Z_5^R \to Z_4$.

7. Show that $\{I, \rho_2\}$ is a normal subgroup of the octic group G.

8. Find $G/\{I, \rho_2\}$.

9. Find a two-element subgroup of the octic group that is not a normal subgroup.

10. Prove Theorem 9.88: If $f : G \to H$ is a homomorphism from group G to group H, and K is a subgroup of H, then $f^{-1}(K)$ is a subgroup of G.

11. Prove Theorem 9.89: If $f : G \to H$ is a homomorphism from group G to group H, and K is a subgroup of G, then $f(K)$ is a subgroup of H.

Express the following permutations as products of disjoint cycles:

12. $\begin{pmatrix} 1 & 2 & 3 & 4 & 5 & 6 & 7 & 8 & 9 \\ 3 & 9 & 4 & 7 & 8 & 6 & 1 & 5 & 2 \end{pmatrix}$

13. $\begin{pmatrix} 1 & 2 & 3 & 4 & 5 & 6 & 7 & 8 & 9 \\ 2 & 4 & 7 & 8 & 3 & 6 & 9 & 1 & 5 \end{pmatrix}$

14. $\begin{pmatrix} 1 & 2 & 3 & 4 & 5 & 6 & 7 \\ 2 & 4 & 1 & 7 & 3 & 6 & 5 \end{pmatrix}$

15. $\begin{pmatrix} 1 & 2 & 3 & 4 & 5 & 6 \\ 3 & 4 & 5 & 6 & 1 & 2 \end{pmatrix}$

Find $\sigma \circ \tau$ when

16. $\sigma = (1, 3, 4)(2, 5)$ and $\tau = (1, 2)(3, 4)(5)$.

17. $\sigma = (1, 2)(3, 4)(5, 6)$ and $\tau = (1, 2)(3, 4)(5, 6)$.

18. $\sigma = (1, 2, 3)(4, 5, 6)$ and $\tau = (1, 2)(3, 4)(5, 6)$.

19. $\sigma = (1, 2, 3, 8)(4, 5, 6)(7)$ and $\tau = (1, 2)(3, 8)(4, 5)(6, 7)$.

20. Prove Theorem 9.91: If H, J, and K are subsets of a group $(G, \circ)$, then $(H \circ J) \circ K = H \circ (J \circ K)$.

21. Prove the converse in Theorem 9.95: For a given subgroup H of a group $(G, \circ)$, if $gH = Hg$ for all $g \in G$, then H is a normal subgroup.

22. An **automorphism** on a group is an isomorphism from the group onto itself. Prove that for any fixed a in a group G, the function $T_a : G \to G$ defined by $T_a(h) = a^{-1}ha$ is an automorphism on G.

23. The automorphisms formed in the previous problem are called **inner automorphisms**. Prove the inner automorphisms of a group form a group under composition.

24. *The set of all elements of a group G of the form $x^{-1}y^{-1}xy$ are called commutators. Prove that the group generated by the set C of all commutators of G is a normal subgroup of G.

25. Prove that if C is a normal subgroup containing the set of all commutators of the group G, then G/C is commutative.

26. (Cayley's Theorem) Let G be a group and for each $\alpha \in G$, define $\varphi_\alpha : G \to G$ by $\varphi_\alpha(g) = ag$ for all $g \in G$.
    (1) Prove that $\varphi_\alpha : G \to G$ is a permutation.
    (2) Prove that if $H = \{\varphi_\alpha : \alpha \in G\}$, then H is a group under composition.
    (3) Prove that G and H are isomorphic, so that every group is isomorphic to a group of permutations.

27. Using Cayley's Theorem, prove that every finite group of order n is isomorphic to a subgroup of the symmetric group S_n.

9.6 Linear Algebra

Let R be the set of real numbers or other field. A **vector space** $(V, R, +, \circ)$ over the set R consists of the set R, which we will call scalars, together with a set V, which we call vectors, that have the following properties:

(a) There is an addition $+$ between the vectors so that $(V, +)$ is a commutative group.

(b) There is a multiplication $\circ$ between elements of R and elements of V such that
  (i) $1 \circ v = v$ for all $v \in V$.
  (ii) $a \circ (b \circ v) = (a \circ b) \circ v$ for all $a, b \in R$, $v \in V$.
  (iii) $a \circ (v + w) = (a \circ v) + (a \circ w)$ for all $a \in R$, $v, w \in V$.
  (iv) $(a + b) \circ v = (a \circ v) + (b \circ v)$ for all $a, b \in R$, $v \in V$.

For simplicity, we shall just refer to the vector space $(V, R, +, \circ)$ as V. Denote the additive identity of $+$ by $\mathbf{0}$.

EXAMPLE 9.105 Let V be the sets of $m \times n$ matrices with real coefficients or scalars and $\circ$ be the usual scalar multiplication discussed in Section 4.3. We showed in the problems of Section 4.3 that V is a vector space. In particular, we are interested in $1 \times n$ matrices called horizontal vectors or simply vectors and $n \times 1$ matrices called vertical vectors. ∎

EXAMPLE 9.106 Let S be a set and V be the set of all functions from S to R. For $f, g : S \to R$, define $(af)(s) = a(f(s))$ and $(f+g)(s) = f(s) + g(s)$ for all $s \in S$. V is a vector space. ∎

EXAMPLE 9.107 Let V be the set of all polynomials of degree n with real coefficients. V is a vector space. ∎

DEFINITION 9.108 Let V be a vector space, and $W \subseteq V$. W is a **subspace** of V if W is a vector space.

Lemma 9.109 Let V be a vector space; then for all $v \in V$, $0 \circ v = \mathbf{0}$.

Proof Left to the reader. ∎

THEOREM 9.110 If $W \subseteq V$, and W is nonempty, then W is a **subspace** of V if for all $w, w' \in W$, and $a, b \in R$, $aw + bw' \in W$.

Proof If W is a **subspace** of V, then $aw, bw' \in W$. Hence, $aw + bw' \in W$. Conversely, if for all $w, w' \in W$, and $a, b \in R$, $aw + bw' \in W$, we first show W is a subgroup. $0w + 0w \in W$, so $\mathbf{0} \in W$. If $w \in W$, then $0w + (-1)w = -w$. Associativity and commutativity follow since $W \subseteq V$, and V has this property. Hence, W is a group. Properties i–iv are also true since $W \subseteq V$. Finally, we need to show that if $a \in R$ and $w \in W$, then $aw \in W$. But $aw + 0w'$ belongs to W, so aw belongs to W. ∎

EXAMPLE 9.111 The vector space W where $W = \{\mathbf{0}\}$ is a subspace. Certainly $a\mathbf{0} + b\mathbf{0} = \mathbf{0}$ for all $a, b \in R$. ∎

EXAMPLE 9.112 Let V be the vector space of all ordered pairs of real numbers (v_1, v_2) and $W = \{v : 2v_1 = v_2\}$. To show that this is a vector space, we must show that if (v_1, v_2) and (v'_1, v'_2) are in W and a is in R then $(v_1, v_2) + (v'_1, v'_2)$ and $a(v_1, v_2)$ are in W. $(v_1, v_2) + (v'_1, v'_2) = (v_1 + v'_1, v_2 + v'_2)$. But $2v_1 = v_2$ and $2v'_1 = v'_2$ so $2v_1 + 2v'_1 = v_2 + v'_2$ or $2(v_1 + v'_1) = v'_1 + v'_2$. Therefore, the sum of the vectors is in W. Also $a(v_1, v_2) = (av_1, av_2)$ and $av_1 = a2v_1 = 2av_1 = 2av_2$ and $a(v_1, v_2)$ is in W. Hence, W is a subspace. ∎

EXAMPLE 9.113 Let V be the vector space of all ordered pairs of real numbers (v_1, v_2) and $W = \{v : v_1 v_2 = 4\}$. Note that $(2, 2)$ belongs to W, but $3 \circ (2, 2) = (6, 6)$ does not belong to W. Hence, W is not a subspace. ∎

EXAMPLE 9.114 If V is the set of all $(v_1, v_2, \ldots v_n)$ for $v_1, v_2, \ldots v_n \in R$, and W is the set of all $(v_1, v_2, \ldots v_n)$ for $v_1, v_2, \ldots v_n \in R$ such that $v_1 = 0$, then W is a subspace of V. If $(v_1, v_2, \ldots v_n)$ and $(v'_1, v'_2, \ldots v'_n)$ are vectors in W, then

$$w = a(v_1, v_2, \ldots v_n) + b(v'_1, v'_2, \ldots v'_n) = (av_1 + bv'_1, av_2 + bv'_2, \ldots, av_n + bv'_n)$$

But since $v_1, v'_1 = 0$, $av_1 + bv'_1 = 0$ and $w \in W$. Hence, W is a subspace of V. ∎

EXAMPLE 9.115 If V is the set of all $(v_1, v_2, \ldots v_n)$ for $v_1, v_2, \ldots v_n \in R$, and W is the set of all $(v_1, v_2, \ldots v_n)$ for $v_1, v_2, \ldots v_n \in R$ such that $v_1 = v_2$, then W is a subspace of V. If $(v_1, v_1, \ldots v_n)$ and $(v'_1, v'_1, \ldots v'_n)$ are vectors in W, then

$$w = a(v_1, v_1, \ldots v_n) + b(v'_1, v'_1, \ldots v'_n) = (av_1 + bv'_1, av_1 + bv'_1, \ldots, av_n + bv'_n)$$

and $w \in W$. ∎

THEOREM 9.116 The intersection of subspaces of a vector space is a subspace.

Proof Let $W_i \subseteq V$ for $i \in I$. Also let $v, w \in W_i$ for all i and a and b be elements R. Since each W_i is a vector space, $av + bw \in W_i$ for all i. Hence, $av + bw \in \bigcap_{i \in I} W_i$. Therefore, $\left(\bigcap_{i \in I} W_i, R \right)$ is a vector space. ∎

DEFINITION 9.117 If V and W are vector spaces, then the function $T : V \to W$ is a **linear transformation** if for all $u, v \in V$ and $a \in R$, $T(u + v) = T(u) + T(v)$ and $T(au) = aT(u)$.

EXAMPLE 9.118 The identity transformation defined by $T(v) = v$ is a linear transformation. ∎

EXAMPLE 9.119 The transformation defined by $T(v) = av$ is a linear transformation. ∎

THEOREM 9.120 If $T : V \to W$ is a linear transformation, then for all $v_i \in V$ and $a_i \in R$, $T\left(\sum_i a_i v_i\right) = \sum_i a_i T(v_i)$.

Proof Left to the reader. ∎

THEOREM 9.121 If V and W are the set of $n \times 1$ and $m \times 1$ horizontal vectors and M is an $m \times n$ matrix, then $v \in V$, $T : V \to W$ is defined by $T(v) = Mv$ and T is a linear transformation.

Proof

$T(u + v) = M(u + v) = Mu + Mv$ by the distributive property of matrices.
$\quad = T(u) + T(v)$

and

$T(au) = M(au) = aMu$ by the linear property of matrices.
$\quad = aT(u)$

Therefore, T is a linear transformation. ∎

Exercises

Let V consist of all ordered pairs of real numbers in the form $v = \begin{bmatrix} v_1 \\ v_2 \end{bmatrix}$. Determine which of the following are subspaces of V.

1. $W = \{v : v_1 = 3v_2\}$
2. $W = \{v : v_1 = v_2^3\}$
3. $W = \{v : v_1 = v_2 + 5\}$

Let V consist of all ordered triples of real numbers in the form $\begin{bmatrix} v_1 \\ v_2 \\ v_3 \end{bmatrix}$.

Determine which of the following are subspaces of V.

4. $W = \{v : v_1 = v_2 - 2v_3\}$
5. $W = \{v : v_1 = v_2 v_3\}$
6. $W = \{v : v_1 = v_2 - v_3 + 5\}$
7. $W = \{v : v_1 = v_2 = 0\}$
8. $W = \{v : v_1 = v_2 = 1\}$

Let $v = \begin{bmatrix} v_1 \\ v_2 \\ v_3 \end{bmatrix}$ and define

$T(v) = \begin{bmatrix} 1 & 3 & 4 \\ 0 & 1 & 0 \\ 2 & 2 & 1 \end{bmatrix} \begin{bmatrix} v_1 \\ v_2 \\ v_3 \end{bmatrix}$.

9. Find $T \begin{bmatrix} 4 \\ 3 \\ 1 \end{bmatrix}$.

10. Find $T \begin{bmatrix} 1 \\ 2 \\ 2 \end{bmatrix}$.

Let $v = \begin{bmatrix} v_1 \\ v_2 \\ v_3 \end{bmatrix}$ and define

$T(v) = \begin{bmatrix} 2 & 0 & 1 \\ 1 & 0 & 1 \\ 3 & 1 & 1 \end{bmatrix} \begin{bmatrix} v_1 \\ v_2 \\ v_3 \end{bmatrix}$.

11. Find $T \begin{bmatrix} 3 \\ 0 \\ 2 \end{bmatrix}$.

12. Find $T \begin{bmatrix} 3 \\ 2 \\ 4 \end{bmatrix}$.

13. Prove that if U and W are subspaces of a vector space V, then $U + W = \{u + v : u \in U, w \in W\}$ is a subspace of V.

14. Prove Lemma 9.109. Let V be a vector space, then for all $v \in V$, $0 \circ v = \mathbf{0}$.

15. Let V be the set of pairs of real numbers. Define addition by $(u_1, v_1) + (u_2, v_2) = (u_1 + u_2, v_1 + v_2)$ and scalar multiplication by $a \circ (u_1, v_1) = (au_1, v_1)$. Does this define a vector space?

16. Let V be the set of pairs of real numbers. Define addition by $(u_1, v_1) + (u_2, v_2) = (3v_1 + 3v_2, -u_1 - u_2)$ and $a \circ (u_1, v_1) = (3av_1, -au_1)$. Show that this defines a vector space or show why it isn't.

17. Let $v_1, v_2, v_3, \ldots v_k$ be vertices in a vector space V. Prove that there is a smallest subspace of V containing $v_1, v_2, v_3, \ldots v_k$. Prove that this vector space is the set of all vectors of the form $a_1 v_1 + a_2 v_2 + a_3 v_3 + \cdots + a_k v_k$ for $a_1, a_2, a_3, \ldots a_k \in R$.

18. Prove Theorem 9.120. If $T : V \to W$ is a linear transformation, then for $v_i \in V$ and $a_i \in R$, where $i = 1$ to n, $T\left(\sum_i a_i v_i\right) = \sum_i a_i T(v_i)$.

19. If W is a subspace of V and $T : V \to V'$ is a linear transformation. Prove that $T(W)$ is a subspace of V'.

20. Let V, W be vector spaces and $T : V \to W$ be a linear transformation. The **null set** of T is the set $\{v : T(v) = \mathbf{0}\}$. Prove that the null set of T is a subspace of V.

GLOSSARY

Chapter 9: Algebraic Structures

Abelian or commutative group (9.4)	If a group G has the property that $a \circ b = b \circ a$ for all a, b in G, then G is called a **commutative** or **Abelian** group.
Abelian or commutative semigroup (9.2)	If $a*b = b*a$ for all a and b in a semigroup with operation $*$, then S with operation $*$ is called an **Abelian** or **commutative semigroup**.
Atom (9.3)	An **atom** of a Boolean algebra is a nonzero element x that has the property that for each y in the algebra, either $xy = x$ or $xy = 0$.
Boolean algebra (9.3)	A **Boolean algebra** is a bounded distributive lattice $(S, +, \cdot)$ that has a unary operator $'$ on the set S such that $x + x' = 1$ and $x \cdot x' = 0$.
Boolean algebra properties (9.3)	A Boolean algebra, which will be denoted by $(S, +, \cdot, ', 1, 0)$, has the following properties for all $x, y, z \in S$: 1. Associativity: $x + (y + z) = (x + y) + z$ and $x \cdot (y \cdot z) = (x \cdot y) \cdot z$ 2. Commutativity: $x + y = y + x$ and $x \cdot y = y \cdot x$ 3. Distributive Laws: $x \cdot (y + z) = (x \cdot y) + (x \cdot z)$ and $x + (y \cdot z) = (x + y) \cdot (x + z)$ 4. Complements: $x + x' = 1$ and $x \cdot x' = 0$ 5. Identities: $x + 0 = x$ and $x \cdot 1 = x$
Bounded (9.3)	A lattice is **bounded** if the set S, considered as a poset, has a least upper bound and a greatest lower bound. The least upper bound is denoted by 1 and the greatest lower bound is denoted by 0. Equivalently, a lattice is bounded if there exist elements $0, 1 \in S$ such that $0 \wedge a = 0$ and $1 \vee a = 1$ for all $a \in S$.
Comparable (9.1)	Two elements a and b of the partially ordered set (poset) $(S, \leq)$ are **comparable** if either $a \leq b$ or $b \leq a$.
Complement (9.3)	For $x \in S$, x' is called the **complement** of x.
Congruence relation (9.2)	Let $(S, \cdot)$ be a semigroup and R be an equivalence relation on S. If R has the property that, if $s_1 R s_2$ and $s_3 R s_4$ then $s_1 s_3 R s_2 s_4$ for all $s_1, s_2, s_3, s_4 \in S$, then R is called a **congruence relation**.
Cycle (9.5)	A cycle is a permutation $\sigma_o = (a_1, a_2, a_3, \ldots, a_k)$ for which $\sigma_o(a_i) = a_{i+1}$ for $i < k$ and $\sigma_o(a_k) = a_1$.
Cyclic semigroup (9.2)	The semigroup $\langle a \rangle = \{a^n : n > 0\} = \{a, a^2, a^3, \ldots\}$ is called a **cyclic semigroup**. More precisely, it is called the cyclic semigroup generated by a.
Disjoint cycles (9.5)	If $\pi = (a_1 a_2 \ldots a_r)$ and $\sigma = (b_1 b_2 \ldots b_n)$ are cycles of S_n, and sets $(a_1 a_2 \ldots a_r)$ and $(b_1 b_2 \ldots b_n)$ are disjoint, we say that π and σ are disjoint cycles.
Distributive lattice (9.3)	A lattice $(S, \vee, \wedge)$ is a **distributive lattice** if for all $a, b \in S$ $a \wedge (b \vee c) = (a \wedge b) \vee (a \wedge c)$ and $a \vee (b \wedge c) = (a \vee b) \wedge (a \vee c)$

Epimorphism (9.5)	The homomorphism $f: G \to H$ is an **epimorphism** if for every $h \in H$, $\exists g \in G$ such that $f(g) = h$.
Factor subgroup (9.5)	If H is a normal subgroup of a group $(G, \circ)$, then cosets of H in G form a group under the operation $(aH)(bH) = (abH)$. This group is called the **factor subgroup** and denoted by G/H.
Finite cyclic group (9.4)	The set $\{g, g^2, \ldots, g^P\}$, where $g^P = 1$, is called a **finite cyclic group**, or more precisely, the finite cyclic group generated by g. It is denoted by $\langle g \rangle$.
Generating set (9.4)	The subgroup A^* is called the group **generated** by A.
Greatest element (9.1)	The least upper bound for the entire poset A (if it exists) is called the **greatest element** of A.
Greatest lower bound (9.1)	The element a is called the **greatest lower bound** (glb) of B if (i) a is a lower bound of B and (ii) if any other element a' of A is a lower bound of B, then $a' \leq a$.
Group (9.4)	A **group** is a set G together with a binary operation (or product) $\circ$ on G that has the properties: 1. $a \circ (b \circ c) = (a \circ b) \circ c$ for all a, b, c in G (the operation $\circ$ on S is associative). 2. There exists an element 1 in G, called the **identity**, which has the property that $a \circ 1 = 1 \circ a = a$ for all a in G. 3. For each element a in G, there exists an element a^{-1}, called the **inverse** of a in G such that $a \circ a^{-1} = a^{-1} \circ a = 1$.
Homomorphism (9.2)	Let $(S, \circ)$ and $(T, \circ)$ be semigroups and $f: S \to T$ be a function such that $f(s \circ s') = f(s) \circ f(s')$. The function f is called a **homomorphism** from S to T.
Homomorphism (9.5)	Let $(G, \circ)$ and $(H, *)$ be groups, where $\circ$ and $*$ are the operations on G and H, respectively. Let $f: G \to H$ be a function. The function f is a **homomorphism** if $f(g \circ g') = f(g) * f(g')$ for all g and g' in G.
Idempotent (9.2)	An element a of semigroup is called an **idempotent** element if $a*a = a$.
Identity/monoid (9.2)	If there is an element I in $(S, *)$ such that $I*a = a*I = a$ for all a in A, then I is called the **identity** of $(S, *)$ and $(S, *)$ is called a **semigroup with identity** or a **monoid**.
Isomorphism (9.5)	The homomorphism $f: G \to H$ is said to be an **isomorphism** if f is one-to-one and onto.
Kernel (9.5)	Let $f: G \to H$ be a homomorphism from the group G to the group H. The **kernel** of G is the set $\{x : x \in G \text{ and } f(x) = 1\} = f^{-1}(\{1\})$, where 1 is the identity of H.
Lattice (9.3)	A poset $(S, \leq)$, which is both an upper semilattice and a lower semilattice, is a **lattice**.
Least element (9.1)	The greatest lower bound for the entire poset A (if it exists) is called the **least element** of A.
Least upper bound (9.1)	The element a is called a **least upper bound** (lub) of B if (i) a is an upper bound of B and (ii) if any other element a' of A is an upper bound of B, then $a \leq a'$.

Left coset (9.4)	For a subgroup H of a group G and any a in G, $a \circ H = \{x : x = a \circ h \text{ for some } h \text{ in } H\}$ is a **left coset** of H in G.
Lower bound (9.1)	An element a of A is a **lower bound** of a subset B of a poset A if $a \leq b$ (or $b \geq a$) for all b in B.
Lower semilattice (9.1)	A poset A for which all two-element subsets have a greatest lower bound in A is called a **lower semilattice** and is denoted by $(A, \wedge)$ or $(A, \circ)$.
Maximal element (9.1)	An element a of subset B of a poset A is a **maximal element** of B if for every element b of B, $b \geq a$ implies $b = a$.
Minimal element (9.1)	An element a of subset B of a poset A is a **minimal element** of B if for every element b of B, $b \leq a$ implies $b = a$.
Minimal generating set (9.4)	If a set A generates a group G and no proper subset of A also generates G, then A is a **minimal generating set** of G.
Minimal generating set of A^* (9.2)	The semigroup A^* is called the semigroup generated by A. If for every proper subset B of A, $B \subset A^*$, then A is called a **minimal generating set** of A^*.
Monomorphism (9.5)	The homomorphism $f: G \to H$ is said to be a **monomorphism** if f is one-to-one; that is, $f(a) = f(b) \to a = b$.
Normal subgroup (9.5)	If H is a subgroup of a group $(G, \circ)$ and has the property that $gHg^{-1} = H$ for all $g \in G$, then H is called a **normal subgroup**.
Octic group or symmetries of the square (9.5)	The eight permutations $D_4 = \{\rho_0, \rho_1, \rho_2, \rho_3, \mu_1, \mu_2, \delta_1, \delta_2\}$ contained in S_4, where $$\rho_0 = \begin{pmatrix} 1 & 2 & 3 & 4 \\ 1 & 2 & 3 & 4 \end{pmatrix} \quad \mu_1 = \begin{pmatrix} 1 & 2 & 3 & 4 \\ 4 & 3 & 2 & 1 \end{pmatrix}$$ $$\rho_1 = \begin{pmatrix} 1 & 2 & 3 & 4 \\ 4 & 1 & 2 & 3 \end{pmatrix} \quad \mu_2 = \begin{pmatrix} 1 & 2 & 3 & 4 \\ 2 & 1 & 4 & 3 \end{pmatrix}$$ $$\rho_2 = \begin{pmatrix} 1 & 2 & 3 & 4 \\ 3 & 4 & 1 & 2 \end{pmatrix} \quad \delta_1 = \begin{pmatrix} 1 & 2 & 3 & 4 \\ 1 & 4 & 3 & 2 \end{pmatrix}$$ $$\rho_3 = \begin{pmatrix} 1 & 2 & 3 & 4 \\ 2 & 3 & 4 & 1 \end{pmatrix} \quad \delta_2 = \begin{pmatrix} 1 & 2 & 3 & 4 \\ 3 & 2 & 1 & 4 \end{pmatrix},$$ form a group called the **symmetries of a square** or the **octic group**.
Orbit (9.5)	If σ is a permutation and R is the equivalence relation defined by aRb when $\sigma^n(a) = b$, then the equivalence classes created are called the **orbits** of σ.
Order of the group (9.4)	If G is a finite group with n elements, then n is called the **order of the group** G.
Partial ordering (9.1)	A relation R on set A is a partial ordering if it is reflexive, antisymmetric, and transitive. The set A is said to be a **partially ordered set** or **poset** with ordering R. This relation is often denoted by $\leq$ (or $\geq$ where $a \geq b$ if $b \leq a$). The partially ordered set A with ordering $\leq$ is denoted by $(A, \leq)$.
Retract (9.2)	The image of a retraction map function is called a **retract**.
Retraction map (9.2)	If $f: S \to S$ is an idempotent, then $f \circ f = f$. Such a function is called a **retraction map**.
Semigroup (9.2)	A set with binary operation $*$ on S such that for all a, b, and c in S, $(a*b)*c = a*(b*c)$ is called a **semigroup** and is denoted by $(S, *)$ or simply S if the operation is understood.

Semilattice (9.2)	A commutative semigroup $(S, *)$ is a **semilattice** if $a*a = a$ for all $a \in S$. A semilattice is a commutative semigroup in which every element is an idempotent.
Subgroup (9.4)	A subset H of a group G is a **subgroup** of G if H with the same operation as G is also a group.
Sublattice (9.3)	A nonempty subset $\overline{S}$ of a lattice $(S, \vee, \wedge)$ is a **sublattice** of S if for all $a, b \in \overline{S}, a \wedge b, a \vee b \in \overline{S}$.
Subsemigroup (9.2)	If $(S, *)$ is a semigroup and $\overline{S} \subseteq S$, then $\overline{S}$ is a **subsemigroup** of S if $*$ is a binary operation on $\overline{S}$. Equivalently, $(\overline{S}, *)$ is a subsemigroup of $(S, *)$ if $\overline{S} \subseteq S$ and for every $a, b \in \overline{S}, a*b \in S$.
Subsemilattice (9.2)	If $(S, *)$ is a semilattice and $\overline{S} \subseteq S$, then $\overline{S}$ is a subsemilattice of S if $*$ is a binary operation on $\overline{S}$. Equivalently, $(\overline{S}, *)$ is a subsemilattice of S if $\overline{S} \subseteq S$ and for every $a, b \in \overline{S}, a*b \in S$.
Total ordering or a chain (9.1)	If every two elements of a partially ordered set $(S, \leq)$ are comparable, then $(S, \leq)$ is a **total ordering** or a **chain**.
Upper bound (9.1)	For a subset B of a poset A, an element a of A is an **upper bound** of B if $b \leq a$ (or $a \geq b$) for all b in B.
Upper semilattice (9.1)	A poset A for which every two-element subset has a least upper bound in A is called an **upper semilattice** and is denoted by $(A, \vee)$ or $(A, +)$.

TRUE-FALSE QUESTIONS

1. In a partially ordered set, all pairs of elements are comparable.
2. Every finite partially ordered set has a minimal element.
3. Every finite partially ordered set has a greatest element.
4. In a partially ordered set A, if B is a subset of A and $a \in A$ is a least upper bound of B, then $a \in B$.
5. If a partially ordered set has a greatest lower bound, then it is unique.
6. In a lower semilattice, every finite set has a least upper bound.
7. In an upper semilattice, every set has a least upper bound.
8. In a lower semilattice A, $a \wedge a = a$ for all $a \in A$.
9. In a lower semilattice A, $a \vee a = a$ for all $a \in A$.
10. Every semilattice is a semigroup.
11. Every semigroup is a semilattice.
12. Every semigroup contains an identity.
13. Every semigroup contains an idempotent.
14. Every finite semigroup contains an idempotent.
15. A semigroup can contain only one idempotent.
16. Every semigroup is contained in a group.
17. Every group is a semigroup.
18. A group can contain only one idempotent.
19. The power set of a set A, together with the operation $\cup$, is a semilattice.
20. The power set of a set A, together with the operation $\cap$, is a group.
21. The power set of a set A, together with the operation $\triangle$, is a group.
22. Every sublattice of a bounded lattice is bounded.
23. Every bounded lattice contains a maximal element.
24. Every bounded lattice contains a greatest element.
25. A finite semigroup contains a maximal group.
26. Every group contains a finite semigroup.
27. A finite lattice is a distributive lattice.
28. A sublattice of a distributive lattice is distributive.
29. The power set of a set A, together with the operations $\cap$ and $\cup$, forms a distributive lattice.
30. The power set of a set A, together with the operations $\cap$ and $\triangle$, forms a distributive lattice.
31. A Boolean algebra is bounded.
32. In a Boolean algebra, the complement of an element is unique.

33. The intersection of a collection of subgroups of a group forms a group.

34. The union of a collection of subgroups of a group forms a group.

35. For every element in a finite group, there is a positive integer n so that $a^n = a$.

36. Every finite group is a cyclic group.

37. Every cyclic group is a finite group.

38. Let G be commutative group. If H and K are subgroups of G, then HK is a subgroup of G.

39. Let G be a group. If H and K are subgroups of G, then HK is a subgroup of G.

40. Let S be a subset of G, and H be the subgroup generated by S. G is cyclic if and only if S contains only one element.

41. A subsemigroup of a group is a group.

42. A subsemigroup of a finite group is a group.

43. A subsemigroup of a commutative group is a group.

44. Let $f : G \to H$ be a homomorphism from a group G to a group H. If $A \subseteq G$, and $f(A) = B$, then A is a subgroup of G if and only if B is a subgroup of H.

45. Let $f : G \to H$ be a homomorphism from a group G to a group H. If B is a subgroup of H, then $f^{-1}(B)$ is a subgroup of G.

46. Let $f : G \to H$ be a homomorphism from a group G to a group H. If A is a subgroup of G, then $f(A)$ is a subgroup of H.

47. Every subgroup of a finite symmetric group is a normal subgroup.

48. Every cyclic subgroup of a finite symmetric group is a normal subgroup.

49. Let $f : G \to H$ be a homomorphism from a group G to a group H. f is a monomorphism if and only if the kernel of f contains only one element.

50. Let $f : G \to H$ be a homomorphism from a group G to a group H. f is a homomorphism if and only if the kernel of f contains only one element.

SUMMARY QUESTIONS

Let $A = \{1, 2, 3, 5, 12, 15, 25, 36\}$. Define the poset $\leq$ by $a \leq b$ if $a \mid b$.

1. Find the Hasse diagram of the poset.

2. Find maximal elements, minimal elements, greatest lower bounds, and least upper bounds of the poset, if they exist.

3. Find all pairs of elements of A that are not comparable.

4. Explain the difference between a semilattice and other semigroups.

5. Give an example of a semilattice that is not finite.

6. Give an example of a semigroup S in which $a^2 = a$ for all a, but S is not a semigroup.

7. Give an example of a lattice that is not bounded.

8. Show that if $(S, \vee, \wedge)$ is finite, then it is bounded.

9. Show that if a lattice has a least element, then it is unique.

10. Let S be a set and G be the power set of S. Explain why $(G, \wedge)$ is not a group.

11. List all groups that are also semilattices.

12. Let $f : (Z, +) \to (Z_5, +)$ be defined by $f(a) = [a]$. What is the kernel of f?

13. Find the subgroup of $(Z, +)$ generated by $\{12, 15\}$.

14. Express the permutation $\begin{pmatrix} 1 & 2 & 3 & 4 & 5 & 6 \\ 2 & 3 & 1 & 4 & 5 & 6 \end{pmatrix}$ as a product of disjoint cycles.

15. Find the inverse of the permutation $\sigma = (1, 4, 3)(2, 5,)$.

ANSWERS TO TRUE-FALSE

1. F 2. T 3. F 4. F 5. T 6. F 7. F 8. T 9. F 10. T 11. F 12. F 13. F 14. T 15. F 16. F 17. T 18. T 19. T 20. F 21. T 22. F 23. T 24. T 25. T 26. T 27. F 28. T 29. T 30. F 31. T 32. T 33. T 34. F 35. T 36. F 37. F 38. T 39. F 40. F 41. F 42. T 43. F 44. F 45. T 46. T 47. F 48. F 49. T 50. F

CHAPTER 10

Number Theory Revisited

10.1 Integral Solutions of Linear Equations

In this chapter we explore some of the more algebraic properties of number theory. We begin with finding solutions of linear equations.

Using the Euclidean algorithm, we can find integral solutions for $ax + by = c$ when they exist. Such equations are called Diophantine Equations. The following theorem was proved previously as Theorem 3.39 and is repeated here for convenience.

THEOREM 10.1 The equation $ax + by = c$, where a, b, and c are integers, has an integral solution (i.e., there exist integers x and y such that $ax + by = c$) if and only if c is divisible by $\gcd(a, b)$.

When c is divisible by $\gcd(a, b)$, a solution of $ax + by = c$ is

$$x_0 = \frac{u \cdot c}{\gcd(a, b)} \qquad y_0 = \frac{v \cdot c}{\gcd(a, b)}$$

where u and v comprise any solution of $\gcd(a, b) = au + bv$. ∎

EXAMPLE 10.2 Find a solution for $85x + 34y = 51$. In Example 7.5 we showed that $\gcd(85, 34) = 17$ and that $(85)(1) + (34)(-2) = 17$. A solution is given by

$$x_0 = \frac{u \cdot c}{\gcd(a, b)} = \frac{1 \cdot 51}{17} = 3$$

and

$$y_0 = \frac{v \cdot c}{\gcd(a, b)} = \frac{(-2) \cdot 51}{17} = -6$$

As a check, we calculate

$$ax_0 + by_0 = 85 \cdot 3 + 34 \cdot (-6) = 255 + (-204) = 51$$

Another way to generate the solution is to use the equation $au + bv = \gcd(a, b)$ directly. Since

$$au + bv = \gcd(a, b)$$

or

$$a \cdot (1) + b \cdot (-2) = 17$$

391

multiplying by 3 we obtain
$$a \cdot (3) + b \cdot (-6) = 51$$
We see there may be more than one solution by observing that
$$85 \cdot 5 + 34 \cdot (-11) = 425 + (-374) = 51$$
when $x = 5$ and $y = -11$. ∎

EXAMPLE 10.3 Solve $252x + 580y = 20$. We showed that $\gcd(252, 580) = 4$ in Example 7.6 and also that $(252)(-23) + (580)(10) = 4$. Multiplying each term by 5, we have $(252)(-115) + (580)(50) = 20$. Hence, $x = -115$ and $y = 50$ is a solution. ∎

We are now able to determine whether a solution exists and to find a specific solution for the equation $ax + by = c$ if a solution exists. The following theorem enables us to find all solutions.

THEOREM 10.4 If a and b are nonzero integers and (x_0, y_0) is a solution of the equation $ax + by = c$, then any other solution (x, y) has the form
$$x = x_0 + \frac{b}{d}t$$
$$y = y_0 - \frac{a}{d}t$$
where t is an arbitrary integer and $d = \gcd(a, b)$.

Proof If (x, y) and (x_0, y_0) are both solutions of $ax + by = c$, then $ax + by = ax_0 + by_0$. Therefore, $ax - ax_0 = by_0 - by$ and $a(x - x_0) = b(y_0 - y)$. Dividing both sides by $d = \gcd(a, b)$, we have $\frac{a}{d}(x - x_0) = \frac{b}{d}(y_0 - y)$. Since, by Theorem 7.8, $\gcd\left(\frac{a}{d}, \frac{b}{d}\right) = 1$, one obtains $\frac{a}{d} \mid (y_0 - y)$ and $\frac{b}{d} \mid (x - x_0)$, say, $x - x_0 = u\left(\frac{b}{d}\right)$ and $(y_0 - y) = v\left(\frac{a}{d}\right)$. Then $\left(\frac{a}{d}\right)u\left(\frac{b}{d}\right) = \left(\frac{b}{d}\right)v\left(\frac{a}{d}\right)$ and, hence, $u = v$ by cancellation. Letting $t = u = v$, we obtain $x = x_0 + \left(\frac{b}{d}\right)t$ and $y = y_0 - \left(\frac{a}{d}\right)t$. We still need to show that the pair $\left(x_0 + \left(\frac{b}{d}\right)t, y_0 - \left(\frac{a}{d}\right)t\right)$ is a solution of $ax + by = c$. But
$$a\left(x_0 + \left(\frac{b}{d}\right)t\right) + b\left(y_0 - \left(\frac{a}{d}\right)t\right) = ax_0 + a\left(\frac{b}{d}\right)t + by_0 - b\left(\frac{a}{d}\right)t$$
$$= ax_0 + by_0$$
$$= c$$
∎

Returning to Example 10.2 we see that the general solution to the equation $85x + 34y = 51$ is $x = 3 + 2t$ and $y = -6 - 5t$. In Example 10.3 the general solution to the equation $252x + 580y = 20$ is $x = -115 + 145t$ and $y = 50 - 63t$.

Exercises

For each of the following equations, find a solution (if it exists):

1. $24x + 81y = 6$
2. $803x + 154y = 33$
3. $73x + 151y = 3$
4. $165x + 418y = 121$
5. $27x + 78y = 12$

6. Obtain the general solution for the equation in Exercise 1 if it exists.
7. Obtain the general solution for the equation in Exercise 2 if it exists.
8. Obtain the general solution for the equation in Exercise 3 if it exists.

9. Obtain the general solution for the equation in Exercise 4 if it exists.

10. Obtain the general solution for the equation in Exercise 5 if it exists.

For each of the following equations, find a solution (if it exists):

11. $23x + 18y = 4$
12. $299x + 533y = 52$
13. $39x + 299y = 27$
14. $272x + 102y = 68$
15. $27x + 180y = 33$

16. Obtain the general solution for each of the equations in Exercise 11 if it exists.

17. Obtain the general solution for each of the equations in Exercise 12 if it exists.

18. Obtain the general solution for each of the equations in Exercise 13 if it exists.

19. Obtain the general solution for each of the equations in Exercise 14 if it exists.

20. Obtain the general solution for each of the equations in Exercise 15 if it exists.

21. Write a procedure in pseudocode for the integral solution of linear equations.

10.2 Solutions of Congruence Equations

In the previous section, we considered equations of the form $ax + by = c$ having integers x and y for solutions. As a special case when $b = 0$, one can find solutions for $ax = c$. We now seek solutions for the congruence $ax \equiv c \pmod{n}$, meaning that we want an integer x so that the integer ax is congruent to c modulo n. Stated in terms of equivalence classes, if $[a]$ and $[c]$ are equivalence classes modulo n, we want an equivalence class $[x]$ so that $[a] \odot [x] = [c]$, where the equality refers to equality for sets.

Given the multiplication tables for Z_5 (see Section 3.6),

$[a] \odot [b]$	[0]	[1]	[2]	[3]	[4]
[0]	[0]	[0]	[0]	[0]	[0]
[1]	[0]	[1]	[2]	[3]	[4]
[2]	[0]	[2]	[4]	[1]	[3]
[3]	[0]	[3]	[1]	[4]	[2]
[4]	[0]	[4]	[3]	[2]	[1]

we can solve the congruence

$$3x \equiv 1 \pmod{5}$$

by considering (modulo 5)

$$[3] \odot [x] = [1]$$

By inspection of the multiplication table, we observe that $[x] = [2]$ will do since $[3] \odot [2] = [1]$. So we could let $x = 2$. Since $[2] = \{\ldots, -8, -3, 2, 7, 12, \ldots\}$, we note that, $x = -3$, $x = 7$, and $x = -8$ are all choices of x satisfying

$$3x \equiv 1 \pmod{5}$$

The next theorem provides some of the important properties of congruences. Parts (a) and (d) are proven in Theorem 3.55. The rest of the proofs are left to the reader.

THEOREM 10.5

(a) If $a \equiv b \pmod{n}$ and $c \equiv d \pmod{n}$, then $a + c \equiv b + d \pmod{n}$ and $ac \equiv bd \pmod{n}$.

(b) If $ac \equiv bc \pmod{n}$ and $\gcd(c, n) = 1$, then $a \equiv b \pmod{n}$.

(c) If $a \equiv b \pmod{n}$, then $a^m \equiv b^m \pmod{n}$ for all positive integers m.

(d) If $a \equiv b \pmod{mn}$, then $a \equiv b \pmod{m}$ and $a \equiv b \pmod{n}$.

(e) For $c \neq 0$, $ac \equiv bc \pmod{n}$ if and only if $a \equiv b \left(\mod \dfrac{n}{\gcd(c, n)} \right)$.

(f) If $a \equiv b \pmod{m}$, $a \equiv b \pmod{n}$, and $\gcd(m, n) = 1$, then $a \equiv b \pmod{mn}$. ∎

The next two theorems give the conditions under which solutions to $ax \equiv c \pmod{n}$ exist and specify the solutions.

THEOREM 10.6 The congruence $ax \equiv c \pmod{m}$ has an integer x as a solution if and only if $\gcd(a, m) \mid c$. All integer solutions are given by

$$x = x_0 + \frac{t \cdot m}{\gcd(a, m)}$$

where t is any integer and for x_0, there is a y_0 so that (x_0, y_0) is a solution of $ax + my = c$.

Proof By definition, $ax \equiv c \pmod{m}$ if and only if $ax - c$ is divisible by m so that there is an integer j such that $ax - c = jm$. This is true if and only if

$$ax + my = c$$

has a solution. The integers a, m, and c are fixed and we want integers x and y so that $ax + my = c$. By Theorem 10.1, $ax + my = c$ has an integral solution if and only if $\gcd(a, m) \mid c$. Also, by this theorem, a solution is given by

$$x_0 = \frac{u \cdot c}{\gcd(a, m)}$$

$$y_0 = \frac{v \cdot c}{\gcd(a, m)}$$

where u and v are selected so that $au + mv = \gcd(a, m)$. By Theorem 10.4 all solutions are given by

$$x = x_0 + \frac{t \cdot m}{\gcd(a, m)}$$

$$y = y_0 - \frac{t \cdot m}{\gcd(a, m)}$$

for any integer t. For this application we need only the solutions for x. Thus, all integral solutions of $ax \equiv c \pmod{m}$ are of the form

$$x = x_0 + \frac{t \cdot m}{\gcd(a, m)}$$

where t may be any integer. ∎

The next theorem gives us the distinct solutions of $ax \equiv c \pmod{m}$. Since there are only a finite number of equivalence classes modulo m, there can be only a finite number of distinct solutions modulo m. These are given in the following theorem. The proof is left to the reader.

THEOREM 10.7 If $\gcd(a, m) \mid c$, then $ax \equiv c \pmod{m}$ has finitely many distinct solutions modulo m. These solutions are given by

$$x_0 + \frac{t \cdot m}{\gcd(a, m)} \text{ modulo } m = \left[\left[x_0 + \frac{t \cdot m}{\gcd(a, m)} \right]\right]_m$$

for $t = 1, 2, 3, \ldots, \gcd(a, m)$, where for x_0 there is a y_0 so that (x_0, y_0) is a solution of $ax + my = c$. ∎

EXAMPLE 10.8 For the congruence

$$35 \cdot x \equiv 14 \pmod{84}$$

since $\gcd(35, 84) = 7$ and $7 \mid 14$, the equation has exactly seven distinct solutions modulo 84, which are of the form

$$x_0 + \frac{84 \cdot t}{7} = x_0 + 12 \cdot t$$

where $t = 1, 2, 3, \ldots, 7$ and (x_0, y_0) is a solution of

$$35x + 84y = 14$$

which is equivalent to

$$5x + 12y = 2$$

By inspection a solution is given by $x_0 = -2$ and $y_0 = 1$. The seven distinct solutions modulo 84 are

t	$x_0 + 12t$
1	$-2 + 12 \cdot 1 = 10$
2	$-2 + 12 \cdot 2 = 22$
3	$-2 + 12 \cdot 3 = 34$
4	$-2 + 12 \cdot 4 = 46$
5	$-2 + 12 \cdot 5 = 58$
6	$-2 + 12 \cdot 6 = 70$
7	$-2 + 12 \cdot 7 = 82$

When $\gcd(a, m) = 1$, there is a unique solution of $ax \equiv c \pmod{m}$. For example, consider

$$6x \equiv 7 \pmod{55}$$

$\gcd(6, 55) = 1$ and certainly 1 divides 7. There will be exactly one solution modulo 55 given by

$$x_0 + \frac{t \cdot m}{\gcd(6, 55)} = x_0 + \frac{1 \cdot 55}{1} = x_0 + 55 \equiv x_0 \pmod{55}$$

where (x_0, y_0) is a solution of $ax + my = c$ or $6x + 55y = 7$. To obtain x_0 and y_0 we begin by backtracking the Euclidean algorithm as shown in the examples following Theorem 7.2 getting $6(-9) + 55(1) = 1 = \gcd(6, 55)$. Multiplying each term by 7, we get $6(-63) + 55(7) = 7$, so $x_0 = -63$ and $x = -63 + 55 = -8$. ∎

EXAMPLE 10.9 Solve the congruence

$$623x \equiv -406 \pmod{84}$$

The integer 623 is greater than the modulus 84 and -406 is negative. Since we want solutions modulo 84, we select integers in the range $0, 1, 2, \ldots, 83$ because these are the possible remainders upon division by 84 and are the simplest representatives of the equivalence classes generated by congruence modulo 84. Using the Division Algorithm,

$$623 = 84 \cdot 7 + 35 \text{ so that } 623 \equiv 35 \pmod{84}$$
$$-406 = 84(-5) + 14 \text{ so that } -406 \equiv 14 \pmod{84}$$

Thus,

$$35x \equiv 14 \pmod{84}$$

is equivalent to the original $623x \equiv -406 \pmod{84}$. We solved the congruence $35x \equiv 14 \pmod{84}$ in a previous example. ■

The proof of the following theorem is left to the reader.

THEOREM 10.10 If $a \equiv b \pmod{n_1}$, $a \equiv b \pmod{n_2}$, ..., $a \equiv b \pmod{n_k}$, and $n = \text{lcm}(n_1, n_2, \ldots, n_k)$, then $a \equiv b \pmod{n}$, and conversely. ■

Exercises

1. For what values of n is $75 \equiv 35 \pmod{n}$?

Find all solutions of the following equations:

2. $4x \equiv 3 \pmod{7}$
3. $27x \equiv 12 \pmod{15}$
4. $28x \equiv 56 \pmod{49}$
5. $24x \equiv 6 \pmod{81}$
6. $91x \equiv 26 \pmod{169}$
7. Prove that if a is an odd integer, then $a^2 \equiv 1 \pmod{8}$.
8. List all of the positive integers that are both congruent to 5 modulo 3 and congruent to 5 modulo 2.
9. Prove that if $a \equiv b \pmod{n_1}$, $a \equiv b \pmod{n_2}$, ..., $a \equiv b \pmod{n_k}$, and $n = \text{lcm}(n_1, n_2, \ldots, n_k)$, then $a \equiv b \pmod{n}$.
10. Prove that if $ac \equiv bc \pmod{n}$ and $\gcd(c, n) = 1$, then $a \equiv b \pmod{n}$.
11. Prove that if $a \equiv b \pmod{n}$, then $a^m \equiv b^m \pmod{n}$ for all positive integers m.
12. Prove that if $a \equiv b \pmod{m}$, $a \equiv b \pmod{n}$, and $\gcd(m, n) = 1$, then $a \equiv b \pmod{mn}$.
13. Prove that for $c \neq 0$, $ac \equiv bc \pmod{n}$ if and only if $a \equiv b \left(\text{mod } \dfrac{n}{\gcd(c, n)} \right)$.

10.3 Chinese Remainder Theorem

In the previous section, we considered equations of the form

$$ax \equiv c \pmod{m}$$

where we were given integers a and c and positive integer m and wanted to find an integer x satisfying the congruence. We now wish to consider solutions of simultaneous congruences:

$$x \equiv a_1 \pmod{m_1}$$
$$x \equiv a_2 \pmod{m_2}$$
$$\vdots$$
$$x \equiv a_n \pmod{m_n}$$

where the m_i are relatively prime in pairs. In other words, we want to find an integer x that leaves a remainder of a_i when divided by m_i, when $\gcd(m_i, m_j) = 1$, if $i \neq j$.

Solutions of simultaneous congruences were considered in ancient times. Often, word problems similar to the following were posed. Suppose that a group of monkeys contemplates a pile of coconuts. If the monkeys put the coconuts in piles of five each, there are four left over. Using piles of four, there are three left over. In piles of seven, there are two left over. In piles of nine, there are six left over. What is the fewest number of coconuts possible?

If x is a possible number of coconuts in the pile, then having four left over from piles of five is expressed as

$$x \equiv 4 \pmod{5}$$

Similarly, the other conditions are

$$x \equiv 3 \pmod{4}$$
$$x \equiv 2 \pmod{7}$$
$$x \equiv 6 \pmod{9}$$

The smallest positive integer x satisfying all four congruences is the required solution. Solutions of such problems are given by the Chinese Remainder Theorem.

THEOREM 10.11 (**Chinese Remainder Theorem**) Let $m_1, m_2, \ldots, m_n$ be pairwise relatively prime, that is, $\gcd(m_i, m_j) = 1$, for all i and j less than or equal to n where $i \neq j$. Then the system of congruences

$$x \equiv a_1 \pmod{m_1}$$
$$x \equiv a_2 \pmod{m_2}$$
$$\vdots$$
$$x \equiv a_n \pmod{m_n}$$

has a solution that is unique modulo the integer $m_1 m_2 \cdots m_n$. Furthermore, if

$$M_j = \frac{\prod_{i=1}^{n} m_i}{m_j}$$

and z_j is a solution of $M_j z_j \equiv a_j \pmod{m_j}$ for each j, then the solution is given by

$$x = \left[\left[\sum_{j=1}^{n} M_j z_j\right]\right]_{m_1 m_2 \cdots m_n}$$

Proof Let x be as defined in the theorem. Then for any k, $1 \leq k \leq n$,

$$x = \left[\left[\sum_{j=1}^{n} M_j z_j\right]\right]_{m_1 m_2 \cdots m_n}$$

so that

$$x \equiv \sum_{j=1}^{n} M_j z_j \pmod{\prod_{i=1}^{n} m_i}$$

$$\equiv \sum_{j=1}^{n} M_j z_j \pmod{m_k}, \text{ by Theorem 10.5(d)}$$

$$\equiv M_k z_k \pmod{m_k}$$

$$\equiv a_k \pmod{m_k}$$

so that x satisfies the n congruences, $x \equiv a_k \pmod{m_k}$ for $1 \leq k \leq n$. If x' also satisfies these n congruences, then

$$x - x' \equiv 0 \pmod{m_i} \quad \text{for } 1 \leq i \leq n$$

Since $\gcd(m_i, m_j) = 1$ for $i \neq j$, we obtain

$$x \equiv x' \pmod{\prod_{i=1}^{n} m_i}$$

and the solution x is unique modulo $\prod_{i=1}^{n} m_i$. ∎

EXAMPLE 10.12 Solve the following set of congruences:

$$x \equiv 5 \pmod{4}, \qquad x \equiv 7 \pmod{11}$$

Since 4 and 11 are relatively prime, there exists an integer, namely 10, such that $(4)(10) \equiv 7 \pmod{11}$ and there exists an integer, namely 3, such that $(11)(3) \equiv 5 \pmod{4}$. Hence, $(4)(10) + (11)(3) = 73$, which is congruent to 29 modulo 44, satisfies the foregoing congruences. ∎

EXAMPLE 10.13 We answer the monkey-coconut question by solving the following set of congruences:

$$x \equiv 4 \pmod{5}$$
$$x \equiv 3 \pmod{4}$$
$$x \equiv 2 \pmod{7}$$
$$x \equiv 6 \pmod{9}$$

We have $M_1 = 4 \cdot 7 \cdot 9 = 252$, $M_2 = 5 \cdot 7 \cdot 9 = 315$, $M_3 = 180$, and $M_4 = 140$. Since 5 and 252 are relatively prime, there exists an integer z_1 such that $252 z_1 \equiv 4 \pmod{5}$ or, equivalently, $2 z_1 \equiv 4 \pmod{5}$ or $z_1 \equiv 2 \pmod{5}$. Hence, z_1 can equal 7.

Since 4 and 315 are relatively prime, there exists an integer z_2 such that $315 z_2 \equiv 3 \pmod{4}$ or equivalently, $3 z_2 \equiv 3 \pmod{4}$. Hence, z_2 can equal 1.

Since 7 and 180 are relatively prime, there exists an integer z_3 such that $180 z_3 \equiv 2 \pmod{7}$ or, equivalently, $5 z_3 \equiv 2 \pmod{7}$. Hence, z_3 can equal 6.

Since 9 and 140 are relatively prime, there exists an integer z_4 such that $140 z_4 \equiv 6 \pmod{9}$ or, equivalently, $5 z_4 \equiv 6 \pmod{9}$. Hence, z_4 can equal 3.

Hence, $x = (7)(252) + (1)(315) + (6)(180) + (3)(140) \pmod{5 \cdot 4 \cdot 7 \cdot 9}$ or $x \equiv 3579 \pmod{1260}$ and $x = 1059$ is the least positive integer solution. ∎

The Chinese Remainder Theorem may be generalized to moduli $m_1, m_2, \ldots, m_n$ that are not relatively prime as follows.

THEOREM 10.14 The system of congruences

$$x \equiv a_1 \pmod{m_1}$$
$$x \equiv a_2 \pmod{m_2}$$
$$\vdots$$
$$x \equiv a_n \pmod{m_n}$$

has a solution if and only if $\gcd(m_i, m_j)$ divides $a_i - a_j$ for all i and j with $1 \le i < j \le n$. When the solution exists, it is unique modulo $\operatorname{lcm}(m_1, m_2, \ldots, m_n)$.

Proof The theorem is obviously true for $n = 1$. To show the technique, we show the proof for $n = 2$. Given the equations

$$x \equiv a_1 \pmod{m_1}$$
$$x \equiv a_2 \pmod{m_2}$$

we have

$$x = a_1 + km_1$$

Substituting this into the second equation, we have

$$a_1 + km_1 \equiv a_2 \pmod{m_2}$$

or

$$km_1 \equiv a_2 - a_1 \pmod{m_2}$$

Since $\gcd(m_1, m_2)$ divides $a_1 - a_2$, there is a solution for k, and $x + km_1$ is a solution to both equations. Since m_1 and m_2 both divide $\operatorname{lcm}((m_1, m_2))$,

$$x + km_1 + \operatorname{lcm}((m_1, m_2))$$

is also a solution. If x_1 and x_2 are both solutions of both congruences, then $x_1 - x_2$ is divisible by both m_1 and m_2 and, therefore, divisible by $\operatorname{lcm}((m_1, m_2))$.

Now assume we have congruences

$$x \equiv a_1 \pmod{m_1}$$
$$x \equiv a_2 \pmod{m_2}$$
$$\vdots$$
$$x \equiv a_k \pmod{m_k}$$
$$x \equiv a_{k+1} \pmod{m_{k+1}}$$

and we have a value $\bar{x}$ so that $\bar{x} \equiv a_i \pmod{m_i}$ for $1 \le i \le k$. Therefore, every solution of $x \equiv a_i \pmod{m_i}$ for $1 \le i \le k$ has the form $\bar{x} + uM_k$ where

$$M_k = \operatorname{lcm}(m_1, m_2, \ldots, m_k)$$

Substitute this value into the last congruence $x \equiv a_{k+1} \pmod{m_{k+1}}$ so we have

$$\bar{x} + uM_k \equiv a_{k+1} \pmod{m_{k+1}}$$

If there is a solution for u, then $\bar{x} + uM_k$ is a solution for all $k + 1$ congruences. There is a solution for this congruence iff $\gcd(M_k, m_{k+1}) | (\bar{x} - +a_{k+1})$. However for all i,

$$\bar{x} + a_{k+1} = \bar{x} - a_i + a_i - a_{k+1}$$

But $\bar{x} - a_i$ is divisible by m_i and $a_i - a_{k+1}$ is divisible by $\gcd(m_i, m_{k+1})$. Therefore, $\gcd(m_i, m_{k+1}) \mid (\bar{x} - a_{k+1})$ for all i. Therefore, $\gcd(M_k, m_{k+1}) \mid (\bar{x} - a_{k+1})$. (To see this, pick the highest power of a prime that divides M_k and m_{k+1}. Show that it

EXAMPLE 10.15

Solve

$$x \equiv 5 \pmod{6}$$
$$x \equiv 3 \pmod{10}$$
$$x \equiv 8 \pmod{15}$$

A solution to

$$x \equiv 5 \pmod{6}$$

has the form $5 + 6t$. Substituting this into the second congruence, we have $5 + 6t \equiv 3 \pmod{10}$. Therefore $6t \equiv 8 \pmod{10}$, and

$$t \equiv 3 \pmod{10}$$

Therefore, $t = 3 + 10u$. Substituting t into the third equation, we have

$$3 + 10u \equiv 8 \pmod{10}$$

Therefore, $10u \equiv 5 \pmod{10}$ and $u \equiv 2 \pmod{15}$. Therefore, $u = 2 + 15v$ for some v, and

$$t = 3 + 10(2 + 15v) = 150v + 23$$

and

$$x = 5 + 6(150v + 23) = 203 + 6 \cdot 10 \cdot 15v$$

so $x = 203$ is a solution. ∎

Exercises

Solve the following sets of simultaneous congruences:

1. $x \equiv 9 \pmod{12}$
   $x \equiv 6 \pmod{25}$

2. $x \equiv 3 \pmod{4}$
   $x \equiv 5 \pmod{9}$
   $x \equiv 10 \pmod{35}$

3. $x \equiv a \pmod{15}$
   $x \equiv b \pmod{16}$

4. $x \equiv 5 \pmod{7}$
   $x \equiv 12 \pmod{15}$
   $x \equiv 18 \pmod{22}$

5. $x \equiv 2 \pmod{13}$
   $x \equiv 5 \pmod{21}$

6. $x \equiv 7 \pmod{17}$
   $x \equiv 9 \pmod{13}$
   $x \equiv 3 \pmod{12}$

7. $x \equiv a \pmod{7}$
   $x \equiv b \pmod{16}$

8. $x \equiv 5 \pmod{9}$
   $x \equiv 13 \pmod{11}$
   $x \equiv 7 \pmod{5}$

9. If the marbles in a bag are lined up in rows of 15 each, there are 4 left in the bag. If the marbles are lined up in rows of 8 each, there are 3 left in the bag. If the marbles are lined up in rows of 23, there are 10 left in the bag. What is the least number of marbles that could have been in the bag initially? How many rows of 15, 8, and 23 were there?

10. Assume that there are two gears, one (gear A) with 25 teeth and the other (gear B) with 54 teeth. Gear A has a "bad" tooth and gear B has a bad section between two teeth. The two gears are meshed together with gear A on the left. Gear A turns clockwise and gear B turns counterclockwise. At the beginning of rotation a tooth of gear A fits exactly with gear B, the bad tooth of gear A is three teeth before the mesh position, and the bad between-teeth section of gear B is 20 sections from reaching the mesh position. How many "teeth" must mesh before the bad tooth first meets the bad between-teeth section? How often after that do the defective parts meet?

If possible, solve each of the following systems of congruences:

11. $x \equiv 21 \pmod{36}$
    $x \equiv 5 \pmod{8}$

12. $x \equiv 8 \pmod{12}$
    $x \equiv 5 \pmod{9}$
    $x \equiv 14 \pmod{15}$

13. $x \equiv 19 \pmod{49}$
    $x \equiv 10 \pmod{14}$

10.4 Properties of the Function ϕ

Let $n = p_1^{\alpha_1} p_2^{\alpha_1} \cdots p_k^{\alpha_k}$ be the prime factorization of n. Every positive divisor of n is either 1 or is divisible by p_i for some i, and every integer relatively prime to n has none of these primes as a factor. Some characteristics of n depend upon the number of integers s, $1 \leq s \leq n$, that do *not* contain any p_i as a factor. For example, if $n = 40 = 2^3 \cdot 5$, then the integers s, $1 \leq s \leq n$, and their factorizations are

1 = 1	**11** = 11	21 = 3 · 7	**31** = 31
2 = 2	12 = 2^2 · 3	22 = 2 · 11	32 = 2^5
3 = 3	**13** = 13	**23** = 23	**33** = 3 · 11
4 = 2^2	14 = 2 · 7	24 = 2^3 · 3	34 = 2 · 17
5 = 5	15 = 3 · 5	25 = 5^2	35 = 5 · 7
6 = 2 · 3	16 = 2^4	26 = 2 · 13	36 = $2^2 \cdot 3^2$
7 = 7	**17** = 17	**27** = 3^3	**37** = 37
8 = 2^3	18 = 2 · 3^2	28 = 2^2 · 7	38 = 2 · 19
9 = 3^2	**19** = 19	**29** = 29	**39** = 3 · 13
10 = 2 · 5	20 = 2^2 · 5	30 = 2 · 3 · 5	40 = 2^3 · 5

The integers highlighted in bold, none of which contain either 2 or 5 as a factor, are those relatively prime to $n = 40$. The number of such s, $1 \leq s \leq n$, relatively prime to 40, is denoted by $\phi(40) = 16$.

DEFINITION 10.16   Let $\phi(n)$ be the number of positive integers less than n that are relatively prime to n; that is, $\phi(n)$ is the number of reduced residues modulo n. ϕ is called **Euler's totient function** or the **Euler ϕ function**.

The foregoing factorization table also shows that

$$\phi(1) = 1 \qquad \phi(5) = 4 \qquad \phi(9) = 6$$
$$\phi(2) = 1 \qquad \phi(6) = 2 \qquad \phi(10) = 4$$
$$\phi(3) = 2 \qquad \phi(7) = 6 \qquad \phi(11) = 10$$
$$\phi(4) = 2 \qquad \phi(8) = 4 \qquad \phi(12) = 4$$

Every positive integer n can be written in terms of the number of relatively prime positive integers less than or equal to each divisor of n. For example, $6 = 2 \cdot 3$ has four divisors: 1, 2, 3, and 6. From the foregoing table,

$$\phi(1) + \phi(2) + \phi(3) + \phi(6) = 1 + 1 + 2 + 2 = 6$$

This property is given by the following theorem.

THEOREM 10.17   **(Gauss)**   If n is a positive integer, then

$$\sum_{d \mid n} \phi(d) = n$$

where the divisors d are positive divisors of n.

Proof   Let d be a positive divisor of n. Let $C(d)$ equal the set of positive integers $1 \leq m \leq n$ where $\gcd(m, n) = d$. $C(d)$ and $C(d')$ are disjoint if $d \neq d'$ because an integer can have only one greatest common divisor with n. But, by Theorem 7.8, $C(d)$ is also the set of positive integers m, $1 \leq m \leq n$ with $\gcd(m/d, n/d) = 1$. But this is the number of positive integers less than n/d and relatively prime to n/d; that is, this is $\phi(n/d)$. Since the union of all of these sets is the set of integers between 1 and n, $n = \sum_{d|n} \phi(n/d)$. But for every d that divides n, there is a corresponding n/d that divides n. Hence, $\sum_{d|n} \phi(n/d) = \sum_{d|n} \phi(d) = n$, which proves the theorem. ∎

EXAMPLE 10.18 Let $n = 12$. The divisors of 12 are 1, 2, 3, 4, 6, and 12. From the preceding table of Euler ϕ values,

$$\phi(1) + \phi(2) + \phi(3) + \phi(4) + \phi(6) + \phi(12) = 1 + 1 + 2 + 2 + 2 + 4 = 12$$

To illustrate the proof of Theorem 10.17, for $d = 1, 2, 3, 4, 6$, and 12, we see that the corresponding values of n/d are, respectively, $n/d = 12, 6, 4, 3, 2$, and 1 so that the two sums mentioned are equal. ∎

We now proceed to determine how to evaluate $\phi(n)$ for any positive integer n. The next three theorems lead to this objective.

THEOREM 10.19 If m and n are relatively prime, then

$$\phi(mn) = \phi(m)\phi(n)$$

Proof Let m and n be relatively prime. An integer is relatively prime to mn if and only if it is relatively prime to both m and n. Let a be relatively prime to m and let $a < m$. Consider the sequence $a, a+m, a+2m, \ldots, a+(n-1)m$. No two of these numbers are congruent modulo n; for if $a + jm \equiv a + km \pmod{n}$, then $n \mid (jm - km)$. Hence, $n \mid m(j-k)$. Since m and n are relatively prime, $n \mid (j-k)$, which is impossible. Therefore, this sequence is a complete residue system modulo n and each of the elements in the sequence is congruent modulo n to a positive integer less than n. Hence, the number of these elements relatively prime to n is equal to $\phi(n)$. Since there are $\phi(m)$ of these sequences, there are $\phi(m)\phi(n)$ numbers that are relatively prime to both m and n, that are less than mn, and that are relative prime to mn. Therefore, $\phi(mn) = \phi(m)\phi(n)$. ∎

For example, let $m = 8$ and $n = 15$. Then $\phi(8) = 4$ since 1, 3, 5, and 7 are the only positive integers less than 8 and relatively prime to 8. Also, $\phi(15) = 8$ since 1, 2, 4, 7, 8, 11, 13, and 14 are the only positive integers less than 15 and relatively prime to 15. Hence, $\phi(120) = \phi(8)\phi(15) = 32$, which may be checked directly.

Because of Theorem 10.19, we say that ϕ is multiplicative for relatively prime factors. We will discuss other multiplicative functions in the next chapter. Next we see how to calculate $\phi(n)$ when n is a power of a single prime.

THEOREM 10.20 If p is a prime number, then $\phi(p^k) = p^k - p^{k-1}$.

Proof The numbers less than or equal to p^k that are not relatively prime to p^k are $p, 2p, 3p, \ldots, (p^{k-1})p$. Since there are p^{k-1} of these integers, then there are $p^k - p^{k-1}$ integers that are relatively prime to p^k. Hence, $\phi(p^k) = p^k - p^{k-1}$. ∎

Corollary 10.21 A positive integer p is prime if and only if $\phi(p) = p - 1$.

Proof If p is prime, it is immediate from Theorem 10.20 that $\phi(p) = p - 1$. On the other hand, if p is not prime, then p has a divisor d different from p and 1. Since by definition, $\phi(p) \leq p - 1$ and d is one of the $p - 1$ positive integers less than p, we have $\phi(p) \leq p - 2$, a contradiction. ∎

Corollary 10.22 $\phi(2^k) = 2^{k-1}$. ∎

The multiplicative property $\phi(mn) = \phi(m)\phi(n)$ for m and n relatively prime and the result $\phi(p^k) = p^k - p^{k-1}$ may be combined to obtain an explicit formula for $\phi(n)$ for any positive integer n using the prime factorization of n. This formula is given in the next theorem whose proof is left as an exercise.

THEOREM 10.23 If n is a positive integer with prime factorization
$$n = p_1^{\alpha_1} p_2^{\alpha_1} \cdots p_t^{\alpha_t}$$
then
$$\phi(n) = \prod_{i=1}^{t} \left[p_i^{\alpha_i - 1}(p_i - 1) \right] = n \prod_{i=1}^{t} \left(1 - \frac{1}{p_i}\right)$$ ∎

EXAMPLE 10.24 Since $n = 40 = 2^3 \cdot 5$, $\phi(40) = 40(1 - 1/2)(1 - 1/5) = 40(1/2)(4/5) = 16$, which agrees with an example at the beginning of this section. Also, in Chapter 2, we saw that $n = 39616304 = 2^4 \cdot 7^2 \cdot 13^3 \cdot 23^1$ so that
$$\phi(39616304) = 2^3(2-1)7^1(7-1)13^2(13-1)23^0(23-1)$$
$$= 8 \cdot 1 \cdot 7 \cdot 6 \cdot 169 \cdot 12 \cdot 1 \cdot 22 = 14990976$$ ∎

There are limitations on the number of integers s, $1 \leq s \leq n$ that are relatively prime to n. One such constraint is given by the next theorem.

THEOREM 10.25 If n is an integer greater than 2, $\phi(n)$ is even.

Proof If $n = 2^k m$ where m is an odd integer and $k > 1$, then $\phi(2^k m) = \phi(2^k)\phi(m) = 2^{k-1}\phi(m)$ and, hence, $\phi(n)$ is even. If $n = p^k m$, where p is an odd prime, and p^k and m are relatively prime, then $\phi(p^k m) = \phi(p^k)\phi(m) = (p^k - p^{k-1})\phi(m)$. But $p^k - p^{k-1} = p^{k-1}(p-1)$ and $p - 1$ is an even number since p is odd. Hence, $\phi(n)$ is even. ∎

Although the following result has been indicated previously in Section 9.5, we state it here formally.

THEOREM 10.26 If n is an integer, then the nonzero reduced residue classes form a group under multiplication modulo n.

Proof Certainly, if a and b are relatively prime to n, then ab is also relatively prime to n and so we have closure. If a is relatively prime to n, then the congruence $ax \equiv 1 \pmod{n}$ has a unique solution and, hence, a has an inverse. ∎

Let p be a prime. Since $\{1, 2, \ldots, p-1\}$ is a set of reduced residues modulo p, $[1], [2], \ldots, [p-1]$ form a group under multiplication as discussed in Section 3.1. The next theorem shows that the product, $[1] \odot [2] \odot \cdots \odot [p-1]$, of all the nonzero residue classes is always the residue class $[p-1] = [-1]$. Stated in terms of congruences, we have equivalently that $1 \cdot 2 \cdots (p-1) \equiv -1 \pmod{p}$.

THEOREM 10.27 (Wilson's Theorem) The positive integer p is a prime if and only if $(p-1)! \equiv -1 \pmod{p}$.

Proof If p is prime, then $p \equiv 0 \pmod{p}$ and $p - 1 \equiv -1 \pmod{p}$. The nonzero residue classes modulo p form a group under multiplication when p is prime so that each residue class is paired with its inverse to yield the product $[1]$. Thus, if $1 \leq u \leq p-1$, then there is a unique integer u^{-1}, $1 \leq u^{-1} \leq p-1$, such that

$u \cdot u^{-1} \equiv 1 \pmod{p}$. Either $u = u^{-1}$ or $u \neq u^{-1}$. Certainly, for $u = 1$, $u^{-1} = 1$ also so that $u^2 = uu^{-1} \equiv 1 \pmod{p}$. If there is an integer a, $1 < a \leq p - 1$, such that $a^2 \equiv 1 \pmod{p}$ also, then $a^2 - 1 = (a-1)(a+1) \equiv 0 \pmod{p}$ and $p \mid (a-1)(a+1)$. Thus, $p \mid (a-1)$ or $p \mid (a+1)$. Since $a - 1 \neq 0$ and $a < p$, we obtain $p \nmid (a-1)$. Thus, $p \mid (a+1)$ implies that $p \leq a + 1$; and since $a \leq p - 1$ implies that $a + 1 \leq p$, we have $p = a + 1$ or $a = p - 1$. Thus, for $1 \leq u \leq p - 1$, only $u = 1$ and $u = p - 1$ have the property that $u^2 \equiv 1 \pmod{p}$. Hence, we obtain $(p-1)! = 1(u_1 u_1^{-1})(u_2 u_2^{-1}) \cdots (u_k u_k^{-1})(p-1) \equiv 1 \cdot 1 \cdot 1 \cdot \cdots \cdot 1 \cdot (p-1) = p - 1 \equiv -1 \pmod{p}$ and where u_j is one of the integers $2, 3 \ldots$, or $(p-2)$, where $k = (p-3)/2$.

If p is not prime, then $p = r \cdot s$ with $1 < r, s < p$. Since $(p-1)!$ contains r as a factor, $(p-1)! \equiv 0 \pmod{r}$ so that $(p-1)! \not\equiv -1 \pmod{r}$. Thus, p must be prime. ∎

For example, let $p = 5$. Then $(p-1)! = 4! = 24 \equiv -1 \pmod{5}$. Notice that the theorem says that the product $(p-1)!$ cannot be congruent to -1 unless p is prime. By the theorem we may check p for primeness by determining whether $(p-1)! \equiv -1 \pmod{p}$; however, this test is not used for large p because the calculation of $(p-1)!$ modulo p is not practical.

We now examine Wilson's theorem from an algebraic point of view. We already know that $Z_p - \{[0]\}$ is a group under multiplication. Thus, every nonzero element of Z_p has a multiplicative inverse. The preceding proof of Wilson's theorem shows that only $[1]$ and $[p-1]$ are their own inverses. Hence, in the product $[1][2][3] \cdots [p-1]$, each of the other elements is paired with its inverse so that $[1][2][3] \cdots [p-1] = [1][p-1] = [p-1]$ or, equivalently,

$$1 \cdot 2 \cdot 3 \cdots (p-1) \equiv p - 1 \pmod{p}$$
$$\equiv -1 \pmod{p}$$

In a cyclic group of even order it is easy to show that there are only two elements that are their own inverses, which in this case are $[1]$ and $[p-1]$.

The ϕ function of this section is named for Léonard Euler, the most prolific writer in mathematics. Many of his theorems appear throughout this book. His works would fill more than 75 large volumes. He was active in virtually every branch of mathematics. In number theory, he did an immense amount of work, including proving several of Fermat's lesser theorems. He is credited with beginning the idea of topology as well as several branches of calculus. It is impossible here to describe the immense amount of important mathematics for which he was responsible. He won the prestigious biennial prize from the Académie des Sciences 12 times. Much of our mathematical notation today is due to Euler. Léonard Euler (1707–1783) was the son of a Lutheran minister in Switzerland. His father was his first teacher and wanted him to enter the ministry. He had the fortune to become the student of Jean Bernoulli, one of Europe's best mathematicians. Since opportunities in Switzerland were limited, he, along with many other mathematicians from Europe, went to the newly organized Academy of St. Petersburg. While in Russia, he lost sight in one eye. Shortly after he arrived, political repression began in Russia. After 14 years, Euler left Russia to head the mathematics division of the Berlin Academy. Apparently, when asked by the queen mother in Germany why he was so shy, he replied that he had just come from a country where he who speaks is hanged.

He was, however, still held in high esteem in Russia. When Russia attacked Germany in 1760, Euler's farm was destroyed by the Russians. When they learned of this, they immediately made good on the loss, and an additional gift was added by the empress. After a disagreement with Frederick II, Euler returned to Russia after 25 years in Germany to accept a generous offer by Catherine the Great. Four years after returning to Russia, he lost sight in his other eye and was completely blind for the last 17 years of his life; however, he did not cease to do mathematics. In 1771 a fire broke out and reached Euler's house. His Swiss servant, Peter Grimes, bravely dashed into the burning house and carried out the blind Euler. Catherine immediately built him a new house. It is said that on September 18, 1783, after spending the afternoon calculating the laws for the ascension of balloons and outlining the calculation for the orbit for the newly discovered Uranus, Euler was playing with his grandchild and smoking his pipe when he suffered a stroke. The pipe dropped from his mouth, he uttered the words "I die," and the career of Euler was ended.

There are several anecdotes about Euler. Reputedly, he could recite Virgil's *Aeneid* line by line, although he had not read it since he was a child. Thièbault relates that Diderot was invited to the Russian court and, being an atheist, began spreading his ideas on atheism. A plot was contrived to silence the guest. Euler, a deeply religious Christian, walked up to Diderot and stated in French that "$(a+b^n)/n = x$; hence, God exists: Reply!" Diderot knew no mathematics and, hence, was silent. Apparently the crowd roared in laughter and Diderot was so embarrassed that he returned to France immediately.

Another story tells how Euler, then blind and elderly, was invited by Princess Daschkoff to attend an address that the princess would give to commence her directorship of the Imperial Academy of Sciences in St. Petersburg. Euler, accompanied by a son and a grandson, traveled with the princess in her personal coach. After the address, in which Euler was highly praised, the princess sat down, intending that Euler would occupy the seat of honor beside her. An arrogant professor, Schtelinn, grabbed the seat before Euler could be led to it. The princess turned to Euler and told him to take any seat and that would be the seat of honor. This act pleased everyone except the arrogant professor.

Exercises

1. Create a table of $\phi(n)$ for $1 \leq n \leq 20$.
2. Create a table of $\phi(n)$ for $20 < n \leq 35$.
3. Find an example of positive integers m and n such that $\phi(mn) \neq \phi(m)\phi(n)$.
4. Find $\phi(86)$.
5. Find $\phi(64)$.
6. Find $\phi(4739)$.
7. Find $\phi(4049)$.
8. Find $\phi(3573)$.
9. Prove that if p is prime and $p > 2$, then $(p-2)! \equiv 1 \pmod{p}$.
10. Prove that $1 \cdot 2 \cdot 3 \cdots \cdots 1007 \equiv 1 \pmod{1009}$.
11. Let
$$S(n) = \sin\left[\pi \cdot \frac{(n-1)!+1}{n}\right]$$
Show that n is prime if and only if $S(n) = 0$.

12. Prove Theorem 10.2.3. If n is a positive integer with prime factorization
$$n = p_1^{\alpha_1} p_2^{\alpha_1} \cdots p_t^{\alpha_t}$$
then
$$\phi(n) = \prod_{i=1}^{t}[p_i^{\alpha_i - 1}(p_i - 1)] = n \prod_{i=1}^{t}\left(1 - \frac{1}{p_i}\right)$$

13. Calculate $\phi(2025)$.
14. If n is composite and $n > 4$, prove that $(n-1)! \equiv 0 \pmod{n}$.
15. Prove that n is prime if and only if n divides $(n-1)! + 1$.

10.5 Order of an Integer

In this section we will determine integers j such that $a^j \equiv 1 \pmod{m}$. In particular, we will be interested in the least such positive integer j.

THEOREM 10.28 (**Euler**) If m is a positive integer and $\gcd(a, m) = 1$, then $a^{\phi(m)} \equiv 1 \pmod{m}$.

Proof Let m be a positive integer and a be relatively prime to m. If $\{x_1, x_2, \ldots, x_k\}$ is a reduced residue system modulo m, then since a and m are relatively prime, $\{ax_1, ax_2, \ldots, ax_k\}$ is also a reduced residue system. Hence, each x_i is congruent to only one ax_j modulo m. Therefore,

$$x_1 x_2 \cdots x_k \equiv ax_1 ax_2 \cdots ax_k \pmod{m}$$

or

$$a^{\phi(m)} x_1 x_2 \cdots x_k \equiv x_1 x_2 \cdots x_k \pmod{m}$$

and since m and $x_1 x_2 \cdots x_k$ are relatively prime, the x_i may be "canceled" giving $a^{\phi(m)} \equiv 1 \pmod{m}$. ∎

For example, let $a = 3$ and $m = 4$ so that $\phi(4) = 2$ and, hence, $3^2 = 9 \equiv 1 \pmod{4}$.

If m is prime in Theorem 10.28, then every positive integer less than m is relatively prime to m so that $\phi(m) = m - 1$. The case for m prime was developed as a corollary to Theorem 10.20. Thus, we then have the following theorem as a special case.

THEOREM 10.29 (**Fermat's Little Theorem**) If p is a prime, then for every integer a such that $0 < a < p$, $a^{p-1} \equiv 1 \pmod{p}$. ∎

For example, if $p = 7$, then $p - 1 = 6$. Then the sixth power of each positive integer less than $p = 7$ should be congruent to 1 modulo 7:

$$1^6 = 1 \equiv 1 \pmod{7}$$
$$2^6 = 64 \equiv 1 \pmod{7}$$
$$3^6 = 729 \equiv 1 \pmod{7}$$
$$4^6 = 4096 \equiv 1 \pmod{7}$$
$$5^6 = 15625 \equiv 1 \pmod{7}$$
$$6^6 = 46656 \equiv 1 \pmod{7}$$
$$7^6 = 117649 \equiv 0 \pmod{7}$$

The converse of Fermat's little theorem is not true. For example, $3^{90} \equiv 1 \pmod{91}$; however, $91 = 7 \cdot 13$ is composite. On the other hand, if p is a positive integer and $0 < a < p$ is such that $a^{p-1} \not\equiv 1 \pmod{p}$, then p cannot be prime. Thus, Fermat's little theorem comprises a partial primeness test since it can be used to show that a positive integer is not prime without finding a nontrivial divisor of p. Composite positive integers n such that $a^{n-1} \equiv 1 \pmod{n}$ for some a, $1 < a < n$, are somewhat primelike; and for this reason, such a composite n is said to be a **pseudoprime** to base a. Thus, $n = 91$ is a pseudoprime to base $a = 3$. However, if we had chosen $a = 2$ instead of $a = 3$, we would have found that $2^{90} \equiv 64 \not\equiv 1 \pmod{91}$, showing that $n = 91$ is not prime. So 91 is a pseudoprime to base 3 but not to base 2.

DEFINITION 10.30 Let n be a positive integer and a be an integer such that $\gcd(a, n) = 1$. The **order** of a modulo n, denoted by $\text{ord}_n a$, is the smallest positive integer k such that $a^k \equiv 1 \pmod{n}$.

10.5 Order of an Integer

THEOREM 10.31 Let n be a positive integer, $\gcd(a, n) = 1$, and $k = \text{ord}_n a$. Then

(a) If $a^m \equiv 1 \pmod{n}$, where m is a positive integer, then $k \mid m$.

(b) $k \mid \phi(n)$.

(c) For integers r and s, $a^r \equiv a^s \pmod{n}$ if and only if $r \equiv s \pmod{k}$.

(d) No two of the integers $a, a^2, a^3, \ldots, a^k$ are congruent modulo k.

(e) If m is a positive integer, then the order of a^m modulo n is $\dfrac{k}{\gcd(k, m)}$.

(f) The order of a^m modulo n is k if and only if m and k are relatively prime.

Proof (a) If $a^m \equiv 1 \pmod{n}$ for a positive integer m, then by the division algorithm, $m = kq + r$ where $0 \le r < k$. Hence, $a^m = a^{kq+r} = a^{kq}a^r$ so that $a^r \equiv 1 \pmod{n}$. But this contradicts the definition of the order of a unless $r = 0$. Hence, $k \mid m$.

(b) Since, by Theorem 10.28, $a^{\phi(n)} \equiv 1 \pmod{n}$, part (a) implies that $k \mid \phi(n)$.

(c) Assume that $r > s$. Since a and n are relatively prime, $a^r \equiv a^s \pmod{n}$ if and only if $a^{r-s} \equiv 1 \pmod{n}$; and, hence, by part (a), k divides $r - s$ and $r \equiv s \pmod{k}$.

(d) This follows directly from part (c).

(e) Let $d = \gcd(k, m)$ so that $k = ud$ and $m = vd$. $(a^m)^{k/\gcd(k,m)} = (a^m)^{ud/d} = a^{um} = a^{uvd} = a^{(ud)v} = a^{kv} \equiv 1 \pmod{n}$. Assume that t is such that $(a^m)^t \equiv 1 \pmod{n}$. Then $a^{mt} \equiv 1 \pmod{n}$ so that $k \mid mt$ because $\text{ord}_n a = k$. Hence, $ud \mid vdt$ and since u and v are relatively prime, $u \mid t$. Since $k = ud$, $(k/d) = k/\gcd(k, m)$ divides t and so, by definition of order, $k/\gcd(k, m)$ is the order of a^m.

(f) This follows directly from part (e). ∎

EXAMPLE 10.32 To illustrate Theorem 10.31, suppose that $n = 14 = 2 \cdot 7$; it now follows that $\phi(n) = (2 - 1)(7 - 1) = 6$. The primary reduced residue system for $n = 14$ is the set $\{1, 3, 5, 9, 11, 13\}$. Consider the following table of primary residues of powers of $a = 5$:

m	$[[a^m]]_n$	m	$[[a^m]]_n$
1	5	8	11
2	11	9	13
3	13	10	9
4	9	11	3
5	3	12	1
6	1	13	5
7	5		

where we see that after $m = 6$, we merely repeat the same pattern. Thus, $k = \text{ord}_{14} 5 = 6$. For $m = 12$, $a^m = 5^{12} \equiv 1 \pmod{14}$ and $k \mid m$, in agreement with Theorem 10.31(a). Also, $\text{ord}_{14} 5 \mid \phi(14)$ since $6 \mid 6$ [Theorem 10.31(b)]. We see that $2 \equiv 8 \equiv 14 \pmod{6}$ and that $5^2 \equiv 5^8 \equiv 5^{14} \equiv 11 \pmod{14}$ [Theorem 10.31(c)].

By inspection of the table, no two of the integers $5^1, 5^2, 5^3, 5^4, 5^5,$ and 5^6 are congruent modulo 14 [Theorem 10.31(d)]. Since $\text{ord}_n b \mid \phi(n)$ for any integer b and since $\phi(n) = 6$ for $n = 14$, the order of every b in $\{1, 3, 5, 9, 11, 13\}$ can quickly be computed as we did for $a = 5$.

b	$\text{ord}_n b$
1	1
3	6
5	6
9	3
11	3
13	2

If $m = 4$, $5^m \equiv 9 \pmod{14}$, but $\text{ord}_{14} 5 / \gcd(\text{ord}_{14} 5, 4) = 6/\gcd(6, 4) = 6/2 = 3$. According to the table of orders, $\text{ord}_{14} 5^4 = 3$ [Theorem 10.31(e)].

Only $b = 3$ and $b = 5$ have order 6 modulo 14. The exponents m in the foregoing table of powers that produce an a^m that is congruent to either 3 or 5 are $m = 1, 5, 7, 11,$ and 13. These are the only such m's that are relatively prime to $n = 14$ [Theorem 10.31(f)]. ∎

THEOREM 10.33 If $\gcd(a, n) = \gcd(b, n) = 1$ and $\text{ord}_n a$ is relatively prime to $\text{ord}_n b$, then $\text{ord}_n(ab) = (\text{ord}_n a) \cdot (\text{ord}_n b)$.

Proof Let $\text{ord}_n a = R$ and $\text{ord}_n b = S$. Then
$$(ab)^{RS} = a^{RS}b^{RS} = (a^R)^S(b^S)^R = 1 \cdot 1 \equiv 1 \pmod{n}.$$

By Theorem 10.31, $\text{ord}_n(ab) \mid RS$. Since R and S are relatively prime, there are integers r and s with $\text{ord}_n(ab) = rs$, $r \cdot w = R$, and $s \cdot x = S$. We now show that $r = R$ and $s = S$. By definition of r and s,
$$(ab)^{rs} = a^{rs}b^{rs} \equiv 1 \pmod{n}$$
$$(a^{rs}b^{rs})^w \equiv 1^w = 1 \pmod{n}$$
$$(a^{rw})^s \cdot (b^{rw})^s \equiv 1 \pmod{n}$$

But since $a^{rw} \equiv 1 \pmod{n}$ and $rw = R$, we have
$$b^{Rs} \equiv 1 \pmod{n}$$

By Theorem 10.31(a), $\text{ord}_n b \mid Rs$ or, equivalently, $S \mid Rs$. Because $\gcd(R, S) = 1$, we have $S \mid s$, but $s \mid S$ also so that $S = s$. Similarly, $R = r$. Thus,
$$\text{ord}_n(ab) = (\text{ord}_n a) \cdot (\text{ord}_n b).$$
∎

EXAMPLE 10.34 If $n = 11$, then all the primary residues are relatively prime to n. The table of orders modulo 11 is

Residue	Order	Residue	Order
1	1	6	10
2	10	7	10
3	5	8	10
4	5	9	5
5	5	10	2

If $a = 3$ and $b = 10$, then $ab = 30 \equiv 8$ (mod 11). Thus, $\text{ord}_{11}(ab) = \text{ord}_{11}(30) = \text{ord}_{11} 8 = 10 = (\text{ord}_{11} 3) \cdot (\text{ord}_{11} 10)$. Of course, $\gcd(3, 11) = \gcd(10, 11) = 1$ and $\text{ord}_{11} 3 = 5$ and $\text{ord}_{11} 10 = 2$ are relatively prime. Note that if $a = 3$ and $c = 7$, then $\text{ord}_{11} 3 = 5$ is not relatively prime to $\text{ord}_{11} 7 = 10$. In this case, $\text{ord}_{11}(ac) = \text{ord}_{11} 21 = \text{ord}_{11} 10 = 2 \neq (\text{ord}_{11} 3) \cdot (\text{ord}_{11} 7) = 50$. ∎

EXAMPLE 10.35   The order of $a = 5$ modulo $n = 14$ was obtained in Example 10.32 by calculating a^m for $m = 1, 2, 3, \ldots, \phi(n)$ until finding that $a^m \equiv 1$ (mod n). Theorem 10.31(b) implies that the order of a modulo n must divide $\phi(n)$; therefore, instead of testing each m, $1 \leq m \leq \phi(n)$, one at a time, test only the m's that divide $\phi(n)$. For $n = 14$, $\phi(n) = 6$, whose only positive divisors are 1, 2, 3, and 6. In this case the work to determine $\text{ord}_{14} 5$ is reduced by only a small amount since in Example 10.32 we tested $m = 4$ and 5 also.

However, for $n = 58$ and $a = 25$, we quickly obtain $\text{ord}_{58} 25 = 7$ using Theorem 10.31(b). $\phi(58) = \phi(2 \cdot 29) = (2 - 1)(29 - 1) = 28 = 2^2 \cdot 7$. The only positive divisors of $2^2 \cdot 7$ are 1, 2, 4, 7, 14, and 28. The following table is easily generated:

m	$[[25^m]]_{58}$
1	25
2	45
4	53
7	1

Therefore, $\text{ord}_{58} 25 = 7$ and we do not need to check $n = 14$ and 28. ∎

The results obtained from Theorems 10.31 and 10.36 lead to a test for primeness called Lucas's primality test.

THEOREM 10.36   (Lucas)   If n is a positive integer and there is an integer a such that

$$a^{n-1} \equiv 1 \quad (\text{mod } n)$$

and

$$a^{\frac{n-1}{p}} \not\equiv 1 \quad (\text{mod } n)$$

for every prime p that divides $n - 1$, then n is prime.

Proof $a^{n-1} \equiv 1$ (mod n) implies that $\gcd(a, n) = 1$ and, by Theorem 10.31(a), that $\text{ord}_n a \mid (n - 1)$. If p is a prime such that $p \mid (n - 1)$, then the congruence $a^{(n-1)/p} \not\equiv 1$ (mod n) implies that $\text{ord}_n a \nmid [(n-1)/p]$ because if $\text{ord}_n a \mid [(n-1)/p]$, it would contradict $a^{\text{ord}_n a} \equiv 1$ (mod n). But $\text{ord}_n a \mid (n - 1)$ and $\text{ord}_n a \nmid [(n-1)/p]$ for all p dividing $n - 1$ imply that $\text{ord}_n a = n - 1$. By Theorem 10.31(b), $\phi(n) = n - 1$. Therefore, by Corollary 10.21, n is prime. ∎

In order to use Lucas's test for testing n, we must be able to factor $n - 1$, which may itself be difficult. Additionally, an appropriate integer a must be found. The integer a of Theorem 10.36 is called a **primitive root** of n. One can show that the Mersenne number $n = 2^{31} - 1$ is prime using Lucas's test with $a = 7$ since $n - 1 = 2 \cdot 3^2 \cdot 7 \cdot 11 \cdot 31 \cdot 151 \cdot 331$. See the exercises at the end of this section.

If $\gcd(a, n) = 1$, Fermat's theorem gives $a^{n-1} \equiv 1$ (mod n) when n is prime. Its generalization, Euler's theorem, gives $a^{\phi(n)} \equiv 1$ (mod n) for any positive integer n.

Aside from these and a few other special cases, the task of evaluating a^e modulo n or, more precisely, calculating $[[a^e]]_n$, the remainder when a^e is divided by n may appear formidable when e is large; for in such cases, just computing a^e and dividing by n is not practical.

In the small illustrative examples given so far, we have just written $e = e_1 + e_2 + \cdots + e_k$ for suitable e_i, calculated $[[a^{e_i}]]_n$, multiplied the results, and reduced the product modulo n. This method works because

$$[[st]]_n = [[\,[[s]]_n \cdot [[t]]_n\,]]_n$$

The e_i were chosen in some ad hoc fashion.

A much more efficient algorithm is similar but expresses the exponent e using a binary representation, that is, a base 2 representation. Thus,

$$e = b_m 2^m + b_{m-1} 2^{m-1} + \cdots + b_1 2^1 + b_0$$
$$= [b_m b_{m-1} \cdots b_1 b_0]_{binary}$$

where $b_i = 0$ or 1 and $b_m = 1$. Thus,

$$a^e = a^{b_m 2^m + b_{m-1} 2^{m-1} + \cdots + b_1 2^1 + b_0}$$

If the exponent is expressed with regrouping to reduce the number of multiplications, we obtain Horner's rule representation

$$e = (\cdots((b_m \cdot 2 + b_{m-1}) \cdot 2 + b_{m-2}) \cdot 2 + \cdots + b_1) \cdot 2 + b_0$$

so that

$$a^e = (\cdots((a^{b_m \cdot 2} \cdot a^{b_{m-1}})^2 \cdot a^{b_{m-2}})^2 \cdots a^{b_1})^2 \cdot a^{b_0}$$

We can calculate $[[a^e]]_n$ by evaluating this last expression "inside out" while reducing each product modulo n. Hence, for $e = [b_m b_{m-1} \cdots b_1 b_0]_{binary}$, start with

$$p_m = [[a]]_n$$

Then, for $k = m - 1, m - 2, \ldots, 2, 1$, and 0, calculate

$$p_k = \begin{cases} [[p_{k+1}^2]]_n & \text{if } b_k = 0 \\ [[p_{k+1}^2 \cdot a]]_n & \text{if } b_k = 1 \end{cases}$$

The final result is $p_0 = [[a^e]]_n$. That is, beginning with $p_m = [[a]]_n$, obtain the next product p_k by squaring the previous product and reducing modulo n when $b_k = 0$ and by squaring the previous product, multiplying by a, and reducing modulo n when $b_k = 1$. The algorithm works because if a is squared k times, the result is a^{2^k}; and if $a^{2^k} b$ is squared j times, the result is $a^{2^{k+j}} b^{2^j}$.

EXAMPLE 10.37

Suppose that we want to evaluate $[[3^{103}]]_{41}$. Since

$$103 = 2^6 + 2^5 + 2^2 + 2^1 + 1$$
$$= 1100111_{binary}$$

and $m = 6$, we obtain

k	b_k	$p_k = [[p_{k+1}^2 \cdot a^{b_k}]]_n$
6	1	$3 \equiv 3$
5	1	$3^2 \cdot 3 = 27$
4	0	$(27)^2 = 729 \equiv 32$
3	0	$(32)^2 = 1024 \equiv 4$
2	1	$(40)^2 \cdot 3 = 4800 \equiv 3$
1	1	$(3)^2 \cdot 3 = 27$
0	1	$(27)^2 \cdot 3 = 2187 \equiv 14$

Therefore, $[[3^{103}]]_{41} = 14$. By an ad hoc method using congruence modulo 41, we obtain

$$3^{10} = 57049 \equiv 9$$
$$3^{50} = (3^{10})^5 \equiv 9^5 = 59049 \equiv 9$$
$$3^{103} = 3^{50} \cdot 3^{50} \cdot 3^3 \equiv 9 \cdot 9 \cdot 27 = 2187 \equiv 14$$ ■

DEFINITION 10.38 Let n be a positive integer and $\gcd(a, n) = 1$. If the order of an element a modulo n is $\phi(n)$, that is, if $\text{ord}_n(a) = \phi(n)$, then we say that a is a **primitive root** of n.

THEOREM 10.39 If a is a primitive root of n, then $\{a, a^2, a^3, \ldots a^{\phi(n)}\}$ is a complete set of reduced residues modulo n. Hence, the reduced residue set is a cyclic group.

Proof By definition of primitive root, a and n are relatively prime. Hence, a^i and n are relatively prime for all $1 \leq i \leq \phi(n)$. Also we know from Theorem 10.39 that the a^i are not congruent. Since there are only $\phi(n)$ positive integers less than n that are relatively prime to n, the set $\{a, a^2, a^3, \ldots a^{\phi(n)}\}$ must be congruent to them. ■

We list without proof some properties of primitive roots:

1. Let p be a prime number. There are exactly $\phi(p-1)$ primitive roots of p.
2. If m and n are relatively prime positive integers greater than 2, then mn has no primitive roots.
3. The integer $2^k m$, where $k > 1$ and m is an odd integer greater than 2, has no primitive roots.
4. There are primitive roots of 2^n if and only if n is a positive integer less than 3.
5. Let p be an odd prime; then p^k has a primitive root for every positive integer k.

Computer Science Application

The following algorithm can be used to determine the number of positive integers less than n which are relatively prime to n.

integer function Euler-phi (integer n)
  integer phi $= n, i, j, k$

  if n mod $2 = 0$ then
    phi = phi div 2

```
        while n mod 2 = 0 do
            n = n div 2
        end-while
    end-if

    if n mod 3 = 0 then
        phi = (2 * phi) div 3
        while n mod 3 = 0 do
            n = n div 3
        end-while
    end-if

    i = 5
    while n >= 5 do
        j = 1
        repeat
            j = j + 1
            k = i mod j
        until k = 0 or j = floor(sqrt(i))
        if k > 0 and n mod i = 0 then
            phi = ((i - 1) * phi) div i
            while n mod i = 0 do
                n = n div i
            end-while
        end-if
        i = i + 2
    end-while

    return phi

end-function
```

Exercises

1. Prove this variation of Fermat's little theorem: If p is prime and $a \not\equiv 0 \pmod{p}$, then $a^{p-1} \equiv 1 \pmod{p}$.

2. Prove that if $\gcd(a, m) = 1$, then the congruence $ax \equiv b \pmod{m}$ has the solution $x \equiv a^{\phi(m)-1}b \pmod{m}$.

3. Let p be prime, $p > 2$, and $J \equiv 0 \pmod{(p-1)}$. Prove that
$$1^J + 2^J + 3^J + \cdots + (p-1)^J \equiv -1 \pmod{p}$$

4. Let $\gcd(m, n) = 1$ and let the sets $\{r_1, r_2, \ldots, r_m\}$ and $\{s_1, s_2, \ldots, s_n\}$ be complete residue systems modulo m and n, respectively. Prove that $\{n \cdot r_i + m \cdot s_j : 1 \leq i \leq m \text{ and } 1 \leq j \leq n\}$ is a set of mn integers that is a complete residue system modulo mn.

5. Fill in the details of a proof of Fermat's little theorem using the binomial theorem and induction on a.

Use the method of exercise 2 to solve these congruences:

6. $5x \equiv 8 \pmod{11}$

7. $7x \equiv 8 \pmod{25}$

8. $9x \equiv 13 \pmod{25}$

9. Prove that for every prime p,
$$(a + b)^p \equiv (a^p + b^p) \pmod{p}$$

10. Prove the converse of Theorem 10.31(a); that is, prove that if n is a positive integer, $\gcd(a, n) = 1$, $k = \text{ord}_n a$, and $k \mid m$, then $a^m \equiv 1 \pmod{n}$.

Determine $\text{ord}_n a$ *for* $1 \leq a \leq n - 1$ *if*

11. $n = 9$

12. $n = 20$

13. $n = 27$

Show that

14. $b^{10} - 1$ is divisible by 11 if b and 11 are relatively prime.

15. $b^{10k} - 1$ is divisible by 11 if $\gcd(b, 11) = 1$.

16. $b^7 - b$ is divisible by 42 for any integer b.

17. $7^4 \equiv 1 \pmod{5}$

18. $7^4 \equiv 1 \pmod{2}$

19. $7^4 \equiv 1 \pmod{10}$

20. $7^{4k} \equiv 1 \pmod{10}$ for any positive integer k

21. What is the last base ten digit of 7^{4000}?

Show that

22. $7^{20} \equiv 1 \pmod{25}$
23. $7^2 \equiv 1 \pmod 4$
24. $7^{20} \equiv 1 \pmod 4$
25. $7^{20} \equiv 1 \pmod{100}$

26. What are the last two base ten digits of 7^{500}?

Calculate the following residues:

27. $[[3^{275}]]_{100}$
28. $[[6^{5000}]]_{1000}$
29. $[[11^{24681}]]_{83}$
30. $[[3497^{100000}]]_{1234}$
31. $[[7^{2 \cdot 3^2 \cdot 7 \cdot 11 \cdot 31 \cdot 331}]]_{2^{31}-1} = [[7^{14221746}]]_{2147483647}$ (The availability of extended integer precision computer software makes the solution of this part more tractable.)

Use Lucas's test to show that the following integers are prime:

32. 37
33. 199

Establish these statements:

34. Prove that if n is an odd positive integer and there is a positive integer a with the properties
  (i) $a^{\frac{n-1}{2}} \equiv -1 \pmod n$, and
  (ii) $a^{\frac{n-1}{p}} \not\equiv 1 \pmod n$ for every odd prime p that divides $(n-1)$, then n is prime.

35. Use the test of Exercise 34 instead of Lucas's test to show that the integers in Exercise 34(i) and (ii) are prime.

36. If $F(n) = 2^{2^n} + 1$ is a Fermat number and there is a positive integer a such that $a^{2^{2^n}} \equiv 1 \pmod{F(n)}$ and $a^{2^{(2^n-1)}} \not\equiv 1 \pmod{F(n)}$, prove that $F(n)$ is prime.

Use Exercise 36 to prove that the following integers are prime:

37. $F(3) = 257$
38. $F(4) = 65537$

GLOSSARY

Chapter 10: Number Theory Revisited

Chinese Remainder Theorem (10.3) — Let $m_1, m_2, \ldots, m_n$ be pairwise relatively prime; that is, $\gcd(m_i, m_j) = 1$, for all i and j less than or equal to n where $i \neq j$. Then the system of congruences

$$x \equiv a_1 \pmod{m_1}$$
$$x \equiv a_2 \pmod{m_2}$$
$$\vdots$$
$$x \equiv a_n \pmod{m_n}$$

has a solution that is unique modulo the integer $m_1 m_2 \cdots m_n$. Furthermore, if

$$M_j = \frac{\prod_{i=1}^{n} m_i}{m_j}$$

and z_j is a solution of $M_j z_j \equiv a_j \pmod{m_j}$ for each j, then the solution is given by

$$x = \left[\left[\sum_{j=1}^{n} M_j z_j\right]\right]_{m_1 m_2 \cdots m_n}$$

Congruence equations (10.2) — Congruence equations are expressions of equality of equivalence classes stated in terms of integers congruent to c modulo n, for example, $ax \equiv c \pmod n$.

Euler's Theorem (10.5) — If m is a positive integer and $\gcd(a, m) = 1$, then $a^{\phi(m)} \equiv 1 \pmod m$.

Euler's totient function or Euler ϕ function (10.4) — $\phi(n)$ is the number of positive integers less than n that are relatively prime to n. ϕ is called **Euler's totient function** or the **Euler ϕ function**.

Fermat's Little Theorem (10.5)	If p is prime, then for every integer a such that $0 < a < p$, $a^{p-1} \equiv 1 \pmod{p}$.	
Lucas's Theorem (10.5)	If n is a positive integer and there is an integer a such that $$a^{n-1} \equiv 1 \pmod{n}$$ and $$a^{\frac{n-1}{p}} \not\equiv 1 \pmod{n}$$ for every p that divides $n-1$, then n is prime.	
Order of a modulo n (10.5)	If n is a positive integer and a is an integer such that $\gcd(a, n) = 1$, then the **order** of a modulo n, denoted by $\text{ord}_n a$, is the smallest positive integer k such that $a^k \equiv 1 \pmod{n}$.	
Pseudoprime (10.5)	Pseudoprimes to base a are composite numbers n such that $a^{n-1} \equiv 1 \pmod{n}$ for some a, $1 < a < n$.	
Wilson's Theorem (10.4)	The positive integer p is a prime if and only if $(p-1)! \equiv -1 \pmod{p}$.	

TRUE-FALSE QUESTIONS

1. The equation $9x + 12y = 14$ has an integral solution.
2. The equation $6x + 9y = 24$ has a unique integral solution.
3. The equation $ax + by = c$ has a unique integral solution if and only if $\gcd(a, b) = c$.
4. If $a \equiv b \pmod{n}$, then $ac \equiv bc \pmod{n}$.
5. If $ac \equiv bc \pmod{n}$, then $a \equiv b \pmod{n}$.
6. $3x \equiv 7 \pmod{11}$ has a unique integral solution.
7. $4x \equiv 2 \pmod{12}$ has a unique integral solution.
8. $4x \equiv 8 \pmod{12}$ has a unique integral solution.
9. $35x \equiv 14 \pmod{84}$ has exactly seven distinct integral solutions.
10. If $a \equiv b \pmod{mn}$, then $a \equiv b \pmod{n}$.
11. If $a \equiv b \pmod{n}$ and $a \equiv b \pmod{m}$, then $a \equiv b \pmod{mn}$.
12. The system of equations
$$x \equiv 7 \pmod{11}$$
$$x \equiv 4 \pmod{14}$$
$$x \equiv 6 \pmod{5}$$
has a unique solution modulo 168.
13. The system of equations
$$x \equiv 3 \pmod{5}$$
$$x \equiv 4 \pmod{7}$$
$$x \equiv 2 \pmod{3}$$
has a unique solution modulo 105.
14. $x = 1059$ is the least positive solution of
$$x \equiv 4 \pmod{5}$$
$$x \equiv 3 \pmod{4}$$
$$x \equiv 6 \pmod{9}$$
$$x \equiv 2 \pmod{7}$$
15.
$$x \equiv 7 \pmod{9}$$
$$x \equiv 1 \pmod{12}$$
$$x \equiv 10 \pmod{21}$$
$$x \equiv 2 \pmod{7}$$
has an integral solution.
16.
$$x \equiv 7 \pmod{9}$$
$$x \equiv 1 \pmod{12}$$
$$x \equiv 10 \pmod{21}$$
$$x \equiv 2 \pmod{6}$$
has an integral solution.
17. $\phi(mn) = \phi(m)\phi(n)$.
18. $\phi(1151) = 1150$.
19. $\phi(28) = 27$.
20. $\phi(16) = 8$.
21. $\phi(56) = 24$.
22. $\phi(567201)$ is even.
23. $\phi(81) = 54$.
24. $p! \equiv 1 \pmod{p}$.
25. $28! \equiv -1 \pmod{29}$.
26. $(25)^{54} \equiv -1 \pmod{81}$.

27. $\text{ord}_{36}((61)^{24}) = \text{ord}_{36}(61)$.

28. $\text{ord}_{81}(29) \mid 54$.

29. $(75)^{29} \equiv (24)^{29} + (51)^{29} \pmod{29}$.

30. n is composite if and only if $n \mid (n-1)!$

SUMMARY QUESTIONS

1. Find a solution for $85x + 68y = 51$ if it exists.
2. Find all solutions to $6x \equiv 5 \pmod{13}$.
3. Find all solutions to $24x \equiv 21 \pmod{9}$.
4. Solve the simultaneous congruence

$$x \equiv 4 \pmod{5}$$
$$x \equiv 3 \pmod{4}$$
$$x \equiv 5 \pmod{7}$$

5. Find $\phi(81)$.
6. Find $11! \pmod{11}$.
7. Find $\phi(1722448)$.
8. Find $11^6 \pmod{7}$.
9. Find $18^{13} - 7^{13} - 11^{13} \pmod{13}$ without using a calculator.
10. Find $\phi(2^{32})$.
11. Find $\text{ord}_8 5$.
12. Find $\text{ord}_8 35$.
13. Find $\sum_{d \mid n} \phi(d)$ where the divisors d are positive divisors of n.
14. For what value of n is $85 \equiv 25 \pmod{n}$?
15. Find the general solution of $7x + 11y = 5$.

ANSWERS TO TRUE-FALSE

1. F **2.** F **3.** F **4.** T **5.** F **6.** T **7.** F **8.** F **9.** T **10.** T **11.** F **12.** F **13.** T **14.** T **15.** F **16.** F **17.** F **18.** T **19.** F **20.** T **21.** T **22.** T **23.** T **24.** F **25.** T **26.** F **27.** F **28.** T **29.** T **30.** T

CHAPTER 11

Recursion Revisited

11.1 Homogeneous Linear Recurrence Relations

In Chapter 4, we began the project of solving recursive functions. By solving a recursive function, we mean that we take a function, say $a(n)$ or a_n, defined on the positive or nonnegative integers, which is described in recursive form, and change it to a function in closed form, where the function a_n is directly expressed as a function of n. For example, the function described in recursive form by

$$a_1 = 1$$
$$a_k = a_{k-1} + k$$

can be defined directly as

$$a_n = \frac{n(n+1)}{2}$$

The method we used was simply to evaluate a_n for the first few values of n and then try to find a pattern from which to find a_n directly in terms of n. In general, the solution of arbitrary recurrence relations is difficult if not virtually impossible. In this chapter we look at a large class of recursive functions whose solutions are known.

We first define a class of recursive functions known as linear recursive functions:

DEFINITION 11.1   A recursive relation of the form
$$a_n = c_1(n)a_{n-1} + c_2(n)a_{n-2} + c_3(n)a_{n-3} + \cdots + c_p(n)an - p + f(n), \quad c_p n \neq 0$$
is called a **linear recurrence relation of order p**.

DEFINITION 11.2   A linear recurrence relation of the form
$$a_n = c_1(n)a_{n-1} + c_2(n)a_{n-2} + c_3(n)a_{n-3} + \cdots + c_p(n)a_{n-p}, \quad c_p n \neq 0$$
is called a **linear homogeneous recurrence relation of order p**.

It is called a linear recurrence relation because each a_i has exponent one. In other words, none of the a_i is raised to any power but one. Thus, $a_n = 3a_{n-1}^3 + 4a_{n-2}$ would not be linear, since a_{n-1} is raised to the third power. However, $a_n = 3n^3 a_{n-1} + na_{n-2}$ is linear. Unfortunately, we must restrict the set of linear recurrence relations even further. We do this by requiring $c_i(n)$ to be a constant for each i.

DEFINITION 11.3

A linear recursive relation of the form

$$a_n = c_1 a_{n-1} + c_2 a_{n-2} + c_3 a_{n-3} + \cdots + c_p a_{n-p} + f(n), \qquad c_p \neq 0$$

with constants c_i for $1 \leq i \leq p$ is called a **linear recurrence relation with constant coefficients of order p**.

We shall be able to discuss some linear recurrences having this form but, even with this restriction, our ability to solve the problem will depend on the choice of $f(n)$. We shall discuss general linear recurrence relations with constant coefficients of order p in the next section, but before solving problems of this type, we must first place one further restriction on our class of linear recurrences.

DEFINITION 11.4

A linear recurrence relation of the form

$$a_n = c_1 a_{n-1} + c_2 a_{n-2} + c_3 a_{n-3} + \cdots + c_p a_{n-p}, \qquad c_p \neq 0$$

with constants c_i for $1 \leq i \leq p$ is called a **linear homogeneous recurrence relation with constant coefficients of order p**.

This is the special case where $f = 0$. One of the examples that we have already seen is the Fibonacci sequence $1, 1, 2, 3, 5, 8, 13, 21, \ldots$, where each element of the sequence, after the first two, is the sum of the two previous elements of the sequence. Recursively this is defined by

$$\text{Fib}(1) = 1, \quad \text{Fib}(2) = 1$$
$$\text{Fib}(n) = \text{Fib}(n-1) + \text{Fib}(n-2) \text{ for } n > 2$$

Thus, in this case $p = 2$, $c_1 = 1$, and $c_2 = 1$. We shall solve this recursive function later in the section.

First, let's consider the case where $p = 1$. Suppose that our recursive function is of the form

$$a_0 = c$$
$$a_n = ba_{n-1} \quad \text{for } n > 0$$

We immediately see that this sequence has the form

$$a_0 = c$$
$$a_1 = bc.$$
$$a_2 = b^2 c$$
$$a_3 = b^3 c$$
$$a_4 = b^4 c$$
$$\vdots$$
$$a_n = b^n c$$

where the value of c is determined by the value assigned to a_0. Thus, if we look at $a_n = ba_{n-1}$, we see that its solution has the form $a_n = b^n c$. We shall see when we look at cases where $n > 1$ that we want solutions of the form $a_n = b^n c$. But suppose that we started with the recursive function

$$a_1 = c$$
$$a_n = ba_{n-1} \quad \text{for } n > 1$$

We immediately see that this sequence has the form

$$a_1 = c$$
$$a_2 = bc.$$
$$a_3 = b^2 c$$
$$a_4 = b^3 c$$
$$a_5 = b^4 c$$
$$\vdots$$
$$a_n = b^{n-1} c$$

Can we still put this in the form $a_n = c'b^n$, if we really think it is important to do so? The answer is yes since $a_n = b^{n-1}c = b^n(c/b)$. If we let $c' = (c/b)$, then $a_n = c'b^n$. Suppose that we had started from the other direction, and for some mysterious reason, having consulted our oracle, assumed that a_n had the form $a_n = r^n$, for some value r. Then since $a_n = ba_{n-1}$, we have $r^n = br^{n-1}$. If we divide both sides by r^{n-1}, we have $r = b$, so that $a_n = b^n$. Further we note that if we multiply both sides of $r^n = br^{n-1}$ by a constant c, we see that $cr^n = bcr^{n-1}$, so that if $a_n = r^n$ is a solution of $a_n = ba_{n-1}$, so is $a_n = cr^n$. Therefore, our general solution is again seen to be of the form cb^n. But then, when we give the value for a_1, we determine the value of c.

For example, if we have

$$a_0 = 4$$
$$a_n = 5a_{n-1} \quad \text{for } n > 1$$

then $a_n = c5^n$. But $a_0 = c5^0 = 4$, so $c = 4$. In general then, the solution of $a_n = ba_{n-1}$ has the solution cb^n where $c = a_0$.

If we have

$$a_1 = 4$$
$$a_n = 5a_{n-1} \quad \text{for } n > 1$$

and $a_n = c5^n$, then $a_1 = c5^1 = 4$, and $c = 4/5$.

The reason for hitting you over the head with the obvious, and making it look obscure, when $p = 1$, is that we now want to consider the case where $p = 2$, so that we have

$$a_n = c_1 a_{n-1} + c_2 a_{n-2}$$

and we again assume that we look for a solution of the form $a_n = r^n$. In this case we have

$$r^n = c_1 r^{n-1} + c_2 r^{n-2}$$

If we now divide both sides of the equation by r^{n-2} we have

$$r^2 = c_1 r + c_2$$

Thus, we can find r by solving the quadratic equation $r^2 = c_1 r + c_2$. This equation is called the **characteristic polynomial** of $a_n = c_1 a_{n-1} + c_2 a_{n-2}$.

For example, suppose $a_n = a_{n-1} + 6a_{n-2}$. Then the characteristic polynomial is the quadratic equation $r^2 = r - 6$ or $r^2 - r - 6 = 0$. Factoring, we have $(r-3)(r+2) = 0$ so that the solutions of the characteristic polynomial are $r = -2$ and $r = 3$. Since solutions of this recurrence relation have the form $a_n = r^n$ where r is a solution of $r^2 - r - 6 = 0$, it follows that $a_n = 3^n$ and $a_n = (-2)^n$ are solutions.

Suppose we find two unequal real solutions to the equation $r^2 = c_1 r + c_2$, say $r = a$ and $r = b$, so we have solutions $a_n = a^n$ and $a_n = b^n$ for the equation $a_n = c_1 a_{n-1} + c_2 a_{n-2}$ so that $a^n = c_1 a^{n-1} + c_2 a^{n-2}$ and $b^n = c_1 b^{n-1} + c_2 b^{n-2}$. But if we multiply $a^n = c_1 a^{n-1} + c_2 a^{n-2}$ by the constant c, then we have $ca^n = c_1 c a^{n-1} + c_2 c a^{n-2}$ so that $a_n = ca^n$ is also a solution of $a_n = c_1 a_{n-1} + c_2 a_{n-2}$.

Further if we add $a^n = c_1 a^{n-1} + c_2 a^{n-2}$ and $b^n = c_1 b^{n-1} + c_2 b^{n-2}$, we have

$$a^n + b^n = c_1 a^{n-1} + c_1 b^{n-1} + c_2 a^{n-2} + c_2 b^{n-2}$$
$$= c_1(a^{n-1} + b^{n-1}) + c_2(a^{n-2} + b^{n-2})$$

so that $a_n = a^n + b^n$ is a solution of $a_n = c_1 a_{n-1} + c_2 a_{n-2}$. Putting this all together, since a constant times a solution of $a_n = c_1 a_{n-1} + c_2 a_{n-2}$ is a solution, then if a^n and b^n are solutions, so are ca^n and db^n for constants c and d. But since the sum of solutions of $a_n = c_1 a_{n-1} + c_2 a_{n-2}$ is a solution, then $ca^n + db^n$ is a solution for constants c and d. Thus, the general solution of $a_n = c_1 a_{n-1} + c_2 a_{n-2}$ has the form $ca^n + db^n$ where c and d are constants and $r = a$ and $r = b$ are solutions of $r^2 = c_1 r + c_2$.

For example, we saw that $a_n = a_{n-1} + 6a_{n-2}$ had solutions $a_n = 3^n$ and $a_n = (-2)^n$. Therefore, $a_n = c3^n + d(-2)^n$ is a general solution of $a_n = a_{n-1} + 6a_{n-2}$. If we wanted to completely determine the solution of $a_n = a_{n-1} + 6a_{n-2}$ as a recursive function, we would have to state the first two values of the sequence. For example, if $a_0 = 1$ and $a_1 = 8$, then $a_0 = c3^0 + d(-2)^0 = c + d = 1$ and $a_1 = c3^1 + d(-2)^1 = 3c - 2d = 8$. Solving the simultaneous equations

$$c + d = 1$$
$$3c - 2d = 8$$

we find that $c = 2$ and $d = -1$, so the solution of the recursive function

$$a_0 = 1$$
$$a_1 = 8$$
$$a_n = a_{n-1} + 6a_{n-2} \quad \text{for } n > 2$$

is $a_n = 2 \cdot 3^n - (-2)^n$.

EXAMPLE 11.5 Solve the recursive function

$$a_1 = 2$$
$$a_2 = 10$$
$$a_n = 5a_{n-1} - 6a_{n-2} \quad \text{for } n > 2$$

We first find the characteristic equation $r^2 = 5r - 6$ or $r^2 - 5r + 6 = 0$. Factoring we have $(r-2)(r-3) = 0$, so that $r = 2$ and $r = 3$. Therefore, the general solution of $a_n = 5a_{n-1} - 6a_{n-2}$ is $a_n = c2^n + d3^n$. But $a_1 = c2^1 + d3^1 = 2c + 3d = 2$ and $a_2 = c2^2 + d3^2 = 4c + 9d = 10$. Solving the simultaneous equations

$$2c + 3d = 2$$
$$4c + 9d = 10$$

we have $c = -2$ and $d = 2$, so $a_n = (-2) \cdot 2^n + 2 \cdot 3^n$. ∎

EXAMPLE 11.6 Solve the Fibonacci sequence defined recursively by

$$\text{Fib}(1) = 1$$
$$\text{Fib}(2) = 1$$
$$\text{Fib}(n) = \text{Fib}(n-1) + \text{Fib}(n-2) \quad \text{for } n > 2$$

Warning: Some books define the Fibonacci sequence differently and will have a different solution.

The characteristic polynomial for $\text{Fib}(n) = \text{Fib}(n-1) + \text{Fib}(n-2)$ for $n > 2$ is $r^2 = r + 1$ or $r^2 - r - 1 = 0$. Using the quadratic formula we find that

$$r = \frac{1 \pm \sqrt{5}}{2}$$

so that the general solution for $\text{Fib}(n) = \text{Fib}(n-1) + \text{Fib}(n-2)$ is

$$\text{Fib}(n) = c\left(\frac{1+\sqrt{5}}{2}\right)^n + d\left(\frac{1-\sqrt{5}}{2}\right)^n$$

To simplify solving our equation, we let $\text{Fib}(0) = 0$. We still have each term of the sequence equal to the sum of the two previous terms where they exist. Solving for c and d, we have

$$\text{Fib}(0) = c\left(\frac{1+\sqrt{5}}{2}\right)^0 + d\left(\frac{1-\sqrt{5}}{2}\right)^0 = c + d = 0$$

and

$$\text{Fib}(1) = c\left(\frac{1+\sqrt{5}}{2}\right)^1 + d\left(\frac{1-\sqrt{5}}{2}\right)^1 = 1$$

Since $c + d = 0$, $c = -d$. Therefore, the second equation may be written as

$$-d\left(\frac{1+\sqrt{5}}{2}\right)^1 + d\left(\frac{1-\sqrt{5}}{2}\right)^1 = 1$$

So that

$$d = -\frac{1}{\sqrt{5}} \quad \text{and} \quad c = \frac{1}{\sqrt{5}}$$

and

$$\text{Fib}(n) = \frac{1}{\sqrt{5}}\left(\frac{1+\sqrt{5}}{2}\right)^n - \frac{1}{\sqrt{5}}\left(\frac{1-\sqrt{5}}{2}\right)^n$$

is the Fibonacci sequence. ∎

Having conquered the case where we have two distinct real roots, we now consider the case where both roots are equal. Thus, given the recursive relation $a_n = c_1 a_{n-1} + c_2 a_{n-2}$, the characteristic function $r^2 = c_1 r + c_2$ has equal roots, say $r = a$. We then claim that $a_n = a^n$ and $a_n = na^n$ are both solutions of $a_n = c_1 a_{n-1} + c_2 a_{n-2}$. We already know that $a_n = a^n$ is a solution, but we need to convince ourselves that $a_n = na^n$ is a solution. Since both roots of $r^2 = c_1 r + c_2$

are a, $r^2 = c_1 r + c_2$ is actually the equation $(r-a)^2 = 0$ or $r^2 = 2ar - a^2$ so that $c_1 = 2a$ and $c_2 = -a^2$. Therefore, our recurrence relation is $a_n = 2aa_{n-1} - a^2 a_{n-2}$. We want to show that $a_n = na^n$ is a solution. If we substitute this into our relation, we have

$$2aa_{n-1} - a^2 a_{n-2} = 2a(n-1)a^{n-1} - a^2(n-2)a^{n-2}$$
$$= 2na^n - 2a^n - na^n + 2a^n$$
$$= na^n$$

so that $a_n = na^n$ does satisfy the recursion relation.

EXAMPLE 11.7 Find the general solution for the recurrence relation $u_n = 6a_{n-1} - 9a_{n-2}$. The characteristic equation is $r^2 = 6r - 9$ or $r^2 - 6r + 9 = 0$. Factoring, we have $(r-3)^2 = 0$ so that $r = 3$ is a double root. Therefore, the general solution is $a_n = c3^n + dn3^n$. ∎

EXAMPLE 11.8 Solve the recursive function

$$a_0 = 2$$
$$a_1 = 6$$
$$a_n = 4a_{n-1} - 4a_{n-2} \quad \text{for } n > 2$$

The characteristic polynomial for $a_n = 4a_{n-1} - 4a_{n-2}$ is $r^2 = 4r - 4$ or $r^2 - 4r + 4 = 0$. Factoring, we have $(r-2)^2 = 0$ so that $r = 2$ is a double root. Therefore, the general solution is $a_n = c2^n + dn2^n$. $a_0 = c2^0 + d(0)2^0 = c = 2$. $a_1 = c2^1 + d(1)2^1 = 2c + 2d = 6$. Therefore, $c = 2$ and $d = 1$. Therefore, $a_n = 2 \cdot 2^n + n2^n = (n+2)2^n$. ∎

The only case remaining is when the two roots of the characteristic polynomial are complex numbers. Before we show this, however, we need to show a property about complex numbers, using properties of trigonometry.

THEOREM 11.9 For angles α and β,

$$(\cos(\alpha) + i\sin(\alpha))(\cos(\beta) + i\sin(\beta)) = (\cos(\alpha + \beta) + i\sin(\alpha + \beta))$$

Proof

$$(\cos(\alpha) + i\sin(\alpha))(\cos(\beta) + i\sin(\beta)) = \cos(\alpha)\cos(\beta) + i^2 \sin(\alpha)\sin(\beta)$$
$$+ i\sin(\alpha)\cos(\beta) + i\sin(\beta)\cos(\alpha)$$
$$= \cos(\alpha)\cos(\beta) - \sin(\alpha)\sin(\beta)$$
$$+ i(\sin(\alpha)\cos(\beta) + \cos(\alpha)\sin(\beta))$$
$$= \cos(\alpha + \beta) + i\sin(\alpha + \beta)$$ ∎

From this theorem it follows that $(\cos(\theta) + i\sin(\theta))^2 = \cos(2\theta) + i\sin(2\theta)$.
The reader is asked to prove the following theorem using induction.

THEOREM 11.10 (**De Moivre**) For a given angle θ, $(\cos(\theta) + i\sin(\theta))^k = \cos(k\theta) + i\sin(k\theta)$. ∎

We need one further property of complex numbers. Consider the complex number $a + bi$ as shown in Figure 11.1.

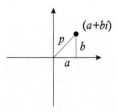

Figure 11.1

By the Pythagorean theorem, $\rho = \sqrt{a^2+b^2}$. By definition, $\cos(\theta) = \frac{a}{\rho}$ and $\sin(\theta) = \frac{b}{\rho}$, so that

$$a + bi = \rho\left(\frac{a}{\rho} + i\frac{b}{\rho}\right) = \rho(\cos(\theta) + i\sin(\theta))$$

Thus, any complex number can be changed to the form $\rho(\cos(\theta) + i\sin(\theta))$. Further, from the foregoing theorem, it follows that $(a+bi)^n = (\rho(\cos(\theta)+i\sin(\theta)))^n = \rho^n(\cos(n\theta) + i\sin(n\theta))$.

Consider the recursion relation $a_n = 2a_{n-1} - 4a_{n-2}$, which has characteristic polynomial $r^2 - 2r + 4 = 0$. Using the quadratic formula we find that

$$r = 1 \pm \sqrt{1-4}$$
$$= 1 \pm \sqrt{3}i$$

As before we assume that our general solution has the form $cr_1^n + dr_1^n$ where r_1 and r_2 are the solutions of the characteristic polynomial. Thus,

$$a_n = c(1+\sqrt{3}i)^n + d(1-\sqrt{3}i)^n$$
$$= 2^n c \left(\frac{1}{2} + \frac{\sqrt{3}}{2}i\right)^n + 2^n d \left(\frac{1}{2} - \frac{\sqrt{3}}{2}i\right)^n$$
$$= 2^n c \left(\cos\left(\frac{\pi}{3}\right) + i\sin\left(\frac{\pi}{3}\right)\right)^n + 2^n d \left(\cos\left(\frac{\pi}{3}\right) - i\sin\left(\frac{\pi}{3}\right)\right)^n$$
$$= 2^n \left(c\left(\cos\left(n\cdot\frac{\pi}{3}\right) + i\sin\left(n\cdot\frac{\pi}{3}\right)\right) + d\left(\cos\left(n\cdot\frac{\pi}{3}\right) - i\sin\left(n\cdot\frac{\pi}{3}\right)\right)\right)$$
$$= 2^n \left((c+d)\cos\left(n\cdot\frac{\pi}{3}\right) + i(c-d)\sin\left(n\cdot\frac{\pi}{3}\right)\right)$$

If we let $k_1 = c + d$ and $k_2 = i(c-d)$, then

$$a_n = 2^n \left(k_1 \cos\left(n\cdot\frac{\pi}{3}\right) + k_2 \sin\left(n\cdot\frac{\pi}{3}\right)\right)$$

At this point a little sleight-of-hand seems to take place since any trace of complex numbers seems to have disappeared. This is indeed the case; for example, suppose that we let $a_0 = 1$ and $a_1 = 4$. Then

$$a_0 = 2^0 \left(k_1 \cos\left(0\cdot\frac{\pi}{3}\right) + k_2 \sin\left(0\cdot\frac{\pi}{3}\right)\right) = k_1\cos(0) + k_2\sin(0) = k_1$$

and $a_0 = k_1 = 1$. Also

$$a_1 = 2^1\left(k_1 \cos\left(1\cdot\frac{\pi}{3}\right) + k_2 \sin\left(1\cdot\frac{\pi}{3}\right)\right)$$
$$= 2\left(\left(\cos\left(\frac{\pi}{3}\right) + k_2\sin\left(\frac{\pi}{3}\right)\right)\right)$$
$$= 2\left(\frac{1}{2} + k_2\frac{\sqrt{3}}{2}\right)$$
$$= 1 + k_2\sqrt{3}$$

and $a_1 = 1 + k_2\sqrt{3} = 4$. Solving for k_2 we have

$$k_2\sqrt{3} = 3$$

so that

$$k_2\sqrt{3} = \frac{3}{\sqrt{3}} = \sqrt{3}$$

and
$$a_n = 2^n\left(\cos\left(n\cdot\frac{\pi}{3}\right) + \sqrt{3}\sin\left(n\cdot\frac{\pi}{3}\right)\right)$$

In general, we see that if the solutions of the characteristic polynomial are $r_1 = a + bi$ and $r_2 = a - bi$, then
$$a_n = \rho^n(k_1\cos(n\theta) + k_2\sin(n\theta))$$
where $\rho = \sqrt{a^2 + b^2}$, $\cos(\theta) = \dfrac{a}{\rho}$, and $\sin(\theta) = \dfrac{b}{\rho}$.

EXAMPLE 11.11 Solve the recursive function given by
$$a_0 = 1$$
$$a_1 = 2$$
$$a_n = 2\sqrt{2}a_{n-1} - 4a_{n-2}$$

The characteristic polynomial is $r^2 - 2\sqrt{2}r + 4$. Using the quadratic formula, we have $r = \sqrt{2} \pm \sqrt{2}i$, so that $\rho = 2$, and $\cos(\theta) = \sin(\theta) = \dfrac{\sqrt{2}}{2}$. Therefore, $\theta = \dfrac{\pi}{4}$ and
$$a_n = 2^n\left(k_1\cos\left(n\cdot\frac{\pi}{4}\right) + k_2\sin\left(n\cdot\frac{\pi}{4}\right)\right)$$

Therefore,
$$a_0 = 2^0\left(k_1\cos\left(0\cdot\frac{\pi}{4}\right) + k_2\sin\left(0\cdot\frac{\pi}{4}\right)\right)$$
$$= k_1\cos(0) + k_2\sin(0)$$
$$= k_1$$

and $a_0 = k_1 = 1$. Also
$$a_1 = 2^1\left(k_1\cos\left(1\cdot\frac{\pi}{4}\right) + k_2\sin\left(1\cdot\frac{\pi}{4}\right)\right)$$
$$= 2\left(\left(\cos\left(\frac{\pi}{4}\right) + k_2\sin\left(\frac{\pi}{4}\right)\right)\right)$$
$$= 2\left(\frac{\sqrt{2}}{2} + k_2\frac{\sqrt{2}}{2}\right)$$
$$= \sqrt{2} + k_2\sqrt{2}$$

and $a_1 = 2 = \sqrt{2} + k_2\sqrt{2}$. Solving for k_2, we have
$$k_2 = \frac{2 - \sqrt{2}}{\sqrt{2}} = \sqrt{2} - 1$$

and
$$a_n = 2^n\left(\cos\left(n\cdot\frac{\pi}{4}\right) + (\sqrt{2}-1)\sin\left(n\cdot\frac{\pi}{4}\right)\right)$$ ∎

We now consider linear recurrence relations of the form
$$a_n = c_1a_{n-1} + c_2a_{n-2} + c_3a_{n-3} + \cdots + c_pa_{n-p}$$
where $p \geq 2$. For example, consider the recurrence relation
$$a_n = -6a_{n-1} - 5a_{n-2} + 24a_{n-3} + 36a_{n-4}$$

This has characteristic polynomial
$$r^4 + 6r^3 + 5r^2 - 24r - 36 = 0$$

Factoring, we have
$$(r-2)(r+2)(r+3)^2 = 0$$

The general solution then has the form
$$a_n = a \cdot (2)^n + b \cdot (-2)^n + c \cdot (-3)^n + d \cdot n(-3)^n$$

We give the following general solution, which will be stated without proof.

THEOREM 11.12 Let the recursive relation
$$a_n = c_1 a_{n-1} + c_2 a_{n-2} + c_3 a_{n-3} + \cdots + c_p a_p$$
have characteristic polynomial
$$r^n - c_1 r^{n-1} - c_2 r^{n-2} - c_3 r^{n-3} - \cdots - c_p r^{n-p} = 0$$
For each r_i that occurs as a root of the characteristic polynomial q_i times so that $(r - r_i)^{q_i}$ is a factor of the characteristic polynomial, then
$$a_{i1} r_i^n + a_{i2} n r_i^n + a_{i3} n^2 r_i^n + a_{i4} n^3 r_i^n + \cdots + a_{iq_i} n^{q_i - 1} r_i^n$$
is included in the general solution of a_n. ∎

EXAMPLE 11.13 Let the recursive relation for a_n have characteristic polynomial
$$(r-1)^3 (r-2)^2 (r-3)^2 = 0$$
Then
$$a_n = a + bn + cn^2 + d \cdot (2)^n + e \cdot n(2)^n + f \cdot 3^n + g \cdot n(3)^n$$
is the general solution of a_n. ∎

The Catalan numbers defined in Section 5.2 are an example of a homogeneous linear recurrence relation, although it does not have constant coefficients. Recall that they are defined by
$$\text{Cat}(0) = 1$$
$$\text{Cat}(n+1) = \frac{2(2n+1)}{(n+2)} \text{Cat}(n)$$

Exercises

State which of the following are linear recurrence relations:

1. $a_n = n^2 a_{n-1} - \sqrt{n} a_{n-2}$
2. $a_n = a_{n-1}^2 - a_{n-2}$
3. $a_n = a_{n-1} + \sqrt{3} a_{n-2} + \sin(n)$
4. $a_n = a_{n-1} + 3 a_{n-2} a_{n-3}$
5. $a_n = a_{n-1} + 3 a_{n-2} - n a_{n-3}$
6. State which of the recurrence relations in the previous problems are homogeneous linear recurrence relations.

Find the general solution for the following recurrence relations:

7. $a_n - 3 a_{n-1} = 0$
8. $a_n + 3 a_{n-1} = 0$
9. $a_n = -a_{n-1} + 6 a_{n-2}$
10. $a_n = a_{n-1} + 3 a_{n-2}$
11. $a_n + 7 a_{n-1} = 0$
12. $a_n - 5 a_{n-1} = 0$
13. $a_n = +2 a_{n-1} + 8 a_{n-2}$
14. $a_n = +3 a_{n-1} - 2 a_{n-2}$

Solve the following recursive functions:

15. $a_0 = 1$
    $a_n = -4 a_{n-1}$ for $n > 0$
16. $a_1 = 1$
    $a_n = 6 a_{n-1}$ for $n > 1$

17. $a_0 = 2$
    $a_1 = 5$
    $a_n = 5a_{n-1} - 6a_{n-2}$
    for $n > 1$

18. $a_0 = 2$
    $a_1 = 4$
    $a_n = 7a_{n-1} - 12a_{n-2}$
    for $n > 1$

30. $a_0 = 5$
    $a_1 = 4$
    $a_n = -16a_{n-2}$
    for $n > 1$

31. $a_0 = 1$
    $a_1 = 0$
    $a_n = -2\sqrt{2}a_{n-1}$
    $\quad - 4a_{n-2}$ for $n > 1$

Solve the following recursive functions:

19. $a_0 = 0$
    $a_1 = 5$
    $a_n = 9a_{n-1} - 20a_{n-2}$
    for $n > 1$

20. $a_0 = 2$
    $a_1 = 5$
    $a_n = 7a_{n-1} - 12a_{n-2}$
    for $n > 1$

32. $a_0 = 2$
    $a_1 = 1 - \sqrt{3}$
    $a_2 = 1 - 2\sqrt{3}$
    $a_n = a_{n-3}$ for $n > 2$

Find the general solution for the following relations:

33. $a_n - 5a_{n-2} + 4a_{n-4} = 0$
34. $a_n - 2a_{n-1} + 2a_{n-2} - 2a_{n-3} + a_{n-4} = 0$
35. $a_n = a_{n-4}$
36. $a_n = +3a_{n-1} - 3a_{n-2} + a_{n-3} - a_{n-4} + 3a_{n-5} - 3a_{n-6} + a_{n-7}$

21. $a_0 = 2$
    $a_1 = 1$
    $a_n = 2a_{n-1} + 2a_{n-2}$
    for $n > 1$

22. $a_0 = 4$
    $a_1 = 2$
    $a_n = -a_{n-1} + a_{n-2}$
    for $n > 1$

23. $a_0 = 3$
    $a_1 = 21$
    $a_n = 6a_{n-1} - 9a_{n-2}$
    for $n > 1$

24. $a_0 = 2$
    $a_1 = 6$
    $a_n = 4a_{n-1} - 4a_{n-2}$
    for $n > 1$

37. Show that
$$F(n) = \binom{n-1}{0} + \binom{n-2}{1} + \binom{n-3}{2} + \cdots$$
satisfies the recurrence relation for the Fibonacci numbers and, hence, is another expression for the Fibonacci numbers.

25. $a_0 = 1$
    $a_1 = -8$
    $a_n = -4a_{n-1} - 4a_{n-2}$
    for $n > 1$

26. $a_0 = -3$
    $a_1 = 1$
    $a_n = 2a_{n-1} - a_{n-2}$
    for $n > 1$

38. The general **Lucas sequence**, L_n, is given by the recursive function
$$L_1 = a$$
$$L_2 = b$$
$$L_n = L_{n-1} + L_{n-2} \text{ for } n > 2$$
where p and q are integers. Note that when $L_1 = L_2 = 1$, we have the Fibonacci sequence Fib(n). Show that $L_n = b\,\text{Fib}(n-1) + a\,\text{Fib}(n-2)$ for all $n > 2$.

27. $a_0 = 2$
    $a_1 = 6$
    $a_n = -2a_{n-1} - a_{n-2}$ for $n > 1$

Solve the following recursive functions:

28. $a_0 = 3$
    $a_1 = 4$
    $a_n = -4a_{n-2}$ for $n > 1$

29. $a_0 = 1$
    $a_1 = 2$
    $a_n = a_{n-1} - a_{n-2}$
    for $n > 1$

39. Using Theorem 11.9 and induction, prove Theorem 11.10: For a given angle θ, $(\cos(\theta) + i\sin(\theta))^k = \cos(k\theta) + i\sin(k\theta)$.

11.2 Nonhomogeneous Linear Recurrence Relations

Consider the recursive function a_n defined by
$$a_1 = c$$
$$a_n = a \cdot a_{n-1} + b \quad \text{for } n \geq 2$$
and $a \neq 1$. We see that
$$a_1 = c$$
$$a_2 = ac + b$$
$$a_3 = a^2c + ab + b$$

$$a_4 = a^3c + a^2b + ab + b$$
$$a_5 = a^4c + a^3b + a^2b + ab + b$$
$$\vdots$$
$$a_n = a^{n-1}c + a^{n-2}b + a^{n-3}b + \cdots + a^2b + ab + b$$
$$= a^{n-1}c + b(a^{n-2} + a^{n-3} + \cdots + a^2 + a + 1)$$
$$= a^{n-1}c + b\left(\frac{a^{n-1} - 1}{a - 1}\right)$$
$$= \left(c + \frac{b}{a-1}\right)a^{n-1} - \frac{b}{a-1}$$
$$= \left(c - \frac{b}{1-a}\right)a^{n-1} + \frac{b}{1-a}$$

Observe that the recursive function has the general form
$$a_n = c_1 a_{n-1} + f(n)$$
where $c_1 = a$ and $f(n) = b$. We would like to solve this problem using some of the techniques we used for homogeneous recurrences. To do this we need the following theorem.

THEOREM 11.14 Let $a_n^0 = P_n$ satisfy the equation
$$a_n^0 = c_1 a_{n-1}^0 + c_2 a_{n-2}^0 + c_3 a_{n-3}^0 + \cdots + c_p a_{n-p}^0$$
and $a_n = Q_n$ be a particular solution of
$$a_n = c_1 a_{n-1} + c_2 a_{n-2} + c_3 a_{n-3} + \cdots + c_p a_{n-p} + f(n)$$
Then $a_n = P_n + Q_n$ is also a solution of
$$a_n = c_1 a_{n-1} + c_2 a_{n-2} + c_3 a_{n-3} + \cdots + c_p a_{n-p} + f(n)$$

Proof Since $a_n^0 = P_n$ satisfies the equation
$$a_n^0 = c_1 a_{n-1}^0 + c_2 a_{n-2}^0 + c_3 a_{n-3}^0 + \cdots + c_p a_{n-p}^0$$
we have
$$P_n = c_1 P_{n-1} + c_2 P_{n-2} + c_3 P_{n-3} + \cdots + c_p P_{n-p}$$
Since $a_n = Q_n$ satisfies the equation
$$a_n = c_1 a_{n-1} + c_2 a_{n-2} + c_3 a_{n-3} + \cdots + c_p a_{n-p} + f(n)$$
we have
$$Q_n = c_1 Q_{n-1} + c_2 Q_{n-2} + c_3 Q_{n-3} + \cdots + c_p Q_{n-p} + f(n)$$
Adding
$$P_n = c_1 P_{n-1} + c_2 P_{n-2} + c_3 P_{n-3} + \cdots + c_p P_{n-p}$$
and
$$Q_n = c_1 Q_{n-1} + c_2 Q_{n-2} + c_3 Q_{n-3} + \cdots + c_p Q_{n-p} + f(n)$$

we have

$$P_n + Q_n = c_1(P_{n-1} + Q_{n-1}) + c_2(P_{n-2} + Q_{n-2}) + \cdots + c_p(P_{n-p} + Q_{n-p}) + f(n)$$

so that $a_n = P_n + Q_n$ satisfies

$$a_n = c_1 a_{n-1} + c_2 a_{n-2} + c_3 a_{n-3} + \cdots + c_p a_{n-p} + f(n)$$ ∎

THEOREM 11.15 Let Q_n be a solution of

$$a_n = c_1 a_{n-1} + c_2 a_{n-2} + c_3 a_{n-3} + \cdots + c_p a_{n-p} + f(n)$$

Then every solution of

$$a_n = c_1 a_{n-1} + c_2 a_{n-2} + c_3 a_{n-3} + \cdots + c_p a_{n-p} + f(n)$$

has the form $P_n + Q_n$ where P_n is a solution of

$$a_n^0 = c_1 a_{n-1}^0 + c_2 a_{n-2}^0 + c_3 a_{n-3}^0 + \cdots + c_p a_{n-p}^0$$

Proof We already know that $P_n + Q_n$ is a solution of

$$a_n = c_1 a_{n-1} + c_2 a_{n-2} + c_3 a_{n-3} + \cdots + c_p a_{n-p} + f(n)$$

Assume that Q_n and R_n are both solutions of

$$a_n = c_1 a_{n-1} + c_2 a_{n-2} + c_3 a_{n-3} + \cdots + c_p a_{n-p} + f(n)$$

Therefore,

$$Q_n = c_1 Q_{n-1} + c_2 Q_{n-2} + c_3 Q_{n-3} + \cdots + c_p Q_{n-p} + f(n)$$

and

$$R_n = c_1 R_{n-1} + c_2 R_{n-2} + c_3 R_{n-3} + \cdots + c_p R_{n-p} + f(n)$$

Subtracting the first equation from the second we have

$$R_n - Q_n = c_1(R_{n-1} - Q_{n-1}) + c_2(R_{n-2} - Q_{n-2}) + \cdots + c_p(R_{n-p} - Q_{n-p})$$

so that $R_n - Q_n$ is a solution of

$$a_n^0 = c_1 a_{n-1}^0 + c_2 a_{n-2}^0 + c_3 a_{n-3}^0 + \cdots + c_p a_{n-p}^0$$

Say $R_n - Q_n = P_n$. Then $R_n = Q_n + P_n$, where P_n satisfies

$$a_n^0 = c_1 a_{n-1}^0 + c_2 a_{n-2}^0 + c_3 a_{n-3}^0 + \cdots + c_p a_{n-p}^0$$

and our theorem is proved. ∎

We know that $a_n = a \cdot a_{n-1}$ has general solution $a_n = k_1 a^n$ where k_1 is a constant. Therefore, if we could find a particular solution Q_n for the recurrence relation $a_n = a \cdot a_{n-1} + b$, then the general solution would be $R_n = Q_n + k_1 a^n$.

Remark 11.16 We shall assume that a particular solution Q_n of

$$a_n = c_1 a_{n-1} + c_2 a_{n-2} + c_3 a_{n-3} + \cdots + c_p a_{n-p} + f(n)$$

has the same form as $f(n)$.

For example, if $f(n) = a \cdot b^n$, then assume that the particular solution Q_n has form $k_2 \cdot b^n$ and if $f(n)$ is a third-degree polynomial, then assume that Q_n has the form $k_2 \cdot n^3 + k_3 \cdot n^2 + k_4 \cdot n + k_5$.

If $a_1 = c$ and $a_n = a \cdot a_{n-1} + b$ for $n > 1$, then $f(n)$ is a constant, so we shall assume that a constant k_2 is a solution. Therefore, substituting k_2 into the equation, we have

$$k_2 = a \cdot k_2 + b$$

so that

$$k_2 = \frac{b}{1-a}$$

Therefore, our recurrence relation has the form

$$a_n = k_1 a^n + \frac{b}{1-a}$$

Since, in our particular problem $a_1 = c$, we have

$$c = k_1 a + \frac{b}{1-a}$$

and

$$k_1 = \frac{1}{a}\left(c - \frac{b}{1-a}\right)$$

so that

$$a_n = \frac{1}{a}\left(c - \frac{b}{1-a}\right)a^n + \frac{b}{1-a}$$
$$= \left(c - \frac{b}{1-a}\right)a^{n-1} + \frac{b}{1-a}$$

which agrees with our solution at the beginning of the section.

We began this section with the recursive function a_n defined by

$$a_1 = c$$
$$a_n = a \cdot a_{n-1} + b$$

where $a \neq 1$. Suppose we let $a = 1$, so we have

$$a_1 = c$$
$$a_n = a_{n-1} + b$$

The equation $a_n = a_{n-1}$ has characteristic polynomial $r - 1 = 0$ so the general solution for $a_n = a_{n-1}$ is

$$a_n = k_1(1)^n = k_1$$

We would like for a particular solution of $a_n = a_{n-1} + b$ to have the form $a_n = k_2$ since $f(n) = b$, a constant, but $a_n = k_2$ satisfies the equation $a_n = a_{n-1}$ so that if $b \neq 0$, it cannot satisfy $a_n = a_{n-1} + b$. Therefore, we shall duplicate our procedure for multiple roots and let $a_n = n \cdot k_2$. If we substitute this into the equation $a_n = a_{n-1} + b$, we have

$$n \cdot k_2 = (n-1)k_2 + b$$

or

$$k_2 = b$$

Therefore,
$$a_n = k_1 + nb$$

Since $a_1 = c$,
$$c = k_1 + b$$

and
$$k_1 = c - b$$

Therefore,
$$a_n = c - b + nb$$
$$= c + (n-1)b$$

Remark 11.17 Assume that we have the recurrence relation
$$a_n = c_1 a_{n-1} + c_2 a_{n-2} + c_3 a_{n-3} + \cdots + c_p a_{n-p} + f(n)$$

If $f(n) = aP(n)$ and the general solution for the homogeneous recurrence relation
$$a_n = c_1 a_{n-1} + c_2 a_{n-2} + c_3 a_{n-3} + \cdots + c_p a_{n-p}$$

has solution $b_1 f(n) + b_2 n f(n) + b_3 n^2 f(n) + \cdots + b_j n^j f(n)$, but $b_j n^j f(n)$ is not a solution, then $Q_n = k n^{j+1} f(n)$.

For example, if $a_n = 4a_{n-1} - 4a_{n-2} + 3 \cdot 2^n$, then the characteristic polynomial is $r^2 - 4r + 4 = 0$, and the homogeneous recurrence relation has the form $a_n = a \cdot 2^n + b \cdot n 2^n$, so we assume the particular solution for $a_n = 4a_{n-1} - 4a_{n-2} + 3 \cdot 2^n$ has the form $k \cdot n^2 2^n$.

EXAMPLE 11.18 Let $a_n = 5a_{n-1} - 4a_{n-2} + 3 \cdot 2^n$. The characteristic polynomial is $r^2 - 5r + 4 = 0$ and the homogeneous recurrence relation has the form $a_n = a + b \cdot 4^n$. Therefore, we assume that Q_n has the form $k \cdot 2^n$. Substituting this into the recurrence relation, we have
$$k \cdot 2^n = 5k \cdot 2^{n-1} - 4k \cdot 2^{n-2} + 3 \cdot 2^n$$

or
$$4k \cdot 2^{n-2} = 10k \cdot 2^{n-2} - 4k \cdot 2^{n-2} + 12 \cdot 2^{n-2}$$

so that
$$4k = 10k - 4k + 12$$

and $k = -6$. Therefore, the general solution of $a_n = 5a_{n-1} - 4a_{n-2} + 3 \cdot 2^n$ is $a_n = a + b \cdot 4^n - 6 \cdot 2^n$. ■

EXAMPLE 11.19 Let $a_n = 5a_{n-1} - 6a_{n-2} + 6 \cdot 3^n$. The characteristic polynomial is $r^2 - 5r + 6 = 0$ and the homogeneous recurrence relation has the form $a_n = a \cdot 2^n + b \cdot 3^n$. Therefore, we assume that Q_n has the form $k \cdot n 3^n$. Substituting this into the recurrence relation, we have
$$kn3^n = 5k(n-1)3^{n-1} - 6k(n-2)3^{n-2} + 6 \cdot 3^n$$

or
$$9kn3^{n-2} = 15k(n-1)3^{n-2} - 6k(n-2)3^{n-2} + 54 \cdot 3^{n-2}$$

so that

$$9kn = 15k(n-1) - 6k(n-2) + 54$$

and $k = 18$. Therefore, the general solution to the recurrence relation $a_n = 5a_{n-1} - 6a_{n-2} + 6 \cdot 3^n$ is $a_n = a \cdot 2^n + b \cdot 3^n + 18n3^n$. ∎

EXAMPLE 11.20 Solve the recursive function

$$a_1 = 1$$
$$a_n = a_{n-1} + n$$

Note that since a_n is the sum of the first n positive integers, we are really finding the formula for the sum of the first n positive integers. The characteristic polynomial is

$$r - 1 = 0$$

so the solution of the homogeneous equation is $a_n = k_1$. Since $f(n) = n$, we would normally assume that the particular solution of $a_n = a_{n-1} + n$ has the form $k_2 n + k_3$. However, since a constant is a solution of the homogeneous equation, we assume the particular solution of $a_n = a_{n-1} + n$ has the form $a_n = k_2 n^2 + k_3 n$. Substituting this into $a_n = a_{n-1} + n$, we have

$$k_2 n^2 + k_3 n = k_2(n-1)^2 + k_3(n-1) + n$$

or

$$k_2 n^2 + k_3 n = k_2(n^2 - 2n + 1) + k_3(n-1) + n$$

Equating coefficients of n, we have

$$k_3 = -2k_2 + k_3 + 1$$

or

$$k_2 = \frac{1}{2}$$

Equating constants, we have

$$k_2 - k_3 = 0$$

so that

$$k_3 = \frac{1}{2}$$

Therefore, the general solution for $a_n = a_{n-1} + n$ has the form

$$a_n = k_1 + \frac{1}{2}n^2 + \frac{1}{2}n$$

Since $a_1 = 1$, we have

$$a_1 = 1 = k_1 + \frac{1}{2} + \frac{1}{2}$$

so that $k_1 = 0$, and

$$a_n = \frac{1}{2}n^2 + \frac{1}{2}n. = \frac{n(n+1)}{2}$$ ∎

EXAMPLE 11.21

Solve the recurrence relation
$$a_n = 3a_{n-1} - 2a_{n-2} + 3\sin\left(\frac{n\pi}{2}\right)$$

for $n \geq 2$. The homogeneous equation is easily solved. The characteristic polynomial is
$$r^2 - 3r + 2 = 0$$

or
$$(r-1)(r-2) = 0$$

so that the general solution for the homogeneous relation is
$$a_n^0 = a + b \cdot 2^n$$

Since $f(n) = 3\sin\left(\frac{n\pi}{2}\right)$ and the solution for homogeneous recurrence relation having the quadratic characteristic polynomial with complex roots $a + bi$ and $a - bi$ is $b_1\cos(n\theta) + b_2\sin(n\theta)$, we shall assume that the particular solution for
$$a_n = 3a_{n-1} - 2a_{n-2} + 3\sin\left(\frac{n\pi}{2}\right)$$

has the form $b_1\cos\left(\frac{n\pi}{2}\right) + b_2\sin\left(\frac{n\pi}{2}\right)$. Substituting this into the recurrence relation we have

$$b_1\cos\left(\frac{n\pi}{2}\right) + b_2\sin\left(\frac{n\pi}{2}\right) = 3b_1\cos\left(\frac{(n-1)\pi}{2}\right) + 3b_2\sin\left(\frac{(n-1)\pi}{2}\right)$$
$$-2\cos b_1\left(\frac{(n-2)\pi}{2}\right) - 2b_2\sin\left(\frac{(n-2)\pi}{2}\right) + 3\sin\left(\frac{n\pi}{2}\right)$$

Since this must be true for all $n \geq 2$, substitute in $n = 2$ to get
$$b_1\cos(\pi) + b_2\sin(\pi) = 3b_1\cos\left(\frac{\pi}{2}\right) + 3b_2\sin\left(\frac{\pi}{2}\right) - 2b_1\cos(0)$$
$$- 2b_2\sin(0) + 3\sin(\pi)$$

or
$$-b_1 = +3b_2 - 2b_1$$

or
$$b_1 = 3b_2$$

If we substitute in $n = 3$, we get
$$b_1\cos\left(\frac{3\pi}{2}\right) + b_2\sin\left(\frac{3\pi}{2}\right) = 3b_1\cos(\pi) + 3b_2\sin(\pi) - 2\cos b_1\left(\frac{\pi}{2}\right)$$
$$-2b_2\sin\left(\frac{\pi}{2}\right) + 3\sin\left(\frac{3\pi}{2}\right)$$

or
$$-b_2 = -3b_1 - 2b_2 - 3$$

or
$$3b_1 + b_2 = -3$$

Substituting $b_1 = 3b_2$ into this equation, we get
$$9b_2 + b_2 = -3$$

or
$$b_2 = -\frac{3}{10}$$
so that
$$b_1 = -\frac{9}{10}$$

and our general equation is $a_n = a + b \cdot 2^n - \frac{9}{10}\cos\left(\frac{n\pi}{2}\right) - \frac{3}{10}\sin\left(\frac{n\pi}{2}\right)$. Note that we have shown that this is the general solution if it exists, since we really only showed that it is true for $n = 2$ and $n = 3$. In general we will assume that the solution exists and use this method. However, to truly show there is a solution, we would have to take the original equation

$$b_1 \cos\left(\frac{n\pi}{2}\right) + b_2 \sin\left(\frac{n\pi}{2}\right) = 3b_1 \cos\left(\frac{(n-1)\pi}{2}\right) + 3b_2 \sin\left(\frac{(n-1)\pi}{2}\right)$$
$$- 2\cos b_1 \left(\frac{(n-2)\pi}{2}\right)$$
$$- 2b_2 \sin\left(\frac{(n-2)\pi}{2}\right) + 3\sin\left(\frac{n\pi}{2}\right)$$

and use trigonometric identities to express all terms in the form $\cos\left(\frac{(n-2)\pi}{2}\right)$ or $\sin\left(\frac{(n-2)\pi}{2}\right)$.

Considering each term at a time we find

$$b_1 \cos\left(\frac{n\pi}{2}\right) = b_1 \cos\left(\frac{(n-2)\pi}{2}\right)\cos(\pi) - b_1 \sin\left(\frac{(n-2)\pi}{2}\right)\sin(\pi)$$
$$= -b_1 \cos\left(\frac{(n-2)\pi}{2}\right)$$

$$b_2 \sin\left(\frac{n\pi}{2}\right) = b_2 \cos\left(\frac{(n-2)\pi}{2}\right)\sin(\pi) + b_2 \sin\left(\frac{(n-2)\pi}{2}\right)\cos(\pi)$$
$$= -b_2 \sin\left(\frac{(n-2)\pi}{2}\right)$$

$$3b_1 \cos\left(\frac{(n-1)\pi}{2}\right) = 3b_1 \cos\left(\frac{(n-2)\pi}{2}\right)\cos\left(\frac{\pi}{2}\right)$$
$$- 3b_1 \sin\left(\frac{(n-2)\pi}{2}\right)\sin\left(\frac{\pi}{2}\right)$$
$$= -3b_1 \sin\left(\frac{(n-2)\pi}{2}\right)$$

$$3b_2 \sin\left(\frac{(n-1)\pi}{2}\right) = 3b_2 \cos\left(\frac{(n-2)\pi}{2}\right)\sin\left(\frac{\pi}{2}\right) + \sin\left(\frac{(n-2)\pi}{2}\right)\cos\left(\frac{\pi}{2}\right)$$
$$= 3b_2 \cos\left(\frac{(n-2)\pi}{2}\right)$$

$$3\sin\left(\frac{n\pi}{2}\right) = 3\cos\left(\frac{(n-2)\pi}{2}\right)\sin(\pi) + 3\sin\left(\frac{(n-2)\pi}{2}\right)\cos(\pi)$$
$$= -3\sin\left(\frac{(n-2)\pi}{2}\right)$$

Substituting these terms into

$$b_1 \cos\left(\frac{n\pi}{2}\right) + b_2 \sin\left(\frac{n\pi}{2}\right) =$$
$$3b_1 \cos\left(\frac{(n-1)\pi}{2}\right) + 3b_2 \sin\left(\frac{(n-1)\pi}{2}\right)$$
$$- 2\cos b_1 \left(\frac{(n-2)\pi}{2}\right) - 2b_2 \sin\left(\frac{(n-2)\pi}{2}\right) + 3\sin\left(\frac{n\pi}{2}\right)$$

we have

$$- b_1 \cos\left(\frac{(n-2)\pi}{2}\right) - b_2 \sin\left(\frac{(n-2)\pi}{2}\right) -$$
$$- 3b_1 \sin\left(\frac{(n-2)\pi}{2}\right) + 3b_2 \cos\left(\frac{(n-2)\pi}{2}\right)$$
$$- 2b_1 \cos\left(\frac{(n-2)\pi}{2}\right) - 2b_2 \sin\left(\frac{(n-2)\pi}{2}\right) - 3\sin\left(\frac{(n-2)\pi}{2}\right)$$

Equating coefficients of $\cos\left(\frac{(n-2)\pi}{2}\right)$, we have

$$-b_1 = 3b_2 - 2b_1 \quad \text{or} \quad b_1 = 3b_2$$

Equating coefficients of $\sin\left(\frac{(n-2)\pi}{2}\right)$, we have

$$-b_2 = -3b_1 - 2b_2 - 3 \quad \text{or} \quad 3b_1 + b_2 = 3$$

which are the same equations we received before. Hence, we know that our solution really exists. ∎

Exercises

Find the general solution for the following recurrence relations:

1. $a_n = 2a_{n-1} + 5$
2. $a_n = -3a_{n-1} + n$
3. $a_n = -a_{n-1} + 12a_{n-2} + 2^n$
4. $a_n = -3a_{n-1} - a_{n-2} + 3^n$
5. $a_n = 4a_{n-1} - 4a_{n-2} + n^2$
6. $a_n = -2a_{n-1} + n^2$
7. $a_n = 4a_{n-1} + 4^n$
8. $a_n = 6a_{n-1} + 9a_{n-2} + (-2)^n$
9. $a_n = -9a_{n-2} + 3^n$
10. $a_n = -4a_{n-1} - 2a_{n-2} + 5$
11. $a_n = 2a_{n-1} + 2^n$
12. $a_n = -6a_{n-1} - 9a_{n-2} + (-3)^n$
13. $a_n = 3_{n-1} - 2a_{n-2} + 5$
14. $a_n = a_{n-1} + 6a_{n-2} + 3^n$
15. $a_n = 2\sqrt{3}a_{n-1} - 4a_{n-2} + 3$
16. $a_n = a_{n-2} + \cos\left(\frac{n\pi}{2}\right)$
17. $a_n = -a_{n-2} + \cos\left(\frac{n\pi}{2}\right)$
18. $a_n = 6a_{n-1} - 12a_{n-2} + 8a_{n-3} + 3^n$
19. $a_n = 6a_{n-1} - 12a_{n-2} + 8a_{n-3} + 2^n$
20. $a_n = -a_{n-3} + n$

Find the following sums using recurrence relations:

21. $1^2 + 2^2 + 3^2 + \cdots + n^2$ Use $a_n = a_{n-1} + n^2$.
22. $1 + 4 + 7 + \cdots + 3n - 2$ Use $a_n = a_{n-1} + 3n - 2$.
23. $2^1 + 2^2 + 2^3 + 2^4 + \cdots + 2^n$ Use $a_n = a_{n-1} + 2^n$.
24. $1 \cdot 2 + 2 \cdot 3 + 3 \cdot 4 + \cdots + n \cdot (n+1)$ $a_n = a_{n-1} + n \cdot (n+1)$

Find the following sums using recurrence relations:

25. $1^3 + 2^3 + 3^3 + \cdots + n^3$ Use $a_n = a_{n-1} + n^3$.
26. $1^2 \cdot 2 + 2^2 \cdot 3 + 3^2 \cdot 4 + \cdots + n^2 \cdot (n+1)$ Use $a_n = a_{n-1} + n^2 \cdot (n+1)$.
27. $1 \cdot 2^0 + 2 \cdot 2^1 + 3 \cdot 2^2 + \cdots + n \cdot 2^{(n-1)}$ Use $a_n = a_{n-1} + n \cdot 2^{(n-1)}$.

11.3 Finite Differences

For students who have had calculus, many of the results of this section may seem familiar. In fact, one may think of finite differences as calculus without limits. Similar to calculus, finite differences have many applications. They are found in such diverse areas as computer science, actuarial science, economics, psychology, and sociology. Uses include error detection, fitting polynomials to data, extrapolation and interpolation, summation of functions, derivation of combinatorial functions, and approximation of area. It is one of the basic tools of numerical analysis. We want to emphasize, however, that calculus is not required for this section. In this section, we introduce finite differences and show some of their elementary properties.

The **first difference** of a function f, denoted by Δf, is defined by

$$\Delta f(x) = f(x+1) - f(x)$$

The **second difference** of f, denoted by $\Delta^2 f$, is defined by

$$\Delta^2 f(x) = \Delta(\Delta f(x))$$

Therefore,

$$\Delta^2 f(x) = \Delta(f(x+1) - f(x))$$
$$= (f(x+2) - f(x+1)) - (f(x+1) - f(x))$$
$$= f(x+2) - 2f(x+1) + f(x)$$

In general the nth difference of f, denoted by $\Delta^n f$, is defined inductively by

$$\Delta^n f(x) = \Delta(\Delta^{n-1} f(x))$$

Later, we shall show a formula for expressing $\Delta^n f(x)$ in terms of $f(x)$. The following table illustrates the difference function:

x	$f(x)$	$\Delta f(x)$	$\Delta^2 f(x)$	$\Delta^3 f(x)$	$\Delta^4 f(x)$
1	1	7	12	6	0
2	8	19	18	6	0
3	27	37	24	6	
4	64	61	30		
5	125	91			
6	216				

Note that in this case, $f(x) = x^3$ and $\Delta^4 f(x) = 0$ for all values of x.

An **operator** is a function that maps functions to functions. Therefore, Δ is an operator. We also define another operator E, by $E(f(x)) = f(x+1)$. Thus,

$$\Delta f(x) = E(f(x)) - f(x))$$
$$= (E - I)(f(x))$$

where we use the notation $(F + G)(u) = F(u) + G(u)$. I is the identity operator. Therefore, $\Delta = (E - I)$ and $E = I + \Delta$.

Using $\Delta = (E - I)$ and letting $E^0 = I$, we have

$$\Delta^n(f(x)) = (E - I)^n(f(x))$$
$$= E^n(f(x)) - n \cdot E^{n-1}(f(x)) + \cdots + (-1)^k \binom{n}{k} \cdot E^{n-k}(f(x))$$
$$+ \cdots + (-1)^n E^0(f(x))$$
$$= f(x + n) - n \cdot f(n - 1) + \cdots + (-1)^k \binom{n}{k}(f(x + n - k))$$
$$+ \cdots (-1)^n f(x)$$

EXAMPLE 11.22 $\Delta^3(f(x)) = f(x + 3) - 3f(x + 2) + 3f(x + 1) - f(x)$ ∎

THEOREM 11.23 The operators Δ and E have the following properties: For a real number a and functions f and g,

(a) $\Delta(f + g) = \Delta(f) + \Delta(g)$; $E(f + g) = E(f) + E(g)$
(b) $\Delta(af) = a\Delta(f)$; $E(af) = aE(f)$
(c) $E\Delta = \Delta E$;
(d) $\Delta(a) = 0$; $E(a) = a$

Proof (a)

$$\Delta((f + g)(x)) = \Delta(f(x) + g(x))$$
$$= f(x + 1) + g(x + 1) - (f(x) + g(x))$$
$$= f(x + 1) - f(x) + g(x + 1) - g(x)$$
$$= \Delta f(x) + \Delta g(x)$$
$$= (\Delta f + \Delta g)(x)$$

and

$$\Delta((af)(x)) = \Delta(af(x))$$
$$= af(x + 1) + af(x)$$
$$= a(f(x + 1) - f(x))$$
$$= a\Delta f(x)$$

Parts (b) and (c) are left to the reader. ∎

EXAMPLE 11.24
$$\Delta(3x^4 + 2x^3 + 5x + 4) = \Delta(3x^4) + \Delta(2x^3) + \Delta(5x) + \Delta(4)$$
$$= 3\Delta(x^4) + 2\Delta(x^3) + 5\Delta(x)$$ ∎

At this point we wish to find $\Delta(x^n)$. Unfortunately, the news here is not so good. We see that

$$\Delta(x) = (x + 1) - x = 1$$

and
$$\Delta(x^2) = (x+1)^2 - x^2$$
$$= x^2 + 2x + 1 - x^2$$
$$= 2x + 1$$

In general
$$\Delta(x^n) = (x+1)^n - x^n$$
$$= x^n + nx^{n-1} + \cdots + \binom{n}{k}x^k \cdots + 1 - x^n$$
$$= nx^{n-1} + \cdots + \binom{n}{k}x^k \cdots + 1$$

which is not terribly pleasant.

We now introduce two properties of the operator Δ, which should look familiar to calculus students. They are called, respectively, the **product rule** and the **quotient rule**.

THEOREM 11.25 For functions f and g,
$$\Delta(fg) = f \cdot \Delta(g) + E(g) \cdot \Delta(f)$$

and
$$\Delta\left(\frac{f}{g}\right) = \frac{g \cdot \Delta(f) - f \cdot (\Delta(g))}{g \cdot (E(g))}$$

Proof
$$\Delta(fg(x)) = \Delta(f(x) \cdot g(x))$$
$$= f(x+1) \cdot g(x+1) - f(x) \cdot g(x)$$
$$= f(x+1) \cdot g(x+1) - f(x) \cdot g(x+1)$$
$$\quad + f(x) \cdot g(x+1) - f(x) \cdot g(x)$$
$$= (f(x+1) - f(x)) \cdot g(x+1) + f(x)$$
$$\quad \cdot (g(x+1) - \cdot g(x))$$
$$= \Delta f(x) \cdot E(g(x)) + f(x) \cdot \Delta g(x)$$
$$= f(x) \cdot \Delta g(x) + Eg(x) \cdot \Delta f(x)$$
$$= (f \cdot \Delta g + Eg \cdot \Delta f)(x)$$

The proof of the second part is left to the reader. ∎

EXAMPLE 11.26
$$\Delta((x^2 + 6x)(2x^2 + 5))$$
$$= (x^2 + 6x)\Delta(2x^2 + 5) + E(2x^2 + 5)\Delta(x^2 + 6x)$$
$$= (x^2 + 6x)(2\Delta(x^2) + \Delta(5)) + E(2x^2 + 5)(\Delta(x^2) + 6\Delta(x))$$
$$= (x^2 + 6x)(4x + 2) + E(2x^2 + 5)(2x + 1 + 6)$$
$$= (x^2 + 6x)(4x + 2) + (2((x+1)^2 + 5)(2x + 7)$$
$$= (x^2 + 6x)(4x + 2) + (2x^2 + 4x + 7)(2x + 7)$$ ∎

EXAMPLE 11.27

$$\Delta\left(\frac{x^2+2}{2x^2+6x+5}\right)$$
$$=\frac{(2x^2+6x+5)\cdot\Delta(x^2+2)-(x^2+2)\cdot\Delta(2x^2+6x+5)}{(2x^2+6x+5)\cdot E(2x^2+6x+5)}$$
$$=\frac{(2x^2+6x+5)\cdot(\Delta(x^2)+\Delta(2))-(x^2+2)\cdot(2\Delta(x^2)+6\Delta(x)+\Delta(5))}{(2x^2+6x+5)\cdot(2(x+1)^2+6(x+1)+5)}$$
$$=\frac{(2x^2+6x+5)\cdot(2x+1)-(x^2+2)\cdot(2(2x+1)+6)}{(2x^2+6x+5)\cdot(2(x^2+2x+1)+6x+11)}$$
$$=\frac{(2x^2+6x+5)\cdot(2x+1)-(x^2+2)\cdot(4x+8)}{(2x^2+6x+5)\cdot(2x^2+10x+13)}$$

∎

We do have a bit more luck finding $\Delta f(x)$ when $f(x) = a^x$, as seen by the following theorem:

THEOREM 11.28 If $f(x) = a^x$, then $\Delta f(x) = a^x(a-1)$. In particular, if $f(x) = 2^x$, then $\Delta f(x) = 2^x$.

Proof $\Delta(a^x) = a^{x+1} - a^x = a^x(a-1)$. ∎

EXAMPLE 11.29

$$\Delta(x^2+3x+4+2^x+4^x) = \Delta(x^2)+3\Delta(x)+\Delta(4)+\Delta(2^x)+\Delta(4^x)$$
$$= 2x+1+3+2^x+3\cdot 4^x$$
$$= 2x+4+2^x+3\cdot 4^x$$

∎

Exercises

1. Find $\Delta^2(x^3)$ evaluated at $x=1$.
2. Find $\Delta^3(2^x)$ evaluated at $x=1$.
3. Find $\Delta^2 E^2(x^4)$ evaluated at $x=4$.
4. Find $\Delta^2 E(n!)$ evaluated at $n=1$.
5. Find $\Delta((3x+2)(x-3))$.
6. Find $\Delta((4x-2)(x+5))$.
7. Find $\Delta\left(\dfrac{x-6}{3x+4}\right)$.
8. Find $\Delta\left(\dfrac{2x+3}{4x+1}\right)$.
9. Show that $\Delta(x!) = x \cdot x!$
10. Prove parts (b), (c), and (d) of Theorem 11.23: For real number a and functions f and g,
    (a) $\Delta(af) = a\Delta(f)$
    (b) $E(af) = aE(f)$
    (c) $E\Delta = \Delta E$
    (d) $\Delta(a) = 0$
    (e) $E(a) = a$
11. Prove the second part of Theorem 11.25:
    $$\Delta\left(\frac{f}{g}\right) = \frac{g\cdot\Delta(f) - f\cdot(\Delta(g))}{g\cdot(E(g))}$$

11.4 Factorial Polynomials

As we have seen, finding finite differences of x^n is really a bit messy, which is unfortunate since polynomial functions are rather popular. To avoid this problem, we develop the factorial polynomial. For $x \geq n \geq 0$, let

$$x^{(n)} = \begin{cases} x(x-1)(x-2)\cdots(x-n+1) & \text{if } n > 0 \\ 1 & \text{if } n = 0 \end{cases}$$

Fortunately,

$$\Delta(x^{(n)}) = (x+1)(x)(x-1)(x-2)\cdots(x-n+2)$$
$$- x(x-1)(x-2)\cdots(x-n+1)$$
$$= (x)(x-1)(x-2)\cdots(x-n+2)((x+1)-(x-n+1))$$
$$= (x)(x-1)(x-2)\cdots(x-n+2)(n)$$
$$= n(x)(x-1)(x-2)\cdots(x-(n-1)+1)$$
$$= nx^{(n-1)}$$

so that we now have a function that has a nice difference. In fact, those who have had calculus probably hoped for such a nice result for x^n. Also notice that $x^{(1)} = x$. Thus, if

$$f(x) = 5x^{(4)} + 6x^{(3)} - 3x^{(2)} + 2x^{(1)} + 7$$

then

$$\Delta f(x) = (5 \cdot 4)x^{(3)} + (6 \cdot 3)x^{(2)} - 3 \cdot 2x^{(1)} + 2$$
$$= 20x^{(3)} + 18x^{(2)} - 6x^{(1)} + 2$$

If $f(x) = a_n x^{(n)} + a_{n-1} x^{(n-1)} + \cdots a_2 x^{(2)} + a_1 x + a_0$ for some n, then f is called a **factorial polynomial**. It is obviously very simple to take the difference of a factorial polynomial. The problem is to get a regular polynomial into the form of a factorial polynomial. We now give an algorithm to perform this task. The reader may observe that this technique is very similar to one of the techniques used to change integers from base 10 to base n in Section 5.6.

We begin with an example. Suppose that we have

$$3x^4 - 19x^3 + 34x^2 - 21x + 5$$

which we wish to change to a factorial polynomial

$$a_4 x^{(4)} + a_3 x^{(3)} + a_2 x^{(2)} + a_1 x + a_0$$

Remember that these polynomials are both the same function. They are only different representations.

If we divide

$$3x^4 - 19x^3 + 34x^2 - 21x - 5$$

by x, we get the quotient $3x^3 - 19x^2 + 34x - 21$ with remainder -5. If we divide

$$a_4 x^{(4)} + a_3 x^{(3)} + a_2 x^{(2)} + a_1 x + a_0$$

by x, we get

$$a_4(x-1)(x-2)(x-3) + a_3(x-1)(x-2) + a_2(x-1) + a_1$$

with remainder a_0. Therefore,

$$a_0 = -5$$

and

$$3x^3 - 19x^2 + 34x - 21 = a_4(x-1)(x-2)(x-3) + a_3(x-1)(x-2) + a_2(x-1) + a_1$$

When

$$3x^3 - 19x^2 + 34x - 21$$

is divided by $x-1$, we get the quotient
$$3x^2 - 16x + 18$$
and remainder -3.
When
$$a_4(x-1)(x-2)(x-3) + a_3(x-1)(x-2) + a_2(x-1) + a_1$$
is divided by $x-1$, we get the quotient
$$a_4(x-2)(x-3) + a_3(x-2) + a_2$$
and remainder a_1. Therefore,
$$a_1 = -3$$
and
$$3x^2 - 16x + 18 = a_4(x-2)(x-3) + a_3(x-2) + a_2$$
When we divide
$$3x^2 - 16x + 18$$
by $x-2$, we get the quotient $3x-10$ and remainder -2. When we divide
$$a_4(x-2)(x-3) + a_3(x-2) + a_2$$
by $x-2$, we get the quotient $a_4(x-3) + a_3$ and remainder a_2. Therefore,
$$a_2 = -2$$
and
$$3x - 10 = a_4(x-3) + a_3$$
Dividing
$$3x - 10$$
by $x-3$, we get 3 and remainder -1.
Dividing $a_4(x-3) + a_3$ by $x-3$, we get a_4 and remainder a_3. Therefore,
$$a_4 = 3$$
and
$$a_3 = -1$$
Our factor polynomial is, therefore,
$$3x^{(4)} - x^{(3)} - 2x^{(2)} - 3x + -5$$

In summation, to find a_0 in the factor polynomial, divide by x and take the remainder. To find a_1 in the factor polynomial, divide the quotient by $x-1$ and take the remainder. Continue this process of dividing the remaining quotient obtained each time by $x-2$, $x-3$, and $x-4$ to find respective remainders a_2, a_3, and a_4.

We now present the algorithm for finding the factorial polynomial $a_n x^{(n)} + a_{n-1} x^{(n-1)} + \cdots + a_2 x^{(2)} + a_1 x^{(1)} + a_0 x$.

Algorithm Factorial Polynomial

Step 1. Given the nth degree polynomial $f(x)$, let $k = 0$;

Step 2. Divide $f(x)$ by $x - k$, getting remainder r and the quotient $q(x)$;

Step 3. Let $a_k = r$ and $f(x) = q(x)$;

Step 4. If $k = n$, the process is finished. Otherwise let $k = k + 1$ and go to Step 2.

End.

The easiest way to perform this process is to use synthetic division. We illustrate the process letting $f(x) = x^4 - 8x^3 + 21x^2 - 6x + 3$.

```
0 | 1   -8    21    -6    3
  |     0     0     0     0
1 | 1   -8    21    -6   [3]
  |     1    -7    14
2 | 1   -7    14   [8]
  |     2   -10
3 | 1   -5   [4]
  |     3
4 | 1  [-2]
     [1]
```

This gives the factorial polynomial $x^{(4)} - 2x^{(3)} + 4x^{(2)} + 8x + 3$.

There are several ways to go from a factorial polynomial to a regular polynomial. One is to simply expand each term and collect terms of like degree. Another method could be to evaluate $n+1$ points with the factorial polynomial, where n is the degree of the polynomial and then fit an nth degree polynomial to these points.

A third method is a reverse form of synthetic division. To show how it works, consider the factorial polynomial

$$3x^{(4)} - x^{(3)} - 2x^{(2)} - 3x - 5$$

Expand each term so we have

$$3x(x-1)(x-2)(x-3) - x(x-1)(x-2) - 2x(x-1) - 3x - 5$$

Split $(x-3)$ in the first term, leaving x with the first term and combining -3 with the second term so that we have

$$3x^2(x-1)(x-2) + (3(-3) - 1)x(x-1)(x-2) - 2x(x-1) - 3x - 5$$

Now we split the $(x-2)$ in the second term, leaving x with the second term and taking -2 to the third term, giving us

$$3x^2(x-1)(x-2) - 10x^2(x-1) + ((-10)(-2) - 2)x(x-1) - 3x - 5$$

Now we split the $(x-1)$ in the third term, leaving x with the third term and taking -1 to the fourth term, giving us

$$3x^2(x-1)(x-2) - 10x^2(x-1) + 18x^2 + (-18 - 3)x - 5$$

or

$$3x^2(x-1)(x-2) - 10x^2(x-1) + 18x^2 - 21x - 5$$

which is our polynomial after stage 1. Note that we would have obtained the same results in

	3	−1	−2	−3	−5
		−9	20	−18	
3	3	−10	18	−21	
2					
1					

using the following process. Note that the numbers in the top row are the coefficients of the factorial polynomial and the numbers in the first column are the integers beginning with 3, one less than the degree of the polynomial. Bring down the 3, just as in synthetic division, and multiply by 3, the first number in the column. But instead of adding to -1, the next number in the row, subtract 9 from -1, getting -10. Now take -10, multiply by 2, the next integer in the column, and subtract from -2, the next number in the row. We get 18. Continue this process by multiplying 18 by 1, and subtracting it from -3, getting -21. The last number in the row is not used and remains unchanged. We can see that the numbers in the third row together with the last number in the first row are the coefficients of the polynomial from stage 1.

We now take the polynomial in the first stage,

$$3x^2(x-1)(x-2) - 10x^2(x-1) + 18x^2 - 21x - 5$$

and repeat the process. We split $x - 2$, leaving x with the first term and taking -2 to the second term, giving us

$$3x^3(x-1) + ((3)(-2) - 10)x^2(x-1) + 18x^2 - 21x - 5$$

We now split $x - 1$ in the second term, leaving x in the second term and taking -1 to the third term, giving us

$$3x^3(x-1) - 16x^3 + (16 + 18)x^2 - 21x - 5$$

or

$$3x^3(x-1) - 16x^3(x-1) + 34x^2 - 21x - 5$$

for the second-stage polynomial. Notice that each stage gives us x to a higher power in the term where the splitting occurs.

Again, returning to our table we repeat the process, except we bring down 3, the first term in the row, multiply by 2, and subtract it from -10, the integer in the second column. This gives us -16. Multiply -16 by 1 and subtract from 18, giving 34. Again -21, the last number in the row, is not used. Again notice that the remaining row, together with the numbers -21 and -5, forms our second-stage polynomial:

	3	−1	−2	−3	−5
		−9	20	−18	
3	3	−10	18	−21	
		−6	16		
2	3	−16	34		
1					

We now take the second-stage polynomial

$$3x^3(x-1) - 16x^3 + 34x^2 - 21x - 5$$

and split $x - 1$ in the first term. This gives us

$$3x^4 - 19x^3 + 34x^2 - 21x - 5$$

which is the polynomial we are seeking.

Again, looking at the table, bring down 3, and multiply by 1. Subtract from -16. This leaves -19 unused. Finally, bring down the 3. Note that the numbers in brackets are the coefficients we seek.

		3	-1	-2	-3	[-5]
			-9	20	-18	
	3	3	-10	18	[-21]	
			-6	16		
	2	3	-16	[34]		
			-3			
	1	3	[-19]			
		[3]				

Having shown the pattern, we leave the algorithm for changing from factorial polynomials to regular polynomials to the reader.

EXAMPLE 11.30 Change the polynomial $x^{(4)} - 2x^{(3)} + 4x^{(2)} + 6x + 3$ to a regular polynomial. Using reverse synthetic division, we have

		1	-2	4	6	[3]
			-3	10	-14	
	3	1	-5	14	[-8]	
			-2	7		
	2	1	-7	[21]		
			-1			
	1	1	[-8]			
		[1]				

giving the polynomial $f(x) = x^4 - 8x^3 + 21x^2 - 8x + 3$. ∎

We have seen that

$$\begin{aligned} x^{(n)} &= x(x-1)(x-2)\cdots(x-(n-1)+1)(x-n+1) \\ &= (x-n+1)x(x-1)(x-2)\cdots(x-(n-1)+1) \\ &= (x-n+1)x^{(n-1)} \end{aligned}$$

Solving for $x^{(n-1)}$, we have

$$x^{(n-1)} = \frac{x^{(n)}}{x-n+1}$$

We now have a way of defining $x^{(n)}$ when n is a negative integer.
Letting $n = 1$, we have

$$x^{(0)} = \frac{x^{(1)}}{x-1+1} = 1$$

Letting $n = 0$, we have

$$x^{(-1)} = \frac{x^{(0)}}{x - 0 + 1} = \frac{1}{x + 1}$$

We now use induction and

$$x^{(n-1)} = \frac{x^{(n)}}{x - n + 1}$$

to show

$$x^{(-m)} = \frac{1}{(x + m)^{(m)}}$$

for all $m \geq 1$, where the denominator is not zero.

We have already shown that

$$x^{(-1)} = \frac{1}{x + 1} = \frac{1}{(x + 1)^{(1)}}$$

Assume that

$$x^{(-k)} = \frac{1}{(x + k)^{(k)}}$$

Then

$$\begin{aligned}
x^{(-(k+1))} &= x^{(-k-1)} \\
&= \frac{x^{(-k)}}{x - (-k) + 1} \\
&= \frac{1}{(x + k)^{(k)}} \cdot \frac{1}{x + k + 1} \\
&= \frac{1}{(x + k)(x + k - 1) \cdots (x + 1)(x + k + 1)} \\
&= \frac{1}{(x + k + 1)(x + k)(x + k - 1) \cdots (x + 1)} \\
&= \frac{1}{(x + k + 1)^{(k+1)}}
\end{aligned}$$

We now need to find $\Delta(x^{-n})$. By definition

$$\begin{aligned}
\Delta(x^{-n}) &= \Delta\left(\frac{1}{(x + n)^{(n)}}\right) \\
&= \frac{1}{(x + n + 1)^{(n)}} - \frac{1}{(x + n)^{(n)}} \\
&= \frac{1}{(x + n + 1)(x + n)(x + n - 1) \cdots (x + 2)} - \frac{1}{(x + n)(x + n - 1) \cdots (x + 1)} \\
&= \frac{x + 1 - (x + n + 1)}{(x + n + 1)(x + n)(x + n - 1) \cdots (x + 1)} \\
&= \frac{-n}{(x + n + 1)^{(n+1)}} \\
&= -nx^{-n-1}
\end{aligned}$$

so for every nonzero integer n, we have $\Delta(x^{(n)}) = nx^{(n-1)}$.

We are now able to generalize one of our theorems on combinations. For a positive integer n, let
$$\binom{x}{n} = \frac{x^{(n)}}{n!}$$
When x is an integer, this is the usual definition of combinations.

Taking the difference, we find that
$$\Delta\binom{x}{n} = \frac{nx^{(n-1)}}{n!}$$
$$= \frac{x^{(n-1)}}{(n-1)!}$$
$$= \binom{x}{n-1}$$
But
$$\Delta\binom{x}{n} = \binom{x+1}{n} - \binom{x}{n}$$
so we have
$$\binom{x}{n-1} = \binom{x+1}{n} - \binom{x}{n}$$
or
$$\binom{x+1}{n} = \binom{x}{n} + \binom{x}{n-1}$$
Letting $y = x + 1$, we have
$$\binom{y}{n} = \binom{y-1}{n} + \binom{y-1}{n-1}$$
in the more familiar form, which we had previously proved only when y was an integer.

DEFINITION 11.31   **Stirling numbers** of the first kind are defined by
$$s_0^{(n)} = 0 \qquad \text{for all } n \geq 1$$
$$s_n^{(n)} = 1 \qquad \text{for all } n \geq 0$$
$$s_k^{(n+1)} = s_{k-1}^{(n)} + ns_k^{(n)}$$

DEFINITION 11.32   **Stirling numbers** of the second kind are defined by
$$S_0^{(n)} = 0 \qquad \text{for all } n \geq 1$$
$$S_n^{(n)} = 1 \qquad \text{for all } n \geq 0$$
$$S_k^{(n+1)} = S_{k-1}^{(n)} + kS_k^{(n)}$$

At this point we show that we can; describe Stirling numbers of the first and second kind in terms of polynomials. Stirling numbers of the first kind are simply the absolute value of coefficients of the polynomial equal to $x^{(n)}$ for $n = 1, 2, 3, \cdots$.

11.4 Factorial Polynomials

THEOREM 11.33 Let

$$x^{(n)} = s_n^{(n)}x^n - s_{n-1}^{(n)}x^{n-1} + s_{n-2}^{(n)}x^{n-2} - \cdots + (-1)^{n-2}s_2^{(n)}x^2$$
$$+ (-1)^{n-1}s_1^{(n)}x + (-1)^n s_0^{(n)}$$

The coefficients $s_i^{(n)}$ are **Stirling numbers of the first kind**.

Proof By definition $s_0^{(n)} = 0$ for all $n \geq 1$, and $s_n^{(n)} = 1$ for all $n \geq 0$. To show the $s_i^{(n)}$ are Stirling numbers of the first kind, we must show that $s_i^{(n+1)} = s_{i-1}^{(n)} + n \cdot s_i^{(n)}$ for $1 \leq i \leq n$.

By definition

$$x^{(n+1)} = (x - n)x^{(n)}$$
$$= x \cdot x^{(n)} - n \cdot x^{(n)}$$

Therefore,

$$\sum_{i=0}^{n+1} (-1)^{n+1-i} s_i^{(n+1)} x^i = x \cdot \sum_{i=0}^{n} (-1)^{n-i} s_i^{(n)} x^i - n \cdot \sum_{i=0}^{n} (-1)^{n-i} s_i^{(n)} x^i$$
$$= \sum_{i=0}^{n} (-1)^{n-i} s_i^{(n)} x^{i+1} - n \cdot \sum_{i=0}^{n} (-1)^{n-i} s_i^{(n)} x^i$$
$$= \sum_{i=1}^{n+1} (-1)^{n-(i-1)} s_{i-1}^{(n)} x^i - n \cdot \sum_{i=0}^{n} (-1)^{n-i} s_i^{(n)} x^i$$

(increasing i by 1 in the first sum)

$$= \sum_{i=1}^{n+1} (-1)^{n-i+1} s_{i-1}^{(n)} x^i + n \cdot \sum_{i=0}^{n} (-1)^{n-i+1} s_i^{(n)} x^i$$

Equating coefficients of x^i, for $1 \leq i \leq n$, we have $s_i^{(n+1)} = s_{i-1}^{(n)} + n s_i^{(n)}$ as desired. ∎

We now find that we get Stirling numbers of the second kind when we look at the coefficients of the factorial polynomials equal to x^n for $n = 1, 2, 3, \cdots$.

THEOREM 11.34 Let

$$x^n = S_n^{(n)} x^{(n)} + S_{n-1}^{(n)} x^{(n-1)} + S_{n-2}^{(n)} x^{(n-2)} + \cdots S_2^{(n)} x^{(2)} + S_1^{(n)} x + S_0^{(n)}$$

The coefficients $S_i^{(n)}$ are **Stirling numbers of the second kind**.

Proof By definition $S_0^{(n)} = 0$ for all $n \geq 1$, and $S_n^{(n)} = 1$ for all $n \geq 0$. We need to show that for $1 \leq i \leq n$,

$$S_i^{(n+1)} = S_{i-1}^n + i \cdot S_i^n$$

By definition,

$$\sum_{i=0}^{n+1} S_i^{(n+1)} x^{(i)} = x^{n+1}$$

$$= x \cdot x^n$$

$$= x \cdot \sum_{i=0}^{n} S_i^{(n)} x^{(i)}$$

$$= (i + x - i) \cdot \sum_{i=0}^{n} S_i^{(n)} x^{(i)}$$

$$= i \cdot \sum_{i=0}^{n} S_i^{(n)} x^{(i)} + (x - i) \cdot \sum_{i=0}^{n} S_i^{(n)} x^{(i)}$$

$$= i \cdot \sum_{i=0}^{n} S_i^{(n)} x^{(i)} + \sum_{i=0}^{n} S_i^{(n)} (x - i) \cdot x^{(i)}$$

$$= i \cdot \sum_{i=0}^{n} S_i^{(n)} x^{(i)} + \sum_{i=0}^{n} S_i^{(n)} x^{(i+1)}$$

$$= i \cdot \sum_{i=0}^{n} S_i^{(n)} x^{(i)} + \sum_{i=1}^{n+1} S_{i-1}^{(n)} x^{(i)}$$

Equating coefficients of x^i, for $1 \leq i \leq n$, we have $S_i^{(n+1)} = i \cdot S_i^{(n)} + S_{i-1}^{(n)}$ as desired. ∎

Exercises

1. Find $\Delta^2(3x^{(3)} - 5x^{(2)} + 4)$.
2. Find $\Delta^4(4x^{(5)} - 3x^{(4)} + 6x^{(3)} + x^{(2)} - 3x + 4)$.
3. Find $\Delta^3(x^{(5)} + 3x^{(4)} - 3x^{(3)} - 2x^{(2)} + 4x - 1)$.
4. Find $\Delta(3x^{(2)} + 2x - 5)(x^{(4)} + 4x^{(3)} + 2x^{(2)})$.
5. Find $\Delta(x^{(3)} - 2x^{(2)} - 4)(x^{(4)} + 2x^{(3)} - 6x^{(2)} + x - 3)$.
6. Find $\Delta\left(\dfrac{x^{(3)} + 6x^{(2)} + 3x}{x^{(5)} + 3x^{(3)}}\right)$.
7. Find $\Delta\left(\dfrac{x^{(4)} - 3x^{(2)} + 3x}{x^{(4)} - 3x^{(3)} + 6x^{(2)}}\right)$.
8. Simplify
$$\binom{y+1}{n+1}\Delta^n u_{x+1} - \binom{y}{n+1}\Delta^n u_{x+1} - \binom{y}{n}\Delta^n u_x.$$
9. Find $\Delta^3((x-3)(x-5)(x-7))$.
10. Find $\Delta^3\binom{x}{8}$.
11. Find $\Delta^4\binom{x}{5}$ evaluated at $x = 0$.
12. Convert $x^4 - 2x^3 + 3x^2 + 5$ to a factorial polynomial.
13. Convert $x^4 - 6x^3 + 4x^2 - 4x + 5$ to a factorial polynomial.
14. Convert $x^4 - x^3 + x^2 - x + 1$ to a factorial polynomial.
15. Convert $x^5 - x^4 + x^3 - x^2 + x - 1$ to a factorial polynomial.
16. Convert $x^{(4)} - x^{(3)} + x^{(2)} - x + 1$ to a regular polynomial.
17. Convert $x^{(4)} + 3x^{(3)} - 4x^{(2)} - 2x + 1$ to a regular polynomial.
18. Convert $4x^{(4)} - 2x^{(3)} + 3x^{(2)} - 4x + 1$ to a regular polynomial.
19. Convert $x^{(5)} - x^{(4)} + x^{(3)} - x^{(2)} + x - 1$ to a regular polynomial.
20. Develop the algorithm for changing from factor polynomials to regular polynomials.
21. Find $\Delta^3\binom{8}{x}$ evaluated at $x = 0$.

11.5 Sums of Differences

We know that

$$\Delta^n(f(x)) = f(x+n) - n \cdot f(n-1) + \cdots + (-1)^k \binom{n}{k}(f(x+n-k)) + \cdots - f(x)$$

Therefore, if x is restricted to integers, any difference relation can be expressed as a recurrence relation. For example,

$$\Delta^3(f(x)) - 3\Delta^2(f(x)) = 2\Delta f(x)$$

can be changed to

$$f(x+3) - 3f(x+2) + 3f(x+1) + f(x) - 3(f(x+2) - 2f(x+1) + f(x))$$
$$= 2(f(x+1) - f(x))$$

so that

$$f(x+3) = 6f(x+2) - 7(f(x+1) - 4f(x)$$

which is in recursive form. Therefore, we can use the results of the first three sections of this chapter to solve this relation.

In this section, however, we take a different approach. We define an operator $\sum$, which may be thought of as the antidifference operator. Therefore, if $\Delta F(x) = f(x)$, then we want $\sum f(x) = F(X)$. Actually, we shall find that the best that we can do is to state that if $\Delta F(x) = f(x)$, then $\sum f(x) = F(X) + c$, where c is a constant. This is because $\Delta(F(x) + c) = \Delta F(x) + \Delta(c) = \Delta F(x)$. Again, for those who have had calculus, this is equivalent to the antiderivative or indefinite integral in calculus.

We must be careful, however, for we cannot assume that because $\Delta(F(x)) = 0$, then $F(x)$ is a constant. Consider, for example, $F(x) = \sin(2\pi x)$:

$$\Delta F(x) = \sin(2\pi(x+1)) - \sin(2\pi x)$$
$$= \sin(2\pi x + 2\pi) - \sin(2\pi x)$$
$$= \sin(2\pi x) - \sin(2\pi x)$$
$$= 0$$

This happens because $\sin(x)$ is a periodic function. However, we shall restrict our functions to polynomials, rational functions, and exponential functions. For all of these functions, if $\Delta(F(x)) = 0$, then $F(x)$ is a constant.

EXAMPLE 11.35 Using the difference formulas in the previous section, we find that if $f(x) = \dfrac{x^{(n+1)}}{n+1}$, then

$$\Delta f(x) = \Delta \frac{(x^{(n+1)})}{n+1}$$
$$= \frac{1}{n+1}\Delta(x^{(n+1)}) \quad \text{by Theorem 11.23}$$
$$= \frac{1}{n+1}(n+1)x^{(n)} \quad \text{since } \Delta(x^{(n)}) = nx^{(n-1)}$$
$$= x^{(n)}$$

so that
$$\sum x^{(n)} = \frac{x^{(n+1)}}{n+1} + c$$

Let
$$\Delta f(x) = 15x^{(4)} + 8x^{(3)} + 6x^{(2)} + 4x^{(1)} + 5$$

Then
$$f(x) = \frac{15x^{(5)}}{5} + \frac{8x^{(4)}}{4} + \frac{6x^{(3)}}{3} + \frac{4x^{(2)}}{2} + 5x + c$$
$$= 3x^{(5)} + 2x^{(4)} + 2x^{(3)} + 2x^{(2)} + 5x + c$$

We further show that if
$$f(x) = \frac{a^n}{a-1}$$

then
$$\Delta f(x) = \Delta\left(\frac{a^x}{a-1}\right)$$
$$= \frac{1}{a-1}\Delta a^x \qquad \text{by Theorem 11.23}$$
$$= \frac{1}{a-1}(a^x)(a-1) \qquad \text{shown in previous section}$$
$$= a^x$$

so that
$$\sum a^x = \frac{a^x}{a-1} + c$$

EXAMPLE 11.36 Let $\Delta f(x) = 15x^{(4)} + 3x^{(2)} + 2^x + 8 \cdot 5^x$. Then
$$f(x) = \frac{15x^{(5)}}{5} + \frac{3x^{(3)}}{3} + 2^x + \frac{8 \cdot 5^x}{5-1} + c$$
$$= 3x^{(5)} + x^{(3)} + 2^x + 2 \cdot 5^x + c$$

Also, let
$$f(x) = \binom{x}{n+1}, \quad \text{then} \quad \Delta f(x) = \binom{x}{n}$$

as shown in the previous section, so that
$$\sum \binom{x}{n} = \binom{x}{n+1} + c$$

EXAMPLE 11.37 Let $\Delta f(x) = 5x^{(4)} + 6 \cdot 4^x + 6 \cdot \binom{10}{5}$. Then
$$f(x) = x^{(5)} + 2 \cdot 4^x + 6 \cdot \binom{10}{6} + c$$

Finally, we consider $f(x) = \frac{1}{1-n}x^{(-n+1)}$:

$$\Delta f(x) = \Delta \frac{x^{(-n+1)}}{1-n}$$

$$= \frac{1}{1-n}\Delta x^{(-n+1)} \qquad \text{by Theorem 11.23}$$

$$= \frac{1}{1-n}(-n+1)x^{(-n+1-1)}$$

$$= x^{(-n)}$$

Therefore,

$$\sum x^{(-n)} = -\frac{x^{(-n+1)}}{-n+1} + c$$

EXAMPLE 11.38 Let $\Delta f(x) = x^{(2)} + 2^x + 5^x + x^{(-3)} + x^{(-7)}$. Then

$$f(x) = \frac{1}{3}x^{(3)} + 2^x + \frac{1}{4}5^x - \frac{1}{2}x^{(-2)} + \frac{1}{6}x^{(-6)} + c \qquad \blacksquare$$

We now consider one more tool, which is called **summation by parts**.

THEOREM 11.39 $\sum f(x)\Delta g(x) = f(x)g(x) - \sum E(g(x))\Delta f(x) + c.$

Proof From Theorem 11.25, we have

$$\Delta(fg) = f \cdot \Delta(g) + E(g) \cdot \Delta(f)$$

Therefore,

$$\sum \Delta(fg) = \sum f \cdot \Delta(g) + \sum E(g) \cdot \Delta(f)$$

so that

$$\sum f \cdot \Delta(g) = (fg) - \sum E(g) \cdot \Delta(f) \qquad \blacksquare$$

We now consider definite sums, which are similar in calculus to definite integrals. Let

$$f(x) = \Delta F(x) = F(x+1) - F(x)$$

Then

$$\sum_{x=1}^{n} f(x) = \sum_{x=1}^{n}(F(x+1) - F(x))$$

$$= F(n+1) - F(1)$$

We denote $F(n+1) - F(1)$ by $F(x)|_1^{n+1}$.

EXAMPLE 11.40 Let $f(x) = x^{(3)} + 3^x + x^{(-4)}$. Then

$$\sum_{x=1}^{7} f(x) = \sum_{x=1}^{7} x^{(3)} + 3^x + x^{(-4)}$$

$$= \left(\frac{x^{(4)}}{4} + \frac{1}{2} 3^x - \frac{1}{3} x^{(-3)} \right) \Big|_{1}^{8}$$

$$= \frac{8^{(4)}}{4} + \frac{1}{2} 3^8 - \frac{1}{3} 8^{(-3)} - \left(\frac{1^{(4)}}{4} + \frac{1}{2} 3 + \frac{1}{3} 3^{(-3)} \right)$$

EXAMPLE 11.41 Find the sum of $1^2 + 2^2 + 3^2 + \cdots + n^2$.

$$1^2 + 2^2 + 3^2 + \cdots + n^2 = \sum_{x=1}^{n} x^2$$

$$= \sum_{x=1}^{n} x^{(2)} + x$$

$$= \left(\frac{x^{(3)}}{3} + \frac{x^{(2)}}{2} \right) \Big|_{1}^{n+1}$$

$$= \frac{1}{6} \left(2 x^{(3)} + 3 x^{(2)} \right) \Big|_{1}^{n+1}$$

$$= \frac{1}{6} \left(2 x^3 - 3 x^2 + x \right) \Big|_{1}^{n+1}$$

$$= \frac{1}{6} (2(n+1)^3 - 3(n+1)^2 + (n+1)) - 0$$

$$= \frac{1}{6} n(n+1)(2n+1)$$

EXAMPLE 11.42

$$\sum_{x=0}^{n} r^x = \frac{r^x}{r-1} \Big|_{0}^{n+1} = \frac{r^{n+1} - 1}{r - 1}.$$

EXAMPLE 11.43

$$\sum_{x=1}^{n} x = \frac{x^{(2)}}{2} \Big|_{1}^{n+1} = \frac{x(x-1)}{2} \Big|_{1}^{n+1} = \frac{(n+1)n}{2}.$$

EXAMPLE 11.44 As mentioned earlier,

$$\Delta(fg) = f \cdot \Delta(g) + E(g) \cdot \Delta(f)$$

Therefore,

$$\sum_{x=1}^{n} \Delta(f(x)g(x)) = \sum_{x=1}^{n} f(x) \cdot \Delta(g(x)) + \sum_{x=1}^{n} E(g(x)) \cdot \Delta(f(x))$$

or

$$(f(x)g(x))|_{1}^{n+1} = \sum_{x=1}^{n} f(x) \cdot \Delta(g(x)) + \sum_{x=1}^{n} g(x+1) \cdot \Delta(f(x))$$

Therefore,

$$\sum_{x=1}^{n} f(x) \cdot \Delta(g(x)) = (f(x)g(x))\Big|_{1}^{n+1} - \sum_{x=1}^{n} g(x+1) \cdot \Delta(f(x))$$

EXAMPLE 11.45

$$\sum_{x=1}^{n} x 3^x = \sum_{x=1}^{n} x \Delta \frac{3^x}{2}$$

$$= \frac{x 3^x}{2}\Big|_{1}^{n+1} - \sum_{x=1}^{n} \frac{3^{x+1}}{2} \cdot \Delta x$$

$$= \frac{(n+1)3^{n+1}}{2} - \frac{3}{2} - \left(\frac{3^{x+1}}{4}\right)\Big|_{1}^{n+1}$$

$$= \frac{(n+1)3^{n+1}}{2} - \frac{3}{2} - \left(\left(\frac{3^{n+2}}{4}\right) - \frac{9}{4}\right)$$

$$= \frac{(2n-1)3^{n+1} + 3}{4}$$

Exercises

1. Find the sum of the first differences of the function $3^x - x^3 + x^2 - x + 5$ from $x = 1$ to $x = 9$.

2. Find the sum of the first differences of the function $2^x - 3x^3 + 2x^2$ from $x = 1$ to $x = 5$.

3. Find $\sum_{x=4}^{8} 2^x - 12x^{(3)}$.

4. Find $\sum_{x=6}^{10} 3^x + 10x^{(4)} - 8x^{(3)}$.

5. Find $\sum_{x=4}^{7} \binom{x}{2} + 4^x + 12x^{(5)}$.

6. Find $\sum_{x=6}^{11} \binom{x}{4} + 3 \cdot 5^x + 14x^{(6)}$.

7. Show that $\sum_{i=1}^{n} 2i = n(n+1)$.

8. Show that $\sum_{i=1}^{n} i^3 = \frac{n^2(n+1)^2}{4}$.

9. Show that $\sum_{i=1}^{n} i(i+1) = \frac{n(n+1)(n+2)}{3}$.

10. Show that $\sum_{i=1}^{n} 4i - 2 = 2n^2$.

11. Using summation by parts find $\sum_{x=1}^{8} (2x+3)3^x$.

12. Using summation by parts find $\sum_{x=2}^{10} 4(3x-6)4^x$.

13. Using summation by parts find $\sum_{x=0}^{8} (x)(x-1)2^x$.

14. Using summation by parts find $\sum_{x=0}^{8} (x)(x-1)3^x$.

15. Using summation by parts find $\sum_{x=0}^{8} x^2 2^x$.

16. Using summation by parts find $\sum_{x=0}^{10} x^{(2)} 2^x$.

17. Using summation by parts find $\sum_{x=0}^{10} x^{(4)} 4^x$.

18. Using summation by parts find $\sum_{x=0}^{9} x^{(4)} x^{(6)}$.

19. Using summation by parts find $\sum_{x=0}^{9} x^{(2)} x^{(8)}$.

20. Prove that $\Delta \sum_{i=c}^{x} f(i) = f(x+1)$.

21. Prove that $\Delta \sum_{i=x}^{c} f(i) = -f(x)$.

GLOSSARY

Chapter 11: Recursion Revisited

Characteristic polynomial (11.1)	The equation $r^n = c_1 r^{n-1} + c_2 r^{n-2} \ldots + c_{k-1} r + c_k$ is called the **characteristic polynomial** of $a_n = c_1 a_{n-1} + c_2 a_{n-2} \ldots + c_{k-1} a_{n-(k-1)} + c_k a_{n-k}$.
De Moivre's Theorem (11.1)	For a given angle θ, $(\cos(\theta) + i \sin(\theta))^k = \cos(k\theta) + i \sin(k\theta)$.
Factorial polynomials (11.4)	If $f(x) = a_n x^{(n)} + a_{n-1} x^{(n-1)} + \cdots + a_2 x^{(2)} + a_1 x + a_0$ for some n, then f is called a **factorial polynomial**.
First difference of a function f (11.3)	The **first difference of a function** f, denoted by Δf, is defined by $\Delta f(x) = f(x+1) - f(x)$.
Linear homogeneous recurrence relation of order p (11.1)	A linear recursive relation of the form $a_n = c_1 a_{n-1} + c_2 a_{n-2} + c_3 a_{n-3} + \cdots + c_p a_{n-p}$ is called a **linear homogeneous recurrence relation of order p**.
Linear homogeneous recurrence relation with constant coefficients of order p (11.1)	A linear recursive relation of the form $a_n = c_1 a_{n-1} + c_2 a_{n-2} + c_3 a_{n-3} + \cdots + c_p a_{n-p}$, $c_p \neq 0$ with constants c_i for $1 \leq i \leq p$ is called a **linear homogeneous recurrence relation with constant coefficients of order p**.
Linear recurrence relation of order p (11.1)	A recursive relation of the form $a_n = c_1(n) a_{n-1} + c_2(n) a_{n-2} + c_3(n) a_{n-3} + \cdots + c_p(n) a_{n-p} + f(n)$ is called a **linear recurrence relation of order p**.
Linear recurrence relation with constant coefficients of order p (11.1)	A linear recursive relation of the form $a_n = c_1 a_{n-1} + c_2 a_{n-2} + c_3 a_{n-3} + \cdots + c_p a_{n-p} + f(n)$, $c_p \neq 0$ with constants c_i for $1 \leq i \leq p$ is called a **linear recurrence relation with constant coefficients of order p**.
Operator (11.3)	An **operator** maps functions to functions.
Product rule (11.3)	$\Delta(fg) = f \cdot \Delta(g) + E(g) \cdot \Delta(f)$
Quotient rule (11.3)	$\Delta\left(\dfrac{f}{g}\right) = \dfrac{g \cdot \Delta(f) - f \cdot (\Delta(g))}{g \cdot (E(g))}$
Second difference of a function f (11.3)	The **second difference of a function** f, denoted by $\Delta^2 f$, is defined by $\Delta^2 f(x) = \Delta(\Delta f(x))$.
Stirling numbers of the first kind (11.4)	**Stirling numbers of the first kind** are defined by $$s_0^{(n)} = 0 \text{ for all } n \geq 1$$ $$s_n^{(n)} = 1 \text{ for all } n \geq 0$$ $$s_k^{(n+1)} = s_{k-1}^{(n)} + n s_k^{(n)}$$
Stirling numbers of the second kind (11.4)	**Stirling numbers of the second kind** are defined by $$S_0^{(n)} = 0 \text{ for all } n \geq 1$$ $$S_n^{(n)} = 1 \text{ for all } n \geq 0$$ $$S_k^{(n+1)} = S_{k-1}^{(n)} + k S_k^{(n)}$$
Summation by parts	$\sum f(x) \Delta g(x) = f(x) g(x) - \sum E(g(x)) \Delta f(x) + c$

TRUE-FALSE QUESTIONS

1. $a_n = 3a_{n-1}^2 - 6a_{n-2} + \sin(n)$ is a recurrence relation.

2. $a_n = 3a_{n-1}^2 - 6a_{n-2} + 3a_{n-3}$ is a linear homogeneous recurrence relation.

3. $a_n = a_{n-1} + 25a_{n-2} + 3$ is a linear homogeneous recurrence relation.

4. $a_n = n^2 a_{n-1} + 2na_{n-2} + 3\sin(n)$ is a linear recurrence relation.

5. $r^2 - 3r + 5$ is the characteristic polynomial for the relation $a_n = -3a_{n-1} + 5a_{n-2}$.

6. $a_n = 3 \cdot 3^n - 5n3^n$ is a solution of the recursive function
$$a_0 = 3$$
$$a_1 = -6$$
$$a_n = -6a_{n-1} - 9a_{n-2}$$

7. The recurrence relation
$$a_n = a_{n-1} - a_{n-2}$$
has no solutions of the form $c_1 a^n + c_2 b^n$ where a and b are real numbers.

8. The recurrence relation
$$a_n = a_{n-1} + a_{n-2}$$
has no solutions of the form $c_1 a^n + c_2 b^n$ where a and b are rational numbers.

9. The recurrence relation $a_n = -2\sqrt{2}a_{n-1} - 4a_{n-2}$ has a general solution of the form
$$2^n \left(k_1 \cos\left(n \cdot \frac{\pi}{4}\right) + k_2 \cos\left(n \cdot \frac{\pi}{4}\right) \right)$$

10. The recurrence relation $a_n = 4a_{n-1} - 5a_{n-2} + 2a_{n-3}$ has a general solution of the form
$$a_n = k_1 + k_2 n + k_3 2^n$$

11. The recurrence relation $a_n = 6a_{n-1} - 12a_{n-2} + 8a_{n-3}$ has a general solution of the form
$$a_n = k_1 2^n + k_2 n 2^n + k_3 n^2 2^n$$

12. The recurrence relation $a_n = 3a_{n-1} - 2a_{n-2} + 3\sin\left(\frac{n\pi}{2}\right)$ has a general solution of the form
$$a_n = k_1 + k_2 2^n + C \cdot \sin\left(\frac{n\pi}{2}\right)$$
where C is a fixed constant.

13. The recurrence relation $a_n = 3a_{n-1} + n$ has a general solution of the form
$$a_n = k_1 3^n + C \cdot n$$
where C is a fixed constant.

14. The recurrence relation $a_n = a_{n-1} + n^2$ has a general solution of the form
$$a_n = k_1 + C \cdot n^2$$
where C is a fixed constant.

15. The recurrence relation $a_n = a_{n-1} + n^2 + n$ has a solution of the form
$$a_n = k_1 \cdot n^3 + k_2 \cdot n^2 + k_3 \cdot n + k_4$$

16. $E = I + \Delta$. 17. $E\Delta = \Delta E$. 18. $\Delta x^n = nx^{n-1}$.

19. $\Delta(fg) = f \cdot \Delta(g) + g \cdot \Delta f$. 20. $\Delta(5^x) = 4 \cdot 5^x$.

21. $E(a) = a$. 22. $\Delta(a) = a$.

23. $\Delta x^{(n)} = nx^{(n-1)}$. 24. $x^{(1)} = x$.

25. $x^{(4)} - 2x^{(3)} + 4x^{(2)} + 6x + 8 = x^4 - 8x^3 + 21x^2 + 6x + 8$.

26. $\Delta \binom{x}{n} = \binom{x-1}{n-1}$.

27. If $x^{(n)}$ is expanded as a regular polynomial, the coefficients of the polynomial are Stirling numbers of the first kind.

28. If x^n is expanded as a factorial polynomial, the coefficients of the polynomial are Stirling numbers of the first kind.

29. $\binom{x}{n}$ is defined only if x is an integer.

30. $\binom{n}{x}$ is defined only if x is an integer.

31. $x^{(n)}$ is defined only if x is an integer.

32. $x^{(n)}$ is defined only if n is an integer.

33. $x^{(n)}$ is defined only if n is a nonnegative integer.

34. $\sum 2^x = 2^x + C$.

35. $\sum x^5 = \frac{1}{6}x^6 + C$.

36. $\sum \binom{x}{5} = \binom{x}{4} + C$.

37. $\sum_{i=1}^{n} \Delta f(x) = f(n) - f(1)$.

38. $\sum_{i=1}^{20} 2^x = 2^{21} - 2$.

39. $x^{(n)} \cdot x^{(m)} = x^{(n+m+1)}$.

40. $\sum x^{(n)} \cdot 2^x = x^{(n)} \cdot 2^x - \sum nx^{(n-1)} \cdot 2^x + C$.

SUMMARY QUESTIONS

1. Solve the recursive function
   $a_0 = 1$
   $a_n = 3a_{n-1}$ for $n > 0$

2. Solve the recursive function
   $a_0 = 1$
   $a_1 = 2$
   $a_n = 3a_{n-1} + 4a_{n-2}$ for $n > 1$

3. Solve the recursive function
   $a_0 = 1$
   $a_1 = 3$
   $a_n = +2a_{n-1} + a_{n-2}$ for $n > 1$

4. $a_0 = 1$
   $a_1 = 4$
   $a_n = -a_{n-1} + a_{n-2}$ for $n > 1$

5. Solve the recursive function
   $a_0 = 1$
   $a_1 = 4$
   $a_n = a_{n-2}$ for $n > 1$

6. Find the general solution for the recurrence relation
   $a_n = a_{n-1} + \sin\left(\frac{n\pi}{2}\right)$

7. $a_n = 4a_{n-1} + 2^n$

8. $a_n = a_{n-1} + n$

9. $a_n = 2a_{n-1} + n$

10. $a_n = -a_{n-2} + 2^n$

11. Find $\Delta^2 2^x$ evaluated at $x = 1$.

12. Find $\Delta((x+1)(2x-3))$.

13. Change $x^{(3)} + 2x^{(2)} + x + 1$ to a regular polynomial.

14. Change $x^3 - 7x^2 + x + 1$ to a factorial polynomial.

15. Find $\sum_{i=1}^{5} x^{(3)} + 3^x$.

ANSWERS TO TRUE-FALSE

1. T 2. F 3. F 4. T 5. F 6. T 7. T 8. T 9. F 10. T
11. T 12. F 13. T 14. F 15. T 16. T 17. T 18. F 19. F
20. T 21. T 22. F 23. T 24. T 25. F 26. F 27. T 28. F
29. F 30. T 31. F 32. T 33. F 34. T 35. F 36. F 37. F
38. T 39. F 40. F

CHAPTER 12

Counting Continued

12.1 Occupancy Problems

In this section, we are concerned with the number of ways to place objects in a box or urn. We note that we have already begun our task. In Section 8.3 we showed that, given n distinguishable objects, the number of ways to place m of them in one box and $n - m$ in another box is equal to

$$C(n, m) = \frac{n!}{m!(n-m)!}$$

More generally, we have shown that, given n distinguishable objects and k distinguishable boxes, the number of ways to place n_i objects in the ith box for all $1 \leq i \leq k$, where $n = n_1 + n_2 + n_3 + \cdots + n_k$, is given by

$$C(n; n_1, n_2, n_3, \cdots, n_k) = \frac{n!}{n_1! n_2! n_3! \cdots n_k!}$$

We note that in both of the preceding cases, a box may be empty. Also note that both the objects and the boxes are distinguishable.

We now consider the number of ways to place n distinguishable objects in k distinguishable boxes, where some of the boxes may be empty. Note that we do not state the number of objects to go in each box. For each object, there are k choices of boxes in which to place the object. Therefore, there are k choices for each of the n objects. By the multiplication counting principle, there are k^n possible ways to place the objects in the boxes. Note that this is also the number of functions from a set S with n objects to a set T with k objects. If one identifies mapping an element s in S to an object t in T with placing object s in box t, it is easy to see why we get the same result.

THEOREM 12.1 There are k^n ways to place n distinguishable objects in k distinguishable boxes, where some of the boxes may be empty. ∎

If none of the boxes may be empty, the problem becomes surprisingly more difficult. Therefore, we shall delay this problem momentarily.

We have already shown in Section 8.6 that the number of ways to place n indistinguishable objects in k distinguishable boxes when one or more of the boxes may be empty is given by

$$C(n + k - 1, n) = C(n + k - 1, k - 1) = \frac{(n + k - 1)!}{n!(k - 1)!}$$

We have also shown that the number of ways to place n indistinguishable objects in k distinguishable boxes when none of the boxes may be empty is given by

$$C(n-1, n-k) = C(n-1, k-1) = \frac{(n-1)!}{(n-k)!(k-1)!}$$

We now count the number of ways of putting n distinguishable balls into k indistinguishable boxes. We first consider the case where none of the boxes may be empty. *This is also the number of ways of partitioning a set of n objects into k disjoint subsets.*

THEOREM 12.2 There are $S_k^{(n)}$ ways of putting n distinguishable balls into k indistinguishable boxes where none of the boxes may be empty, and where $\{S_k^{(n)} : 0 \leq k \leq n\}$ is the set of **Stirling numbers of the second kind**.

Proof We must show that if $f(n, k)$ is the number of ways of putting n distinguishable balls into k indistinguishable boxes where none of the boxes may be empty, then $f(n, n) = 1$ for all $n \geq 0$, $f(n, 0) = 0$ for all $n \geq 1$, and $f(n+1, k) = f(n, k-1) + kf(n, k)$. For $n \geq 1$, there is obviously only one way to put n balls in n boxes so that no box is empty. Therefore, $f(n, n) = 1$ for all $n \geq 1$. The case where $n = 0$ is a bit tricky, but it seems reasonable to assume that there is only one way to put 0 objects in 0 boxes. Therefore, we let $f(0, 0) = 1$.

For $n \geq 1$, there is obviously no way in which to place 0 objects in n boxes and leave no box empty. Thus, $f(n, 0) = 0$ for all $n \geq 1$.

To show that $f(n+1, k) = f(n, k-1) + kf(n, k)$, assume that we have placed n objects in k boxes and now wish to place the $n+1$st object in a box. There are two possibilities. Place the object in a box by itself, or place the object in a box that is already occupied. If we place the object in a box by itself, then the n objects have all been placed in the other boxes. Since there are $k-1$ of these boxes, n objects were placed in $k-1$ boxes. There are $f(n, k-1)$ ways in which this could have occurred. If we place the $n+1$st object in a box already occupied, then n objects were placed in k boxes. This can be done in $f(n, k)$ ways. The $n+1$st object can then be placed in any of the k boxes, so there are $k \cdot f(n, k)$ ways in which the balls can be placed in the boxes if all k boxes were already occupied. Therefore, there are $f(n, k-1) + kf(n, k)$ ways to place $n+1$ objects in k boxes, and

$$f(n+1, k) = f(n, k-1) + kf(n, k)$$

Therefore, $f(n, k) = S_k^{(n)}$ where $\{S_k^{(n)} : 0 \leq k \leq n\}$ is the set of Stirling numbers of the second kind. ∎

We can generate a table of Stirling numbers of the second kind similar to the Pascal triangle, using $S_k^{(n+1)} = S_{k-1}^{(n)} + kS_k^{(n)}$ as follows:

Triangle for Stirling numbers of the second kind

n	$S_0^{(n)}$	$S_1^{(n)}$	$S_2^{(n)}$	$S_3^{(n)}$	$S_4^{(n)}$	$S_5^{(n)}$	$S_6^{(n)}$	$S_7^{(n)}$	$S_8^{(n)}$	$S_9^{(n)}$	$S_{10}^{(n)}$
0	1										
1	0	1									
2	0	1	1								
3	0	1	3	1							
4	0	1	7	6	1						
5	0	1	15	25	10	1					
6	0	1	31	90	65	15	1				
7	0	1	63	301	350	140	21	1			
8	0	1	127	966	1701	1050	266	28	1		
9	0	1	255	3025	7770	6951	2646	462	36	1	
10	0	1	511	9330	34105	42525	22827	5880	750	45	1

THEOREM 12.3 We now remove the restriction that none of the boxes be empty. The possibilities are (1) all n objects are placed in one box, which can be done in $S_1^{(n)}$ ways; (2) all n objects are placed in two boxes, which can be done in $S_2^{(n)}$ ways,...; (k) all n objects are placed in k boxes, which can be done in $S_k^{(n)}$ ways,.... Adding these possibilities we have

$$S_1^{(n)} + S_2^{(n)} + S_3^{(n)} + \cdots + S_k^{(n)}$$

ways of placing n objects in k boxes where boxes may be empty. ■

Using the fact that $S_k^{(n)}$ is the number of ways that n distinguishable objects may be placed in k indistinguishable boxes when none of the boxes may be empty, we are now able to determine the number of ways that n distinguishable objects may be placed in k distinguishable boxes when none of the boxes may be empty. Essentially, we are changing from unordered boxes to ordered boxes. In changing from ordered boxes to unordered boxes, we would divide the number of ordered boxes by $k!$, as we did in changing from permutations to combinations. In changing from unordered boxes to ordered boxes, we therefore multiply by $k!$, so that there are $k!S_k^{(n)}$ ways that n distinguishable objects may be placed in k distinguishable boxes when none of the boxes may be empty.

Previously, we have counted the number of functions from a set S, with n objects, to a set T, with k objects. We have also counted the number of one-to-one functions from a set S, with n objects to a set T, with k objects. We can now count the number of onto functions from a set S, with n objects to a set T, with k objects. If we think of the k objects in T as boxes, then we can identify mapping an object s in S to an element t in T with placing object s in box t. Since the function is onto, no box will be empty. *Therefore, there are $k!S_k^{(n)}$ onto functions.*

We can now form the following table:

Placing of n objects in k boxes	Number of ways
(1) objects-distinguishable, boxes-distinguishable, n_i objects in the ith box	$C(n; n_1, n_2, n_3, \cdots, n_k)$
(2) objects-distinguishable, boxes-distinguishable, boxes may be empty	k^n
(3) objects-distinguishable, boxes-distinguishable, boxes may not be empty	$k!S_k^{(n)}$
(4) objects-indistinguishable, boxes-distinguishable, boxes may be empty	$C(n+k-1, n)$
(5) objects-indistinguishable, boxes-distinguishable, boxes may not be empty	$C(n-1, k-1)$
(6) objects-distinguishable, boxes-indistinguishable, boxes may be empty	$S_1^{(n)} + S_2^{(2)} + \cdots + S_k^{(n)}$
(7) objects-distinguishable, boxes-indistinguishable, boxes may not be empty	$S_k^{(n)}$

Having completed our round with objects and boxes, we now consider the number of ways of dividing n objects into k cycles, where no two of the k cycles have an element in common. A cycle of m objects has the form $a_1 a_2 a_3 \cdots a_m$, where a_m is considered to be next to a_1 so that

$$a_1 a_2 a_3 \cdots a_j \cdots a_m = a_j \cdots a_m a_1 a_2 a_3 \cdots a_{j-1}$$

Therefore,

$$a_1 a_2 a_3 = a_3 a_1 a_2 = a_2 a_3 a_1$$

We may picture an n-cycle as n people seated around a circular table. In Chapter 8 we found that there are $(m-1)!$ ways to seat m people around a circular table. Therefore, given m objects, there are $(m-1)!$ ways to create an m-cycle.

We now want to find the number of possible ways of dividing n objects into k cycles, where no two of the k cycles have an element in common.

THEOREM 12.4 The number of possible ways of dividing n objects into k cycles, where no two of the k cycles have an element in common is equal to $s_k^{(n)}$, where $\{s_k^{(n)} : 0 \leq k \leq n\}$ is the set of **Stirling numbers of the first kind**.

Proof Let $f(n, k)$ be the number of possible ways of dividing n objects into k cycles, where no two of the k cycles have an element in common. To show that $f(n, k) = s_k^{(n)}$, we need first to show that $f(n, 0) = 0$ for all $n \geq 1$. But if $n \geq 1$, it is impossible to form 0 cycles and use all n objects. Therefore, $f(n, 0) = 0$ for all $n \geq 1$. We next show that $f(n, n) = 1$ for all $n \geq 0$. If $n \geq 1$, we are creating n 1 cycles. There is obviously one way to do this since the order that the cycles appear is unimportant. If $n = 0$, it seems somewhat reasonable to consider that there is only one way to create no cycles using no objects. Simply don't do it. Therefore, $f(n, n) = 1$ for all $n \geq 0$. Next we have to show that

$$f(n+1, k) = f(n, k-1) + nf(n, k)$$

Every arrangement of $n + 1$ objects either places the $n + 1$st object in a cycle by itself, or it places it in a cycle with other objects. If it places the $n + 1$st object in a cycle by itself, then the other n objects are placed in $k - 1$ cycles. There are $f(n, k-1)$ ways of doing this. If the $n + 1$st object is placed in one of the other cycles, say one containing m objects, then the $n + 1$st object can be placed after any of the m objects in the cycle. Therefore, there are m different cycles that can be formed if the $n + 1$st object is placed in a cycle, containing m objects. Assume that there are cycles $c_1, c_2, \ldots, c_k$ where c_i is an n_i cycle and $n_1 + n_2 + \cdots + n_k = n$. Then if a_{n+1} is placed in cycle c_i, it creates n_i new cycles. Therefore, there are $n_1 + n_2 + \cdots + n_k = n$ new cycles that can be created from a fixed set of k cycles containing n elements by adding the $n + 1$st element. Since there are $f(n, k)$ ways of placing n element in k cycles, there are $nf(n, k)$ ways of creating k cycles with $n + 1$ objects if the $n + 1$st object is added to a cycle. Therefore, there are

$$f(n, k-1) + nf(n, k)$$

ways of creating k cycles with $n + 1$ objects and $f(n, k) = s_k^{(n)}$. ∎

We can generate a table of Stirling numbers of the first kind in a manner that is similar to those of the table of Stirling numbers of the second kind, using $s_k^{(n+1)} = s_{k-1}^{(n)} + ns_k^{(n)}$ as follows.

Triangle for Stirling numbers of the first kind

n	$s_0^{(n)}$	$s_1^{(n)}$	$s_2^{(n)}$	$s_3^{(n)}$	$s_4^{(n)}$	$s_5^{(n)}$	$s_6^{(n)}$	$s_7^{(n)}$	$s_8^{(n)}$	$s_9^{(n)}$	$s_{10}^{(n)}$
0	1										
1	0	1									
2	0	1	1								
3	0	2	3	1							
4	0	6	11	6	1						
5	0	24	50	35	10	1					
6	0	120	274	225	85	15	1				
7	0	720	1764	1624	735	175	21	1			
8	0	5040	13068	13132	6769	1960	322	28	1		
9	0	40320	109584	118124	67284	22449	4536	546	36	1	
10	0	362880	1026576	1172700	723680	269325	63273	9450	870	45	1

Exercises

How many ways can we place 7 objects in 3 boxes if

1. The objects are distinguishable, the boxes are distinguishable, and the boxes may be empty?
2. The objects are distinguishable, the boxes are distinguishable, and the boxes may not be empty?
3. The objects are indistinguishable, the boxes are distinguishable, and the boxes may be empty?
4. The objects are indistinguishable, the boxes are distinguishable, and the boxes may not be empty?
5. The objects are distinguishable, the boxes are indistinguishable, and the boxes may be empty?
6. The objects are distinguishable, the boxes are indistinguishable, and the boxes may not be empty?

How many ways can we place 10 objects in 4 boxes if

7. The objects are distinguishable, the boxes are distinguishable, and the boxes may be empty?
8. The objects are distinguishable, the boxes are distinguishable, and the boxes may not be empty?
9. The objects are indistinguishable, the boxes are distinguishable, and the boxes may be empty?
10. The objects are indistinguishable, the boxes are distinguishable, and the boxes may not be empty?
11. The objects are distinguishable, the boxes are indistinguishable, and the boxes may be empty?
12. The objects are distinguishable, the boxes are indistinguishable, and the boxes may not be empty?
13. Complete the next two rows of the triangle for Stirling numbers of the second kind.
14. Complete the next two rows of the triangle for Stirling numbers of the first kind.

How many ways can we place 12 objects in 4 boxes if

15. The objects are distinguishable, the boxes are distinguishable, and the boxes may be empty?
16. The objects are distinguishable, the boxes are distinguishable, and the boxes may not be empty?
17. The objects are indistinguishable, the boxes are distinguishable, and the boxes may be empty?
18. The objects are indistinguishable, the boxes are distinguishable, and the boxes may not be empty?
19. The objects are distinguishable, the boxes are indistinguishable, and the boxes may be empty?
20. The objects are distinguishable, the boxes are indistinguishable, and the boxes may not be empty?

How many ways can we place 14 objects in 6 boxes if

21. The objects are distinguishable, the boxes are distinguishable, and the boxes may be empty?
22. The objects are distinguishable, the boxes are distinguishable, and the boxes may not be empty?
23. The objects are indistinguishable, the boxes are distinguishable, and the boxes may be empty?
24. The objects are indistinguishable, the boxes are distinguishable, and the boxes may not be empty?
25. The objects are distinguishable, the boxes are indistinguishable, and the boxes may be empty?
26. The objects are distinguishable, the boxes are indistinguishable, and the boxes may not be empty?

Let $S(n,k) = \frac{1}{k!} \sum_{i=0}^{k} (-1)^i \binom{k}{i} (k-i)^n$.

27. Show $S(n, 0) = 0$ for all $n \geq 1$.
28. Show $S(n, k)$ satisfies the recursive relation $S(n+1, k) = S(n, k-1) + kS(n, k)$.

12.2 Catalan Numbers

At this point we look at three problems and see that, if viewed appropriately, they are the same problem.

■ 12.2.1 Problem 1

A man goes to work 10 blocks east and 10 blocks north of where he lives. There is a river running diagonally from where he lives to where he walks. Unfortunately there are no bridges across the river between his home and work so that when he reaches the diagonal he also reaches the river. Hence, he can reach the diagonal but must then turn right. If the streets are laid out in square block grids so there are north-south streets every block and east-west streets every block, how many ways can he get to work without going out of the way or getting wet? Examples of routes that he can take are given in Figure 12.1.

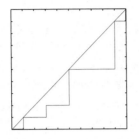
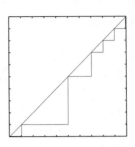
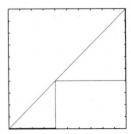

Figure 12.1

If we denote his route by using an E when he goes a block east and an N when he goes a block north, then his first route is denoted

ENEENEENNNEEEENNNNEN

His second route is denoted

ENEEEENNNNEENNENENEN

and his third route is denoted

EEEENNNNEEEEEENNNNNN

Notice that in each case the route must start with an E and, at any point along the route, there must be as many Es as Ns or he goes into the river.

■ 12.2.2 Problem 2

In how many ways can a sequence of ten 1's and ten -1's be added together so that if S_k is the sum of the first k elements in the sequence, then $S_k \geq 0$ for all $1 \leq k \leq 20$? Possible sums are

$$1, -1, 1, 1, -1, 1, 1, -1, -1, -1, 1, 1, 1, 1, -1, -1, -1, -1, 1, -1$$

$$1, -1, 1, 1, 1, 1, -1, -1, -1, -1, 1, 1, -1, -1, 1, -1, 1, -1, 1, -1$$

and

$$1, 1, 1, 1, -1, -1, -1, -1, 1, 1, 1, 1, 1, 1, -1, -1, -1, -1, -1, -1$$

Again we see that at any point along any of the sequences, there must be as many 1's as -1's.

■ 12.2.3 Problem 3

Given 11 numbers with binary operations between them, how many different ways can 10 pairs of parentheses be inserted? This seems to be the same type of problem since, as we go from left to right, we must have at least as many left parentheses as right parentheses. But the problem is where to put them. It turns out that it is better to use left parentheses and the first 10 numbers rather than left and right parentheses, since there can never be more numbers than left parentheses. For example, consider

$$3 + 4 - 8 \times 2 \div 4 - 7 + 4 \times 5\hat{\,}6 - 2 \times 4$$

Corresponding to

$$1, -1, 1, 1, -1, 1, 1, -1, -1, -1, 1, 1, 1, 1, -1, -1, -1, -1, 1, -1$$

we place a left parenthesis when there is a 1 and a letter when there is a -1 so that we have

$$(3 + ((4 - ((8 \times 2) \div 4)) - ((((7 + 4) \times 5)\hat{\,}6) - (2 \times 4))))$$

The left parentheses are put in to pair up with the right parentheses. Again consider

$$1, -1, 1, 1, 1, 1, -1, -1, -1, -1, 1, 1, -1, -1, 1, -1, 1, -1, 1, -1$$

which corresponds to

$$(3 + ((((4 - 8) \times 2) \div 4) - ((7 + 4) \times (5\verb|^|6 - (2 \times 4))))))$$

and finally

$$1, 1, 1, 1, -1, -1, -1, -1, 1, 1, 1, 1, 1, 1, -1, -1, -1, -1, -1, -1$$

which corresponds to

$$((((3 + 4) - 8) \times 2) \div (((((4 - 7) + 4) \times 5)\verb|^|6) - 2) \times 4))$$

We have now stalled long enough. It is time to solve the problem. We first solve it for $n = 10$ since we have beaten this example to death and then find a formula for arbitrary n. Let us return to the man walking to work. First, consider all ways that he could walk to work wet or dry as long as he does not go out of his way. He must walk 10 blocks north and 10 blocks east in some order. Since we are considering all cases, including wading the river, the 10 Ns and 10 Es can be in any order. If we consider the string of 10 Ns and 10 Es as consisting of 20 blanks, then we must choose 10 and place Es in them. Thus, the total number of ways to walk to work is the number of ways that 10 blanks can be chosen out of 20 to place the Es. But this is $C(10, 20)$ or $\binom{20}{10} = \frac{20!}{10!10!}$. We are now going to use a common trick in counting. We are going to find the number of ways that our hero gets wet and subtract them from the total. Put perhaps more clearly, we find the number of ways that do not work and subtract them from the total number to find the number of ways that our hero can safely get to work. Our hero ends up in the river if at any point along his route he goes more blocks north than east. In other words, at some point he has one more N than E and into the river he goes. For example, in the string

$$EENNNEENENENEENNNENE \qquad (12.1)$$

the N indicates the point where he goes into the river. At the time he hits the river we have used 2 Es and 3 Ns, leaving 8 Es and 7 Ns. We are now going to change each E in the string after N to an N and each N to an E, so that we have

$$EENNNNNENENENNEEENEN \qquad (12.2)$$

In the new path (which does not get him to the office) there are 2 Es and 3 Ns followed by 8 Ns and 7 Es for a total of 9 Es and 11 Ns. This makes sense because there was one more N when he hit the river so there was one more E in the rest of the original path. When we changed the rest of the original path, there was one more N in the rest of the new path. Thus, there are 2 more Ns than Es. Let's try it again. In the string

$$ENENENENNNENENEENENNE \qquad (12.3)$$

he goes in the river when there are 5 Ns and 4 Es. After N, we have 5 Ns and 6 Es. We again change each E in the string after N to an N and each N to an E, so that we have 6 Ns and 5 Es after N:

$$ENENENENNNENENNNENEEN \qquad (12.4)$$

The new string, therefore, has a total of 9 Es and 11 Ns.

Conversely, if we have a string with 9 Es and 11 Ns, then at some point in the string, there must be more Ns than Es. We select the first such point and change the N where this occurs into N. We again change the rest of the string, replacing each N with an E and each E with an N. For example, given the string

$$ENENEENENENNNENENNNE \tag{12.5}$$

which has 9 Es and 11 Ns, we mark with $\mathbf{N}$ the first place where there are more Ns than Es. At this point, there are 6 Es and 7 Ns in the string up to and including $\mathbf{N}$. The remainder of the string has 3 Es and 4 Ns. We change each N with an E and each E with an N after $\mathbf{N}$ to get

$$ENENEENENEN\mathbf{N}NENEENN \tag{12.6}$$

and we have 4 Es and 3 Ns after $\mathbf{N}$. There is a total of 10 Es and 10 Ns. This is again a path that lands our hero in the river, and it occurs at point $\mathbf{N}$. This makes sense since we started with 9 Es and 11 Ns and the string up to and including $\mathbf{N}$ has one more N than E. Therefore, the remainder of the string, after $\mathbf{N}$, also has one more N than E. When we exchange each N with an E and each E with an N after $\mathbf{N}$, we get one more E than N. Therefore, in the entire new string we will always have an even number of Es and Ns and the front string including $\mathbf{N}$ will always have one more N than E and N is the extra block north that puts our hero in the river. We see then that each string that puts our hero in the river can be changed to a string with 9 Es and 11 Ns and, conversely, each string with 9 Es and 11 Ns can be changed to a string that puts our hero in the river. Thus, the number of strings putting our hero in the river is the number of ways we can form a string with 9 Es and 11 Ns. Thus, we are picking 9 places to put Es out of 20 places and there are $C(20, 9) = \binom{20}{9} = \frac{20!}{9!11!}$ ways to do this. Therefore, the number of ways to get to work and stay dry is equal to the total number of ways minus the number of ways that the hero gets wet. We shall call this C_{10}, so $C_{10} = \frac{20!}{10!10!} - \frac{20!}{9!11!}$. We could calculate this, but a true mathematician never uses numbers when he can use letters, so we shall attempt to work with letters and avoid the possibility of arithmetic errors.

Assume that there are n blocks east and n blocks north, so that in our earlier example $n = 10$. We are now using a string of length $2n$, where half of them are Es and half of them are Ns. As before, we never want our path to have more Ns than Es or its back in the river. As before, we consider all possible ways of getting to the office, wet or dry, as long as we don't go out of the way. We are picking n places to put Es out of $2n$ possible places. As before, there are $C(2n, n) = \binom{2n}{n} = \frac{(2n)!}{n!n!}$ ways to do this. Next we again find the paths *not* to take, that is, the paths where the hero goes in the water. Using the same procedure as before, we find that the number of these paths is equal to the number of strings with $n - 1$ occurrences of E and $n + 1$ occurrences of N. The number of these is the number of ways to pick $n - 1$ places to put E out of $2n$ possible places. This is equal to $C(2n, n-1) = \binom{2n}{n-1} = \frac{(2n)!}{(n-1)!(n+1)!}$, so that C_n, the number of safe ways to get to work, is equal to $\frac{(2n)!}{n!n!} - \frac{(2n)!}{(n-1)!(n+1)!}$. It is now safe to calculate, so we have

$$\begin{aligned} C_n &= \frac{(2n)!}{n!n!} - \frac{(2n)!}{(n-1)!(n+1)!} \\ &= \frac{(2n)!}{n!n!} - \left(\frac{n}{n+1}\right)\left(\frac{(2n)!}{n!n!}\right) \\ &= \left(1 - \left(\frac{n}{n+1}\right)\right)\left(\frac{(2n)!}{n!n!}\right) \\ &= \left(\frac{1}{n+1}\right)\left(\frac{(2n)!}{n!n!}\right) \\ &= \left(\frac{1}{n+1}\right)\binom{2n}{n} \end{aligned}$$

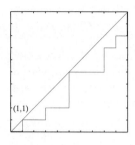

Figure 12.2

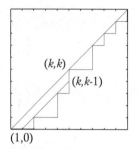

Figure 12.3

Before abandoning this adventure, we again return to look at the case $n = 10$. We now look at a different way of counting the number of good paths, that is, C_n. Now consider the case the path our hero reaches the river by going 1 mile east and then 1 mile north. We shall consider our hero's home to be located at the origin $(0, 0)$, so that he reaches the point $(1, 1)$ and then continues his journey to the office. This is shown in Figure 12.2.

How many of these paths are there? Obviously there is only one way to stay dry and get to point $(1, 1)$. To go from point $(1, 1)$ to point $(10, 10)$, that is, the office, we see that really this is the same as the original problem except we have to go 9 blocks east and 9 blocks north without landing in the river instead of going 10 blocks each way. Thus, there are C_9 ways of doing it. Since $C_0 = 1$, we say there are $C_0 \cdot C_9$ ways of getting to the office and passing through $(1, 1)$. This will fit with our notation later. Now suppose that on the way to work our hero reaches the river the first time at point (k, k), where $1 < k < 10$. We first calculate how many paths there are from $(0, 0)$ to (k, k) without reaching the river before reaching (k, k). Any such path must begin by going east 2 blocks since otherwise he would reach the river at $(1, 1)$. He must finish the path by going north for 2 blocks, since otherwise he would reach the river at $(k - 1, k - 1)$. We see in Figure 12.3 that if we draw a line parallel to the river and 1 mile south between the points $(1, 0)$ and $(k, k - 1)$, then the path does not cross this line. Hence, we can use any path that begins at $(1, 0)$ and goes $k - 1$ north and $k - 1$ blocks east to get to $(k, k - 1)$ **without crossing the new diagonal**. There are C_{k-1} ways of doing this. Therefore, there are C_{k-1} paths from $(0, 0)$ to (k, k) without reaching the river before reaching (k, k). Now consider paths which may be taken from (k, k) to the office without landing in the river. But this is just the problem of going north $10 - k$ blocks and east $10 - k$ blocks without landing in the river and is equal to C_{10-k}. Since there are C_{k-1} ways of getting to (k, k) and C_{10-k} ways of getting from (k, k) to the office, there are $C_{k-1} \cdot C_{10-k}$ ways of getting to the office by reaching the river for the first time at (k, k). Furthermore, for each path there is only one k since the place where it reaches the river for the first time is unique. Consider the case where our hero does not reach the river before he reaches the office. In other words, $k = 10$. We use the same argument as before to show that there are C_{k-1}, that is, C_9 ways of getting to the office, but in this case he is there, so there is no second part to the journey. Again for consistency of notation and since $C_0 = 1$, we shall write this as $C_9 \cdot C_0$ ways of getting to the office without reaching the river. If we sum the paths for all possible values of $1 \leq k \leq 10$, we have all possible ways of getting from home to the office. Therefore,

$$C_{10} = \sum_{k=1}^{10} C_{k-1} \cdot C_{10-k}$$

As we can see, there is nothing special in our argument about using 10. If we use n in place of 10, we would have

$$C_n = \sum_{k=1}^{n} C_{k-1} \cdot C_{n-k}$$

Figure 12.4

Figure 12.5

Given an $(n + 2)$-sided convex polygon, we wish to divide the polygon into triangles by drawing lines between vertices, without any of the lines crossing. How many ways can we do this? We shall let T_n be this number. For example, when $n = 4$, a possible triangulation is shown in Figure 12.4. For convenience we shall label the vertices $v_0, v_a, v_1, v_2, v_3, \ldots, v_n$. Suppose we draw a line from v_0 to v_1. How many triangles can we then draw? See Figure 12.5. We now have the polygon

Figure 12.6

Figure 12.7

$v_0v_1v_2v_3\ldots v_n$ left to triangulate. Since it has $n+1$ sides, there are T_{n-1} ways to triangulate this polygon. Now suppose we triangulate our $n+2$-sided polygon by drawing a line from v_0 to v_k where $1 < k < n$ and we allow no line to be drawn from v_0 to any vertex v_i where $i < k$ so that k is the smallest integer for which there is a line from v_0 to the vertex with that subscript. But then there must be a line from v_k to v_a to form triangle $v_k v_a v_0$ This leaves the polygon $v_a v_1 v_2 v_3 \ldots v_k$ left to triangulate. Since there are $k+1$ sides to this polygon, there are T_{k-1} ways to triangulate it. See Figure 12.6.

There is also the polygon $v_0 v_k v_{k+1} v_{k+2} v_{k+3} \ldots v_n$. It has $n-k+2$ sides so that it can be triangulated in T_{n-k} ways. Therefore, the polygon to the right of the line from v_0 to v_k can be triangulated in T_{k-1} ways and the polygon to the left of the line from v_0 to v_k can be triangulated in T_{n-k} ways, so the whole polygon with the line from v_0 to v_k can be triangulated in $T_{k-1} \cdot T_{n-k}$ ways. We now have to consider where the first line is drawn from v_0 to v_n. This should be interpreted as meaning that there is no line drawn from v_0 to any other vertex. But then there must be a line from v_n to v_a and the polygon $v_a v_1 v_2 v_3 \ldots v_n$ must be triangulated. See Figure 12.7.

But this polygon has $n+1$ sides and so can be triangulated in T_{k-1} ways. Since summing over k from 1 to n produces all possible triangulations of our $n+2$-sided triangle and letting $T_0 = 1$, we have

$$T_n = \sum_{k=1}^{n} T_{k-1} \cdot T_{n-k}$$

But this is the definition of Catalan numbers, so $T_n = C_n$.

Exercises

1. Compute C_5.
2. Compute C_6.
3. Compute C_7.
4. During intermission, 20 people are lined up at a drink machine. The drinks are 50 cents each. The machine is out of change but can return change once it receives some. If 10 people have the correct change and 10 have only dollar bills, how many ways can the people line up so that everyone gets correct change?
5. How many ways can 15 men and 15 women form a line if, at any point in the line, there are at least as many women as men?
6. If 12 numbers in a list are to be multiplied together, how many ways can this be done without rearranging the numbers?
7. Suppose we have n integers. We will place them in order in a stack but may pop the stack at any time. When the number is popped, it is placed in a sequence. Thus, suppose $n = 3$, so we have 1, 2, 3. We may push 1 and then pop it, push 2 and then pop it, and finally push 3 and then pop it. This gives us the sequence 1, 2, 3. We could, however, push 1, push 2, pop 2, pop 1, push 3, and then pop it. This gives the sequence 2, 1, 3. Given n integers, how many sequences may we create?

12.3 General Inclusion-Exclusion and Derangements

We have already discussed the principle of inclusion-exclusion involving three sets or less. At this point we want to discuss the principle of inclusion-exclusion involving any finite number of finite sets. We begin with the following theorem.

THEOREM 12.5 Let $A_1, A_2, A_3, \ldots, A_n$ be a collection of finite sets. The number of elements in $A_1 \cup A_2 \cup A_3 \cup \ldots \cup A_n$ is given by the formula

$$|A_1 \cup A_2 \cup A_3 \cup \ldots \cup A_n| = \sum_i |A_i| - \sum_{i<j} |A_i \cap A_j| + \sum_{i<j<k} |A_i \cap A_j \cap A_k| -$$
$$\cdots + (-1)^{n+1} |A_1 \cap A_2 \cap A_3 \cap \ldots \cap A_n|$$

Proof We want to show that each element of $A_1 \cup A_2 \cup A_3 \cup \ldots \cup A_n$ is counted exactly once on the right side of the preceding equation. Assume an element a is contained in exactly p of the sets $A_1, A_2, A_3, \ldots, A_n$. In $\sum_i |A_i|$ a is counted p times. In $\sum_{i<j} |A_i \cap A_j|$, a is counted whenever two sets are selected in which a is contained. There are $\binom{p}{2}$ ways to select these two sets. Therefore, in $\sum_{i<j} |A_i \cap A_j|$, a is counted $\binom{p}{2}$ times. In $\sum_{i<j<k} |A_i \cap A_j \cap A_k|$, a is counted whenever three sets are selected in which a is contained. There are $\binom{p}{3}$ ways to select these three sets. Therefore, in $\sum_{i<j<k} |A_i \cap A_j \cap A_k|$, a is counted $\binom{p}{3}$ times. In the sum of elements in all possible intersection of r sets, for $r \leq p$, a is counted $\binom{p}{r}$ times. Therefore, in

$$\sum_i |A_i| - \sum_{i<j} |A_i \cap A_j|$$
$$+ \sum_{i<j<k} |A_i \cap A_j \cap A_k| - \cdots + (-1)^{n+1} |A_1 \cap A_2 \cap A_3 \cap \ldots \cap A_n|$$

a is counted $p - \binom{p}{2} + \binom{p}{3} - \cdots + (-1)^{i+1} \binom{p}{i} + \cdots + (-1)^{p+1}$ times. But

$$0 = (1-1)^p$$
$$= 1 - p + \binom{p}{2} - \binom{p}{3} + \cdots + (-1)^p$$

so that

$$1 = p - \binom{p}{2} + \binom{p}{3} - \cdots + (-1)^{i+1} \binom{p}{i} + \cdots + (-1)^{p+1}$$

and a is counted exactly once. ∎

Using this theorem, we now state the theorem that is known as the inclusion-exclusion theorem.

THEOREM 12.6 (**Inclusion-Exclusion**) Let $A_1, A_2, A_3, \ldots, A_n$ be a collection of finite sets. The number of elements in $A_1' \cap A_2' \cap A_3' \cap \ldots \cap A_n'$ is given by the formula

$$|A_1' \cap A_2' \cap \ldots \cap A_n'| = |U| - \sum_i |A_i| + \sum_{i<j} |A_i \cap A_j| - \sum_{i<j<k} |A_i \cap A_j \cap A_k|$$
$$+ \cdots + (-1)^n |A_1 \cap A_2 \cap \ldots \cap A_n|$$

Proof From set theory, we know that

$$A_1' \cap A_2' \cap A_3' \cap \ldots \cap A_n' = (A_1 \cup A_2 \cup A_3 \cup \ldots \cup A_n)'$$
$$= U - (A_1 \cup A_2 \cup A_3 \cup \ldots \cup A_n)$$

so that

$$|A_1' \cap A_2' \cap A_3' \cap \ldots \cap A_n'| = |U| - |A_1 \cup A_2 \cup A_3 \cup \ldots \cup A_n|$$

and substituting in

$$|A_1 \cup A_2 \cup A_3 \cup \ldots \cup A_n| = \sum_i |A_i| - \sum_{i<j} |A_i \cap A_j| + \sum_{i<j<k} |A_i \cap A_j \cap A_k|$$
$$- \cdots + (-1)^{n+1}|A_1 \cap A_2 \cap A_3 \cap \ldots \cap A_n|$$

we have the desired result. ∎

EXAMPLE 12.7

How many positive integers less than 2003 are divisible by 2, 3, 5, or 7? How many positive integers less than 2003 are not divisible by 2, 3, 5, or 7? There are $\lfloor \frac{2002}{2} \rfloor = 1001$ integers divisible by 2, $\lfloor \frac{2002}{3} \rfloor = 666$ integers divisible by 3, $\lfloor \frac{2002}{5} \rfloor = 400$ integers divisible by 5, and $\lfloor \frac{2002}{7} \rfloor = 286$ integers divisible by 7. Taking two at a time, we have $\lfloor \frac{2002}{6} \rfloor = 333$ integers divisible by 2 and 3, $\lfloor \frac{2002}{10} \rfloor = 200$ integers divisible by 2 and 5, $\lfloor \frac{2002}{15} \rfloor = 133$ integers divisible by 3 and 5, $\lfloor \frac{2002}{14} \rfloor = 143$ integers divisible by 2 and 7, $\lfloor \frac{2002}{21} \rfloor = 95$ integers divisible by 3 and 7, and $\lfloor \frac{2002}{35} \rfloor = 57$ integers divisible by 5 and 7. Taking three as a time, we have $\lfloor \frac{2002}{30} \rfloor = 66$ integers divisible by 2, 3, and 5, $\lfloor \frac{2002}{42} \rfloor = 47$ integers divisible by 2, 3, and 7, $\lfloor \frac{2002}{70} \rfloor = 28$ integers divisible by 2, 5, and 7, and $\lfloor \frac{2002}{105} \rfloor = 19$ integers divisible by 3, 5, and 7. Finally, there are $\lfloor \frac{2002}{210} \rfloor = 9$ integers divisible by 2, 3, 5, and 7. Therefore, there are

$$1001 + 666 + 400 + 286 - (333 + 200 + 133 + 143 + 95 + 57)$$
$$+ 66 + 47 + 28 + 19 - 9$$

or $2353 - 961 + 160 - 9 = 1543$ integers are divisible by 2, 3, 5, or 7. There are

$$2002 - 1543 = 459$$

integers not divisible by 2, 3, 5, or 7. ∎

We now use our theorem on inclusion-exclusion to count derangements.

DEFINITION 12.8

A **derangement** on n distinct ordered symbols is a permutation in which none of the n symbols is left fixed. The number of derangements on n distinct ordered symbols is denoted by D_n.

For convenience, we shall denote the n symbols by $1, 2, 3, \ldots, n$. $D_1 = 0$, since a permutation on one element must leave it fixed. For $n = 2$, the only derangement is the permutation

$$\begin{pmatrix} 1 & 2 \\ 2 & 1 \end{pmatrix}$$

so that $D_2 = 1$. For $n = 3$, the derangements are

$$\begin{pmatrix} 1 & 2 & 3 \\ 2 & 3 & 1 \end{pmatrix} \text{ and } \begin{pmatrix} 1 & 2 & 3 \\ 3 & 1 & 2 \end{pmatrix}$$

so that $D_3 = 2$. There are nine derangements when $n = 4$. The reader is asked to find them.

12.3 General Inclusion-Exclusion and Derangements

We shall also try to count the derangements when $n = 4$ using counting techniques. Let U be the set of all permutations on the set $\{1, 2, 3, 4\}$. Therefore, $|U| = 4!$. Let A_1 be the set of permutations leaving the symbol 1 fixed, A_2 be the set of permutations leaving the symbol 2 fixed, A_3 be the set of permutations leaving the symbol 3 fixed, and A_4 be the set of permutations leaving the symbol 4 fixed. The set $A'_1 \cap A'_2 \cap A'_3 \cap A'_4$ is then the set leaving none of the symbols fixed. By the inclusion-exclusion theorem,

$$|A'_1 \cap A'_2 \cap A'_3 \cap A'_4| = |U| - \sum_{i \leq 4} |A_i| + \sum_{i < j \leq 4} |A_i \cap A_j|$$

$$- \sum_{i < j < k \leq 4} |A_i \cap A_j \cap A_k| + |A_1 \cap A_2 \cap A_3 \cap A_4|$$

Let $A_{ij} = A_i \cap A_j$, $A_{ijk} = A_i \cap A_j \cap A_k$, and $A_{1234} = A_1 \cap A_2 \cap A_3 \cap A_4$. Thus, A_{ij} is the set of all permutations leaving i and j fixed, A_{ijk} is the set of all permutations leaving i, j, and k fixed, and A_{1234} contains only the identity permutation, which leaves all four symbols fixed. To find $\sum_{i \leq 4} |A_i|$, we need to find the number of permutations leaving i fixed.

For a specific case, consider A_1. Each permutation in A_1 leaves 1 fixed and permutes 2, 3, and 4. Therefore, A_1 contains 3! and, by a similar argument, A_i contains 3! elements. Therefore,

$$|A_1| + |A_2| + |A_3| + |A_4| = 4 \times 3! = 4! = 4! \times \frac{1}{1!}$$

We now find

$$\sum_{i < j \leq 4} |A_i \cap A_j| = \sum_{i < j \leq 4} |A_{ij}|$$

Consider the specific case $|A_{12}|$. To belong to A_{12}, a permutation must leave 1 and 2 fixed and permute 3 and 4. There are 2! ways to do this. Thus, $|A_{12}| = 2$. Similarly $|A_{ij}| = 2$ for all $1 \leq i < j \leq 4$. Now the question becomes, how many A_{ij} are there? Since there are four elements from which to select i and j, there are $\binom{4}{2}$ sets of the form A_{ij}. Therefore,

$$\sum_{i < j \leq 4} |A_{ij}| = \binom{4}{2} 2! = \frac{4!}{2!2!} 2! = 4! \frac{1}{2!}$$

We next find

$$\sum_{i < j < k \leq 4} |A_i \cap A_j \cap A_k| = \sum_{i < j < k \leq 4} |A_{ijk}|$$

First consider A_{123}. Since 1, 2, and 3 are left fixed, so is 4. Therefore, A_{123} contains only 1 element. Similarly A_{ijk} contains only 1 element for $1 \leq i < j < k \leq 4$. Since we select i, j, and k from 4 elements, there are $\binom{4}{3}$ sets of the form A_{ijk}. Therefore,

$$\sum_{i < j < k \leq 4} |A_{ijk}| = \frac{4!}{3!1!} \times 1 = 4! \frac{1}{3!}$$

Finally, as previously mentioned, A_{1234} contains only 1 element, but

$$1 = \frac{4!}{4!} = 4! \times \frac{1}{4!}$$

Putting all of this together, we have

$$|A_1' \cap A_2' \cap A_3' \cap A_4'| = 4! - 4! \times \frac{1}{1!} + 4!\frac{1}{2!} - 4!\frac{1}{3!} + 4! \times \frac{1}{4!}$$

$$= 4!\left(1 - \frac{1}{1!} + \frac{1}{2!} - \frac{1}{3!} + \frac{1}{4!}\right)$$

More generally, we have the following theorem.

THEOREM 12.9 For $n > 1$, D_n the number of derangements of the n symbols $1, 2, 3, \ldots, n$ is given by

$$D_n = n!\left(1 - \frac{1}{1!} + \frac{1}{2!} - \frac{1}{3!} + \cdots + (-1)^n\frac{1}{n!}\right)$$

Proof To prove this we look at $A_1 \cap A_2 \cap A_3 \cap \ldots \cap A_k$ for $1 \leq k \leq n$. Every element of $A_1 \cap A_2 \cap A_3 \cap \ldots \cap A_k$ leaves $1, 2, 3, \ldots k$ fixed, so it permutes $n - k$ symbols. Therefore, $|A_1 \cap A_2 \cap A_3 \cap \ldots \cap A_k| = (n-k)!$. Similarly for any fixed $m_1, m_2, m_3, \ldots m_k$,

$$|A_{m_1} \cap A_{m_2} \cap A_{m_3} \cap \ldots \cap A_{m_k}| = (n-k)!$$

Since we are selecting k symbols out of n, there are $\binom{n}{k}$ sets of the form

$$A_{m_1} \cap A_{m_2} \cap A_{m_3} \cap \ldots \cap A_{m_k}$$

Therefore,

$$\sum |A_{m_1} \cap A_{m_2} \cap A_{m_3} \cap \ldots \cap A_{m_k}| = \binom{n}{k} \times (n-k)!$$

$$= \frac{n!}{k!(n-k)!} \times (n-k)!$$

$$= n! \times \frac{1}{k!}$$

Again, let U be the set of all permutations on n objects, so $|U| = n!$. We then have

$$D_n = |A_1' \cap A_2' \cap A_3' \cap \ldots \cap A_n'|$$

$$= |U| - \sum_i |A_i| + \sum_{i<j} |A_i \cap A_j|$$

$$- \sum_{i<j<k} |A_i \cap A_j \cap A_k| + \cdots + (-1)^n |A_1 \cap A_2 \cap \ldots \cap A_n|$$

$$= n! - \frac{1}{1!}n! + \frac{1}{2!}n! - \frac{1}{3!}n! + \cdots + (-1)^n\frac{1}{n!}n!$$

$$= n!\left(1 - \frac{1}{1!} + \frac{1}{2!} - \frac{1}{3!} + \cdots + (-1)^n\frac{1}{n!}\right) \blacksquare$$

It is shown in calculus that the Taylor expansion for e^{-1} is given by

$$e^{-1} = 1 - \frac{1}{1!} + \frac{1}{2!} - \frac{1}{3!} + \frac{1}{4!} - \frac{1}{5!} + \cdots$$

Therefore, $D_n \approx n!e^{-1}$.

EXAMPLE 12.10 An incompetent waiter is serving 10 distinct meals to 10 customers. How many ways can he serve the meals so that no person gets the proper meal? This is the derangement of 10 objects, so it is equal to

$$10!\left(1 - \frac{1}{1!} + \frac{1}{2!} - \frac{1}{3!} + \frac{1}{4!} - \frac{1}{5!} + \frac{1}{6!} - \frac{1}{7!} + \frac{1}{8!} - \frac{1}{9!} + \frac{1}{10!}\right) \blacksquare$$

EXAMPLE 12.11

An 8×8 permutation matrix is a matrix that contains exactly one 1 in each row and each column, with all of the other entries being 0. How many 8×8 permutation matrices are there that have all 0s along their main diagonal? This is simply the matrices representing permutations on 8 symbols that are derangements. Thus, there are

$$8! \left(1 - \frac{1}{1!} + \frac{1}{2!} - \frac{1}{3!} + \frac{1}{4!} - \frac{1}{5!} + \frac{1}{6!} - \frac{1}{7!} + \frac{1}{8!}\right)$$

such matrices. ■

EXAMPLE 12.12

In chess, two rooks (castles) can attack each other only if they are on the same row or the same column. Therefore, nonattacking rooks are those that are not in the same row or column. It is easily seen that eight nonattacking rooks may be placed on an 8×8 chessboard. How many different ways may eight nonattacking rooks be placed on an 8×8 chessboard? Since placing eight nonattacking rooks is really equivalent to placing 1s in a permutation matrix, and there are 8! permutations on 8 objects, there are then 8! different 8×8 permutation matrices.

If one wants to avoid thinking about permutation matrices, let a rook placed in the ith row, jth column be identified with the ordered pair (i, j). If this is done for each rook, then we have 8 ordered pairs. These form a permutation in which the first element of each of the ordered pairs is mapped to the second element. This is the standard representation of a function as a set of ordered pairs. ■

EXAMPLE 12.13

How many different ways may 8 nonattacking rooks be placed on an 8×8 chessboard if no rook is on the same row and column? Thus, a rook on row i cannot be in column i. This is equivalent to the number of permutation matrices with no 1s (and, hence, all 0s) on the main diagonal. As a permutation, it means that no element is mapped to itself. Hence, there are

$$8! \left(1 - \frac{1}{1!} + \frac{1}{2!} - \frac{1}{3!} + \frac{1}{4!} - \frac{1}{5!} + \frac{1}{6!} - \frac{1}{7!} + \frac{1}{8!}\right)$$

different ways of placing the rooks. ■

Exercises

1. Prove that for any positive integer n,

$$1 - \binom{n}{1} + \binom{n}{2} - \binom{n}{3} + \binom{n}{4} - \cdots + (-1)^n \binom{n}{1} = 0$$

2. A teacher has 25 students. She gives a pop quiz and then asks the student to exchange papers so that no students grade their own paper. How many ways can this be done?

3. A student takes an exam in which the student is to match 15 questions with the unique answer for that question. If no two questions have the same answer, in how many ways can the student get all of the questions wrong?

4. Let $A = \{1, 2, 3, 4, 5, 6, 7, 8, 9\}$ and $B = \{-1, -2, -3, -4, -5, -6, -7, -8, -9\}$. Each of the elements in the first set is to be paired with a unique element in the second set, so the pairing is a $1 - 1$ correspondence. The elements in each pair is added. In how many ways can this be done so that no pair has sum zero?

The illiterate mailman has a letter for each of 7 customers in a given area. He delivers exactly one letter to each customer

5. In how many ways can he deliver the letters so no one person gets his or her own mail?

6. In how many ways can he deliver the letters so at least one person gets his or her own mail?

7. In how many ways can he deliver the letters so that exactly one person gets his or her own mail?

8. In how many ways can he deliver the letters so that exactly one person does not get his or her own mail?

9. Let A have 12 elements, and φ be a permutation of the elements of A. Let $S = \{\theta : \theta$ is a permutation on A and $\theta(a) \neq \varphi(a)$ for all $a \in A\}$. Find $|S|$.

Seven gentlemen attend a party and each checks his hat. In how many ways can the hats be returned by an incompetent hat checker so that

10. When the hats are returned, no receives his own hat back?

11. When the hats are returned, exactly one person receives his own hat back?

12. When the hats are returned, at least one person receives his own hat back?

13. When the hats are returned, at least two people receive their own hat back?

How many permutations are there of $\{1, 2, 3, 4, 5, 6, 7, 8, 9\}$ such that

14. No number is left fixed by any the permutation; that is, no number is mapped to itself by the permutation?

15. Not one of the even numbers is left fixed by the permutation?

16. At least one even number is left fixed by the permutation?

17. Exactly four numbers are left fixed by the permutation?

18. Using the previous problem for patterns, determine the number of permutations on n elements leaving exactly k elements fixed.

19. Find the number of positive integers ≤ 2300 that are relatively prime to 700.

20. Find the number of positive integers ≤ 5460 that are relatively prime to 700.

12.4 Rook Polynomials and Forbidden Positions

In the previous examples we examined nonattacking rooks on an 8×8 chessboard and on an 8×8 chessboard where a rook could not appear in the same row and column. We may think of the squares where the rook could not be placed as **forbidden positions**. We now want to consider chessboards or parts of a chessboard that have forbidden positions where the rooks cannot be placed. When an $n \times n$ chessboard is used with forbidden positions, this is also referred to as **permutations with forbidden positions** because of the identification of permutations with nonattacking rooks. We should probably mention, before all those who do not play chess skip to the next section or toss the book away, that we are using rooks and a chessboard because it is a good way to illustrate what we plan to do rather than have any real value in playing chess.

We begin by introducing the rook polynomial.

DEFINITION 12.14 Let C be an arbitrary board with m squares, which is part of an $n \times n$ chessboard for some positive integer n. For $0 \leq k \leq m$, the integer $r_k(C)$ is the number of ways that k rooks can be placed on C in nonattacking position. The **rook polynomial** $R(x, C)$ on C is the generating function for the numbers $r_k(C)$, so that

$$R(x, C) = r_0(C) + r_1(C)x + r_2(C)x^2 + r_3(C)x^3 + \cdots + r_m(C)x^m$$

EXAMPLE 12.15 Let C be the board shown in Figure 12.8. There is always one way of placing 0 rooks on the board, so $r_0(C) = 1$ for any board C. In this case there are three ways of placing one rook on C, since it can be placed on any square of C. There is one way of placing two rooks so they are in nonattacking position. Therefore,

$$R(x, C) = 1 + 3x + x^2$$

∎

Figure 12.8

EXAMPLE 12.16

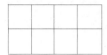

Figure 12.9

Let C be the board shown in Figure 12.9. There are 8 blocks, so there are 8 ways to place one rook on C. To place two rooks in nonattacking position, there are 4 choices for placing a rook in the top row and then 3 places for placing a rook on the second row, since the second rook can be placed on any square except the one directly below the one on which the first rook is placed. Thus, there are 12 possible ways for placing two nonattacking rooks. Therefore,

$$R(x, C) = 1 + 8x + 12x^2$$

EXAMPLE 12.17

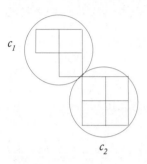

Figure 12.10

Let C be the board shown in Figure 12.10. To make our task easier, we observe that if we consider C consisting of two parts, C_1 and C_2 as shown, then since neither shares a common row or column with the other, the placement of rooks in one part does not affect the placement of rooks in the other part. To place 1 rook in C, one must either place a rook in C_1 or a rook in C_2. There are 3 ways to place a rook in C_1 and 4 ways to place a rook in C_2, so there are 7 ways to place a rook in C. To place 2 nonattacking rooks in C, we can place both rooks in C_1, 1 in C_1 and 1 in C_2, or both in C_2. There is 1 way to place 2 rooks in C_1. There are 3 ways to place a rook in C_1 and 4 ways to place a rook in C_2, making 12 ways to place 1 rook in C_1 and 1 in C_2. There are 2 ways to place 2 nonattacking rooks in C_2. Thus, there are $1 + 12 + 2 = 15$ ways to place 2 nonattacking rooks in C. To place 3 rooks in C, one may place 2 rooks in C_1 and 1 in C_2 or place 2 rooks in C_2 and 1 in C_1. There is 1 way to place 2 nonattacking rooks in C_1 and 4 ways to place 1 rook in C_2 giving 4 ways to place 2 rooks in C_1 and 1 in C_2. There are 3 ways to place 1 rook in C_1 and 2 ways to place 2 nonattacking rooks in C_2, so there are 6 ways to place 2 rooks in C_2 and 1 in C_1. This gives a total of $4 + 6 = 10$ ways to place 3 rooks in C. Finally, to place four nonattacking rooks in C, one must place 2 nonattacking rooks in C_1 and 2 in C_2. There is 1 way to place 2 nonattacking rooks in C_1 and 2 ways of placing 2 nonattacking rooks in C_2. Therefore, there are 2 ways placing 4 nonattacking rooks in C. Thus, the rook polynomial of C,

$$R(x, C) = 1 + 7x + 15x^2 + 10x^3 + 2x^4$$

EXAMPLE 12.18

Let C be the board shown in Figure 12.11. As in the previous example, we observe that if we consider C consisting of two parts, C_1 and C_2 as shown, then since neither shares a common row or column with the other, the placement of rooks in one part does not affect the placement of rooks in the other part. There are 5 ways to place a rook in C_1 and 8 ways to place a rook in C_2. Therefore, there are 13 ways to place a rook in C. Note that $r_1(C) = r_1(C_1) + r_1(C_2)$. However, since $r_0(C_1) = r_0(C_2) = 1$, we can let

$$r_1(C) = r_1(C_1)r_0(C_2) + r_0(C_1)r_1(C_2)$$

Figure 12.11

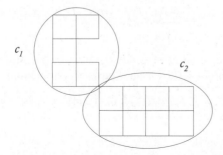

To select 2 nonattacking rooks we may select 2 rooks from C_1, select 1 rook from C_1 and 1 from C_2, or select 2 rooks from C_2. There are 4 ways to select 2 nonattacking rooks from C_1. There are 5 ways to select 1 rook from C_1 and 8 ways to select 1 rook from C_2. Therefore, there are $5 \times 8 = 40$ ways to select 1 rook from C_1 and 1 from C_2. There are 12 ways to select 2 nonattacking rooks from C_2. Therefore, there are $5 + 40 + 12 = 56$ ways to select 2 nonattacking rooks from C. From our method of selection we see that

$$r_2(C) = r_2(C_1) + r_1(C_1)r_1(C_2) + r_2(C_2)$$

which we can write as

$$r_2(C) = r_2(C_1)r_0(C_2) + r_1(C_1)r_1(C_2) + r_0(C_1)r_2(C_2)$$

Next, to get 3 nonattacking rooks from C, we could try to get 3 nonattacking rooks from C_1 but there is no way of getting them. We could get 1 rook from C_1 and 2 nonattacking rooks from C_2. There are 5 ways to get 1 rook from C_1 and 12 ways to get 2 nonattacking rooks from C_2. This gives us $5 \times 12 = 60$ ways to get 3 nonattacking rooks by getting 1 rook from C_1 and 2 nonattacking rooks from C_2. We could also get 2 nonattacking rooks from C_1 and 1 rook from C_2. There are 4 ways to get 2 nonattacking rooks from C_1 and 8 ways to get 1 rook from C_2. Therefore, there are $4 \times 8 = 32$ ways to get 3 nonattacking rooks by getting 2 nonattacking rooks from C_1 and 1 rook from C_2. Again, we could try to get 3 nonattacking rooks from C_2 but there is no way of getting them. Therefore, $r_3(C) = 60 + 32 = 92$ ways to get 3 nonattacking rooks from C. From our method of selection, we could also write

$$r_3(C) = r_2(C_1)r_1(C_2) + r_1(C_1)r_2(C_2)$$

Finally, to get 4 nonattacking rooks from C, we must get 2 nonattacking rooks from C_1 and 2 nonattacking rooks from C_2. There are 4 ways to get 2 nonattacking rooks from C_1 and 12 ways to get 2 nonattacking rooks from C_2. Thus, there are $4 \times 12 = 48$ ways of getting 4 nonattacking rooks from C. We could also write

$$r_4(C) = r_2(C_1)r_2(C_2)$$

Thus,

$$R(x, C) = 1 + 13x + 56x^2 + 92x^3 + 48x^4$$

Also

$$\begin{aligned}R(x, C) &= r_0(C_1)r_0(C_2) + (r_1(C_1)r_0(C_2) + r_0(C_1)r_1(C_2))x \\ &\quad + (r_2(C_1)r_0(C_2) + r_1(C_1)r_1(C_2) + r_0(C_1)r_2(C_2))x^2 \\ &\quad + (r_2(C_1)r_1(C_2) + r_1(C_1)r_2(C_2))x^3 + r_2(C_1)r_2(C_2)x^4 \\ &= (r_0(C_1) + r_1(C_1)x + r_2(C_1)x^2)(r_0(C_2) + r_1(C_2)x + r_2(C_2)x^2) \\ &= R(x, C_1)R(x, C_2)\end{aligned}$$

Therefore, we could have gotten the rook polynomial for C by taking the product of the rook polynomial for C_1 and the rook polynomial for C_2. Indeed

$$R(x, C_1) = 1 + 5x + 4x^2$$

and

$$R(x, C_2) = 1 + 8x + 12x^2$$

and their product

$$\begin{aligned} R(x, C_1)R(x, C_2) &= (1 + 5x + 4x^2)(1 + 8x + 12x^2) \\ &= 1 + 13x + 56x^2 + 92x^3 + 48x^4 \\ &= R(x, C) \end{aligned}$$

■

Using the previous example for guidance, we have the following theorem.

THEOREM 12.19 If a board C consists of two parts C_1 and C_2, which have no row or column in common, then $R(x, C) = R(x, C_1)R(x, C_2)$.

Proof Consider the coefficient $r_k(C)$ of x^k in $R(x, C)$. To select k nonattacking rooks in C, we must select l of them from C_1 and $k - l$ of them from C_2. There are $r_l(C_1)$ ways of selecting l nonattacking rooks from C_1 and $r_{k-l}(C_2)$ ways of selecting $k - l$ nonattacking rooks from C_2. This gives $r_l(C_1)r_{k-l}(C_2)$ ways of selecting l nonattacking rooks from C_1 and $k - l$ nonattacking rooks from C_2. If we let l take on all value from 0 to k, we have

$$\begin{aligned} r_k(C) = r_0(C_1)r_k(C_2) + r_1(C_1)r_{k-1}(C_2) \\ + \cdots + r_l(C_1)r_{k-l}(C_2) + \cdots + r_k(C_1)r_0(C_2) \end{aligned}$$

But this is the coefficient of x^k in the product

$$\begin{aligned} R(x, C_1)R(x, C_2) = (r_0(C_1) + r_1(C_1)x + \cdots + r_m(C_1)x^m)(r_0(C_2) \\ + r_1(C_2)x + \cdots + r_m(C_2)x^m) \end{aligned}$$

■

EXAMPLE 12.20 Again consider the board in Figure 12.12.

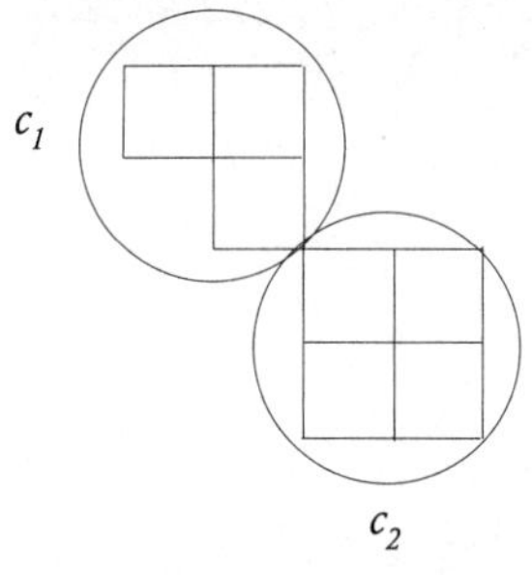

Figure 12.12

$$R(x, C_1) = 1 + 3x + x^2$$

and

$$R(x, C_2) = 1 + 4x + 2x^2$$

Thus,

$$\begin{aligned} R(x, C) &= R(x, C_1)R(x, C_2) \\ &= (1 + 3x + x^2)(1 + 4x + 2x^2) \\ &= 1 + 7x + 15x^2 + 10x^3 + 2x^4 \end{aligned}$$

■

We have yet another way of finding $R(x, C)$ by breaking C down into boards. In this case the boards may have common rows and columns.

THEOREM 12.21 Given a board C, let s be a square of C. Let C_s be the board C with the square s eliminated. Let $C_s^{\#}$ be the board C with the entire row and column containing s eliminated. Then $R(x, C) = xR(x, C_s^{\#}) + R(x, C_s)$.

Proof Let k be a positive integer. If k nonattacking rooks are placed on C and C has m squares, then either a rook is placed on square s or it is not. If it is, then there are $k - 1$ rooks to be selected from $C_s^{\#}$, which can be done in $r_{k-1}(C_s^{\#})$ ways. If a rook is not placed on square s, then k rooks must be selected from C_s. This may be done in $r_k(C_s)$ ways. Therefore,

$$r_k(C) = r_{k-1}(C_s^{\#}) + r_k(C_s)$$

and since $r_0(C) = r_0(C_s) = 1$,
$$\sum_{k=0}^{m} r_k(C)x^k = \sum_{k=1}^{m} r_{k-1}(C_s^{\#})x^k + \sum_{k=0}^{m} r_k(C_s)x^k$$

But
$$\sum_{k=1}^{m} r_{k-1}(C_s^{\#})x^k = \sum_{k=0}^{m-1} r_k(C_s^{\#})x^{k+1}$$
$$= x \sum_{k=0}^{m-1} r_k(C_s^{\#})x^k$$
$$= x \sum_{k=0}^{m} r_k(C_s^{\#})x^k$$

since $r_m(C_s^{\#}) = 0$. It must equal 0, since $C_s^{\#}$ has less than m squares. Therefore,
$$R(x, C) = xR(x, C_s^{\#}) + R(x, C_s)$$

EXAMPLE 12.22 Let C be the board shown in Figure 12.13 where s is the square specified. Then C_s is the board shown in Figure 12.14 and $C_s^{\#}$ is the board shown in Figure 12.15.

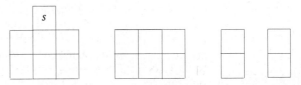

Figure 12.13   Figure 12.14   Figure 12.15

The rook polynomials are
$$R(x, C_s) = 1 + 6x + 6x^2$$
and
$$R(x, C_s^{\#}) = 1 + 4x + 2x^2$$
Therefore,
$$R(x, C) = xR(x, C_s^{\#}) + R(x, C_s)$$
$$= x(1 + 4x + 2x^2) + 1 + 6x + 6x^2$$
$$= 1 + 7x + 10x^2 + 2x^3$$

EXAMPLE 12.23 Let C be the board shown in Figure 12.16 where s is the square specified. Then C_s is the board shown in Figure 12.17 and $C_s^{\#}$ is the board shown in Figure 12.18.
$$R(x, C_s) = R(x, C_s')R(x, C_s'')$$

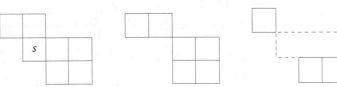

Figure 12.16   Figure 12.17   Figure 12.18

where C'_s is the upper left rectangle and C''_s is the lower right square

$$R(x, C'_s) = 1 + 2x$$

and

$$R(x, C''_s) = 1 + 4x + 2x^2$$

Therefore,

$$R(x, C_s) = (1 + 2x)(1 + 4x + 2x^2)$$
$$= 1 + 6x + 10x^2 + 4x^3$$

It is easily seen that

$$R(x, C_s^\#) = 1 + 3x + 2x^2$$

Therefore,

$$R(x, C) = xR(x, C_s^\#) + R(x, C_s)$$
$$= x(1 + 3x + 2x^2) + 1 + 6x + 10x^2 + 4x^3$$
$$= 1 + 7x + 13x^2 + 6x^3$$

∎

Since we have spent so much time looking at rook polynomials, it seems reasonable at this point to find some use for them. It may be recalled that when looking at permutation matrices, each one had exactly one 1 in each row and column and 0s elsewhere. If we identity rooks with the 1's, then a permutation on n elements can be identified with n nonattacking rooks on an $n \times n$ chessboard. If we consider a permutation ϕ on n elements as a function, when we can consider it as a set of ordered pairs $\{(k, \phi(k)) : k = 1, 2, 3, \ldots n\}$. If we identify $(k, \phi(k))$ with a square on a chessboard at row k, column $\phi(k)$, then again we return to n nonattacking rooks on an $n \times n$ chessboard. We, therefore, can identity permutations on n elements with n nonattacking rooks on an $n \times n$ chessboard and will feel free to shift back and forth.

Suppose that a student is trying to take five different courses in five different class periods. If we view this as a chessboard, then we can use marked out areas of the board to represent the class periods when a particular course is not offered. Let ϕ map each course to the period that it is taken. As before, we can think of ϕ as a permutation. We then have values for each k that $\phi(k)$ cannot equal. For example, suppose that $n = 5$, and $\phi(1)$ cannot equal 2, 3, or 5; $\phi(2)$ cannot equal 1, 2, or 3; $\phi(3)$ cannot equal 1, 3, or 5; $\phi(4)$ cannot equal 1, 2, or 5; and $\phi(5)$ cannot equal 1, 3, or 4. The permutations that satisfy these restrictions are called the **permissible arrangements** on the board. The forbidden area or forbidden positions are those that are shaded in the chessboard in Figure 12.19.

Figure 12.19

It is obvious, and possibly even true, that the only permissible arrangements are

$$\begin{pmatrix} 1 & 2 & 3 & 4 & 5 \\ 1 & 5 & 4 & 3 & 2 \end{pmatrix} \text{ and } \begin{pmatrix} 1 & 2 & 3 & 4 & 5 \\ 1 & 4 & 2 & 3 & 5 \end{pmatrix}$$

Suppose, however, that the squares in the forbidden area are relatively small. For example, suppose that we have the board with shaded forbidden area shown in Figure 12.20.

It would be easier if we could work with the forbidden area, which is much smaller. Assume that we have an $n \times n$ board with a relatively small forbidden area A. Let A_i represent the set of all permutations ϕ such that $\phi(i)$ is in the

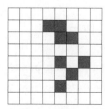

Figure 12.20

forbidden area, A. Then we are seeking $|A'_1 \cap A'_2 \cap \ldots \cap A'_n|$. But

$$|A'_1 \cap A'_2 \cap \ldots \cap A'_n| = n! - \sum_i |A_i| + \sum_{i<j} |A_i \cap A_j|$$
$$- \sum_{i<j<k} |A_i \cap A_j \cap A_k| + \cdots + (-1)^n |A_1 \cap A_2 \cap \ldots \cap A_n|$$

so we need to find the values of the terms on the right side of the equation. Consider $|A_i|$ for a fixed i. Let n_i be the number of squares of the forbidden area in row i. Then for $\phi(i)$ to belong to A_i there are n_i choices for $\phi(i)$, but for $k \neq i$, $\phi(k)$ can be any value as long as ϕ is a permutation. Therefore, there are $(n-1)!$ choices for the other values of ϕ. Hence, there are $n_i(n-1)!$ permutations in A_i. Adding, we have

$$\sum_i |A_i| = \sum_i n_i(n-1)! = \left(\sum_i n_i \right)(n-1)!$$

But $\left(\sum_i n_i \right) = r_1(A)$, the number of squares in the forbidden area A. Therefore,

$$\sum_i |A_i| = r_1(A)(n-1)!$$

Now consider $\sum_{i<j} |A_i \cap A_j|$. Let i and j be fixed. For ϕ to belong to $A_i \cap A_j$, $\phi(i)$ and $\phi(j)$ must both belong to A, and they must be in different columns. Let n_{ij} be the number of ways that $\phi(i)$ and $\phi(j)$ can both belong to A and not be in the same column. There are, therefore, n_{ij} ways to pick $\phi(i)$ and $\phi(j)$. For $k \neq i, j$ the value $\phi(k)$ can be any value as long as ϕ is a permutation. Therefore, there are $(n-2)!$ ways of selecting the other value of ϕ. Therefore, $A_i \cap A_j$ contains $n_{ij}(n-2)!$ permutations. Summing over i and j, we have

$$\sum_{i<j} |A_i \cap A_j| = \sum_{i<j} n_{ij}(n-2)! = \left(\sum_{i<j} n_{ij} \right)(n-2)!$$

But $\sum_{i<j} n_{ij}$ is simply the number of ways that we can place two nonattacking rooks in A. Therefore, $\sum_{i<j} n_{ij} = r_2(A)$, and

$$\sum_i |A_i| = r_2(A)(n-2)!$$

Similarly, to find $\sum_{i_1 < i_2 < \cdots < i_m} |A_{i_1} \cap A_{i_2} \cap \cdots \cap A_{i_m}|$, let $n_{i_1 i_2 \ldots i_m}$ be the number of ways that $\phi(i_1), \phi(i_2), \ldots, \phi(i_m)$ can all belong to A and not be in the same column. There are, therefore, $n_{i_1 i_2 \ldots i_m}$ ways to pick $\phi(i_1), \phi(i_2), \ldots, \phi(i_m)$. For $k \neq i_1, i_2, \ldots, i_m$, the value $\phi(k)$ can be any value as long as ϕ is a permutation. Therefore, there are $(n-m)!$ ways of selecting the other value of ϕ and $A_{i_1} \cap A_{i_2} \cap \cdots \cap A_{i_m}$ contains $n_{i_1 i_2 \ldots i_m}(n-m)!$ permutations. Summing, over $i_1, i_2, \ldots, i_m$, we have

$$\sum_{i_1 < i_2 < \cdots < i_m} |A_{i_1} \cap A_{i_2} \cap \cdots \cap A_{i_m}| = \sum_{i_1 < i_2 < \cdots < i_m} n_{i_1 i_2 \ldots i_m}(n-m)!$$
$$= \left(\sum_{i_1 < i_2 < \cdots < i_m} n_{i_1 i_2 \ldots i_m} \right)(n-m)!$$

But $\sum_{i_1 < i_2 < \cdots < i_m} n_{i_1 i_2 \ldots i_m}$ is simply the number of ways that we can place m nonattacking rooks in A. Therefore, $\sum_{i_1 < i_2 < \cdots < i_m} n_{i_1 i_2 \ldots i_m} = r_m(A)$, and

$$\sum_{i_1 < i_2 < \cdots < i_m} |A_{i_1} \cap A_{i_2} \cap \cdots \cap A_{i_m}| = r_m(A)(n-m)!$$

12.4 Rook Polynomials and Forbidden Positions

Therefore,
$$|A'_1 \cap A'_2 \cap \ldots \cap A'_n| = n! - r_1(A)(n-1)! + r_2(A)(n-2)! + \cdots + (-1)^n r_n(A)$$
where the rook polynomial for A, $R(x, A) = \sum_{k=0}^{n} r_k(A) x^k$.

EXAMPLE 12.24 Given the board shown in Figure 12.21 where the shaded area A is the forbidden area, find the number of permissible arrangements. We want to find
$$|A'_1 \cap A'_2 \cap A'_3 \cap A'_4 \cap A'_5 \cap A'_6| = 6! - r_1(A) \cdot 5! + r_2(A \cdot) \cdot 4! + \cdots + (-1)^6 r_6(A)$$

We already know the rook polynomial for A is
$$R(x, A) = 1 + 7x + 15x^2 + 10x^3 + 2x^4$$
so the number of permissible arrangements is given by
$$|A'_1 \cap A'_2 \cap A'_3 \cap A'_4 \cap A'_5 \cap A'_6| = 6! - 7 \cdot 5! + 15 \cdot 4! - 10 \cdot 3! + 2 \cdot 2!$$
$$= 184$$

Figure 12.21

■

EXAMPLE 12.25 Given the board shown in Figure 12.22 where the shaded area A is the forbidden area, find the number of permissible arrangements. We want to find
$$|A'_1 \cap A'_2 \cap \cdots \cap A'_7| = 7! - r_1(A) \cdot 6! + r_2(A \cdot) \cdot 5! + \cdots + (-1)^7 r_7(A)$$

and so we again need the rook polynomial for A. Note that on the board in Figure 12.23, the shaded area B has the same rook polynomial as A. The shaded area B consists of the areas B_1, B_2, and B_3, which have no rows or columns in common. Therefore,

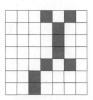

Figure 12.22

$$R(x, A) = R(x, B)$$
$$= R(x, B_1) \cdot R(x, B_2) \cdot R(x, B_3)$$

The polynomials
$$R(x, B_1) = 1 + 4x + 2x^2$$
$$R(x, B_2) = 1 + 3x$$
$$R(x, B_3) = 1 + 2x$$

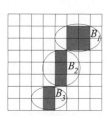

Figure 12.23

so that
$$R(x, A) = (1 + 4x + 2x^2)(1 + 3x)(1 + 2x)$$
$$= 1 + 9x + 28x^2 + 34x^3 + 12x^4$$

Therefore,
$$|A'_1 \cap A'_2 \cap \cdots \cap A'_7| = 7! - 9 \cdot 6! + 28 \cdot 5! - 34 \cdot 4! + 12 \cdot 3!$$
$$= 1176$$

so there are 1176 permissible arrangements. ■

Computer Science Application

A number of algorithms can easily be written to compute factorials, Fibonacci numbers, Catalan numbers, Stirling numbers, permutations, combinations, derangements, and so on. In almost all cases, the algorithms are difficult to implement because of the hazards of integer overflow that are associated with such large numbers. In these cases, special data structures are needed to represent integers with large numbers of digits. An array of integers could easily be used to represent large positive integers of up to 100 digits, each component containing a single digit. If we assume that A and B are two such "big integer" arrays, the following algorithms demonstrate how easily addition and multiplication can be performed.

function add_big(big integer A, big integer B, big integer C)

    integer carry $= 0$

    for $j = 100$ to 1 by -1 do
      $k = A[j] + B[j] +$ carry
      carry $= k/10$ (integer division)
      $C[j] = k$ mod 10
    end-for

    if carry $= 1$ then
      print "Integer Overflow"
      stop execution
    end-if

    return

end-function

function multiply_big(big integer A, big integer B, big integer C)

    for $i = 1$ to 200 do
      $C[i] = 0$
    end-for

    $n = 200$
    for $i = 100$ to 1 by -1 do
      $m = n$
      for $j = 100$ to 1 by -1 do
        $C[m] = C[m] + A[j] * B[i]$
        $m = m - 1$
      end-for
      $n = n - 1$
    end-for

    for $i = 200$ to 2 by -1 do
      $C[i - 1] = C[i - 1] + C[i]/10$ (integer division)
      $C[i] = C[i]$ mod 10
    end-for

    if $C[1] > 9$ then
      print "Integer Overflow"
      stop execution
    end-if

    return

end-function

In both functions parameter C is passed back to the caller with the result of the operation, leaving A and B unchanged. In the multiply function C is an array of 200 components, giving it the capacity to store the product even if both A and B have 100-digit values.

12.4 Rook Polynomials and Forbidden Positions

Exercises

1. Find the rook polynomial for the board in Figure 12.24.

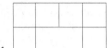

Figure 12.24

2. Find the rook polynomial for the board in Figure 12.25.

Figure 12.25

3. Find the rook polynomial for the board in Figure 12.26.

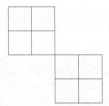

Figure 12.26

4. Find the rook polynomial for the board in Figure 12.27.

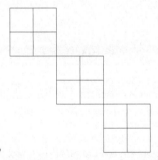

Figure 12.27

5. Find the rook polynomial for the board in Figure 12.28.

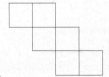

Figure 12.28

6. Find the rook polynomial for the board in Figure 12.29.

Figure 12.29

7. Find the rook polynomial for the board in Figure 12.30.

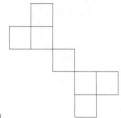

Figure 12.30

8. Find the rook polynomial for the board in Figure 12.31.

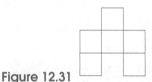

Figure 12.31

9. Find the rook polynomial for the board in Figure 12.32.

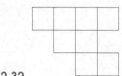

Figure 12.32

10. Find the rook polynomial for the board in Figure 12.33.

Figure 12.33

11. Find the rook polynomial for the board in Figure 12.34.

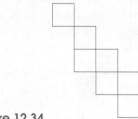

Figure 12.34

12. Find the rook polynomial for the board in Figure 12.35.

Figure 12.35

13. Given the board in Figure 12.36 where the shaded area *A* is the forbidden area, find the number of permissible arrangements.

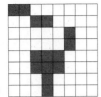

Figure 12.36

14. Given the board in Figure 12.37 where the shaded area *A* is the forbidden area, find the number of permissible arrangements.

Figure 12.37

15. Given the board in Figure 12.38 where the shaded area *A* is the forbidden area, find the number of permissible arrangements.

Figure 12.38

16. Given the board in Figure 12.39 where the shaded area *A* is the forbidden area, find the number of permissible arrangements.

Figure 12.39

17. Given the board in Figure 12.40 where the shaded area *A* is the forbidden area, find the number of permissible arrangements.

Figure 12.40

18. Given the board in Figure 12.41 where the shaded area *A* is the forbidden area, find the number of permissible arrangements.

Figure 12.41

19. Find the number of permissible arrangements on a 4×4 board where the squares along both diagonals are forbidden.

20. On a six-man football team, player *A* cannot play positions 1 and 3, player *B* cannot play positions 2 and 4, player *C* cannot play position 6, player *D* cannot play positions 5 and 6, player *E* cannot play positions 5 and 6, and player *F* cannot play positions 1 and 3. Find the number of ways in which the positions can be assigned.

21. Six men are renting a house with six bedrooms. The first man will not use bedrooms 4 and 6. The second man will not use bedrooms 4, 5, and 6. The third man will not use bedroom 3. The fourth man will not use bedrooms 1 and 2. The fifth man will not use bedrooms 1 and 2. The sixth man will not use bedrooms 1, 2, and 3. In how many ways can the bedrooms be assigned?

22. The football coach has bought six new cars for his six top players. He has purchased a Volvo, a Mercedes, a BMW, a Lexus, a Lincoln, and Cadillac. The first player will not drive a Cadillac or a Lincoln. The second player will not drive a Cadillac or Lincoln. The third player will not drive a Lexus. The fourth player will not drive a Mercedes. The fifth player will not drive a Volvo or BMW. The sixth player will not drive a Volvo or BMW. In how many ways can the coach give out the cars?

23. An $n \times n$ **latin square** is a square in which the numbers in each row and each column are a permutation of the first n positive integers. Thus, each number appears once in each row and each column. Suppose the first two rows of numbers in a 4×4 latin square are 1, 4, 3, 2 and 4, 1, 2, 3. How many different third rows are possible? *Hint*: in the first row of the board, black out the position where 1 has occurred, and so on.

24. Suppose the first two rows of numbers in a 4×4 latin square are 4, 3, 1, 2 and 3, 4, 2, 1. How many different third rows are possible?

25. Suppose the first two rows of numbers in a 5×5 latin square are 1, 3, 2, 4, 5 and 2, 1, 3, 5, 4. How many different third rows are possible.

26. Suppose the first two rows of numbers in a 5×5 latin square are 1, 2, 3, 4, 5 and 3, 4, 5, 2, 1. How many different third rows are possible?

27. Suppose the first two rows of numbers in a 5×5 latin square are 1, 2, 3, 4, 5 and 2, 3, 4, 5, 1. How many different third rows are possible?

GLOSSARY

Chapter 12: Counting Continued

Catalan numbers (12.2)	The Catalan number for the positive integer n is given by the formula $$Cat(n) = \frac{(2n)!}{(n+1)!(n!)}$$
Derangement (12.3)	A **derangement** on n distinct ordered symbols is a permutation in which none of the n symbols is left fixed. The number of derangements on n distinct ordered symbols is denoted by D_n.
Forbidden positions (12.4)	On a chessboard, **forbidden positions** are the squares where the rook cannot be placed.
Inclusion-exclusion (12.3)	Let $A_1, A_2, A_3, \ldots, A_n$ be a collection of finite sets. The number of elements in $A'_1 \cap A'_2 \cap A'_3 \cap \ldots \cap A'_n$ is given by the formula $$\left|A'_1 \cap A'_2 \cap A'_3 \cap \ldots \cap A'_n\right| = |U| - \sum_i |A_i| + \sum_{i<j} |A_i \cap A_j| - \sum_{i<j<k} |A_i \cap A_j \cap A_k| + \cdots (-1^n) |A_1 \cap A_2 \cap \ldots \cap A_n|$$
Number of ways of dividing n distinguishable objects into k distinguishable, boxes that may not be empty (12.1)	$k! S_k^{(n)}$
Number of ways of dividing n indistinguishable objects into k distinguishable boxes that may be empty (12.1)	$C(n+k-1, n)$
Number of ways of dividing n objects into k cycles (12.1)	$s_k^{(n)} = \{s_k^{(n)} : 0 \leq k \leq n\}$, the set of Stirling numbers of the first kind.
Number of ways of partitioning a set of n distinguishable objects into k indistinguishable boxes that may not be empty (12.1)	$S_k^{(n)} = \{S_k^{(n)} : 0 \leq k \leq n\}$, the set of **Stirling numbers of the second kind**.
Number of ways of placing n distinguishable objects in k distinguishable boxes, where some of the boxes may be empty (12.1)	k^n

Number of ways of placing n distinguishable objects in k indistinguishable boxes, where some of the boxes may be empty (12.1)	$S_1^{(n)} + S_2^{(n)} + S_3^{(n)} + \cdots + S_k^{(n)}$
Number of ways of placing n indistinguishable objects in k distinguishable boxes, where some of the boxes may not be empty (12.1)	$C(n-1, k-1)$
Number of ways to take n distinguishable objects and k distinguishable boxes and place n_i objects in the ith box for all $1 \le i \le k$, where $n = n_1 + n_2 + \cdots n_k$ (12.1)	$C(n; n_1, n_2, n_3, \cdots n_k) = \dfrac{n!}{n_1 \cdot n_2 \cdot n_3 \cdots \cdot n_k}$
Number of ways to take n distinguishable objects and place m of them into one box and $n-m$ into the other (12.1)	$C(n, m) = \dfrac{n!}{m!(n-m)!}$
Permissible arrangements (12.4)	Permutations that avoid forbidden positions called the **permissible arrangements** on the board.
Permutations with forbidden positions (12.4)	**Permutations with forbidden positions** are permutations that are not permissible arrangements.
Rook polynomial (12.4)	Let C be an arbitrary board with m squares, which is part of an $n \times n$ chessboard for some positive number n. For $0 \le k \le m$, the integer $r_k(C)$ is the number of ways that k rooks can be placed on C in nonattackable positions. The **rook polynomial** $R(x, C)$ on C is the generating function for the numbers $R(x, C)$, so that $R(x, C) = r_0(C) + r_1(C)x + r_2(C)x^2 + r_3(C)x^3 + \cdots + r_m(C)x^m$.

TRUE-FALSE QUESTIONS

1. Stirling numbers of the first kind satisfy the relation $S_k^{(n+1)} = S_{k-1}^{(n)} + kS_k^{(n)}$.

2. $S_0^{(n)} = s_0^{(n)} = 0$ for all $n \ge 0$.

3. $S_n^{(n)} = s_n^{(n)} = 1$ for all $n \ge 0$.

4. $S_1^{(n)} = 1$ for all $n \ge 1$.

5. $s_n^{(n)} = 1$ for all $n \ge 1$.

6. The number of ways 10 distinguishable objects may be placed in 5 distinguishable boxes where the boxes may be empty is $5! \cdot S_5^{(10)}$.

7. The number of ways 10 distinguishable objects may be placed in 5 indistinguishable boxes where the boxes may not be empty is $S_5^{(10)}$.

8. The number of ways 10 indistinguishable objects may be placed in 5 distinguishable boxes where the boxes may be empty is $\binom{14}{10}$.

9. The number of ways 10 indistinguishable objects may be placed in 5 distinguishable boxes where the boxes may not be empty is $\binom{10}{5}$.

10. The Catalan number $C_{10} = 16796$.

11. The number of outcomes of tossing 12 coins where there are never more heads than tails occurring as they are tossed is $\dfrac{12!}{6!6!}$.

12. $|(A_1 \cup A_2 \cup A_3 \cup \cdots \cup A_n)'| = |U| - \sum_i |A_i| + \sum_{i<j} |A_i \cap A_j| + \cdots + (-1)^n |A_1 \cap A_2 \cap \cdots \cap A_n|$.

13. The number of positive integers less than 500 not divisible by 2, 4, or 6 is
$499 - 249 - 124 - 83 + 62 + 41 + 20 - 10$

14. The number of permutations of 4 elements where no element is mapped to itself is 9.

15. The number of derangements on n elements is equal to $\dfrac{n!}{e}$.

16. If C is the board [board], C_1 is the board [board], and C_2 is the board [board], then $R(x, C) = R(x, C_1) R(x, C_2)$.

17. If C is the board [board], C_1 is the board [board], and C_2 is the board [board], then $R(x, C) = R(x, C_1) R(x, C_2)$.

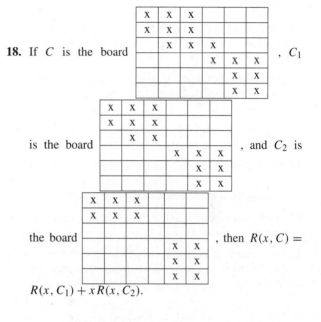

18. If C is the board [board], C_1 is the board [board], and C_2 is the board [board], then $R(x, C) = R(x, C_1) + x R(x, C_2)$.

19. If $C = $ [board], then $R(x, C) = 1 + 7x + 8x^2 + 6x^3$.

20. If the area that is not blank in the board [board] is the forbidden area, then the number of permissible arrangements is given by $5! - 8 \cdot 4! + 20 \cdot 3! - 16 \cdot 2! + 4$.

SUMMARY QUESTIONS

1. How many ways can 4 objects be placed in 2 boxes if the objects are distinguishable, the boxes are distinguishable, and the boxes may be empty?

2. Two candidates Smith and Jones are running for office. Each receives n votes. In the counting, Smith is never behind Jones. In how many ways can the votes be counted?

3. How many ways can 6 objects be placed in 3 boxes if the objects are distinguishable, the boxes are distinguishable, and the boxes may not be empty?

4. How many derangements are there of the symbols 1, 2, 3, 4, 5?

5. How many ways can 12 objects be placed in 6 boxes if the objects are indistinguishable, the boxes are distinguishable, and the boxes may be empty?

6. A student takes a test on which 25 questions are to be matched with 25 answers. How many ways can the student answer the questions so that he does not get one answer right?

7. How many ways can 8 objects be placed in 4 boxes if the objects are indistinguishable, the boxes are distinguishable, and the boxes may not be empty?

8. How many ways can 6 objects be placed in 4 boxes if the objects are distinguishable, the boxes are indistinguishable, and the boxes may be empty?

9. Find the number of positive integers ≤2000 that are relatively prime to 500.

10. How many ways can 7 objects be placed in 3 boxes if the objects are distinguishable, the boxes are indistinguishable, and the boxes may not be empty?

11. How many positive integers less than or equal to 1000 are not divisible by 2, 3, 5, or 7?

12. A student takes a test on which 15 questions are to be matched with 15 answers. How many ways can the student answer the questions so that he gets exactly one answer right?

13. Find $R(X, C)$ for the board

14. Find $R(X, C)$ for the board

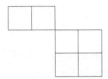

Given the board below where the shaded area is the forbidden area, find the number of permissible arrangements.

ANSWERS TO TRUE-FALSE

1. F **2.** F **3.** T **4.** T **5.** F **6.** F **7.** T **8.** T **9.** F **10.** T **11.** F **12.** T **13.** F **14.** T **15.** F **16.** F **17.** T **18.** T **19.** F **20.** T

CHAPTER 13

Generating Functions

13.1 Defining the Generating Function (Optional)

This section presents an abstract definition of the generating function. Anyone not desiring this degree of abstraction may skip to the next section without loss of continuity.

At this point we define generating functions. We shall see that they are a very powerful tool in solving recurrence relations and in counting.

Let $a_0, a_1, a_2, a_3, a_4, \ldots, a_n, \ldots$ be a sequence of real numbers. For the moment, we shall write it as $\langle a_0, a_1, a_2, a_3, a_4, \ldots, a_n, \ldots \rangle$. Addition of two sequences is defined by

$$\langle a_0, a_1, a_2, a_3, a_4, \ldots, a_n, \ldots \rangle + \langle b_0, b_1, b_2, b_3, b_4, \ldots, b_n, \ldots \rangle$$
$$= \langle a_0 + b_0, a_1 + b_1, a_2 + b_2, a_3 + b_3, a_4 + b_4, \ldots, a_n + b_n, \ldots \rangle$$

For a real number c, we shall define

$$c \circ \langle a_0, a_1, a_2, a_3, a_4, \ldots, a_n, \ldots \rangle = \langle ca_0, ca_1, ca_2, ca_3, ca_4, \ldots, ca_n, \ldots \rangle$$

and

$$x \circ \langle a_0, a_1, a_2, a_3, a_4, \ldots, a_n, \ldots \rangle = \langle 0, a_0, a_1, a_2, a_3, a_4, \ldots, a_n, \ldots \rangle$$

so that x shifts every element in the sequence one place to the right and places a 0 in the vacant space. We shall define

$$x^{k+1} \circ \langle a_0, a_1, a_2, a_3, a_4, \ldots, a_n, \ldots \rangle$$
$$= x \circ (x^k \circ \langle a_0, a_1, a_2, a_3, a_4, \ldots, a_n, \ldots \rangle)$$

so that

$$x \circ \langle 1, 0, 0, 0, 0, \ldots \rangle = \langle 0, 1, 0, 0, 0, \ldots \rangle$$
$$x^2 \circ \langle 1, 0, 0, 0, 0, \ldots \rangle = x \circ \langle 0, 1, 0, 0, 0, \ldots \rangle$$
$$= \langle 0, 0, 1, 0, 0, \ldots \rangle$$
$$x^3 \circ \langle 1, 0, 0, 0, 0, \ldots \rangle = x \circ (x^2 \circ \langle 1, 0, 0, 0, 0, \ldots \rangle)$$
$$= x \circ \langle 0, 0, 1, 0, 0, \ldots \rangle$$
$$= \langle 0, 0, 0, 1, 0, \ldots \rangle$$

It can be shown inductively that $x^k \circ \langle 1, 0, 0, 0, 0, \ldots \rangle$ is a sequence with k zeros followed by a one and then zeros in the remainder of the sequence. Since $\langle 1, 0, 0, 0, 0, \ldots \rangle$ is the sequence $1, 0, 0, 0, \ldots$, we shall identify it with the number 1. Thus,

$$x = x \cdot 1 = x \circ \langle 1, 0, 0, 0, 0, \ldots \rangle = \langle 0, 1, 0, 0, 0, \ldots \rangle$$
$$x^2 = x^2 \cdot 1 = x^2 \circ \langle 1, 0, 0, 0, 0, \ldots \rangle = \langle 0, 0, 1, 0, 0, \ldots \rangle$$
$$x^3 = x^3 \cdot 1 = x^3 \circ \langle 1, 0, 0, 0, 0, \ldots \rangle = \langle 0, 0, 0, 1, 0, 0, \ldots \rangle$$
$$x^4 = x^4 \cdot 1 = x^4 \circ \langle 1, 0, 0, 0, 0, \ldots \rangle = \langle 0, 0, 0, 0, 1, 0, 0, \ldots \rangle$$

and $x^k = x^k \circ \langle 1, 0, 0, 0, 0, \ldots \rangle$ is a sequence with k zeros followed by a one and then zeros in the remainder of the sequence. We see that

$$\langle a_0, a_1, a_2, a_3, a_4, 0, 0, 0, 0, \ldots \rangle = a_0 \circ \langle 1, 0, 0, 0, 0, 0, \ldots \rangle$$
$$+ a_1 \circ \langle 0, 1, 0, 0, 0, 0, \ldots \rangle$$
$$+ a_2 \circ \langle 0, 0, 1, 0, 0, 0, \ldots \rangle$$
$$+ a_3 \circ \langle 0, 0, 0, 1, 0, 0, \ldots \rangle$$
$$+ a_4 \circ \langle 0, 0, 0, 0, 1, 0, \ldots \rangle$$
$$= a_0 + a_1 x + a_2 x^2 + a_3 x^3 + a_4 x^4$$

and

$$\langle a_0, a_1, a_2, a_3, a_4, \ldots, a_n, 0, 0, 0, 0, \ldots \rangle$$
$$= a_0 + a_1 x + a_2 x^2 + a_3 x^3 + a_4 x^4 + \cdots + a_n x^n$$

In general,

$$\langle a_0, a_1, a_2, a_3, a_4, \ldots, a_n, \ldots \rangle = a_0 + a_1 x + a_2 x^2 + a_3 x^3 + a_4 x^4 + \cdots + a_n x^n \cdots$$

The expression $a_0 + a_1 x + a_2 x^2 + a_3 x^3 + a_4 x^4 + \cdots + a_n x^n \cdots$ is the **generating function** of the sequence $a_0, a_1, a_2, a_3, \ldots$.

In Section 13.5, we use a somewhat different type of generating function called the exponential function. Above, we let

$$\langle 1, 1, 1, 1, 1, 1, \ldots \rangle = 1 + x + x^2 + x^3 + x^4 + x^5 + \cdots$$
$$\langle a_0, a_1, a_2, a_3, a_4, \cdots \rangle = a_0 + a_1 x + a_2 x^2 + a_3 x^3 + a_4 x^4 + \cdots$$

and

$$\langle a_0, a_1, a_2, a_3, \ldots, a_n, 0, 0, 0, 0, \ldots \rangle = a_0 + a_1 x + a_2 x^2 + a_3 x^3 + \cdots + a_n x^n$$

For the exponential generating function, we let

$$\langle 1, 1, 1, 1, 1, 1, \ldots \rangle = 1 + \frac{x}{1!} + \frac{x^2}{2!} + \frac{x^3}{3!} + \frac{x^4}{4!} + \frac{x^5}{5!} + \frac{x^6}{6!} + \cdots$$

We then have

$$\langle a_0, a_1, a_2, a_3, a_4, \cdots \rangle = a_0 + a_1 \frac{x}{1!} + a_2 \frac{x^2}{2!} + a_3 \frac{x^3}{3!} + a_4 \frac{x^4}{4!} + \cdots$$

and

$$\langle a_0, a_1, a_2, a_3, \ldots, a_n, 0, 0, 0, 0, \ldots \rangle = a_0 + a_1 \frac{x}{1!} + a_2 \frac{x^2}{2!} + a_3 \frac{x^3}{3!} + \cdots + a_n \frac{x^n}{n!}$$

13.2 Generating Functions and Recurrence Relations

In this section we show how generating functions can be used to solve recurrence relations. In the next three sections, we shall find that they are a powerful tool for developing counting techniques. We shall call $a_0 + a_1 x + a_2 x^2 + a_3 x^3 + a_4 x^4 + \cdots + a_n x^n \cdots$ the **generating function** of the sequence $a_0, a_1, a_2, a_3, \ldots$.

If $f(x)$ and $g(x)$ are a polynomials, we shall define $\dfrac{f(x)}{g(x)}$ by

$$\frac{f(x)}{g(x)} = a_0 + a_1 x + a_2 x^2 + a_3 x^3 + a_4 x^4 + \cdots + a_n x^n \cdots$$

if and only if

$$f(x) = g(x)(a_0 + a_1 x + a_2 x^2 + a_3 x^3 + a_4 x^4 + \cdots + a_n x^n \cdots)$$

DEFINITION 13.1 If

$$\frac{f(x)}{g(x)} = a_0 + a_1 x + a_2 x^2 + a_3 x^3 + a_4 x^4 + \cdots + a_n x^n \cdots$$

then

$$a_0 + a_1 x + a_2 x^2 + a_3 x^3 + a_4 x^4 + \cdots + a_n x^n \cdots$$

is called the **expansion** of $\dfrac{f(x)}{g(x)}$.

THEOREM 13.2 For all m,

$$\frac{1}{(1-ax)^m} = 1 + \binom{m}{1} ax + \binom{m+1}{2} a^2 x^2 + \binom{m+2}{3} a^3 x^3 + \cdots + \binom{m+n-1}{n} a^n x^n + \cdots$$

Proof Using induction, we first show for $m = 1$ that

$$\frac{1}{(1-ax)} = 1 + ax + a^2 x^2 + a^3 x^3 + \cdots + a^n x^n + \cdots$$

but this is equivalent to showing that

$$1 = (1-ax)(1 + ax + a^2 x^2 + a^3 x^3 + \cdots + a^n x^n + \cdots)$$

But

$$(1-ax)(1 + ax + a^2 x^2 + a^3 x^3 + \cdots + a^n x^n + \cdots)$$
$$= 1 + ax + a^2 x^2 + \cdots + a^n x^n + \cdots - ax(1 + ax + a^2 x^2 + \cdots + a^n x^n + \cdots)$$
$$= 1 + ax + a^2 x^2 + \cdots + a^n x^n + \cdots - ax - a^2 x^2 - \cdots - a^n x^n - \cdots$$
$$= 1$$

and the theorem is true for $m = 1$.

Assume the theorem is true for $m = k$ so that

$$\frac{1}{(1-ax)^k} = 1 + \binom{k}{1}ax + \binom{k+1}{2}a^2x^2 + \binom{k+2}{3}a^3x^3 + \cdots + \binom{k+n-1}{n}a^nx^n + \cdots$$

We want to prove that

$$\frac{1}{(1-ax)^{k+1}} = 1 + \binom{k+1}{1}ax + \binom{k+2}{2}a^2x^2 + \binom{k+3}{3}a^3x^3 + \cdots + \binom{k+n}{n}a^nx^n + \cdots$$

Since $\dfrac{1}{(1-ax)^{k+1}} = \dfrac{1}{(1-ax)} \dfrac{1}{(1-ax)^k}$, we have $\dfrac{1}{(1-ax)^k} = (1-ax) \dfrac{1}{(1-ax)^{k+1}}$.

Therefore, the equation for the generating function for $\dfrac{1}{(1-ax)^{k+1}}$ given previously is correct if and only if, when multiplied by $(1-ax)$, we get the generating function for $\dfrac{1}{(1-ax)^k}$.

But

$$(1-ax)\left(1 + \binom{k+1}{1}ax + \binom{k+2}{2}a^2x^2 + \binom{k+3}{3}a^3x^3 + \cdots + \binom{k+n}{n}a^nx^n + \cdots\right)$$

$$= 1 + \binom{k+1}{1}ax + \binom{k+2}{2}a^2x^2 + \binom{k+3}{3}a^3x^3 + \cdots + \binom{k+n}{n}a^nx^n + \cdots$$

$$- (ax)\left(1 + \binom{k+1}{1}ax + \binom{k+2}{2}a^2x^2 + \binom{k+3}{3}a^3x^3 + \cdots + \binom{k+n}{n}a^nx^n + \cdots\right)$$

$$= 1 + \binom{k+1}{1}ax + \binom{k+2}{2}a^2x^2 + \binom{k+3}{3}a^3x^3 + \cdots + \binom{k+n}{n}a^nx^n + \cdots$$

$$- ax - \binom{k+1}{1}a^2x^2 - \binom{k+2}{2}a^3x^3 - \cdots - \binom{k+n-1}{n-1}a^nx^n - \cdots$$

$$= 1 + \binom{k}{1}ax + \binom{k+1}{2}a^2x^2 + \binom{k+2}{3}a^3x^3 + \cdots + \binom{k+n-1}{n}a^nx^n + \cdots$$

since by Pascal's identity

$$\binom{k+n-1}{n} + \binom{k+n-1}{n-1} = \binom{k+n}{n}$$

so that

$$\binom{k+n}{n} - \binom{k+n-1}{n-1} = \binom{k+n-1}{n}$$

Therefore,

$$(1-ax)^{k+1}\left(1 + \binom{k+1}{1}ax + \binom{k+2}{2}a^2x^2 + \binom{k+3}{3}a^3x^3 + \cdots + \binom{k+n}{n}a^nx^n + \cdots\right)$$

$$= (1-ax)^k\left(1 + \binom{k}{1}ax + \binom{k+1}{2}a^2x^2 + \binom{k+2}{3}a^3x^3 + \cdots + \binom{k+n-1}{n}a^nx^n + \cdots\right)$$

But by our induction hypothesis,

$$(1-ax)^k\left(1 + \binom{k}{1}ax + \binom{k+1}{2}a^2x^2 + \binom{k+2}{3}a^3x^3 + \cdots + \binom{k+n-1}{n}a^nx^n + \cdots\right) = 1$$

and we are finished. ∎

From this theorem we conclude that

$$\frac{1}{(1-ax)^2} = 1 + 2ax + 3a^2x^2 + 4a^3x^3 + \cdots + (n+1)a^nx^n + \cdots$$

and
$$\frac{1}{(1-ax)^3} = 1 + 3ax + 6a^2x^2 + 10a^3x^3 + \cdots + \frac{(n+1)(n+2)}{2}a^n x^n + \cdots$$

It is shown in calculus that any fraction of the form
$$\frac{Ax + B}{(1 - Cx)(1 - Dx)}$$
where C and D are distinct is equal to
$$\frac{a}{1 - Cx} + \frac{b}{1 - Dx}$$
for some constants a and b. This is called **partial fraction decomposition**.

Further, any fraction of the form
$$\frac{Ax^2 + Bx + C}{(1 - Dx)(1 - Ex)(1 - Fx)}$$
where D, E, and F are distinct, is equal to
$$\frac{a}{1 - Dx} + \frac{b}{1 - Ex} + \frac{c}{1 - Fx}$$
for constants a, b, and c. Also if D and E are distinct, then
$$\frac{Ax^2 + Bx + C}{(1 - Dx)^2(1 - Ex)} = \frac{a}{1 - Dx} + \frac{b}{(1 - Dx)^2} + \frac{c}{1 - Ex}$$
for constants a, b, and c, and if E and F are distinct, then
$$\frac{Ax^3 + Bx^2 + Cx + D}{(1 - Ex)^3(1 - Fx)} = \frac{a}{1 - Ex} + \frac{b}{(1 - Ex)^2} + \frac{c}{(1 - Ex)^3} + \frac{d}{1 - Fx}$$
for constants a, b, c, and d.

Suppose we want to expand the generating function
$$\frac{15x}{1 + 3x - 4x^2} = \frac{15x}{(1 + 4x)(1 - x)}$$

Let
$$\frac{15x}{(1 + 4x)(1 - x)} = \frac{a}{1 + 4x} + \frac{b}{1 - x}$$
for constants a and b. Multiplying both sides by $(1 + 4x)(1 - x)$, we have
$$15x = a(1 - x) + b(1 + 4x)$$

Letting $x = 1$, we have
$$15(1) = a(1 - 1) + b(1 + 4)$$
and solving, we have
$$b = 3$$

Letting $x = -\frac{1}{4}$, we have
$$15\left(-\frac{1}{4}\right) = a\left(1 - \frac{1}{4}\right) + b\left(1 + 4\left(-\frac{1}{4}\right)\right)$$

and solving, we have
$$a = -5$$
so that
$$\frac{15x}{(1+4x)(1-x)} = \frac{-5}{1+4x} + \frac{3}{1-x}$$

Using the formula
$$\frac{1}{(1-ax)} = 1 + ax + a^2x^2 + a^3x^3 + \cdots + a^nx^n + \cdots$$

for $a = -4$ and $a = 1$, we have
$$\frac{1}{1+4x} = 1 + (-4)x + (-4)^2x^2 + (-4)^3x^3 + \cdots + (-4)^nx^n + \cdots$$

and
$$\frac{1}{(1-x)} = 1 + x + x^2 + x^3 + \cdots + x^n + \cdots$$

so that
$$\frac{-5}{1+4x} + \frac{3}{1-x} = -5(1 + (-4)x + (-4)^2x^2 + (-4)^3x^3 + \cdots + (-4)^nx^n + \cdots)$$
$$+ 3(1 + x + x^2 + x^3 + \cdots + x^n + \cdots)$$
$$= -2 + 23x - 77x^2 + 323x^3 + \cdots + (-5(-4)^n + 3)x^n + \cdots$$

We have one additional theorem to prove about generating functions that we shall use later.

THEOREM 13.3 For all $n \geq 0$, $\dfrac{1-x^{n+1}}{1-x} = 1 + x + x^2 + x^3 + \cdots + x^n$.

Proof
$$(1-x^{n+1})\frac{1}{1-x} = (1-x^{n+1})(1 + x + x^2 + \cdots + x^n + \cdots)$$
$$= (1)(1 + x + x^2 + \cdots + x^n + \cdots)$$
$$- x^{n+1}(1 + x + x^2 + \cdots + x^n + \cdots)$$
$$= (1 + x + x^2 + \cdots + x^n + x^{n+1} + \cdots)$$
$$- (x^{n+1} + x^{n+2} + x^{n+2} + \cdots)$$
$$= 1 + x + x^2 + x^3 + \cdots + x^n \qquad \blacksquare$$

We shall now consider how to use the generating function to solve recurrence relations. We begin with a simple example to illustrate the method. Consider the recurrence relation $a_n = 3a_{n-1}$ or $a_n - 3a_{n-1} = 0$. Let
$$f(x) = a_0 + a_1x + a_2x^2 + a_3x^3 + \cdots + a_nx^n + \cdots$$
and multiply both sides by $3x$, remembering that x shifts every element to the right, so that
$$3xf(x) = 3a_0x + 3a_1x^2 + 3a_2x^3 + 3a_3x^4 + \cdots + 3a_{n-1}x^n + 3a_nx^{n+1} + \cdots$$

Subtracting $3xf(x)$ from $f(x)$ we have

$$f(x) - 3xf(x) = a_0 + (a_1 - 3a_0)x + (a_2 - 3a_1)x^2$$
$$+ (a_3 - 3a_2)x^3 + \cdots + (a_n - 3a_{n-1})x^n + \cdots$$

But $a_n - 3a_{n-1} = 0$ for all $n \geq 1$, so that

$$f(x) - 3xf(x) = a_0$$

Thus

$$(1 - 3x)f(x) = a_0$$

and

$$f(x) = \frac{a_0}{(1 - 3x)}$$
$$= a_0(1 + 3x + 3^2 x^2 + 3^3 x^3 + \cdots + 3^n x^n + \cdots)$$
$$= a_0 + 3a_0 x + 3^2 a_0 x^2 + 3^3 a_0 x^3 + \cdots + 3^n a_0 x^n + \cdots$$

Therefore, $a_n = 3^n a_0$. Notice that we multiplied by $3x$ so that when we subtracted we would get $a_n - 3a_{n-1}$, which we knew would equal 0 because of the definition of the recurrence relation.

EXAMPLE 13.4 Solve the recursive function

$$a_0 = 5$$
$$a_k = a_{k-1} + 3$$

We can also write the second equation as $a_k - a_{k-1} - 3 = 0$.

Again let

$$f(x) = a_0 + a_1 x + a_2 x^2 + a_3 x^3 + \cdots + a_n x^n + \cdots$$

Since the coefficient of a_{k-1} is 1, this time we want to find

$$xf(x) = a_0 x + a_1 x^2 + a_2 x^3 + a_3 x^4 + \cdots + a_n x^{n+1} + \cdots$$

We also want a generating function that has 3 in each term, but this is

$$\frac{3}{1-x} = 3(1 + x + x^2 + x^3 + \cdots + x^n + \cdots)$$
$$= 3 + 3x + 3x^2 + 3x^3 + \cdots + 3x^n + \cdots$$

Now

$$f(x) - xf(x) - \frac{3}{1-x} = a_0 - 3 + (a_1 - a_0 - 3)x + (a_2 - a_1 - 3)x^2$$
$$+ \cdots + (a_n - a_{n-1} - 3) + \cdots$$

But $a_k - a_{k-1} - 3 = 0$ for all $k \geq 1$, so that

$$f(x) - xf(x) - \frac{3}{1-x} = a_0 - 3 = 2$$

Solving for $f(x)$, we have

$$f(x) = \frac{3}{(1-x)^2} + \frac{2}{1-x}$$
$$= 3(1 + 2x + 3x^2 + \cdots + (n+1)x^n + \cdots) + 2(1 + x + x^2 + \cdots x^n \cdots)$$
$$= 5 + 8x + 11x^2 + 14x^3 + \cdots + (3n + 5)x^n + \cdots$$

so that
$$a_n = 3n + 5$$

EXAMPLE 13.5 Solve the recursive function
$$a_0 = 1$$
$$a_1 = 4$$
$$a_k = a_{k-1} + 6a_{k-2}$$

As before, we let
$$f(x) = a_0 + a_1 x + a_2 x^2 + a_3 x^3 + \cdots + a_n x^n + \cdots$$

Since the coefficient of a_{k-1} is 1 and the coefficient of a_{k-2} is 6, we want
$$xf(x) = a_0 x + a_1 x^2 + a_2 x^3 + a_3 x^4 + \cdots + a_n x^{n+1} + \cdots$$

and
$$6x^2 f(x) = 6a_0 x^2 + 6a_1 x^3 + 6a_2 x^4 + 6a_3 x^5 + \cdots + 6a_n x^{n+2} + \cdots$$

Therefore,
$$f(x) - xf(x) - 6x^2 f(x) =$$
$$a_0 + (a_1 - a_0)x + (a_2 - a_1 - 6a_0)x^2$$
$$+ (a_3 - a_2 - 6a_1)x^3 + \cdots + (a_n - a_{n-1} - 6a_{n-2})x^n + \cdots$$

But $a_n x^n - a_{n-1} - 6a_{n-2} = 0$ for all $n \geq 1$, so that
$$f(x) - xf(x) - 6x^2 f(x) = a_0 + (a_1 - a_0)x$$
$$= 1 + 3x$$

Solving for $f(x)$, we have
$$f(x) = \frac{1 + 3x}{1 - x - 6x^2}$$
$$= \frac{1 + 3x}{(1 - 3x)(1 + 2x)}$$

Letting
$$\frac{1 + 3x}{(1 - 3x)(1 + 2x)} = \frac{a}{(1 - 3x)} + \frac{b}{(1 + 2x)}$$

and multiplying both sides by $(1 - 3x)(1 + 2x)$ we have
$$1 + 3x = a(1 + 2x) + b(1 - 3x)$$

Letting $x = -\frac{1}{2}$, we have
$$1 + \left(-\frac{3}{2}\right) = a\left(1 + 2\left(-\frac{1}{2}\right)\right) + b\left(1 - 3\left(-\frac{1}{2}\right)\right)$$

so that
$$b = -\frac{1}{5}$$

Letting $x = \frac{1}{3}$, we have

$$1 + 1 = a\left(1 + 2\left(\frac{1}{3}\right)\right) + b\left(1 - 3\left(\frac{1}{3}\right)\right)$$

so that

$$a = \frac{6}{5}$$

Therefore, we have

$$f(x) = \left(\frac{1}{5}\right)\left(\frac{6}{(1-3x)} - \frac{1}{(1+2x)}\right)$$
$$= \left(\frac{6}{5}\right)(1 + 3x + 3^2x^2 + 3^3x^3 + \cdots + 3^nx^n + \cdots)$$
$$- \left(\frac{1}{5}\right)1 + (-2)x + (-2)^2x^2 + (-2)^3x^3 + \cdots + (-2)^nx^n + \cdots$$

so that

$$a_n = \left(\frac{1}{5}\right)(6 \cdot (3)^n - (-2)^n)$$

∎

EXAMPLE 13.6   Solve the recursive function

$$a_0 = 1$$
$$a_k = 3a_{k-1} + 4^n$$

As before, we let

$$f(x) = a_0 + a_1x + a_2x^2 + a_3x^3 + \cdots + a_nx^n + \cdots$$

Since the coefficient of $a_{k-1} = 3$, we want

$$3xf(x) = 3a_0x + 3a_1x^2 + 3a_2x^3 + 3a_3x^4 + \cdots + 3a_nx^{n+1} + \cdots$$

Since

$$\frac{1}{(1-4x)} = (1 + 4x + 4^2x^2 + 4^3x^3 + \cdots + 4^nx^n + \cdots)$$

this will provide 4^nx^n. Therefore,

$$f(x) - 3xf(x) - \frac{1}{(1-4x)}$$
$$= a_0 - 1 + (a_1 - 3a_0 - 4)x + (a_2 - 3a_1 - 4^2)x^2$$
$$+ (a_3 - 3a_2 - 4^2)x^3 + \cdots + (a_n - a_{n-1} - 4^n)x^n + \cdots$$

Since $a_n - 3a_{k-1} - 4^n = 0$ for all $n \geq 1$, we have

$$f(x) - 3xf(x) - \frac{1}{(1-4x)} = a_0 - 1 = 0$$

Solving for $f(x)$ we have

$$f(x) = \frac{1}{(1-4x)(1-3x)} = 0$$

Let
$$\frac{1}{(1-4x)(1-3x)} = \frac{a}{(1-4x)} + \frac{b}{(1-3x)}$$
Multiplying both sides by $(1-4x)(1-3x)$, we have
$$1 = (1-3x)a + (1-4x)b$$
Letting $x = \frac{1}{3}$, we have
$$1 = \left(1 - 3 \cdot \frac{1}{3}\right)a + \left(1 - 4 \cdot \frac{1}{3}\right)b$$
so that
$$b = -3$$
Letting $x = \frac{1}{4}$, we have
$$1 = \left(1 - 3 \cdot \frac{1}{4}\right)a + \left(1 - 4 \cdot \frac{1}{4}\right)b$$
so that
$$a = 4$$
Therefore, we have
$$f(x) = \frac{4}{(1-4x)} - \frac{3}{(1-3x)}$$
$$= 4(1 + 4x + 4^2x^2 + 4^3x^3 + \cdots + 4^nx^n + \cdots)$$
$$- 3(1 + 3x + 3^2x^2 + 3^3x^3 + \cdots + 3^nx^n + \cdots)$$
so that
$$a_n = 4 \cdot 4^n - 3 \cdot 3^n = 4^{n+1} - 3^{n+1}$$ ∎

EXAMPLE 13.7 Solve the recursive function
$$a_0 = 3$$
$$a_n = 2a_{n-1} + n$$
We can also write the second equation as $a_n - 2a_{n-1} - n = 0$.
Again let
$$f(x) = a_0 + a_1x + a_2x^2 + a_3x^3 + \cdots + a_nx^n + \cdots$$
Since the coefficient of a_{k-1} is -2, this time we want to find
$$2xf(x) = 2a_0x + 2a_1x^2 + 2a_2x^3 + 2a_3x^4 + \cdots + 2a_nx^{n+1} + \cdots$$
We also want a generating function that has n in each term, but this is
$$\frac{x}{(1-x)^2} = x(1 + 2x + 3x^2 + \cdots + nx^{n-1} + \cdots)$$
$$= x + 2x^2 + 3x^3 + \cdots + nx^n + \cdots$$
Now
$$f(x) - 2xf(x) - \frac{x}{(1-x)^2} = a_0 + (a_1 - 2a_0 - 1)x + (a_2 - 2a_1 - 2)x^2$$
$$+ \cdots + (a_n - 2a_{n-1} - n) + \cdots$$

But $a_k - 2a_{k-1} - k = 0$ for all $k \geq 1$, so that

$$f(x) - 2xf(x) - \frac{x}{(1-x)^2} = a_0 = 3$$

Solving for $f(x)$, we have

$$f(x) = \frac{x}{(1-2x)(1-x)^2} + \frac{3}{(1-2x)}$$

$$= \frac{3 - 5x + 3x^2}{(1-2x)(1-x)^2}$$

$$= \frac{A}{(1-x)} + \frac{B}{(1-x)^2} + \frac{C}{(1-2x)}$$

so that

$$3 - 5x + 3x^2 = A(1-x)(1-2x) + B(1-2x) + C(1-x)^2$$
$$= A(1 - 3x + 2x^2) + B(1 - 2x) + C(1 - 2x + x^2)$$

Equating coefficients of x^2, x, and 1, respectively, we have

$$3 = 2A + C$$
$$-5 = -3A - 2B - 2C$$
$$3 = A + B + C$$

Solving, we have $A = -1$, $B = -1$, and $C = 5$. Therefore,

$$f(x) = \frac{-1}{(1-x)} - \frac{1}{(1-x)^2} + \frac{5}{(1-2x)}$$

$$= -(1 + x + x^2 + \cdots + x^n + \cdots)$$
$$- (1 + 2x + 3x^2 + \cdots + (n+1)x^n + \cdots)$$
$$+ 5(1 + 2x + (2x)^2 + \cdots + (2x)^n \cdots)$$

and

$$a_n = -n - 2 + 5(2^n)$$ ∎

EXAMPLE 13.8   We again provide a solution to Example 12 from the Shier collection. The problem is to find the number of spanning trees of $K_{2,n}$. We already know that t_n, the number of spanning trees, satisfies the recurrence relations

$$t_1 = 1$$
$$t_n = 2t_{n-1} + 2^{n-1}$$

Let $T(x) = t_0 + t_1 x + t_2 x^2 + \ldots$

$$xT(x) = t_0 x + t_1 x^2 + t_2 x^3 + \ldots$$

$$\frac{x}{1-2x} = x + 2x^2 + 2^3 x^3 + \ldots$$

so that $T(x) - 2xT(x) = \frac{x}{1-2x}$.

Solving for $T(x)$, we have $T(x) = \frac{x}{(1-2x)^2} = x + 2(2)x^2 + 2^2(3)x^3 + \cdots + (n+1)2^n x^{n+1}$ and $t_n = n2^{n-1}$. ∎

Chapter 13 Generating Functions

Exercises

Express the following as the sum of partial fractions:

1. $\dfrac{1}{(x+4)(x+3)}$

2. $\dfrac{2x-3}{(x-2)(x-3)}$

3. $\dfrac{x^2+2x+3}{(x-1)^2(x-2)}$

4. $\dfrac{x}{(x-1)(x-2)(x-3)}$

5. $\dfrac{3x^2+1}{(x-1)^2(x-2)^2}$

6. $\dfrac{2x+1}{(x-4)(x+2)}$

7. $\dfrac{2x-3}{(x-2)(x+4)}$

8. $\dfrac{x^2+2x+3}{(x+1)(x-2)^2}$

9. $\dfrac{x^2}{(x-1)(x+3)(x-5)}$

10. $\dfrac{3x^2+1}{(x-1)^3}$

Expand the following generating functions:

11. $\dfrac{1}{x+4}$

12. $\dfrac{1}{(x-2)^2}$

13. $\dfrac{1}{(x-1)^3}$

14. $\dfrac{x}{(x-1)(x-2)}$

15. $\dfrac{3x^2+1}{(x-1)^2(x-2)}$

16. $\dfrac{1}{x-3}$

17. $\dfrac{x}{(x-1)^2}$

18. $\dfrac{x}{(x-1)^3}$

19. $\dfrac{x}{(x+1)(x-2)(x+3)}$

20. $\dfrac{2x^2+6}{(x-1)^2(x+2)^2}$

Use generating functions to solve the following recursive functions:

21. $a_0 = 1$
    $a_n = 2a_{n-1} + 3^n$ for $n > 0$

22. $a_0 = 2$
    $a_n = 3a_{n-1} + n$ for $n > 0$

23. $a_0 = 8$
    $a_1 = 16$
    $a_n = 2a_{n-1} + 3a_{n-2}$ for $n > 1$

24. $a_0 = 1$
    $a_1 = 0$
    $a_n = 4a_{n-1} - 4a_{n-2}$ for $n > 1$

25. $a_0 = 4$
    $a_n = 2a_{n-1} - 3$ for $n > 0$

26. $a_0 = 1$
    $a_n = 2a_{n-1}$ for $n > 0$

27. $a_0 = 1$
    $a_n = 4a_{n-1} + 2n$ for $n > 0$

28. $a_0 = 1$
    $a_1 = 3$
    $a_n = 5a_{n-1} - 6a_{n-2}$ for $n > 1$

29. $a_0 = 1$
    $a_1 = 3$
    $a_n = 7a_{n-1} - 10a_{n-2}$ for $n > 1$

30. $a_0 = 1$
    $a_n = 2a_{n-1} + n - 1$ for $n > 0$

31. $a_0 = 1$
    $a_n = a_{n-1} + n$ for $n > 0$

32. $a_0 = 2$
    $a_n = 3a_{n-1} + 2^n$ for $n > 0$

33. $a_0 = 1$
    $a_n = 2a_{n-1} + 2^n + n$ for $n > 0$

34. $a_0 = 1$
    $a_1 = 8$
    $a_n = 6a_{n-1} - 9a_{n-2} + 2^n$ for $n > 1$

13.3 Generating Functions and Counting

As one might guess from the title of this section, we now explore the use of generating functions for counting. This is really a beautiful technique that one can really appreciate when and if one becomes accustomed to using it. As in the last section, we shall consider generating functions to be formal expressions and not be concerned with many of the concepts that one encounters in calculus, such as convergence, Taylor's series, integration, and derivatives.

We begin by looking at $(1+x)^n$. We know, by the binomial theorem, that

$$(1+x)^n = \binom{n}{0}x^0 + \binom{n}{1}x^1 + \binom{n}{2}x^2 + \cdots + \binom{n}{n}x^n$$

so that the coefficient of x^k is the number of ways that k objects can be selected from n objects without order. Therefore, the generating function given by $(1+x)^n$ is the generating function for combinations of n objects taken k at a time.

By Theorem 13.2,

$$\frac{1}{(1-x)^m} = 1 + \binom{m}{1}x + \binom{m+1}{2}x^2 + \binom{m+2}{3}x^3 + \cdots + \binom{m+n-1}{n}x^n + \cdots$$

so that the coefficient of x^k is the number of combinations of k objects selected from m objects with repetition. Therefore, the generating function $\frac{1}{(1-x)^m}$ is the generating function used to determine the number of combinations of m objects taken k at a time with repetition.

Consider now the product

$$(1_a + x_a)(1_b + x_b)(1_c + x_c + x_c^2)(1_d + x_d + x_d^2)$$

where the subscripts are only to keep track of where each x comes from rather than to affect its value. Thus, x_a, x_b, x_c, and x_d are all simply x, with a label to keep track of it. If we let $x_i^0 = 1_i$ for $i = a, b, c,$ and d, and look at the x^4 term, we see that it is the sum

$$x_a^0 x_b^0 x_c^2 x_d^2 + x_a^0 x_b^1 x_c^2 x_d^1 + x_a^1 x_b^0 x_c^2 x_d^1 + x_a^1 x_b^1 x_c^2 1 x_d^0$$
$$+ x_a^0 x_b^1 x_c^1 x_d^2 + x_a^1 x_b^0 x_c^1 x_d^2 + x_a^1 x_b^1 x_c^0 x_d^2 + x_a^1 x_b^1 x_c^1 x_d^1$$

Each term has the sum of its exponents equal to four. Thus, looking at the exponents of the terms, we see that we have all possible solutions of the equation

$$e_a + e_b + e_c + e_d = 4$$

for $0 \le e_a \le 1$, $0 \le e_b \le 1$, $0 \le e_c \le 2$, and $0 \le e_d \le 2$. In general, for $0 \le r \le 6$, the coefficient of x^r is the number of solutions of

$$e_a + e_b + e_c + e_d = r$$

for $0 \le e_a \le 1$, $0 \le e_b \le 1$, $0 \le e_c \le 2$ and $0 \le e_d \le 2$.

Suppose we have a set A containing 1 object of type a, 1 object of type b, 2 objects of type c, and 2 objects of type d. Let x_i^j represent j objects of type i. Thus, 1_a represents 0 elements of type a, x_b represents 1 element of type b, and x_c^2 represents 2 elements of type c. For example, $x_a^1 x_b^1 x_c^2 1 x_d^0$ represents selecting 1 object of type a, 1 object of type b, and 2 objects of type c. Then the coefficient of x^4 is the number of ways of selecting 4 objects from the 4 types in A. It may helpful to think of $1_c + x_c + x_c^2$ as meaning that 0, 1, or 2 objects of type c may be selected.

Similarly, leaving off the labels, in the product

$$(1+x)(1+x+x^2)(1+x+x^2+x^3)(1+x+x^2+x^3+x^4+x^5)$$

the coefficient of x^r represents the number of ways of selecting r objects from a set containing 1 element of type a, 2 elements of type b, 3 elements of type c, and 5 elements of type d. As we have seen, it can also represent the number of solutions of

$$e_a + e_b + e_c + e_d = r$$

for $0 \le e_a \le 1$, $0 \le e_b \le 2$, $0 \le e_c \le 3$ and $0 \le e_d \le 5$.

EXAMPLE 13.9 An urn contains 4 red, 5 blue, and 2 green balls. (a) Find the number of ways 7 balls may be drawn from the urn. (b) Find the number of ways 7 balls may be drawn if at least 1 red ball and 2 blue balls must be drawn.

498 Chapter 13 Generating Functions

For part (a), the generating function is
$$(1 + x + x^2 + x^3 + x^4)(1 + x + x^2 + x^3 + x^4 + x^5)(1 + x + x^2)$$
and the number of ways 7 balls may be drawn is the coefficient of x^7 for this generating function.

For part (b) since at least 1 red ball must be drawn, the corresponding polynomial is $(x + x^2 + x^3 + x^4)$, which represents 1, 2, 3, or 4 red balls being drawn. Similarly, since at least 2 blue balls may be drawn, the corresponding polynomial is $(x^2 + x^3 + x^4 + x^5)$, which represents 2, 3, 4, or 5 red balls being drawn. Thus, the generating polynomial is
$$(x + x^2 + x^3 + x^4)(x^2 + x^3 + x^4 + x^5)(1 + x + x^2)$$
and the number of ways of selecting 7 balls is the coefficient of x^7 in the expansion of the generating function. ∎

EXAMPLE 13.10 Suppose that an urn contains 3 red, 8 green, 9 orange, and 2 white balls. How many ways can 12 balls be selected from the urn if at least 1 red ball must be selected, an even number of green balls must be selected, and an odd number of orange balls must be selected? Since at least 1 red ball must be selected, 1, 2, or 3 balls must be selected, and the polynomial representing the red balls is
$$x + x^2 + x^3$$

Since an even number of green balls must be selected, 0, 2, 4, 6, or 8 green balls must be selected, and the polynomial representing the green balls is
$$1 + x^2 + x^4 + x^6 + x^8$$

An odd number of orange balls must be selected so that 1, 3, 5, 7, or 9 orange balls must be selected, and the polynomial representing the orange balls is
$$x + x^3 + x^5 + x^7 + x^9$$

Therefore, the generating function for this problem is
$$(x + x^2 + x^3)(1 + x^2 + x^4 + x^6 + x^8)(x + x^3 + x^5 + x^7 + x^9)(1 + x + x^2)$$
and the answer is the coefficient of x^{12}. ∎

We now introduce the generating functions from the previous section to help us find coefficients for some of our polynomial generating functions.

EXAMPLE 13.11 Find the coefficient of x^{24} in the $(x^3 + x^4 + x^5 + x^6 + \cdots)^4$.
$$(x^3 + x^4 + x^5 + x^6 + \cdots)^4 = (x^3)^4(1 + x + x^2 + x^3 + x^4 + \cdots)^4$$
$$= (x^{12})\left(\frac{1}{1-x}\right)^4 \quad \text{by Theorem 13.2}$$
$$= (x^{12})\frac{1}{(1-x)^4}$$
$$= (x^{12})\left(1 + 4x + \binom{5}{2}x^2 + \cdots + \binom{4+n-1}{n}x^n + \cdots\right)$$
by Theorem 13.2

Therefore, to find the coefficient of x^r in this generating function we need to find the coefficient of x^{r-12} in the generating function
$$1 + 4x + \binom{5}{2}x^2 + \cdots + \binom{4+n-1}{n}x^n + \cdots$$

so that the coefficient of x^r is
$$\binom{4+r-12-1}{r-12} = \binom{r-9}{r-12}$$
and the coefficient of x^{24} is
$$\binom{24-9}{24-12} = \binom{15}{12}$$

EXAMPLE 13.12

Find the number of ways 12 objects can be selected from five types of objects if at most 2 objects of the first three types may be chosen and an unlimited number of objects of the other two types may be chosen.

The generating function is
$$(1+x+x^2)^3(1+x+x^2+x^3+\cdots)^2$$
But by Theorem 13.3,
$$(1+x+x^2)^3 = \left(\frac{1-x^3}{1-x}\right)^3$$
Therefore,
$$(1+x+x^2)^3(1+x+x^2+x^3+\cdots)^2 = \left(\frac{1-x^3}{1-x}\right)^3 (1+x+x^2+x^3+\cdots)^2$$
$$= \frac{(1-x^3)^3}{(1-x)^3} \cdot \frac{1}{(1-x)^2}$$
$$= (1-x^3)^3 \cdot \frac{1}{(1-x)^5}$$
Since
$$(1-x^3)^3 = 1 - 3x^3 + 3x^6 - x^9$$
and
$$\frac{1}{(1-x)^5} = 1 + 5x + \binom{6}{2}x^2 + \cdots + \binom{5+n-1}{n}x^n \cdots$$
by the product rule of polynomials the coefficient of x^{12} in
$$(1 - 3x^3 + 3x^6 - x^9)\left(1 + 5x + \binom{6}{2}x^2 + \cdots + \binom{5+n-1}{n}x^n \cdots\right)$$
is
$$1 \cdot \binom{5+12-1}{12} - 3\binom{5+9-1}{9} + 3\binom{5+6-1}{6} - \binom{5+3-1}{3}$$

EXAMPLE 13.13

In how many ways can 20 objects be selected from five types of objects if the first type can be selected only in multiples 5, the second can be selected only in multiples of 3, at most 4 of the fourth kind can be selected, at least 3 of the fourth kind can be selected, and at most 2 of the fifth kind can be selected?

The generating function is
$$(1 + x^5 + x^{10} + \cdots) \cdot (1 + x^3 + x^6 + \cdots)$$
$$\times (1 + x + \cdots + x^4) \cdot (x^3 + x^4 + \cdots)(1 + x + x^2)$$

which is equal to

$$\frac{1}{1-x^5} \cdot \frac{1}{1-x^3} \cdot \frac{1-x^5}{1-x} \cdot \frac{x^3}{1-x} \cdot \frac{1-x^3}{1-x} = \frac{x^3}{(1-x)^3}$$

so that the generating function is

$$x^3 \left(1 + 3x + \binom{4}{2}x^2 + \cdots \binom{3+n-1}{n}x^n + \cdots\right)$$

The coefficient of x^r is $\binom{r-1}{r-3}$ and the coefficient of x^{20} is $\binom{19}{17}$. ∎

EXAMPLE 13.14 Find a generating function whose nth coefficient gives the number of nonnegative integral solutions of $e_1 + 4e_2 + 5e_3 + 3e_4 = n$. This is equivalent to finding the number of ways of selecting n objects when objects of type two are selected 4 at a time, objects of type three are selected 5 at a time, and objects of type four are selected 3 at a time. Therefore, the generating function is

$$(1 + x + x^2 + \cdots) \cdot (1 + x^4 + x^8 + \cdots)$$
$$\cdot (1 + x^5 + x^{10} + \cdots) \cdot (1 + x^3 + x^6 + \cdots)$$

but this is equal to

$$\frac{1}{(1-x)(1-x^4)(1-x^5)(1-x^3)}$$

∎

Exercises

Find generating functions whose rth coefficient gives the number of solutions of

1. $e_1 + e_2 + e_3 + e_4 = r$ where $0 \le e_2 \le 2$ and $0 \le e_4 \le 4$.
2. $e_1 + e_2 + e_3 = r$ where $0 \le e_1 \le 3$ and e_3 is even.
3. $e_1 + e_2 + e_3 + e_4 = r$ where e_1 is odd, e_2 is even, and $e_4 \ge 4$.
4. $e_1 + e_2 + e_3 + e_4 = r$ where $e_i \ge i$ for all i.
5. $e_1 + 2e_2 + 3e_3 + 4e_4 = r$.
6. $e_1 + e_2 + e_3 + e_4 = r$ where $e_1 \ge 2$ and $0 \le e_3 \le 3$.
7. $e_1 + e_2 + e_3 = r$ where $0 \le e_i \le 3$ for all i.
8. $e_1 + e_2 + e_3 + e_4 = r$ where e_1 is odd, e_2 and e_3 are even, and $e_4 \ge 2$.
9. $e_1 + e_2 + e_3 + e_4 = r$ where $e_i \ge 3$ for all i.
10. $e_1 + 2e_2 + e_3 + 2e_4 = r$.

Find a generating function used for finding the number of ways of selecting r items out of

11. 4 red, 3 blue, 6 orange, and 2 green balls.
12. 2 red, 5 green, 4 orange, and 3 green balls, if at least 1 ball of each kind must be selected.
13. 6 red, 12 black, 7 white, and 10 blue balls, if at least 4 black balls must be selected, and an even number of blue balls must be selected.
14. 6 red, 12 black, 7 white, and 10 blue balls, if at least 1 ball of each kind must be selected, an even number of red balls must be selected, and an odd number of white balls must be selected.
15. 6 red, 5 blue, 4 orange, and 3 green balls.
16. 7 red, 5 green, 8 orange, and 4 white balls, if an odd number of green and red balls must be selected and an even number of orange and white balls must be selected.
17. 5 red, 4 purple, 6 white, and 8 black balls if at least 2 black balls, 1 purple ball, and 3 red balls must be selected, and an even number of white balls must be selected.
18. 9 red, 7 black, 6 white, and 11 blue balls if at least 2 balls of each kind must be selected, an odd number of red balls must be selected, and an odd number of white balls must be selected.

19. Find the generating function used to determine the number of ways to get k cents, using pennies, nickels, dimes, and quarters.

20. In a small town of 50 registered voters, each person either votes twice or stays home, except for the mayor who votes either three or five times. Use a generating function to determine the number of ways n votes may be cast.

21. For $n \geq 8$, n cents postage can always be supplied using 5- and 3-cent stamps. Use a generating function to describe the number of ways n cents postage can be supplied using only 5- and 3-cent stamps.

22. Find a generating function to determine how many ways k pieces of chocolate can be distributed in 6 boxes if an even number of chocolates are placed in the first box, an odd number of chocolates in the second box, at least 3 in the third box, at most 3 in the fourth box, and any number in the fifth and sixth boxes.

23. Use a generating function to determine the number of ways k dice may thrown for a sum of n.

24. Find the coefficient of x^7 in the generating function formed by the expansion of $f(x) = (x + x^2 + x^3)^5$.

25. Find the coefficient of x^{10} in the generating function formed by the expansion of $f(x) = \dfrac{x^4}{(1-x)^5}$.

26. Find the coefficient of x^{17} in the generating function $f(x) = \dfrac{x^4}{1 - x^5}$.

27. Find the coefficient of x^{19} in the generating function $f(x) = \dfrac{x^4}{1 - x^5}$.

28. Find the coefficient of x^{20} in the generating function
$$f(x) = (x^4 + x^8 + x^{16} + \cdots)^3$$

29. Find the coefficient of x^{12} in the generating function
$$f(x) = (x^2 + x^3 + x^4 + \cdots)^4$$

30. Find the coefficient of x^{10} in the generating function
$$f(x) = (x^2 + x^3 + x^4 + x^5)(1 + x^5 + x^{10})$$

31. Find the coefficient of x^{12} in the generating function
$$f(x) = \dfrac{x+3}{1 - 2x + x^2}.$$

32. Find the coefficient of x^{12} in the generating function
$$f(x) = \dfrac{x^3 - 3x^2}{(1-x)^4}.$$

33. Find the coefficient of x^{12} in the generating function $f(x) = (1 - 3x)^{-4}$.

34. Find the coefficient of x^{12} in the generating function
$$f(x) = \dfrac{1}{(1 - x^3)^4}.$$

Use generating functions to find the number of ways to select 10 balls from 20 red, 20 white, and 20 blue balls if

35. At least one ball of each color is selected.

36. An even number of red balls and an even number of blue balls are chosen.

37. Use a generating function to determine how many ways there are to distribute 15 toys among 5 children, if no child can get more than 4 toys.

38. Use a generating function to determine how many ways there are to place 16 chocolates in a box if there are 4 types of chocolates and at most 4 of each of the first 3 types are selected.

39. Use a generating function to determine how many ways there are to distribute 16 identical objects into 5 distinct boxes if each box contains not less than 2 or more that 6 objects.

40. Let $F(x)$ be the generating function whose coefficients give the Fibonacci numbers. Using the recurrence relation for $F(x)$, show that
$$F(x) = \dfrac{x}{(1 - (x + x^2))}$$
Use this to expand the generating function $F(x)$. Show that the coefficient of x^{n+1} is
$$\binom{n}{0} + \binom{n-1}{1} + \binom{n-2}{2} + \cdots$$
so that
$$\text{Fib}(n + 1) = \binom{n}{0} + \binom{n-1}{1} + \binom{n-2}{2} + \cdots$$

13.4 Partitions

In this section we use generating functions to describe the number of partitions of a set of n indistinguishable objects into a given number of indistinguishable boxes. This is equivalent to the number of ways of partitioning the integer n into collections of a given number of integers whose sum is n, where the collections are not distinguished by order.

For example, 3 may be expressed as the sums $1+1+1$ and $1+2$. The integer 4 may be expressed as $1+1+1+1$, $1+1+2$, $1+3$, $2+2$, and 4. The ways in which 4 may be expressed as the sum of two integers are $1+3$ and $2+2$. There are two of these, so there are two ways that 4 indistinguishable objects can be placed in 2 indistinguishable boxes such that no box is empty. The ways in which 4 may expressed as the sum of two or fewer integers are $1+3$ and $2+2$, and 4. There are three of these, so there are three ways that 4 indistinguishable objects can be placed in 2 indistinguishable boxes such that some of the boxes may be empty.

We first consider the number of ways of placing n indistinguishable objects in n indistinguishable boxes, where some of boxes may be empty. Equivalently we are finding the number of ways of partitioning the integer n into collections of n or fewer integers whose sum is n. This is the number of nonnegative integral solutions of

$$e_1 + 2e_2 + 3e_3 + \cdots + ne_n = n$$

If n were fixed, we could find the number of solutions as the coefficient of x^n in the generating function

$$\frac{1}{(1-x)(1-x^2)(1-x^3)\cdots(1-x^n)}$$

We saw in the last example of the previous section that this is equivalent to finding the number of ways of selecting n objects when objects of type one are selected one at a time, objects of type two are selected two at a time, and objects of type three are selected three at a time, and so on. This is what we are doing, since each one selected contributes 1 to the sum of exponents adding up to n in x^n, each two selected contributes 2 to the sum adding up to n, and so on. As n increases, we need more terms in our product. Since we want our generating function to find the number of solutions for all n, we need the generating function

$$\frac{1}{(1-x)(1-x^2)(1-x^3)\cdots(1-x^k)\cdots}$$

and we have the following theorem:

THEOREM 13.15 The generating function whose nth coefficient is the number of ways of placing n indistinguishable objects in n indistinguishable boxes, where some of boxes may be empty, or equivalently, the number of ways of partitioning the integer n into collections of n or fewer integers whose sum is n is

$$\frac{1}{(1-x)(1-x^2)(1-x^3)\cdots(1-x^k)\cdots}$$ ∎

EXAMPLE 13.16 Show that any integer can be written in the form

$$2^0 + 2^1 + 2^2 + 2^3 + \cdots + 2^k$$

This shows that every positive integer can be written in binary, which is rather critical when using integers on a computer.

(a) We first show that, for all n,

$$(1-x)(1+x)(1+x^2)\cdots(1+x^{2^m})\cdots$$
$$= (1-x^{2^n})(1+x^{2^n})(1+x^{2^{n+1}})(1+x^{2^{n+2}})\cdots$$

We show this by induction on n. It is certainly true for $n = 0$. Assume that it is true for $n = k$, so that

$$(1-x)(1+x)(1+x^2)\cdots(1+x^{2^m})\cdots$$
$$= (1-x^{2^k})(1+x^{2^k})(1+x^{2^{k+1}})(1+x^{2^{n+2}})\cdots$$

but

$$(1-x^{2^k})(1+x^{2^k}) = (1-x^{2^{k+1}})$$

so we have

$$(1-x)(1+x)(1+x^2)\cdots(1+x^{2^m})\cdots$$
$$= (1-x^{2^{k+1}})(1+x^{2^{k+1}})(1+x^{2^{n+2}})\cdots$$

so the statement is true for $n = k+1$, and, therefore, it is true for all n.

(b) We next show that $(1-x)(1+x)(1+x^2)\cdots(1-x^{2^n})\cdots = 1$. By part (a), for any $k \geq 1$, on the left side of the equation, the coefficient of x^k is 0. Therefore, the generating function is equal to $1 + 0x + 0x^2 + \cdots = 1$.

(c) By part (b), it follows that

$$(1+x)(1+x^2)\cdots(1+x^{2^n})\cdots = \frac{1}{1-x} = 1 + x + x^2 + x^3 + \cdots$$

(d) The left-hand side of the equation in part (c) is the generating function for all sums of unique powers of 2, and the right side is the generating function for all positive integers. Therefore, each positive integer is a sum of unique powers of 2. ∎

EXAMPLE 13.17 Consider the number of ways of partitioning the integer n into collections of n or fewer distinct positive integers. Thus, each integer can only appear in a given sum once. For example, 4 may be expressed as 4, and $1 + 3$. In each expression of the number n as the sum of distinct integers, each integer less than or equal to n will either appear once or not at all. Therefore, the generating function is

$$(1+x)(1+x^2)(1+x^3)\cdots(1+x^k)\cdots$$

∎

EXAMPLE 13.18 The number of ways of partitioning the integer n into collections of n or fewer distinct positive integers is equal to the number of ways of partitioning the integer n into collections of n or fewer odd integers whose sum is n.

We first find a generating function used to determine the number of ways of partitioning the integer n into collections of n or fewer odd integers whose sum is n. Using the same argument as for Theorem 13.15, we find that the generating function is

$$\frac{1}{(1-x)(1-x^3)(1-x^5)\cdots(1-x^{2k-1})\cdots}$$

From the previous example the generating function used for determining the number of ways of partitioning the integer n into collections of n or fewer distinct integers is equal to

$$(1+x)(1+x^2)(1+x^3)\cdots(1+x^k)\cdots$$

But

$$(1+x^k) = \frac{1-x^{2k}}{1-x^k}$$

so

$$(1+x)(1+x^2)(1+x^3)\cdots(1+x^k)\cdots$$
$$= \frac{1-x^2}{1-x} \cdot \frac{1-x^4}{1-x^2} \cdot \frac{1-x^6}{1-x^3} \cdot \frac{1-x^8}{1-x^4} \cdot \frac{1-x^{10}}{1-x^5}\cdots$$

But the kth term in the numerator cancels with the $2k$th term in the denominator leaving only the $(2k+1)$th terms in the denominator, so we have

$$(1+x)(1+x^2)(1+x^3)\cdots(1+x^k)\cdots = \frac{1}{1-x}\frac{1}{1-x^3}\frac{1}{1-x^5}\frac{1}{1-x^7}\frac{1}{1-x^9}\cdots$$

which is the generating function used for determining the number of ways of partitioning the integer n into collections of n or fewer odd integers whose sum is n. ∎

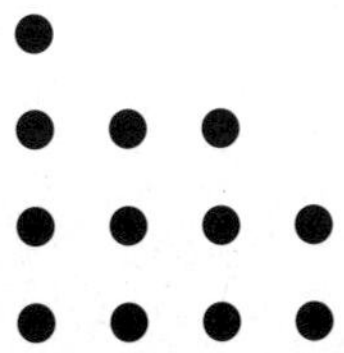

Our next goal is to describe the number of ways to place n indistinguishable objects in k indistinguishable boxes. Before doing so, however, we need a simple, but convenient, tool called a **Ferrer's diagram**. We shall, however, use a modified version, which seems simpler, although less algebraic in description. In our version, we begin by describing a distribution of objects into boxes using rows in increasing size. For example, if we are placing 12 objects in 4 boxes, this can be shown as in Figure 13.1, where, looking at the rows, we are placing 1 object in the first box, 3 objects in the second box, 4 objects in the third box, and 4 objects in the fourth box. If we look at the columns, we would find that we have 4 objects in the first box, 3 objects in the second box, 3 objects in the third box, and two objects in the fourth box. We can see that in looking at the columns, the boxes are in decreasing size. Since the boxes are indistinguishable, this is irrelevant. In particular, we see that for every distribution using rows, we get a distribution using columns, and conversely. Hence, there is a one-to-one correspondence between the distributions using the rows with boxes in increasing order and using columns with boxes in decreasing order. Consider an arbitrary distribution of 12 objects into 4 boxes. For example, suppose we have 1 object in the first box, 2 objects in the second box, 4 objects in the third box, and 6 objects in the fourth box. The Ferrer's diagram is shown in Figure 13.2.

Figure 13.1

Figure 13.2

If we now look at the distribution using the columns, we see we have 4 objects in the first box, 3 objects in the second box, 2 objects in the third and fourth boxes, and 1 object in the fifth and sixth boxes. However, since we began with 4 boxes using the columns, when considering the rows, we cannot exceed 4 objects in each box using the columns. Conversely, if we do not use more than 4 objects in each box using the columns, then we are using four or less boxes using the rows. Furthermore, there will be 4 nonempty boxes using the rows only if there is one box using the columns with exactly 4 objects in it. Therefore, there is a one-to-one correspondence between distributions of 12 objects in 4 boxes, where some may be empty, and distributions of 12 objects with at most 4 objects in each box. Furthermore, there is a one-to-one correspondence between distributions of 12 objects in 4 boxes, where none are empty, and distributions of 12 objects, where at least 1 box contains 4 objects. This same argument is true with n objects and k boxes giving us the following theorem.

THEOREM 13.19 The number of ways that n objects may be placed in k boxes, where some boxes may be empty, is equal to the number of distributions of n objects into boxes, where there is not more than k objects in each box. Furthermore, the number of ways that n objects may be placed in k boxes, where no box may be empty, is equal to the number of distributions of n objects into boxes, where there is not more than k objects in each box and at least one box contains k objects. ∎

The equivalent theorem in terms of partitioning the integer n is given next.

THEOREM 13.20 The number of ways of partitioning the integer n into collections of k or fewer integers is equal to the number of ways of partitioning the integer n into collections of n or fewer integers where no integer in the collection exceeds k. Furthermore, the number of ways of partitioning the integer n into collections of exactly k integers is equal to the number of ways of partitioning the integers into collections of n or fewer nonnegative integers where the largest integer in each collection is k. ∎

Since the number of ways of partitioning the integer n into collections of k or fewer integers where no integer in the collection exceeds k is equal to the number of solutions of nonnegative integral solutions of

$$e_1 + 2e_2 + 3e_3 + \cdots + ke_k = n$$

the generating function is

$$(1 + x + x^2 + \cdots)(1 + x^2 + x^4 + \cdots)(1 + x^3 + x^6 + \cdots) \cdots (1 + x^k + x^{2k} + \cdots)$$

or

$$\frac{1}{1-x} \cdot \frac{1}{1-x^2} \cdot \frac{1}{1-x^3} \cdots \frac{1}{1-x^k}$$

and we have the following theorem.

THEOREM 13.21 The generating function used for determining the number of ways of partitioning the integer n into collections of k or fewer nonnegative integers is

$$(1 + x + x^2 + \cdots)(1 + x^2 + x^4 + \cdots)(1 + x^3 + x^6 + \cdots) \cdots (1 + x^k + x^{2k} + \cdots)$$

or

$$\frac{1}{1-x} \cdot \frac{1}{1-x^2} \cdot \frac{1}{1-x^3} \cdots \frac{1}{1-x^k}$$ ∎

We now give the equivalent theorem for partitions of indistinguishable objects.

THEOREM 13.22 The generating function used for determining the number of ways of placing n indistinguishable objects in k boxes where some may be empty is

$$(1 + x + x^2 + \cdots)(1 + x^2 + x^4 + \cdots)(1 + x^3 + x^6 + \cdots) \cdots (1 + x^k + x^{2k} + \cdots)$$

$$\frac{1}{1-x} \cdot \frac{1}{1-x^2} \cdot \frac{1}{1-x^3} \cdots \frac{1}{1-x^k}$$ ∎

We next consider the case for the number of ways the integer n is partitioned into collections of k integers whose sum is n. By Theorem 13.20, this is the number of ways of partitioning the integers into collections of n or fewer nonnegative integers where the largest integer in each collection is k. The generating function for this is

$$(1 + x + x^2 + \cdots)(1 + x^2 + x^4 + \cdots)$$
$$\cdots (1 + x^{k-1} + x^{2(k-1)} + \cdots)(x^k + x^{2k} + \cdots)$$

This is equal to

$$x^k(1 + x + x^2 + \cdots)(1 + x^2 + x^4 + \cdots)$$
$$\times (1 + x^3 + x^6 + \cdots) \cdots (1 + x^k + x^{2k} + \cdots)$$

or

$$\frac{x^k}{(1-x)(1-x^2)(1-x^3)\cdots(1-x^k)}$$

Another way to see this is to take the generating function used for determining the number of ways an integer n is partitioned into collections of k or fewer integers whose sum is n and subtract the generating function used for determining the number of ways an integer n is partitioned into collections of $k-1$ or fewer integers whose sum is n. This should give the generating function when integer n is partitioned into collections of exactly k integers whose sum is n. Doing this we have

$$(1+x+x^2+\cdots)(1+x^2+x^4+\cdots)\cdots(1+x^{k-1}+x^{2(k-1)}+\cdots)(1+x^k+x^{2k}+\cdots)$$
$$-(1+x+x^2+\cdots)(1+x^2+x^4+\cdots)$$
$$\cdots(1+x^{k-1}+x^{2(k-1)}+\cdots)$$
$$=(1+x+x^2+\cdots)(1+x^2+x^4+\cdots)$$
$$\cdots(1+x^{k-1}+x^{2(k-1)}+\cdots)((1+x^k+x^{2k}+\cdots)-1)$$
$$=(1+x+x^2+\cdots)(1+x^2+x^4+\cdots)$$
$$\cdots(1+x^{k-1}+x^{2(k-1)}+\cdots)((x^k+x^{2k}+\cdots).$$

Thus, we have the following theorem.

THEOREM 13.23 The generating function used for determining the number of ways an integer n is partitioned into collections of exactly k nonnegative integers whose sum is n is equal to

$$x^k(1+x+x^2+\cdots)(1+x^2+x^4+\cdots)$$
$$\times(1+x^3+x^6+\cdots)\cdots(1+x^k+x^{2k}+\cdots)$$

or

$$\frac{x^k}{(1-x)(1-x^2)(1-x^3)\cdots(1-x^k)}$$ ∎

Equivalently, we have a theorem.

THEOREM 13.24 The generating function used for determining the number of ways n indistinguishable objects can be place in k boxes, such that no box is empty, is equal to

$$x^k(1+x+x^2+\cdots)(1+x^2+x^4+\cdots)$$
$$\times(1+x^3+x^6+\cdots)\cdots(1+x^k+x^{2k}+\cdots)$$

or

$$\frac{x^k}{(1-x)(1-x^2)(1-x^3)\cdots(1-x^k)}$$ ∎

Exercises

1. Find the number of partitions of the integer 7.

2. Find the number of partitions of the integer 5.

3. Find the generating function to determine the number of ways 10 indistinguishable balls can be placed in 6 indistinguishable boxes.

4. Find the generating function to determine the number of ways to partition n into distinct odd integers.

5. Find the generating function to determine the number of ways to partition n into distinct even integers.

6. Assume that an athlete participates in 3 events. In each event he can win 1, 2, 3, 4, or 5 points. Find the generating function to determine how many different ways he can accumulate n points.

7. Use Ferrer's diagram to show that the number of partitions of an integer into parts of even size is equal to the number of partitions of the integer into parts so that each part occurs an even number of times.

8. Use Ferrer's diagram to show that the number of partitions of n into 3 parts is equal to the number of partitions of $2n$ into 3 parts of size less than n. What about the number of partitions of n into 4 parts and the number of partitions of $3n$ into 4 parts of size less than n? Can this be generalized?

9. Use Ferrer's diagram to show that the number of partitions of n is equal to the number of partitions of $2n$ into n parts.

10. Use Ferrer's diagram to show that the number of partitions of $2n + k$ into $n + k$ parts is the same for all k.

11. If accidents are classified according to the day of the week on which they occur, find a generating function used to describe the number of different ways we can classify n accidents.

13.5 Exponential Generating Functions

In this section we use generating functions of the form

$$a_0 + a_1 \frac{x}{1!} + a_2 \frac{x^2}{2!} + a_3 \frac{x^3}{3!} + \cdots + a_n \frac{x^n}{n!}$$

We first let

$$e^x = e(x) = 1 + \frac{x}{1!} + \frac{x^2}{2!} + \frac{x^3}{3!} + \frac{x^4}{4!} + \frac{x^5}{5!} + \frac{x^6}{6!} + \cdots$$

The only property of this function, which we shall need, is given in the following theorem. The proof of this is left to the reader.

THEOREM 13.25 For every nonnegative integer n

$$(e^x)^n = e^{nx}$$ ∎

We know that

$$(1+x)^n = \binom{n}{0} + \binom{n}{1}x + \binom{n}{2}x^2 + \cdots + \binom{n}{n}x^n$$

so that $(1+x)^n$ is the generating function for combinations using ordinary generating functions. However,

$$(1+x)^n = \binom{n}{0} + \binom{n}{1}x + \binom{n}{2}x^2 + \cdots + \binom{n}{n}x^n$$

$$= \frac{n!}{0!(n)!} + \frac{n!}{1!(n-1)!}x + \frac{n!}{2!(n-2)!}x^2 + \cdots + \frac{n!}{k!(n-k)!}x^k + \cdots + \frac{n!}{0!(n)!}x^n$$

$$= \frac{n!}{n!} + \frac{n!}{(n-1)!} \cdot \frac{x}{1!} + \frac{n!}{(n-2)!} \cdot \frac{x^2}{2!} + \cdots + \frac{n!}{(n-k)!} \cdot \frac{x^k}{k!} + \cdots + \frac{n!}{0!} \cdot \frac{x^n}{n!}$$

$$= P(n,0) + P(n,1)\frac{x}{1!} + P(n,2)\frac{x^2}{2!} + \cdots + P(n,k)\frac{x^k}{k!} + \cdots + P(n,n)\frac{x^n}{n!}$$

so that $(1+x)^n$ is the exponential generating function for permutations.

Since we have just expressed $(1+x)^n$ as an exponential generating function whose coefficients are permutations instead of combinations, as they were with ordinary generating functions, one might guess that we are now concerned with counting objects where there is order. One would be correct.

Suppose that we have an unlimited number of red, green, white, and blue objects. We know that the generating function used to describe the number of ways to select n objects when there are at least 2 green ones and 3 blue ones is given by

$$(1 + x + x^2 + \cdots)(x^2 + x^3 + x^4 + \cdots)(1 + x + x^2 + \cdots)(x^3 + x^4 + x^5 + \cdots)$$

For example, if we are selecting 12 objects, then $x^2 x^3 x^4 x^3$ represents picking 2 red, 3 green, 4 white, and 3 blue objects. Suppose, however, that instead of picking red, green, white, and blue objects, we were picking the letters r, g, w, and b and deciding how many words we could make from them. Instead of just wanting the one result—2 r, 3 g, 4 w, and 3 b—we want the number

$$\frac{12!}{2!3!4!2!}$$

which is the number of ways we can rearrange the 2 r, 3 g, 4 w, and 3 b to get words. Notice that if instead of using the foregoing product of ordinary generating functions, we had used

$$\left(1 + \frac{x}{1!} + \frac{x^2}{2!} + \cdots\right)\left(\frac{x^2}{2!} + \frac{x^3}{3!} + \frac{x^4}{4!} + \right)$$

$$\left(1 + \frac{x}{1!} + \frac{x^2}{2!} + \cdots\right)\left(\frac{x^3}{3!} + \frac{x^4}{4!} + \frac{x^5}{5!} + \cdots\right)$$

then instead of getting $x^2 x^3 x^4 x^3$ as one of the terms forming x^{12}, we would have gotten

$$\frac{x^2}{2!}\frac{x^3}{3!}\frac{x^4}{4!}\frac{x^3}{3!} = \frac{x^{12}}{2!3!4!2!} = \frac{12!}{2!3!4!2!}\frac{x^{12}}{12!}$$

The coefficient of $\frac{x^{12}}{12!}$ is $\frac{12!}{2!3!4!2!}$. Thus, if we want the number of ways that we can rearrange the objects, rather than just select them, we should use exponential generating functions instead of ordinary generating functions.

The coefficient of x^n, using ordinary generating functions, then gives the total number of ways n objects may be selected. The coefficient of $\frac{x^n}{n!}$, using exponential generating functions, gives the number of ways n objects may be selected and rearranged.

Remember, as discussed in Section 8.6,

$$P(n; r_1, r_2, \cdots, r_k) = \frac{n!}{r_1!r_2!\cdots r_k!} = C(n; r_1, r_2, \cdots, r_k)$$

is both the number of arrangements of n objects where there are r_i objects of type i and

$$r_1 + r_2 + \cdots + r_k = n$$

and the number of partitionings of a set of n distinguishable objects into k distinguishable boxes, where r_i objects go into the ith box and

$$r_1 + r_2 + \cdots + r_k = n$$

Hence, we may use exponential generating functions for both types of problems.

13.5 Exponential Generating Functions

EXAMPLE 13.26 How many sequences of length n can be formed from the integers 1, 2, 3, and 4 if there must be at least one 1, two 2's, an odd number of 3's, and an even number of 4's?

If we were simply selecting the objects, the generating function would be given by

$$(x + x^2 + x^3 \cdots)(x^2 + x^3 + x^4 + \cdots)(x + x^3 + x^5 \cdots)(x^2 + x^4 + x^6 + \cdots)$$

However, since we want the number of sequences, we are concerned with order, so the exponential generating function is given by

$$\left(\frac{x}{1!} + \frac{x^2}{2!} + \cdots\right)\left(\frac{x^2}{2!} + \frac{x^3}{3!} + \frac{x^4}{4!} + \cdots\right)$$
$$\left(\frac{x}{1!} + \frac{x^3}{3!} + \frac{x^5}{5!} + \cdots\right)\left(\frac{x^2}{2!} + \frac{x^4}{4!} + \frac{x^6}{6!} + \cdots\right)$$

∎

EXAMPLE 13.27 Find the generating function used for determining the number of ways n people can be placed in three rooms with at least 2, but not more than 10 people, in a room. Following the guidelines for regular exponential functions, the exponential generating function is given by

$$f(x) = \left(\frac{x^2}{2!} + \frac{x^3}{3!} + \cdots + \frac{x^{10}}{10!}\right)\left(\frac{x^2}{2!} + \frac{x^3}{3!} + \cdots + \frac{x^{10}}{10!}\right)\left(\frac{x^2}{2!} + \frac{x^3}{3!} + \cdots + \frac{x^{10}}{10!}\right)$$

or

$$f(x) = \left(\frac{x^2}{2!} + \frac{x^3}{3!} + \cdots + \frac{x^{10}}{10!}\right)^3$$

∎

EXAMPLE 13.28 Suppose that we want to find the number of different arrangements of k objects chosen from n types of objects, where there is an unlimited number of each type of object. We already know from our study of permutations with unlimited repetitions that the answer is n^k. See Section 8.7. However, let's try our luck with exponential generating functions. The exponential generating function is

$$f(x) = \left(1 + \frac{x}{1!} + \frac{x^2}{2!} + \frac{x^3}{3!} + \cdots\right)^n$$

But

$$e^x = 1 + \frac{x}{1!} + \frac{x^2}{2!} + \frac{x^3}{3!} + \frac{x^4}{4!} + \frac{x^5}{5!} + \frac{x^6}{6!} + \cdots$$

so that

$$(e^x)^n = \left(1 + \frac{x}{1!} + \frac{x^2}{2!} + \frac{x^3}{3!} + \cdots\right)^n$$

But

$$(e^x)^n = e^{nx} = \left(1 + \frac{nx}{1!} + \frac{(nx)^2}{2!} + \frac{(nx)^3}{3!} + \cdots\right)$$

so that the coefficient of $\frac{x^r}{k!}$ is n^k as expected.

∎

The proof of the following theorem is left to the reader.

THEOREM 13.29

(a) $e^x - 1 = \dfrac{x}{1!} + \dfrac{x^2}{2!} + \dfrac{x^3}{3!} + \cdots$

(b) $\dfrac{e^x + e^{-x}}{2} = 1 + \dfrac{x^2}{2!} + \dfrac{x^4}{4!} + \dfrac{x^6}{6!} + \cdots$

(c) $\dfrac{e^x - e^{-x}}{2} = \dfrac{x}{1!} + \dfrac{x^3}{3!} + \dfrac{x^5}{5!} + \dfrac{x^7}{7!} + \cdots$

EXAMPLE 13.30 Find the exponential generating function to describe how many ways n guests may be placed in three rooms, if at least one guest must be in the first room, an odd number of guests must be in the second room, and an even number of guests must be in the third room. The generating function is given by

$$f(x) = \left(\dfrac{x}{1!} + \dfrac{x^2}{2!} + \dfrac{x^3}{3!} + \cdots\right)\left(\dfrac{x}{1!} + \dfrac{x^3}{3!} + \dfrac{x^5}{5!} + \cdots\right)\left(1 + \dfrac{x^2}{2!} + \dfrac{x^4}{4!} + \dfrac{x^6}{6!} + \cdots\right)$$

$$= (e^x - 1)\dfrac{e^x - e^{-x}}{2} \cdot \dfrac{e^x + e^{-x}}{2}$$

$$= (e^x - 1)\dfrac{e^{2x} - e^{-2x}}{4}$$

$$= \dfrac{1}{4}(e^{3x} - e^{-x} - e^{2x} + e^{-2x})$$

so that the coefficient of $\dfrac{x^n}{n!}$ is $\dfrac{1}{4}((3)^n - (-1)^n - (2)^n + (-2)^n)$. ∎

We now have the tools to find the number of ways to place n distinguishable objects into k distinguishable boxes where no box is empty, and the number of ways to place n distinguishable objects into k indistinguishable boxes where no box is empty.

THEOREM 13.31 The number of ways to place n distinguishable objects into k distinguishable boxes where no box is empty and $1 \leq k \leq n$ is

$$T_k^{(n)} = \sum_{i=1}^{k} (-1)^i \binom{k}{i}(k-i)^n$$

Proof The exponential generating function for placing n distinguishable objects into k distinguishable boxes where no box is empty is

$$f(x) = \left(\dfrac{x}{1!} + \dfrac{x^2}{2!} + \dfrac{x^3}{3!} + \cdots\right)^k$$

But

$$\left(\dfrac{x}{1!} + \dfrac{x^2}{2!} + \dfrac{x^3}{3!} + \cdots\right)^k = (e^x - 1)^k \qquad \text{by Theorem 13.29}$$

$$= \sum_{i=1}^{k} \binom{k}{i}(-1)^i e^{(k-i)x} \qquad \text{by the binomial theorem}$$

By definition,

$$e^{(k-i)x} = 1 + \frac{(k-i)x}{1!} + \frac{(k-i)^2 x^2}{2!} + \frac{(k-i)^3 x^3}{3!} + \cdots$$

$$= \sum_{n=0}^{\infty} \frac{(k-i)^n x^n}{n!}$$

so that

$$f(x) = \sum_{i=1}^{k} \binom{k}{i}(-1)^k \sum_{n=0}^{\infty} \frac{(k-i)^n x^n}{n!}$$

$$= \sum_{n=0}^{\infty} \frac{x^n}{n!} \sum_{i=1}^{k} \binom{k}{i}(-1)^k (k-i)^n$$

and the coefficient of $\frac{x^n}{n!}$ is

$$T_k^{(n)} = \sum_{i=1}^{k} \binom{k}{i}(-1)^i (k-i)^n \qquad \blacksquare$$

We are now able to prove the following theorem.

THEOREM 13.32 The number of ways to place n distinguishable objects into k indistinguishable boxes where no box is empty and $1 \leq k \leq n$ is

$$S_k^{(n)} = \frac{1}{k!} \sum_{i=1}^{k} (-1)^i \binom{k}{i} (k-i)^n$$

where $\{S_k^{(n)} : 0 \leq k \leq n\}$ is the set of **Stirling numbers of the second kind**.

Proof Since $T_k^{(n)}$ is the number of ways to place n distinguishable objects into k distinguishable boxes, to change from distinguishable sets to indistinguishable sets we need to divide by $k!$, since all $k!$ of the arrangements of the k boxes are now identified. Therefore,

$$S_k^{(n)} = \frac{1}{k!} T_k^{(n)} = \frac{1}{k!} \sum_{i=1}^{k} (-1)^i \binom{k}{i} (k-i)^n \qquad \blacksquare$$

Exercises

Set up the appropriate generating function to determine how many 10-digit code words can be formed from 0, 1, and 2 if

1. The integer 0 must appear at least twice in the code.

2. The integer 1 can appear at most three times, and the integer 2 at most four times.

3. The integer 1 can appear an even number of times and the integer 0 an odd number of times.

Twelve balloons are to be tied to 12 lampposts along a parade route. Set up the appropriate exponential generating function used to determine how many ways red, blue, yellow, and green balloons can be arranged if

4. Every color must appear at least once.

5. At most 4 red, 3 yellow, 7 green, and 6 blue balloons may be used.

6. At least 2 red balloons, at most 3 green balloons, between 1 and 5 green balloons, and an even number of blue balloons may be used.

7. Find the exponential generating function that generates the sequence of factorials, 0!, 1!, 2!, 3!, ...

8. Find the exponential generating function that is used to describe the number of ways of coloring an 8×8 chessboard using the colors red, white, and blue, if an even number of squares must be colored red.

9. Use an exponential generating function to find the number of n-digit numbers with all digits odd, if each of 1 and 3 occur at least once.

10. Use an exponential generating function to find the number of n-digit numbers with all digits odd if all digits occur at least once and 5 occurs an even number of times.

11. Use an exponential generating function to find the number of ways to place 30 distinguishable people in 3 distinguishable hotel rooms if at least one person must be in each room.

12. Use an exponential generating function to find the number of ways to place 30 distinguishable people in 3 distinguishable hotel rooms if an even number of people must be in each room.

13. Use an exponential generating function to find the number of ways to place 25 objects from four different types in 25 labeled places to be filled by these objects, one per position, if there an even number of the first type, an odd number of the second type, and at least one of the third and fourth types.

14. Use an exponential generating function to determine how many ways there are to distribute 9 different pieces of candy to 4 children if each of the first two children must receive a piece of candy.

15. Determine the generating function used to count the number of arrangements of pennies, nickels, dimes, and quarters totaling n cents.

16. Show that $\frac{e^n}{(1-x)^n}$ is the exponential generating function used to describe the number of ways to choose some possibly empty subset of k objects to be distributed into n boxes where order of the subset is counted.

17. Find the exponential generating function used to describe the number of ways to partition n into k parts if order is important.

18. Prove Theorem 13.25: $e^{nx} = (e^x)^n$ for every nonnegative integer n, using induction.

19. Prove Theorem 13.29:

(a) $e^x - 1 = \frac{x}{1!} + \frac{x^2}{2!} + \frac{x^3}{3!} + \cdots$

(b) $\frac{e^x + e^{-x}}{2} = 1 + \frac{x^2}{2!} + \frac{x^4}{4!} + \frac{x^6}{6!} + \cdots$

(c) $\frac{e^x - e^{-x}}{2} = \frac{x}{1!} + \frac{x^3}{3!} + \frac{x^5}{5!} \frac{x^7}{7!} + \cdots$

GLOSSARY

Chapter 13: Generating Functions

Exponential generating function for permutations (13.5)	$(1+x)^n = P(n,0) + P(n,1)\frac{x}{1!} + P(n,2)\frac{x^2}{2!} + \cdots + P(n,k)\frac{x^k}{k!} + \cdots P(n,n)\frac{x^n}{n!}$
Ferrer's diagram (13.4)	A Ferrer's diagram of a partition is an arrangement of n dots on a square grid where a part i in the partition is represented by placing p_i dots in a row. A new partition is then formed by examining the columns instead of the rows.
Generating function (13.1)	The expression $a_0 + a_1 x + a_2 x^2 + a_3 x^3 + a_4 x^4 + \cdots a_n x^n + \cdots$ is the **generating function** of the sequence $a_0, a_1, a_2, a_3, \ldots$.
Exponential generating functions (13.5)	**Exponential generating functions** are of the form $a_0 + a_1 \frac{x}{1!} + a_2 \frac{x^2}{2!} + a_1 \frac{x^3}{3!} + \cdots + a_n \frac{x^n}{n!}$.

Number of ways to place n distinguishable objects into k distinguishable boxes where no box is empty (13.5)	The number of ways to place n distinguishable objects into k distinguishable boxes where no box is empty and $1 \le k \le n$ is $$T_k^{(n)} = \sum_{i=1}^{k} (-1)^i \binom{k}{i} (k-i)^n.$$
Number of ways to place n distinguishable objects into k indistinguishable boxes where no box is empty (13.5)	The number of ways to place n distinguishable objects into k indistinguishable boxes where no box is empty and $1 \le k \le n$ is $S_k^{(n)} = \frac{1}{k!}\sum_{i=1}^{k}(-1)^i \binom{k}{i}(k-i)^n$ where $\{S_k^{(n)} : 0 \le k \le n\}$ is the set of Stirling numbers of the second kind.
Partial fraction decomposition (13.2)	Any fraction of the form $\frac{Ax+B}{(1-Cx)(1-Dx)}$ where C and D are distinct is equal to $\frac{a}{1-Cx} + \frac{b}{1-Dx}$ for constants a and b. This is called **partial fraction decomposition**.
Partitions (13.4)	Let A and I be sets and let $\langle A \rangle = \{A_1 : i \in I\}$, with I nonempty, be a set of nonempty subsets of A. The set $\langle A \rangle$ is called a **partition** of A if both of the following are satisfied: a. $A_i \cap A_j = \emptyset$ for all $i \ne j$. b. $A = \bigcup_{i \in I} A_i$; that is, $a \in A$ if and only if for some $i \in I$.

TRUE-FALSE QUESTIONS

1. $\dfrac{1}{1+x} = 1 - x + x^2 - x^3 + \cdots + (-1)^{n+1}x^n$.

2. $\dfrac{1}{(1-x^2)^2} = 1 + 2x^2 + 3x^4 + 4x^6 + \cdots + (n+1)x^{2n} + \cdots$.

3. $\dfrac{x}{1-x} = x + x^2 + x^3 + \cdots + x^n + \cdots$.

4. $\dfrac{1-x^n}{1-x} = 1 + x + x^2 + x^3 + \cdots + x^{n-1}$.

5. $\dfrac{Ax^2 + Bx + Cx^n}{(1-Dx)^2(1-Ex)}$ may be expressed in the form $\dfrac{a}{(1-Dx)^2} + \dfrac{b}{(1-Ex)}$.

6. $\dfrac{1}{(1-x)(1+x)} = 1 + 2x + 3x^2 + \cdots + (n+1)x^n + \cdots$.

7. To solve the recurrence function
$$a_0 = 1$$
$$a_1 = 2$$
$$a_k = 2a_{k-1} + 3a_{k-2}$$
we use the generating function $f(x) - 2xf(x) - 3x^2 f(x)$.

8. To solve the recurrence function
$$a_0 = 1$$
$$a_k = 2a_{k-1} + 2^n$$
we use the generating function $f(x) - 2xf(x) - \dfrac{1}{1-2x}$.

9. To solve the recurrence function
$$a_0 = 1$$
$$a_k = -3a_{k-1} + n$$
we use the generating function $f(x) + 3xf(x) - \dfrac{1}{(1-x)^2}$.

514 Chapter 13 Generating Functions

10. To solve the recurrence function
$$a_0 = 1$$
$$a_k = -3a_{k-1} + 3$$
we use the generating function $f(x) + 3xf(x) - \dfrac{3}{(1-x)}$.

11. The generating function used to determine the number of combinations of m objects taken k at a time with repetition is $(1+x)^m$.

12. The generating function that represents the number of solutions of $e_1 + e_2 + e_3 = r$ for $0 \le e_1 \le 2$, $1 \le e_2 \le 4$, $0 \le e_3 \le 3$ is $(1 + x + x^2)(x + x^2 + x^3 + x^4)(1 + x + x^2 + x^3)$.

13. The generating function for determining the number of ways n objects can be selected from four sets of objects where an even number of the first set must be selected, an odd number of the second must be selected, at least two of the third set must be selected, and at most 10 of the last set must be selected is
$$(1 + x^2 + x^4 + \cdots)(x + x^3 + x^5 + \cdots)$$
$$\times (x^2 + x^3 + x^4 + \cdots) \cdot \dfrac{1 - x^{11}}{1 - x}$$

14. The generating function whose nth coefficient gives the number of nonnegative integral solutions of $e_1 + 2e_2 + 3e_3$ is
$$(x + x^2 + x^3 + \cdots)(x^2 + x^3 + x^4 + \cdots)$$
$$\times (x^3 + x^4 + x^5 + \cdots)$$

15. The generating function used for determining the number of ways of partitioning the integer n into collections of k or fewer nonnegative integers is
$$\dfrac{1}{1-x} \cdot \dfrac{1}{1-x^2} \cdots \dfrac{1}{1-x^k}$$

16. The generating function used to determine the number of ways of placing n indistinguishable objects in k boxes where some may be empty is
$$\dfrac{1}{1-x} \cdot \dfrac{1}{1-x^2} \cdots \dfrac{1}{1-x^k}$$

17. The number of partitions of n is equal to the number of partitions of $2n$ into n parts.

18. $\dfrac{e^x + e^{-x}}{2} = \dfrac{x}{1!} + \dfrac{x^3}{3!} + \dfrac{x^5}{5!} + \cdots$

19. The number of ways to place 5 distinguishable objects into 3 distinguishable boxes where no box is empty is the Stirling number $S_3^{(5)}$.

20. The exponential generating function for describing how many ways n guests may be placed in four rooms if at least one guest must be in each room, an even number must be in the third room, and an odd number in the fourth room is
$$(e^x - 1)^2 \left(\dfrac{e^{2x} - e^{-2x}}{4} \right)$$

SUMMARY QUESTIONS

1. Expand $\dfrac{1}{(1-x)^2}$.

2. Express $\dfrac{x}{(x-1)}$ as the sum of partial fractions.

3. Expand the generating function $\dfrac{2x}{(x-1)(x+2)}$.

4. Solve the relation
$$a_0 = 1$$
$$a_n = 3a_{n-1} + n, n \ge 1$$

5. Solve the relation
$$a_0 = 1$$
$$a_n = 2a_{n-1} + 3^n, n \ge 1$$

6. Solve the relation
$$a_0 = 1$$
$$a_1 = -2$$
$$a_n = 5a_{n-1} - 6a_{n-2}, n \ge 2$$

7. Find a generating function whose rth coefficient gives the number of solutions of $e_1 + e_2 + e_3 + e_4 = r$ where e_2 is odd and $e_4 \ge 4$.

8. Find a generating function used for finding the number of ways of selecting r items out of 3 red, 7 blue, 4 orange, and 5 green balls.

9. Find the coefficient of x^6 in the generating function $\dfrac{1}{(1-x^2)^3}$.

10. Find the number of partitions of the integer 6.

11. Find the generating function to determine the number of ways 8 indistinguishable balls can be placed in 4 indistinguishable boxes.

12. Find the coefficient of x^9 in the generating function $f(x) = (x + x^2 + x^3 + x^4)^3$.

13. Find the coefficient of x^{16} in the generating function $f(x) = \dfrac{x^3}{(1-x)^4}$.

14. Find a generating function whose rth coefficient gives the number of solutions of $e_1 + 2e_2 + 3e_3 + e_4$ where e_2 is odd.

15. Find the coefficient of x^{16} in the generating function $f(x) = \frac{x+2}{1-2x-3x^2}$.

ANSWERS TO TRUE-FALSE

1. F 2. T 3. T 4. T 5. F 6. T 7. T 8. T 9. F
10. T 11. F 12. T 13. T 14. F 15. T 16. T
17. T 18. F 19. F 20. F

CHAPTER 14
Graphs Revisited

14.1 Algebraic Properties of Graphs

In this section we will discuss relationships between graphs such that if two graphs have a given relationship, then properties possessed by one graph are necessarily possessed by the other. We shall present the material in this section with the understanding that most of the concepts can be generalized to multigraphs and pseudographs.

DEFINITION 14.1 A function f from the graph $G(V, E)$ to the graph $G'(V', E')$ is called a **homomorphism** from G to G', denoted $f : G \to G'$, if it has the following properties:
(a) If $e \in E$, then $f(e) \in E'$. ($f(E) \subseteq E'$)
(b) If $v \in V$, then $f(v) \in V'$. ($f(V) \subseteq V'$)
(c) If vertices u and v are incident to edge e in G, then $f(u)$ and $f(v)$ are incident to $f(e)$ in G'.

Proofs of the following theorems are left to the reader.

THEOREM 14.2 If f is a homomorphism from G to G', then $f(G)$ is the subgraph $(f(V), f(E))$ of G'. ∎

THEOREM 14.3 If a graph G is connected and f is a homomorphism, then $f(G)$ is connected. ∎

THEOREM 14.4 If a graph G is complete and f is a homomorphism, then $f(G)$ is complete. ∎

We shall see in the exercises that many properties of G are not invariant in $f(G)$.

DEFINITION 14.5 The homomorphism $f : G \to G'$ is an **isomorphism** if $f : V \to V'$ and $f : E \to E'$ are one-to-one correspondences. If $f : G \to G'$ is an isomorphism, then G and G' are **isomorphic**.

Thus, an isomorphism is really a relabeling of the vertices and edges of V that preserves the homomorphic property so that if vertices u and v are incident to edge e in G, then $f(u)$ and $f(v)$ are incident to $f(e)$ in G'. Invariants of isomorphisms are given in the problems.

It becomes obvious that virtually all properties of graphs are invariant under isomorphisms. Hence, it is easiest to show that two graphs are not isomorphic by showing that one graph has a property that the other graph does not have.

DEFINITION 14.6 If a graph $G(V, E)$ has an edge $e = \{v_1, v_2\}$ and graph $G'(V', E')$ is derived from $G(V, E)$ by adding a new vertex v to V and replacing the edge $\{v_1, v_2\}$ with the edges $\{v_1, v\}$ and $\{v, v_2\}$, then $G'(V', E')$ is an **extension** of $G(V, E)$. If there are graphs $G_1, G_2, G_3, \ldots G_n$ such that G_{i+1} is an extension of G_i, then G_n can be **derived** from G_1.

If a graph $G'(V', E')$ is an extension of $G(V, E)$, then one of the edges of V has had a new vertex placed in the middle of it and the original edge is now divided into two new edges that connect the vertices incident to the old edge to the new vertex.

DEFINITION 14.7 The graphs G and G' are **homeomorphic** if there is a graph G'' such that both G and G' can be derived from G''.

EXAMPLE 14.8 The graph in Figure 14.1 is an extension of the graph in Figure 14.2.

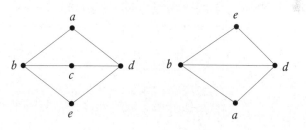

Figure 14.1   Figure 14.2

EXAMPLE 14.9 The graph in Figure 14.3 can be derived from the graph in Figure 14.4.

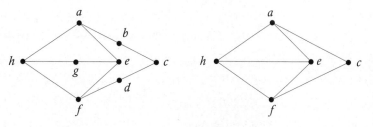

Figure 14.3   Figure 14.4

EXAMPLE 14.10 The graph in Figure 14.5 can be derived from the graph in Figure 14.6.

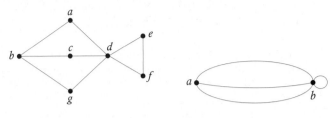

Figure 14.5   Figure 14.6

EXAMPLE 14.11 The graph in Figure 14.7 is homeomorphic to the graph in Figure 14.8.

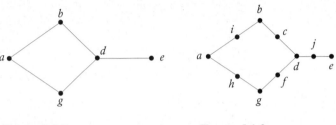

Figure 14.7   Figure 14.8

The following definitions are being developed primarily for the discussion of planar graphs in the next section. They also show more examples of how properties of graphs may remain invariant for relationships between graphs.

If a graph $G'(V', E')$ is an extension of $G(V, E)$, then a new vertex has been added that has degree 2. The degrees of all other vertices are unchanged. Hence, we have the following theorem.

THEOREM 14.12 If graphs G and G' are homeomorphic, then they both have the same number of vertices of odd degree.

From this theorem, Theorem 6.45, and Theorem 6.49, we immediately get the following theorem.

THEOREM 14.13 If graphs G and G' are homeomorphic, G has a Euler cycle (proper path) if and only if G' has a Euler cycle (proper path).

If G' is a subgraph of G, denote this by $G' \preceq G$.

DEFINITION 14.14 Let $G = G(V, E)$ be a graph and $G_1, G_2, G_3, \ldots, G_n$ be subgraphs of G. The subgraph G' of G is the **union** of $G_1, G_2, G_3, \ldots, G_n$, denoted by $\bigcup_{i=1}^{n} G_i$ if

(1) A vertex $v \in G'$ if and only if $v \in G_i$ for some $1 \leq i \leq n$.
(2) An edge $e \in G'$ if and only if $e \in G_i$ for some $1 \leq i \leq n$.

DEFINITION 14.15 Let $G = G(V, E)$ be a graph and $G_1, G_2, G_3, \ldots, G_n$ be subgraphs of G. The subgraph G' of G is the **intersection** of $G_1, G_2, G_3, \ldots, G_n$, denoted by $\bigcap_{i=1}^{n} G_i$ if

(1) A vertex $v \in G'$ if and only if $v \in G_i$ for all $1 \leq i \leq n$.
(2) An edge $e \in G'$ if and only if $e \in G_i$ for all $1 \leq i \leq n$.

14.1 Algebraic Properties of Graphs

DEFINITION 14.16 Let $G = G(V, E)$ be a graph and $G_1, G_2, G_3, \ldots, G_n$ be subgraphs of G. The subgraphs $G_1, G_2, G_3, \ldots, G_n$ are **pairwise disjoint** if for all $1 \leq i < j \leq n$, $G_i \cap G_j = \emptyset$.

The proofs of the following theorems are left to the reader.

THEOREM 14.17 If G_1 and G_2 are distinct components of a graph G, then G_1 and G_2 are pairwise disjoint. ∎

THEOREM 14.18 A graph G is the union of pairwise disjoint components. ∎

DEFINITION 14.19 Let $G(V, E)$ be a graph. The **complement** of G denoted by $G^c(V, E')$ is the graph such that for all $u, v \in V$ there is an edge between u and v in G^c if and only if there is no edge in E between u and v in G.

DEFINITION 14.20 A subgraph $G'(V', E')$ is a **spanning graph** for the graph $G = (V, E)$ if $V' = V$.

EXAMPLE 14.21 Given the graph in Figure 14.9, the graphs in Figures 14.10, 14.11, and 14.12 are spanning graphs

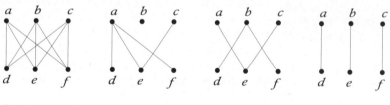

Figure 14.9   Figure 14.10   Figure 14.11   Figure 14.12

whereas the graphs in Figure 14.13, 14.14, and 14.15 are not spanning graphs.

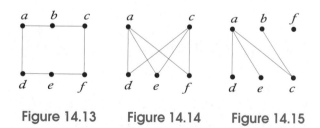

Figure 14.13   Figure 14.14   Figure 14.15   ∎

The spanning tree was defined in Chapter 6 and is repeated here for convenience. It was also shown that every connected graph has a spanning tree.

DEFINITION 14.22 A tree is a **spanning tree** for the graph G if it is a spanning graph of G.

The proof of the following theorem is left to the reader.

THEOREM 14.23 If $T(V, E')$ is a spanning tree of the graph $G = (V, E)$, then for any cycle $v_0 v_1 v_2 v_3 v_4, \ldots v_0$, at least one of the edges must belong to $E - E'$. ∎

Suppose that a company constructs a communications network as shown in Figure 14.16.

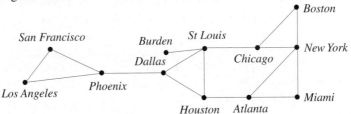

Figure 14.16

It is important to the company to know how many cities are disconnected from each other if certain lines are down. In particular, it is important to know how many lines must be down before cities are disconnected.

DEFINITION 14.24 A set of edges C of a connected graph $G = (V, E)$ is a **cut set** if the removal of the edges of C causes the graph to no longer be connected, but the removal of any proper subset of C leaves the set connected. If the set C consists of a single edge, then the edge in this set is called a **cut edge**.

Another way of describing a cut set is that it is a minimal set of edges that causes the set to be disconnected if it is removed.

EXAMPLE 14.25 In the graph shown in Figure 14.17 e_5 and e_6 are cut edges. ∎

EXAMPLE 14.26 In the graph shown in Figure 14.18, $\{\{v_1, v_2\}, \{v_5, v_6\}\}$ and $\{\{v_2, v_3\}, \{v_6, v_7\}\}$ are cut sets.

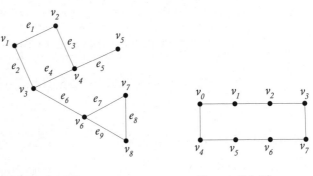

Figure 14.17   **Figure 14.18** ∎

The proof of the following theorems is left to the reader.

THEOREM 14.27 If $T(V, E')$ is a spanning tree of the graph $G = (V, E)$ and C is a cut set for G, then $C \cap E' \neq \varnothing$. ∎

THEOREM 14.28 An edge e of a graph G is a cut edge of G if and only if it is not an edge in a cycle of G. ∎

Another important concern might be how many cities would be disconnected if the communication relay in a particular city ceased to function. Since cities are represented by vertices, we are asking what would happen if a vertex of the graph were removed.

14.1 Algebraic Properties of Graphs

DEFINITION 14.29 A connected graph $G = (V, E)$ has a **cut vertex** or **articulation point** $a \in V$ if the removal of a and its incident edges causes the graph to no longer be connected.

DEFINITION 14.30 A graph $G(V, E)$ is **biconnected** if it contains no articulation points.

THEOREM 14.31 A vertex a of a graph $G = (V, E)$ is an articulation point if and only if there exist distinct vertices v and w such that every path from v to w is through a.

Proof We first prove the "if" part of the theorem. Assume that every path from v to w is through a. Then if a is removed, there is no path from v to w and G is no longer connected. Hence, a is an articulation point.

To prove the "only if" part of the theorem, we use the contrapositive method of proof. Assume that for every pair of distinct vertices v and w there is a path that does not pass through a; then if a is removed, there is still a path from v to w for all $v, w \in V$, and G is still connected. Hence, a is not an articulation point. ∎

EXAMPLE 14.32 In the graph shown in Figure 14.19, v_3, v_4, and v_6 are cut vertices. ∎

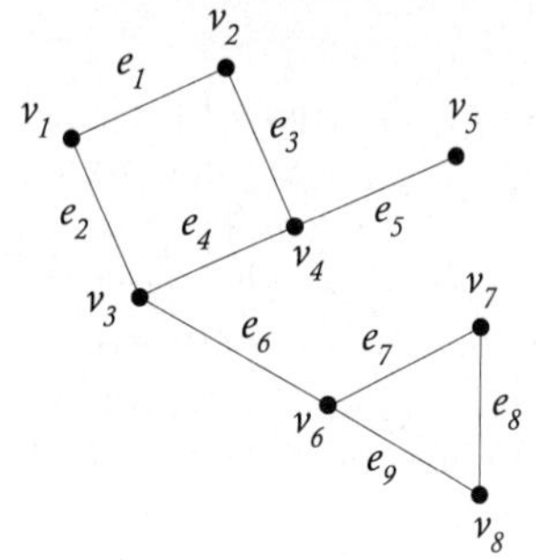

Figure 14.19

THEOREM 14.33 For a connected graph $G = (V, E)$, define a relation R on E by $e_1 R e_2$ if $e_1 = e_2$ or there is a simple cycle in G containing e_1 and e_2 as edges. The relation R is an equivalence relation.

Proof The reflexive and symmetric properties are obvious. Suppose $e_1 R e_2$ and $e_2 R e_3$. If either $e_1 = e_2$ or $e_2 = e_3$, then the result is obvious. Otherwise there is a cycle containing e_1 and e_2 as edges, and a cycle containing e_2 and e_3 as edges. If either cycle contains both e_1 and e_3 as edges, we are finished. If not, these are distinct cycles that have the edge e_2 and perhaps other edges in common. Let $v_0 v_1 v_2 v_3 \cdots v_m v_0$ be the cycle containing the edge e_1 and let $v'_0 v'_1 v'_2 v'_3 \cdots v'_n v'_0$ be the cycle containing the edge e_3. Let $v_i v_{i+1} v_{i+2} \cdots v_j$ and $v'_k v'_{k+1} v'_{k+2} \cdots v'_l$ be the longest paths in the cycles $v_0 v_1 v_2 v_3 \cdots v_m v_0$ and $v'_0 v'_1 v'_2 v'_3 \cdots v'_n v'_0$, respectively, such that e_1 is an edge of $v_i v_{i+1} v_{i+2} \cdots v_j$, e_3 is an edge of $v'_k v'_{k+1} v'_{k+2} \cdots v'_l$, both $v_i = v'_k$, and $v_j = v'_l$ but the paths have no other vertex in common. Then the path

$$v_i v_{i+1} v_{i+2} \cdots v_j v'_{l-1} v'_{l-2} \cdots v'_k$$

is a simple cycle containing both e_1 and e_3 as edges. Therefore, R is transitive. ∎

DEFINITION 14.34 For each equivalence class E_i for the equivalence relation R, let V_i be the vertices incident to edges of E_i and $G_i(V_i, E_i)$ be the subgraph of $G(V, E)$ with vertices V_i and edges E_i. The subgraph $G_i(V_i, E_i)$ is a **bicomponent** of $G(V, E)$.

THEOREM 14.35 If (a, b) and (c, d) are distinct edges of bicomponent $G_i(V_i, E_i)$, then there is a simple cycle in $G_i(V_i, E_i)$ containing (a, b) and (c, d) as edges.

Proof If (a, b) and (c, d) are distinct edges of bicomponent $G_i(V_i, E_i)$, then there is a simple cycle in $G(V, E)$ containing (a, b) and (c, d) as edges. But if there is an edge (c, d) of this cycle not in E_i, then there are edges in two different equivalence classes contained in a simple cycle, which contradicts the construction of the equivalence classes. ∎

The proof of the following theorem is left to the reader.

THEOREM 14.36 If bicomponent $G_i(V_i, E_i)$ consists of a single edge e_i, then e_i is a cut edge of G. ∎

THEOREM 14.37 If every two distinct edges are edges of a common simple cycle in a graph $G(V, E)$, then $G(V, E)$ is biconnected.

Proof Assume every two distinct edges are edges of a common simple cycle. If $G(V, E)$ has a single edge e_i, then the result is immediate. Otherwise, let u and v be vertices in G and a be a vertex on the path from u to v. Let $\{c, a\}$ and $\{a, b\}$ be distinct edges on the path. Then $\{c, a\}$ and $\{a, b\}$ are edges of a common simple cycle so there is a path from b to c that does not pass through a. Hence, there is a path from u to v that does not pass through a. By Theorem 14.28, a cannot be an articulation point. Hence, $G(V, E)$ is biconnected. ∎

Corollary 14.38 The subgraph $G_i(V_i, E_i)$ is biconnected. ∎

THEOREM 14.39 If $G_i(V_i, E_i)$ and $G_j(V_j, E_j)$ are bicomponents of $G(V, E)$, then $V_i \cap V_j$ contains at most one vertex.

Proof If $V_i \cap V_j$ contains two vertices a and b, then there are edges $\{a, c\}$ and $\{b, d\}$ in E_i. By definition of E_i, there is a simple cycle in $G_i(V_i, E_i)$ containing $\{a, c\}$ and $\{b, d\}$. Hence, there is a path in $G_i(V_i, E_i)$ from a to b. Similarly, there is a path from b to a in $G_j(V_j, E_j)$. But since $G_i(V_i, E_i)$ and $G_j(V_j, E_j)$ have no edge in common, the path from a to b in $G_i(V_i, E_i)$ followed by the path from b to a in $G_j(V_j, E_j)$ is a cycle, and this cycle contains a simple cycle. But then there are edges from both $G_i(V_i, E_i)$ and $G_j(V_j, E_j)$ in a common simple cycle, which contradicts the construction of $G_i(V_i, E_i)$ and $G_j(V_j, E_j)$. ∎

THEOREM 14.40 A vertex a is an articulation point if and only if for some $i \ne j$ it is a vertex in $V_i \cap V_j$ for bicomponents $G_i(V_i, E_i)$ and $G_j(V_j, E_j)$.

Proof If a is an articulation point, then there exist edges $\{a, b\}$ and $\{a, c\}$ such that there is no path from b to c that does not pass through a. (See proof of Theorem 14.37.) Hence, $\{a, b\}$ and $\{a, c\}$ are not edges of a common cycle so they belong to different bicomponents, say $G_i(V_i, E_i)$ and $G_j(V_j, E_j)$. Therefore, $a \in V_i \cap V_j$. Conversely, if $a \in V_i \cap V_j$, then there is an edge $\{a, b\}$ in E_i and an edge $\{a, c\}$ in E_j. But there is no other path from b to c except through a. Otherwise $\{a, b\}$ and $\{a, c\}$ would be edges of a common simple cycle, which contradicts the construction of E_i and E_j. Hence, a is an articulation point. ∎

THEOREM 14.41 $G(V, E)$ is biconnected if and only if every two distinct edges in $G(V, E)$ are edges of a common simple cycle in $G(V, E)$.

Proof Half of the theorem is a restatement of Theorem 14.37. Assume there are two distinct edges in $G(V, E)$ that are not edges of a common simple cycle in $G(V, E)$. Then they are in different bicomponents. Hence, the path between them must pass through a vertex that is common to two different bicomponents and, by the previous theorem, is an articulation point so that $G(V, E)$ is not biconnected. ∎

THEOREM 14.42 If $G_i(V_i, E_i)$ is a bicomponent and $G_i(V_i, E_i) \neq G(V, E)$, then there exists at least one distinct bicomponent $G_j(V_j, E_j)$ such that $V_i \cap V_j$ contains exactly one vertex.

Proof By Theorem 14.39, for all distinct bicomponents V_i and V_j, $V_i \cap V_j$ contains at most one vertex. However, if $a \in V_i$ and $b \notin V_i$, then there exists a path from a to b and the last vertex v along the path that is in V_i is in the intersection of V_i and V_j for some j. ∎

Exercises

In each of the following pairs of graphs, find a homomorphism from the first onto the second if it exists.

1.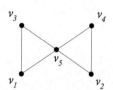

6.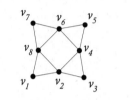

For each of the following pairs of graphs, describe an isomorphism or show the graphs are not isomorphic because an invariant is not preserved.

2.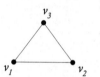

7.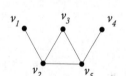

3.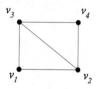

8.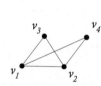

4.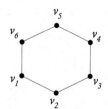

9.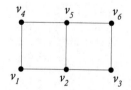

5.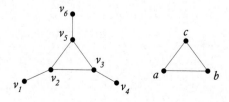

10.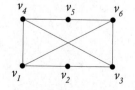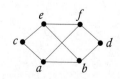

11.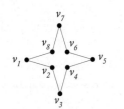

17.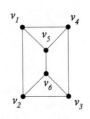

12.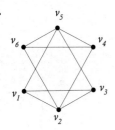

18.

13.

19.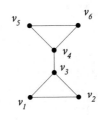

14.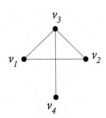

20.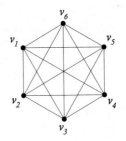

15.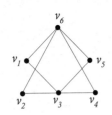

21.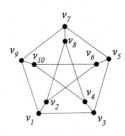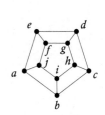

16.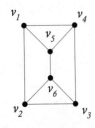

22.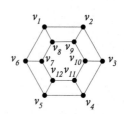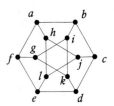

14.1 Algebraic Properties of Graphs

23.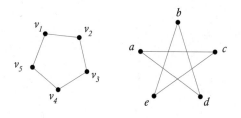

24.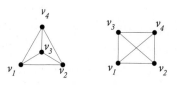

Which of the following graphs can be derived from the graph in Figure 14.20?

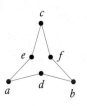

Figure 14.20

25.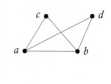

26.

27.

28.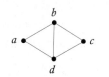

Which of the following graphs can be derived from the graph in Figure 14.21?

Figure 14.21

29.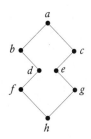

30.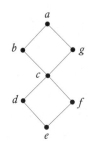

31.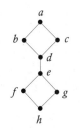

32.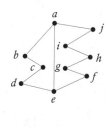

Which of the following pairs of graphs are homeomorphic?

33.

34.

35.

36.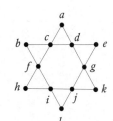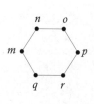

37.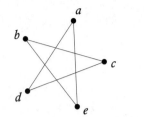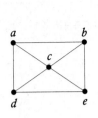

43.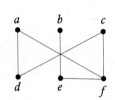

38.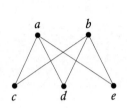

44.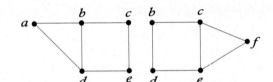

Find the union and intersection of the following collections of graphs.

39.

45.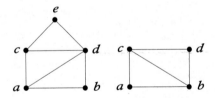

40.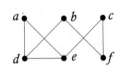

46.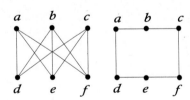

41.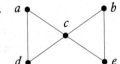

47.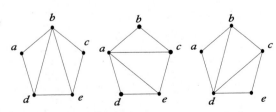

42.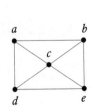

48.

49.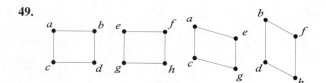

57.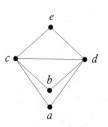

50.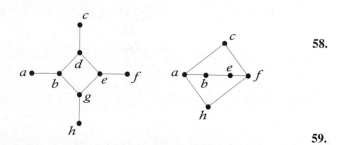

58.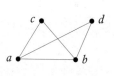

Find the complement of the following graphs.

59.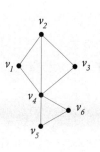

51. 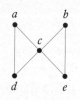 **52.**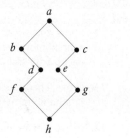

60.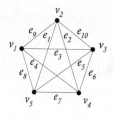

53. 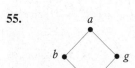 **54.**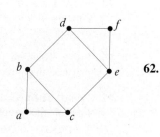

61.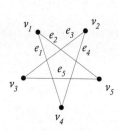

55. 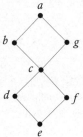 **56.**

62.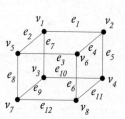

Given the graph G in Figure 14.22, which of the following are spanning graphs?

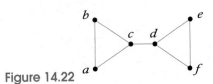

Figure 14.22

63.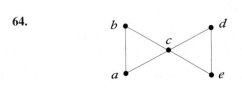

64.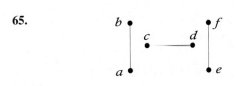

65.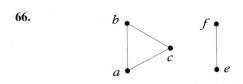

66.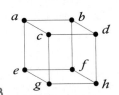

Given the graph G in Figure 14.23, which of the following are spanning graphs?

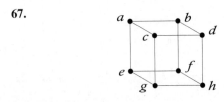

Figure 14.23

67.

68.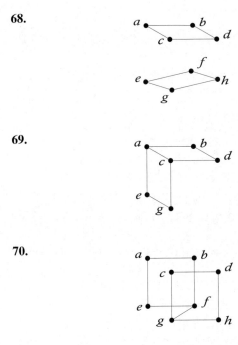

69.

70.

71. Show that any homomorphism from the graph in Figure 14.24 onto a graph G is an isomorphism.

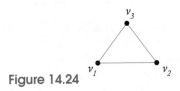

Figure 14.24

72. Show that if G and G' are graphs, $f : G \to G'$ is a homomorphism, and G is connected, then it is not necessarily true that G' is connected.

73. Prove Theorem 14.3: If a graph G is connected and f is a homomorphism, then $f(G)$ is connected.

74. Prove Theorem 14.4: If a graph G is complete and f is a homomorphism, then $f(G)$ is complete.

75. Prove Theorem 14.17: If G_1 and G_2 are distinct components of a graph G, then G_1 and G_2 are pairwise disjoint.

76. Prove Theorem 14.18: A graph G is the union of pairwise disjoint components.

77. Prove Theorem 14.23: If $T(V, E')$ is a spanning tree of the graph $G = (V, E)$, then for any cycle $v_0 v_1 v_2 v_3 v_4, \ldots v_0$ in G, at least one of the edges must belong to $E - E'$.

78. Prove that if G is connected and f is a homeomorphism, then $f(G)$ is connected.

79. Show that if $G(V, E)$ and $G'(V', E')$ are graphs, $f: G \to G'$ is a homomorphism, and $v \in V$ has degree n, then $f(v)$ does not necessarily have degree n in $f(G)$. What can be said about the degree of $f(v)$ as a vertex of $f(G)$? What can be said about the degree of $f(v)$ as a vertex of G'?

80. Show that if graphs G and G' are isomorphic, then they have the same number of edges and vertices.

81. Show that if f is an isomorphism from G to G', then G is connected if and only if G' is connected.

82. Show that if f is an isomorphism from G to G' and H is a component of G, then $f(H)$ is a component of G'.

83. Show that if f is an isomorphism from G to G' and $v \in V$ has degree n in G, then $f(v)$ has degree n in G'.

84. Show that if f is an isomorphism from G to G' and G has an Euler cycle, then G' has an Euler cycle.

85. Show that if f is an isomorphism from G to G' and G has a proper Euler path, then G' has a proper Euler path.

86. Show that if f is an isomorphism from G to G' and G is a bipartite graph, then G' is a bipartite graph.

87. Show that if f is an isomorphism from G to G' and G is a complete bipartite graph K_{mn}, then G' is a complete bipartite graph K_{mn}.

88. Prove that the intersection of a graph $G = (V, E)$, and its complement graph is the graph consisting only of the vertices of G with no edges.

89. Prove that the union of a graph $G = (V, E)$ and its complement graph is a complete graph.

90. Show that if f is an isomorphism from G to G' and e is a cut edge of G, then $f(e)$ is a cut edge of G'.

91. A graph G is **self-complementary** if it is isomorphic to its complement (see Definition 14.19). Give an example of a graph that is self-complementary.

92. What is the relationship between the matrix of a graph and the adjacency matrix of its complement?

93. Show that for a graph G, a graph G_1 is isomorphic to G if and only if G_1^c is isomorphic to G^c.

94. *Let G and G' be graphs and let B and B' be their respective adjacency matrices. Prove that G is isomorphic to G' if and only if there is a permutation P such that $B' = PBP^t = PBP^{-1}$.

95. Find three cut sets in the graph shown in Figure 14.25. Find an articulation point. Are there any cut edges?

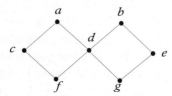

Figure 14.25

96. Find three cut sets in the graph shown in Figure 14.26. Find any articulation points. Find any cut edges.

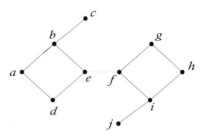

Figure 14.26

97. Find three cut sets in the graph shown in Figure 14.27.

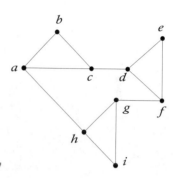

Figure 14.27

Find any articulation points. Find any cut edges.

98. Prove Theorem 14.28: An edge e of a graph G is a cut edge of G if and only if it is not an edge in a cycle of G.

99. Prove Theorem 14.27: If $T(V, E')$ is a spanning tree of the graph $G = (V, E)$ and C is a cut set for G, then $C \cap E' \neq \emptyset$.

100. Prove Theorem 14.36: If bicomponent $G_i(V_i, E_i)$ has a single edge e_i, then e_i is a cut edge of G.

101. Given the following definition: Let $G = G(V, E)$ be a graph and $G_1, G_2, G_3, \ldots, G_n$ be subgraphs of G. The subgraph G' of G is the **intersection** of $G_1, G_2, G_3, \ldots, G_n$, denoted by $\bigcap_{i=1}^{n} G_i$ if

   1. For all i, $G' \preceq G_i$.
   2. If for all i, $G'' \preceq G_i \preceq G$, then $G'' \preceq G'$.

   Show that this definition is equivalent to the one for intersection given in this section.

102. Given the following definition: Let $G = G(V, E)$ be a graph and $G_1, G_2, G_3, \ldots, G_n$ be subgraphs of G. The subgraph G' of G is the **union** of $G_1, G_2, G_3, \ldots, G_n$, denoted by $\bigcup_{i=1}^{n} G_i$ if

   1. For all i, $G_i \preceq G'$.
   2. If for all i, $G_i \preceq G'' \preceq G$, then $G' \preceq G''$.

   Show that this definition is equivalent to the one for union given in this section.

14.2 Planar Graphs

An integrated circuit chip consists of layers of miniature circuits placed together in a wafer. It is critical that the wires of the miniature circuits not cross except at designated connections. If we consider the designated connections to be vertices of a graph, then we are concerned with constructing a graph in which the edges do not cross. It is important to note that we are concerned with whether the graph *can* be drawn without the edges crossing. For example, the graph shown in Figure 14.28 can also be drawn as shown in Figure 14.29. The graph shown in Figure 14.30 can also be drawn as shown in Figure 14.31.

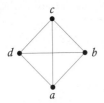
Figure 14.28

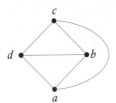
Figure 14.29

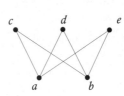
Figure 14.30

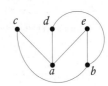
Figure 14.31

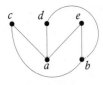

Figure 14.32

If the three house–three utilities problem of Section 6.1 had consisted of three houses and two utilities or two houses and three utilities, we could solve the problem since, in either case, we can represent the problem with the graph in Figure 14.32 so that the lines do not cross.

Before tackling the three house–three utilities problem, we need some definitions and some tools. We begin by giving a name to the graphs with the properties we are considering.

DEFINITION 14.43   A **planar graph** is a graph that can be drawn on a plane so that none of the edges cross. A graph that is not planar is called **nonplanar**.

Think of a graph as a figure drawn on a piece of paper. If a graph is planar and is drawn so that none of the lines cross, and if we can cut along the edges, we will cut the graph into pieces, including the outside piece that is left. These pieces are called **faces**. Notice that the boundary of each face is a cycle.

DEFINITION 14.44   A face of a planar graph is a maximal section of a plane such that any two points within the section may be connected by a curve that does not cross an edge.

We now prove a theorem, called Euler's formula, which will help us determine whether or not a graph is planar.

THEOREM 14.45   If G is a connected planar graph with v vertices, e edges, and f faces, then $v - e + f = 2$.

Proof   We prove this using induction on the number of edges. If $e = 0$, then there is one vertex, no edge, and one face, which is the entire plane. Therefore, we have $1 - 0 + 1 = 2$, and the theorem is true if $e = 0$. For good measure we note that if there is one edge, then there are two vertices and one face. Therefore, we have $2 - 1 + 1 = 2$, and the theorem is true for $e = 1$.

Assume that the theorem is true for any connected planar graph with k edges and that we have a connected planar graph G_{k+1} with $k+1$ edges. We now want to remove an edge so that we have a connected planar graph G_k with k edges, so that the formula $v - e + f = 2$ is true for G_k and then try to find a way to argue that the formula will also be true for the connected planar graph G_{k+1}.

First assume that the graph G_{k+1} has no cycles. Start on a path and continue until a vertex is reached for which there is no other edge out. This must happen if graph G_{k+1} has no cycles. This is a vertex of degree 1. We remove that vertex and its incident edge. This does not change the number of faces. Also, the graph is still planar, connected, and has k edges, so the formula $v - e + f = 2$ is true. Since the number of edges is decreased by 1 and the number of vertices is decreased by 1, then $v - e + f$ remains unchanged by the removal so that $v - e + f = 2$ for the connected planar graph G_{k+1}.

Next assume that the graph G_{k+1} has a cycle. Remove an edge e_i in the cycle with incident vertices u_i and v_i but leave the vertices. From Theorem 14.28, we know that there is still a path from u_i to v_i so that the graph is still connected. It is also planar and has k edges so that it satisfies the formula $v - e + f = 2$. Since e_i is in a cycle, it separates two faces. Hence, its removal removes a face. We, therefore, have removed one edge and one face. Therefore, we have not changed the value of $v - e + f$ and the formula $v - e + f = 2$ is true for G_{k+1}. ∎

We are now ready to show that the graph for the three house–three utilities problem, which is the complete bipartite graph $K_{3,3}$, cannot be drawn without the edges crossing.

THEOREM 14.46 The complete bipartite graph $K_{3,3}$ is *not* planar.

Proof We use a reductio ad absurdum argument, so we assume that $K_{3,3}$ is planar. If $K_{3,3}$ is planar, then since there are nine edges and six vertices, $6 - 9 + f = 2$ so $f = 5$. Let A and B be the disjoint three-element sets of vertices forming the set V of vertices of $K_{3,3}$. If we start a path in one of the disjoint sets, say A, and do not repeat edges, we must go to a vertex in B, return to a vertex in A, again return to a vertex in B, and finally return to a vertex in A before we can complete a cycle. Each cycle in $K_{3,3}$ has a path whose length is at least 4. Therefore, each face is determined by a cycle of at least four edges. Therefore, if we find the sum of the edges of all the faces, it must be greater than $4f$. But each edge will be counted at most two times, since an edge can only be an edge for two faces. Therefore, the sum of the edges of all the faces must be less than $2e$. Putting these two inequalities together, we have $4f \leq 2e$, so $4f \leq 18$. But this contradicts the fact that $f = 5$. Therefore, we have a contradiction and $K_{3,3}$ is *not* planar. ∎

We next want to show that the complete graph K_5 shown in Figure 14.33 is not planar. Before we do this, however, we need the following lemma (which is a theorem that is used to help prove another theorem).

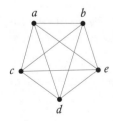

Figure 14.33

Lemma 14.47 In any connected planar graph G, with at least three vertices, $3v - e \geq 6$.

Proof If the graph G has no cycles, then the number of edges is less than the number of vertices $e \leq v$. Also since $v \geq 3$, $2v - 6 \geq 0$ so $e \leq v + 2v - 6 = 3v - 6$ or $3v - e \geq 6$. If the graph G has a cycle, we again sum the edges on the boundaries of the faces. Since the boundary of each face has at least three edges, the sum of the edges of the faces must be greater than $3f$. But, as in the previous proof, each edge is only on the boundary of at most two faces so the sum of the faces of the edges must be less than $2e$. Therefore, $3f \leq 2e$ and $2 = v - e + f$ so $6 = 3v - 3e + 3f \leq 3v - 3e + 2e = 3v - e$ and $3v - e \geq 6$. ∎

THEOREM 14.48 The complete graph K_5 is not planar.

Proof Since K_5 has five vertices and 10 edges, $3v - e = 3 \cdot 5 - 10 = 5$, so by the previous lemma, K_5 is not planar. ∎

Using the previous lemma, we can also prove another theorem that is used in the next section to show that a planar graph can be colored using five colors or less.

THEOREM 14.49 Every planar graph G contains a vertex of degree 5 or less.

Proof As mentioned in Chapter 2, since for each edge there are two vertices, each edge contributes 2 to the sum of the degrees of the vertices. Therefore, the sum of the degrees of the vertices is equal to $2e$. If there are at most two vertices in G, obviously *all* vertices have degree 5 or less. Therefore, assume there are three or more vertices. Then by the previous lemma $e \leq 3v - 6$, so $2e \leq 6v - 12$. But if every vertex has degree 6 or more, then the sum of the degrees of the vertices is greater than or equal to $6v$. The sum of the degrees of the vertices is equal to $2e$ so $2e \geq 6v$. This is impossible since $2e \leq 6v - 12$. Hence, there is a vertex of degree 5 or less. ∎

The proof of the following theorem is left to the reader.

THEOREM 14.50 If two connected graphs are homeomorphic, then they are both planar or both nonplanar. ∎

From this theorem we immediately have the following theorem.

THEOREM 14.51 Any graph that is homeomorphic to either $K_{3,3}$ or K_5 is not planar. ∎

Since any graph that contains a subgraph homeomorphic to either $K_{3,3}$ or K_5 is not planar, we have proven one part of the following theorem. We accept the remainder of the theorem without proof.

THEOREM 14.52 **(Kuratowski's Theorem)** A graph is planar if and only if it does not contain a subgraph homeomorphic to either $K_{3,3}$ or K_5. ∎

EXAMPLE 14.53 Show that the graph in Figure 14.34 is planar. Since this graph consists of two components, we can separate them to the form shown in Figure 14.35, which is certainly planar.

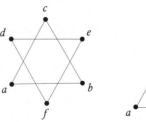
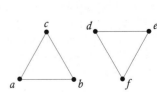

Figure 14.34   Figure 14.35

14.2 Planar Graphs

EXAMPLE 14.54 Show that the graph in Figure 14.36 is planar. By moving vertex d so that we have the graph in Figure 14.37, we have greatly simplified the graph. By then moving vertex c, we have the graph in Figure 14.38, which is obviously planar.

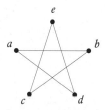

Figure 14.36

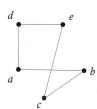

Figure 14.37

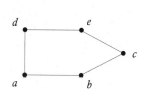

Figure 14.38

EXAMPLE 14.55 Show that the Petersen graph in Figure 14.39 is not planar. This graph can be shown to have a subgraph homeomorphic to $K_{3,3}$. The only hard part is to pick the right sets of vertices. If e, f, and g are chosen as the top vertices and j, a, and i are chosen as the bottom vertices and we connect each top vertex to each bottom vertex, we have the graph in Figure 14.40. We see that vertex c is not needed, so the vertex and all the edges incident to it are removed. All of the other edges are still present. We can remove points h, b, and d forming edges $\{f, j\}$, $\{a, g\}$, $\{e, i\}$, respectively, and still have a graph in Figure 14.41 that is homeomorphic to the previous graph. But this can be redrawn as $K_{3,3}$. Therefore, the Petersen graph has a subgraph homeomorphic to $K_{3,3}$ and so it is not planar.

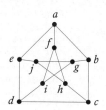

Figure 14.39

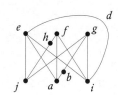

Figure 14.40

Figure 14.41

Exercises

For each graph that follows, determine whether or not it is planar. Give reasons why or why not.

1.

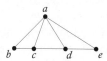

2.

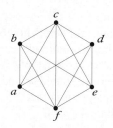

3.

4.

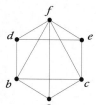

5.

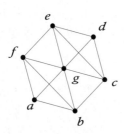

6. For each planar graph in the previous problems verify that $v - e + f = 2$.

For each graph that follows, determine whether or not it is planar. Give reasons why or why not.

7.

8.

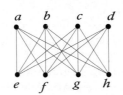

9.

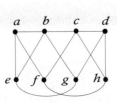

10.

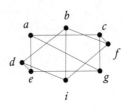

11.

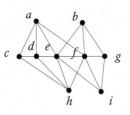

12. For each planar graph in Exercises 7–11, verify that $v - e + f = 2$.

13. If a planar graph has 12 vertices, each of degree 3, how many faces and edges does the graph have?

14. If a planar graph has vertices with degrees 2, 2, 2, 3, 3, 3, 4, 4, 4, and 5, respectively, how many edges and faces does it have?

15. Find the value of $v - e + f$ in a graph with

   (a) two components

   (b) three components

   (c) k components

16. If G is a planar graph with n vertices and every face of G is bounded by a 3-cycle, find the number of edges and faces in terms of n.

17. Prove that any graph with four vertices is planar.

18. Prove that any graph with five vertices, one of which has degree 2, is planar.

19. Using induction, prove that a tree is a planar graph.

20. Prove Theorem 14.50: If two connected graphs are homeomorphic, then they are both planar or both nonplanar.

14.3 Coloring Graphs

One of the most famous unsolved problems for many years was the four-color problem. It states that if one is drawing a map, then it may be drawn, using only four colors, so that any two countries sharing a common border are a different color. It was well known that five colors were sufficient and that three colors were not sufficient to color a map. The minimum number of colors needed if the map was drawn on a doughnut had been solved, but the four-color problem remained unsolved. It has now been solved, but not perhaps in the most satisfying way. The method required using a computer to check out a finite but still large number of cases that have not been solved using other methods. While this method is certainly an outstanding achievement and should be sufficient to stop most people from looking for counterexamples, it would be nice if someone could someday find a more elegant proof. Perhaps one of the main contributions of this proof is that it has forced us to broaden our concept of what constitutes a mathematical proof.

14.3 Coloring Graphs

Figure 14.42

Figure 14.43

Consider the map in Figure 14.42. We see that if the left-side face is painted red, then we can alternate painting the central adjoining faces blue and green. We can again paint the right face red since it does not adjoin the left face as we see in Figure 14.43.

In transferring the map to a graph, the most natural transition is to use edges as boundaries and faces as countries, so that the map in Figure 14.44 becomes the graph in Figure 14.45.

Rather than work with this graph, however, it is better to make one change. It is more convenient to work with the **dual graph** G' of a planar graph G, which is formed by changing the faces (or countries) to vertices and constructing an edge between two vertices if the faces are adjacent. Hence, vertices in the dual graph G' are adjacent if and only if the faces in the original graph G are adjacent. Thus, the dual of the graph becomes the graph shown in Figure 14.46. The outside edge would be included as a vertex only if it were to be colored.

Figure 14.44   Figure 14.45   Figure 14.46

EXAMPLE 14.56

Figure 14.47

Figure 14.48

Figure 14.49

Figure 14.50

The dual graph of the graph G in Figure 14.47 is the graph G' of Figure 14.48 so, in this case, G is isomorphic to G'. ∎

Since the faces of G have become vertices in G', we paint the vertices in G'. Our problem has now changed from painting countries of a map so that no two countries that have a common boundary to painting the vertices of a graph so no two adjacent vertices have the same color.

At this point, we shall take a new form of attack. For a graph, if we have a fixed number of colors, we shall try to determine how many ways, using these colors, we can paint the vertices so no adjacent vertices are the same color. This concept was discovered by G. D. Birkhoff [8]

For example, suppose we have five colors and want to paint the graph in Figure 14.49. Then there are five colors to paint the first vertex but only four colors to paint the second vertex since its color must be different from the first vertex. Hence, there are $5 \cdot 4 = 20$ ways of coloring the graph. Notice that this is the counting method that we previously used for permutations and other cases where we used the multiplication counting principle in Section 8.1. Suppose we have four colors and want to color the graph G in Figure 14.50. We have four colors for painting vertex a. There are then three colors to paint vertex b. There are just two colors to paint vertex c, since it is adjacent to both a and b so that its color must be different both of them. There are also two colors to paint vertex d since it may be the same color as b but must be a different color from adjacent vertices a and c. Therefore, we have $4 \cdot 3 \cdot 2 \cdot 2 = 48$ ways of coloring G. It is important not to paint both vertices b and d before counting the colors for vertex c, since we would have to break our counting into the cases where b and d were the same color and where b and d were different colors. Suppose we have λ colors to paint the graph G. We have λ colors for painting a, $\lambda - 1$ colors for painting b, $\lambda - 2$ colors for painting c, and $\lambda - 2$ colors for painting d. Thus, there are $(\lambda)(\lambda - 1)(\lambda - 2)(\lambda - 2)$ ways of coloring G.

We denote this by $C_G(\lambda)$. Thus, $C_G(\lambda)$ is the number of ways of coloring G using λ colors. Note that $C_G(0) = C_G(1) = C_G(2) = 0$.

This leads us to the following definition. In this definition we shall state that $C_G(\lambda)$ is a polynomial, which seems intuitively likely, and prove it formally later.

DEFINITION 14.57 Let G be a graph. A **coloring** of G is a painting of the vertices of G such that no two adjacent vertices are the same color. Let $C_G(\lambda)$ denote the number of ways the graph G can be painted using λ colors so that no two vertices are the same color, that is, the number of colorings of G. For a fixed graph G, $C_G(\lambda)$ is a polynomial function of λ, called the **chromatic polynomial** of G. The **chromatic number** of a graph is the smallest number of colors that can be used to color the graph. It is the smallest positive number n so that $C_G(n) \neq 0$.

EXAMPLE 14.58 Let G be the graph consisting of five isolated vertices, so that there are no edges and, hence, no adjacent vertices. If we paint this graph with λ colors, we have λ choices of color of each vertex so that there are $\lambda \cdot \lambda \cdot \lambda \cdot \lambda \cdot \lambda = \lambda^5$ ways of coloring the graph and $C_G(\lambda) = \lambda^5$. Obviously if G is a graph consisting of k isolated vertices, then $C_G(\lambda) = \lambda^k$. The chromatic number is 1. ∎

EXAMPLE 14.59 Let G be the graph K_5, so that each of the five vertices is adjacent to all of the other four vertices. If we paint this graph with λ colors, we have λ choices for the color of the first vertex. Since the second vertex is adjacent to the first colored vertex, we have $\lambda - 1$ choices for colors for the next vertex. The third vertex is adjacent to each of the first two colored vertices, so there are $\lambda - 2$ choices for colors for this vertex. The fourth vertex is adjacent to each of the first three colored vertices, so there are $\lambda - 3$ choices for colors for this vertex. Finally, the fifth vertex is adjacent to each of the first four colored vertices, so there are $\lambda - 4$ choices for colors for this vertex. Thus, $C_G(\lambda) = (\lambda)(\lambda - 1)(\lambda - 2)(\lambda - 3)(\lambda - 4)$. Notice that this graph cannot be colored in four colors since $C_G(4) = 0$, but since G is not a planar graph, this doesn't contradict the fact that a planar graph can be colored with four colors. Using the special case of K_5 above as our guide, we find that if $G = K_n$ then $C_G(\lambda) = (\lambda)(\lambda - 1)(\lambda - 2)(\lambda - 3) \cdots (\lambda - (n - 1))$. Thus, the chromatic number for G is n. ∎

The painting of one component of a graph does not restrict the painting of another component since no vertex of one component is adjacent to the vertex of another component. Therefore, we have the following theorem.

THEOREM 14.60 If $G = G_1 \cup G_2 \cup G_3 \cup \cdots \cup G_n$, where $G_1, G_2, G_3, \ldots, G_n$ are the components of G, then $C_G(\lambda) = C_{G_1}(\lambda) \cdot C_{G_2}(\lambda) \cdot C_{G_3}(\lambda) \cdot \cdots \cdot C_{G_n}(\lambda)$. ∎

From this theorem it follows that $C_G(\lambda) = 0$ if and only if $C_{G_i}(\lambda) = 0$ for some $1 \leq i \leq n$. Hence, we have the following corollary.

Corollary 14.61 If it requires k colors to color G, then it requires k colors to color one of its components. ∎

Because of these results we will only consider the coloring of connected graphs.

At this point, from a given graph $G(V, E)$, we wish to form two special graphs. Let $e = \{a, b\}$ be an edge of G. Let $G_{\hat{e}}$ be the graph G with the edge e removed, but the vertices a and b remaining. Let $G_{/e}$ be the graph $G_{\hat{e}}$ with the vertices a and b identified.

(Note that the graph $G_{/e}$ is really the homomorphic image of the graph $G_{\hat{e}}$ where the homomorphism $f : G_{\hat{e}} \to G_{/e}$ is defined for the vertices of $G_{\hat{e}}$ by $f(a) = f(b)$

and $f(v) = v$ for all v in $V - \{a, b\}$. For the edges, $f(\{u, v\}) = \{f(u), f(v)\}$. Also note that $f = f \circ f$ since f is the identity map of the vertices and edges of $G_{/e}$. A function with this property is called a retraction map.)

EXAMPLE 14.62 Let G be the graph in Figure 14.51. Then $G_{\hat{e}}$ is the graph in Figure 14.52 and $G_{/e}$ is the graph in Figure 14.53.

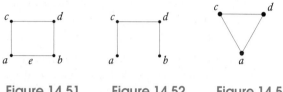

Figure 14.51   Figure 14.52   Figure 14.53

EXAMPLE 14.63 Let G be the graph in Figure 14.54. Then $G_{\hat{e}}$ is the graph in Figure 14.55 and $G_{/e}$ is the graph in Figure 14.56.

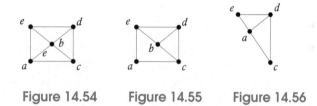

Figure 14.54   Figure 14.55   Figure 14.56

We now use these special graphs to prove a very useful theorem:

THEOREM 14.64 For any planar graph G with edge e,
$$C_{G_{\hat{e}}}(\lambda) = C_G(\lambda) + C_{G_{/e}}(\lambda)$$

Proof Suppose that we have a coloring of $G_{\hat{e}}$. If the color of the vertex a is not the same as the color of the vertex b, then this is also a coloring of G and conversely. If the color of a is the same as the color of the vertex b, then we have a coloring of $G_{/e}$ and conversely. Therefore, each coloring of $G_{\hat{e}}$ is either a coloring of G or $G_{/e}$ but not both, and the number of colorings of $G_{\hat{e}}$ is equal to the number of colorings of G plus the number of colorings of $G_{/e}$ or $C_{\hat{e}}(\lambda) = C_G(\lambda) + C_{/e}(\lambda)$. ∎

It is intuitively obvious that for any planar graph G with n vertices, $C_G(\lambda)$ is an nth degree polynomial since, when we constructed $C_G(\lambda)$ in our examples, for each vertex we formed $\lambda - k$ for some integer $0 \le k \le n$. In fact we used this assumption when defining chromatic polynomials. We shall now prove it formally, using the previous theorem and induction on the number of edges.

THEOREM 14.65 For any planar graph G with n vertices, $C_G(\lambda)$ is an nth degree polynomial.

Proof Let m be the number of edges in a planar graph G with n vertices. If $m = 0$, then by Example 14.58, $C_G(\lambda) = \lambda^n$, which is an nth-degree polynomial. Assume the theorem is true for all planar graphs with n vertices and k edges. Remember k is fixed but n is not. Let G have n vertices and $k + 1$ edges. By Theorem 14.64, $C_G(\lambda) = C_{G_{\hat{e}}}(\lambda) - C_{G_{/e}}(\lambda)$. But $G_{\hat{e}}$ has n vertices and k edges and $G_{/e}$ has $n - 1$ vertices and k edges. Also they are both planar. Hence, by the induction hypothesis, $C_{G_{\hat{e}}}(\lambda)$ and $C_{G_{/e}}(\lambda)$ are polynomials of degrees n and $n - 1$, respectively. Therefore, $C_G(\lambda) = C_{G_{\hat{e}}}(\lambda) - C_{G_{/e}}(\lambda)$ is a polynomial of degree n. ∎

From this theorem, we easily get the next theorem.

THEOREM 14.66 For any nonempty planar connected graph G the constant term of $C_G(\lambda)$ is 0. If G has two or more vertices, then the sum of the coefficients of $C_G(\lambda)$ is 0.

Proof Since a nonempty planar graph G cannot be colored by 0 colors of paint, $C_G(0) = 0$. If G has two or more vertices then, since G is connected, there are two adjacent vertices and G cannot be colored by one color of paint. Therefore, $C_G(1) = 0$. But $C_G(1)$ is the sum of the coefficients of $C_G(\lambda)$, so the sum of the coefficients of $C_G(\lambda)$ is 0. ∎

The equation $C_{G_{\hat{e}}}(\lambda) = C_G(\lambda) + C_{G_{/e}}(\lambda)$ can be used in two ways. If we let our graph be $G_{\hat{e}}$, then we can find $C_{G_{\hat{e}}}(\lambda)$ is expressed as the sum of $C_G(\lambda)$, and $C_{G_{/e}}(\lambda)$. But G has an edge added to $G_{\hat{e}}$ and, thus, increases the number of edges, and $G_{/e}$, which identifies two points in $G_{\hat{e}}$ and decreases the number of points. Eventually we will get to complete graphs. Theoretically this will get us an answer but not necessarily the simplest way.

For example, suppose we want to find the chromatic polynomial of the graph in Figure 14.57. If we let $e = \{a, b\}$ and $G_{\hat{e}}$ be the graph above, then G is the graph in Figure 14.58 and $G_{/e}$ is the graph in Figure 14.59.

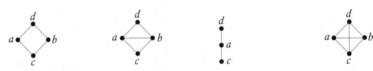

Figure 14.57   Figure 14.58   Figure 14.59   Figure 14.60

Therefore, $C_{G_{\hat{e}}}(\lambda) = C_G(\lambda) + G_{/e}$. If f is the edge $\{c, d\}$ and $G = G'_{\hat{f}}$, then G' is the graph in Figure 14.60, $G_{/f}$ is the graph in Figure 14.61, and $C'_{G'_{\hat{f}}}(\lambda) = C_{G'}(\lambda) + C_{G'_{/f}}(\lambda)$. But G' is the graph K_4 so

$$C_{G'}(\lambda) = \lambda(\lambda - 1)(\lambda - 2)(\lambda - 3)$$

and $G_{/f}$ is the graph K_3 so

$$C_{G_{/f}}(\lambda) = \lambda(\lambda - 1)(\lambda - 2)$$

Figure 14.61

Therefore,
$$C_G(\lambda) = C'_{G'_{\hat{f}}}(\lambda) = C_{G'}(\lambda) + C_{G_{/f}}(\lambda)$$
$$= \lambda(\lambda - 1)(\lambda - 2)(\lambda - 3) + \lambda(\lambda - 1)(\lambda - 2)$$
$$= \lambda(\lambda - 1)(\lambda - 2)(\lambda - 2)$$

Figure 14.62

Let $g = \{c, d\}$ and $G_{/e}$ be the graph $G''_{\hat{g}}$. Then G'' is the graph in Figure 14.62, $G_{/f}$ is the graph in Figure 14.63, and $C'_{G''_{\hat{g}}}(\lambda) = C_{G''}(\lambda) + C_{G_{/g}}(\lambda)$. But G'' is the graph K_3 so

$$C_{G''}(\lambda) = \lambda(\lambda - 1)(\lambda - 2)$$

$G_{/g}$ is the graph K_2 so

$$C_{G_{/f}}(\lambda) = \lambda(\lambda - 1)$$

Figure 14.63

Therefore,
$$C_{G_{/e}} = C'_{G''_{\hat{g}}}(\lambda) = C_{G''}(\lambda) + C_{G_{/g}}(\lambda)$$
$$= \lambda(\lambda - 1)(\lambda - 2) + \lambda(\lambda - 1)$$
$$= \lambda(\lambda - 1)(\lambda - 1)$$

Thus,

$$C_{G_{\hat{e}}}(\lambda) = C_G(\lambda) + G_{/e}$$
$$= \lambda(\lambda-1)(\lambda-2)(\lambda-2) + \lambda(\lambda-1)(\lambda-1)$$
$$= \lambda(\lambda-1)[(\lambda-2)(\lambda-2) + (\lambda-1)]$$
$$= \lambda(\lambda-1)(\lambda^2 - 3\lambda + 3)$$

The second method of using the equation $C_{G_{\hat{e}}}(\lambda) = C_G(\lambda) + C_{G_{/e}}(\lambda)$ is recursive, where we use $C_G(\lambda) = C_{G_{\hat{e}}}(\lambda) - C_{G_{/e}}(\lambda)$. We eliminate edges and identify points so we reduce the numbers of edges and points until we can, if necessary, get isolated points. Consider the same graph in Figure 14.64. If we let G be this graph and remove the edge $\{a, d\}$, then $G_{\hat{e}}$ is the graph in Figure 14.65 and $G_{/e}$ is the graph in Figure 14.66.

Figure 14.64

Figure 14.65

Figure 14.66

Using λ paints to color $G_{\hat{e}}$, we see that we can use λ colors to paint vertex a and $\lambda - 1$ to color each of the other vertices since each may be colored any color but that of the preceding vertex. Thus,

$$C_{G_{\hat{e}}}(\lambda) = \lambda(\lambda-1)^3$$

Graph $G_{/e} = K_3$ so that

$$C_{G_{/e}}(\lambda) = \lambda(\lambda-1)(\lambda-2)$$

and $C_{G_{\hat{e}}}(\lambda) - C_{G_{/e}}(\lambda)$,

$$C_G(\lambda) = C_{G_{\hat{e}}}(\lambda) - C_{G_{/e}}(\lambda)$$
$$= \lambda(\lambda-1)^3 + \lambda(\lambda-1)(\lambda-2)$$
$$= \lambda(\lambda-1)((\lambda-1)^2 - (\lambda-2))$$
$$\lambda(\lambda-1)(\lambda^2 - 3\lambda + 3)$$

which fortunately is the same answer we calculated before.

The four-color problem is equivalent to stating that for every planar graph G, $C_G(4) \neq 0$. As previously mentioned, the four-color problem remained a famous unsolved problem for some time. It had the two basic ingredients for being a famous unsolved problem. The statement of the problem was so simple that it could be explained to anyone, and it could not be solved. The problem was first formulated in 1852 by Francis Gutherie, a student of Augustus De Morgan. Some mathematicians have devoted their lives to trying to solve this problem. There have been many "proofs" of the four-color problems by both amateur and professional mathematicians. In 1880, it was assumed that the problem had been solved by Alfred Bay Kempe [47]. However, in 1890, P. J. Heawood [41] found an error in Kempe's proof. He was, however, able to modify Kempe's proof to prove that every planar map could be colored with five colors.

THEOREM 14.67 Every planar graph G can be colored using only five colors.

Proof As previously discussed, we shall assume that G is connected. We shall also assume that G has been drawn so that none of its edges cross. This proof uses induction on the number of vertices. If there is only one vertex (in fact five or less), then it can certainly be colored using only five colors. Assume that any graph with k vertices can be colored using only five colors. Let G be a graph with $k + 1$ vertices. From Theorem 14.49, we know that there is a vertex of degree 5 or less. Let v be this vertex. Let G' be the subgraph of G with v and all of its incident edges removed. Since G' has only k vertices, by our induction hypothesis, G' can be colored using only five colors. If the degree of v is less than 5, then it is adjacent to four vertices or less in G and, hence, may be colored a color different than that

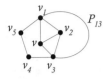

Figure 14.67

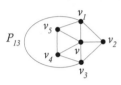

Figure 14.68

of its adjacent vertices and we are finished. Hence, assume the degree of v is 5. Therefore, v is adjacent to five vertices in G, say v_1, v_2, v_3, v_4, and v_5. If any of these vertices are the same color, then only four colors are used to paint them and we again choose a color not used, use it to paint v, and we are finished. Therefore, assume that v_1, v_2, v_3, v_4, and v_5 are all different colors. We shall let 1, 2, 3, 4, and 5 be the colors of v_1, v_2, v_3, v_4, and v_5, respectively. For convenience, assume that v_1, v_2, v_3, v_4, and v_5 occur in clockwise order around v. Beginning at vertex v_1 in G', construct the subgraph G_{13} of G' as follows: The set of vertices V_{13} of G_{13} consists of v_1 and all vertices of G' that can be connected to v_1 using paths that pass only through vertices of color 1 or 3. First assume $v_3 \notin V_{13}$. By construction of G_{13}, there is no vertex of G' having color 1 or 3 that is not in V_{13} and is adjacent to a vertex in V_{13}. Therefore, we may interchange the colors 1 and 3 in G_{13}, leaving all other vertices alone. Now v_1 and v_3 have the same color so by our previous discussion we can now color G. If $v_3 \in V_{13}$, then there is a path from v_1 to v_3 that passes only through vertices of color 1 or 3. Call this path P_{13}. Then $vv_1 P_{13} v_3 v$ is a cycle and has form shown in Figure 14.67 or in Figure 14.68. In the first form, v_2 is inside the cycle and v_4 is outside.

In the second form, v_4 is inside the cycle and v_2 is outside. In either case there is no path from v_2 to v_4 in G' through vertices of color 2 or 4. The reason is that G is planar so edges do not cross, and to get from v_2 to v_4 requires going through a vertex in the cycle $vv_1 P_{13} v_3 v$ and all these vertices have color 1 or 3 except possibly v, which is not in G'. Therefore, we can go through the same procedure as before forming G_{24} in place of G_{13} and color v. ∎

> ### Computer Science Application
>
> An algorithm in which, at each step, a decision is made that appears to be good without regard for future consequences is usually referred to as a *greedy algorithm*. A greedy algorithm for coloring the vertices of a graph in such a way that adjacent vertices have different colors is easy to program, though the number of colors required may not be minimal. The following algorithm assumes the adjacency matrix A of an undirected graph G on n nodes has been constructed.
>
> function coloring($n \times n$ matrix A, integer array V)
>
> ```
>     integer array colors[n]
>
>     V[1] = 1
>     for i = 2 to n do
>         colors = 0
>         for j = 1 to i − 1 do
>             if A[i, j] = 1 then
>                 colors[V[j]] = 1
>             end-if
>         end-for
>         k = 1
>         while colors[k] = 1 do
>             k = k + 1
>         end-while
>         V[i] = k
>     end-for
>
>     return
>
> end-function
> ```

Vertex 1 is assigned color 1. Each vertex thereafter is assigned color k, where k is the smallest color not assigned to an adjacent vertex already colored. The array V will be returned to the caller with the color assignments for each vertex of G (i.e., vertex i is assigned the color $V[i]$).

Exercises

Find the dual of the following graphs:

1.

2.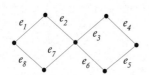

3.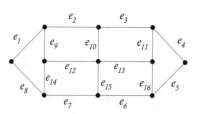

4.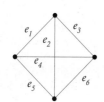

5.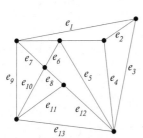

6.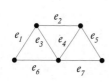

7.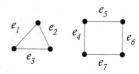

8.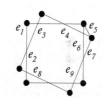

9.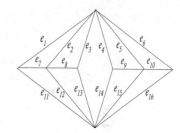

10.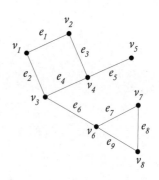

By inspection, what is the chromatic number of each of the following graphs?

11.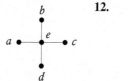

12.

542 Chapter 14 Graphs Revisited

13.

14.

15.

16.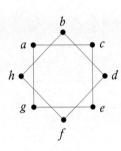

17.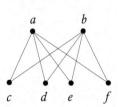

18.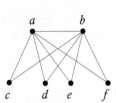

19.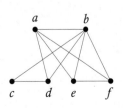

20.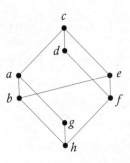

21. Show that bipartite graph $K_{m,n}$ can always be colored with two colors. What is the chromatic polynomial of $K_{m,n}$?

22. What is the chromatic number of a hypercube?

Using Theorem 14.64, find the chromatic polynomial for each of the following:

23. 24.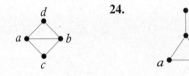

25. 26.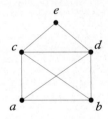

27.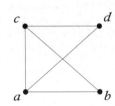

28.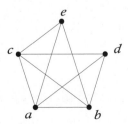

29.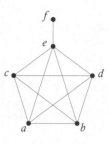

30.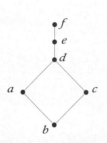

31.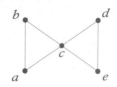

32.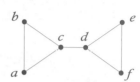

33. Prove that any planar graph with n vertices that is isomorphic to its dual has $2n - 2$ edges.

34. Prove that if a connected graph with at least one edge has chromatic number 2, then it is a bipartite graph.

14.4 Hamiltonian Paths and Cycles

In 1857, the mathematician William Rowan Hamilton invented a game. There are various versions of what really happened. In one version, he described the game in a letter to a friend. In another version he actually developed the game and sold it to a toy manufacturer. In any case, the game apparently consisted of a dodecahedron, which is a solid with faces consisting of 12 congruent regular pentagons. At each of the 20 corners or vertices of the solid, a hole was drilled and a peg placed in the hole, which represented a city. Using a string, one was supposed to find a path among the cities so that each city was visited exactly once and the traveler returned to the original city. The dodecahedron can be flattened out as shown in Figure 14.69.

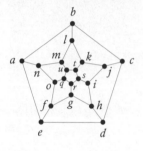

Figure 14.69

The problem then becomes finding a cycle in the graph such that each vertex, except the starting vertex, is visited exactly once. Because of this game, any cycle on a graph that has this property is called a Hamiltonian cycle. This is in contrast to the Euler cycle in which each edge is used exactly once. In some ways the Euler cycle and the Hamiltonian cycle may seem similar, but we shall see that the Hamiltonian cycle is much more complex.

We formally describe a Hamiltonian path and a Hamiltonian cycle as follows.

DEFINITION 14.68 Let G be a graph. A **Hamiltonian path** is a simple path that passes through every vertex of G. A **Hamiltonian cycle** is a simple cycle that passes through every vertex of G.

To show that the Hamiltonian cycle deserves the name, we show that the graph for Hamilton's game indeed has a Hamiltonian cycle. One such cycle is shown in Figure 14.70.

EXAMPLE 14.69 The graph in Figure 14.71 has the Hamiltonian cycle in Figure 14.72.

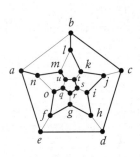
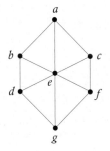

Figure 14.70   Figure 14.71   Figure 14.72

EXAMPLE 14.70 An n-hypercube n for $n \geq 3$ has a Hamiltonian cycle. It is given by the Gray code for n described in Section 6.7. Thus, the Hamiltonian cycle for a 4-hypercube is

$$\begin{array}{cccc} 1 & 1 & 1 & 1 \\ 1 & 1 & 1 & 0 \\ 1 & 1 & 0 & 0 \\ 1 & 1 & 0 & 1 \\ 1 & 0 & 0 & 1 \\ 1 & 0 & 0 & 0 \\ 1 & 0 & 1 & 0 \\ 1 & 0 & 1 & 1 \\ 0 & 0 & 1 & 1 \\ 0 & 0 & 1 & 0 \\ 0 & 0 & 0 & 0 \\ 0 & 0 & 0 & 1 \\ 0 & 1 & 0 & 1 \\ 0 & 1 & 0 & 0 \\ 0 & 1 & 1 & 0 \\ 0 & 1 & 1 & 1 \\ 1 & 1 & 1 & 1 \end{array}$$

EXAMPLE 14.71 The complete graph K_n for $n \geq 3$ has a Hamiltonian cycle. Let $v_1, v_2, v_3, \ldots, v_n$ be the vertices of K_n. Since there is an edge between any two vertices, there is always an edge from v_i to v_{i+1} and, finally, from the last vertex v_n back to v_1.

Before examining our next example we need the following theorem.

THEOREM 14.72 In a Hamiltonian cycle, for a given vertex, there are exactly two edges of the cycle that are incident to that vertex.

Proof As we follow the cycle, for each vertex V, there is an edge to the cycle and an edge from the cycle. If there were another edge of the cycle incident to V, then the cycle would return to V and V would again appear in the cycle, which is a contradiction of the definition of a Hamiltonian cycle. Thus, there are exactly two edges incident to V in a Hamiltonian cycle.

Note that as a result of the previous theorem, any graph with a vertex of degree 1 cannot be Hamiltonian.

EXAMPLE 14.73 The Peterson graph in Figure 14.73 has a Hamiltonian path but not a Hamiltonian cycle. The path is shown in Figure 14.74.

To show that the Peterson graph does not have a Hamiltonian cycle requires a certain amount of trial and error. We try to construct a Hamiltonian cycle and always

14.4 Hamiltonian Paths and Cycles

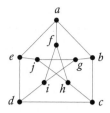

Figure 14.73

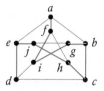

Figure 14.74

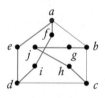

Figure 14.75

reach a dead end. To begin, we know that the star in the center must be connected to the outside pentagon so assume without loss of generality that $\{a, f\}$ is an edge in the cycle. From f the cycle must go to either i or h, and by symmetry, it doesn't matter which, so assume $\{f, i\}$ is in the cycle. By the previous theorem, $\{f, h\}$ cannot be in the cycle since there would be three edges incident to f. Therefore, $\{j, h\}$ and $\{h, c\}$ must be in the cycle since they are the only possible edges incident to h. If $\{i, d\}$ is in the cycle, then since $\{f, i\}$ is in the cycle, $\{i, g\}$ cannot be in the cycle since by the previous theorem, only two edges can be incident to i. Therefore, $\{j, g\}$ and $\{b, g\}$ must be in the cycle since there are the only two edges that can be incident to g. Since $\{j, h\}$ and $\{j, g\}$ are in the cycle, $\{j, e\}$ cannot be in the cycle or there would be three edges incident to j. Therefore, $\{d, e\}$ and $\{e, a\}$ must be in the cycle so that there are two edges incident to e, and we have the following parts of the cycle in Figure 14.75.

Since $\{i, d\}$ and $\{d, e\}$ are in the cycle, $\{d, c\}$ cannot be in the cycle or there would be three edges incident to d, so $\{c, b\}$ must be in the cycle so that there are two edges incident to c. Since $\{b, g\}$ and $\{c, b\}$ are in the cycle, $\{a, b\}$ cannot be in the cycle, and we have the graph in Figure 14.76 with no further edges to put in the cycle. Thus, this path cannot produce a Hamiltonian cycle.

If we return to where we had $\{a, f\}$, $\{f, i\}$, $\{j, h\}$, and $\{h, c\}$ in the cycle and $\{f, h\}$ not in the cycle, since $\{i, d\}$ cannot be in the cycle, $\{i, g\}$ must be in the cycle so that there are two edges incident to i. Edges $\{e, d\}$ and $\{d, c\}$ must be in the cycle so that there are two edges incident to d. Therefore, since $\{h, c\}$ and $\{d, c\}$ are in the cycle, $\{b, c\}$ cannot be in the cycle or there would be three edges incident to c. Therefore, $\{a, b\}$ and $\{g, b\}$ must be in the cycle so that two edges are incident to b. Since $\{i, g\}$ and $\{g, b\}$ must be in the cycle, $\{j, g\}$ cannot be in the cycle or there would be three edges incident to g. At this point we have the graph in Figure 14.77.

The edge $\{e, j\}$ must be in the cycle so that there are two edges incident to j. Since $\{e, d\}$ and $\{e, j\}$ must be in the cycle, $\{e, a\}$ cannot be in the cycle. This gives us the graph in Figure 14.78 with no edges left, which is not a Hamiltonian cycle. Since we have exhausted all possibilities, there is no Hamiltonian cycle for the Peterson graph.

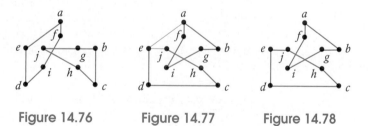

Figure 14.76 Figure 14.77 Figure 14.78

Previously, we found that for a graph to have a Hamiltonian cycle, the degree of each vertex must be two or more. It is also obvious that a graph must be connected to have a Hamiltonian cycle. The following theorem provides us with more information. The proof is left to the reader.

THEOREM 14.74 If the graph G has a cut edge, then G cannot have a Hamiltonian cycle. If the components in the graph formed by the removal of the cut edge have Hamiltonian cycles, then G has a Hamiltonian path. ■

As we have seen, there are very nice, necessary, and sufficient conditions for a graph to have an Euler cycle. Unfortunately, no one has discovered necessary and sufficient conditions for a graph to have a Hamiltonian cycle. The following theorems, while far from ideal, do give us some conditions for showing that graphs have a Hamiltonian cycle. Obviously the more edges that a graph with a fixed number of vertices has, the higher the degree of the vertices and the more likely there is a cycle through all of the vertices without repeating a vertex. Intuitively, if one vertex has low degree, then this must be compensated for by others having higher degree. This is reflected in the following theorem.

THEOREM 14.75 If $G(V, E)$ is a connected graph with $n \geq 3$ vertices and if for every pair of distinct nonadjacent vertices $u, v \in V$, $\deg(u) + \deg(v) \geq n$, then G has a Hamiltonian cycle.

Proof Let $v_1 v_2 v_3 \ldots v_m$ be a maximal simple path in G. Let $a = \deg(v_1)$ and $b = \deg(v_m)$. We first want to show that by possibly rearranging the vertices in the path, we can form a simple cycle. If v_1 and v_m are adjacent, then $v_1 v_2 v_3 \ldots v_m v_1$ is a cycle and we have found our cycle. If v_1 and v_m are not adjacent, then $a + b \geq n$. We want to show that there are vertices v_i and v_{i+1} in the path so that v_1 is adjacent to v_{i+1} and v_m is adjacent to v_i. If we can do this, then we start with the path $v_1 v_2 v_3 \ldots v_i v_{i+1} \ldots v_m$. We remove the edge between v_i and v_{i+1}. We begin our new path with v_{i+1}, then follow it with v_1 since v_{i+1} and v_1 are adjacent. Then continue the path to get $v_{i+1} v_1 v_2 v_3 \ldots v_i v_m$ since v_i and v_m are adjacent as shown in Figure 14.79.

Figure 14.79 $v_1 \; v_2 \; \cdots \; v_k \; (v_{k+1}) \; v_{k+2} \; v_{k+3} \; \cdots \; (v_m)$

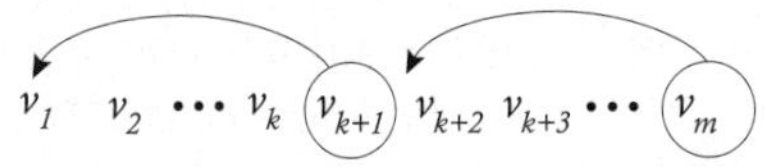

We then travel in reverse along the path $v_m v_{m-1} v_{m-2} \ldots v_{i+2} v_{i+1}$ and we have our cycle. $v_{i+1} v_1 v_2 v_3 \ldots v_i v_m v_{m-1} v_{m-2} \ldots v_{i+2} v_{i+1}$.

We must now show that there are vertices v_i and v_{i+1} in the path so that v_1 is adjacent to v_{i+1} and v_m is adjacent to v_i. Assume there is not. Now vertex v_1 is not adjacent to any vertex that is not in the path, for if v_1 is adjacent to w and w is not in the path, then $w v_1 v_2 v_3 \ldots v_m$ is a simple path, which contradicts the assumption that $v_1 v_2 v_3 \ldots v_m$ is a maximal simple path in G, then there are a vertices in the path $v_1 v_2 v_3 \ldots v_m$ that are adjacent to v_1. Similarly, there are b vertices in the path $v_1 v_2 v_3 \ldots v_m$ that are adjacent to v_m. If we include v_1, there are $a + 1$ vertices in the path $v_1 v_2 v_3 \ldots v_m$ that are equal to or adjacent to v_1. There are b vertices in the path $v_1 v_2 v_3 \ldots v_m$ that are adjacent to v_m. If for each v_i in the path adjacent to v_m, v_{i+1} is not adjacent to v_1. Therefore, there are $a + 1$ vertices in the path equal to or adjacent to v_1 and there are b vertices in the path that are not equal to or adjacent to v_1. Thus, there are $a + b + +1$ distinct vertices in the path $v_1 v_2 v_3 \ldots v_m$, which is impossible. Thus, our assumption was incorrect and there are vertices v_i and v_{i+1} in the path so that v_1 is adjacent to v_{i+1} and v_m is adjacent to v_i. Thus, we have our cycle. For simplicity, assume we have relabeled our vertices so that our cycle is $v_1 v_2 v_3 \ldots v_m v_1$.

We now show that this cycle contains every vertex of V. If not and v' is not equal to any of the v_i since G is connected, there is a path from v' to one of the v_i and there is a vertex w not in the path $v_1 v_2 v_3 \ldots v_m$ that is adjacent to one of the v_j. But then

$$w v_j v_{j+1} v_{j+2} \ldots v_m v_1 v_2 v_3 \ldots v_{j-1}$$

is a simple path longer than $v_1 v_2 v_3 \ldots v_m$, which is a contradiction. Hence, our cycle is $v_1 v_2 v_3 \ldots v_m v_1$ is a Hamiltonian cycle. ∎

From this theorem, the following corollary follows immediately. The corollary is older and better known than the theorem itself.

Corollary 14.76 If $G(V, E)$ is a connected graph with $n \geq 3$ vertices and if for every $v \in V$, $\deg(v) \geq \frac{n}{2}$, then G has a Hamiltonian cycle. ∎

In proving Theorem 14.75, we only used the fact that the sum of the degrees of vertices v_1 and v_m of the maximal path $v_1 v_2 v_3 \ldots v_m$ is greater than the number of vertices. The degrees of the other vertices in the path were not material as long as they had degree 2 or greater. In the following theorem we shall see that we need only assume that the sum of the degrees of vertices v_1 and v_m of the maximal path $v_1 v_2 v_3 \ldots v_m$ is greater than the number of vertices.

THEOREM 14.77 If $G(V, E)$ is a connected graph with $n \geq 3$ vertices and u and v are nonadjacent vertices of G such that $\deg(u) + \deg(v) \geq n$, then the graph G^e consisting of G together with the added edge $e = \{u, v\}$ has a Hamiltonian cycle if and only if G has a Hamiltonian cycle.

Proof If G has a Hamiltonian cycle, then certainly G^e has a Hamiltonian cycle, since the same cycle can be used. Conversely, assume that G^e has a Hamiltonian cycle. If edge e does not appear in the cycle, then certainly G has a Hamiltonian cycle. If edge e does appear in the cycle, then the Hamiltonian cycle in G^e may be written in the form $u v_1 v_2 v_3 \cdots v_m v u$. But then $u v_1 v_2 v_3 \cdots v_m v$ is a Hamiltonian path in G and, hence, a maximal path and also $\deg(u) + \deg(v) \geq n$. By the proof in the previous theorem, the vertices in this path can be rearranged to form a cycle in G. Since this cycle contains every vertex of G, it is a Hamiltonian cycle. ∎

At this point we wish to begin with a graph G and add edges to nonadjacent vertices u and v of G such that

$$\deg(u) + \deg(v) \geq n$$

until we can no longer do so. When we have finished, we shall call this the closure of G.

DEFINITION 14.78 Let G be a graph with n vertices. The **closure** of G, denoted by $cl(G)$, is the graph obtained from G by recursively adding edges to nonadjacent vertices u and v of G for which $\deg(u) + \deg(v) \geq n$ until we can no longer do so.

Thus, $cl(G)$ has the property that for any two nonadjacent vertices u and v of $cl(G)$ such that $\deg(u) + \deg(v) \geq n$, there is an edge between them. We need to show that $cl(G)$ is well defined, which means that if we obtain the graph G' from G by recursively adding edges to nonadjacent vertices u and v of G such that $\deg(u) + \deg(v) \geq n$ until we can no longer do so and also obtain the graph G'' from G by recursively adding edges to nonadjacent vertices u and v of G such that $\deg(u) + \deg(v) \geq n$ until we can no longer do so, but using a possibly different recursion, then $G' = G''$. To show this let $e'_1, e'_2, e'_3, \ldots, e'_l$ be the edges recursively added to G to construct G' and $e''_1, e''_2, e''_3, \ldots, e''_m$ be the edges recursively added to G to construct G' if the edges are not the same, say there is an edge in G' that is not in G''. Let e'_j be the first edge recursively added to G' that is not in G'' and let $e'_j = \{u, v\}$. Then with edges $e'_1, e'_2, e'_3, \ldots, e'_{j-1}$ added to G, $\deg(u) + \deg(v) \geq n$. But since e'_j is the first edge recursively added to G' that is not in G'', $e'_1, e'_2, e'_3, \ldots, e'_{j-1}$ are in G''. Therefore, $\deg(u) + \deg(v) \geq n$ in G'' but there is no edge between u and v, which is a contradiction. Therefore, $G' = G''$ and cl(G) is uniquely defined.

Notice that for any graph G, cl(cl(G)) = cl(G).

EXAMPLE 14.79 If G is the graph in Figure 14.80, then cl(G) is the graph in Figure 14.81.

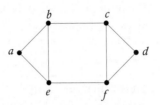

Figure 14.80   Figure 14.81

EXAMPLE 14.80 If G is the graph in Figure 14.82, then cl(G) is the graph in Figure 14.83.

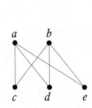

Figure 14.82   Figure 14.83

EXAMPLE 14.81 If G is the complete bipartite graph $K_{m,m}$ for $m \geq 1$, then cl(G) is the complete graph K_{m+m}.

Using induction and Theorem 14.77, the reader is asked to prove the following theorem.

THEOREM 14.82 A graph G has a Hamiltonian cycle if and only if cl(G) has one.

As a result of the previous theorem and example, we see that for $m \geq 1$, the complete bipartite graph $K_{m,m}$ has a Hamiltonian cycle.

Exercises

Find a Hamiltonian cycle for each of the following graphs if it exists.

1.

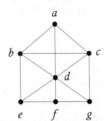

2.

3.

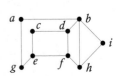

4.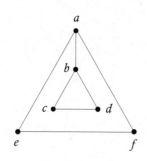

10.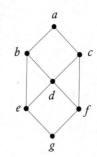

5.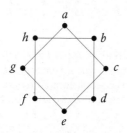

11.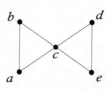

6.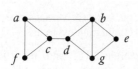

12.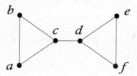

7.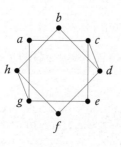

13.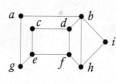

8.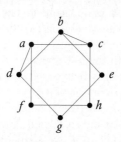

14.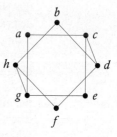

9.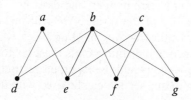

15.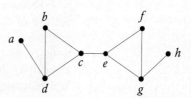

16.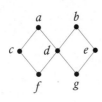

17.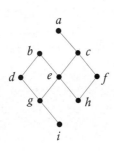

18.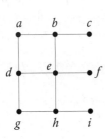

19.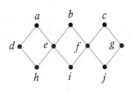

20.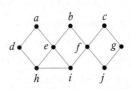

21. Prove Theorem 14.74: If the graph G has a cut edge, then G cannot have a Hamiltonian cycle. If the components in the graph formed by the removal of the cut edge have Hamiltonian cycles, then G has a Hamiltonian path.

22. Draw a graph with six vertices that has a Hamiltonian cycle but not a Euler cycle.

23. Draw a graph with six vertices that has a Euler cycle but not a Hamiltonian cycle.

24. Using Theorem 14.77 and induction, prove Theorem 14.82: A graph G has a Hamiltonian cycle if and only if $cl(G)$ has one.

14.5 Weighted Graphs and Shortest Path Algorithms

So far, in looking at graphs, we have been concerned only with vertices and edges that can be traversed. At this point, we will not only be concerned with getting from point A to point B, but also doing it the best way. The first question, of course, is what is meant by the best way. It could be the cheapest way, the safest way, the shortest way, the way that uses the least energy, or some other measure. To determine this best way, we assign a weight or measure to each edge. If we are trying to find the shortest distance, by interstate, between two cities, then the cities would be represented by vertices and the weight assigned to the edges would be the distance between the cities. If we are trying to find the cheapest way to fly from one city to another, the weight assigned to the edges between the vertices representing the cities would be the cost of flying directly from one city to another. If there is no direct flight between cities, then there is no edge between the corresponding vertices. Although, as we have seen, the weights or measures assigned to the edges may have many different meanings, to simplify the discussion, we shall refer to the weights as distances and the best way of getting from point A to point B as the shortest path between A and B. This is simply terminology and does not in any way limit the use of the theory or the generality of the results. We will make one important restriction, however. We shall assume that the weight or measure, which we now call distance, assigned to an edge between two distinct points must be positive. There are algorithms that allow for negative weights and, with certain restrictions, the Floyd-Warshall algorithm, which is given in this section can be used with negative weights.

In the remainder of this section we shall use the symbol ∞. For ease of discussion we shall consider all integers to be less than ∞ so that $\min(a, \infty) = a$ for every nonnegative integer a, and $\min(\infty, \infty) = \infty$. We shall also consider $a + \infty = \infty + \infty = \infty$. These are just notational conveniences.

To help in our notation, we use the following definition.

DEFINITION 14.83 Let $A = (a_1, a_2, a_3, \ldots, a_n)$ and $B = (b_1, b_2, b_3, \ldots, b_n)$ be row matrices where each of the a_i and b_i are nonnegative integers or ∞. Then
$$A \wedge B = (\min(a_1, b_1), \min(a_2, b_2), \min(a_3, b_3), \ldots, \min(a_n, b_n))$$

DEFINITION 14.84 Let c be a number and $A = (a_1, a_2, a_3, \ldots, a_n)$ be a row matrix. Then
$$c + A = c + (a_1, a_2, a_3, \ldots, a_n) = (c + a_1, c + a_2, c + a_3, \ldots, c + a_n)$$

The first algorithm we shall describe is called Dijkstra's algorithm. There are several versions of Dijkstra's algorithm. First, there is the original version and the improved version, which is more efficient. In addition, there is an optional feature that allows us not only to find the length of the shortest path but also to find the path. This is done by a pointer that, for each vertex along the shortest path, indicates the previous vertex in the path. Thus, if we find the length of the shortest path from A and B, we may begin at B and backtrack along the path until we reach A, and we have found our path. We shall demonstrate the improved version of Dijkstra's algorithm with the path option.

Before beginning, however, we wish to point out a theorem that may seem intuitively obvious. The proof is left to the reader.

THEOREM 14.85 Let $a = v_1$ and $b = v_n$. If $v_1, v_2, \ldots, v_i, v_{i+1}, \ldots, v_j, \ldots, v_n$ is a shortest path between a and b, then $v_i, v_{i+1}, \ldots, v_j$, the part of this path between v_i and v_j, is a shortest path between v_i and v_j. ■

We begin with the statement of Dijkstra's first algorithm and then give an example of how to use it. In this algorithm, we seek the shortest distance from v_1 to v_n. We begin with v_1 and find the distance from v_1 to each of the adjacent vertices. We select the vertex that is the shortest distance from v_1 say v_i and find the distance from v_1 to each vertex adjacent to v_i by passing through v_i. If this distance is less than the current distance assigned to any of the vertices, it replaces the current distance. Again the vertex closest to v_1, but not previously chosen, is selected and the process is repeated.

■ Dijkstra's Algorithm (1)

Given a weighted graph, this algorithm gives the shortest distance from v_1 to v_n. Associate an ordered pair $(\infty, 0)$ with each vertex. The first coordinate of $v_i(m, v_r)$ will denote the distance assigned from v_1 to v_i and the second will denote the preceding vertex in the path from v_1 to v_i.

1. Begin at $v_1(\infty, 0)$, change it to $v_1(0, 0)$, and make it permanent. All other vertices will be temporary at this point.

2. When a vertex $v_k(m, v_r)$ becomes permanent, for each vertex v_j adjacent to v_k, add m to the distance from v_k to v_j. If this value is less than the current distance assigned to v_j, replace the current distance with this sum and replace the second coordinate with v_k.

3. Find the minimum of the distances assigned to the temporary vertices. Make the first vertex with that distance permanent.
4. If v_n is not permanent, return to (2).
5. If v_n is permanent, the distance assigned to v_n is the shortest distance from v_1 to v_n.
6. To find the path, begin at v_n and find its previous vertex in the path (the second coordinate). For each v_j in the path, find its previous vertex in the path until v_1 is reached. Reversing these vertices gives the path.

EXAMPLE 14.86

Let the graph in Figure 14.84 be a weighted graph, where we seek the shortest distance from A to F. Setting the ordered pair for each vertex to $(\infty, 0)$, our graph will now have the form shown in Figure 14.85.

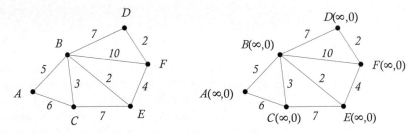

Figure 14.84   Figure 14.85

We shall begin at A and start constructing paths to the other vertices. As a vertex is reached, the first component of the ordered pair will indicate the length of the shortest path to the vertex at that time, and the second component will indicate the previous vertex in the shortest path. Until a path is reached, the first component will remain ∞ and the second will remain 0. We shall indicate that a vertex is permanent by placing it in boldface. Using step 1 in the algorithm, we have the graph shown in Figure 14.86.

Since vertices B and C are adjacent to A, we use step 2 to assign the value $(5, A)$ to the ordered pair for B and $(6, A)$ to the ordered pair for C. (We actually make a change only if the new distances are shorter but since the previous distances to B and C were ∞, this is no concern.) Using step 3 we take the shortest of these temporary values that we have assigned, which is in this case is the distance 5 to A, and make $B(5, A)$ permanent. Thus, we now have Figure 14.87.

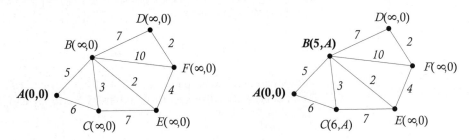

Figure 14.86   Figure 14.87

Returning to step 2, we now consider the temporary vertices C, D, E, and F adjacent to B. In each case, we add the distance from A to B to the distance from B to the given vertex. Thus, for C, we have $5 + 3 = 8$. For D, we have $5 + 7 = 12$. For

E, we have $5+2 = 7$. For F, we have $5+10 = 15$. **Since the new distance to C is not less than the distance it is already assigned, we do not change $C(6, A)$**. The new distances to D, E, and F are all smaller and they got these values by taking the path from B, so we change them to $D(12, B)$, $E(7, B)$, and $F(15, B)$. Using step 3, we now take the minimum of the distances assigned to the temporary vertices, so we take $\min\{6, 12, 15, 7\} = 6$, and since C has this distance, we make $C(6, A)$ permanent. Thus, we have Figure 14.88.

We now take the new permanent vertex C. Step 2 produces no change. Using step 3, we make $E(7, B)$ permanent. Thus, we have Figure 14.89.

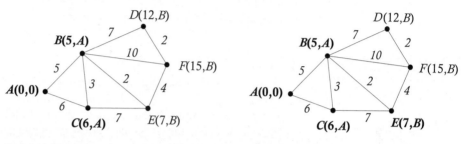

Figure 14.88   **Figure 14.89**

We now take the new permanent vertex E and using step 2, we change $F(15, B)$ to $F(11, E)$. Using step 3, we make $F(11, E)$ permanent. Thus, we have Figure 14.90.

Figure 14.90

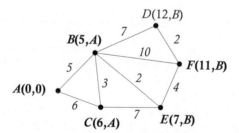

Once F becomes permanent, we are finished and 11 is the shortest distance from A to F. If we had run out of adjacent vertices to a permanent vertex before reaching F, then there would be no solution since there would have been no path from A to F. To find the shortest path, we see that F is preceded by E, E is preceded by B and B is preceded by A. Thus, the shortest path is $ABEF$. ∎

We now state Dijkstra's second algorithm, which uses matrices to find the shortest distance between two given vertices. Let n be the number of vertices in the graph. The matrix W is an $n \times n$ matrix in which the value of $W(i, j)$ is the distance between vertices v_i and v_j if they are adjacent and ∞ otherwise. $W(i)$ will denote the ith row of W. We also have $1 \times n$ matrices P, T, Sum, and $pred$ and variable V. $pred(i)$ is the subscript of the vertex, which is the predecessor of v_i in the shortest path to v_i. $P(i)$ is the permanent distance from v_1 to v_i when v_i is selected as a permanent vertex. $T(i)$ is the temporary distance from v_1 to v_i. The variable V keeps track of the last permanent vertex. Sum is simply used to temporarily store $P(V) + W(V)$.

■ Dijkstra's Algorithm (2)

1. Set $P(1) = 0$, $T(1) = \infty$, and $W_{j1} = \infty$ for $1 \leq j \leq n$. Let $V = 1$.
2. Add $P(V)$ to each element of $W(V)$. More precisely, let $Sum = P(V) + W(V)$. For $Sum(j) < T(j)$, let $pred(j) = V$. Then let $T = T \wedge Sum$, and for the least i such that $T(i) = \min\{T(j) : 1 \leq j \leq n\}$ let $P(i) = T(i)$, $T(i) = \infty$, and $W_{ji} = \infty$ for $1 \leq j \leq n$. Let $V = i$.
3. If $i \neq n$, return to step 2.
4. If $i = n$, then $P(i)$ is the shortest distance from v_1 to v_n.
5. To find the path, use $pred(n)$ to find the vertex that precedes n. For each v_j in the path, use $pred(j)$ to find its previous vertex in the path until v_1 is reached. The reverse of these vertices is the shortest path.

EXAMPLE 14.87

Determine the shortest path from v_1 to v_5 in the graph shown in Figure 14.91.

We shall demonstrate Dijkstra's algorithm using matrices, and we will also show the corresponding graph. We shall associate ordered pairs similarly defined to those in the previous example so that we have Figure 14.92.

Instead of using the previous vertex v_i in the path of a given vertex for the second coordinate of that vertex, we shall just use i to represent v_i. Thus, instead of having $v_j(m, v_i)$ where m is the assigned length of the path from v_1 to v_j and v_i is the previous vertex in the path, we shall just use $v_j(m, i)$.

Initially we set $pred(i) = 0$ for $1 \leq i \leq 5$, since at this point, no vertex has a predecessor. Set $P(1) = 0$, $T(1) = \infty$, and $W_{j1} = \infty$ for $1 \leq j \leq 5$. This makes v_1 a permanent vertex giving us Figure 14.93.

We add $P(1) = 0$ to each element of

$$W(1) = (W_{11}, W_{12}, W_{13}, W_{14}, W_{15}) = (\infty, 2, \infty, 6, \infty)$$

More precisely, let $Sum = P(1) + W(1)$. For $Sum(j) < T(j)$, let $pred(j) = 1$. Then let $T = T \wedge Sum$, so that $T = (\infty, 2, \infty, 6, \infty)$. This simply assigns distances to the vertices adjacent to v_1. For the least i such that $T(i) = \min\{T(j) : 1 \leq j \leq 5\}$, let $P(i) = T(i)$, $T(i) = \infty$, and $W_{ji} = \infty$ for $1 \leq j \leq 5$. This selects v_2 as the second permanent vertex. Let $V = 2$. (The variable V keeps track of the last permanent variable.) Thus, $T = (\infty, \infty, \infty, 6, \infty)$, $P = (0, 2, \infty, \infty, \infty)$, and $pred = (0, 1, 0, 1, 0)$. The matrix

$$W = \begin{bmatrix} \infty & \infty & \infty & 6 & \infty \\ \infty & \infty & 4 & \infty & 8 \\ \infty & \infty & 0 & 6 & 2 \\ \infty & \infty & 6 & 0 & 4 \\ \infty & \infty & 2 & 4 & 0 \end{bmatrix}$$

This gives us the graph in Figure 14.94.

We now add $P(2) = 2$ to each element of

$$W(2) = (W_{21}, W_{22}, W_{23}, W_{24}, W_{25}) = (\infty, \infty, 4, \infty, 8)$$

More precisely, let $Sum = P(2) + W(2)$. This gives the distance from v_1 to each vertex adjacent to v_2 on the path passing through v_2. For $Sum(j) < T(j)$, let $pred(j) = 2$. Then let $T = T \wedge Sum$, so that $T = (\infty, \infty, 6, 6, 10)$ and for the least i such that $T(i) = \min\{T(j) : 1 \leq j \leq 5\}$, let $P(i) = T(i)$, $T(i) = \infty$, and $W_{ji} = \infty$ for $1 \leq j \leq 5$. Thus, v_3 is selected as a permanent vertex. Let $V = 3$. Thus, $T = (\infty, \infty, \infty, 6, 10)$, $P = (0, 2, 6, \infty, \infty)$, and $pred = (0, 1, 2, 1, 2)$.

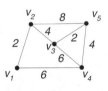

Figure 14.91

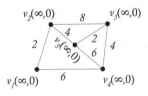

Figure 14.92

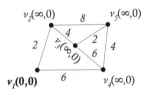

Figure 14.93

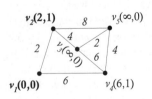

Figure 14.94

The matrix

$$W = \begin{bmatrix} \infty & \infty & \infty & 6 & \infty \\ \infty & \infty & \infty & \infty & 8 \\ \infty & \infty & \infty & 6 & 2 \\ \infty & \infty & \infty & 0 & 4 \\ \infty & \infty & \infty & 4 & 0 \end{bmatrix}$$

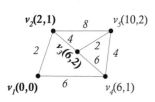

Figure 14.95

Thus, we have the graph in Figure 14.95.

We now add $P(3) = 6$ to each element of

$$W(3) = (W_{31}, W_{32}, W_{33}, W_{34}, W_{35}) = (\infty, \infty, \infty, 6, 2)$$

More precisely, let $Sum = P(3) + W(3)$. This gives the distance from v_1 to each vertex adjacent to v_3 on the path passing through v_3. For $Sum(j) < T(j)$, let $pred(j) = 3$. Then let $T = T \wedge Sum$, so that $T = (\infty, \infty, \infty, 6, 8)$ and for the least i such that $T(i) = \min\{T(j) : 1 \leq j \leq 5\}$, let $P(i) = T(i)$, $T(i) = \infty$, and $W_{ji} = \infty$ for $1 \leq j \leq 5$. This selects v_4 as a permanent variable. Thus, $T = (\infty, \infty, \infty, \infty, 10)$, $P = (0, 2, 6, 6, \infty)$, and $pred = (0, 1, 2, 1, 3)$, and

$$W = \begin{bmatrix} \infty & \infty & \infty & \infty & \infty \\ \infty & \infty & \infty & \infty & 8 \\ \infty & \infty & \infty & \infty & 2 \\ \infty & \infty & \infty & \infty & 4 \\ \infty & \infty & \infty & \infty & 0 \end{bmatrix}$$

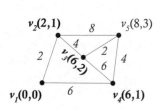

Figure 14.96

Thus, we have the graph in Figure 14.96.

Finally, we add $P(4) = 6$ to each element of

$$W(4) = (W_{41}, W_{42}, W_{43}, W_{44}, W_{45}) = (\infty, \infty, \infty, \infty, 4)$$

More precisely, let $Sum = P(4) + W(4)$. This gives the distance from v_1 to each vertex adjacent to v_4 on the path passing through v_4. For $Sum(j) < T(j)$, let $pred(j) = 4$. Then let $T = T \wedge Sum$, so that $T = (\infty, \infty, \infty, \infty, 8)$ and for the least i such that $T(i) = \min\{T(j) : 1 \leq j \leq 5\}$, let $P(i) = T(i)$, $T(i) = \infty$, and $W_{ji} = \infty$ for $1 \leq j \leq 5$. This selects v_5 as the final permanent variable. Thus, $T = (\infty, \infty, \infty, \infty, \infty)$, $P = (0, 2, 6, 6, 8)$, and $pred = (0, 1, 2, 1, 3)$, and

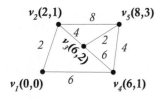

Figure 14.97

$$W = \begin{bmatrix} \infty & \infty & \infty & \infty & \infty \\ \infty & \infty & \infty & \infty & \infty \\ \infty & \infty & \infty & \infty & \infty \\ \infty & \infty & \infty & \infty & \infty \\ \infty & \infty & \infty & \infty & \infty \end{bmatrix}$$

Thus, we have the graph in Figure 14.97 and we are finished. The length of the path is 8. Tracing the path as in the first example, we see that it is $v_1 v_2 v_3 v_5$. ∎

We now consider the Floyd-Warshall algorithm. This algorithm has the advantage of being both time and space efficient. It also finds the shortest distance between any two vertices. This algorithm is very similar to Warshall's algorithm we used to determine all paths in a graph by determining the matrix $\hat{A}$ where $\hat{A}_{ij} = 1$ if and only if there is a path from v_i to v_j. In the previous case, we first found the set of all 1-paths. We then found the set of all 2-paths where the middle vertex is v_1 and combined them with the paths of length 1. We then found all path of length 3 or less passing through v_1 and/or v_2 and/or v_3 (if any) and continued until we had found all paths of length n or less passing through any of the n vertices.

This time, instead of just finding the paths from v_i to v_j, at each stage, we find the length of the path and compare it to the previously found shortest path from v_i

to v_j (if any) and replace the previous value in the ith row, jth column with the new length if we have found a shorter path from v_i to v_j. At this point, it is obvious that we have two lengths. One is the length determined by adding the lengths of each edge in the path. The other is the normal length of a path, which is the number of edges in the path. To (hopefully) avoid confusion we shall refer to the length of an edge as the weight of an edge, as we did previously for Dijkstra's algorithms.

We give two algorithms for computing $\hat{A}$. The first is the handier if computing by hand. The second is better if using a computer and is given at the end of the section.

■ Floyd-Warshall Algorithm 1

1. Look at column 1 of A. If there is a positive integer in the jth row of that column, add that integer to row 1 to form A_j^1. Let A_j be the jth row of A and replace the jth row of A with $A_j^1 \wedge A_j$. Call this matrix $A^{(1)}$.

2. Look at column 2 of the matrix $A^{(1)}$ constructed in step 1. If there is a positive integer in the jth row of that column, add that integer to row 2 to form A_j^2. Let A_j be the jth row of A and replace the jth row of $A^{(1)}$ with $A_j^2 \wedge A_j$. Call this matrix $A^{(2)}$.

3. Look at column 3 of the matrix $A^{(2)}$ constructed in step 2. If there is a positive integer in the jth row of that column, add that integer to row 3 to form A_j^3. Let A_j be the jth row of A and replace the jth row of $A^{(2)}$ with $A_j^3 \wedge A_j$. Call this matrix $A^{(3)}$.

4. Continue this process of looking at the next column, say the ith column, in the previously constructed matrix and, if there is a positive integer in the jth row of that column, add that integer to the row corresponding to the to form A_j^i. Let A_j be the jth row of A and replace the jth row of $A^{(i-1)}$ with $A_j^i \wedge A_j$. Call this matrix $A^{(i)}$.

5. Continue until all columns have been examined.

EXAMPLE 14.88 Given the graph in Figure 14.98, let A be the matrix where A_{ij} is the weight of the edge $\{v_i, v_j\}$ if the edge exists and ∞ otherwise. In this particular case

$$A = \begin{bmatrix} 0 & 2 & \infty & 3 & \infty \\ 2 & 0 & 4 & \infty & 1 \\ \infty & 4 & 0 & 6 & 2 \\ 3 & \infty & 6 & 0 & 3 \\ \infty & 1 & 2 & 3 & 0 \end{bmatrix}$$

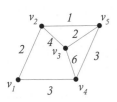

Figure 14.98

We want to find weights of the 2-paths where the middle vertex is v_1. We begin with the first column. Find the first row in column 1 in which there is a positive integer. In this case there is a 2 in row 2. Thus, there is an edge from v_2 to v_1 of weight 2. If there is an integer m in row 1 in position $(1, j)$, then there is an edge from v_1 to v_j of weight m. Then $2 + m$ is the weight of a 2-path from v_2 to v_j. Thus, if we add 2 to each element in row 1, we get a row, which we consider as a row matrix, call it $A_2^1 = (2, 4, \infty, 5, \infty)$, so that $A_2^1(j)$ is the weight of a 2-path from v_2 to v_j. If we let $A_2 = (2, 0, 4, \infty, 1)$ be row 2 of A considered a row matrix, then $A_2(j)$ is the weight of a 1-path from v_2 to v_j. Since we want the shortest path for each j, we replace row 2 of A with $A_2^1 \wedge A_2 = (2, 4, \infty, 5, \infty) \wedge (2, 0, 4, \infty, 1) = (2, 0, 4, 5, 1)$. Similarly, there is a 3 in row 4 of the first column. Thus, there is an edge from v_4 to v_1 of weight 3. If there is an integer m in row 1 in position $(1, j)$, then there is an edge from v_4 to v_j of weight m. Then $3+m$ is the weight of a 2-path from v_4 to v_j. Thus, if we add 3 to each element in row 1, we get a row, which we consider a row matrix, call it $A_4^1 = (3, 5, \infty, 6, \infty)$ so that $A_4^1(j)$ is the weight of a 2-path from v_4 to v_j. If we let $A_4 = (3, \infty, 6, 0, 3)$ be row 4 of A considered as a row matrix, then $A_4(j)$ is the weight of a 1-path from v_4 to v_j. Since we want the shortest path for each j, we replace row 4 of A with $A_4^1 \wedge A_4 = (3, 5, \infty, 6, \infty) \wedge (3, \infty, 6, 0, 3) = (3, 5, 6, 0, 3)$. Since there are no other rows in column 1 with positive integers, we have finished the first phase and we have

$$A^{(1)} = \begin{bmatrix} 0 & 2 & \infty & 3 & \infty \\ 2 & 0 & 4 & 5 & 1 \\ \infty & 4 & 0 & 6 & 2 \\ 3 & 5 & 6 & 0 & 3 \\ \infty & 1 & 2 & 3 & 0 \end{bmatrix}$$

We now use the second column of $A^{(1)}$. If there is a k in row i of column 2, then there is an edge from v_i to v_2 of weight k. If there is an integer m in row j in position $(2, j)$, then there is an edge from v_2 to v_j of weight m. Then $k+m$ is the weight of a path from v_i to v_j. Thus if we add k to each element in row 2, we get a row, which we consider a row matrix, call it A_i^2, so that $A_i^2(j)$ is the length of a 2-path from v_i to v_j. If we let A_i be row i of $A^{(1)}$ considered a row matrix, then $A_i(j)$ is the length of a path from v_i to v_j. Since we want the shortest path for each j, we replace row i of $A^{(1)}$ with $A_i^2 \wedge A_i$. There is a 2 in row 1, a 4 in row 3, a 5 in row 4, and a 1 in row 5 of the second column. We use this process on each of these rows to get

$$A^{(2)} = \begin{bmatrix} 0 & 2 & 6 & 3 & 3 \\ 2 & 0 & 4 & 5 & 1 \\ 6 & 4 & 0 & 6 & 2 \\ 3 & 5 & 6 & 0 & 3 \\ 3 & 1 & 2 & 3 & 0 \end{bmatrix}$$

We now consider the third column of $A^{(2)}$. For each position $(3, j)$ of $A^{(2)}$ which is a positive integer, we add this value to the third row of $A^{(2)}$ to form A_j^3 and let the jth row of $A^{(2)}$ be A_j. We then replace the jth row of $A^{(2)}$ with $A_j^3 \wedge A_j$ and we have

$$A^{(3)} = \begin{bmatrix} 0 & 2 & 6 & 3 & 3 \\ 2 & 0 & 4 & 5 & 1 \\ 6 & 4 & 0 & 6 & 2 \\ 3 & 5 & 6 & 0 & 3 \\ 3 & 1 & 2 & 3 & 0 \end{bmatrix}$$

We now consider the fourth column of $A^{(3)}$. For each position $(4, j)$ of $A^{(3)}$ which is a positive integer, we add this value to the fourth row of $A^{(3)}$ to form A_j^4 and let the jth row of $A^{(3)}$ be A_j. We then replace the jth row of $A^{(3)}$ with $A_j^4 \wedge A_j$ and we have

$$A^{(4)} = \begin{bmatrix} 0 & 2 & 6 & 3 & 3 \\ 2 & 0 & 4 & 5 & 1 \\ 6 & 4 & 0 & 6 & 2 \\ 3 & 5 & 6 & 0 & 3 \\ 3 & 1 & 2 & 3 & 0 \end{bmatrix}$$

Finally, we consider the fifth column of $A^{(4)}$. For each position $(5, j)$ of $A^{(4)}$, which is a positive integer, we add this value to the fifth row of $A^{(4)}$ to form A_j^5

and let the jth row of $A^{(4)}$ be A_j. We then replace the jth row of $A^{(4)}$ with $A_j^5 \wedge A_j$ and we have

$$A^{(5)} = \begin{bmatrix} 0 & 2 & 5 & 3 & 3 \\ 2 & 0 & 3 & 4 & 1 \\ 5 & 3 & 0 & 5 & 2 \\ 3 & 4 & 5 & 0 & 3 \\ 3 & 1 & 2 & 3 & 0 \end{bmatrix}$$

■

The second method uses the same process expressed earlier in pseudocode using the same notational conveniences for ∞ as before. The only difference is that earlier we did not bother with some cases where A_{ik} was not a positive integer.

■ Floyd-Warshall Algorithm 2

For $k = 1$ to n
    For $i = 1$ to n
        For $j = 1$ to n
            $A_{ij} = A_{ij} \wedge (A_{ik} + A_{kj})$

Exercises

Use Dijkstra's algorithm to find the shortest distance between v_0 and v in the following graphs.

1.

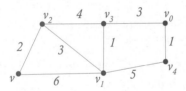

2.

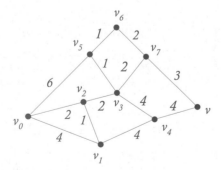

3.

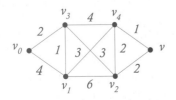

4.

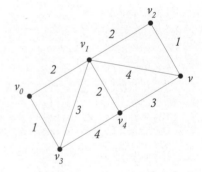

5.

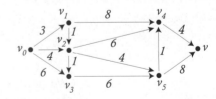

6. Use the Floyd-Warshall algorithm to find the shortest distance between the points for the graph in problem (1).

7. Use the Floyd-Warshall algorithm to find the shortest distance between the points for the graph in problem (2).

8. Use the Floyd-Warshall algorithm to find the shortest distance between the points for the graph in problem (3).

9. Use the Floyd-Warshall algorithm to find the shortest distance between the points for the graph in problem (4).

10. Use the Floyd-Warshall algorithm to find the shortest distance between the points for the graph in problem (5).

Use Dijkstra's algorithm to find the shortest distance between v_0 and v in the following graphs:

11.

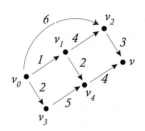

14.

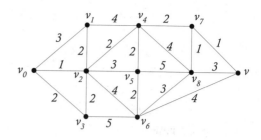

15.

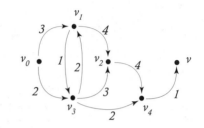

12.

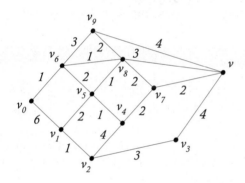

13.

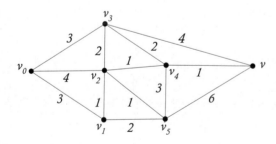

16. Use the Floyd-Warshall algorithm to find the shortest distance between the points for each graph in problem (11).

17. Use the Floyd-Warshall algorithm to find the shortest distance between the points for each graph in problem (12).

18. Use the Floyd-Warshall algorithm to find the shortest distance between the points for each graph in problem (13).

19. Use the Floyd-Warshall algorithm to find the shortest distance between the points for each graph in problem (14).

20. Use the Floyd-Warshall algorithm to find the shortest distance between the points for each graph in problem (15).

21. Prove that Dijkstra's first algorithm finds the length of a shortest path between vertices in a connected weighted graph.

22. Prove that Dijkstra's first algorithm uses $O(n^2)$ additions and comparisons to find a shortest path for a graph with n vertices.

GLOSSARY

Chapter 14: Graphs Revisited

Bicomponent (14.1) — Let R be the relation defined such that aRb if $a = b$ or if there is a simple cycle containing a and b. For each equivalence class E_i for the equivalence relation R, let V_i be the vertices incident to edges of E_i and $G_i(V_i, E_i)$ be the subgraph of $G(V,E)$ with vertices V_i and edges E_i. The subgraph $G_i(V_i, E_i)$ is a **bicomponent** of $G(V,E)$.

Biconnected (14.1)	A graph $G(V, E)$ is **biconnected** if it contains no articulation points (cut vertices).
Chromatic number (14.3)	The **chromatic number** of a graph is the smallest number of colors that can be used to color the graph. It is the smallest positive integer n so that $C_G(n) \neq 0$.
Chromatic polynomial (14.3)	For a fixed graph G, $C_G(\lambda)$ is a polynomial function of λ, called the **chromatic polynomial** of G. $C_G(\lambda)$ is the number of ways of coloring G with λ colors.
Closure (14.4)	Let G be a graph with n vertices. The **closure** of G, denoted by cl(G), is the graph obtained from G by recursively adding edges to nonadjacent vertices u and v of G for which $\deg(u) + \deg(v) \geq n$ until we can no longer do so.
Coloring (14.3)	If G is a graph, a **coloring** of G is a painting of the vertices of G such that no two vertices are the same color. $C_G(\lambda)$ denotes the number of ways the graph G can be painted using λ colors so that no two vertices are the same color, that is, the number of colorings of G.
Complement (14.1)	Let $G(V, E)$ be a graph. The **complement** of G denoted by $G^C(V, E')$ is the graph such that for all $u, v \in V$ there is an edge between u and v in G^C if and only if there is no edge in E between u and v in G.
Cut edge (14.1)	If the cut set C consists of a single edge, then the edge in this set is called a **cut edge**.
Cut set (14.1)	A set of edges C of a connected graph $G = (V, E)$ is a **cut set** if the removal of the edges of C causes the graph to no longer be connected, but the removal of any proper subset of C leaves the set connected.
Cut vertex/ articulation point (14.1)	A connected graph $G = (V, E)$ has a **cut vertex** or **articulation point** $a \in V$ if the removal of a and its incident edges causes the graph to no longer be connected.
Derived (14.1)	If there are graphs $G_1, G_2, G_3, \ldots G_n$ such that G_{i+1} is an extension of G_i, then G_n can be **derived** from G_1.
Dijkstra's algorithm(1) (14.5)	Given a weighted graph, **Dijkstra's algorithm** gives the shortest distance between two vertices.
Dijkstra's algorithm(2) (14.5)	**Dijkstra's second algorithm** uses matrices to find the shortest distance between two vertices.
Dual graph (14.3)	A **dual graph** G' of a planar graph G is formed by changing the faces to vertices and constructing an edge between two vertices if the faces are adjacent.
Extension (14.1)	If a graph $G(V, E)$ has an edge $e = \{v_1, v_2\}$ and graph $G'(V', E')$ is derived from $G(V, E)$ by adding a new vertex v to V and replacing the edge $\{v_1, v_x\}$ with the edges $\{v_1, v\}$ and $\{v, v_2\}$, then $G'(V', E')$ is an **extension** of $G(V, E)$.
Face of a planar graph (14.2)	A **face of a planar graph** is a maximal section of a plane such that any two points within the section may be connected by a curve that does not cross an edge.
Floyd-Warshall algorithm 1 (14.5)	The first **Floyd-Warshall algorithm** also finds the shortest distance between two vertices, but it has the advantage of being both time and space efficient.
Floyd-Warshall algorithm 2 (14.5)	The second **Floyd-Warshall algorithm** uses the same process but does not bother with some cases where A_{ik} is not a positive integer.

Hamiltonian cycle (14.4)	G has a **Hamiltonian cycle** if there is a simple cycle that passes through every vertex of G.
Hamiltonian path (14.4)	G has a **Hamiltonian path** if there is a simple path that passes through every vertex of G.
Homeomorphic (14.1)	The graphs G and G' are **homeomorphic** if there is a graph G'' such that both G and G' can be derived from G''.
Homomorphism (14.1)	A function f from the graph $G(V, E)$ to the graph $G'(V', E')$ is called a **homomorphism** from G to G', denoted by $f : G \to G'$, if it has the following properties: a. If $e \in E$, then $f(e) \in E'$, $(f(E) \subseteq E')$. b. If $v \in V$, then $f(v) \in V'$, $(f(V) \subseteq V')$. c. If vertices u and v are incident to edge e in G, then $f(u)$ and $f(v)$ are incident to $f(e)$ in G'.
Intersection (14.1)	Let $G = G(V, E)$ be a graph and $G_1, G_2, G_3, \ldots G_n$, be subgraphs of G. The subgraph G' of G is the **intersection** of $G_1, G_2, G_3, \ldots G_n$, denoted by $\bigcap_{i=1}^{n} G_i$ if 1. A vertex $v \in G'$ if and only if $v \in G_i$ for all $1 \leq i \leq n$. 2. An edge $e \in G'$ if and only if $e \in G_i$ for all $1 \leq i \leq n$.
Isomorphism/isomorphic (14.1)	The homomorphism $f : G \to G'$ is an **isomorphism** if $f : V \to V'$ and $f : E \to E'$ are one-to-one correspondences. If $f : G \to G'$ is an isomorphism, then G and G' are **isomorphic**.
Kuratowski's theorem (14.2)	A graph is planar if and only if it does not contain a subgraph homeomorphic to either $K_{3,3}$ or K_5.
Pairwise disjoint (14.1)	Let $G = G(V, E)$ be a graph and $G_1, G_2, G_3, \ldots G_n$ be subgraphs of G. The subgraphs $G_1, G_2, G_3, \ldots G_n$ are **pairwise disjoint** if for all $1 \leq i < j \leq n$, $G_i \cap G_j = \phi$.
Peterson graph (14.4)	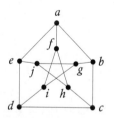
Planar graph (14.2)	A **planar graph** is a graph that can be drawn on a plane so that none of the edges cross. A graph that is not planar is called **nonplanar**.
Spanning graph (14.1)	A subgraph $G'(V', E')$ is a **spanning graph** for the graph $G = (V, E)$ if $V' = V$.
Spanning tree (14.1)	A tree is a **spanning tree** for the graph G if it is a spanning graph of G.
Union (14.1)	If $G = G(V, E)$ be a graph and $G_1, G_2, G_3, \ldots G_n$, be subgraphs of G. The subgraph G' of G is the **union** of $G_1, G_2, G_3, \ldots G_n$, denoted by $\bigcup_{i=1}^{n} G_i$ if 1. A vertex $v \in G'$ if and only if $v \in G_i$ for some $1 \leqslant i \leqslant n$. 2. An edge $e \in G'$ if and only if $e \in G_i$ for some $1 \leqslant i \leqslant n$.

TRUE-FALSE QUESTIONS

1. There is a homomorphism from K_3 to K_2.
2. There is a homomorphism from $K_{2,2}$ to K_2.
3. $K_{2,2}$ is homeomorphic to K_3.
4. $K_{3,2}$ is homeomorphic to $K_{3,3}$.
5. A graph is never isomorphic to its complement.
6. Let G_1 and G_2 be subgraphs of a graph G. $(G_1 \cup G_2)^c = G_1^c \cap G_2^c$.
7. The complement of a graph G is a spanning graph of G.
8. The complement of a tree is never a tree.
9. Every graph has a spanning tree.
10. If G_1 and G_2 are homeomorphic and G_1 is biconnected, then G_2 is biconnected.
11. For a graph G having n vertices, $G \cup G' = K_n$.
12. If two graphs are isomorphic, then they are homeomorphic.
13. If two graphs are homeomorphic, then they are isomorphic.
14. If a graph G has an Euler cycle and G is homeomorphic to G', then G' has an Euler cycle.
15. If a graph G has a Hamiltonian cycle and G is homeomorphic to G', then G' has a Hamiltonian cycle.
16. If there is a homomorphism from a graph G onto a graph G' and G is biconnected, then G' is biconnected.
17. If there is a homomorphism from a graph G onto a graph G' and G has an Euler cycle, then G' has an Euler cycle.
18. If there is a homomorphism from a graph G onto a graph G' and G has a Hamiltonian cycle, then G' has a Hamiltonian cycle.
19. If there is a homomorphism from a graph G onto a graph G' and G is a tree, then G' is a tree.
20. If a graph G is a tree and G is homeomorphic to G', then G' is a tree.
21. A vertex is an articulation point if and only if it is contained in the intersection of two bicomponents.
22. The intersection of two bicomponents may contain more than one articulation point.
23. The intersection of two bicomponents may contain an edge.
24. A bicomponent is biconnected.
25. A bicomponent of G is not properly contained in any biconnected subgraph of a graph G.
26. A graph contains a planar spanning graph.
27. A planar graph is connected.
28. Every subgraph of a planar graph is planar.
29. Every graph homeomorphic to a planar graph is planar.
30. A 3-hypercube is planar.
31. K_4 is planar.
32. $K_{4,4}$ is planar.
33. If a graph can be colored with four colors, then it is planar.
34. The chromatic polynomial for K_5 is λ^5.
35. $\lambda(\lambda-1)(\lambda-2)(\lambda-3)(\lambda-4)$ cannot be the chromatic polynomial for a planar graph.
36. Every planar graph G with n vertices has an nth-degree chromatic polynomial with λ as a factor.
37. If a graph has a Euler cycle, then it has a Hamiltonian cycle.
38. If a graph has a Hamiltonian cycle, then it has a Euler cycle.
39. An n-hypercube has a Hamiltonian cycle.
40. An $n \times m$ grid has a Hamiltonian path.
41. If a graph has a Hamiltonian cycle, then it is biconnected.
42. If a graph has a Hamiltonian cycle, then it has no cut edges.
43. For $n \geq 2$, $K_{n,n}$ has a Hamiltonian cycle.
44. $\text{cl}(K_{5,6}) = K_{11}$.
45. Dijkstra's algorithm gives the minimal distance between any two vertices.
46. The shortest path determined by Dijkstra's algorithm is unique.
47. The Floyd-Warshall algorithm does not indicate the shortest path.
48. Dijkstra's algorithm continues until all vertices are made permanent.
49. If a graph has a Hamiltonian cycle, then it has a proper Hamiltonian path.
50. If a graph has a proper Hamiltonian path, then it has a Hamiltonian cycle.

SUMMARY QUESTIONS

1. Find a graph with 6 vertices that can be derived from K_3.
2. Find a homomorphism from K_4 to K_3 if it exists.
3. Find a spanning tree for $K_{3,3}$.
4. Is the graph whose edges form a regular pentagon homeomorphic to $K_{2,2}$?
5. Show that K_4 is planar.
6. Show that K_6 is not planar.
7. Find the chromatic polynomial for the 4-hypercube.
8. Show that K_5 has a Hamiltonian cycle.
9. Show that

has a Hamiltonian cycle.

10. Use the Floyd-Warshall algorithm to find the shortest distance between the points for the graph in

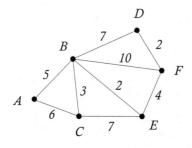

11. Use Dijkstra's algorithm to find the shortest distance between v_1 and v_5

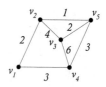

12. Find the cut edges of the graph

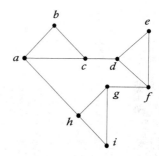

13. Give an example of a cycle with a Euler path that does not have a Hamiltonian path.
14. Give an example of a cycle with a Hamiltonian path that does not have a Euler path.
15. Find any articulation points in the graph

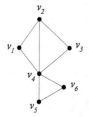

ANSWERS TO TRUE-FALSE

1. F 2. T 3. T 4. F 5. F 6. T 7. F 8. F 9. F 10. T 11. T
12. T 13. F 14. T 15. T 16. F 17. F 18. F 19. F 20. T
21. T 22. F 23. F 24. T 25. T 26. T 27. F 28. T 29. T
30. T 31. T 32. F 33. F 34. F 35. T 36. T 37. F 38. F
39. T 40. T 41. T 42. T 43. T 44. F 45. F 46. F 47. T
48. F 49. T 50. F

CHAPTER 15

Trees

15.1 Properties of Trees

In Section 6.3, a tree was defined to be a connected graph that contains no cycles. The following theorem summarizes the properties of trees collected from Section 6.3.

THEOREM 15.1 The following statements are equivalent:

(a) The graph G is a tree.
(b) The graph G is connected and $v = e + 1$, where G has v vertices and e edges.
(c) For every pair of distinct vertices a and b, there is a unique path from a to b.
(d) G is acyclic (has no cycles) and $v = e + 1$.

Proof Parts (a) and (b) are shown equivalent in Theorem 6.37 and Theorem 6.38 of Section 6.3. Parts (a) and (c) are shown equivalent in Theorem 6.40 and Theorem 6.41. Part (d) follows directly from parts (a) and (b). To show that part (a) follows from part (d), assume G is acyclic. Then the components $G_1, G_2, G_3, \ldots, G_m$ of G are all trees. For each G_i, $v_i = e_i + 1$ where G_i has v_i vertices and e_i edges. Therefore, $\sum_{i=1}^m v_i = \sum_{i=1}^m e_i + m$. But $v = \sum_{i=1}^m v_i$ and $e = \sum_{i=1}^m e_i$ so $v = e + m$. Therefore, $m = 1$, so there is only one component, and G is connected. Thus, G is a tree. ∎

In Chapter 6, we stated that a **directed tree** T is a loop-free directed graph whose underlying graph is a tree such that if there is a path from a vertex a to a vertex b, it is unique. A directed tree T is a **rooted directed tree** if there is a unique vertex v_0 such that $\text{indeg}(v_0) = 0$ and there is a path from v_0 to every other vertex in T. As we did for trees, we show several equivalent descriptions of rooted directed trees. In the following proof, if p and q are paths, then pq denotes the path p followed by the path q.

THEOREM 15.2 The following statements are equivalent for a directed graph G:

(a) G is a rooted directed tree.
(b) G has a unique element v_0 such that for any vertex a of G, there is a unique directed path from v_0 to a.

(c) The underlying graph for G is connected and G has a unique element v' such that $\text{indeg}(v') = 0$ and $\text{indeg}(a) = 1$ for every other vertex a of G.

(d) The graph underlying G is connected and G has a unique element v_0 such that for any vertex a of G, there is a unique path from v_0 to a.

Proof $(a) \to (b)$ Since G is a rooted directed tree, there is a directed path from the root v_0 to any given vertex a, and since it is a directed tree, the path is unique. The root is unique since if v_0' had the same property, then there is a path p from v_0 to v_0' and a path p' from v_0' to v_0. But then the path $pp'p$ is another path from v_0 to v_0' and the path would not be unique.

$(b) \to (c)$ Let v' be the unique element v_0 of (b). Then $\text{indeg}(v') = 0$ for if $\text{indeg}(v') \neq 0$, then there is an edge (a, v') for some vertex a. But there is also a path p from v' to a, and $p(a, v')p$ is also a path from p to a and the path from p to a is not unique. The vertex v' is the only vertex that has an indegree equal to 0. If there another vertex v with $\text{indeg}(v) = 0$, then there would be no path from v' to v, which contradicts (b). Every other edge has indegree 1, for if vertex v has indegree greater than 1, then there is a directed edge from a to v and a directed edge from b to v for distinct vertices a and b. But there are paths p from v' to a, and q from v' to b, so that $p(a, v)$ and $q(b, v)$ are distinct paths from v' to v, which contradicts (b). The underlying graph of G is connected since, given vertices a and b, there is a path from v' to a and from v' to b, and, hence, in the underlying graph, there is a path from a to v' and a path from v' to b, so there is a path from a to b.

$(c) \to (d)$ Certainly $v_0 = v'$. By (c), there is a path from v_0 to a for any vertex a in the graph. Assume there are two paths from v_0 to a for some vertex a. Let b be the first vertex such that the two paths are the same from b to a. It may be a itself. Then $\text{indeg}(b) > 1$, which contradicts (c).

$(d) \to (a)$ Assume that in the underlying graph G', there are two undirected paths from v^* to a vertex a. If both are directed paths from v^* to a, this contracts the hypothesis. If one is a directed path p from a to v^* and the other is a directed path q from v^* to a, then qpq is also a path from v^* to a, which contradicts the uniqueness of the path. The only other possibility is that one of the connected paths from v^* to a is not a directed path. Therefore, there must be "an arrow going the wrong way" in one path. More precisely there is an edge (b, c) and an edge (d, c) for some vertex c in one of the paths. But, since there is a directed path p' from v^* to b and a directed path p'' from v^* to d, there are distinct paths $p'(b, c)$ and $p''(d, c)$ from v^* to c, which is again a contradiction. Therefore, the undirected path from v^* to a vertex a is unique. The undirected paths from any vertex a to any vertex b must be unique, for if there were two paths from a to b, then there would be two distinct paths from v^* to a to b, or in other words, two distinct paths from v^* to b, which we have already shown is not true. Therefore, G is a directed tree, and since v^* satisfies the definition of a root, G is a rooted directed tree. ■

For the remainder of this book, only rooted directed trees will be considered; hence, all directed trees will be understood to be rooted directed trees.

Parts of the following definitions are repeated from Section 6.3 for convenience. We warn the reader that these definitions vary in different texts.

DEFINITION 15.3 The **level** of a vertex v in a directed tree is the length of the path from the root of the tree to v. The **height** of a directed tree is the length of the longest path from the root to a leaf.

DEFINITION 15.4 A *m*-ary **directed tree** is a directed tree such that $\text{outdeg}(v) \leq m$ for every vertex v in the tree. Thus, a parent has at most m children. A **full *m*-ary directed tree** is a directed tree such that $\text{outdeg}(v) = m$ for every vertex v in the tree that is not a leaf, and every leaf is at the same level. Thus, each parent has exactly m children.

DEFINITION 15.5 A *m*-ary directed tree of height h is **balanced** (**complete** or **nearly full**) if the level of every leaf is either h or $h - 1$.

THEOREM 15.6 In a full *m*-ary directed tree with n vertices and i internal vertices,

$$n = mi + 1$$

Solving for i,

$$i = \frac{n-1}{m}$$

Proof Every vertex except the root is the child of an internal vertex. Since there are i internal vertices and each internal vertex has m children, there are im children. When the root is added, there are a total of $mi + 1$ vertices. ∎

THEOREM 15.7 In a full *m*-ary directed tree with n vertices, i internal vertices, and l leaves,

$$l = (m-1)i + 1$$

Solving for i,

$$i = \frac{l-1}{m-1}$$

Proof By the previous theorem, we have $n = mi + 1$ vertices. Subtracting the i internal vertices, we have $l = n - i = mi + 1 - i = (m-1)i + 1$ leaves. ∎

THEOREM 15.8 A full *m*-ary directed tree of height h has $\frac{m^{h+1}-1}{m-1}$ vertices and m^h leaves. In particular a full binary directed tree of height h has $2^{h+1} - 1$ vertices and 2^h leaves.

Proof Since the number of vertices at any level is m times larger than the previous level, this forms a geometric sequence, and a full *m*-ary directed tree of height h has

$$1 + m + m^2 + m^3 + \cdots + m^{h-1} + m^h = \frac{m^{h+1}-1}{m-1}$$

vertices. Since only those at level h are leaves, there are m^h leaves. ∎

The proof of the following theorem is left to the reader.

THEOREM 15.9

(a) For a full *m*-ary tree with height h and l leaves, $h = \log_m(l)$.
(b) For a *m*-ary tree with height h and l leaves, $h \geq \log_m(l)$.
(c) For a full binary tree with height h and v vertices, $h = \log_2(v+1) - 1$.
(d) For a binary tree with height h and v vertices, $h \geq \log_2(v+1) - 1$. ∎

Recall that a function f from the graph $G(V, E)$ to the graph $G'(V', E')$ is called a **homomorphism** from G to G', denoted $f : G \to G'$, if it has the following properties:

DEFINITION 15.10

(a) If $e \in E$, then $f(e) \in E'$ ($f(E) \subseteq E'$).
(b) If $v \in V$, then $f(v) \in V'$ ($f(V) \subseteq V'$).
(c) If vertices u and v are incident to edge e in G, then $f(u)$ and $f(v)$ are incident to $f(e)$ in G'.

Also recall that the homomorphism $f : G \to G'$ is an **isomorphism** if $f : V \to V'$ and $f : E \to E'$ are one-to-one correspondences. If $f : G \to G'$ is an isomorphism, then G and G' are **isomorphic**.

DEFINITION 15.11

Two rooted binary trees $T(E, V)$ and $T'(E', V')$ are **isomorphic** if there exists a graph isomorphism f from T to T' such that
(1) v_i is a left child of v_j if and only if $f(v_i)$ is a left child of $f(v_j)$.
(2) v_i is a right child of v_j if and only if $f(v_i)$ is a right child of $f(v_j)$.
(3) f maps the root r of T to the root r' of T'.

For those who like to count things, we have the following theorem.

THEOREM 15.12 The number of nonisomorphic rooted binary trees with n vertices is the Catalan number C_n.

Proof Let I_n be the number of isomorphic rooted binary trees with n vertices. If $n = 0$, we have the empty tree and define $I_0 = 1$. If $n = 1$, we have only one vertex and obviously only one tree, so $I_1 = 1$. If $n = 2$, we have the two rooted binary trees shown in Figure 15.1. If $n = 3$, we have the five rooted binary trees shown in Figure 15.2.

Assume that $n > 3$ and choose k so $1 \leq k \leq n$. Let T_n denote a tree with n vertices and let $T_{n,k}$ be a tree with n vertices defined as follows: Since one vertex is the root, assume that there is a right subtree T_{k-1} with $k - 1$ vertices and a left subtree T_{n-k} with $n - k$ vertices shown in Figure 15.3.

Figure 15.1

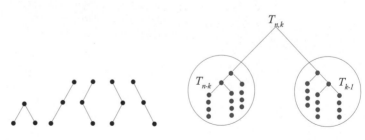

Figure 15.2   **Figure 15.3**

Then, by definition, the number of ways that T_{k-1} can be formed is I_{k-1} and the number of ways that T_{n-k} can be formed is I_{n-k}. Thus, the number of ways $T_{n,k}$ can be formed is $I_{k-1} \cdot I_{n-k}$. As we sum over k, we create all possible ways that the tree T_n can be formed. Thus,

$$I_n = \sum_{k=1}^{n} I_{k-1} \cdot I_{n-k}$$

but this is the definition of the Catalan number C_n. Thus,

$$I_n = C_n = \frac{1}{n+1} \cdot \frac{(2n)!}{n! \cdot n!}$$ ∎

Exercises

Find the nonisomorphic rooted binary trees with n vertices when

1. $n = 2$ **2.** $n = 3$ **3.** $n = 4$

How many nonisomorphic rooted binary trees with n vertices are there when

4. $n = 5$?   **5.** $n = 6$?   **6.** $n = 8$?

7. $n = 10$?   **8.** $n = 20$?

In a full m-ary tree with height h, how many leaves, how many vertices, and how many internal vertices are there when

9. $m = 2$ and $h = 5$?   **10.** $m = 3$ and $h = 4$?

11. $m = 2$ and $h = 8$?   **12.** $m = 4$ and $h = 3$?

13. $m = 1$ and $h = 10$?

In the tree in Figure 15.4,

(i) What is the height of the rooted tree?

(ii) What is the level of vertex e?

(iii) What is the level of vertex g?

(iv) What is the level of vertex a?

(v) What is the parent of i?

(vi) What are the children of b?

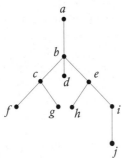

Figure 15.4

14. If vertex d is selected as the root?

15. If vertex f is selected as the root?

16. If vertex c is selected as the root?

17. If vertex j is selected as the root?

18. If vertex b is selected as the root?

In the tree in Figure 15.5,

(i) What is the height of the rooted tree?

(ii) What is the level of vertex e?

(iii) What is the level of vertex g?

(iv) What is the level of vertex a?

(v) What is the parent of i?

(vi) What are the children of b?

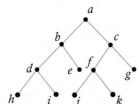

Figure 15.5

19. If vertex d is selected as the root?

20. If vertex f is selected as the root?

21. If vertex c is selected as the root?

22. If vertex j is selected as the root?

23. If vertex b is selected as the root?

Which of the following trees are balanced? Which are full?

24.   **25.**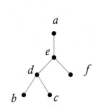

26. 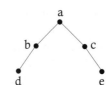   **27.**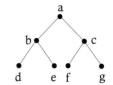

28.   29.

30.   31.

32.   33.

If a full binary tree has 32 leaves,

34. What is the height of the tree?
35. How many vertices does it have?

If a full binary tree has 128 leaves,

36. What is the height of the tree?
37. How many vertices does it have?

Prove Theorem 15.9

38. For a full m-ary tree with height h and l leaves, $h = \log_m(l)$.

39. For an m-ary tree with height h and l leaves, $h \geq \log_m(l)$.

40. For a full binary tree with height h and v vertices, $h = \log_2(v+1) - 1$.

41. For a binary tree with height h and v vertices, $h \geq \log_2(v+1) - 1$.

42. Prove that a tree with two or more vertices is a bipartite graph.

43. Prove that a connected graph is a tree if and only if the addition of a new edge between vertices of the graph always causes the graph to have one cycle.

44. Prove or disprove that a connected graph is a tree if and only if every edge is a cut edge.

45. Prove or disprove that a connected graph is a tree if and only if every vertex is a cut vertex.

46. In a rooted tree, which vertices are cut vertices?

47. Let G be a connected graph. Given necessary and sufficient conditions on G so that if any edge is removed from G, then G is a tree.

48. Let $T(V, E)$ be a rooted tree. Define a relation $\leq$ on V by $a \leq b$ if $a = b$ or a is a descendant of b. Show that the relation $\leq$ is a partial ordering. Define the multiplication $a \cdot b$ by $a \cdot b = \text{lub}(a, b)$ so that $(V, \cdot)$ is an upper semilattice. Describe $a \cdot b$ in terms of paths from a and b to the root.

49. Given the multiplication on V in the previous exercise, for a given vertex a of V, define $\hat{a} : V \to V$ by $\hat{a}(v) = av$. Prove that $\hat{a}$ is a semigroup homomorphism.

15.2 Binary Search Trees

The rooted binary tree, which we shall simply refer to as a binary tree, provides an excellent method of arranging data so that any particular piece of data can be found or easily discovered to be missing. Obviously the most inefficient method to search is simply to go through each item and, if it is missing, go through all of the data to find this out. The binary search tree avoids this. The only requirement is that there be some sort of linear order attached to the data. It may, for example, be in alphabetical order or numerical order. The linear order may be on a tag, pointer, file, or some other key that reveals the data. Our only concern is that there be some sort of linear order that, for simplicity, we shall assume to be names in alphabetical order.

A binary search tree is, first of all, a binary tree with a name (or other key) at each node. We shall discuss how to construct a binary search tree and, in doing so, we shall pretty much indicate how to use the tree. Suppose we want to store the names Peterson, Johnson, Smith, Holt, Spenser, Russell, Dover, Martin, Wilson, and Lyman. We begin with Peterson and place him at the root of the tree. Since Johnson, the next name, occurs before Peterson alphabetically, we make Johnson the left child of Peterson as shown in Figure 15.6. Smith, the next name occurs after Peterson alphabetically, so Smith becomes the right child of Peterson as shown in Figure 15.7.

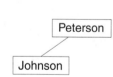

Figure 15.6

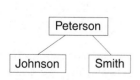

Figure 15.7

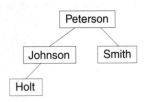

Figure 15.8

We now consider Holt. Since Holt is before (or less than) Peterson alphabetically, we come down to the left child, Johnson, and since Holt is before Johnson alphabetically, we make Holt the left child of Johnson as shown in Figure 15.8.

Next, we consider Spenser. Since Spenser is after Peterson alphabetically, we go down to the right child, Smith. Since Spenser is after Smith alphabetically, Spenser is made the right child of Smith as shown in Figure 15.9.

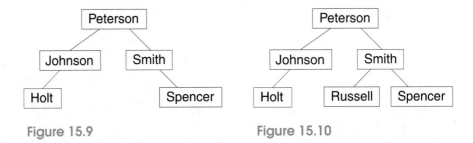

Figure 15.9   Figure 15.10

Now we consider Russell. Since Russell is after Peterson in the alphabet, we go down to the right child, Smith. Since Russell is before Smith in the alphabet, we make Russell the left child of Smith as shown in Figure 15.10.

Hopefully, by now, a pattern is beginning to emerge. Assume we have a name to put on the tree. When we come to a vertex of the tree, if there is a name there, we go to the left child if the name we wish to store is alphabetically before the name at that vertex, and we go to the right child if the name we wish to store is alphabetically after the name at that vertex. If there is no name at that vertex, we place the name we wish to store at that vertex.

Now, using this process, let's finish the list. The next name is Dover. It is before Peterson alphabetically, so we go to the left child, Johnson. It is before Johnson alphabetically, so we go to the left child, Holt. It is before Holt alphabetically, so we go to the left child. Since there is no name there, we make Dover the left child as shown in Figure 15.11.

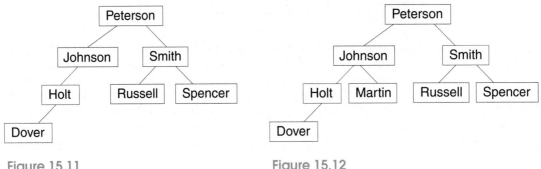

Figure 15.11   Figure 15.12

The next name we wish to store is Martin. Martin is before Peterson in the alphabet, so we go to the left child, Johnson. It is after Johnson in the alphabet so we go to the right child, and since there is no name there, we make Martin the right child as shown in Figure 15.12.

The next name we wish to store is Wilson. Since it is after Peterson alphabetically, we go to its right child, Smith. Since it is after Smith alphabetically, we go to the right child, Spenser. Since it is after Spenser alphabetically and there is no name there, we make it the right child of Spenser as shown in Figure 15.13.

Figure 15.13

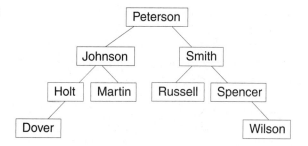

Last, but not least, we put Lyman in his place. Since Lyman is before Peterson alphabetically, we go to the left child, Johnson. Since Lyman is after Johnson in the alphabet, we go to its right child, Martin. Since Lyman is before Martin in the alphabet, we go to its left child. There is no name there, so we make Lyman the left child of Martin.

It is probably more precise, when reaching a left (or right child) that has not been defined, to say "if the left (or right) child does not exist, create the vertex and store the name" rather than to say "if there is no name there, then store the name." However, we are more concerned with the concept rather than the technical details. At this point we give an algorithm for inserting a name in a search tree, which is really creating a search tree except for placing a name at the root vertex. We shall speak more generally, using < and >, so that it applies to any ordering.

■ **Insert(item)**

1. Start at root vertex.
2. If item < vertex object, go to left child.
3. If item > vertex object, go to right child.
4. Repeat steps 2 and 3 until vertex reached is undefined.
5. If the vertex reached is undefined, define vertex and insert item.

Once the method of creating the search tree is described, it is easy to see how to search for an item in the tree. We basically use the same method, but instead of just checking to see if the name is greater than or less than the name at the vertex, we also check to see it is the same as the name at the vertex. If it is, we are finished. If it isn't, then we use the same path as before. If we reach a vertex that has not been defined, then we know that the name isn't stored in the tree. For example, suppose in the tree above, we searched for the name Jenkins. Since it is "less than Peterson," we go to the left child, Johnson. Since it is "less than Johnson," we go to the left child, Holt. Since is "greater than Holt" (no one is really greater than Holt), we go to the right child. But it is undefined, so the name is not on the list. The following is an algorithm for searching for a name in a search tree.

■ **Search(item)**

1. Start at root vertex.
2. If item < vertex object, go to left child.
3. If item > vertex object, go to right child.

4. If item = vertex object, name is located, take appropriate action, and exit.

5. Repeat steps 2, 3, and 4 until vertex reached is undefined.

6. If the vertex reached is undefined and the name is not in the tree, take appropriate action and exit.

It is obvious that the search path for an item cannot exceed the height of the path. Therefore, by Theorem 15.9, if the tree is full, the worst case has order $O(\ln(n))$, where n is the number of items on the tree. Hence, if a tree is balanced, the worst case has order $O(\ln(n))$.

Finally we consider how to delete a vertex from the tree. This is simply one method. There is no claim that it is the best method. Also, not being total fools, we shall omit any technical details with regard to actually implementing this method.

We summarize this method in the following algorithm.

■ Algorithm Delete(v_0)

1. If v_0 has no child, simply remove it.

2. If v_0 has one child, remove v_0 and replace it with its child.

3. If v_0 has two children, find the right child v_1 of v_0 and then find the left child (if it exists) of v_1. Continue taking the left child of each vertex found until there is a vertex v with no left child. Replace v_0 with v and let the right child of v be the left child of the parent of v.

EXAMPLE 15.13 Consider the tree in Figure 15.14. If x were removed, we would have the tree in Figure 15.15. If k were removed, we would have the tree in Figure 15.16.

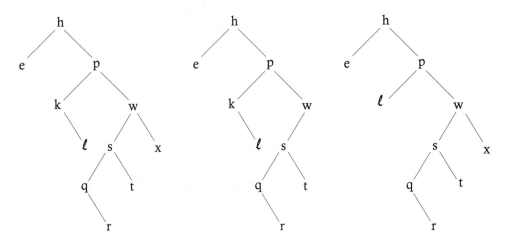

Figure 15.14   Figure 15.15   Figure 15.16

In our original tree, if we delete h, we go to the right child p and then to the left child k. Since k has no left child, it is the replacement we seek, so we replace h with k and make l the left child of p. This gives us the tree in Figure 15.17.

If, instead, we had removed p, we would go right to w, then left to s, then left to q. Since q has no left child, it is the desired replacement vertex. Therefore, we replace p with q and let r be the left child of s, so that we have the tree shown in Figure 15.18.

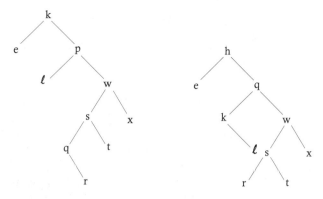

Figure 15.17   Figure 15.18

Computer Science Application

Recall from Section 6.3 that the height of a tree is the length of the longest path from the root of the tree to a leaf. In the following algorithm we assume that the tree is made up of nodes that consist of a left link, a right link, and a content field. T is assumed to be a pointer to the root of a binary tree.

integer function height(TreeNodePointer T)

   integer HeightOfLeftSubtree, HeightOfRightSubtree

   if T = null then
     return -1
   end-if

   HeightOfLeftSubtree = height(T->leftlink)
   HeightOfRightSubtree = height(T->rightlink)

   if HeightOfLeftSubtree > HeightOfRightSubtree then
     return HeightOfLeftSubtree + 1
   else
     return HeightOfRightSubtree + 1
   end-if

end-function

When T = null the tree is empty. In each of the recursive invocations of the height function the parameter is a pointer to the root of a subtree. The recursion is easy to express since the height of any tree is simply the larger of the two subtree heights plus 1. A function that returns the number of nodes on a binary tree is done in similar fashion by noting that the number of nodes on a tree is the sum of the nodes of the two subtrees plus 1.

integer function node-count(TreeNodePointer T)

   if T = null then
     return 0
   end-if

   return node-count(T->leftlink) + node-count(T->rightlink) + 1

end-function

Exercises

Given the tree in Figure 15.19,

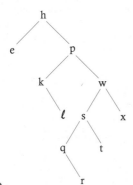

Figure 15.19

1. Insert the letter j.
2. Insert the letter u.
3. Insert the letter m.
4. How many comparisons are necessary to determine that i is missing?
5. How many comparisons are necessary to determine that u is missing?
6. How many comparisons are necessary to determine that f is missing?

Given the tree in Figure 15.20,

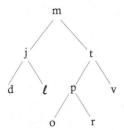

Figure 15.20

7. Insert the letter e.
8. Insert the letter n.
9. Insert the letter u.
10. How many comparisons are necessary to determine that n is missing?
11. How many comparisons are necessary to determine that u is missing?
12. How many comparisons are necessary to determine that b is missing?
13. Describe the binary tree with five vertices that would give a worst-case solution for searching.

14. Construct the binary tree for the following names, stored in alphabetical order, if the data are entered in the following order: Howard, James, Aaron, Spencer, Jackson, Madison, Venus, Calhoun, Barnes, McDuff, and Peterson.

15. Construct the binary tree for the presidents of the United States, stored in alphabetical order, if they are entered in the order in which they served their first term.

16. Construct the binary tree for the following numbers, stored in numerical order, if they are entered in the following order: 25, 35, 15, 27, 48, 36, 22, 44, 18, 30, 42, 11, 39, 32.

17. Construct the binary tree for the following letters, stored in alphabetical order, if they are entered in the following order: $m, a, r, i, g, o, l, d, s, t, u, l, i, p$.

18. Construct the binary tree for the following names, stored in alphabetical order, if the data are entered in the following order: Benson, Chevrolet, Chrysler, Dodge, Mercury, Nash, and Oldsmobile.

19. Construct a binary tree of length 4 for the letters of the English alphabet.

Given the tree in Figure 15.21,

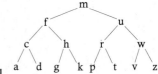

Figure 15.21

20. Delete a.
21. Delete f.
22. Delete m.
23. Delete u.
24. Delete r.

Given the tree in Figure 15.22,

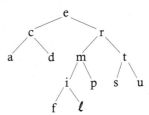

Figure 15.22

25. Delete a.
26. Delete e.
27. Delete r.
28. Delete m.
29. Delete t.

30. Write a procedure for the algorithm Insert(item) in pseudocode or an assigned language.

31. Write a procedure for the algorithm Search(item) in pseudocode or an assigned language.

32. Write a procedure for the algorithm Delete(item) in pseudocode or an assigned language.

15.3 Weighted Trees

In this section we look at two problems and find that they essentially have the same solution. In the computer, all letters and other symbols are stored as strings of 1's and 0's. For example, when symbols are stored as ASCII symbols, they consists of strings of length 7 plus an eighth symbol that may be used for parity check. Let A be the set of all the symbols that we wish to store. It is always possible to store the set A as strings of 1's and 0's of some fixed length n. This has a tremendous advantage when it is necessary to "decode" the string back to symbols in A. The first n bytes (1's and 0's in the string) form the first symbol of A, the second n bytes form the second symbol, and so on.

If the amount of data is large, however, it is often desirable to perform a compactification of the data, which means to store it using less space. The obvious way to do this is to vary the length of the strings used to store the data so that shorter strings are used to store symbols that are used more often and longer strings for those used less often. Here we have a problem, for if strings representing different symbols have different lengths, how do we know when one symbol string ends and another begins? For example, if b is represented by 101, e by 11, a by 10, q by 1011, and w by 110, does the string 1011110 represent bea or qw?

We define a **uniquely decipherable code** for a language as a set such that every string in the language can be expressed uniquely as the concatenation of elements of C. In this case, the strings of 1's and 0's representing the elements of A will be our code. If these strings form a uniquely decipherable code, then at least when we separate strings into the elements representing A, we know that the representation is unique and, hence, words that we have decoded are correct. We wish to improve on this somewhat. We define a code C to be a **prefix code** if it has the property that an element of a code cannot be the beginning string of another element of the code. Thus, when we have read a string of 1's and 0's that represent a symbol of A, we know that this is the symbol we seek rather than the beginning of another symbol. The next 1 or 0 must, therefore, begin a string for the next symbol we are decoding. For example, if the strings 111, 1011, 1001, 110, and 01 represent a, e, i, o, and u, respectively, in a five-letter alphabet, then if we read 110 as the first three digits in a string, we know this represents the letter o since no other letter begins with these three digits. Similarly, it next three digits read are 011, we know that 01 represents the letter u, and 1 begins the next letter.

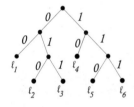

Figure 15.23

Consider the following binary tree with leaves t_1, t_2, t_3, t_4, t_5, and t_6 in Figure 15.23. If we consider a path from the root of the tree to any leaf, each edge along the path must be to a right child or a left child of the previous vertex. If it is to a left child, we denote the edge by 0, and if it is to a right child, we denote the edge by 1. Since the path to the leaf v_1 is to a left child, then to another left child, we denote the path to v_1 by 00. Similarly the paths to v_2, v_3, v_4, v_5, and v_6, respectively, are denoted by 010, 011, 10, 110, and 111. We shall call the string corresponding to a given leaf, the **leaf path code**. Note that the strings corresponding to these leaves form a prefix code. The string corresponding to each leaf is unique since the path to the leaf is unique. More generally we have the following theorem. The proof is left to the reader.

THEOREM 15.14 In any binary tree, the leaf path codes for the leaves in the tree are a prefix code. ∎

Suppose that we have a binary tree and, for each symbol in our alphabet A, which we desire to represent with a binary string, there is a leaf on the tree. We assign to each letter of A, the leaf path code of the corresponding leaf on the binary tree. Thus, if the letters a, e, i, o, u, and w were assigned, respectively, to the letters v_1, v_2, v_3, v_4, v_5, and v_6 in the above tree, then a, e, i, o, u, and w would be assigned the strings 00, 010, 011, 10, 110, and 111, respectively.

In our original problem we wanted symbols that occurred more often to have shorter strings. Using the method just described, this means that we want symbols that occur more often to have shorter paths from the root, or equivalently, be at lower levels. We need some sort of measure to determine how well we have achieved our goal of giving more frequently occurring letters shorter strings so that we minimize our code during compactification. For each letter a_i in our alphabet of symbols A, assume that it occurs with relative frequency f_i and that the binary string representing it has length l_i. Then

$$w = l_1 f_1 + l_2 f_2 + l_3 f_3 + l_4 f_4 + \cdots + l_n f_n = \sum_{i=1}^{n} l_i f_i$$

is called the **weight of the code**. When the leaves of a weighted tree represent the letters of the code, the weight of the leaf is the frequency the letter occurs, and the distance from the root to the vertex is also the length of the binary code for the symbol at the leaf. Then we define the **weight of the tree** to be

$$w = l_1 f_1 + l_2 f_2 + l_3 f_3 + l_4 f_4 + \cdots + l_n f_n = \sum_{i=1}^{n} l_i f_i$$

More generally, the **weight of a tree** is

$$w = l_1 w_1 + l_2 w_2 + l_3 w_3 + l_4 w + \cdots + l_n w_n = \sum_{i=1}^{n} l_i w_i$$

where w_i is the weight assigned to a vertex v_i and l_i is the length of the path from root to the vertex. This should not be confused with weighted graphs, including trees, where weights are assigned to the edges of the graphs. These are discussed in Section 15.6.

The lower the weight of the tree, the better we have achieved our goal. Thus, to find the best code for minimizing our data, we want to find the code with minimal weight. The process for obtaining this tree is called **Huffman's algorithm**. The code assigned to the symbols by their leaf path codes is called **Huffman's code**.

Before giving the algorithm, let's see how it works. Suppose we have the following letters and their frequencies:

symbol	frequency
a	7
e	10
w	3
k	5
m	8
#	2
%	1

We begin by placing the frequencies in increasing order, so we have (1, 2, 3, 5, 7, 8, 10). We then form a tree using the two letters with lowest frequencies as vertices and the sum of the frequencies as their parent. We place a 0 on the edge to the left child and a 1 on the edge to the right child so we have the tree in Figure 15.24.

In the frequency list, we also replace the values of the two lowest frequencies with their sum, so we now have (3, 3, 5, 7, 8, 10). If this sum had been greater than one or more of the other entries in the frequency list, we would have rearranged them so they were still in increasing order. We again pick the two vertices or trees with the lowest number on the frequency list and form a tree with these vertices or trees as children and their sum as parent. In this case it is the tree with value 3 and the letter w with value 3 as children and their sum 6 as parent. We place a 0 on the edge to the left child and a 1 on the edge to the right child, so we have the graph in Figure 15.25.

In the frequency list, we also replace the values of the two lowest frequencies with their sum, 6, and place the values in increasing order, so we now have (5, 6, 7, 8, 10). In all cases, the values of vertices or trees used as children have their values removed from the frequency list and from the tree being constructed. We again pick the two vertices or trees with the lowest number on the frequency list and form a tree with these vertices or trees as children and their sum as parent. In this case it is the vertex k, with value 5, and the tree with value 6. We place a 0 on the edge to the left child and a 1 on the edge to the right child, so we have the tree in Figure 15.26.

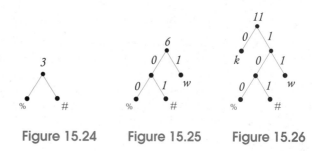

Figure 15.24   Figure 15.25   Figure 15.26

In the frequency list, we also replace the values of the two lowest frequencies with their sum, 11, and place the values in increasing order, so we now have (7, 8, 10, 11). Continuing this process we have the following trees and frequency lists: Figure 15.27 (10, 11, 15) and Figure 15.28 (15, 21).

Finally, we connect the two remaining trees, placing a 0 on the edge to the left child and a 1 on the edge to the right child, so we have the tree in Figure 15.29.

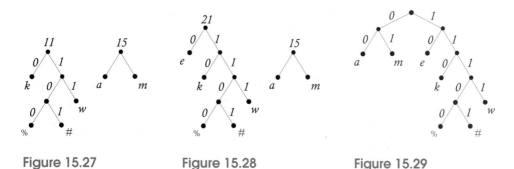

Figure 15.27   Figure 15.28   Figure 15.29

We don't need to the place the sum at the root. Notice that Huffman's code numbers for a, e, w, k, m, #, and % are 00, 10, 1111, 01, 110, 11101, and 11100, respectively.

The following is Huffman's algorithm for constructing a tree for symbols $s_1, s_2, \ldots, s_n$ having respective frequencies $f_1, f_2, \ldots, f_n$.

■ Huffman's Algorithm $((s_1, f_1), (s_2, f_2), \ldots, (s_n, f_n))$

1. Place the frequencies in increasing order in a frequency table.
2. If f_i and f_j are the two lowest frequencies, form a binary tree with s_i and s_j as the children, with s_i to the left, and $f_i + f_j$ as the frequency of the parent. Place a 0 on the edge to the left child and a 1 on the edge to the right child.
3. Remove f_i and f_j from the frequency table and replace with $f_i + f_j$.
4. Again place the frequencies in increasing order.
5. Remove the two lowest frequencies from the frequency table and form a tree with the symbols or trees corresponding to these frequencies as children, with the smaller tree or symbol on the left, and the sum of their frequencies as the frequency of the parent. If either of the children is a tree, remove its frequency label from the tree.
6. Replace the value of the two lowest frequencies with their sum.
7. Repeat steps 4–7 until the frequency table has only one value.
8. Remove the frequency from the root of the tree and from the table.
9. Find the Huffman code for an item by forming a string of 1's and 0's on the edges of the path from the root to the given item.

Earlier we mentioned a second problem with the same solution. Suppose that we have n items of data and we know the relative frequency that each item is used. We want to store them as leaves on a binary tree so that the most commonly used items are the easiest to retrieve. But this means that the most commonly used items will have the shortest paths from the root. This is exactly what we were doing previously, since the most commonly used symbols had the shortest binary strings and, hence, had the shortest path from the root.

EXAMPLE 15.15 Assume that we have the following letters and their frequencies:

symbol	frequency
A	15
B	25
C	10
D	30
E	15
F	5

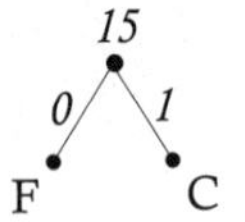

Figure 15.30

Figure 15.31

Construct a binary tree so that the most frequently used are the closest to the root. More precisely find the tree so that the **weight of the tree**

$$w = l_1 f_1 + l_2 f_2 + l_3 f_3 + l_4 f_4 + \cdots + l_n f_n = \sum_{i=1}^{n} l_i f_i$$

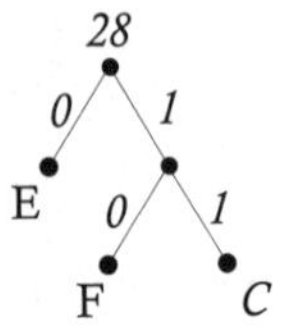

is minimal, where l_i and f_i are the level and frequency of a given item.

We first place the frequencies in order getting the frequency table (5, 10, 13, 17, 25, 30). Using the above algorithm, we have the trees and frequency lists: Figure 15.30 (13, 15, 17, 25, 30); Figure 15.31 (17, 25, 28, 30); Figure 15.32 (28, 30, 42); Figure 15.33 (42, 58).

Finally we combine the two trees to form the desired tree in Figure 15.34.

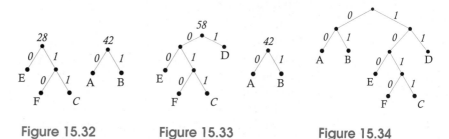

Figure 15.32   Figure 15.33   Figure 15.34

We now need to show that Huffman's tree is the tree with minimum weight. We begin with a series of lemmas.

Lemma 15.16 Given a set of n symbols and their frequency, there is a binary tree of minimum weight with the symbols as leaves.

Proof There is only a finite number of binary trees with n leaves. For each one, compute

$$w = l_1 f_1 + l_2 f_2 + l_3 f_3 + l_4 f_4 + \cdots + l_n f_n = \sum_{i=1}^{n} l_i f_i$$

and pick the tree that gives the smallest value. ∎

Lemma 15.17 In a tree with minimum weight, at the maximal level of the tree, leaves occur in pairs; that is, wherever there is a left child, there is a right child, and conversely.

Figure 15.35

Proof Assume at the maximal level there is a single child, say a left child with symbol s, but no right child so that we have the tree in Figure 15.35. Then remove the left child and label the parent s. This produces a tree with less weight, which contradicts our assumption that the tree has minimal weight. ∎

Lemma 15.18 In a tree with minimum weight, the two symbols with smallest frequency occur at the maximal level.

Proof If this is not true, then we have a symbol a at level l_i and frequency f_i and a symbol b at level l_j and frequency f_j where $l_i > l_j$ and $f_i > f_j$. There two symbols contribute weight $l_i f_i + l_j f_j$ to the total weight of the tree. If the two symbols are interchanged, then the two symbols contribute weight $l_i f_j + l_j f_i$ to the tree. But

$$l_i f_i + l_j f_j - (l_i f_j + l_j f_i) = l_i f_i + l_j f_j - l_i f_j - l_j f_i$$
$$= (l_i - l_j)(f_i - f_j) > 0$$

since $l_i - l_j > 0$ and $f_i - f_j > 0$. Therefore, $l_i f_i + l_j f_j > l_i f_j + l_j f_i$ so that switching symbols contributes less weight to the weight of the tree. But this contradicts the assumption that we have a tree of minimal weight. ∎

Lemma 15.19 There is a tree with minimum weight, in which the two symbols with smallest frequency have the same parent.

Proof By Lemma 15.17, every leaf at the maximal level has a sibling. By Lemma 15.18, both of the symbols with minimum weight, say a and b, are at the same level. If they are not siblings, then since they are at the same level exchange, a's current sibling for b and the weight of the tree are not affected. ∎

We are now ready for the big theorem.

THEOREM 15.20 For a given set of symbols with corresponding frequencies, Huffman's tree is a tree with minimum weight.

Figure 15.36

Proof We prove this theorem using induction on the number of symbols. If we have two symbols, say a and b, then Huffman's tree shown in Figure 15.36 is certainly a tree with minimum weight. Therefore, the theorem is true for $n = 2$. Assume the statement is true for a tree with n symbols. Let T_{n+1} be a tree with minimum weight containing $n + 1$ symbols. Assume that a and b, with frequencies f_a and f_b, have the lowest frequency and, by Lemma 15.18, may be assumed to be siblings at the greatest level $l+1$. Now remove a and b from the tree and assign their parent the symbol c with frequency $f_a + f_b$. Call this tree T_n. Since there are only n symbols on this tree, by induction, there is a Huffman tree with minimum weight with these symbols. Call this tree T'_n. Since T'_n has minimal weight, the weight of T'_n is less than or equal to the tree T_n. Now replace c with frequency $f_a + f_b$, by a parent with children a and b to form tree T'_{n+1}. In other words, reverse the process we used to get from tree T_{n+1} to tree T_n. But this is the construction for the Huffman tree with $n + 1$ symbols, since to construct the Huffman tree with $n + 1$ symbols, we would begin with a and b and form a tree with parent having frequency $f_a + f_b$, so that if we think of c representing this tree, the construction of the Huffman tree T'_n is the continuation of the Huffman tree T'_{n+1}.

We now have four trees; the Huffman trees T'_n and T'_{n+1}, where T'_n is a tree with minimal weight and trees T_n and T_{n+1} where T_{n+1} is a tree with minimal weight. We get from tree T'_n to T'_{n+1} and from tree T_n to T_{n+1} by replacing c, with frequency $f_a + f_b$, with vertices a, with frequency f_a, b, with frequency f_b, and their parent. We now consider how much this change affects the weight of the tree. In changing from T_n to T_{n+1}, we change from vertex c with frequency and length l, for a total weight of

$$l(f_a + f_b)$$

to vertex a, with frequency f_a and length $l+1$, and b, with frequency f_b and length $l+1$, for a total weight of

$$f_a(l+1) + f_b(l+1) = l(f_a + f_b) + f_a + f_b$$

Thus, the increase of $f_a + f_b$ in the weight of the tree. Similarly, in changing from T'_n to T'_{n+1} the increase in the weight of the tree is also $f_a + f_b$. Since the weight of T'_n is less than or equal to the weight of T_n and the increase in weight in changing to T'_{n+1} and T_{n+1} is the same, then the weight of T'_{n+1} is less than or equal to the weight of T_{n+1}. Therefore, T'_{n+1} is a minimal weight tree. ∎

Exercises

1. Given the following symbols and their codes,

symbol	code
c	111
d	101
e	01
i	11001
s	0010
u	1101

    (a) Encode the word *discuss*.

    (b) Encode the word *suede*.

    (c) Decode the word 1010111111100110101101.

    (d) Decode the word 11001001000101101010010.

2. Given the following symbols and their codes,

symbol	code
a	111
c	101
e	01
i	11011
g	0010
h	1001
m	0001
s	10001
t	11001

(a) Encode the word *mathematics*.
(b) Encode the word *mammas*.
(c) Decode the word 111001010011111000111001.
(d) Decode the word 100101110110010100111001.

3. Given the following symbols with their frequencies and three sets of codes,

symbol	frequency	code 1	code 2	code 3
c	7	111	010	111
d	8	101	11	101
e	10	01	001	001
i	2	11001	101	10001
s	9	0010	011	011
u	4	1101	000	100011

Find the weight of each code and select the most desirable one for minimal storage space.

4. Given the following symbols with their frequencies and three sets of codes,

symbol	frequency	code 1	code 2	code 3
a	8	111	1111	111
c	4	101	1110	011
e	13	01	1100	101011
i	9	11011	1001	11011
g	5	0010	1011	001
h	2	1001	1000	000
m	3	0001	1010	100
s	7	10001	0011	10111
t	11	11001	0111	10110

Find the weight of each code and select the most desirable one for minimal storage space.

5. Given the following symbols with their frequencies,

symbol	frequency
a	8
b	6
c	3
r	4

(a) Construct the Huffman tree.
(b) Determine the Huffman code.
(c) Find the weight of the code.
(d) Encode the word *crab*.
(e) Encode the word *bar*.
(f) Decode 1100111.
(g) Decode 10011110.

6. Given the following symbols with their frequencies,

symbol	frequency
a	8
b	5
e	10
k	1
r	3

(a) Construct the Huffman tree.
(b) Determine the Huffman code.
(c) Find the weight of the code.
(d) Encode the word *break*.
(e) Encode the word *barber*.
(f) Decode 11011011000.
(g) Decode 11110110111101101.

7. Given the following symbols with their frequencies,

symbol	frequency
a	8
d	7
e	12
k	1
r	4
s	6

(a) Construct the Huffman tree.
(b) Determine the Huffman code.
(c) Find the weight of the code.
(d) Encode the word *eraser*.
(e) Encode the word *darker*.
(f) Decode 10010110001100.
(g) Decode 0011011001111001.

8. Given the following symbols with their frequencies,

symbol	frequency
c	7
e	12
g	3
h	2
o	9
r	4
s	5
t	8
u	1

(a) Construct the Huffman tree.
(b) Determine the Huffman code.
(c) Find the weight of the code.
(d) Encode the word *ghost*.
(e) Encode the word *choose*.
(f) Decode 100010010000111101.
(g) Decode 10001001110111101.
(h) Decode 111000100001001100001.

9. Given the following symbols with their frequencies,

symbol	frequency
a	8
c	4
e	13
i	9
g	5
h	2
m	3
s	7
t	11

(a) Construct the Huffman tree.

(b) Determine the Huffman code.

(c) Find the weight of the code.

10. Given the following symbols with their frequencies,

symbol	frequency	symbol	frequency
a	19	m	7
d	8	n	9
e	20	o	12
i	11	r	11
g	3	s	16
h	1	t	14
l	5		

(a) Construct the Huffman tree.

(b) Determine the Huffman code.

(c) Find the weight of the code.

(d) Encode the words *moonlightandroses*.

(e) Encode the word *roosters*.

11. Given the following symbols with their frequencies,

symbol	frequency	symbol	frequency
a	20	n	7
d	9	o	13
e	25	r	15
f	4	s	11
g	8	t	5
h	2	u	14
l	6	w	1
m	10		

(a) Construct the Huffman tree.

(b) Determine the Huffman code.

(c) Find the weight of the code.

(d) Encode the words *eastofthesunandwestofthemoon*.

(e) Encode the word *sweater*.

12. Given the following symbols with their frequencies,

symbol	frequency	symbol	frequency
a	25	n	12
b	10	o	20
e	30	r	9
g	7	s	11
h	6	t	15
i	18	u	3
l	8	y	13

(a) Construct the Huffman tree.

(b) Determine the Huffman code.

(c) Find the weight of the code.

(d) Encode the words *lightsout*.

13. Prove Theorem 15.14: In any binary tree, the leaf path codes for the leaves in the tree are a prefix code.

15.4 Traversing Binary Trees

In this section we describe three different ways of traversing a binary tree so the item at each vertex is read exactly once. We shall concentrate on trees that represent arithmetic expressions, although these transversals have other uses. First, trees representing arithmetic expressions are constructed, and then the methods of traversing the trees are shown.

Trees that represent arithmetic expression have the property that all internal vertices are binary operators. First, consider the expression $A + B$. The arithmetic expression tree for this expression is shown in Figure 15.37.

We shall use a top-down approach, which we shall demonstrate by example. Consider the expression $(A + B) \times (C + D)$. Since this expression is the product of $A + B$ and $C + D$, we first form the tree in Figure 15.38. Since $A + B$ is the sum of A and B, and $C + D$ is the sum of C and D, we form the tree in Figure 15.39.

Figure 15.37

Figure 15.38

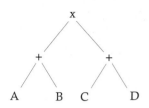

Figure 15.39

Figure 15.40

Each arithmetic expression that still contains a binary operation is really a binary operation on two arithmetic expressions (which may contain no binary operations). Beginning at the root, if the expression at a vertex has the form $E_1 \diamond E_2$ where $\diamond$ is a binary operation on expressions E_1 and E_2, replace this vertex with the tree in Figure 15.40.

Repeat this process with vertices E_1 and E_2. Continue until there are no vertices that are arithmetic expressions containing binary operations. We now show three ways to traverse an arithmetic expression tree. In each of these, we shall use the command **process**(n), where n is a node. We shall leave this term undefined, since its use will depend on what we want to achieve during the transversal of the vertices. For our purposes it is sufficient to consider the command as simply printing the symbol. We begin with the **inorder transversal**. In this transversal, we reverse the process that was used to create the tree. Given the tree in Figure 15.40 where $\diamond$ is a binary operation on expressions E_1 and E_2 we process (or print) this as $E_1 \diamond E_2$. The result is to form an expression in **infix form**, as described in Section 5.5 of Chapter 5.

The following algorithm describes the procedure for the inorder traversal. We shall abbreviate the left child of vertex v by left(v) and the right child of vertex v by right(v).

■ Algorithm inorder (*root*)

1. If left(*root*) exists, then inorder (left(*root*)).
2. Process (*root*).
3. If right(*root*) exists, then inorder (right(*root*)).

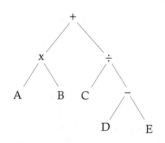

Figure 15.41

For example, consider the tree in Figure 15.41. Beginning with inorder(+), the first instruction calls inorder(×). The first instruction of inorder(×) then calls inorder(A). Since A has no left child, we proceed to step 2 of inorder(A) and process (print) A. We then proceed to step 3 of inorder (A), and since inorder (A) has no right child, inorder (A) is finished and we return to step 2 of inorder(×). Here we process (print) × and in step 3, call inorder(B). It may be observed that when we called inorder(A), the result was simply to process (print) the leaf A and return to the parent. Henceforth, when we have inorder(L), where L is a leaf, we shall simply abbreviate the three-step algorithm by saying **process leaf L**. Using this notation, we print B and return to inorder(×). At this point we have processed (printed)

$$A \times B$$

We have now completed inorder(×), so we return to inorder(+). (Also since we have surely beaten the concept to death by now, we shall simply say process, but keep in mind that this may well mean print.) We have now reached step 2 of inorder(+), so we process + and proceed to step 3 of inorder(+). In this step we call inorder(÷). In the first step of inorder(÷), we call inorder(C), so we process leaf C and return to step 2 of inorder(÷). We now process ÷, so we have now processed

$$A \times B + C \div$$

Now we return to step 3 of inorder(÷), and call inorder(−). In step 1 of inorder(−), we call inorder(D) and process leaf D. We then return to step 2 of inorder(−) and process −. We then go to step 3 of inorder(−) and call inorder(E). We process leaf E and then return to inorder(−). However, we have completed inorder(−), so we return to inorder(÷). But this also completes so we return to inorder(+). Finally, this completes inorder(+), and we are finished. We have processed

$$A \times B + C \div D - E$$

Figure 15.42

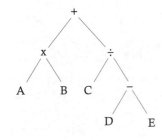

Figure 15.43

This expression is now in **infix form**, as described in Section 5.5. Unfortunately the expression now has no parentheses and, hence, as simply a printed expression, has no meaning. It may be recalled from Section 5.5, that this was the reason for switching to **prefix (Polish)** or **postfix (reverse Polish)** expressions. Fortunately, these are produced by the next two transversals.

We conclude our discussion of the infix transversal with the observation that if the infix transversal is applied to a binary search tree, it processes the items in alphabetical order.

We now consider **preorder transversal**. In this transversal, given the tree in Figure 15.42 where $\diamond$ is a binary operation on expressions E_1 and E_2. We process (or print) this as $\diamond E_1 E_2$. The result is to form an express in **prefix form**. This form is commonly used in the study of logic. It has the advantage that parentheses are not needed.

The following algorithm describes the procedure for the preorder traversal. As before, we shall abbreviate the left child of vertex v by left(v) and the right child of vertex v by right(v).

■ Algorithm preorder (*root*)

1. Process (*root*).
2. If left(*root*) exists, then preorder (left(*root*)).
3. If right(*root*) exists, then preorder (right(*root*)).

Again, we consider the tree in Figure 15.43. Beginning with preorder(+), the first instruction processes +. Step 2 calls preorder(×). The first instruction of preorder(×) processes ×. The second instruction then calls preorder(A). In step of preorder(A), we process A. At this point, we have processed

$$+ \times A$$

Since A has no left root, we proceed to step 3 of preorder(A). Since there is no right child, we return to preorder(×). We again observe, as in the case of inorder(A), that when we called preorder(A), the result was simply to process (print) the leaf A and return to the parent. Henceforth, when we have preorder(L), where L is a leaf, we shall simply abbreviate the three-step algorithm by saying **process leaf L**. We are now at step 3 of preorder(×), so we call preorder(B). We process leaf B and return to preorder(×). This concludes preorder(×), so we return to preorder(+). At this point, we have processed

$$+ \times AB$$

We are now at step 3 of preorder(+), so we call preorder(÷). The first step of preorder(÷) processes ÷. The second step calls preorder(C). We process leaf C and return to step 3 of preorder(÷). This instruction calls preorder(−). The first step of preorder(−) processes −. Step 2 calls preorder(D). We process leaf D and return to step 3 of preorder(−). At this point, we have processed

$$+ \times AB \div C - D$$

Step 3 of preorder(−) calls preorder(E). We process leaf E and return to preorder(−). But preorder(−) is finished so we return to preorder(÷). However, preorder(÷) is also finished, so we return to preorder(+). Finally preorder(+) is finished and we have processed

$$+ \times AB \div C - DE$$

which is expressed in **prefix form** or **Polish form** as described in Section 5.5 of Chapter 5.

Figure 15.44

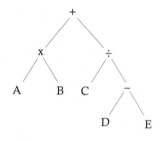

Figure 15.45

Finally, we consider **postorder transversal**. In this transversal, given the tree in Figure 15.44 where ◇ is a binary operation on expressions E_1 and E_2, we process (or print) this as $E_1 E_2 \diamond$. The result is to form an expression in **postfix form or reverse Polish form**, as described in Section 5.5. Anyone who has used a Hewlett-Packard calculator is familiar with postfix or reverse Polish notation. It also has the advantage that parentheses are not needed, which is the main reason for its use in a calculator.

The following algorithm describes the procedure for the postorder transversal. As before, we shall abbreviate the left child of vertex v by left(v) and the right child of vertex v by right(v).

■ **Algorithm postorder (*root*)**
1. If left(*root*) exists, then postorder(left(*root*)).
2. If right(*root*) exists, then postorder(right(*root*)).
3. Process(*root*).

Again, we consider the tree in Figure 15.45. Beginning with postorder(+), the first instruction calls postorder(×). The first instruction of postorder(×) calls postorder(A). Since A has no left root, we proceed to step 2 of postorder(A). Since there is no right child, we proceed to step 3 of postorder(A), where we process A and return to postorder(×). We again observe, as in the case of inorder(A) and preorder (A), that when we called postorder(A), the result was simply to process (print) the leaf A and return to the parent. Henceforth, when we have postorder(L), where L is a leaf, we shall simply abbreviate the three-step algorithm by saying **process leaf L**. We are now at step 2 of postorder(×), so we call postorder(B). We process leaf B and return to step 3 of postorder(×). We process × and, since this concludes postorder(×), we return to postorder(+). At this point, we have processed

$$AB\times$$

We are now at step 2 of postorder(+), so we call postorder(÷). The first step of postorder(÷) calls postorder(C). We process leaf C and return to step 2 of postorder(÷). This instruction calls postorder(−). The first step of postorder(−) calls postorder(D). We process leaf D and return to step 2 of postorder(−). The second step of postorder(−) calls postorder(E). We now process E and return to step 3 of postorder(−). We process − and return to step 3 of postorder(÷). At this point, we have processed

$$AB \times CDE-$$

In step 3 of postorder(÷), we process ÷ and return to step 3 of postorder(+). We process + and are finished. We have processed

$$AB \times CDE - \div +$$

and have, thus, transformed the expression to **postfix form**.

Once we can transverse a tree, it is a simple matter to determine if two binary trees are isomorphic. As we transverse the two trees simultaneously, we simply check to see if one tree has a vertex whenever the other tree has a vertex. As we transverse the two trees simultaneously, processing the vertices will mean checking both trees to see if the corresponding vertices are there. It seems reasonable to use prefix transversal to transverse the trees since there is no need to worry about children of vertices if one of the parent vertices is not there. When we process a vertex V, we will assign N to the variable $exists(V)$ if the vertex V is not there and Y to the variable $exists(V)$ if the vertex V is there. We then check the corresponding vertices and, if $exists(V)$ is Y for one vertex and N, for the corresponding vertex, then the two trees are not isomorphic.

We shall have a variable called *Iso*, which accepts the values T and F. It is initially assigned the value T, but if at any point it is seen that the trees are not isomorphic, it is assigned the value F. We then have the following algorithm for two trees T_1, with root r_1, and T_2, with root r_2.

Let $Iso = T$.

■ Algorithm Isomorphic Binary Tree (r_1, r_2)

1. Process r_1 and r_2
2. If $exists(r_1) = Y$ and $exists(r_2) = N$ or $exists(r_1) = N$ and $exists(r_2) = Y$, then $Iso = F$.
3. If $Iso = F$, then conclude Isomorphic Binary Tree(r_1, r_2).
4. If $exists(r_1) = Y$ and $exists(r_2) = Y$, then Isomorphic Binary Tree(left(r_1), left(r_2)).
5. If $exists(r_1) = Y$ and $exists(r_2) = Y$, then Isomorphic Binary Tree(right(r_1), right(r_2)).

Lets try this algorithm out on the following trees T_1 and T_2, respectively, shown in Figure 15.46.

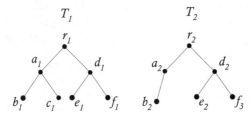

Figure 15.46

We process r_1 and r_2. Since $exists(r_1) = Y$ and $exists(r_2) = Y$, we proceed to step 3 of Isomorphic Binary Tree(r_1, r_2). Here Isomorphic Binary Tree(left(r_1), left(r_2)) is called, so we process a_1 and a_2. Since $exists(a_1) = Y$ and $exists(a_2) = Y$, we proceed to step 3 of Isomorphic Binary Tree(a_1, a_2). Here Isomorphic Binary Tree(left(a_1), left(a_2)) is called, so we process b_1 and b_2. Since $exists(b_1) = Y$ and $exists(b_2) = Y$, we proceed to step 3 of Isomorphic Binary Tree(a_1, a_2). Here Isomorphic Binary Tree(left(b_1), left(b_2)) is called, so we process left(b_1) and left(b_2). Since $exists(\text{left}(b_1)) = N$ and $exists(\text{left}(b_2)) = N$, we return to step 4 of Isomorphic Binary Tree(a_1, a_2), which calls Isomorphic Binary Tree(right(a_1), right(a_2)). We process c_1 and right(a_2). Since $exists(c_1) = Y$ and $exists(\text{right}(a_2)) = N$, $iso = F$, and no further relevant processing can take place. Hence, the two trees are not isomorphic.

Computer Science Application

The order in which the nodes of a binary search tree are transversed is not always important, and in those cases there is no advantage in selecting one of preorder, inorder, or postorder over the other two. There are occasions, however, when the order does matter. The following function deletes the nodes of binary search tree one at a time but in such a way that the links to the subtrees are not destroyed prematurely.

```
function DisposeOfTree(TreeNodePointer T)

    if T = null then
        return
    end-if

    DisposeOfTree(T->leftlink)
    DisposeOfTree(T->rightlink)

    delete the node T is pointing to

end-function
```

The visit at each node is the actual deallocation of the root of the tree that the parameter is pointing to. Preorder and inorder transversal would cause one or both of the links T->leftlink and T->rightlink to be deleted before all of the nodes accessible through them had been deleted. Inorder transversal should be used whenever the nodes of the tree need to transversed in ascending order of key field. Preorder transversal would have to be used to make a copy of a binary search tree.

Exercises

1. Given the tree in Figure 15.47,

Figure 15.47

(a) Give the inorder form of the vertices of the tree.

(b) Give the preorder form of the vertices of the tree.

(c) Give the postorder form of the vertices of the tree.

2. Given the tree in Figure 15.48,

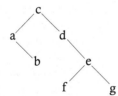

Figure 15.48

(a) Give the inorder form of the vertices of the tree.

(b) Give the preorder form of the vertices of the tree.

(c) Give the postorder form of the vertices of the tree.

3. Given the tree in Figure 15.49,

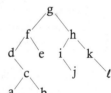

Figure 15.49

(a) Give the inorder form of the vertices of the tree.

(b) Give the preorder form of the vertices of the tree.

(c) Give the postorder form of the vertices of the tree.

4. Given the tree in Figure 15.50,

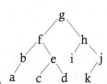

Figure 15.50

(a) Give the inorder form of the vertices of the tree.

(b) Give the preorder form of the vertices of the tree.

(c) Give the postorder form of the vertices of the tree.

5. Given the tree in Figure 15.51,

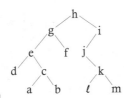

Figure 15.51

   (a) Give the inorder form of the vertices of the tree.
   (b) Give the preorder form of the vertices of the tree.
   (c) Give the postorder form of the vertices of the tree.

6. Given the tree in Figure 15.52,

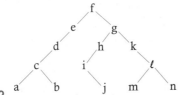

Figure 15.52

   (a) Give the inorder form of the vertices of the tree.
   (b) Give the preorder form of the vertices of the tree.
   (c) Give the postorder form of the vertices of the tree.

7. Given the infix expression $(a + b) \times c$,
   (a) Find a binary tree to represent this expression.
   (b) Give the prefix form of this expression.
   (c) Give the postorder form of this expression.

8. Given the infix expression $(a + b) \times (c + d)$,
   (a) Find a binary tree to represent this expression.
   (b) Give the prefix form of this expression.
   (c) Give the postorder form of this expression.

9. Given the infix expression $((a + b) \times c) \div (c + d)$,
   (a) Find a binary tree to represent this expression.
   (b) Give the prefix form of this expression.
   (c) Give the postorder form of this expression.

10. Given the infix expression $(a \div b) + (c \times (d + e))$,
    (a) Find a binary tree to represent this expression.
    (b) Give the prefix form of this expression.
    (c) Give the postorder form of this expression.

11. Given the infix expression

    $(a \times (b + (d - (e + f)))) \div ((g + h) \times (i + j))$

    (a) Find a binary tree to represent this expression.
    (b) Give the prefix form of this expression.
    (c) Give the postorder form of this expression.

12. Given the infix expression $(a + b) \times ((((e - f) + g) \div h) - i)$,
    (a) Find a binary tree to represent this expression.
    (b) Give the prefix form of this expression.
    (c) Give the postorder form of this expression.

13. Given the infix expression $(((a + b) - c) \times d) \div (e - (f \times (g + h)))$,
    (a) Find a binary tree to represent this expression.
    (b) Give the prefix form of this expression.
    (c) Give the postorder form of this expression.

14. Find a binary tree such that the inorder transversal gives *acbde* and the postorder transversal gives *abced*.

15. Find a binary tree such that the inorder transversal gives *acbde* and the postorder transversal gives *acedb*.

16. Find a binary tree such that the inorder transversal gives *acbdefg* and the preorder transversal gives *dcbfeg*.

17. Find a binary tree such that the inorder transversal gives *acbdefg* and the preorder transversal gives *edcabfg*.

18. Find a tree with at least three vertices such that the infix expression is the same as the prefix notation.

19. Find a tree with at least three vertices such that the infix expression is the same as the postfix notation.

20. Construct three nonisomorphic binary trees that have *abcd* as the postfix expression.

21. Construct three nonisomorphic binary trees that have *abcd* as the preorder expression.

22. Prove that if the inorder and postorder expressions for a tree are given, then the construction of the binary tree is unique.

23. If the arithmetic operations are all binary, which of the following are true for the representation tree?
    (a) The binary tree is balanced.
    (b) The binary tree is full.
    (c) A leaf can be an arithmetic expression.
    (d) An internal vertex must be an arithmetic expression.

24. Write an algorithm that places parentheses in the proper place in the infix arithmetic expression for the inorder transversal of a tree.

25. Give necessary and sufficient conditions for an expression to be a postfix arithmetic expression.

26. Is the exponential arithmetic operation binary or unary? Why? Give an example of a unary arithmetic operation.

27. Describe where and how the isomorphic binary tree algorithm will show that the trees in Figure 15.53 are not isomorphic.

28. Describe where and how the isomorphic binary tree algorithm will show that the trees in Figure 15.54 are not isomorphic.

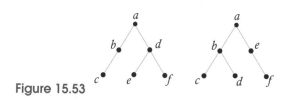

Figure 15.53

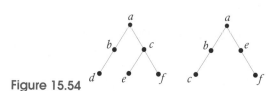

Figure 15.54

15.5 Spanning Trees

In Section 6.3, we defined a tree T to be a **spanning tree** of a subgraph of G such that every vertex in G is a vertex in T. It was shown that every connected graph has a spanning tree.

At this point we discuss two methods of creating a spanning tree. The first is the breadth-first search method and the second is the depth-first search method. We begin with the breadth-first search method. In this method, given a graph G, we select any vertex v_0 of the graph and designate it as the root of the tree T. For each vertex v of G that is adjacent to v_0, we add the vertex and v the edge $\{v, v_0\}$ to the tree T. These are the vertices of level 1. We then take each vertex v_i of level 1, and for each vertex v_j of G that is adjacent to that v_i and has not already been selected, add v_j and the edge $\{v_i, v_j\}$ to the tree T. The vertices added in this step are the edges of level 2. We continue this process until there are no vertices of G left to add to the tree. By construction, T is a tree. If the distance from v_0 to a vertex v of G is n, then it will be added to the tree at level n. Hence, T is certainly a spanning tree.

In constructing an algorithm, we need to know when a vertex and an edge have been added to a tree, both to describe the tree and to eliminate the possibility of using a vertex more than once so that we no longer have a tree. We also need to know the level at which a vertex is added, since it is more efficient to use only the vertices selected at the last level to determine the leaves at the next level. It is also necessary to be able to select vertices one at a time for determining new vertices to add to the tree. There are several ways of performing some of these tasks, including using arrays and queues, so that we will not be specific in determining the "next" vertex to use. If desired, one can just assume that the vertices of G have been preordered. We shall let V and E denote the vertices and edges of G, and let V^T and E^T denote the vertices and edges of T. When a vertex v is added to the tree, $L(v)$ will denote the level at which it is added.

■ **Algorithm Breadth-First Tree Search (G)**

1. Select an arbitrary element v_0 of G. Let $v_0 \in V^T$ and $L(v_0) = 0$.

2. For all $v \in V - V^T$ such that v is adjacent to v_0, let $v \in V^T$, $\{v_0, v\} \in E^T$, and $L(v) = 1$.

3. Let $i = 1$.

4. Select $v_j \in V^T$ such that $L(v_j) = i$.

5. Select $v \in V - V^T$. If v is adjacent to v_j, let $v \in V^T$, $\{v_0, v\} \in E^T$ and $L(v) = i + 1$.
6. Continue step 5 until all elements of $V - V^T$ have been examined. (Note that $V - V^T$ is constantly changing.)
7. Repeat steps 4, 5, and 6 until all v_j such that $L(v_j) = i$ have been selected.
8. Let $i = i + 1$.
9. Repeat steps 4–8 until $V = V^T$.

EXAMPLE 15.21

Consider the graph in Figure 15.55. Suppose v_0 is selected as the first vertex. Then $L(v_0) = 0$, and $v_0 \in V^T$. Since v_1 is adjacent to v_0, let $v_1 \in V^T$, $\{v_0, v_1\} \in E^T$, and $L(v_1) = 1$. Also v_2 is adjacent to v_0, so let $v_2 \in V^T$, $\{v_0, v_2\} \in E^T$, and $L(v_2) = 1$. Finally v_3 is adjacent to v_0, so let $v_3 \in V^T$, $\{v_0, v_3\} \in E^T$, and $L(v_3) = 1$. At this point, we have the tree in Figure 15.56.

We now consider the vertices v such that $L(v) = 1$. Starting with $L(v_1) = 1$, we find vertices that have not been used that are adjacent to $L(v_1)$. Since v_5 is adjacent to v_1, let $v_5 \in V^T$, $\{v_1, v_5\} \in E^T$, and $L(v_5) = 2$. There are no further unused vertices adjacent to v_1, so we go to v_2. Since v_4 is adjacent to v_2, let $v_4 \in V^T$, $\{v_2, v_4\} \in E^T$, and $L(v_4) = 2$. We have now used all the vertices so we have finished and have the tree in Figure 15.57.

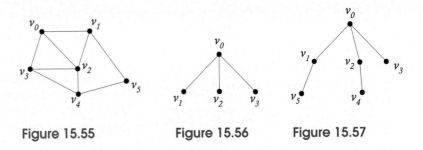

Figure 15.55   Figure 15.56   Figure 15.57

EXAMPLE 15.22

Consider the graph in Figure 15.58.

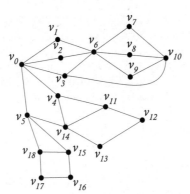

Figure 15.58

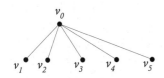

Figure 15.59

As before we pick v_0 as the initial vertex. Then $L(v_0) = 0$, and $v_0 \in V^T$. Since v_1, v_2, v_3, v_4, v_5 are all adjacent to v_0, for $1 \le i \le 5$, let $v_i \in V^T$ $\{v_0, v_i\} \in E^T$, and $L(v_i) = 1$. This gives us the tree in Figure 15.59.

We now start at level 1 with v_1. At this level, we have v_6 adjacent to v_1, v_{10} is adjacent to v_3, v_{11} and v_{14} are adjacent to v_4, and v_{15} and v_{18} are adjacent to v_5. Therefore, $v_6, v_{10}, v_{11}, v_{14}, v_{15}, v_{18} \in V^T$, $L(v_6) = L(v_{10}) = L(v_{11}) = $

$L(v_{14}) = L(v_{15}) = L(v_{18}) = 2$, and $\{v_1, v_6\}, \{v_3, v_{10}\}, \{v_4, v_{11}\}, \{v_4, v_{14}\}, \{v_5, v_{15}\},$ $\{v_5, v_{18}\} \in E^T$. This gives us the tree in Figure 15.60.

We now start at level 2. At this level we have v_7, v_8, v_9 adjacent to v_6, v_{12} adjacent to v_{11}, v_{13} adjacent to v_{14}, v_{16} adjacent to v_{15}, and v_{17} adjacent to v_{18}. Therefore, $v_7, v_8, v_9, v_{12}, v_{13}, v_{16}, v_{17} \in V^T$, $\{v_6, v_7\}, \{v_6, v_8\}, \{v_6, v_9\}, \{v_{11}, v_{12}\},$ $\{v_{14}, v_{13}\}, \{v_{16}, v_{13}\}, \{v_{17}, v_{18}\} \in E^T$, and $L(v_7) = L(v_8) = L(v_9) = L(v_{12}) = L(v_{13}) = L(v_{16}) = L(v_{17}) = 3$. This gives us the tree in Figure 15.61.

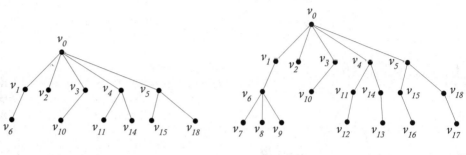

Figure 15.60   **Figure 15.61**   ∎

In the breadth-first search, we first found all vertices adjacent to a given vertex before proceeding to the next level. In the depth-first search, we try to form as long a path for the tree as possible. When a path reaches a dead end so we have created a leaf, we back up to the parent of the leaf and again try to form a path. We return to a parent only after trying all possible paths from a child of the parent. In terms of the tree being created, we are trying to reach the deepest or largest level possible. Our algorithm will begin at a given vertex of the graph, which we shall designate as the root. We select a vertex, which is adjacent to the root and forms an edge of the tree. We then select a vertex, which is adjacent to the previously selected vertex and forms a new edge. As we continue this process, we need to mark each vertex as it is selected, so that each vertex will be used only once. If at any vertex v we select another vertex w and find that vertex w has already been added to the tree, then the edge $\{v, w\}$ between these two vertices cannot be added to the tree. We shall call such an edge a **back edge** and we shall also keep track of these for future use. Before declaring an edge to be a back edge, we do have to be careful to check that the vertex w is not the parent of v, since in this case, the edge $\{v, w\}$ is already an edge of the tree. Essentially we avoid considering the parent of v when selecting a vertex w as a possible new vertex. If $\{v, w\}$ is a back edge, we remain at v, and select a new adjacent vertex if possible. By definition, any edge of the graph is eventually either an edge of the tree or a back edge.

In our algorithm, we call the set of edges to the tree TREE EDGE, and the set of back edges we shall call BACK EDGE. At the beginning, we shall assume that all vertices are labeled n for "new" and when it is added to the tree, we shall change the label to u for "used."

■ **Algorithm Depth-First Tree Search (G)**

1. Label each vertex of G with the symbol n.
2. Select an arbitrary element v_0 of G and let this be the root of the tree.
3. Change the label of v_0 from n to u and let $v = v_0$.
4. While there is a vertex adjacent to v that has not been selected, do the following:

    (a) Select a vertex w that is adjacent to v.

(b) If w has label n, add (v, w) to TREE EDGE, change the label of w to u, let $w = v$, and repeat step 4.

(c) If w has label u, and w is not the parent of v, add (v, w) to BACK EDGE, and repeat step 4.

5. If $v \neq v_0$, let $v = parent(v)$ and repeat step 4.
6. End.

Note that as long as any vertex v has a vertex w adjacent to it that has not been used, we extend the path from v to w. It is only when we can go no further that we get to step 5 and back up to the parent of the vertex v.

EXAMPLE 15.23

Consider the graph in Figure 15.62.

We arbitrarily select a as the root. We change the label of a from n to u. Since b is adjacent to a, and b has label n, we add edge $\{a, b\}$ to TREE EDGE and change the label of b to u as shown in Figure 15.63.

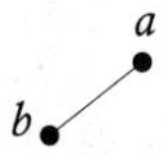

From b, we now go to d, since it is adjacent to b. The vertex d has label n, so we add edge $\{b, d\}$ to TREE EDGE, and change the label of d to u as shown in Figure 15.64.

At this point we select a vertex adjacent to d. We can select a, f, or g. The selection will determine the form of the tree so that the depth-first search does not produce a unique tree. Suppose we next select g. The vertex g has label n, so we add edge $\{d, g\}$ to TREE EDGE, and change the label of g to u as shown in Figure 15.65.

From vertex g, we select vertex f, since it is adjacent to g. The vertex f has label n, so we add edge $\{g, f\}$ to TREE EDGE and change the label of f to u as shown in Figure 15.66.

From vertex f, we select vertex d, since it is adjacent to f. However, vertex d already has label u, so we add edge $\{d, f\}$ to BACK EDGE as shown in Figure 15.67.

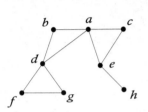

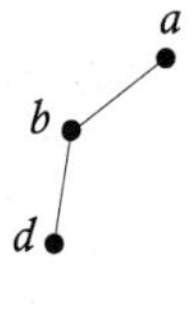

Figure 15.62

Figure 15.63

Figure 15.64

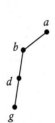

Figure 15.65

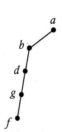

Figure 15.66

Figure 15.67

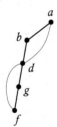

Figure 15.68

Since there are no other vertices to check that are adjacent to f, except the parent, we return to vertex g. (We shall no longer mention the parent as a possible adjacent vertex.) But there are also no other vertices to check that are adjacent to g, so we return to vertex d. The only vertex to check at d is a, but vertex a already has label u, so we add edge $\{d, f\}$ to BACK EDGE, and return to b as shown in Figure 15.68.

But there are no other adjacent vertices to consider at b, so we return to a. Note that if it had been possible to reach every vertex by a path from b, we would not have returned to a until the entire tree had been completed. Having returned to a, we can select either c or e. Suppose we select vertex c. Since vertex c has label n, we add edge $\{a, c\}$ to TREE EDGE and change the label of c to u as shown in Figure 15.69.

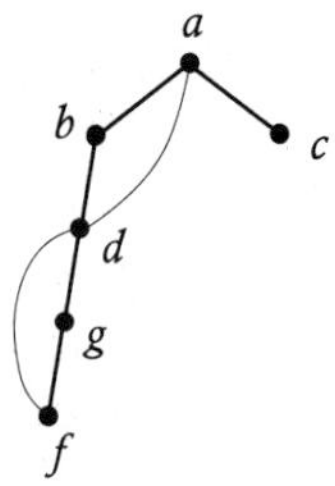

Figure 15.69

Since e is adjacent to c and has label n, add edge $\{c, e\}$ to TREE EDGE, and change the label of e to u as shown in Figure 15.70. Now assume we select vertex a, since it is adjacent to e. But vertex a already has label u, so we add edge $\{e, a\}$ to BACK EDGE, and return to e as shown in Figure 15.71.

Since h is adjacent to e, add edge $\{e, h\}$ to TREE EDGE, and change the label of h to u as shown in Figure 15.72.

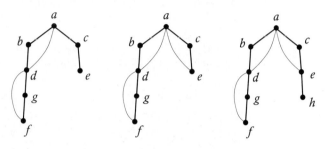

Figure 15.70   Figure 15.71   Figure 15.72

There are no further vertices to check at h, so we return to e. But there are no further vertices to check at e, so we return to c. Since there are no further vertices to check at c, we return to a. There are no further vertices to check at a, so we are finished. ∎

EXAMPLE 15.24

Find the spanning tree for the graph in Figure 15.73. Assume that all the vertices have been labeled n. Let v_0 be the first vertex selected. Add edges (v_0, v_5), (v_5, v_{18}), (v_{18}, v_{17}), (v_{17}, v_{16}), and (v_{16}, v_{15}) to TREE EDGE and change the labels of $v_0, v_5, v_{18}, v_{17}, v_{16}$, and v_{15} to u. At vertex v_{15} we find that v_5 and v_{18} are both adjacent to v_{15}, but they have already been labeled u, so edges (v_{15}, v_{18}) and (v_{15}, v_5) are added to BACK EDGE. We now return to v_{16}, then v_{17}, then v_{18}, and finally v_5. At v_5, we add (v_5, v_{14}), (v_{14}, v_{13}), (v_{13}, v_{12}), (v_{12}, v_{11}), and (v_{11}, v_4) to TREE EDGE and change the labels of $v_{14}, v_{13}, v_{12}, v_{11}$, and v_4 to u. Vertices v_0 and v_{14} are both adjacent to v_4, but they have already been labeled u, so edges (v_4, v_0), and (v_4, v_{14}) are both added to BACK EDGE. We return to vertex v_{11}, where we find v_{14} adjacent to v_{11}. But v_{14} has already been labeled u, so we add edge (v_{11}, v_{14}) to BACK EDGE. We now backtrack until we reach v_0. At this point we have the tree and back edges shown in Figure 15.74.

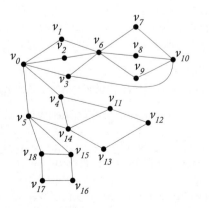 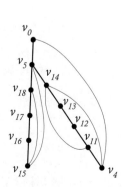

Figure 15.73   Figure 15.74

At v_0, we select a vertex, say v_3, adjacent to v_0, so we add (v_0, v_3) to TREE EDGE and change the label of v_3 to u. We continue this process, adding (v_3, v_6), (v_6, v_9), (v_9, v_{10}), and (v_{10}, v_8) to TREE EDGE and change the labels of v_6, v_9, v_{10}, and v_8 to u. We are now at vertex v_8. Vertex v_6 is adjacent to v_8, but v_6 has already been labeled u, so we add edge (v_8, v_6) to BACK EDGE. Returning to v_{10}, we find v_7 adjacent to v_{10}, so we add (v_{10}, v_7) to TREE EDGE, and changing the label of v_7 to u. The vertex v_7 is adjacent to v_6, but v_6 has already been labeled u, so we add edge (v_7, v_6) to BACK EDGE. Returning to v_{10}, we find v_3 adjacent to v_{10}. But v_3 has already been labeled u, so we add edge (v_{10}, v_3) to BACK EDGE. We now backtrack to vertex v_6. Vertex v_1 is adjacent to v_6, so we add (v_6, v_1) to TREE EDGE, and changing the label of v_1 to u. At v_1, we find v_0 adjacent to v_1, and since v_0 has already been labeled u, so we add edge (v_1, v_0) to BACK EDGE. Returning to v_6, we find v_2 is adjacent to v_6, so we add (v_6, v_2) to TREE EDGE and change the label of v_2 to u. At v_2, we find v_0 adjacent to v_2, and since v_0 has already been labeled u, we add edge (v_2, v_0) to BACK EDGE. We have now finished the tree and in Figure 15.75, we show the tree and its back edges in.

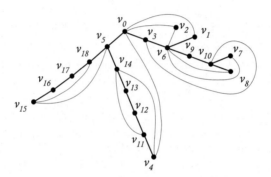

Figure 15.75

If a graph is not connected, then the algorithm for the depth-first spanning tree will not span the entire graph but only the component containing the vertex selected as the root. For this reason, the depth-first spanning tree can be used to test whether or not a graph is connected. Simply run the algorithm and if any vertices are not reached, then the graph is not connected. The algorithm can also be used to construct a tree for each component of a graph that is not connected. Simply select a vertex as the root of the first tree as before and construct a spanning tree. Then select one of the vertices that was not reached and construct a second tree. Continue this process until no vertices remain. The forest of spanning trees constructed in this manner is called a **spanning forest**.

The depth-first spanning tree can also be used to test a graph for articulation points. To show how this is done, we need the following theorem.

THEOREM 15.25 If T is a depth-first spanning tree of a graph $G(V, E)$ and $\{a, b\}$ is an edge of $G(V, E)$, then a is a descendant of b or conversely.

Proof If $\{a, b\}$ is an edge of T, the result is immediate. If not, then one of the vertices, say a, must be placed on the tree first. But since b has not been placed on the tree, using step 4, the depth-first search will continue searching from a until it finds b, and b will be a descendant of a. ∎

THEOREM 15.26 Let T be a depth-first spanning tree of a graph $G(V, E)$. A vertex of $a \in V$ is an articulation point of $G(V, E)$ if and only if either (1) the vertex a is the root of the tree T and has more than one child or (2) the vertex a is not the root of T and there exists a child c such that there is no back edge between c or any of its descendants and a proper ancestor of a.

Proof (1) If a is the root of T and has only one child c, then there is a path from any vertex in T and c. Therefore, given two vertices v and w, neither of which is a, there is a path from v to c to w that does not pass through a, and a is not an articulation point. If root a has two children, c and c', then if the depth-first search begins with c, it can reach c' only by returning to a, and there is no path from c to c' that does not pass through a. (2) Assume there exists a child c of a such that there is no back edge between c or any of its descendants and a proper ancestor of a. Let p be the parent of a. Since there is no back edge between c or any of its descendants and a proper ancestor of a, by the previous theorem, every edge from a or one of its descendants must go to a or one of its descendants. Thus, every path beginning at c must end at c or one of its descendants or pass through a. Therefore, the only path from c to p is through a, and a is an articulation point. Conversely, assume that for every child c of a, there is a back edge from c to a proper ancestor of a. Let c and c' be children of a, and b and b' be proper ancestors of a such that $e = \{d, b\}$ and $e = \{d', b'\}$ are back edges where d and d' are descendants of c and c', respectively. Thus, there is a path $p_{c,b}$ from c to d to b and a path $p_{c',b'}$ from c' to d' to b'. Since there is a path p from b to b' that does not pass through a, path $p_{c,b}$ followed by path p and the reverse of path $p_{c',b'}$ is a path from c to c' that does not pass through a. Let v and w be vertices of $G(V, E)$. If v and w are both descendants of a, then there are paths, which do not pass through a, from v and w to c and c', respectively, where c and c' are children of a. Since there is also a path from c and c', which does not pass through a, then there is a path from v to c to c' to w that does not pass through a. If v is a descendant of a and w is not, since there is a path from v to child c of a, a path $p_{c,b}$ from c to b where b is a proper ancestor of a, and a path along the tree from b to w, and none of these paths contain a, there is again a path from v to w that does not contain a. If neither v nor w is a descendant of a, then there is a path along the tree from v to w that does not contain a. Thus, there is always a path from v to w that does not contain a, and a is not an articulation point. ∎

To find the articulation points of a graph G, we want to take advantage of the previous theorem and find out if there exists a child c of T such that there is no back edge between c or any of its descendants and a proper ancestor of a. We first need some sort of ordering of the vertices so that if a vertex a is a proper descendant of a vertex b, then the value assigned to a is greater than the value assigned to b. We shall do this by counting the vertices as they are first reached on the tree. Thus, if c is the nth vertex reached on the tree, we shall assign it a count value n, which we shall denote by $CV(c)$. Using this ordering, we now want to assign to a vertex c the lowest value of any vertex a such that there is a back edge between a and c or any of its descendants. We shall denote this value by $BE(c)$ and call it the BE value of c. We noted previously that in the depth-first search, as long as any vertex v has a vertex w adjacent to it that has not been used, we extend the path from v to w. It is only when we can go no further that we leave step 4. We wish for our new algorithm to do the same except that it assigns a count value $CV(w)$ as the new vertex w is added. We also temporarily assign $BE(w) = CV(w)$. We now begin a bottom-up approach for determining the BE value for a vertex w. When we reach a leaf l of the tree, we compare its BE value and the count value or CV values of all of the vertices adjacent to l, except its parent, and assign to $BE(l)$ the minimum of these values. When we return to a parent p, we compare its current CV value with the BE value of the child c from which the algorithm returned. If $BE(c) > CV(p)$, then no descendant of c has a back edge with a proper ancestor of p, and by the previous theorem, p is an articulation point.

We now assign to $BE(p)$ the minimum of its current value and the value of $BE(c)$. Finally when there are no more new edges from p, we let $BE(p)$ be the minimum of its current value and the CV values of all of the vertices such that there is a back edge from p to these vertices. We continue until the depth-first search is completed.

Note that if, when we are searching the tree and reach an articulation point a, we remove the edges of the tree that have not already been removed, whose vertices are a and its descendants, but leaving vertex a, we have the spanning tree for a bicomponent of the graph G.

We formally express the procedure in the following algorithm. We shall use the recursive property of letting an algorithm call itself. We shall also have a counter i originally set at 0. As before, we shall assume each vertex begins with the label n.

■ **Algorithm DFT Articulation Point Search (v)**

1. Label v with the symbol u.
2. i gets value $i + 1$.
3. Set $BE(v) = CV(v) = i$.
4. While there is a vertex adjacent to v that has not been selected, do the following:

    (a) Select a vertex w that is adjacent to v.
    (b) If w has label n,
        (i) Add (v, w) to TREE EDGE (to form bicomponents).
        (ii) Let $v = \text{parent}(w)$.
        (iii) Call DFT articulation point search (w).
        (iv) If $BE(w) \geq CV(v)$, then v is an articulation point. The edges below v are removed that span a component.
        (v) Set $BE(v) = \min(BE(v), BE(w))$.
    (c) If w has label u and w is not the parent(w), $BE(v) = \min(BE(v), CV(w))$.

5. End.

EXAMPLE 15.27   Again consider the graph in Figure 15.76 and the corresponding tree in Figure 15.77.

As the path *abdgf* of the tree is formed, we define $BE(a) = CV(a) = 1$, $BE(b) = CV(b) = 2$, $BE(d) = CV(d) = 3$, $BE(g) = CV(g) = 4$, and $BE(f) = CV(f) = 5$. At vertex f, we find a back edge from f to d, and set $BE(f) = CV(d) = 3$. Returning to vertex g, since f is the child of g, $BE(g) = \min(BE(f), BE(g)) = \min(4, 3) = 3$. Since $BE(g) = 3 \geq CV(d)$, and g is a child of d, d is an articulation point and the tree in Figure 15.78 formed by edges $\{f, g\}$ and $\{d, g\}$ span the bicomponent in Figure 15.79.

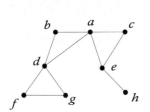

Figure 15.76

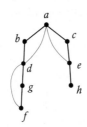

Figure 15.77

Figure 15.78

Figure 15.79

Figure 15.80

Figure 15.81

Returning to vertex d, as g is a child of d, we first take $BE(d) = \min((BE(g), BE(d)) = \min(3, 3) = 3$. Then, since there is a back edge from d to a, set $BE(d) = CV(a) = 1$. Returning to b, since d is the child of b, set $BE(b) = \min(BE(d), BE(b)) = \min(1, 2) = 1$. Since $BE(b) \geq CV(a)$, a is an articulation point (which we already knew since it is a root with two children), and the tree in Figure 15.80 formed by the edges $\{d, b\}$ and $\{b, a\}$ span the bicomponent in Figure 15.81.

Returning to vertex a, since b is a child of a, we compare their BE values and find that $BE(a) = \min((BE(a), BE(b)) = \min(1, 1) = 1$, so $BE(a)$ remains unchanged. As the path $aceh$ of the tree is formed set $BE(c) = CV(c) = 6$, $BE(e) = CV(e) = 7$, and $BE(h) = CV(h) = 8$. Since vertex h has no back edges, $BE(h)$ remains unchanged. Since $BE(h) \geq CV(e)$ and h is a son of e, e is an articulation point and edge $\{e, h\}$ is a spanning tree for the bicomponent in Figure 15.82.

Returning to e, since h is the son of e, we now have $BE(e) = \min(BE(e), BE(h)) \min(7, 8) = 7$, so $BE(e)$ remains unchanged. Since there is a back edge from e to a, $BE(e) = CV(a) = 1$. $BE(c) = \min(BE(c), BE(e)) = \min(1, 6) = 1$. Since $BE(e) \geq CV(a)$, as expected a is again found to be an articulation point and the tree in Figure 15.83 formed by the edges $\{a, c\}$ and $\{c, e\}$ spans the bicomponent in Figure 15.84.

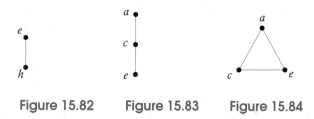

Figure 15.82 Figure 15.83 Figure 15.84

Finally, we return to vertex a and we are finished. ∎

A set of n vertices is **labeled** if each of them is assigned a unique integer between 1 and n. Thus, we have the vertices $v_1, v_2, v_3, \ldots, v_n$. If we consider all graphs whose n vertices are labeled, how many spanning trees can be formed? In other words, how many spanning trees are there for n labeled vertices? The answer to this is given by Cayley's tree formula. Note that this is the number of spanning trees for K_n.

THEOREM 15.28 (**Cayley's tree formula**) The number of spanning trees for n labeled vertices is n^{n-2}.

Proof We show this by showing that each spanning tree can be represented by a unique sequence $a_1, a_2, a_3, \ldots, a_{n-2}$ of length $n-2$ such that for $1 \leq k \leq n-2$, a_k can be any integer between 1 and n, and conversely for any sequence $a_1, a_2, a_3, \ldots, a_{n-2}$ of length $n-2$ such that for $1 \leq k \leq n-2$, and $1 \leq a_k \leq n$, there corresponds a unique spanning tree. If we can do this then, we are really asking how many ways we can fill $n-2$ blanks when there are n choices for each blank and there are n^{n-2} ways. Rather than attempt a formal proof, we shall state an algorithm for producing a sequence from a spanning tree and an algorithm for producing a spanning tree from a sequence. We shall then show by example that if we start with a spanning tree, use the algorithm to produce the sequence, and then apply the algorithm to the sequence to produce the spanning tree, we again have the same spanning tree. ∎

We begin with the algorithm for converting spanning trees to sequences.

■ Tree to Sequence Algorithm (T) for $n \geq 3$

1. To pick a_1, select leaf v_j with smallest value j. There exists a unique k so that $\{v_j, v_k\}$ is an edge in the tree. Remove this edge and let $a_1 = k$.
2. Assume a_{i-1} has been picked. To pick a_i, from the remaining tree, select leaf v_j with smallest value j. There exists a unique k so that $\{v_j, v_k\}$ is an edge in the tree. Remove this edge and let $a_i = k$.
3. Continue until a_{n-2} has been selected.

EXAMPLE 15.29

Let T be the tree in Figure 15.85. The vertex v_2 is the leaf with the smallest subscript, so remove edge $\{v_2, v_1\}$ and let $a_1 = 1$ as shown in Figure 15.86. Since v_3 is now the leaf with the smallest subscript in the remaining tree, remove edge $\{v_3, v_4\}$ and let $a_2 = 4$ as shown in Figure 15.87.

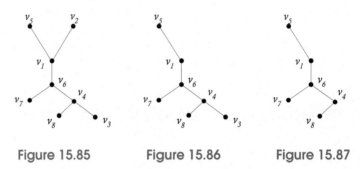

Figure 15.85   **Figure 15.86**   **Figure 15.87**

Since v_5 is now the leaf with the smallest subscript in the remaining tree, remove edge $\{v_5, v_1\}$ and let $a_3 = 1$ as shown in Figure 15.88. Since v_1 is the leaf with the smallest subscript in the remaining tree, remove edge $\{v_1, v_6\}$ and let $a_4 = 6$ as shown in Figure 15.89. Now that v_7 is the leaf with the smallest subscript in the remaining tree, remove edge $\{v_7, v_6\}$ and let $a_5 = 6$ as shown in Figure 15.90.

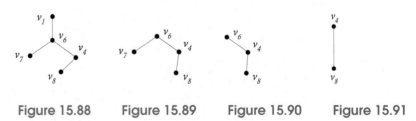

Figure 15.88   **Figure 15.89**   **Figure 15.90**   **Figure 15.91**

Finally, since v_6 is the leaf with the smallest subscript in the remaining tree, remove edge $\{v_6, v_4\}$ and let $a_6 = 4$. Only the edge in Figure 15.91 remains. The sequence is, therefore,

$$1, 4, 1, 6, 6, 4$$

Note that each vertex appears once less than its degree. This is because the edge is not counted where the vertex is removed as a leaf or else is left as a vertex in the edge that remains. ■

EXAMPLE 15.30

Let T be the tree in Figure 15.92. Since v_2 is the leaf with the smallest subscript remove edge $\{v_2, v_9\}$ and let $a_1 = 9$ as shown in Figure 15.93. Since v_3 is the leaf with the smallest subscript in the remaining tree, remove edge $\{v_3, v_8\}$ and let $a_2 = 8$ as shown in Figure 15.94.

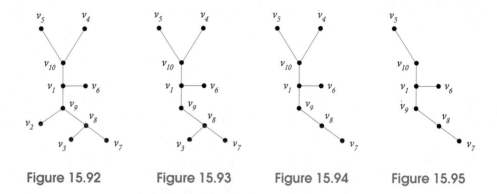

Figure 15.92 **Figure 15.93** **Figure 15.94** **Figure 15.95**

Since v_4 is the leaf with the smallest subscript in the remaining tree, remove edge $\{v_4, v_{10}\}$ and let $a_3 = 10$ as shown in Figure 15.95. Since v_5 is the leaf with the smallest subscript in the remaining tree, remove edge $\{v_5, v_{10}\}$ and let $a_4 = 10$ as shown in Figure 15.96. Since v_6 is the leaf with the smallest subscript in the remaining tree, remove edge $\{v_6, v_1\}$ and let $a_5 = 1$ as shown in Figure 15.97.

Since v_7 is the leaf with the smallest subscript in the remaining tree, remove edge $\{v_7, v_8\}$ and let $a_6 = 8$ as shown in Figure 15.98. Since v_8 is the leaf with the smallest subscript in the remaining tree, remove edge $\{v_8, v_9\}$ and let $a_7 = 9$ as shown in Figure 15.99. Finally, since v_9 is the leaf with the smallest subscript in the remaining tree, remove edge $\{v_9, v_1\}$ and let $a_8 = 1$. Only the edge in Figure 15.100 remains.

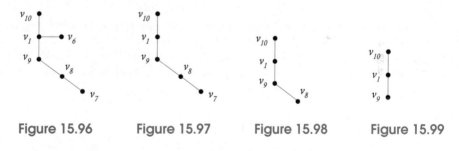

Figure 15.96 **Figure 15.97** **Figure 15.98** **Figure 15.99**

Figure 15.100

The sequence is, therefore,

$$9, 8, 10, 10, 1, 8, 9, 1$$

We now need the algorithm for going the other way:

■ **Sequence to Tree Algorithm** $(a_1, a_2, a_3, \ldots, a_{n-2})$ for $n \geq 3$

1. For each v_i, if i occurs in the sequence n_i times, let $\deg(v_i) = n_i + 1$.
2. Read a_1 and form the edge $\{v_{a_1}, v_j\}$ where j is the smallest label so that $\deg(v_j) = 1$.
3. Reduce both $\deg(v_{a_1})$ and $\deg(v_j)$ by 1.
4. Assuming a_{i-1} has been read, then read a_i and form the edge $\{v_{a_i}, v_j\}$ where j is the smallest label so that $\deg(v_j) = 1$.
5. Reduce both $\deg(v_{a_i})$ and $\deg(v_j)$ by 1.
6. After a_{n-2} has been read, create an edge between the two remaining vertices of degree 1.

EXAMPLE 15.31 We now take the sequence

$$1, 4, 1, 6, 6, 4$$

and use it to retrieve the original tree in Example 15.29.

Since 1, 4, and 6 all occur twice in the sequence,

$$\deg(v_1) = \deg(v_4) = \deg(v_6) = 3$$

Since 2, 3, 5, 7, and 8 do not occur at all,

$$\deg(v_2) = \deg(v_3) = \deg(v_5) = \deg(v_7) = \deg(v_8) = 1$$

Figure 15.101

We shall denote these by the 8-tuple $(3, 1, 1, 3, 1, 3, 1, 1)$, which we shall call the degree 8-tuple.

We now read $a_1 = 1$ and since v_2 has the smallest subscript of any vertex of degree 1, we create the edge $\{v_1, v_2\}$ shown in Figure 15.101. We let $\deg(v_1) = 2$ and $\deg(v_2) = 0$, so the degree 8-tuple is $(2, 0, 1, 3, 1, 3, 1, 1)$. Now read $a_2 = 4$ and since v_3 has the smallest subscript of any vertex of degree 1, we create the edge $\{v_3, v_4\}$ giving the graph in Figure 15.102.

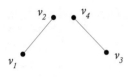

Figure 15.102

We let $\deg(v_4) = 2$ and $\deg(v_3) = 0$, so the degree 8-tuple is $(2, 0, 0, 2, 1, 3, 1, 1)$. The second edge has been slanted because I cheated and looked at the tree I'm trying to create. Now read $a_3 = 1$ and since v_5 has the smallest subscript of any vertex of degree 1, we create the edge $\{v_1, v_5\}$ giving the graph in Figure 15.103.

We let $\deg(v_1) = 1$ and $\deg(v_5) = 0$, so the degree 8-tuple is $(1, 0, 0, 2, 0, 3, 1, 1)$. Now read $a_4 = 6$ and since v_1 has the smallest subscript of any vertex of degree 1, we create the edge $\{v_1, v_6\}$ giving the graph in Figure 15.104.

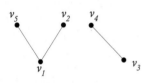

Figure 15.103

We let $\deg(v_1) = 0$ and $\deg(v_6) = 2$, so the degree 8-tuple is $(0, 0, 0, 2, 0, 2, 1, 1)$. Now read $a_5 = 6$ and since v_7 has the smallest subscript of any vertex of degree 1, we create the edge $\{v_7, v_6\}$ giving the graph in Figure 15.105.

We let $\deg(v_7) = 0$ and $\deg(v_6) = 1$, so the degree 8-tuple is $(0, 0, 0, 2, 0, 1, 0, 1)$. Now read $a_6 = 4$ and since v_6 has the smallest subscript of any vertex of degree 1, we create the edge $\{v_4, v_6\}$ giving the graph in Figure 15.106.

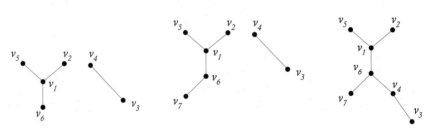

Figure 15.104   Figure 15.105   Figure 15.106

We let $\deg(v_4) = 1$ and $\deg(v_6) = 0$, so the degree 8-tuple is $(0, 0, 0, 1, 0, 0, 0, 1)$. Finally, since we have read all of the sequence and $\deg(v_4) = \deg(v_8) = 1$, we form the edge $\{v_4, v_8\}$ and we have the desired tree of Example 15.29.   ∎

EXAMPLE 15.32 We now take the sequence

$$9, 8, 10, 10, 1, 8, 9, 1$$

and use it to retrieve the original tree in Example 15.30.

Since 10, 9, 8, and 1 all occur twice in the sequence,

$$\deg(v_{10}) = \deg(v_9) = \deg(v_8) = \deg(v_1) = 3$$

15.5 Spanning Trees **601**

Figure 15.107

Figure 15.108

Since 2, 3, and 5 do not occur at all,

$$\deg(v_2) = \deg(v_3) = \deg(v_4) = \deg(v_5) = \deg(v_6) = \deg(v_7) = 1$$

We shall denote these by the 10-tuple $(3, 1, 1, 1, 1, 1, 1, 3, 3, 3)$, which we shall call the degree 10-tuple.

We now read $a_1 = 9$ and since v_2 has the smallest subscript of any vertex of degree 1, we create the edge $\{v_9, v_2\}$. as shown in Figure 15.107.

We let $\deg(v_9) = 2$ and $\deg(v_2) = 0$, so the degree 10-tuple is $(3, 0, 1, 1, 1, 1, 1, 3, 2, 3)$. Now read $a_2 = 8$ and since v_3 has the smallest subscript of any vertex of degree 1, we create the edge $\{v_8, v_3\}$ as shown in Figure 15.108. We let $\deg(v_8) = 2$ and $\deg(v_3) = 0$, so the degree 10-tuple is $(3, 0, 0, 1, 1, 1, 1, 2, 2, 3)$. Now read $a_3 = 10$ and since v_4 has the smallest subscript of any vertex of degree 1, we create the edge $\{v_{10}, v_4\}$ as shown in Figure 15.109.

We let $\deg(v_{10}) = 2$ and $\deg(v_4) = 0$, so the degree 10-tuple is $(3, 0, 0, 0, 1, 1, 1, 2, 2, 2)$. Now read $a_4 = 10$ and since v_5 has the smallest subscript of any vertex of degree 1, we create the edge $\{v_{10}, v_5\}$ as shown in Figure 15.110. We let $\deg(v_{10}) = 1$ and $\deg(v_5) = 0$, so the degree 10-tuple is $(3, 0, 0, 0, 0, 1, 1, 2, 2, 1)$. Now read $a_5 = 1$ and since v_6 has the smallest subscript of any vertex of degree 1, we create the edge $\{v_1, v_6\}$ as shown in Figure 15.111.

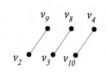 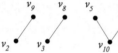

Figure 15.109   Figure 15.110   Figure 15.111

We let $\deg(v_1) = 2$ and $\deg(v_6) = 0$, so the degree 10-tuple is $(2, 0, 0, 0, 0, 0, 1, 2, 2, 1)$. Now read $a_6 = 8$ and since v_7 has the smallest subscript of any vertex of degree 1, we create the edge $\{v_8, v_7\}$ as shown in Figure 15.112.

We let $\deg(v_8) = 1$ and $\deg(v_7) = 0$, so the degree 10-tuple is $(2, 0, 0, 0, 0, 0, 0, 1, 2, 1)$. Now read $a_7 = 9$ and since v_8 has the smallest subscript of any vertex of degree 1, we create the edge $\{v_8, v_9\}$ as shown in Figure 15.113.

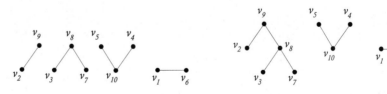

Figure 15.112   Figure 15.113

We let $\deg(v_9) = 1$ and $\deg(v_8) = 0$, so the degree 10-tuple is $(2, 0, 0, 0, 0, 0, 0, 0, 1, 1)$. Now read $a_8 = 1$ and since v_9 has the smallest subscript of any vertex of degree 1, we create the edge $\{v_1, v_9\}$ as shown in Figure 15.114.

Figure 15.114

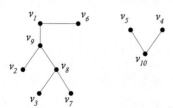

We let deg(v_1) = 1 and deg(v_9) = 0, so the degree 10-tuple is (1, 0, 0, 0, 0, 0, 0, 0, 0, 1). Finally, since we have read all of the sequence and deg(v_1) = deg(v_{10}) = 1, we form the edge $\{v_1, v_{10}\}$ and we have the desired tree of Example 15.30. ∎

The previous theorem gives the number of distinct spanning trees on n labeled vertices but does not consider a specific graph with these vertices. When a particular graph is considered, the number of distinct spanning trees on this graph is reduced, since only edges of the graph can be used. We conclude this section with a method of determining how many distinct spanning trees there are on graph with n labeled vertices.

DEFINITION 15.33 Let G be a graph with n labeled vertices, $v_1, v_2, v_3, \ldots, v_n$. The **degree matrix** of G is the $n \times n$ matrix D defined by $D_{ij} = 0$ if $i \neq j$ and D_{ii} is the degree of vertex v_i.

With this definition, we can now state the following rather amazing theorem, which we shall give without proof.

THEOREM 15.34 (Kirchoff's Matrix-Tree Formula) Let G be a connected graph with labeled vertices. Let $K = D - A$, where A is the adjacency matrix of G, and D is the degree matrix of G. The number of spanning trees of G is equal to the value of any cofactor of K. ∎

EXAMPLE 15.35 Find the number of spanning trees of the graph in Figure 15.115. Since deg(v_1) = deg(v_3) = 2 and deg(v_2) = deg(v_4) = 3,

$$C = \begin{bmatrix} 2 & 0 & 0 & 0 \\ 0 & 3 & 0 & 0 \\ 0 & 0 & 2 & 0 \\ 0 & 0 & 0 & 3 \end{bmatrix}$$

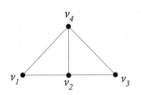

Figure 15.115

A is the matrix

$$\begin{bmatrix} 0 & 1 & 0 & 1 \\ 1 & 0 & 1 & 1 \\ 0 & 1 & 0 & 1 \\ 1 & 1 & 1 & 0 \end{bmatrix}$$

and

$$K = C - A = \begin{bmatrix} 2 & 0 & 0 & 0 \\ 0 & 3 & 0 & 0 \\ 0 & 0 & 2 & 0 \\ 0 & 0 & 0 & 3 \end{bmatrix} - \begin{bmatrix} 0 & 1 & 0 & 1 \\ 1 & 0 & 1 & 1 \\ 0 & 1 & 0 & 1 \\ 1 & 1 & 1 & 0 \end{bmatrix} = \begin{bmatrix} 2 & -1 & 0 & -1 \\ -1 & 3 & -1 & -1 \\ 0 & -1 & 2 & -1 \\ -1 & -1 & -1 & 3 \end{bmatrix}$$

The cofactor of K_{11} is

$$\det \begin{bmatrix} 3 & -1 & -1 \\ -1 & 2 & -1 \\ -1 & -1 & 3 \end{bmatrix} = 3 \det \begin{bmatrix} 2 & -1 \\ -1 & 3 \end{bmatrix}$$

$$- (-1) \det \begin{bmatrix} -1 & -1 \\ -1 & 3 \end{bmatrix} + (-1) \det \begin{bmatrix} -1 & 2 \\ 2 & -1 \end{bmatrix}$$

$$= 3(5) - 4 - 3 = 8.$$

And there are eight spanning trees. ∎

EXAMPLE 15.36 We again provide a solution to Example 8.12 from the Shier collection. The problem is to find the number of spanning trees of $K_{2,n}$. Using Kirchoff's matrix-tree formula, and listing the vertices as $A, B, 1, 2, 3 \ldots n$, we have

$$C = \begin{bmatrix} n & & & & & & \\ & n & & & & & \\ & & 2 & & & 0 & \\ & & & 2 & & & \\ & & & & 2 & & \\ & & 0 & & & 2 & \\ & & & & & & \ddots \\ & & & & & & & 2 \end{bmatrix}$$

and

$$A = \begin{bmatrix} 0 & 0 & 1 & 1 & 1 & 1 & \cdots & 1 \\ 0 & 0 & 1 & 1 & 1 & 1 & \cdots & 1 \\ 1 & 1 & 0 & 0 & 0 & 0 & \cdots & 0 \\ 1 & 1 & 0 & 0 & 0 & 0 & \cdots & 0 \\ 1 & 1 & 0 & 0 & 0 & 0 & \cdots & 0 \\ 1 & 1 & 0 & 0 & 0 & 0 & \cdots & 0 \\ \vdots & \vdots & \vdots & \vdots & \vdots & \vdots & \ddots & 0 \\ 1 & 1 & 0 & 0 & 0 & 0 & 0 & 0 \end{bmatrix}$$

so that $K = C - A = A =$

$$\begin{bmatrix} n & 0 & -1 & -1 & -1 & -1 & \cdots & -1 \\ 0 & n & -1 & -1 & -1 & -1 & \cdots & -1 \\ -1 & -1 & 2 & 0 & 0 & 0 & \cdots & 0 \\ -1 & -1 & 0 & 2 & 0 & 0 & \cdots & 0 \\ -1 & -1 & 0 & 0 & 2 & 0 & \cdots & 0 \\ -1 & -1 & 0 & 0 & 0 & 2 & \cdots & 0 \\ \vdots & \vdots & \vdots & \vdots & \vdots & \vdots & \ddots & 0 \\ -1 & -1 & 0 & 0 & 0 & 0 & 0 & 2 \end{bmatrix}$$

The cofactor of $K_{11} =$

$$\begin{bmatrix} n & -1 & -1 & -1 & -1 & \cdots & -1 \\ -1 & 2 & 0 & 0 & 0 & \cdots & 0 \\ -1 & 0 & 2 & 0 & 0 & \cdots & 0 \\ -1 & 0 & 0 & 2 & 0 & \cdots & 0 \\ -1 & 0 & 0 & 0 & 2 & \cdots & 0 \\ \vdots & \vdots & \vdots & \vdots & \vdots & \ddots & 0 \\ -1 & 0 & 0 & 0 & 0 & 0 & 2 \end{bmatrix}$$

Multiply the first row by 2 so we have

$$\frac{1}{2} \begin{bmatrix} 2n & -2. & -2 & -2 & -2 & \cdots & -2 \\ -1 & 2 & 0 & 0 & 0 & \cdots & 0 \\ -1 & 0 & 2 & 0 & 0 & \cdots & 0 \\ -1 & 0 & 0 & 2 & 0 & \cdots & 0 \\ -1 & 0 & 0 & 0 & 2 & \cdots & 0 \\ \vdots & \vdots & \vdots & \vdots & \vdots & \ddots & 0 \\ -1 & 0 & 0 & 0 & 0 & 0 & 2 \end{bmatrix}$$

then add rows 2 through $n+1$ to the first row getting

$$\frac{1}{2}\begin{bmatrix} n & 0. & 0 & 0 & 0 & \cdots & 0 \\ -1 & 2 & 0 & 0 & 0 & \cdots & 0 \\ -1 & 0 & 2 & 0 & 0 & \cdots & 0 \\ -1 & 0 & 0 & 2 & 0 & \cdots & 0 \\ -1 & 0 & 0 & 0 & 2 & \cdots & 0 \\ \vdots & \vdots & \vdots & \vdots & \vdots & \ddots & 0 \\ -1 & 0 & 0 & 0 & 0 & 0 & 2 \end{bmatrix}$$

The determinant of this triangulated matrix is $n2^n$. Multiplying by $\frac{1}{2}$ we get the answer $n2^{n-1}$. ■

Exercises

1. Given the graph in Figure 15.116 and assuming that the vertices have been ordered as labeled,

Figure 15.116

   (a) Find a spanning tree by removing an edge from each cycle.
   (b) Find the breadth-first spanning tree.
   (c) Find the depth-first spanning tree.

2. Given the graph in Figure 15.117 and assuming that the vertices have been ordered as labeled,

Figure 15.117

   (a) Find a spanning tree by removing an edge from each cycle.
   (b) Find the breadth-first spanning tree.
   (c) Find the depth-first spanning tree.

3. Given the graph in Figure 15.118 and assuming that the vertices have been ordered as labeled,
   (a) Find a spanning tree by removing an edge from each cycle.
   (b) Find the breadth-first spanning tree.
   (c) Find the depth-first spanning tree.

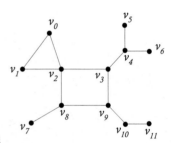

Figure 15.118

4. Given the graph in Figure 15.119 and assuming that the vertices have been ordered as labeled,

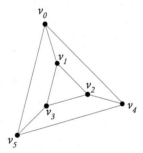

Figure 15.119

   (a) Find a spanning tree by removing an edge from each cycle.
   (b) Find the breadth-first spanning tree.
   (c) Find the depth-first spanning tree.

5. Given the graph in Figure 15.120 and assuming that the vertices have been ordered as labeled,
   (a) Find a spanning tree by removing an edge from each cycle.
   (b) Find the breadth-first spanning tree.
   (c) Find the depth-first spanning tree.

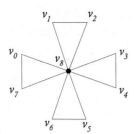

Figure 15.120

6. Given the graph in Figure 15.121 and assuming that the vertices have been ordered as labeled,

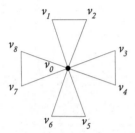

Figure 15.121

(a) Find a spanning tree by removing an edge from each cycle.

(b) Find the breadth-first spanning tree.

(c) Find the depth-first spanning tree.

7. Given the graph in Figure 15.122 and assuming that the vertices have been ordered as labeled,

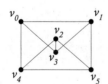

Figure 15.122

(a) Find a spanning tree by removing an edge from each cycle.

(b) Find the breadth-first spanning tree.

(c) Find the depth-first spanning tree.

8. Given the graph in Figure 15.123, and assuming that the vertices have been ordered as labeled,

(a) Find a spanning tree by removing an edge from each cycle.

(b) Find the breadth-first spanning tree.

(c) Find the depth-first spanning tree.

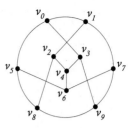

Figure 15.123

9. Given the graph K_5,

(a) Find a spanning tree by removing an edge from each cycle.

(b) Find the breadth-first spanning tree.

(c) Find the depth-first spanning tree.

10. Given the graph $K_{3,4}$,

(a) Find a spanning tree by removing an edge from each cycle.

(b) Find the breadth-first spanning tree.

(c) Find the depth-first spanning tree.

11. For the Petersen graph,

(a) Find a spanning tree by removing an edge from each cycle.

(b) Find a breadth-first spanning tree.

(c) Find a depth-first spanning tree.

12. Using the DFT articulation point search algorithm, find the articulation points and bicomponents for the graph in Figure 15.124.

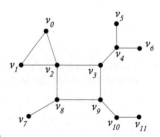

Figure 15.124

13. Using the DFT articulation point search algorithm, find the articulation points and bicomponents for the graph in Figure 15.125.

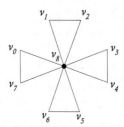

Figure 15.125

Using the DFT articulation point search algorithm, find the articulation points and bicomponents for the following graphs:

14.

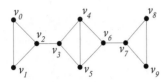

15.

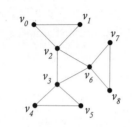

16.

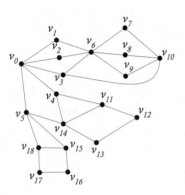

17. If G is a graph that is a simple cycle, describe the breadth-first and depth-first spanning trees for G. If G has n edges, how many spanning trees are there?

18. If G is a graph that is a cycle but not necessarily a simple cycle, is the description the same? Why? Can G have an articulation point?

19. Prove that an edge is in every spanning tree of a graph G if and only if it is a cut edge.

20. Find necessary and sufficient conditions on G so that the depth-first tree and the breadth-first tree are equal.

21. Show that the depth-first spanning tree for K_n for $n \geq 3$ is a simple path but the breadth-first spanning tree is not.

22. Find a so that if $n \geq a$, then K_n has two spanning trees with no edge in common.

23. Prove that if E' is a cut set of a graph G, then every spanning tree contains an edge of E'.

24. How many spanning trees does K_3 have? How many spanning trees does K_4 have? How many spanning trees does K_n have?

25. How many spanning trees does $K_{2,n}$ have?

26. If G is a graph with v vertices and e edges, how many edges can be removed without causing the remaining graph to be disconnected?

27. Let T and T' be distinct spanning trees of a graph G. Let e be an edge of T that is not an edge of T' and T_e be T with the edge e removed. Prove that there is an edge e' of T' that may be added to T_e so that the resulting graph is again a spanning tree of G.

Using the tree to sequence algorithm, determine the corresponding sequence for each of the following trees:

28.

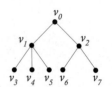

29.

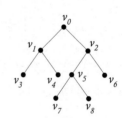

30.

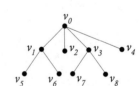

31.

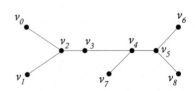

32.

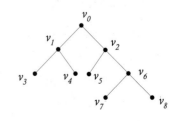

33.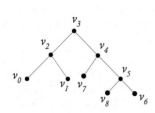

34.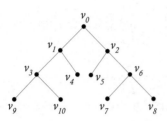

35.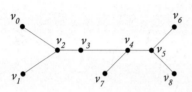

Using the sequence to tree algorithm, determine the corresponding trees for the following sequences:

36. 1, 2, 3, 3, 2, 3

37. 1, 3, 3, 5, 5, 4, 4

38. 1, 2, 2, 2, 5, 4, 5, 6

Using the sequence to tree algorithm, determine the corresponding trees for the following sequences:

39. 1, 4, 2, 3, 2, 3

40. 1, 4, 3, 3, 4, 4, 4

41. 1, 5, 2, 5, 3, 6, 5, 5

Using Kirchoff's Matrix-Tree Formula, find the number of spanning trees of each of the following graphs:

42.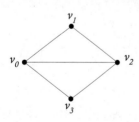

43.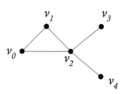

44.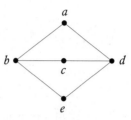

45.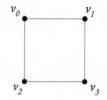

46.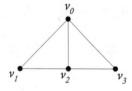

47.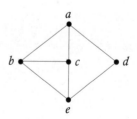

48. How many spanning trees does $K_{n,n}$ have?

15.6 Minimal Spanning Trees

In the previous chapter we defined a weighted graph to be a graph such that each edge had a positive number, called a weight, attached to it. This weight could represent distance, cost, or numerous other items. For example, if the vertices represented cities, the weight could be the distance between two cities or the cost of flying from one city to the other. The **weight** of a spanning tree of a weighted graph G is the sum of the weights assigned to the edges of the spanning tree. A **minimal spanning tree** is a spanning tree of G such that the weight of T is less than or equal to the weight of any other spanning tree of G.

We shall show two ways of constructing a minimal spanning tree of a weighted graph G. The first of these is called Kruskal's algorithm. It consists of simply picking the edges with least weight, such that no cycle occurs, and forming a tree $T(E, V)$.

■ Kruskal's Algorithm

1. Choose an edge e of least weight from the graph G with the property it does not already belong to E and that its addition to E will not create a cycle in T.
2. Add this edge to the set of edges E.
3. Continue until there are no more edges with this property.

EXAMPLE 15.37 Consider the weighted graph in Figure 15.126. We first choose the edge $\{b, d\}$ and add it to E. We can then select the edge $\{d, e\}$, so that

$$E = \{\{b, d\}, \{d, e\}\}$$

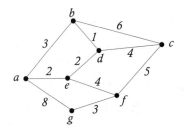

Figure 15.126

We then add the edge $\{a, e\}$, so that

$$E = \{\{b, d\}, \{d, e\}, \{a, e\}\}$$

We cannot add $\{a, b\}$ since it would create a cycle, so we add $\{f, g\}$ and

$$E = \{\{b, d\}, \{d, e\}, \{a, e\}, \{f, g\}\}$$

We can next add $\{e, f\}$ and

$$E = \{\{b, d\}, \{d, e\}, \{a, e\}, \{f, g\}, \{e, f\}\}$$

Finally, we add $\{c, d\}$, and

$$E = \{\{b, d\}, \{d, e\}, \{a, e\}, \{f, g\}, \{e, f\}\}, \{c, d\}$$

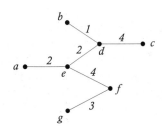

Figure 15.127

We can add no further edges, so we have the tree in Figure 15.127. ■

THEOREM 15.38 Kruskal's algorithm produces a minimal spanning tree.

Proof By construction, Kruskal's algorithm produces a minimal spanning tree. Let

$$e_1, e_2, e_3, e_4, \ldots, e_n$$

be the edges of T in the order they are picked by Kruskal's algorithm. Let k be the largest integer such that $e_1, e_2, e_3, e_4, \ldots, e_k$ are edges of a minimal spanning tree, and let T' be that tree. If $k \neq n$, then attach edge e_{k+1} to T'. Now T' with the new edge attached contains a cycle. Remove e from this cycle where $e \neq e_i$, for $1 \leq i \leq k + 1$ and let T'' be the new tree created. The weight of e cannot be less than the weight of e_{k+1}, since if it were, it would already have been picked by Kruskal's algorithm or it would have produced a cycle so that the edges in $\{e_1, e_2, e_3, e_4, \ldots, e_k, e\}$ cannot be part of the edges of a tree. But this is impossible since this set of edges is part of the tree T'. Therefore, since T'' was created from T' by replacing e with an edge that is no heavier, the weight of T'' is less than or equal to the weight of T'. Therefore, T'' is a minimal spanning tree having $\{e_1, e_2, e_3, e_4, \ldots, e_k, e_{k+1}\}$ as part of its set of edges. But this contradicts the definition of k. Therefore, $k = n$, and T is a minimal spanning tree. ∎

The second and final algorithm we present for finding a minimal spanning tree for a graph G is called Prim's algorithm. One principal difference with this algorithm is that we always have a tree, to which we add edges until we have a spanning tree. Using Prim's algorithm, we begin by selecting a vertex v_0 from G and then select the edge $e_1 = \{v_1, v_0\}$, incident to v_0, with least weight and form the tree T_1. We next select the edge e_2 with least weight, such that one vertex of the edge is a vertex of T_1, and the other is not, and add it to the tree, so that we now have a tree T_2 with two edges. Once we have formed the tree T_k, we then form the tree T_{k+1} by selecting the edge e_{k+1} with least weight, such that one vertex of the edge is a vertex of T_k, and the other is not, and add it to the tree. We continue until the tree contains all of the vertices of G.

We now formally state the algorithm:

■ Prim's Minimum Spanning Tree Algorithm

1. Select a vertex v_0 from G, and select edge e_1 with least weight from those having v_0 as a vertex, to form T_1.
2. Given tree T_k, with edges $e_1, e_2, e_3, \ldots, e_k$, if there is a vertex not in T_k, select the edge with minimal weight that is adjacent to an edge of T_k and has a vertex not in T_k. Add this edge to tree T_k, to form T_{k+1}.
3. Continue until there are no vertices of G that are not in the tree.
4. End.

EXAMPLE 15.39 Again consider the tree in Figure 15.128. If we begin at vertex a, we first select the edge $\{a, e\}$, since it has minimal weight. Therefore, T_1 consists of the edge $\{a, e\}$. We now select edge $\{e, d\}$, since it is the edge of minimal weight with only one vertex in T_1, and T_2 is the tree in Figure 15.129.

Next, we select edge $\{d, b\}$, since it is the edge of minimal weight with only one vertex in T_2, and T_3 is the tree in Figure 15.130. We now have a choice, since there are two vertices with the same weight with only one vertex in T_3. Suppose we select $\{d, c\}$, so T_4 is the tree in Figure 15.131.

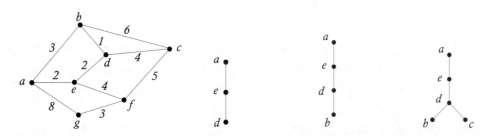

Figure 15.128   Figure 15.129   Figure 15.130   Figure 15.131

Next we select $\{e, f\}$, since it is the edge of minimal weight with only one vertex in T_4, and T_5 is the tree in Figure 15.132.

Figure 15.132

Finally, we select the edge $\{f, g\}$ since it is the edge of minimal weight with one vertex in T_4 and the other vertex g, since it is the only remaining vertex. We now have the tree T_6, which is our spanning tree. ∎

We next prove that this algorithm does indeed produce a minimal spanning tree.

THEOREM 15.40 For a given weighted graph G, the tree produced by Prim's Minimum Spanning Tree Algorithm is a minimal spanning tree.

Proof It is immediate from the construction of the tree that it is a spanning tree. The proof for showing that it is a minimal spanning tree is similar to the proof for Kruskal's algorithm. Let

$$e_1, e_2, e_3, e_4, \ldots, e_n$$

be the edges of T in the order they are picked by Prim's Minimum Spanning Tree Algorithm. Let k be the largest integer such that $e_1, e_2, e_3, e_4, \ldots, e_k$ are edges of a minimal spanning tree, and let T' be that tree. Let T_k be the tree formed by edges $e_1, e_2, e_3, e_4, \ldots, e_k$. If $k \neq n$, then attach edge e_{k+1} to T'. Now T' with the new edge attached contains a cycle. There is another edge e in the cycle such that only one of its vertices is in the tree T_k. Remove e from $T' \cup \{e_{k+1}\}$, forming a tree T'', with edge e replaced by e_{k+1}. Since e_{k+1} is an edge of least weight with only one vertex in T_k, the weight of e_{k+1} is less than or equal to the weight of e. Therefore, the weight of T'' is less than or equal to the weight of T', so that T'' is also a minimal spanning tree, with edges $e_1, e_2, e_3, e_4, \ldots, e_k, e_{k+1}$, but this contradicts the definition of k. Therefore, $k = n$, and Prim's Minimum Spanning Tree Algorithm generates a minimal spanning tree. ∎

Prim's algorithm can also be accomplished using matrices. For a graph G with n vertices, we need an $(n+1) \times (n+1)$ matrix. We first find the $n \times n$ weighted matrix W, where W_{ij} is the weight of the edge from v_i to v_j if there is an edge from v_i to v_j and 0 if there is no edge. We then add row $n+1$ and column $n+1$, which are temporarily blank. We shall call the new matrix W^*.

Formally we state the following algorithm.

■ **Prim's Algorithm with Matrices**

1. Create weighted matrix W.

2. Add additional row and column to create W^*.

3. In row 1 of W^*, place $*$ in the last column. In column 1, change all of the numbers to 0 and place a U in the last row.

4. Select the smallest number such that the row of the number has $*$ in column $n+1$ and the column of the number does not have a U in row $n+1$.

5. If the number selected is in row i and column j, place $*$ in the last column of row j, change the remaining numbers in column j to 0, place U in row $n+1$ of column j, and add edge (v_i, v_j) to the spanning tree.

6. Continue steps 4 and 5 until no further numbers can be selected.

EXAMPLE 15.41

Assume that we have the graph in Figure 15.133, so that the matrices W and W^* are, respectively,

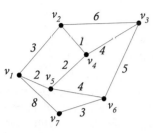

Figure 15.133

$$\begin{bmatrix} 0 & 3 & 0 & 0 & 2 & 0 & 8 \\ 3 & 0 & 6 & 1 & 0 & 0 & 0 \\ 0 & 6 & 0 & 4 & 0 & 5 & 0 \\ 0 & 1 & 4 & 0 & 2 & 0 & 0 \\ 2 & 0 & 0 & 2 & 0 & 4 & 0 \\ 0 & 0 & 5 & 0 & 4 & 0 & 3 \\ 8 & 0 & 0 & 0 & 0 & 3 & 0 \end{bmatrix} \text{ and } \begin{bmatrix} 0 & 3 & 0 & 0 & 2 & 0 & 8 \\ 3 & 0 & 6 & 1 & 0 & 0 & 0 \\ 0 & 6 & 0 & 4 & 0 & 5 & 0 \\ 0 & 1 & 4 & 0 & 2 & 0 & 0 \\ 2 & 0 & 0 & 2 & 0 & 4 & 0 \\ 0 & 0 & 5 & 0 & 4 & 0 & 3 \\ 8 & 0 & 0 & 0 & 0 & 3 & 0 \end{bmatrix}$$

We first select row 1, which means that we select the first vertex, v_1. We set each entry in the first column of W^* equal to 0 since we do not wish to return to v_1, and place a U in the eighth $(n+1)$st row of column 1, to indicate that column 1, is not to be used again. We place a $*$ in row 1 to indicate we may select values from this row since this row represents values of edges from v_1 to adjacent vertices. Thus, we have

$$\begin{bmatrix} 0 & 3 & 0 & 0 & 2 & 0 & 8 & * \\ 0 & 0 & 6 & 1 & 0 & 0 & 0 & \\ 0 & 6 & 0 & 4 & 0 & 5 & 0 & \\ 0 & 1 & 4 & 0 & 2 & 0 & 0 & \\ 0 & 0 & 0 & 2 & 0 & 4 & 0 & \\ 0 & 0 & 5 & 0 & 4 & 0 & 3 & \\ 0 & 0 & 0 & 0 & 0 & 3 & 0 & \\ U & & & & & & & \end{bmatrix}$$

We select the smallest nonzero value in this row (so we select the edge with smallest weight adjacent to v_1). In this case, the value is 2, in column 5. Thus, we select (v_1, v_5) for our first edge. We set the other values in the column equal to 0 and set the value in the eighth row of that column equal to U so it will not be used again and we will not return to v_5. We place a $*$ in the eighth column of row 5 to indicate that elements in row 5 may now be used as well as row one since these rows represent edges from v_1, and v_5 to adjacent vertices.

$$\begin{bmatrix} 0 & 3 & 0 & 0 & \underline{2} & 0 & 8 & * \\ 0 & 0 & 6 & 1 & 0 & 0 & 0 & \\ 0 & 6 & 0 & 4 & 0 & 5 & 0 & \\ 0 & 1 & 4 & 0 & 0 & 0 & 0 & \\ 0 & 0 & 0 & 2 & 0 & 4 & 0 & * \\ 0 & 0 & 5 & 0 & 0 & 0 & 3 & \\ 0 & 0 & 0 & 0 & 0 & 3 & 0 & \\ U & & & & U & & & \end{bmatrix}$$

We now select the smallest nonzero value in either row 1 or row 5. In this case, the value is 2 in column 4, row 5. Thus, we select (v_5, v_4) for our next edge. We set the other values in the column equal to 0 and set the value in the eighth row of that column equal to U so it will not be used again and we will not return to v_4. We place $*$ in the eighth row of row 4 to indicate that elements in it may now be used as well as row 1 and row 5, since these rows represent edges from v_1, v_4, and

v_5 to adjacent vertices.

$$\begin{bmatrix} 0 & 3 & 0 & 0 & 2 & 0 & 8 & * \\ 0 & 0 & 6 & 0 & 0 & 0 & 0 & \\ 0 & 6 & 0 & 0 & 0 & 5 & 0 & \\ 0 & 1 & 4 & 0 & 0 & 0 & 0 & * \\ 0 & 0 & 0 & \underline{2} & 0 & 4 & 0 & * \\ 0 & 0 & 5 & 0 & 0 & 0 & 3 & \\ 0 & 0 & 0 & 0 & 0 & 3 & 0 & \\ U & & & U & U & & & \end{bmatrix}$$

We now select the smallest nonzero value in either row 1, 4, or 5. In this case, the value is 1 in column 2, row 4. Thus, we select (v_4, v_2) for our next edge. We set the other values in the column equal to 0 and set the value in the eighth row of that column equal to U. We place $*$ in the eighth column of row 2 to indicate that elements in it may now be used as well as rows 1, 4, and 5, since these rows represent edges from $v_1, v_2, v_4,$ and v_5 to adjacent vertices.

$$\begin{bmatrix} 0 & 0 & 0 & 0 & 2 & 0 & 8 & * \\ 0 & 0 & 6 & 0 & 0 & 0 & 0 & * \\ 0 & 0 & 0 & 0 & 0 & 5 & 0 & \\ 0 & \underline{1} & 4 & 0 & 0 & 0 & 0 & * \\ 0 & 0 & 0 & 2 & 0 & 4 & 0 & * \\ 0 & 0 & 5 & 0 & 0 & 0 & 3 & \\ 0 & 0 & 0 & 0 & 0 & 3 & 0 & \\ U & U & & U & U & & & \end{bmatrix}$$

Using the same procedures, we choose the 4 in row 4, column 3, and the 4 in row 5, column 6, giving us edges respectively edges (v_4, v_3) and (v_5, v_6), and matrices

$$\begin{bmatrix} 0 & 0 & 0 & 0 & 2 & 0 & 8 & * \\ 0 & 0 & 0 & 0 & 0 & 0 & 0 & * \\ 0 & 0 & 0 & 0 & 0 & 5 & 0 & * \\ 0 & 1 & \underline{4} & 0 & 0 & 0 & 0 & * \\ 0 & 0 & 0 & 2 & 0 & 4 & 0 & * \\ 0 & 0 & 0 & 0 & 0 & 0 & 3 & \\ 0 & 0 & 0 & 0 & 0 & 3 & 0 & \\ U & U & U & U & U & & & \end{bmatrix} \text{ and } \begin{bmatrix} 0 & 0 & 0 & 0 & 2 & 0 & 8 & * \\ 0 & 0 & 0 & 0 & 0 & 0 & 0 & * \\ 0 & 0 & 0 & 0 & 0 & 0 & 0 & * \\ 0 & 1 & 4 & 0 & 0 & 0 & 0 & * \\ 0 & 0 & 0 & 2 & 0 & \underline{4} & 0 & * \\ 0 & 0 & 0 & 0 & 0 & 0 & 3 & * \\ 0 & 0 & 0 & 0 & 0 & 0 & 0 & \\ U & U & U & U & U & U & & \end{bmatrix}$$

Finally we select the 3 in row 6, column 7, giving us edge (v_6, v_7) and the matrix

$$\begin{bmatrix} 0 & 0 & 0 & 0 & 2 & 0 & 0 & * \\ 0 & 0 & 0 & 0 & 0 & 0 & 0 & * \\ 0 & 0 & 0 & 0 & 0 & 0 & 0 & * \\ 0 & 1 & 4 & 0 & 0 & 0 & 0 & * \\ 0 & 0 & 0 & 2 & 0 & 4 & 0 & * \\ 0 & 0 & 0 & 0 & 0 & 0 & \underline{3} & * \\ 0 & 0 & 0 & 0 & 0 & 0 & 0 & \\ U & U & U & U & U & U & & \end{bmatrix}$$

and we have the spanning tree in Figure 15.134.

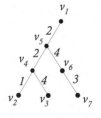

Figure 15.134

Exercises

Given the weighted graphs in Figures 15.135 and 15.136, find the minimum spanning tree using

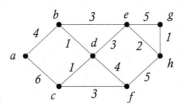

Figure 15.135

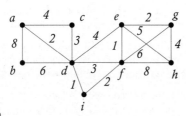

Figure 15.136

1. Kruskal's algorithm.
2. Prim's algorithm.
3. Prim's algorithm with matrices.

Given the weighted graphs in Figures 15.137 and 15.138, find the minimum spanning tree using

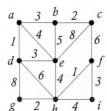

Figure 15.137

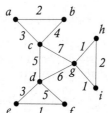

Figure 15.138

4. Kruskal's algorithm.
5. Prim's algorithm.
6. Prim's algorithm with matrices.

Given the weighted graphs in Figures 15.139 and 15.140, find the minimum spanning tree using

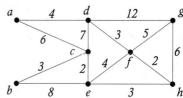

Figure 15.139

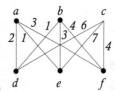

Figure 15.140

7. Kruskal's algorithm.
8. Prim's algorithm.
9. Prim's algorithm with matrices.

Given the weighted graphs in Figures 15.141 and 15.142, find the minimum spanning tree using

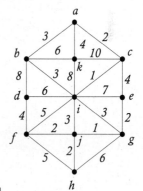

Figure 15.141

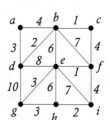

Figure 15.142

10. Kruskal's algorithm.
11. Prim's algorithm.
12. Prim's algorithm with matrices.

GLOSSARY

Chapter 15: Trees

Articulation point (15.5)	A connected graph $G = (V, E)$ has a **articulation point** or cut edge $a \in V$ if the removal of a and its incident edges causes the graph to no longer be connected.
Back edge (15.5)	If in trying to create a spanning tree we encounter a vertex v, select another vertex w, and find that vertex w has already been added to the tree, then the edge $\{v, w\}$ between these two vertices cannot be added to the tree and we call such an edge a **back edge**.
Balanced (complete or nearly full)	An m-ary directed tree of height h is **balanced (completely** or **nearly full)** if the level of every leaf is either h or $h + 1$.
Binary search tree (15.2)	A **binary search tree** is a binary tree with a name (or other key) at each node and a process for finding nodes of the tree.
Breadth-first tree search algorithm (15.5)	The **Breadth-First Tree Search algorithm** is a method of creating a spanning tree in which we find all vertices adjacent to a given vertex before proceeding to the next level.
Degree matrix of G (15.5)	If we let G be a graph with n labeled vertices, $v_1, v_2, v_3, \ldots v_n$, then the **degree matrix of** G is the $n \times n$ matrix D defined by $D_{ij} = 0$ if $i \neq j$ and D_{ij} is the degree of the vertex v_i.
Depth-first tree search algorithm (15.5)	The **Depth-First Tree Search algorithm** is a method of creating a spanning tree in which we try to form as long a path as possible before returning to an articulation point.
DFT articulation point search algorithm (15.5)	The **DFT Articulation Point Search algorithm** is a search to find articulation points and bicomponents in a graph.
Directed tree (15.1)	A **directed tree** T is a loop-free graph whose underlying graph is a tree such that if there is a path from vertex a to vertex b, it is unique.
Full m-ary directed tree (15.1)	A **full m-ary directed tree** is a directed tree such that $\text{outdeg}(v) = m$ for every vertex v in the tree that is not a leaf, and every leaf is at the same level. (Thus, each parent has exactly m children.)
Height of a directed tree (15.1)	The **height of a directed tree** is the length of the longest path from the root to a leaf.
Homomorphism (15.1)	A function f from the graph $G(V, E)$ to the graph $G'(V', E')$ is called a **homomorphism** from G to G', denoted by $f : G \to G'$ if it has the properties a. If $e \in E$, then $f(e) \in E' (f(E) \subseteq f(E'))$. b. If $v \in V$, then $f(v) \in V' (f(V) \subseteq f(V'))$. c. If vertices u and v are incident to edge e in G, then $f(u)$ and $f(v)$ are incident to $f(e)$ in G'.
Huffman's algorithm (15.3)	**Huffman's algorithm** is the process for finding the best code for minimizing our data to find the code with minimal weight.
Huffman's code (15.3)	**Huffman's code** is the code assigned to the symbols by their leaf path codes.
Inorder algorithm (15.4)	The **inorder algorithm** describes the procedure for the inorder transversal of a binary tree.

Isomorphic (15.1)	The homomorphism $f : G \to G'$ is an **isomorphism** if $f : V \to V'$ and $f : E \to E'$ are one-to-one correspondences. If $f : G \to G'$ is an isomorphism, then G and G' are **isomorphic**.
Isomorphic Binary Tree algorithm (15.4)	The **Isomorphic Binary Tree algorithm** is used to determine whether two binary trees are isomorphic.
Isomorphic trees (15.1)	Two rooted binary trees $T(E, V)$ and $T'(E', V')$ are **isomorphic** if there exists a graph isomorphism f from T to T' such that 1. v_i is a left child of v_j if and only if $f(v_i)$ is a left child of $f(v_j)$. 2. v_i is a right child of v_j if and only if $f(v_i)$ is a right child of $f(v_j)$. 3. f maps the root r of T to the root r' of T'.
Kirchoff's Matrix-Tree Formula (15.5)	**Kirchoff's Matrix-Tree Formula** finds the number of spanning trees of G using the cofactor of a matrix.
Kruskal's algorithm (15.6)	**Kruskal's algorithm** constructs a minimal spanning tree of a weighted graph G by picking the edges with least weight, such that no cycle occurs, and forming a tree $T(E, V)$.
Labeled (15.5)	A set of n vertices is **labeled** if each of them is assigned a unique integer between 1 and n. This gives us the vertices $v_1, v_2, v_3, \ldots v_n$.
Leaf path code (15.3)	The string of 0's and 1's corresponding to a given leaf from the root is called the **leaf path code**.
Level of a vertex (15.1)	The **level of a vertex** v in a directed tree is the length of the path from the root of the tree to v.
m-ary directed tree (15.1)	An m**-ary directed tree** is a directed tree such that outdeg$(v) \leqslant m$ for every vertex v in the tree. (Thus a parent has at most m children.)
Minimal spanning tree (15.6)	A **minimal spanning tree** is a spanning tree such that the weight T is less than or equal to the weight of any other spanning tree of G.
Postorder algorithm (15.4)	The **postorder algorithm** is used to perform the postorder transversal.
Prefix code (15.3)	A **prefix code** of a code C has the property that an element of a code cannot be the beginning string of another element of the code.
Preorder algorithm (15.4)	The **preorder algorithm** is used to perform the preorder transversal.
Preorder transversal (15.4)	**Preorder transversal** on a tree is a binary operation on expressions E_1 and E_2 to form an expression in prefix form.
Prim's Algorithm with Matrices (15.6)	**Prim's Algorithm with Matrice** creates a minimal spanning tree using matrices.
Prim's Minimum Spanning Tree Algorithm (15.6)	**Prim's Minimum Spanning Tree algorithm** constructs a minimal spanning tree by adding edges of minimal weight to an existing tree until we have a spanning tree.
Rooted directed tree (15.1)	A directed tree T is a **rooted directed tree** if there is a unique vertex v_0 such that indeg$(v_0) = 0$ and there is a path from v_0 to every other vertex in T.

Sequence to Tree algorithm (15.5)	A **Sequence to Tree algorithm** produces a spanning tree from a sequence.
Spanning forest (15.5)	A **spanning forest** is a graph consisting of a spanning tree for each component in a given graph.
Spanning tree (15.5)	A tree T is a **spanning tree** of a subgraph G if every vertex in G is a vertex in T.
Tree to Sequence algorithm (15.5)	A **Tree to Sequence algorithm** produces a sequence from a spanning tree.
Uniquely decipherable code (15.3)	A **uniquely decipherable code** for a language is a set such that every string in the language can be expressed uniquely as the concatenation of elements of a code C.
Weight of a code (15.3)	If every letter a_i in our alphabet of symbols A occurs with relative frequency f_i and has a binary string of length l_i with weight $w = l_1 f_1 + l_2 f_2 + l_3 f_3 + l_4 f_4 + \cdots + l_n f_n = \sum_{i=1}^{n} l_i f_i$ then w is the **weight of the code**.
Weight of a leaf (15.3)	The **weight of a leaf** is the frequency of the occurrence of a letter represented by the leaf.
Weight of a spanning tree (15.6)	The **weight of a spanning tree** of a weighted graph G is the sum of the weights assigned to the edges of the spanning tree.
Weight of a tree (15.3)	The **weight of a tree** is given by $w = l_1 f_1 + l_2 f_2 + l_3 f_3 + l_4 f_4 + \cdots + l_n f_n = \sum_{i=1}^{n} l_i f_i$ or $w = l_1 w_1 + l_2 w_2 + l_3 w_3 + l_4 w_4 + \cdots + l_n w_n = \sum_{i=1}^{n} l_i w_i$ where l_i is the distance from the root to v_i.

TRUE-FALSE QUESTIONS

1. A graph G is a tree if and only if every edge of G is a cut edge.

2. Once a root is selected for a tree, the direction of each directed tree in the directed tree is determined.

3. Two rooted directed binary trees are isomorphic if and only if they are isomorphic as directed graphs.

4. The following tree

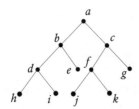

   is balanced.

5. The tree in Problem 4 is full.

6. If i is selected as the root in the tree in Problem 4, then the tree has length 5.

7. If e is selected as the root in the tree in Problem 4, then c has level 3.

8. Any binary tree can be changed to a balanced tree if the proper root is selected.

9. The number of binary trees isomorphic to the tree in Problem 4 is $\dfrac{22!}{11! \cdot 11!}$.

10. The height of a full binary tree with 31 vertices is 4.

11. Every tree that is a directed graph is a rooted directed tree.

12. The shape of a binary search tree depends on the order in which the data are entered.

13. The efficiency of a binary search tree depends on the shape of the tree.

14. The number of steps necessary to locate an item in a search tree depends on the width of the tree.

15. The number of steps necessary to locate an item in a search tree depends on the length of the tree.

16. A Huffman code is uniquely decipherable.

17. For a fixed frequency table, there is a unique Huffman tree.
18. For a given Huffman tree, the code produced is unique.
19. For a fixed frequency table, all Huffman trees have the same weight.
20. In a tree with minimum weight, every leaf has a sibling.
21. For any message to be encoded, the Huffman tree will always give the most compact encoding.
22. In a Huffman tree, two leaf symbols with least frequency have the same parent.
23. In a Huffman tree, two leaf symbols with least frequency always have the same parent.
24. Given a Huffman code, it is possible to reconstruct the tree.
25. A Huffman tree is always balanced.
26. In the tree

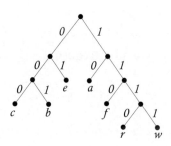

the code for letter f is 110 and the code for the letter b is 001.

27. In the tree in Problem 26, it is assumed that c occurs more frequently than a.
28. In the tree in Problem 26, r and w are assumed to have the least frequency.
29. The weight of the tree in Problem 26 is uniquely determined by the tree.
30. The prefix form of an expression is also called the reverse Polish form.
31. The expression produced using the infix algorithm for the tree

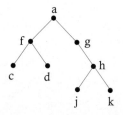

is *cfdagjhk*.

32. The expression produced using the prefix algorithm for the tree

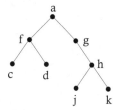

is *cdfjkhga*.

33. There exists a transversal tree with five unique vertices such that the expression produced using the prefix algorithm is the same as the one produced by the postfix algorithm.
34. There exists a transversal tree with five unique vertices such that the expression produced using the infix algorithm is the same as the one produced by the postfix algorithm.
35. There exists a transversal tree with five unique vertices such that the expression produced using the infix algorithm is the same as the one produced by the prefix algorithm.
36. If the vertices for a graph have been preordered, then the spanning tree produced by the breadth-first tree algorithm is unique.
37. The spanning tree produced by the depth-first tree algorithm is unique.
38. If a vertex v is a descendant of a vertex w using the DFT Articulation Point Search algorithm, then $BE(v)$ is always less than or equal to $BE(w)$.
39. If a vertex v is a descendant of a vertex w using the DFT Articulation Point Search algorithm, then $BE(v)$ is less than or equal to $BE(w)$ if and only if w is an articulation point.
40. If the root of a spanning tree has more than one child, then it is an articulation point.
41. The corresponding sequence for the tree in the following figure

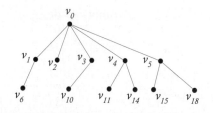

using the tree to sequence algorithm is 1, 0, 2, 3, 4, 4, 5, 5, 0, 0.

42. There are eight spanning trees for the graph

43. If G is a connected graph, A is its adjacency matrix, and K is Kirchoff's matrix, then $K_{i,j} = -A_{i,j}$ for all $i \neq j$.

44. If a connected graph G contains an articulation point, then every bicomponent contains an articulation point of G.

45. Using Prim's Minimum Spanning Tree algorithm, the next edge selected is the edge of minimal weight that has not been selected and will not form a cycle.

46. Kruskal's algorithm and Prim's algorithm always produce the same minimal spanning tree.

47. Using Kruskal's algorithm, it is more difficult to detect cycles.

48. Prim's algorithm with matrices is the simplest algorithm for avoiding cycles.

49. All minimal spanning trees for a fixed graph have the same weight.

50. All minimal spanning trees for a fixed graph are isomorphic.

SUMMARY QUESTIONS

1. In a full m-ary tree with height h how many leaves, how many vertices, and how many internal vertices are there when $m = 4$ and $h = 5$.

2. In the tree

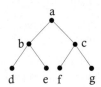

what is the level of d if c is selected as a root?

3. In the previous tree, what is the parent of a if c is selected as a root?

4. In the previous tree, what is the length of the tree if e is selected as a root?

5. Given the following symbols and their codes,

Symbol	code
c	101
y	11
n	1101
u	1011
o	001
t	111
r	1110

encode the word *country*.

6. Given the following symbols with their frequencies

Symbol	frequency
a	8
y	2
e	9
m	4
n	7
i	6
o	3

construct the Huffman tree.

7. Using the Huffman tree in Problem 6, determine the Huffman code.

8. Given the tree

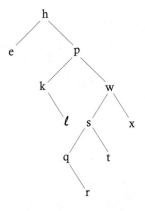

insert the letter n.

9. Given the tree in Problem 8, remove the letter w.

10. Given the tree

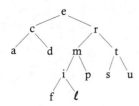

give the inorder transversal of the tree.

11. Given the tree in Problem 10, give the postorder transversal of the tree.

12. Given the tree

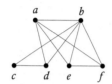

find the breadth-first spanning tree.

13. Given the tree in Problem 12, find the depth-first spanning tree.

14. Assume we have the graph

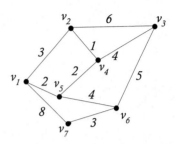

Find the minimum spanning tree using Kruskal's algorithm.

15. Given the graph in Problem 14, find the minimum spanning tree using Prim's algorithm.

ANSWERS TO TRUE-FALSE

1. T 2. T 3. F 4. T 5. F 6. F 7. T 8. F 9. F 10. T 11. F
12. T 13. T 14. F 15. F 16. T 17. F 18. T 19. T 20. T
21. F 22. T 23. F 24. T 25. F 26. T 27. F 28. T 29. F
30. F 31. T 32. F 33. F 34. T 35. T 36. T 37. F 38. F
39. F 40. T 41. F 42. T 43. T 44. T 45. F 46. F 47. T
48. T 49. T 50. F

CHAPTER 16

Networks

16.1 Networks and Flows

A network may be pictured as a system where some product is transported from one point to another through the system. This product may be people, electricity, natural gas, oil, or any other various products. An example would be an oil pipeline system, where oil flows between various points in the system, and our terminology will be consistent with this concept. Using this concept, we may think of the network as a directed graph where the edges are oil pipes between the points, represented by vertices, in the system. There is a positive integer $c(e)$ associated with each edge $e = (v_i, v_j)$, called the capacity of e. If there is no edge between two vertices, then we consider the capacity to be zero. In our example of oil networks, the capacity may be associated with the amount of oil that may pass through the pipe (edge). Before we define a network, we limit our definition of a directed graph. We shall not allow loops, since we will be concerned only with transporting a product between distinct points. Also, if there is an edge from v_i to v_j, then there is no edge from v_j to v_i. Thus, we consider only a product flowing in one direction. We also require that the directed graph be connected, since, if there is a path from a to z, we are interested only in the component containing a and z. If there is no path between a and z, then there is nothing to determine. A directed graph, with these restrictions, is often called a simple connected directed graph. We shall simply call it a directed graph, with the understanding that it has these restrictions.

We shall also have a special vertex a, called the source, and another special vertex z, called the sink. Vertex a has indegree 0, so that nothing flows into the source. Vertex z has outdegree 0, so that nothing flows out of the sink. Thus, the product is shipped from a and has destination z. More precisely, we define a network as follows.

DEFINITION 16.1 A **network** is a directed graph (G, V, E) together with a weight function $c : E \to N$ and designated vertices a, z such that

(i) $\text{indeg}(a) = 0$

(ii) $\text{outdeg}(z) = 0$

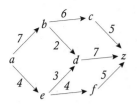

Figure 16.1

For example, the graph in Figure 16.1 is an example of a network where the number on each edge is the capacity.

Into this network, which we envision as an oil pipeline, we introduce the concept of flow. We may think of this as the amount of oil that passes through the pipes in the oil pipelines. Thus, for each edge e, we have a value $f(e)$, which is the flow through that particular edge or pipe. Obviously, the flow through a pipe cannot exceed the capacity of the pipe. We also require that, with the exception of a and z, the flow into a vertex be equal to the flow out of the vertex. This is called the **conservation of flow**. Let in(v) be the set of edges for which v is the terminal vertex and out(v) be the set of edges for which v is the initial vertex. Thus, out(v) is the set of edges out of v and in(v) is the set of edges into v. We, therefore, have the following definition.

DEFINITION 16.2

A **flow** in a network is a function $f : E \to N \cup \{0\}$ such that

(i) for all $e \in E$, $0 \leq f(e) \leq c(e)$

(ii) for all $v \in V$ such that $v \neq a, z$, $\sum_{e \in \text{in}(v)} f(e) = \sum_{e \in \text{out}(v)} f(e)$

Assume that we have a fixed flow. Let $flow(a) = \sum_{e \in \text{out}(a)} f(e)$ so that $flow(a)$ is the flow out of the source vertex a. Let $flow(z) = \sum_{e \in \text{in}(z)} f(e)$ so that $flow(z)$ is the flow into of the sink vertex z. Figure 16.2 is an example of the flow of a network, where the first element in the ordered pair on each edge is the capacity and the second is the flow.

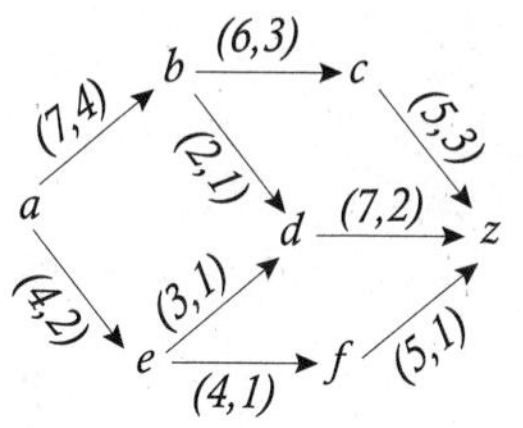

On each edge, the flow is less than the capacity. Notice, for example, that at vertex b, the flow into b is 4 and the flow out of b is 4. Thus, they are the same and we have conservation of flow at b. This is also true at all of the other vertices except a and z. We want to use the conservation of flow to prove something that seems intuitively obvious. The flow out of a is equal to the flow into z, that is, $flow(a) = flow(z)$, which we state later as a theorem. First we look at a specific network.

Figure 16.2

Notice that in Figure 16.2, that $flow(a) = flow(z) = 6$. Let S be the set of vertices $\{b, c, d\}$ and $T = V - S$ be the set of vertices $\{a, e, f, z\}$. If we sum the flow into S, we see from the following table that 9, the flow out, minus 9, the flow in, is equal to 0.

vertex	flow in	edge	flow out	edge
b	4	(a, b)	3	(b, c)
			1	(b, d)
c	3	(b, c)	3	(c, z)
d	1	(e, d)	2	(d, z)
	1	(b, d)		
sum	9		9	

If we add a to S, we then have that 15, the flow out minus 9, the flow in equals $6 = flow(a)$.

vertex	flow in	edge	flow out	edge
b	4	(a,b)	3	(b,c)
			1	(b,d)
c	3	(b,c)	3	(c,z)
d	1	(e,d)	2	(d,z)
	1	(b,d)		
a	0		4	(a,b)
			2	(a,e)
sum	9		15	

Notice that the edges (a,b), (b,c), and (b,d), appear in both the flow in column and the flow out columns. That is because both vertices of each edge are in S. These edges cancel each other out in the subtraction so if we remove them from both sides, we have

vertex	flow in	edge	flow out	edge
b				
c			3	(c,z)
d	1	(e,d)	2	(d,z)
a	0			
			2	(a,e)
sum	1		7	

so again, flow out minus flow in is equal to 6. But since we have eliminated the edges that have both vertices in S, the flow out is the sum of the flow of edges going from S to T, and the flow in is the sum of the flow of edges going from T to S. We shall be discussing these same concepts in the following theorem.

THEOREM 16.3 For any fixed flow f,

$$\text{flow}(a) = \sum_{e \in \text{out}(a)} f(e) = \sum_{e \in \text{in}(z)} f(e) = \text{flow}(z).$$

Proof Let S be a subset of V containing a, but not z, and $T = V - S$. We know by the conservation of flow that for $v \neq a, z$, $\sum_{e \in \text{in}(v)} f(e) = \sum_{e \in \text{out}(v)} f(e)$. Therefore, summing over the vertices in $S - \{a\}$, we have

$$\sum_{v \in S-\{a\}} \sum_{e \in \text{in}(v)} f(e) = \sum_{v \in S-\{a\}} \sum_{e \in \text{out}(v)} f(e)$$

or

$$\sum_{v \in S-\{a\}} \sum_{e \in \text{out}(v)} f(e) - \sum_{v \in S-\{a\}} \sum_{e \in \text{in}(v)} f(e) = 0$$

Therefore, if we also include a, we have

$$\sum_{v \in S} \sum_{e \in \text{out}(v)} f(e) - \sum_{v \in S} \sum_{e \in \text{in}(v)} f(e) = \sum_{e \in \text{out}(a)} f(e)$$

Let (S, T) denote the set of all edges from S to T, that is, $(S, T) = \{e :$ the initial vertex of e is in S and the terminal vertex of e is in $T\}$. Similarly, let (T, S) denote the set of all edges from T to S. Again consider

$$\sum_{v \in S} \sum_{e \in \text{out}(v)} f(e) - \sum_{v \in S} \sum_{e \in \text{in}(v)} f(e)$$

If an edge e has both its initial and its terminal vertices in S, then e appears in both sums and, hence, cancels out. If we cancel these, on the left side, we only have edges that go from S to T, and on the right side, we only have edges that go from T to S. Thus, we have

$$\sum_{v \in S} \sum_{e \in \text{out}(v)} f(e) - \sum_{v \in S} \sum_{e \in \text{in}(v)} f(e) = \sum_{e \in (S,T)} f(e) - \sum_{e \in (T,S)} f(e)$$

Therefore,

$$\sum_{e \in (S,T)} f(e) - \sum_{e \in (T,S)} f(e) = \sum_{e \in \text{out}(a)} f(e)$$

so that given any set S of vertices containing a, but not z, the flow out of S minus the flow into S is equal to $flow(a)$, the flow out of a as seen above. Now let $S = V - \{z\}$ so that $T = \{z\}$. Then $\sum_{e \in (S,T)} f(e)$ is simply the flow into z so that

$$\sum_{e \in (S,T)} f(e) = \sum_{e \in \text{in}(z)} f(e) = flow(z)$$

and $\sum_{e \in (T,S)} f(e) = 0$ since it is the flow out of t. Therefore,

$$\sum_{e \in (S,T)} f(e) - \sum_{e \in (T,S)} f(e) = \sum_{e \in \text{in}(z)} f(e) = flow(z).$$

and $flow(a) = flow(z)$. ∎

In proving the theorem, we also proved the following corollary.

Corollary 16.4 Let S be a subset of V containing a, but not z, and $T = V - S$. Then

$$\sum_{e \in (S,T)} f(e) - \sum_{e \in (T,S)} f(e) = flow(a) = flow(z) \qquad \blacksquare$$

DEFINITION 16.5 The **value of a flow** f, denoted by $val(f)$, is equal to $flow(a) = flow(z)$.

DEFINITION 16.6 Let S be a subset of V and $T = V - S$. Then $\{e : e \in (S, T)\}$ is called a **cut**. If $a \in S$ and $z \in T$, then the cut is called an **a-z cut**.

DEFINITION 16.7 The **capacity of a cut**, $C(S, T) = \sum_{e \in (S,T)} c(e)$.

DEFINITION 16.8 A flow $\hat{f}$ on a network is a **maximal flow** if $val(\hat{f}) \geq val(f)$ for every possible flow f on the network.

DEFINITION 16.9 An $a - z$ cut (S, T) is a **minimal cut** if $C(S, T)$ is less than or equal to the capacity of any other $a - z$ cut.

THEOREM 16.10 Let S be a subset of V containing a, but not z, and $T = V - S$. Then

$$val(f) \leq C(S, T)$$

Proof

$$val(f) = \sum_{e \in (S,T)} f(e) - \sum_{e \in (T,S)} f(e)$$
$$\leq \sum_{e \in (S,T)} f(e)$$
$$\leq \sum_{e \in (S,T)} c(e)$$
$$= C(S, T) \qquad \blacksquare$$

The following corollaries follow immediately from the previous theorem.

Corollary 16.11 If $val(f) = C(S, T)$ for some flow f and $a - z$ cut (S, T), then f is a maximal flow and C is a minimal cut. $\blacksquare$

Corollary 16.12 For some flow f and $a - z$ cut (S, T), $val(f) = C(S, T)$ if and only if $f(e) = c(e)$ for all $e \in (S, T)$ and $f(e) = 0$ for all $e \in (T, S)$. $\blacksquare$

We now seek to find ways to find the maximal flow of a network. Consider the network in Figure 16.3. We can easily find a maximal flow (it need not be unique) as shown in Figure 16.4.

In this case we know that the flow is maximal because the flow out of a cannot exceed the sum of capacity of the edges out of a.

Consider the network in Figure 16.5. It may appear that this network has maximal flow, since there does not appear to be any directed path we can take for which we can increase the flow. However note that the network in Figure 16.6 has a greater flow and, in fact, a maximal flow.

Please note that if the flow from the source is equal to the sum of the capacity of the edges out of the source, or if the flow into the sink is equal to the sum of the capacity of the edges into the sink, then the flow is maximal. However, the flow can be maximal without either of these occurring. For example, the network in Figure 16.7 is a maximal flow.

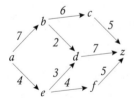

Figure 16.3

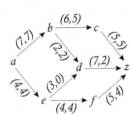

Figure 16.4

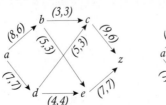

Figure 16.5

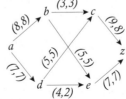

Figure 16.6

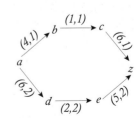

Figure 16.7

So how do we find maximal networks? We do it by forming paths from a to z, ignoring the direction of the edges. We shall call such paths **chains**. Consider again the network with flow shown in Figure 16.5. One such path is

$$a \xrightarrow{(8,6)} b \xrightarrow{(3,3)} c \xrightarrow{(9,6)} z$$

Obviously there is no way to increase the flow on this path since the edge from b to c is filled to capacity. The same is true for chain

$$a \xrightarrow{(8,6)} b \xrightarrow{(5,3)} e \xrightarrow{(7,7)} z$$

Consider, however, the chain

$$a \xrightarrow{(8,6)} b \xrightarrow{(5,3)} e \xleftarrow{(4,4)} d \xrightarrow{(5,3)} c \xrightarrow{(9,6)} z$$

If we increase the value by 2, when the arrow is following the chain and decrease the value by 2 when the arrow is going in the opposite direction, we then get

$$a \xrightarrow{(8,8)} b \xrightarrow{(5,5)} e \xleftarrow{(4,2)} d \xrightarrow{(5,5)} c \xrightarrow{(9,8)} z$$

which increases the flow by 2. The first question we should probably ask is, "Why choose 2?" Obviously we want to increase the flow as much as possible. The flow, however, cannot exceed the capacity for any given edge. This limits us to 2. Also, although it was not a problem this time, if an edge is opposite to the flow of the chain, we cannot then reduce the flow of the edge by more that its current flow, or it would have negative flow. Thus, if there had been no other limit, then

$$e \xleftarrow{(4,4)} d$$

would have limited our change in flow to 4. The second question would probably be, "What does this do to the conservation of flow?" The answer is that conservation of flow is preserved. Consider, for example changing

$$a \xrightarrow{(8,6)} b \xrightarrow{(5,3)} e$$

to

$$a \xrightarrow{(8,8)} b \xrightarrow{(5,5)} e$$

The flow out of b is increased by the same amount as the flow into b. So the net flow through b is unchanged. In the change from

$$b \xrightarrow{(5,3)} e \xleftarrow{(4,4)} d$$

to

$$b \xrightarrow{(5,5)} e \xleftarrow{(4,2)} d$$

the flow from b to e is increased by the same amount that the flow from d to e is decreased, so that the net flow to e is unchanged. Finally, consider the change from

$$e \xleftarrow{(4,4)} d \xrightarrow{(5,3)} c$$

to

$$e \xleftarrow{(4,2)} d \xrightarrow{(5,5)} c$$

The flow from d to e is reduced by the same amount that the flow from d to c is increased. So the net flow out of d is unchanged.

This process for increasing the flow, called augmenting the flow, is fairly simple. Form a chain from a to z. If possible, increase the flow by finding the largest amount that we can add to each of the edges going the same direction as the chain without the flow exceeding the capacity and that we can subtract from each of the edges going the opposite direction without producing negative flow. Since the final edge into z is going the same direction as the chain, the flow into z is increased. Similarly the edge coming from a is going the same direction as the change so, as expected, the flow from a is increased. It is expected because $flow(a) = flow(z)$. Since the capacity is finite, and we are increasing the flow by integral amounts, we eventually arrive at the point when we can no longer augment the flow. When this occurs, we have maximized the flow.

So far we have shown how to augment an already existing flow. One might wonder if there is a way to start at the beginning. The answer is fairly simple. Just begin with the flow where the flow for each edge is 0 and begin augmenting.

We now state a systematic algorithm for finding a maximum flow. With each vertex, we have an ordered pair. The first is the predecessor of the vertex in our chain so we can find our way back. The second is the slack, or the amount that each edge along the path can be increased if the orientation of the edge is the same as the chain (which we will call properly oriented) or decreased if the edge is not properly oriented. Put more simply, if less accurately, the slack of a given vertex is the largest amount that the flow along the chain can be increased up to that vertex and still have a flow. We also have a set S, which consists of all vertices that have not been used in our attempt to find a chain to z. If S is empty before we get to z, then we no longer have a vertex that we have not checked on our quest for z, so we cannot get to z, there are no more augmentations, and the flow is maximized.

■ Algorithm (Ford-Fulkerson) Maximum Flow

1. Set the predecessor of each vertex equal to—and the slack of each vertex equal to the symbol—(unlabeled). A vertex is labeled when the slack is no longer the symbol—. Set the slack of a equal to ∞, so that it will not be a restriction on the slack of other vertices. Let $S = \{a\}$.

2. If S is empty, the flow is maximized. If S is not empty, pick an element from S and remove it. Set it equal to v.

3. If w is not labeled, (v, w) is an edge and $f((v, w)) < c((v, w))$. Let the slack of w be the minimum of $c((v, w)) - f((v, w))$ and the slack of v. Set the predecessor of w to v. If $w \neq z$, add w to S.

4. If w is not labeled, (w, v) is an edge and $f((v, w)) > 0$. Let the slack of w be the minimum of $f((v, w))$ and the slack of v. Set the predecessor of w to v. Add w to S.

5. If z is labeled, use the predecessor function to get back to a, and for each edge on the chain, add the slack of z to the flow of each edge that is properly oriented and subtract the slack of z from each edge that is not properly oriented. Return to step 1.

6. Return to step 2.

We still need to prove that the algorithm indeed produces a maximal flow, but first let's see if we can figure out what's going on.

EXAMPLE 16.13   Find the maximal flow for the network in Figure 16.8.

In step 1, we set the predecessor and the slack equal to the symbol—for each vertex, set the slack for a equal to ∞, and let $S = \{a\}$. We then have the network in Figure 16.9.

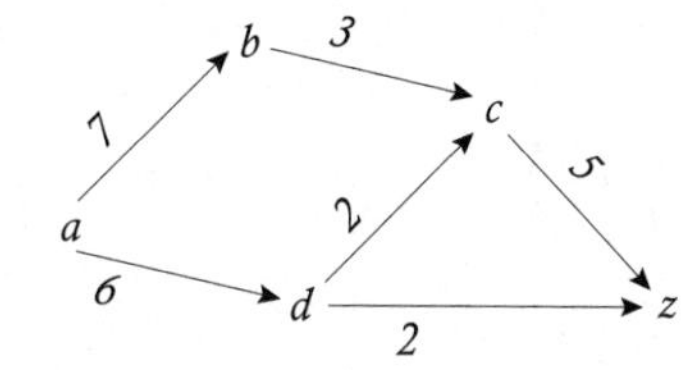

16.1 Networks and Flows

Figure 16.8

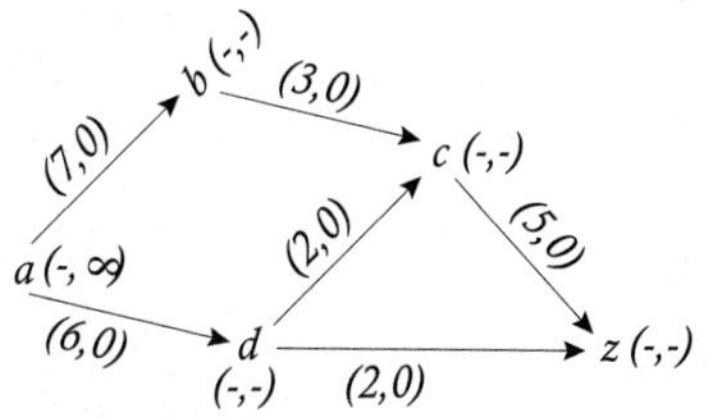

Figure 16.9

In step 2, we check that S is not empty and select a from S. In step 3, we set the slack of b equal to $\min(7, \infty) = 7$ and set the predecessor of b equal to a. We place b in S. We also set the slack of d equal to $\min(6, \infty) = 6$ and set the predecessor of d equal to a. We place d in S. Steps 4 and 5 do not apply, so we return to step 2.

We check that S is not empty and then select a vertex, say b, from S. The set S now contains only d. In step 3, we set the slack of c equal to $\min(3, 7) = 3$ and set the predecessor of c equal to b. We place c in S so that $S = \{c, d\}$. Again steps 4 and 5 do not apply, so we return to step 2.

We check that S is not empty and then select a vertex, say c, from S. The set S again contains only d. In step 3, we set the slack of z equal to $\min(3, 5) = 3$ and set the predecessor of z equal to c. We do not place z in S. In step 4, we would label d, but it is already labeled.

In step 5, we see that z is labeled, and, using the predecessor function, we find that we have the chain

$$a \xrightarrow{(7,0)} b \xrightarrow{(3,0)} c \xrightarrow{(5,0)} z$$

and adding 3, the slack of z to the flow of each edge, we have

$$a \xrightarrow{(7,3)} b \xrightarrow{(3,3)} c \xrightarrow{(5,3)} z$$

giving us the network in Figure 16.10.

We now return to step 1 where we again reset the labels and predecessors and let $S = \{a\}$. In step 2, we select a from S. In step 3, we set the slack of b equal to $\min(\infty, 7-3) = 4$ and set the predecessor of b equal to a. We place b in S. We also set the slack of d equal to 6 and set the predecessor of d equal to a. We place d in S. Steps 4 and 5 do not apply, so we return to step 2. Choose b from S. Since the flow from b to c is equal to the capacity from b to c, we cannot label c. We select d from S, and continue the process so that we have the chain

$$a \xrightarrow{(6,0)} d \xrightarrow{(2,0)} z$$

and adding 2, the slack of z to the flow of each edge, we have

$$a \xrightarrow{(6,2)} d \xrightarrow{(2,2)} z$$

giving us Figure 16.11.

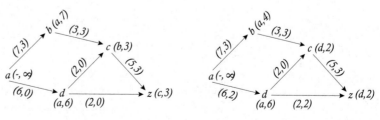

Figure 16.10   **Figure 16.11**

We now return to step 1 and repeat the process until we have the chain

$$a \xrightarrow{(6,2)} d \xrightarrow{(2,0)} c \xrightarrow{(5,3)} z$$

and adding 2, the slack of z to the flow of each edge, we have

$$a \xrightarrow{(6,4)} d \xrightarrow{(2,2)} c \xrightarrow{(5,5)} z$$

giving us Figure 16.12.

We now return to step 1 where we again reset the labels and predecessors and let $S = \{a\}$. In step 2, we select a from S. As before, in step 3, we set the slack of b equal to 4 and set the predecessor of b equal to a. We place b in S. We also set the slack of d equal to $6 - 4 = 2$ and place d in S. We return to step 2. Choose b from S. Since the flow from b to c is equal to the capacity from b to c, we cannot label c, so we eventually return to step 2. Choose d from S. Since the flow from d to both c and z is equal to their capacity, we cannot label c and z. We again return to step 2, but S is empty, so we are finished. The final flow is shown in Figure 16.13.

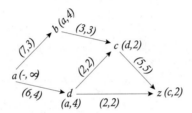

Figure 16.12   Figure 16.13

EXAMPLE 16.14 Find the maximal flow for the network in Figure 16.14.

This time we will give fewer details on the first few passes. The first pass we label $a(-, \infty)$, $b(a, 9)$, $d(a, 8)$, $c(b, 4)$, $e(b, 6)$, and $z(c, 4)$. This gives us the $a - z$ path

$$a \xrightarrow{(9,0)} b \xrightarrow{(4,0)} c \xrightarrow{(8,0)} z$$

and since the slack of z is 4, we add 4 to each flow, giving

$$a \xrightarrow{(9,4)} b \xrightarrow{(4,4)} c \xrightarrow{(8,4)} z$$

so that we have Figure 16.15.

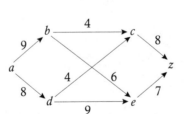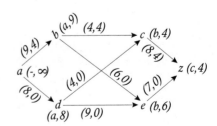

Figure 16.14   Figure 16.15

The second pass we label $a(-, \infty)$, $b(a, 5)$, $d(a, 8)$, $c(d, 4)$, $e(d, 8)$, and $z(e, 7)$. This gives us the $a - z$ path

$$a \xrightarrow{(8,0)} d \xrightarrow{(9,0)} e \xrightarrow{(7,0)} z$$

and since the slack of z is 7, we add 7 to each flow, giving

$$a \xrightarrow{(8,7)} d \xrightarrow{(9,7)} e \xrightarrow{(7,7)} z$$

so that we have Figure 16.16.

The third pass we label $a(-, \infty)$, $b(a, 5)$, $d(a, 1)$, $c(d, 1)$, $e(d, 1)$, and $z(c, 1)$. This gives us the $a - z$ path

$$a \xrightarrow{(8,7)} d \xrightarrow{(4,0)} c \xrightarrow{(8,4)} z$$

and since the slack of z is 1, we add 1 to each flow, giving

$$a \xrightarrow{(8,8)} d \xrightarrow{(4,1)} c \xrightarrow{(8,5)} z$$

so that we have Figure 16.17.

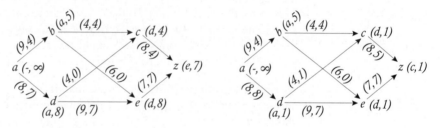

Figure 16.16 **Figure 16.17**

It now starts to get interesting so we go into more detail. We label $b(a, 5)$ but cannot label d, placing b in S. Selecting b from S, we cannot label c, but we label $e(b, 5)$, placing c in S. Selecting e from S, we cannot label z, but, since d has not been labeled, we can label d. Since (d, e) is not properly oriented, we let the slack of d equal $\min(7, 5) = 5$. We set the predecessor of d equal to e and place d in S. Selecting d from S, the only choice we have is to label $c(d, 3)$ and place c in S. We select c from S and label $z(c, 3)$. This gives us the $a - z$ path

$$a \xrightarrow{(9,4)} b \xrightarrow{(6,0)} e \xleftarrow{(9,7)} d \xrightarrow{(4,1)} c \xrightarrow{(8,5)} z$$

and since the slack of z is 3, we add 3 to each flow, except for edge (d, e). Since it is not properly oriented, we subtract 3 from the flow of this edge giving

$$a \xrightarrow{(9,7)} b \xrightarrow{(6,3)} e \xleftarrow{(9,4)} d \xrightarrow{(4,4)} c \xrightarrow{(8,8)} z$$

and we have Figure 16.18.

We now begin the final pass. We begin with $S = \{a\}$. We remove a from S label $b(a, 2)$, and place b in S. We remove b from S, label $e(b, 2)$, label $c(b, 2)$, and place c, e in S. We remove e from S, label $d(e, 2)$, since the slack of d is equal to the minimum of the flow of (d, e) and the slack of d. We then place d in S.

Next we remove d from S but cannot perform any of the remaining steps so we return to step 2. Next we remove c from S but cannot perform any of the remaining steps so we return to step 2. But S is empty so we are finished. We now have Figure 16.19.

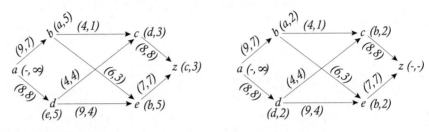

Figure 16.18   Figure 16.19

THEOREM 16.15 The maximum flow algorithm produces a maximum flow for the network.

Proof Let S be the set of all vertices that were labeled during the last pass of the algorithm and $T = V - S$. The set S is not empty since $a \in S$. If e is an edge from S to T, say $e = (s, t)$, so that t is unlabeled, then $f(e) = c(e)$, since otherwise t could be labeled. If e is an edge from T to S, say $e = (t, s)$, so that t is unlabeled, then $f(e) = 0$, since otherwise t could be labeled. Therefore, by Corollaries 16.11 and 16.12, f is a maximal flow and (S, T) is a minimal cut. ∎

From Corollary 16.12 we know that if $val(f) = C(S, T)$ for some flow f and $a - z$ cut (S, T), then f is a maximal flow and C is a minimal cut. We have shown in the previous algorithm how to find a minimum cut (S, T) so that f is a maximal flow, $val(f) = C(S, T)$. Putting these together, we have the following theorem.

THEOREM 16.16 On a given network N, the flow f on N is a maximal flow if and only if there exists a cut (S, T) such that $val(f) = C(S, T)$. ∎

Notice that in Example 16.13, S consists of all of the vertices that we could label in the last pass, so $S = \{a, b, d\}$ and $T = \{c, z\}$.
In Example 16.14, $S = \{a, b, c, d, e\}$ and $T = \{z\}$.

Exercises

In Figure 16.20,

1. Verify the conservation of flow at b, c, and d.

2. Find $val(f)$, the value of the flow.

3. Find the value of $C(S, T)$ when $S = \{a, b, c, d\}$.

4. Find the value of $C(S, T)$ when $S = \{a, d, e\}$.

5. Find the value of $C(S, T)$ when $S = \{a, b, d\}$.

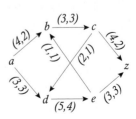

Figure 16.20

In Figure 16.21,

6. Verify the conservation of flow at b, d, and e.
7. Find $val(f)$, the value of the flow.
8. Find the value of $C(S, T)$ when $S = \{a, b, c, d\}$.
9. Find the value of $C(S, T)$ when $S = \{a, b, d, e\}$.
10. Find the value of $C(S, T)$ when $S = \{a, b, e\}$.

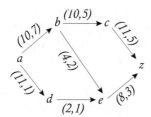

Figure 16.21

11. Complete the flow in the network in Figure 16.22 such that there is conservation of flow.

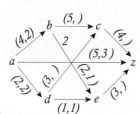

Figure 16.22

12. Complete the flow in the network in Figure 16.23 such that there is conservation of flow.

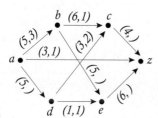

Figure 16.23

13. Find the maximum flow in the flow network in Figure 16.24.

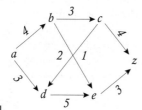

Figure 16.24

14. Find the maximum flow in the flow network in Figure 16.25.

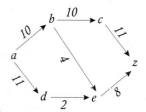

Figure 16.25

15. Find the maximum flow in the flow network in Figure 16.26.

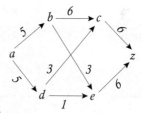

Figure 16.26

16. Find the maximum flow in the flow network in Figure 16.27.

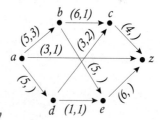

Figure 16.27

17. Find the minimal cut for the flow network in Exercise 13.
18. Find the minimal cut for the flow network in Exercise 14.
19. Find the minimal cut for the flow network in Exercise 15.
20. Find the minimal cut for the flow network in Exercise 16.

16.2 Matching

We immediately get a chance to use our theory of network flows. Recall that a graph $G = (V, E)$ is a bipartite graph if V can expressed as the disjoint union of nonempty sets, say $V = A \cup B$, such that every edge has the form $\{a, b\}$ where $a \in A$ and $b \in B$. Thus every edge connects a vertex in A to a vertex in B and no vertices both in A or both in B are connected. A subset M of E is a **matching** if no two edges have a common vertex. Thus, no two edges are incident. If $\{a, b\}$ is an edge in a matching, then both a and b have **matching edges**. The vertices a and b are **matched**. For example, in the following bipartite graph in Figure 16.28, the matching edges, denoted by heavy lines in Figure 16.28, are a matching.

One example would be matching people with jobs. One set of vertices is a set of people and the other set is a set of jobs. The edges of the graph indicate which people are capable of doing particular jobs. Another example would be matching boys with girls. The boys form one set and the girls the other. An edge would indicate that a pair are compatible. (We shall assume that this is a bipartite graph without further comment.)

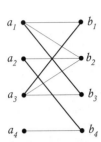

Figure 16.28

DEFINITION 16.17 A matching M on a bipartite graph $G = (V, E)$ is **maximal** if no other matching on G contains more edges.

DEFINITION 16.18 A matching M on a bipartite graph $G = (V, E)$, where $V = A \cup B$, is **complete** if for every $a \in A$, there exists $b \in B$ such that $\{a, b\} \in M$.

EXAMPLE 16.19 The graph in Figure 16.29 is maximal but not complete. ∎

Obviously our next task is to determine how to find a maximal matching. To help us, we change our bipartite graph into a directed graph, where edges are from vertices in A to vertices in B. We also add two new vertices a and z and edges from a to each vertex in A and edges from each vertex in B to z. Therefore, the graph in Figure 16.30 is changed to the one in Figure 16.31.

We have changed a bipartite graph into a network. Such a network is called a **matching network**. For each $a_i \in A$, let e_i be the edge from a to a_i. Let the capacity $c(e_i) = 1$ for all e_i. Similarly, for each $b_i \in B$, let ε_i be an edge from b_i to z, and let the capacity $c(\varepsilon_i) = 1$ for all ε_i. For an edge k_{ij} from a_i to b_j we let $c(k_{ij}) = |A|+1$, so that it is greater than the number of vertices in A. The reason will be explained later. Notice that since the flow into each a_i cannot exceed 1, the flow from any edge k_{ij} from a_i to b_j must be either 1 or 0 **regardless of the capacity** $c(k_{ij})$. For the edge k_{ij} from a_i to b_j, we define the flow $f(k_{ij})$ equal to 1 if there is a matching edge between a_i and b_j and equal to 0 otherwise. The value $val(f)$ of the flow is then equal to the number of edges in the matching between A and B. A maximal flow occurs when there is a maximal matching. We have now converted the maximal matching problem into a maximal flow problem.

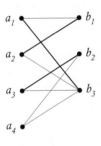

Figure 16.29

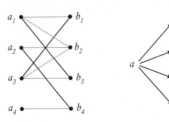

Figure 16.30   Figure 16.31

16.2 Matching

Using the maximal flow algorithm developed in the last section, we must start with an edge e_i to a vertex a_i, which is not a matching edge, since otherwise $f(e_i) = c(e_i) = 1$. We then take a directed edge k_{ij} over to a vertex b_j of B. If b_j has not been matched, then we create a matching edge between a_i and b_j, and we have augmented the flow. If b_j has been matched, then we must return by an edge that is not properly oriented. Hence, its flow must be 1, so we continue our chain back to A on an edge that is a matching edge. Again returning to B, we must select an edge that is not a matching edge. We continue this chain until we reach a vertex in B that is not matched. On the chain, we must add 1 to the flow of the properly oriented edges from A to B and subtract 1 from the edges from B to A that are not properly oriented. The result on the chain is to change to a matching edge, any edge that was not a matching edge and to remove matching edges where there were matching edges. This adds one matching edge to the bipartite graph and so increases the matching by 1. Using the maximal flow algorithm in this manner enables us to achieve a maximal matching.

We can begin with no matching edges as required by the maximum flow algorithm, in which case we use the algorithm to draw matching edges and then augment them as necessary to find a maximal matching. In our examples, we shall assume we have some matching and then augment it until we have a maximal matching.

Once we have developed the technique, we see that we really don't have to worry about a and z. We simply begin with a vertex in A that does not have a matching edge and continue back and forth until we end at a vertex in B that does not have a matching edge.

EXAMPLE 16.20

Given the graph in Figure 16.32, develop a maximal matching. We begin at a_4, since it does not have a matching edge. We follow the directed edge to b_4, since (a_4, b_4) is not a matching edge. We then continue our chain back to a_2, since (a_2, b_4) is a matching edge. From a_2, we continue our chain to b_1, since (a_2, b_1) is not a matching edge. From b_1, we continue our chain back to a_3, since (a_3, b_1) is a matching edge. From a_3, we continue our chain to b_3, since (a_3, b_3) is not a matching edge. From b_3, we continue our chain back to a_1, since (a_1, b_3) is a matching edge. From a_1, we continue our chain to b_2, since (a_1, b_2) is not a matching edge. From b_2, there is no matching edge, so we have completed our chain, which is $a_4 \rightarrow b_4 \leftarrow a_2 \rightarrow b_1 \leftarrow a_3 \rightarrow b_3 \leftarrow a_1 \rightarrow b_2$. We then change matching edges to nonmatching edges and conversely to produce the graph in Figure 16.33.

Since the number of edges in the matching equals the number of vertices in A, we now have a maximal matching and, in fact, a complete matching.

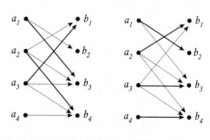

Figure 16.32   Figure 16.33

EXAMPLE 16.21

Given the graph in Figure 16.34, develop a maximal matching. We begin at a_5, since it does not have a matching edge. We follow the directed edge to b_5, since (a_5, b_5) is not a matching edge. We then continue our chain back to a_3, since (a_3, b_5) is a matching edge. From a_3, we continue our chain to b_4, since (a_3, b_4) is not a matching edge. We then continue our chain back to a_1, since (a_1, b_4) is a matching edge. From a_1, we continue our chain to b_3, since (a_1, b_3) is not a matching edge. We then continue our chain back to a_2, since (a_2, b_3) is a matching edge. From a_2, we continue our chain to b_2, since (a_2, b_2) is not a matching edge. From b_2, there is no matching edge, so we have completed our chain, which is

$$a_5 \to b_5 \leftarrow a_3 \to b_4 \leftarrow a_1 \to b_3 \leftarrow a_2 \to b_2$$

We then change matching edges to nonmatching edges and conversely to produce the graph in Figure 16.35.

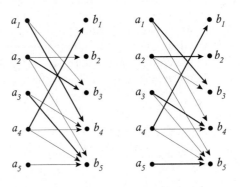

Figure 16.34   Figure 16.35

Since the number of edges in the matching equals the number of vertices in A, we now have a maximal matching and, in fact, a complete matching. ∎

DEFINITION 16.22   For a subset X of A, let $R(A) = \{b : b \in B$ and b is adjacent to a vertex in $A\}$. $R(A)$ is called the range of A.

This is consistent with the definition of range in the discussion of relations given in Chapter 2. We are now able to state necessary and sufficient conditions for a matching to be complete.

THEOREM 16.23   **(Hall's Theorem)** A bipartite graph $G(E, V)$, where $V = A \cup B$, has a complete matching if and only if for each subset $X \subseteq A$, $|X| \leq |R(X)|$.

Proof   Clearly if $G(E, V)$ is complete, then it is necessary that $|X| \leq |R(X)|$ for every subset X of A, since otherwise we could not have a matching of all the elements of X and, hence, no matching of the elements of A. In particular, every element of A is adjacent to an element of B, and $|A| \leq |B|$.

Conversely, assume that $|X| \leq |R(X)|$ for every subset X of A. By Theorem 16.16 we know there is an (S, T) cut such that $val(f) = C(S, T)$. We already know that $val(f)$ is the number of edges in the maximal matching, so $C(S, T)$ is the number of edges in the maximal matching. Let $|A| = n$. Then $C(S, T) \leq n$, since the number of elements in a maximal matching cannot exceed the number of vertices in A. Remember that $C(S, T)$ is the sum of the capacities of all the edges between S and T. We know that this cut cannot contain an edge k_{ij} between a vertex a_i in A and a vertex b_j in B, since $c(k_{ij}) = n + 1$. Therefore, if $a_i \in S$, then $R(\{a\}) \subseteq S$. It follows that $R(A \cap S) \subseteq S$. There is an edge between every vertex in $R(A \cap S)$ and z. Further, each of these edges is in the cut (S, T), since $z \in T$. Therefore, each of these edges contributes 1 to $C(S, T)$. A is the disjoint union of $A \cap S$ and $A \cap T$. Hence, $n = |A| = |A \cap S| + |A \cap T|$. There is an edge between a and every element of $A \cap T$. Since $a \in S$, each of these edges is in the cut (S, T). Therefore, each edge contributes 1 to $C(S, T)$. We conclude that

$$C(S, T) \geq |R(A \cap S)| + |A \cap T|$$
$$\geq |A \cap S| + |A \cap T| \quad \text{since, by hypothesis, } |A \cap S| \leq |R(A \cap S)|$$
$$= |A| = n$$

But remember that $C(S, T)$ is the number of edges in the maximal matching. Therefore, $C(S, T) \leq n$, and it follows that $C(S, T) = n$, so the maximal matching is complete. ∎

We now present a different way of representing the steps in finding a maximal matching. This time we use matrices. We begin with a variation of the incidence matrix. Since the only edges are from A to B, only that portion of the matrix of A where the row corresponds to an element of A and the column corresponds to an element of B is needed. Therefore, we place the vertices in A along the left side and we place the vertices in B along the top. As before, we place a 1 in the ith row, jth column if there is an edge from a_i to b_j. We begin by placing brackets around each 1 in the ith row, jth column if there is a matching edge from a_i to b_j. We illustrate this process for the matching in Example 16.20. We then place a # at the end of row with a_4 to indicate that there is no in [1] in that row and hence no matching edge for a_4. We consider this row to be labeled and will consider any row to be labeled that has a symbol in the last column of that row. Likewise, any column is labeled that has a symbol in the bottom row. The matrix at this point is

	b_1	b_2	b_3	b_4	
a_1	0	1	[1]	0	
a_2	1	0	1	[1]	
a_3	[1]	0	1	1	
a_4	0	0	0	1	#

Since the row containing a_4 is labeled, wherever there is a 1 without a bracket in that row, we place a_4 at the bottom of the column containing each 1. The numbers at the bottom and at the right are to help us find our way back. Going to the 1 in column b_4 is equivalent to going from a_4 to b_4 on a nonmatching edge. We now select the row in the b_4 column containing [1], which occurs in a_2 and place b_4 in the right column of that row. This is equivalent to taking a matching edge from b_4 to a_2. At this point we have

	b_1	b_2	b_3	b_4	
a_1	0	1	[1]	0	
a_2	1	0	1	[1]	b_4
a_3	[1]	0	1	1	
a_4	0	0	0	1	#
				a_4	

We now find each occurrence of a 1 in row a_2 and place an a_2 at the bottom of each column. In this case we have 1's in columns b_1 and b_3. Thus, we have two choices for an unmatched edge. In columns b_1 and b_3, we find the rows containing [1]. These are rows a_3 and a_1, respectively. We then place b_1 in the right column of row a_3 and b_3 in the right column of row a_1. This gives us

	b_1	b_2	b_3	b_4	
a_1	0	1	[1]	0	b_3
a_2	1	0	1	[1]	b_4
a_3	[1]	0	1	1	b_1
a_4	0	0	0	1	#
	a_2		a_2	a_4	

In row a_1 we seek a 1 in one of the columns. This occurs in column b_2. We then place a_1 at the bottom of this column. In row a_3 we seek a 1 in one of the columns. This occurs in columns b_3 and b_4, but these columns are already labeled. In column b_2, there is no [1], so we have found an unmatched edge in B and, therefore, the chain we are seeking. We now have the matrix

	b_1	b_2	b_3	b_4	
a_1	0	1	[1]	0	b_3
a_2	1	0	1	[1]	b_4
a_3	[1]	0	1	1	b_1
a_4	0	0	0	1	#
	a_2	a_1	a_2	a_4	

Backtracking, using the edge markers at the bottom and extreme right, we see that we have the chain

$$a_4 \to b_4 \leftarrow a_2 \to b_3 \leftarrow a_1 \to b_2$$

Note that this is *not* the same chain we got in Example 16.20. It is shorter. Changing matched edges to unmatched edges and unmatched edges to matched edges as in Example 16.20 gives us the matching in Figure 16.36.

To get the new matrix we simply take each 1 in our journey and change it to [1], and take each [1] in our journey and change it to 1. This gives us

	b_1	b_2	b_3	b_4
a_1	0	[1]	1	0
a_2	1	0	[1]	1
a_3	[1]	0	1	1
a_4	0	0	0	[1]

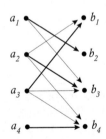

Figure 16.36

Using the procedure above for the matching in Example 16.21, we begin with the matrix

	b_1	b_2	b_3	b_4	b_5
a_1			1	[1]	
a_2			[1]		1
a_3				1	[1]
a_4	[1]				1
a_5					1

and finish with the matrix

	b_1	b_2	b_3	b_4	b_5	
a_1			1	[1]		b_4
a_2		1	[1]		1	b_3
a_3				1	[1]	b_5
a_4	[1]			1	1	
a_5					1	#
	a	a_2	a_1	a_3	a_5	

Backtracking, using the edge markers at the bottom and extreme right, we see that we have the chain

$$a_5 \to b_5 \leftarrow a_3 \to b_4 \leftarrow a_1 \to b_3 \leftarrow a_2 \to b_2$$

which is the same result given in Example 16.21. The matrix of the new matching is

	b_1	b_2	b_3	b_4	b_5	
a_1			[1]	1		b_4
a_2		[1]	1		1	b_3
a_3				[1]	1	b_5
a_4	[1]			1	1	
a_5					[1]	#
	a	a_2	a_1	a_3	a_5	

Computer Science Application

Recall that an undirected graph G is bipartite if all edges of G go between two sets of nodes. If G has n vertices and A is the $n \times n$ adjacency matrix of G, the following algorithm determines if G is bipartite or not.

boolean function bipartite($n \times n$ matrix A)

    integer array Group[$n + 1$]
    integer queue Q

    Group = 0
    Group[1] = 1
    $k = 1$
    while k is not equal to $n + 1$ do
        for $i = 1$ to n do
            if $A[k, i] = 1$ then
                if Group[i] = Group[k] then
                    return false
                else if Group[i] = 0 then
                    insert i to the rear of Q
                    Group[i] = (Group[k] mod 2) + 1
                end-if
            end-if
        end-for
        if Q is empty then
            $k = 2$
            while Group[k] is not equal to 0 do
                $k = k + 1$
            end-while

```
        else
            take integer from front of Q and place into k
        end-if
    end-while
    return true
end-function
```

The expression (Group[k] mod 2) + 1 produces 1 when Group[k] is 2, and 2 when Group[k] is 1. If the graph is bipartite, the Group array will contain 1's and 2's in the first n components, giving a division of the vertices into two distinct groups. Group[$n+1$] is always 0 and provides the necessary control of the major while-loop.

Exercises

Find a maximal matching for the following bipartite graphs in the following figures using methods described in this chapter. Is this matching complete?

Find a maximal matching for the following bipartite graphs using methods described in this chapter. Is this matching complete?

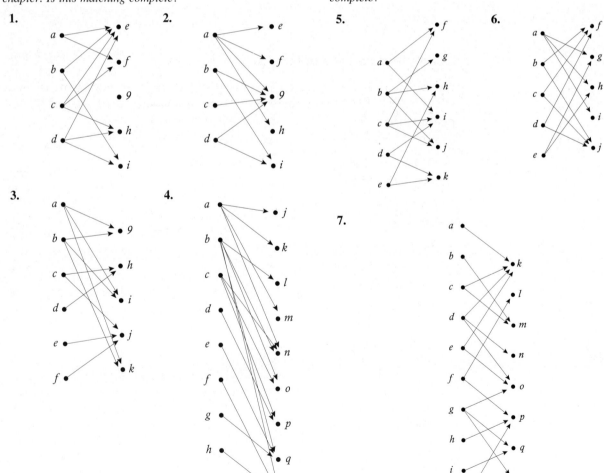

8. Six people are seeking a job. Person a is qualified for jobs A, D, and F. Person b is qualified for job C. Person c is qualified for jobs B and D. Person d is qualified for jobs A and E. Person e is qualified for jobs B and D. Person f is qualified for job E. Find a matching that will give the most people a job for which they are qualified. Does everyone get a job?

9. Five girls want a date for the dance. Alice likes Charles and Edward; Betty likes Albert and Dan; Clara likes Charles and Edward; Donna likes Dan; Elizabeth likes Bart and Edward. Match as many girls as possible with dates that they like. Is it possible for each girl to date a boy whom she likes?

10. Amy is on committees a and b. Bruce is on committees a and c. Charles is on committee b. Denise is on committees b, c, and e. Ellen is on committees d and e. The president of the student government wants to pick five members for the student council, each from a different committee. Is this possible?

11. Six friends are going to a costume party. Each wants to select a different costume. Ann likes the clown costume, Ben likes the ghost and Frankenstein costumes, Courtney likes the clown, and Western costumes. Dennis likes the ghost, Dracula, and Frankenstein costumes. Ellen likes the clown, Western, and Sherlock Holmes costumes. Frank likes the Dracula and clown costumes. Is it possible for each person to have a distinct costume that he or she likes?

12. Write a formal algorithm for finding the maximal matching M on a bipartite graph $G = (V, E)$ without using matrices.

13. Write a formal algorithm for finding the maximal matching M on a bipartite graph $G = (V, E)$ using matrices.

16.3 Petri Nets

In this section, we will continue to consider bipartite graphs with directed edges. However, unlike the previous section, we shall allow directed edges in both directions. Let $G(V, E)$ be a directed bipartite graph, where $V = P \cup T$. The set P is called the set of **places** and the set T is called the set of **transitions**. E is the set of directed edges between P and T. We also include a set of functions M such that each $\mu \in M$ is a function that assigns a nonnegative integer to each element of P. The function μ is called the **marking** of G. The directed graph with these properties is called a **Petri net**.

Petri nets primarily are used to model concurrent processes. They are used to model components of a computer, parallel processing, robotics, and even to describe music structures (see Haus [40]). In general, Petri nets are used to detect flaws in the design of a system although they obviously have other uses. It has many of the properties of a flowchart and also of a finite state machine, which will defined in a later chapter. For details, see Peterson [79].

Each place in the Petri net is denoted by a circle, and each transition is denoted by a vertical line. If p is a place, then $\mu(p)$ is called the marking of p. The marking of the set G is shown on the graph by placing large black dots called **tokens** in the circles representing the places. If the circle of a place p is blank, then there are no tokens in p so that $\mu(p) = 0$.

Consider the graph in Figure 16.37. In this Petri net, p_1 and p_2 are places and t_1 is a transition. Place p_1 contains 1 token and p_2 contains no tokens.

A **firing** of a transition t is the removal of a token from each place p_i, such that there is a directed edge from p_i to t, and the addition of a token to each place p_j, such that there is a directed edge from t to p_j. For example, the firing of t_1 in the above example results in the Petri net in Figure 16.38.

In the Petri net in Figure 16.39, the firing of the transition t_1 results in the Petri net in Figure 16.40.

In the Petri net in Figure 16.41, the firing of the transition t_1 results in the Petri net in Figure 16.42.

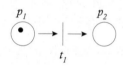

Figure 16.37

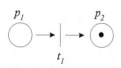

Figure 16.38

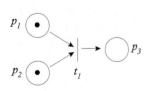

Figure 16.39

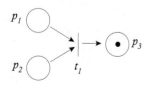

Figure 16.40

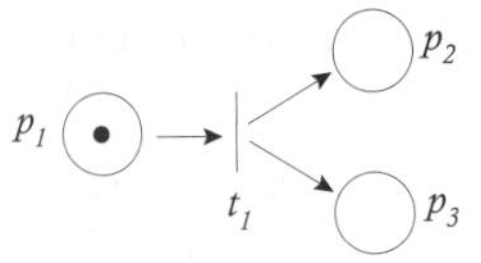

640 Chapter 16 Networks

Figure 16.41

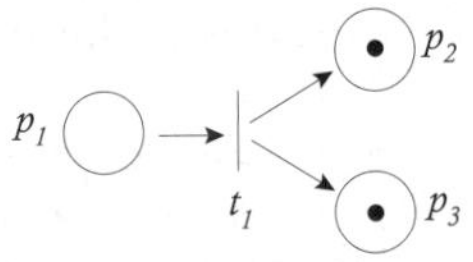

Figure 16.42

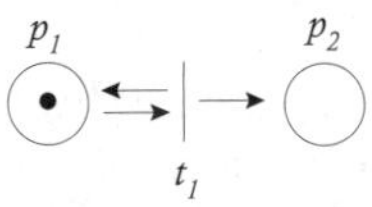

Figure 16.43

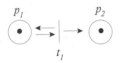

Figure 16.44

In the Petri net in Figure 16.43, the firing of the transition t_1 results in the Petri net in Figure 16.44, since there is a directed edge from t_1 back to p_1.

A transition t may fire only when each place p_i, such that there is a directed edge from p_i to t, contains a token. When this occurs, the transition t is said to be **enabled**. The transition t_1 in the Petri net shown in Figure 16.45 cannot fire since there is no token in p_2.

In some cases, we may want more than one token to be removed or added when a transition is fired. If more than one token is to be removed from a place p when the transition t is fired, the directed edge from p to t is labeled with the number of tokens to be removed from p. If there is no label, then the number is assumed to be 1. If more than one token is to be added to a place p when the transition t is fired, the directed edge from t to p is labeled with the number of tokens to be added to p. If there is no label, then the number is again assumed to be 1. If n tokens are to removed from p when the transition t is fired, then there must be n tokens in p if t is enabled. In the Petri net in Figure 16.46, the firing of t_1 results in the Petri net in Figure 16.47. The transition t_1 cannot fire again, since there are not enough tokens in p_1.

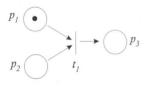

Figure 16.45

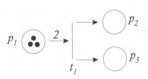

Figure 16.46

The firing of a transition is assumed to be instantaneous and may occur at any time that the transition is enabled. The order in which transitions are fired is limited only by the necessity of enough tokens being available so that the transition is enabled. It is quite possible for more than one transition to be enabled at any given time, and these may fire in any given order. For example, if we are modeling several processors that are sharing data and peripherals, it is uncertain when one of the processors may need a printer or data. The Petri net is used to make sure that, even within this randomness, the system will work properly.

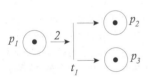

Figure 16.47

The **state** s of a Petri net is the number of tokens in each place, as determined by a function $\mu_i \in M$. Thus, the state of a Petri net is the same as the marking of a Petri net. The term *state* is used when considering automata and other finite state machines. Since μ_i completely determines the state, we shall refer to the state as μ_i. The firing of a transition changes the state as determined by another function μ_j of M. Thus, the new state depends on the current state of the Petri net and the transition fired. Thus, $\mu_j = \delta(\mu_i, t)$ where $\delta(\mu_i, t)$ is the state produced when the Petri net is in state μ_i. If at any point there is no transition that is enabled to fire, then the state function δ is no longer defined and the process described by the Petri net ends.

Consider the Petri net in Figure 16.48. When transition t_1 is fired, the Petri net is in the state shown in Figure 16.49. If transition t_4 is fired, we then have the Petri net in the state shown in Figure 16.50.

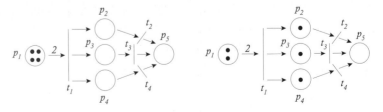

Figure 16.48

Figure 16.49

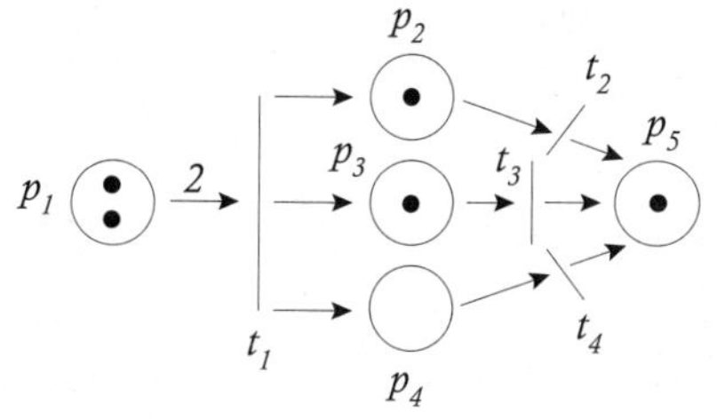

Figure 16.50

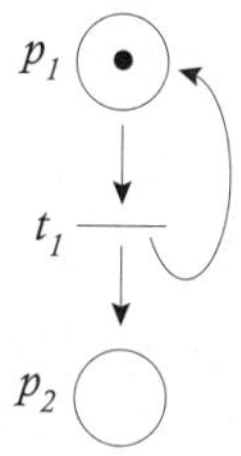

Figure 16.51

It is obvious that eventually all of the tokens will end in p_5 and the process will terminate. In the Petri net shown in Figure 16.51, the process never ends since t_1 is always enabled.

The Petri net in Figure 16.52 performs the computation $(a+b) \times (c+d)$. Either t_1 or t_2 may be fired first; however, t_3 is not enabled until both t_1 and t_2 are fired.

Figure 16.52

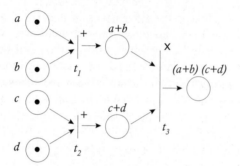

A state μ_j is **immediately reachable** from a state μ_i if the firing of some transition t, while in state μ_i, produces state μ_j. More formally state, μ_j is immediately reachable from a state μ_i if there exists a transition t such that $\mu_j = \delta(\mu_i, t)$. A state μ_j is **reachable** from a state μ_i if, beginning in state μ_i, the firing of a sequence of transitions produces state μ_j. More formally, μ_{i_m} is **reachable** from a state μ_{i_1} if there exists states $\mu_{i_2}, \mu_{i_3}, \mu_{i_4}, \ldots, \mu_{i_{m-1}}$ and transitions $t_1, t_2, t_3, \ldots, t_{m-1}$ such that $\mu_{i_{k+1}} = \delta(\mu_{i_k}, t_k)$ for $k = 1, 2, \ldots, m-1$. Thus, the Petri net in the state shown in Figure 16.53 is immediately reachable from the Petri net in the state shown in Figure 16.54 since it is produced by the firing of t_1.

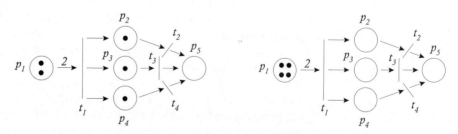

Figure 16.53

Figure 16.54

The Petri net in the state shown in Figure 16.55 is reachable from the Petri net in the state shown in Figure 16.56 since it is produced by the firing of t_1 followed by t_4.

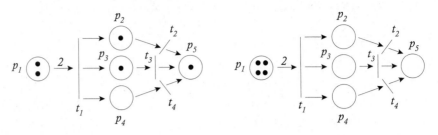

Figure 16.55   Figure 16.56

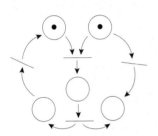

Figure 16.57

A Petri net is **live** if for any state μ and any transition t, there exists a state μ' that is reachable from μ such that the transition t is enabled in μ'. Thus, regardless of the current state, there is a sequence of firing of transitions, beginning at that state, so that any given transition may fire. It is easily seen that the Petri net in Figure 16.57 is live.

One definition of a Petri net being deadlocked is that it is not live. Therefore, there is a state so that if the Petri net is in that state, one or more transitions can never be fired. We shall call this a **partial deadlock**. We define a Petri net to be **deadlocked** if there is a state where no transition can be fired. Thus, there exists a state μ such that $\delta(\mu, t)$ is undefined for every transition t. We consider a standard example where two processors share at least two resources, for example, a printer and a storage area, each of which cannot be used by both processors at the same time. However, each processor must use both the printer and the storage area to perform the task. If one person has access to the printer and the other has access to the storage area, they cannot complete their task. Also they cannot release the resource that they have so the entire system is deadlocked. This is shown in Figure 16.58.

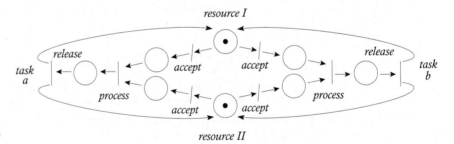

Figure 16.58

Another example, which will be described in the problems, is the **dining philosophers' problem**.

Another problem considered when using a Petri net when sharing files is the mutual exclusion problem. Suppose it is not desirable for two people to access data at the same time. For example, it would not be desirable for one person to read the data while another is changing it. This problem is solved by mutual exclusion. Until one person is through with the data and has modified and returned it, the other person cannot access the data. This is illustrated in Figure 16.59. Notice that the single token in place c prevents the other party from accessing it until the token is returned.

A Petri net is **safe** if each place contains only one token at a time. When a Petri net is safe, there is only one token or no token in each place so that recording the token count in a place is a binary operation. In most control models, a safe Petri net is desirable, since a token present can denote a process is enabled and no token present denotes that a process is disabled.

Figure 16.59

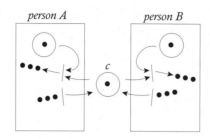

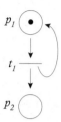

Figure 16.60

A Petri net is **bounded** if the tokens in any place cannot exceed some integer k. If a Petri net is bounded, it is possible to control overflow problems. A safe Petri net is obviously bounded. The Petri net in Figure 16.60 is not bounded.

In some Petri nets, conditions are each assigned a capacity and the number of tokens in the condition cannot exceed the capacity.

A Petri net is **conservative** if the total number of tokens in all the places is always a constant. When this occurs, the number of tokens into each transition must equal the numbers out of the transition. If tokens represent resources, a conservative Petri net ensures that resources are neither created nor reduced. A conservative Petri net is obviously bounded. The Petri net in Figure 16.61 is conservative.

Figure 16.61

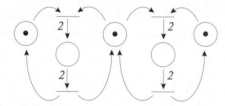

Exercises

In the following Petri nets, which

1. Are safe?
2. Are conservative?
3. Are bounded?
4. Have every state reachable?
5. Are live?
6. Transitions are currently enabled?
7. Transitions can be enabled in some state reachable from the current state?

(i)

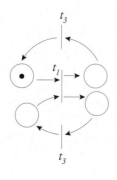

(ii)

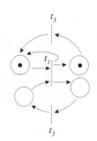

(iii)

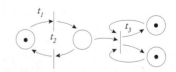

(iv)

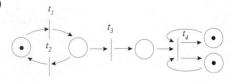

(v)

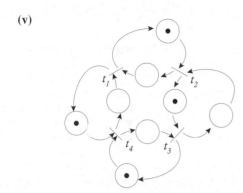

(vi)

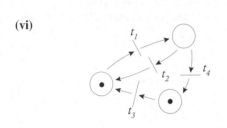

(vii)

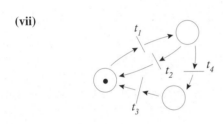

In the Petri nets in (c)–(vii)

8. Find all possible states of the Petri net.
9. Find the state when t_1 has been fired.
10. Find the state when t_1 followed by t_2 has been fired.
11. Find any state in which there is a deadlock.
12. Find any state in which there is a partial deadlock.
13. Find an example of a Petri net in which two transitions are enabled but firing either of them disables the other.
14. Find an example of a Petri net that is safe but not live.
15. Find a Petri net that has a bound of 4.
16. Find an example of a Petri net that is bounded but not safe.
17. Find an example of a Petri net that is conservative but not safe.
18. Find a Petri net which is equivalent to the one in Figure 16.58 that is not deadlocked.
19. Can a conservative Petri net be deadlocked?
20. Find a Petri net that is partially disabled but not disabled.
21. In the dining philosophers' problem, five philosophers sit down at a table to eat and meditate. At any given time they may do one or the other. They are eating Chinese food with chopsticks so that when they eat, they need two chopsticks. When they meditate, they need no chopsticks. There is one chopstick between each two philosophers, or a total of five chopsticks. Obviously if each philosopher simultaneously picks up the right chopstick, everybody starves. Design a Petri net to model this problem that will not deadlock and so that when a philosopher enters the eat state, he has a chopstick on each side of him to use.

GLOSSARY

Chapter 16: Networks

Bounded Petri net (16.3)	A Petri net is **bounded**, if the tokens in any one place cannot exceed some integer k.
Capacity of a cut (16.1)	The **capacity of cut**, $C(S, T) = \sum_{e \in (S,T)} c(e)$.
Chains (16.1)	**Chains** are created by forming paths from a source a to a sink z, ignoring the direction of the edges.
Complete matching (16.2)	A matching M on a bipartite graph $G = (V, E)$, where $V = A \cup B$ is **complete** if for every $a \in A$, there exists $b \in B$ such that $\{a, b\} \in M$.
Conservative Petri net (16.3)	A Petri net is **conservative** if the total number of tokens in all the places is always a constant.
Cut / $a - z$ cut (16.1)	Let S be subset of V and $T = V - S$. Then $\{e : e \in (S, T)\}$ is called a **cut**. If $a \in S$ and $z \in T$, then the **cut** is called an $a - z$ **cut**.

Deadlocked (16.3)	A Petri net is **deadlocked** if there is a state where no transition can be fired.
Dining philosopher's problem (16.3)	In the **dining philosopher's problem**, five philosophers sit down at a table to eat and meditate. At any given time they may do one or the other. They are eating Chinese food with chopsticks so that when they eat, they need two chopsticks. When they meditate, they need no chopsticks. There is one chopstick between each two philosophers, or a total of five chopsticks. Obviously, if every philosopher simultaneously picks up the right chopstick, everybody starves. How can this problem be solved so that when a philosopher enters the eat state, he has a chopstick on each side of him to use?
Enabled transition (16.3)	Transition t is **enabled** if for every directed edge from p_i to t, p_i contains the necessary tokens for t to fire.
Firing of a transition (16.3)	The **firing of a transition** t is the removal of a token or tokens from each place p_i, such that there is a directed edge from p_i to t_i, and the addition of a token or tokens to each place p_j, such that there is a directed edge from t to p_j.
Flow (16.1)	A **flow** in a network is a function $f: E \to N \cup \{0\}$ such that i. for all $e \in E, 0 \leqslant f(e) \leqslant c(e)$ ii. for all $v \in V$ such that $v \neq a, z$, $\sum_{e \in in(v)} f(e) = \sum_{e \in out(v)} f(e)$
Ford-Fulkerson Maximal Flow algorithm (16.1)	The **Ford-Fulkerson Maximal Flow algorithm** finds the maximal flow for a network.
Hall's theorem (16.2)	A bipartite graph $G(E, V)$, where $V = A \cup B$, has a complete matching if and only if for each subset $X \subseteq A, |X| \leq |R(X)|$.
Immediately reachable (16.3)	A state μ_j is **immediately reachable** from a state μ_i if the firing of some transition t, while in state μ_i, produces state μ_j.
Marking (16.3)	A marking is a function μ that assigns a nonnegative integer to each element of P. The function μ is called the **marking of** G.
Marking of p (16.3)	If p is a place in a Petri net, then $\mu(p)$ is called the **marking of p**.
Matched vertices (16.2)	If $\{a, b\}$ is an edge in a matching, then vertices a and b are **matched**.
Matching (16.2)	Given a bipartite graph $G(V, E)$, a subset M of E is a **matching** if no two edges have a common vertex. (Thus, no two edges are incident.)
Matching edges (16.2)	If $\{a, b\}$ is an edge in a matching, then both a and b have **matching edges**.
Maximal flow (16.1)	A flow $\hat{f}$ on a network is called a **maximal flow** if $val(\hat{f}) \geqslant val(f)$ for every possible flow f on the network.
Maximal matching (16.2)	A matching M on a bipartite graph $G = (V, E)$ is **maximal** if no other matching on G contains more edges.
Minimal cut (16.1)	An $a - z$ cut (S, T) is a **minimal cut** if $C(S, T)$ is less than or equal to the capacity of any other $a - z$ cut.
Mutual exclusion (16.3)	Until one person is through with the data and has modified and returned it, another person cannot access the data.

Network (16.1)	A **network** is a directed graph $G(G, V, E)$ together with a weight function $c : E \to N$ and designated vertices a, z such that **i.** indeg $(a) = 0$ **ii.** outdeg $(z) = 0$
Partial deadlock (16.3)	A Petri net **partial deadlock** is defined as one that is not live. There is a state so that if the Petri net is in that state, one or more transitions can never be fired.
Range (16.2)	Given a matching for a subset X of A, let $R(X) = \{b : b \in B \text{ and } b \text{ is adjacent to a vertex in } X\}$. $R(X)$ is called the **range** of X.
Reachable from a state (16.3)	State μ_{i_m} is **reachable from a state** μ_{i_1} if there exists states $u_{i_2}, u_{i_3}, u_{i_4}, \ldots, u_{i_{m-1}}$ and transitions $t_1, t_2, t_3, \ldots t_{m-1}$ such that $\mu_{i_{k+1}} = \delta(\mu_{i_k}, t_k)$ for $k = 1, 2, \ldots, m-1$.
Safe Petri net (16.3)	Each place can contain only one token at a time.
Set of directed edges (16.3)	If $G(V, E)$ is a directed bipartite graph, where $V = P \cup T$, then the set E is called the **set of directed edges** between P and T.
Sink vertex (16.1)	If we have a path flowing from vertex a to vertex z, and nothing flows out of z, it is called the **sink vertex**.
Source vertex (16.1)	If we have a path flowing from vertex a to vertex z and nothing flows into a, it is called the **source vertex**.
State of a Petri net (16.3)	The **state** of a Petri net is the number of tokens in each place. It represents the marking of the Petri net.
Tokens (16.3)	The marking of the set G is accomplished by placing large black dots called **tokens** in the circles representing the places.
Value of a flow (16.1)	The **value of a flow** f, denoted by $val(f)$, is equal to $flow(a) = flow(z)$.

TRUE-FALSE QUESTIONS

1. The flow $f(e)$ of an edge e in a network must be positive.
2. The capacity $c(e)$ of an edge e in a network must be positive.
3. The capacity of edges into a vertex must equal the capacity of edges out of a vertex.
4. The capacity of an $a - z$ cut is the same for every $a - z$ cut in a network.
5. The flow out of an $a - z$ cut is the same for every $a - z$ cut in a network.
6. The flow out of a cut is the same for every cut in a network.
7. If a flow f in a network is maximal, then $f(e) = c(e)$ for every edge out of a.
8. The flow out of a cut is always equal to the flow into a cut.
9. For every flow f, there is a cut such that the flow is equal to the capacity of the cut.
10. A flow for a network is maximal if and only if it is equal to the capacity of some cut.
11. For a given network, there exist a flow f and a cut (S, T) such that $val(f) = C(S, T)$.
12. If the flow through a network is minimal, then the flow is unique.
13. If $C(S, T)$ is a minimal cut, then $f(e) = 0$ for all $e \in (T, S)$.
14. There exists a cut (S, T) such that $c(e) = f(e)$ for all $e \in (S, T)$.
15. In the $a - z$ path $a \xrightarrow{(5,1)} b \xrightarrow{(6,3)} e \xleftarrow{(9,7)} d \xrightarrow{(4,1)} c \xrightarrow{(8,3)} z$, the slack of z is 2.
16. The set S produced, using the maximum flow algorithm, gives the minimal cut (S, T).
17. In an $a - z$ path, if the slack of z is 4, then the flow can be increased by 4.

18. In an $a - z$ path, if the slack of z is 4, then $f(e)$ is increased by 4 for every edge in the path.

19. The maximum flow of the following network

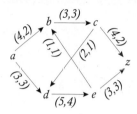

is 5.

20. The minimal cut for the network in the previous problem is (S, T) where $S = \{a\}$.

21. If a matching is complete, then it is maximal.

22. The bipartite graph $G = (V, E)$ where $V = A \cup B$ has a complete matching if $|A| \leq |B|$.

23. The bipartite graph $G = (V, E)$ where $V = A \cup B$ has a complete matching if each vertex in A is adjacent to two vertices in B.

24. A conservative Petri net is safe.

25. A live Petri net is safe.

26. A live Petri net is bounded.

27. In a live Petri net, every state is reachable from every other state.

28. If from some state every other state is reachable, then the Petri net is live.

29. If state u_i is reachable from u_j and u_j is reachable from u_k, then u_i is reachable from u_k.

30. A state is never immediately reachable from itself.

SUMMARY QUESTIONS

1. Find the maximum flow in the flow network in

2. Complete the flow in the following network such that there is conservation of flow.

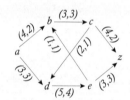

3. Verify the conservation of flow at b and c.

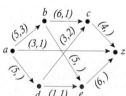

4. For the network in Problem 3, find the value of the flow.

5. Find the minimal cut for the following network.

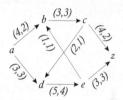

6. Six people are seeking a job. Person a is qualified for jobs A, B, and D. Person b is qualified for job E. Person c is qualified for jobs A and D. Person d is qualified for jobs C, D and E. Person e is qualified for jobs C and E. Person f is qualified for jobs A and B. Does everyone get a job?

7. Show the Petri net after t_1 followed by t_3 has been fired in the Petri net

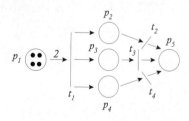

8. Is the Petri net in Problem 7 bounded?

9. Is the Petri net in Problem 7 conservative?

10. Is the Petri net in Problem 7 live?

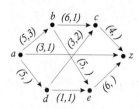

ANSWERS TO TRUE-FALSE

1. F 2. T 3. F 4. F 5. T 6. F 7. F 8. F 9. F 10. T 11. T
12. F 13. T 14. T 15. F 16. F 17. T 18. F 19. F 20. F
21. T 22. F 23. F 24. F 25. F 26. F 27. F 28. F 29. T
30. F

CHAPTER 17

Theory of Computation

17.1 Regular Languages

In Example 9.17, we introduced the following definitions:

DEFINITION 17.1 Let A be a finite collection of symbols. This set of symbols is called an **alphabet**. For example, A could be a subset of the English alphabet or it might simply consist of the set $\{0, 1\}$. A **string** or **word** of symbols of A has the form $a_1 a_2 a_3 a_4 \ldots a_n$ where $a_i \in A$.

Thus, if $A = \{0, 1\}$, then 1, 100, 101, 11011, and 0001 would all be strings of symbols of A. In addition we include an empty string denoted by λ, which has no symbols in it.

DEFINITION 17.2 Let A^* denote the set of all strings of A. Define the binary operation $\circ$ called **concatenation** on A^* as follows: If $a_1 a_2 a_3 a_4 \ldots a_n$ and $b_1 b_2 b_3 b_4 \ldots b_m \in A^*$, then

$$a_1 a_2 a_3 a_4 \ldots a_n \circ b_1 b_2 b_3 b_4 \ldots b_m = a_1 a_2 a_3 a_4 \ldots a_n b_1 b_2 b_3 b_4 \ldots b_m$$

Thus, if $A = \{0, 1\}$, then $11011 \circ 100010 = 11011100010$. In particular, if ω is a string in A^*, then $\lambda \circ \omega = \omega \circ \lambda = \omega$, so that a string followed or preceded by the empty string simply gives the original string.

DEFINITION 17.3 Let A^* denote the set of all strings of A. A subset L of A^* is called a **language**.

If A is the set of letters in the English alphabet, then L could be the set of words in the English language. If A is the set of letters in the Greek alphabet, then L could be the set of words in the Greek language. If A is the set of symbols used in a computer language, then L could be the set of words in that language.

If A is the set $\{1, 0\}$, then the following are languages:

$$L_1 = \{0, 1, 00, 11, 000, 111, \ldots\}$$
$$L_2 = \{w : w \in A^* \text{ and contains exactly one 1}\}$$
$$L_3 = \{w : w \in A^* \text{ and contains exactly two 1's}\}$$
$$L_4 = \{w : w \in A^* \text{ and contains at least two 1's}\}$$

If A is the set $\{a, b, c\}$, then the following are languages:

$$L_1 = \{acb, aacbb, aaacbbb, \ldots\}$$
$$L_2 = \{w : w = a^n b^n \text{ for } n \geq 1\}$$
$$L_3 = \{w : w \in a^n b^m \text{ for } m, n \geq 1\}$$
$$L_4 = \{w : w \in A^* \text{ and contains no consecutive identical letters}\}$$

DEFINITION 17.4

Let S be a subset of A^*. Then S^* is the set of all strings or words formed by concatenating words from S; that is, $S^* = \{w_1 w_2 \ldots w_n : w_i \in S\}$. The symbol $*$ is called the Kleene star and is named after the mathematician and logician Steven Cole Kleene.

Note that A^* is consistent with this definition.

DEFINITION 17.5

Let A be an alphabet. An **expression** in A is a string of symbols that represents a subset of A^*. A **regular expression** uses the symbols $*$, $\vee$, λ, $(,)$, and $\mathbf{a}$ for each element $\mathbf{a}$ in A. The class of **regular expressions** is defined by the following rules:

(i) The symbol λ is a regular expression and for every $a \in A$, the symbol $\mathbf{a}$ is a regular expression.

(ii) If w_1 and w_2 are regular expressions, then $w_1 w_2$, $w_1 \vee w_2$, w_1^*, and (w_1) are regular expressions.

(iii) There are no regular expressions that are not generated by (i) and (ii).

Each expression corresponds to a set with the following correspondence £ defined by

$$£(\lambda) = \lambda$$
$$£(\mathbf{a}) = \{a\} \text{ for all } a \in A$$
$$£(w_1 \vee w_2) = £(w_1) \cup £(w_2)$$
$$£(w_1 w_2) = £(w_1) \circ £(w_2)$$
$$£(w_1^*) = (£(w_1))^*$$

so that

$£(\mathbf{aa}^*)\quad = \{a\} \circ \{a\}^* \quad\quad\quad\quad = \{a, aa, aaa, aaaa, aaaa, \ldots\}$
$£(\mathbf{a(b \vee c)d}) = \{a\} \circ \{\{b\} \cup \{c\}\} \circ \{d\} = \{abd, acd\}$
$£((\mathbf{a \vee b})^*) \ = \{a \cup b\}^* \quad\quad\quad\quad\quad = \{\lambda, a, b, ab, ba, abb, aba, \ldots\}$
$\quad\quad\quad\quad\quad\quad\quad\quad\quad\quad\quad\quad\quad\quad\ = $ all strings consisting of 0 or more a's and b's
$£(\mathbf{ab}^*\mathbf{c})\quad = \{a\} \circ \{b\}^* \circ \{c\} \quad\quad = \{ac, abc, abbc, abbbc, \ldots\}$
$£(\mathbf{a \vee b \vee c}) = \{a\} \cup \{b\} \cup \{c\} \quad\quad = \{a, b, c\}$

DEFINITION 17.6

The class R of **regular languages** over A has the following properties:
(i) The emptyset $\emptyset \in R$, and if $a \in A$, then $\{a\} \in R$.
(ii) If s_1 and $s_2 \in R$, then $s_1 \cup s_2$, $s_1 \circ s_2$, $s_1^* \in R$.
(iii) Only sets formed using (i) and (ii) belong to R.

Although it will not be shown here, the intersection of regular sets is a regular set and the complement of a regular set is a regular set.

DEFINITION 17.7

Let S and T be subsets of A^*. If $S = T^*$, then S is **generated** by T.

If S is generated by T, then every element of S is the finite concatenation of elements of T. The elements in the concatenation forming a given word are not necessarily unique, however. If $T = \{a, b\}$, then T^* consists of the empty word and of all possible finite strings of the symbols a and b. If $\overline{T} = \{a, ab, b\}$, then $T^* = \overline{T}^*$ but, although every string in T^* can be expressed uniquely as the concatenation of elements of T, this is not true of elements of $\overline{T}$ since the expression **abab** can be expressed as **(a)(b)(a)(b)**, and also as **(a)(b)(ab)** and **(ab)(ab)**.

DEFINITION 17.8

Let S and C be subsets of A^*. If $S = C^*$, then every string in S can be expressed as the concatenation of elements of C, and we say that C is a **Code**. A code C is **uniquely decipherable** if every string in S can be uniquely expressed as the concatenation of elements of C.

Therefore, $\{ba, ab, c\}$, $\{ade, ddbee, dfc, ddd\}$, and $\{a, b, c, de\}$ are uniquely decipherable codes whereas $\{a, ab, bc, c\}$, $\{ab, abc, cde, de\}$, and $\{a, bc, abc\}$ are not uniquely decipherable codes.

EXAMPLE 17.9

A function on a set S is called a **retraction map** if $f \circ f = f$. The image of f is called a **retract**. Let R be a retract of a retraction map $f : A^* \to A^*$. Tom Head has shown that a retract is generated by a finite code that has the property that each word of the code contains an element of A that is contained only once in that word and does not occur in any other word of the code. Such codes are called **key codes**. Therefore, $\{a, b, c\}$, $\{a, be, ce\}$, $\{ade, ddbee, caa\}$, and $\{ad, bdde, cede\}$ are key codes that generate retracts, since a, b, and c each occur in exactly one word and occur only once in that word. However, $\{ba, ab, c\}$, $\{ab, abc, cde\}$, and $\{a, ab, bc, d\}$ are codes but do not generate retracts. The intersection of retracts is not necessarily a retract. It is left to the reader to show that the intersection of the sets $\{ac, bc, d\}^*$ and $\{a, cb, cd\}^*$ is generated by the code described by the expression **ac(bc)*d**. ∎

DEFINITION 17.10

Let A be an alphabet. A code $C \subseteq A^*$ is called a **prefix code** if for all words $u, v \in C$, if $u = vw$ for $w \in A^*$, then $u = v$ and $w = \lambda$. This means that one word in a code cannot be the beginning string of another word in the code. A code $C \subseteq A^*$ is called a **suffix code** if for all words $u, v \in C$, if $u = wv$ for $w \in A^*$, then $u = v$ and $w = \lambda$. This means that one word in a code cannot be the final string of another word in the code.

The set $\{a, ab, ac\}$ is a code but it is not a prefix code since a is the initial string of both ab and ac. It is, however, a suffix code. The set $\{a, ba, ca\}$ is a prefix code, but it is not a suffix code since a is the final string of both ba and ca. The set $\{ad, ab, ac\}$ is both a prefix and a suffix code.

Exercises

1. Find regular sets corresponding to the following expressions. If the set is infinite, list ten elements in the set.
   (a) a(b ∨ c ∨ d)a
   (b) a*b*c
   (c) (a∨b)(c∨d)
   (d) (ab*λ)∨(cd)*
   (e) a(bc)*d

2. Find regular sets corresponding to the following expressions. If the set is infinite, list ten elements in the set.
   (a) bc(bc)*
   (b) (a∨b* ∨ λ)(c∨d*)
   (c) (a∨bc∨d)*
   (d) (a∨b)(c∨d)b
   (e) a*(b∨c∨d*)

3. Find regular expressions that correspond to the following regular sets.
   (a) $\{ab, ac, ad\}$
   (b) $\{ab, ac, bb, bc\}$
   (c) $\{a, ab, abb, abbb, abbbb, \dots\}$
   (d) $\{ab, abab, ababab, abababab, ababababab, \dots\}$
   (e) $\{ab, abb, aab, aabb\}$

4. Find regular expressions that correspond to the following regular sets:
   (a) $\{ab, acb, adb\}$
   (b) $\{ab, abb, abbb, abbbb, \dots\}$
   (c) $\{ad, ae, af, bd, be, bf, cd, ce, cf\}$
   (d) $\{abcd, abcbcd, abcbcbcd, abcbcbcbcd, \dots\}$
   (e) $\{abcd, abef, cdcd, cdef\}$

5. Let $A = \{a, b, c\}$.
   (a) Give a regular expression for the set of all elements of A^* containing exactly two b's.
   (b) Give a regular expression for the set of all elements of A^* containing exactly two b's and two c's.
   (c) Give a regular expression for the set of all elements of A^* containing two or more b's.
   (d) Give a regular expression for the set of all elements of A^* beginning and ending with a and containing at least one b and one c.
   (e) Give a regular expression for the set of all elements of A^* consisting of one or more a's, followed by one or more b's, and then one or more c's.

6. Let $A = \{a, b\}$.
   (a) Give a regular expression for the set of all elements of A^* containing exactly two b's or exactly two a's.
   (b) Give a regular expression for the set of all elements of A^* containing an even number of b's.
   (c) Give a regular expression for the set of all elements of A^* beginning and ending with a and containing at least one b.
   (d) Give a regular expression for the set of all elements of A^* such that the number of a's in each string is divisible by 3 and the number of b's is divisible by 5.
   (e) Give a regular expression for the set of all elements of A^* such that the length of each string is divisible by 3.

7. Which of the following are uniquely decipherable codes?
   (a) $\{ab, ba, a, b\}$
   (b) $\{ab, acb, accb, acccb, \dots\}$
   (c) $\{a, b, c, bd\}$
   (d) $\{ab, ba, a\}$
   (e) $\{a, ab, ac, ad\}$

8. Which of the following expressions describe uniquely decipherable codes?
   (a) ab*
   (b) ab*∨ baaa
   (c) ab*c ∨ baaac
   (d) (a ∨ b)(b ∨ a)
   (e) (a ∨ b ∨ λ)(b ∨ a∨ λ)

9. Which of the following are uniquely decipherable codes? Which are suffix codes?
   (a) $\{ab, ba, \}$
   (b) $\{ab, abc, bc\}$
   (c) $\{a, b, c, bd\}$
   (d) $\{aba, ba, c\}$
   (e) $\{ab, acb, accb, acccb\}$

10. Which of the following expressions describe prefix codes? Which describe suffix codes?

    (a) ab*
    (b) ab*c
    (c) a*bc*
    (d) (a ∨ b)(b ∨ a)
    (e) a*b

11. Which of the following codes are uniquely decipherable? Justify your answer.

    (a) 1110, 111, 1011, 11, 0101
    (b) 10, 110, 1110, 0101
    (c) 010, 101, 0101, 1010, 10101
    (d) 0101, 1001, 011, 11, 10011, 00101
    (e) 101, 1001, 10001, 11011, 11001, 00111

12. Which of the following codes are uniquely decipherable? Justify your answer.

    (a) 110, 1110, 11110, 111110, 10
    (b) 1011, 1010, 1100, 0101, 0011, 1100, 1111, 0000
    (c) 101, 010, 11, 10101, 1011, 1010
    (d) 110, 101, 1001, 11101, 0101, 0110, 0111, 10001
    (e) 11011, 101, 010100, 11, 1011, 1110

13. Which of the following codes are prefix codes?

    (a) 101, 1101, 0101, 001, 1001, 11101, 0110, 01001
    (b) 1101, 0011, 1010, 1110, 1010, 0101, 1100, 1111
    (c) 101, 1100, 01010, 10101, 0011, 0110
    (d) 1011, 10101, 10110, 0101, 01100, 111000
    (e) 1101, 10001, 1001, 01110, 10110, 110110

14. Which of the codes in Exercise 1 are prefix codes?

15. Which of the codes in Exercise 2 are prefix codes?

16. (a) Which of the codes in Exercise 1 are suffix codes?

    (b) Which of the codes in Exercise 2 are suffix codes?

    (c) Which of the codes in Exercise 3 are suffix codes?

17. A code is an **infix code** if no string in the code is a substring of another string in the code. Is a code that is a suffix code and a prefix code always an infix code? Prove or give a counterexample.

18. Is a code that is an infix code always a prefix code and a suffix code? Prove or give a counterexample.

19. Which of the codes in Exercise 1 are infix codes?

20. Which of the codes in Exercise 2 are infix codes?

21. Which of the codes in Exercise 3 are infix codes?

22. Determine whether there is a prefix code with code word lengths of 1, 2, 3, 3.

23. Determine whether there is a prefix code with code word lengths of 1, 3, 3, 3, 4, 5, 5, 5.

24. Determine whether there is an infix code with code word lengths of 1, 3, 3, 3.

25. Prove that a key code is both a prefix code and a suffix code.

17.2 Automata

An **automaton** is basically a device that recognizes or accepts certain elements of A^*, where A is a finite alphabet. Different automata (the plural of *automaton*) recognize or accept different elements of A^*. The subset of elements of A^* accepted by an automaton M is a language called the **language over A accepted by** M and denoted $M(L)$.

For a given finite alphabet, an automaton consists of a set S of states and a function $F : A \times S \to S$, called the **transition function**. The set S contains an element s_0 and one or more acceptance states. The automaton M is denoted by (A, S, s_0, T, F) where A is the alphabet, S is the set of states, s_0 is the initial or starting state, T is the set of acceptance states, and F is the transition function. Note that the input of F is a letter of A and a state belonging to S. The output is a state of S (possibly the same one). If the automaton is in state s and "reads" the letter a, then (a, s) is the input for F and $F(a, s)$ is the next state. We can think of an automaton "beginning" at the initial state s_0. If the automaton "reads" a letter a of A, then it "moves" to state $s = F(a, s_0)$. If the automaton now "reads" a letter of A, say b, it then moves to state $s' = F(a, s)$. Therefore, as the automaton continues to "read" letters of the alphabet, it "moves" from one state to another. Let M be the automaton with alphabet $A = \{a, b\}$, set of states $S = \{s_0, s_1, s_2\}$, and F defined by the table

F	s_0	s_1	s_2
a	s_1	s_1	s_2
b	s_2	s_2	s_1

Suppose M "reads" the letter a, followed by the letter b and the letter a. Since the automaton begins in state s_0, and the letter read is a, and $F(a, s_0) = s_1$, the automaton is now in state s_1. The next letter read is b and $F(b, s_1) = s_2$. Finally, the last letter a is read and, since $F(a, s_2) = s_2$, the automaton remains in state s_2. We may also state F as a set of rules as follows:

If in state s_0 and a is read, go to state s_1.
If in state s_1 and a is read, go to state s_1.
If in state s_2 and a is read, go to state s_2.
If in state s_0 and b is read, go to state s_2.
If in state s_1 and b is read, go to state s_2.
If in state s_2 and b is read, go to state s_1.

This automaton is best shown by pictorially by a **state diagram**, which is essentially a directed graph where the states are represented by the vertices and a directed edge from s to s' is labeled with a letter, say a, of the alphabet A if $F(a, s) = s'$. A directed arrow labeled with the letter a will be called an **a-arrow**. Therefore, the preceding automaton may be represented pictorially as seen in Figure 17.1.

(Note that although the state diagram in Figure 17.1 is a directed graph, the figures representing some automata may not truly be directed graphs, as we have defined them, since there may be more than one directed edge from one state to another. Such figures are more precisely called **multigraphs** or **labeled digraphs**.)

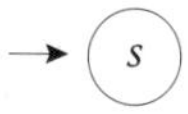

The automaton "reads" a word or string $a_0 a_1 a_2 \ldots a_n$ of A^* by first reading a_0, then reading a_1, and continuing until it has read a_n. As previously mentioned, certain states are designated as acceptance states. An automaton **accepts** or **recognizes** $a_0 a_1 a_2 \ldots a_n$ if after reading a_n, the automaton finishes in an acceptance state. If s is a starting state, then its vertex is denoted by the diagram shown in Figure 17.2.

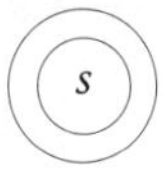

If s is an acceptance state, its vertex is denoted by the diagram shown in Figure 17.3.

For example, the automaton with the state diagram shown in Figure 17.4 has initial state s_0 and acceptance state s_3.

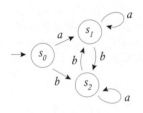

Figure 17.1

Figure 17.2

Figure 17.3

Figure 17.4

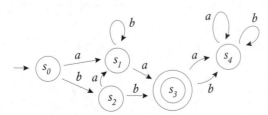

It accepts the word *baa* since after reading *b*, it is in state s_2. After reading *a*, it is in state s_1. After reading the second *a*, it in state s_3, which is an acceptance state. One can see that it also accepts *abbba* and *bb*, but not *bbb*, *abab*, or *abb*. As mentioned earlier, the set of all words accepted by an automaton is called the language accepted by the automaton.

EXAMPLE 17.11 Consider the automaton with the state diagram shown in Figure 17.5.

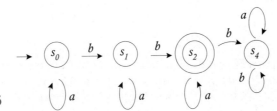

Figure 17.5

It obviously accepts the word *bb*. In each state, since there is a loop for *a* so that if *a* is read, then the state does not change. This enables us to read as many *a*'s as desired before reading another *b*. Thus, the automaton reads *aababaaa*, *baabab*, *baaab*, *babaaa*, *aabaabaa*, and in fact we can read any word in the language described by the regular expression **a*ba*ba***. Since $A = \{a, b\}$, this language can also be described as the set of all words containing exactly two *b*'s. The state s_4 is an example of a **sink state**. Once the automaton is in the sink state, it can never get out again. ∎

EXAMPLE 17.12 Consider the automaton with state diagram shown in Figure 17.6, which we simplify as shown in Figure 17.7 to decrease the number of arrows.

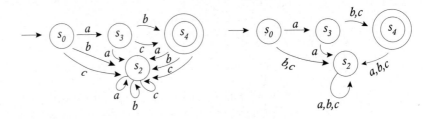

Figure 17.6   **Figure 17.7**

This automaton obviously accepts only the words *ab* and *ac*. This language may be described by the regular expression **a(b∨c)**. ∎

EXAMPLE 17.13 Consider the automaton with the state diagram shown in Figure 17.8.

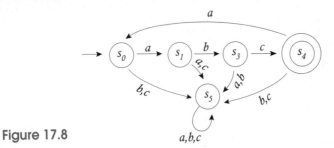

Figure 17.8

Every word must begin with *ab* and end with *c*. However, the loop *aabc* may be repeated as often as desired since this loop begins and ends at s_4. Therefore, the regular expression for this automaton is **abc(aabc)***.

EXAMPLE 17.14

Consider the automaton with the state diagram shown in Figure 17.9.

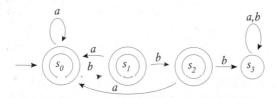

Figure 17.9

In this automaton, if three consecutive *b*'s are read, then the automaton is in state s_3, which is a sink state. This is the only way to get to s_3 and every other state is an acceptance state. Thus, the language accepted by this automaton consists of all words that do not have three consecutive *b*'s.

The automata that we have been discussing are often called **deterministic automata** since in every state and for every value of the alphabet that is read, there is one and only one state. In other words, it is because $F : A \times S \to S$ is a function. It is often convenient to relax the rules so that F is no longer a function but instead a relation. If we again consider F as a set of rules, given $a \in A$ and $s \in S$, the rules may allow advancement to each of several states or there may not be a rule that allows it to go to any state. In the latter case, the automaton is "hung up" and can proceed no further. An automaton for which F is not necessarily a function is called a **nondeterministic automaton**. For example, consider the following state diagram shown in Figure 17.10.

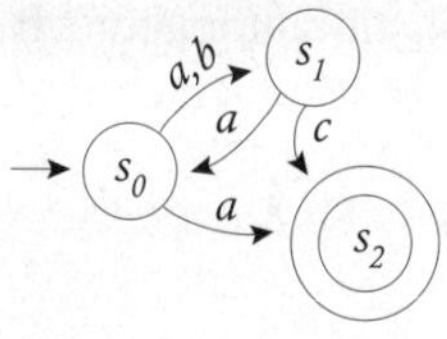

Figure 17.10

If in state s_0, *a* is read, then the automaton may be in either state s_1 or s_2. If, however, the automaton is in state s_1 and *b* is read, the automaton "hangs up," since there is no *b* arrow out of state s_1. Note that in our definition the set of deterministic automata is contained in the set of nondeterministic automata.

EXAMPLE 17.15

It is easily seen that the automaton with the state diagram shown in Figure 17.11 accepts the language with regular expression **ab*c**.

Figure 17.11

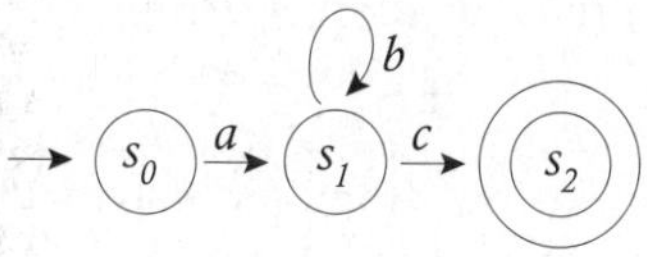

EXAMPLE 17.16

The automaton with the state diagram shown in Figure 17.12 accepts the language with regular expression **a∨b**.

EXAMPLE 17.17

The automaton with the state diagram shown in Figure 17.13 accepts the language with regular expression **aa*bb***.

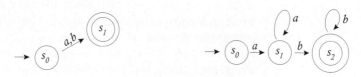

Figure 17.12   Figure 17.13

EXAMPLE 17.18 The automaton with the state diagram shown in Figure 17.14 accepts the language consisting of strings with at least two a's.

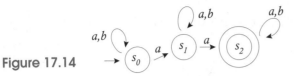

Figure 17.14

It is easier to use nondeterministic automata since we do not have to include nonrelevant edges. In particular, we shall see in some of the later constructions that it is much simpler to use nondeterministic automata. Obviously any language accepted by a deterministic automaton is accepted by a nondeterministic automaton since the set of deterministic automata is a subset of the set of nondeterministic automata. In the following theorem, however, we shall see that any language accepted by a nondeterministic automaton is also accepted by a deterministic automaton.

THEOREM 17.19 For each nondeterministic automaton, there is an equivalent deterministic automaton that accepts the same language.

Instead of a formal proof, we shall demonstrate how to construct a deterministic automaton that accepts the language accepted by a nondeterministic automaton. If S is the set of states for the nondeterministic automaton, we shall use elements of $\mathcal{P}(S)$, that is, the subset of S, as states for the deterministic automaton that we are constructing. Consider the nondeterministic automaton N shown in Figure 17.15.

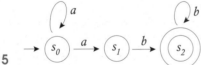

Figure 17.15

Let $\{s_0\}$ be the starting state. Construct an a-arrow from $\{s_0\}$ to the set of all states so that there is an a-arrow from s_0 to that state. In this case there is an a-arrow from s_0 to s_0 and an a-arrow from s_0 to s_1, so we construct an a-arrow from $\{s_0\}$ to $\{s_0, s_1\}$. There is no b-arrow from s_0 to any state. In this case we construct a b-arrow from $\{s_0\}$ to the empty set $\emptyset$. We now consider the state $\{s_0, s_1\}$. We construct an a-arrow from $\{s_0, s_1\}$ to the set of all states such that there is an a-arrow from either s_0 or s_1 to that state. In this case we construct an a-arrow from $\{s_0, s_1\}$ to itself. We construct a b-arrow from $\{s_0, s_1\}$ to the set of all states such that there is a b-arrow from either s_0 or s_1 to that state. In this case we construct a b-arrow from $\{s_0, s_1\}$ to $\{s_2\}$. Since there are no a-arrows or b-arrow from any state in the empty set to any other state, we construct an a-arrow and a b-arrow from the empty set to itself. We now turn our attention to $\{s_2\}$. Since there is no a-arrow from s_2 to any other state, we construct an a-arrow from $\{s_2\}$ to the empty set. Since the only b-arrow from s_2 is to itself, we construct a b-arrow from $\{s_2\}$ to itself. The terminal state(s) consist of all sets that contain an element of the terminal set of N. In this case $\{s_2\}$ is the only terminal state. We have now completed the state diagram in Figure 17.16, which is easily seen to be the state diagram of a deterministic automaton. This automaton also reads the same language as N, namely, the language described by the expression $\mathbf{a^*abb^*}$.

Figure 17.16

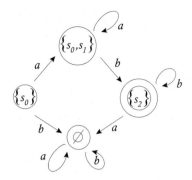

In general we have the following procedure for constructing a deterministic automaton from a nondeterministic automaton.

1. Begin with the state $\{s_0\}$ where s_0 is the start state of the nondeterministic automaton.
2. For each $a_i \in A$, construct an a_i arrow from $\{s_0\}$ to the set of all states such that there is an a_i-arrow from s_0 to that state.
3. For each newly constructed set of states S_j and for each $a_i \in A$, construct an a_i arrow from S_j to the set of all states such that there is an a_i arrow from an element of S_j to that state.
4. Continue this process until no new states are created.
5. Make each set of states S_j, which contains an element of the terminal set of the nondeterministic automaton, into a terminal state.

EXAMPLE 17.20 Given the nondeterministic automaton in Figure 17.17, we construct the corresponding deterministic automaton.

Since there is an a-arrow from s_0 to s_1, construct an a-arrow from $\{s_0\}$ to $\{s_1\}$. Since there is no b-arrow from s_0 to any other state, construct a b-arrow from s_0 to the empty set $\emptyset$. There is an a-arrow from s_1 to s_3 so construct an a-arrow from $\{s_1\}$ to $\{s_3\}$. There is a b-arrow from s_1 to s_0 and s_1 to s_2, so construct a b-arrow from $\{s_1\}$ to $\{s_0, s_2\}$. Since there are no arrows from s_3 to another state, construct an a-arrow and a b-arrow from s_3 to the empty set. There are a-arrows from s_0 to s_1, s_2 to s_1, and s_2 to s_3, so construct an a-arrow from $\{s_0, s_2\}$ to $\{s_1, s_3\}$. There are no b-arrows from either s_0 or s_2 to another state, so construct a b-arrow from $\{s_0, s_2\}$ to the empty set. There is an a arrow from s_1 to s_3 so construct an a arrow from $\{s_1, s_3\}$ to $\{s_3\}$. There are b-arrows from s_1 to s_0 and s_1 to s_2, so construct a b-arrow from $\{s_1, s_3\}$ to $\{s_0, s_2\}$. This completes the deterministic automaton in Figure 17.18.

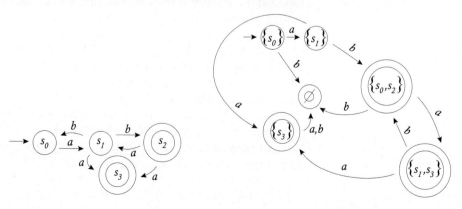

Figure 17.17 **Figure 17.18**

We unite our study of regular languages and automata theory with the following theorem.

THEOREM 17.21 (Kleene's Theorem) A language is regular if and only if it is accepted by an automaton.

Proof We first show that a regular language can the read by an automaton. The language consisting of the empty set is accepted by the following automaton.

The finite set of elements $\{a_1, a_2, a_3, \ldots, a_n\}$ is accepted by the following automaton.

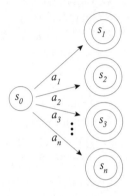

We next need to show we can find an automaton for the concatenation of two languages. Given automata $M_1 = (A, S, s_0, T, F)$, which accepts L, and $M_2 = (A, S', s_0', T', F')$, which accepts L', we construct an automaton $M_3 = (A, S'', s_0'', T'', F'')$, which accepts LL'. Essentially we place M_2 after M_1. Let $s_0'' = s_0$, $S'' = S \cup S'$, $T'' = T'$ and $F'' = F \cup F''$ together with additional rules given as follows: If there is an a-arrow from a state s in S to a state in T, add an arrow from s to s_0''. It is easily seen that this automaton accepts LL'.

Next we must show that if automata $M_1 = (A, S, s_0, T, F)$, which accepts L, then there is an automaton that accepts $M_2 = (A, S, s_0, T, F')$, which accepts L^*. This automaton has the same starting state, and acceptance states. We add the following states to F. If there is an a-arrow from a state s to an acceptance state t, then add an a-arrow from s to the starting state s_0. It is easily seen that this is the desired automata.

Finally, we need to show that given machines $M_1 = (A, S, s_0, T, F)$ and $M_2 = (A, S', s_0', T', F')$, accepting languages L and L', there is a machine accepting $L \cup L'$. The machine is $M = (A, S \times S', s_0 \times s_0', T', F')$. Basically we read letters and move from one pair of states to another until one of the states is a final state, at which point we accept the word. We define F' more specifically as follows: We begin in state $s_0 \times s_0'$. If we are in state $s_1 \times s_2$, read a and there are a-arrows from s_1 to s_1' in M_1 and s_2 to s_2' in M_2, then we define an a-arrow from $s_1 \times s_2$ to $s_1' \times s_2'$. We continue reading letters until we reach $s_n \times s_m$ where either s_n or s_m is an acceptance state in its respective automaton. When this occurs, $s_n \times s_m$ is an acceptance state in M. Thus, we continue reading letters until the word is accepted in one of the two machines.

We now show how, given an automaton, we can construct the language accepted by this automaton. To perform this construction, we shall use transition graphs, which read strings of letters of A rather than letters of A. We shall also have a rule that reads the empty state, which means that it changes states without reading anything. We shall assume there is only one acceptance state. If there is more than one, form a new acceptance state and use the empty word to get from the old acceptance states to the new and only acceptance state.

We wish to eliminate states and create words until we have

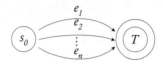

which accepts the expression $e_1 \vee e_2 \vee e_3 \vee \cdots \vee e_n$. To eliminate s_i we use the following rules until we get the desired figure:

1. If

$$\rightarrow \underset{s_i}{\bigcirc} \overset{a,b,c}{\circlearrowright} \rightarrow$$

occurs, replace it with

$$\rightarrow \underset{s_i}{\bigcirc} \overset{(a \vee b \vee c)^*}{\circlearrowright} \rightarrow$$

2. If the diagram

$$\rightarrow s_{i-1} \overset{a}{\rightarrow} \underset{s_i}{\bigcirc} \overset{b}{\circlearrowright} \overset{c}{\rightarrow} s_{i+1}$$

occurs, then replace it with

$$\rightarrow s_{i-1} \overset{ab^*c}{\rightarrow} s_{i+1}$$

3. If the diagram

$$\overset{w}{\rightarrow} s_i \overset{\overset{a}{\underset{c}{b}}}{\rightrightarrows} s_{i+1}$$

occurs, then replace it with

$$\xrightarrow{w} (s_i) \xrightarrow{avbvc} (s_{i+1})$$

4. If the diagram

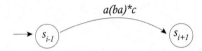

occurs, then replace it with

$$\longrightarrow (s_{i-1}) \xrightarrow{a(ba)^*c} (s_{i+1})$$

5. If the diagram

$$\longrightarrow (s_{i-1}) \xrightarrow{a} (s_i) \begin{array}{c} \xrightarrow{b} (s_{i+1}) \\ \xrightarrow{c} (s_{i+2}) \end{array}$$

occurs, then replace it with the diagram

$$\longrightarrow (s_{k-1}) \begin{array}{c} \xrightarrow{ab} (s_{i+1}) \\ \xrightarrow{ac} (s_{i+2}) \end{array}$$

∎

Exercises

1. Which of the following words are accepted by the automaton in Figure 17.19?

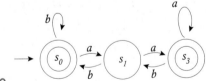

Figure 17.19

(a) *abba*

(b) *aabbb*

(c) *babab*

(d) *aaabbb*

(e) *bbaab*

2. Which of the following words are accepted by the automaton in Figure 17.20?

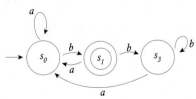

Figure 17.20

(a) *aaabb*

(b) *abbbabbb*

(c) *bababa*

(d) *aaabab*

(e) *bbbabab*

3. Write an expression for the language accepted by the automaton in Figure 17.21.

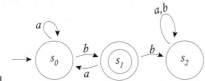

Figure 17.21

4. Write an expression for the language accepted by the automaton in Figure 17.22.

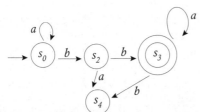

Figure 17.22

5. Write an expression for the language accepted by the automaton in Figure 17.23.

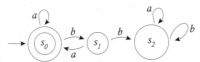

Figure 17.23

6. Write an expression for the language accepted by the automaton in Figure 17.24.

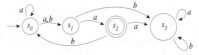

Figure 17.24

7. Find a deterministic automaton that accepts the language expressed by **aa*bb*cc***.
8. Find a deterministic automaton that accepts the language expressed by **(a*ba*ba*b)***.
9. Find a deterministic automaton that accepts the language expressed by **(a*(ba)*bb*a)***.
10. Find a deterministic automaton that accepts the language expressed by **(a*b)∨(b*a)***.
11. Find a nondeterministic automaton that accepts the language expressed by **aa*bb*cc***.
12. Find a nondeterministic automaton that accepts the language expressed by **(a*b)∨(c*b)∨(ac)***.
13. Find a nondeterministic automaton that accepts the language expressed by **(a∨b)*(aa∨bb)(a∨b)***.
14. Find a nondeterministic automaton that accepts the language expressed by **((aa*b)∨bb*a)ac***.
15. Find a deterministic automaton that accepts the same language as the nondeterministic automaton in Figure 17.25.

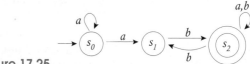

Figure 17.25

16. Find a deterministic automaton that accepts the same language as the nondeterministic automaton in Figure 17.26.

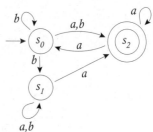

Figure 17.26

17. Find a deterministic automaton that accepts the same language as the nondeterministic automaton in Figure 17.27.

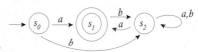

Figure 17.27

18. Find a deterministic automaton that accepts the same language as the nondeterministic automaton in Figure 17.28.

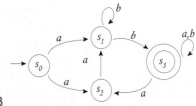

Figure 17.28

17.3 Finite State Machines with Output

Previously, we defined a deterministic automaton, as a device that accepts or recognizes words of A^*. We begin with a basic review and then use this to extend our machine to a Moore machine and to a Mealy machine. An automaton consists of a finite set of states S together with a function $F : A \times S \to S$ called a transition function. If the automaton is in a given state s and "reads" a letter a of A, then it "moves" to state $F(a, s)$ or remains in state s if $F(a, s) = s$. Thus, the function F is used for transitions to different states of S as it "reads" elements of A. The set S contains a state s_0 called the initial state and a subset of one or more acceptance states.

As we have seen, given a string $w = a_1 a_2 a_3 \ldots a_n \in A^*$, the automaton begins in the initial state and "reads" the first letter a_1 of w. It then "moves" to state $s' = F(a_1, s_0)$. From this state, it reads the next letter a_2 and moves to $F(a_2, s')$. It continues this process until it "reads" a_n. If at this point the automaton is in an acceptance state, the word is "accepted" by the automaton.

We have also seen that an automaton is best shown as a diagram called a finite state diagram. The states are denoted by circles containing the label of the state. The initial state s_0 is preceded by $\to$. Acceptance states are represented as double circles. If $F(a, s_i) = s_j$, this is represented by

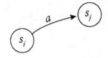

The finite automaton represented by the diagram

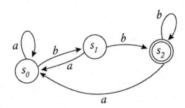

has alphabet $A = \{a, b\}$, states $\{s_0, s_1, s_2\}$, and acceptance state s_2. It accepts words bb, abb, and any word ending in bb.

Formally an automaton M is denoted by (A, S, s_0, T, F), where T is the set of acceptance states. Thus, in the previous example, $A = \{a, b\}$, $S = \{s_0, s_1, s_2\}$, $T = \{s_2\}$, and F is given by the table

F	s_0	s_1	s_2
a	s_0	s_0	s_0
b	s_1	s_2	s_2

An automaton only accepts or rejects words. We now introduce another type of finite state machine that produces output. These machines are called finite automata with output or finite state transducers. These machines no longer have acceptance states, so they no longer directly accept or reject words. They simply read input and produce output.

The first machine we introduce is called a **Moore machine** or **Moore automaton**, created by E. F. Moore [68]. It again has a finite set of states including a starting state. It contains two alphabets A and Σ. The first is the alphabet of input characters read by the machine and the second is the alphabet of output characters produced by the machine. The Moore machine retains the transition function $F: A \times S \to S$ of the finite state automata. In addition it contains a function $\phi: S \to \Sigma$. In the operation of a Moore machine, the output is first produced using the function ϕ before input is read. Then the transition function F is used to read the input and change states. Again imitating the finite state automaton, the Moore machine reads each element of a string w of characters of A until it has read the entire string. In the process it produces output consisting of a string of characters of Σ. Since the Moore machine produces output $\phi(s_0)$ before the first input character is read and produces output from the last state reached before the transition function tries and fails to read input, the output string contains one more character than the input string. Also since $\phi(s_0)$ is always executed first, each output string begins with $\phi(s_0)$. When we say a Moore machine **reads** a symbol a of the alphabet A, we simply mean that the letter a is used as input for the function F. Similarly, output is simply the value of ϕ in a given state s_i. We shall say that the machine **prints** the value $\phi(s_0)$, although the output may be used for an entirely different purpose. One may envision a Moore machine reading input from a tape and printing the output on the tape or on another tape.

Formally a Moore machine M is denoted by $(A, \Sigma, S, s_0, F, \phi)$. As with the finite state automaton, the Moore machine is best illustrated using a finite state diagram. As before, if $F(a, s_i) = s_j$, this is represented by

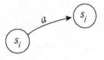

If $\phi(s_i) = z$, this is represented by

In the diagram

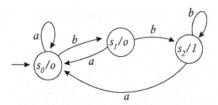

$A = \{a, b\}$, $\Sigma = \{0, 1\}$, $S = \{s_0, s_1, s_2\}$, F is given by the table

F	s_0	s_1	s_2
a	s_0	s_0	s_0
b	s_1	s_2	s_2

and ϕ is given by the table

s	$\phi(s)$
s_0	0
s_1	0
s_2	1

Given the input string bb, the machine first prints the value $\phi(s_0) = 0$. It then reads b and travels to state $F(b, s_0) = s_1$. It then prints $\phi(s_1) = 0$. Next it reads b and travels to state $F(b, s_1) = s_2$. Finally, it reads $\phi(s_2) = 1$. Since there is no more input, operations cease. The result is the output string 001. The input string *aabba* produces the output string 000010. The input string *abb* produces the output string 0001.

Note that the Moore machine we have produced is actually the finite automaton at the beginning of the section except that we have added ϕ with the property that $\phi(s_i) = 0$ if s_i is not an acceptance state and $\phi(s_i) = 1$ if s_i is an acceptance state. When we do this, the last character printed will be 1 if and only if the input is accepted by the finite automaton. Thus, since the output for *bb* and *abb* are 001 and 0001, respectively, *bb* and *abb* are accepted by the automaton. Using this procedure we can "duplicate" any finite automaton with a Moore machine where a word is accepted only if the last character output is 1. It may also be observed that whenever a 1 appears in the output, the initial string of input that has been read at that point is accepted by the finite automaton since the state at that point is an acceptance state. For example, in the preceding example input *abbabbb* produces output 00010011, so *abb*, *abbabb*, and *abbabbb* are all accepted by the automaton. In this particular example, since a word is accepted only if it ends in *bb*, the number of 1's tells us how many times the string *bb* occurs in the input. In general, the number of 1's in the output of a Moore machine that "duplicates" a finite automaton is the number of initial strings of the input that is accepted by the finite automaton.

EXAMPLE 17.22

The automaton

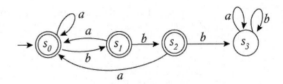

accepts all words that do not contain three consecutive *b*'s. The corresponding Moore machine is

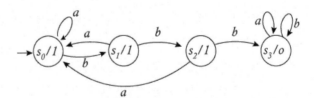

Input *abbaa* produces output 111111 so every substring of *abbaa* is accepted by the automaton. The first 1 in the output indicates that the "empty string" λ, a string containing no input, is accepted. The input *ababbbaa* produces output 111111000. Until an initial substring contains three consecutive *b*'s it is accepted by the automaton. All further initial substrings are not accepted since they contain three consecutive *b*'s. ∎

EXAMPLE 17.23

Let $Z_5 = \{\overline{0}, \overline{1}, \overline{2}, \overline{3}, \overline{4},\}$ be the set of integers modulo 5, where the "sum" of two integers is found by adding the numbers and finding the remainder of this sum when divided by 5. Therefore, $\overline{3} + \overline{4} = \overline{2}$ and $\overline{2} + \overline{3} = \overline{0}$. The following Moore machine gives a sum of initial strings of elements of Z_5. Thus, the input $\overline{2}\,\overline{1}\,\overline{4}\,\overline{0}\,\overline{3}\,\overline{2}\,\overline{1}$ produces output $\overline{0}\,\overline{2}\,\overline{3}\,\overline{2}\,\overline{2}\,\overline{0}\,\overline{2}\,\overline{3}$. ∎

EXAMPLE 17.24 A unit delay machine delays the appearance of a bit in s string by one bit. Hence, the appearance of a character in the output is preceded by one in the input. The following machine is a unit delay machine. ∎

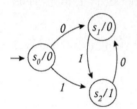

So far, we have primarily shown that a Moore machine may be used to "duplicate" a finite automaton. This is only one of the uses of a Moore machine. However, any task performed by a Moore machine can be performed by another machine called a Mealy machine and conversely. In most cases the task is more easily shown using a Mealy machine.

The **Mealy machine**, created by George H. Mealy [64], also contains an output function, however, the input is an edge rather than a state. Since the edge depends on the state and the input, the output function δ "reads" a letter of $a \in A$ and the current state and prints out a character of the output alphabet. Hence, δ is a function from $S \times A$ to Σ. More formally a Mealy machine is a six-tuple $M_e = (A, \Sigma, S, s_0, F, \delta)$ where $A, \Sigma, S, s_0,$ and F are the same as in the Moore machine and $\delta : S \times A \to \Sigma$. The Mealy machine is also best illustrated using a finite state diagram. We shall denote the output by placing it on the edge so that

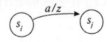

corresponds to $F(a, s_i) = s_j$ and $\delta(a, s_i) = b$. Note that, unlike the Moore machine, the output occurs after the input is read. Hence, for every letter of input, there is a character of output.

Consider the Mealy machine

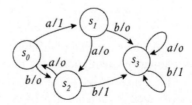

The functions F and δ are given by tables

F	s_0	s_1	s_2	s_3
a	s_1	s_2	s_0	s_3
b	s_2	s_3	s_3	s_3

and

δ	s_0	s_1	s_2	s_3
a	1	0	0	0
b	0	0	1	1

Given the input string $aaabb$, a is read, 1 is printed, and the machine moves to state s_1. The second a is read, 0 is printed, and the machine moves to state s_2. The third a is read, 0 is printed, and the machine returns to state s_0. The letter b is read, 0 is printed, and the machine returns to state s_2. Finally, b is read, 1 is printed, and the machine reaches state s_3. Thus, input $aaabb$ produces output 10001.

Note that the Mealy machine

simply converts every a in the string to x, every b to y, and every c to z. Thus, $aabbcca$ is converted to $xxyyzzx$.

EXAMPLE 17.25 The **1's complement** of a binary string converts each 1 in the string to a 0 and each 0 to a 1. It is given by the state diagram

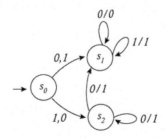

∎

EXAMPLE 17.26 If 1 is added to the 1's complement of a binary string of length n, we obtain the **2's complement** used to express the negative of an integer if we discard any number carried over beyond n digits. Thus, $1111 + 1 = 0000$. See Section 5.7.

The following Mealy machine adds 1 to a binary string in this fashion. The input string must be read in backward and the output is printed out backward so the unit digit is read first. The state diagram

describes the Mealy machine. In this diagram, s_1 is the state reached if there is no 1 to carry when adding the digits. s_2 is the state reached if there is a 1 to carry when adding the digits. Let 1101 be the number in reverse. (Hence, the actual number is 1011.) First, input 1 is read. The output is 0 and the machine is in state s_2. (This corresponds to $1 + 1 = 10$ so 0 is output and 1 is carried.) Now input 1 is read. The output is 0 and the machine remains in state s_2. (This corresponds to $1 + 1 = 10$ so 0 is output and 1 is carried.) Next 0 is input. The output is 1 and the machine moves to state s_1. (This corresponds to $1 + 0 = 1$ so 1 is output and nothing is carried.) Finally, 1 is input. The output is 1 and the machine remains in state s_1. (This corresponds to $1 + 0 = 1$ so 1 is output and nothing is carried.) Thus, the output is 0011 and the number is 1100. ∎

EXAMPLE 17.27

The Mealy machine M_+ adds two integers. The signed integer m is subtracted from the signed integer n by adding n to the 2's complement of m. Thus, M_+ can also be used for subtraction by first using the machine in the previous example to find the 2's complement of the number to be subtracted. Assume $a_n, a_{n-1}, \ldots a_2, a_1$ and $b_n, b_{n-1}, \ldots b_2, b_1$ are the two strings to be added. We again assume that the two strings to be added are read in reverse so the first two digits to be input are a_1 and b_1, followed by a_2 and $b_2, \ldots,$ followed by a_n and b_n. We shall consider the pair of digits to be input as ordered pairs, so that (a_1, b_1) is the first element of input. M_+ is the machine

The machine is in state s_0 when no 1 has been carried in adding the previous input and is in state s_1 when a 1 has been carried in the addition. Assume that 0101 and 1101 are added. First, $(1, 1)$ is read, so the machine moves to s_1 and prints 0. Next $(0, 0)$ is read, so the machine moves to s_0 and prints 1. Then $(1, 1)$ is read, so the machine returns to s_1 and prints 0. Finally, $(1, 0)$ is read, so the machine remains at s_1 and prints 0. Note that the 1, if it exists, which is carried from adding the last two digits, is discarded. Thus, the sum of 0101 and 1101 is 0010. ∎

Earlier in this section, we implied that Moore machines and Mealy machines were equivalent in the sense that every Moore machine could be duplicated by a Mealy machine and conversely. More specifically, given a Moore machine, there is a Mealy machine that will produce output equivalent to the Moore machine when given the same input. Conversely, given a Mealy machine, there is a Moore machine that will produce the output equivalent to the Mealy machine when given the same input.

We first need to specify what we mean by equivalent output, since a Mealy machine always has one less symbol of output than the Moore machine. A string of output of a Mealy machine is **equivalent** to a string of output of a Moore machine if it is equal to the substring of the Moore machine excluding the first symbol $\phi(s_0)$. Thus, if the Moore machine produced output 010010101, the equivalent output from the Mealy machine would be 10010101.

The transformation from the Moore machine to an equivalent Mealy machine is the simplest. With the transition

in a Moore machine, with input c, a_0 will be printed, the machine will move to state s_i and a_1 will next be printed. In the transition

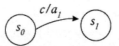

of a Mealy machine, the machine will move to state s_i with input c and a_1 will be printed. Since we disregard a_0 in our definition of equivalent output, we have begun the same output. Assume that we have the transition

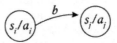

in a Moore machine and a_i has already been printed. Input b will produce output a_j as the next output, when the machine is in state s_i. The machine moves to state s_j. The corresponding transition in the Mealy machine is

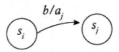

which produces the same transition and output. The Mealy machine corresponding to the Moore machine

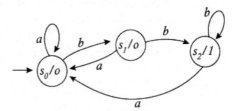

is

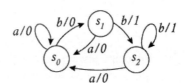

In transforming a Mealy machine to a Moore machine, we have to consider the problem in which arrows into a given state produce different output. Consider the following example:

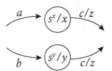

In a Moore machine, the state s_i produces unique output so it cannot produce both x and y as output. We solve this by making two copies of s.

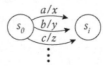

One will produce x as output and the other y as output as follows. Obviously both machines produce output x with input a and output y with input b. For simplicity, we shall simplify s^x/x to s/x and s^y/y to s/y noting that they are different states.

In general, for each state s, except the starting state, in a Mealy machine and for each output symbol z, we shall produce a copy s/z of the state s. This may result in some overkill since in the preceding example, if the output symbols were x, y, and z, we would not have needed state s/z since there was no arrow entering s with output z. We begin with initial state s_0 and give it an arbitrary output variable x_0 from the set of output variables since it is not used in producing output equivalent to the Mealy machine. If we have

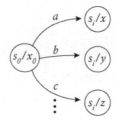

in the Mealy machine, we replace it with

in the Moore machine. For other states, we replace

with

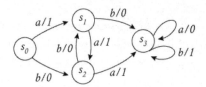

We produce the same output at each step for both machines. Thus, the machine equivalent to

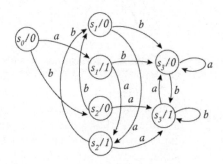

is

Exercises

1. Let the Moore machine $M_o = (A, \Sigma, S, s_0, F, \phi)$ be given by the diagram

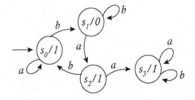

Describe A, Σ, and S. Find tables for F and ϕ.

2. Let the Moore machine $M_o = (A, \Sigma, S, s_0, F, \phi)$ be given by the diagram

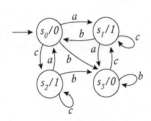

Describe A, Σ, and S. Find tables for F and ϕ.

Let the Moore machine $M_o = (A, \Sigma, S, s_0, F, \phi)$ be given by the diagram

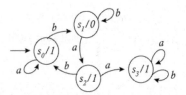

3. Find the output with input *abbabab*.
4. Find the output with input *bbaaba*.
5. Find the output with input *bbbaaa*.
6. Find the output with input λ, the empty word.

Let the Moore machine $M_o = (A, \Sigma, S, s_0, F, \phi)$ be given by the diagram

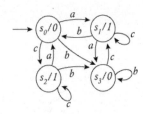

7. Find the output with input *abccbac*.
8. Find the output with input *bbaacc*.
9. Find the output with input *aabbcca*.
10. Find the output with input λ, the empty word.
11. Find the Moore machine that duplicates the finite automaton

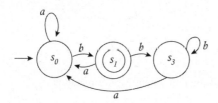

12. Find the Moore machine that duplicates the finite automaton

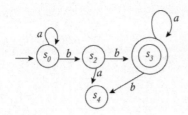

13. Find the Moore machine that duplicates the finite automaton

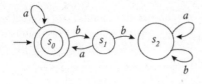

14. Find the Moore machine that duplicates the finite automaton

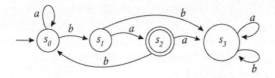

15. Let the Mealy machine $M_e = (A, \Sigma, S, s_0, F, \delta)$ be given by the diagram

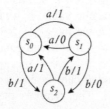

Describe A, Σ, and S. Find tables for F and δ.

16. Let the Mealy machine $M_e = (A, \Sigma, S, s_0, F, \delta)$ be given by the diagram

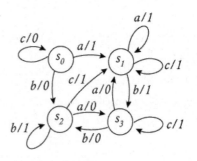

Describe A, Σ, and S. Find tables for F and δ.

Let the Mealy machine $M_e = (A, \Sigma, S, s_0, F, \delta)$ be given by the diagram

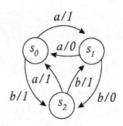

17. Find the output with input *abaabbab*.
18. Find the output with input *bbaaba*.
19. Find the output with input *aabbaaa*.
20. Find the output with input λ, the empty word.

Let the Mealy machine $M_e = (A, \Sigma, S, s_0, F, \delta)$ be given by the diagram

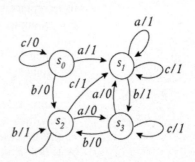

21. Find the output with input *abcccbab*.
22. Find the output with input *bbaabc*.
23. Find the output with input *aaccbba*.

672 Chapter 17 Theory of Computation

24. Given the Moore machine $M_o = (A, \Sigma, S, s_0, F, \phi)$

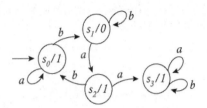

find the equivalent Mealy machine.

25. Given the Moore machine $M_o = (A, \Sigma, S, s_0, F, \phi)$

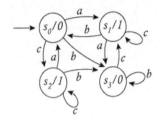

find the equivalent Mealy machine.

26. Given the Mealy machine $M_e = (A, \Sigma, S, s_0, F, \delta)$

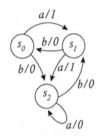

find the equivalent Moore machine.

27. Given the Mealy machine $M_e = (A, \Sigma, S, s_0, F, \delta)$

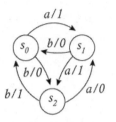

find the equivalent Moore machine.

28. Given the Mealy machine $M_e = (A, \Sigma, S, s_0, F, \delta)$

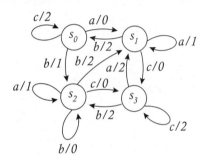

find the equivalent Moore machine.

29. Explain how Mealy machines can be used to subtract a binary number from another binary number.

17.4 Grammars

A **grammar** is a set of rules that is used to construct a language. These rules allow us to replace symbols or strings of symbols with other symbols or strings of symbols. For example, suppose that we begin with a word *add*, and that we can replace *add* with $A + B$, and that both A and B can be replaced with any nonnegative integer less than 10. Using this rule, we can replace A with 3 and B with 2 to get $3 + 2$. If we replace A with 5 and B with 6, we get $5 + 6$.

If we add a further rule that A can be replaced by $A + B$, we can start with replacing *add* with $A + B$. If we then replace A with $A + B$, we get $A + B + B$. Notice that A could have been replaced by either $A + B$ or by an integer, so there is not necessarily any uniqueness in replacing a symbol. Also both of the Bs in $A + B + B$ do not have to be replaced by the same value. If we replace the A with 3, the first B with 5, and the second B with 7, we have $3 + 5 + 7$. We could also have again replaced A with $A + B$ to get $A + B + B + B$ and continue to build sums of arbitrary length.

Note that in the foregoing rules, *add*, A and B can be replaced by other symbols while $+$ and the integers cannot be replaced. The symbols that can be replaced by other symbols are called *nonterminal symbols* and the symbols that cannot be replaced by other symbols are called *terminal symbols*. The rules that tell us how to replace symbols are called *productions*. We denote the production (or rule) that tells us that *add* can be replaced $A + B$ by

$$add \to A + B$$

Thus, the productions for our first example above are

$$add \to A + B$$
$$A \to 0 \qquad B \to 0$$
$$A \to 1 \qquad B \to 1$$
$$A \to 2 \qquad B \to 2$$
$$\vdots \qquad \vdots$$
$$A \to 9 \qquad B \to 9$$

When we add the rule that A can be replaced by $A + B$, we then have the following productions:

$$add \to A + B \qquad A \to A + B$$
$$A \to 0 \qquad B \to 0$$
$$A \to 1 \qquad B \to 1$$
$$A \to 2 \qquad B \to 2$$
$$\vdots \qquad \vdots$$
$$A \to 9 \qquad B \to 9$$

A grammar is formally defined as follows.

DEFINITION 17.28 A **formal grammar** or **phrase structure grammar** consists of a finite set of **nonterminal symbols** N, a finite set of **terminal symbols** T, an element $S \in N$, called the **start symbol**, and a finite set of productions P, which is a relation in $(N \cup T)^*$ such that each first element in an ordered pair of P contains a symbol from N and at least one first element in some ordered pair is the start symbol S. If W and W' are elements of $(N \cup T)^*$, $W = uvw$, $W' = uv'w$, and $v \to v'$ is a production, this is denoted by $W \Rightarrow W'$. If

$$W_1 \Rightarrow W_2 \Rightarrow W_3 \Rightarrow \cdots \Rightarrow W_n$$

for $n \geq 0$, then W_n is **derived** from W_1. This is denoted by $W_1 \Rightarrow_n^* W_n$. A formal grammar Γ is denoted by the 4-tuple (N, T, S, P). The set of all strings of elements of T that may be generated by the set of productions P is called the **language generated by the grammar** Γ and is denoted by $\Gamma(L)$.

To generate a word from the grammar Γ, we keep using productions to derive new strings until we have a string consisting only of terminal elements.

Thus, in our final example above,

$$N = \{add, A, B\}$$
$$T = \{+, 0, 1, 2, 3, 4, 5, 6, 7, 8, 9\}$$
$$S = add$$

and

$$P = \{(add, A + B),$$
$$(A, A + B), (A, 0), (A, 1), \ldots, (A, 9), (B, 0), (B, 1), \ldots, (B, 9)\}$$

where we will denote $(add, A + B)$ by $add \to A + B$, $(A, A + B)$ by $A \to A + B$, and so on. The language generated by Γ is the set of all formal expressions of finite sums of nonnegative integers less than 10.

EXAMPLE 17.29 In the grammar described previously, derive the expression $3 + 2 + 4$.

Begin with the production

$$add \to A + B$$

to derive

$$A + B$$

Then use the production

$$A \to A + B$$

to derive

$$A + B + B$$

Then use the productions

$$A \to 3 \quad B \to 2 \quad B \to 4$$

to derive

$$3 + 2 + 4 \qquad \blacksquare$$

EXAMPLE 17.30 Suppose we want a grammar that derives arithmetic expressions for the set of integers $\{0, 1, 2, 3, 4, 5, 6, 7, 8, 9\}$. Thus, the language generated by the grammar is the set of all finite arithmetic expressions for the set of integers $\{0, 1, 2, 3, 4, 5, 6, 7, 8, 9\}$. Examples would be $3 \times (5 + 4)$ and $(4 + 5) \div (3\hat{\ }2)$, where $\hat{\ }$ denotes exponent. We obviously want to exclude expressions such as $3 + \times 6$ and $3 + \div 6 \times 4 - 5$. Let the set $N = \{S, A, B\}$ and $T = \{+, -, \times, \div, \hat{\ }, 0, 1, 2, 3, 4, 5, 6, 7, 8, 9, (,)\}$. We will need the following productions:

$$\begin{array}{rclcrcl}
S & \to & (A + B) & \quad & B & \to & (A + B) \\
S & \to & (A - B) & \quad & B & \to & (A - B) \\
S & \to & (A \times B) & \quad & B & \to & (A \times B) \\
S & \to & (A \div B) & \quad & B & \to & (A \div B) \\
S & \to & (A\hat{\ }B) & \quad & B & \to & (A\hat{\ }B) \\
A & \to & (A + B) & \quad & A & \to & 0 \\
A & \to & (A - B) & \quad & \vdots & & \vdots \\
A & \to & (A \times B) & \quad & A & \to & 9 \\
A & \to & (A \div B) & \quad & B & \to & 0 \\
A & \to & (A\hat{\ }B) & \quad & \vdots & & \vdots \\
 & & & \quad & B & \to & 9 \\
\end{array}$$

We will use the grammar to derive the arithmetic expression $((2 + 3) \times (4 + 5))$. We begin with the production

$$S \to (A \times B)$$

17.4 Grammars

We then use the productions
$$A \to (A + B)$$
and
$$B \to (A + B)$$
to derive
$$((A + B) \times (A + B))$$
The productions
$$A \to 2 \quad \text{and} \quad B \to 3$$
give us
$$((2 + 3) \times (A + B))$$
Finally, we use the productions
$$A \to 4 \quad \text{and} \quad B \to 5$$
to derive
$$((2 + 3) \times (4 + 5))$$

We next use the grammar to derive the arithmetic expression $((3\hat{\ }2) \div (5 + 7))$. We begin with the production
$$S \to (A \div B)$$
We then use the productions
$$A \to (A\hat{\ }B) \quad \text{and} \quad B \to (A + B)$$
to derive
$$((A\hat{\ }B) \div (A + B))$$
The productions
$$A \to 3 \quad \text{and} \quad B \to 2$$
give us
$$((3\hat{\ }2) \div (A + B))$$
Finally we use the productions
$$A \to 5 \quad \text{and} \quad B \to 7$$
to derive
$$((3\hat{\ }2) \div (5 + 7))$$
■

EXAMPLE 17.31

In a similar manner, we may form arithmetic expression in postfix notation. Let the set $N = \{S, A, B\}$ and
$$T = \{+, -, \times, /, \hat{\ }, 0, 1, 2, 3, 4, 5, 6, 7, 8, 9\}$$
We will need the following productions:

$$
\begin{array}{llll}
S \to AB+ & A \to AB+ & B \to AB+ & A \to 0 \\
S \to AB- & A \to AB- & B \to AB- & \vdots \\
S \to AB\times & A \to AB\times & B \to AB\times & A \to 9 \\
S \to AB\div & A \to AB\div & B \to AB\div & B \to 0 \\
S \to AB\hat{\ } & A \to AB\hat{\ } & B \to AB\hat{\ } & \vdots \\
& & & B \to 9
\end{array}
$$

Consider the expression $32 + 47 + \times$. Since our integers are all less than 10, $32+$ represents the integer symbol 3, followed by the integer symbol 2 and the $+$ symbol. To construct this expression, we begin with the production

$$S \rightarrow AB\times$$

We then use the productions

$$A \rightarrow A + B \quad \text{and} \quad B \rightarrow A + B$$

to derive

$$AB + AB + \times$$

The productions

$$A \rightarrow 2 \quad \text{and} \quad B \rightarrow 3$$

give us

$$23 + AB + \times$$

Finally, we use the productions

$$A \rightarrow 4 \quad \text{and} \quad B \rightarrow 7$$

to derive

$$23 + 47 + \times$$

■

EXAMPLE 17.32

A grammar may also be used to derive proper sentences. These sentences are proper in the sense that they are grammatically correct, although they may not have any meaning. Suppose we want a grammar that will derive the following statements, among others:

Joe chased Fred.
The large dog leaped over the old fence.
Tom drove slowly into the sunset.

Before actually stating the grammar, let's decide on its structure. Each of our sentences has a noun phrase (noun p), a verb phrase (verb p), and another noun phrase. In addition the last two sentences have a preposition (prep). Therefore, let the first production be

$$S \rightarrow \langle \text{noun p} \rangle \langle \text{verb p} \rangle \langle \text{prep} \rangle \langle \text{noun p} \rangle$$

In our example, the most general form of a noun phrase is an article followed by an adjective and then a noun. Therefore, let the next production be

$$\langle \text{noun phrase} \rangle \rightarrow \langle \text{art} \rangle \langle \text{adj} \rangle \langle \text{noun} \rangle$$

where "art" represents article and "adj" represents adjective.

The most general form of a verb phrase is a verb followed by an adverb. Therefore, let the next production be

$$\langle \text{verb p} \rangle \rightarrow \langle \text{adv} \rangle \langle \text{verb} \rangle$$

where "adv" represents adverb.

At this point, we know that the terminal set $T = \{$Joe, chased, Fred, the, large, dog leaped, over, the, old, fence, Tom, drove, slowly, into, the, sunset$\}$. The nonterminal set $N = \{S, \langle \text{noun p} \rangle, \langle \text{verb p} \rangle, \langle \text{art} \rangle, \langle \text{adj} \rangle, \langle \text{noun} \rangle, \langle \text{adv} \rangle, \langle \text{verb} \rangle, \langle \text{prep} \rangle\}$.

We next need productions that will assign values to ⟨art⟩, ⟨adj⟩, ⟨noun⟩, ⟨adv⟩, and ⟨verb⟩. In some of our sentences we do not need ⟨art⟩, ⟨adjective⟩, ⟨prep⟩, and ⟨adv⟩. To solve this problem, we include the productions

$$\langle art \rangle \to \lambda \qquad \langle adj \rangle \to \lambda \qquad \langle adv \rangle \to \lambda \qquad \langle prep \rangle \to \lambda$$

By assigning these symbols to the empty set, we simply erase them when they are not needed. The remainder of our productions consists of the following:

⟨art⟩ → the	⟨noun⟩ → Fred	⟨noun⟩ → fence
⟨adj⟩ → large	⟨noun⟩ → dog	⟨adv⟩ → slowly
⟨adj⟩ → old	⟨noun⟩ → Tom	⟨verb⟩ → chased
⟨noun⟩ → Joe	⟨noun⟩ → sunset	⟨verb⟩ → leaped
⟨verb⟩ → drove	⟨prep⟩ → over	⟨prep⟩ → into

To derive the sentence "Joe chased Fred," we begin with

$$S \to \langle noun\ p \rangle \langle verb\ p \rangle \langle prep \rangle \langle noun\ p \rangle$$

to derive

$$\langle noun\ p \rangle \langle verb\ p \rangle \langle prep \rangle \langle noun\ p \rangle$$

Using the production

$$\langle noun\ p \rangle \to \langle article \rangle \langle adjective \rangle \langle noun \rangle$$

we derive

$$\langle art \rangle \langle adj \rangle \langle noun \rangle \langle verb\ p \rangle \langle prep \rangle \langle noun\ p \rangle$$

Using

$$\langle art \rangle \to \lambda \qquad \langle adj \rangle \to \lambda$$

we derive

$$\langle noun \rangle \langle verb\ p \rangle \langle prep \rangle \langle noun\ p \rangle$$

Repeating the process for the second ⟨noun phrase⟩, we derive

$$\langle noun \rangle \langle verb\ p \rangle \langle prep \rangle \langle noun \rangle$$

Using

$$\langle verb\ p \rangle \to \langle adv \rangle \langle verb \rangle$$

we derive

$$\langle noun \rangle \langle adv \rangle \langle verb \rangle \langle prep \rangle \langle noun \rangle$$

Using

$$\langle adv \rangle \to \lambda \qquad \langle prep \rangle \to \lambda$$

we derive

$$\langle noun \rangle \langle verb \rangle \langle noun \rangle$$

Using

$$\langle noun \rangle \to Joe \qquad \langle noun \rangle \to Fred \qquad \langle verb \rangle \to chased$$

we derive "Joe chased Fred."

To derive the sentence "The large dog leaped over the old fence," we again begin with

$$S \to \langle noun\ p \rangle \langle verb\ p \rangle \langle prep \rangle \langle noun\ p \rangle$$

to derive

$$\langle\text{noun p}\rangle\langle\text{verb p}\rangle\langle\text{prep}\rangle\langle\text{noun p}\rangle$$

Using the production

$$\langle\text{noun p}\rangle \rightarrow \langle\text{art}\rangle\langle\text{adj}\rangle\langle\text{noun}\rangle$$

we derive

$$\langle\text{art}\rangle\langle\text{adj}\rangle\langle\text{noun}\rangle\langle\text{verb p}\rangle\langle\text{prep}\rangle\langle\text{noun p}\rangle$$

Using

$$\langle\text{art}\rangle \rightarrow \text{the} \quad \langle\text{adj}\rangle \rightarrow \text{large} \quad \langle\text{noun}\rangle \rightarrow \text{dog}$$

we derive

$$\text{the large dog}\langle\text{verb p}\rangle\langle\text{prep}\rangle\langle\text{noun p}\rangle$$

Using

$$\langle\text{verb p}\rangle \rightarrow \langle\text{adv}\rangle\langle\text{verb}\rangle$$

we derive

$$\text{the large dog}\langle\text{adv}\rangle\langle\text{verb}\rangle\langle\text{prep}\rangle\langle\text{noun p}\rangle$$

Using

$$\langle\text{adv}\rangle \rightarrow \lambda \quad \langle\text{verb}\rangle \rightarrow \text{leaped}$$

we derive

$$\text{the large dog leaped }\langle\text{prep}\rangle\langle\text{noun p}\rangle$$

Using

$$\langle\text{prep}\rangle \rightarrow \text{over}$$

we derive

$$\text{the large dog leaped over}\langle\text{noun p}\rangle$$

Using the production

$$\langle\text{noun p}\rangle \rightarrow \langle\text{art}\rangle\langle\text{adj}\rangle\langle\text{noun}\rangle$$

we derive

$$\text{the large dog leaped over}\langle\text{art}\rangle\langle\text{adj}\rangle\langle\text{noun}\rangle$$

Using

$$\langle\text{art}\rangle \rightarrow \text{the} \quad \langle\text{adj}\rangle \rightarrow \text{old} \quad \langle\text{noun}\rangle \rightarrow \text{fence}$$

we derive

$$\text{the large dog leaped over the old fence}$$

Derivation of the last sentence is left to the reader. ■

17.4 Grammars

DEFINITION 17.33 For each production $P \to w_1 w_2 w_3 \ldots w_n$ the **corresponding tree** is shown in Figure 17.29.

Thus, the corresponding tree for $S \to A + B$ is the tree in Figure 17.30.

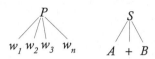

Figure 17.29   Figure 17.30

DEFINITION 17.34 If the corresponding trees of the productions used to derive a given expression are connected, they form a tree with root S, called the **parse tree** or the **derivation tree**. The leaves of the tree, when read left to right, form the expression.

EXAMPLE 17.35 In Example 17.29, we used productions to derive $3 + 2 + 4$.

To construct the tree, begin with the first production used:

$$add \to A + B$$

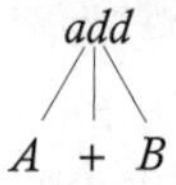

Figure 17.31

to form the corresponding tree in Figure 17.31.

Then use the corresponding tree in Figure 17.32 of the production

$$A \to A + B$$

Figure 17.32

to form the tree in Figure 17.33.

Then use the corresponding trees of the next productions

$$A \to 3 \quad B \to 2 \quad B \to 4$$

to form the parse tree in Figure 17.34.

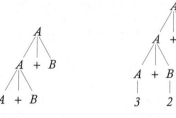

Figure 17.33   Figure 17.34

EXAMPLE 17.36 In Example 17.31, to derive $((2 + 3) \times (4 + 5))$, we use the productions

$$S \to (A \times B) \quad A \to (A + B) \quad A \to 2 \quad B \to 3$$
$$B \to (A + B) \quad A \to 4 \quad B \to 5$$

Therefore, the parse tree is the tree in Figure 17.35.

EXAMPLE 17.37

In Example 17.32, to derive the sentence "The large dog leaped over the old fence," we use the productions

$S \to$ ⟨noun p⟩⟨verb p⟩⟨prep⟩⟨noun p⟩ ⟨noun p⟩ $\to$ ⟨art⟩⟨adj⟩⟨noun⟩
⟨noun p⟩ $\to$ ⟨art⟩⟨adj⟩⟨noun⟩ ⟨adj⟩ $\to$ large ⟨noun⟩ $\to$ dog
⟨verb p⟩ $\to$ ⟨adv⟩⟨verb⟩ ⟨adv⟩ $\to \lambda$ ⟨verb⟩ $\to$ leaped
⟨prep⟩ over ⟨art⟩ $\to$ the ⟨art⟩ $\to$ the
⟨adj⟩ $\to$ old ⟨noun⟩ $\to$ fence

Thus, the parse tree is shown in Figure 17.36.

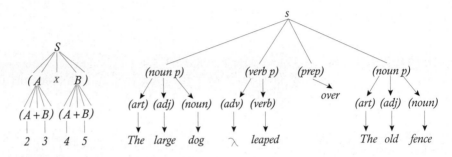

Figure 17.35   Figure 17.36

In all of the grammars in this section, the productions have been of the form $A \to W$, where A is a nonterminal symbol. Therefore, the production can be used everywhere that A appears, regardless of its position in an expression. Such grammars are called **context-free grammars**. If a grammar has a production of the form $aAb \to W$ where A is a nonterminal and either a or b is not the empty word, then this production can only be used when a is on the left side of A and b is on the right side. It, therefore, cannot be used whenever A appears and so it is dependent on the context in which A appears. Such a grammar is called a **context-sensitive grammar**. ∎

In the following examples, we consider context-free grammars that generate more abstract languages:

EXAMPLE 17.38

Let $\Gamma = (N, T, S, P)$ be the grammar defined by $N = \{S, A, B\}$, $T = \{a, b\}$, and P be the set of productions

$$S \to AB \quad A \to a \quad B \to Bb \quad B \to \lambda$$

Using the production $S \to AB$, we derive AB. Next using the productions $A \to a$ and $B \to \lambda$, we derive a. If we use the productions

$$S \to AB \quad A \to a \quad B \to Bb \quad B \to \lambda$$

in order, we derive ab. We can also generate $abbb$, $abbbb$, and $abbbbb$. In fact, we can generate ab^n for all nonnegative integers n. Hence, the expression for the language generated by Γ is ab^*. ∎

EXAMPLE 17.39

Let $\Gamma' = (N, T, S, P)$ be the grammar defined by $N = \{S, A\}$, $T = \{a, b\}$, and P be the set of productions

$$S \to aAb \quad A \to aAb \quad A \to \lambda$$

Using the productions $S \to aAb$ and $A \to \lambda$, we derive ab. Using the productions

$$S \to aAb \quad A \to aAb \quad A \to \lambda$$

in order, we derive *aabb* or a^2b^2. Using the productions

$$S \to aAb \quad A \to aAb \quad A \to aAb \quad A \to ab$$

in order, we derive *aaabbb* or a^3b^3. It is easily seen that the language generated by Γ' is $\{a^nb^n : n \text{ is a positive integer}\}$. Note that this is not the same as a^*b^* since this would also include a^mb^n where m and n are not equal. ∎

EXAMPLE 17.40 Let $\Gamma'' = (N, T, S, P)$ be the grammar defined by $N = \{S, A, B\}$, $T = \{a, b\}$, and P be the set of productions

$$S \to ABABABA \quad A \to Aa \quad A \to \lambda \quad B \to b$$

It can be shown that the expression for the language generated by Γ'' is $a^*ba^*ba^*ba^*$. This is the language consisting of all words containing exactly three b's. ∎

In Example 17.39 we generated the language $\{a^nb^n : n \text{ is a positive integer}\}$. Intuitively, we can see that this is not a regular language since the only way that we can generate an infinite regular language using a finite alphabet is with the Kleene star ∗. In this case the only possibility is a^*b^* but, as mentioned earlier, this doesn't work since this would also include a^mb^n where m and n are not equal.

One might ask if there is a particular type of grammar that generates only regular languages. The answer is yes, as we shall now show.

DEFINITION 17.41 A context-free grammar $\Gamma = (N, T, S, P)$ is called a **regular grammar** if every production $p \in P$ has the form $n \to w$ where the string w contains at most one nonterminal symbol and it occurs at the end of the string, if at all.

Therefore, w could be of the form $aacA$, ab, or bA, where a, b, and c are terminals and A is a nonterminal. However, w could not be of the form aAb, aAB, or Aa. The production $n \to abcA$ could be replaced by the productions

$$n \to aB \quad B \to bC \quad C \to cA$$

It is also possible w could contain no terminal and one nonterminal so we have $B \to C$, but if this is followed by $C \to tC$, where t is a terminal, we can then combine the two productions to get $B \to tC$. Hence, it is no restriction to require each production to be one of the following forms:

$$A \to aB \quad B \to b \quad C \to \lambda$$

where A, B, and C are nonterminal elements, and a and b are terminal elements.

The following theorem which unites regular grammars, regular languages, and automata is stated without proof.

THEOREM 17.42 A language is regular if and only if it is generated by a regular grammar. ∎

Exercises

1. Using the grammar in Example 17.38, construct a parse tree for *abbb*.

2. Using the grammar in Example 17.39, construct a parse tree for *aaabbb*.

3. Using the grammar in Example 17.40, construct a parse tree for *babaab*.

4. Find the language generated by the grammar $\Gamma = (N, T, S, P)$ defined by $N = \{S, A, B\}$, $T = \{a, b\}$ and

the set of productions P given by

$$S \to AB \quad A \to aA \quad A \to \lambda \quad B \to Bb \quad B \to \lambda$$

5. Find the language generated by the grammar $\Gamma = (N, T, S, P)$ defined by $N = \{S, A, B\}$, $T = \{a, b\}$ and the set of productions P given by

$$S \to aB \quad B \to bA \quad A \to aB \quad B \to b$$

6. Find the language generated by the grammar $\Gamma = (N, T, S, P)$ defined by $N = \{S, A, B\}$, $T = \{a, b\}$ and the set of productions P given by

$$S \to aA \quad B \to aA \quad S \to bB \quad A \to aB$$
$$B \to bB \quad A \to bA \quad B \to b \quad A \to a$$

7. Find the language generated by the grammar $\Gamma = (N, T, S, P)$ defined by $N = \{S, A, B, C\}$, $T = \{a, b\}$ and the set of productions P given by

$$S \to C \quad A \to aB \quad C \to bC \quad B \to bB$$
$$C \to aA \quad B \to aA \quad A \to bA \quad B \to \lambda$$

17.5 Turing Machines

A Turing machine is another theoretical machine. It seems very crude, yet it is incredibly powerful with regard to the functions that it can perform. The Turing machine has an input table A and a set of states S. It has two special states, the start state s_0 and the halt state h. The Turing machine begins in the start state and halts if it reaches the halt state. It also contains a tape that is infinitely long on the right together with an output alphabet P. The input table and the output table may be the same. There is no loss of generality if they are assumed to be the same. The tape contains a string of squares on which letters of the alphabet may be printed or erased. Only a finite number of squares may contain letters of the input or output alphabet. All of the squares to the right of the last square containing a letter of the alphabets are assumed to be blank. The Turing machine also contains a head that can read a blank or a letter of the input alphabet that is on the square of the tape in front of it. After reading the input and depending on the state the machine is in, the head can print or erase on the square in front of it and then move left or right by one square or remain where it is. We shall use # for the blank. The following Turing machine is in state q_1 and is reading letter a.

The Turing machine has a set of rules that tells the machine what to do. An example of a rule is (s_i, a, s_j, b, R). This rule says that if the machine is in state s_i and reads a, it changes to state s_j, prints b, and moves one square to the right. The rule $(s_i, a, s_j, \#, L)$ says that if the machine is in state s_i and reads a, it changes to state s_j, erases the letter b, and moves one square to the left.

We shall assume that a Turing machine always begins at the leftmost square on the tape. Suppose we wish to go to an end of a string on the tape where the input and output characters on the string are from $A = \{a, b\}$. To do this let the states be $S = \{s_0, s_1, h\}$ and the rules be (s_0, a, s_1, a, R), (s_0, b, s_1, b, R), (s_1, a, s_1, a, R), (s_1, b, s_1, b, R), and $(s_1, \#, h, \#, \#, \#)$. One can easily see that the machine simply reads letters and moves to the right until there is no letter to read so it halts. We shall call this routine **go-end**.

We next show you to move the machine to the right n steps. We again assume both the input and output alphabet is $A = \{a, b\}$. Let $S = \{s_0, s_1, \ldots, s_n\}$. We use the following rules:

$$(s_0, a, s_1, a, R), (s_1, a, s_2, a, R), \ldots, (s_{n-1}, a, s_n, a, R), (s_0, b, s_1, b, R),$$
$$(s_1, b, s_2, b, R), \ldots, (s_{n-1}, b, s_n, b, R)$$

It is easily seen that if we begin in state s_1, then each application of a rule regardless of the letter read moves the machine right and increases the state. After n steps, the head has been moved to the right n squares and the machine is in state s_{n+1}. We call this subroutine **go-right** (n). Similarly, we define **go-left** (n).

We next consider a subroutine that inserts a letter at a given point in the string. Assume the alphabet is $A = \{a, b, c\}$. Say that we have string $a_1, a_2 \ldots a_i a_{i+1} \ldots a_n$ and want to replace it with $a_1, a_2 \ldots a_i c a_{i+1} \ldots a_n$ so that the string $a_{i+1} \ldots a_n$ must be moved one square to the right. First we use **go-right** (i) to place the head where the letter c is to be placed. Assume we are in state s_x when we read this square. We are going to need a state for each letter in the alphabet. Thus, we shall need s_a, s_b, s_c. The process is rather simple. When we print c, we need to remember a_{i+1} so we can print it in the next square. We do this by entering $s_{a_{i+1}}$ after we have printed c and then moving to the right. Thus, the rule is $(s_i, a_{i+1}, s_{a_{i+1}}, c, R)$. In state $s_{a_{i+1}}$, we print a_{i+1} in the square occupied by a_{i+2} and again move right. Each time we print a letter, we enter the state corresponding to the letter destroyed and in this way remember this letter so it can be printed in the next square. Finally, we reach a blank square, print a_n, and then use go-left (n) to return to the beginning of the string. We call this subroutine **insert** (n).

Turing machines are extremely powerful. **Church's thesis** states that an algorithm that can be performed by a computer can be performed by a Turing machine. We shall limit ourselves to one property of the Turing machine. We shall show how to use a Turing machine to recognize a regular language. We know that an automaton recognizes a regular language, so what we do is program the Turing machine to imitate an automaton. An automaton reads a word beginning with the first letter and reads from left to right until it has reached the last letter. Our Turing machine will do the same. When the word has been read, we need for the Turing machine to know that this has occurred so that it will stop. We begin with the machine ready to read the first letter. We won't care what is printed out, so assume we print a # in each square as it is read. Each time a letter is read we wish for the machine to move one square right so the next letter is read.

We are now ready to imitate an automaton. If the symbol

occurs in an automaton, we shall imitate it with the rule $(s_i, a, s_j, \#, R)$. For every acceptance state s of the automaton, we will add a rule $(s, \#, h, \#, \#)$ so that if the word is accepted by the automaton, it will also end up in state s of the Turing machine, read the # in the front of the word, and halt. Thus, the Turing machine programmed in this manner accepts the same words as the automaton it is imitating.

EXAMPLE 17.43   Consider the automaton

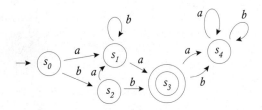

The corresponding Turing machine consists of the rules. Assume **insert** (#) has placed a blank at the beginning of the word.

$(s_0, a, s_1, \#, R), (s_0, b, s_2, \#, R), (s_1, a, s_3, \#, R), (s_1, b, s_1, \#, R), (s_2, a, s_1, \#, R),$
$(s_2, b, s_3, \#, R), (s_3, a, s_4, \#, R), (s_3, b, s_4, \#, R), (s_4, a, s_4, \#, R), (s_4, b, s_4, \#, R),$
$(s_3, \#, h, \#, \#)$

The string aba is accepted by the automaton by first beginning in state s_0, reading a, and moving to state s_1. This is imitated in the Turing machine by the rule $(s_0, a, s_1, \#, R)$ leaving #ba on the tape. Next the automaton while in state s_1 reads b and remains in state s_1. This is imitated in the Turing machine by the rule $(s_1, b, s_1, \#, R)$, leaving #a on the tape. Next the automaton while in state s_1 reads a and ends up in the acceptance state s_3. This is imitated in the Turing machine by the rule $(s_1, a, s_3, \#, R)$, leaving # on the tape. Then the rule $(s_3, \#, h, \#, \#)$ results in the Turing machine accepting the word.

The languages accepted by a Turing machine are called **recursively enumerable**. We have shown that they include regular languages. They also include context free languages. ∎

Exercises

1. Describe a Turing machine **delete** (c) that will delete the letter c from a string and close the gap.

2. Describe a Turing machine that reads all words consisting of a string of zeroes followed by an equal number of ones.

3. Describe a Turing machine **go-left** (n) that will go left n squares.

4. Construct an automaton that reads $a^*ba^*ba^*$ and use it to construct a Turing machine that reads $a^*ba^*ba^*$.

5. Construct an automaton that reads a^*bc and use it to construct a Turing machine that reads a^*bc.

6. Construct an automaton that reads aa^*bb^* and use it to construct a Turing machine that reads aa^*bb^*.

7. Construct an automaton that reads $(aa^*bb^*)^*$ and use it to construct a Turing machine that reads aa^*bb^*.

8. Design a Turing machine that accepts the language of all strings in a and b that have the same number of a's and b's.

GLOSSARY

Chapter 17: Theory of Computation

a-arrow (17.2)	A directed arrow labeled with the letter a is called an **a-arrow**.
Accept or recognize a string (17.2)	An automaton **accepts** or **recognizes a string** $a_0 a_1 a_2 \ldots a_n$ if after reading a_n, the automaton finishes in an acceptance state.
Alphabet (17.1)	An **alphabet** is a finite collection of symbols.
Automaton (17.2)	An **automaton** is a device that recognizes or accepts certain elements of A^*, where A is a finite alphabet.
Class of regular expressions (17.1)	The **class of regular expressions** is defined by the following rules: i. The symbol λ is a regular expression and for every $a \in A$, the symbol a is a regular expression. ii. If w_1 and w_2 are regular expressions, then $w_1 w_2$, $w_1 \vee w_2$, w_1^*, and (w_1) are regular expressions. iii. There are no regular expressions that are not generated by (i) and (ii).
Code (17.1)	If S and C are subsets of A^* and $S = C^*$, then every string in S can be expressed as the concatenation of elements in C, and we can say that C is a **Code** for S.

Concatenation (17.1)	Let A^* be the set of all strings of A. **Concatenation** is a binary operation $\circ$ on A defined as follows: If $a_1a_2a_3a_4\ldots a_n$, $b_1b_2b_3b_4\ldots b_m \in A^*$, then $a_1a_2a_3a_4\ldots a_n \circ b_1b_2b_3b_4\ldots b_m = a_1a_2a_3a_4\ldots a_nb_1b_2b_3b_4\ldots b_m$.
Context-free grammar (17.3)	If a grammar has a production of the form $A \to W$, where A is a nonterminal symbol, and the production can be used everywhere that A appears, regardless of its position in an expression, this grammar is called a **context-free grammar**.
Context-sensitive grammar (17.3)	If a grammar is dependent on the context in which A appears, it is called a **context-sensitive grammar**.
Corresponding tree (17.3)	For each production $P \to w_1w_2w_3\ldots w_n$, the **corresponding tree** is $$\begin{array}{c} P \\ \swarrow \swarrow \downarrow \searrow \\ w_1\ w_2\ w_3\ \cdots\ w_n \end{array}$$
Derived (17.3)	If W and W' are elements of $(N \cup T)^*$, $W = uvw$, $W' = uv'w$, and $v \to v'$ is a production denoted by $W \Rightarrow W'$. If $W_1 \Rightarrow W_2 \Rightarrow W_3 \Rightarrow \cdots \Rightarrow W_n$ for n $\geq$ 0, then W_n is **derived** from W_1. This is denoted by $W_1 \Rightarrow_n^* W_n$.
Deterministic automata (17.2)	In a **deterministic automata** for every state and for every value of the alphabet that is read, there is one and only one state.
Expression (17.1)	If A is an alphabet, an **expression** in A is a string of symbols that represents a subset of A^*.
Formal grammar or phrase structure grammar (17.3)	A **formal grammar** or **phrase structure grammar** consists of a finite set of nonterminal symbols N, a finite set of terminal symbols T, an element $S \in N$, called the *start symbol*, and a finite set of productions P, which is a relation in $(N \cup T)^*$ such that each first element in an ordered pair of P contains a symbol from N and at least one first element in some ordered pair is the start symbol S.
Generation of sets (17.1)	If S and T are subsets of A^*, and $S = T^*$, then S is **generated** by T.
Grammar (17.3)	A **grammar** is a set of rules that is used to construct a language.
Key codes (17.1)	**Key codes** are generated by a finite code and have the property that each word of the code has an element of A that is contained only once in that word and does not occur in any other word of the code.
Kleene star (17.1)	If S is a subset of A^*, then S^* is the set of all strings or words formed by concatenating words from S, including the empty word. The symbol $*$ is called the **Kleene star**.
Kleene's theorem (17.2)	**Kleene's theorem** states that a language is regular if and only if it is accepted by an automaton.
Language (17.1)	If A^* is the set of all strings of A, then a subset L of A^* is called a **language**.
Language generated by the grammar (17.3)	A formal grammar Γ is denoted by the 4-tuple (N, T, S, P). The set of all strings of elements of T that may be generated by the set of productions P is called the **language generated by the grammar** Γ and is denoted by $\Gamma(L)$.
Language over A accepted by M (17.2)	The subset of elements of A^* accepted by an automaton M is a language called the **language over A accepted by M** and denoted by $M(L)$.

Nondeterministic automaton (17.2)	A **nondeterministic automaton** is an automaton on (A, S, S_0, T, F) for which F is not necessarily a function.
Nonterminal symbols (17.3)	The symbols that can be replaced by other symbols in a grammar are called **nonterminal symbols**.
Parse tree or the derivation tree (17.3)	If the corresponding trees of the productions used to derive a given expression are connected, they form a tree with root S, called the **parse tree** or the **derivation tree**. The leaves of the tree, when read left to right, form the expression.
Prefix code (17.1)	A code $C \subseteq A^*$ is called a **suffix code** if for all words $u, v \in C$, if $u = wv$ for $w \in A$, then $u = v$ and $w = \lambda$. This means that one word in a code cannot be the final string of another word in the code.
Productions (17.3)	**Productions** are the rules that tell us how to replace other symbols in a grammar.
Regular expression (17.1)	If A is an alphabet, a **regular expression** is an expression that uses the symbols $*, \vee, \lambda, (,)$, and a for each element a in A.
Regular grammar (17.3)	A context-free grammar $\Gamma = (N, T, S, P)$ is called a **regular grammar** if every production $p \in P$ has the form $n \to w$ where the string w contains at most one nonterminal symbol and it occurs at the end of the string, if at all.
Regular language (17.1)	The class R is a **regular language** if it has the following properties: i. the empty word $\lambda \in R$, and if $a \in R$, then $\{a\} \in R$. ii. If s_1 and $s_2 \in R$, then $s_1 \cup s_2, s_1 \circ s_2, s_1^* \in R$. iii. Only sets formed by using (i) and (ii) belong to R.
Regular language (17.3)	A regular language is one generated by regular grammar.
Retract (17.1)	The image of f in a retraction map is called a **retract**.
Retraction map (17.1)	A function $f : S \to S$ is called a **retraction map** if $f \circ f = f$.
Sink state (17.2)	The *sink state* is a state from which an automaton can never exit.
Start/goal symbol (17.3)	The *start* symbol is a special nonterminal designated as the one from which all strings are derived.
State diagram (17.2)	A **state diagram** is a directed graph where the states are represented by the vertices, and a directed edge from s to s' is labeled with a letter, say a, of the alphabet A if $F(a, s) = s'$.
String or word of symbols (17.1)	A **string** or **word of symbols** of A has the form $a_1 a_2 a_3 a_4 \ldots a_n$ where $a_i \in A$.
Suffix code (17.1)	If A is an alphabet, a code $C \subseteq A^*$ is called a **prefix code** if for all words $u, v \in C$, if $u = vw$ for $w \in A$, then $u = v$ and $w = \lambda$. This means that one word in a code cannot be the beginning string of another word in the code.
Terminal symbols (17.3)	The symbols in a grammar that cannot be replaced by other symbols are called **terminal symbols**.
Transition function (17.2)	For a given finite alphabet and an automaton consisting of a set S of states, a function $F : A \times S \to S$ is called the **transition function**.
Uniquely decipherable (17.1)	A code C is uniquely decipherable if every string in S can be uniquely expressed as the concatenation of elements of C.

TRUE-FALSE QUESTIONS

1. The expression **ab*** describes a prefix code.
2. The expression **a*b** describes a prefix code.
3. The expression **b*ab*** describes a uniquely decipherable code.
4. A prefix code is uniquely decipherable.
5. The expression **ab*c** describes a code that is both a prefix code and a suffix code.
6. The automaton

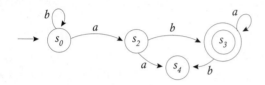

   is a deterministic automaton.
7. The automaton in Problem 6 accepts the word *bbbabab*.
8. **bb*abaa*** describes the language accepted by the automaton in Problem 6.
9. Every language accepted by a deterministic automaton is accepted by a nondeterministic automaton.
10. Every language accepted by a nondeterministic automaton is accepted by a deterministic automaton.
11. The language $\{aab, aaabb, aaaabbb, \ldots, a^{101}b^{100}\}$ is accepted by an automaton.
12. The language $\{aab, aaabb, aaaabbb, \ldots, a^{n+1}b^n, \ldots\}$ is accepted by an automaton.
13. $(\mathbf{a} \vee \mathbf{b} \vee \mathbf{c})^*$ and $(\mathbf{a}^*\mathbf{b}^*\mathbf{c}^*)^*$ are expressions for the same set.
14. $(\mathbf{a} \vee \mathbf{b})^*$ and $\mathbf{a}^* \vee \mathbf{b}^*$ are expressions for the same set.
15. If a code is an infix code, then it is a prefix code and a suffix code.
16. If a code is a prefix code and a suffix code, then it is an infix code.
17. A phrase structure grammar is a context-free grammar.
18. A context-free grammar is a phrase structure grammar.
19. A context-sensitive grammar is a phrase structure grammar.
20. A regular grammar is a context-free grammar.
21. A context-free grammar is a regular grammar.
22. A language is regular if and only if it is generated by a regular grammar.
23. A language is regular if and only if it is accepted by a deterministic automata.
24. The set $\{a^n b^n : n \text{ is a nonnegative integer}\}$ is generated by a regular grammar.
25. The set $\{a^n c b^n : n \text{ is a nonnegative integer}\}$ is generated by a context-free grammar.
26. Let $\Gamma = (N, T, S, P)$ be the grammar defined by $N = \{S, A\}$, $T = \{a, b\}$ and P be the set of productions

$$S \to aAb \qquad A \to aBb$$
$$B \to aAb \qquad A \to \lambda$$

   The word $a^3 b^3$ is generated by Γ.
27. The grammar Γ defined in Problem 26 generates the language $\{a^n b^n : n \text{ is a nonnegative integer}\}$.
28. Let $\Gamma = (N, T, S, P)$ be the grammar defined by $N = \{S, A\}$, $T = \{a, b\}$ and P be the set of productions

$$S \to aAb \qquad A \to aBb \qquad B \to aAb$$
$$A \to \lambda \qquad B \to \lambda$$

   The word $a^2 b^2$ is generated by Γ.
29. The grammar Γ defined in Problem 28 generates the language $\{a^n b^n : n \text{ is a nonnegative integer}\}$.
30. Let $\Gamma = (N, T, S, P)$ be the grammar defined by $N = \{S, A\}$, $T = \{a, b\}$ and P be the set of productions

$$S \to A \qquad A \to aBb \qquad B \to aAb$$
$$A \to \lambda \qquad B \to \lambda$$

   Γ generates the language $\{a^n b^n : n \text{ is a nonnegative integer}\}$.

SUMMARY QUESTIONS

1. Explain the difference between a regular language and a regular set.
2. Find a regular set corresponding to $(\mathbf{a} \vee \mathbf{b} \vee \mathbf{c})^*$.
3. Determine if $(\mathbf{a} \vee \mathbf{b})(\mathbf{a} \vee \mathbf{c})$ is uniquely decipherable.
4. Let $A = \{a, b\}$. Give a regular expression for the set of all elements with as least two *a*'s followed by as least three *b*'s.
5. Construct the automaton that accepts the language $(\mathbf{aa}^*\mathbf{bb}^*)^*$.

6. Construct a deterministic automaton that accepts the same language as the automaton

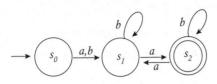

7. Find the language generated by the grammar with productions
$S \to AB, A \to aaA, A \to \lambda, B \to Bbbb, B \to \lambda$

8. Find the language generated by the grammar with productions
$S \to A, S \to B, A \to aaA,$
$A \to \lambda, B \to Bbbb, B \to \lambda$

9. State the rules for the Turing machine that accepts the same language as the automaton

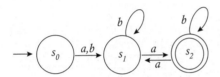

10. Write an expression for the language accepted by the automaton

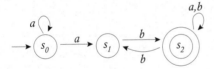

11. Explain why the language generated by productions
$S \to abB, B \to baA, A \to aA, A \to bB, A \to a, B \to b$ generates a regular language.

12. Describe the language accepted by the automaton

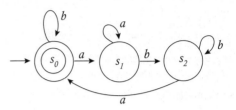

13. Construct the Moore machine that accepts the same language as

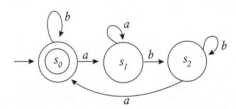

14. Construct the Mealy machine corresponding to the Moore machine in Problem 13.

15. Explain why a prefix code is a uniquely decipherable code.

ANSWERS TO TRUE-FALSE

1. F 2. T 3. F 4. T 5. T 6. F 7. F 8. F 9. T 10. T
11. T 12. F 13. T 14. F 15. T 16. F 17. F 18. T 19. T
20. T 21. F 22. T 23. T 24. F 25. T 26. T 27. F 28. T
29. F 30. T

CHAPTER 18
Theory of Codes

18.1 Introduction

When we think of codes, we often think of secret codes. Some of us may remember having secret decoder rings when we were young. Certainly transmission of codes, which theoretically can be understood only by the person who is supposed to receive them, is important for military purposes and to ensure privacy on networks and other places where information may be available to unauthorized personnel. Along with the creation of these codes, there is also the breaking of the code. This is the intercepting and decoding of messages by unauthorized personnel. This is serious work for the military and others who are interested in securing the integrity of their code and breaking the codes of others. It is also an increasingly popular hobby. Thus, there is a contest between those who break codes and those who try to create codes that cannot be broken. Perhaps the most important case of codebreaking was the breaking of the German enigma code during World War II. For details see Kahn [51] and [50]. The branch of code theory concerned with all this is called cryptology.

In this chapter, however, we will be concerned with a different type of code. As in Section 15.3, we define a code to be a representation of a set of symbols as strings of 1's and 0's. This set of symbols usually includes the letters of the alphabet, the numbers on the keyboard, and, often, control symbols. These codes are representations of binary strings for use by computers in the storage and transmission of data. There are several properties that we would like our codes to have. Unfortunately, we shall find that some of these properties are not mutually compatible.

The most important property of a code is that, when a message is expressed as a binary string consisting of the concatenation of elements of the code, this concatenation is unique. If a message is decoded, there must be no problem in deciding which letters the code represents. We call such a code a uniquely decipherable code (see Definition 17.8).

There are several ways to achieve this goal. One way is to encode all symbols with binary strings of the same length. Such a code is called a **block code**. For example, if each symbol is encoded using 8 bits, then we would know that at the end of each eight bits, we would have a code string representing a symbol of the transmitted message. The block code is particularly useful if it is necessary to limit the length of the code for each symbol or letter sent. Another method for constructing a uniquely decipherable code is to use a **prefix code**. This was defined in Definition 17.10 and discussed in the section on weighted trees. We defined a code C to be a prefix code if it has the property that an element of a code cannot be the beginning string of another element of the code. Thus, when we have read a string of 1's and 0's that represents the symbol of A, we know this the moment the string for A is completed. A prefix code is also called an **instantaneous code**.

One type of prefix code is the **comma code**. Each symbol is encoded into a string consisting of a string of 1's followed by a 0 at the end. Thus, the set of strings of the code has the form {0, 10, 110, 11110, 111110, ...}. This code has the obvious disadvantage that the elements of the code are going to be very long and take up a lot of storage space.

Often it is desirable to compress data to minimize storage space or transmission time. The most efficient code, with regard to minimizing space, is the **Huffman code**. Encoding with the Huffman code is described in Section 15.3. The Huffman code also has the advantage of being an instantaneous code.

A well-known example of a code that minimizes transmission time is the **Morse code**. In both the Huffman and Morse codes, letters or symbols that would occur most often are shorter. In Morse code, letters are separated by "spaces" and words by three "spaces." In this case, spaces are units of time.

		Morse Code			
A	.—	J	.———	S	...
B	—...	K	—.—	T	—
C	—.—.	L	.—..	U	..—
D	—..	M	——	V	...—
E	.	N	—.	W	.——
F	..—.	O	———	X	—..—
G	——.	P	.——.	Y	—.——
H	....	Q	——.—	Z	——..
I	..	R	.—.		

When transmitting data, errors may occur during transmission. We refer to anything that may produce errors with the vague term *noise*. For example, data received from a distant spacecraft such as Voyager and Galileo could certainly have problems with various kinds of noise. In some cases, we may only be interested in detecting an error. This could be the case if the data could be retransmitted. Codes that have the property that they can detect errors are called **error-detecting codes**.

In other cases, where data cannot be transmitted, such as data from a distant spacecraft, we want enough information about the data so we can not only detect the error but also correct it. Codes that have the property to correct errors are called **error-correcting codes**. It might seem reasonable to always use error-correcting codes. The problem with error-correcting codes, or even error-detecting codes, is that more information must be included for these codes, so they are less efficient with regard to minimizing space. Unfortunately, with both error-correcting and error-detecting codes, we can never be sure that errors are corrected or detected. The problem is with multiple errors. It is certainly possible to correct or detect the error if there is only one. The best that we can do in general is reduce the probability that errors will occur undetected or uncorrected. Again we have the problem that the more that we reduce the probability, the more information we have to send and the less efficient our code will be.

Before beginning our discussion of error-correcting and error-detecting codes, however, we want to include one more type of code. There is a classic example that demonstrates the use for this code. Suppose we have a rotating disk, which is divided into sectors, and a series of brushes or laser beams sending back digital information about how far the disk has rotated. If the binary strings recording the numbering of adjacent sectors are substantially different, in the sense that there are a lot of changes of the individual digits in going from one sector to the next, then a reading taken just as the sector was changing could produce a number totally different from the number of either of the sectors. In this case, it is desirable to number the sectors so that the binary string determining the sector has only one digit change between adjacent sectors. A code that has this property is the **Gray code**. Its construction is discussed in Section 6.7.

The first error-detecting method we consider is the parity bit. We demonstrate this method using ASCII code. This code is a block code that uses 7 bits, so that every symbol is a string of seven 1's and 0's. An eighth bit is then added that makes the number of 1's even. Therefore, if the code for a transmitted string is received with a single error, the number of 1's will be odd, and the receiver will know that an error has occurred. Unfortunately, if two errors occur, they will not be detected since the sum of the 1's will again be even. Assume the probability of an error in transmission is 0.01 both for a 1 changing to a 0 and for a 0 changing to a 1. Further assume that the probability for an error is the same regardless of the location of the error and whether the error is changing 1 to 0 or conversely. We also assume that the occurrence of one error does not affect the probability of another occurring.

We know from the binomial theorem for probability (Theorem 8.103) that the probability for exactly one error is $\binom{8}{1}(.01)(.99)^7$, which is approximately 0.07. However, the probability of exactly two errors is $\binom{8}{2}(.01)^2(.99)^6$, which is approximately 0.002 and substantially smaller.

Since three errors would be detected, the probability of more than two errors going undetected is less than the probability of four or more errors occurring, since any odd number of errors would be detected. This probability is almost negligible in comparison to that of one or less errors occurring.

Consider the code that is encoded by simply repeating each string to be encoded a given number of times. For example, if each string of code to be encoded is repeated once, then 10110 would be encoded as 1011010110. If each string to be encoded is repeated twice, then 10110 would be encoded as 101101011010110. If each string to be encoded is repeated once, then we have an error-detecting code. If an error occurs, then the corresponding positions will not be the same. For example, if the encoded string is 1111110101110111011, then an error has occurred in the third and last bit. We cannot correct the error since we don't know which error occurred in which copy of the string. If the strings to be encoded are repeated twice, the best we can do is error detection. If we have three copies of the string, then we can correct the code for a single error. If there is a difference in bits at corresponding positions in the string, then we accept the value that occurs twice. For example, if our string has length 4 and we receive 110110011101, then in the second position we receive two 1's and a 0. Thus, we assume that the correct value is 1, and that the correct string, which was encoded, was 1101. Obviously, a problem occurs if an error occurs more than once in the same position in the string. If a string is repeated so that we have n copies of the string, then the error correction gives the right result as long as an error occurs in the same position in the repetition less than $\lfloor \frac{n}{2} \rfloor$ times.

ASCII Code with Parity Bit

Code	Symbol	Code	Symbol	Code	Symbol	Code	Symbol
00000000	NUL	10100000	SP	11000000	@	01100000	`
10000001	SOH	00100001	!	01000001	A	11100001	a
10000010	STX	00100010	"	01000010	B	11100010	b
00000011	ETX	10100011	#	11000011	C	01100011	c
10000100	EOT	00100100	$	01000100	D	11100100	d
00000101	ENQ	10100101	%	11000101	E	01100101	e
00000110	ACK	10100110	&	11000110	F	01100110	f
10000111	BEL	00100111	'	01000111	G	11100111	g
10001000	BS	00101000	(	01001000	H	11101000	h
00001001	HT	10101001	)	11001001	I	01101001	i
00001010	LF	10101010	*	11001010	J	01101010	j
10001011	VT	00101011	+	01001011	K	11101011	k
00001100	FF	10101100	,	11001100	L	01101100	l
10001101	CR	00101101	−	01001101	M	11101101	m
10001110	SO	00101110	.	01001110	N	11101110	n
00001111	SI	10101111	/	11001111	O	01101111	o
10010000	DLE	00110000	0	01010000	P	11110000	p
00010001	DC1	10110001	1	11010001	Q	01110001	q
00010010	DC2	10110010	2	11010010	R	01110010	r
10010011	DC3	00110011	3	01010011	S	11110011	s
00010100	DC4	10110100	4	11010100	T	01110100	t
10010101	NAK	00110101	5	01010101	U	11110101	u
10010110	SYN	00110110	6	01010110	V	11110110	v
00010111	ETB	10110111	7	11010111	W	01110111	w
00011000	CAN	10111000	8	11011000	X	01111000	x
10011001	EM	00111001	9	01011001	Y	11111001	y
10011010	SUB	00111010	:	01011010	Z	11111010	z
10011011	ESC	10111011	;	11011011	[	01111011	{
10011100	FS	00111100	<	01011100	\	11111100	—
00011101	GS	10111101	=	11011101	]	01111101	}
00011110	RS	10111110	>	11011110	^	01111110	~
10011111	US	00111111	?	01011111	_	11111111	DEL

18.2 Generator Matrices

Beginning at this point, we shall assume that all strings in the code have a fixed length n and we shall treat these strings as vectors or $1 \times n$ matrices. Therefore, we shall have addition of vectors. However, we shall define addition mod 2, so that $1 + 1 = 0$. Thus,

$$11110001 + 10100111 = 01010110$$

The codes that we encounter in this section are called **linear codes**. These codes are also known as **group codes**. As previously mentioned, we shall consider the strings of a code C to be binary strings of length n, which we shall also consider as vectors or $1 \times n$ matrices. If B_n is the set of all binary strings of length n, then C is a subset of B_n. The reader is asked to prove that B_n, with the preceding addition, forms a group under addition. Each string in B_n is its own inverse, so in a sense, adding and subtracting are the same thing. If in some cases it may seem that we should be subtracting instead of adding, it doesn't matter. A code C is a **linear code** if C is a subgroup of B_n. If you have taken linear algebra, you may observe that C is really a linear vector space, and many of the properties of linear algebra would be helpful and illuminating here. However, in this development, the only properties we shall use regarding C are that C is a group, and that an element of C is a vector and, hence, can be multiplied by a matrix of the proper size. We shall also use the distributive law of matrices, which states that

$$A(B + C) = AB + AC$$

for all matrices A, B, and C where multiplication is defined. The reader may recall that if $u = (u_1, u_2, u_3, \ldots, u_n)$ and $v = (v_1, v_2, v_3, \ldots, v_n)$, then the **inner product** of u and v, denoted by $u \bullet v$ is equal to

$$u_1v_1 + u_2v_2 + u_3v_3 + \cdots + u_nv_n$$

The **weight** of a string of code c, denoted by $wt(c)$, is the number of 1's in the string. For example, if $c = 1011010$, then $wt(c) = 4$.

Suppose we have a $k \times n$ matrix G such that the first k columns and rows form the $k \times k$ identity matrix I_k, and all the columns are distinct. Thus, G has the form $[I_k | A_{n-k}]$. For example,

$$\begin{bmatrix} 1 & 0 & 0 & 1 & 0 & 1 \\ 0 & 1 & 0 & 1 & 1 & 0 \\ 0 & 0 & 1 & 0 & 1 & 1 \end{bmatrix}$$

is such a matrix. G is called a **generator matrix**. Consider the rows of the generator matrix as vectors or strings of a code. Call this set of strings S. For example, in the matrix, preceding

$$S = \{100101, 010110, 001011\}$$

Let C be the code created by taking all vectors that are finite sums of strings in S. The reader is asked to prove that C is a subgroup of B_n. In our example, we get 110011 by adding the first two strings in S, so 110011 would be in C. The group C is **generated** by the set S. It is also a minimal set generating C, since no elements of S are sums of other elements of S. We denote this by saying that $C = S^*$. The code C with form $[I_k | A_{n-k}]$ (i.e., generated by the rows of $[I_k | A_{n-k}]$) is called an $[n, k]$-code.

THEOREM 18.1 The $[n, k]$-code C contains 2^k strings.

Proof The first k bits of a string in C determine the elements of C. The positions where 1's occur in the first k bits of a string in C indicate which strings in S were added together. For example, if 1's occurred in the first and third bits of a string in C, then this string was created by adding the first and third rows of G. Since there are 2^k distinct ways of forming the first k bits, there are 2^k strings in C. ∎

If we wish to transmit message strings of length k, we encode them by multiplying them on the right by G. Thus, if $w = w_1w_2w_3 \cdots w_k$ or $(w_1, w_2, w_3, \ldots, w_k)$, then we encode it as the string wG. In our example, we would encode 110 or $(1, 1, 0)$ as

$$(1, 1, 0) \begin{bmatrix} 1 & 0 & 0 & 1 & 0 & 1 \\ 0 & 1 & 0 & 1 & 1 & 0 \\ 0 & 0 & 1 & 0 & 1 & 1 \end{bmatrix} = (1, 1, 0, 0, 1, 1)$$

or 110011. We note that the message string is the first three bits of the encoded string. In general the message string of length k will be the first k bits of the encoded string, since the identity matrix I_k, forming the first k columns of G, simply repeats the original string. Thus, we can decode by simply taking the first k bits of the encoded string. Also note that when we multiplied $(1, 1, 0)$ by G previously, we formed

$$(1, 1, 0, 0, 1, 1) = 1 \cdot (1, 0, 0, 1, 0, 1) + 1 \cdot (0, 1, 0, 1, 1, 0) + 0 \cdot (0, 0, 1, 0, 1, 1)$$

where for any vector v, $1 \cdot v = v$ and $0 \cdot v = (0, 0, 0, 0, 0, 0)$. Thus the encoded string is a sum of the vectors in S and, hence, is in C, since C is a group. In general, if $S = (s_1, s_2, s_3, \ldots, s_k)$ is the set of rows of the generator matrix and $w = (w_1, w_2, w_3, \ldots, w_k)$ is a message code, then the encoded string is

$$w_1 s_1 + w_2 s_2 + w_3 s_3 + \cdots + w_k s_k$$

This is a sum of the strings in S, since each w_i is either 1 or 0 and, hence, is in C since C is the group generated by S.

We have noted that G has the form $[I_k | A_{n-k}]$ and that I_k relays the message. What does A_{n-k} do? Let us again look at our example. If the message string is (w_1, w_2, w_3), then the encoded string is

$$(w_1, w_2, w_3) \begin{bmatrix} 1 & 0 & 0 & 1 & 0 & 1 \\ 0 & 1 & 0 & 1 & 1 & 0 \\ 0 & 0 & 1 & 0 & 1 & 1 \end{bmatrix} = (w_1, w_2, w_3, w_1 + w_2, w_2 + w_3, w_1 + w_3)$$

Thus, the fourth bit in the encoded string must be $w_1 + w_2$, the fifth bit must be $w_2 + w_3$, and the sixth bit must be $w_1 + w_3$. Therefore, if the encoded string is $(w_1, w_2, w_3, w_4, w_5, w_6)$, then $w_4 = w_1 + w_2$, $w_5 = w_2 + w_3$, and $w_6 = w_1 + w_3$. If any encoded string after it is transmitted does not satisfy these equations, we know there is an error in transmission. For example, if we received the encoded string 101100 then, since $w_1 = 1$, $w_2 = 0$, and $w_3 = 1$, we must have $w_4 = 1 = 1 + 0$, $w_5 = 0 = 0 + 1$, and $w_6 = 0 = 1 + 1$. Since the equation for w_5 is not correct, we know that an error has occurred. Thus, the matrix A_{n-k} serves as a check on the accuracy of the transmission, much as the parity check did earlier. In general if we have the encoded string $w_1 w_2 w_3 \ldots w_k \ldots w_n$, and

$$G = \begin{bmatrix} 1 & 0 & 0 & \cdots & 0 & A_{1,k+1} & A_{1,k+2} & \cdots & A_{1,n} \\ 0 & 1 & 0 & \cdots & 0 & A_{2,k+1} & A_{2,k+2} & \cdots & A_{2,n} \\ 0 & 0 & 1 & \ddots & \vdots & A_{3,k+1} & A_{3,k+2} & \cdots & A_{3,n} \\ \vdots & \vdots & \vdots & \ddots & 0 & \vdots & \vdots & & \ddots \\ 0 & 0 & 0 & 0 & 1 & A_{k,k+1} & A_{k,k+2} & & A_{k,n} \end{bmatrix}$$

then for $i > k$, $w_i = w_1 A_{1,i} + w_2 A_{2,i} + w_3 A_{3,i} + \cdots + w_k A_{k,i}$ and the encoded string must satisfy these $n - k$ equations.

The next problem is to correct an error, given that a single error has occurred. The method below is known as using **coset leaders**. We illustrate this method with our example where

$$G = \begin{bmatrix} 1 & 0 & 0 & 1 & 0 & 1 \\ 0 & 1 & 0 & 1 & 1 & 0 \\ 0 & 0 & 1 & 0 & 1 & 1 \end{bmatrix}$$

We know

$$S = \{100101, 010110, 001011\}$$

so that

$$C = \{000000, 100101, 010110, 001011, 110011, 011101, 101110, 111000\}$$

We now form cosets of C in B_n as described in Section 9.4. The first coset is C itself. To form the next coset, we select an element of B_n, which has minimum weight and is not in C. For example, we could choose $b_1 = 100000$. The coset

$$b_1 + C = \{100000, 000101, 110110, 101011, 010011, 111101, 001110, 011000\}$$

We again select an element of B_n which has minimum weight and is not in either of the previous cosets. For example, we could select $b_2 = 010000$. The coset

$$b_2 + C = \{010000, 110101, 000110, 011011, 100011, 001101, 111110, 101000\}$$

Continuing in this process, we have the following table of B_n divided into cosets, where the element of least weight is listed first. The elements in the first column are called coset leaders.

000000	100101	010110	001011	110011	011101	101110	111000
100000	000101	110110	101011	010011	111101	001110	011000
010000	110101	000110	011011	100011	001101	111110	101000
001000	101101	011110	000011	111011	010101	100110	110000
000100	100001	010010	001111	110111	011001	101010	111100
000010	100111	010100	001001	110001	011111	101100	111010
000001	100100	010111	001010	110010	011100	101111	111001
100010	000111	110100	101001	010001	111111	001100	011010

In the last coset, we had to use a string of weight 2. We could have selected either 100010, 010001, or 001100. Frankly, we were happy just to find one. The process works as follows: When an encoded string is received, we find it in the table. For example, assume that we receive 110110. We then look at the first column of the row containing 110110 to find 100000 so we assume that the error occurred in the first bit since 100000 was added to each element of C to get this coset. We look at the first row of the column containing 110110 to find 010110, which we assume is the correct string of C. Again, suppose we get the string 001010. We then look to the first column of the row containing 001010 to find 000001 so we assume that the error occurred in the sixth bit. We look at the first row of the column containing 001010 to find 001011, which we assume is the correct string of C.

We now look for an easier way to determine errors. Let v and w be vectors (or strings) of length n. The vector v is **orthogonal** to the vector w if their dot product, $v \bullet w = 0$. Given a code C, the **dual code** of C, denoted by $C^\perp$, is the set of all strings of B_n that are orthogonal to every string in C. It is left to the reader to show that $C^\perp$ is a subgroup of B_n. If code C is an $[n, k]$-code so that it generated by k strings, then $C^\perp$ is generated by $n - k$ strings. We shall not prove this, since it requires a knowledge of linear algebra, but simply accept it as true. We shall, however, prove the following theorem.

THEOREM 18.2 Let C be a group code and $C^\perp$ be its dual code. A string t is in $C^\perp$ if and only if it is orthogonal to each string in S, the set of generators of C.

Proof Obviously if t is in $C^\perp$, then it is orthogonal to each string in S, since it is orthogonal to every string in C and $S \subseteq C$. Assume that $S = \{s_1, s_2, s_3, \ldots s_k\}$ and t is orthogonal to s_i for all $s_i \in S$, so that $t \bullet s_1 = 0$ for all $s_i \in S$. Every element of C is of the form

$$w_1 s_1 + w_2 s_2 + w_3 s_3 + \cdots + w_k s_k$$

where each w_1 is either 1 or 0. From the linear property of matrices we know that

$$t \bullet (w_1 s_1 + w_2 s_2 + \cdots + w_k s_k) = w_1(t \bullet s_1) + w_2(t \bullet s_2) + \cdots + w_k(t \bullet s_k)$$
$$= 0 + 0 + 0 + \cdots + 0$$
$$= 0$$
∎

Recall that in our example

$$G = \begin{bmatrix} 1 & 0 & 0 & 1 & 0 & 1 \\ 0 & 1 & 0 & 1 & 1 & 0 \\ 0 & 0 & 1 & 0 & 1 & 1 \end{bmatrix} = [I_3 | A_3]$$

We now define

$$G^\perp = \begin{bmatrix} 1 & 1 & 0 & 1 & 0 & 0 \\ 0 & 1 & 1 & 0 & 1 & 0 \\ 1 & 0 & 1 & 0 & 0 & 1 \end{bmatrix} = [A_3^t | I_3]$$

where A_3^t is the transpose of A_3 and is obtained by changing the rows of A_3 to columns. The matrix $G^\perp$ is called the **parity check matrix**.

By checking each possibility in our example, we can show that the inner product of any row of G with any row of $G^\perp$ is equal to 0. From this we know that $G^\perp r_i^t = 0$ where r_i^t is the transpose of the ith row of G. By definition of transpose, r_i^t is the ith row of G, changed to a column. We place it in column form so we can multiply it on the right by the matrix $G^\perp$:

$$G^\perp(w_1 r_1^t + w_2 r_2^t + w_3 r_3^t) = w_1 G^\perp r_1^t + w_2 G^\perp r_2^t + w_3 G^\perp r_3^t$$
$$= 0 + 0 + 0$$
$$= 0$$

Thus, if we multiply $G^\perp$ and the transpose of any element of C, we get 0.

In general if

$$G = [I_k | A_{n-k}] = \begin{bmatrix} 1 & 0 & 0 & \cdots & 0 & A_{1,k+1} & A_{1,k+2} & \cdots & A_{1,n} \\ 0 & 1 & 0 & \cdots & 0 & A_{2,k+1} & A_{2,k+2} & \cdots & A_{2,n} \\ 0 & 0 & 1 & \ddots & \vdots & A_{3,k+1} & A_{3,k+2} & \cdots & A_{3,n} \\ \vdots & \vdots & \vdots & \ddots & 0 & \vdots & \vdots & & \ddots \\ 0 & 0 & 0 & 0 & 1 & A_{k,k+1} & A_{k,k+2} & & A_{k,n} \end{bmatrix}$$

then

$$G^\perp = [A_{n-k}^t | I_{n-k}] = \begin{bmatrix} A_{1,k+1} & A_{2,k+1} & \cdots & A_{k,k+1} & 1 & 0 & 0 & \cdots & 0 \\ A_{1,k+2} & A_{2,k+2} & \cdots & A_{k,k+2} & 0 & 1 & 0 & \cdots & 0 \\ A_{1,k+3} & A_{2,k+3} & \cdots & A_{k,k+3} & 0 & 0 & 1 & \ddots & \vdots \\ \vdots & \vdots & \ddots & \vdots & \vdots & \vdots & & \ddots & 0 \\ A_{1,n} & A_{2,n} & & A_{k,n} & 0 & 0 & 0 & & 1 \end{bmatrix}$$

The inner product of the ith row of G with the jth row of $G^\perp$ is equal to

$$0 + 0 + \cdots 0 + A_{i,j} + 0 + \cdots 0 + A_{i,j} + \cdots + 0 + 0 = 0$$

so that in the general case $G^\perp r_i^t = 0$ where r_i^t is the transpose of the ith row of G. Using the same argument as previously, if we multiply $G^\perp$ and the transpose of any element of C, we get 0.

We also get another rather remarkable result. If two elements b_1 and b_2 of B_n are in the same coset where we form cosets of B_n using the group C as we did previously, then

$$G^\perp b_1^t = G^\perp b_2^t$$

To show this, we use the fact that if b_1 and b_2 are in the same coset, then $b_1 = b_2 + c$ for some $c \in C$. Therefore, $b_1^t = b_2^t + c^t$ and

$$\begin{aligned}
G^\perp b_1^t &= G^\perp (b_2^t + c^t) \\
&= G^\perp b_2^t + G^\perp c^t \\
&= G^\perp b_2^t + 0 \\
&= G^\perp b_2^t
\end{aligned}$$

since $G^\perp c^t = 0$ for all $c \in C$.

Since $G^\perp b^t$ is the same for all b in a coset, we can select any b in the coset and determine this value. Thus, in the preceding table we can add this common value of the image of $G^\perp$ to each row, since the elements in each row determine a coset. We choose the coset leader since it is the simplest and place the value of its image in the second column. These value are called **syndromes**. We already know that the first syndrome is $\begin{bmatrix} 0 \\ 0 \\ 0 \end{bmatrix}$.

We find that

$$\begin{bmatrix} 1 & 1 & 0 & 1 & 0 & 0 \\ 0 & 1 & 1 & 0 & 1 & 0 \\ 1 & 0 & 1 & 0 & 0 & 1 \end{bmatrix} \begin{bmatrix} 1 \\ 0 \\ 0 \\ 0 \\ 0 \\ 0 \end{bmatrix} = \begin{bmatrix} 1 \\ 0 \\ 1 \end{bmatrix}$$

so $\begin{bmatrix} 1 \\ 0 \\ 1 \end{bmatrix}$ is the second syndrome, and

$$\begin{bmatrix} 1 & 1 & 0 & 1 & 0 & 0 \\ 0 & 1 & 1 & 0 & 1 & 0 \\ 1 & 0 & 1 & 0 & 0 & 1 \end{bmatrix} \begin{bmatrix} 0 \\ 1 \\ 0 \\ 0 \\ 0 \\ 0 \end{bmatrix} = \begin{bmatrix} 1 \\ 1 \\ 0 \end{bmatrix}$$

so $\begin{bmatrix} 1 \\ 1 \\ 0 \end{bmatrix}$ is the third syndrome. Continuing, we have the following table.

000000	000	100101	010110	001011	110011	011101	101110	111000
100000	101	000101	110110	101011	010011	111101	001110	011000
010000	110	110101	000110	011011	100011	001101	111110	101000
001000	011	101101	011110	000011	111011	010101	100110	110000
000100	100	100001	010010	001111	110111	011001	101010	111100
000010	010	100111	010100	001001	110001	011111	101100	111010
000001	001	100100	010111	001010	110010	011100	101111	111001
100010	111	000111	110100	101001	010001	111111	001100	011010

Having completed the table, suppose that we receive the transmitted string 101100. We then multiply its transpose by $G^{\perp}$ getting

$$\begin{bmatrix} 1 & 1 & 0 & 1 & 0 & 0 \\ 0 & 1 & 1 & 0 & 1 & 0 \\ 1 & 0 & 1 & 0 & 0 & 1 \end{bmatrix} \begin{bmatrix} 1 \\ 0 \\ 1 \\ 1 \\ 0 \\ 0 \end{bmatrix} = \begin{bmatrix} 0 \\ 1 \\ 0 \end{bmatrix}$$

so we know that 101100 is in row 6. The coset leader is 000010, the element in the leftmost column of the row containing 101100. The element of C, in the top row of the column, containing 101100, is 101110. By the way in which the table was constructed we know that $101100 = 101110 + 000010$, so we assume that the transmitted message 101100 should have been 101110 and there was an error in the fifth bit.

This method is much quicker, since we only have to multiply the transpose of the transmitted message by $G^{\perp}$, find the row containing the syndrome, and find the transmitted message. The coset leader for that message is the error, and the element of C in the first row of the column containing the transmitted message is the corrected message.

We can, however, make the process even quicker, and we need only the first two columns of the preceding table. Suppose we receive the transmitted message 110000. We multiply its transpose by $G^{\perp}$ getting

$$\begin{bmatrix} 1 & 1 & 0 & 1 & 0 & 0 \\ 0 & 1 & 1 & 0 & 1 & 0 \\ 1 & 0 & 1 & 0 & 0 & 1 \end{bmatrix} \begin{bmatrix} 1 \\ 1 \\ 0 \\ 0 \\ 0 \\ 0 \end{bmatrix} = \begin{bmatrix} 0 \\ 1 \\ 1 \end{bmatrix}$$

so the syndrome is $\begin{bmatrix} 0 \\ 1 \\ 1 \end{bmatrix}$ and the coset leader is 001000. Since the coset leader tells us the error occurred in the third bit, if we add 001000 to 110000, we get 111000, the corrected code. Thus, our method is simple. Multiply the transpose of the transmitted message by $G^{\perp}$ to find the syndrome. Find the coset leader for that syndrome and add it to the transmitted message to get the corrected code. Notice that we use only the first two columns of the table. At present, however, we do have one problem.

If we should receive the transmitted message 101001 so the syndrome is $\begin{bmatrix} 1 \\ 1 \\ 1 \end{bmatrix}$, there are three strings of weight 2 in the corresponding row. Remember 100010 was chosen arbitrarily. Any of these are equally likely to be the error so our correction using syndromes here is hopeless. We are also trying to correct a string with two errors instead of one.

Exercises

1. Which of the following are proper generator matrices? If not, why not?

   (a) $\begin{bmatrix} 1 & 0 & 0 & 1 & 1 & 1 \\ 0 & 0 & 1 & 0 & 1 & 1 \\ 0 & 1 & 0 & 1 & 1 & 0 \end{bmatrix}$ (b) $\begin{bmatrix} 1 & 0 & 0 & 1 & 1 & 0 \\ 0 & 1 & 0 & 0 & 1 & 1 \\ 0 & 0 & 1 & 1 & 1 & 0 \end{bmatrix}$

   (c) $\begin{bmatrix} 1 & 0 & 0 & 1 & 1 & 1 \\ 0 & 1 & 0 & 1 & 0 & 1 \\ 0 & 0 & 1 & 1 & 1 & 0 \end{bmatrix}$ (d) $\begin{bmatrix} 1 & 0 & 0 & 1 & 0 & 1 \\ 0 & 1 & 0 & 0 & 0 & 1 \\ 0 & 0 & 1 & 1 & 0 & 0 \end{bmatrix}$

2. Given the generator matrix

   $\begin{bmatrix} 1 & 0 & 0 & 1 & 1 & 1 \\ 0 & 1 & 0 & 0 & 1 & 1 \\ 0 & 0 & 1 & 1 & 1 & 0 \end{bmatrix}$

   (a) Find $G^{\perp}$.
   (b) Encode 111, 011, 101, 110.
   (c) Decode 111011, 110100, 010010, 011110.
   (d) Use $G^{\perp}$ to determine which strings in (c) are proper encodings.

3. Given the generator matrix

   $\begin{bmatrix} 1 & 0 & 0 & 0 & 1 & 1 & 1 \\ 0 & 1 & 0 & 0 & 1 & 1 & 0 \\ 0 & 0 & 1 & 0 & 1 & 0 & 0 \\ 0 & 0 & 0 & 1 & 0 & 1 & 1 \end{bmatrix}$

   (a) Find $G^{\perp}$.
   (b) Encode 1111, 0101, 1001, 1010.
   (c) Decode 1111110, 0111010, 0111101, 1011110.
   (d) Use $G^{\perp}$ to determine which strings in (c) are proper encodings.

4. Given the generator matrix

   $\begin{bmatrix} 1 & 0 & 0 & 0 & 1 & 1 \\ 0 & 1 & 0 & 0 & 1 & 0 & 1 \\ 0 & 0 & 1 & 0 & 1 & 1 & 0 \\ 0 & 0 & 0 & 1 & 1 & 0 & 0 \end{bmatrix}$

   (a) Find $G^{\perp}$.
   (b) Construct the table of cosets with coset leaders.
   (c) Use this table to correct errors (if any) in the transmissions of 1111100, 1111000, 0110101, and 1011000.

5. Given the generator matrix

   $\begin{bmatrix} 1 & 0 & 0 & 0 & 1 & 1 \\ 0 & 1 & 0 & 1 & 0 & 1 \\ 0 & 0 & 1 & 1 & 1 & 0 \end{bmatrix}$

   (a) Find $G^{\perp}$.
   (b) Construct the table of cosets with coset leaders.
   (c) Use this table to correct errors (if any) in the transmissions of 011001, 111000, 110111, and 101100.

6. Given the generator matrix

   $\begin{bmatrix} 1 & 0 & 0 & 1 & 1 & 0 \\ 0 & 1 & 0 & 0 & 1 & 1 \\ 0 & 0 & 1 & 0 & 1 & 1 \end{bmatrix}$

   (a) Find $G^{\perp}$.
   (b) Find the syndromes for the cosets.
   (c) Use these syndromes to correct the errors (if any) for the transmitted strings 111101, 111001, 110101, 101001.

7. Given the generator matrix

$$\begin{bmatrix} 1 & 0 & 0 & 1 & 0 & 1 \\ 0 & 1 & 0 & 0 & 1 & 1 \\ 0 & 0 & 1 & 1 & 1 & 1 \end{bmatrix}$$

   (a) Find $G^\perp$.

   (b) Find the syndromes for the cosets.

   (c) Use these syndromes to correct the errors (if any) for the transmitted strings 111101, 111001, 110010, and 101001.

8. Explain what happens using the generator matrix in Exercise 7, if there are two or more errors.

9. Using the generator matrix in Exercise 5, find other strings than the rows of $G^\perp$ that are in $C^\perp$ and construct a parity check matrix using these strings.

10. Using the generator matrix in Exercise 4, find other strings than the rows of $G^\perp$ that are in $C^\perp$, and construct a parity check matrix using these strings.

11. Prove that B_n, the set of all binary strings of length n, is a group under the addition defined in this chapter.

12. Let C be the code created by taking all vectors that are finite sums of strings in S. Prove that C is a subgroup of B_n under addition.

13. Prove that the **dual code** of C, denoted by $C^\perp$, is a subgroup of B_n under addition.

18.3 Hamming Codes

At the end of the last section, we saw that there was difficulty trying to correct the code for certain strings, since not all coset leaders had weight 1. This is remedied by using a matrix called a **Hamming matrix** as a generating matrix. Before looking at the Hamming matrix G_H, we first look at the parity check matrix $G_H^\perp$. Let $G_H^\perp$ be a matrix with r rows such that the columns consist of all possible strings of length r except the string consisting of all 0's. We shall assume that $r \geq 3$. There are $2^r - 1$ such strings so $G_H^\perp$ is an $r \times n$ matrix where $n = 2^r - 1$. We use the columns of weight 1 as the final r columns, forming the identity matrix so that $G_H^\perp$ has the form $[A^t | I_r]$ where A^t is an $r \times (n-r)$ matrix. The Hamming matrix G_H is the $(n-r) \times n$ matrix of the form $[I_{n-r} | A]$ where A is an $(n-r) \times r$ matrix. The code generated by the rows of the Hamming matrix is called the **Hamming code**.

For example, let $r = 3$, and $G_H^\perp$ be the matrix

$$\begin{bmatrix} 1 & 1 & 0 & 1 & 1 & 0 & 0 \\ 0 & 1 & 1 & 1 & 0 & 1 & 0 \\ 1 & 0 & 1 & 1 & 0 & 0 & 1 \end{bmatrix}$$

Then G_H is the matrix

$$\begin{bmatrix} 1 & 0 & 0 & 0 & 1 & 0 & 1 \\ 0 & 1 & 0 & 0 & 1 & 1 & 0 \\ 0 & 0 & 1 & 0 & 0 & 1 & 1 \\ 0 & 0 & 0 & 1 & 1 & 1 & 1 \end{bmatrix}$$

To study Hamming matrices, we need the concept of distance between two strings and their relation to the weight of the strings. We begin with a theorem about weights of strings.

THEOREM 18.3 For strings c and c', $wt(c + c') \leq wt(c) + wt(c')$.

Proof Let $c = c_1 c_2 c_3 \ldots c_n$ and $c' = c'_1 c'_2 c'_3 \ldots c'_n$. If $c_i + c'_i = 1$ then either $c_i = 1$ or $c'_i = 1$. Therefore, for every 1 occurring in $c + c'$, there is a 1 occurring in either c or c'. ∎

The **Hamming distance** or simply **distance** between two strings of code c and c', having the same length, is the number of corresponding bits in the string where one string has the digit 1 and the other has the digit 0. We shall denote the distance function by $\delta(c, c')$. For example, if $c = 101011$ and $c' = 110010$, then $\delta(c, c') = 3$, since the two strings differ in the second, third, and sixth positions. Obviously, the greater the distance between two strings, the more errors that can be made in transmitting one string without accidentally getting another in the code. Since we have called δ a distance function we should show that it has the basic properties of a distance function.

THEOREM 18.4 The Hamming distance function has the following properties:

(a) For strings c and c', $\delta(c, c') = 0$ if and only if $c = c'$.
(b) For strings c and c', $\delta(c, c') = \delta(c', c)$.
(c) For strings c, c' and c'', $\delta(c, c'') \leq \delta(c, c') + \delta(c', c'')$.

Proof Parts (a) and (b) follow directly and are left to the reader. For part (c), we note that for strings c and c'', $wt(c + c'') = \delta(c, c'')$. To see this we note that if $c = c_1 c_2 c_3 \ldots c_n$ and $c'' = c''_1 c''_2 c''_3 \ldots c''_n$, then $c_i + c''_i$ contributes 1 to $wt(c + c'')$ if and only if $c_i = 0$ and $c''_i = 1$ or $c_i = 1$ and $c''_i = 0$. But this true if and only if c_i and c''_i are different, which contributes 1 to $\delta(c, c'')$. We also note that for any string c', the string $c' + c'$ consists only of 0's. We shall call this string $\mathbf{0}$. By the definition of addition, $\mathbf{0} + c = c$. for every string c. Therefore,

$$\delta(c, c'') = wt(c + c'')$$
$$= wt(c + \mathbf{0} + c'')$$
$$= wt(c + c' + c' + c'')$$
$$\leq wt(c + c') + wt(c' + c'')$$
$$= \delta(c, c') + \delta(c', c'')$$

■

It is important to know the minimal distance between any two strings in a code. If C is a code, then the **minimal distance** of a code C, denoted by $D(C)$, is the smallest distance between any two strings in C. The following theorem gives us an important measure of the number of errors that can be corrected or detected using the code.

THEOREM 18.5 For a code C,

(a) If $D(C) = k + 1$, then up to k errors can be detected using the code.
(b) If $D(C) = 2k + 1$, then up to k errors can be corrected using the code.

Proof (a) If $D(C) = k + 1$ and $c \in C$, then c differs from any other string in the code in at least $k + 1$ places. Therefore, if c is transmitted and has k or fewer errors, it cannot possibly be another string in the code, and an error is detected.

(b) If string c is transmitted as c' with k or fewer errors, we have $\delta(c, c') \leq k$. If $\delta(c', c'') \leq k$, for some string c'' in C, then $\delta(c, c') + \delta(c', c'') \leq 2k$. But $\delta(c', c'') \leq \delta(c, c') + \delta(c', c'')$ and $\delta(c, c'') \geq 2k + 1$, giving a contradiction. Therefore, c' can be corrected to c, the only string whose distance from c' is less than $k + 1$. ■

We want to determine $D(C)$, the smallest distance between any two strings in C. To do this, however, we first need the following theorem.

THEOREM 18.6 $D(C)$ is equal to $W(C) = \min\{wt(c) : c \in C \text{ and } c \neq \mathbf{0}\}$.

Proof By definition of $D(C)$, there exists $c, c' \in C$ such that $\delta(c, c') = D(C)$. But $\delta(c, c'') = wt(c + c'')$, and, since $c + c'' \in C$, $W(C) \leq wt(c + c'')$. Therefore

$W(C) \leq D(C)$. Conversely, for $c \in C$, $wt(c) = wt(c+\mathbf{0}) = \delta(c,\mathbf{0}) \geq D(C)$. Therefore, $W(C) \geq D(C)$. Thus, $W(C) = D(C)$. ∎

We now show that for a Hamming code C, $W(C) \geq 3$. There is no $c \in C$ of weight 1. If there were and c were a string with all 0's, except for a 1 in the jth place, then since c is orthogonal to every row of $G_H^\perp$, the jth column of $G_H^\perp$ would consist of all 0's, which contradicts the construction of $G_H^\perp$. Also there is no $c \in C$ of weight 2. If there were, then c would be a string with all 0's except for two 1's, say in the ith and jth positions. Again since c is orthogonal to every row of $G_H^\perp$, the ith and jth columns of every row in $G_H^\perp$ would have to be either both 1's or both 0's. But then the ith and jth columns of $G_H^\perp$ would have to be the same, which again contradicts the construction of $G_H^\perp$. Therefore, $W(C) \geq 3$ and C can be used for error correction for a single error.

We now want to show that there is one element of weight 1 in each coset except for C itself. If we do this, it will greatly simplify the problem of decoding. By Theorem 18.1, an $[n,k]$-code C contains 2^k strings. Since G_H has the form $[I_{n-r}|A]$, the Hamming code C is an $[n, n-r]$-code and so C contains 2^{n-r} elements. B_n contains 2^n elements. Therefore, there are

$$\frac{2^n}{2^{n-r}} = 2^r$$

cosets including C. The strings in C have length $2^r - 1$. Therefore, there are $2^r - 1$ strings of weight 1. We now have to show that no coset contains two strings of weight 1. Assume that s and s' are both strings of weight 1 in the same coset. Then, by definition of a coset, $s = s' + c$ for some $c \in C$. Therefore, $c = s + s'$, so by Theorem 18.3,

$$wt(c) \leq wt(s) + wt(s') \leq 1 + 1 = 2$$

But $wt(c) \geq 3$, so this is obviously impossible. Therefore, each coset contains exactly one string of weight 1, except for C which contains a string of weight 0.

We return to our example where $G_H^\perp$ is the matrix

$$\begin{bmatrix} 1 & 1 & 0 & 1 & 1 & 0 & 0 \\ 0 & 1 & 1 & 1 & 0 & 1 & 0 \\ 1 & 0 & 1 & 1 & 0 & 0 & 1 \end{bmatrix}$$

and G_H is the matrix

$$\begin{bmatrix} 1 & 0 & 0 & 0 & 1 & 0 & 1 \\ 0 & 1 & 0 & 0 & 1 & 1 & 0 \\ 0 & 0 & 1 & 0 & 0 & 1 & 1 \\ 0 & 0 & 0 & 1 & 1 & 1 & 1 \end{bmatrix}$$

Since every coset contains a coset leader of weight 1, consider the string 0010000. If we multiply it by $G_H^\perp$ we have

$$\begin{bmatrix} 1 & 1 & 0 & 1 & 1 & 0 & 0 \\ 0 & 1 & 1 & 1 & 0 & 1 & 0 \\ 1 & 0 & 1 & 1 & 0 & 0 & 1 \end{bmatrix} \begin{bmatrix} 0 \\ 0 \\ 1 \\ 0 \\ 0 \\ 0 \\ 0 \end{bmatrix} = \begin{bmatrix} 0 \\ 1 \\ 1 \end{bmatrix}$$

so the syndrome is $\begin{bmatrix} 0 \\ 1 \\ 1 \end{bmatrix}$, which is the third column of $G_H^\perp$. In fact, if a 1 occurs in the jth digit of a string of weight 1, the syndrome, when the transpose of the string is multiplied by $G_H^\perp$, is the jth column in $G_H^\perp$. Therefore, whenever we receive a transmitted message string and multiply it by $G_H^\perp$, if the transmission is correct, we get a syndrome with all 0's. If there is a single error, we get one of the columns of $G_H^\perp$, since the transmitted message string has to be in one of the cosets, so there is a coset leader of weight 1. Therefore, if the ith column of $G_H^\perp$ is the syndrome, we know the coset leader has a 1 in the ith column, so the error is in the ith column or ith bit of the string. For example, suppose we get the transmitted message string 1110110. We multiply its transpose by $G_H^\perp$, getting

$$\begin{bmatrix} 1 & 1 & 0 & 1 & 1 & 0 & 0 \\ 0 & 1 & 1 & 1 & 0 & 1 & 0 \\ 1 & 0 & 1 & 1 & 0 & 0 & 1 \end{bmatrix} \begin{bmatrix} 1 \\ 1 \\ 1 \\ 0 \\ 1 \\ 1 \\ 0 \end{bmatrix} = \begin{bmatrix} 1 \\ 1 \\ 0 \end{bmatrix}$$

which is the second row of $G_H^\perp$. Therefore, the error is in the second bit, and the message should have been 1010110.

In the remainder of this section, we will quickly explore other codes. The first of these is the Golay code. Hamming codes were discovered independently by Hamming in 1950 and Golay in 1949. We will not try to explain why they are called Hamming codes. It is a long and involved story. For details see Thompson [106]. However, there is a Golay code. It was in Golay's 1949 paper. This is the (23, 12, 7) model published by Golay. This means that it is a (23, 12) generating matrix with a minimum distance of 7 between the strings of C. This **Golay code** has generator $G = [I_{11}|A]$ where

$$A = \begin{bmatrix} 1 & 0 & 0 & 1 & 1 & 1 & 0 & 0 & 0 & 1 & 1 & 1 \\ 1 & 0 & 1 & 0 & 1 & 1 & 0 & 1 & 1 & 0 & 0 & 1 \\ 1 & 0 & 1 & 1 & 0 & 1 & 1 & 0 & 1 & 0 & 1 & 0 \\ 1 & 0 & 1 & 1 & 1 & 0 & 1 & 1 & 0 & 1 & 0 & 0 \\ 1 & 1 & 0 & 0 & 1 & 1 & 1 & 0 & 1 & 1 & 0 & 0 \\ 1 & 1 & 0 & 1 & 0 & 1 & 1 & 1 & 0 & 0 & 0 & 1 \\ 1 & 1 & 0 & 1 & 1 & 0 & 0 & 1 & 1 & 0 & 1 & 0 \\ 1 & 1 & 1 & 0 & 0 & 1 & 0 & 1 & 0 & 1 & 1 & 0 \\ 1 & 1 & 1 & 0 & 1 & 0 & 1 & 0 & 0 & 0 & 1 & 1 \\ 1 & 1 & 1 & 0 & 0 & 0 & 0 & 1 & 1 & 0 & 1 \\ 0 & 1 & 1 & 1 & 1 & 1 & 1 & 1 & 1 & 1 & 1 \end{bmatrix}$$

This matrix has a geometrical interpretation using five lines in the plane (see Thompson). It is more easily studied by using the extended generating (24, 12, 8) matrix $G = [I_{12}|A]$ where

$$A = \begin{bmatrix} 0 & 1 & 1 & 1 & 1 & 1 & 1 & 1 & 1 & 1 & 1 \\ 1 & 1 & 1 & 0 & 1 & 1 & 1 & 0 & 0 & 0 & 1 & 0 \\ 1 & 1 & 0 & 1 & 1 & 1 & 0 & 0 & 0 & 1 & 0 & 1 \\ 1 & 0 & 1 & 1 & 1 & 0 & 0 & 0 & 1 & 0 & 1 & 1 \\ 1 & 1 & 1 & 1 & 0 & 0 & 0 & 1 & 0 & 1 & 1 & 0 \\ 1 & 1 & 1 & 0 & 0 & 0 & 1 & 0 & 1 & 1 & 0 & 1 \\ 1 & 1 & 0 & 0 & 0 & 1 & 0 & 1 & 1 & 0 & 1 & 1 \\ 1 & 0 & 0 & 0 & 1 & 0 & 1 & 1 & 0 & 1 & 1 & 1 \\ 1 & 0 & 0 & 1 & 0 & 1 & 1 & 0 & 1 & 1 & 1 & 0 \\ 1 & 0 & 1 & 0 & 1 & 1 & 0 & 1 & 1 & 1 & 0 & 0 \\ 1 & 1 & 0 & 1 & 1 & 0 & 1 & 1 & 1 & 0 & 0 & 0 \\ 1 & 0 & 1 & 1 & 0 & 1 & 1 & 1 & 0 & 0 & 0 & 1 \end{bmatrix}$$

(see Hill [43]). This is produced by rearranging the Golay generating matrix and adding a parity bit. The symmetry of this matrix makes it much easier to study. It is easily seen that $G^\perp = [A|I_{12}]$ since $A = A^t$. Golay introduced several other codes, including a (4096, 244, 8) code, which was used by the Voyager spacecraft to transmit images of Jupiter, Uranus, and Neptune.

For a given string s, of length n, let $\bar{s} = s+\mathbf{1}$, where $\mathbf{1}$ is a string of length n consisting of all 1's. Given a set of strings S, let $Plot(S) = \{ss : s \in S\} \cup \{s\bar{s} : s \in S\}$. Given the set

$$S = \{0000, 0011, 1100, 1111\}$$

let $S_1 = Plot(S)$, $S_2 = Plot(S_1)$, $S_3 = Plot(S_2)$, and $S_n = Plot(S_{n-1})$. A plot is a construction created by M. Plotkin [81]. The codes generated by sets $S_1, S_2, S_3, \ldots$ are called **Reed-Muller codes**. The set S_3 is a (64, 32, 16) code that was used for error correction on images transmitted by the Mariner 9 spacecraft. More specifically, the Reed-Muller matrix used is called a Hadamard matrix, which is defined next.

Consider matrices defined recursively as follows:

$$A_1 = [0]$$

$$A_{2n} = \begin{bmatrix} A_n & A_n \\ A_n & \overline{A_n} \end{bmatrix}$$

where the matrix $\overline{A}_n$ is defined by $\overline{A}_{ij} = \overline{A_{ij}}$ for $1 \leq i, j \leq n$. Therefore,

$$A_2 = \begin{bmatrix} 0 & 0 \\ 0 & 1 \end{bmatrix}$$

and

$$A_4 = \begin{bmatrix} 0 & 0 & 0 & 0 \\ 0 & 1 & 0 & 1 \\ 0 & 0 & 1 & 1 \\ 0 & 1 & 1 & 0 \end{bmatrix}$$

Let H_n denote the matrix that results from A_n by replacing 0 with 1 and 1 with -1, for each A_{ij} for $1 \leq i, j \leq n$. Thus,

$$H_2 = \begin{bmatrix} 1 & 1 \\ 1 & -1 \end{bmatrix}$$

The matrix H_n has the property $H_n H_n^t = nI$, where I is the identity matrix. Matrices having this property are called **Hadamard matrices**. The code generated is called the **Hadamard code**.

Computer Science Application

The *Hamming distance* between two character strings of the same length is defined as the number of corresponding characters that mismatch. The following brute force algorithm searches a character string S of length m for the first occurrence of a substring whose Hamming distance from a specified character string P of length n ($n <= m$) is at most k.

int StringSearch(character string S, character string P, integer k)

```
n = length of string P
if n <= k then
   return 1
end-if

for i = 1 to m - n + 1
   count = 0
   j = 1
```

```
        while j <= n and count <= k do
            if P[j] is not equal to S[i + j − 1] then
                count = count + 1
            end-if
            j = j + 1
        end-while
        if count <= k then
            return i
        end-if
    end-for
    return 0
end-function
```

The integer returned by the function is the index of the first character of the substring found in S with the desired Hamming distance from P. If no such substring is found, 0 is returned. When $k = 0$, the algorithm becomes a simple string search.

Exercises

1. Given that $G_H^\perp$ is the matrix
$$\begin{bmatrix} 1 & 0 & 1 & 1 & 1 & 0 & 0 \\ 1 & 1 & 1 & 0 & 0 & 1 & 0 \\ 1 & 1 & 0 & 1 & 0 & 0 & 1 \end{bmatrix}$$
find G_H.

2. Given that $G_H^\perp$ is the matrix
$$\begin{bmatrix} 1 & 0 & 1 & 1 & 1 & 0 & 0 \\ 0 & 1 & 1 & 1 & 0 & 1 & 0 \\ 1 & 1 & 0 & 1 & 0 & 0 & 1 \end{bmatrix}$$
find G_H.

3. Find the distance between strings 110010101 and 010101111.

4. Find the distance between strings 110011001 and 111100001.

5. Find three strings orthogonal to 110011001.

6. Find three strings orthogonal to 010101111.

7. Given the generator matrix
$$\begin{bmatrix} 1 & 0 & 0 & 0 & 1 & 1 & 1 \\ 0 & 1 & 0 & 0 & 0 & 1 & 1 \\ 0 & 0 & 1 & 0 & 1 & 1 & 0 \\ 0 & 0 & 0 & 1 & 1 & 0 & 1 \end{bmatrix}$$

   (a) Find $G_H^\perp$.

   (b) Using only $G_H^\perp$, correct errors (if any) in the transmitted messages 1110110, 1011001, 1101010, and 1111110.

8. Given the generator matrix
$$\begin{bmatrix} 1 & 0 & 0 & 0 & 1 & 1 & 1 \\ 0 & 1 & 0 & 0 & 1 & 1 & 0 \\ 0 & 0 & 1 & 0 & 0 & 1 & 1 \\ 0 & 0 & 0 & 1 & 1 & 0 & 1 \end{bmatrix}$$

   (a) Find $G_H^\perp$.

   (b) Using only $G_H^\perp$, correct errors (if any) in the transmitted messages 1111110, 1010010, 1101100, and 1111101.

9. For each of the Hamming matrices in the matrices in Exercises 7 and 8, if 1001010 is transmitted as 1001001, what is the corrected code? How does this affect the original word encoded?

10. Which of the following are Hamming matrices?
$$\begin{bmatrix} 1 & 0 & 0 & 0 & 1 & 1 & 1 \\ 0 & 1 & 0 & 0 & 0 & 1 & 1 \\ 0 & 0 & 1 & 0 & 1 & 1 & 0 \end{bmatrix} \quad \begin{bmatrix} 1 & 0 & 0 & 0 & 1 & 1 & 1 \\ 0 & 1 & 0 & 0 & 0 & 1 & 1 \\ 0 & 0 & 1 & 0 & 1 & 1 & 1 \\ 0 & 0 & 0 & 1 & 1 & 0 & 1 \end{bmatrix}$$
$$\begin{bmatrix} 1 & 0 & 0 & 0 & 1 & 1 & 1 \\ 0 & 1 & 0 & 0 & 0 & 1 & 1 \\ 0 & 0 & 1 & 0 & 1 & 1 & 0 \\ 0 & 0 & 0 & 1 & 0 & 0 & 0 \end{bmatrix}$$

11. Prove Theorem 18.4: The Hamming distance function has the following properties:

    (a) For strings c and c', $\delta(c, c') = 0$ if and only if $c = c'$.

    (b) For strings c and c', $\delta(c, c') = \delta(c', c)$.

12. Construct the Reed-Muller code S_1.

13. Construct the Hadamard matrices H_2, H_4, H_8, and H_{16}.

GLOSSARY

Chapter 18: Theory of Codes

$[n, k]$-code (18.2)	The code C with form $[I_k \mid A_{n-k}]$ is called an $[n, k]$-code.
Block code (18.1)	**Block code** is encoded with binary strings of the same length.
Comma code (18.1)	**Comma code** is a type of prefix code in which each symbol is encoded into a string consisting of a string of 1's followed by a 0 at the end.
Coset leaders (18.2)	**Coset leaders** are used to correct an error, given that a single error has occurred that produces a string in that coset.
Dual code (18.2)	The **dual code** of C, denoted by $C^\perp$, is the set of all strings of B_n that are orthogonal to every string in C.
Error-detecting codes (18.1)	**Error-detecting codes** are those that can detect errors in transmission.
Generator matrix (18.2)	G is called a **generator matrix** if we have a $k \times n$ matrix G such that the first k columns and rows form the $k \times k$ identity matrix I_k, and all the columns are distinct. G has the form $[I_k \mid A_{n-k}]$.
Group generated by a set (18.2)	If C is the code created by taking all the vectors that are finite sums of strings in S, then the **group C is generated by the set S**.
Hamming code (18.3)	The code generated by the rows of the Hamming matrix is called the **Hamming code**.
Hamming distance function properties (18.3)	The **Hamming distance function properties** are: a. For strings c and c', $\delta(c, c') = 0$ if and only if $c = c'$. b. For strings c and c', $\delta(c, c') = \delta(c', c)$. c. For strings c, c', and c'', $\delta(c, c'') \leq \delta(c, c') + \delta(c', c'')$.
Hamming distance or distance between two strings of code (18.3)	The **Hamming distance or distance between two strings of code c and c'**, having the same length, is the number of corresponding bits in the string where one string has the digit 1 and the other has the digit 0. This is denoted by $\delta(c, c')$.
Hamming matrix (18.3)	The **Hamming matrix** G_H is the $(n-r) \times n$ matrix of the form $[I_{n-r} \mid A]$ where A is an $(n-r) \times n$ matrix and G_H has the form $[A^t \mid I_r]$.
Inner product (18.2)	If $u = (u_1, u_2, u_3, \ldots u_n)$ and $v = (v_1, v_2, v_3, \ldots v_n)$, then the **inner product** of u and v, denoted by $u \bullet v$, is equal to $u_1 v_1 + u_2 v_2 + u_3 v_3 + \cdots + u_n v_n$.
Linear codes or group codes (18.2)	**Linear code** or **group codes** are codes that are subgroups under addition of the group B_n of linear strings of length n.
Minimal distance of a code (18.3)	The **minimal distance of a code**, denoted by $D(C)$, is the smallest distance between any two strings in C.
Orthogonal vectors (18.2)	The **vector v is orthogonal** to the vector w if their dot product $u \bullet v = 0$.
Parity check matrix (18.2)	The **parity check matrix**, $G^\perp$, is a matrix used to detect an error in code transmitted.
Prefix code (18.1)	A **prefix code** has the property that an element of a code cannot be at the beginning of another element of the code.
Syndromes (18.2)	The syndrome of a received code word is the product of the parity check matrix, the transpose of an element in the corresponding coset.
Weight of a string of code (18.2)	The **weight of a string of code c**, denoted by $wt(c)$, is the number of 1's in the string.

TRUE-FALSE QUESTIONS

1. A block code is a uniquely decipherable code.
2. A comma code is a prefix code.
3. A prefix code is a comma code.
4. ASCII code is a block code.
5. Morse code is a block code.
6. ASCII code is an error-detecting code.
7. Any subset of the set B_n of all binary strings of length n is a linear code.
8. For any element c of the group code C, $wt(c+c) = 0$.
9. If $G = \begin{bmatrix} 1 & 0 & 0 & 0 & 1 & 1 \\ 0 & 1 & 0 & 1 & 0 & 1 \\ 0 & 0 & 1 & 1 & 1 & 0 \end{bmatrix}$ is a generator matrix, then $(1,1,1,0,0,0) \in C$.
10. If $G = \begin{bmatrix} 1 & 0 & 0 & 0 & 1 & 1 \\ 0 & 1 & 0 & 1 & 1 & 1 \\ 0 & 0 & 1 & 1 & 0 & 1 \end{bmatrix}$ is a generator matrix, then $(1,1,0,0,0,0) \in C$.
11. If $G = \begin{bmatrix} 1 & 0 & 0 & 0 & 1 & 1 \\ 0 & 1 & 0 & 1 & 1 & 1 \\ 0 & 0 & 1 & 1 & 0 & 1 \end{bmatrix}$ is a generator matrix, then the order of C is 2^6.
12. If $G = \begin{bmatrix} 1 & 0 & 0 & 0 & 1 & 1 \\ 0 & 1 & 0 & 1 & 1 & 1 \\ 0 & 0 & 1 & 1 & 0 & 1 \end{bmatrix}$ is a generator matrix, then the encoding of $(1,1,0)$ is $(1,1,0,1,0,1)$.
13. If $G = \begin{bmatrix} 1 & 0 & 0 & 0 & 1 & 1 \\ 0 & 1 & 0 & 1 & 1 & 1 \\ 0 & 0 & 1 & 1 & 0 & 1 \end{bmatrix}$ is a generator matrix, then $G^\perp = \begin{bmatrix} 1 & 0 & 0 & 0 & 1 & 1 \\ 0 & 1 & 0 & 1 & 1 & 0 \\ 0 & 0 & 1 & 1 & 1 & 1 \end{bmatrix}$.

14. Let $G = \begin{bmatrix} 1 & 0 & 0 & 0 & 1 & 1 \\ 0 & 1 & 0 & 1 & 1 & 1 \\ 0 & 0 & 1 & 1 & 0 & 1 \end{bmatrix}$ be a generator matrix and $v \in C^\perp$. v is orthogonal to every row in G.
15. If C is a linear code, then $C^\perp$ is a group.
16. An $[n, n-r]$ Hamming code contains 2^r cosets in B_n.
17. Let $G = \begin{bmatrix} 1 & 0 & 0 & 1 & 0 & 1 \\ 0 & 1 & 0 & 1 & 1 & 0 \\ 0 & 0 & 1 & 0 & 1 & 1 \end{bmatrix}$ be a generator matrix. The syndrome for the coset containing $(0,0,0,1,0,0)$ is $\begin{bmatrix} 0 \\ 1 \\ 1 \end{bmatrix}$.
18. The rows of the parity check matrix generate $C^\perp$.
19. Let $G = \begin{bmatrix} 1 & 0 & 0 & 1 & 0 & 1 \\ 0 & 1 & 0 & 1 & 1 & 0 \\ 0 & 0 & 1 & 0 & 1 & 1 \end{bmatrix}$ be a generator matrix. Without error correction, 110010 is decoded as 110.
20. Let $G = \begin{bmatrix} 1 & 0 & 0 & 1 & 0 & 1 \\ 0 & 1 & 0 & 1 & 1 & 0 \\ 0 & 0 & 1 & 0 & 1 & 1 \end{bmatrix}$ be a generator matrix. With error correction, 110010 is decoded as 110.
21. Let $G = \begin{bmatrix} 1 & 0 & 0 & 1 & 0 & 1 \\ 0 & 1 & 0 & 1 & 1 & 0 \\ 0 & 0 & 1 & 0 & 1 & 1 \end{bmatrix}$ be a generator matrix. 011010 is a proper encoding.
22. For a code C, if $D(C) = k$, up to k errors can be detected using the code.
23. For a code C, if $D(C) = k+1$, up to k errors can be corrected using the code.
24. Let C be a Hamming code. If $c \in C$ and $c \neq 0$, the $wt(c) \geq 3$.
25. For any linear code, $D(C) = W(C)$.

ANSWERS TO TRUE-FALSE

1. T 2. T 3. F 4. T 5. F 6. T 7. F 8. T 9. T 10. F
11. F 12. F 13. F 14. T 15. T 16. T 17. F 18. T 19. T
20. T 21. F 22. F 23. F 24. T 25. T

CHAPTER 19
Enumeration of Colors

19.1 Burnside's Theorem

In the beginning of this section we consider the number of ways to paint the vertices of a regular n-polygon with m colors. Since $m = 1$ does not produce anything particularly interesting, we shall begin with two colors, say, red and blue. We shall also begin with a square, so we are trying to color the vertices of a square with two colors. We shall refer to a coloring of the vertices simply as a **coloring**. Thus, painting the upper-left vertex red and the others blue is a coloring. If we do not allow the square to move, then we have two choices for each of the four vertices and there are $2^4 = 16$ ways to color the square. To complicate the problem, we shall allow the square to be rotated and we shall also allow the square to be flipped over about an axis between opposite vertices or the midpoints of opposite edges. We shall refer to the first class of actions as **rotations** and the second class as **reflections**. All colorings that can be obtained from a given coloring of the vertices of a square using rotations and reflections are considered equivalent. Thus, the colorings in Figure 19.1 are equivalent since each can be obtained from the others by rotation.

This is also true of the colorings in Figure 19.2.

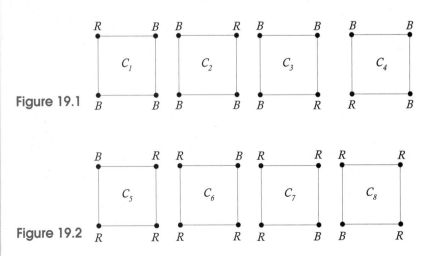

Figure 19.1

Figure 19.2

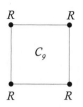

Figure 19.3

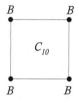

Figure 19.4

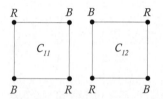

Figure 19.5

The coloring in Figure 19.3 is equivalent only with itself since any rotation or reflection produces the same coloring. In a similar manner we see the coloring in Figure 19.4 is equivalent only to itself.

The colorings in Figure 19.5 are equivalent since either is obtained from the other by a 90° rotation and no further colorings are produced with reflections and rotations.

Finally, we see that the sets of colorings in Figure 19.6 are equivalent since each can be obtained from the other by a rotation.

Actually, it may be noted that for the 2-coloring of a square, reflections are not needed, since all possible equivalent colorings can be obtained from each other by rotations alone.

Recall in Section 9.4, rotations and reflections of the square were shown to be described by a subgroup of S_4, the group of permutations on four elements. This group is called the **octic group** or the **group of symmetries of the square**. Given the square in Figure 19.7,

$$\delta_1 = \begin{pmatrix} 1 & 2 & 3 & 4 \\ 3 & 2 & 1 & 4 \end{pmatrix} \text{ and } \delta_2 = \begin{pmatrix} 1 & 2 & 3 & 4 \\ 1 & 4 & 3 & 2 \end{pmatrix}$$

are reflections which interchange vertices 1 and 3 and vertices 2 and 4, respectively. Permutations

$$\phi_1 = \begin{pmatrix} 1 & 2 & 3 & 4 \\ 4 & 3 & 2 & 1 \end{pmatrix} \text{ and } \phi_2 = \begin{pmatrix} 1 & 2 & 3 & 4 \\ 2 & 1 & 4 & 3 \end{pmatrix}$$

reflect the square around the horizontal line through the center of the square and around the vertical line through the center of the square respectively. Permutations ρ_1, ρ_2, and ρ_3 rotate the square clockwise by 90°, 180°, and 270°, respectively, where

$$\rho_1 = \begin{pmatrix} 1 & 2 & 3 & 4 \\ 2 & 3 & 4 & 1 \end{pmatrix} \quad \rho_2 = \begin{pmatrix} 1 & 2 & 3 & 4 \\ 3 & 4 & 1 & 2 \end{pmatrix} \quad \rho_3 = \begin{pmatrix} 1 & 2 & 3 & 4 \\ 4 & 1 & 2 & 3 \end{pmatrix}$$

I is the identity permutation

$$\begin{pmatrix} 1 & 2 & 3 & 4 \\ 1 & 2 & 3 & 4 \end{pmatrix}$$

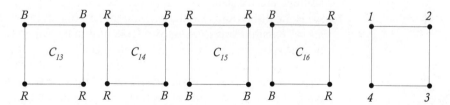

Figure 19.6 **Figure 19.7**

Note that in the group of symmetries of a regular n-sided polygon, there are always $2n$ transformations. There are always n rotations. If n is odd, there are n reflections, each about an axis between a vertex and the center of the opposite side. If n is even, there are $\frac{n}{2}$ reflections about axes between opposite vertices and $\frac{n}{2}$ reflection centers of opposite sides of the polygon. In each case we get a n rotations and n reflections.

Returning to 2-colors of the square, for each permutation σ we can produce a permutation $\hat{\sigma}$ between colorings, where $\hat{\sigma}(C_i) = C_j$ if C_j is the coloring produced when the square with coloring C_i is transformed by permutation σ. For example, $\hat{\rho}_1(C_1) = C_2$ and $\hat{\rho}_1(C_{11}) = C_{12}$. To keep notation simple, we shall write $\hat{\sigma}$ as σ also and use σ both as a permutation of vertices and a permutation of colors.

Earlier we discussed equivalent colorings. We now create an equivalence relation such that equivalent colorings are indeed equivalent. Define the relation R on the set of colorings by $C_i R C_j$ if there is a permutation σ in the cyclic group such that $\sigma(C_i) = C_j$. We now show that R is an equivalence relation. Certainly $C_i R C_i$ for coloring C_i, since $I(C_i) = C_i$. If $C_i R C_j$, then there is a permutation σ such that $\sigma(C_i) = C_j$. But since σ^{-1} belongs to the octic group and $\sigma^{-1}(C_j) = C_i$, then $C_j R C_i$. If $C_i R C_j$ and $C_j R C_k$, then there exist permutations σ and σ' such that $\sigma(C_i) = C_j$ and $\sigma(C_j) = C_k$. Therefore, $\sigma'\sigma(C_i) = C_k$ and since $\sigma'\sigma$ is also a permutation in the group of symmetries of the square, we have $C_i R C_k$. Therefore, R is an equivalence relation. This equivalence relation can be created in the same manner for any set K of colorings on a set S and group of permutations of S. Thus, we have the following theorem.

THEOREM 19.1 Let K be the set of colorings on a set S and G a group of permutations of S. Then the relation R on K is defined by $C_i R C_j$ if there is a permutation $\sigma \in G$ such that $\sigma(C_i) = C_j$. The relation R is an equivalence relation. ∎

Returning to 2-coloring of the square, when we first mentioned colorings being equivalent we meant that one could be produced from the other by means of rotation and reflections. When we discuss equivalent colorings in the above theorem, we mean that one coloring can be produced from the other by means of a permutation. Since each permutation can be considered as a rotation or reflection, it should be obvious that the equivalences are really the same.

We need to remember that our goal is to find the number of "different" colorings where colors are the "same" if they are in the same equivalence class. Therefore, what we are really seeking is the number of equivalence classes, since this is the number of "different" colorings. For example, consider a round table seating n people. Suppose we consider two seatings to be the same if everyone has the same person to their right and the same person to their left. Therefore, if the table is rotated, then the seatings are the same. We do not consider reflections, since flipping the table over spills the dishes and also the person to the right and to the left will be reversed. There are $n!$ possible ways to seat n people at a table. Since two seatings are equivalent, or the same, if one is a rotation of the other and there are n rotations, the number of different seatings is equal to the number of equivalence classes. But the number of equivalence classes is equal to $n!$, the number of ways to seat n people at a table, divided by n, the number of seatings in each equivalence class and therefore is equal to $\frac{n!}{n} = (n-1)!$. This works because each rotation produces a different member of the equivalent class. In case anyone is wondering where the colors went, we can consider each person as representing a different color, so we are really considering an n-edged polygon where each vertex has a different color.

Bouncing back again to our 2-coloring of a square, we notice that not every permutation produces a different coloring. For example, $\rho_2(C_{11}) = C_{11} = \delta_1(C_{11}) = \delta_2(C_{11})$. In fact we already know that all of our equivalence classes are not the same size. Therefore, a new line of attack is needed.

For a fixed coloring C, let $G_C = \{\sigma : \sigma(C) = C\}$, so that G_C is the set of all permutations in G leaving C fixed. The set G_C is called the **stabilizer** of C.

THEOREM 19.2 For a fixed coloring C, the stabilizer G_C is a subgroup of G, the group of permutations.

Proof To show G_C is a group, we first show that if $\sigma_i, \sigma_j \in G_C$, then $\sigma_i \circ \sigma_j \in G_C$. But $\sigma_i(C) = C$ and $\sigma_j(C) = C$, so

$$\sigma_i \circ \sigma_j(C) = \sigma_i(\sigma_j(C)) = \sigma_i(C) = C$$

and $\sigma_i \circ \sigma_j \in G_C$. Obviously the identity permutation I belongs to G_C, since $I(C_i) = C_i$ for every coloring C_i. If $\sigma \in G_C$, then $\sigma(C) = C$ and hence $\sigma^{-1}(C) = C$, since

$$C = I(C) = \sigma^{-1} \circ \sigma(C) = \sigma^{-1}(\sigma(C)) = \sigma^{-1}(C)$$

Therefore, $\sigma^{-1} \in G_C$. Therefore, G_C is a group. ∎

Let C and C' be in the same equivalence class. The number of permutations in G that map C to C' is equal to $|G_C|$. To prove this let $\sigma_0(C) = C'$ and $\sigma(C) = C$. Then

$$\sigma_0 \circ \sigma(C) = \sigma_0(\sigma(C)) = \sigma_0(C) = C'$$

Thus, if σ_0 is fixed, for every σ such that $\sigma(C) = C$, we have a permutation $\sigma_0 \circ \sigma$ that maps C to C'. For each σ such that $\sigma(C) = C$, let $f(\sigma) = \sigma_0 \circ \sigma$. Then f maps every permutation mapping C to itself to a permutation mapping of C to C'. Furthermore, if σ_1 and σ_2 both map C to itself and $\sigma_0 \circ \sigma_1 = \sigma_0 \circ \sigma_2$, then

$$\sigma_0^{-1} \circ (\sigma_0 \circ \sigma_1) = \sigma_0^{-1} \circ (\sigma_0 \circ \sigma_2)$$

so

$$(\sigma_0^{-1} \circ \sigma_0) \circ \sigma_1 = (\sigma_0^{-1} \circ \sigma_0) \circ \sigma_2$$

giving

$$I \circ \sigma_1 = I \circ \sigma_2$$

and

$$\sigma_1 = \sigma_2$$

Therefore, f is one-to-one. Let σ map C to C'. Then

$$f(\sigma_0^{-1} \circ \sigma) = \sigma_0 \circ (\sigma_0^{-1} \circ \sigma) = (\sigma_0 \circ \sigma_0^{-1}) \circ \sigma = I \circ \sigma = \sigma$$

so that f is onto. Therefore, f is a one-to-one correspondence, so there are the same number of permutations that map C to C' as there are permutations that map C to itself. Therefore, if C_1 and C_2 are both in E, the same equivalence class as C, then the number of permutations mapping C to C_1 is equal to the number of permutations mapping C to C_2, namely $|G_C|$. This number is called the **multiplicity** of the equivalence class E. It follows then that $|G|$, the number of permutations in G, is equal to $|G_C| \cdot |E|$ where $|E|$ is the number of colorings in E or

$$|G_C| \cdot |E| = |G|$$

We can also write this as

$$|G| = \sum_{C \in E} |G_C|$$

so that if N is the number of equivalence classes then

$$N \cdot |G| = \sum_E \sum_{C \in E} |G_C| = \sum_{C \in K} |G_C|$$

where K is the set of colorings. Therefore, the set of distinct colorings, that is, the number N of equivalence classes, is given by

$$N = \frac{\sum_{C \in K} |G_C|}{|G|} = \frac{1}{|G|} \sum_{C \in K} |G_C|$$

For each permutation σ, let $\varphi(\sigma)$ equal the number of colorings that σ leaves fixed. Note that if for each color we find the number of permutations leaving that color fixed and find this sum over all of the colors, we get the same result as finding the number of colors that each permutation leaves fixed. Summing over the permutations, therefore,

$$\sum_{C \in K} |G_C| = \sum_{\sigma \in G} \varphi(\sigma)$$

and

$$N = \frac{1}{|G|} \sum_{C \in K} |G_C| = \frac{1}{|G|} \sum_{\sigma \in G} \varphi(\sigma)$$

We summarize this in the following theorem.

THEOREM 19.3 (Burnside's Counting Lemma) If K is the number of colorings over a set S, G is the a group of permutations on a set S, and N is the number of equivalence classes (i.e., distinct colorings), then

$$N = \frac{1}{|G|} \sum_{C \in K} |G_C| = \frac{1}{|G|} \sum_{\sigma \in G} \varphi(\sigma) \qquad \blacksquare$$

Exercises

1. Determine the group of symmetries of an equilateral triangle.
2. Determine the group of symmetries of a regular pentagon.
3. Determine elements of the group of symmetries of a regular octagon.
4. Determine elements of the group of symmetries of a tetrahedron.
5. If a rod is decorated with five bands using two colors, how many ways can the rod be colored? (Remember the rod can be turned over.)
6. How many ways are there to color an equilateral triangle with two colors?
7. How many ways are there to color a rectangle that is not a square?

19.2 Polya's Theorem

Having taken the first step, we now develop more theory to help us in our color counting. We develop Polya's theorem, which gives us a generating function for finding the number of colorings of an unoriented figure using a given number of colors. Actually it is the last part of Burnside's equation that we shall find most helpful. If we can find a simple method of determining how many colors are left fixed by each permutation, then we can simply sum them and divide by the order of G. In Section 9.4, we found that every permutation can be expressed as the product of disjoint cycles. Returning to our 2-coloring of a square,

$$\phi_1 = \begin{pmatrix} 1 & 2 & 3 & 4 \\ 4 & 3 & 2 & 1 \end{pmatrix}$$

can be expressed as

(14)(23)

19.2 Polya's Theorem

In a cycle, if the coloring is to remain unchanged, the color of each vertex must be the same color as the next vertex to which it is moved. Therefore, every vertex in a cycle must have the same color. If ϕ_1 is to leave a coloring unchanged, then vertices 1 and 4 must be the same color, and vertices 2 and 3 must be the same color. Therefore, vertices 1 and 4 may be red or blue, and vertices 2 and 3 may be red or blue. Thus, there are four possible colorings that are left unchanged by ϕ_1. They are C_9, C_{10}, C_{14}, and C_{16}. Since for each cycle there are two choices of color, in general the number of colorings left fixed by a permutation σ is $2^{Cy(\sigma)}$ where $Cy(\sigma)$ is the number of cycles of σ, including the cycles of length 1. The identity permutation I may be written as $(1)(2)(3)(4)$. Since it has four cycles, it leaves 2^4 colorings fixed. In the following table, we give each permutation, its expression as cycles, the cycles structure, the number of cycles, and the colorings left fixed. To determine cycle structure, we use c_m^n to denote that there are n cycles of length m. Thus, the cycle structure of ϕ_1 is c_2^2, meaning that ϕ_1 is expressed as two 2-cycles.

Permutation	Cycle structure	No. of cycles	Colorings left fixed	No. of colorings
$\delta_1 = (13)(2)(4)$	$c_1^2 c_2$	3	$C_2, C_4, C_6, C_8, C_9, C_{10}, C_{11} C_{12}$	8
$\delta_2 = (1)(3)(24)$	$c_1^2 c_2$	3	$C_1, C_3, C_5, C_7, C_9, C_{10}, C_{11} C_{12}$	8
$\phi_1 = (14)(23)$	c_2^2	2	$C_9, C_{10}, C_{14} C_{16}$	4
$\phi_2 = (12)(34)$	c_2^2	2	$C_9, C_{10}, C_{13} C_{15}$	4
$\rho_1 = (1234)$	c_4	1	C_9, C_{10}	2
$\rho_2 = (13)(24)$	c_2^2	2	$C_9, C_{10}, C_{11} C_{12}$	4
$\rho_3 = (1432)$	c_4	1	C_9, C_{10}	2
$I = (1)(2)(3)(4)$	c_1^4	4	$C_1, C_2, C_3, C_4, \ldots, C_{16}$	16
Total	$c_1^4 + 2c_1^2 c_2 + 3c_2^2 + 2c_4$			48

As previously mentioned, to find the number of colorings left fixed by a permutation σ, we only need to find $2^{Cy(\sigma)}$, where $Cy(\sigma)$ is the number of cycles of σ. We then take the sum over all the permutations to find $\sum_{\sigma \in G} \varphi(\sigma)$. Another method is to take the cyclic form and let each cycle have the value 2. For example $\delta_2 = c_1^2 c_2$ leaves $2^2 \cdot 2 = 8$ colorings fixed. We would again take the sum to find $\sum_{\sigma \in G} \varphi(\sigma)$. A third method is to add the cyclic forms. As shown in the table, we have

$$c_1^4 + 2c_1^2 c_2 + 3c_2^2 + 2c_4$$

Now if we let each cycle have value 2, we have

$$2^4 + 2 \cdot 2^2 \cdot 2 + 3 \cdot 2^2 + 2 \cdot 2 = 48$$

So there are 48 total colorings left fixed by the permutations. If we now divide this by 8, the number of permutations, we have

$$\frac{1}{|G|} \sum_{\sigma \in G} \varphi(\sigma) = \frac{48}{8} = 6$$

distinct colorings.

Suppose we have three colors. Using similar arguments in the table, we could take $3^{Cy(\sigma)}$ where $Cy(\sigma)$ is the number of cycles σ to find the number of colorings left fixed by the permutation σ and then sum these to find $\sum_{\sigma \in G} \varphi(\sigma)$. We could then again divide by 8 to have the number of distinct colorings. Another method, since we already have

$$c_1^4 + 2c_1^2 c_2 + 3c_2^2 + 2c_4$$

is to set each cycle equal to 3 so that sum of the colorings left fixed by the permutations is

$$3^4 + 2 \cdot 3^2 \cdot 3 + 3 \cdot 3^2 + 2 \cdot 3 = 168$$

Now dividing by 8, we have

$$\frac{1}{|G|} \sum_{\sigma \in G} \varphi(\sigma) = \frac{168}{8} = 21$$

different colorings. If we have m colors, then

$$\frac{1}{|G|} \sum_{\sigma \in G} \varphi(\sigma) = \frac{m^4 + 2 \cdot m^2 \cdot m + 3 \cdot m^2 + 2 \cdot m}{8}$$

Since, in general, we want $\frac{1}{|G|} \sum_{\sigma \in G} \varphi(\sigma)$, which we get from

$$\frac{1}{|G|}(c_1^4 + 2c_1^2 c_2 + 3c_2^2 + 2c_4)$$

we let

$$P_G(c_1, c_2, c_3, c_4) = \frac{1}{|G|}(c_1^4 + 2c_1^2 c_2 + 3c_2^2 + 2c_4)$$

where c_i is a cycle of length i, and we call $P_G(c_1, c_2, c_3, c_4)$ the **cycle index**.

EXAMPLE 19.4

It was previously mentioned that for 2-coloring the square, only the rotations were really needed. Suppose we let $G = \{I, \rho_1, \rho_2, \rho_3\}$. We would then have the following table:

Permutation	Cycle structure	No. of cycles	Colorings left fixed	No. of colorings
$\rho_1 = (1234)$	c_4	1	C_9, C_{10}	2
$\rho_2 = (13)(24)$	c_2^2	2	$C_9, C_{10}, C_{11}C_{12}$	4
$\rho_3 = (1432)$	c_4	1	C_9, C_{10}	2
$I = (1)(2)(3)(4)$	c_1^4	4	$C_1, C_2, C_3, C_4, \ldots, C_{16}$	16
Total	$c_1^4 + c_2^2 + 2c_4$			24

Then

$$\frac{1}{|G|} \sum_{\sigma \in G} \varphi(\sigma) = \frac{24}{4} = 6$$

and, as before, we have six distinct colorings. Again we see that if we set each cycle equal to 2 in

$$P_G(c_1, c_2, c_3, c_4) = \frac{1}{4}(c_1^4 + c_2^2 + 2c_4)$$

we have

$$\frac{1}{4}(2^4 + 2^2 + 2 \cdot 2) = 6$$

and we again get six distinct colorings. ■

In general we see that if the group of permutations G has $|G|$ elements, we are coloring with m colors and the sum of the cycles is $a_1 c_1^{e_1} + a_2 c_1^{e_2} + a_3 c_1^{e_3} + \cdots + a_n c_n^{e_n}$. Then the cycle index

$$P_G(c_1, c_2, c_3, \ldots, c_n) = \frac{1}{|G|} (a_1 c_1^{e_1} + a_2 c_1^{e_2} + a_3 c_1^{e_3} + \cdots + a_n c_n^{e_n})$$

and the number of distinct colorings is $P_G(m, m, m, \ldots, m)$.

At this point, we shift from counting patterns to looking at patterns of the colorings that are preserved by particular permutations. Again we consider the 2-coloring of the square. Consider the cycle structure $c_1^2 c_2$ of $\delta_1 = (13)(2)(4)$. We first look at the cycle structure c_1^2, which represents two 1 cycles. We have four choices of colors for the two cycles so the coloring is left fixed; the vertices in both cycles may be colored red, those in one cycle may be colored red and the other blue, or they may both be colored blue. There is obviously only one way to color vertices in both cycles red and one way to color them blue. If vertices in one cycle are colored red and the other blue, we have two choices for choosing the blue vertices and the others must be colored red. We denote this color pattern as $bb + 2br + rr$, where the $+$ sign may be read as "or." For convenience we shall denote this pattern by $b^2 + 2br + r^2$. Note that algebraically, this is $(b+r)^2$. The other cycle is c_2, which represents one 2-cycle. Since it is a single cycle of length 2, both vertices must be colored red or both blue. We denote this by $bb + rr$ or $b^2 + r^2$. Since the pattern for two of the vertices is $(b+r)^2$ and for the other two vertices, the pattern is $b^2 + r^2$, the pattern for colorings left fixed by δ_1 may be described by $(b+r)^2(b^2 + r^2)$.

Assume that we have a cycle c_m^n. This is the product of n cycles of length m. Let R represent coloring a cycle red and B represent coloring a cycle blue. Thus, $R^j B^{n-j}$ represents coloring j cycles red and $n - j$ colors blue. The number of ways of coloring j cycles red and $n - j$ colors blue is equal to the number of ways of choosing j cycles out of n to color red. This is equal to $C(n, j) = \binom{n}{j}$. We shall, thus, let $\binom{n}{j} R^j B^{n-j}$ represent the colorings with j cycles red and $n - j$ cycles blue. But this is the coefficient of $R^j B^{n-j}$ in $(R + B)^n$. Thus, we can represent the colorings of the cycle c_m^n by $(R + B)^n$. Since each coloring of a cycle consists of coloring m vertices, we let $R = r^m$ and $B = b^m$. Thus, our patterns for c_m^n are described by $(r^m + b^m)^n$. This gives us the following set of colorings, for each permutation, which we will call the **inventory** of colorings, for the permutation:

Permutation	Cycle structure	Inventory of fixed colorings
$\delta_1 = (13)(2)(4)$	$c_1^2 c_2$	$(r+b)^2(r^2+b^2)$
$\delta_2 = (1)(3)(24)$	$c_1^2 c_2$	$(r+b)^2(r^2+b^2)$
$\phi_1 = (14)(23)$	c_2^2	$(r^2+b^2)^2$
$\phi_2 = (12)(34)$	c_2^2	$(r^2+b^2)^2$
$\rho_1 = (1234)$	c_4	(r^4+b^4)
$\rho_2 = (13)(24)$	c_2^2	$(r^2+b^2)^2$
$\rho_3 = (1432)$	c_4	(r^4+b^4)
$I = (1)(2)(3)(4)$	c_1^4	$(r+b)^4$
Total	$c_1^4 + 2c_1^2 c_2 + 3c_2^2 + 2c_4$	$8r^4 + 8r^3 b + 16 r^2 b^2 + 8rb^3 + 8b^4$

The total of the inventories of the coloring inventory is obtained by expanding the individual inventories and adding them.

Earlier it was shown that for a coloring C, $|G|$, the number of permutations in G, is equal to $|G_C| \cdot |E|$ where $|E|$ is the number of colorings in the equivalent class E containing C and $|G_C|$ is the number of permutations leaving C fixed. Since any permutation maps a coloring to another coloring with the same cycle structure, in summing our inventory we have counted the colorings equivalent to C and described by the same structure $|G|$ times. Thus, to find the inventory of nonequivalent color patterns, we define the total inventory of fixed colorings by $|G| = 8$, so that we have

$$\frac{1}{8}(8r^4 + 8r^3b + 16r^2b^2 + 8rb^3 + 8b^4) = r^4 + r^3b + 2r^2b^2 + rb^3 + b^4$$

for our pattern inventory.

To get our total pattern inventory, we obtained the fixed colorings from the cycle structure of each permutation and then added them. We could first add the cycle structures as in column 2 in the preceding table and then convert cycles to colorings to get the same result. Thus, for the sum

$$c_1^4 + 2c_1^2c_2 + 3c_2^2 + 2c_4$$

we get the total pattern inventory

$$(r^4 + b^4) + 2(r+b)^2(r^2+b^2)(r+b)^2 + 3(r^2+b^2)^2 + 2(r^4+b^4)$$

which, if we expand and collect like terms, is equal to

$$8r^4 + 8r^3b + 16r^2b^2 + 8rb^3 + 8b^4$$

Thus, if the sum of the cycles is $a_1c_1^{e_1} + a_2c_1^{e_2} + a_3c_1^{e_3} + \cdots + a_nc_n^{e_n}$, then the inventory of nonequivalent color patterns is found to be

$$P_G(r+b, r^2+b^2, \ldots, r^n+b^n) = \frac{1}{|G|}(a_1(r+b)^{e_1} + a_2(r^2+b^2)^{e_2} + \cdots + a_n(r^n+b^n)^{e_n})$$

If three colors red (r), white (w), and blue (b) were used, it can be shown that the inventory of nonequivalent color patterns is

$$P_G(r+w+b, r^2+w^2+b^2, \ldots, r^n+w^n+b^n)$$

In general, we have the following theorem.

THEOREM 19.5 (Polya's Counting Theorem) Given a set S, cycle index $P_G(c_1, c_2, \ldots, c_n)$, and colors $k_1, k_2, k_3, \ldots, k_m$, the inventory of nonequivalent color patterns is

$$P_G\left(\sum_{i=1}^n k_i, \sum_{i=1}^n k_i^2, \ldots, \sum_{i=1}^n k_i^m\right)$$ ∎

EXAMPLE 19.6

Again consider the 2-coloring of the square, where only the rotations are used. Also let $G = \{I, \rho_1, \rho_2, \rho_3\}$. We then have the table

Permutation	Cycle structure	Inventory of fixed colorings
$\rho_1 = (1234)$	c_4	$(r^4 + b^4)$
$\rho_2 = (13)(24)$	c_2^2	$(r^2 + b^2)^2$
$\rho_3 = (1432)$	c_4	$(r^4 + b^4)$
$I = (1)(2)(3)(4)$	c_1^4	$(r + b)^4$
Total	$c_1^4 + c_2^2 + 2c_4$	$4r^4 + 4r^3b + 8r^2b^2 + 4rb^3 + 4b^4$

and

$$P_G(r+b, r^2+b^2, r^3+b^3, r^4+b^4) = \frac{1}{4}((r+b)^4 + (r^2+b^2)^2 + 2(r^4+b^4))$$
$$= r^4 + r^3b + 2r^2b^2 + rb^3 + b^4$$

so that the inventory of nonequivalent color patterns is the same as previously where the entire symmetric group was used. ∎

EXAMPLE 19.7

Let S be the vertices of a triangle and G the group of symmetries of a triangle. Find the inventory of nonequivalent color patterns. Let $\rho_1 = (123)$, and $\rho_2 = (132)$ be rotations of the vertices. Let $\delta_1 = (1)(23)$, $\delta_2 = (13)(2)$, and $\delta_3 = (12)(3)$ be reflections and $I = (1)(2)(3)$ be the identity. We then have the table

Permutation	Cycle structure	Inventory of fixed colorings
$\rho_1 = (123)$	c_3	$(r^3 + b^3)$
$\rho_2 = (132)$	c_3	$(r^3 + b^3)$
$\delta_1 = (1)(23)$	$c_1 c_2$	$(r + b)(r^2 + b^2)$
$\delta_2 = (13)(2)$	$c_1 c_2$	$(r + b)(r^2 + b^2)$
$\delta_2 = (12)(3)$	$c_1 c_2$	$(r + b)(r^2 + b^2)$
$I = (1)(2)(3)$	c_1^3	$(r + b)^3$
Total	$c_1^3 + 3c_1c_2 + 2c_3$	$6r^3 + 6r^2b + 6rb^2 + 6b^3$

so that

$$P_G(r+b, r^2+b^2, r^3+b^3) = \frac{1}{6}((r+b)^3 + 3(r^2+b^2)(r+b) + (r^3+b^3))$$
$$= r^3 + r^2b + rb^2 + b^3$$

∎

Computer Science Application

Another interesting coloring problem involves the squares of an $n \times n$ matrix. A *chromatic rectangle* is formed when the four corners of a rectangle on the matrix have the same color. An *ncr-board* is a matrix with no chromatic rectangles. It is known that a 10×10 ncr-board exists using only three colors, but no square matrix of larger dimension can be 3-colored to produce one. The following algorithm returns the number of chromatic rectangles on an $n \times n$ matrix. The colors of the squares are represented by integers.

integer function chromatic-rectangles($n \times n$ matrix A)

    integer Counter $= 0$

    for $i = 1$ to $n - 1$ do
      for $j = 1$ to $n - 1$ do
        for $k = i + 1$ to n do
          for $l = j + 1$ to n do
            if $A[i, j] = A[i, l]$ and $A[i, j] = A[k, j]$ and $A[i, j] = A[k, l]$ then
              Counter $=$ Counter $+ 1$
            end-if
          end-for
        end-for
      end-for
    end-for

    return Counter

end-function

If all the squares of A are assigned the same color, the function returns the number of rectangles within the matrix. If the dimension of A is 10×10 the total number of rectangles is 2025. While it is known that the 100 squares of such a matrix can be colored with 3 colors in such a way that none of these 2025 rectangles are chromatic, it is an extremely challenging exercise to produce such a coloring.

Exercises

In the following problems, it is assumed that all rotations and reflections are possible.

1. A bracelet contains five beads. If they are to be colored using two different colors, how many colorings are there? How many 3-colorings are there?

2. In how many ways can the squares on the top of a 3×3 chessboard be colored?

3. How many ways are there to color an equilateral triangle with three colors?

4. How many ways are there to color a cube with two colors?

5. How many ways are there to color a regular pentagon with two colors?

6. How many ways are there to color a regular pentagon with three colors?

7. How many ways are there to color a regular octagon with two colors?

8. Find the pattern inventory for the 2-coloring of a bracelet with seven beads.

9. Find the pattern inventory for the 2-coloring of a rectangle that is not a square.

10. Find the pattern inventory for the 3-coloring of an equilateral triangle.

11. Find the pattern inventory for the 2-coloring of a pentagon.

12. Find the pattern inventory for the 2-coloring of a tetrahedron.

13. Find the pattern inventory for the 2-coloring of the squares on the top of a 3×3 chessboard.

14. Find the pattern inventory for the 2-coloring of an octagon.

15. Ten bowling pins are placed in a triangle in the usual manner. Find the pattern inventory for the 2-coloring of the these pins with no restrictions on movement. Find the pattern inventory if only rotation is allowed.

GLOSSARY

Chapter 19: Enumeration of Colors

Burnside's counting lemma (19.1)	If K is the number of colorings over a set S, G is a group of permutations on a set S, and N is the number of equivalence classes (distinct colorings), then $N = \frac{1}{|G|} \sum_{C \in K} |G_C| = \frac{1}{|G|} \sum_{\sigma \in G} \varphi(\sigma)$.
Coloring (19.1)	A **coloring** is a coloring of the vertices of a regular n-polygon.
Cycle index (19.2)	If $P_G(c_1, c_2, c_3, c_4) = \frac{1}{|G|}(c_1^4 + 2c_1^2 c_2 + 3c_2^2 + 2c_4)$ where c_i is a cycle of length I, then $P_G(c_1, c_2, c_3, c_4)$ is the **cycle index**.
Inventory of colorings (19.2)	Patterns for cycle c_m^n are described by $(r^m + b^m)^n$. This is used to give a set of colorings, using red and blue, which is called the **inventory of colorings** for a permutation.
Multiplicity of an equivalence class E (19.1)	If C_1 and C_2 are both in E, the same equivalence class as C, then the number of permutations mapping C to C_1 is equal to the number of permutations mapping C to C_2, namely $|G_C|$. This number is called the **multiplicity of the equivalence class E**.
Octic group or group of symmetries of the square (19.1)	The **octic group** is the group of permutations of the four vertices of a square, formed by rotating or reflecting the square.
Polya's counting theorem (19.2)	Given a set S, cycle index $P_G(c_1, c_2, \ldots, c_n)$, and colors $k_1, k_2, k_3, \ldots, k_m$, the inventory of nonequivalent color patterns is $$P_G\left(\sum_{i=1}^{n} k_i, \sum_{i=1}^{n} k_i^2, \ldots, \sum_{i=1}^{n} k_i^m\right)$$
Stabilizer set (19.1)	The set G_C is called the **stabilizer** of C if it is the set of all permutations in G leaving C fixed.

TRUE-FALSE QUESTIONS

1. In the group of symmetries of a regular n-sided polygon, if n is even there are n rotations and n reflections.

2. In the group of symmetries of a regular n-sided polygon, if n is odd there are n rotations and n reflections.

3. If permutations σ and τ leave a given coloring C fixed, then so does permutation $\sigma \circ \tau$.

4. The group of symmetries of a triangle is the group S_3.

5. All equivalence classes of an equivalence relation on a group contain the same number of elements.

6. The number of elements in the stabilizer G_c is the same for each coloring C.

7. When two colors are used, the permutation $(13)(2)(4)$ leaves 9 colors fixed.

8. When three colors are used, the permutation $(13)(2)(4)$ leaves 27 colors fixed.

9. Using two colors, there are 6 distinct colorings of the vertices of a square.

10. The pattern inventory is the sum of the inventory of fixed colorings.

ANSWERS TO TRUE-FALSE

1. T **2.** T **3.** T **4.** T **5.** F **6.** F **7.** F **8.** T **9.** T **10.** F

CHAPTER 20
Rings, Integral Domains, and Fields

20.1 Rings and Integral Domains

In the sections on lattices and Boolean algebras, we had two binary operations and described certain relationships between them. We are going to do so again. This time we generalize the integers. In fact we shall denote the operations $+$ and $\cdot$. Thus, we combine many of the properties of the integers and call any set with two operations that satisfies these properties a ring. Once we have defined a ring, we shall then devote some time in trying to get back to the integers.

DEFINITION 20.1

A **ring** is a nonempty set R together with binary operations called multiplication and addition, denoted by $\cdot$ and $+$, respectively, which satisfies the following conditions:

(1) R is closed under addition; that is, if $x \in R$ and $y \in R$, then $x + y \in R$.
(2) Addition in R is associative; that is, $x + (y + z) = (x + y) + z$ for all x, y, and z in R.
(3) R has an additive identity 0 such that $x + 0 = 0 + x = x$ for all x in R.
(4) R contains an additive inverse $-x$ for each element x in R such that $x + (-x) = -x + x = 0$.
(5) Addition in R is commutative; that is, $x + y = y + x$ for all x and y in R.
(6) R is closed under multiplication; that is, if $x \in R$ and $y \in R$, then $x \cdot y \in R$.
(7) Multiplication in R is associative; that is, $x \cdot (y \cdot z) = (x \cdot y) \cdot z$ for all x, y, and z in R.
(8) The following distributive laws hold for all x, y, and z in R:
$$x \cdot (y + z) = (x \cdot y) + (x \cdot z)$$
and
$$(y + z) \cdot x = (y \cdot x) + (z \cdot x)$$

If there is an element 1 in R such that $1 \cdot r = r \cdot 1 = r$ for all r in R, then R is called a **ring with unity**. If $r \cdot r' = r' \cdot r$ for all r and r' in R, then R is called a **commutative ring**.

Note that a ring R is a group under addition and a semigroup under multiplication. Examples of rings are the integers, real numbers, rational numbers, and complex numbers under the usual addition and multiplication. Other examples include the set of $n \times n$ matrices for a fixed integer n, and the set of polynomials, both with the normal addition and multiplication. The matrices are the only ring of those given previously that is not a commutative ring.

DEFINITION 20.2   An **integral domain** is a commutative ring with unity not equal to 0 such that $ab = 0$ implies that $a = 0$ or $b = 0$.

The integers, the rational numbers, and the real numbers are examples of integral domains. The set of 2×2 matrices is not an integral domain since

$$\begin{bmatrix} 0 & 1 \\ 0 & 0 \end{bmatrix} \begin{bmatrix} 1 & 0 \\ 0 & 0 \end{bmatrix} = \begin{bmatrix} 0 & 0 \\ 0 & 0 \end{bmatrix}$$

so that the product of two nonzero matrices may be the zero matrix.

Integral domains have the multiplicative cancellation property, which means that if $ab = ac$ and $a \neq 0$, then $a = b$. This is important since rings do not necessarily have inverses so we cannot multiply both sides of the equation $ab = ac$ by a^{-1} to get $a = b$. The proof of the following theorem is left to the reader.

THEOREM 20.3   Let R be a commutative ring with unity. R is an integral domain if and only if $ab = ac$ implies that $b = c$ for all b, c, and nonzero a in R.   ■

At this point let us examine Z_n with regard to being a ring, or an integral domain. We can easily show that Z_n is always a ring. We have already shown that it satisfies all of the laws except the distributive laws, and that is easily shown. If we look at Z_6, however, we see that $[3] \odot [2] = [0]$ and, hence, Z_6 is not an integral domain. In fact, if n is not prime, say $n = pq$, then $[p] \odot [q] = [0]$ and Z_n is not an integral domain.

If n is a prime, then given a nonzero integer $a \not\equiv 0 \pmod{n}$ we know, by Theorem 10.6, that $ax \equiv 1 \pmod{n}$ has a solution say a'. Hence, $[a] \odot [a'] = [1]$ and every element has an inverse. Therefore, if $ab = 0$ and $a \neq 0$, then $b = a^{-1}ab = a^{-1}0 = 0$. Hence, if n is prime, Z_n is an integral domain.

If n is not prime, consider the subset $R = \{[x] : x \text{ is relatively prime to } n\}$. It is easily shown that this set forms a group under multiplication. The product of two integers relatively prime to n is relatively prime to n. The identity, 1, is relatively prime to n; and if b is relatively prime to n, then $bx \equiv 1 \pmod{n}$ has a unique solution so that $[b]$ has an inverse. R is not even a semigroup under addition, however, since the sum of two numbers relatively prime to n is not necessarily relatively prime to n.

DEFINITION 20.4   Let R and R' be rings and let $f : R \to R'$ be a function from R into R'. f is said to be a **ring homomorphism** if and only if

$$f(a + b) = f(a) + f(b)$$
$$f(a \cdot b) = f(a) \cdot f(b)$$

for all $a, b \in R$. Addition and multiplication are those defined in the respective rings. If ring homomorphism $f : R \to R'$ is one-to-one, it is said to be a **monomorphism**. If a ring homomorphism $f : R \to R'$ is onto R', it is said to be an **epimorphism**. A ring homomorphism $f : R \to R'$ is said to be an

isomorphism provided that f is both one-to-one and onto R'. Normally, when describing a homomorphism from a ring R with unity to a ring R' with unity, it is required that the multiplicative identity of R be mapped to the multiplicative identity of R'.

For example, consider the ring of integers Z and the ring

$$Z_7 = \{[0], [1], [2], [3], [4], [5], [6]\}$$

of integers modulo 7. Suppose that $f : Z \to Z_7$ is defined by $f(a) = [a]$; that is, $f(a)$ is the equivalence class of integers congruent to a modulo 7. If each of a and b is an integer, then $f(ab) = [ab] = [a][b] = f(a)f(b)$ and $f(a+b) = [a+b] = [a]+[b] = f(a)+f(b)$ using the definition of multiplication and addition of equivalence classes of Z_7. Thus, f is a homomorphism. f is clearly an epimorphism because $f(i) = [i]$ for $0 \le i \le 6$. f is not one-to-one because $7 \equiv 0 \pmod{7}$ and, consequently, $f(7) = [7] = [0] = f(0)$. Therefore, f is neither a monomorphism nor an isomorphism. The same argument may be applied, of course, to rings Z and Z_n for any positive integer $n > 1$.

Rings that are isomorphic have the same algebraic structure and differ only in the naming of their elements.

THEOREM 20.5 In a ring R, $a \cdot 0 = 0$ for all a in R.

Proof The proof is left to reader. ∎

DEFINITION 20.6 A subset R' of a ring R is called a **subring** of R if R' is a ring with the same operations.

For example, the integers are a subring of the rational numbers. The rational numbers are a subring of the real numbers. The real numbers are a subring of the complex numbers. The set of $n \times n$ matrices of integers form a subring of the set of $n \times n$ matrices of rational numbers. One can easily check that $\{[0], [2], [4]\}$ is a subring of Z_6.

At this point we shall define a field, although we shall really use it in a later chapter.

DEFINITION 20.7 A **field** is a commutative ring with unity not equal to 0 such that every nonzero element has a multiplicative inverse.

The rational numbers, real numbers, and complex numbers are fields. If we consider just the positive reals and positive rationals, we have an integral domain. In many cases, an integral domain may not be a field, but it is a subring of a field. We saw earlier that for a prime number n, every element of Z_n has an inverse, so that Z_n is a field.

The following theorem gives us the relationship between an integral domain and a field.

THEOREM 20.8 A field is an integral domain. A finite integral domain is a field.

Proof If F is a field and $ab = 0$ for $a, b \in F$, and if $a \ne 0$, then

$$a^{-1}ab = a^{-1}0 = 0$$

But $a^{-1}ab = b$, so $b = 0$, and F is an integral domain. Conversely, if D is a finite integral domain with elements $0, 1, a_1, a_2, a_3, \ldots, a_n$ and $a = a_i$ for some $0 \leq i \leq n$, then $a \cdot 0, a \cdot 1, a \cdot a_1, a \cdot a_2, a \cdot a_3, \ldots, a \cdot a_n$ are all distinct elements. Since if $a \cdot a_j = a \cdot a_k$, then $a \cdot (a_j - a_k) = 0$, so $(a_j - a_k) = 0$, and $a_j = a_k$. Therefore, $a \cdot 0, a \cdot 1, a \cdot a_1, a \cdot a_2, a \cdot a_3, \ldots, a \cdot a_n$ are the complete set of elements of D, so $a \cdot a_k = 1$ for some k for $0 \leq k \leq n$, so a has an inverse. Therefore, D is a field. ∎

Note that the integers are an integral domain but are not a field. Hence, all integral domains are not fields. However, we shall now show that every integral domain is embedded in a field. By this we mean that given an integral domain A, there is a field F and a monomorphism from A to F.

Let A be an integral domain. In particular, A could be the set of integers Z. Consider the set of ordered pairs

$$P = \{(a, b) : (a, b) \in A \times A \text{ and } b \neq 0\}$$

and define the relation $\sim$ on P as follows.

DEFINITION 20.9   If each of (a, b) and (c, d) is in P, then $(a, b) \sim (c, d)$ if and only if $ad = bc$.

For example, if $A = Z$, the equivalence class $[(2, 3)]$ contains these ordered pairs: $(2, 3), (4, 6), (6, 9), \ldots, (-2, -3), (-4, -6), \ldots$, which correspond to the rational number representations $2/3, 4/6, 6/9, \ldots, (-2)/(-3), (-4)/(-6), \ldots$. All of these are different ways of expressing the same rational number, $[(2, 3)]$.

THEOREM 20.10   The relation $\sim$ on P is an equivalence relation.

Proof   The proof is straightforward and is left to the reader. ∎

Notation   Let the equivalence class containing $(a, b) \in P$, namely $[(a, b)]$, be denoted by a/b. Let F denote the set of equivalence classes on P.

DEFINITION 20.11   Given elements a, b, c, and d of A, let addition be defined on F by $a/b + c/d = (ad + bc)/bd$, and let multiplication on F be defined by $(a/b)(c/d) = ac/bd$.

The next theorem shows that the definitions of addition and multiplication on F are independent of the representatives of the equivalence classes used. When the definitions are independent of the representatives, the operations are well defined.

THEOREM 20.12

(a) Addition in F is well defined.

(b) Multiplication in F is well defined.

Proof   Let $a/b = a'/b'$ and $c/d = c'/d'$. We have immediately that $ab' = a'b$ and $cd' = c'd$.

(a) By definition of addition we have $a/b + c/d = (ad + bc)/bd$ and $a'/b' + c'/d' = (a'd' + b'c')/b'd'$. We need to show that $(ad + bc)/bd = (a'd' + b'c')/b'd'$ or, equivalently, that $b'd'(ad + bc) = bd(a'd' + b'c')$, but

$$b'd'(ad + bc) = b'add' + cd'bb'$$
$$= a'bdd' + c'dbb'$$
$$= bd(a'd' + b'c')$$

where the second equality holds because $ab' = a'b$ and $cd' = c'd$.

(b) The definition of multiplication gives both $(a/b)(c/d) = ac/bd$ and $(a'/b')(c'/d') = a'c'/b'd'$. It is necessary to prove that $ac/bd = a'c'/b'd'$ or, equivalently, that $acb'd' = a'c'bd$. But $acb'd' = ab'cd' = a'bc'd = a'c'bd$. ∎

THEOREM 20.13 The set of equivalence classes F is a commutative ring with additive identity $0/1$ and multiplicative identity $1/1$.

Proof The proof is straightforward and is left to the reader. ∎

Lemma 20.14 For a and b in A, $a/b = 0/1$ if and only if $a = 0$.

Proof Clearly, $a/b = 0/1$ if and only if $a = a(1) = b(0) = 0$. ∎

THEOREM 20.15 The commutative ring F is a field.

Proof If $a/b \neq 0$, then $a \neq 0$ and $b/a \in F$. But
$$(a/b)(b/a) = ab/ab = 1/1 = 1$$
Hence, every nonzero element of F has an inverse, and F is a field. ∎

Let $f : A \to F$ be defined by $f(a) = a/1$. $f(ab) = ab/1 = (a/1)(b/1) = f(a)f(b)$ and $f(a+b) = (a+b)/1 = (a(1)+b(1))/1 = a/1+b/1 = f(a)+f(b)$. Also, f is one-to-one for if $f(a) = f(b)$, then $a/1 = b/1$ and $a = a1 = b1 = b$. Hence, f is an monomorphism and the elements in A may be identified with the elements $\{a/1 : a \in A\}$ in F. Thus, A may be considered to be a subring of F.

THEOREM 20.16 The mapping $f : A \to F$ defined by $f(a) = a/1$ is a monomorphism and we say that the integral domain A is **embedded** in the field F or that F contains A. ∎

It is easily shown that F is the smallest field in which A can be embedded.

DEFINITION 20.17 F is called the **fraction field** of A. If A is the set of integers Z, then F is the set of **rational numbers** usually denoted by Q.

In this section further algebraic structures of the integers are developed and the integers are determined algebraically up to isomorphism. Notice that this development does not imply the existence of the integers but only their algebraic properties if they do exist.

DEFINITION 20.18 A subset I of a ring R is called an **ideal** of R if
(a) I is a subring of R.
(b) If x is in I and r is in R, then $x \cdot r$ and $r \cdot x$ are in I.
An ideal is a **proper ideal** if $I \neq \{0\}$ and $I \neq R$.

For example, in the ring of integers, the set of all multiples of a fixed integer p forms an ideal.

DEFINITION 20.19 Let R be a commutative ring. An ideal I of R is called a **principal ideal** generated by a if I consists of all products of a by elements of R, that is, $I = \langle a \rangle = \{ar : r \in R\}$.

THEOREM 20.20 Every nonempty ideal I of the ring of integers is a principal ideal.

Proof For $I \neq \{0\}$, let p be the least positive integer in I and let m belong to I. By the division algorithm, $m = pq + r$ where $0 \leq r < p$. Since $r = m - pq$, r is in I. Thus, since $r < p$ and p is the smallest positive integer in I, $r = 0$. Hence, every integer in I is a multiple of p and $I = \langle p \rangle$. ∎

Note that the proof of this theorem depends upon the well-ordering principle and the division algorithm, both of which are available for integers.

EXAMPLE 20.21 Consider the ring Z of integers and the two principal ideals generated by the integers 8 and 12:

$$\langle 8 \rangle = \{8r : r \in Z\}$$
$$= \{\ldots, -24, -16, -8, 0, 8, 16, 24, \ldots\}$$
$$\langle 12 \rangle = \{12s : s \in Z\}$$
$$= \{\ldots, -24, -12, 0, 12, 24, \ldots\}$$

The intersection $\langle 8 \rangle \cap \langle 12 \rangle$ is the set

$$\{\ldots, -48, -24, 0, 24, 48, \ldots\}$$

which is the principal ideal generated by the integer 24. Note that 24 is the least common multiple of 8 and 12. In general,

$$\langle a \rangle \cap \langle b \rangle = \langle \mathrm{lcm}(a, b) \rangle$$

Results pertinent to this example are given in the next theorem. The proof is left to the reader. ∎

THEOREM 20.22 If s and t are nonzero integers and $\langle s \rangle$ and $\langle t \rangle$ are the corresponding principal ideals in the ring Z, then

(a) If $\langle s \rangle \subseteq \langle t \rangle$, then $t \mid s$.
(b) $\langle s \rangle \cap \langle t \rangle = \langle u \rangle$ where $u = \mathrm{lcm}(s, t)$. ∎

EXAMPLE 20.23 If $\langle a, b \rangle$ is the smallest ideal containing a and b, then $\langle a, b \rangle = \langle \gcd(a, b) \rangle$. For example,

$$\langle 8, 12 \rangle = \{-16, -12, -8, -4, 0, 4, 8, 12, 16, \ldots\}$$
$$= \langle \gcd(8, 12) \rangle$$

since $\gcd(8, 12) = 4$ and

$$\langle 4 \rangle = \{\ldots, -8, -4, 0, 4, 8, \ldots\}$$

Let I be the smallest ideal containing integers a and b. Then every element of I is of the form $am + bn$, where m and n are integers. The ideal I is said to be generated by a and b. By Theorem 20.22, I is generated by the smallest positive number of this form. But this number is the greatest common divisor of a and b, as mentioned earlier. This idea may be extended to an ideal generated by any finite number of positive integers. The ideal is again generated by the greatest common divisor of these numbers. ∎

Chapter 20 Rings, Integral Domains, and Fields

The proof of the following theorems is left to the reader.

THEOREM 20.24 An I ideal of R with unity is equal to R if and only if 1, the multiplicative identity of R, is in I. ∎

THEOREM 20.25 A field has no proper ideals. ∎

DEFINITION 20.26 An ideal I of a commutative ring R is a **prime ideal** if $ab \in I$ implies that either $a \in I$ or $b \in I$.

THEOREM 20.27 In the ring of integers an ideal $\langle a \rangle$ is a prime ideal if and only if a is a prime number.

Proof Assume that a is not prime, say $a = rs$ where r and s are both integers greater than 1. Then a is in $\langle a \rangle$ but r and s are not. If a is prime and $rs \in \langle a \rangle$, then a divides rs; hence, by Theorem 3.44, a divides r or a divides s. Hence, either r is in $\langle a \rangle$ or s is in $\langle a \rangle$. ∎

DEFINITION 20.28 An integral domain D is a **principal ideal domain** if every ideal in D is a principal ideal.

We have already shown that the integral domain Z of integers is a principal ideal domain.

DEFINITION 20.29 If A is a commutative ring with unity, let $A^{\#}$ denote the set $\{a \in A : \text{there exists } b \in A \text{ with } ab = 1\}$. The subset $A^{\#}$ is a group under multiplication called the *group of units* of A. Each element of $A^{\#}$ is called a *unit* of A. In a ring with unity, an element s is *irreducible* if it is nonzero, is not a unit, and cannot be expressed as a product of two nonunits.

Units are just divisors of 1. In the integral domain Z, $ab = 1$ only if $a = b = 1$ or $a = b = -1$ so that 1 and -1 are units in Z. In a field, every nonzero element is a unit because $a \cdot a^{-1} = 1$ for $a \neq 0$.

EXAMPLE 20.30 The set $Z_6 = \{[0], [1], [2], [3], [4], [5]\}$ is a commutative ring with unity $[1]$ and zero $[0]$. The multiplication table in Z_6 is

$\odot$	[0]	[1]	[2]	[3]	[4]	[5]
[0]	[0]	[0]	[0]	[0]	[0]	[0]
[1]	[0]	[1]	[2]	[3]	[4]	[5]
[2]	[0]	[2]	[4]	[0]	[2]	[4]
[3]	[0]	[3]	[0]	[3]	[0]	[3]
[4]	[0]	[4]	[2]	[0]	[4]	[2]
[5]	[0]	[5]	[4]	[3]	[2]	[1]

From the table we see that $[3] \odot [2] = [0]$, but $[3] \nmid [0]$ and $[2] \nmid [0]$ so that $[3]$ and $[2]$ are nonzero divisors of $[0]$. Thus, Z_6 is not an integral domain. ∎

In the previous example it is seen that the cancellation property does not hold since $[3] \odot [1] = [3] = [15] = [3] \odot [5]$ and, although $[3] \neq [0]$, we have $[1] \neq [5]$. The units of Z_6 correspond to integers relative prime to 6; therefore, as the multiplication foregoing indicates, the units of Z_6 are $[1]$ and $[5]$.

In the ring $Z_5 = \{[0], [1], [2], [3], [4]\}$, with unity $[1]$ and zero $[0]$, every nonzero member of Z_5 is a unit since it is relatively prime to the prime 5. Hence, every element has an inverse so that Z_5 is a field. See Example 3.62 for operations tables for Z_5.

Exercises

1. Prove that if F is a field, then the equation $ax = b$ has a unique solution in the field for $a \neq 0$.

2. An element a of a ring R is a zero divisor if there exists a $b \neq 0$ in R such that $ab = 0$. Prove or disprove that the sum of two elements that are not zero divisors is not a zero divisor.

3. Prove or disprove that the product of two elements in a ring R that are not zero divisors is not a zero divisor.

4. What are the zero divisors in Z_{10}?

5. What are the zero divisors in Z_7?

6. Prove Theorem 20.25: A field has no proper ideals.

7. Prove Theorem 20.3: Let R be a commutative ring with unity. R is an integral domain if and only if $ab = ac$ implies that $b = c$ for all b, c, and nonzero a in R.

8. Let M be the ring of 2×2 matrices of the form $\begin{bmatrix} a & b \\ c & 0 \end{bmatrix}$. Prove that the set of all matrices of the form $\begin{bmatrix} a & b \\ 0 & 0 \end{bmatrix}$ is an ideal of M.

9. Prove that the sum of ideals I and J of a commutative ring R defined by $I + J = \{i + j \mid i \in I \text{ and } j \in J\}$ is an ideal of R.

10. Let $A = \langle 14 \rangle$ and $B = \langle 16 \rangle$. Find $A \cap B$ and $A + B$.

11. Describe the smallest ideal of the integers containing 6, 9, and 12.

12. Prove that if I is an ideal of a ring R and $-1 \in I$, then $I = R$.

13. Let $f : R \to R'$ be a ring homomorphism. Prove that $f(R)$ is a ring.

14. Let $f : R \to R'$ be a ring homomorphism. Prove that if R is a field, then $f(R)$ is a field.

15. Let $f : R \to R'$ be a ring homomorphism. If R is an integral domain, is $f(R)$ an integral domain?

16. Let $f : R \to R'$ be a ring homomorphism. Prove that

$$\{x \in R \mid f(x) = 0\}$$

is an ideal of R.

17. Prove or disprove that the zero divisors of a commutative ring R form an ideal of R.

18. Does the set of polynomials with integer coefficients form an integral domain?

19. An element a of a ring R is called an idempotent if $a^2 = a$. Prove that if R is an integral domain, then the only idempotents are 0 and 1.

20. Prove Theorem 20.24: An I ideal of R is equal to R if and only if 1, the multiplicative identity of R, is in I.

21. Prove that for all a, b in a ring R, $a(-b) = -(ab) = (-a)b$ and $(-a)(-b) = ab$.

20.2 Integral Domains

DEFINITION 20.31

An integral domain D is said to be a *unique factorization domain* provided that

(a) If an element of D is not zero and not a unit, then it can be factored into a product of a finite number of irreducibles.

(b) If an element of D has factorizations $p_1 \cdots p_r$ and $q_1 \cdots q_s$ as products of irreducibles, then $r = s$ and the q_j can be renumbered so that the p_i and q_i differ only by a unit for all i; that is, $p_i = a_i q_i$ for some unit a_i.

728 Chapter 20 Rings, Integral Domains, and Fields

We already know that the integers are a unique factorization domain. We have defined a prime integer as one that has no nontrivial factors. In other words, a prime integer is irreducible. An alternative definition for a prime integer is that p is prime if and only if $p \mid ab$ implies that $p \mid a$ or $p \mid b$. These definitions are equivalent for the integers but not for arbitrary integral domains. For example, let A be the set of all complex numbers of the form

$$a + b\sqrt{5}i$$

It is easily shown that A is an integral domain and that

$$21 = 3 \cdot 7 = (1 + 2\sqrt{5}i)(1 - 2\sqrt{5}i)$$

All of these factor are irreducible and, hence, A is not a unique factorization domain. However, none of them is a prime using the alternative definition.

DEFINITION 20.32 A commutative ring A with unity is an *ordered ring* if and only if there exists a nonempty subset, A^+, of A, called the subset of positive elements of A, such that
(a) If $a, b \in A^+$, then $a + b \in A^+$.
(b) If $a, b \in A^+$, then $a \cdot b \in A^+$.
(c) Given any $a \in A$, one and only one of the following alternatives holds for a:
   (i) $a \in A^+$
   (ii) $a = 0$
   (iii) $-a \in A^+$

A commutative ring with unity that has such a set A^+ is said to satisfy the trichotomy axiom. If $a \in A^+$, then we say that $a > 0$. If $-a \in A^+$, then we say that $a < 0$.

The proof of the following theorem is left to the reader.

THEOREM 20.33 Every ordered ring is an integral domain. Given any $a \neq 0$, $a^2 > 0$. In particular, $1^2 > 0$ and, hence, $1 > 0$. ∎

DEFINITION 20.34 An ordered integral domain A is called **well ordered** if and only if any nonempty subset S of A^+ has a first element; that is, there exists $s \in S$ such that if $t < s$, then $t \notin S$.

We have already seen that the integers form a well-ordered integral domain.

THEOREM 20.35 If A is a well-ordered integral domain, then there is no element c of A such that $0 < c < 1$.

Proof Let S be the subset of all c in A such that $0 < c < 1$. If S is not empty, then there exists a least element s in S. But s^2 is in A; and since $s > 0$, $s^2 > 0$. Since $s < 1$, $0 < s$ implies that $s^2 < s$ and $0 < s^2 < s$. Therefore, $s^2 \in S$ and $s^2 < s$, which contradicts the fact that s is the least element of S. Hence, S is empty. ∎

DEFINITION 20.36 In an ordered integral domain I, let $\overline{n+1} = \overline{n} + 1$, where 1 is the multiplicative identity.

THEOREM 20.37 In any ordered integral domain A, the following are equivalent for A^+:

1. First principle of induction
2. Well-ordering principle
3. Second principle of induction

Proof (1) → (2) Let $S = \{\bar{n} \in A^+ : \text{if } \bar{n} \in T \text{ where } T \text{ is a subset of } A^+, \text{ then } T$ has a least element$\}$. We want to show that $S = A^+$. Obviously, $1 \in S$, for if $1 \in T$, then 1 is the least element of T. Assume that $k \in S$ and let $k + 1 \in T$. We wish to show that T has a least element. If $1 \in T$, then T has a least element and we are finished. Assume that $1 \notin T$. Let $T' = \{s : s + 1 \in T\}$. Since $k \subset T'$, T' has a least element, say r. By definition, $r + 1 \in T$. Also, $r + 1$ is the least element in T, for, if $u \in T$ and $u < r + 1$, then since $u = v + 1$ for some $v \in T'$, we obtain $v < r$, which is a contradiction. Hence, T has a least element, $k + 1 \in S$ and $S = A^+$.

(2) → (3) Assume that S is a subset of A^+ with the properties that

(a) $1 \in S$;
(b) If $k \in A^+$ and $s \in S$ for all $s < k$, then $k \in S$.

We want to show $S = A^+$. Choose S with the preceding properties. Let $T = \{\bar{n} : \bar{n} \in A^+ \text{ and } \bar{n} \notin S\}$. We want to show that T is empty. If T is not empty, by the well-ordering principle, T contains a least element, say t, and $t \neq 1$ because 1 is in S and t is not. Hence, $t = r + 1$ for some r. The set of all $x \in A^+$ such that $x < t$ is nonempty since r is in it. All of these elements belong to S since t is the least positive integer not belonging to S. Hence, by part (b), $t \in S$ which is a contradiction; therefore, T is empty.

(3) → (1). Let S be a subset of A^+. Let $p(1)$ be the statement "$\bar{n} \in S$ implies that $(n + 1) \in S$" and $p(2)$ be the statement "If for all $m < \bar{n}$ with $m \in S$, then $\bar{n} \in S$." Obviously, $p(1)$ implies $p(2)$. The second principle of induction may be stated: "If $1 \in S$ and $p(2)$, then $S = A^+$" and the first principle of induction may be stated: "If $1 \in S$ and $p(1)$, then $S = A^+$." Since $1 \in S$ and $p(1)$ implies that $1 \in S$ and $p(2)$, then the statement "$1 \in S$ and $p(2)$ implies that $S = A^+$" implies the statement "$0 \in S$ and $p(1)$ implies that $S = A^+$)." (This result is true because $(p \to q) \implies ((q \to r) \to (p \to r))$.) ∎

THEOREM 20.38 Any two well-ordered, ordered integral domains are isomorphic and, therefore, are isomorphic to Z.

Proof Let A be a well-ordered, ordered integral domain and N the set of positive integers. Let I be the multiplicative identity of A. Define $f : N \to A^+$ by $f(1) = I$ and if $f(k) = K$, then $f(k + 1) = K + I$. It is easily shown that $f(N)$ satisfies the first principle of induction and, hence, $f(N) = A^+$. We need to show that $f(m + n) = f(m) + f(n)$ and $f(m \cdot n) = f(m)f(n)$.

Using induction on n, we first show that $f(m + n) = f(m) + f(n)$. For $n = 1$, $f(m + 1) = f(m) + I$ by definition of f, so $f(m + 1) = f(m) + f(1)$. Assume that $f(m + k) = f(m) + f(k)$. Then

$$f(m + (k + 1)) = f((m + k) + 1)$$
$$= f(m + k) + I$$
$$= f(m) + f(k) + I$$
$$= f(m) + f(k + 1)$$

and

$$f(m + n) = f(m) + f(n)$$

for all $m, n \in N$.

We now show that $f(m \cdot n) = f(m)f(n)$ by induction on n. For $n = 1$,
$$f(m \cdot 1) = f(m)$$
$$= f(m) \cdot I$$
$$= f(m)f(1)$$
Assume that $f(m \cdot k) = f(m)f(k)$. Then
$$f(m \cdot (k+1)) = f((m \cdot k) + m)$$
$$= f(m \cdot k) + f(m)$$
$$= f(m)f(k) + f(m)$$
$$= f(m)f(k) + f(m)I$$
$$= f(m)(f(k) + I)$$
$$= f(m)f(k+1)$$
and $f(m \cdot n) = f(m)f(n)$ for all $m, n \in N$. Therefore, f is an isomorphism. This isomorphism is easily extended to an isomorphism from Z to A. ∎

DEFINITION 20.39 An integral domain is called a **minimal domain** if and only if it has no subdomains except itself.

A minimal domain may be found by taking the intersection of all subdomains of an integral domain. The proof of the following theorem is left to the reader.

THEOREM 20.40 Any two ordered minimal integral domains are isomorphic. They are isomorphic to the integers and, hence, are well ordered. ∎

Exercises

1. Prove that the intersection of all subdomains of an integral domain is an integral domain and use this to show that every integral domain contains a minimal integral domain.
2. Prove that any two ordered minimal integral domains are isomorphic. Prove that they are isomorphic to the integers and hence are well ordered.
3. In an ordered integral domain prove that for any nonzero element a, $a^2 > 0$. In particular $1 > 0$ (Theorem 20.33).

20.3 Polynomials

In this section we formally define polynomial forms and study some of their algebraic properties. Polynomial functions $f(x)$ and $g(x)$ are equal provided that $f(b) = g(b)$ for all b in the common domain. Polynomial forms f and g are equal provided that corresponding powers of x have equal coefficients. For polynomials over the integers, the two concepts are the same; however, such is not the case over all commutative rings with unity. We will reconcile the two concepts of equality at the end of this section. Most of the following discussions are in terms of commutative rings with unity and integral domains, but we will usually be using the ring of integers or the rational numbers. To aid in understanding, one may wish to think of A as being the integers in much of the following development.

In the following, we see that we are really defining polynomials as generating functions.

20.3 Polynomials

DEFINITION 20.41 Let A be a commutative ring with unity and let S be the set of all sequences $(a_0, a_1, a_2, \ldots)$ of elements of A such that if $f \in S$, then there exists an integer N_f so that $a_j = 0$ for all $j > N_f$. If $f \in S$, then f is said to be a **polynomial** or a **polynomial form** over A.

For example,
$$(1, 0, 0, 0, 0, \ldots)$$
and
$$(1, 1, 1, 0, 0, 0, \ldots)$$
are in S, but
$$(1, 1, 1, 1, 1, 1, 1, 1, \ldots)$$
and
$$(1, 0, 1, 0, 1, 0, 1, 0, 1, \ldots)$$
are not in S. If $A = Z$, then
$$(0, -5, 2, 0, 0, \ldots)$$
and
$$(3, 7, 5, 8, 0, 0, \ldots)$$
are polynomials in S.

DEFINITION 20.42 Let A be a commutative ring and let $f = (a_i)^* = (a_0, a_1, a_2, \ldots)$ and $g = (b_i)^* = (b_0, b_1, b_2, \ldots)$ be in S, the set of polynomials over A. Define the **sum** of f and g to be the sequence $f + g = (a_i + b_i)^* = (a_0 + b_0, a_1 + b_1, \ldots)$ and the **product** of f and g to be the sequence $fg = (c_k)^*$, where $c_k = \sum_{i+j=k} a_i b_j$.

THEOREM 20.43 For $f, g \in S$, the set of polynomials over a commutative ring A with unity, $f + g \in S$ and $fg \in S$.

Proof The proof for $f + g$ is left to the reader. To show that $fg \in S$, note that for any $k > N_f + N_g + 1$, if $i + j = k$ and $i \leq N_f$, then $j > N_g + 1$, and if $j \leq N_g$, then $i > N_f + 1$. Hence, $c_k = 0$ for any $k > N_f + N_g + 1$ since either $a_i = 0$ or $b_j = 0$ in the sum c_k. ∎

The proof of the following theorem is left to the reader.

THEOREM 20.44 If S is the set of polynomials over a commutative ring A with unity, then S is also a commutative ring with unity. Its unit is $(1, 0, 0, 0, \ldots)$ and its zero element is $(0, 0, 0, \ldots)$. ∎

DEFINITION 20.45
(a) Let A be a commutative ring with unity, and let $f \in S$ and $f = (a_i)^*$. If $f \neq 0$, let $\deg(f)$ equal the largest integer k, such that $a_k \neq 0$. The function $\deg(f)$ is called the **degree** of f. S is called the **ring of polynomials** over A. The set S is denoted by $A[x]$. Any element of $A[x]$ is called a **polynomial** over A. Any polynomial that is of degree 0 or is equal to zero is called a **constant**.

(b) Let $f = (a_0, a_1, a_2, \ldots)$ belong to $A[x]$. The terms a_i of the sequence are called the **coefficients** of f. If $f \neq 0$ and $n = \deg(f)$, then a_n is called the **leading coefficient** of f. If $a_n = 1$, then f is called a **monic** polynomial. If

> $f \neq 0$ has the property that any greatest common divisor of all of its nonzero coefficients is a unit, then f is called a **primitive** polynomial.
>
> (c) Two elements f and g of $A[x]$ are **equal**, written $f = g$, if their respective coefficients are equal; that is, if $f = (a_0, a_1, a_2, \ldots)$ and $g = (b_0, b_1, b_2, \ldots)$, then $f = g$ if and only if $a_i = b_i$ for every nonnegative integer i.
>
> (d) The polynomial f **divides** the polynomial g provided that there is a polynomial h such that $fh = g$. In this case, we say that f and h are **factors** of g.

For example, if
$$f = (0, 1, 1, 0, 1, 0, 0, \ldots)$$
then $\deg(f) = 4$. If
$$g = (1, 0, 0, 0, \ldots)$$
then $\deg(g) = 0$ and g is a constant polynomial. The zero polynomial
$$(0, 0, 0, \ldots)$$
has no degree. It is left to the reader to show that equality is an equivalence relation on $A[x]$.

THEOREM 20.46 Let A be a commutative ring with unity, let $A[x]$ be the ring of polynomials over A, and let f and g be in $A[x]$.

(a) If $f, g \neq 0$, then $\deg(f + g) \leq \max(\deg(f), \deg(g))$.

(b) Either $fg = 0$ or $\deg(fg) \leq \deg(f) + \deg(g)$.

(c) If A is an integral domain, then either $fg = 0$ or $\deg(fg) = \deg(f) + \deg(g)$.

(d) If A is an integral domain, then so is $A[x]$.

Proof (a) Let $f = (a_0, a_1, a_2, \ldots)$ and $g = (b_0, b_1, b_2, \ldots)$, where $n = \deg(f)$ and $m = \deg(g)$. Since $a_i = 0$ for $i > n$ and $b_j = 0$ for $j > m$, if $k > \max(m, n)$, then $a_k + b_k = 0$ and $\deg(f + g) \leq \max(\deg(f), \deg(g))$.

(b) Let $n = \deg(f)$, $m = \deg(g)$, and $fg = (c_0, c_1, c_2, \ldots)$ so that $c_k = \sum_{i+j=k} a_i b_j$. The product $a_i b_j = 0$ when $i > n$ since $a_i = 0$. If $i \leq n$, and $i + j > m + n$, then $j > m + (n - i)$ so $b_j = 0$ and, in either case, $c_k = 0$. Hence, $\deg(fg) \leq \deg(f) + \deg(g)$.

(c) By part (b), $\deg(fg) \leq \deg(f) + \deg(g)$. However, since a_n and b_m are not 0, then $c_{n+m} = a_n b_m$ is not 0. Hence, $\deg(fg) \geq \deg(f) + \deg(g)$ and we have equality.

(d) Let f and g be nonzero elements of $A[x]$ with degree n and m, respectively. Since A is an integral domain, $\deg(fg) = \deg(f) + \deg(g)$ and fg is not 0. ∎

THEOREM 20.47 There is a monomorphism from A into $A[x]$, the ring of polynomials over A, which makes the image of A a subring of $A[x]$. If A is an integral domain, then any unit of $A[x]$ corresponds to a unit of A via the monomorphism.

Proof Define the function $\phi : A \to A[x]$ as follows: If $a \in A$, let $\phi(a) = (a, 0, 0, 0, \ldots)$. If a and b are in A, then

$$\phi(a+b) = (a+b, 0, 0, 0, 0, \ldots)$$
$$= (a, 0, 0, 0, 0, \ldots) + (b, 0, 0, 0, 0, \ldots)$$
$$= \phi(a) + \phi(b)$$
$$\phi(a \cdot b) = (a \cdot b, 0, 0, 0, \ldots)$$
$$= (a, 0, 0, 0, \ldots) \cdot (b, 0, 0, 0, \ldots)$$
$$= \phi(a) \cdot \phi(b)$$

Also, $\phi(1) = (1, 0, 0, 0, \ldots) = 1$. Hence, ϕ is a homomorphism from A into $A[x]$.

If $(c, 0, 0, 0, \ldots) = \phi(c) = 0 = (0, 0, 0, 0, \ldots)$, then $c = 0$. If $\phi(a) = \phi(b)$, then $\phi(a) - \phi(b) = 0$; that is, $\phi(a - b) = 0$ and $a - b = 0$. Therefore, $a = b$ and ϕ is one-to-one.

Recall that a unit of a ring is an element that has an inverse; that is, if b is a unit there exists an element b^{-1} so that $b \cdot b^{-1} = 1$.

Assume that A is an integral domain and f is a unit in $A[x]$. Let g be its inverse, so $fg = 1 = (1, 0, 0, 0, \ldots)$. But $0 = \deg(fg) = \deg(f) + \deg(g)$, which implies that $\deg(f)$ and $\deg(g)$ are both 0. Thus, $f = (a, 0, 0, 0, \ldots)$ and $g = (b, 0, 0, 0, \ldots)$ for some $a, b \in A$. Since

$$\phi(a \cdot b) = fg = (1, 0, 0, 0, \ldots) = \phi(1)$$

and since ϕ is one-to-one, $ab = 1$ and a and b are units in A.

Because the monomorphic image $\phi(A)$ in $A[x]$ is a ring and is isomorphic with A, we identify A with $\phi(A)$ as a subring of $A[x]$ and may refer to A as a subring of $A[x]$. ∎

DEFINITION 20.48 The Kronecker δ, δ_{ij}, for integers i and j, is defined by

$$\delta_{ij} = \begin{cases} 1 & \text{if } i = j \\ 0 & \text{if } i \neq j \end{cases}$$

For example, if $f = (a_0, a_1, a_2, \ldots)$ where $a_i = \delta_{i3}$, then

$$f = (0, 0, 0, 1, 0, 0, \ldots)$$

If $g = \sum_{j=1}^{5} (\delta_{ij})^*$, then

$$g = (0, 1, 1, 1, 1, 1, 0, 0, \ldots)$$

THEOREM 20.49 If $x = (0, 1, 0, 0, 0, \ldots) = (c_i)^*$ where $c_i = \delta_{i1}$, then for each $k > 0$, $x^k = (a_0, a_1, a_2, \ldots)$, where $a_i = \delta_{ik}$.

Proof The theorem is clearly true for $k = 1$. Assume that the theorem holds for $k = n$. Since $x^{n+1} = x \cdot x^n$, simple multiplication shows the ith term in the sequence x^{n+1} to be $\delta_{i,n+1}$. ∎

If $a \in A$ and $f = (a_0, a_1, a_2, \ldots)$, then $af = (aa_0, aa_1, aa_2, \ldots)$. In particular, if $a \in A$, then $ax^k = (0, 0, 0, \ldots, a, 0, 0, 0, \ldots)$, where a occurs in the $(k+1)$st place.

Let $f \in A[x]$ be of degree n. Then
$$\begin{aligned}
f &= (a_0, a_1, a_2, \ldots, a_n, 0, 0, \cdots) \\
&= (a_0, 0, 0, \ldots) + (0, a_1, 0, \ldots) + \cdots + (0, 0, \ldots, 0, a_n, 0, 0, \ldots) \\
&= a_0(1, 0, 0, \ldots) + a_1(0, 1, 0, 0, \ldots) + \cdots + a_n(0, 0, \ldots, 0, 1, 0, 0, \ldots) \\
&= a_0 + a_1 x + a_2 x^2 + \cdots + a_n x^n
\end{aligned}$$

DEFINITION 20.50 Let A be a commutative ring with unity and let $A[x]$ be the set of polynomials over A. The symbol x is called an **indeterminate** over A. For every
$$f = a_0 + a_1 x + a_2 x^2 + \cdots + a_n x^n \in A[x]$$
there is associated a function from A to A
$$f(x) = a_0 + a_1 x + a_2 x^2 + \cdots + a_n x^n$$
or
$$f(x) = a_n x^n + \cdots + a_2 x^2 + a_1 x + a_0$$
called a **polynomial function**. Let $A(x) = \{f(x) : f \in S\}$. The degree of $f(x)$ is the same as the degree of the corresponding polynomial $f \in A[x]$. The elements f of $A[x]$ are called **polynomial forms** to distinguish them more easily from the polynomial functions.

Define $\theta : A[x] \to A(x)$ by $\theta(f) = f(x)$. It is easily shown that the map θ, which maps elements of $A[x]$ into the corresponding polynomial functions in $A(x)$, is a homomorphism. Since θ is onto, it is an epimorphism as well. Thus, all the properties of $A[x]$ preserved by homomorphisms are imbued upon the commutative ring of polynomial functions $A(x)$. We shall show shortly that if A is an infinite integral domain, then $\theta : A[x] \to A(x)$ is an isomorphism.

DEFINITION 20.51 Let A be a commutative ring with unity. If
$$f(x) = a_n x^n + \cdots + a_2 x^2 + a_1 x + a_0$$
and
$$g(x) = b_n x^n + \cdots + b_2 x^2 + b_1 x + b$$
where some of the a_i and b_i in A may be 0 including a_n or b_n, then
$$f(x) + g(x) = (a_n + b_n)x^n + \cdots + (a_1 + b_1)x + (a_0 + b_0)$$
and
$$f(x)g(x) = c_m x^m + \cdots + c_2 x^2 + c_1 x + c_0$$
where $c_k = \sum_{i+j=k} a_i b_j$. Define $f(x) = g(x)$ if and only if $f(b) = g(b)$ for all $b \in A$. A **solution** of the equation $f(x) = 0$ is an element $a \in A$ such that $f(a) = 0$.

From the preceding discussion, we have the following theorem for polynomial functions over an integral domain A.

THEOREM 20.52 If $f(x)$ and $g(x)$ are polynomial functions over an integral domain A, $f(x)$ has degree n, and $g(x)$ has degree m, then

(a) $f(x) + g(x)$ has degree less than or equal to $\max\{m, n\}$.

(b) $f(x)g(x)$ has degree $m + n$. ∎

Two polynomial forms f and g were defined to be equal if and only if their coefficients were equal. Two polynomial functions $f(x)$ and $g(x)$ were defined to be equal provided that $f(b) = g(b)$ for all $b \in A$. In the next section, we shall show for a polynomial f of degree n that $f(x) = 0$ has at most n solutions. At this point we state the theorem without proof and use it to prove a much weaker result.

THEOREM 20.53 If f is a polynomial of degree n over an infinite integral domain and $f(x)$ is the corresponding polynomial function, then $f(x) = 0$ has at most n solutions. ∎

Corollary 20.54 Let

$$f(x) = a_n x^n + \cdots + a_2 x^2 + a_1 x + a_0$$

be a polynomial function over an infinite integral domain A. If $f(a) = 0$ for all $a \in A$, then $a_0 = a_1 = a_2 = \cdots = a_n = 0$. ∎

THEOREM 20.55 Let f and g be polynomials over an infinite integral domain A. Then $f = g$ if and only if the corresponding polynomial functions $f(x)$ and $g(x)$ have the property that $f(b) = g(b)$ for every $b \in A$.

Proof If f and g are polynomials over the integral domain A such that $f = g$, then clearly $f(b) = g(b)$ for all $b \in A$. Let

$$f(x) = a_n x^n + \cdots + a_2 x^2 + a_1 x + a_0$$

and

$$g(x) = b_n x^n + \cdots + b_2 x^2 + b_1 x + b$$

be polynomial functions of degrees n and m, respectively, and $f(c) = g(c)$ for every $c \in A$. Let f and g be the respective corresponding polynomials. Then the polynomial function $h(x) = c_s x^s + \cdots + c_2 x^2 + c_1 x + c$, where $s = \max(n, m)$ and $c_i = a_i - b_i$, has the property $h(c) = f(c) - g(c) = 0$ for every $c \in A$. But by the previous corollary, if $h(c) = 0$ for all c in A, then, $h(x)$ must be the zero polynomial. Therefore, $c_i = a_i - b_i = 0$ or $a_i = b_i$ for $i \geq 1$, and $f = g$. ∎

From the previous theorem, we immediately have the following theorem that was discussed earlier.

THEOREM 20.56 Let A be an infinite integral domain and define $\theta : A[x] \to A(x)$ by $\theta(f) = f(x)$. The function θ is an isomorphism. ∎

In the following example, we show that it is necessary in the preceding corollary and theorem that A be infinite.

EXAMPLE 20.57 The set $Z_5 = \{[0], [1], [2], [3], [4]\}$ of integers modulo 5 is a field and, therefore, by Theorem 20.8, is an integral domain. By Fermat's theorem, if $a \not\equiv 0 \pmod{p}$ for p prime, then $a^{p-1} \equiv 1 \pmod{p}$. Then $a^p \equiv a \pmod{p}$ for any integer a, even if $a \equiv 0 \pmod{p}$. For $p = 5$, we have $a^5 \equiv a \pmod{5}$ for all $a \in Z$. This congruence is equivalent to $[a]^5 = [a]$ or to $[a]^5 - [a] = [0]$ for any integer a. Thus, if $g(x) = x^5 - x$, then $g(x) = [0]$ holds for any x in Z_5. But if h is the zero polynomial, then $h(x) = [0]$ for any x in Z_5. The polynomials g and h are not equal, but $g(b) = h(b)$ for every b in the integral domain Z. Thus, over a finite integral domain, equality of polynomial forms is not equivalent to equality of polynomial functions. On the other hand, since the set of integers is infinite, the concepts of equality are the same for polynomial forms and polynomial functions over Z. ∎

Exercises

1. Prove Theorem 20.44.
2. Show that equality of S in Definition 20.45 is an equivalence relation.
3. Prove or disprove: Let f and g be polynomials over a commutative ring A. If $f, g \neq 0$, then
$$\deg(f + g) = \max(\deg(f), \deg(g))$$
4. Find two polynomials from the set of polynomials over Z_{12} (i.e., the coefficients of the polynomials are from Z_{12}) such that $\deg(fg) < \deg(f) + \deg(g)$.
5. Prove or disprove: In the set of polynomial functions over Z_6, if $f(x) = g(x)$ for all x, then the polynomials are equal.

20.4 Algebra and Polynomials

In this section, we discuss special subrings and subfields of the complex numbers. These consist of the Gaussian integers, algebraic integers, and algebraic numbers.

In this section we will discuss further properties of polynomials and also find extension fields of a given field F so that polynomials which cannot be factored in F can be factored in the extension field. Recall that a unit is a divisor of 1; that is, u is a unit provided that there is an element u' such that $uu' = 1$. Recall also that, in a field, every nonzero element is a unit.

In the following material we build up algebraic structure that we will then apply to polynomials. All rings will be assumed to be commutative.

DEFINITION 20.58 A field F is called a **subfield** of F' if it is contained in F' and has the same operations. The field F' is called an **extension field** of F. More generally, a field F' is an extension field of F if there is an isomorphism $\eta : F \to E$ where E is a subfield of F'.

DEFINITION 20.59 If A is an integral domain, a nonzero element a of A is said to be **irreducible** if when $a = bc$, then either b or c is a unit and is said to be **reducible** otherwise. Let $A[x]$ denote the set of polynomials with coefficients in A. A polynomial f in $A[x]$ with positive degree is called **irreducible** if whenever $f = g \cdot h$, then either g or h is an element of A and is said to be **reducible** otherwise.

Thus, if a polynomial f is reducible, then there are polynomials g and h of positive degree such that $f = g \cdot h$.

We shall see that the definitions of irreducible for integral domains and for polynomials are the same if A is a field or if the polynomials are restricted to monic polynomials.

DEFINITION 20.60 A polynomial f in $A[x]$ is called **prime** if it has degree greater than 0 and if whenever $f \mid g \cdot h$, then either $f \mid g$ or $f \mid h$. More generally, in an integral domain, an element p is prime if $p \mid ab$ implies that $p \mid a$ or $p \mid b$.

In the set of integers, prime integers and irreducible integers are the same; however, such is not true in some rings of polynomials.

THEOREM 20.61 In an integral domain, a prime is always irreducible.

Proof Assume that p is a prime and $p = ab$; then by definition, $p \mid a$ or $p \mid b$. Assume that $p \mid a$; then $a = mp$ for some m. Then $p = mpb = mbp$, so $mb = 1$ and b is a unit. Hence, p is irreducible. ∎

DEFINITION 20.62 Let $A^{\emptyset}$ denote the nonzero elements of the integral domain A. Then A is a **Euclidean domain** if there exists a function $\phi : A^{\emptyset} \to N$, where N is the set of positive integers, such that
(1) If $a, b \in A^{\emptyset}$, then $\phi(a) \leq \phi(ab)$ or equivalently, if $a \mid c$, then $\phi(a) \leq \phi(c)$.
(2) If $a, b \in A^{\emptyset}$, then there exists $q, r \in A^{\emptyset}$ such that $a = bq + r$ and either $r = 0$ or $\phi(r) < \phi(b)$.
In a Euclidean domain A, $\phi(a)$, for $a \in A^{\emptyset}$, is called the **norm** of a.

THEOREM 20.63 Let F be a field. Then $F[x]$ is a Euclidean domain using the norm $\phi(f) = \deg(f)$ for $f \in F[x]$.

Proof Let $A^{\emptyset} = F[x] - \{(0)^*\}$ and define $\phi : A^{\emptyset} \to N$ by $\phi(f) = deg(f)$ when $f \neq 0$. Theorem 20.46 establishes part (a). For part (b), let $f \in F[x]$ be a polynomial of degree n and g a polynomial of degree k. Let S be the set of polynomials of the form $f - q \cdot g$. If any element of S is 0, we are finished; otherwise, select one, say r, with least degree and q the corresponding quotient. This element has degree less than n, for if the leading coefficient of b is v, select a polynomial of degree $n - k$ with leading coefficient c so that vc is the leading coefficient of a. Then $f - q \cdot g$ has degree less than n. Similarly, if $\deg(r)$ is not less than $\deg(b)$, then

$$r = q' \cdot f + r'$$

where $\deg(r')$ is less than $\deg(r)$ so that

$$f - q \cdot g + r = q' \cdot g + r'$$

and

$$f - (q + q')g = r'$$

where $\deg(r') < \deg(r)$. But this contradicts the definition of r. Hence, $\deg(r) < \deg(b)$. ∎

The following theorem is a special case of Theorem 20.89, which will be proved later.

THEOREM 20.64 Let F be a field and $p \in F[x]$ be a polynomial of degree n. Then for any $a \in F$, $x - a$ is a factor of p if and only if $p(a) = 0$.

Proof Certainly if $x - a$ is a factor of p, then $p(a) = 0$. Conversely assume that $p(a) = 0$. By the previous theorem, $p = (x - a) \cdot q + r$ where $r \in F$. Therefore, $p(a) = (a - a) \cdot q + r = r$, and $r = 0$. Therefore $(x - a)$ is a factor of p. ∎

Using the two previous theorems and induction we have the following theorem.

THEOREM 20.65 Let F be a field and $p \in F[x]$ be a polynomial of degree n. Then $p(x) = 0$ has at most n solutions. ∎

Since an integral domain can be embedded in a field, we have the following corollary.

Corollary 20.66 Let A be a an integral domain and $p \in A[x]$ be a polynomial of degree n. Then $p(x) = 0$ has at most n solutions. ∎

THEOREM 20.67 Every Euclidean domain A is a principal ideal domain.

Proof Let I be a nonzero ideal of A. Let d be any nonzero element of I with minimal norm value, and let $p \in I$. But $p = dq + r$, where the norm of r is less than the norm of d. Since $r \in I$, d has minimal norm, and the norm of r is less than the norm of d, $r = 0$, and $p = dq$. Hence, d generates I. ∎

Corollary 20.68 Let F be a field; then $F[x]$ is a principal ideal domain. ∎

DEFINITION 20.69

An ideal I is **maximal** in a ring R if for any ideal J, $I \subseteq J \subseteq R$ implies that $I = J$ or $J = R$.

THEOREM 20.70 An ideal I of a principal ideal domain is maximal if and only if p, the generator of I, is irreducible over F.

Proof Assume that I is maximal and p is not irreducible, say $p = rs$. Then p is in the ideal J generated by r (or s) and $I \subset J$, which is a contradiction.
Conversely, if p is irreducible and I is not maximal, say $I \subset J$ and r generates J, then $p = rs$ for some s and this is impossible since p is irreducible. ∎

THEOREM 20.71 Let I be an ideal contained in a ring R. Define a relation $\sim$ on R by $a \sim b$ if and only if $a - b \in I$. The relation $\sim$ is an equivalence relation.

Proof The relation $\sim$ has the following properties:

(a) Reflexivity: $a \sim a$ since $0 \in I$.
(b) Symmetry: If $a \sim b$, then $a - b \in I$; hence, $b - a = -(a - b) \in I$ since I is a group under addition. So $b \sim a$.
(c) Transitivity: Let $a \sim b$ and $b \sim c$ so that $a - b \in I$ and $b - c \in I$. Hence,

$$(a - b) + (b - c) = a - c \in I$$

and $a \sim c$. ∎

THEOREM 20.72 If $a \sim b$ and $c \sim d$, then $a + c \sim b + d$ and $ac \sim bd$.

Proof If $a \sim b$ and $c \sim d$, then $a - b \in I$ and $c - d \in I$, then

$$(a - b) + (c - d) = (a + c) - (b + d) \in I$$

and $a + c \sim b + d$. Also, $c(a - b) = ac - bc \in I$ and $b(c - d) = bc - ad \in I$ since I is an ideal. Hence, $(ac - bc) + (bc - bd) = ac - bd \in I$ and $ac \sim bd$. ∎

We can now define addition and multiplication between elements of the equivalence relation. If $a + I$ denotes the equivalence class containing a and if $b + I$ the equivalence class containing b, then we define

$$(a + I) + (b + I) = (a + b) + I$$

and

$$(a + I) \cdot (b + I) = a \cdot b + I$$

When we can multiply and add equivalence classes, then as mentioned before, the relation is called a **congruence relation**. The ring of equivalence classes is denoted by R/I. Note that $0 + I$ is the zero element of the ring and $1 + I$ is the multiplicative identity. The equivalence class $a + I$ is called a **left coset** of I, in particular, the one generated by a. Hence, the distinct left cosets of I are just the equivalence classes in R/I.

THEOREM 20.73 Let R be a commutative ring with unity and I be an ideal of R. Then R/I is a field if and only if I is a maximal ideal.

Proof Assume that I is maximal and let $a + I \in R/I$, where $a \notin I$. Let $M = \{ra + b \mid r \in R \text{ and } b \in I\}$. It is straightforward to show that M is an ideal. Since $a \notin I$, one obtains $M \subset I$ and $M = R$. Hence, $1 \in M$ and there exists b in R and c in I so that $1 = ab + c$. Thus, $1 + I = ab + I = (a + I)(b + I)$ and $a + I$ has an inverse. Hence, R/I is a field.

Conversely, assume that R/I is a field. Assume that there exists an ideal J such that $I \subset J \subseteq R$. Consider $a + I$ where a is in J but not I. Then there exists a' in R so that $a' + I$ is the inverse of $a + I$, making $aa' + I = 1 + I$. Furthermore, since J is an ideal, $aa' \in J$. Also, $1 = aa' + c$ for some c in I. Since aa' and c are in J, 1 is in J. But if 1 is in J, then $r1$ is in J for all r and $J = R$. ∎

THEOREM 20.74 Let R be a commutative ring with identity. Then R/I is an integral domain if and only if I is a prime ideal.

Proof If R/I is an integral domain and ij is an element of I, then $ij + I = 0 + I$. Hence, $(i + I)(j + I) = 0 + I$. But since R/I is an integral domain, either $i + I$ or $j + I$ equals $0 + I$. Hence, either $i \in I$ or $j \in I$ and I is a prime ideal.

Conversely, if I is a prime ideal and $(i + I)(j + I) = 0$, then $ij + I = 0$ and $ij \in I$. But since I is a prime ideal, $i \in I$ or $j \in I$; therefore, $i + I = 0$ or $j + I = 0$ and I is a integral domain. ∎

THEOREM 20.75 Every maximal ideal is a prime ideal.

Proof The proof follows immediately from the results of Theorems 20.74 and 20.75. ∎

THEOREM 20.76 In a principal ideal domain A, every irreducible element is a prime.

Proof Let p be an irreducible element of A and let $\langle p \rangle$ be the principal ideal generated by p. Since p is irreducible, by Theorem 20.70, $\langle p \rangle$ is maximal, and hence by Theorem 20.75, is a prime ideal. Therefore, if $p \mid (ab)$, then $ab \in \langle p \rangle$. Thus, $a \in \langle p \rangle$ or $b \in \langle p \rangle$ so that $p \mid a$ or $p \mid b$ and p is prime. ∎

Combining Corollary 20.68 and Theorem 20.76, we obtain the following result.

Corollary 20.77 If F is a field, then in $F[x]$ every irreducible element is a prime. ∎

In the next theorem, we have an analogy to the unique factorization of integers into products of primes.

THEOREM 20.78 If F is a field, then every nonunit polynomial in $F[x]$ can be uniquely factored into a product of irreducible polynomials in the sense that factors are unique up to multiplication by a unit.

Proof Using the second form of induction on the degree of the polynomial, we show that every nonconstant polynomial of degree n can be factored into irreducible elements. For $n = 1$ we have a polynomial of the form $ax + b$, which is already irreducible because if not, then $ax + b = fg$ but $\deg(ax+b) = \deg(f) + \deg(g) \geq 2$, an obvious contradiction. Assume that $k > 1$ and the theorem holds for true for all $n < k$ and let f have degree k. If f is irreducible, we are finished. If not, then f factors into polynomials g and h, each with positive degree less than f. Hence, g and h both factor into irreducible elements; therefore, f also factors into irreducible factors.

To show that the factors are unique we again use induction on the number of factors in one reduction. For $n = 1$, the result is obvious. Assume that the uniqueness holds for $n = k$ factors. Assume that f factors into $k + 1$ irreducible elements $g_1, g_2, \ldots, g_{k+1}$ and into m irreducible elements $g'_1, g'_2, \ldots, g'_m$. Select the factor g_1 from the $k + 1$ irreducible elements. Since an irreducible element is a prime by Corollary 20.77 and g_1 divides f, g_1 must divide one of the m irreducible elements in the second factorization, say g'_j. But since g'_j is irreducible, g_1 and g'_j differ by a unit. Hence, divide both factorizations by g_1, leaving the first factorization of f with k irreducible factors $g_2, \ldots, g_{k+1}$ and the second factorization with $m - 1$ irreducible factors $g'_1, \ldots, g'_{j-1}, g'_{j+1}, \ldots, g'_m$, and a possible unit factor. By induction, the factorization of f/g_1 is unique and so the factorization of f is unique. ∎

Recall that a polynomial $f = a_0 + a_1 x + \cdots + a_n x^n$ is primitive if $\gcd(a_0, a_1, \ldots, a_n)$ is a unit.

THEOREM 20.79 (Gauss) Let A be an integral domain. The product of two primitive polynomials in $A[x]$ is a primitive polynomial.

Proof Let $f = a_0 + a_1 x + \cdots + a_n x^n$ and $g = b_0 + b_1 x + \cdots + b_m x^m$ and let p be an irreducible element of A. Since f is primitive, there exists an a_i such that p does not divide a_i. Let a_r be the first such coefficient. Similarly, there is a b_j such that p does not divide b_j. Let b_s be the first such coefficient. Let $h = c_0 + c_1 x + \cdots + c_{m+n} x^{m+n}$ be the product of f and g. Then

$$c_{r+s} = a_0 b_{r+s} + \cdots + a_{r-1} b_{s+1} + a_r b_s + a_{r+1} b_{s-1} + \cdots + a_{r+s} b_0$$

But p divides $a_0, \ldots, a_{r-1}$, so p divides $a_0 b_{r+s} + \cdots + a_{r-1} b_{s+1}$; and p divides $b_0, \ldots, b_{s-1}$, so p divides $a_{r+1} b_{s-1} + \cdots + a_{r+s} b_0$. But p does not divide $a_r b_s$ so p does not divide h. Hence, h is primitive. ∎

It might seem that a polynomial that is irreducible in the integers might not be irreducible if we allow factorization into polynomials with rational coefficients; however, the following theorem shows that this limitation is not true.

THEOREM 20.80 Let A be a unique factorization domain, that is, an integral domain having unique factorization into primes, and let F be its field of fractions. If $p \in A[x]$ and if p is reducible in $F[x]$, then it is reducible in $A[x]$. If p is primitive in $A[x]$, then it is reducible in $F[x]$ if and only if it is reducible in $A[x]$.

Proof Let $p \in A[x]$ be reducible in $F[x]$, say $p = fg$. Let u be the least common multiple of the denominators of the coefficients of f and g; then $u \cdot p = f'g'$, where f' and g' are in $A[x]$. But $f' = af''$ and $g' = bg''$, where f'' and g'' are primitive; and $p = cp'$, where c is primitive. Hence, $uc \cdot p' = ab \cdot f''g''$. But since $f''g''$ is also primitive, it is easily shown that two expressions for a primitive polynomial differ only by a unit. Thus, $ucv = ab$, where v is a unit. Hence, $ucp' = ucvf''g''$. So

$$up = ucvf''g''$$

or

$$p = cvf''g''$$

and p has been factored in $A[x]$.

Obviously, if a polynomial factors in $A[x]$, then it factors in $F[x]$, but if p is not primitive in $A[x]$, then it is reducible in $A[x]$ because it is the product of an integer and a polynomial; however, in $F[x]$ it may be irreducible since an integer is a unit. ∎

THEOREM 20.81 An integral domain A is a unique factorization domain if and only if $A[x]$ is.

Proof Let F be the field of fractions of A and p be a polynomial in $A[x]$. Then p factors into irreducibles in $F[x]$ and, hence, p factors into irreducibles in $A[x]$, where each irreducible of the factorization in $F[x]$ is equal to a corresponding irreducible of the factorization in $A[x]$ multiplied by an element of F.

Let d equal the greatest common divisor of the coefficients of p so that $p = dp'$, where p' is a primitive polynomial. Since d factors uniquely, we need only show p' factors uniquely; hence we will assume that p is primitive. Since, by Theorem 20.78, $F[x]$ is a unique factorization domain, let

$$p' = r_1 r_2 \cdots r_n = s_1 s_2 \cdots s_n$$

Assume that by rearrangement we have r_i and s_i differing only by a nonzero element of F. Say that $r_i = (a/b)s_i$ so $br_i = as_i$, but since r_i and s_i are primitive, a and b differ by a unit in $A[x]$ and the factorization is unique.

Conversely, since A is a subset of $A[x]$, if $A[x]$ is a unique factorization domain, then every element of A must factor uniquely and A is a unique factorization domain. ∎

THEOREM 20.82 Let A be a unique factorization domain; then two polynomials in $A[x]$ have a greatest common divisor.

Proof Let p and q be polynomials in $A[x]$. Since $A[x]$ is a unique factorization domain, factor p and q into irreducibles. Then $\gcd(p, q)$ is the product of the least powers of irreducibles occurring as factors in both p and q. ∎

THEOREM 20.83 Let A be a principal ideal domain and I be an ideal generated by p and q. If d generates I, then $d = \gcd(p, q)$ and d may be written in the form $up + vq$.

Proof Certainly, d divides p and q. Since $d \in I$, d may be written in the form $up + vq$. If c divides p and c divides q, then c divides $up + vq$ and c divides d. Thus, $d = \gcd(p, q)$. ∎

Beginning with an integral domain A, such as the integers, we may of course find the fraction field of A, say F, and form $F[x]$. Since $A[x]$ is an integral domain, we may also form the fraction field of $A[x]$, essentially generating elements of the form p/q, where p and q belong to $A[x]$. Since $F[x]$ is also an integral domain, we may form the fraction field of $F[x]$, which essentially consists of elements of the form p/q where p and q belong to $F[x]$. This field is called the **field of rational functions over the field** F.

DEFINITION 20.84 An element $c \in C$, the set of complex numbers, is an **algebraic integer** if it is a zero of some monic polynomial p in $Z[x]$, that is, $p(c) = 0$.

DEFINITION 20.85 An element a of an extension field E of a field F is **algebraic** over F if $f(a) = 0$ for some nonzero $f \in F[x]$.

DEFINITION 20.86 An element $c \in C$ that is algebraic over Q, the field of rational numbers, is an **algebraic number**; that is, there is a polynomial $p \in Q[x]$ such that $p(c) = 0$.

DEFINITION 20.87 Let A be an integral domain. An element $a \in A$ is a **root** of a polynomial p, if $p(a) = 0$.

In the following we show abstractly how, given a field and an irreducible polynomial over that field, we can find an extension field with a root of the irreducible polynomial. We then compare this extension field to one that would be formed by a field that has been extended by a root of the irreducible polynomial in the complex numbers.

THEOREM 20.88 Let F be a field and p be an irreducible polynomial over F. Then F is isomorphic to a subfield of $F[x]/\langle p \rangle$ and there is an element f of $F[x]/\langle p \rangle$ such that $p(f) = 0$.

Proof By Theorems 20.70 and 20.73, $F[x]/\langle p \rangle$ is a field. The field F can be identified with a subfield F' of $F[x]/\langle p \rangle$ as follows: Let $g : F \to F'$ be defined by $g(a) = a + \langle p \rangle$. The function g is a monomorphism (i.e., one-to-one) for if $g(a) = g(b)$, then $a + \langle p \rangle = b + \langle p \rangle$ and p divides $a - b$. Since p is a polynomial and a and b are in F, $a - b = 0$ and $a = b$.

Let $f = x + \langle p \rangle$ and $p = a_0 + a_1 x + \cdots + a_n x^n$. Then

$$p(f) = a_0 + a_1(x + \langle p \rangle) + \cdots + a_n(x + \langle p \rangle)^n$$
$$= a_0 + a_1 x + \cdots + a_n x^n + \langle p \rangle$$
$$= p + \langle p \rangle$$
$$= 0 + \langle p \rangle$$
$$= 0$$

in $F[x]/\langle p \rangle$. ∎

THEOREM 20.89 Let p be an irreducible polynomial in $F[x]$ and $p(a) = 0$ in an extension field of F. If $g(a) = 0$ for g in $F[x]$, then p divides g.

Proof It is easily shown that $I = \{g : g \in F[x] \text{ and } g(a) = 0\}$ is an ideal. This ideal is obviously generated by p since p is irreducible. Hence, if $g \in I$, that is, $g(a) = 0$, then g must be a multiple of p. ∎

THEOREM 20.90 Let F be a field and $p \in F[x]$ be an irreducible polynomial. If K is an extension field of F containing a root a of p, that is, $p(a) = 0$, then $F[a]$ is isomorphic to $F[x]/\langle p \rangle$.

Proof Define $h : F[x]/\langle p(x) \rangle \to F[a]$ by $h(f + \langle p \rangle) = f(a)$. It is obvious that h is a homomorphism. If $h(f + \langle p \rangle) = h(g + \langle p \rangle)$, then $f(a) = g(a)$ and $(f - g)(a) = 0$ Hence, by the Theorem 20.89, p divides $f - g$ and $f + \langle p \rangle = g + \langle p \rangle$ and h is one-to-one. Let $f \in F[a]$; then $h(f + \langle p \rangle) = f(a)$, h is onto, and consequently, h is an isomorphism. ∎

Let F be the field of real numbers and $p = x^2 + 1$, and let K be an extension of F containing i, which is a root of $p(x)$, that is, $i^2 + 1 = 0$. By Theorem 20.89, $F[x]/\langle p \rangle$ is isomorphic to $F[i]$. Since $p(x)$ has degree 2, we need only look at $f \in F[i]$ of degree 1. Hence, every element of $F[i]$ will be of the form $a + bi$ for real numbers a and b and we have created the complex numbers.

The subring consisting of elements of $F[i]$ where a and b are both integers is called the set of **Gaussian integers**. This may also be considered as the smallest subring of the complex numbers containing the integers and i. It is obvious that 5 is not an irreducible in this ring since $5 = (1 - 2i)(1 + 2i)$. Let the norm n_G on the set of Gaussian integers be defined by $n_G(a + bi) = a^2 + b^2$. Then

$$\begin{aligned} n_G((a+bi)(c+di)) &= n_G((ac - bd) + (ad + bc)i) \\ &= (ac - bd)^2 + (ad + bc)^2 \\ &= a^2c^2 - 2abcd + b^2d^2 + a^2d^2 + 2abcd + b^2c^2 \\ &= (a^2 + b^2)(c^2 + d^2) \\ &= n_G(a + bi)n_G(c + di) \end{aligned}$$

so that n_G is a multiplicative homomorphism from the Gaussian integers to the integers. Since $n_G(1 - 2i) = n_G(1 + 2i) = 5$, then $1 - 2i$ and $1 + 2i$ are irreducibles in the ring of Gaussian integers. The product of irreducible Gaussian integers $2 + i$ and $2 - i$ is also 5. Hence, we do not have unique factorization into irreducibles. Note that the irreducibles in this example are not primes.

As another example, let F be the field of rational numbers and $p(x) = x^2 - 2$; then $F[x]/\langle p \rangle = F[\sqrt{2}]$. Again only polynomials of degree 1 need be considered, and $F[x]/\langle p \rangle$ is the field of all elements of the rational numbers of the form $a + b\sqrt{2}$ where a and b are rational numbers. We may consider the subring consisting only of elements of $F[x]/\langle p \rangle$ of the form $a + b\sqrt{2}$ where a and b are integers. $(3 + \sqrt{2})(3 - \sqrt{2}) = 7$ so that 7 is not an irreducible in this set.

We have assumed that, beginning with the rational numbers, all of the algebraic integers are a subset of the complex numbers. Under this assumption, we prove the following theorem.

THEOREM 20.91 The set A of all algebraic numbers over the rational numbers Q is a subfield of the complex numbers.

Proof Let a and b be nonzero algebraic numbers. If a is not rational, then there exists an irreducible polynomial p with degree 2 or greater such that a is a root of p. Let $F = Q[x]/\langle p \rangle$. If b is not in F, then there exists an irreducible polynomial in F, say q, of degree greater than 2 so that b is a root of q. Let $G = F[x]/\langle q \rangle$. Then G is a field containing a and b and so contains $a + b$, $a - b$, ab, and a/b. Since G is contained in A, these elements are in A and A is a field. ∎

Exercises

1. Prove that the set of Gaussian integers form a subring of the set of complex numbers.

2. Prove or disprove that the set of Gaussian integers is an integral domain.

3. Prove that $\sin(x)$ cannot be expressed as a polynomial.

4. Prove or disprove: The product of two prime polynomials is a prime polynomial.

5. Prove or disprove: The sum of two prime polynomials is a prime polynomial.

6. Find the generator of the smallest principal ideal containing the polynomials $x^2 - 4$ and $x^2 + 4x + 4$ where the polynomials are over the integers.

7. Prove or disprove that every Gaussian integer is an algebraic number.

8. Find an ideal of the integers that is not a maximal ideal.

9. Describe the smallest subfield of the complex numbers containing the roots of $x^2 - 5 = 0$.

10. Describe the smallest subfield of the complex numbers containing the roots of $x^2 + 5 = 0$.

11. Describe the smallest subfield of the complex numbers containing the roots of $x^3 - 1 = 0$.

12. Describe the smallest subfield of the complex numbers containing the roots of $x^4 - 1 = 0$.

GLOSSARY

Chapter 20: Rings, Integral Domains, and Fields

Addition in a field (20.1)	Given elements $a, b, c, d \in A$, **addition in a field** is defined on F by $\frac{a}{b} + \frac{c}{d} = \frac{ad+bc}{bd}$.
Algebraic (20.4)	An element a of an extension field E of a field F is **algebraic** over F if $f(a) = 0$ for some nonzero $f \in F[x]$.
Algebraic integer (20.4)	An element $c \in C$, the set of complex numbers, is an **algebraic integer** if it is a zero of some monic polynomial p in $Z[x]$, that is $p(c) = 0$.
Algebraic number (20.4)	An element $c \in C$ that is algebraic over Q, the field of rational numbers, is an **algebraic number**; that is, there is a polynomial $p \in Q[x]$ such that $p(c) = 0$.
Coefficients (20.3)	Let $f = (a_0, a_1, a_2, \ldots)$ belong to $A[x]$. The terms a_i in the sequence are called the **coefficients** of f.
Commutative ring (20.1)	If $r \cdot r' = r' \cdot r$ for all r and r' in R, then R is called a **commutative ring**.
Congruence relation (20.4)	A **congruence relation** in a ring is an equivalence relation in which we can multiply and add equivalence classes.
Constant (20.3)	Any polynomial that is of degree 0 or is equal to zero is called a **constant**.
Degree of f (20.3)	Let A be a commutative ring with unity, and let $f \in S$ and $f = (a_i)^*$. If $f \neq 0$, $\deg(f)$ equals the largest integer k, such that $a_k \neq 0$. The integer is called the **degree of f**.
Degree of $f(x)$ (20.3)	If $A(x) = \{f(x) : f \in S\}$, then the **degree of $f(x)$** is the same as the degree of the corresponding polynomial $f \in A[x]$.
Divides (20.3)	The polynomial f **divides** the polynomial g provided that there is a polynomial h such that $fh = g$.
Embedded in a field (20.1)	If the mapping $f : A \to F$ defined by $f(a) = \frac{a}{1}$ is a monomorphism, then the integral domain A is **embedded in the field** F.
Epimorphism (20.1)	If ring homomorphism $f : R \to R'$ is onto, it is said to be an **epimorphism**.

Equal elements (20.3)	Two elements f and g of $A[x]$ are **equal** if their respective coefficients are equal.
Euclidean domain (20.4)	Let A^ϕ denote the nonzero elements of the integral domain A. Then A is an **Euclidean domain** if there exists a function $\phi : A^\phi \to N$, where N is the set of positive integers, such that 1. If $a, b \in A^\phi$, then $\phi(a) \leq \phi(ab)$ or equivalently, if $a\mid c$, then $\phi(a) \leq \phi(c)$. 2. If $a, b \in A^\phi$, then there exist $q, r \in A^\phi$ such that $a = bq + r$ and either $r = 0$ or $\phi(r) < \phi(b)$.
Extension field (20.4)	A field F' is an **extension field** of F if there is an isomorphism $\eta : F \to E$ where E is a subfield of F'.
Factors of g (20.3)	If polynomial f divides polynomial g such that $fh = g$, then f and h are **factors of g**.
Field (20.1)	A **field** is a commutative ring with unity not equal to 0 such that every nonzero element has a multiplicative inverse.
Field of rational functions over F (20.4)	The **field of rational numbers over the field F** is the fraction field of integral domain $F[x]$.
Fraction field (20.1)	F is called the **fraction field** of A if the mapping $f : A \to F$ defined by $f(a) = \dfrac{a}{1}$ is a monomorphism.
Gauss' theorem (20.4)	If A is an integral domain, the product of two primitive polynomials in $A[x]$ is a primitive polynomial.
Gaussian integers (20.4)	The subring consisting of elements of the field of complex numbers $F[i]$ where a and b are both integers is called the set of **Gaussian integers**.
Group of units (20.1)	If A is a commutative ring with unity, let A^* denote the set $\{a \in A : \exists b \in A \text{ with } ab = 1\}$. The subset A^* is a group under multiplication called the **group of units** of A. Each element of A^* is called a **unit** of A. In a ring with unity, an element s is **irreducible** if it is nonzero, is not a unit, and cannot be expressed as a product of two nonunits.
Group of units (20.1)	If A is a commutative ring with unity, and $A^\#$ denotes the set $\{a \in A : \text{ there exists } b \in A \text{ with } ab = 1\}$, then the subset $A^\#$ is a group under multiplication called the **group of units** of A.
Ideal (20.1)	A subset I of a ring R is called an **ideal** of R if a. I is a subring of R. b. If $x \in R$ and $r \in R$, then $x \cdot r, r \cdot x \in I$.
Ideal maximal in a ring (20.4)	An ideal I is **maximal** in a ring R if for any ideal J, $I \subseteq J \subseteq R$ implies that $I = J$ or $J = R$.
Indeterminate symbol over A (20.3)	If A is a commutative ring with unity and $A[x]$ is the set of polynomials over A, then the symbol x is called an **indeterminate symbol over A**.
Integral domain (20.1)	An **integral domain** is a commutative ring with unity not equal to 0 such that $ab = 0$ implies that $a = 0$ or $b = 0$.
Irreducible (20.1)	In a ring with unity, an element s is **irreducible** if it is nonzero, is not a unit, and cannot be expressed as a product of two nonunits.
Isomorphism (20.1)	A ring homomorphism $f : R \to R'$ is said to be an **isomorphism** provided that f is both one-to-one and onto.

Kronecker delta (20.3)	The Kronecker δ, δ_{ij}, for integers i and j, is defined by $$\delta_{ij} = \begin{cases} 1 \text{ if } i = j \\ 0 \text{ if } i \neq j \end{cases}$$
Leading coefficient (20.3)	If $f \neq 0$ and $n = \deg(f)$, then a_n is called the **leading coefficient** of f.
Minimal domain (20.2)	An integral domain is called a **minimal domain** if and only if it has no subdomains except itself.
Monic polynomial (20.3)	If $a_n = 1$, then f is called a **monic polynomial**.
Monomorphism (20.1)	If ring homomorphism $f : R \to R'$ is one-to-one, it is said to be a **monomorphism**.
Norm of a (20.4)	In an Euclidean domain A, $\phi(a)$, for which $a \in A^\phi$, is called the **norm** of a where $\phi : A \to N$.

Ordered ring (20.2)	A commutative ring A with unity is an **ordered ring** if and only if there exists a nonempty subset $A+$ of A called the subset of positive elements of A such that
	 a. If $a, b \in A^+$, then $a + b \in A^+$.
	 b. If $a, b \in A^+$, then $a \cdot b \in A^+$.
	 c. Given any $a \in A$, one and only one of the following alternatives is true for a:
	    (i) $a \in A^+$
	    (ii) $a = 0$
	    (iii) $-a \in A^+$

Polynomial forms (20.3)	The elements f of $A[x]$ are called **polynomial forms**.
Polynomial function (20.3)	$f(x) = a_n x^n + \cdots + a_2 x^2 + a_1 x + a_0$ is called a **polynomial function**.
Polynomial over A (20.3)	Let A be a commutative ring with unity. Any element of $A[x]$ is called a **polynomial over A**.
Prime ideal (20.1)	An ideal I of a commutative ring R is a **prime ideal** if $ab \in I$ implies that either $a \in I$ or $b \in I$.
Prime polynomial (20.4)	A polynomial f in $A[x]$ is called **prime** if it has degree greater than 0 and if whenever $f \mid g \cdot h$, then either $f \mid g$ or $f \mid h$.
Primitive polynomial (20.3)	If $f \neq 0$ has the property that any greatest common divisor of all of its nonzero coefficients is a unit, then f is called a **primitive polynomial**.
Principal ideal (20.1)	If R is a commutative ring, an ideal I of R is called a **principal ideal** generated by a if I consists of all products of a by elements of R; that is, $I = \langle a \rangle = \{ar : r \in R\}$.
Principal ideal domain (20.1)	An integral domain D is a **principal ideal domain** if every ideal in D is a principal ideal.
Product of f and g (20.3)	If A is a commutative ring and let $f = (a_i)^* = (a_0, a_1, a_2, \ldots)$ and $g = (b_i)^* = (b_0, b_1, b_2, \ldots)$ belong to S, the set of polynomials over A, then the **product of f and g** is the sequence $fg = (c_k)^*$ where $c_k = \sum_{i+j=k} a_i b_j$.
Proper ideal (20.1)	An ideal I is a **proper ideal** if $I \neq \{0\}$ and $I \neq R$.
Reducible and irreducible elements (20.4)	If A is an integral domain, a nonzero element a of A is said to be **irreducible** if when $a = bc$ then b or c is a unit and is said to be **reducible** otherwise.

Reducible and irreducible polynomials (20.4)	If $A[x]$ denotes the set of polynomials with coefficients in A, then the polynomial f in $A[x]$ with positive degree is **irreducible** if whenever $f = g \cdot h$, then either g or h is an element of A and is **reducible** otherwise.
Ring (20.1)	A **ring** is a nonempty set R together with binary operations called multiplication and addition, denoted by $\cdot$ and $+$, respectively, which satisfies the following conditions: 1. R is closed under addition. 2. Addition in R is associative. 3. R has an additive identity. 4. R contains an additive inverse. 5. Addition in R is commutative. 6. R is closed under multiplication. 7. Multiplication in R is associative. 8. The following distributive laws hold for all x, y, and z in R: $$x \cdot (y + z) = (x \cdot y) + (x \cdot z) \text{ and } (y + z) \cdot x = (y \cdot x) + (z \cdot x).$$
Ring homomorphism (20.1)	If R and R' are rings and $f : R \to R'$ is a function from R into R', then f is said to be a **ring homomorphism** if and only if $f(a + b) = f(a) + f(b)$ and $f(a \cdot b) = f(a) \cdot f(b)$ for all $a, b \in R$.
Ring with unity (20.1)	If there is an element 1 in R such that $1 \cdot r = r \cdot 1 = r$ for all r in R, then R is called a **ring with unity**.
Root of a polynomial (20.4)	If A is an integral domain, an element $a \in A$ is a **root of a polynomial** p if $p(a) = 0$.
Subfield (20.4)	A field F is called a **subfield** of F' if it is contained in F' and has the same operations.
Subring (20.1)	A subset R' of a ring R is called a **subring** of R if R' is a ring with the same operations.
Sum of f and g (20.3)	If A is a commutative ring and let $f = (a_i) = (a_0, a_1, a_2, \ldots)$ and $g = (b_i) = (b_0, b_1, b_2, \ldots)$ be in S, the set of polynomials over A, then the **sum of f and g** is the sequence $f + g = (a_i + b_i)^* = (a_0 + b_0, a_1 + b_1, \ldots)$.
Unique factorization domain (20.2)	An integral domain D is said to be a **unique factorization domain** provided that a. If an element of D is not zero and not a unit, then it can be factored into a product of a finite number of irreducibles. b. If an element of D has factorization $p_1 \cdots p_r$ and $q_1 \cdots q_s$ as products of irreducibles, then $r = s$ and the q_j can be renumbered so that the p_i and q_i differ only by a unit for all I; that is, $p_i = a_i q_i$ for some unit a_i.
Well ordered (20.2)	An ordered integral domain A is called **well ordered** if and only if any nonempty subset S of A^+ has a first element; that is, there exists $s \in S$ such that if $t < s$, then $t \notin S$.

TRUE-FALSE QUESTIONS

1. The set of all subsets of a given set S together with the operations $\cap$ and $\cup$ form a ring.
2. The set of all subsets of a given set S together with the operations $\cap$ and Δ form a ring.
3. Z_n is an integral domain if and only if n is prime.
4. Z_n is a field if and only if n is prime.
5. The set of $m \times m$ matrices with real coefficients together with matrix multiplication and addition form an integral domain.
6. All infinite integral domains are fields.
7. The rational numbers are an ideal of the real numbers.
8. The multiples of 4 are an ideal of the ring of integers.
9. All infinite rings are commutative.
10. All finite rings are commutative.
11. A field contains no proper integral domains.
12. All proper ideals of the ring of integers are principal ideals.
13. All proper ideals of the ring of integers are prime ideals.
14. In the ring of integers, the smallest ideal containing integers a, b, and c is the principal ideal generated by $\text{lcm}(a, b, c)$.
15. The ring of integers has no proper ideal containing two distinct positive primes.
16. The intersection of two ideals is an ideal.
17. The ring of integers contains no finite proper ideals.
18. The union of two ideals is an ideal.
19. Every integral domain is a field or may be embedded in a field.
20. The zero divisors of a ring form an ideal.
21. In Z_n, every element is either a unit or a zero divisor.
22. The units of a ring form a subring.
23. An element of Z_n is a unit if and only if it is relatively prime to n.
24. Let $F : R \to R'$ be a ring homomorphism. If R is an integral domain, then the image of R is an integral domain.
25. Let $F : R \to R'$ be a ring homomorphism. If the image of R is a field, then R is a field.
26. The ring of integers is a minimal domain.
27. Let set of Gaussian integers is an integral domain.
28. Let set of Gaussian integers is an ordered ring.
29. An ordered ring contains only one unit.
30. A field cannot be an ordered ring.
31. An integral domain contains exactly two idempotents.
32. The set of Gaussian integers is a field.
33. Let A be an integral domain and $\theta : A[x] \to A(x)$ be defined by $\theta(f) = f(x)$. θ is an isomorphism.
34. If $f(x)$ and $g(x)$ are polynomial functions over a commutative ring A, $f(x)$ has degree m and $g(x)$ has degree n, then $f(x)g(x)$ has degree $m + n$.
35. If $f(x)$ and $g(x)$ are polynomial functions over a commutative ring A, $f(x)$ has degree m and $g(x)$ has degree n, then $f(x) + g(x)$ has degree less than or equal to $\max(m, n)$.
36. Let f and g be polynomials over an integral domain A, then $f = g$ if and only if $f(b) = g(b)$ for every $b \in A$.
37. In a principal ideal domain, every irreducible element is a prime.
38. In a principal ideal domain every prime is irreducible.
39. Let R be a commutative ring with unity. R/I is a field if and only if I is a prime ideal.
40. Every maximal ideal is a prime ideal.
41. Every prime ideal is maximal.
42. Every rational number is an algebraic number.
43. Every algebraic number is a real number.
44. Let F be the field of rational numbers and $p(x) = x^2 + 2$. $F[x]/\langle p \rangle = F[\sqrt{2}]$.
45. Let F be the field of rational numbers and $p(x) = x^3 - 1$. $F[x]/\langle p \rangle = F[i\sqrt{3}]$.
46. Every complex number $a + bi$ such that $a^2 + b^2 = 1$ is an algebraic integer.
47. The set of algebraic numbers is a subfield of the complex numbers.
48. The set of Gaussian integers is a subring of the algebraic numbers.
49. The integer 5 is irreducible in the ring of algebraic numbers.
50. The Gaussian integers are a unique factorization domain.

ANSWERS TO TRUE-FALSE

1. F 2. T 3. T 4. T 5. F 6. F 7. F 8. T 9. F 10. F 11. F
12. T 13. F 14. F 15. T 16. T 17. T 18. F 19. T 20. F
21. T 22. F 23. T 24. F 25. F 26. T 27. T 28. F 29. F
30. F 31. T 32. F 33. F 34. F 35. T 36. F 37. T 38. T
39. F 40. T 41. T 42. T 43. F 44. F 45. T 46. F 47. T
48. T 49. F 50. F

CHAPTER 21

Group and Semigroup Characters

21.1 Complex Numbers

In the preceding chapter, we used the complex numbers and developed then using roots of polynomial functions. At this point, we formally introduce them from a different point of view.

DEFINITION 21.1 Let R be the set of real numbers and C the set
$$\{(r, s) : r, s \in R\} = R \times R$$
so that every element of C is an ordered pair of real numbers. Define addition on C by
$$(a, b) + (c, d) = (a + c, b + d)$$
and multiplication by
$$(a, b) \cdot (c, d) = (ac - bd, ad + bc)$$
Then C is called the set of **complex numbers**.

It is straightforward to show that C is a field. The mapping $f : R \to C$ defined by $f(a) = (a, 0)$ is a monomorphism and is an isomorphism into the subfield $\{(a, 0) : a \in R\}$. Therefore, we speak of the real number field R as being contained in the complex number field and we say that R is a **subfield** of C. Further, $(0, 1) \cdot (0, 1) = (-1, 0) = -1$. If we define $i = (0, 1)$, then $i^2 = -1$. Hence, we may write (a, b) as $(a, 0) + (b, 0) \cdot (0, 1) = a + bi$. Thus, the definitions of addition and multiplication become
$$(a + bi) + (c + di) = (a + c) + (b + d)i$$
and
$$(a + bi) \cdot (c + di) = (ac - bd) + (ad + bc)i$$

For a complex number $z = a + bi = (a, b)$, a is called the **real part** and b is called the **imaginary part** and we write $real(z) = real(a + bi) = a$ and

$\text{im}(z) = \text{im}(a + bi) = b$. Just as we often plot ordered pairs (a, b) of real numbers on a Cartesian coordinate system (in the plane), we plot the complex number $a + bi$ in the plane as (a, b). In this context, the horizontal axis is called the real axis and the vertical axis is called the imaginary axis. Thus, the real part of $a + bi$ [that is, $\text{real}(a + bi) = a$] is measured along the horizontal axis and the imaginary part of $a+bi$ [that is, $\text{im}(a+bi) = b$] is measured along the vertical axis. When the Cartesian coordinate system is used in this manner, it is referred to as the **complex plane**.

DEFINITION 21.2 Let C be the set of complex numbers. For $c = a + bi$ in C, $|c| = \sqrt{a^2 + b^2}$ is called the **absolute value** or **length** of c.

On the complex plane $|c|$ is the distance from the point $c = a + bi$ to the origin. It is easily shown that $|c| |d| = |cd|$ and $|c + d| \leq |c| + |d|$ for complex numbers c and d. Let r be a complex number which is an nth root of 1, that is, $r^n = 1$. Then, $|r|^n = |r^n| = |1| = 1$; so r has length 1 and is on the **unit circle**, the circle of all points on the complex plane whose distance from the origin is 1.

An important property of complex numbers is given by the following theorem, which we will state without proof.

FUNDAMENTAL THEOREM OF ALGEBRA (Gauss) For any polynomial p of positive degree with coefficients in $\mathbb{R}$, the set of real numbers, or in C, the set of complex numbers, there is an element b in C such that $p(b) = 0$.

Thus, in particular, every polynomial over Z has at least one root or zero in the set of complex numbers.

21.2 Group Characters

DEFINITION 21.3 A **group character** on a commutative group G is a homomorphism from G into the nonzero complex numbers with the operation of multiplication.

DEFINITION 21.4 The **conjugate** of the complex number $c = a + bi$ is the number $a - bi$, which is denoted by $\text{conj}(c)$.

For the remainder of this section we will assume that all groups forming the domain of the group characters are commutative and finite. Let χ be a group character. Thus, there exists $g \in G$ such that $\chi(g) \neq 0$. Since G is finite, there exists n so that $g^n = 1$. Hence, $\chi(g^n) = (\chi(g))^n = 1$, $\chi(g)$ is an nth root of 1, and the element g is mapped onto the unit circle. Thus, a group character maps G into the unit circle.

The proof of the following theorem is left to the reader.

THEOREM 21.5 If $\chi(g) = a + bi$, then $\chi(g^{-1})$ is the conjugate of $\chi(g)$, that is, $\chi(g^{-1}) = a - bi$. ∎

DEFINITION 21.6 A **primitive nth root** r of a complex number a is an nth root of a such that $r^k \neq a$ for $0 < k < n$.

DEFINITION 21.7 The group character χ_1 defined by $\chi_1(g) = 1$ for all g in the group G is called the **principal character** of the group G. The **character group** of G consists of the set of group characters of G with the operation $(\chi \cdot \chi')(g) = \chi(g)\chi'(g)$ for group characters χ and χ'.

DEFINITION 21.8 Let $G_1, G_2, \ldots, G_m$ be subgroups of a group G. Then G is the **direct sum** of $G_1, G_2, \ldots, G_m$, denoted $G_1 \oplus G_2 \oplus \cdots \oplus G_m$, if every element of G may be uniquely written in the form $a_1 a_2 \cdots a_m$ where $a_k \in G_k$.

THEOREM 21.9 Every finite Abelian group is the direct sum of cyclic subgroups. Thus, if $G = G_1 \oplus G_2 \oplus G_3 \oplus \cdots \oplus G_m$ for cyclic groups G_i with generators g_i of order $k(i)$, then each $g \in G$ is uniquely expressed in the form

$$g = g_1^{j(1)} g_2^{j(2)} g_3^{j(3)} \cdots g_m^{j(m)}$$

with $0 \leq j(i) < k(i)$.

Proof We show that there exists an integer m so that

$$G = G_1 \oplus G_2 \oplus G_3 \oplus \cdots \oplus G_m$$

and each element of G can be written uniquely in the form $g_1^{j(1)} g_2^{j(2)} g_3^{j(3)} \cdots g_m^{j(m)}$ where $0 \leq j(i) < k(i)$. Let $g_1 \in G$ have order $k(1)$ and G_1 be the subgroup generated by g_1. If $G_1 = G$, then $m = 1$ and we are finished. If not, select g_2 in $G - G_1$ and let G_2 be the group generated by g_2. Then $g_2^n = 1$ for some n since G is finite. Hence, some power of g_2 is in G_1. Let $k(2)$ be the smallest power of g_2 so that $g_2^{k(2)}$ is in G_1. If $G_1 \cup G_2 = G_2$, then relabel with $G_1 = G_2$ and start over. If $G_1 \cup G_2 \neq G_2$, then it is easily shown that every element in $G_1 \cup G_2$ can be uniquely written in the form $g_1^j g_2^s$, where $0 \leq j < k(1), 0 \leq s < k(2)$. If we select g_3 in $G - (G_1 \oplus G_2)$ and continue the process, we obtain $G_1 G_2 G_3 \cdots G_r G_{r+1} \cdots$; and since G is finite, there exists an m so that $G = G = G_1 \oplus G_2 \oplus G_3 \oplus \cdots \oplus G_m$ and each element of G can be written uniquely in the form $g_1^{j(1)} g_2^{j(2)} g_3^{j(3)} \cdots g_m^{j(m)}$, where $0 \leq j(i) < k(i)$. ∎

THEOREM 21.10 The character group of a group G is isomorphic to G.

Proof Let $G = G_1 \oplus G_2 \oplus \cdots \oplus G_m$, where g_i generates G_i; and let a_i be a primitive n_ith root of unity, where G_i has order n_i. For each $g \in G$, define χ_g as by

$$\chi_g(b) = a_1^{j(1) \cdot h(1)} a_2^{j(2) \cdot h(2)} \cdots a_m^{j(m) \cdot h(m)}$$

where members g and b of G have the direct sum representations

$$g = g_1^{j(1)} g_2^{j(2)} g_3^{j(3)} \cdots g_m^{j(m)}$$
$$b = g_1^{h(1)} g_2^{h(2)} g_3^{h(3)} \cdots g_m^{h(m)}$$

with $0 \leq j(i) < n_i$ and $0 \leq h(i) < n_i$. It is straightforward to show that χ_g is a homomorphism from G into the nth roots of unity for any $g \in G$. Further, if $g \neq g'$, then $\chi_g \neq \chi_{g'}$ so there are $\prod_{i=1}^m n_i = n$ group characters, where G has n elements. Define the mapping η from G into the character group of G as follows:

$$\eta(g) = \eta\left(g_1^{j(1)} g_2^{j(2)} g_3^{j(3)} \cdots g_m^{j(m)}\right) = \left(\chi_{g_1}\right)^{j(1)} \left(\chi_{g_2}\right)^{j(2)} \cdots \left(\chi_{g_m}\right)^{j(m)}$$

η is easily shown to be an isomorphism. ∎

Further, if A is a matrix where $A_{ij} = \chi_{g_i}(g_j)$, then

$$A = \begin{bmatrix} a_1 & 1 & 1 & \cdots & 1 \\ 1 & a_2 & 1 & \cdots & 1 \\ 1 & 1 & a_3 & \cdots & 1 \\ & & \vdots & & \\ 1 & 1 & 1 & \cdots & a_k \end{bmatrix}$$

EXAMPLE 21.11 Consider the set $G_5 = \{[1], [2], [3], [4]\}$, which consists of the nonzero elements of Z_5. Since 5 is prime, the reduced residue classes are just the nonzero classes in Z_5, and the nonzero classes corresponding to the reduced residues form a group under multiplication. Thus, G_5 is a group under multiplication. The multiplication table for G_5 is

$\odot$	[1]	[2]	[3]	[4]
[1]	[1]	[2]	[3]	[4]
[2]	[2]	[4]	[1]	[3]
[3]	[3]	[1]	[4]	[2]
[4]	[4]	[3]	[2]	[1]

To make the following construction of the group characters clearer, we will use the notation found in the proof of Theorem 21.10.

The complex number

$$e^{2\pi t i/n} = \cos\left(\frac{2\pi t}{n}\right) + i \sin\frac{2\pi t}{n}$$

is on the unit circle because, clearly,

$$\left|e^{2\pi t i/n}\right| = \sqrt{\cos^2\left(\frac{2\pi t}{n}\right) + \sin^2\left(\frac{2\pi t}{n}\right)} = 1$$

For any integer t, $e^{2\pi t i/n}$ is an nth root of unity since

$$\left(e^{2\pi t i/n}\right)^n = e^{2\pi t i} = \cos(2\pi t) + i \sin(2\pi t) = 1$$

In fact, the set

$$\left\{e^{2\pi t i/n} : t = 0, 1, 2, \ldots, (n-1)\right\}$$

is the set of the n distinct complex nth roots of unity.

For $n = 4$, the order of G_5, the foregoing formula a gives the four fourth roots of unity.

t	$e^{(2\pi t i)/4}$
0	1
1	i
2	-1
3	$-i$

Calculating $1^4 = i^4 = (-1)^4 = (-i)^4 = 1$, we verify that $1, i, -1$, and $-i$ are fourth roots of unity.

If we let $g_1 = [3]$, then

$$g_1^2 = [3]^2 = [4]$$
$$g_1^3 = [3]^3 = [2]$$
$$g_1^4 = [3]^4 = [1] = g_1^0$$

Hence, $k(1) = 4$ and $G_1 = G$ so that $m = 1$ in the notation of the proof of Theorem 21.10. For $g_1 = [3]$, select the fourth root $r_1 = i$. Since $r_1^2 = -1$, $r_1^3 = -i$, and $r_1^4 = 1$, $r_1 = i$ is primitive. Using this primitive root, we can define a character for every element g of G_5. The following table gives the definition of all the group characters of G_5:

$j(1)$	$g = g_1^{j(1)}$	$\chi_g(b) = \chi_g(g_1^{h(1)})$
0	[1]	$r_1^{0 \cdot h(1)} = i^0$
3	[2]	$r_1^{3 \cdot h(1)} = i^{3h(1)}$
1	[3]	$r_1^{1 \cdot h(1)} = i^{h(1)}$
2	[4]	$r_1^{2 \cdot h(1)} = i^{2h(1)}$

Therefore, the four group characters of the character group of G_5 are given in the next table, where $b = g_1^{h(1)}$:

		$b =$	[1]	[2]	[3]	[4]
		$h(1) =$	0	3	1	2
g	$\chi_g(b)$					
[1]	$\chi_{[1]}(b)$	=	1	1	1	1
[2]	$\chi_{[2]}(b)$	=	1	i	$-i$	-1
[3]	$\chi_{[3]}(b)$	=	1	$-i$	i	-1
[4]	$\chi_{[4]}(b)$	=	1	-1	-1	1

Clearly, $\chi_{[1]} = 1$ is the principal character.

To illustrate the homomorphic nature of group characters, consider $\chi_{[2]}$:

$$\chi_{[2]}([3][4]) = \chi_{[2]}([2]) = i$$
$$\chi_{[2]}([3]) = -i$$
$$\chi_{[2]}([4]) = -1$$
$$\chi_{[2]}([3])\chi_{[2]}([4]) = (-i)(-1) = i$$

The other 15 products may be checked for $\chi_{[2]}$ similarly.

Using the preceding table of group character definitions, it is easy to construct the multiplication table for the character group:

$\cdot$	$\chi_{[1]}$	$\chi_{[2]}$	$\chi_{[3]}$	$\chi_{[4]}$
$\chi_{[1]}$	$\chi_{[1]}$	$\chi_{[2]}$	$\chi_{[3]}$	$\chi_{[4]}$
$\chi_{[2]}$	$\chi_{[2]}$	$\chi_{[4]}$	$\chi_{[1]}$	$\chi_{[3]}$
$\chi_{[3]}$	$\chi_{[3]}$	$\chi_{[1]}$	$\chi_{[4]}$	$\chi_{[2]}$
$\chi_{[4]}$	$\chi_{[4]}$	$\chi_{[3]}$	$\chi_{[2]}$	$\chi_{[1]}$

∎

Exercises

1. Prove Theorem 21.5.

2. Prove that if b is an element of a commutative group G and $b \neq e$, the identity element of G, then there is a group character χ such that $\chi(b) \neq 1$.

3. If g and g' are elements of a finite commutative group G such that $g \neq g'$, then prove that $\chi_g \neq \chi_{g'}$.

4. Complete the proof of Theorem 21.10.

21.3 Semigroup Characters

DEFINITION 21.12 Let S be a finite commutative semigroup. A **semigroup character** on S is a homomorphism from S into the complex numbers with the operation multiplication. As in the case with groups, we will assume that in this section all semigroups used are commutative and finite.

For a finite commutative semigroup S, let $s \in S$. If $|\chi(s)| > 1$, then
$$\left|\chi(s^2)\right| > |\chi(s)|$$
and
$$\left|\chi(s^3)\right| > \left|\chi(s^2)\right|, \ldots, \left|\chi(s^{k+1})\right| > \left|\chi(s^k)\right|, \ldots$$
which is impossible since S is finite. Similarly, if
$$0 |\chi(s)| < 1$$
then
$$0 < \left|\chi(s^2)\right| < |\chi(s)|$$
and
$$0 < \left|\chi(s^3)\right| < \left|\chi(s^2)\right| \cdots$$
$$0 < \left|\chi(s^{k+1})\right| < \left|\chi(s^k)\right| < \cdots < \left|\chi(s^2)\right| < |\chi(s)| < 1$$
which is again impossible. Hence, either $|\chi(s)| = 0$ or $|\chi(s)| = 1$ for each s in S.

DEFINITION 21.13 A subsemigroup I of a commutative semigroup S is an **ideal** of S if $i \in I$ and $s \in S$ implies that $s \cdot i \in I$. The ideal I is a **prime ideal** if $S - I$ is a semigroup.

The proof of the following theorem is left to the reader.

THEOREM 21.14 If S is a finite commutative semigroup and χ is a semigroup character on S, then $I = \{s : |\chi(s)| = 0\}$ is prime ideal. Conversely, if I is a prime ideal in S, then there exists a nonzero character χ_I such that $\chi_I(i) = 0$ if and only if i is in I. If I is a prime ideal, then any group character on $S - I$ can be extended to a semigroup character on S. ∎

At this point we consider the set Z_n of residue classes modulo n. Recall that the set of elements of Z_n consisting of equivalence classes containing integers that are relatively prime to n is called the set of reduced residue classes. The set of reduced residue classes forms a group under class multiplication. We denote this group by G_n. The set Z_n is a semigroup under class multiplication. The set $Z_n - G_n$ is a semigroup prime ideal since the product of any integer and an integer not relatively prime to n is not relatively prime to n. Hence, we can form characters on Z_n that are group characters on G_n and map all of the elements of $Z_n - G_n$ to 0. These characters are called **Dirichlet characters**. We know that the number of Dirichlet characters modulo n is equal to the number of group characters on G_n. Since the set of group characters of G_n is isomorphic to G_n and there are $\phi(n)$ elements in G_n, there are $\phi(n)$ Dirichlet characters modulo n and we have the following theorem.

THEOREM 21.15 There exist $\phi(n)$ distinct Dirichlet characters modulo the positive integer n. ∎

Lemma 21.16 Let G be a finite group, g' an arbitrary element of G, and $G' = \{x : x = g'g \text{ for } g \text{ in } G\}$. Then $G = G'$.

Proof Assume that $G' = \{x : x = g'g \text{ for } g \text{ in } G\}$. Certainly $G' \subseteq G$. Let g_1 and g_2 be elements of G. Then $g'g_1$ and $g'g_2$ are in G'. If $g'g_1 = g'g_2$, then $g_1 = (g')^{-1}g'g_1 = g_2$. Hence, there are as many elements in G' as in G; and since G is finite, $G = G'$. ∎

THEOREM 21.17 Let G be a finite commutative group of order n. If χ is a group character, then

$$\sum_{g \in G} \chi(g) = \begin{cases} n & \text{if } \chi = \chi_1 \\ 0 & \text{if } \chi \neq \chi_1 \end{cases}$$

Proof Since $\chi_1(g) = 1$ for all g in G, the first part is obvious. If $\chi \neq \chi_1$ and g' is an element of G, such that $\chi(g') \neq 1$, then

$$\sum_{g \in G} \chi(g) = \sum_{g \in G} \chi(g'g) \quad \text{by Lemma 21.16}$$

$$= \sum_{g \in G} \chi(g')\chi(g)$$

$$= \chi(g') \sum_{g \in G} \chi(g)$$

so that

$$[1 - \chi(g')] \sum_{g \in G} \chi(g) = 0$$

and since $1 - \chi(g') \neq 0$, $\sum_{g \in G} \chi(g) = 0$. ∎

THEOREM 21.18 If G is a finite commutative group of order n and $g \in G$, then

$$\sum_{\chi} \chi(g) = \begin{cases} n & \text{if } g = e \\ 0 & \text{if } g \neq e \end{cases}$$

where e is the group identity element and where the sum is over all group characters χ.

Proof If $g = e$, then $\chi(e) = 1$ for all χ since every χ is a homomorphism. Thus, the sum equals n. Assume that $g \neq e$ and χ_a is a character such that $\chi_a(g) \neq 1$. Such a character exists since $g = g_1^{j(1)} g_2^{j(2)} g_3^{j(3)} \cdots g_m^{j(m)}$ with $0 \leq j(i) < k(i)$, as shown in the proof of Theorem 21.10. Further, since $g \neq e$, then $j(s) \neq 0$ for some s. We can then let $\chi_a(g)$ be $\chi_{g_s}(g)$ since $\chi_{g_s}(g) = r_s^{j_s} \neq 1$, where r_s is a $k(s)$th primitive root of unity. Then

$$\sum_\chi \chi(g) = \sum_\chi (\chi_a \chi)(g) \quad \text{by Lemma 21.16 applied to character groups}$$

$$= \sum_\chi \chi_a(g)\chi(g)$$

$$= \chi_a(g) \sum_\chi \chi(g)$$

and

$$(1 - \chi_a(g)) \sum_\chi \chi(g) = 0$$

Since $\chi_a(g) \neq 1$, we get $\sum_\chi \chi(g) = 0$. ∎

The next theorem gives two useful orthogonality properties of group characters. The proof is left to the reader.

THEOREM 21.19 If G is a finite commutative group of order n, $g_i, g_j \in G$, and both χ_a and χ_b are group characters of G, then

(a) $\sum_\chi \chi(g_i^{-1})\chi(g_j) = \begin{cases} n & \text{if } g_i = g_j \\ 0 & \text{if } g_i \neq g_j \end{cases}$

(b) $\sum_{g \in G} \chi_a(g) \chi_b^{-1}(g) = \begin{cases} n & \text{if } \chi_a = \chi_b \\ 0 & \text{if } \chi_a \neq \chi_b \end{cases}$ ∎

THEOREM 21.20 Assume that the group G has order n and σ is the $n \times n$ matrix with $\sigma_{ij} = \chi_i(g_j)$, where $\chi_1, \chi_2, \chi_3, \ldots, \chi_n$ are the characters of G. If σ' is the matrix with $\sigma'_{ij} = \text{conj}(\sigma_{ji})$, then

$$\sigma'\sigma = \begin{bmatrix} n & 0 & 0 & \cdots & 0 \\ 0 & n & 0 & \cdots & 0 \\ 0 & 0 & n & \cdots & 0 \\ & & \vdots & & \\ 0 & 0 & 0 & \cdots & n \end{bmatrix} = \sigma\sigma'$$

Proof Let $C = [C_{ij}] = \sigma'\sigma$. Then

$$C_{ij} = \sum_{k=1}^n \sigma'_{ik}\sigma_{kj}$$

$$= \sum_{k=1}^n \text{conj}(\sigma_{ki})\sigma_{kj}$$

$$= \sum_{k=1}^n \text{conj}(\chi_k(g_i))\chi_k(g_j)$$

$$= \sum_{k=1}^n \chi_k(g_i^{-1})\chi_k(g_j)$$

and the first equality of the theorem follows from Theorem 21.19. The proof of the second equality of the theorem is left to the reader. ∎

THEOREM 21.21 There are $\phi(m)$ distinct Dirichlet characters modulo the positive integer m. Each Dirichlet character is totally multiplicative and has the property that $\chi([a+m]) = \chi([a])$, where $[a]$ is the equivalence class of a modulo m, or, equivalently, $\chi(a+m) = \chi(a)$.

Note: Since addition and multiplication in Z_m are the same for any member of an equivalence class in Z_m, we often write $\chi(a+m)$ instead of $\chi([a+m])$. Thus, χ is extended to have domain Z. It is in this extended sense that we speak of each character as being totally multiplicative.

Proof Since each Dirichlet character is a unique extension of a group character from the group of reduced residue classes modulo m, then the first part follows from the fact that the number of group characters of a group is equal to the number of elements in the group. Since each "integer" in the image of a Dirichlet character really represents a congruence class modulo m, $[a] = [a+m]$ and $\chi([a+m]) = \chi([a])$.

The character χ is totally multiplicative since if a and b are relatively prime to m, then $[a]$ and $[b]$ are in the reduced residue group and χ is a homomorphism on this group. If either a or b is not relatively prime to m, then neither is the product and so $\chi([a])\chi([b])$ and $\chi([ab])$ are both 0 or, equivalently, $\chi(a)\chi(b)$ and $\chi(ab)$ are both 0. ∎

Exercises

1. Prove Theorem 21.14.
2. Prove Theorem 21.19.
3. Complete the proof of Theorem 21.20.

GLOSSARY

Chapter 21: Group and Semigroup Characters

Absolute value or length (21.1)	Let C be the set of complex numbers. For $c = a + bi$ in C, $\|c\| = \sqrt{a^2 + b^2}$ is called the **absolute value** or **length** of c.
Character group (21.2)	The **character group of G** consists of the set of group characters of G with the operation $(\chi \cdot \chi')(g) = \chi(g)\chi'(g)$ for group characters χ and χ'.
Conjugate (21.2)	The **conjugate** of the complex number $c = a + bi$ is the number $a - bi$, which is denoted by conj(c).
Direct sum (21.2)	Let $G_1, G_2, G_3, \ldots, G_m$ be subgroups of a group G. Then G is the **direct sum** of $G_1, G_2, G_3, \ldots, G_m$, denoted by $G_1 \oplus G_2 \oplus \cdots \oplus G_m$, if every element of G may be uniquely written in the form $a_1 a_2 \cdots a_m$ where $a_k \in G_k$.
Dirichlet characters (21.3)	Characters on Z_n, which are group characters on G_n and map all of the elements of $Z_n - G_n$ to 0, where G_n is the set of reduced residue classes, are called **Dirichlet characters**.
Group characters (21.2)	A **group character** on a commutative group G is a homomorphism from G into the nonzero complex numbers with the operation of multiplication.
Ideal (21.3)	A subsemigroup I of a commutative semigroup S is an **ideal** of S if $i \in I$ and $s \in S$ implies that $s \cdot i \in I$.
Prime ideal (21.3)	The ideal I is a **prime ideal** if $S - I$ is a semigroup.

Primitive nth root r (21.2)	A **primitive nth root** r of a complex number a is an nth root of a such that $r^k \neq a$ for $0 < k < n$.
Principal character of a group (21.2)	The group character χ_1 defined by $\chi_1(g) = 1$ for all g in the group G is called the **principal character of the group G**.
Semigroup character (21.3)	Let S be a finite commutative semigroup. A **semigroup character** on S is a homomorphism from S into the complex numbers with the operation multiplication.
Unit circle (21.1)	The **unit circle** is the circle of all points on the complex plane whose distance from the origin is 1.

TRUE-FALSE QUESTIONS

1. Let $g \in G$ and $\chi(g) > 1$. G is an infinite group.
2. Every finite group is the direct sum of cyclic groups.
3. A finite Abelian group is isomorphic to its character group.
4. Let g and g' be elements of a finite commutative group. $\chi_g = \chi_{g'}$ if and only if $g = g'$.
5. Let G be a finite commutative group and $g \in G$. $\sum_\chi \chi(g) = 1$ if and only g is the identity of G.
6. If G is a finite commutative group and $g \in G$, then $\sum_{g \in G} \chi(g) = 0$.
7. There are n distinct Dirichlet characters modulo n.
8. The semigroup characters on a group G are the same as the group characters.
9. Let S be a semigroup. For all $s \in S$, and all χ, $\chi(s^2) = \chi(s)$.
10. The multiples of 3 form a prime ideal in the semigroup of integers with the operation multiplication.

ANSWERS TO TRUE-FALSE

1. T **2.** F **3.** T **4.** T **5.** F **6.** F **7.** F **8.** F **9.** T **10.** T

CHAPTER 22
Applications of Number Theory

22.1 Application: Pattern Matching

Pattern matching is a task that appears frequently in computer applications. A simple example of pattern matching is in word processing computer programs when one is writing or editing a document and wants to locate a certain word or phrase, say, the word "effect" in a document. If the document is more than a few paragraphs, then visually searching is tedious, time consuming, and error-prone. A more complex example is when one is presented with a two-dimensional rectangular image or picture represented within the computer application as a two-dimensional array of zero or one bit that indicates black or white. Such a digital image may be of size 1024×1024 bits and can be displayed on a computer display screen. In this case, the task is to search for a certain irregular shape or particular pattern of black or white, say a cross, ✠.

The "brute force" search method simply compares the pattern with every possibility. In the example of searching for the word "effect," we would compare the six characters in "effect" with every sequence of six characters in the document text. This action would require at worst six letters to be compared with every sequence of six contiguous characters in the document. For short patterns and short texts, this method may be quite efficient; however, the text usually is relatively long and the pattern may be long, making the brute force method impractical. By an efficient method we mean one that requires a relatively "short" time to complete or one that requires relatively "few" comparisons or computations. More specifically, we want a search algorithm that uses a minimum of storage, whose number of required comparisons/computations is linearly related to the lengths of the pattern and the search text, and, perhaps for some applications, that executes in "real time" (accepts the search text one character at a time without having to know the entire text beforehand and requires the same computation time for each character of text). Aside from being efficient, it is desirable that an algorithm be extensible to many different types of pattern-searching contexts. We present here a rather general pattern-matching algorithm due to Richard M. Karp and Michael O. Rabin [52] that relies upon choosing random primes. This algorithm has these desirable properties and is an example from a class of probabilistic methods.

All numbers and characters (digits and alphabetic letters) are represented in a computer as sequences of zeros and ones. For example, the representation of the letter "e" using ASCII 8-bit encoding is the bit sequence

$$[0\ 1\ 1\ 0\ 0\ 1\ 0\ 1]$$

which is treated as the base 2 integer (see Section 1.9)

$$0 \cdot 2^7 + 1 \cdot 2^6 + 1 \cdot 2^5 + 0 \cdot 2^4 + 0 \cdot 2^3 + 1 \cdot 2^2 + 0 \cdot 2^1 + 1 \cdot 2^0$$

The word "effect" may be represented as

$$[01100101][01100110][01100110][01100101][01100011][01110110]$$

which could be considered to be a single, long bit string and could be treated as the base 2 integer

$$0 \cdot 2^{47} + 1 \cdot 2^{46} + 1 \cdot 2^{45} + \cdots + 1 \cdot 2^1 + 0 \cdot 2^0$$

The bit string [01100101] for "e" could be represented in base 16 or hexadecimal as 65_{hex} or in base 8 or octal as 145_{octal}. In an actual implementation of the algorithm, a base besides 2 may be more natural; however, in the description of the Karp and Rabin algorithm, we will consider the "pattern" and "text" to be strings of bits.

Let X be the pattern and Y be the text so that

$$X = [x_1\ x_2\ x_3\ \cdots\ x_n] \quad \text{where } x_i = 0 \text{ or } 1$$

$$Y = [y_1\ y_2\ y_3\ \cdots\ y_m] \quad \text{where } y_i = 0 \text{ or } 1$$

The pattern has length n bits and the text has length m bits, where $m \geq n$. Let $Y(i)$ be a substring of n consecutive bits of Y starting at position i, $1 \leq i \leq m - n + 1$, so that

$$Y(i) = [y_i\ y_{i+1}\ y_{i+2}\ \cdots\ y_{i+n-1}]$$

Clearly, we have a match between the pattern and a part of the text for any i for which $X = Y(i)$. Note that X and $Y(i)$ may be considered to be elements of $\{0, 1\}^n = \{0, 1\} \times \{0, 1\} \times \cdots \times \{0, 1\}$, the cross product of the set $\{0, 1\}$ taken n times or the set of all n-tuples of $\{0, 1\}$. Let $N(X)$ and $N(Y(i))$ be the base 2 integers having these representations:

$$N(X) = x_1 2^{n-1} + x_2 2^{n-2} + \cdots + x_{n-1} 2^1 + x_n$$

$$N(Y(i)) = y_i 2^{n-1} + y_{i+1} 2^{n-2} + \cdots + y_{i+n-2} 2^1 + y_{i+n-1}$$

where we have a match if and only if $N(X) = N(Y(i))$.

Comparing $N(X)$ and $N(Y(i))$ requires just as much work as comparing X and $Y(i)$. We need to "contract" the information inherent in $N(X)$ and $N(Y(i))$ to a more manageable amount so that the comparison requires a more manageable amount of work. One way of doing this contraction is to map the integers $N(X)$ and $N(Y(i))$ into smaller integers using modular arithmetic. Let p be a prime in the range $1 \leq p \leq M$ where M is some suitably large positive integer that will be specified later. Let N_p be the mapping from the set of bit strings of length n into the set of primary residues modulo p, $\{0, 1, 2, \ldots, (p-1)\}$, defined by

$$N_p(W) = [[N(W)]]_p$$

so that $N_p(X) = [[N(X)]]_p$ and $N_p(Y(i)) = [[N(Y(i))]]_p$. In a typical application, n may be around 200 and p may be between 2^{16} and 2^{64}. Many computer architectures can compare 16- to 64-bit integers efficiently. The function N_p produces a "proxy" number $N_p(W)$ for $N(W)$ that will be used in comparisons instead of W or of $N(W)$. Such a function is sometimes called a **fingerprint** function and $N_p(W)$ is said to be the fingerprint of W. Note that in this case, the fingerprint function N_p is not one-to-one so that it is possible to have two bit strings W and V of length n with the same fingerprint, that is, with $N_p(W) = N_p(V)$. If $N_p(W) = N_p(V)$, but $W \neq V$, we say that there is a false match. We will resolve this apparent difficulty presently.

If $N(W)$ is reduced modulo p to $N_p(W)$ starting from $N(W)$ for each new W, a great deal of time may be spent in this conversion. Fortunately, it will be possible to compute the sequence $N_p(Y(1))$, $N_p(Y(2))$, $N_p(Y(3))$, ... without such computational overhead. The reduction of $N(W)$ to $N_p(W)$ may be accomplished piecemeal. We note that

$$N(W) = w_1 2^{n-1} + w_2 2^{n-2} + \cdots + w_{n-1} 2^1 + w_n$$
$$= (\cdots((w_1 \cdot 2 + w_2) \cdot 2 + w_3) \cdot 2 + \cdots + w_{n-1}) \cdot 2 + w_n$$

where in this last form the order of the operations of multiplication and addition is precisely specified. According to Theorem 3.55, if $a \equiv b \pmod{p}$, then b may be substituted for a in any expression involving multiplication, addition, and subtraction without changing the modulo p residue of the expression. We say that an operation $a \odot b$ for integers a and b is performed modulo p provided that after the operation is finished, the result is reduced modulo p to its least nonnegative residue; that is, substitute the primary residue $[[a \odot b]]_p$ for $a \odot b$. For brevity, we will sometimes write $a \odot_p b$ instead of $[[a \odot b]]_p$ in order to emphasize that the reduction modulo p is to occur after the operation is performed. For example,

$$\text{substitute } w_1 \cdot_p 2 \text{ for } w_1 \cdot 2$$

and then

$$\text{substitute } (w_1 \cdot_p 2) +_p w_2 \text{ for } (w_1 \cdot_p 2) + w_2$$

giving in the end

$$N_p(W) = (\cdots((w_1 \cdot_p 2 +_p w_2) \cdot_p 2 +_p w_3) \cdot_p 2 +_p \cdots +_p w_{n-1}) \cdot_p 2 +_p w_n$$

Thus, reduction of $N(W)$ to its primary residue modulo p, $N_p(W)$, may be done one operation at a time beginning with the innermost nested parentheses.

We still need to be able to compare X and $Y(i)$ for each i by comparing the proxy integers $N_p(X)$ and $N_p(Y(i))$ for each i; thus, unless $N_p(Y(i))$ can be calculated efficiently for each i, the overall search algorithm may not be practical. It is the case, however, that $N_p(Y(i+1))$ can be obtained from $N_p(Y(i))$ and y_{i+1} using only a few arithmetical operations without completely recalculating $N_p(Y(i+1))$ from scratch. Since

$$N(Y(i+1)) = (N(Y(i)) - 2^{n-1} \cdot y_i) \cdot 2 + y_{i+n}$$

we have, using only operations modulo p,

$$N_p(Y(i+1)) = (N_p(Y(i)) -_p [[2^{n-1}]]_p \cdot_p y_i) \cdot_p 2 +_p y_{i+n}$$

A formula such as this last one which allows the "next" value to be computed easily from the current value is called an **update formula**. Thus, we say that $N_p(Y(i))$ admits an easy update.

The basic idea of the Karp and Rabin algorithm is to choose a prime p at random from a suitable set $1 \le p \le M$ and to compute the fingerprint $N_p(X)$ of the pattern bit string X. Next, compare $N_p(X)$ with the fingerprint $N_p(Y(i))$ of every sequence $Y(i)$ of n contiguous bits in the text bit string Y for $i = 1, 2, 3, \ldots, m - n + 1$. If it happens that $N_p(X) = N_p(Y(k))$ for some k, then we have evidence that $X = Y(k)$ and we stop. Ordinarily, if the algorithm is applied with only one prime p, we would check to see if $X = Y(k)$ because N_p is not one-to-one. If $X \ne Y(k)$, then we would continue with $i = k + 1$; however, as we shall soon see, the use of several primes and several fingerprint functions exemplify the essence of the probabilistic methodology.

PATTERN SEARCH ALGORITHM (Karp and Rabin) Suppose that we are given a pattern bit string X of length n, a text bit string Y of length m with $m \ge n$, and a positive integer M which will depend upon m and n. Let $S = \{p : p \text{ is prime and } 1 \le p \le M\}$. For a prime p in S, let $N_p : \{0, 1\}^n \to \{0, 1, 2, \ldots, (p-1)\}$ be the fingerprint function. Choose k primes $p_1, p_2, \ldots, p_k$ at random from the set $\{1, 2, \ldots, M\}$ by choosing integers at random from $\{1, 2, 3, \ldots, M\}$, testing them for being prime, and stopping when k primes have been found. The search proceeds as follows.

For each i, $1 \le i \le m - n + 1 = t$, compare $N_{p(j)}(X)$ with $N_{p(j)}(Y(i))$ for each prime $p_j = p(j)$ for $1 \le j \le k$ either until there is an i such that

$$N_{p(j)}(X) = N_{p(j)}(Y(i)) \text{ for all } j, 1 \le j \le k$$

at which time we declare that a "match" has occurred, or until $i = t$ with no "match" having occurred.

Note that a complete match requires that all k fingerprint functions produce a match. The other important feature is that the primes $p_1, p_2, \ldots, p_k$ are chosen at random from a large enough set. It is still possible for a false match to occur, namely, for $N_{p(j)}(X) = N_{p(j)}(Y(i))$ for all j but with $X \ne Y(i)$; however, the nub of the matter is that the probability or chance of a false match can be made arbitrarily small. This last characteristic embodies what is meant by a probabilistic algorithm.

EXAMPLE 22.1

In order to fix the ideas of this algorithm, we will consider a smaller problem in this example than that for which the method is intended to be applied. Suppose that

$$X = [1001\ 1011] \text{ where } n = 8$$

and

$$Y = [1011011001101100 \cdots 0] \text{ where } m = 300$$

Let $t = m - n + 1 = 293$ and $M = nt^2 = 686792$. Suppose that $k = 2$ and we have obtained these primes at random:

$$p_1 = 47 \text{ and } p_2 = 31$$

We first obtain the fingerprints of X by computing the primary residues

$$N_{p(1)}(X) = [[N(X)]]_{p(1)} = [[155_{ten}]]_{47} = 14$$
$$N_{p(2)}(X) = [[N(X)]]_{p(2)} = [[155_{ten}]]_{31} = 0$$

Starting with $i = 1$, we calculate the fingerprints of $Y(i)$ relative to the primes 47 and 31 and check for matches. Matches with either of the two fingerprints are indicated by an asterisk ($*$).

i	$Y(i)$	$N(Y(i))$	$N_{47}(Y(i))$		$N_{31}(Y(i))$		Match?
1	[1011 0110]	182	41		27		No
2	[0110 1100]	108	14	*	15		No
3	[1101 1001]	217	29		0	*	No
4	[1011 0011]	179	38		24		No
5	[0110 0110]	102	8		9		No
6	[1100 1101]	205	17		19		No
7	[1001 1011]	155	14	*	0	*	Yes

The first match occurs at position 7 in the bit string Y. We should note that because some of the numbers are small in this example, we performed direct calculations of the primary residues. In an actual computer implementation, we would need to utilize the modular operations and the updating formula described above. ∎

Following Karp and Rabin, we now derive a bound on the probability of a false match when using the algorithm. First, we need two concepts: probability and a number theoretic function.

Let T be a nonempty set of "possibilities," and any one of these possibilities may be chosen. A subset A of T contains the possibilities in which we are interested. Choose one element of T. The probability of choosing an element of A is defined to be

$$\text{PROB}(A) = \frac{\text{Number of elements in } A}{\text{Number of elements in } T}$$

Under these circumstances, we say that the element is chosen at random from the set T (any element of T is as likely to be chosen as any other element) and that event A has occurred if the element chosen is in A. The probability of event A is given by the foregoing formula. For example, suppose that T is the set of faces of a cubical die so that $T = \{1, 2, 3, 4, 5, 6\}$ and we roll the die once (choose a side at random). Let A be the event (set of possible outcomes) of having a prime number of spots. $A = \{2, 3, 5\}$. Then $\text{PROB}(A) = 3/6 = 0.5$. Evidently, the probability of any event is between 0 and 1 inclusive.

Let $\pi(w)$ be the number of primes less than or equal to w. Thus, $\pi(1) = 0$, $\pi(2) = 1$, $\pi(5) = 3$, and $\pi(20) = 8$.

We state without proof two results from J. B. Rosser and L. Schoenfeld [94].

THEOREM 22.2 If $w \geq 29$, then

$$p_1 p \cdots p_{\pi(w)} > 2^w$$

where $p_1, p_2, p_3, \ldots, p_{\pi(w)}$ are the primes less than or equal to w. ∎

THEOREM 22.3 If $w \geq 17$, then

$$\frac{w}{\log(w)} \leq \pi(w) \leq \frac{1.25506w}{\log(w)}$$
∎

THEOREM 22.4 If $w \geq 29$ and $b \leq 2^w$, then b has fewer than $\pi(w)$ prime divisors.

Proof Assume that $w \geq 29$ and $b \leq 2^w$, but that b has at least $\pi(w)$ prime divisors. Let $d_1 < d_2 < \cdots < d_r$ be these prime divisors of b so that $r \geq \pi(w)$. Let $p_1 < p_2 < \cdots < p_k$ be the first k primes for any k. Then

$$2^w \geq b \qquad \text{given}$$
$$b \geq d_1 d_2 d_3 \cdots d_r \qquad \text{since } d_1 d_2 \cdots d_r \mid b$$
$$d_1 d_2 \cdots d_r \geq p_1 p_2 \cdots p_r \qquad \text{since } d_i \geq p_i \text{ for each } i$$
$$p_1 p_2 \cdots p_r \geq p_1 p_2 \cdots p_{\pi(w)} \qquad \text{since } r \geq \pi(w)$$
$$p_1 p_2 \cdots p_{\pi(w)} > 2^w \qquad \text{by Theorem 22.2}$$

But $2^w > 2^w$ is a contradiction. Thus, b must have fewer than $\pi(w)$ prime divisors. ∎

As mentioned before, this search algorithm has the possibility of terminating with a false match. The algorithm would not be of much use if a false match occurred often. The next theorem says that the chances of this algorithm producing a false match are no larger than a known value.

THEOREM 22.5 If the algorithm is applied to a search instance of X and Y with related characteristics n, m, t, M, k, and S, then the probability of a false match is less than or equal to

$$t \left(\frac{\pi(n)}{\pi(M)} \right)^k \quad \text{when } n \geq 29$$

Proof Assume that $n \geq 29$, $1 \leq r \leq t$, and $X \neq Y(r)$. Since $0 \leq N(X) < 2^n$ and $0 \leq N(Y(r)) < 2^n$, then $|N(X) - N(Y(r))| < 2^n$. By Theorem 22.4, the number of primes dividing $|N(X) - N(Y(r))|$ is less than $\pi(n)$; therefore, for any j, $1 \leq j \leq k$, since the prime p_j is chosen at random from S which has $\pi(M)$ elements,

$$\text{PROB}\left(p_j \text{ divides } |N(X) - N(Y(r))|\right) \leq \frac{\pi(n)}{\pi(M)}$$

Since the k primes $p_1, p_2, \ldots, p_k$ are randomly and independently chosen, it is a property of probability of the intersection of these k events that

$$\text{PROB}\left(p_j \text{ divides } |N(X) - N(Y(r))| \text{ for all } j\right) \leq \left(\frac{\pi(n)}{\pi(M)} \right)^k$$

Since there are t mutually exclusive instances of r, it is a property of probability that all the primes $p_1, p_2, \ldots, p_k$ divide $|N(X) - N(Y(r))|$ for some r is less than or equal to $t \cdot \left(\frac{\pi(n)}{\pi(M)} \right)^k$. ∎

THEOREM 22.6 Under the hypotheses of Theorem 22.5, if $M = nt^2$ with $n \geq 29$, then the probability of a false match is less than or equal to

$$(1.25506)^k \, t^{-(2k-1)} \, (1 + 0.6 \log(t))^k$$

Proof Theorem 22.5 implies that

$$\text{PROB (false match)} \leq t \cdot \left(\frac{\pi(n)}{\pi(M)} \right)^k$$

Theorem 22.3 implies that

$$\pi(n) \leq c \frac{n}{\log(n)} \quad \text{where } c = 1.25506$$

and that
$$\frac{nt^2}{\log(nt^2)} \leq \pi(nt^2)$$

Thus, we have

$$\text{PROB (false match)} \leq t \cdot \left[\frac{c \cdot n}{\log(n)} \cdot \frac{\log(nt^2)}{nt^2}\right]^k$$

$$= t \cdot \left[\frac{c}{t^2} \cdot \left(1 + 2 \cdot \frac{\log(t)}{\log(n)}\right)\right]^k$$

But for $n \geq 29$, $\log(n) \geq \log(29)$; therefore, the conclusion of the theorem holds since $\frac{2}{\log(n)} \leq \frac{2}{\log(29)} \leq 0.6$ by direct calculation. ∎

To illustrate Theorem 22.6, suppose that we have a string-matching problem with the pattern X being of length $n = 200$, which corresponds to 25 8-bit ASCII-encoded characters, and with the text string Y being of length 32000 corresponding to 4000 ASCII characters. Then $t = 31801$ and $M = nt^2 < 200(32000)^2 = 2^{11} \cdot 10^8 \approx 2^{11} \cdot 2^{26.6} < 2^{38}$ so that the base 2 computer representations of the primes p_i and corresponding residues will have fewer than 38 bits. If $k = 4$, then the probability of a false match will be less than about $2 \cdot 10^{-28}$.

The Karp and Rabin algorithm is a real-time algorithm requiring the same computation time for each text character. It also requires a constant amount of storage depending only upon n, k, and t. The algorithm can be proved to have a small probability of error. In addition, the method easily generalizes to apply to multidimensional problems. Other "fingerprinting" functions besides N_p may be appropriate.

We note that Karp and Rabin give other similar but different probabilistic pattern-searching algorithms in the reference cited earlier. For comparison purposes, information about some efficient nonprobabilistic searching algorithms may be found in (1) "Fast Pattern Matching in Strings," by D. E. Knuth, J. H. Morris, and V. R. Pratt [57]; (2) "A Fast String Searching Algorithm," by R. S. Boyer and J. S. Moore [9]; (3) by Jack Purdum "Pattern Matching Alternatives: Theory vs. Practice" [86]; and (4) "A Very Fast Substring Search Algorithm," by Daniel M. Sunday [105].

Exercises

1. Prove that if W and V are bit strings of length n, then $|N(W) - N(V)| < 2^n$.

2. In Theorem 22.2, the product of all primes less than or equal to w was shown to be greater than 2^w as long as $w \geq 29$. Determine which positive integers $w < 29$ have the property of the conclusion of this theorem.

3. In Theorem 22.3, two inequalities were shown to hold for every positive integer $w \geq 17$. By direct computation, determine which positive integers $w < 17$ also satisfy these inequalities.

4. For a pattern X of length $n = 100$ and a text of length $m = 20000$, use Theorem 22.6 to determine an upper bound on the probability of a false match for $k = 1, 2, 3,$ and 4.

5. Assume that the pattern length n bits is given, that the number of primes k has been given, that M is such that all fingerprinting primes and their residues have fewer than 32 bits, and that the updating described in the discussion will be used. Determine the number of 32-bit computer words of information storage that are needed to implement the Karp and Rabin algorithm.

22.2 Application: Hashing Functions

In computer applications the situation frequently occurs in which there are n sets of information $I_1, I_2, \ldots, I_n$ where each set is "named" with a key, say $k_1, k_2, \ldots, k_n$, respectively. It is desirable to be able to find the information I_j quickly when the key k_j is given. The keys may be considered to be integers even though they may be finite sequences of alphabetic characters. Searching lists of keys may require too much time; moreover, if a search for information using the keys must be conducted many times, the computational work may be considerable. A compiler is a computer program that converts another computer program written in a high-level computer language such as FORTRAN or Pascal into a machine language program. During the process of "compiling," the compiler must produce tables and find items with the associated information in the tables many times. The important idea here is that the "lookup" table is created once but is used many times. Since there may be many such tables, one also needs to use the least storage feasible.

One solution to the "location" problem is to use a **hash function** $h : K \to \{0, 1, \ldots, m\}$ where $K = \{k_1, k_2, \ldots, k_n\}$ and where $m + 1 \geq n$. We then provide $m + 1$ computer storage locations where the n sets of information are kept. Thus, to find I_j, we calculate $h(k_j)$ and go to the location $h(k_j)$ where I_j is kept. It is generally difficult to find an ideal hash function, that is, one where $h(k_r) \neq h(k_s)$ when $k_r \neq k_s$ and where $m + 1 = n$. Hash functions that are often used map K into a set much larger than n elements and are not one-to-one. For h not one-to-one, provision must be made for what to do when there is collision, that is, when k_r and k_s are distinct keys but $h(k_r) = h(k_s)$. Donald E. Knuth in Volume 3 of *The Art of Computer Programming* [56] discusses various choices for hash functions and methods of dealing with collisions. G. Jaeschke in "Reciprocal Hashing: A Method for Generating Minimal Perfect Hashing Functions" [49] describes a particular type of minimum hashing function, that is, a hash function that is one-to-one and uses minimum storage (**minimum perfect hashing function**). Jaeschke's hashing function requires that one be able to map the keys $\{k_1, k_2, \ldots, k_n\}$, which may not be relatively prime to one another, to a set of integers $\{f(k_1), f(k_2), \ldots, f(k_n)\}$ that are pairwise relatively prime. He proved the following theorem.

THEOREM 22.7 Let $K = \{k_1, k_2, \ldots, k_n\}$ be a set of n distinct positive integers. There exist two integers D and E such that if $f(x)$ is defined by $f(x) = Dx + E$, then the members of the set of positive integers $\{f(k_1), f(k_2), \ldots, f(k_n)\}$ are pairwise relatively prime. Thus, the integers $Dk_1 + E, Dk_2 + E, \ldots,$ and $Dk_n + E$ are pairwise relatively prime.

Proof The proof is longer than we can present here (see Jaeschke's 1981 paper [49]); however, we will consider a special case later that will be adequate for our purposes. ∎

Jaeschke's main result is the next theorem, where $\lfloor w \rfloor$ is the greatest positive integer less than or equal to w.

THEOREM 22.8 If the set $K = \{k_1, k_2, \ldots, k_n\}$ has distinct positive integer keys, then there exist integers C, D, and E such that

$$h(x) = \left[\left[\left\lfloor \frac{C}{Dx+E} \right\rfloor \right] \right]_n \equiv \left\lfloor \frac{C}{Dx+E} \right\rfloor \pmod{n}$$

where $x \in K$, is a minimum perfect hashing function.

Proof Let $k_1 < k_2 < \cdots < k_n$ and let the integers D and E be given by Theorem 22.7 so that $f(k_j) = Dk_j + E$ and $\gcd(f(k_i), f(k_j)) = 1$ for $i \neq j$. Since D and E may be chosen so that $f(k_j) > n$ for each j, it is possible to choose n integers $a_1, a_2, \ldots, a_n$ such that $a_i \not\equiv a_j \pmod{n}$ and such that

$$(i-1) \cdot (Dk_i + E) \leq a_i < i \cdot (Dk_i + E)$$

Then, by the version of the Chinese remainder theorem applicable to nonrelatively prime moduli (Theorem 10.14), there is a number C such that

$$C \equiv a_1 \pmod{n(Dk_1 + E)}$$
$$C \equiv a_2 \pmod{n(Dk_2 + E)}$$
$$\vdots$$
$$C \equiv a_n \pmod{n(Dk_n + E)}$$

Therefore, there are integers q_i such that $C = q_i[n(Dk_i + E)] + a_i$ for all i. Consequently,

$$\left\lfloor \frac{C}{Dk_i + E} \right\rfloor = q_i n + (i-1) \equiv i - 1 \pmod{n}$$

This result implies that the function h defined in the theorem has the property $h(k_i) \neq h(k_j)$ when $i \neq j$ so that h is one to one. Further, the range of the function h is $\{0, 1, \ldots, (n-1)\}$. ∎

Jaeschke gives algorithms for computing C, D, and E; however, the methods are exhaustive in nature. On the other hand, C. C. Chang and J. C. Shieh in "Pairwise Relatively Prime Generating Polynomials and Their Applications" [17] give a way to calculate a required polynomial of the form $f(x) = Dx + 1$ where $E = 1$. The proof of the next lemma is left to the reader.

Lemma 22.9 If a and b are positive integers, $a > b$, and d is a multiple of $a - b$, then $d(a-b)$ and $da + 1$ are relatively prime. ∎

THEOREM 22.10 Let $K = \{k_1, k_2, \ldots, k_n\}$ be a set of n distinct positive integers with $k_i < k_{i+1}$. If $\{t_1, t_2, \ldots, t_s\} = \{k_i - k_j : 1 \leq j < i \leq n\}$ is the set of $s = n(n-1)/2$ differences, then $D = w \cdot \mathrm{lcm}(t_1, t_2, \ldots, t_s)$, where w is any positive integer, has the property that $Dk_1 + 1, Dk_2 + 1, \ldots, Dk_n + 1$ are pairwise relatively prime.

Proof Since $\gcd(a, b) = \gcd(a - b, a)$ when a and b are integers for which both $\gcd(a, b)$ and $\gcd(a - b, b)$ are defined, we obtain $\gcd(Dk_i + 1, Dk_j + 1) = \gcd(D(k_i - k_j), Dk_i + 1)$ for $i > j$, where D is as given in the theorem. By definition of D, D is a multiple of $(k_i - k_j)$; therefore, Lemma 22.9 gives $\gcd(Dk_i + 1, Dk_j + 1) = 1$ for $i > j$. ∎

We note that the proof of Theorem 22.10 requires only that D be a multiple of all of $t_1, t_2, \ldots, t_s$. Ordinarily, one would probably choose as small a D as possible ($w = 1$); however, the proof of Theorem 22.8 requires a D such that $f(k_i) = Dk_i + 1 > n$. Such a D can always be obtained by selecting w large enough.

EXAMPLE 22.11 Suppose that $K = \{3, 6, 7, 12\}$. Then

$$\{k_i - k_j : 1 \leq j < i \leq 4\} = \{6-3, 7-3, 12-3, 7-6, 12-6, 12-7\}$$
$$= \{3, 4, 9, 1, 6, 5\}$$

so that $D = \text{lcm}(3, 4, 9, 1, 6, 5) = 180$. Thus, $f(x) = 180x + 1$ and

i	k_i	$f(k_i) = Dk_i + 1$
1	3	$180 \cdot 3 + 1 = 541$
2	6	$180 \cdot 6 + 1 = 1081$
3	7	$180 \cdot 7 + 1 = 1261$
4	12	$180 \cdot 12 + 1 = 2161$

By inspection, any two of 541, 1081, 1261, and 2161 are relatively prime. ∎

In order to implement Jaeschke's reciprocal hashing we must be able to compute the integer C. Fortunately, C. C. Chang and J. C. Shieh in "A Fast Algorithm for Constructing Reciprocal Hashing Functions" [16] provide an algorithm for computing the integer C.

In the rest of this section we use both the greatest integer function $\lfloor w \rfloor$ described earlier and the ceiling function $\lceil w \rceil$, which is the least positive integer greater than or equal to w. Thus, $\lceil 5.23 \rceil = 6$ and $\lceil 8 \rceil = 8$.

THEOREM 22.12 If

(a) $m_1, m_2, \ldots, m_n$ are pairwise relatively prime integers,

(b) $m_i > n$ for all i, $1 \leq i \leq n$,

(c) $M = \prod_{j=1}^{n} m_j$,

(d) $M_i = n \prod_{j \neq i} m_j = \dfrac{nM}{m_i}$ for all i,

(e) b_i is such that $M_i b_i \equiv n \pmod{nm_i}$, and

(f) $N_i = \left\lceil \dfrac{(i-1)m_i}{n} \right\rceil$ for each i,

then

1.
$$\left\lfloor \dfrac{\sum_{j=1}^{n} M_j b_j N_j}{m_i} \right\rfloor \equiv (i-1) \pmod{n} \text{ for all } i$$

2.
$$C = \left[\left[\sum_{j=1}^{n} M_j b_j N_j \right] \right]_{nM}$$

is the smallest positive integer such that $\left\lfloor \dfrac{C}{m_i} \right\rfloor \equiv (i-1) \pmod{n}$ for all i.

Proof Let $a_i = N_i \cdot n$. Then a_i is the smallest multiple of n such that $(i-1) \cdot m_i \leq a_i < i \cdot m_i$ for all i and $a_i \not\equiv a_j \pmod{n}$ when $i \neq j$. The sum $\sum_{j=1}^{n} M_j b_j N_j$ is a solution of the n congruences

$$x \equiv a_1 \pmod{nm_1}$$
$$x \equiv a_2 \pmod{nm_2}$$
$$\vdots$$
$$x \equiv a_n \pmod{nm_n}$$

because if $j \neq i$, then M_j contains the factor nm_i; so $M_j b_j N_j \equiv 0 \pmod{nm_i}$ and

$$\sum_{j=1}^{n} M_j b_j N_j \equiv M_i b_i N_i \pmod{nm_i}$$
$$\equiv n N_i \equiv a_i \pmod{nm_i}$$

Therefore,

$$W_i = \frac{\sum_{j=1}^{n} M_j b_j N_j}{m_i} = \frac{\sum_{j \neq i}^{n} M_j b_j N_j}{m_i} + \frac{M_i b_i N_i}{m_i} = n J_i + \frac{M_i b_i N_i}{m_i}$$

for an appropriate integer J_i. Since $M_i b_i \equiv n \pmod{nm_i}$, there exists an integer t_i such that $M_i b_i = t_i(nm_i) + n$ so that

$$W_i = nJ_i + t_i n N_i + \frac{nN_i}{m_i}$$

Thus, because $(i-1) \cdot m_i \leq a_i < i \cdot m_i$,

$$\lfloor W_i \rfloor \equiv \left\lfloor \frac{nN_i}{m_i} \right\rfloor \equiv \left\lfloor \frac{a_i}{m_i} \right\rfloor \equiv (i-1) \pmod{n}$$

which is part (1).

Let $C = \left[\left[\sum_{j=1}^{n} M_j b_j N_j\right]\right]_{nM}$. We need to show the congruence $\left\lfloor \frac{C}{m_i} \right\rfloor \equiv (i-1)$ $\pmod{n}$ for $1 \leq i \leq n$. By the division algorithm there is an integer J such that $\sum_{j=1}^{n} M_j b_j N_j = J(nM) + C$. For $1 \leq i \leq n$, again using the division algorithm, there are integers q_i and r_i with $0 \leq r_i < m_i$ such that

$$\sum_{j=1}^{n} M_j b_j N_j = q_i m_i + r_i$$

so that

$$\left\lfloor \frac{\sum_{j=1}^{n} M_j b_j N_j}{m_i} \right\rfloor = q_i$$

Therefore, by the result of part (1), there is an integer t_i such that

$$q_i = (i-1) + t_i n$$
$$q_i m_i = (i-1)m_i + t_i n m_i$$
$$q_i m_i + r_i = (i-1)m_i + t_i n m_i + r_i$$
$$J(nM) + C = (i-1)m_i + t_i n m_i + r_i$$

so that

$$\frac{C}{m_i} = (i-1) + \left(t_i - \frac{JM}{m_i}\right)n + \frac{r_i}{m_i}$$

and

$$\left\lfloor \frac{C}{m_i} \right\rfloor = (i-1) + \left(t_i - \frac{JM}{m_i}\right)n$$

$$\left\lfloor \frac{C}{m_i} \right\rfloor \equiv (i-1) \pmod{n}$$

It is left to the reader to show that if C' is any other positive integer such that $\left\lfloor \frac{C'}{m_i} \right\rfloor \equiv (i-1) \pmod{n}$ for all i, $1 \le i \le n$, then $C' > C$. ∎

Finally, we combine the last several results into the following theorem.

THEOREM 22.13 Let $K = \{k_1, k_2, \ldots, k_n\}$ be a set of n distinct positive integers; and let D and E be the integers, and $f(x) = Dx + E$ the function, described in Theorems 22.8 and 22.10. Also, assume that C is given by part (2) of Theorem 22.12 where $m_i = f(k_i)$. Then

$$h(x) = \left[\left[\left\lfloor \frac{C}{Dx + E} \right\rfloor\right]\right]_n$$

is a minimum perfect hashing function.

Proof If $m_i = f(k_i)$, Theorem 22.12 implies that

$$h(k_i) = \left[\left[\left\lfloor \frac{C}{f(k_i)} \right\rfloor\right]\right] = (i-1) \text{ for } 1 \le i \le n$$

Thus, h is one-to-one and onto $\{0, 1, \ldots, (n-1)\}$. ∎

EXAMPLE 22.14 We now continue Example 22.11 where $K = \{k_1, k_2, k_3, k_4\} = \{3, 6, 7, 12\}$ and $f(x) = 180x + 1$. We found that $\{f(k_1), f(k_2), f(k_3), f(k_4)\} = \{541, 1081, 1261, 2161\}$. Let $m_i = f(k_i)$. The m_i's are pairwise relatively prime and $m_i > n = 4$. Using Theorem 22.12, we obtain the following table:

i	m_i	M_i	b_i	N_i	$M_i b_i N_i$	$h(k_i)$
1	541	11782990804	235	0	0	0
2	1081	5896945444	12	273	19318393274544	1
3	1261	5055192724	172	631	548650176721168	2
4	2161	2949837124	1303	1621	6230536829339212	3

For example, to find b_3 we solve $5055192724 \cdot b_3 \equiv 4 \pmod{5044}$, which is equivalent to solving $1263798181 \cdot b_3 \equiv 1 \pmod{1261}$ or to solving $22 \cdot b_3 \equiv 1 \pmod{1261}$. Therefore, we want b_3 and y such that $22b_3 + 1261y = 1$. Using the Euclidean algorithm and back substitution, we obtain the result $22 \cdot 172 + (-3) \cdot 1261 = 1$. Thus,

$$C = \left[\left[\sum_{j=1}^{n} M_j b_j N_j\right]\right]_{nM} = [[6798505399334924]]_{nM} = 3183904723300$$

Next we use C, f, and h to hash $k_3 = 7$.

$$h(k_3) = h(7) = \left[\left[\left\lfloor \frac{3183904723300}{1261} \right\rfloor\right]\right]_4 = [[2524904618]]_4 = 2 \qquad \blacksquare$$

Jaeschke's reciprocal hashing function generates large integers. The computational effort necessary to implement it should be compared to the resources required for using a nonminimal and nonperfect hashing function: (1) possible large memory requirements, (2) dealing with collisions, and (3) using search algorithms to find items in a table.

Exercises

1. If a and b are integers for which both $gcd(a, b)$ and $gcd(a - b, a)$ are defined, prove that $gcd(a, b) = gcd(a - b, a)$.

For each of these sets, find a polynomial $f(x) = Dx + 1$ that maps the set one-to-one and onto a set of pairwise relatively prime integers:

2. $\{12, 14, 15, 18\}$ 3. $\{5, 15, 20, 25\}$

4. Let $a_1, a_2, \ldots, a_n$ be n distinct positive integers. Prove or disprove: $gcd(a_1, a_2, \ldots, a_n) = 1$ if and only if $gcd(a_i, a_j) = 1$ for all $i \neq j$.

5. In the proof of Theorem 22.12, show that $a_i = N_i n$ is the smallest multiple of n such that $(i - 1) \cdot m_i \leq a_i < i \cdot m_i$.

6. Verify the numbers in the table of Example 22.14.

7. For $K = \{12, 14, 15, 18\}$ of Exercise 2

8. $K = \{5, 15, 20, 25\}$ of Exercise 3

compute the integer C of Theorem 22.12 and, therefore, the minimum perfect reciprocal hashing function of Theorem 22.13. Verify that $h(k_i) = i - 1$ for each i.

22.3 Application: Cryptography

Governments, companies, and individuals have a need to send messages in such a way that only the intended recipient is able to read the messages. Armies send battle orders, banks wire fund transfers, and individuals make purchases using credit or debit cards. The messages are sent over telephone lines, via radio and satellite, or through computer networks and are subject to being intercepted by, perhaps hostile, third parties. To prevent a third party from knowing or modifying the messages, they are scrambled in such a way that the original is obscured. The scrambling process is called encryption or encipherment and the unscrambling process is called decryption or decipherment. The original message is called the plaintext and the scrambled message is the ciphertext. See Figure 22.1.

Of course, both the sender and recipient must have agreed upon the scrambling and unscrambling procedure. This procedure usually consists of a general method along with a "key." The key provides specific information that, with the general method, allows easy encryption by the sender and easy decryption by the recipient; however, without the key, anyone who intercepts the transmission finds it computationally impractical to recover the original message even though they know the general method. For example, message recovery would require, say, a million years using the fastest computer available or would require examining all possible messages through an exhaustive trial-and-error search. An attempt to decrypt the plaintext from the ciphertext without benefit of the general method or the key is called cryptanalysis. The mechanism is illustrated in Figure 22.1 where E is the encryption function and D is the decryption function. Thus, $D(E(M)) = M$. Cryptography is the discipline that deals with the study of methods of encryption and the cryptanalysis of those methods.

Figure 22.1

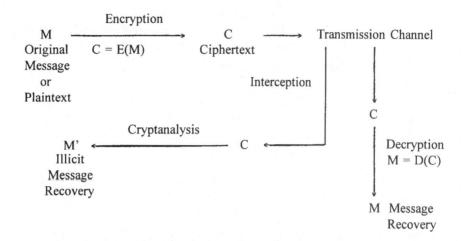

Encryption and decryption are usually done by computers; therefore, a message M is first converted to an integer or a series of integers in order to allow manipulation using computer arithmetic. Thus, we will think of M as an integer. The integer message M is then usually divided into blocks $M_1, M_2, M_3, \ldots, M_k$ so that each block is an integer M_i with $0 \le M_i < n$, where n is previously chosen. The encryption is then done block by block as $E(M_i) = C_i$ and decryption proceeds similarly to $D(C_i) = M_i$. For example, one could use 8-bit ASCII computer codes for each alphabetic character as follows:

$$
\begin{aligned}
M &\quad 01001101_{binary} = 77_{ten} \\
O &\quad 01001111_{binary} = 79_{ten} \\
N &\quad 01001110_{binary} = 78_{ten} \\
E &\quad 01001010_{binary} = 69_{ten} \\
Y &\quad 01011001_{binary} = 89_{ten}
\end{aligned}
$$

Then M_i could correspond to groups of two letters, giving

$$"MO" \to M_1 = 01001101\,01001111_{binary} = 19791_{ten}$$

so that $0 \le M_i < n = 2^{16} = 65536$. Of course, the entire message could be considered to be a single string of binary digits which could then be blocked into a sequence of blocks of any convenient size.

R. L. Rivest, A. Shamir, and L. Adleman in "A Method for Obtaining Digital Signatures and Public Key Cryptosystems" [91] introduce a general method of encryption, known as the RSA method after the authors. The method has many desirable characteristics.

Let p and q be two large primes, each having, say, 100 digits each and let $n = p \cdot q$. Two integers e and d that are related to p, q, and n will be used to encrypt and decrypt. Both e and d will be determined later. To encrypt a message $M = M_i$ we calculate

$$C = E(M) = [\![M^e]\!]_n$$

and to decrypt we calculate

$$M = D(C) = [\![C^d]\!]_n$$

Since the products M^e and C^d are reduced modulo n, both ciphertext and plaintext blocks are integers in the range 0 to $(n-1)$. The key to encrypt is the integer pair (n, e) and the key to decrypt is the integer pair (n, d).

First, choose d to be a "large" integer relatively prime to the product $(p-1) \cdot (q-1)$. Then determine the integer e, $0 \leq e < n$, that is, the unique solution (Theorem 3.66) of the congruence

$$d \cdot x \equiv 1 \pmod{(p-1)(q-1)}$$

Therefore, since $\phi(n) = \phi(p)\phi(q) = (p-1)(q-1)$,

$$d \cdot e \equiv 1 \pmod{\phi(n)}$$

THEOREM 22.15 If

(a) $n = p \cdot q$, where p and q are distinct primes,
(b) $\gcd(d, \phi(n)) = 1$,
(c) $d \cdot e \equiv 1 \pmod{\phi(n)}$,
(d) for $0 \leq J < n$, define functions

$$E(J) = \left[\left[J^e\right]\right]_n \text{ and } D(J) = \left[\left[J^d\right]\right]_n$$

then, for $0 \leq J < n$,

$$D(E(J)) = J \text{ and } E(D(J)) = J$$

Proof Let $0 \leq J < n$. Since, $A \equiv B \pmod{n}$ implies that $[[A]]_n = [[B]]_n$, we have

$$D(E(J)) = D\left(\left[\left[J^e\right]\right]_n\right) = \left[\left[\left[\left[J^e\right]\right]_n^d\right]\right]_n$$

$$= \left[\left[(J^e)^d\right]\right]_n = \left[\left[J^{ed}\right]\right]_n$$

$$= \left[\left[J^1\right]\right]_n = J$$

because $d \cdot e \equiv 1 \pmod{\phi(n)}$ and because $r \equiv s \pmod{\phi(n)}$ if and only if $x^r \equiv x^s \pmod{n}$ by Theorem 10.31. The result $E(D(J)) = J$ is proved similarly. ∎

EXAMPLE 22.16 Suppose that we decide to use 8-bit ASCII encoding for alphabetic characters with a block size of one character or letter. We will need $n \geq 2^8 = 256$. Let $p = 41$ and $q = 73$ so that $n = p \cdot q = 2993$. $\phi(n) = 40 \cdot 72 = 2880 = 2^6 \cdot 3^2 \cdot 5$. Let $d = 217 = 31 \cdot 7$ so that $\gcd(217, 2880) = 1$. Using the Euclidean algorithm, we obtain $e = 1513$ as a solution to $217x \equiv 1 \pmod{2880}$. Then *"MONEY"* is encrypted in the following table using the RSA method with key $(n, e) = (2993, 1513)$:

i	Block	M_i	$E(M_i) = \left[\left[M_i^e\right]\right]_n = C_i$
1	"M"	77	1683
2	"O"	79	79
3	"N"	78	2560
4	"E"	69	872
5	"Y"	89	2571

We have used the algorithm of Section 3.6 to calculate M_i^e modulo n efficiently. One may show also that $D(C_i) = M_i$ in every case using the same algorithm with key $(n, d) = (2993, 217)$. The small size of n in this example does not yield a secure encryption function E because small n may be easily factored. In addition, secure blocking will have more than one letter in a block since otherwise we would have a simple substitution cipher that yields easily to direct cryptanalysis. ∎

The RSA method is intended to be used as follows. Both the two primes p and q giving $n = pq$ and the integer d relatively prime to $\phi(n)$ are kept secret; however, the integers n and e are made public, and the general method is public. Therefore, anyone knowing the encryption key (n, e) may encrypt a message M, resulting in an encrypted result C that only someone with the key (n, d) may read. Everyone could have his own set of RSA cryptosystem keys. Each person's public key, (n, e), could be listed, say, in a register as telephone numbers now are. If a person, say Alice, wants to send a secure message M_A to Bob, then Alice would look up Bob's public key, (n_B, e_B) in the public registry. (In this literature, the two parties are often named Bob and Alice.) Next, Alice would encrypt her message M_A as follows:

$$E_B(M_A) = [[M_A^{e_B}]]_{n_B} = C_A$$

The ciphertext C_A is then transmitted to Bob by any convenient method. Anyone encountering C_A besides Bob would be unable to reconstruct M_A in a reasonable amount of time. When Bob receives the enciphered message C_A, he recovers the original using his private key (n_B, d_B) as follows:

$$D_B(C_A) = [[C_A^{d_B}]]_{n_B} = M_A$$

The feature of RSA that is different from typical cryptosystems is that part of the key is public. Such methods are called public-key cryptosystems. They were introduced by W. Diffie and M. E. Hellman in "New Directions in Cryptography" [26].

Because $D(E(J)) = J$ and $E(D(J)) = J$ in Theorem 22.15, the RSA cryptosystem has another very useful characteristic also proposed by Diffie and Hellman. Anyone can obtain Bob's public key (n_B, e_B) and send messages to Bob; therefore, if Alice sends Bob a message, how can Bob be sure that the message came from Alice and not someone else? Alice can compose a signature message S_A containing her name and other identifying documentation. Next, Alice uses her own **secret deciphering key** (n_A, d_A)

$$D_A(S_A) = [[S_A^{d_A}]]_{n_A} = T_A$$

and includes T_A along with or as part of the main ciphertext C_A of M_A. When Bob gets Alice's transmission, he isolates T_A and applies Alice's **public enciphering key** (n_A, e_A), giving

$$E_A(T_A) = E_A(D_A(S_A)) = S_A$$

Since, presumably, only Alice could have produced a message T_A that could be converted to S_A with Alice's public key (n_A, e_A), the associated message M_A must have been authentic. For such an application, the instrument of authentication must be message dependent and signer dependent. A simple way to implement this authentication for a message M from Alice to Bob consisting of blocks $M_1, M_2, \ldots, M_k$ would be for Alice to generate a signature S consisting of blocks $S_1, S_2, \ldots, S_k$ that are transformed to $T_i = D_A(S_i)$ using Alice's decryption key. We could even let $S_i = M_i$ so that $T_i = D_A(M_i)$. The super block pairs $T_1M_1, T_2M_2, \ldots, T_kM_k$ are enciphered using Bob's public key. The integers $T_1M_1, T_2M_2, \ldots, T_kM_k$ may need to be reblocked to fit Bob's modulus n_B, say as $J_1, J_2, \ldots, J_m$ where $0 \le J_i < n_B$. Thus, we have

$$U_i = E_B(J_i)$$

Upon receiving the sequence $U_1, U_2, \ldots, U_k$, Bob uses his secret key to obtain

$$J_i = D_B(U_i)$$

which may be unblocked to give again $T_1 M_1, T_2 M_2, \ldots, T_k M_k$. Then Bob checks the signatures with

$$E_A(T_i) = M_i$$

assuming that we used $S_i = M_i$. Since T_i giving $M_i = E_A(T_i)$ could have been produced only by Alice, the message is authentic.

The security of the RSA cryptosystem depends upon the difficulty of factoring $n = p \cdot q$, thereby obtaining p and q. Since e is public, d could then be determined if p and q were known by solving $ex \equiv 1 \pmod{(p-1)(q-1)}$. The security also depends upon the difficulty of finding indices. For a known M, $C = [[M^e]]_n$ may be calculated using public knowledge. A cryptanalyst would need to solve $M \equiv C^d \pmod{n}$ for $d = \text{ind}_C M$ modulo n, which is also difficult. Also, p and q are randomly chosen large primes to make factoring n difficult. Moreover, d is randomly chosen relatively prime to $\phi(n)$ and large to make a quick exhaustive search over small d unproductive.

Traditional cryptosystems require that both parties be in possession of the same key but no one else. For example, the United States Data Encryption Standard (DES) uses a 56-bit key. (See *Cryptography: A New Dimension in Computer Data Security* by C. H. Meyer and S. M. Matyas [65].) The use of DES for high-volume message traffic may be preferable to using the RSA method since DES has less computational overhead. The question is: How can the parties communicate the key to one another? Diffie and Hellman in their 1976 paper [26] proposed the following method of exchanging keys.

Suppose that p is a prime and g is a primitive root of p. p and g may be publicly known. Suppose that Alice and Bob want to share the same key. Alice chooses a random integer a such that $1 \leq a \leq p-2$ and Bob chooses a random integer b such that $1 \leq b \leq p-2$. a and b are kept secret and are known only to Alice and Bob, respectively. Alice computes

$$A = [[g^a]]_p$$

and transmits A to Bob in any convenient way. On the other hand, Bob computes

$$B = [[g^b]]_p$$

and transmits B to Alice in any way. Alice computes

$$K_1 = [[B^a]]_p$$

and Bob computes

$$K_2 = [[A^b]]_p$$

But $K_1 = K_2$ because

$$[[A^b]]_p = [[(g^a)^b]]_p = [[(g^b)^a]]_p = [[B^a]]_p$$

Thus, Bob and Alice now have the same key, $K = K_1 = K_2$.

Since B, A, p, and g are perhaps all known, a cryptanalyst can break the cipher by solving $A \equiv g^a \pmod{p}$ for a or by solving $B \equiv g^b \pmod{p}$ for b; that is, one must be able to compute $\text{ind}_g A$ or $\text{ind}_g B$ modulo p, both of which are computationally difficult. The complexity of this calculation is on the order of $\sqrt{p}$ according to the discussion of Section 3.14; therefore, p should be a large prime and $p - 1$ should have at least one large prime factor.

For general information about cryptography from ancient times, see David Kahn's *The Codebreakers: The Story of Secret Writing* [50]. See M. E. Hellman's "The Mathematics of Public-Key Cryptography" [42] for a discussion of public-key cryptography. The RSA method was patented by Rivest, Shamir, and Adleman in 1983 (U.S. Patent 4,405,829) (which expired in the fall of 2000).

Computer Science Application

The computation of raising one number to a power modulo another number is known as *modular exponentiation*. Modular exponentiation is an essential operation in the encryption and decryption of text in the RSA public-key cryptosystem. In the computation of a^b mod n, the integers a, b, and n may need to be very large in order for the system to be secure. The following algorithm returns a^b mod n by using the binary representation of b along with the identities

$$a^{2c} \bmod n = (a^c)^2 \bmod n,$$
$$a^{2c+1} \bmod n = a(a^c)^2 \bmod n.$$

integer function modular-exp(a, b, n)

$d = 1$
$e = b$

while $e > 0$ do
    push e mod 2 to stack S
    $e = e/2$ (integer division)
end-while

while stack S is not empty do
    pop stack S into x
    $d = (d * d) \bmod n$
    if x mod 2 = 1 then
        $d = (d * a) \bmod n$
    end-if
end-while

return d

end-function

By computing powers like x^{15} as $x^0 x^1 x^2 x^4 x^8$, the number of multiplications is greatly reduced, especially when the exponent is large. When used in the context of RSA cryptosystems, special functions to handle arithmetic with large integers are needed.

Exercises

1. In Example 22.16, verify that $D(J) = [[J^{217}]]_{2993}$ decrypts the message

   1683, 79, 2560, 872, 2571

2. Generate an RSA cryptosystem using $n = 83 \cdot 107$ by finding a public key (n, e) and private key (n, d).

   (a) Use these keys to encipher the message "*HELP*."

   (b) Let the RSA keys of Example 22.16 belong to Alice and the keys for $n = 83 \cdot 107$ belong to Bob. Use the method of signatures described in this section to send the signed message "*MONEY*" from Alice to Bob.

(c) Using Bob's private and Alice's public keys, decipher the message of part (b) and verify that it came from Alice.

3. In the RSA cryptosystem with (n, e) public, show that if $\phi(n)$ can be found, then the cryptosystem may be compromised by calculating d.

4. Consider the RSA cryptosystem where $n = p \cdot q$ and $\phi(n)$ are known but the primes p and q are not known.

(a) Show that $p + q$ may be expressed in terms of n and $\phi(n)$.

(b) Show that if $q > p$, then $q - p = \sqrt{(p+q)^2 - 4n}$.

(c) Determine p and q.

(d) Refer to Example 22.16 and assume only that $n = 2993$ and $\phi(n) = 2880$. Then use the method above to find p and q.

GLOSSARY

Chapter 22: Applications of Number Theory

DES (22.3)	The United States Data Encryption Standard (DES) has a 56-bit key.
Hash function (22.2)	Let $k_1, k_2, \ldots, k_n$ be keys to n sets of information. Then **hash function** is defined as $h : K \rightarrow \{0, 1, 2, \ldots, m\}$ where $K = \{k_1, k_2, \ldots, k_n\}$ and where $m + 1 \geqslant n$.
Jaeschke's hashing function (22.2)	**Jaeschke's hashing function** requires one to be able to map the keys $\{k_1, k_2, \ldots, k_n\}$, which may not be relatively prime to one another, to a set of integers $\{f(k_1), (k_2), \ldots, f(k_n)\}$ that are pairwise relatively prime.
Minimum perfect hashing function (22.2)	A **minimum perfect hashing function** is a hashing function that is one-to-one and uses minimum computer storage.
Public enciphering key/deciphering key (22.3)	Keys are provided so that anyone knowing the keys and the method may encrypt a message. However, an additional key is needed to decipher the message.
RSA (22.3)	RSA is a general method of encryption named after R. L. Rivest, A. Shamir, and L. Adleman.

TRUE-FALSE QUESTIONS

1. It is possible to have two bit string patterns with the same fingerprint.

2. Using the Karp and Rabin pattern search algorithm, it is possible to eliminate false "matches" by selecting the primes randomly.

3. The Karp and Rabin pattern search algorithm requires the same computational time for each text character.

4. A hashing function is used to locate sets of data that are reused.

5. A hashing function is one-to-one.

6. Given a set of n positive distinct integers, $\{k_1, k_2, k_3, \ldots, k_n,\}$ it is possible to find integers $C, D,$ and E so that $h(k_i) = \left[\left[\left\lfloor \dfrac{C}{Dk_i + E} \right\rfloor\right]\right] = i - 1$ for $1 \leq i \leq n$.

7. Encryption is the process of decoding a message.

8. Cryptanalysis is decrypting a message using the key.

9. Using the RSA method for encryption, it is possible to encrypt a message without being able to decode a message.

10. When receiving and decrypting a message using the RSA method, since it is a public-key cryptosystem, it is impossible to determine if the proper person is sending the message.

ANSWERS TO TRUE-FALSE

1. T **2.** F **3.** T **4.** T **5.** F **6.** T **7.** F **8.** F **9.** T **10.** F

BIBLIOGRAPHY

[1] L. M. Adleman, "A Subexponential Algorithm for the Discrete Logarithm Problem with Applications to Cryptography," *Proc. 20th IEEE Found. Comput. Sci. Symp.* (1979), 55–60.

[2] W. R. Alford, A. Granville, and C. Pomerance, "There Are Infinitely Many Carmichael Numbers," *Ann. Math.*, Volume 140 (1994), 703–722.

[3] J. A. Anderson and E. F. Wilde, "A Generalization of a Mathematical Curiosity," *J. Recreational Math.*, Volume 29, No. 1 (1998), 1–7.

[4] G. E. Andrews, *Number Theory*, Dover, New York, 1994.

[5] E. T. Bell, *The Development of Mathematics*, 2nd ed., McGraw-Hill, New York, 1945.

[6] E. T. Bell, *Men of Mathematics*, Simon & Schuster, New York, 1965.

[7] C. Berge, *Principles of Combinatorics*, Academic Press, New York, 1971.

[8] G. D. Birkhoff, "A Determinant Formula for the Number of Ways of Coloring a Map," *Ann. of Math.*, 14 (1912), 42–46.

[9] R. S. Boyer and J. S. Moore, "A Fast String Searching Algorithm," *Commun. ACM*, Volume 20, No. 10 (1977), 762–772.

[10] R. P. Brent, "Irregularities in the Distribution of Primes and Twin Primes," *Math. Comput.*, Volume 29 (1975), 43–56.

[11] R. P. Brent and J. M. Pollard, "Factorization of the Eighth Fermat Number," *Math. of Comput.*, Volume 36, No. 154 (1981), 627–630.

[12] D. M. Burton, *Elementary Number Theory*, 3rd ed., Wm. C. Brown, Dubuque, Iowa, 1994.

[13] F. Cajori, *A History of Mathematics*, 2nd ed., Macmillan, New York, 1961.

[14] R. D. Carmichael, "On Composite Numbers P Which Satisfy the Fermat Congruence $a^{P-1} \equiv 1 \bmod P$," *Amer. Math. Monthly*, Volume 19, No. 2 (1912), 22–27.

[15] W. Chan, R. Anderson, P. Beame, S. Burns, F. Modugno, and D. Notkin, "Model Checking Large Software Specifications," *Proceedings of the 4th ACM SIGSOFT Symposium on the Foundations of Software Engineering* (1996), 156–166.

[16] C. C. Chang and J. C. Shieh, "A Fast Algorithm for Constructing Reciprocal Hashing Functions," *Proc. Internat. Symp. New Directions Comput.* (1985), 232–236.

[17] C. C. Chang and J. C. Shieh, "Pairwise Relatively Prime Generating Polynomials and Their Applications," *Proc. Internat. Workshop Discrete Algorithms Complexity*, November (1989), 137–139.

[18] Nan-xian Chen, "Modified Möbius Inversion Formula and Its Application to Physics," *Phys. Rev. Lett.*, Volume 64, No. 11 (1990), 1193–1195.

[19] L. W. Cohen and G. Ehrlich, *The Structure of the Real Number System*, Van Nostrand Reinhold, New York, 1963.

[20] H. Cohn, *Advanced Number Theory*, Dover, New York, 1980.

[21] D. Coppersmith, "Cryptography," *IBM J. Res. Develop.*, Volume 31, No. 2 (1987), 244–248.

[22] R. Crandall, J. Doenias, C. Norrie, and J. Young, "The Twenty-Second Fermat Number Is Composite," *Math. Comput.*, Volume 64 (1995), 863–868.

[23] T. Dantzig, *Number: The Language of Science*, 4th ed., Macmillan, New York, 1954.

[24] H. Davenport, *The Higher Arithmetic*, 6th ed., Cambridge University Press, New York, 1992.

[25] L. E. Dickson, *Studies in the Theory of Numbers*, Chelsea, New York, 1957.

[26] W. Diffie and M. E. Hellman, "New Directions in Cryptography," *IEEE Trans. Inform. Theory*, Volume IT-22, No. 6 (1976), 644–654.

[27] H. W. Eves, *In Mathematical Circles, Quadrants I and II*, Prindle, Weber & Schmidt, Boston, 1969.

[28] C. V. Eynden, *Elementary Number Theory*, McGraw-Hill, New York, 1987.

[29] G. S. Fishman and L. R. Moore, "A Statistical Evaluation of Multiplicative Congruential Random Number Generators with Modulus $2^{31} - 1$," *J. Amer. Statist. Assoc.*, Volume 77 (1982), 129–136.

[30] G. S. Fishman and L. R. Moore III, "An Exhaustive Analysis of Multiplicative Congruential Random Generators with Modulus $2^{31} - 1$," *SIAM J. Sci. Statist. Comput.*, Volume 7, No. 1 (1986), 24–45.

[31] M. Gardner, "Through the Looking Glass," by Lewis Carroll, from *The Annotated Alice*, Bramhall House, New York, 1960.

[32] C. W. Gear, *Computer Organization and Programming*, 2nd ed., McGraw-Hill, New York, 1974.

[33] J. Gilbert and L. Gilbert, *Elements of Modern Algebra*, 4th ed. PWS, Boston, 1995.

[34] E. Grosswald, *Representations of Integers as Sums of Squares*, Springer-Verlag, New York, 1985.

[35] G. H. Hardy and E. M. Wright, *An Introduction to the Theory of Numbers*, 5th ed., Clarendon Press, Oxford, 1979.

[36] D. Harel, "Statecharts: A Visual Formalism for Complex Systems CS84–05," Department of Applied Mathematics, The Weizmann Institute of Science, February 1984.

[37] D. Harel, "Statecharts: A Visual Formalism for Complex Systems," *Science of Computer Programming*, 8 (1987), 231–274.

[38] D. Harel, A. Pnueli, J. P. Schmidt, and R. Sherman, "On the Formal Semantics of Statecharts," *Proceedings of the 2nd IEEE Symposium of Logic in Computer Science Ithaca. N. Y. June 22–24*. IEEE Press, New York (1987), 54–56.

[39] D. Harel and E. Gery, "Executable Object Modeling with Statecharts," *IEEE Computer*, Volume 30, No. 7, July (1997), 31–42.

[40] Goffredo Haus and A. Sametti, "ScoreSnyth: A System for the Synthesis of Music Scores Based on Petri Nets and a Music Algebra," *Readings in Computer-Generated Music*, Denis Baggi, ed., pp. 53–77, IEEE Computer Press Society, Los Alamitos, California, 1991.

[41] P. J. Heawood, "Map-Color Theorem," *Quart. J. Math.*, 23 (1890), 332–339.

[42] M. E. Hellman, "The Mathematics of Public-Key Cryptography," *Sci. Amer.*, Volume 241, August (1979), 146–157.

[43] Raymond Hill, *A First Course in Coding Theory*, Oxford University Press, New York, 1993.

[44] F. S. Hillier and G. Lieberman, *Operations Research*, 2nd ed., Holden-Day, San Francisco, 1974.

[45] Ian Horrocks, *Constructing the User Interface with Statecharts*, Addison-Wesley, New York, 1988.

[46] John M. Howie, *Automata and Languages*, Oxford University Press, New York, 1991.

[47] Alfred Bay Kempe, "On the Geographical Problem of the Four Colors," *Amer. J. Math.*, 2 (1879), 193–200.

[48] A. E. Ingham, *The Distribution of Prime Numbers*, Cambridge University Press, New York, 1992.

[49] G. Jaeschke, "Reciprocal Hashing: A Method for Generating Minimal Perfect Hashing Functions," *Commun. ACM*, Volume 24, No. 12 (1981), 829–833.

[50] D. Kahn, *The Codebreakers: The Story of Secret Writing*, Macmillan, New York, 1968.

[51] David Kahn, *Kahn on Codes*, Macmillan Publishing Company, New York, 1983.

[52] R. M. Karp and M. O. Rabin, "Efficient Randomized Pattern-Matching Algorithms," *IBM J. Res. Develop.*, Volume 31, No. 2, March (1987), 249–260.

[53] A. Ya. Khinchin, *Continued Fractions*, The University of Chicago Press, Chicago, 1964.

[54] H. A. Klein, *The Science of Measurement: A Historical Survey*, Dover, New York, 1988.

[55] M. Kline, *Mathematical Thought from Ancient to Modern Times*, Oxford University Press, New York, 1972.

[56] D. E. Knuth, *The Art of Computer Programming*, 2nd ed., Volumes 2 and 3, Addison-Wesley, Reading, Mass., 1981 and 1973.

[57] D. E. Knuth, J. H. Morris, and V. R. Pratt, "Fast Pattern Matching in Strings," *SIAM J. Comput.*, Volume 6, No. 2, June (1977), 323–350.

[58] J. C. Lagarias, V. S. Miller, and A. M. Odlyzko, "Computing $\pi(x)$: The Meissel-Lehmer Method," *Math. Comput.*, Volume 44 (1985), 537–560.

[59] P. L'Ecuyer, "Efficient and Portable Combined Random Number Generators," *Commun. ACM*, Volume 31 (1988), 742–774.

[60] A. K. Lenstra and H. W. Lenstra, Jr., editors, *The Development of the Number Field Sieve*, Springer-Verlag, New York, 1993.

[61] W. J. LeVeque, *Elementary Theory of Numbers*, Dover, New York, 1990.

[62] N. Leveson, E. Heimdahl, H. Hildreth, and D. Reese, "Requirements Specification for Process Control Systems," *IEEE Transactions on Software Engineering*, Volume 20, No. 9, September (1994), 684–707.

[63] G. Marsaglia, "Random Numbers Fall Mainly in the Planes," *Proc. Nat. Acad. Sci. U.S.A.*, Volume 61 (1968), 25–28.

[64] George H. Mealy, "A Method for Synthesizing Sequential Circuits," *Bell System Technical Journal*, Volume 34 (1955), 1045–1070.

[65] C. H. Meyer and S. M. Matyas, *Cryptography: A New Dimension in Computer Data Security*, Wiley, New York, 1982.

[66] R. P. Millane, "A Product Form of the Möbius Transform," *Phys. Lett. A*, Volume 162 (1992), 213–214.

[67] R. P. Millane, "Möbius Transform Pairs," *J. Math. Phys.*, Volume 34 (1993), 875–877.

[68] Edward F. Moore, "Gedanken-Experiments on Sequential Machines," *Automata Studies, Annals of Mathematical Studies*, No. 32 (1956), 129–133.

[69] O. Neugebauer, *The Exact Sciences in Antiquity*, 2nd ed., Dover, New York, 1969.

[70] O. Neugebauer and A. Sachs, *Mathematical Cuneiform Texts*, American Oriental Society, New Haven, Conn., 1945.

[71] J. R. Newman, *The World of Mathematics*, Simon & Schuster, New York, 1956.

[72] B. W. Ninham et al., "Möbius, Mellin, and Mathematical Physics," *Physica A*, Volume 186 (1992), 441–481.

[73] I. Niven and H. S. Zuckerman, *An Introduction to the Theory of Numbers*, 4th ed., Wiley, New York, 1980.

[74] J. M. H. Olmsted, *The Real Number System*, Appleton-Century-Crofts, New York, 1962.

[75] G. Ottewell, *The Astronomical Companion*, Astronomical Workshop, Furman University, Greenville, S.C., 1979.

[76] R. E. A. C. Paley, "On Orthogonal Matrices," *J. of Math. Phys.*, Volume 12 (1933), 311–320.

[77] B. K. Parady, J. R. Smith, and S. Zarantonello, "Largest Known Twin Primes," *Math. Comput.*, Volume 55 (1990), 381–382.

[78] S. K. Park and K. W. Miller, "Random Number Generators: Good Ones Are Hard to Find," *Commun. ACM*, Volume 31 (1988), 1192–1201.

[79] James L. Peterson, *Petri Net Theory and the Modeling of Systems*, Prentice Hall, Upper Saddle River NJ, 1981.

[80] R. L. Plackett and J. P. Burman, "The Design of Optimum Multifactorial Experiments," *Biometrika*, Volume 33 (1946), 305–325.

[81] M. Plotkin, "Binary Codes with Specified Minimal Distance," *IEEE Trans. Info. Theory,* 6, (1960), 445–450.

[82] S. Pohlig and M. Hellman, "An Improved Algorithm for Computing Logarithms over $GF(p)$ and Its Cryptographic Significance," *IEEE Trans. Inform. Theory*, Volume IT-24, No. 1 (1978), 106–110.

[83] J. M. Pollard, "Theorems on Factorization and Primality Testing," *Proc. Cambridge Philos. Soc.*, Volume 76 (1974), 521–528.

[84] J. M. Pollard, "A Monte Carlo Method for Factorization," *Bit* (1975), 331–334.

[85] C. Pomerance, *Lecture Notes on Primality Testing and Factoring*, The Mathematical Association of America, Washington, D.C., 1984.

[86] J. Purdum, "Pattern Matching Alternatives: Theory vs. Practice," *Comput. Language*, November (1987), 33–44.

[87] M. Rabin, "Probabilistic Algorithm for Testing Primality," *J. Number Theory*, Volume 12 (1980), 128–138.

[88] F. Reid and K. Honeycutt, "A Digital Clock for Sidereal Time," *Sky & Telescope*, July (1976), 59–63.

[89] S. Y. Ren and J. D. Dow, "Generalized Möbius Transforms for Inverse Problems," *Phys. Lett. A*, Volume 154 (1991), 215–216.

[90] P. Ribenboim, *The Little Book of Big Primes*, Springer-Verlag, New York, 1991.

[91] R. L. Rivest, A. Shamir, and L. Adleman, "A Method for Obtaining Digital Signatures and Public Key Cryptosystems," *Commun. ACM*, Volume 21, No. 2 (1978), 120–126.

[92] N. Robbins, *Beginning Number Theory*, Wm. C. Brown, Dubuque, Iowa, 1993.

[93] K. H. Rosen, *Elementary Number Theory and Its Applications*, 3rd ed., Addison-Wesley, Reading, Mass., 1993.

[94] J. B. Rosser and L. Schoenfeld, "Approximate Formulas for Some Functions of Prime Numbers," *Illinois J. Math.*, Volume 6 (1962), 64–94.

[95] A. Schimmel, *The Mystery of Numbers*, Oxford University Press, New York, 1993.

[96] M. R. Schroeder, *Number Theory in Science and Communication*, 2nd ed., Springer-Verlag, Berlin, 1986.

[97] D. Shanks, *Solved and Unsolved Problems in Number Theory*, Chelsea, New York, 1985.

[98] D. Shier, "Spanning Trees: Let Me Count the Ways." Mathematics Magazine 73 (2000), 376–381.

[99] R. W. Sinnott, *Sky & Telescope*, January (1989), 80–82.

[100] D. E. Smith, *History of Mathematics*, Dover, New York, 1958.

[101] D. E. Smith, *A Source Book in Mathematics*, Dover, New York, 1959.

[102] R. Solovay and V. Strassen, "A Fast Monte-Carlo Test for Primality," *SIAM J. Comput.*, Volume 6, No. 1 (1977), 84–85. "Erratum: A Fast Monte-Carlo Test for Primality," *SIAM J. Comput.*, Volume 7, No. 1 (1978), 118.

[103] B. M. Stewart, *Theory of Numbers*, 2nd ed., Macmillan, New York, 1964.

[104] J. K. Strayer, *Elementary Number Theory*, PWS Publishing Company, Boston, 1993.

[105] D. M. Sunday, "A Very Fast Substring Search Algorithm," *Commun. ACM*, Volume 33, No. 8 (1990), 132–142.

[106] Thomas M. Thompson, *From Error-Correcting Codes through Sphere Packings to Simple Groups*, Mathematical Association of America, 1984.

[107] J. A. Todd, "A Combinatorial Problem," *J. Math. Phys.*, Volume 12 (1933), 321–333.

[108] B. L. van der Waerden, *Science Awakening*, Volume I, 3rd ed., Oxford University Press, New York, 1971.

[109] A. E. Western and J. C. Miller, *Tables of Indices and Primitive Roots*, Royal Society Mathematical Tables, Volume 9, Cambridge University Press, Cambridge, 1968.

A N S W E R S

Chapter 1—Section 1.1

1. no, it is a question
3. yes, false
5. yes, false
7. no, it is a command
9. yes, false
11. $\sim r \wedge \sim q$
13. $\sim (r \wedge q)$
15. $\sim p \vee q$
17. $\sim (q \wedge r)$

19.
Case	q	r	$\sim$	r	$\wedge$	$\sim$	q
1	T	T	F	T	**F**	F	T
2	T	F	T	F	**F**	F	T
3	F	T	F	T	**F**	T	F
4	F	F	T	F	**T**	T	F
					*		

21.
Case	q	r	$\sim$	(r	$\wedge$	q)
1	T	T	**F**	T	T	T
2	T	F	**T**	F	F	T
3	F	T	**T**	T	F	F
4	F	F	**T**	F	F	F
			*			

23.
Case	p	q	$\sim$	p	$\vee$	q
1	T	T	F	T	**T**	T
2	T	F	F	T	**F**	F
3	F	T	T	F	**T**	T
4	F	F	T	F	**T**	F
					*	

25.
Case	q	r	$\sim$	(q	$\wedge$	r)
1	T	T	**F**	T	T	T
2	T	F	**T**	T	F	F
3	F	T	**T**	F	F	T
4	F	F	**T**	F	F	F
			*			

27. $((p \wedge (q \wedge r) \vee (\sim p \wedge s)))$
29. $q \wedge (p \wedge s)$
31. $(p \vee q) \wedge \sim s$
33. $(p \vee q) \wedge \sim (p \wedge q)$, $p \underline{\vee} q$
35. This game is not very difficult or I will not play chess.
37. This game is very difficult and I play chess, and it takes time to play chess.
39. Great Danes are large dogs and I do not have a small house or a great Dane.
41. Great Danes are large dogs and I have a great Dane, or I have a small house and not a great Dane.

43.
Case	p	q	$\sim$	p	$\vee$	$\sim$	q
1	T	T	F	T	**F**	F	T
2	T	F	F	T	**T**	T	F
3	F	T	T	F	**T**	F	T
4	F	F	T	F	**T**	T	F
					*		

45.
Case	p	q	r	(p	$\wedge$	q)	$\wedge$	r
1	T	T	T	T	T	T	**T**	T
2	T	T	F	T	T	T	**F**	F
3	T	F	T	T	F	F	**F**	T
4	T	F	F	T	F	F	**F**	F
5	F	T	T	F	F	T	**F**	T
6	F	T	F	F	F	T	**F**	F
7	F	F	T	F	F	F	**F**	T
8	F	F	F	F	F	F	**F**	F
							*	

47.
Case	p	q	r	(p	$\wedge$	($\sim$	q)	$\vee$	$\sim$	r)
1	T	T	T	T	**F**	F	T	F	F	T
2	T	T	F	T	**T**	F	T	T	T	F
3	T	F	T	T	**T**	T	F	F	F	T
4	T	F	F	T	**T**	T	F	T	T	F
5	F	T	T	F	**F**	F	T	F	F	T
6	F	T	F	F	**F**	F	T	T	T	F
7	F	F	T	F	**F**	T	F	F	F	T
8	F	F	F	F	**F**	T	F	T	T	F
					*					

49.
Case	p	q	r	(p	$\wedge$	r)	$\vee$	(q	$\wedge$	$\sim$	r
1	T	T	T	T	T	T	**T**	T	F	F	T
2	T	T	F	T	F	F	**T**	T	T	T	F
3	T	F	T	T	T	T	**T**	F	F	F	T
4	T	F	F	T	F	F	**F**	F	F	T	F
5	F	T	T	F	F	T	**F**	T	F	F	T
6	F	T	F	F	F	F	**T**	T	T	T	F
7	F	F	T	F	F	T	**F**	F	F	F	T
8	F	F	F	F	F	F	**F**	F	F	T	F
							*				

51. Cabbages have one head and two heads are better than one, and a cabbage can use a computer and I don't like cabbage.

53. The statement is not true: she likes to quote Milton and he likes Bogart movies and Bogart quotes Milton, or they don't go to movies.

55.

Case	p	q	r	p	∧	(q	∨	~	r)
1	T	T	T	T	**T**	T	T	F	T
2	T	T	F	T	**T**	T	T	T	F
3	T	F	T	T	**F**	F	F	F	T
4	T	F	F	T	**T**	F	T	T	F
5	F	T	T	F	**F**	T	T	F	T
6	F	T	F	F	**F**	T	T	T	F
7	F	F	T	F	**F**	F	F	F	T
8	F	F	F	F	**F**	F	T	T	F

*

57.

Case	p	q	r	~	(p	∧	r)	∨	(~	q	∧	r)
1	T	T	T	F	T	T	T	**F**	F	F	F	T
2	T	T	F	T	T	F	F	**T**	F	F	F	F
3	T	F	T	F	T	T	T	**T**	T	T	T	T
4	T	F	F	T	T	F	F	**T**	T	T	F	F
5	F	T	T	T	F	F	T	**T**	F	F	F	T
6	F	T	F	T	F	F	F	**T**	F	F	F	F
7	F	F	T	T	F	F	T	**T**	T	T	T	T
8	F	F	F	T	F	F	F	**T**	T	T	F	F

*

59.

Case	p	q	r	(p	∧	r)	∨	(p	∧	~	q)
1	T	T	T	T	T	T	**T**	T	F	F	T
2	T	T	F	T	F	F	**F**	T	F	F	T
3	T	F	T	T	T	T	**T**	T	T	T	F
4	T	F	F	T	F	F	**T**	T	T	T	F
5	F	T	T	F	F	T	**F**	F	F	F	T
6	F	T	F	F	F	F	**F**	F	F	F	T
7	F	F	T	F	F	T	**F**	F	F	T	F
8	F	F	F	F	F	F	**F**	F	F	T	F

*

61.

Case	p	q	r	(~	q	∧	r)	∨	~	(p	∧	r)
1	T	T	T	F	T	F	T	**F**	F	T	T	T
2	T	T	F	F	T	F	F	**T**	T	T	F	F
3	T	F	T	T	F	T	T	**T**	F	T	T	T
4	T	F	F	T	F	F	F	**T**	T	T	F	F
5	F	T	T	F	T	F	T	**T**	T	F	F	T
6	F	T	F	F	T	F	F	**T**	T	F	F	F
7	F	F	T	T	F	T	T	**T**	T	F	F	T
8	F	F	F	T	F	F	F	**T**	T	F	F	F

*

63.

Case	p	q	r	~	(~	p	∧	(q	∨	~	r))
1	T	T	T	**T**	F	T	F	T	T	F	T
2	T	T	F	**T**	F	T	F	T	T	T	F
3	T	F	T	**T**	F	T	F	F	F	F	T
4	T	F	F	**T**	F	T	F	F	T	T	F
5	F	T	T	**F**	T	F	T	T	T	F	T
6	F	T	F	**F**	T	F	T	T	T	T	F
7	F	F	T	**T**	T	F	F	F	F	F	T
8	F	F	F	**F**	T	F	T	F	T	T	F

*

65.

Case	p	q	r	(p	∨	q)	∧	(q	∨	r)
1	T	T	T	T	F	T	**F**	T	F	T
2	T	T	F	T	F	T	**F**	T	T	F
3	T	F	T	T	T	F	**T**	F	T	T
4	T	F	F	T	T	F	**F**	F	F	F
5	F	T	T	F	T	T	**F**	T	F	T
6	F	T	F	F	T	T	**T**	T	T	F
7	F	F	T	F	F	F	**F**	F	T	T
8	F	F	F	F	F	F	**F**	F	F	F

67.

Case	p	q	r	~	(p	∨	q)	∨	~	(~	(p	∨	r)	∨	~	(q	∨	r))
1	T	T	T	T	T	F	T	**T**	F	T	T	F	T	T	T	T	F	T
2	T	T	F	T	T	F	T	**T**	F	T	T	F	F	F	T	T	F
3	T	F	T	F	T	T	F	**F**	T	T	F	T	T	T	F	F	T	T
4	T	F	F	F	T	T	F	**F**	T	T	F	T	F	F	F	F	F
5	F	T	T	F	F	T	T	**F**	T	T	T	F	T	T	T	F	T
6	F	T	F	F	F	T	T	**F**	F	T	F	F	F	T	T	F
7	F	F	T	T	F	F	F	**T**	T	F	T	T	F	T	T	F	T
8	F	F	F	T	F	F	F	**T**	F	T	F	F	F	T	T	F	F

69. (a) On the ranch they do not raise cattle or they do not raise sheep.

(b) $p \wedge q$, $\sim p \vee \sim q$

(c)

Case	p	q	~ (p ∧ q)		Case	p	q	~ p ∨ ~ q
1	T	T	**F** T		1	T	T	F T **F** F T
2	T	F	**T** F		2	T	F	F T **T** T F
3	F	T	**T** F		3	F	T	T F **T** F T
4	F	F	**T** F		4	F	F	T F **T** T F

71. (a) 16, 2^k

(b) In each case increase to one more variable by placing the previous table with T in front of each row followed by placing the previous table with F in front of each row.

(c)

Case	p	q	r	s
1	T	T	T	T
2	T	T	T	F
3	T	T	F	T
4	T	T	F	F
5	T	F	T	T
6	T	F	T	F
7	T	F	F	T
8	T	F	F	F
9	F	T	T	T
10	F	T	T	F
11	F	T	F	T
12	F	T	F	F
13	F	F	T	T
14	F	F	T	F
15	F	F	F	T
16	F	F	F	F

(d) To increase from k to $k+1$ variables, place the table for k variables with all T's for the $k+1$st variable, then repeat the table for k variables with all F's for the $k+1$st variable.

73. $2^4, 2^8, 2^{16}, 2^{2^n}$

Chapter 1—Section 1.2

1. $r \to (p \wedge q)$
3. $(r \to q) \wedge (\sim r \to \sim p)$
5. $(p \wedge q) \to r$
7. $(p \to q) \wedge (\sim p \to r)$
9. $\sim (p \vee r) \to (q \wedge \sim s)$
11. $\sim r \to ((\sim (p \leftrightarrow s)) \vee q)$
13. If he loves purple ties and he is popular, then he has strange friends.
15. If he loves purple ties, then he is popular or he has strange friends.
17. It is not true that if he is successful then he is popular.
19. He is popular if and only if he is successful and rich.
21. Castles are drafty and kings live in castles if and only if, if barns are not drafty then kings live in castles or cattle live in barns but not both.
23. If kings live in castles and cattle live in barns, then castles are not drafty or barns are not drafty.
25. If x is less than 5 and greater than 1, then $x = 2$, and $x \neq 6$.
27. It is not true that if x is less than 5 then $x = 2$, and if x is greater than 1 then $x = 6$.
29. true
31. true
33. false
35. true
37. true

39.
Case	p	q	r	(p	→	q)	→	r
1	T	T	T	T	T	T	**T**	T
2	T	T	F	T	T	T	**F**	F
3	T	F	T	T	F	F	**T**	T
4	T	F	F	T	F	F	**T**	F
5	F	T	T	F	T	T	**T**	T
6	F	T	F	F	T	T	**F**	F
7	F	F	T	F	T	F	**T**	T
8	F	F	F	F	T	F	**F**	F

*

41.
Case	p	q	r	(q	→	(p	∧	r))	↔	((q	→	p)	∧	(q	→	r))
1	T	T	T	T	T	T	T	T	**T**	T	T	T	T	T	T	T
2	T	T	F	T	F	T	F	F	**T**	T	T	T	T	F	F	F
3	T	F	T	F	T	T	T	T	**T**	F	T	T	T	F	T	T
4	T	F	F	F	T	T	F	F	**T**	F	T	T	T	F	T	F
5	F	T	T	T	F	F	F	T	**T**	T	F	F	F	T	T	T
6	F	T	F	T	F	F	F	F	**T**	T	F	F	F	F	F	F
7	F	F	T	F	T	F	F	T	**T**	F	T	F	F	T	F	T
8	F	F	F	F	T	F	F	F	**T**	F	T	F	F	T	F	F

*

43.
Case	p	q	r	(p	→	q)	→	(q	→	r)
1	T	T	T	T	T	T	**T**	T	T	T
2	T	T	F	T	T	T	**F**	T	F	F
3	T	F	T	T	F	F	**T**	F	T	T
4	T	F	F	T	F	F	**T**	F	T	F
5	F	T	T	F	T	T	**T**	T	T	T
6	F	T	F	F	T	T	**F**	T	F	F
7	F	F	T	F	T	F	**T**	F	T	T
8	F	F	F	F	T	F	**T**	F	T	F

*

45.
Case	p	q	r	(p	∨	r)	→	(p	∧	q)
1	T	T	T	T	T	T	**T**	T	T	T
2	T	T	F	T	T	F	**T**	T	T	T
3	T	F	T	T	T	T	**F**	T	F	F
4	T	F	F	T	T	F	**F**	T	F	F
5	F	T	T	F	T	T	**F**	F	F	T
6	F	T	F	F	F	F	**T**	F	F	T
7	F	F	T	F	T	T	**F**	F	F	F
8	F	F	F	F	F	F	**T**	F	F	F

*

47.
Case	p	q	r	((p	→	q)	∧	∼	(r	∨	p))	→	(∼	p	∨	∼	q)
1	T	T	T	T	T	T	F	F	T	T	T	**T**	F	T	F	F	T
2	T	T	F	T	T	T	F	F	F	T	T	**T**	F	T	F	F	T
3	T	F	T	T	F	F	F	F	T	T	T	**T**	F	T	T	T	F
4	T	F	F	T	F	F	F	F	F	T	T	**T**	F	T	T	T	F
5	F	T	T	F	T	T	F	F	T	T	F	**T**	T	F	T	F	T
6	F	T	F	F	T	T	T	T	F	F	F	**T**	T	F	T	F	T
7	F	F	T	F	T	F	F	F	T	T	F	**T**	T	F	T	T	F
8	F	F	F	F	T	F	T	T	F	F	F	**T**	T	F	T	T	F

*

49.
Case	p	q	r	(p	∧	∼	(q	∨	∼	r))	↔	(p	→	q)
1	T	T	T	T	F	F	T	T	F	T	**F**	T	T	T
2	T	T	F	T	F	F	T	T	T	F	**F**	T	T	T
3	T	F	T	T	T	T	F	F	F	T	**F**	T	F	F
4	T	F	F	T	F	F	F	T	T	F	**T**	T	F	F
5	F	T	T	F	F	F	T	T	F	T	**F**	F	T	T
6	F	T	F	F	F	F	T	T	T	F	**F**	F	T	T
7	F	F	T	F	F	T	F	F	F	T	**F**	F	T	F
8	F	F	F	F	F	F	F	T	T	F	**F**	F	T	F

*

51.
Case	p	q	r	((p	→	q)	∧	(q	→	∼	r))	→	(r	→	p)
1	T	T	T	T	T	T	F	T	F	F	T	**T**	T	T	T
2	T	T	F	T	T	T	T	T	T	T	F	**T**	F	T	T
3	T	F	T	T	F	F	F	F	T	F	T	**T**	T	T	T
4	T	F	F	T	F	F	F	F	T	T	F	**T**	F	T	T
5	F	T	T	F	T	T	F	T	F	F	T	**T**	T	F	F
6	F	T	F	F	T	T	T	T	T	T	F	**T**	F	T	F
7	F	F	T	F	T	F	F	F	T	F	T	**F**	T	F	F
8	F	F	F	F	T	F	F	F	T	T	F	**T**	F	T	F

*

53.

Case	p	q	r	(~	(p	∧	~	r)	∨	q)	→	(q	∨	r)
1	T	T	T	T	T	F	F	T	T	T	**T**	T	T	T
2	T	T	F	F	T	T	T	F	T	T	**T**	T	T	F
3	T	F	T	T	T	F	F	T	T	F	**T**	F	T	T
4	T	F	F	F	T	T	T	F	F	F	**T**	F	F	F
5	F	T	T	T	F	F	F	T	T	T	**T**	T	T	T
6	F	T	F	T	F	F	T	F	T	T	**T**	T	T	F
7	F	F	T	T	F	F	F	T	T	F	**T**	F	T	T
8	F	F	F	T	F	F	T	F	T	F	**F**	F	F	F
											*			

55.

Case	p	q	r	(~	(p	→	q)	→	(q	→	r)
1	T	T	T	F	T	T	T	**T**	T	T	T
2	T	T	F	F	T	T	T	**T**	T	F	F
3	T	F	T	T	T	F	F	**T**	F	T	T
4	T	F	F	T	T	F	F	**T**	F	T	F
5	F	T	T	F	F	T	T	**T**	T	T	T
6	F	T	F	F	F	T	T	**T**	T	F	F
7	F	F	T	F	F	T	F	**T**	F	T	T
8	F	F	F	F	F	T	F	**T**	F	T	F

57.

Case	p	q	r	((p	∨	r)	→	q)	→	((p	→	q)	∨	(p	→	r))
1	T	T	T	T	T	T	T	T	**T**	T	T	T	T	T	T	T
2	T	T	F	T	T	T	T	T	**T**	T	T	T	T	T	F	F
3	T	F	T	T	T	T	F	F	**T**	T	F	F	T	T	T	T
4	T	F	F	T	T	F	F	T	**T**	T	F	F	F	T	F	F
5	F	T	T	F	T	T	T	T	**T**	F	T	T	T	T	F	T
6	F	T	F	F	F	F	T	T	**T**	F	T	T	T	T	F	F
7	F	F	T	F	T	T	F	F	**T**	F	T	F	T	F	T	T
8	F	F	F	F	F	F	T	F	**T**	F	T	F	T	F	T	F
									*							

59. true

61. false

63. If either r or s is false, then p and q are false. If r and s are both true, then p and q can be either true for false.

65.

p	r	s
T	T	T
F	T	T
F	F	F

The letter q can take any value.

67.

Case	p	q	r	((p	|	q)	|	r)	↔	(p	|	(q	|	r))
1	T	T	T	T	F	T	T	T	**T**	T	T	T	F	T
2	T	T	F	T	F	T	T	F	**F**	T	F	T	T	F
3	T	F	T	T	T	F	F	T	**T**	T	F	F	T	T
4	T	F	F	T	T	F	T	F	**F**	T	T	F	T	F
5	F	T	T	F	T	T	F	T	**F**	F	T	T	F	T
6	F	T	F	F	T	T	T	F	**T**	F	T	T	T	F
7	F	F	T	F	T	F	F	T	**F**	F	T	F	F	T
8	F	F	F	F	T	F	T	F	**T**	F	T	F	T	F

69.

Case	p	q	r	(q	↓	r)	↓	(p	∧	r)
1	T	T	T	T	F	T	**F**	T	T	T
2	T	T	F	T	F	F	**T**	T	F	F
3	T	F	T	F	F	T	**F**	T	T	T
4	T	F	F	F	T	F	**F**	T	F	F
5	F	T	T	T	F	T	**T**	F	F	T
6	F	T	F	T	F	F	**T**	F	F	F
7	F	F	T	F	F	T	**T**	F	F	T
8	F	F	F	F	T	F	**F**	F	F	F

71.

Case	p	q	r	(p	∨	q)	↓	(q	→	r)
1	T	T	T	T	T	T	**F**	T	T	T
2	T	T	F	T	T	T	**F**	T	F	F
3	T	F	T	T	T	F	**F**	F	T	T
4	T	F	F	T	T	F	**F**	F	T	F
5	F	T	T	F	T	T	**F**	T	T	T
6	F	T	F	F	T	T	**F**	T	F	F
7	F	F	T	F	F	F	**F**	F	T	T
8	F	F	F	F	F	F	**F**	F	T	F

Chapter 1—Section 1.3

1.

Case	p	q	~	(p ∧ q)	~p	∨	~q
1	T	T	**F**	T	F	**F**	F
2	T	F	**T**	F	F	**T**	T
3	F	T	**T**	F	T	**T**	F
4	F	F	**T**	F	T	**T**	T
			*			*	

3.

Case	p	q	r	p	∨	(q ∧ r)	p ∨ q	∧	(p ∨ r)
1	T	T	T	T	**T**	T	T	**T**	T
2	T	T	F	T	**T**	F	T	**T**	T
3	T	F	T	T	**T**	F	T	**T**	T
4	T	F	F	T	**T**	F	T	**T**	T
5	F	T	T	F	**T**	T	T	**T**	T
6	F	T	F	F	**F**	F	T	**F**	F
7	F	F	T	F	**F**	F	F	**F**	T
8	F	F	F	F	**F**	F	F	**F**	F
					*			*	

5.
$\sim (p \to q) \equiv \sim (\sim p \lor q)$ by exercise 4
$\equiv \sim (\sim p) \land \sim q$ De Morgan's law
$\equiv p \land \sim q$ double negation

7. $\sim (p \land s) \to (\sim s \land p) \equiv \sim\sim (p \land s) \lor (\sim s \land p)$
  problem 4
$\equiv (p \land s) \lor (\sim s \land p)$
  double negation
$\equiv (p \land s) \lor (p \land \sim s)$
  commutive law
$\equiv p \land (s \lor \sim s)$
  distributive law
$\equiv p \land \mathbf{T}$ property of **T**
$\equiv p$ property of **T**

9. If he is a Martian, then he has six legs.

11. If he has money, then he is popular.

13. If I am literary, then I read Shakespeare.

15. If he enjoys downhill skiing, then he must have snow.

17. If I am able to wake up in the morning, then I have drunk three cups of coffee.

19. If one is truly educated, then one knows Latin.

21. If I am a good citizen, then I vote.

23. If I am not a good citizen, then I don't vote.

25. If I pay this loan, then I won't have to leave town.

27. (a) Any statement has the same truth table as itself.
    (b) $u \equiv v$ and $v \equiv u$ both state that u and v have the same truth tables.
    (c) If statements u and v have the same truth table and statements v and w have the same truth table, then statements u and w will have the same truth table.

29. logically equivalent

Case	p	q	r	p → (q ∧ r)			(p → q) ∧ (p → r)		
1	T	T	T	T	**T**	T	T	**T**	T
2	T	T	F	T	**F**	F	T	**F**	F
3	T	F	T	T	**F**	F	F	**F**	T
4	T	F	F	T	**F**	F	F	**F**	F
5	F	T	T	F	**T**	T	T	**T**	T
6	F	T	F	F	**T**	F	T	**T**	T
7	F	F	T	F	**T**	F	T	**T**	T
8	F	F	F	F	**T**	F	T	**T**	T
					*			*	

31. not logically equivalent

Case	p	q	r	p ∧ q	→	r	(p → r) ∧ (q → r)		
1	T	T	T	T	**T**	T	T	**T**	T
2	T	T	F	T	**F**	F	F	**F**	F
3	T	F	T	F	**T**	T	T	**T**	T
4	T	F	F	F	**T**	F	F	**F**	T
5	F	T	T	F	**T**	T	T	**T**	T
6	F	T	F	F	**T**	F	T	**F**	F
7	F	F	T	F	**T**	T	T	**T**	T
8	F	F	F	F	**T**	F	T	**T**	T
					*			*	

33. logically equivalent

Case	p	q	r	p ∧ q	→	r	(p → r) ∨ (q → r)		
1	T	T	T	T	**T**	T	T	**T**	T
2	T	T	F	T	**F**	F	F	**F**	F
3	T	F	T	F	**T**	T	T	**T**	T
4	T	F	F	F	**T**	F	F	**T**	T
5	F	T	T	F	**T**	T	T	**T**	T
6	F	T	F	F	**T**	F	T	**T**	F
7	F	F	T	F	**T**	T	T	**T**	T
8	F	F	F	F	**T**	F	T	**T**	T
					*			*	

35. not logically equivalent

Case	p	q	r	~	((p ∧ q) → r)	~p	∨	~q	∨	r
1	T	T	T	**F**	T	T	T	F	F	F	T	T
2	T	T	F	**T**	T	F	F	F	F	F	F	F
3	T	F	T	**F**	F	T	T	F	T	T	T	T
4	T	F	F	**F**	F	T	F	F	T	T	T	F
5	F	T	T	**F**	F	T	T	T	T	F	T	T
6	F	T	F	**F**	F	T	F	T	T	F	T	F
7	F	F	T	**F**	F	T	T	T	T	T	T	T
8	F	F	F	**F**	F	T	F	T	T	T	**T**	F
				*						*		

37. logically equivalent

Case	p	q	r	p ∧ (q ∨ r)			(p ∧ q) ∨ (p ∧ r)		
1	T	T	T	T	**F**	F	T	**F**	T
2	T	T	F	T	**T**	T	T	**T**	F
3	T	F	T	T	**T**	T	F	**T**	T
4	T	F	F	T	**F**	F	F	**F**	F
5	F	T	T	F	**F**	F	F	**F**	F
6	F	T	F	F	**F**	T	F	**F**	F
7	F	F	T	F	**F**	T	F	**F**	F
8	F	F	F	F	**F**	F	F	**F**	F
					*			*	

39. logically true

Case	p	q	r	p ∧ q	→	p
1	T	T	T	T	**T**	T
2	T	T	F	T	**T**	T
3	T	F	T	F	**T**	T
4	T	F	F	F	**T**	T
5	F	T	T	F	**T**	F
6	F	T	F	F	**T**	F
7	F	F	T	F	**T**	F
8	F	F	F	F	**T**	F
					*	

41. logically true

Case	p	q	r	(p ∨ q)	∧	~p	→	q
1	T	T	T	T	F	F	**T**	T
2	T	T	F	T	F	F	**T**	T
3	T	F	T	T	F	F	**T**	F
4	T	F	F	T	F	F	**T**	F
5	F	T	T	T	T	T	**T**	T
6	F	T	F	T	T	T	**T**	T
7	F	F	T	F	F	T	**T**	F
8	F	F	F	F	F	T	**T**	F
							*	

43.

Case	p	q	r	(((p → q) ∧ (q → r)) ∧ r) → p						
1	T	T	T	T	T	T	T	T	**T**	T
2	T	T	F	T	F	F	F	F	**T**	T
3	T	F	T	F	F	T	F	T	**T**	T
4	T	F	F	F	F	F	F	F	**T**	T
5	F	T	T	T	T	T	T	T	**F**	F
6	F	T	F	T	F	F	F	F	**T**	F
7	F	F	T	T	T	T	T	T	**F**	F
8	F	F	F	T	T	F	F	F	**T**	F
									*	

45.

Case	p	T	$\to$	p
1	T	T	**T**	T
2	F	T	**F**	F

47. logically true

Case	p	F	$\to$	p
1	T	F	**T**	T
2	F	F	**T**	F

49. logically false

case	p	q	$(((p \leftrightarrow q)$	$\wedge$	$(p \underline{\vee} q))$
1	T	T	T	**F**	F
2	T	F	F	**F**	T
3	F	T	F	**F**	T
4	F	F	T	**F**	F

51. logically true

case	p	$(\sim p$	$\to$	F)	$\to$	p
1	T	F	T	F	**T**	T
2	F	T	F	F	**T**	F

53.
$$p \to (q \vee r) \equiv \sim p \vee (q \vee r)$$
$$\equiv (\sim p \vee \sim p) \vee q \vee r)$$
$$\equiv (\sim p \vee q) \vee (\sim p \vee r)$$
$$\equiv (p \to q) \vee (p \to r)$$

55.
$$(p \wedge q) \to r \equiv \sim (p \wedge q) \vee r)$$
$$\equiv (\sim p \vee \sim q) \vee r$$
$$\equiv (\sim p \vee \sim q) \vee (r \vee r)$$
$$\equiv (\sim p \vee r) \vee (\sim q \vee r)$$
$$\equiv (p \to r) \vee (q \to r)$$

Chapter 1—Section 1.4

1.

Case	p	q	r	$p \to q$	$q \to r$	$p \to r$	
1	T	T	T	T	T	T	*
2	T	T	F	T	F	F	
3	T	F	T	F	T	T	
4	T	F	F	F	T	F	
5	F	T	T	T	T	T	*
6	F	T	F	T	F	T	
7	F	F	T	T	T	T	*
8	F	F	F	T	T	T	*

3.

Case	r	w	$\sim w \to (r \wedge \sim r)$	w	
1	T	T	T	T	*
2	T	F	F	F	
3	F	T	T	T	*
4	F	F	F	F	

5. invalid

Case p q r | $p \to q$ $q \to r$ $q \vee r$ p

Case	p	q	r	$p \to q$	$q \to r$	$q \vee r$	p	
1	T	T	T	T	T	T	T	*
2	T	T	F	T	F	T	T	
3	T	F	T	F	T	T	T	
4	T	F	F	F	T	F	T	
5	F	T	T	T	T	T	F	**
6	F	T	F	T	F	T	F	
7	F	F	T	T	T	T	F	**
8	F	F	F	T	T	F	F	

7. valid

Case	p	q	r	$p \to q$	$p \to r$	$\sim (p \wedge q)$	$\sim p$	
1	T	T	T	T	T	F	F	
2	T	T	F	T	F	F	F	
3	T	F	T	F	T	T	F	
4	T	F	F	F	F	T	F	
5	F	T	T	T	T	T	T	*
6	F	T	F	T	T	T	T	*
7	F	F	T	T	T	T	T	*
8	F	F	F	T	T	T	T	*

9. valid

Case	p	q	r	$\sim r$	$p \to r$	$q \to r$	$\sim (p \wedge q)$	
1	T	T	T	F	T	T	F	
2	T	T	F	T	F	F	F	
3	T	F	T	F	T	T	T	
4	T	F	F	T	F	T	T	
5	F	T	T	F	T	T	T	
6	F	T	F	T	T	F	T	
7	F	F	T	F	T	T	T	
8	F	F	F	T	T	T	T	*

11. invalid

Case	p	q	r	$p \to q$	$q \to r$	$q \vee r$	p	
1	T	T	T	T	T	T	T	*
2	T	T	F	T	F	T	T	
3	T	F	T	F	T	T	T	
4	T	F	F	F	T	F	T	
5	F	T	T	T	T	T	F	**
6	F	T	F	T	F	T	F	
7	F	F	T	T	T	T	F	**
8	F	F	F	T	T	F	F	

13. valid. Try to prove that it is invalid. Thus, the hypotheses are true and the conclusion is false.

$$\begin{array}{ll} s \vee r & T \\ t \to r & T \\ \underline{s \to w} & T \\ r \vee w & F \end{array}$$

Since $r \vee w$ is false, r is false and w is false. Since w is false, s must be false to make $s \to w$ true. Since r is false, t must be false to make $t \to r$ true. But since r and s are both false, $s \vee r$ must be false. It is not possible to have the hypotheses true and the conclusion false, so the argument is valid.

15. **invalid**. Try to prove that it is invalid. Thus, the hypotheses are true and the conclusion is false.

$$\begin{array}{ll} p \to q & T \\ \sim q \to \sim s & T \\ s \to t & T \\ t \vee q & T \\ \hline p \vee s & F \end{array}$$

Since $p \vee s$ is false, p and s must be false. Therefore, t can be true, making $t \vee q$ and $s \to t$ true. Whether q is true or false, $p \to q$ and $\sim q \to \sim s$ are true. Therefore, we have the hypotheses true and the conclusion is false.

17. **valid**. For the conclusion to be false t must be false. q must be true. Therefore, for $\sim (s \wedge q)$ to be true, s must be false. But then $s \vee t$ is false.

19. **invalid**. If p is true s is false, r is false, and q is false, then the hypotheses are all true, but the conclusion is false.

21.
1. $\sim (s \wedge t)$ given
2. $\sim w \to t$ given
3. $\sim s \vee \sim t$ 1 and De Morgan's law
4. $s \to \sim t$ 3 and Theorem 1.3(h)
5. $\sim t \to w$ 2 and contrapositive
6. $s \to w$ syllogism

23.
1. $\sim (\sim p \vee q)$ given
2. $\sim z \to \sim s$ given
3. $(p \wedge \sim q) \to s$ given
4. $\sim z \vee r$ given
5. $\sim \sim p \wedge \sim q$ 1 and De Morgan's Law
6. $p \wedge \sim q$ 5 and double negation
7. s 3, 6, and modus ponens
8. $s \to z$ 2 and contrapositive
9. z 7, 8, and modus ponens
10. $z \to r$ 4 and Theorem 1.3(h)
11. r 9, 10, and modus ponens

25. $p \equiv q$ if and only if they have the same truth tables
if and only if in each case, p and q are both true or both false
if and only if in each case, $p \leftrightarrow q$ is true
if and only if $p \leftrightarrow q$ is a tautology

$p \leftrightarrow q$ is a tautology
if and only if $(p \to q) \wedge (q \to p)$ is a tautology
if and only if $(p \to q) \wedge (q \to p)$ is true in each case
if and only if $(p \to q)$ and $(q \to p)$ are true in each case
if and only if $(p \to q)$ and $(q \to p)$ are tautologies

27. yes

p	q	p	$\to$	$(p \vee q)$
T	T	T	T	**T**
T	F	F	T	**T**
F	T	T	T	**T**
F	F	F	T	**F**

29. yes

p	q	$((p \to q)$	$\wedge$	$\sim q)$	$\to$	$\sim p$
T	T	T	F	F	T	F
T	F	F	F	T	T	F
F	T	T	F	F	T	T
F	F	T	T	T	T	T

31.
$p \vee q$ given
$q \to r$ given
$\sim s \to \sim r$ given
$s \to t$ given
$\sim t$ given
$\sim s$ (4), (5), and modus tollens
$\sim r$ (3), (6), modus ponens
$\sim q$ (2), (7), and modus tollens
p (1), (8), and case elimination

33. b-brush teeth a-dentist is angry f-drives too fast

$b \to a$ given
$a \to f$ given
$\sim f$ given
$\sim a$ (2), (3), and modus tollens
b (1), (4), and modus tollens

Conclusion: I brush my teeth.

35. m-Martians v-vegetarians r-rodents s-squirrels

$m \to v$ given
$v \to \sim r$ given
$s \to r$ given
$\sim r \to \sim s$ 3), contrapositive
$m \to \sim r$ (1), (2), syllogism
$m \to \sim s$ (4), (5), syllogism

Conclusion: Martians do not eat squirrels.

37. s-sharks g-golfers p-Gary Player d-dangerous fish

$s \to \sim g$ given
$p \to g$ given
$\sim s \to \sim d$ given
$g \to \sim s$ (1), contrapositive
$p \to \sim s$ (2), (4), syllogism
$p \to \sim d$ (3), (5), syllogism

Conclusion: Gary Player is not a dangerous fish.

39. b-bears c-can count h-honest p-politicians w-worthy people

$b \to \sim c$	given
$\sim h \to p$	given
$w \to c$	given
$h \to w$	given
$\sim p \to h$	(2), contrapositive
$\sim p \to w$	(2), (5), syllogism
$\sim p \to c$	(3), (6), syllogism
$c \to \sim b$	(1), contrapositive
$\sim p \to \sim b$	(7), (8), syllogism
$b \to p$	(9), contrapositive

Conclusion: Bears can become politicians.

41. b-wear bow tie p-attend party c-live on the coast s-act silly i-look intelligent g-chew gum o-bassoon player

$b \to p$	given
$\sim c \to \sim s$	given
$i \to b$	given
$p \to s$	given
$o \to \sim g$	given
$\sim g \to i$	given
$o \to i$	(5), (6), syllogism
$o \to b$	(3), (7), syllogism
$o \to p$	(1), (8), syllogism
$o \to s$	(4), (9), syllogism
$s \to c$	(2), contrapositive
$o \to c$	(10), (11), syllogism

Conclusion: Bassoon players live on the coast.

43.

$p \to q$	assume
$\sim q$	assume
$\sim q \to \sim p$	(1), contrapositive
$\sim p$	(2), (3), modus ponens

Hence, from $p \to q$ and $\sim q$, we can conclude $\sim p$.

45. Assume that there is a finite number of primes, say, $a_1, a_2, \ldots, a_n$. Consider the number $a_1 a_2 \cdots a_n + 1$. It is not divisible by any of the primes so that it is a prime. But this is a contradiction. Therefore, there is an infinite number of primes.

Chapter 1—Section 1.5

1.

Case	p	q	$p \downarrow q$	$\sim p$
1	T	T	T F T	F
2	T	F	T F T	F
3	F	T	F T F	T
4	F	F	F T F	T
			*	*

3.

Case	p	q	$(p \downarrow q) \downarrow (p \downarrow q)$	$p \vee q$
1	T	T	T F T **T** T F T	**T**
2	T	F	T F F **T** T F F	**T**
3	F	T	F F T **T** F F T	**T**
4	F	F	F T F **F** F T F	**F**
			*	*

5. $(p \wedge q \wedge r) \vee (p \wedge \sim q \wedge r) \vee (\sim p \wedge q \wedge \sim r) \vee (\sim p \wedge \sim q \wedge \sim r)$

7. $(p \wedge q \wedge \sim r) \vee (p \wedge \sim q \wedge r) \vee (p \wedge \sim q \wedge \sim r) \vee (\sim p \wedge \sim q \wedge \sim r)$

9. $(p \wedge q \wedge r) \vee (p \wedge \sim q \wedge r) \vee (p \wedge \sim q \wedge \sim r) \vee (\sim p \wedge \sim q \wedge r)$

11. $(\sim p \vee q \vee \sim r) \wedge (p \vee \sim q \vee \sim r) \wedge (p \vee q \vee r)$

13. $(\sim p \vee \sim q \vee \sim r) \wedge (\sim p \vee q \vee \sim r) \wedge (p \vee \sim q \vee \sim r)$

15. $(\sim p \vee \sim q \vee r) \wedge (p \vee \sim q \vee r) \wedge (p \vee q \vee r)$

17. $((p \downarrow (q \downarrow q)) \downarrow (p \downarrow (q \downarrow q)) \downarrow ((p \downarrow (q \downarrow q)) \downarrow (p \downarrow (q \downarrow q)))) \downarrow ((q \downarrow (p \downarrow p)) \downarrow (q \downarrow (p \downarrow p))) \downarrow ((q \downarrow (p \downarrow p)) \downarrow (q \downarrow (p \downarrow p)))$

19. $((p|p)|(p|p))|(q|q)$

21. $((((p|(q|q))|(p|(q|q)))|((p|(q|q))|(p|(q|q))))|(((p|p)|q)|((p|p)|q))|((p|p)|q)|((p|p)|q)))$

Chapter 1—Section 1.6

1. $r \vee (p \wedge \sim q)$

3. $(p \wedge \sim r) \vee (\sim r \wedge s) \vee (\sim q \wedge r \wedge \sim s)$

5. $(r \wedge \sim s) \vee (\sim r \wedge s)$

7.

	q	q	$\sim q$	$\sim q$
p	×	×		
$\sim p$	×	×	×	
	r	$\sim r$	$\sim r$	r

$q \vee (\sim p \wedge \sim r)$

9.

	q	q	$\sim q$	$\sim q$	
p	×				s
p			×	×	$\sim s$
$\sim p$			×	×	$\sim s$
$\sim p$	×				s
	r	$\sim r$	$\sim r$	r	

$(\sim q \wedge \sim s) \vee (q \wedge r \wedge s)$

11.

	q	q	$\sim q$	$\sim q$	
p		×	×		s
p					$\sim s$
$\sim p$					$\sim s$
$\sim p$	×	×	×	×	s
	r	$\sim r$	$\sim r$	r	

$(\sim p \wedge s) \vee (\sim r \wedge s)$

Chapter 1—Section 1.7

1. $pq(q + r')$

3. $(pq'r)' + q'r + pq'r'$

5. $pr + pq'(p + r + s')$

7. $q'r + (pr)'$

9. $pqr + p'qr' + p'qr + p'q'r'$

11. $(p'q')' + q'r$

13. $(pq + r) + (q + r)'$

15. $(p + q)((qr')' + (p' + r))$

17. $((pq)' + r)'pq$

19. $((pq)' + qr)((rs)' + qr)$

21.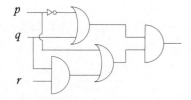

23.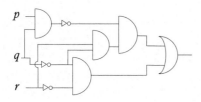

25.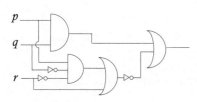

27.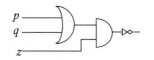

29.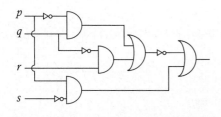

31.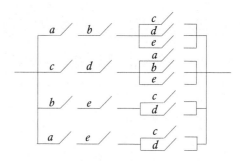

33.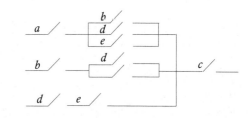

35.
p	q	r	
1	1	1	1
1	1	0	1
1	0	1	0
1	0	0	1
0	1	1	1
0	1	0	1
0	0	1	0
0	0	0	0

$pqr + pqr' + pq'r' + p'qr' + p'qr$

	q	q	q'	q'	
p	×	×	×		
p'	×	×			$q + pr'$
	r	r'	r'	r	

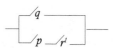

37.
p	q	r	
1	1	1	1
1	1	0	0
1	0	1	1
1	0	0	0
0	1	1	0
0	1	0	0
0	0	1	0
0	0	0	0

$pqr + pq'r$

	q	q	q'	q'	
p	×			×	
p'					pr
	r	r'	r'	r	

___/p___/r___

39.

p	q	r	s	
1	1	1	1	0
1	1	1	0	0
1	1	0	1	0
1	1	0	0	0
1	0	1	1	1
1	0	1	0	1
1	0	0	1	1
1	0	0	0	0
0	1	1	1	0
0	1	1	0	0
0	1	0	1	0
0	1	0	0	0
0	0	1	1	1
0	0	1	0	0
0	0	0	1	0
0	0	0	0	0

$pq'rs + pq'rs' + pq'r's + p'q'rs$

	q	q	$\sim q$	$\sim q$	
p			×	×	s
p				×	$\sim s$
$\sim p$					$\sim s$
$\sim p$				×	s
	r	$\sim r$	$\sim r$	r	

$pq's + pq'r + q'rs$

41.

43.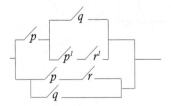

45.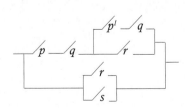

47.

	q	q	q'	q'
p		×		×
p'	×		×	×
	r	r'	r'	r

$pqr' + p'q' + p'r + q'r$

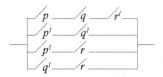

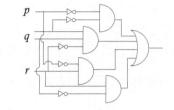

49.

	q	q	$\sim q$	$\sim q$	
p	×	×	×	×	s
p		×			$\sim s$
$\sim p$	×	×			$\sim s$
$\sim p$	×	×	×	×	s
	r	$\sim r$	$\sim r$	r	

$s + p'q + qr'$

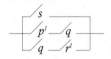

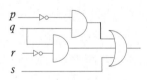

Chapter 2—Section 2.1

1. $\{-9, -8, -7, -6, -5, -4, -3, -2, -1, 0, 1, 2, 3, 4, 5, 6, 7, 8, 9\}$

3. $\{a, e, i, o, u\}$

5. $\{x : 0 < x \leq 24 \text{ and } x \text{ is a multiple of } 3\}$

7. $\{x : x \text{ is a state that begins with the letter "O"}\}$

9. $\varnothing, \{a\}$

11. $\varnothing, \{a\}, \{b\}, \{c\}, \{a, b\}, \{a, c\}, \{b, c\}, \{a, b, c\}$

13. $\varnothing$

15. true

17. false

19. false
21. true
23. true
25. true
27. true
29. true
31. false
33. true
35. 2
37. 4
39. 3
41. Either way is a contradiction.

Chapter 2—Section 2.2

1. $A \cup C = \{1, 2, 3, 4, 5, 6, 7, 8, 10\}$
3. $A \cap (B \cup C) = \{1, 2, 3, 4, 5, 6, 7\} \cap \{2, 4, 5, 6, 7, 8, 9, 10\} = \{2, 4, 5, 6, 7\}$
5. $(A \cap B)' = \{4, 5, 6, 7\}' = \{1, 2, 3, 8, 9, 10\}$
7. $A \triangle B = \{1, 2, 3, 4, 5, 6, 7, 8, 9, 10\} - \{4, 5, 6, 7\} = \{1, 2, 3, 8, 9, 10\}$
9. $A - C = \{1, 2, 3, 4, 5, 6, 7\} - \{2, 4, 6, 8, 10\} = \{1, 3, 5, 7\}$
11. $A \cap B \cap C' = \{4, 5, 6, 7\} \cap \{1, 3, 5, 7, 9\} = \{5, 7\}$
13. $(A - \varnothing) \cup (A - A) = A \cup \varnothing = A$
15. $C - A = \{2, 4, 6, 8, 10\} - \{1, 2, 3, 4, 5, 6, 7\} = \{8, 10\}$
17. $B \times B = \{(a, a), (a, b), (b, a), (b, b)\}$
19. $\{(1, 1), (1, 2), (1, 3), (2, 1), (2, 2), (2, 3), (3, 1), (3, 2), (3, 3)\}$
21. $\varnothing$
23. $\mathcal{P}\{\varnothing, \{\varnothing\}\} = \{\varnothing, \{\varnothing\}, \{\{\varnothing\}\}, \{\varnothing, \{\varnothing\}\}\}$
25. F
27. T
29. F
31. T
33. F
35. T
37. $x \in (A \cup B)' \Leftrightarrow x \notin (A \cup B)$
    definition of complement
    $\Leftrightarrow \sim (x \in (A \cup B))$
    definition of $\notin$
    $\Leftrightarrow \sim ((x \in A) \vee (x \in B))$
    definition of union
    $\Leftrightarrow \sim (x \in A) \wedge \sim (x \in B)$
    De Morgan's law for logic
    $\Leftrightarrow (x \notin A') \wedge (x \notin B')$
    definition of complement
    $\Leftrightarrow x \notin A' \cap B'$
    definition of intersection

39. $\{\varnothing\}$
41. $\{\varnothing, \{\varnothing\}, \{\varnothing, \{\varnothing\}\}\}$
43. $\{\varnothing, \{\varnothing\}, \{\varnothing, \{\varnothing\}\}, \{\varnothing, \{\varnothing\}, \{\varnothing, \{\varnothing\}\}\}, \{\{\varnothing, \{\varnothing\}, \{\varnothing, \{\varnothing\}\}, \{\varnothing, \{\varnothing\}, \{\varnothing, \{\varnothing\}\}\}\}\}$

   A does belong to the power set. A does belong to its successor. In general the power set is not the successor.

45. no
47. yes
49. Let $A = \{a\}$, and $B = \{b\}$, then the power set of A is $\{\varnothing, \{a\}\}$ and the power set of B is $\{\varnothing, \{b\}\}$. The union of the power sets of A and B is $\{\varnothing, \{a\}, \{b\}, \}$. $A \cup B = \{a, b\}$ so its power set is $\{\varnothing, \{a\}, \{b\}, \{a, b\}\}$. Hence, they are not equal.
51. A and B are both empty.
53. A is a subset of B.
55. (a) Any element of C must be a multiple of 2 to belong to $A_\square$, a multiple of 3 to belong to $A_\triangle$, and a multiple of 5 to belong to $A_\bigcirc$. Hence, any element of C must be a multiple of 30. Therefore, $C = \{x : x = 30k \text{ for some positive integer } k\}$.

    (b) Any given integer i does not belong to A_{i+1}. Hence, no positive integer belongs to all of the A_i and $\bigcap_{i \in I} A_i = \varnothing$.

Chapter 2—Section 2.3

1.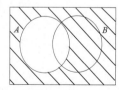

3.

A-12 Answers

5.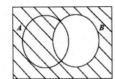

7.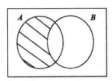

9.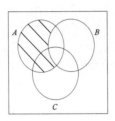

11.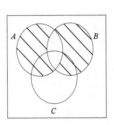

13.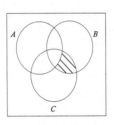

15.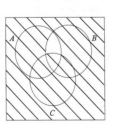

17.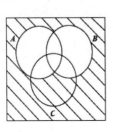

19.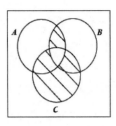

21.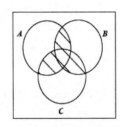

23. A'

25. $((A \cap B) \cup (A \cap C) \cup (B \cap C))'$

27. $A \cup B \cup C - (A \cap B) - (A \cap C) - (B \cap C)$

29. $(A \cup B - C) \triangle (A \cap B)$

31. $(B \cup C)$

33.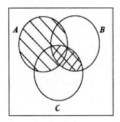

35. They are not the same.

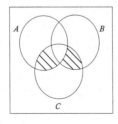 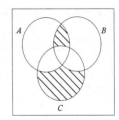

37. They are the same.

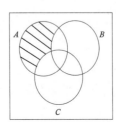 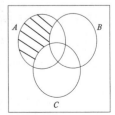

39. They are not the same.

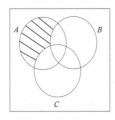 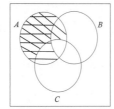

41. They are not the same

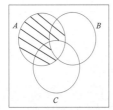 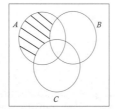

43. $(A \triangle B) \cup B = (A \cap B') \cup (B \cap A') \cup B$
$ = (A \cap B') \cup B$
 since $(B \cap A') \subseteq B$
$ = (A \cup B) \cap (B' \cup B)$
 De Morgan's law
$ = (A \cup B) \cap \mathbf{U}$
 complement property
$ = (A \cup B)$ identity property

45. $(A \cap B \cap C) \cup (A \cap B' \cap C) \cup (B - (A \cap C)) =$
$(B \cap A \cap C) \cup (B' \cap A \cap C) \cup (B - (A \cap C)) =$ commutivity
$((B \cup B') \cap (A \cap C)) \cup (B - (A \cap C))$ $\quad =$ distributive law
$(\mathbf{U} \cap (A \cap C)) \cup (B - (A \cap C))$ $\quad =$ complement law
$(A \cap C) \cup (B - (A \cap C))$ $\quad =$ idempotent law
$(A \cap C) \cup (B \cap (A \cap C)')$ $\quad =$ definition
$((A \cap C) \cup B) \cap ((A \cap C) \cup (A \cap C)')$ $\quad =$ distributive law
$((A \cap C) \cup B) \cap \mathbf{U}$ $\quad =$ complement law
$(A \cap C) \cup B)$ $\quad$ idempotent

47. The expression is $(A \cap B' \cap C) \cup (A \cap B \cap C) \cup (A' \cap B' \cap C) \cup (A' \cap B \cap C) \cup (A' \cap B \cap C')$.

The Karnaugh map is

	B	B	B'	B'
A	x			x
A'	x	x		x
	C	C'	C'	C

The simplified expression is $C \cup (A' \cap B)$.

49. The expression is $(A' \cap B' \cap C') \cup (A \cap B' \cap C') \cup (A \cap B' \cap C) \cup (A \cap B \cap C') \cup (A \cap B \cap C) \cup (A' \cap B \cap C') \cup (A' \cap B \cap C)$.

The Karnaugh map is

	B	B	B'	B'
A	x	x	x	x
A'	x	x	x	
	C	C'	C'	C

The simplified expression is $A \cup C' \cup B = A \cup B \cup C'$.

Chapter 2—Section 2.4

1. $x \cdot x = x \cdot x + 0$ identity law
$ = x \cdot x + x \cdot x'$ complement law
$ = x \cdot (x + x')$ distributive law
$ = x \cdot 1$ complement law
$ = x$ identity law

3. $x \cdot (x + y) = (x + 0) \cdot (x + y)$ identity law
$ = x + (0 \cdot y)$ distributive law
$ = x + 0$ null law
$ = x$ identity law

5. $1 + 0 = 1$ identity law
$1 \cdot 0 = 0$ identity law

Therefore, 0 satisfies the complement law for 1, and, by the unique complement law, it is the only element that does so.

7. $(x + y + z') \cdot (x + y' + z)$

9. $xy' + (z'y)'$

11. $((x + y) \cdot 0) + (1 \cdot x) + z$

13. $(a_1, a_2, a_3, \ldots, a_n) \cdot (b_1, b_2, b_3, \ldots, b_n) = (a_1 \cdot b_1, a_2 \cdot b_2, a_3 \cdot b_3, \ldots, a_n \cdot b_n)$ and $(a_1, a_2, a_3, \ldots, a_n) + (b_1, b_2, b_3, \ldots, b_n) = (a_1 + b_1, a_2 + b_2, a_3 + b_3, \ldots, a_n + b_n)$ where $\cdot$ and $+$ are described in Problem 12.

Unity is 30, zero is 1.

15. If $xy = xz$ and $x'y = x'z$, then $xy + x'y = xz + x'z$. By the distributive law, $(x + x') \cdot y = (x + x') \cdot z$. But $x + x' = 1$, so $1 \cdot y = 1 \cdot z$ and $y = z$.

17. If $xy' = 0$, then $xy' + xy = 0 + xy = xy$. But

$xy' + xy = x(y + y')$ by the distributive law
$ = x \cdot 1$ by the complement law
$ = x$ by the identity law

so $xy = x$. Conversely, if $xy = x$, then $(xy) \cdot y' = xy'$. But

$(xy) \cdot y' = x(y \cdot y')$ by the associative law
$ = x \cdot 0$ by the complement law
$ = 0$ by the identity law

so $xy' = 0$.

19. If (i) $0 + a = a$ for all a and (ii) $0' + a = a$ for all a, then

$$\begin{aligned} 0' &= 0 + 0' & \text{by (i)} \\ &= 0' + 0 & \text{by the commutative law} \\ &= 0 & \text{by (ii)} \end{aligned}$$

Hence, 0 is uniquely defined by its properties. The proof that 1 is uniquely defined by its properties is the dual of the proof that 0 is uniquely defined by its properties.

Chapter 2—Section 2.5

1. range = $\{1, 2, 4, 5\}$ domain = $\{a, c, d\}$
3. range = R, domain = $\{x : x \in R \text{ and } x \geq 0\}$
5. range = $\{x : -4 \leq x \leq 4\}$ = domain
7. $R^{-1} = \{(7, 1), (6, 4), (6, 5), (8, 2)\}$. $S^{-1} = \{(10, 6), (11, 6), (10, 7), (13, 8)\}$.
9. $S^{-1} \circ S = \{(6, 6), (6, 7), (7, 6), (7, 7), (8, 8)\}$
11. $T \circ (S \circ R) = \{(1, \triangle), (4, \triangle), (5, \triangle), (2, *), (2, \bigcirc)\}$
13. $(T \circ S) \circ R = \{(1, \triangle), (4, \triangle), (5, \triangle), (2, *), (2, \bigcirc)\}$
15. $B^{-1} = \{(a, v), (e, w), (i, x), (o, y), (u, z)\}$
17. $B^{-1} \circ A = \{(b, v), (c, w), (d, x), (f, y), (g, z)\}$
19. $\{(x, y) : y = 3x^2 + 15\}$
21. $\{(x, y) : y = \frac{1}{3}x\}$
23. $\{(a, a), (b, b), (c, c), (d, d), (e, e)\}$
25. $\{(a, a), (a, b), (b, a), (b, c), (c, b), (b, d), (d, b), (c, e), (e, c), (e, d), (d, e), (b, b), (c, c), (d, d), (e, e)\}$
27. $\{(a, a), (a, b), (b, c), (b, d), (c, e), (e, d), (c, b), (a, c), (a, d), (b, e), (b, b), (c, d), (c, c)\}$
29. U
31. S
33. $S \cup T = \{(a, a), (a, b), (b, c), (b, d), (c, e), (e, d), (c, a), (b, a), (e, e), (d, e), (c, b)\}$
35. $U \triangle S = \{(b, b), (e, e), (b, a), (c, b), (c, c), (d, d), (a, c), (b, d), (c, e), (e, d)\}$
37. Let R and S be symmetric relations and $(a, b) \in R \cap S$. Since R is symmetric, $(b, a) \in R$. Since S is symmetric, $(b, a) \in S$. Therefore, $(b, a) \in R \cap S$ and $R \cap S$ is symmetric.
39. $\{(a, c), (c, a), (b, c), (c, b)\}$
41. $\{(a, a), (b, b), (c, c), (d, d), (e, e), (a, b), (b, a), (c, b), (b, c)\}$
43. $\{(a, a), (b, b), (c, c), (d, d), (e, e), (a, b)\}$
45. true
47. false
    Let $A = \{a, b\}$, $R = \{(a, a), (b, b), (a, b)\}$, and $S = \{(a, a), (b, b), (b, a)\}$. Then $R - S = \{(a, b)\}$ is not reflexive.
49. true
51. false. Let $R = \{(b, c), (c, b)\}$ and $S = \{(b, a), (a, b)\}$, then $(a, c) \in R \circ S$, but $(c, a) \notin R \circ S$.
53. true
55. false. Let $A = \{a, b\}$, $R = \{(a, a), (b, b), (a, b)\}$, and $S = \{(a, a), (b, b), (b, a)\}$. Then $R \cup S = \{(a, a), (b, b), (a, b), (b, a)\}$ is not antisymmetric.
57. true
59. true
61. false. Let $R = \{(c, a)(b, b)\}$ and $S = \{(a, b), (b, c), (a, c)\}$, then R and S are transitive, but $R \circ S = \{(a, b), (b, a), (a, a)\}$ is not transitive.
63. $R = \{(a, b), (b, c), (a, c)\}$ and $S = \{(a, c)\}$, then R and S are transitive, but $R \triangle S = \{(a, b), (b, c)\}$ is not transitive.

65.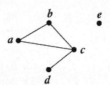

67.

69.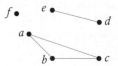

71.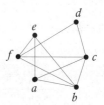

73. $V = \{a, b, c, d, e\}$, $E = \{\{a, e\}, \{b, e\}, \{c, e\}, \{d, e\}\}$

$R = \{(a, e), (e, a), (b, e), (e, b), (c, e), (e, c), (d, e), (e, d)\}$

75. $V = \{a, b, c, d, e, f\}$, $E = \{\{a,c\}, \{b,c\}, \{b,e\}, \{c,e\}, \{c,d\}, \{a,d\}\}$

  $R = \{(a,c), (c,a), (c,b), (b,c), (b,e), (e,b), (c,e), (e,c), (c,d), (d,c), (a,d), (d,a)\}$

77. $V = \{a, b, c, d\}$, $E = \{\{a,b\}, \{a,d\}, \{b,d\}, \{c,d\}, \{a,c\}, \{b,c\}\}$,

  $R = \{(a,b), (b,a), (a,d), (d,a), (b,d), (d,b), (c,d), (d,c), (a,c), (c,a), (b,c), (c,b)\}$

79. $V = \{a, b, c, d, e,\}$, $E = \{\{a,b\}, \{c,d\}, \{d,e\}, \{c,e\}\}$

  $R = \{(a,b), (b,a), (c,d), (d,c), (d,e), (e,d), (c,e), (e,c)\}$

81.

83.

85.

87.

89. $V = \{a, b, c, d, e\}$, $E = \{(a,e), (b,e), (c,e), (d,e)\}$

91. $V = \{a, b, c, d, e,\}$, $E = \{(a,b), (b,a), (c,a), (d,c), (c,d), (e,e)\}$

93. $V = \{a, b, c, d\}$, $E = \{(d,b), (d,a), (c,d), (b,c)\}$

95. $V = \{a, b, c, d, e,\}$, $E = \{(a,b), (b,c), (c,d), (d,e), (e,a), (b,d), (b,e), (c,e), (e,d), (a,c)\}$

97. Show that $(T \circ S) \circ R \subseteq T \circ (S \circ R)$. Let $(a, d) \in (T \circ S) \circ R$, then there exists $b \in B$ such that $(a, b) \in R$ and $(b, d) \in T \circ S$. Since $(b, d) \in T \circ S$, there exists $c \in C$ so that $(b, c) \in S$ and $(c, d) \in T$. Since $(a, b) \in R$ and $(b, c) \in S$, $(a, c) \in S \circ R$. Since $(a, c) \in S \circ R$ and $(c, d) \in T$, $(a, d) \in T \circ (S \circ R)$. Thus, $T \circ (S \circ R) \subseteq T \circ (S \circ R)$.

Chapter 2—Section 2.6

1. yes
3. no
5. no
7. no
9. yes
11.
13.
15.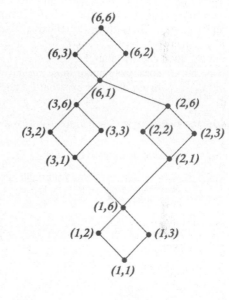
17.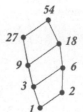
19. $\{(a,a), (b,b), (c,c), (d,d), (e,e), (d,b), (d,a), (d,e), (c,e), (c,a), (c,b), (e,a), (e,b)\}$
21. $\{(a,a), (b,b), (c,c), (d,d), (e,e), (e,a), (e,b), (e,c), (e,d), (d,a), (d,b), (d,c)\}$
23. c, e
25. a, e
27. b

29. d

31. d, e, f

33. none

35. b, c

37. a, b

39. b, d

41. d, e

43. No two distinct elements are comparable.

45. Let x be an atom of B and y belong to B. $xy \leq x$; therefore, $xy = 0$ or $xy = x$.

47. Assume $a \preceq b$ and $b \preceq a$. If $a_1 \leq b_1$ and $b_1 \leq a_1$, then $a_1 = b_1$. Assume for all $i < k$, $a_i = b_i$. Since $a \preceq b$, $a_k < b_k$ or a_k does not exist but b_k does. Since $b \preceq a$, $b_k < a_k$ or b_k does not exist but a_k does. Therefore, $a_k = b_k$. Therefore, $a = b$. Therefore, $\preceq$ is antisymmetric. Trivially $\preceq$ is reflexive. If $a \preceq b$ and $b \preceq c$, then for some k, $a_k < b_k$ or a_k does not exist but b_k does and for some j, $b_j < c_j$ or b_j does not exist but c_j does. Combining these we have for some m, $a_m < c_m$ or a_m does not exist but c_m does.

Chapter 2—Section 2.7

1. no, not reflexive or transitive

3. yes, $[p] = \{q : q$ has the same truth table as $p\}$

5. yes, $[n] = \{n, -n\}$

7. no, not reflexive

9. no, not transitive

11. Yes, equivalence class is $\{1, 2, 3\}$

13. yes, $[1] = A$

15. no, not transitive

17. yes, $[p] = \{q : p$ is written in the same computer language as $q\}$

19. $\{\square, \bigcirc, z\}, \{w, h\}, \{t, \bigstar\}, \{\triangle\}$

21. $[1] = \{1, -1\}$, $[2] = \{2, -2\}$, $[0] = \{0\}$, $[3] = \{3\}$, $[-\frac{1}{2}] = \{\frac{1}{2}, -\frac{1}{2}\}$, $[4] = \{4\}$.

23. If $(a, b) \in R$, then $(b, a) \in R^{-1}$. Hence, if $R = R^{-1}$, then R is symmetric. Conversely, if $(b, a) \in R^{-1}$, then $(a, b) \in R$. If R is symmetric, then $(b, a) \in R$ and $R^{-1} \subseteq R$. Similarly, $R \subseteq R^{-1}$. Hence, $R = R^{-1}$.

25. no

27. no

29. $A \times A$

31. no, 8 is not in A

33. yes, $R = \{(1, 1), (1, 7), (7, 1), (7, 7), (3, 3)(3, 5), (5, 3), (5, 5), (2, 2), (2, 4), (4, 2), (4, 4), (6, 6)\}$

35. yes. Show R is reflexive. $(a, b)R(a, b)$ if and only if $ab = ab$. Show R is symmetric. Assume $(a, b)R(c, d)$. But $(a, b)R(c, d)$ if and only if $ad = bc$. Also $(c, d)R(a, b)$ if and only if $cb = ad$. Therefore, $(c, d)R(a, b)$. Show R is transitive. Let $(a, b)R(c, d)$ and $(c, d)R(e, f)$. Therefore, $ad = bc$ and $cf = ed$. From the first equation $adf = bcf$. Substituting $cf = ed$ into this equation we have $adf = bed$. Dividing by d, we have $af = be$ and $(a, b)R(e, f)$. Elements related to $(1, 2)$ are $(2, 4), (3, 6), (4, 8), \ldots$.

37. false. Not reflexive for strings less than 4.

39. A string certainly contains the same number of 0's as itself. Hence, the relation is reflexive. If a string u has the same number of 0's as a string v, then the string v has the same number of 0's as a string u. Hence, the relation is symmetric. If a string u has the same number of 0's as the string v, and the string v has the same number of 0's as the string w, then the string u has the same number of 0's as the string w. Hence, the relation is transitive.

Chapter 2—Section 2.8

1. The domain of f is R. The range of f is $\{y : y \geq 4\}$.

3. The domain of f is $\{x : x > 2\}$. The range of f is $\{y : y > 0\}$.

5. The domain of f is $R - \{2, -2\}$. The range of f is $\{y : y \leq -\frac{1}{4}\} \cup \{y : y > 0\}$.

7. not a function

9. function

11. not a function unless the domain is restricted

13. $f(g(x)) = \sqrt{(x^2 + 3)^2 + 2}$ $g(f(x)) = x^2 + 5$

15. $f(x) = \dfrac{7}{x}$, $g(x) = x^2 + 3$

17. $f(x) = \sqrt{x}$, $g(x) = \dfrac{1}{x}$, $h(x) = x^2 - 2$

19. $y = x^{\frac{1}{3}}$

21. none

23. 1-1, onto, has inverse

25. onto

27. Let $a \in f^{-1}(B_1)$. Then $f(a) \in B_1$. Since $B_1 \subseteq B_2$, $f(a) \in B_2$. Therefore, $a \in f^{-1}(B_2)$, and $f^{-1}(B_1) \subseteq f^{-1}(B_2)$.

29. Since $B_1 \subseteq B_1 \cup B_2$, by part (b) $f^{-1}(B_1) \subseteq f^{-1}(B_1 \cup B_2)$. Similarly, $f^{-1}(B_2) \subseteq f^{-1}(B_1 \cup B_2)$. Therefore, $f^{-1}(B_1) \cup f^{-1}(B_2) \subseteq f^{-1}(B_1 \cup B_2)$. Conversely, let $a \in f^{-1}(B_1 \cup B_2)$. Then $f(a) = b$ where $b \in B_1 \cup B_2$. By definition of union $b \in B_1$ and $a \in f^{-1}(B_1)$ or $b \in B_2$ and $a \in f^{-1}(B_2)$. Therefore, $a \in f^{-1}(B_1) \cup f^{-1}(B_2)$ and $f^{-1}(B_1 \cup B_2) \subseteq f^{-1}(B_1) \cup f^{-1}(B_2)$.

31. Since $B_1 \cap B_2 \subseteq B_1$, by part (b) $f^{-1}(B_1 \cap B_2) \subseteq f^{-1}(B_1)$. Similarly, $f^{-1}(B_1 \cap B_2) \subseteq f^{-1}(B_2)$. Therefore, $f^{-1}(B_1 \cap B_2) \subseteq f^{-1}(B_1) \cap f^{-1}(B_2)$.

33. Let $A_1 = \{a_1\}$ and $A_2 = \{a_2\}$. Let $B = \{b\}$. Let $A = A_1 \cup A_2$ and define $f : A \to B$ by $f(a_1) = f(a_2) = b$. Then $f(A_1 \cap A_2) = f(\emptyset) = \emptyset$, but $f(A_1) \cap f(A_2) = \{b\}$.

35. Assume f is not one-to-one. Then there exists $a_1, a_2 \in A$ such that $f(a_1) = f(a_2) = b$. Let $X = \{a_1\}$ and $Y = \{a_2\}$. Then $f(X \cap Y) \neq f(X) \cap f(Y)$.

37. If f is not onto, then there exists $b \in B$, such that there is not in $f(A)$. Therefore, $ff^{-1}(A \cup \{b\}) \neq A \cup \{b\}$. On the other hand, suppose f is onto. By definition of f^{-1}, it is always true that $ff^{-1}(W) \subseteq W$. Let $w \in W$. Since f is onto, there exists $a \in A$ such that $f(a) = w$, so that $a \in f^{-1}(W)$. Therefore, $f(a) \in ff^{-1}(W)$, so $w \in ff^{-1}(W)$, and $W \subseteq ff^{-1}(W)$. Therefore, $W = ff^{-1}(W)$.

39. Assume f and g are one-to-one and $(f \circ g)(a) = (f \circ g)(a') = c$. Hence, $f(g(a)) = f(g(a'))$. Let $b = g(a)$ and $b' = g(a')$. Therefore, $f(b) = f(b')$. Since f is one-to-one, $b = b'$ and $g(a) = g(a')$. Since g is one-to-one, $a = a'$. Therefore, $f \circ g$ is one-to-one.

41. $(f \circ g)^{-1} = \{(c, a) : (a, c) \in f \circ g\}$
$= \{(c, a) : \text{There exists } b \in B$
$\quad\quad \text{so } (a, b) \in g \text{ and } (b, c) \in f$
$= \{(c, a) : \text{There exists } b \in B$
$\quad\quad \text{so } (b, a) \in g^{-1} \text{ and } (c, b) \in f^{-1}$
$= g^{-1} \circ f^{-1}$

43. Assume g is not one-to-one, then $g(a) = g(b)$ and $a \neq b$. Hence, $f(g(a)) = f(g(b))$ with $a \neq b$, and $f \circ g$ is not one-to-one.

Chapter 3—Section 3.1

1. $3^2 + 4^2 \geq 5^2$

3. $|3 - 1| \leq 7$

5. John likes Sue better than Mary.

7. If $2^2 = (-2)^2$, then $-2 = 2$.

9. $0 \leq (-3)^2 \leq 4$

11. There exists x, and there exists y, such that $x^2 + y^2 \geq 25^2$.

13. For every r, there is an x such that $|x - 1| \leq r$.

15. John likes Sue better than anyone.

17. For all x and for all y, if $y^2 = x^2$, then $x = y$.

19. For all x, there exists a and there exists b such that $a \leq x^2 \leq b$.

21. Domain of discourse for x is the set of all dogs. The domain for y is the set of all days. $R(x, y)$: x has his y.

$\forall x \exists y R(x, y)$.

23. Domain of discourse—people. $T(x, y) : x$ plays better tennis than y.

$\forall x T(x, \text{Fred})$.

25. Domain of discourse—golfers. $V(x, y) : x$ will eventually be defeated by a better y.

$\forall x \exists y V(x, y)$.

27. Domain of discourse—integers. $T(n, x, y, z) : x^n = y^n + z^n$.

$\forall n \exists x \exists y \exists z T(n, x, y, z)$.

29. Domain of discourse—athletes in the world. $W(x, y) :$ x is better than y.

$\forall y W(he, y)$.

31. true

33. true

35. false

37. The first statement is true. Since it is true that 7 is odd and
$$7 \text{ is odd} \to (\exists k)(7 = 2k + 1)$$
we may conclude that $(\exists k)(7 = 2k + 1)$.

39.
1. $\forall x P(x) \vee \forall x Q(x)$	given	
2. $P(a) \vee \forall x Q(x)$ for arbitrary a	1, universal instantiation	
3. $P(a) \vee Q(b)$ for arbitrary b	1, universal instantiation	
4. $P(a) \vee Q(a)$	3, since b is arbitrary, choose $b = a$	
5. $\forall x (P(x) \vee Q(x))$	3, universal generalization	

41. valid

43. invalid

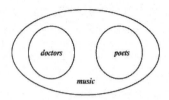

45. valid

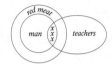

47. invalid

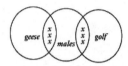

49. valid

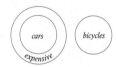

51. Someone does not like Charlie.

53. Some men are not mortal.

55. $\forall x \; x^2 \neq 9$

57. $\forall x \forall y \exists z \exists n \; x^n + y^n \neq z^n$

Chapter 3—Section 3.2

1.
$b + a = c + a$	given
$(b + a) + (-a) = (c + a) + (-a)$	I2
$b + (a + (-a)) = (c + a) + (-a)$	associative law
$b + 0 = c + 0$	I7 additive inverse
$b = a$	I6 additive identity

3. Since $-a$ is the inverse of a, we have $a + (-a) = (-a) + a$. But this also satisfies the requirement for a to be the inverse of $-a$. Therefore, $a = -(-a)$.

5. $a \cdot (-b) + (-a) \cdot (-b) = (-b) \cdot a + (-b) \cdot (-a)$
by the commutative law
$= (-b) \cdot (a + (-a))$
by the distributive law
$= (-b) \cdot 0$
by definition of inverse
$= 0$ by part (b)

Similarly, $(-a) \cdot (-b) + a \cdot (-b) = 0$, so $(-a) \cdot (-b)$ is the inverse of $a \cdot (-b)$. But $a \cdot b$ is the inverse of $a \cdot (-b)$. Therefore, $a \cdot b = (-a) \cdot (-b)$.

7. If $a + b = a$, then $a + b = a + 0$ since 0 is the additive. Hence, $b + a = 0 + a$, since addition is commutative. Therefore, $b = 0$, by the cancellation theorem (Theorem 3.8(a)).

9. If a is positive, then $a \geq 1$ and $a \cdot a \geq a \cdot 1$ by Theorem 3.11. If $a = 0$, then $a^2 = 0 \cdot 0 = 0$, and $a^2 = a$, so $a^2 \geq a$. If a is negative, then $0 > a$. But by Theorem 3.12, $a^2 \geq 0$. By Theorem 3.10, $a^2 \geq a$. (Note that when we state that $a \geq 1$, we are assuming that there is no integer between 0 and 1. Although this seems obvious, we do not have the necessary axioms to prove it here. It is proven in a later section.)

11. If a and b are odd, then $a = 2m + 1$ and $b = 2n + 1$ for some m, n.

$ab = (2m + 1)(2n + 1)$
by the definition of an odd integer
$= 2m(2n + 1) + 1(2n + 1)$
by the distributive law
$= 4mn + 2m + 1(2n + 1)$
by the distributive law
$= 4mn + 2m + 2n + 1$
by definition of multiplicative identity
$= 2(2mn + m + n) + 1$
by definition of inverse

so that ab is an odd number.

13. Not true. $-1 \geq -3$ and $-1 \geq -2$ but 1 is not greater than 6.

15. If the inverse n of an odd integer m were even, then the inverse of the even number n would be the odd integer m, which contradicts the results of the previous problem.

17. There is no largest positive integer. If n were the largest integer, then $n + 1$ is a larger positive integer, which is a contradiction.

19. If a and b are negative, then $-a$ and $-b$ are positive and $(-a)(-b) = ab$ is positive, which is a contradiction.

21. Not true for $a = 1$.

23. Since 1 is positive, m and n are both positive or both negative. If m and n are both positive and $m = 1$, then $n = mn = 1m = 1$. If m and n are both greater than 1, then mn is greater than 1, a contradiction. If m and n are both negative, then $-m$ and $-n$ are both positive and by the preceding argument $-m = -n = 1$, so that $m = n = -1$.

25. $(a+b)(a^2 - ab + b^2) = (a+b)a^2 + (a+b)(-ab)$
    $\qquad\qquad\qquad\qquad\quad + (a+b)b^2 \quad$ distributive law
    $\qquad\qquad\qquad = aa^2 + ba^2 + a(-ab) + b(-ab) + ab^2$
    $\qquad\qquad\qquad\quad + bb^2 \quad$ distributive law
    $\qquad\qquad\qquad = a^3 + ba^2 - a^2b - bab + ab^2 + b^3$
    $\qquad\qquad\qquad\quad$ property of $-$
    $\qquad\qquad\qquad = a^3 + a^2b - a^2b - ab^2 + ab^2 + b^3$
    $\qquad\qquad\qquad\quad$ associative law
    $\qquad\qquad\qquad = a^3 + 0 + 0 + b^3$
    $\qquad\qquad\qquad\quad$ property of inverse
    $\qquad\qquad\qquad = a^3 + b^3$
    $\qquad\qquad\qquad\quad$ property of identity

27. Assume $(a+b)$ is not even. Then one term is even and the other is odd. Say a is even and b is odd. Then $a = 2m$ and $b = 2n + 1$ for some m and n. $(a^2 - ab + b^2)$ must be even. But $a^2 = 4m^2$, $ab = 2m(2n+1) = 4mn + 2m$ and $b^2 = (2n+1) = 4n^2 + 4n + 1$ so that
$$(a^2 - ab + b^2) = 4m^2 + 4mn + 2m + 4n^2 + 4n + 1$$
$$= 2(2m^2 + 2mn + m + 2n^2 + 2n) + 1$$
and $(a^2 - ab + b^2)$ is odd, which is a contradiction.

29. Assume $a + b$ is not odd. Then it is even. But then $a^3 + b^3$ is even. Hence, $a^3 + b^3$ is not odd. Hence, using the contrapositive we have proven that if $(a^3 + b^3)$ is odd then $a + b$ is odd.

31. Not true. $6 = 3 \cdot 2$

Chapter 3—Section 3.3

1. For $n = 1$, we have $1 = \dfrac{1(3-1)}{2}$, so the statement is true. Assume that the statement is true for $n = k$, so that $1 + 4 + 7 + \cdots + (3k - 2) = \dfrac{k(3k-1)}{2}$. We now want to prove that the statement is true for $n = k + 1$, so that
$$1 + 4 + 7 + \cdots + (3k - 2) + (3(k+1) - 2)$$
$$= \dfrac{(k+1)(3(k+1) - 1)}{2}$$
$$= \dfrac{(k+1)(3k+2)}{2}$$
If we add $(3(k+1) - 2) = 3k + 1$ to both sides of the statement for $n = k$, we have
$$1 + 4 + 7 + \cdots + (3k-2) + (3(k+1) - 2)$$
$$= \dfrac{k(3k-1)}{2} + 3k + 1$$
$$= \dfrac{3k^2 + 5k + 2}{2}$$
$$= \dfrac{(k+1)(3k+2)}{2}$$
and we have proven the statement for $n = k + 1$.

3. For $n = 1$, we have $1 = 2 - 1$, so the statement is true for $n = 1$. Assume true for $n = k$ so $1 + 2 + 2^2 + \cdots + 2^{k-1} = 2^k - 1$. We want to prove true for $n = k + 1$ so $1 + 2 + 2^2 + \cdots + 2^{k-1} + 2^k = 2^{k+1} - 1$. Adding 2^k to each side of the equation for $n = k$, $1 + 2 + 2^2 + \cdots + 2^{k-1} + 2^k = 2^k - 1 + 2^k$. But $2^k - 1 + 2^k = 2 \cdot 2^k - 1 = 2^{k+1} - 1$. Thus, we have proven the statement is true for $n = k + 1$.

5. For $n = 1$, we have $a = \dfrac{a - ar}{1 - r} = a$, so the statement is true. Assume true for $n = k$, so we have
$$a + ar + ar^2 + \cdots + ar^{k-1} = \dfrac{a - ar^k}{1 - r}$$
We want to prove
$$a + ar + ar^2 + \cdots + ar^{k-1} + ar^k = \dfrac{a - ar^{k+1}}{1 - r}$$
Adding ar^k to both sides of the equation we have
$$a + ar + ar^2 + \cdots + ar^{k-1} + ar^k$$
$$= \dfrac{a - ar^k}{1 - r} + ar^k$$
$$= \dfrac{a - ar^k}{1 - r} + \dfrac{ar^k(1 - r)}{1 - r}$$
$$= \dfrac{a - ar^k + ar^k - ar^{k+1}}{1 - r}$$
$$= \dfrac{a - ar^{k+1}}{1 - r}$$
Thus, we have proven the statement is true for $n = k + 1$.

7. For $n = 1$, we have $1^3 = \dfrac{1^2 2^2}{4}$, so the statement is true. Assume true for $n = k$, so we have $1^3 + 2^3 + 3^3 + \cdots + k^3 = \dfrac{k^2(k+1)^2}{4}$. We want to prove the statement is true for $n = k + 1$, so we have
$$1^3 + 2^3 + 3^3 + \cdots + k^3 + (k+1)^3$$
$$= \dfrac{(k+1)^2(k+2)^2}{4}$$
Adding $(k+1)^3$ to both sides of the statement for $n = k$, we have
$$1^3 + 2^3 + 3^3 + \cdots + k^3 + (k+1)^3$$
$$= \dfrac{k^2(k+1)^2}{4} + (k+1)^3$$
$$= \dfrac{k^2(k+1)^2 + 4(k+1)^3}{4}$$
$$= \dfrac{(k+1)^2(k^2 + 4(k+1))}{4}$$
$$= \dfrac{(k+1)^2(k+2)^2}{4}$$
showing the statement for $n = k + 1$.

9. For $n = 1$, we have
$$1(1+1) = \frac{1(1+1)(1+2)}{3}$$
or $2 = 2$, so the statement is true. Assume true for $n = k$, so we have
$$1 \cdot 2 + 2 \cdot 3 + \cdots + k(k+1) = \frac{k(k+1)(k+2)}{3}$$
We want to prove the statement is true for $n = k+1$, so we have
$$1 \cdot 2 + 2 \cdot 3 + \cdots + k(k+1) + (k+1)(k+2)$$
$$= \frac{(k+1)(k+2)(k+3)}{3}$$
Adding $(k+1)(k+2)$ to both sides of the equation for $n = k$, we have
$$1 \cdot 2 + 2 \cdot 3 + \cdots + k(k+1) + (k+1)(k+2)$$
$$= \frac{k(k+1)(k+2)}{3} + (k+1)(k+2)$$
but
$$\frac{k(k+1)(k+2)}{3} + (k+1)(k+2)$$
$$= \frac{k(k+1)(k+2) + 3(k+1)(k+2)}{3}$$
$$= \frac{(k+3)(k+1)(k+2)}{3}$$
showing the statement for $n = k+1$. So the statement is true.

11. For $n = 1$, we have $2 = 2(1)^2 = 2$ and the statement is true for $n = 1$.

Assuming true for $n = k$, we have
$$2 + 6 + 10 + \cdots (4k-2) = 2k^2$$
We want to prove that
$$2 + 6 + 10 + \cdots (4k-2) + (4k+2) = 2(k+1)^2$$
Adding $(4k+2)$ to both sides of the equation in the assumption, we have
$$2 + 6 + 10 + \cdots (4k-2) + (4k+2) = 2k^2 + (4k+2)$$
$$= 2k^2 + 4k + 2$$
$$= 2(k^2 + 2k + 1)$$
$$= 2(k+1)^2$$
and we have proven the statement for $n = k+1$.

13. $3^2 > 2 \cdot 3 + 1$, so the statement is true for $n = 3$. Assume true for $n = k$, so that $k^2 > 2 \cdot k + 1$. We need to show that
$$(k+1)^2 > 2(k+1) + 1 = 2 \cdot k + 3$$
or
$$k^2 + 2k + 1 > 2(k+1) + 1 = 2 \cdot k + 3$$

Since $k \geq 3$, certainly $2k + 1 > 2$. Combining this with $k^2 > 2 \cdot k + 1$, we have
$$k^2 + 2k + 1 > 2 \cdot k + 1 + 2 = 2 \cdot k + 3$$

15. $2^1 > 1!$, $2^2 > 2!$, $2^3 > 3!$. Therefore, $2^n > n!$, for $n = 1, 2, 3$. For $n \geq 4$, $2^n \leq n!$. Prove this using induction. $2^4 = 16 \leq 24 = 4!$. Therefore, true for $n = 4$. Assume true for $n = k$, so $2^k \leq k!$. We need to show that $2^{k+1} \leq (k+1)!$. If we multiply the left side of $2^k \leq k!$ by 2 and the left side by $k+1$, we get the desired result. This is possible since $2 \leq k+1$.

17. For $n = 1$, we have $7 - 2$ divisible by 5 so the statement is true for $n = 1$. Assume it is true for $n = k$, so that $7^k - 2^k$ is divisible by 5. Say $7^k - 2^k = 5m$. We want to show that $7^{k+1} - 2^{k+1}$ is divisible by 5.
$$7^{k+1} - 2^{k+1} = 7(7^k - 2^k) + 7 \cdot 2^k - 2^{k+1}$$
$$= 7(7^k - 2^k) + (7-2) \cdot 2^k$$
$$= 5 \cdot 7 \cdot m + 5 \cdot 2^k$$
$$= 5(7m + 2^k)$$
and the statement is true for $n = k+1$.

19. For $n = 1$, we have $1^3 + 2 \cdot 1 = 3$ divisible by 3. Assume $k^3 + 2k$ is divisible by 3, say, $k^3 + 2k = 3m$. We need to show that $(k+1)^3 + 2(k+1)$ is divisible by 3.
$$(k+1)^3 + 2(k+1) = k^3 + 3k^2 + 3k + 1 + 2k + 2$$
$$= k^3 + 2k + 3k^2 + 3k + 3$$
$$= 3m + 3(k^2 + k + 1)$$
$$= 3(m + k^2 + k + 1)$$
so that $(k+1)^3 + 2(k+1)$ is divisible by 3. Thus, the statement is true for $n = k+1$.

21. For $n = 8$, we have $8 = 3(1) + 5(1)$. Assume true for $n = k$, so $k = 3r + 5s$. If $r \geq 3$, then $k+1 = 3(r-3) + 5(s+2)$. If $r < 3$, then $s \geq 1$ so $k+1 = 3(r+2) + 5(s-1)$. Either way we have expressed $k+1$ in the proper form.

23. For $n = 6$, $6! = 720 > 216 = 6^3$. Therefore, the statement is true for $n = 6$. Assume the statement is true for $n = k$ so $k! > k^3$. We need to show that $(k+1)! > (k+1)^3$. Multiplying both sides of $k! > k^3$ by $(k+1)$, we have $(k+1)! > k^3(k+1) = k^3 + k^4$. From exercise 2, we have $k^2 > 2k + 1$, so $k^4 > (2k+1)^2 = 4k^2 + 4k + 1$, so that
$$(k+1)! > k^3 + k^4$$
$$> k^3 + 4k^2 + 4k + 1$$
$$> k^3 + 3k^2 + 3k + 1$$
$$= (k+1)^3$$
and the statement is true for $n = k+1$.

25. Suppose $w \geq x$ for all $w \in W$. Let $T = \{t : t = w - x + 1 \text{ for some } w \in W.\}$. Each integer in T is positive. Therefore, T contains a least element, say, $t_0 = w_0 - x + 1$. Then w_0 is the least element of W.

27. (a) For $n = 1$, $A_1 \cap A = A_1 \cap A$. Assume true for $n = k$, so that $\left(\bigcup_{i=1}^{k} A_i\right) \cap A = \bigcup_{i=1}^{k} (A_i \cap A)$. We want to prove that $\left(\bigcup_{i=1}^{k+1} A_i\right) \cap A = \bigcup_{i=1}^{k+1} (A_i \cap A)$.

$$\left(\bigcup_{i=1}^{k+1} A_i\right) \cap A = \left(\bigcup_{i=1}^{k} A_i \cup A_{k+1}\right) \cap A$$
by definition of union
$$= \left(\left(\bigcup_{i=1}^{k} A_i\right) \cap A\right) \cup (A_{k+1} \cap A)$$
by distributive law
$$= \left(\bigcup_{i=1}^{k} (A_i \cap A)\right) \cup (A_{k+1} \cap A)$$
by induction hypothesis
$$= \bigcup_{i=1}^{k+1} (A_i \cap A).$$
by definition of union

(b) For $n = 1$, $A_1' = A_1'$ so the equation is true. Assume true for $n = k$, so we have $\left(\bigcup_{i=1}^{k} A_i\right)' = \bigcap_{i=1}^{k} A_i'$. We want to prove that $\left(\bigcup_{i=1}^{k+1} A_i\right)' = \bigcap_{i=1}^{k+1} A_i'$. By De Morgan's law,

$$\left(\bigcup_{i=1}^{k+1} A_i\right)' = \left(\bigcup_{i=1}^{k} A_i \cup A_{k+1}\right)' = \left(\bigcup_{i=1}^{k} A_i\right)' \cap A_{k+1}'$$

Using the induction hypothesis, we have

$$\left(\bigcup_{i=1}^{k} A_i\right)' \cap A_{k+1}' = \bigcap_{i=1}^{k} A_i' \cap A_{k+1}' = \bigcap_{i=1}^{k+1} A_i'$$

Therefore, the statement is true for $n = k + 1$.

29. Using induction on m, for $m = 1$ we have $a^{m+1} = a^m \cdot a$, which is true by definition. Assume true for $n = k$, so we have $a^{m+k} = a^m \cdot a^k$. We need to prove that $a^{m+k+1} = a^m \cdot a^{k+1}$. But

$$\begin{aligned} a^{m+k+1} &= a^{m+k} \cdot a & \text{by definition} \\ &= (a^m \cdot a^k) \cdot a & \text{by induction hypothesis} \\ &= a^m \cdot (a^k \cdot a) & \text{by associativity} \\ &= a^m \cdot a^{k+1}. & \text{by definition} \end{aligned}$$

showing the statement for $n = k + 1$.

31. For $n = 1$, we have $a^m = a^m$, so the equality is true for $n = 1$. Assume true for $n = k$, so $a^{mk} = (a^m)^k$. Then, using the induction hypothesis,

$$a^{m(k+1)} = a^{mk+m} = a^{mk} \cdot a^m = (a^m)^k \cdot a^m = (a^m)^{k+1}$$

so the statement is true for $n = k + 1$.

33. Assume that T satisfies (a) and (b) of the second order of induction. Let T' be the set of all integers for which T is not true. If T' is nonempty, then by the well-ordering theorem, T' contains a least positive integer, say, n. But since n is the least element in T', for all m such that $m < n$, T is true. This contradicts (b) so T' is empty, and T is true for all n.

35. Assume T is not empty. Then by the well-ordering theorem, it has a least element, say, n. But by part 2, there is $m < n$, which satisfies T. This is a contradiction, so T is empty.

37. Using induction, for $n = 1$ we have $\cos(\theta) + i\sin(\theta) = \cos(\theta) + i\sin(\theta)$, so the statement is true for $n = 1$. Assume the statement is true for k so that $(\cos(\theta) + i\sin(\theta))^k = \cos(k\theta) + i\sin(k\theta)$. We need to prove that $(\cos(\theta) + i\sin(\theta))^{k+1} = \cos((k+1)\theta) + i\sin((k+1)\theta)$.

$$\begin{aligned} (\cos(\theta) + i\sin(\theta))^{k+1} &= (\cos(\theta) + i\sin(\theta))^k(\cos(\theta) + i\sin(\theta)) \\ &= (\cos(k\theta) + i\sin(k\theta))(\cos(\theta) + i\sin(\theta)) \\ &= \cos(k\theta)\cos(\theta) - \sin(k\theta)\sin(\theta) \\ &\quad + i(\sin(\theta)\cos(k\theta) \\ &\quad + \sin(k\theta)\cos(\theta)) \\ &= \cos(k\theta + \theta) + i\sin(k\theta + \theta) \\ &= \cos((k+1)\theta) + i\sin((k+1)\theta) \end{aligned}$$

39. For $n = 2$, there is obviously only one way to multiply the numbers together. Hence, the statement is true for $n = 2$. Assume true for $n = k$ so that there are $k - 1$ multiplications necessary. Assume there are $k + 1$ numbers to be multiplied. For a fixed set of multiplications of the $k + 1$ numbers, multiply the first two numbers to be multiplied together. There are now k numbers, so they require $k - 1$ multiplications. Replacing the first product with its two factors, we have k multiplications as desired.

Chapter 3—Section 3.4

1. 54, 27, 18, 9, 6, 3, 2, 1

3. 72, 36, 18, 9, 24, 12, 6, 3, 8, 4, 2, 1

5. 74, 37, 2, 1

7. $47 = 1 \cdot 47 + 0 \quad q = 1, r = 0$

9. $43 = 5 \cdot 8 + 3 \quad q = 5, r = 3$

11. $\gcd(54, 27) = 27$, $\text{lcm}(54, 27) = 54$, $\gcd(54, 27) \cdot \text{lcm}(54, 27) = 1458$

13. $\gcd(6, 15) = 3$, $\text{lcm}(6, 15) = 30$, $\gcd(6, 15) \cdot \text{lcm}(6, 15) = 90$

15. $\gcd(33, 1) = 1$, $\text{lcm}(33, 1) = 33$, $\gcd(33, 1) \cdot \text{lcm}(33, 1) = 33$

17. 0 might be a possible solution for lcm$(a, 0)$. 0 is a multiple of a and 0. If any number is a multiple of a and 0, then it is a multiple of 0.

19. gcd(a, b) divides a and b, so it divides $a - b$. Therefore, gcd$(a, b)|$ gcd$(a - b, b)$. Conversely, if any integer divides $a - b$ and b, then it divides a. Therefore, gcd$(a - b, b) |$ gcd(a, b). Therefore, gcd$(a - b, b) =$ gcd(a, b).

21. $-a$ and $-b$ are positive. $-a \mid -b$ and $-b \mid -a$. Therefore, $-a = -b$ and $a = b$.

23. $bc = mac$ for some m. Therefore, $b = ma$ and $a \mid b$.

25. We first prove that if a^2 is even, then a is even or, conversely, if a is odd, then a^2 is odd. Let $a = 2k + 1$. Then $a^2 = 4k^2 + 4k + 1 = 2(2k^2 + 2k) + 1$ is odd. If a is even, then $a = 2k$ for some k so $a^2 = 4k^2$ and a^2 is divisible by 4.

27. If $a \mid b$, then $b = am$ for some m. Therefore, $b^2 = a^2 m^2$ and $a^2 \mid b^2$.

Chapter 3 — Section 3.5

1. $7 \cdot 2^3 \cdot 13$

3. $4899 = 4900 - 1 = 69 \cdot 71 = 3 \cdot 23 \cdot 71$

5. 523

7. Since 71 and 23 are both primes, gcd$(71, 23) = 1$ and
$$\text{lcm}(71, 23) = 71 \cdot 23 = 1633$$

9. Since $n! \mid (n+2)!$, gcd$(n!, (n+2)!) = n!$ and lcm$(n!, (n+2)!) = (n+2)!$.

11. 5 and 7, 11 and 13, 17 and 19.

13. Unless one of the primes is 2, the sum $a^2 + b^2$ is never prime, since it is divisible by 2.

15. $479001603 = 12! + 3$ and $479001603 = 12! + 7$.

17. Let p, $p + 2$, and $p + 4$ be three consecutive odd numbers. p has one of the forms $3m$, $3m + 1$, or $3m + 2$. If it has form $3m$, then it is divisible by 3. If it has form $3m + 1$, then $p + 2$ is divisible by 3. If it has form $3m + 2$, then $p + 4$ is divisible by 3. Since one of the three numbers must be divisible by 3, they cannot all be prime.

19. If $0 \leq b(i) \leq a(i)$, then $p_i^{b(i)} | p_i^{a(i)}$ for i, so $b|a$. If $b|a$, then $p_i^{b(i)}|a$. By the unique factorization of an integer into primes, we know that $p_i^{b(i)}$ is part of the unique factorization of a, so $p_i^{b(i)}|p_i^{a(i)}$.

21. No. gcd$(8, 26) = 2$.

23. No. gcd$(327, 443) = 3$.

25. If $a = p_1^{k_1} p_2^{k_2} p_3^{k_3} \cdots p_n^{k_n}$, then $a^2 = p_1^{2k_1} p_2^{2k_2} p_3^{2k_3} \cdots p_n^{2k_n}$. If $a^2 \mid b^2$, then $p_i^{2k_i} \mid b^2$ for all $1 \leq i \leq n$ and so $p_i^{k_i} \mid b$ for all $1 \leq i \leq n$ and $a \mid b$.

27. Not necessarily since 23 divides 667.

29. No. $13^2 - 13 - 1 = 155$.

Chapter 3 — Section 3.6

1. Since $37 = 4 \cdot 9 + 1$, $[[37]]_4 = 1$.

3. Since $48 = 2 \cdot 23 + 2$, $[[48]]_{23} = 2$.

5. Since $33 = 5 \cdot 6 + 3$, $[[33]]_6 = 3$.

7. Since $49 = 2 \cdot 17 + 15$, $[[49]]_{17} = 15$.

9. Since 6 is a factor of 8!, $[[8!]]_6 = 0$.

11. $[[1^2 + 2^2 + 3^2 + \cdots + 25^2]]_2 = [[1^2 + 0^2 + 1^2 + 0^2 + \cdots + 1^2]]_2 = [[13]]_2 = 1$

13. $4^2 \equiv 3^2 \pmod{7}$, however, 4 is not congruent to 3.

15.
+	[0]	[1]	[2]	[3]
[0]	[0]	[1]	[2]	[3]
[1]	[1]	[2]	[3]	[0]
[2]	[2]	[3]	[0]	[1]
[3]	[3]	[0]	[1]	[2]

17.
+	[0]	[1]	[2]	[3]	[4]	[5]	[6]
[0]	[0]	[1]	[2]	[3]	[4]	[5]	[6]
[1]	[1]	[2]	[3]	[4]	[5]	[6]	[0]
[2]	[2]	[3]	[4]	[5]	[6]	[0]	[1]
[3]	[3]	[4]	[5]	[6]	[0]	[1]	[2]
[4]	[4]	[5]	[6]	[0]	[1]	[2]	[3]
[5]	[5]	[6]	[0]	[1]	[2]	[3]	[4]
[6]	[6]	[0]	[1]	[2]	[3]	[4]	[5]

19. [2]

21. [2]

23. [3]

25. [1], [3]

27. [4]

29. [6]

31. If $a \equiv b \pmod{mn}$, then $a - b$ is equal to kmn for some k. Therefore, since $a - b$ is equal to $(kn)m$, we have $a \equiv b \pmod{m}$. Similarly, $a \equiv b \pmod{n}$.

33. We already know that there is one and only one r for $0 \leq r < n$ such that $a \equiv r \pmod{n}$. We must show that r is also relatively prime to n. Since $a = nq + r$ for some q, if r is not relatively prime to n, then there exists an integer m such that $n = cm$ and $r = dm$ for integers c and d. Therefore,
$$a = cmq + dm = (cq + d)m$$
so that m is also a factor of a, which contradicts the assumption that a and n are relatively prime.

35. 32

37. 78

39. 776

41. 535

Chapter 4—Section 4.1

1. $f \circ g = \begin{pmatrix} 1 & 2 & 3 \\ 1 & 2 & 3 \end{pmatrix}, g \circ f = \begin{pmatrix} 1 & 2 & 3 \\ 1 & 2 & 3 \end{pmatrix},$
$f^{-1} = \begin{pmatrix} 1 & 2 & 3 \\ 3 & 1 & 2 \end{pmatrix}, g^{-1} = \begin{pmatrix} 1 & 2 & 3 \\ 2 & 3 & 1 \end{pmatrix}$

3. $f \circ g = \begin{pmatrix} 1 & 2 & 3 & 4 \\ 1 & 2 & 3 & 4 \end{pmatrix}, g \circ f = \begin{pmatrix} 1 & 2 & 3 & 4 \\ 1 & 2 & 3 & 4 \end{pmatrix},$
$f^{-1} = \begin{pmatrix} 1 & 2 & 3 & 4 \\ 3 & 2 & 4 & 1 \end{pmatrix}, g^{-1} = \begin{pmatrix} 1 & 2 & 3 & 4 \\ 4 & 2 & 1 & 3 \end{pmatrix}$

5. $\dfrac{10!}{8!} = 10 \cdot 9 = 90$

7. $\dfrac{n!}{(n-2)!} = n \cdot (n-1)$

9. $\lfloor 1.001 \rfloor = 1$

11. $\lceil -4.01 \rceil = -4$

13. 630630

15. -2

17. -3

19. $4, 11, 30, 67, 128$

21. $1, 1, 1, 2, 2$

23. $6, 11, 18, 27, 38$

25. $-4, -2, -1, -1, 0$

27. $A_n = 3 + 5(n-1)$

29. $A_n = \left(2(n-1) + \left\lfloor \dfrac{1}{n} \right\rfloor\right)(3 + 1 + 2 + 3 \cdots n) = \left(2(n-1) + \left\lfloor \dfrac{1}{n} \right\rfloor\right)\left(3 + \dfrac{n(n+1)}{2}\right)$

31. $A(n) = 2n^2$

33. $A(n) = (-1)^{n+1} \left\lfloor \dfrac{n+1}{2} \right\rfloor$

Chapter 4—Section 4.2

1. $\begin{bmatrix} 1 & -4 & -3 & 12 \\ 39 & 5 & 21 & 8 \\ -10 & 5 & 0 & -20 \end{bmatrix}$

3. $\begin{bmatrix} -2 & 3 & -5 & 0 \\ -6 & 9 & -15 & 0 \\ -8 & 12 & -20 & 0 \\ -10 & 15 & -25 & 0 \end{bmatrix}$

5. $\begin{bmatrix} 10 & -2 \\ -12 & 14 \end{bmatrix}$

7. $A^t = \begin{bmatrix} -2 & 6 & 2 \\ 4 & 1 & 3 \\ 7 & 5 & -4 \end{bmatrix}$

9. $A^t A = \begin{bmatrix} 44 & 4 & 8 \\ 4 & 26 & 21 \\ 8 & 21 & 58 \end{bmatrix}$

11. $BA = \begin{bmatrix} -5 & 13 \\ 69 & 41 \end{bmatrix}$

13. $A^t B = \begin{bmatrix} -26 & 60 \\ -2 & 47 \end{bmatrix}$

15. $B^t A^t = \begin{bmatrix} -16 & 18 \\ 15 & 52 \end{bmatrix}$

17. $B - A = \begin{bmatrix} 7 & -1 \\ -9 & 4 \end{bmatrix}$

19. $2B - 3A = \begin{bmatrix} 17 & -4 \\ -25 & 3 \end{bmatrix}$

21. $BA = \begin{bmatrix} 3 & -8 & 7 \\ 21 & -36 & 69 \\ 10 & 0 & 50 \end{bmatrix}$

23. $B^t = \begin{bmatrix} -1 & 3 & 10 \\ 2 & 9 & 0 \end{bmatrix}$

25. $(BA)^t = \begin{bmatrix} 3 & 21 & 10 \\ -8 & -36 & 0 \\ 7 & 69 & 50 \end{bmatrix}$

27. Let $A = \begin{bmatrix} 0 & 0 \\ 0 & 1 \end{bmatrix}$ and $B = \begin{bmatrix} 1 & 0 \\ 0 & 0 \end{bmatrix}$. Then $AB = \begin{bmatrix} 0 & 0 \\ 0 & 0 \end{bmatrix}$ but $A, B \neq \begin{bmatrix} 0 & 0 \\ 0 & 0 \end{bmatrix}$.

29. (a) For each $1 \le i \le m$, $1 \le j \le n$, $A_{ij} + B_{ij} = B_{ij} + A_{ij}$, so $A + B = B + A$.

(b) For each $1 \le i \le m$, $1 \le j \le n$, $A_{ij} + (B_{ij} + C_{ij}) = (A_{ij} + B_{ij}) + C_{ij}$ so $A + (B + C) = (A + B) + C$.

31. For each $1 \le i \le m$, $1 \le j \le n$, $(r+s)A_{ij} = rA_{ij} + sA_{ij}$ so $(r+s)A = rA + sA$.

33. For each $1 \le i \le m$, $1 \le j \le n$, let $C_{ij} = -A_{ij} + B_{ij}$, so $A_{ij} + C_{ij} = A_{ij} + -A_{ij} + B_{ij} = B_{ij}$ and $A + C = B$.

35. Let R and S be relations with representations M and N, respectively. Let $A = M \vee N$. If $(a_i, a_j) \in R \cup S$, then $(a_i, a_j) \in R$ or $(a_i, a_j) \in S$. Therefore, $M_{ij} = 1$

or $N_{ij} = 1$, so $A_{ij} = M_{ij} \vee N_{ij} = 1$. Conversely, if $A_{ij} = M_{ij} \vee N_{ij} = 1$, then $M_{ij} = 1$ or $N_{ij} = 1$, so that $(a_i, a_j) \in R$ or $(a_i, a_j) \in S$. Therefore, $(a_i, a_j) \in R \cup S$. Therefore, $M_{ij} \vee N_{ij}$ is the representation matrix for $R \cup S$.

37. Let $(a_i, c_j) \in S \circ R$. Then there exists b_k so that $(a_i, b_k) \in R$ and $(b_k, c_j) \in S$. The representation matrix M for R has 1 in the ith row, jth column since $(a_i, b_k) \in R$ and the representation matrix N for S has 1 in the ith row, jth column since $(b_k, c_j) \in R$.

Since $(M \odot N)_{i,j} = \bigvee_{m=1}^{n} M_{im} \wedge N_{mj}$, it is equal to 1 since $M_{ik} \wedge N_{kj} = 1$.

Conversely, if $\bigvee_{m=1}^{n} M_{im} \wedge N_{mj} = 1$, then $M_{ik} \wedge N_{kj} = 1$ for some k. But then $(a_i, b_k) \in R$ and $(b_k, c_j) \in S$, so $(a_i, c_j) \in S \circ R$. Therefore, $(M \odot N)$ is the representation matrix for $S \circ R$.

39.
$$S \circ R = \begin{bmatrix} 1 & 0 & 0 & 0 \\ 0 & 0 & 0 & 1 \\ 0 & 0 & 0 & 0 \\ 1 & 1 & 0 & 0 \\ 1 & 1 & 0 & 0 \end{bmatrix}$$

41.
$$R^{-1} \circ S^{-1} = \begin{bmatrix} 1 & 0 & 0 & 1 & 1 \\ 0 & 0 & 0 & 1 & 1 \\ 0 & 0 & 0 & 0 & 0 \\ 0 & 1 & 0 & 0 & 0 \end{bmatrix}$$

43.
$$T \circ S = \begin{bmatrix} 0 & 1 & 0 & 0 \\ 0 & 1 & 0 & 0 \\ 0 & 0 & 1 & 1 \\ 0 & 0 & 0 & 0 \end{bmatrix}$$

45.
$$R = \begin{bmatrix} 1 & 0 & 0 & 0 & 0 \\ 0 & 1 & 1 & 0 & 0 \\ 1 & 1 & 0 & 0 & 1 \\ 0 & 0 & 0 & 1 & 0 \end{bmatrix}$$

$$S = \begin{bmatrix} 1 & 1 & 0 & 0 & 0 \\ 0 & 0 & 0 & 1 & 0 \\ 1 & 0 & 1 & 0 & 1 \\ 0 & 0 & 0 & 0 & 0 \end{bmatrix}$$

$$R^{-1} = \begin{bmatrix} 1 & 0 & 1 & 0 \\ 0 & 1 & 1 & 0 \\ 0 & 1 & 0 & 0 \\ 0 & 0 & 0 & 1 \\ 0 & 0 & 1 & 0 \end{bmatrix}$$

47.
$$S \circ R^{-1} = \begin{bmatrix} 1 & 1 & 1 & 0 & 1 \\ 1 & 0 & 1 & 1 & 1 \\ 0 & 0 & 0 & 1 & 0 \\ 0 & 0 & 0 & 0 & 0 \\ 1 & 0 & 1 & 0 & 1 \end{bmatrix}$$

49. U, V

51. U

53.
$$S = \begin{bmatrix} 1 & 1 & 0 & 0 & 0 \\ 0 & 0 & 1 & 1 & 0 \\ 1 & 0 & 0 & 0 & 1 \\ 0 & 0 & 0 & 0 & 0 \\ 0 & 0 & 0 & 1 & 0 \end{bmatrix} \quad T = \begin{bmatrix} 0 & 1 & 0 & 0 & 0 \\ 1 & 0 & 1 & 1 & 0 \\ 0 & 1 & 0 & 0 & 0 \\ 0 & 0 & 0 & 0 & 1 \\ 0 & 0 & 0 & 0 & 1 \end{bmatrix}$$

$$U = \begin{bmatrix} 1 & 1 & 1 & 0 & 0 \\ 1 & 1 & 1 & 0 & 0 \\ 1 & 1 & 1 & 0 & 0 \\ 0 & 0 & 0 & 1 & 0 \\ 0 & 0 & 0 & 0 & 1 \end{bmatrix} \quad V = \begin{bmatrix} 0 & 1 & 1 & 0 & 0 \\ 1 & 1 & 1 & 0 & 0 \\ 1 & 1 & 0 & 0 & 0 \\ 0 & 0 & 0 & 1 & 0 \\ 0 & 0 & 0 & 0 & 1 \end{bmatrix}$$

55.
$$S \cap T = \begin{bmatrix} 0 & 1 & 0 & 0 & 0 \\ 0 & 0 & 1 & 1 & 0 \\ 0 & 0 & 0 & 0 & 0 \\ 0 & 0 & 0 & 0 & 0 \\ 0 & 0 & 0 & 0 & 0 \end{bmatrix} \quad U \cap V = \begin{bmatrix} 0 & 1 & 1 & 0 & 0 \\ 1 & 1 & 1 & 0 & 0 \\ 1 & 1 & 0 & 0 & 0 \\ 0 & 0 & 0 & 1 & 0 \\ 0 & 0 & 0 & 0 & 1 \end{bmatrix}$$

57. $\{(1, a), (1, c), (2, b), (3, c), (4, a), (4, b)\}$

59. $\{(1, a), (1, c), (2, b), (3, a), (3, c), (4, b)\}$

61. $\{(a, c), (a, e), (b, b), (b, d), (c, a), (c, c), (c, e), (d, b), (d, d), (e, a), (e, c)\}$

63. $\{(a, a), (a, e), (b, b), (b, d), (c, a), (c, c), (d, b), (d, d), (e, a), (e, c), (e, e)\}$

65.

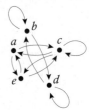

67.

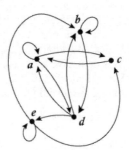

69.

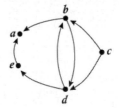

71. $$\begin{bmatrix} 1 & 1 & 1 & 0 & 1 \\ 0 & 1 & 1 & 0 & 1 \\ 1 & 1 & 1 & 1 & 1 \\ 1 & 0 & 0 & 1 & 0 \\ 1 & 1 & 1 & 0 & 1 \end{bmatrix}$$

Chapter 4—Section 4.3

1. (a) 7, (b) 11, (c) countably infinite

3. 1

5. 3

7. $\begin{array}{ccccc} 1 & 2 & 3 & 4 & 5 & \cdots \\ \downarrow & \downarrow & \downarrow & \downarrow & \downarrow \\ 0 & 1 & 2 & 3 & 4 & \cdots \end{array}$, $a(n) = n - 1$

9. $\begin{array}{ccccccc} 1 & 2 & 3 & 4 & 5 & \cdots & 10 & 11 & \cdots \\ \downarrow & \downarrow & \downarrow & \downarrow & \downarrow & & \downarrow & \downarrow \\ 0 & 1 & -1 & 2 & -2 & \cdots & 5 & -5 & \cdots \end{array}$

$a(n) = \begin{cases} \frac{n}{2} & \text{if } n \text{ is even} \\ -\frac{n-1}{2} & \text{if } n \text{ is odd} \end{cases}$

11. Since A is countable, there is a one-to-one correspondence a from N to A. If ϕ is a 1-1 correspondence from A to B, then ϕa is a one-to-one correspondence from N to B, and B is countably infinite.

13. Since Z, the set of integers, has been shown to be countably infinite, define a from Z to the desired set by $a(i) = 3i$.

Chapter 4—Section 4.4

1. $\{2, 4, \}$

3. $\emptyset$

5. By Theorem 4.38, the set of positive rationals is countably infinite. Since there is a one-to-one correspondence between the positive rationals and the negative rationals, the negative rationals are countable. Therefore, their union is countable by Theorem 4.39. Since $\{0\}$ is countable, its union with the nonzero rationals is countable. Therefore, the set of rational numbers is countable.

7. Using induction, the statement is obviously true for $n = 1$. Assume $Z_1 \times Z_2 \times Z_3 \times \cdots \times Z_k$ is countable. $Z_1 \times Z_2 \times Z_3 \times \cdots \times Z_k \times Z_{k+1} = (Z_1 \times Z_2 \times Z_3 \times \cdots \times Z_k) \times Z_{k+1}$ and since both $Z_1 \times Z_2 \times Z_3 \times \cdots \times Z_k$ and Z_{k+1} are countable, $Z_1 \times Z_2 \times Z_3 \times \cdots \times Z_k \times Z_{k+1}$ is countable.

9. The two previous exercises prove that kth-degree polynomials with $k \leq n$ are countable. Since the set of all polynomials with degree n or less is a finite union of kth-degree polynomials with $k \leq n$, this set is countable.

11. If a set A is countably infinite, it has the form $a_1, a_2, a_3, \ldots$ for some one-to-one correspondence a from Z to A. Define $\phi : A \to A$ by $\phi(a_i) = a_{i+1}$. This maps A into itself.

13. By mapping each element a of A into $\{a\}$ we define a one-to-one mapping from A to its power set. Therefore, $A \preceq \mathcal{P}(A)$. We have already shown there is no one-to-one mapping from A onto its power set. Hence, $A \prec \mathcal{P}(A)$.

15. If a set S is finite, then it cannot be in one-to-one correspondence with itself. If it is infinite, then it contains a countably infinite subset T. There is a one-to-one correspondence ϕ from T to a subset of itself. Define $\theta : S \to S$ by $\theta(s) = \phi(s)$ if $s \in T$ and $\theta(s) = s$ if $s \notin T$. This is a one-to-one correspondence from S to a subset of itself.

Chapter 5—Section 5.1

1. T is the transpose of A.

   Procedure Transpose (A, m, n, T)
   For $i = 1$ to m
     For $j = 1$ to n,
       $T_{ji} = A_{ij}$
     Endfor
   Endfor
   End

3. $\begin{bmatrix} -3 & -5 & 10 \\ 11 & -3 & 26 \\ 3 & -11 & 31 \end{bmatrix}$

5. $\begin{bmatrix} -18 \end{bmatrix}$

7. $\begin{bmatrix} -4 & -9 \\ 9 & -1 \end{bmatrix}$

9. $\begin{bmatrix} 29 & 23 & -17 \\ 23 & 51 & 5 \\ -17 & 5 & 21 \end{bmatrix}$

11. $AB = \begin{bmatrix} 0 & -5 \\ 6 & -11 \end{bmatrix}$

13. $AB^t = \begin{bmatrix} 1 & 2 \\ 4 & 5 \end{bmatrix} \begin{bmatrix} 4 & -2 \\ 1 & -3 \end{bmatrix} = \begin{bmatrix} 6 & -8 \\ 21 & -23 \end{bmatrix}$

15. $(AB)^t = \begin{bmatrix} 0 & -5 \\ 6 & -11 \end{bmatrix}^t = \begin{bmatrix} 0 & 6 \\ -5 & -11 \end{bmatrix}$

17. $A - B = \begin{bmatrix} -3 & 1 \\ 6 & 8 \end{bmatrix}$

19. $$5A = \begin{bmatrix} 5 & 10 \\ 20 & 25 \end{bmatrix}$$

21. $$\begin{bmatrix} -31 & 2 \\ -14 & -12 \end{bmatrix}$$

23. $$\begin{bmatrix} 1 & 2 \\ 0 & -4 \\ -5 & 0 \end{bmatrix}$$

25. $$\begin{bmatrix} 3 & 11 & 6 \\ -8 & -16 & 0 \\ 5 & -15 & -30 \end{bmatrix}$$

27. Procedure MaxInt$(a_1, a_2, a_3, \ldots, a_n)$
 (finds maximum value a_k)
 Let $k = 1$
 For $i = 2$ to n
  If $a_i \geq a_k$ let $k = i$
 Endfor
 End

29. Procedure Max-MinInt$(a_1, a_2, a_3, \ldots, a_n)$
 (finds minimum value a_k and maximum value a_j)
 Let $k = 1$
 Let $j = 1$
 For $i = 2$ to n
  If $a_i \leq a_k$ let $k = i$
  If $a_i \geq a_j$ let $j = i$
 Endfor
 End

31. Procedure Reflexive Test (A, n)
 Let $T = 1$
 For $i = 1$ to n
  If $A_{ii} = 0$ let $T = 0$
 Endfor
 End

33. First determine procedure for adding 1 to an integer n.

 Procedure Add (a, b, m, n)
  (adds b and a where b has length m, a has length n, $m \leq n$ and $b \leq a$. Assume $b = b_1 b_2 b_3 \cdots b_n$ and $a = a_1 a_2 a_3 \cdots a_n$ and $s = s_0 s_1 s_2 s_3 \cdots s_n$ where $b_i = 0$ for $i < n - m$.) Also assume single-digit integers can be added.
  Let $c = 0$.
   For $i = 1$ to n
    Let $d = \left\lfloor \dfrac{a_{n-i+1} + b_{n-i+1} + c}{10} \right\rfloor$
    Let $s_{n-i+1} = a_{n-i+1} + b_{n-i+1} + c - 10d$
    Let $c = d$
   Endfor
  Let $s_0 = c$
 End

35. Procedure Multiply (a, b, m, n)
  (Assume $b = b_m \cdots b_3 b_2 b_1$ and $a = a_n \cdots a_3 a_2 a_1$, $t = t_{m+1} \cdots t_3 t_2 t_1$, $p = p_{m+n} \cdots p_3 p_2 p_1$, $s = s_{m+n} \cdots s_3 s_2 s_1$)
  For $k = 1$ to $m+n$ set $p_k = 0$
  For $i = 1$ to n
   Set $c = 0$.
   For $j = 1$ to m
    $d = \left\lfloor \dfrac{a_i b_j + c}{10} \right\rfloor$
    $t_j = a_i b_j + c - 10d$
    $c = d$
   Endfor
   $t_{m+1} = c$
   For $k = 1$ to $m+n$
    For $k < i - 1$, $s_k = 0$
    For $k \geq i - 1$, $s_{k+i-1} = t_k$
   Endfor
   $p = s + p$
  Endfor
 End

37. Procedure Biadd (a, b)
  (adds b and a in binary where b has length m, a has length n, $m \leq n$, and $b \leq a$. Assume $b = b_1 b_2 b_3 \cdots b_n$ and $a = a_1 a_2 a_3 \cdots a_n$ and $s = s_0 s_1 s_2 s_3 \cdots s_n$ where $b_i = 0$ for $i < n - m$.) Also assume single-digit integers can be added.
  Set $c = 0$.
   For $i = 1$ to n
    Let $d = \left\lfloor \dfrac{a_{n-i+1} + b_{n-i+1} + c}{10} \right\rfloor$ where 10 is in binary.
    Let $s_{n-i+1} = a_{n-i+1} + b_{n-i+1} + c - 10d$
    Let $c = d$
   Endfor
  $s_0 = c$
 End

39. Assume the length of the truth table is $m = 2^n$ for some integer n. Let p_i and q_i be the values of p and q in case i. Let s_i be the truth value of $p \wedge q$ in case i.

 Procedure And (a, b, m, s)
  For $i = 1$ to m
   If $p_i = T$ and $q_i = T$, then $s_i = T$.
   If $p_i = T$ and $q_i = F$, then $s_i = F$.
   If $p_i = F$ and $q_i = T$, then $s_i = F$.
   If $p_i = F$ and $q_i = F$, then $s_i = F$.
  Endfor
 End

41. Assume the length of the truth table is $m = 2^n$ for some integer n. Let p_i and q_i be the values of p and q in case i. Let s_i be the truth value of $p \rightarrow q$ in case i.

 Procedure Implies (a, b, m, s)
  For $i = 1$ to m
   If $p_i = T$ and $q_i = T$, then $s_i = T$.

If $p_i = T$ and $q_i = F$, then $s_i = F$.
If $p_i = F$ and $q_i = T$, then $s_i = T$.
If $p_i = F$ and $q_i = F$, then $s_i = T$.
Endfor
End

Chapter 5—Section 5.2

1. $f(1) = 3 \cdot 3 = 9$ $f(2) = 3 \cdot 9 = 27$ $f(3) = 3 \cdot 27 = 81$ $f(4) = 3 \cdot 81 = 243$

3. $f(1) = 2^2 = 4$ $f(2) = 4^2 = 16$ $f(3) = 16^2 = 256$ $f(4) = 256^2 = 65536$

5. $f(1) = 2^1 = 2$ $f(2) = 2^2 = 4$ $f(3) = 2^4 = 16$ $f(4) = 2^{16} = 65536$

7. $f(1) = 0 + 3 = 3$ $f(2) = 1 + 6 = 7$ $f(3) = 3 + 9 = 12$ $f(4) = 6 + 12 = 18$

9. $f(1) = \frac{4}{1} = 4$ $f(2) = \frac{4}{4} = 1$ $f(3) = \frac{1}{9}$ $f(4) = \frac{1}{9} \div 16 = \frac{1}{144}$

11. $f(2) = 2 \cdot 3 - 1 = 5$ $f(3) = 2 \cdot 5 - 3 = 7$ $f(4) = 2 \cdot 7 - 5 = 9$ $f(5) = 2 \cdot 9 - 7 = 11$

13. $f(2) = 2^2 - 1^2 + 2^2 = 7$ $f(3) = 7^2 - 2 + 3^2 = 56$ $f(4) = 56^2 - 7 + 16 = 3145$ $f(5) = 3145^2 - 56 + 25 = 9{,}890{,}994$

15. $f(2) = \frac{1}{-1} = -1$ $f(3) = \frac{-1}{1} = -1$ $f(4) = \frac{-1}{-1} = 1$ $f(5) = \frac{1}{-1} = -1$

17. $f(2) = 2! - 1! = 1$ $f(3) = 1! - 2! = -1$ $f(4)$ undefined

19. $f(2) = \lfloor \frac{10+20}{2!} \rfloor = 15$ $f(3) = \lfloor \frac{20+15}{3!} \rfloor = 5$ $f(4) = \lfloor \frac{15+5}{4!} \rfloor = 0$ $f(5) = \lfloor \frac{5+0}{5!} \rfloor = 0$

21. $f(0) = 1$ $f(1) = 2$ $f(2) = 2^2$ $f(3) = 2^3$ $f(n) = 2^n$,

23. $f(0) = 1$ $f(1) = \frac{1}{1} = 1$ $f(2) = \frac{1}{2}$ $f(3) = \frac{1}{2 \cdot 3}$ $f(n) = \frac{1}{n!}$,

25. $f(0) = 1$ $f(1) = 5$ $f(2) = 5^2$ $f(3) = 5^3$ $f(n) = 5^n$.

27. $f(0) = 1$ $f(1) = 2$ $f(2) = 3$ $f(3) = 4$ $f(n) = n + 1$,

29. $f(0) = 1$ $f(1) = -3$ $f(2) = (-3)^2$ $f(3) = (-3)^3$ $f(n) = (-3)^n$,

31. $f(0) = 1$ $f(1) = -1$ $f(2) = -1$ $f(3) = 1$ $f(4) = -1$ $f(5) = -1$ $f(6) = 1$ $f(n) = (-1)^{\lfloor\lfloor n+1 \rfloor\rfloor_3 + 1}$ where $\lfloor\lfloor n+1 \rfloor\rfloor_3$ is the remainder when $n+1$ is divided by 3.

33. $a_0 = 3^0 - 0 \cdot 3^1 = 1$. $a_1 = 3^1 - 3^2 = -6$. Need to show
$$3^k - k3^{k+1} = 6(3^{k-1} - (k-1)3^k) - 9(3^{k-2} - (k-2)3^{k-1})$$
but
$$6(3^{k-1} - (k-1)3^k) - 9(3^{k-2} - (k-2)3^{k-1})$$
$$= 6 \cdot 3^{k-1} - 6(k-1)3^k - 9(3^{k-2}) + 9(k-2)3^{k-1}$$
$$= 2 \cdot 3^k - 2k \cdot 3^{k+1} + 6 \cdot 3^k - 3^k + k3^{k+1} - 6 \cdot 3^k$$
$$= 3^k - 2 \cdot 3^{k+1}$$

35. $a_0 = \frac{1 - r^1}{1 - r} = 1$. Substituting into the recursive function, we have
$$\frac{1 - r^{k+1}}{1 - r} = \frac{1 - r^k}{1 - r} + r^k$$
but
$$\frac{1 - r^k}{1 - r} + r^k = \frac{1 - r^k + r^k(1 - r)}{1 - r}$$
$$= \frac{1 - r^{k+1}}{1 - r}$$

37. $a_1 = 1^2 \cdot 2^2 = 4$. Substituting into the recursive function, we have
$$k^2(k+1)^2 = (k-1)^2 k^2 + 4k^3$$
but
$$(k-1)^2 k^2 + 4k^3 = k^2((k-1)^2 + 4k)$$
$$= k^2(k^2 - 2k + 1 + 4k)$$
$$= k^2(k^2 + 2k + 1)$$
$$= k^2(k+1)^2$$

39. $a_0 = 2(-3)^0 + 5 \cdot 2^0 = 7$. $a_1 = 2(-3)^1 + 5 \cdot 2^1 = 4$. Substituting into the recursive function, we have
$$2(-3)^k + 5 \cdot 2^k$$
$$= -(2(-3)^{k-1} + 5 \cdot 2^{k-1}) + 6(2(-3)^{k-2} + 5 \cdot 2^{k-2})$$
$$= -(2(-3)^{k-1} + 5 \cdot 2^{k-1}) - 4(-3)^{k-1} + 15 \cdot 2^{k-1}$$
$$= -2(-3)^{k-1} - 4(-3)^{k-1} - 5 \cdot 2^{k-1} + 15 \cdot 2^{k-1}$$
$$= -6 \cdot (-3)^{k-1} + 10 \cdot 2^{k-1}$$
$$= 2(-3)^k + 5 \cdot 2^k$$

41. $a_1 = \frac{1}{\sqrt{5}}\left(\left(\frac{1+\sqrt{5}}{2}\right)^1 - \left(\frac{1-\sqrt{5}}{2}\right)^1\right) = \frac{1}{\sqrt{5}} \frac{2\sqrt{5}}{2} = 1$

$a_2 = \frac{1}{\sqrt{5}}\left(\left(\frac{1+\sqrt{5}}{2}\right)^2 - \left(\frac{1-\sqrt{5}}{2}\right)^2\right)$
$= \frac{1}{\sqrt{5}}\left(\left(\frac{1+2\sqrt{5}+5}{4}\right) - \left(\frac{1-2\sqrt{5}+5}{4}\right)\right)$
$= \frac{1}{\sqrt{5}}\left(\frac{4\sqrt{5}}{4}\right) = 1$

$\dfrac{1}{\sqrt{5}}\left(\left(\dfrac{1+\sqrt{5}}{2}\right)^k - \left(\dfrac{1-\sqrt{5}}{2}\right)^k\right)$

$= \dfrac{1}{\sqrt{5}}\left(\left(\dfrac{1+\sqrt{5}}{2}\right)^{k-1} - \left(\dfrac{1-\sqrt{5}}{2}\right)^{k-1}\right)$

$+ \dfrac{1}{\sqrt{5}}\left(\left(\dfrac{1+\sqrt{5}}{2}\right)^{k-2} - \left(\dfrac{1-\sqrt{5}}{2}\right)^{k-2}\right)$

$= \dfrac{1}{\sqrt{5}}\left(\left(\dfrac{1+\sqrt{5}}{2}\right)^{k-1} + \left(\dfrac{1+\sqrt{5}}{2}\right)^{k-2}\right)$

$- \dfrac{1}{\sqrt{5}}\left(\left(\dfrac{1-\sqrt{5}}{2}\right)^{k-1} + \left(\dfrac{1-\sqrt{5}}{2}\right)^{k-2}\right)$

$= \dfrac{1}{\sqrt{5}}\left(\dfrac{1+\sqrt{5}}{2}\right)^{k-2}\left(\dfrac{1+\sqrt{5}}{2}+1\right)$

$- \dfrac{1}{\sqrt{5}}\left(\dfrac{1-\sqrt{5}}{2}\right)^{k-2}\left(\dfrac{1-\sqrt{5}}{2}+1\right)$

$= \dfrac{1}{\sqrt{5}}\left(\dfrac{1+\sqrt{5}}{2}\right)^{k-2}\left(\dfrac{1+\sqrt{5}+2}{2}\right)$

$- \dfrac{1}{\sqrt{5}}\left(\dfrac{1-\sqrt{5}}{2}\right)^{k-2}\left(\dfrac{1-\sqrt{5}+2}{2}\right)$

$= \dfrac{1}{\sqrt{5}}\left(\dfrac{1+\sqrt{5}}{2}\right)^{k-2}\left(\dfrac{3+\sqrt{5}}{2}\right)$

$- \dfrac{1}{\sqrt{5}}\left(\dfrac{1-\sqrt{5}}{2}\right)^{k-2}\left(\dfrac{3-\sqrt{5}}{2}\right)$

$= \dfrac{1}{\sqrt{5}}\left(\dfrac{1+\sqrt{5}}{2}\right)^{k-2}\left(\dfrac{1+\sqrt{5}}{2}\right)^2$

$- \dfrac{1}{\sqrt{5}}\left(\dfrac{1-\sqrt{5}}{2}\right)^{k-2}\left(\dfrac{1-\sqrt{5}}{2}\right)^2$

$= \dfrac{1}{\sqrt{5}}\left(\dfrac{1+\sqrt{5}}{2}\right)^k - \dfrac{1}{\sqrt{5}}\left(\dfrac{1-\sqrt{5}}{2}\right)^k$

Chapter 5—Section 5.3

1. $2^8 - 1 = 255$
3. (i) 5 (ii) 4
5. (i) 18 (ii) 9
7. $2 \cdot 3 \cdot 3 \cdot 3 = 54$
9. no
11. yes
13. yes
15. no
17. yes
19. yes
21. yes
23. no
25. yes
27. yes
29. 4
31. 2
33. For $n=1$, $1! = 1 = 1^1$, so the statement is true. Assume that the statement is true for $n = k$, so $k! \leq k^k \leq (k+1)^k$ since the expansion of $(k+1)^k$ has k^k as one of its terms. Multiplying both sides of $k! \leq (k+1)^k$ by $(k+1)$, we get $(k+1)! \leq (k+1)^{k+1}$.

35. If $f(n) = \Theta(g(n))$, then $f(n) \leq mg(n)$ for some $n \geq s$. If $h(n) = \Theta(k(n))$, then $h(n) \geq pk(n)$ for some p and for $n \geq t$. Therefore, $\dfrac{f(n)}{h(n)} \leq mp\dfrac{g(n)}{k(n)}$ for $n \geq \max(s, t)$. By symmetry $\dfrac{g(n)}{k(n)} \leq q\dfrac{f(n)}{h(n)}$ for some q. Therefore, $\dfrac{f(n)}{h(n)} = \Theta\left(\dfrac{g(n)}{k(n)}\right)$.

37. If $f(n) = O(g(n))$, then there exists k_1, m_1 such that $f(n) \leq k_1 g(n)$ for $n > m_1$. If $g(n) = O(h(n))$, then there exists k_2, m_2 such that $g(n) \leq k_2 h(n)$ for $n > m_2$. Letting $M = \max(m_1, m_2)$, we have $f(n) \leq k_1 g(n) \leq k_1 k_2 h(n)$ for $n > M$, so $f(n) = O(h(n))$.

39. Let $P(m)$ be the statement, "If r and m are positive integers, $r \leq m$ and $n > 1$, then $n^r \leq n^m$." The statement is certainly true if $m = 1$. Assume that $P(k)$ is true, so that if $r \leq k$, then $n^r \leq n^k$. We now want to prove $P(k+1)$ is true, so that if $r \leq k+1$, then $n^r \leq n^{k+1}$. Since $1 < n$, $n^k \cdot 1 \leq n^k \cdot n$. Therefore, $n^k \leq n^{k+1}$. Assume $r \leq k+1$. If $r < k+1$, then $r \leq k$ so $n^r \leq n^k \leq n^{k+1}$. If $r = k+1$, then $n^r \leq n^{k+1}$. Therefore, $P(k+1)$ is true.

41. Obviously $f(n) = \Theta f(n)$. Since $f(n) = \Theta(g(n))$ may be interpreted as $f(n) = O(g(n))$ and $g(n) = O(f(n))$, symmetry is obvious. Transitivity follows from Problem 37.

43. $f(n) = n\left|\sin\left(\dfrac{n\pi}{4}\right)\right|$, $g(n) = n\left|\cos\left(\dfrac{n\pi}{4}\right)\right|$

45. $O(M(n)) = n^n$

Chapter 5—Section 5.4

Arrangements for quicksort and mergesort are condensed and do not directly follow algorithms.

Answers **A-29**

1. The following arrangements occur.

    $7, 11, 4, 0, 3, 1, 9, 4, 2, 8 \Rightarrow 0, 11, 4, 7, 3, 1, 9, 4, 2, 8$
    $\Rightarrow 0, 1, 4, 7, 3, 11, 9, 4, 2, 8$
    $\Rightarrow 0, 1, 2, 7, 3, 11, 9, 4, 4, 8$
    $\Rightarrow 0, 1, 2, 3, 7, 11, 9, 4, 4, 8$
    $\Rightarrow 0, 1, 2, 3, 4, 11, 9, 7, 4, 8$
    $\Rightarrow 0, 1, 2, 3, 4, 4, 9, 7, 11, 8$
    $\Rightarrow 0, 1, 2, 3, 4, 4, 7, 9, 11, 8$
    $\Rightarrow 0, 1, 2, 3, 4, 4, 7, 8, 11, 9$
    $\Rightarrow 0, 1, 2, 3, 4, 4, 7, 8, 9, 11$

3. The following arrangements occur.

    7, 11, 4, 0, 3, 1, 9, 4, 2, 8
    7, 2, 4, 0, 3, 1, 4,   8,   11, 9
    1, 2, 4, 0, 3, 4,   7,   8,   11, 9
    1, 2, 3, 0,   4, 4,   7,   8,   11, 9
    0,   1, 2, 3,   4, 4,   7,   8,   11, 9
    0,   1, 2, 3,   4, 4,   7,   8,   9, 11

5. The following arrangements occur.

    7, 11
    4, 7, 11
    0, 4, 7, 11
    0, 3, 4, 7, 11
    0, 1, 3, 4, 7, 11
    0, 1, 3, 4, 7, 9, 11
    0, 1, 3, 4, 4, 7, 9, 11
    0, 1, 2, 3, 4, 4, 7, 9, 11
    0, 1, 2, 3, 4, 4, 7, 8, 9, 11

7. The following arrangements occur.

    12, 50, −1, −10, 10, 11, 52, 30, 2, 8, −12
    $\Rightarrow$ −12, 12, 50, −1, −10, 10, 11, 52, 30, 2, 8
    $\Rightarrow$ −12, −10, −1, 12, 50, 2, 10, 11, 52, 30, 8
    $\Rightarrow$ −12, −10, −1, 2, 12, 50, 8, 10, 11, 52, 30
    $\Rightarrow$ −12, −10, −1, 2, 8, 10, 11, 12, 50, 30, 52
    $\Rightarrow$ −12, −10, −1, 2, 8, 10, 11, 12, 30, 50, 52

9. The following arrangements occur.

    12, 50, −1, −10, 10, 11, 52, 30, 2, 8, −12
    12, 50, −1, −10, 10|11, 52, 30, 2, 8, −12
    12, 50| −1, −10, 10|11, 52, 30, 2, 8, −12
    12|50| −1, −10, 10|11, 52, 30, 2, 8, −12
    12, 50| −1| −10, 10|11, 52, 30, 2, 8, −12
    12, 50| −10, −1, 10|11, 52, 30, 2, 8, −12
    −10, −1, 10, 12, 50|11, 52, 30| −12, 2, 8
    −10, −1, 10, 12, 50|11|52, 30| −12, 2, 8
    −10, −1, 10, 12, 50|11|52|30| −12, 2, 8
    −10, −1, 10, 12, 50|11|30, 52| −12, 2, 8
    −10, −1, 10, 12, 50|11, 30, 52| −12, 2, 8
    −10, −1, 10, 12, 50|11, 30, 52| −12, 2, 8
    −10, −1, 10, 12, 50| −12, 2, 8, 11, 30, 52
    −12, −10, −1, 2, 8, 10, 11, 12, 30, 50, 52

11. The following arrangements occur.

    $x, a, c, y, p, z, f, t, m, y, u, b, d, n, s, r$
    $\Rightarrow a, x, c, y, p, z, f, t, m, y, u, b, d, n, s, r$
    $\Rightarrow a, b, c, y, p, z, f, t, m, y, u, x, d, n, s, r$
    $\Rightarrow a, b, c, d, p, z, f, t, m, y, u, x, y, n, s, r$
    $\Rightarrow a, b, c, d, f, z, p, t, m, y, u, x, y, n, s, r$
    $\Rightarrow a, b, c, d, f, m, p, t, z, y, u, x, y, n, s, r$
    $\Rightarrow a, b, c, d, f, m, n, t, z, y, u, x, y, p, s, r$
    $\Rightarrow a, b, c, d, f, m, n, p, z, y, u, x, y, t, s, r$
    $\Rightarrow a, b, c, d, f, m, n, p, r, y, u, x, y, t, s, z$
    $\Rightarrow a, b, c, d, f, m, n, p, r, s, u, x, y, t, y, z$
    $\Rightarrow a, b, c, d, f, m, n, p, r, s, t, x, y, u, y, z$
    $\Rightarrow a, b, c, d, f, m, n, p, r, s, t, u, y, x, y, z$
    $\Rightarrow a, b, c, d, f, m, n, p, r, s, t, u, x, y, y, z$

13. The following arrangements occur.

    $x, a, c, y, p, z, f, t, m, y, u, b, d, n, s, r$
    $n, a, c, d, p, b, f, m \ \ , r \ \ y, u, z, y, x, s, t$
    $f, a, c, d, b, \ \ m \ n, p, \ \ r \ \ , u, z, y, x, s, t$
    $a, b, c, d, \ \ f, \ \ m \ n, p, \ \ r \ \ , u, z, y, x, s, t$
    $a, b, c, d, \ \ f, \ \ m \ n, p, \ \ r \ s, \ \ t, \ \ z, x, y, u$
    $a, b, c, d, \ \ f, \ \ m \ n, p, \ \ r \ s, \ \ t, \ \ u, \ \ x, y, z$

15. The following arrangements occur.

    a, x
    a, c, x
    a, c, x, y
    a, c, p, x, y
    a, c, p, x, y, z
    a, c, f, p, x, y, z
    a, c, f, p, t, x, y, z
    $a, c, f, m, p, t, x, y, z$
    $a, c, f, m, p, t, x, y, y, z$
    $a, c, f, m, p, t, u, x, y, y, z$
    $a, b, c, f, m, p, t, u, x, y, y, z$
    $a, b, c, d, f, m, p, t, u, x, y, y, z$
    $a, b, c, d, f, m, n, p, t, u, x, y, y, z$
    $a, b, c, d, f, m, n, p, s, t, u, x, y, y, z$
    $a, b, c, d, f, m, n, p, r, s, t, u, x, y, y, z$

17. The following arrangements occur.
    Johnson, Brown, Black, Jackson, Murphy, Smith, Jones
    Black, Johnson, Brown, Jackson, Jones, Murphy, Smith
    Black, Brown, Johnson, Jackson, Jones, Murphy, Smith
    Black, Brown, Jackson, Johnson, Jones, Murphy, Smith

19. The following arrangements occur.
    Johnson, Brown, Black | Jackson, Murphy, Smith, Jones
    Johnson | Brown, Black | Jackson, Murphy, Smith, Jones
    Johnson | Brown | Black | Jackson, Murphy, Smith | Jones
    Johnson | Black, Brown | Jackson, Murphy, Jones, Smith
    Black, Brown, Johnson | Jackson, Jones, Murphy, Smith
    Black, Brown, Jackson, Johnson, Jones Murphy, Smith

21. $9, 8, 7, 6, 5, 4, 3, 2, 1$ (number of comparisons always the same for n objects)

23. $1, 2, 3, 4, 5, 6, 7, 8, 9$

25. $1, 2, 3, 4, 5, 6, 7, 8, 9$

27. language dependent

29. language dependent

31. language dependent

33. $2D(\frac{n}{2})(\log_2(\frac{n}{2}) + A) + Dn$
    $= Dn(\log_2 n - \log_2 2 + A) + Dn$
    $= Dn(\log_2 n - 1 + A) + Dn$
    $= Dn(\log_2 n + A)$

35. $\dfrac{Q_{2^m}}{2^m} = \dfrac{cQ_{2^{m-1}}}{2 \cdot 2^{m-1}} + D$
    Recursive formula divided by $n = 2^m$

    $= \left(\dfrac{c}{2}\right)^2 \dfrac{Q_{2^{m-2}}}{2^{m-2}} + \dfrac{c}{2}D + D$
    Recursive formula repeated

    $= \left(\dfrac{c}{2}\right)^3 \dfrac{Q_{2^{m-3}}}{2^{m-3}} + \left(\left(\dfrac{c}{2}\right)^2 + \dfrac{c}{2} + 1\right)D$
    Recursive formula repeated

    $\vdots \qquad \vdots$

    $= \left(\dfrac{c}{2}\right)^m Q_1 + \left(\left(\dfrac{c}{2}\right)^{m-1} + \cdots \right.$
    $\left. + \left(\dfrac{c}{2}\right)^2 + \dfrac{c}{2} + 1\right)D$
    Recursive formula repeated

    $= \dfrac{c^{\log_2 n}}{2^{\log_2 n}} Q_1 + \left(\left(\dfrac{c}{2}\right)^{m-1} + \cdots \right.$
    $\left. + \left(\dfrac{c}{2}\right)^2 + \dfrac{c}{2} + 1\right)D$
    $m = \log_2 n$ since $n = 2^m$

    $= \dfrac{c^{\log_2 n}}{n} Q_1 + \left(\dfrac{1 - \left(\dfrac{c}{2}\right)^m}{1 - \dfrac{c}{2}}\right)D$
    Sum of geometric series

    $= \dfrac{n^{\log_2 c}}{n} Q_1 + \left(\dfrac{1 - \left(\dfrac{c}{2}\right)^m}{1 - \dfrac{c}{2}}\right)D$
    Since $c^{\log_2 n} = n^{\log_2 c}$
    (take $\log_2$ of both sides)

    $= \dfrac{n^{\log_2 c}}{n}\left(A + \dfrac{2D}{2-c}\right) + \left(\dfrac{1 - \dfrac{n^{\log_2 c}}{n}}{1 - \dfrac{c}{2}}\right)D$
    Sub in value of Q_1

    $= \dfrac{n^{\log_2 c}}{n} A + \dfrac{2D}{2-c} \dfrac{n^{\log_2 c}}{n} + \dfrac{2D}{2-c}$
    $\quad - \dfrac{2D}{2-c} \dfrac{n^{\log_2 c}}{n}$

    $= \dfrac{n^{\log_2 c}}{n} A + \dfrac{2D}{2-c}$

    Therefore
    $$\dfrac{Q_n}{n} = \dfrac{n^{\log_2 c}}{n} A + \dfrac{2D}{2-c}$$
    and
    $$Q_n = n^{\log_2 c} A + \dfrac{2D}{2-c} n$$

Chapter 5—Section 5.5

1. $a2\hat{\;}b2\hat{\;}+$
3. $a2\hat{\;}b + cd2\hat{\;}+$
5. $a2\hat{\;}bc + \times$
7. $ab + c - 3\hat{\;}$
9. $ab - ab + \times$
11. $(a + b) \div (c - d)$
13. $3 \times a \times b + 4 \times c \times d\hat{\;}2$
15. $(a\hat{\;}2 + b) \times (c + d\hat{\;}2)$
17. $(a + b)\hat{\;}2 + c - (c + d\hat{\;}2)$
19. $((a + b)\hat{\;}2 \div (c + d))$
21. $+\hat{\;}a2\hat{\;}b2$
23. $\times + \hat{\;}a2b + c\hat{\;}d2$
25. $\times\hat{\;}a2 + bc$
27. $\hat{\;} - + abc3$
29. $\times - ab + ab$
31. $(a + b) \div (c - d)$

33. $3ab + 4cd + 5ad$

35. $(a\char`\^2 + b)(c + d\char`\^2)$

37. $(a + b + c)(c + d + e)$

39. $((a + b\char`\^2) \div (c + d\char`\^2))\char`\^2$

41. (a) Add parentheses to the expression until it is fully parenthesized.

   (b) Beginning with the innermost parentheses, remove the pair of parentheses and move the corresponding operator to replace the left parenthesis. If there is more than one innermost parentheses, begin with the right one.

   (c) Moving out to the next pair of parentheses, again remove the parentheses and move the corresponding operator to replace the left parenthesis corresponding to the operation.

   (d) Continue step (3) until all parentheses are removed.

Chapter 5—Section 5.6

1. 1, 10, 11, 100, 101, 110, 111, 1000, 1001, 1010, 1011, 1100, 1101, 1110, 1111

   10000, 10001, 10010, 10011, 10100, 10101, 10110, 10111, 11000, 11001,

3. 101101

5. 11110011

7. 100111000

9. $99 = 2 \cdot 49 + \underline{1}$ $49 = 2 \cdot 24 + \underline{1}$ $24 = 2 \cdot 12 + \underline{0}$ $12 = 2 \cdot 6 + \underline{0}$
   $6 = 2 \cdot 3 + \underline{0}$ $3 = 2 \cdot 1 + \underline{1}$ $1 = 2 \cdot 0 + \underline{1}$ 1100011

11. $186 = 2 \cdot 93 + \underline{0}$ $93 = 2 \cdot 46 + \underline{1}$ $46 = 2 \cdot 23 + \underline{0}$ $23 = 2 \cdot 11 + \underline{1}$
    $11 = 2 \cdot 5 + \underline{1}$ $5 = 2 \cdot 2 + \underline{1}$ $2 = 2 \cdot 1 + \underline{0}$ $1 = 2 \cdot 0 + \underline{1}$ 10111010

13. 11

15. 36

17. 53

19. $2 + 4 + 16 = 22$

21. $1 + 4 + 16 + 32 = 53$

23. $3F$

25. 178

27. $147E$

29. $297 = 16 \cdot 18 + \underline{9}$ $18 = 16 \cdot 1 + \underline{2}$ $1 = 16 \cdot 0 + \underline{1}$ 129

31. $861 = 16 \cdot 53 + \underline{13}$ $53 = 16 \cdot 3 + \underline{5}$ $3 = 16 \cdot 0 + \underline{3}$ $35D$

33. 78

35. 1274

37. 53731

39. $4 \cdot 1 + 10 \cdot 16 + 3 \cdot 16^2 = 932$

41. $10 \cdot 1 + 3 \cdot 16 + 1 \cdot 16^2 + 13 \cdot 16^3 = 53562$

43. 10111.111

45. 1010110.001

47. $110100.01\overline{0011}$

49. 22 is 10110 in binary. $.625 \cdot 2 = \underline{1}.25$ $.25 \cdot 2 = \underline{0}.5$ $.5 \cdot 2 = \underline{1}.0$ 10110.101

51. 16 is 10000 in binary. $.53125 \cdot 2 = \underline{1}.0625$ $.0625 \cdot 2 = \underline{0}.125$ $.125 \cdot 2 = \underline{0}.25$ $.25 \cdot 2 = \underline{0}.5$ $.5 \cdot 2 = \underline{1}$ 10000.10001

53. 14.75

55. 4.5625

57. 13.3125

59. $11100 \Rightarrow 4 + 8 + 16 = 28$ $.1010 \Rightarrow \frac{1}{2} + \frac{1}{8} = .625$ 28.625

61. $1101.1011 \Rightarrow 1 + 4 + 8 = 13$ $.1011 \Rightarrow \frac{1}{2} + \frac{1}{8} + \frac{1}{16} = .6875$ 13.6875

63. $24.A$

65. 250.28

67. $147E.59$

69. $349 \Rightarrow 15D$ $.796875 \cdot 16 = \underline{12}.75$ $.75 \cdot 16 = \underline{12}$ $15D.CC$

71. $1997 \Rightarrow 7CD$ $.35 \cdot 16 = \underline{5}.6$ $.6 \cdot 16 = \underline{9}.6$ $7CD.5\overline{9}$

73. 92.2265625

75. 1199.767578125,

77. 5070.821289063

79. $11F \Rightarrow 15 + 16 + 16^2 = 287$ $.E33 \Rightarrow \frac{14}{16} + \frac{3}{16^2} + \frac{3}{16^3} = .887451172$ 287.887451172

81. $237F \Rightarrow 15 + 7 \cdot 16 + 3 \cdot 16^2 + 2 \cdot 16^3 = 9087$ $.755 \Rightarrow \frac{7}{16} + \frac{5}{16^2} + \frac{5}{16^3} = .458251953$ 9087.458251953

83. $5.A$

85. $1B6.6F8$

87. $3D4.228$

89. 25.78

91. $6AA.CDA$

93. 101100.01001011

95. $101011111000.010010001100$

97. $1100000101001010.001011010100$

99. $111100011010.111000111101$

101. $1111101000110111.001011100101$

Chapter 5—Section 5.7

1. $11111111 - 10001001 + 1 = 10001010$

3. $11111111 - 00110111 + 1 = 11001001$

5. $11111111 - 01100100 + 1 = 10011100$

7. $11111111 - 00111011 + 1 = 11000101$,

9. 01001001

11. 11011011

13. $48 \Rightarrow 00110000$

15. $58 \Rightarrow 00111010 \quad -58 \Rightarrow 11111111 - 00111010 + 1 = 11000110$

17. 00100110

19. $00110101 + 10001010 = 10111111$

21. 00100111

23. $01011100 + 10001001 = 11100101$

25. $43 \Rightarrow 00101011 \quad 28 \Rightarrow 00011100 \quad 43 + 28 \Rightarrow 01000111$

27. $43 \Rightarrow 00101011 \quad 67 \Rightarrow 01000011$
$43 - 67 \Rightarrow 00101011 - 01000011 = 00101011 + 10111101 = 11101000$

29. $29 \Rightarrow 00011101 \quad 61 \Rightarrow 00111101 \quad 29 + 61 = 01011010$

31. $13 \Rightarrow 00001101 \quad 87 \Rightarrow 01010111$
$13 - 87 \Rightarrow 00001101 - 01010111 = 00001101 + 10101001 = 10110110$

33. $FFFF - 723A + 1 = 8DC6$

35. $FFFF - FA12 + 1 = 05EE$

37. $FFFF - 23CF + 1 = DC31$

39. $FFFF - FF13 + 1 = 00ED$

41. $270 \Rightarrow 010E$

43. $17 \Rightarrow 0011 \quad -17 \Rightarrow FFEF$

45. $48 \Rightarrow 0030$

47. $8858 \Rightarrow 229A, \quad -8858 \Rightarrow DD66$

49. $3C14 + 89AC = C5C0$

51. $B127 - C1AB = B127 + 3E55 = EF7C$

53. $4713 + 0F23 = 5636$

55. $37B4 - AA44 = 37B4 + 556C = 8D70 \text{ overflow}$

57. $936 \Rightarrow 03A8 \quad 258 \Rightarrow 0102 \quad 936 + 258 \Rightarrow 03A8 + 0102 = 04AA$

59. $8836 \Rightarrow 2284 \quad 19923 \Rightarrow 4DD3 \quad 8836 - 19923 \Rightarrow 2284 - 4DD3 = 2284 + b22D = D4B1$

61. $829 \Rightarrow 033D \quad 1499 \Rightarrow 05DB \quad 829 + 1499 \Rightarrow 033D + 05DB = 0918$

63. $13 \Rightarrow 000d \quad 87 \Rightarrow 0057 \quad 13 - 87 \Rightarrow 000d - 0057 = 000d + FFa9 = FFB6$

65. -2147483648 to 2147483647

Chapter 5—Section 5.8

1. $5 \cdot 1 - 3 \cdot 2 = -1$

3. $3 \cdot 5 - 4 \cdot 2 = 7$

5. $3 \cdot 5 - 4 \cdot 2 = 7 \quad 3 \cdot 6 - 2 \cdot 9 = 0$

7. $6 \cdot \begin{vmatrix} -11 & 2 \\ -1 & 1 \end{vmatrix} - (-4) \cdot \begin{vmatrix} 1 & 2 \\ 3 & 1 \end{vmatrix} + (-5) \cdot \begin{vmatrix} 1 & -11 \\ 3 & -1 \end{vmatrix}$
$= 6 \cdot ((-11) \cdot 1 - 2(-1)) + 4(1 \cdot 1 - 2 \cdot 3)$
$\quad -5((-1) \cdot 1 - 3(-11))$
$= 6 \cdot (-9) + 4(-5) - 5(32) = -234$

9. $1 \cdot \begin{vmatrix} -1 & 2 \\ 2 & 1 \end{vmatrix} - (-6) \cdot \begin{vmatrix} -6 & 2 \\ 3 & 1 \end{vmatrix} + 3 \cdot \begin{vmatrix} -6 & -1 \\ 3 & 2 \end{vmatrix}$
$= ((-1) \cdot 1 - 2 \cdot 2) + 6((-6) \cdot 1 - 2 \cdot 3)$
$\quad + 3((-6) \cdot 2 - (-1) \cdot (3))$
$= -5 + 6(-12) + 3 \cdot (-9) = -104$

11. $\begin{vmatrix} 2 & 3 \\ 5 & 7 \end{vmatrix} = 2 \cdot 7 - 5 \cdot 3 = -1$

$\begin{bmatrix} 2 & 3 \\ 5 & 7 \end{bmatrix}^{-1} = \begin{bmatrix} \frac{7}{-1} & \frac{-3}{-1} \\ \frac{-5}{-1} & \frac{2}{-1} \end{bmatrix} = \begin{bmatrix} -7 & 3 \\ 5 & -2 \end{bmatrix}$

13. $\begin{vmatrix} 2 & -5 \\ -4 & 10 \end{vmatrix} = 0$. Hence, the inverse does not exist.

15. $\begin{vmatrix} 4 & -1 & 3 \\ -3 & 1 & -3 \\ 2 & 0 & 1 \end{vmatrix} = 4 \cdot \begin{vmatrix} 1 & -3 \\ 0 & 1 \end{vmatrix} - (-1)$
$\cdot \begin{vmatrix} -3 & -3 \\ 2 & 1 \end{vmatrix} + 3 \cdot \begin{vmatrix} -3 & 1 \\ 2 & 0 \end{vmatrix} = 1$

$\begin{bmatrix} 4 & -1 & 3 \\ -3 & 1 & -3 \\ 2 & 0 & 1 \end{bmatrix}^{-1}$

$$= \begin{bmatrix} \begin{vmatrix} 1 & -3 \\ 0 & 1 \end{vmatrix} & -\begin{vmatrix} -1 & 3 \\ 0 & 1 \end{vmatrix} & \begin{vmatrix} -1 & 3 \\ 1 & -3 \end{vmatrix} \\ -\begin{vmatrix} -3 & -3 \\ 2 & 1 \end{vmatrix} & \begin{vmatrix} 4 & 3 \\ 2 & 1 \end{vmatrix} & -\begin{vmatrix} 4 & 3 \\ -3 & -3 \end{vmatrix} \\ \begin{vmatrix} -3 & 1 \\ 2 & 0 \end{vmatrix} & -\begin{vmatrix} 4 & -1 \\ 2 & 0 \end{vmatrix} & \begin{vmatrix} 4 & -1 \\ -3 & 1 \end{vmatrix} \end{bmatrix}$$

$$= \begin{bmatrix} 1 & 1 & 0 \\ -3 & -2 & 3 \\ -2 & -2 & 1 \end{bmatrix}$$

17. $\begin{bmatrix} 1 & 1 & -5 \\ 0 & 1 & -2 \\ -2 & 0 & 7 \end{bmatrix}^{-1} = \frac{1}{10} \begin{bmatrix} 5 & -25 & 5 \\ 1 & 7 & -1 \\ -1 & 3 & 1 \end{bmatrix}$

19. The matrix of coefficients is $\begin{bmatrix} 2 & 5 \\ 1 & 3 \end{bmatrix}$. Its inverse is $\begin{bmatrix} 3 & -5 \\ -1 & 2 \end{bmatrix}$. Thus, since

$$\begin{bmatrix} 2 & 5 \\ 1 & 3 \end{bmatrix} \begin{bmatrix} x_1 \\ x_2 \end{bmatrix} = \begin{bmatrix} 14 \\ 8 \end{bmatrix}$$

$$\begin{bmatrix} 3 & -5 \\ -1 & 2 \end{bmatrix} \begin{bmatrix} 2 & 5 \\ 1 & 3 \end{bmatrix} \begin{bmatrix} x_1 \\ x_2 \end{bmatrix} = \begin{bmatrix} 3 & -5 \\ -1 & 2 \end{bmatrix} \begin{bmatrix} 14 \\ 8 \end{bmatrix}$$

and

$$\begin{bmatrix} x_1 \\ x_2 \end{bmatrix} = \begin{bmatrix} 2 \\ 2 \end{bmatrix}$$

so $x_1 = 2$ and $x_2 = 2$.

21. The matrix of coefficients is $\begin{bmatrix} 4 & -1 & 3 \\ -3 & 1 & -3 \\ -1 & -2 & 6 \end{bmatrix}$.

$\begin{vmatrix} 4 & -1 & 3 \\ -3 & 1 & -3 \\ -1 & -2 & 6 \end{vmatrix} = 0$. Hence, there are no inverse and no solution.

23. The matrix of coefficients is $\begin{bmatrix} 7 & -7 & 3 \\ 4 & -3 & 2 \\ 2 & -2 & 1 \end{bmatrix}$. Its inverse is $\begin{bmatrix} 1 & 1 & -5 \\ 0 & 1 & -2 \\ -2 & 0 & 7 \end{bmatrix}$.

Since

$$\begin{bmatrix} 1 & 1 & -5 \\ 0 & 1 & -2 \\ -2 & 0 & 7 \end{bmatrix} \begin{bmatrix} x_1 \\ x_2 \\ x_3 \end{bmatrix} = \begin{bmatrix} 14 \\ 16 \\ 24 \end{bmatrix}$$

$$\begin{bmatrix} 1 & 1 & -5 \\ 0 & 1 & -2 \\ -2 & 0 & 7 \end{bmatrix} \begin{bmatrix} 1 & 1 & -5 \\ 0 & 1 & -2 \\ -2 & 0 & 7 \end{bmatrix} \begin{bmatrix} x_1 \\ x_2 \\ x_3 \end{bmatrix}$$

$$= \begin{bmatrix} 1 & 1 & -5 \\ 0 & 1 & -2 \\ -2 & 0 & 7 \end{bmatrix} \begin{bmatrix} 14 \\ 16 \\ 24 \end{bmatrix}$$

and

$$\begin{bmatrix} x_1 \\ x_2 \\ x_3 \end{bmatrix} = \begin{bmatrix} -90 \\ -32 \\ 140 \end{bmatrix}$$

Therefore, $x_1 = 10$, $x_2 = 16$, $x_3 = -8$.

25. Assume that A' and A'' are both inverses of A. Then

$$A' = A'I = A'(AA'') = (A'A)A'' = IA'' = A''$$

so the inverse is unique.

27. $I = I \cdot I$. Therefore, $\det(I) = \det(I \cdot I) = \det(I) \cdot \det(I)$. Hence, $\det(I) = 1$ or $\det(I) = 0$. If $\det(I) = 0$, since $\det(A) = \det(A \cdot I) = \det(A) \cdot \det(I)$, we would have $\det(A) = 0$ for all A. Since it is not easy to show this is not true, we take a different approach. Let I_n be the $n \times n$ identity matrix. We show $\det(I_n) = 1$ using induction on n. Certainly, $\det(I_1) = 1$. Assume $\det(I_k) = 1$. But by construction of determinant, $\det(I_{k+1}) = 1 \cdot \det(I_k) = 1 \cdot 1 = 1$.

29. If $\det(A) \neq 0$, then we can construct the inverse. If $\det(A) = 0$, then A cannot have an inverse. If it did, $\det(A) \cdot \det(A^{-1}) = \det(I) = 1$. But if $\det(A) = 0$, then $\det(I) = 0$, which is a contradiction.

31. Assume A and B have inverses, A^{-1} and B^{-1}, respectively. Then

$$(AB)(B^{-1}A^{-1}) = A(BB^{-1})A^{-1}$$
$$= AIA^{-1} = AA^{-1} = 1$$

and

$$(B^{-1}A^{-1})(AB) = B^{-1}(A^{-1}A)B$$
$$= B^{-1}IB = B^{-1}B = 1$$

33. Assume B is the inverse of A. Then $\sum_{k=1}^{n} A_{ik}B_{kj} = \sum_{k=1}^{n} B_{jk}A_{ki} = \delta_{ij}$ where $\delta_{ij} = 1$ if $i = j$ and 0 otherwise. $\sum_{k=1}^{n} A_{ik}^{t}B_{kj}^{t} = \sum_{k=1}^{n} A_{ki}B_{jk} = \sum_{k=1}^{n} B_{jk}A_{ki} = \delta_{ij}$

Similarly, $\sum_{k=1}^{n} B_{ik}^{t}A_{kj}^{t} = \delta_{ij}$. Therefore, the inverse of A^t is the transpose of A^{-1}.

35. If AB is the zero matrix, then $AB = 0 = A0$. Therefore, $A^{-1}AB = A^{-1}A0$ and $B = 0$. Hence, if B is not 0, then AB is not 0.

37. If AB has an inverse, then $\det(AB) \neq 0$. But $\det(AB) = \det(A)\det(B)$ so $\det(A) \neq 0$. Hence, A has an inverse. (Try to prove without using determinant.)

39. $\begin{bmatrix} 1 & 1 & 7 \\ 2 & 2 & 4 \\ 3 & 3 & 9 \end{bmatrix}$ has no inverse because its determinant is 0. The determinant of a matrix is always 0 if any row (column) is a multiple of any other row (column).

41. $\begin{bmatrix} 1 & 1 & 1 \\ 2 & 2 & 2 \\ 3 & 1 & 7 \end{bmatrix}$ has no inverse because its determinant is 0. The determinant of a matrix is always 0 if any row (column) is a multiple of any other row (column).

Chapter 6—Section 6.1

1. path, simple path, length 5
3. path, not simple, length 7
5. path, not simple, length 6
7. not path
9. not cycle
11. not cycle
13. not cycle
15. cycle, simple, length 5
17.
19.
21. If a vertex v occurs more than once in a cycle, remove the vertices between two occurrences of v along with one of the two occurrences of v. The remaining path is still a cycle. Continue until no two vertices occur more than once in the cycle.
23. A relation on A is reflexive if and only if every element of a is related to itself. Since the elements of A are the vertices in the graph, there is an edge from every vertex to itself if and only if there is a loop at every vertex. A graph with loops is not reflexive unless there is a loop at every vertex.
25. $1 + 2 + 3 + \cdots + (n-1)$
27. The maximal number of edges is equal to the number of vertices in one of the disjoint sets multiplied by the number of vertices in the other disjoint set since there is an edge from each vertex in one set to each vertex in the other set. The number of vertices in each set must be equal. For simplicity assume there are m elements in each set, so there are m^2 edges. If k vertices are removed from one set and placed in the other set, then there are $m + k$ vertices in one set and $m - k$ elements in the other set. There are then $(m+k)(m-k) = m^2 - k^2$ edges, which is less than m^2 edges. Thus, there are $n/2$ vertices in each set and $n^2/4$ edges.
29. Assume there are n vertices. The possible degrees of the vertices are $0, 1, 2, \ldots, n-1$. If one vertex has degree $n-1$, then there is an edge from that vertex to every other vertex. Hence, there is no vertex of degree 0. Thus, there are $n-1$ values for the degrees and n vertices. Thus, two vertices must have the same degree. This is an example of the pigeonhole principle, which is found in Chapter 8.
31. mn

Chapter 6—Section 6.2

1. $V = \{a, b, c, d, e\}$. Directed edges (a, a), (b, a), (c, b), (d, c), (e, c). a has indegree 2 and outdegree 1. b has indegree 1 and outdegree 1. c has indegree 2 and outdegree 1. d and e both have indegree 0 and outdegree 1. d and e are sources.
3. $V = \{a, b, c, d, e, f\}$. Directed edges (a, f), (b, f), (c, f), (d, f), (e, f), (a, b), (b, c), (c, d), (d, e). a has indegree 0 and outdegree 2. b, c, and d have indegree 1 and outdegree 2. e has indegree 1 and outdegree 1. f has indegree 5 and outdegree 0. a is a source and f is a sink.
5. $V = \{a, b, c, d, e\}$. Directed edges (a, b), (b, a), (d, c), (c, d), (d, e), (a, e), (b, e), (c, e). Vertices a, b, c, d have indegree 1 and outdegree 2. Vertex e has indegree 4 and outdegree 0. Vertex e is a sink.
7. $V = \{a, b, c, d, e\}$. Directed edges (a, b), (b, c), (c, d), (d, e), (e, a), (b, e), (b, d), (c, e), (d, a). a has indegree 2 and outdegree 1. b has indegree 1 and outdegree 3. c has indegree 1 and outdegree 2. d has indegree 2 and outdegree 2. e has indegree 3 and outdegree 1.
9. [figure]
11. [figure]
13. $V = \{a, b, c, d\}$. Directed edges (a, b), (b, b), (b, c), (d, b), (b, d), (d, c), (a, d). a has indegree 0 and outdegree 2. b has indegree 3 and outdegree 3. c has indegree 2 and outdegree 0. d has indegree 2 and outdegree 2. a is a source. c is a sink.
15. $V = \{a, b, c, d, e\}$. Directed edges (a, b), (a, e), (a, d), (b, a), (b, e), (b, c), (c, d), (c, e), (d, e). a, b have indegree 1 and outdegree 3. c has indegree 1 and outdegree 2. d has indegree 2, outdegree 1. e has indegree 4 and outdegree 0. e is a sink.

17. A directed cycle is a directed path that begins and ends at the same vertex.

19. connected

21. not connected

23. length 2 *adc*, no longer paths, *adc* maximal length, no cycles.

25. length 2 *cab*, length 3 *cabc*, length 4 *cabcd*, length 5 *bcabcd*, simple path of maximal length *bca*. Longest simple cycle *bcab*.

27. not connected

29. connected

31. length 2 *cab*, length 3 *cabc*, length 4 *cabca*, length 5 *cabcab*, simple path of maximal length *abc*. Longest simple cycle *abca*.

33. Yes, if directed edges between two vertices for which there is no directed edge in the other direction are eliminated.

35. No. The sum of the indegrees must equal the sum of the outdegrees.

Chapter 6—Section 6.3

1. yes

3. no

5.
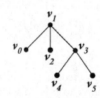

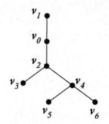

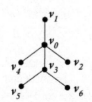

7.

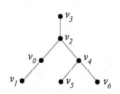

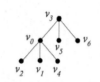

9. yes

11. yes

13. no

15.
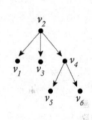

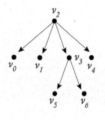

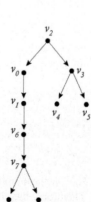

17.

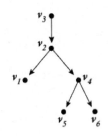

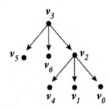

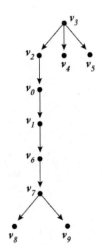

19. no

21. no

23. no

25. v_0, v_3, v_6

27. 2

29. 3

31. no, v_0 has outdegree 3.

33. v_0, v_2, v_3

35. 3

37. 4

39. Answer depends on the individual.

41. Using induction on the number of vertices, if $v = 1$, then $e = 0$. Therefore, the theorem is true if $v = 1$. Assume the theorem is true if $v = k$. Let T be a tree with $k + 1$ vertices. Remove a leaf and its incident edge. This leaves a tree with k vertices. Therefore, by the induction hypothesis, for this tree, $v = e + 1$. If we add the leaf and edge back on, we increase the number of edges and vertices by one, so we still have $v = e + 1$ for the tree with $k + 1$ vertices.

43. Using induction on the number of vertices, the theorem is certainly true if the theorem is equal to 1. Assume the theorem is true for $n = k$, so the sum of the degree of the vertices is $2k - 1$. Assume we have a tree with $k + 1$ edges. Remove one edge that is incident to a leaf and remove the leaf. We have reduced the sum of the degree of the vertices by two and the number of edges by 1. We now have a tree with k vertices, so the sum of the degree of the vertices is $2k - 1$. Therefore, our original tree had the sum of the degree of the vertices $2k - 1 + 2 = 2(k + 1) - 1$, and the theorem is true for all n.

Conversely, assume a connected graph has the sum of the degrees of its vertices equal to $2k - 1$. If it is not a tree, we can remove an edge from each cycle until we get a tree. We do not remove any vertices but we have decreased the value of the sum of the vertices by two for each edge removed. Therefore, the tree cannot have degree $2k - 1$, which is a contradiction.

45. Assume that we have a tree with two vertices a and b of degree 3. There is a path from a to b. There are two edges from a and two edges from b that are not on the path. These edges can all be extended to form maximal edges that end in a leaf. Hence, there are four distinct leaves. Assume there is an edge from a to b, and each of the other edges from a and b are incident to leaves. Then there are exactly four leaves.

47. Since a tree has no cycles, if any edge from a to b is removed, then there is no other path from a to b since the edge from a to b is not part of a cycle. Hence, the graph is no longer connected. Conversely, if an edge from a to b in a graph is removed and the graph is still connected, then there is another path from a to b and the edge from a to b is the edge of a cycle. Hence, the graph is not a tree.

Chapter 6—Section 6.4

1. The drawing connecting colors on opposite sides is

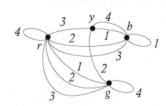

From this we can remove

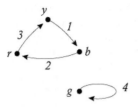

leaving

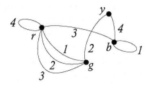

From this figure, we get

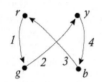

This gives us the solution

front	back	left	right
y	b	r	g
b	r	g	y
r	y	b	r
g	g	y	b

3. The drawing connecting colors on opposite sides is

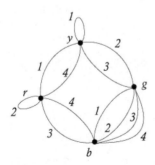

From this we can remove

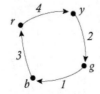

leaving

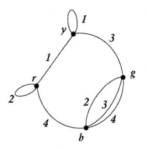

From this figure, we get

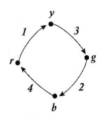

This gives us the solution

front	back	left	right
g	b	r	y
y	g	g	b
b	r	y	g
r	y	b	r

5. no solution

Chapter 6—Section 6.5

1. (a) connected and all vertices have even degree.

3. (c) exactly two edges have odd degree.

5. (a), (c)

7. (b)

9. Assume that G is connected and has exactly two vertices of odd degree. Let a and b be the vertices of odd degree. If there is no edge between a and b, add one. Now each vertex has even degree so the new graph has an Euler cycle. In this cycle, the edge $\{a, b\}$ is traversed. Say that in traversing the cycle we pass from a to b. If we begin with this edge and traverse the cycle, we see that when the edge is removed and we begin the path at b and follow the path of the cycle, we have an Euler path from b to a. If there is an edge between a and b, remove it. The new graph has an Euler cycle if it is still connected. Let the Euler cycle begin at a and end at a. If we traverse this cycle and then follow the removed edge from a to b, we have an Euler path

from a to b. If it is no longer connected, then the new graph has an Euler cycle for the component containing a, which begins and ends at a, and an Euler cycle for the component containing b, which begins and ends at b. Traverse the Euler cycle at a, pass on the removed edge from a to b, then traverse the Euler cycle from b to b. This gives an Euler path from a to b.

Assume G has an Euler path. Say it begins at a and ends at b. After the first edge on the path that leaves a, for each edge on the path that leads to a, there must be an edge that leaves a. Therefore, a must have odd degree. Similarly, b must have odd degree. At any other vertices, for each edge on the path that leads to that vertex, there must be an edge that leaves the vertices. Therefore, the vertex has even degree.

11. If the graph is strongly connected, then certainly for every vertex v every vertex is reachable from v and v is reachable from every other vertex.

Conversely, assume there is a v with the property given. Let a and b be arbitrary vertices. Since there is a path from a to v and a path from v to b, there is a path from a to b and the graph is strongly connected.

Chapter 6—Section 6.6

1. $\begin{bmatrix} & e_1 & e_2 & e_3 & e_4 & e_5 & e_6 \\ v_1 & 1 & 0 & 0 & 1 & 1 & 0 \\ v_2 & 1 & 1 & 0 & 0 & 0 & 0 \\ v_3 & 0 & 1 & 1 & 0 & 0 & 0 \\ v_4 & 0 & 0 & 0 & 0 & 1 & 1 \\ v_5 & 0 & 0 & 1 & 1 & 0 & 1 \end{bmatrix}$

3. $\begin{bmatrix} & e_1 & e_2 & e_3 & e_4 & e_5 & e_6 \\ v_1 & 1 & 0 & 0 & 1 & 0 & 0 \\ v_2 & 1 & 1 & 1 & 0 & 0 & 0 \\ v_3 & 0 & 1 & 0 & 0 & 1 & 1 \\ v_4 & 0 & 0 & 1 & 1 & 1 & 0 \\ v_5 & 0 & 0 & 0 & 0 & 0 & 1 \end{bmatrix}$

5. $\begin{bmatrix} & e_1 & e_2 & e_3 & e_4 & e_5 & e_6 & e_7 & e_8 & e_9 \\ v_1 & 1 & 1 & 1 & 0 & 0 & 0 & 0 & 0 & 0 \\ v_2 & 0 & 0 & 0 & 1 & 1 & 1 & 0 & 0 & 0 \\ v_3 & 0 & 0 & 0 & 0 & 0 & 0 & 1 & 1 & 1 \\ v_4 & 1 & 0 & 0 & 1 & 0 & 0 & 1 & 0 & 0 \\ v_5 & 0 & 1 & 0 & 0 & 1 & 0 & 0 & 1 & 0 \\ v_6 & 0 & 0 & 1 & 0 & 0 & 1 & 0 & 0 & 1 \end{bmatrix}$

7. $\begin{bmatrix} & e_1 & e_2 & e_3 & e_4 & e_5 & e_6 & e_7 & e_8 & e_9 & e_{10} \\ v_1 & 1 & 1 & 1 & 1 & 0 & 0 & 0 & 0 & 0 & 0 \\ v_2 & 0 & 0 & 0 & 1 & 1 & 1 & 1 & 0 & 0 & 0 \\ v_3 & 0 & 0 & 1 & 0 & 0 & 0 & 1 & 1 & 1 & 0 \\ v_4 & 0 & 1 & 0 & 0 & 0 & 1 & 0 & 0 & 1 & 1 \\ v_5 & 1 & 0 & 0 & 0 & 1 & 0 & 0 & 1 & 0 & 1 \end{bmatrix}$

9. $\begin{bmatrix} & v_1 & v_2 & v_3 & v_4 & v_5 \\ v_1 & 0 & 1 & 0 & 1 & 1 \\ v_2 & 1 & 0 & 1 & 0 & 0 \\ v_3 & 0 & 1 & 0 & 0 & 1 \\ v_4 & 1 & 0 & 0 & 0 & 1 \\ v_5 & 1 & 0 & 1 & 1 & 0 \end{bmatrix}$

11. $\begin{bmatrix} & v_1 & v_2 & v_3 & v_4 & v_5 \\ v_1 & 0 & 1 & 0 & 1 & 0 \\ v_2 & 1 & 0 & 1 & 1 & 0 \\ v_3 & 0 & 1 & 0 & 1 & 1 \\ v_4 & 1 & 1 & 1 & 0 & 0 \\ v_5 & 0 & 0 & 1 & 0 & 0 \end{bmatrix}$

13. $\begin{bmatrix} & v_1 & v_2 & v_3 & v_4 & v_5 & v_6 \\ v_1 & 0 & 0 & 0 & 1 & 1 & 1 \\ v_2 & 0 & 0 & 0 & 1 & 1 & 1 \\ v_3 & 0 & 0 & 0 & 1 & 1 & 1 \\ v_4 & 1 & 1 & 1 & 0 & 0 & 0 \\ v_5 & 1 & 1 & 1 & 0 & 0 & 0 \\ v_6 & 1 & 1 & 1 & 0 & 0 & 0 \end{bmatrix}$

15. $\begin{bmatrix} & v_1 & v_2 & v_3 & v_4 & v_5 \\ v_1 & 0 & 1 & 1 & 1 & 1 \\ v_2 & 1 & 0 & 1 & 1 & 1 \\ v_3 & 1 & 1 & 0 & 1 & 1 \\ v_4 & 1 & 1 & 1 & 0 & 1 \\ v_5 & 1 & 1 & 1 & 1 & 0 \end{bmatrix}$

17.

19.

21. $A = \begin{bmatrix} & v_1 & v_2 & v_3 & v_4 & v_5 \\ v_1 & 0 & 1 & 1 & 0 & 0 \\ v_2 & 1 & 0 & 1 & 0 & 0 \\ v_3 & 1 & 1 & 0 & 0 & 0 \\ v_4 & 0 & 0 & 0 & 0 & 1 \\ v_5 & 0 & 0 & 0 & 1 & 0 \end{bmatrix}$

23. The 3-paths are given by

$$A^{\odot 3} = \begin{bmatrix} & v_1 & v_2 & v_3 & v_4 & v_5 \\ v_1 & 1 & 1 & 1 & 0 & 0 \\ v_2 & 1 & 1 & 1 & 0 & 0 \\ v_3 & 1 & 1 & 1 & 0 & 0 \\ v_4 & 0 & 0 & 0 & 0 & 1 \\ v_5 & 0 & 0 & 0 & 1 & 0 \end{bmatrix}$$

The 3-paths are v_1 to v_1, v_2, v_3, v_2 to v_1, v_2, v_3, v_3 to v_1, v_2, v_3, v_4 to v_5, v_5 to v_4.

25.
$$A^{\odot 2} = \begin{bmatrix} 1 & 1 & 1 & 1 & 1 & 0 \\ 1 & 1 & 1 & 1 & 1 & 1 \\ 1 & 1 & 1 & 1 & 1 & 1 \\ 1 & 1 & 1 & 1 & 1 & 1 \\ 1 & 1 & 1 & 1 & 1 & 1 \\ 0 & 1 & 1 & 1 & 1 & 1 \end{bmatrix}$$

The two paths are a to a, a to b, a to c, a to d, a to e, b to every other vertex, c to every other vertex, d to every other vertex, e to every other vertex, and f to every other vertex except a.

27.
$$A = \begin{bmatrix} 0 & 0 & 1 & 1 & 0 \\ 0 & 0 & 1 & 1 & 0 \\ 1 & 1 & 0 & 1 & 1 \\ 1 & 1 & 1 & 0 & 1 \\ 0 & 0 & 1 & 1 & 0 \end{bmatrix}$$

29.
$$A^{\odot 3} = \begin{bmatrix} 0 & 0 & 1 & 1 & 0 \\ 0 & 0 & 1 & 1 & 0 \\ 1 & 1 & 0 & 0 & 1 \\ 1 & 1 & 0 & 0 & 1 \\ 0 & 0 & 1 & 1 & 1 \end{bmatrix}.$$

The three paths are v_1 to v_3, v_4, v_2 to v_3, v_4, v_3 to v_1, v_2, v_5, v_4 to v_1, v_2, v_5, v_5 to v_3, v_4, v_5.

31.
$$A^{\odot 2} = \begin{bmatrix} 1 & 1 & 1 & 1 & 1 \\ 1 & 1 & 1 & 1 & 1 \\ 1 & 1 & 1 & 1 & 1 \\ 1 & 1 & 1 & 1 & 1 \\ 1 & 1 & 1 & 1 & 1 \end{bmatrix}$$

There are two paths between any two vertices.

33.
$$A = \begin{bmatrix} 0 & 1 & 1 & 0 & 0 \\ 1 & 0 & 1 & 0 & 0 \\ 1 & 1 & 0 & 0 & 0 \\ 0 & 0 & 0 & 0 & 1 \\ 0 & 0 & 0 & 1 & 0 \end{bmatrix} \quad A^{\odot 2} = \begin{bmatrix} 1 & 1 & 1 & 0 & 0 \\ 1 & 1 & 1 & 0 & 0 \\ 1 & 1 & 1 & 0 & 0 \\ 0 & 0 & 0 & 1 & 0 \\ 0 & 0 & 0 & 0 & 1 \end{bmatrix}$$

$$A^{\odot 3} = \begin{bmatrix} 1 & 1 & 1 & 0 & 0 \\ 1 & 1 & 1 & 0 & 0 \\ 1 & 1 & 1 & 0 & 0 \\ 0 & 0 & 0 & 0 & 1 \\ 0 & 0 & 0 & 1 & 0 \end{bmatrix} \text{ and } A^{\odot 4} = A^{\odot 2}.$$

$$\hat{A} = \begin{bmatrix} 1 & 1 & 1 & 0 & 0 \\ 1 & 1 & 1 & 0 & 0 \\ 1 & 1 & 1 & 0 & 0 \\ 0 & 0 & 0 & 1 & 1 \\ 0 & 0 & 0 & 1 & 1 \end{bmatrix}$$

$R = \{(v_1, v_1), (v_1, v_2), (v_1, v_3), (v_2, v_1), (v_2, v_2),$
$(v_2, v_3), (v_3, v_1), (v_3, v_2), (v_3, v_3), (v_4, v_4),$
$(v_5, v_5), (v_4, v_5), (v_5, v_4)\}$

35.
$$A = \begin{bmatrix} 0 & 0 & 1 & 1 \\ 1 & 0 & 0 & 0 \\ 0 & 1 & 0 & 0 \\ 1 & 0 & 0 & 0 \end{bmatrix} \quad A^{\odot 2} = \begin{bmatrix} 1 & 1 & 0 & 0 \\ 0 & 0 & 1 & 1 \\ 1 & 0 & 0 & 0 \\ 0 & 0 & 1 & 1 \end{bmatrix}$$

$$A^{\odot 3} = \begin{bmatrix} 1 & 0 & 1 & 1 \\ 1 & 1 & 0 & 0 \\ 0 & 0 & 1 & 1 \\ 1 & 1 & 0 & 0 \end{bmatrix}$$

$$A + A^{\odot 2} + A^{\odot 3} = \begin{bmatrix} 1 & 1 & 1 & 1 \\ 1 & 1 & 1 & 1 \\ 1 & 1 & 1 & 1 \\ 1 & 1 & 1 & 1 \end{bmatrix} \quad R = A \times A$$

37.
$$A = \begin{bmatrix} 0 & 1 & 1 & 0 & 0 \\ 1 & 0 & 1 & 0 & 0 \\ 1 & 1 & 0 & 0 & 0 \\ 0 & 0 & 0 & 0 & 1 \\ 0 & 0 & 0 & 1 & 0 \end{bmatrix}$$

After the first round, we have

$$\begin{bmatrix} 0 & 1 & 1 & 0 & 0 \\ 1 & 1 & 1 & 0 & 0 \\ 1 & 1 & 1 & 0 & 0 \\ 0 & 0 & 0 & 0 & 1 \\ 0 & 0 & 0 & 1 & 0 \end{bmatrix}$$

After the second round, we have

$$\begin{bmatrix} 1 & 1 & 1 & 0 & 0 \\ 1 & 1 & 1 & 0 & 0 \\ 1 & 1 & 1 & 0 & 0 \\ 0 & 0 & 0 & 0 & 1 \\ 0 & 0 & 0 & 1 & 0 \end{bmatrix}$$

The third round produces nothing new. After the fourth round, we have

$$\begin{bmatrix} 1 & 1 & 1 & 0 & 0 \\ 1 & 1 & 1 & 0 & 0 \\ 1 & 1 & 1 & 0 & 0 \\ 0 & 0 & 0 & 0 & 1 \\ 0 & 0 & 0 & 1 & 1 \end{bmatrix}$$

Finally, we have

$$\begin{bmatrix} 1 & 1 & 1 & 0 & 0 \\ 1 & 1 & 1 & 0 & 0 \\ 1 & 1 & 1 & 0 & 0 \\ 0 & 0 & 0 & 1 & 1 \\ 0 & 0 & 0 & 1 & 1 \end{bmatrix}$$

$R = \{(v_1, v_1), (v_1, v_2), (v_1, v_3), (v_2, v_1), (v_2, v_2),$
$(v_2, v_3), (v_3, v_1), (v_3, v_2), (v_3, v_3), (v_4, v_4), (v_5, v_5),$
$(v_4, v_5), (v_5, v_4)\}$

39.
$$A = \begin{bmatrix} 0 & 0 & 1 & 1 \\ 1 & 0 & 0 & 0 \\ 0 & 1 & 0 & 0 \\ 1 & 0 & 0 & 0 \end{bmatrix}$$

After the first round, we have
$$\begin{bmatrix} 0 & 0 & 1 & 1 \\ 1 & 0 & 1 & 1 \\ 0 & 1 & 0 & 0 \\ 1 & 0 & 1 & 1 \end{bmatrix}$$

After the second round, we have
$$\begin{bmatrix} 0 & 0 & 1 & 1 \\ 1 & 0 & 1 & 1 \\ 1 & 1 & 1 & 1 \\ 1 & 0 & 1 & 1 \end{bmatrix}.$$

After the third round, we have
$$\begin{bmatrix} 1 & 1 & 1 & 1 \\ 1 & 1 & 1 & 1 \\ 1 & 1 & 1 & 1 \\ 1 & 1 & 1 & 1 \end{bmatrix}$$

The transitive closure is $A \times A$.

41.
$$A = \begin{bmatrix} 0 & 1 & 1 & 0 & 0 \\ 0 & 0 & 0 & 0 & 0 \\ 0 & 0 & 0 & 0 & 0 \\ 0 & 0 & 0 & 0 & 1 \\ 0 & 0 & 0 & 1 & 0 \end{bmatrix}$$

After three rounds, we have no change. After the fourth round, we have
$$\begin{bmatrix} 0 & 1 & 1 & 0 & 0 \\ 0 & 0 & 0 & 0 & 0 \\ 0 & 0 & 0 & 0 & 0 \\ 0 & 0 & 0 & 0 & 1 \\ 0 & 0 & 0 & 1 & 1 \end{bmatrix}$$

After the fifth round, we have
$$\begin{bmatrix} 0 & 1 & 1 & 0 & 0 \\ 0 & 0 & 0 & 0 & 0 \\ 0 & 0 & 0 & 0 & 0 \\ 0 & 0 & 0 & 1 & 1 \\ 0 & 0 & 0 & 1 & 1 \end{bmatrix}$$

and we are finished.
$$R = \{(v_1, v_3), (v_1, v_2), (v_4, v_4), (v_4, v_5),$$
$$(v_5, v_4), (v_5, v_5)\}$$

43. The edges have identical columns.

45. By definition of adjacency matrix, the theorem is true when $n = 1$.

Assume true for $n = k$. Therefore, there is a k-path from v_i to v_j if and only if $A_{ij}^{\odot k} = 1$. $A_{ij}^{\odot k+1} = \bigvee_{m=1}^{n} (A_{im}^{\odot k} \wedge A_{mj}) = 1$. Now
$$A_{ij}^{\odot k+1} = \bigvee_{m=1}^{n} (A_{im}^{\odot k} \wedge A_{mj}) = 1$$

if and only if there exists m so that $A_{im}^{\odot k} = 1$ and $A_{mj} = 1$. But using the induction hypothesis, this is true if and only if there is a k-path from i to m and a 1-path from m to j. This is true if and only if there is a $k + 1$-path from i to j.

47. There is a path from v_i to v_j if and only if there is a k-path from v_i to v_j for some $1 \leq k \leq m$ if and only if $A_{ij}^{\odot k} = 1$ for some $1 \leq k \leq m$ if and only if $\hat{A}_{ij} = 1$.

49. If the tree $T(V, E)$ is not a directed tree, then the transitive closure is $V \times V$. If $T(V, E)$ is a directed tree, then aRb if a is an ancestor of b.

Chapter 6—Section 6.7

1. 1

3. 2

5. 4

7. 5

9. 4

11. See table below.

1	1	1	1	1	1		1	0	0	1	1	1		0	0	1	1	1	1		0	1	0	1	1	1
1	1	1	1	1	0		1	0	0	1	1	0		0	0	1	1	1	0		0	1	0	1	1	0
1	1	1	1	0	0		1	0	0	1	0	0		0	0	1	1	0	0		0	1	0	1	0	0
1	1	1	1	0	1		1	0	0	1	0	1		0	0	1	1	0	1		0	1	0	1	0	1
1	1	1	0	0	1		1	0	0	0	0	1		0	0	1	0	0	1		0	1	0	0	0	1
1	1	1	0	0	0		1	0	0	0	0	0		0	0	1	0	0	0		0	1	0	0	0	0
1	1	1	0	1	0		1	0	0	0	1	0		0	0	1	0	1	0		0	1	0	0	1	0
1	1	1	0	1	1		1	0	0	0	1	1		0	0	1	0	1	1		0	1	0	0	1	1
1	1	0	0	1	1		1	0	1	0	1	1		0	0	0	0	1	1		0	1	1	0	1	1
1	1	0	0	1	0		1	0	1	0	1	0		0	0	0	0	1	0		0	1	1	0	1	0
1	1	0	0	0	0		1	0	1	0	0	0		0	0	0	0	0	0		0	1	1	0	0	0
1	1	0	0	0	1		1	0	1	0	0	1		0	0	0	0	0	1		0	1	1	0	0	1
1	1	0	1	0	1		1	0	1	1	0	1		0	0	0	1	0	1		0	1	1	1	0	1
1	1	0	1	0	0		1	0	1	1	0	0		0	0	0	1	0	0		0	1	1	1	0	0
1	1	0	1	1	0		1	0	1	1	1	0		0	0	0	1	1	0		0	1	1	1	1	0
1	1	0	1	1	1		1	0	1	1	1	1		0	0	0	1	1	1		0	1	1	1	1	1

13.

	111	110	100	101	001	000	101	011
11	11111	11110	11100	11101	11001	11000	11101	11011
10	10111	10110	10100	10101	10001	10000	10101	10011
00	00111	00110	00100	00101	00001	00000	00101	00011

15. Two vertices in a hypercube are adjacent if and only if the strings differ by a single symbol in the string. Thus, if one of the vertices contains an even number of ones, the other must contain an odd number of ones.

Chapter 7—Section 7.1

1. **1**, 2, 3, **4**, 5, **6**, 7, **8**, 9, **10**, 11, **12**, 13, **14**, **15**, **16**, 17, **18**, 19, **20**, **21**, **22**, 23, **24**, **25**, **26**, **27**, **28**, 29, **30**, 31, **32**, 33, **34**, **35**, **36**, 37, **38**, **39**, **40**, 41, **42**, 43, **44**, **45**, 46, 47, **48**, **49**, **50**, 51, **52**, 53, **54**, **55**, **56**, 57, **58**, 59, **60**, 61, **62**, **63**, **64**, **65**, **66**, 67, **68**, **69**, **70**, 71, **72**, 73, **74**, **75**, **76**, 77, **78**, 79, **80**, **81**, **82**, 83, **84**, **85**, **86**, 87, **88**, 89, **90**, **91**, **92**, **93**, **94**, **95**, **96**, 97, **98**, **99**, **100**, 101, **102**, 103, **104**, **105**, **106**, 107, **108**, 109, **110**, **111**, **112**, 113, **114**, **115**, **116**, **117**, **118**, **119**, **120**, **121**, **122**, **123**, **124**, **125**, **126**, **127**, **128**, **129**, **130**, 131, **132**, **133**, **134**, **135**, **136**, 137, **138**, 139, **140**, **141**, **142**, **143**, **144**, **145**, **146**, **147**, **148**, 149, **150**, 151, **152**, **153**, **154**, **155**, **156**, 157, **158**, **159**, **160**, 161, **162**, 163, **164**, **165**, **166**, 167, **168**, **169**, **170**, 171, **172**, 173, **174**, **175**, **176**, **177**, **178**, 179, **180**, 181, **182**, **183**, **184**, **185**, **186**, **187**, **188**, **189**, **190**, 191, **192**, 193, **194**, **195**, **196**, 197, **198**, 199, **200**

3. $37 \cdot 13$

5. $37 \cdot 47$

7. $47 \cdot 83$

Chapter 7—Section 7.2

1. $45^2 - 32^2 = 1001$, so 1001 is not a prime.

3. $70^2 - 7^2 = 4851$ so 4851 is not prime.

5. $90^2 - 7^2 = 8051$, so 8051 is not prime.

7. $90^2 - 13^2 = 7931$, so 7931 is not prime.

Chapter 7—Section 7.3

1. $75 = 8 \cdot 9 + 3$

3. $81 = 9 \cdot 9 + 0$

5. $75 = 25 \cdot 3 + 0$. Therefore, 25 is the gcd of 75 and 25.

7. $621 = 437 \cdot 1 + 184 \quad 437 = 184 \cdot 2 + 69 \quad 184 = 69 \cdot 2 + 46 \quad 69 = 46 \cdot 1 + 23 \quad 46 = 23 \cdot 2 + 0$. Therefore, 23 is the gcd of 621 and 437.

9. $822 = 436 \cdot 1 + 386 \quad 436 = 386 \cdot 1 + 50 \quad 386 = 50 \cdot 7 + 36 \quad 50 = 36 \cdot 1 + 14 \quad 36 = 14 \cdot 2 + 8 \quad 14 = 8 \cdot 1 + 6 \quad 8 = 6 \cdot 1 + 2 \quad 6 = 2 \cdot 3 + 0$. Therefore, 2 is the gcd of 822 and 436.

11. $\operatorname{lcm}(27, 18) = \dfrac{27 \cdot 18}{9} = 54$

13. $\operatorname{lcm}(377, 289) = \dfrac{377 \cdot 289}{1} = 108953$

15. $83 = 17 \cdot 4 + 15 \quad 17 = 15 \cdot 1 + 2 \quad 15 = 2 \cdot 7 + 6$
Therefore, 1 is the gcd of 83 and 17.

$$\begin{aligned} 1 &= 15 - 2 \cdot 7 \\ &= 15 - (17 - 15) \cdot 7 \\ &= 8 \cdot 15 - 7 \cdot 17 \\ &= 8 \cdot (83 - 17 \cdot 4) - 7 \cdot 17 \\ &= 8 \cdot 83 - 39 \cdot 17 \end{aligned}$$

17. $25 = 15 \cdot 1 + 10 \quad 15 = 10 \cdot 1 + 5 \quad 10 = 5 \cdot 2 + 0$. Therefore, 5 is the gcd of 25 and 15.

$$\begin{aligned} 5 &= 15 - 10 \\ &= 15 - (25 - 15) \\ &= 2 \cdot 15 + (-1) \cdot 25 \end{aligned}$$

19. $324 = 216 \cdot 1 + 108. \quad 216 = 108 \cdot 2 + 0$. Therefore, 108 is the gcd of 216 and 324.

$$1 \cdot 324 + (-1) \cdot 216 = 108.$$

21. If $ax + by = \gcd(a, b)$, then $\dfrac{a}{\gcd(a,b)} x + \dfrac{b}{\gcd(a,b)} y = 1$.

Any divisor of $\dfrac{a}{\gcd(a,b)}$ and $\dfrac{b}{\gcd(a,b)}$ must also divide 1.

Therefore, $\dfrac{a}{\gcd(a,b)}$ and $\dfrac{b}{\gcd(a,b)}$ are relatively prime.

23. Procedure Euclidean Algorithm (a, b)
$r_0 = \max(a, b)$
$r_1 = \min(a, b)$
Let $i = 0$
  While $r_{i+1} > 0$
    Use Procedure Division Algorithm(r_i, r_{i+1}) to find q, r_{i+2} so that $r_i = q r_{i+1} + r_{i+2}$
    $d = r_{i+1}$
    $i = i + 1$
  Endwhile
End

25. (a) $F(0) + 2 = (2^{2^0} + 1) + 2 = 2 + 2 + 1 = 2^{2^0} + 1 = F(1)$ Therefore, the statement is true for $m = 0$. Assume $F(m) = F(0) \cdot F(1) \cdot F(2) \cdot F(3) \cdot \cdots \cdot F(m-1) + 2$.

$$F(0) \cdot F(1) \cdot F(2) \cdot F(3) \cdot \ldots \cdot F(m-1) \cdot F(m) + 2$$
$$= (F(m) - 2) \cdot F(m) + 2$$
$$= (2^{2^m} + 1 - 2) \cdot (2^{2^m} + 1) + 2$$
$$= (2^{2^m} - 1) \cdot (2^{2^m} + 1) + 2$$
$$= (2^{2^m})^2 - 1 + 2$$
$$= 2^{2^m + 2^m} + 1$$
$$= 2^{2^{m+1}} + 1$$
$$= F(m+1)$$

(b) Assume $n < m$. Then $F(m) = F(0) \cdot F(1) \cdot F(2) \cdot F(3) \cdot \ldots \cdot F(m-1) + 2$ where $F(n)$ is one of the factors on the right-hand side. If an integer greater than 1 divides $F(n)$ and $F(m)$, then it must also divide 2. Therefore, it must equal 2. But this is impossible since both $F(n)$ and $F(m)$ are odd. Therefore, $F(n)$ and $F(m)$ are relatively prime.

27. Assume $a + nb$ and b are not relatively prime. Then there exist $d \neq 1$ so that $a + nb = ud$ for some u and $b = vd$ for some v.
$$a = a + nb - nb$$
$$= ud - nvd$$
$$= d(u - nv)$$
so that $d \mid a$, so a and b are not relatively prime.

29. Let $ax + by \in S$, and $d = \gcd(a, b)$. $d \mid a$, so $a = dv$ for some v. $d \mid b$, so $a = dw$ for some w. $ax + by = dvx + dwy = d(vx + wy)$ so $d \mid (ax + by)$.

31. Assume $a \mid c$ and $b \mid c$ and $\gcd(a, b) = 1$. There exists u and v such that $au + bv = 1$. Therefore, $auc + bvc = c$. Since $b \mid c$, $ab \mid auc$. Since $a \mid c$, $ab \mid bvc$. Therefore, $ab \mid c$.

Chapter 7—Section 7.4

1. $[3; 2, 1, 3]$, $[3; 2, 1, 2, 1]$

$$\frac{37}{11} = 3 + \frac{4}{11}$$
$$= 3 + \frac{1}{\frac{11}{4}}$$
$$= 3 + \frac{1}{2 + \frac{3}{4}}$$
$$= 3 + \frac{1}{2 + \frac{1}{\frac{4}{3}}}$$
$$= 3 + \frac{1}{2 + \frac{1}{1 + \frac{1}{3}}}$$

$$\frac{37}{11} = 3 + \frac{4}{11}$$
$$= 3 + \frac{1}{\frac{11}{4}}$$
$$= 3 + \frac{1}{2 + \frac{3}{4}}$$
$$= 3 + \frac{1}{2 + \frac{1}{\frac{4}{3}}}$$
$$= 3 + \frac{1}{2 + \frac{1}{1 + \frac{1}{3}}}$$
$$= 3 + \frac{1}{2 + \frac{1}{1 + \frac{1}{2 + \frac{1}{1}}}}$$

3. $[-1; 1, 6, 1, 3, 1, 5, 1, 1, 1, 2]$, $[-1; 1, 6, 1, 3, 1, 5, 1, 1, 1, 1]$,

$$\frac{-257}{2003} = -1 + \frac{1746}{2003} = -1 + \frac{1746}{2003} = -1 + \frac{1746}{2003}$$

$$= -1 + \frac{1746}{2003} = -1 + \frac{1}{\frac{2003}{1746}} = -1 + \frac{1}{1 + \frac{257}{1746}}$$

$$= -1 + \cfrac{1}{1 + \cfrac{1}{\frac{1746}{257}}} = -1 + \cfrac{1}{1 + \cfrac{1}{6 + \frac{204}{257}}}$$

$$= -1 + \cfrac{1}{1 + \cfrac{1}{6 + \cfrac{1}{\frac{257}{204}}}}$$

$$= -1 + \cfrac{1}{1 + \cfrac{1}{6 + \cfrac{1}{1 + \frac{53}{204}}}}$$

$$= -1 + \cfrac{1}{1 + \cfrac{1}{6 + \cfrac{1}{1 + \cfrac{1}{\frac{204}{53}}}}}$$

$$= -1 + \cfrac{1}{1 + \cfrac{1}{6 + \cfrac{1}{1 + \cfrac{1}{3 + \frac{45}{53}}}}}$$

$$= -1 + \cfrac{1}{1 + \cfrac{1}{6 + \cfrac{1}{1 + \cfrac{1}{3 + \cfrac{1}{\frac{53}{45}}}}}}$$

$$= -1 + \cfrac{1}{1 + \cfrac{1}{6 + \cfrac{1}{1 + \cfrac{1}{3 + \cfrac{1}{1 + \frac{8}{45}}}}}}$$

$$= -1 + \cfrac{1}{1 + \cfrac{1}{6 + \cfrac{1}{1 + \cfrac{1}{3 + \cfrac{1}{1 + \frac{1}{\frac{45}{8}}}}}}}$$

$$= -1 + \cfrac{1}{1 + \cfrac{1}{6 + \cfrac{1}{1 + \cfrac{1}{3 + \cfrac{1}{1 + \frac{1}{5 + \frac{5}{8}}}}}}}$$

$$= -1 + \cfrac{1}{1 + \cfrac{1}{6 + \cfrac{1}{1 + \cfrac{1}{3 + \cfrac{1}{1 + \frac{1}{5 + \frac{1}{8}{5}}}}}}}$$

$$= -1 + \cfrac{1}{1 + \cfrac{1}{6 + \cfrac{1}{1 + \cfrac{1}{3 + \cfrac{1}{1 + \cfrac{1}{5 + \frac{1}{1 + \frac{3}{5}}}}}}}}$$

$$= -1 + \cfrac{1}{1 + \cfrac{1}{6 + \cfrac{1}{1 + \cfrac{1}{3 + \cfrac{1}{1 + \cfrac{1}{5 + \frac{1}{1 + \frac{1}{5}{3}}}}}}}}$$

$$= -1 + \cfrac{1}{1 + \cfrac{1}{6 + \cfrac{1}{1 + \cfrac{1}{3 + \cfrac{1}{1 + \cfrac{1}{5 + \frac{1}{1 + \frac{1}{1 + \frac{2}{3}}}}}}}}}$$

$$= -1 + \cfrac{1}{1 + \cfrac{1}{6 + \cfrac{1}{1 + \cfrac{1}{3 + \frac{1}{B}}}}}$$

$$= -1 + \cfrac{1}{1 + \cfrac{1}{6 + \cfrac{1}{1 + \cfrac{1}{3 + \frac{1}{C}}}}}$$

where $B = 1 + \cfrac{1}{5 + \cfrac{1}{1 + \frac{1}{1 + \frac{1}{\frac{3}{2}}}}}$

$C = 1 + \cfrac{1}{5 + \cfrac{1}{1 + \cfrac{1}{1 + \frac{1}{1 + \frac{1}{2}}}}}$

and we add

$$-1 + \cfrac{1}{1 + \cfrac{1}{6 + \cfrac{1}{1 + \cfrac{1}{3 + \frac{1}{D}}}}}$$

where

$D = 1 + \cfrac{1}{5 + \cfrac{1}{1 + \cfrac{1}{1 + \frac{1}{1 + \frac{1}{1+1}}}}}$

to get the second representation.

5. $[-1; 1, 7, 1, 4]$, $[-1; 1, 7, 1, 3, 1]$,

$\dfrac{-5}{44} = -1 + \dfrac{39}{44}$ $\qquad = -1 + \dfrac{1}{\frac{44}{39}}$

$= -1 + \cfrac{1}{1 + \frac{5}{39}}$ $\qquad = -1 + \cfrac{1}{1 + \frac{1}{\frac{39}{5}}}$

$= -1 + \cfrac{1}{1 + \cfrac{1}{7 + \frac{4}{5}}}$ $\qquad = -1 + \cfrac{1}{1 + \cfrac{1}{7 + \frac{1}{\frac{5}{4}}}}$

$= -1 + \cfrac{1}{1 + \cfrac{1}{7 + \frac{1}{1 + \frac{1}{4}}}}$

Where we add

$-1 + \cfrac{1}{1 + \cfrac{1}{7 + \cfrac{1}{1 + \frac{1}{3 + \frac{1}{1}}}}}$

to get the second representation.

7.
$$3 + \cfrac{1}{5+\frac{1}{2}} = 3 + \cfrac{1}{\frac{11}{2}}$$
$$= 3 + \frac{2}{11}$$
$$= \frac{35}{11}$$

9.
$$-10 + \cfrac{1}{1 + \cfrac{1}{4+\frac{1}{3}}} = -10 + \cfrac{1}{1 + \cfrac{1}{\frac{13}{3}}}$$
$$= -10 + \cfrac{1}{1 + \frac{3}{13}}$$
$$= -10 + \cfrac{1}{\frac{16}{13}}$$
$$= -10 + \frac{13}{16}$$
$$= -\frac{147}{16}$$

11.
$$2 + \cfrac{1}{5 + \frac{1}{3}} = 2 + \cfrac{1}{\frac{16}{3}}$$
$$= 2 + \frac{3}{16}$$
$$= \frac{35}{16}$$

13. $[2; 4, 4, 4, 4, 4, 2 + \sqrt{5}]$

$$2 + \sqrt{5} - 2 = 2 + \cfrac{1}{\cfrac{1}{\sqrt{5}-2}} = 2 + \cfrac{1}{\sqrt{5}+2}$$
$$= 2 + \cfrac{1}{4 + (\sqrt{5}-2)}$$
$$= 2 + \cfrac{1}{4 + \cfrac{1}{\cfrac{1}{(\sqrt{5}-2)}}}$$
$$= 2 + \cfrac{1}{4 + \cfrac{1}{(\sqrt{5}+2)}}$$
$$= 2 + \cfrac{1}{4 + \cfrac{1}{4 + \cfrac{1}{(\sqrt{5}+2)}}}$$
$$= 2 + \cfrac{1}{4 + \cfrac{1}{4 + \cfrac{1}{4 + \cfrac{1}{(\sqrt{5}+2)}}}}$$
$$= 2 + \cfrac{1}{4 + \cfrac{1}{4 + \cfrac{1}{4 + \cfrac{1}{4 + \cfrac{1}{(\sqrt{5}+2)}}}}}$$
$$= 2 + \cfrac{1}{4 + \cfrac{1}{4 + \cfrac{1}{4 + \cfrac{1}{4 + \cfrac{1}{4 + \cfrac{1}{(\sqrt{5}+2)}}}}}}$$

15. $[3; 7, 15, 1, 292, 1, 1.736\ldots]$ answer will depend on the accuracy of the estimate of π.

17. Prove the statement for $n = 1$. $[t_0; t_1] = t_0 + \cfrac{1}{t_1}$ and $[t_0; t_1 + b] = t_0 + \cfrac{1}{t_1 + b}$ and, hence, $[t_0; t_1] > [t_0; t_1 + b]$.

Assume the statement is true for $n = k$, that is, assume

$[t_0; t_1, t_2 \cdots t_k] > [t_0; t_1, t_2 \cdots t_k + b]$ if k is odd

$[t_0; t_1, t_2 \cdots t_k] < [t_0; t_1, t_2 \cdots t_k + b]$ if k is even

Consider the statement for $n = k + 1$

$[t_0; t_1, t_2 \cdots t_k, t_{k+1}] = [t_0 t_1; t_2 \cdots, [t_k; t_{k+1}]]$ and

$[t_0; t_1, t_2 \cdots t_k, t_{k+1} + b] = [t_0; t_1; t_2 \cdots [t_k; t_{k+1} + b]]$
But $[t_k; t_{k+1}] > [t_k; t_{k+1} + b]$ so that $[t_k; t_{k+1}] = [t_k; t_{k+1} + b] + b'$.

Therefore, $[t_0 t_1; t_2 \cdots, [t_k; t_{k+1}]] = [t_0 t_1; t_2 \cdots, [t_k; t_{k+1} + b] + b']$. If $k + 1$ is even, then k is odd and, by the induction hypothesis,

$[t_0; t_1, t_2 \cdots t_k, t_{k+1} + b]$
$= [t_0; t_1; t_2 \cdots [t_k; t_{k+1} + b]]$
$> [t_0 t_1; t_2 \cdots, [t_k; t_{k+1} + b] + b']$
$= [t_0; t_1, t_2 \cdots t_k, t_{k+1}]$

If $k + 1$ is odd, then k is even and, by the induction hypothesis,

$[t_0; t_1, t_2 \cdots t_k, t_{k+1} + b]$
$= [t_0; t_1; t_2 \cdots [t_k; t_{k+1} + b]]$
$< [t_0 t_1; t_2 \cdots, [t_k; t_{k+1} + b] + b']$
$= [t_0; t_1, t_2 \cdots t_k, t_{k+1}]$

Chapter 7—Section 7.5

1. The proof is by induction. Since $q_0 = 1$, $q_0 > 0$. Assume $q_m > 0$ for all $m < k$. But $q_k = q_{k-1} t_k + q_{k-2}$. By the induction hypothesis, q_{k-1} and q_{k-2} are greater than 0. By definition of finite continued fraction, t_k is positive. Therefore, q_k is positive.

3.

k	p_k	q_k	$x - (p_k/q_k)$
0	2	1	$0.236\cdots$
1	9	4	$-0.0139\cdots$
2	38	17	$7.73\cdots \times 10^{-4}$
3	161	72	$-4.31\cdots \times 10^{-5}$
4	682	305	$2.40\cdots \times 10^{-6}$
5	2899	1292	$-1.33\cdots \times 10^{-7}$

5.

k	p_k	q_k	$x - (p_k/q_k)$
0	3	1	$0.141\cdots$
1	22	7	$-1.26\cdots \times 10^{-3}$
2	333	106	$8.32\cdots \times 10^{-5}$
3	355	113	$-2.66\cdots \times 10^{-7}$
4	103993	33102	$5.77\cdots \times 10^{-10}$
5	104348	33215	$-3.31\cdots \times 10^{-10}$

7. (a) $q_0 = 1$. Therefore, $q_0 \geq 0$. Assume $q_k \geq k$. $q_1 = t_1 \geq 1$.
$$q_2 = q_1 t_2 + q_0$$
$$\geq t_2 + q_0$$
$$\geq 2$$
Assume $q_k \geq k \geq 2$.
$$q_{k+1} = q_k t_{k+1} + q_{k-1}$$
$$\geq k \cdot 1 + q_{k-1}$$
$$\geq k + k - 1$$
$$\geq k + 1$$

9. By Theorem 7.11,
$$\frac{q_k}{q_{k-1}} = \frac{q_{k-1} t_k + q_{k-2}}{q_{k-1}}$$
$$= t_k + \frac{q_{k-2}}{q_{k-1}}$$
$$= t_k + \frac{1}{t_{k-1} + \frac{q_{k-3}}{q_{k-2}}}$$
$$= t_k + \frac{1}{t_{k-1} + \frac{1}{t_{k-2} \cdots}}$$

11. $p_{-2} = 0$,
$q_{-2} = 1$

Chapter 8—Section 8.1

1. $\left\lfloor \frac{699}{5} \right\rfloor = 139$, $\left\lfloor \frac{699}{3} \right\rfloor = 233$, $\left\lfloor \frac{699}{15} \right\rfloor = 46$

3. (a) $53 \cdot 52 \cdot 51 \cdot 50 = 7027800$
   (b) $8 \cdot 52 \cdot 51 \cdot 50 = 1060800$
   (c) $45 \cdot 52 \cdot 51 \cdot 50 = 5967000$
   (d) $20 \cdot 33 \cdot 32 \cdot 31 + 33 \cdot 32 \cdot 31 \cdot 30 = 1636800$ or $33 \cdot 32 \cdot 31 \cdot 50 = 1636800$

5. $10^4 - 10 \cdot 9 \cdot 8 \cdot 7 = 4960$

7. $2^{30} = 1073741824$

9. $2 \cdot (26)^2 + 2 \cdot (26)^3 = 36504$

11. (a) $2^4 = 16$, (b) 1, (c) $2^4 - 1 = 15$

13. $(10)^9$

15. $3^6 = 729$

17. $699 - 87 = 612$

19. $26^2 + 26^3 = 18252$

21. $26^2 + 26^3 - 25^2 - 25^3$

23. $2 \cdot 2^5 = 64$

25. $7 \cdot 10^9$

27. $9^4 = 6561$ assuming leading zeroes are allowed.

29. 128^7

31. $128^7 - 7 \cdot 128^6$

33. $2 \cdot 4!$

Chapter 8—Section 8.2

1. $120 + 110 - 80 = 150$ $200 - 150 = 50$

3. 50

5. 70

7. $57 + 36 - 5 = 88$

9. $40 + 26 - 13 = 53$

11. $100 - 14 = 86$

13. $2 \cdot 8! = 80640$ $9! - 2 \cdot 8!$

15. $31 + 10 - 3 = 38$ are squares or cubes so $1000 - 38 = 962$ are not squares or cubes.

17. $667 + 400 + 286 - 133 - 95 - 57 + 19 = 1087$

19. $500 + 400 + 333 - 100 - 66 - 166 + 33 = 934$

21.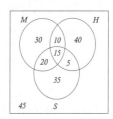

(a) 155
(b) $30 + 40 + 35 = 105$
(c) $30 + 10 + 40 = 80$
(d) $200 - 35 = 165$
(e) 80

23. $9 \cdot 10^4 - 6 \cdot 7^4 = 75594$

25. Let x be the number studying all three languages.

$35 + 42 + 43 - 17 - 15 - 13 + x = 80$. $x = 5$.

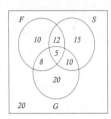

(a) 38 **(b)** 45 **(c)** 30 **(d)** 40 **(e)** 15

Chapter 8—Section 8.3

1. (a) $P(8, 5) = \dfrac{8!}{3!} = 6720$

(b) $P(11, 8) = \dfrac{11!}{3!} = 6652800$

(c) $C(12, 7) = \dfrac{12!}{7!5!} = 792$

(d) $C(14, 2) = \dfrac{14!}{2!12!} = 91$

(e) $C(14, 12) = \dfrac{14!}{2!12!} = 91$

3. $P(7, 3) = 210$

Find those less than 450. Those beginning with 4 must have 2 or 3 for the second digit. There are 5 choices for the third digit. Therefore, there are 10 numbers that begin with 4. For those whose first digit is less than 4, there are 2 choices for the first digit, 6 for the second, and 5 for the third or 60 numbers. This makes a possibility of 70 numbers.

Find those that are even. There are 4 choices for the last digit, 6 for the first digit, and 5 for the second digit. This makes a total of 120 numbers. The last two digits must be divisible by 4. The possibilities are 24, 28, 32, 36, 48, 52, 56, 64, 68, 84, 92, 96. Thus, there are 12 of them. For each of these, there are 5 choices for the first digit. Therefore, there are 60 numbers.

5. $18 \cdot 17 \cdot 16 = 4896$

7. (a) $10! = 3628800$ **(b)** $2^5 \cdot 5! = 3840$

9. (a) $2 \cdot P(8, 4) = 3360$

(b) $7 \cdot 6 \cdot P(7, 3) = 8820$

(c) There are 10 ways to position the 7 and 8. Thus, there are $10 \cdot 8 \cdot 7 \cdot 6 \cdot 5 = 16800$ possible numbers.

(d) There are 22 two-digit numbers divisible by 4 that can serve as the last two digits. Thus, there are $22 \cdot 8 \cdot 7 \cdot 6 \cdot 5 = 36960$ numbers.

(e) $90 \cdot 7 \cdot 6 \cdot 5 = 18900$

(f) $\binom{6}{2} \cdot 8 \cdot 7 \cdot 6 \cdot 5 = 50400$

11. $\dfrac{5! \cdot 5!}{10} = 1440$

13. $\binom{15}{6}\binom{20}{7} = 387987600$

15. $\binom{52}{13}$

17. $4 \cdot \binom{13}{7} \cdot \binom{39}{6} = 22394644272$

19. $4 \cdot \binom{13}{9} \cdot \binom{39}{4} = 235237860$

21. 58656 including full houses or 54912 excluding full houses.

23. 252

25. $\binom{10}{4} = 210$, $2^{10} - \binom{10}{10} - \binom{10}{9} - \binom{10}{8} = 968$

27. $\binom{30}{10} = 30045015$

29. $\binom{20}{9} = 167960$

31. $a^8 + 8a^7b + 28a^6b^2 + 56a^5b^3 + 70a^4b^4 + 56a^3b^5 + 28a^2b^6 + 8ab^7 + b^8$

33. $\binom{13}{8} \cdot 3^8 \cdot (-4)^5 = -8646663168$

35. $\binom{10}{3} \cdot (-3)^7 = -262440$

Chapter 8—Section 8.4

1. 21453

3. 21534

5. $\{a, c, d, e\}$

7. $\{1, 3\}$

9. $\{1, 5\}$

11. $\{1, 5, 6\}$

13. $\{1, 3, 4, 6\}$

15. $xyz, xzy, yxz, yzx, zxy, zyx$

17. $\{a, b, c, d\}, \{a, b, c, e\}, \{a, b, d, e\}, \{a, c, d, e\}, \{b, c, d, e\}$

19. $abcdefghijklmnopqrstuvwxyz, zyxwvutsrqponmlkjihgfedcba$

21. $\{abcde\}, \{vwxyz\}$

23. For $i = 1$ to $C(n, r)$
    Use Procedure Combination$(n.r)$ to generate the ith
    r combination of n items
    Use Procedure Generate Permutation to order all
    permutations of these r items
    generated by .Procedure Combination$(n.r)$
    Endfor

Chapter 8—Section 8.5

1. (a) $\dfrac{3}{36}$, (b) $\dfrac{6}{36}$, (c) $\dfrac{4}{36}$,
   (d) $\dfrac{4}{36}$, (e) $\dfrac{6}{36}$,
   $\dfrac{6}{36}, \dfrac{18}{36}$

3. (a) $\dfrac{12}{36}$, (b) $\dfrac{16}{36}$,
   (c) $\dfrac{30}{36}$, (d) $\dfrac{22}{36}$

5. (a) $\dfrac{100}{301}$, (b) $\dfrac{14}{301}$,
   (c) $\dfrac{129}{301}$, (d) $\dfrac{172}{301}$

7. $\dfrac{15120}{59049}$

9. (a) $\dfrac{53}{250}$, (b) $\dfrac{105}{250}$,
   (c) $\dfrac{97}{250}$, (d) $\dfrac{23}{250}$

11. (a) $\dfrac{54912}{2598960}$, (b) $\dfrac{36}{2598960}$,
    (c) $\dfrac{4}{2598960}$, (d) $\dfrac{5148}{2598960}$
    (e) $\dfrac{9216}{2598960}$

13. $\dfrac{1}{3} \cdot \dfrac{4}{9} + \dfrac{1}{3} \cdot \dfrac{6}{9} + \dfrac{1}{3} \cdot \dfrac{5}{7} = \dfrac{345}{567}$

15. $\dfrac{1000 + 900 + 3168 - 100 - 271 - 252 + 19}{9000} = \dfrac{4464}{9000}$

17. $1 - \dfrac{8 \cdot 9 \cdot 9 \cdot 9}{9000}$, $1 - \dfrac{2 \cdot 8 \cdot 9 \cdot 9 \cdot 9 - 7 \cdot 8 \cdot 8 \cdot 8}{9000}$

Chapter 8—Section 8.6

1. $\dfrac{8!}{3! \cdot 2!}$

3. $\dfrac{6!}{3!}$

5. $\dfrac{26!}{8! \cdot 6! \cdot 9! \cdot 3!}$

7. $\dfrac{35!}{7! \cdot 7! \cdot 7! \cdot 7! \cdot 7!}$

9. $\dfrac{24!}{(4!)^6}$

11. $\dfrac{12!}{3! \cdot 6! \cdot 3!} \cdot 2^6 \cdot 3^3 = 31933440$

13. $\dfrac{18!}{4! \cdot 5! \cdot 6! \cdot 3!} \cdot 2^4 \cdot 4^5 \cdot 5^3$

15. $\dfrac{11!}{2! \cdot 4! \cdot 2! \cdot 3!} 4^2 \cdot 2^4 \cdot 5^3 = 2217600000$

Chapter 8—Section 8.7

1. $\dfrac{21!}{4!17!} = 5985$

3. $\dfrac{12!}{9!3!} = 220$

5. $\dfrac{14!}{4!10!} = 1001$

7. 8^3

9. $\dfrac{12!}{2!10!} = 66$

11. $\dfrac{23!}{4!19!}$

13. $\dfrac{31!}{12!19!}$

15. $\dfrac{20!}{8!12!}$

17. $\dfrac{13!}{10!3!} \cdot \dfrac{13!}{9!4!} \cdot \dfrac{13!}{8!5!} \cdot \dfrac{13!}{12!1!}$

19. 4^{12}, $\dfrac{12!}{3!3!3!3!}$

Chapter 8—Section 8.8

1. 41

3. If one person has 29 mutual friends, then everyone has 29 mutual friends. If not, then the number of mutual friends that the people can have ranges from 0 to 28, so there are 29 possibilities. But there are 30 people, so at least 2 must have the same number of friends.

5. 51

7. 25

9. Let $a_1, a_2, a_3, \ldots, a_n$ be a sequence of n positive integers. Let $s_1 = a_1$, $s_2 = a_1 + a_2$, $s_3 = a_1 + a_2 + a_3, \ldots, s_n = a_1 + a_2 + a_3 + \cdots + a_n$. Each s_i is equivalent to r_i modulo n, where $0 \leq r_i \leq n - 1$. If any of the r_i is equal to 0, then s_i is divisible by n and we are done. If not, then there are $n - 1$ values for the r_i and n of the r_i. Hence, two of the r_i must have the same value, say, r_j and r_k. Assume $j < k$, then $a_{j+1} + a_{j+2} + \cdots + a_k$ is divisible by n.

11. $26^2 + 1$

13. 11

15. $\left\lceil \dfrac{170}{14} \right\rceil = 13$

17. Consider the pairs of numbers

$$\{1, 2\}, \{3, 4\}, \{5, 6\}, \ldots \{2n - 1, 2n\}$$

There are n pairs of numbers, so if we have $n + 1$ numbers, two must belong to the same pair. These two numbers differ by 1.

19. Since

$$0 + 1 + 2 + 3 + \cdots + n - 1 = \dfrac{n(n-1)}{2}$$

there are only $n - 1$ distinct integers whose sum is less than $\dfrac{n(n-1)}{2}$.

Therefore, m is the sum of at most $n - 1$ distinct integers. There are only $n - 1$ distinct integers and n boxes, so two boxes must contain integers that are not distinct. These boxes contain the same number of balls.

Chapter 8—Section 8.9

1. $\dfrac{3}{4}$

3. $\dfrac{6}{18}$

5. $\dfrac{1}{2}, \dfrac{1}{2}$

7. $\dfrac{\frac{1}{8}}{1 - \frac{1}{8}} = \dfrac{1}{7}$

9. $\dfrac{45}{57} = \dfrac{15}{19}$

11. $\dfrac{19}{6}$

13.
value	2	3	4	5	6	7	8	9	11	12	13	14	15	16	17
no. of ways	1	6	6	6	3	4	6	3	2	4	6	3	1	2	3
product	2	18	24	30	18	28	48	27	22	48	78	42	15	32	51

$$\dfrac{483}{56}$$

15.
value	2	3	4	5	6	7	8	9	10	11	12
no. of ways	1	2	3	4	5	6	5	4	3	2	1
product	2	6	12	20	30	42	40	36	30	22	12

$$\dfrac{252}{36}$$

17. $\dfrac{-50000000 \cdot .33 + 10000000}{50000} - 13$ cents

19. $\binom{5}{3}(.6)^3(.4)^2 = .3456$

21. $\dfrac{\binom{7}{4}}{2^7} = \dfrac{35}{2^7}$

23. $\binom{10}{2}(.1)^2(.9)^8 \approx .1937$

25. $\dfrac{\binom{13}{2}\binom{39}{8}}{\binom{52}{10}} \approx .3$

27. $\dfrac{36}{38}$

29. $\dfrac{36}{38}$

31. 3, 2.1

Chapter 8—Section 8.10

1. $p(A) = .45$ $.p(B) = .25$ $p(C) = .30$ $.p(E \mid A) = .05$ $.p(E \mid B) = .1$ $p(E \mid C) = .1$

$p(B \mid E)$

$$= \dfrac{.p(E \mid B) \cdot p(B)}{p(E \mid A) \cdot p(A) + p(E \mid B) \cdot p(B) + p(E \mid C) \cdot p(C)}$$

$$= \dfrac{(.1) \cdot (.25)}{(.05) \cdot (.45) + (.1) \cdot (.25) + (.02) \cdot (.30)}$$

$$= \dfrac{.025}{.0535}$$

$$\approx .47$$

3. $p(A) = p(B) = p(C) = \dfrac{1}{3}$ $p(H \mid A) = 1.0$ $p(H \mid B) = .5$ $p(H \mid C) = .7$

$p(B \mid H)$

$$= \dfrac{p(H \mid B) \cdot p(B)}{p(H \mid A) \cdot p(A) + p(H \mid B) \cdot p(B) + p(H \mid C) \cdot p(C)}$$

$$= \dfrac{.3 \cdot \frac{1}{3}}{1 \cdot \frac{1}{3} + .5 \cdot \frac{1}{3} + .7 \cdot \frac{1}{3}}$$

$$= \dfrac{.5}{2.2}$$

$$\approx .23$$

5. $p(A) = p(B) = p(C) = \frac{1}{3}$ $p(P \mid A) = .4$ $p(P \mid B) = .3$ $p(H \mid C) = .6$

$p(A \mid P)$
$= \frac{p(P \mid A) \cdot p(A)}{p(P \mid A) \cdot p(A) + p(P \mid B) \cdot p(B) + p(P \mid C) \cdot p(C)}$

$= \frac{.4 \cdot \frac{1}{3}}{.4 \cdot \frac{1}{3} + .3 \cdot \frac{1}{3} + .6 \cdot \frac{1}{3}}$

$= \frac{4}{13}$

$\approx .31$

7. $p(G) = .25$ $p(A) = .5$ $p(B) = .25$ $p(W \mid G) = .1$ $p(W \mid A) = .2$

$p(W \mid B) = .6$
$p(G \mid W) = \frac{(.25) \cdot (.1)}{(.25) \cdot (.1) + (.5) \cdot (.2) + (.25) \cdot (.3)}$
$= \frac{.025}{.02}$
$= .125$

Chapter 8—Section 8.11

Let $M = \begin{bmatrix} .2 & .5 & .3 \\ .3 & .6 & .1 \\ .2 & .3 & .5 \end{bmatrix}$ $M^2 = \begin{bmatrix} .25 & .49 & .26 \\ .26 & .54 & .2 \\ .23 & .43 & .34 \end{bmatrix}$

$M^3 = \begin{bmatrix} .249 & .497 & .254 \\ .254 & .514 & .232 \\ .243 & .475 & .282 \end{bmatrix}$

$M^4 = \begin{bmatrix} .2497 & .4989 & .2514 \\ .2514 & .5050 & .2436 \\ .2475 & .4911 & .2614 \end{bmatrix}$

1. $\begin{bmatrix} 1 & 0 & 0 \end{bmatrix} \begin{bmatrix} .25 & .49 & .26 \\ .26 & .54 & .2 \\ .23 & .43 & .34 \end{bmatrix} = \begin{bmatrix} \underline{.25} & .49 & .26 \end{bmatrix}$

answer .25

3. $\begin{bmatrix} 1 & 0 & 0 \end{bmatrix} \begin{bmatrix} .2497 & .4989 & .2514 \\ .2514 & .5050 & .2436 \\ .2475 & .4911 & .2614 \end{bmatrix} = \begin{bmatrix} .2497 & \underline{.4989} & .2514 \end{bmatrix}$

answer .4989

Let $M = \begin{bmatrix} .3 & .4 & .3 \\ .4 & .2 & .4 \\ .2 & .5 & .3 \end{bmatrix}$ $M^2 = \begin{bmatrix} .31 & .35 & .34 \\ .28 & .40 & .32 \\ .32 & .33 & .35 \end{bmatrix}$

$M^3 = \begin{bmatrix} .301 & .364 & .335 \\ .308 & .352 & .34 \\ .298 & .369 & .333 \end{bmatrix}$

$M^4 = \begin{bmatrix} .3029 & .3607 & .3364 \\ .3012 & .3636 & .3352 \\ .3036 & .3595 & .3369 \end{bmatrix}$

5. $\begin{bmatrix} 1 & 0 & 0 \end{bmatrix} \begin{bmatrix} .301 & .364 & .335 \\ .308 & .352 & .34 \\ .298 & .369 & .333 \end{bmatrix} = \begin{bmatrix} .301 & .364 & \underline{.335} \end{bmatrix}$

answer .335

Let $M = \begin{bmatrix} .7 & .2 & .1 \\ .2 & .6 & .2 \\ .1 & .1 & .8 \end{bmatrix}$ $M^2 = \begin{bmatrix} .54 & .27 & .19 \\ .28 & .42 & .30 \\ .17 & .16 & .67 \end{bmatrix}$

$M^3 = \begin{bmatrix} .451 & .289 & .26 \\ .31 & .338 & .352 \\ .218 & .197 & .585 \end{bmatrix}$

7. $\begin{bmatrix} 1 & 0 & 0 \end{bmatrix} \begin{bmatrix} .451 & .289 & .26 \\ .31 & .338 & .352 \\ .218 & .197 & .585 \end{bmatrix} = \begin{bmatrix} .451 & .289 & \underline{.26} \end{bmatrix}$

answer .26

9. $\begin{bmatrix} 0 & 0 & 1 \end{bmatrix} \begin{bmatrix} .451 & .289 & .26 \\ .31 & .338 & .352 \\ .218 & .197 & .585 \end{bmatrix} = \begin{bmatrix} .218 & .197 & \underline{.585} \end{bmatrix}$

answer .585

$M = \begin{bmatrix} .6 & .4 \\ .3 & .7 \end{bmatrix}$ $M^2 = \begin{bmatrix} .48 & .52 \\ .39 & .61 \end{bmatrix}$

$M^3 = \begin{bmatrix} .444 & .556 \\ .417 & .583 \end{bmatrix}$

$M^4 = \begin{bmatrix} .4332 & .5668 \\ .4251 & .5749 \end{bmatrix}$ $M^5 = \begin{bmatrix} .42996 & .57004 \\ .42753 & .57247 \end{bmatrix}$

11. $\begin{bmatrix} 0 & 1 \end{bmatrix} \begin{bmatrix} .444 & .556 \\ .417 & .583 \end{bmatrix} = \begin{bmatrix} .417 & \underline{.583} \end{bmatrix}$

answer .583

$M = \begin{bmatrix} .6 & .4 \\ .3 & .7 \end{bmatrix}$ $M^2 = \begin{bmatrix} .48 & .52 \\ .39 & .61 \end{bmatrix}$

13. $\begin{bmatrix} .5 & .5 \end{bmatrix} \begin{bmatrix} .48 & .52 \\ .39 & .61 \end{bmatrix} = \begin{bmatrix} .435 & \underline{.565} \end{bmatrix}$

answer .565

15. Form the following table in which the numbers on the side and top are the engines remaining.

	4	3	2	1 or 0
4	$(.75)^4$	$\binom{4}{1}(.75)^3(.25)$	$\binom{4}{2}(.75)^2(.25)^2$	$\binom{4}{3}(.75)(.25)^3 + (.25)^4$
3	0	$(.75)^3$	$\binom{3}{1}(.75)^2(.25)$	$\binom{3}{2}(.75)(.25)^2 + (.25)^3$
2	0	0	$(.75)^2$	$\binom{2}{1}(.75)(.25) + (.25)^2$
1 or 0	0	0	0	1

we have

$$M = \begin{bmatrix} .3164 & .4219 & .2109 & .0508 \\ 0 & .4219 & .4219 & .1562 \\ 0 & 0 & .5625 & .4375 \\ 0 & 0 & 0 & 1 \end{bmatrix}$$

$$M^2 = \begin{bmatrix} .10011 & .31149 & .36336 & .22504 \\ 0 & .178 & .41532 & .40668 \\ 0 & 0 & .31641 & .68359 \\ 0 & 0 & 0 & 1 \end{bmatrix}$$

$$M^3 = \begin{bmatrix} .031674 & .17365 & .35692 & .43775 \\ 0 & .075098 & .30871 & .61619 \\ 0 & 0 & .17798 & .82202 \\ 0 & 0 & 0 & 1 \end{bmatrix}$$

$$M^4 = \begin{bmatrix} 1 & 0 & 0 & 0 \end{bmatrix} \begin{bmatrix} .010022 & .086628 & .28071 & .62264 \\ 0 & .031684 & .20534 & .76298 \\ 0 & 0 & .10011 & .89989 \\ 0 & 0 & 0 & 1 \end{bmatrix}$$

$$= \begin{bmatrix} .010022 & .086628 & .28071 & .62264 \end{bmatrix}$$

answer $.010022 + .086628 + .28071 = .37736$

Chapter 9—Section 9.1

1.

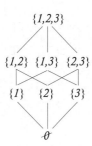

3.

5. (a) greatest element f

(b) least element a

(c) maximal element f, minimal element a

(d) both

7. (a) greatest element g

(b) least element does not exist

(c) maximal element g, minimal elements a, b, c, d

(d) upper

9. (a) greatest element i

(b) least element a

(c) maximal element i, minimal element a

(d) both

11. there is exactly one maximal element.

13. there is exactly one minimal element.

15. Prove $a \wedge (b \wedge c) = (a \wedge b) \wedge c$

b	$\geq (a \wedge b)$	$\geq (a \wedge b) \wedge c$	definition of glb
c	$\geq (a \wedge b) \wedge c$		definition of glb
$\therefore (b \wedge c)$	$\geq (a \wedge b) \wedge c$		definition of glb
a	$\geq (a \wedge b)$	$\geq (a \wedge b) \wedge c$	definition of glb
$\therefore a \wedge (b \wedge c)$	$\geq (a \wedge b) \wedge c$		definition of glb

Similarly, $(a \wedge b) \wedge c \geq a \wedge (b \wedge c)$ and $(a \wedge b) \wedge c = a \wedge (b \wedge c)$.

Prove $a = a \wedge a$

$a \geq a \wedge a$ by definition of greatest lower bound.

Since $a \geq a$ by definition of greatest lower bound, $a \wedge a \geq a$.

Therefore, $a = a \wedge a$.

Prove $a \wedge b = b \wedge a$

$a \geq a \wedge b$ by definition of greatest lower bound.

Similarly, $b \geq b \wedge a$.

Therefore, by definition of greatest lower bound, $a \wedge b \geq b \wedge a$.

Similarly, $b \wedge a \geq a \wedge b$, and $a \wedge b = b \wedge a$.

17. Prove by induction on the number of elements in the subset of the semilattice having a least upper bound. Certainly, a single element has a least upper bound. By definition of semilattice, it is also true for any two elements. Assume any k elements have a least upper bound. If there are only k elements in the semilattice, we are done. Otherwise let $a_1, a_2, a_3, \ldots a_k, a_{k+1}$ be elements of the semilattice and σ_k be the least upper bound of $a_1, a_2, a_3, \ldots a_k$. Then $\sigma_k \vee a_{k+1}$ is the least upper bound of $a_1, a_2, a_3, \ldots a_k, a_{k+1}$. Certainly, $a_i \leq \sigma_k \vee a_{k+1}$ for $1 \leq i \leq k+1$. Assume $a_i \leq b$ for $1 \leq i \leq k+1$. Since $a_i \leq b$ for $1 \leq i \leq k$, $\sigma_k \leq b$. Therefore, since $a_{k+1} \leq b$ and $\sigma_k \leq b$, $\sigma_k \vee a_{k+1} \leq b$. Therefore, $\sigma_k \vee a_{k+1}$ is the least upper bound of $a_1, a_2, a_3, \ldots a_k, a_{k+1}$.

19. Reflexive is obvious. If $A \leq B$ and $B \leq A$, then $A_{ij} \leq B_{ij}$ and $B_{ij} \leq A_{ij}$ $A_{ij} = B_{ij}$ for all $1 \leq i, j \leq n$ and $A = B$. Hence, it is antisymmetric. If $A \leq B$ and $B \leq C$, then $A_{ij} \leq B_{ij}$ for all $1 \leq i, j \leq n$ and $B_{ij} \leq C_{ij}$ for all $1 \leq i, j \leq n$. Therefore, $A_{ij} \leq C_{ij}$ for all $1 \leq i, j \leq n$ and $A \leq C$. Hence, it is transitive.

21. Since $A_{ij} \wedge B_{ij} \leq A_{ij}, B_{ij}$ for all $1 \leq i, j \leq n$, $A \wedge B$ is a lower bound for A and B. Assume D is a lower bound of A and B. If $A_{ij} \wedge B_{ij} = 1$, then either $A_{ij} = 0$ or $B_{ij} = 0$, and $D_{ij} = 0$. Therefore, $D_{ij} \leq A_{ij} \wedge B_{ij}$ for all $1 \leq i, j \leq n$, and $D \leq A \wedge B$.

23. The greatest element is the matrix U where $U_{ij} = 1$ for all $1 \leq i, j \leq n$. The least element is the matrix L where $L_{ij} = 0$ for all $1 \leq i, j \leq n$.

Chapter 9—Section 9.2

1. $\{\{a\}, \{a, b, c\}\}, \{\emptyset, \{a\}, \{a, b\}, \{a, b, c\}\}$,

    $\{\{a\}, \{b\}, \{a, b\}, \{a, b, c\}\}$

3. We need only show that the product of two integers of the form $4k + 1$ has the same form. Let $a = 4m + 1$, and $b = 4n + 1$ be integers of this form. The product
$$ab = (4m + 1)(4n + 1)$$
$$= 16mn + 4m + 4 + 1$$
$$= 4(4mn + m + n) + 1$$
   also has the form $4k + 1$, where $k = 4mn + m + n$.

5. Since $[a] \oplus [b] = [a + b]$ and $[a] \odot [b] = [ab]$ for all positive integers a and b, we have closure.

    Since
$$([a] \oplus [b]) \oplus [c] = [a + b] \oplus [c]$$
$$= [(a + b) + c]$$
$$= [a + (b + c)]$$
$$= [a] \oplus [b + c]$$
$$= ([a] \oplus ([b]) \oplus [c])$$
    addition is associative.

    Since
$$([a] \odot [b]) \odot [c] = [ab] \odot [c]$$
$$= [(ab)c]$$
$$= [a(bc)]$$
$$= [a] \odot [bc]$$
$$= ([a] \odot ([b]) \odot [c])$$
    multiplication is associative.

7. Let a and b be finite products of elements of $\{a_1, a_2, a_3, \ldots, a_k\}$. Then ab is certainly a finite product of elements of $\{a_1, a_2, a_3, \ldots, a_k\}$. Since A^* is a subset of S, we have associativity. Therefore, A^* is a subsemigroup of S. Any subgroup that contains A must contain all finite products of elements of A and, hence, must contain A^*.

9. Since it has been shown in Chapter 2 that composition is associative, we need only show that if f and g are one-to-one correspondences, then so is $f \circ g$. If $(f \circ g)(x) = (f \circ g)(y)$, then $f(g(x)) = f(g(y))$ and $g(x) = g(y)$, since f is one-to-one. But then $x = y$, since g is one-to-one. Therefore, $f \circ g$ is one-to-one. Let $z \in S$. Since f is onto, there exists $y \in S$ such that $f(y) = z$. Since g is onto, there exists $x \in S$ such that $g(x) = y$. Therefore,
$$(f \circ g)(x) = f(g(x))$$
$$= f(y)$$
$$= z$$
    and $f \circ g$ is onto.

11. Since $|x|$ is positive, its absolute value is again $|x|$, and f is a retraction. The retract is the set of nonnegative reals.

    Since $\lfloor x \rfloor$ is an integer, $g(\lfloor x \rfloor) = \lfloor x \rfloor$ and g is a retraction. The retract is the set of integers.

    Since $\lceil x \rceil$ is an integer, $h(\lceil x \rceil) = \lceil x \rceil$ and h is a retraction. The retract is the set of integers.

13. no

15. Let A and B be $n \times n$ be matrices with nonzero determinant. Since
$$\det(AB) = \det(A)\det(B)$$
    then $\det(AB)$ is not zero. Therefore, the set of $n \times n$ matrices with nonzero determinant is closed under multiplication. Therefore, they form a subsemigroup of the set of $n \times n$ matrices. Since $\det(I) = 1$, where I is the multiplicative identity, the set of $n \times n$ matrices with nonzero determinant forms, a monoid. If $\det(A) = 0$ and $\det(B) = 0$, then $\det(AB) = 0$. Therefore, the set of $n \times n$ matrices with zero determinant is closed under multiplication. Therefore, they also from a semigroup.

17. Assume $I * a = a = a * I$ for all $a \in S$ and $e * a = a = a * e$ for all $a \in S$. Therefore,
$$e = I * e \quad \text{since } I \text{ is an identity}$$
$$= I \quad \text{since } e \text{ is an identity}$$

19. $([a] \circ [b]) \circ [c] = [ab] \circ [c] = [abc] = [a] \circ [bc] = [a] \circ ([b] \circ [c])$.

Chapter 9—Section 9.3

1. No

3. No

5. No

7. The sublattice obtained by removing e is isomorphic to the first lattice in Exercise 6.

9. yes (a), (b) (c) (e)

11. yes

13. yes

15. $a \leq a, a \wedge b \leq a$, therefore, $a \vee (a \wedge b) \leq a$. Also $a \leq a \vee (a \wedge b)$. Therefore, $a \vee (a \wedge b) = a$.

17. The miniterms are $p \wedge q \wedge r$, $p \wedge q \wedge \sim r$, $p \wedge \sim q \wedge r$, $p \wedge \sim q \wedge r$, $\sim p \wedge q \wedge r$, $\sim p \wedge q \wedge \sim r$, $\sim p \wedge \sim q \wedge r$, $\sim p \wedge \sim q \wedge \sim r$. The intersection of any two produces a contradiction and, hence, is equal to F, the zero element.

19. b, c, d

21. Since there are 2^n subsets of a set with n elements, this problem is a direct result of the previous problem.

Chapter 9—Section 9.4

1.

$\oplus$	[0]	[1]	[2]	[3]
[0]	[0]	[1]	[2]	[3]
[1]	[1]	[2]	[3]	[0]
[2]	[2]	[3]	[0]	[1]
[3]	[3]	[0]	[1]	[2]

$\odot$	[0]	[1]	[2]	[3]
[0]	[0]	[0]	[0]	[0]
[1]	[0]	[1]	[2]	[3]
[2]	[0]	[2]	[0]	[2]
[3]	[0]	[3]	[2]	[1]

3. [2]

5. [3]

7. yes

9. Let $a = 3m$ and $b = 3n$. Then
$$a + b = 3k + 3n = 3(k + n)$$
so that the sum of a and b is a multiple of 3. Certainly, $0 = 3 \cdot 0$ belongs to the set. The inverse of $3k$ is $(-3)k$. Hence, it is a subgroup. The cosets are $3Z$, $1 + 3Z$, and $2 + 3Z$.

11. By Problem 5, Section 9.2, it is a semigroup. Since
$$[0] + [a] = [0 + a] = [a]$$
[0] is the identity. Since $[a] + [6 - a] = [6] = [0]$, each element has an inverse. Hence, Z_6 is a group under addition. It is not a group under multiplication since [0], [2], and [3] have no inverses.

13. Prove $a^{m+n} = a^m \cdot a^n$ Prove for nonnegative integer n using induction on n. Trivially true for $n = 0$. Assume $a^{m+k} = a^m \cdot a^k$.
$$a^{m+k+1} = a^{m+k} \cdot a$$
$$= a^m \cdot a^k \cdot a$$
$$= a^m \cdot a^{k+1}$$
Assume n is negative.
$$a^{m+n} \cdot a^{-n} = a^{m+n-n} = a^m$$
Therefore,
$$a^{m+n} \cdot a^{-n} \cdot a^n = a^{m+n-n} = a^m a^n$$
and
$$a^{m+n} = a^m a^n$$

15. Prove $(a^{-n})^{-1} = a^n$. $a^n \cdot a^{-n} = a^{n-n} = 0$. Therefore, (a^{-n}) is the inverse of a^n so $(a^{-n})^{-1} = a^n$.

17. Assume $a \cdot a = a$, then $a^{-1} \cdot a \cdot a = a^{-1} \cdot a$ so $1 \cdot a = 1$ and $a = 1$.

19. Certainly, if H is a subgroup of G, then for all $a, b \in H$, $ab \in H$. Conversely, if H is finite, $b^n = 1$ for some n. Therefore, $b^{n-1} = b^{-1}$. Since $b^{n-1} \in H$, $ab^{n-1} = ab^{-1} \in H$, and H is a subgroup.

21. Let H and K be subgroups of G. Let $a, b \in H \cap K$. Then $a, b \in H$ and $a, b \in K$. Therefore, $ab^{-1} \in H$ and $ab^{-1} \in K$. Hence, $ab^{-1} \in H \cap K$ and $H \cap K$ is a subgroup.

23. Let a and b belong to the center of G and $g \in G$. Then $abg = agb = gab$, so ab belongs to the center of G. $b^{-1}g = b^{-1}(g^{-1})^{-1} = (g^{-1}b)^{-1} = (bg^{-1})^{-1} = (g^{-1})^{-1}b^{-1} = gb^{-1}$. Therefore, b^{-1} commutes with every element of G. Hence, b^{-1} belongs to the center of G and G is a group.

Chapter 9—Section 9.5

1. $f(1) = f(1 \cdot 1) = f(1) \cdot f(1)$, so that $f(1)$ is an idempotent. Hence, it is the identity of H, since the only idempotent in H is the identity.

3. If $f(h) = f(h')$, then $ah = ah'$. Therefore, $a^{-1}ah = a^{-1}ah'$ and $h = h'$. Therefore, f is one-to-one. Let $h \in H$. $f(a^{-1}h) = aa^{-1}h = h$. Hence, f is onto.

5. Assume $f : G \to H$ is a monomorphism, and let $k \in \ker(f)$. Therefore, $f(k) = 1$. But since f is a monomorphism, $k = 1$. Therefore, $\ker(f) = \{1\}$. Conversely, assume $\ker(f) = \{1\}$. Assume $f(a) = f(b)$. Therefore, $f(a)(f(b))^{-1} = f(ab^{-1}) = 1$ and $ab^{-1} \in \ker(f) = \{1\}$. Therefore, $ab^{-1} = 1$ and $a = b$. Hence, $f : G \to H$ is a monomorphism.

7. Since ρ_2 and 1 commute with all elements of the octet group, for any element g of the octet group
$$g\{1, \rho_2\} = \{g, g\rho_2\} = \{g, \rho_2 g\} = \{1, \rho_2\}g$$
and $\{1, \rho_2\}$ is a normal subgroup.

9. Let $H = \{1, \delta_1\}$. $\rho_1\{1, \delta_1\} = \{\rho_1, \phi_1\}$ and $\{1, \delta_1\}\rho_1 = \{\rho_1, \phi_2\}$ so H is not normal.

11. We have already shown that $\phi(1)$ is the identity of H and, hence, of $\phi(K)$. Let $h, h' \in \phi(K)$. Then $h = \phi(g)$ and $h' = \phi(g')$ for some $g, g' \in H$.
$$h \cdot h' = \phi(g)\phi(g') = \phi(g \cdot g')$$
so $h \cdot h' \in \phi(K)$. $\phi(g^{-1}) \cdot \phi(g) = \phi(g \cdot g^{-1}) = \phi(1)$ so if $h \in \phi(K)$, then $h^{-1} \in \phi(K)$. Hence, $\phi(K)$ is a subgroup of H.

13. $(1, 2, 4, 8)(3, 7, 9, 5)(6)$

15. $(1, 3, 5)(2, 4, 6)$

17. I

19. $(1, 3)(2)(4, 6, 7)(5)(8)$.

21. Assume $gH = Hg$ for all $g \in G$. Let $h \in H$. Therefore, $hg \in Hg$. Since $Hg = gH$, there exists h' so that $hg = gh'$ and $h = gh'g^{-1} \in gHg^{-1}$. Therefore, $H \subseteq gHg^{-1}$. Assume $ghg^{-1} \in gHg^{-1}$. Since $Hg = gH$, there exists h' so that $gh = gh'$ and $h' = ghg^{-1}$. Therefore, $ghg^{-1} \in H$ and $gHg^{-1} \subseteq H$. Hence, $gHg^{-1} = H$.

23. The set of inner automorphisms is associative since any set of functions from a set to itself is associative under composition.
$$(T_1 \circ T_a)k = T_1(T_a(k))$$
$$= T_1(a^{-1}ka)$$
$$= 1^{-1}(a^{-1}ka)1$$
$$= a^{-1}ka$$
$$= T_a(k)$$

So $T_1 \circ T_a = T_a$. Similarly, $(T_a \circ T_1) = T_a$. Hence, T_1 is the identity.
$$(T_a \circ T_{a^{-1}})(k) = T_a(T_{a^{-1}}(k))$$
$$= T_a((a^{-1})^{-1}ka^{-1})$$
$$= a^{-1}aka^{-1}a$$
$$= k$$
$$= T_1(k)$$

Therefore, $T_a \circ T_{a^{-1}} = T_1$. Similarly, $T_{a^{-1}} \circ T_a = T_1$ and $T_{a^{-1}} = T_a^{-1}$. Hence, the inner automorphisms form a group.

25. Let $a, b \in G$. $aba^{-1}b^{-1} \in C$ if $aba^{-1}b^{-1}C = 1C$. Hence, $aCbCa^{-1}Cb^{-1}C = 1C$. But elements of a group x and y of a group commute if $xyx^{-1}y^{-1} = 1$. Therefore, aC and bC commute.

27. By Cayley's theorem, G is isomorphic to a group of permutations from G onto itself. Since G has n elements, G is isomorphic to a group of permutations on n elements. Hence, G is isomorphic to a subgroup of S_n.

Chapter 9—Section 9.6

1. Yes. Let (v_1, v_2) be a vector. If $v_1 = 3v_2$ and $v_3 = 3v_4$, then $v_1 + v_3 = 3(v_2 + v_4)$ closed under addition. For $a(v_1, v_2) = (av_1, av_2)$, $av_1 = 3v_2$ so closed under scalar multiplication.

3. No. Let (v_1, v_2) and be (v_3, v_4) vectors. If $v_1 = v_2 + 5$ and $v_3 = v_4 + 5$, then $v_1 + v_3 = v_2 + 5 + v_4 + 5 = v_2 + v_4 + 10$. So not closed under vector addition.

5. No. $av_1 = av_2av_3$, so $v_1 = av_2v_3$, which does not satisfy the equation so not closed under scalar multiplication.

7. Yes. If $(0, 0, v_3)$ and $(0, 0, v_3')$ are vectors, then $(0, 0, v_3) + (0, 0, v_3') = (0, 0, v_3 + v_3')$ is also in the set so there is closure under addition. Also $a(0, 0, v_3) = (0, 0, av_3)$ is in the set so there is closure under scalar multiplication.

9. $T\left(\begin{bmatrix} 4 \\ 3 \\ 1 \end{bmatrix}\right) = \begin{bmatrix} 1 & 3 & 4 \\ 0 & 1 & 0 \\ 2 & 2 & 1 \end{bmatrix} \begin{bmatrix} 4 \\ 3 \\ 1 \end{bmatrix} = \begin{bmatrix} 17 \\ 3 \\ 15 \end{bmatrix}$

11. $T\left(\begin{bmatrix} 3 \\ 0 \\ 2 \end{bmatrix}\right) = \begin{bmatrix} 2 & 0 & 1 \\ 1 & 0 & 1 \\ 3 & 1 & 1 \end{bmatrix} \begin{bmatrix} 3 \\ 0 \\ 2 \end{bmatrix} = \begin{bmatrix} 8 \\ 5 \\ 11 \end{bmatrix}$

13. Let $u, u' \in U$ and $v, v' \in V$, $(u + v), (u' + v') \in U + V$. $(u + v) + (u' + v') = (u + u') + (v + v') \in U + V$. Therefore, there is closure under addition. $a(u + v) = (au + av) \in U + V$. Therefore, there is closure under scalar multiplication and $U + V$ is a subspace.

15. Let $a = 0$. $0 \circ (u_1, v_1) = \mathbf{0}$, but by this definition, $0 \circ (0, 1) = (0, 1)$. Therefore, not a vector space.

17. The intersection W of all subspaces containing $v_1, v_2, \cdots v_n$ is the smallest subspace containing $v_1, v_2, \cdots v_n$. Certainly, $a_1v_1 + a_2v_2 + \cdots + a_nv_n$ belongs to any vector space containing $v_1, v_2, \cdots v_n$. Hence, the set U of all vectors of the form $a_1v_1 + a_2v_2 + \cdots + a_nv_n$ is contained in W.

$$a_1v_1 + a_2v_2 + \cdots + a_nv_n + a_1'v_1 + a_2'v_2 + \cdots + a_n'v_n$$
$$= (a_1 + a_1')v_1 + (a_2 + a_2')v_2 + \cdots + (a_n + a_n')v_n$$
$$= a_1''v_1 + a_2''v_2 + \cdots + a_n''v_n$$

where $a_i'' = a_i + a_i'$ is in U. Similarly,

$$a(a_1v_1 + a_2v_2 + \cdots + a_nv_n + a_1'v_1 + a_2'v_2 + \cdots + a_nv_n)$$
$$= aa_1v_1 + aa_2v_2 + \cdots + aa_nv_n$$
$$= a_1''v_1 + a_2''v_2 + \cdots + a_n''v_n$$

where $a_i'' = aa_i$ is in U. Hence, U is a subspace containing $v_1, v_2, \cdots v_n$, and since $U \subseteq W$, $U = W$.

19. Let $u, v \in T(W)$, then $u = T(w)$ and $v = T(w')$ for some $w, w' \in W$. Since W is a vector space, $a_1w + a_2w' \in W$ for all $a_1, a_2 \in R$. Therefore, $T(a_1w + a_2w') \in T(W)$. But $T(a_1w + a_2w') = a_1T(w) + a_2T(w')$ so $a_1T(w) + a_2T(w') \in T(W)$ and $T(W)$ is a subspace.

Chapter 10—Section 10.1

1. $81 = 24 \cdot 3 + 9$. $24 = 2 \cdot 9 + 6$ $9 = 6 + 3$. Therefore, $\gcd(81, 24) = 3$. Backtracking, we have $(-10) \cdot 24 + 3 \cdot 81 = 3$. Thus, $(-20) \cdot 24 + 6 \cdot 81 = 12$, so
$$x = -20 \quad y = 6$$

3. $151 = 2 \cdot 73 + 5$ $73 = 5 \cdot 14 + 3$ $5 = 3 + 2$ $3 = 2 + 1$. Therefore, $\gcd(151, 73) = 1$. Backtracking, we have $73 \cdot 60 + 151(-29) = 1$. Multiplying by 3, we have $73 \cdot 180 + 151(-87) = 1$. So we have
$$x = 180 \quad y = -87$$

5. $78 = 2 \cdot 27 + 24$ $27 = 24 + 3$. Therefore, $\gcd(78, 27) = 3$. Backtracking, we have $27 \cdot (3) + 78(-1) = 3$. Multiplying by 4, we have $27 \cdot (12) + 78(-4) = 3$. So we have
$$x = 12 \quad y = -4$$

7. $x = 15 + 14t \quad y = -78 - 73t$

9. $x = -55 + 38t \quad y = 22 - 15t$

11. $23 = 18 + 5$ $18 = 3 \cdot 5 + 3$ $5 = 3 + 2$ $3 = 2 + 1$. Therefore, $\gcd(23, 18) = 1$. Backtracking, we have $23(-7) + 18 \cdot 9 = 1$. Multiplying by 4, we have $23(-28) + 18 \cdot 36 = 4$. So we have
$$x = -28, y = 36$$

13. $299 = 7 \cdot 39 + 26$ $39 = 26 + 13$. Therefore, $\gcd(299, 39) = 13$. Since 27 is not divisible by 13, there is no solution.

15. $180 = 27 \cdot 6 + 18$ $27 = 18 + 9$. Therefore, $\gcd(180, 27) = 9$. Since 33 is not divisible by 9, there is no solution.

17. $x = -64 + 41t$ $y = 36 - 23t$

19. $x = -2 + 3t$ $y = 6 - 8t$

21. Procedure Integral Solution(m, n, c)
      //Solution for $x \cdot m + y \cdot n = c$//
      $t = m$
      $u = n$
      $x = 0$
      $x' = 1$
      $y = 0$
      $y' = 1$
      While $u \neq 0$
        $q = \lfloor t \div u \rfloor$
        $t = u$
        $u = t - qu$
        $x = x'$
        $x' = x - qx'$
        $y = y'$
        $y' = y - qy'$
      Endwhile
      $d = u$
      let $e = c \div d$
      If $e \neq \lfloor e \rfloor$ no solution
        Otherwise $x = ex$
                  $y = ey$
      Endif
    End

Chapter 10—Section 10.2

1. Each value of n must divide $75 - 35 = 40$. Hence, $n = 1, 2, 4, 5, 8, 10, 20, 40$.

3. 1 is a solution since $27 \cdot 1 - 12 = 15$ is divisible by 15. All solutions have the form
$$1 + k\frac{15}{3} = 1 + 5k$$
The distinct solutions modulo 7 are 1, 6, 11.

5. $81 = 24 \cdot 3 + 9$ $24 = 2 \cdot 9 + 6$ $9 = 6 + 3$. Therefore, 3 is the greatest common divisor of 81 and 24. Backtracking, we have $24 \cdot (-10) + 81 \cdot 3 = 3$. Therefore,
$$24 \cdot (-20) + 81 \cdot 6 = 6$$
and $24 \cdot (-20) \equiv 6 \pmod{81}$. All solutions have the form
$$-20 + k\frac{81}{3} = -20 + 27k$$
The distinct solutions modulo 81 are 7, 34, 61.

7. If a is odd, then $a = 2m + 1$ for some integer m. Therefore,
$$a^2 = (2m + 1)^2 = 4m^2 + 4m + 1$$
and
$$a^2 - 1 = 4m^2 + 4m = 4m(m + 1)$$
But either m or $m + 1$ is even. Therefore, $a^2 - 1$ is divisible by 8.

9. If $a \equiv b \pmod{n}$ where $n = \text{lcm}(n_1, n_2, n_3, \ldots, n_k)$, then $a - b$ is divisible by $\text{lcm}(n_1, n_2, n_3, \ldots, n_k)$, so $a - b$ is divisible by n_i for $1 \leq i \leq k$. Therefore, $a \equiv b \pmod{n_i}$. Conversely, assume $a \equiv b \pmod{n_i}$, for $1 \leq i \leq k$. Then $a - b$ is divisible for n_i for $1 \leq i \leq k$. By definition of lcm, $a - b$ is divisible by $\text{lcm}(n_1, n_2, n_3, \ldots, n_k)$, and $a \equiv b \pmod{n}$.

11. Prove by induction on m. Certainly, $a \equiv b \pmod{n}$. So true for $m = 1$. Assume true for $n = k$, so $a^k \equiv b^k \pmod{n}$. By Theorem 3.55, if $a \equiv b \pmod{n}$ and $a^k \equiv b^k \pmod{n}$, then $a^{k+1} \equiv b^{k+1} \pmod{n}$, and so true for $n = k + 1$.

13. If $a \equiv b \left(\mod \dfrac{n}{\gcd(c, n)}\right)$, then $a - b = k\dfrac{n}{\gcd(c, n)}$ for some k. Therefore, $ac - bc = k\dfrac{n}{\gcd(c, n)}c$. Since $\gcd(c, n) | c$, $\dfrac{c}{\gcd(c, n)}$ is an integer, say q. Therefore, $ac - bc = kqn$ and $ac - bc \mod n$. Conversely, if $ac \equiv bc \mod n$, then $ac - bc = kn$ for some k. Since $(a - b)c = kn$, $(a - b)\dfrac{c}{\gcd(c, n)} = \dfrac{n}{\gcd(c, n)}$. Since $\dfrac{c}{\gcd(c, n)}$ and $\dfrac{n}{\gcd(c, n)}$ are relatively prime, $\dfrac{n}{\gcd(c, n)}$ divides $(a - b)$, so $a \equiv b \left(\mod \dfrac{n}{\gcd(c, n)}\right)$.

Chapter 10—Section 10.3

1. Since $25 \cdot 9 \equiv 9 \pmod{12}$ and $12 \cdot 13 \equiv 6 \pmod{25}$,
$$x \equiv 25 \cdot 9 + 12 \cdot 13 \pmod{300} \equiv 81 \pmod{300}$$

3. Since $16 \cdot a \equiv a \pmod{15}$ and $-15 \cdot b \equiv b \pmod{16}$,
$$x \equiv 16a - 15b \pmod{240}$$

5. Since $21 \cdot 10 \equiv 2 \pmod{13}$ and $13 \cdot 2 \equiv 5 \pmod{21}$,
$$x \equiv 21 \cdot 10 + 13 \cdot 2 \pmod{273} \equiv 236 \pmod{273}$$

7. Since $-48 \cdot a \equiv a \pmod{7}$ and $49 \cdot b \equiv b \pmod{16}$,
$$x \equiv -48a + 49b \pmod{112}$$

9. $x \equiv 4 \pmod{15}$, $x \equiv 3 \pmod 8$, and $x \equiv 10 \pmod{23}$. Since $184 \cdot 1 \equiv 4 \pmod{15}$, $345 \cdot 3 \pmod 8$, and $120 \cdot 2 \equiv 10 \pmod{23}$,

$$x = 184 \cdot 1 + 345 \cdot 3 + 120 \cdot 2 = 1459 \text{ marbles}$$

97 rows of 15, 182 rows of 8, and 63 rows of 23.

11. $x \equiv 21 \pmod{36}$ implies $x = 21 + k \cdot 36$ for some k. Thus, $21 + k \cdot 36 \equiv 5 \pmod 8$, and there is a k' such that $21 + k \cdot 36 = 5 + k' \cdot 8$ or $8k' - 36k = 16$. Since $\gcd(8, 36) \mid 16$, there is a solution $k' = -16$ and $k = -4$. $x = (21 + (-4) \cdot 36) \pmod{72} = 21 \pmod{72}$.

13. No solution since $\gcd(49, 14)$ does not divide $19 - 10$.

Chapter 10—Section 10.4

1. | n | 1 2 3 4 5 6 7 8 9 10 11 12 13 14 15 16 17 18 19 20 |
    |---|---|
    | $\phi(n)$ | 1 1 2 2 4 2 6 4 6 4 10 4 12 6 8 8 16 6 18 8 |

3. $m = 4$ $n = 2$. $\phi(8) = 4$, $\phi(2) = 1$, $\phi(4) = 2$

5. $\phi(64) = \phi(2^6) = 2^5 = 32$

7. $\phi(4049) = 4048$ since 4049 is prime.

9. If p is a prime, by Wilson's Theorem, $(p-1)! \equiv 1 \bmod (p)$. Therefore, $(p-2)!(p-1) \equiv 1 \bmod (p)$. But $(p-1) \equiv -1 \bmod (p)$. Therefore, $(p-2)! \equiv -1 \bmod (p)$.

11. $S(n) = \sin\left[\pi \cdot \dfrac{(n-1)! + 1}{n}\right] = 0$ if and only if $\dfrac{(n-1)! + 1}{n} = k$ if and only if $(n-1)! + 1 = kn$ if and only if $(n-1)! \equiv -1 \bmod (n)$ if and only if n is prime.

13. $\phi(2025) = \phi(25)\phi(81) = 20 \cdot 54 = 1080$

15. By Wilson's Theorem, p is prime if and only if $(p-1)! \equiv -1 \pmod p$ if and only if $(p-1)! + 1 \equiv 0 \pmod p$ if and only if $(p-1)! + 1$ is divisible by p.

Chapter 10—Section 10.5

1. If a is not equal to 0 modulo p, then $a \equiv b \pmod p$, where $1 \le b \le p - 1$. By Fermat's Little Theorem, $b^{p-1} \equiv 1 \pmod p$. But $a^{p-1} \equiv b^{p-1} \pmod p$. Therefore, $a^{p-1} \equiv 1 \pmod p$.

3. Since $J \equiv 0 \pmod{p-1}$, $J = k(p-1)$ for some integer k. Let $1 \le a \le p-1$. Since

$$a^{p-1} \equiv 1 \pmod p, \; a^J = a^{k(p-1)} = (a^{(p-1)})^k \equiv 1^k \equiv 1 \pmod p$$

Therefore,

$$1^J + 2^J + \cdots + (p-1)^J \equiv 1 + 1 + \ldots + 1 \pmod p$$
$$\equiv p - 1 \pmod p$$
$$\equiv -1 \pmod p$$

5. Using induction, the theorem is certainly true when $a = 1$. Assume $k^{p-1} \equiv 1 \pmod p$. Hence, $k^p \equiv k \pmod p$. $(k+1)^p = k^p + pk^{p-1} + \cdots + \binom{p}{r}k^{p-r} + \cdots + 1$. But $\binom{p}{r}$ is divisible by p for $1 \le r \le k-1$ since p is in the numerator and cannot cancel with anything in the denominator. Therefore, $(k+1)^p \equiv k^p + 1 \pmod p \equiv k + 1 \pmod p$, and $(k+1)^{p-1} \equiv 1 \pmod p$.

7.
$$x \equiv 7^{\phi(25)-1} \cdot 8 \pmod{25}$$
$$\equiv 7^{19} \cdot 8 \pmod{25}$$
$$\equiv 19 \pmod{25}$$

9. Since $a^p \equiv a \pmod p$, $b^p \equiv b \pmod p$, and $(a+b)^p \equiv a + b \pmod p$, then $(a+b)^p \equiv a^p + b^p \pmod p$.

11. | a | 1 | 2 | 4 | 5 | 7 | 8 |
    |---|---|---|---|---|---|---|
    | $ord_n a$ | 1 | 6 | 3 | 6 | 3 | 2 |

13. | a | 1 | 2 | 4 | 5 | 7 | 8 | 10 | 11 | 13 |
    |---|---|---|---|---|---|---|---|---|---|
    | $ord_n a$ | 1 | 18 | 9 | 18 | 9 | 6 | 3 | 18 | 9 |

a	14	16	17	19	20	22	23	25	26
$ord_n a$	18	9	6	3	18	9	18	9	2

15. If b and 11 are relatively prime, then $b^{10} \equiv 1 \pmod{11}$. $(b^{10})^k \equiv 1 \pmod{11}$ and $b^{10k} - 1$ is divisible by 11.

17. $7^{\phi(5)} = 7^4 \equiv 1 \pmod 5$

19. Since 2 and 5 are relatively prime, $7^4 \equiv 1 \pmod{10}$.

21. The last base ten digit of 7^{4000} is 1.

23. $7^2 = 49 \equiv 1 \pmod 4$

25. Since 4 and 25 are relatively prime, $7^{20} \equiv 1 \pmod{100}$.

27. $\phi(100) = 40$. Therefore,

$$[[3^{275}]]_{100} = [[3^{6 \cdot 40 + 35}]]_{100}$$
$$= [[3^{6 \cdot 40} 3^{35}]]_{100}$$
$$= [[3^{35}]]_{100}$$
$$= [[(3^5)^7]]_{100}$$
$$= [[(43)^7]]_{100}$$
$$= [[((43)^2)^3 \cdot 43]]_{100}$$
$$= [[((49)^3 \cdot 43)]]_{100}$$
$$= [[((49) \cdot 43)]]_{100}$$
$$= [[7]]_{100}$$
$$= 7$$

29. $\phi(83) = 82$. Therefore,

$$[[11^{24681}]]_{83} = [[11^{300 \cdot 82 + 81}]]_{83} = [[11^{300 \cdot 82 \cdot 1} 11^{81}]]_{83}$$
$$= [[11^{81}]]_{83} = [[(11^2)^{40} \cdot 11]]_{83}$$
$$= [[(38)^{40} \cdot 11]]_{83} = [[(33)^{20} \cdot 11]]_{83}$$
$$= [[((10)^{10}) \cdot 11]]_{83} = [[(17)^5 \cdot 11]]_{83}$$
$$= [[40^2 \cdot 17 \cdot 11]]_{83} = [[23 \cdot 17 \cdot 11]]_{83}$$
$$= [[59 \cdot 11]]_{83} = 68$$

31. 535044134

33. $3^{198} \equiv 1 \pmod{199}$, $3^{99} \equiv -1 \pmod{199}$, $3^{66} \equiv 106 \pmod{199}$, $3^{18} \equiv 125 \pmod{199}$. Therefore, 199 is prime.

35. $2^{18} \equiv -1 \pmod{37}$, $2^{12} \equiv 26 \pmod{37}$. Therefore, 37 is prime.

$3^{99} \equiv -1 \pmod{199}$, $3^{66} \equiv 106 \pmod{199}$, $3^{18} \equiv 125 \pmod{199}$. Therefore, 199 is prime.

37. $3^{256} \equiv 1 \pmod{257}$, $3^{128} \equiv -1 \pmod{257}$. Therefore, 257 is prime.

Chapter 11—Section 11.1

1. yes

3. yes

5. yes

7. $r - 3 = 0$. Therefore, $r = 3$ and $a_n = c_1 \cdot 3^n$.

9. $r^2 = -r + 6$. Therefore, $r^2 + r - 6 = 0$. Solving, we have $r = -3$ and $r = 2$. Therefore, $a_n = c_1 \cdot (-3)^n + c_2 \cdot 2^n$.

11. $r + 7 = 0$. Therefore, $r = -7$ and $a_n = c_1 \cdot (-7)^n$.

13. $r^2 = 2r + 8$. Therefore, $r^2 - 2r - 8 = 0$. Solving, we have $r = 4$ and $r = -2$. Therefore, $a_n = c_1 \cdot 4^n + c_2 \cdot (-2)^n$.

15. $r = -4$ and $a_n = c_1 \cdot (-4)^n$. Therefore, $a_0 = 1 = c_1 \cdot (-4)^0$ and $c_1 = 1$. Therefore, $a_n = (-4)^n$.

17. $r^2 = 5r - 6$. Therefore, $r^2 - 5r + 6 = 0$. Solving we have $r = 3$ and $r = 2$, and $a_n = c_1 \cdot 3^n + c_2 \cdot 2^n$. $a_0 = 2 = c_1 + c_2$ $a_1 = 5 = 3 \cdot c_1 + 2 \cdot c_2$. Solving, we have $c_1 = c_2 = 1$ and $a_n = 3^n + 2^n$.

19. $r^2 = 9r - 20$. Therefore, $r^2 - 9r + 20 = 0$. Solving we have $r = 5$ and $r = 4$, and $a_n = c_1 \cdot 5^n + c_2 \cdot 4^n$. $a_0 = 0 = c_1 + c_2$, $a_1 = 5 = 5 \cdot c_1 + 4 \cdot c_2$. Solving, we have $c_1 = 5$, $c_2 = -5$ and $a_n = 5 \cdot 5^n - 5 \cdot 4^n$.

21. $r^2 = 2r + 2$. Therefore, $r^2 - 2r - 2 = 0$. Solving we have $r = \pm\sqrt{3}$, and $a_n = c_1 \cdot (1 + \sqrt{3})^n + c_2(1 - \sqrt{3})^n$. $a_0 = 2 = c_1 + c_2$, $a_1 = 1 = c_1 \cdot (1 + \sqrt{3}) + c_2 \cdot (1 - \sqrt{3})$. Solving, we have $c_1 = \dfrac{6 - \sqrt{3}}{2}$, $c_2 = \dfrac{6 + \sqrt{3}}{2}$, and $a_n = \dfrac{6 - \sqrt{3}}{2} \cdot (1 + \sqrt{3})^n + \dfrac{6 + \sqrt{3}}{2}(1 - \sqrt{3})^n$.

23. $r^2 = 6r - 9$. Therefore, $r^2 - 6r + 9 = 0$. Solving, we have $r = 3, 3$ and $a_n = c_1 \cdot 3^n + c_2 n \cdot 3^n$. $a_0 = 3 = c_1$, $a_1 = 21 = 3c_1 + 3c_2$. Solving, we have $c_1 = 3$, $c_2 = 4$, and $a_n = 3 \cdot 3^n + 4n \cdot 3^n$.

25. $r^2 = -4r - 4$. Therefore, $r^2 + 4r + 4 = 0$. Solving, we have $r = -2, -2$ and $a_n = c_1 \cdot (-2)^n + c_2 n \cdot (-2)^n$. $a_0 = 1 = c_1$, $a_1 = -8 = -2c_1 - 2c_2$. Solving, we have $c_1 = 1$, $c_2 = 3$, and $a_n = (-2)^n + 3n \cdot (-2)^n$.

27. $r^2 = -2r - 1$. Therefore, $r^2 + 2r + 1 = 0$. Solving, we have $r = -1, -1$ and $a_n = c_1 \cdot (-1)^n + c_2 \cdot (-1)^n$. $a_0 = 2 = c_1$, $a_1 = 6 = -c_1 - c_2$. Solving, we have $c_1 = 2$, $c_2 = -8$, and $a_n = 2 \cdot (-1)^n - 8n \cdot (-1)^n$.

29. $r^2 = r - 1$. Therefore, $r^2 - r + 1 = 0$. Solving, we have $r = \dfrac{1}{2} \pm \dfrac{\sqrt{3}}{2} i$. Therefore, $\cos(\theta) = \dfrac{1}{2}$, $\sin(\theta) = \dfrac{\sqrt{3}}{2}$ and $\theta = \dfrac{\pi}{3}$ and $\rho = 1$. Therefore,

$$a_n = (c_1 \cos\left(\tfrac{n\pi}{3}\right) + c_2 \sin\left(\tfrac{n\pi}{3}\right)$$
$$a_0 = 1 = c_1$$
$$a_1 = 2$$
$$= \tfrac{1}{2} c_1 + \tfrac{\sqrt{3}}{2} c_2$$
$$= \tfrac{1}{2} + \tfrac{\sqrt{3}}{2} c_2$$

Solving, $c_2 = \sqrt{3}$. Therefore, $a_n = (1 \cos\left(\tfrac{n\pi}{3}\right) + \sqrt{3} \sin\left(\tfrac{n\pi}{3}\right))$.

31. $r^2 = -2\sqrt{2} r - 4$. Therefore, $r^2 + 2\sqrt{2} r + 4 = 0$. Solving, we have

$$r = -\sqrt{2} \pm \sqrt{2} i$$
$$= 2\left(-\dfrac{\sqrt{2}}{2} \pm \dfrac{\sqrt{2}}{2}\right) i$$

Therefore, $\cos(\theta) = -\dfrac{\sqrt{2}}{2}$, $\sin(\theta) = \dfrac{\sqrt{2}}{2}$ and $\theta = \dfrac{3\pi}{4}$ and $\rho = 2$. Therefore,

$$a_n = 2^n \left(c_1 \cos\left(\dfrac{3n\pi}{4}\right) + c_2 \sin\left(\dfrac{3n\pi}{4}\right)\right)$$

$a_0 = 1 = c_1$. $a_1 = 0 = -\dfrac{\sqrt{2}}{2} + \dfrac{\sqrt{2}}{2} c_2$. Therefore, $c_2 = 1$. Therefore,

$$a_n = 2^n \left(\cos\left(\dfrac{3n\pi}{4}\right) + \sin\left(\dfrac{3n\pi}{4}\right)\right).$$

33. $r^4 - 5r^2 + 4 = 0$. Solving, we have $r = \pm 1, \pm 2$. Therefore,
$$a_n = c_1 + c_2(-1)^n + c_3 \cdot 2^n + c_4(-2)^n$$

35. $r^4 = 1$. Solving, we have $r = 1, -1, i, -i$. Therefore,
$$a_n = c_1 + c_2 \cdot (-1)^n + c_3 \cos\left(\frac{n\pi}{2}\right) + c_4 \sin\left(\frac{n\pi}{2}\right)$$

37.
$$F(n) = \binom{n-1}{0} + \binom{n-2}{1} + \binom{n-3}{2}$$
$$+ \binom{n-4}{3} + \cdots$$

$$F(1) = \binom{0}{0} = 1$$

$$F(2) = \binom{1}{0} = 1$$

$$F(k) = \binom{k-1}{0} + \binom{k-2}{1} + \binom{k-3}{2}$$
$$+ \binom{k-4}{3} + \cdots + \binom{k-i}{i-1} + \cdots$$

$$F(k-1) = \binom{k-2}{0} + \binom{k-3}{1} + \binom{k-4}{2}$$
$$+ \binom{k-5}{3} + \cdots + \binom{k-i+1}{i-1} + \cdots$$

$$F(k-2) = \binom{k-3}{0} + \binom{k-4}{1} + \binom{k-5}{2}$$
$$+ \binom{k-6}{3} + \cdots + \binom{k-i+2}{i-1} + \cdots$$

Using $\binom{n}{r} = \binom{n-1}{r-1} + \binom{n-1}{r}$, $\binom{k-i}{i-1}$
$$= \binom{k-i-1}{i-2} + \binom{k-i-1}{i-1}$$

$$F(k) = \binom{k-1}{0} + \sum_{i=2}^{\infty} \binom{k-i}{i-1}$$

$$F(k-1) = \binom{k-2}{0} + \sum_{i=2}^{\infty} \binom{k-i-1}{i-1}$$

$$F(k-2) = \sum_{i=1}^{\infty} \binom{k-i-2}{i-1} = \sum_{i=1}^{\infty} \binom{k-i-1}{i-2}$$

$$F(k-1) + F(k-2) = \binom{k-2}{0} + \sum_{i=2}^{\infty} \binom{k-i-1}{i-1}$$
$$+ \sum_{i=2}^{\infty} \binom{k-i-1}{i-2}$$
$$= \binom{k-2}{0} + \sum_{i=2}^{\infty} \binom{k-i}{i-1}$$
$$= \binom{k-1}{0} + \sum_{i=2}^{\infty} \binom{k-i}{i-1}$$
$$= F(k)$$

39. Using induction, for $n = 1$, $(\cos(\theta) + i\sin(\theta))^1 = (\cos(1 \cdot \theta)) + i(\sin(1 \cdot \theta))$.

Assume $(\cos(\theta) + i\sin(\theta))^k = \cos(k\theta) + i\sin(k\theta)$.

$$(\cos(\theta) + i\sin(\theta))^{k+1} = (\cos(\theta) + i\sin(\theta))^k$$
$$\times (\cos(\theta) + i\sin(\theta))$$
$$= (\cos(k\theta) + i\sin(k\theta))$$
$$\times (\cos(\theta) + i\sin(\theta))$$
$$= (\cos((k+1)\theta) + i\sin$$
$$\times ((k+1)\theta))$$

Chapter 11—Section 11.2

1. $r = 2$. Therefore, $a_n = c_1 \cdot 2^n + k$. $k = 2k + 5$. Therefore, $k = -5$ and $a_n = c_1 \cdot 2^n - 5$.

3. $r^2 = -r + 12$ or $r^2 + r - 12 = 0$. Solving, $r = -4, 3$. Therefore,
$$a_n = c_1 \cdot 3^n + c_2 \cdot (-4)^n + k \cdot 2^n. \ k \cdot 2^n$$
$$= -k \cdot 2^{n-1} + 12k \cdot 2^{n-2} + 2^n$$
or
$$4k \cdot 2^{n-2} = -2k \cdot 2^{n-2} + 12k \cdot 2^{n-2} + 4 \cdot 2^{n-2}$$
Solving, $k = -\frac{2}{3}$ and
$$a_n = c_1 \cdot 3^n + c_2 \cdot (-4)^n - \frac{2}{3} \cdot 2^n$$

5. $r^2 = 4r - 4$ or $r^2 - 4r + 4 = 0$. Solving, $r = 2, 2$. Therefore,
$$a_n = c_1 \cdot 2^n + c_2 n \cdot 2^n + k_1 n^2 + k_2 n + k_3$$
$$k_1 n^2 + k_2 n + k_3 = 4[k_1(n-1)^2 + k_2(n-1) + k_3]$$
$$- 4[k_1(n-2)^2 + k_2(n-2) + k_3] + n^2$$
$$= 4[k_1 n^2 - 2k_1 n + k_1 + k_2 n - k_2 + k_3]$$
$$- 4[k_1 n^2 - 4k_1 n + 4k_1 + k_2 n - 2k_2$$
$$+ k_3] + n^2$$

Equating coefficients of n^2, $k_1 = 1$

Equating coefficients of n, $k_2 = -8k_1 + 4k_2 + 16k_1 - 4k_2$ so $k_2 = 8$

Equating constants, $k_3 = 4k_1 - 4k_2 + 4k_3 - 16k_1 + 8k_2 - 4k_3$ so $k_3 = 20$

$$a_n = c_1 \cdot 2^n + c_2 n \cdot 2^n + n^2 + 8n + 20$$

7. $r = 4$. Therefore, $a_n = c_1 \cdot 4^n + k_1 n 4^n$
$$k_1 n 4^n = 4k_1(n-1)4^{n-1} + 4^n$$
so $k_1 = 1$ and
$$a_n = c_1 \cdot 4^n + n4^n$$

9. $r^2 = -9$. Solving, $r = \pm 3i$. Therefore,
$$a_n = 3^n c_1 \left(c_1 \cos\left(\frac{n\pi}{2}\right) + c_2 \sin\left(\frac{n\pi}{2}\right)\right) + k3^n$$
$k3^n = -9k3^{(n-2)} + 3^n$ or $k3^n = -k3^n + 3^n$. Solving, $k = \frac{1}{2}$ so
$$a_n = 3^n c_1 \left(c_1 \cos\left(\frac{n\pi}{2}\right) + c_2 \sin\left(\frac{n\pi}{2}\right)\right) + \frac{1}{2}3^n$$

11. $r = 2$. Therefore,
$$a_n = c_1 \cdot 2^n + kn2^n$$
$kn2^n = 2k(n-1)2^{n-1} + 2^n$ or $kn2^{n-1} = k(n-1)2^n + 2^n$. Solving, $k = 1$. Therefore,
$$a_n = c_1 \cdot 2^n + n \cdot 2^n$$

13. $r^2 = 3r - 2$ or $r^2 + 3r + 2 = 0$. Solving, $r = -1, -2$. Therefore,
$$a_n = c_1 + c_2 \cdot 2^n + kn$$
$kn = 3k(n-1) - 2k(n-2) + 5$. Solving, $k = -5$ and
$$a_n = c_1 + c_2 \cdot 2^n - 5n$$

15. $r^2 = 2\sqrt{3}r - 4$ or $r^2 - 2\sqrt{3}r + 4 = 0$. Solving, $r = 2\dfrac{\sqrt{3} \pm 2i}{2}$. Therefore, $\rho = 2, \theta = \dfrac{\pi}{6}$ and $a_n = 2^n\left(c_1 \cos\left(\dfrac{n\pi}{6}\right) + c_2 \sin\left(\dfrac{n\pi}{6}\right)\right) + k$
$k = 2\sqrt{3}k - 4k + 3$. Solving, $k = \dfrac{3}{5 - 2\sqrt{3}}$ and
$$a_n = 2^n\left(c_1 \cos\left(\dfrac{n\pi}{6}\right) + c_2 \sin\left(\dfrac{n\pi}{6}\right)\right) + \dfrac{3}{5 - 2\sqrt{3}}$$

17. $r^2 = -1$. Solving, $r = \pm i$.
$$a_n = c_1 \cos\left(\dfrac{n\pi}{2}\right) + c_2 \sin\left(\dfrac{n\pi}{2}\right) + k_1 n \cos\left(\dfrac{n\pi}{2}\right)$$
$$+ k_2 n \sin\left(\dfrac{n\pi}{2}\right) k_1 n \cos\left(\dfrac{n\pi}{2}\right)$$
$$+ k_2 n \sin\left(\dfrac{n\pi}{2}\right)$$
$$= k_1(n-1) \cos\left(\dfrac{(n-2)\pi}{2}\right)$$
$$+ k_2(n-1) \sin\left(\dfrac{(n-2)\pi}{2}\right) + \cos\left(\dfrac{n\pi}{2}\right)$$
Letting $n = 0$, $0 = -k_1 \cos(-\pi) + \cos(0)$, so $k_1 - 1$.
Letting $n = 1$, $k_2 \sin\left(\dfrac{\pi}{2}\right) = 0$, so $k_2 = 0$. Therefore,
$$a_n = c_1 \cos\left(\dfrac{n\pi}{2}\right) + c_2 \sin\left(\dfrac{n\pi}{2}\right) - n \cos\left(\dfrac{n\pi}{2}\right)$$

19. $r^3 = 6r^2 - 12r + 8$ or $r^3 - 6r^2 + 12r - 8 = 0$. Solving, $r = 2, 2, 2$. Therefore,
$$a_n = c_1 2^n + c_2 n 2^n + c_3 n^2 2^n + kn^3 2^n$$
$kn^3 2^n = 6k(n-1)^3 2^{n-1} - 12k(n-2)^3 2^{n-2}$
$\qquad + 8k(n-3)^3 2^{n-3} + 2^n$

or
$$kn^3 2^n = 3k(n-1)^3 2^n - 3k(n-2)^3 2^n$$
$$+ k(n-3)^3 2^n + 2^n$$
Therefore,
$$kn^3 = 3k(n-1)^3 - 3k(n-2)^3 + k(n-3)^3 + 1$$

Letting $n = 0$,
$$0 = -3k + 24k - 27k + 1$$
Solving, $k = \dfrac{1}{6}$. Therefore,
$$a_n = c_1 2^n + c_2 n 2^n + c_3 n^2 2^n + \dfrac{1}{6} n^3 2^n$$

21. $r = 1$. Therefore,
$$a_n = c_1 + k_1 n + k_2 n^2 + k_3 n^3$$
$k_1 n + k_2 n^2 + k_3 n^3 = k_1(n-1) + k_2(n-1)^2$
$\qquad + k_3(n-1)^3 + n^2$
$\qquad = k_1(n-1) + k_2(n^2 - 2n + 1)$
$\qquad + k_3(n^3 - 3n^2 + 3n - 1) + n^2$

Checking coefficients of n^3, $k_3 = k_3$
Checking coefficients of n^2, $k_2 = k_2 - 3k_3 + 1$, so $k_3 = \dfrac{1}{3}$
Checking coefficients of n, $k_1 = k_1 - 2k_2 + 3k_3$, so $k_2 = \dfrac{1}{2}$
Letting $n = 0$, $0 = -k_1 + k_2 - k_3$, so $k_1 = \dfrac{1}{6}$. Therefore,
$$a_n = c_1 + \dfrac{1}{6}n + \dfrac{1}{2}n^2 + \dfrac{1}{3}n^3$$
$a_n = 1 = c_1 + \dfrac{1}{6} + \dfrac{1}{2} + \dfrac{1}{3}$ so $c_1 = 0$.
$$a_n = \dfrac{1}{6}n + \dfrac{1}{2}n^2 + \dfrac{1}{3}n^3$$
$$= \dfrac{n(n+1)(2n+1)}{6}$$

23. $r = 1$. Therefore,
$$a_n = c + k \cdot 2^n$$
$k \cdot 2^n = k \cdot 2^{n-1} + 2^n$ or $2k \cdot 2^{n-1} = k \cdot 2^{n-1} + 2 \cdot 2^n$
so $2k = k + 2$ or $k = 2$. Therefore,
$$a_n = c + 2 \cdot 2^n$$
$$= c + 2^{n+1}$$
$a_1 = 2 = c + 2^2$ so $c = -2$ and
$$a_n = 2^{n+1} - 2$$

25. $r = 1$. Therefore,
$$a_n = c + k_1 n + k_2 n^2 + k_3 n^3 + k_4 n^4$$
$k_1 n + k_2 n^2 + k_3 n^3 + k_4 n^4$
$= k_1(n-1) + k_2(n-1)^2 + k_3(n-1)^3$
$\quad + k_4(n-1)^4 + n^3$
$= k_1(n-1) + k_2(n^2 - 2n + 1)$
$\quad + k_3(n^3 - 3n^2 + 3n - 1)$
$\quad + k_4(n^4 - 4n^3 + 6n^2 - 4n + 1) + n^3$

Checking coefficients of n^4, $k_4 = k_4$
Checking coefficients of n^3, $k_3 = k_3 - 4k_4 + 1$, so $k_4 = \dfrac{1}{4}$
Checking coefficients of n^2, $k_2 = k_2 - 3k_3 + 6k_4$, so $k_3 = \dfrac{1}{2}$

Checking coefficients of n, $k_1 = k_1 - 2k_2 + 3k_3 - 4k_4$ so $k_2 = \frac{1}{4}$

Letting $n = 0$, $0 = -k_1 + k_2 - k_3 + k_4$ so $k_1 = 0$

$$a_n = c + \frac{1}{4}n^2 + \frac{1}{2}n^3 + \frac{1}{4}n^4$$

$a_1 = 1 = c + \frac{1}{4} + \frac{1}{2} + \frac{1}{4}$ so $c = 0$ and

$$a_n = \frac{1}{4}n^2 + \frac{1}{2}n^3 + \frac{1}{4}n^4$$
$$= \frac{n^2(n+1)^2}{4}$$

27. $r = 1$. Therefore,

$$a_n = c + k_1 n \cdot 2^{n-1} + k_2 \cdot 2^{n-1}$$
$$k_1 n \cdot 2^{n-1} + k_2 \cdot 2^{n-1} = k_1(n-1) \cdot 2^{n-2}$$
$$+ k_2 \cdot 2^{n-2} + n \cdot 2^{n-1}$$

so

$$2k_1 n \cdot 2^{n-2} + 2k_2 \cdot 2^{n-2}$$
$$= k_1(n-1) \cdot 2^{n-2} + k_2 \cdot 2^{n-2} + 2n \cdot 2^{n-2}$$

Hence,

$$2k_1 n + 2k_2 = k_1(n-1) + k_2 + 2n$$

Checking coefficients of n, $2k_1 = k_1 + 2$ so $k_1 = 2$

Letting $n = 0$, $2k_2 = -k_1 + k_2$ so $k_2 = -2$. Therefore,

$$a_n = c + 2n \cdot 2^{n-1} - 2 \cdot 2^{n-1}$$
$$= c + (n-1)2^n$$

$a_1 = 1 = c + 0 = c$. Therefore,

$$a_n = 1 + (n-1)2^n$$

Chapter 11—Section 11.3

1. $\Delta^2(x^3) = \Delta(3x^2 + 3x + 1) = 3(2x+1) + 3 = 6x + 6$. At $x = 1$, $\Delta^2 x^3 = 12$.

3.
$$\Delta^2 E^2(x^4) = E^2 \Delta^2(x^4)$$
$$= E^2 \Delta(4x^3 + 6x^2 + 4x + 1)$$
$$= E^2(4(3x^2 + 3x + 1) + 6(2x+1) + 4)$$
$$= E^2(12x^2 + 24x + 14)$$
$$= 12(x+2)^2 + 24(x+2) + 14$$

At $x = 4$,
$$\Delta^2 E^2(x^4) = 12 \cdot 6^2 + 24 \cdot 6 + 14 = 590$$

at $n = 1$, $\Delta^2 E(n!) = 14$

5. $\Delta((3x+2)(x-3)) = (3x+2) + (x-2) \cdot 3 = 6x - 4$

7.
$$\Delta\left(\frac{x-6}{3x+4}\right) = \left(\frac{(3x+4) - (x-6) \cdot 3}{(3x+4)(3(x+1)+4)}\right)$$
$$= \left(\frac{22}{(3x+4)(3x+7)}\right)$$

9.
$$\Delta(x!) = (x+1)! - x!$$
$$= (x+1) \cdot x! - x!$$
$$= x \cdot x! + x! - x!$$
$$= x \cdot x!$$

11.
$$\Delta \frac{f(x)}{g(x)}$$
$$= \frac{f(x+1)}{g(x+1)} - \frac{f(x)}{g(x)}$$
$$= \frac{f(x+1)g(x) - f(x)g(x+1)}{g(x+1)g(x)}$$
$$= \frac{f(x+1)g(x) - f(x)g(x) + f(x)g(x) - f(x)g(x+1)}{g(x+1)g(x)}$$
$$= \frac{(f(x+1) - f(x))g(x) - f(x)(g(x+1) - g(x))}{g(x+1)g(x)}$$
$$= \frac{\Delta f(x) \cdot g(x) - f(x) \cdot \Delta g(x)}{g(x+1)g(x)}$$

Chapter 11—Section 11.4

1. $\Delta^2(3x^{(3)} - 5x^{(2)} + 4) = \Delta(9x^{(2)} - 10x) = 18x - 10$

3.
$$\Delta^3(x^{(5)} + 3x^{(4)} - 3x^{(3)} - 2x^{(2)} + 4x - 1)$$
$$= \Delta^2(5x^{(4)} + 12x^{(3)} - 9x^{(2)} - 4x + 4)$$
$$= \Delta(20x^{(3)} + 36x^{(2)} - 18x - 4)$$
$$= 60x^{(2)} + 72x - 18$$

5.
$$\Delta((x^{(3)} - 2x^{(2)} - 4)(x^{(4)} + 2x^{(3)} - 6x^{(2)} + x - 3))$$
$$= (x^{(3)} - 2x^{(2)} - 4)(4x^{(3)} + 6x^{(2)} - 12x + 1)$$
$$+ ((x+1)^{(4)} + 2(x+1)^{(3)} - 6(x+1)^{(2)}$$
$$+ x - 2)(3x^{(2)} - 4x)$$

7. $\Delta \frac{x^{(4)} - 3x^{(2)} + 3x}{x^{(4)} - 3x^{(3)} + 6x^{(2)}} =$
$$\frac{(x^{(4)} - 3x^{(3)} + 6x^{(2)})(4x^{(3)} - 6x + 3) - (x^{(4)} - 3x^{(2)} + 3x)(4x^{(3)} - 9x^{(2)} + 12x)}{(x^{(4)} - 3x^{(3)} + 6x^{(2)})((x+1)^{(4)} - 3(x+1)^{(3)} + 6(x+1)^{(2)})}$$

9.
$$\Delta^3((x-3)(x-5)(x-7)) = \Delta^3(x^{(3)})$$
$$= \Delta^2(3x^2 + 3x + 1)$$
$$= \Delta(6x + 6)$$
$$= 6!$$

11.
$$\Delta^4 \binom{x}{5} = \Delta^3 \binom{x}{4} = \Delta^2 \binom{x}{3} = \Delta \binom{x}{2} = \binom{x}{1} = x$$

At $x = 0$, $\Delta^4 \binom{x}{5} = 0$.

13. For $x^4 - 6x^3 + 4x^2 - 4x + 5$

	0	1	−6	4	−4	5
			0	0	0	0
1		1	−6	4	−4	[5]
				1	−5	−1
2		1	−5	−1	[−5]	
				2	−6	
3		1	−3	[−7]		
				3		
4		1	[0]			
			[1]			

The factor polynomial is $x^{(4)} - 7x^{(2)} - 5x + 5$.

15. For $x^5 - x^4 + x^3 - x^2 + x - 1$

0	1	−1	1	−1	1	−1
		0	0	0	0	0
1	1	−1	1	−1	1	[−1]
		1	0	1	0	
2	1	0	1	0	[1]	
		2	4	10		
3	1	2	5	[10]		
		3	15			
4	1	5	[20]			
		4				
5		[9]				
	[1]					

and the factorial polynomial is $x^{(5)} + 9x^{(4)} + 20x^{(3)} + 10x^{(2)} + x - 1$.

17. For the factorial polynomial $x^{(4)} + 3x^{(3)} - 4x^{(2)} - 2x + 1$

	1	3	−4	−2	[1]
		−3	0	4	
3	1	0	−4	[2]	
		−2	2		
2	1	−2	[−2]		
		−1			
1	1	[−3]			
	[1]				

and the polynomial is $x^4 - 3x^3 - 2x^2 + 2x + 1$.

19. For the factorial polynomial $x^{(5)} - x^{(4)} + x^{(3)} - x^{(2)} + x - 1$

	1	−1	1	−1	1	[−1]
		−4	15	−32	33	
4	1	−5	16	−33	[34]	
		−3	16	−32		
3	1	−8	32	[65]		
		−2	10			
2	1	−10	[42]			
		−1				
1	1	[−11]				
	[1]					

and the polynomial is $x^5 - 11x^4 + 42x^3 - 65x^2 + 34x - 1$.

21.
$$\Delta^3 \binom{8}{x}$$
$$= \Delta^2 \left[\binom{8}{x+1} - \binom{8}{x}\right]$$
$$= \Delta\left[\binom{8}{x+2} - \binom{8}{x+1} - \binom{8}{x+1} + \binom{8}{x}\right]$$
$$= \Delta\left[\binom{8}{x+2} - 2\binom{8}{x+1} + \binom{8}{x}\right]$$
$$= \binom{8}{x+3} - 3\binom{8}{x+2} + 3\binom{8}{x+1} - \binom{8}{x}$$

At $x = 0$,
$$\Delta^3\binom{8}{x} = \binom{8}{3} - 3\binom{8}{2} + 3\binom{8}{1} - \binom{8}{0}$$
$$= 56 - 3 \cdot 28 + 3 \cdot 8 - 1$$
$$= -5$$

Chapter 11—Section 11.5

1. $\sum_{x=1}^{9} \Delta(3^x - x^3 + x^2 - x + 5)$
$$= (3^x - x^3 + x^2 - x + 5)\Big|_1^{10}$$
$$= (3^{10} - 10^3 + 10^2 - 10 + 5) - (3 - 1 + 1 - 1 + 5)$$
$$= 58137$$

3. $\sum_{x=4}^{8} 2^x - 12x^{(3)}$
$$= 2^x - 3x^{(4)}\Big|_4^9$$
$$= (2^9 - 3 \cdot 9^{(4)}) - (2^4 - 3 \cdot 4^{(4)})$$
$$= 512 - 3 \cdot (9 \cdot 8 \cdot 7 \cdot 6) - (16 - 3 \cdot 4 \cdot 3 \cdot 2 \cdot 1)$$
$$= -8504$$

5. $\sum_{x=4}^{7} \binom{x}{2} + 4^x + 12x^{(5)}$
$$= \binom{x}{3} + \tfrac{1}{3} \cdot 4^x + 2x^{(6)}\Big|_4^8$$
$$= \left(\binom{8}{3} + \tfrac{1}{3} \cdot 4^8 + 2 \cdot 8^{(6)}\right) - \left(\binom{4}{3} + \tfrac{1}{3} \cdot 4^4 + 2 \cdot 4^{(6)}\right)$$
$$= (56 + \tfrac{1}{3} \cdot 65536 + 2 \cdot 8 \cdot 7 \cdot 6 \cdot 5 \cdot 4 \cdot 3)$$
$$- \left(4 + \tfrac{1}{3} \cdot 256\right)$$
$$= (62221.33\cdots) - 89.33\cdots$$
$$= 62132$$

7. $\sum_{i=1}^{n} 2i = i^{(2)}\Big|_1^{n+1}$
$$= (n+1)^{(2)} - 1^{(2)}$$
$$= n(n+1)$$

Answers **A-61**

9.
$$\sum_{i=1}^{n} i(i+1) = \sum_{i=1}^{n} i^2 + i$$
$$= \sum_{i=1}^{n} i^{(2)} + 2i$$
$$= \frac{i^{(3)}}{3} + i^{(2)} \Big|_{1}^{n+1}$$
$$= \frac{(n+1)(n)(n-1)}{3} + (n+1)(n)$$
$$= \frac{(n+1)(n)(n+2)}{3}$$

11.
$$\sum_{x=1}^{8}(2x+3) \cdot 3^x = (2x+3) \cdot \tfrac{1}{2}3^x \Big|_{1}^{9} - \tfrac{3}{2} \sum_{x=1}^{8} 3^x \cdot 2$$
$$= (2x+3) \cdot \tfrac{1}{2}3^x \Big|_{1}^{9} - \tfrac{3}{2} 3^{x+1} \Big|_{1}^{9}$$
$$= (2x+3) \cdot \tfrac{1}{2}3^x \Big|_{1}^{9} - \tfrac{3}{2} \cdot 3^x \Big|_{1}^{9}$$
$$= x 3^x \Big|_{1}^{9}$$
$$= 9 \cdot 3^9 - 3$$
$$= 177144$$

13.
$$\sum_{x=0}^{8} x(x-1) \cdot 2^x$$
$$= \sum_{x=0}^{8} x^{(2)} \cdot 2^x$$
$$= x^{(2)} \cdot 2^x \Big|_{0}^{9} - \sum_{x=0}^{8} 2x \cdot 2^{x+1}$$
$$= x^{(2)} \cdot 2^x \Big|_{0}^{9} - 4 \sum_{x=0}^{8} x \cdot 2^x$$
$$= x^{(2)} \cdot 2^x \Big|_{0}^{9} - 4((x \cdot 2^x) \Big|_{0}^{9} - \sum_{x=0}^{8} 2^{x+1})$$
$$= x^{(2)} \cdot 2^x \Big|_{0}^{9} - 4(x \cdot 2^x) \Big|_{0}^{9} + 8 \sum_{x=0}^{8} 2^x$$
$$= x^{(2)} \cdot 2^x \Big|_{0}^{9} - 4(x \cdot 2^x) \Big|_{0}^{9} + 8(2^x) \Big|_{0}^{9}$$
$$= 9 \cdot 8 \cdot 2^9 - 4 \cdot 9 \cdot 2^9 + 8 \cdot 2^9 - 8$$
$$= 44 \cdot 2^9 - 8$$
$$= 22520$$

15.
$$\sum_{x=0}^{8} x^2 \cdot 2^x$$
$$= \sum_{x=0}^{8} (x^{(2)} + x) \cdot 2^x$$
$$= (x^{(2)} + x) \cdot 2^x \Big|_{0}^{9} - \sum_{x=0}^{8} (2x+1) 2^{x+1}$$
$$= (x^{(2)} + x) \cdot 2^x \Big|_{0}^{9} - 2 \sum_{x=0}^{8} (2x+1) \cdot 2^x$$
$$= (x^{(2)} + x) \cdot 2^x \Big|_{0}^{9} - 2((2x+1) \cdot 2^x \Big|_{0}^{9} - 4 \sum_{x=0}^{8} 2^{x+1})$$
$$= (x^{(2)} + x) \cdot 2^x \Big|_{0}^{9} - 2(((2x+1) \cdot 2^x) \Big|_{0}^{9} + 8 \sum_{x=0}^{8} 2^x)$$
$$= (x^{(2)} + x) \cdot 2^x \Big|_{0}^{9} - 2(((2x+1) \cdot 2^x) \Big|_{0}^{9} + 8 \cdot 2^x \Big|_{0}^{9})$$
$$= ((x^{(2)} + x) - (4x+2) + 8) 2^x \Big|_{0}^{9}$$
$$= (9 \cdot 8 - 27 + 6) 2^9 - 6$$
$$= 26106$$

17.
$$\sum_{x=0}^{10} x^{(4)} \cdot 4^x$$
$$= \tfrac{1}{3} x^{(4)} \cdot 4^x \Big|_{0}^{11} - \sum_{x=0}^{10} 4 x^{(3)} \cdot 4^{x+1}$$
$$= \tfrac{1}{3} x^{(4)} \cdot 4^x \Big|_{0}^{11} - \tfrac{1}{3} \sum_{x=0}^{10} 4 x^{(3)} \cdot 4^{x+1}$$
$$= \tfrac{1}{3} x^{(4)} \cdot 4^x \Big|_{0}^{11} - \tfrac{16}{3} \sum_{x=0}^{10} x^{(3)} \cdot 4^x$$
$$= \tfrac{1}{3} x^{(4)} \cdot 4^x \Big|_{0}^{11} - \tfrac{16}{9} x^{(2)} 4^x \Big|_{0}^{11} + \tfrac{16}{3} \sum_{x=0}^{10} x^{(2)} \cdot 4^{x+1}$$
$$= \tfrac{1}{3} x^{(4)} \cdot 4^x \Big|_{0}^{11} - \tfrac{16}{9} x^{(2)} 4^x \Big|_{0}^{11}$$
$$+ \tfrac{64}{9} x^{(2)} 4^x \Big|_{0}^{11} - \tfrac{512}{9} \sum_{x=0}^{10} x 4^x$$
$$= \tfrac{1}{3} x^{(4)} \cdot 4^x \Big|_{0}^{11} - \tfrac{16}{9} x^{(2)} 4^x \Big|_{0}^{11}$$
$$+ \tfrac{64}{9} x^{(2)} 4^x \Big|_{0}^{11} - \tfrac{512}{27} x 4^x \Big|_{0}^{11} + \tfrac{2048}{27} \sum_{x=0}^{10} 4^x$$
$$= \tfrac{1}{3} x^{(4)} \cdot 4^x \Big|_{0}^{11} - \tfrac{16}{9} x^{(2)} 4^x \Big|_{0}^{11}$$
$$+ \tfrac{64}{9} x^{(2)} 4^x \Big|_{0}^{11} - \tfrac{512}{27} x 4^x \Big|_{0}^{11} + \tfrac{2048}{81} 4^x \Big|_{0}^{11}$$
$$= 4^{11} (\tfrac{11 \cdot 10 \cdot 9 \cdot 8}{3} - \tfrac{16}{9}(11 \cdot 10 \cdot 9) + \tfrac{64}{9}(11 \cdot 10)$$
$$- \tfrac{512}{27} \cdot 11 + \tfrac{2048}{81}) - \tfrac{2048}{81}$$
$$= 6203013120$$

19.
$$\sum_{x=0}^{9} x^{(2)} x^{(8)}$$
$$= \tfrac{1}{9} x^{(2)} x^{(9)} \Big|_{0}^{10} - \tfrac{1}{9} \sum_{x=0}^{9} x(x+1)^{(9)}$$
$$= \tfrac{1}{9} x^{(2)} x^{(9)} - \tfrac{2}{9 \cdot 10} x(x+1)^{(10)} \Big|_{0}^{10} + \tfrac{2}{9 \cdot 10} \sum_{x=0}^{9} (x+2)^{(10)}$$
$$= \tfrac{1}{9} x^{(2)} x^{(9)} - \tfrac{2}{9 \cdot 10} x(x+1)^{(10)} + \tfrac{2}{9 \cdot 10 \cdot 11} (x+2)^{(11)} \Big|_{0}^{10}$$
$$= x^{(9)} (\tfrac{1}{9} x^{(2)} - \tfrac{2}{9 \cdot 10} x(x+1) + \tfrac{2}{9 \cdot 10 \cdot 11}(x+1)(x+2)) \Big|_{0}^{10}$$
$$= 10^{(9)}(\tfrac{1}{9} \cdot 10^{(2)} - \tfrac{2}{9 \cdot 10} \cdot 10 \cdot 11 + \tfrac{2}{9 \cdot 10 \cdot 11} \cdot 11 \cdot 12))$$
$$= 10^{(9)}(\tfrac{1}{9} \cdot 10 \cdot 9 - \tfrac{2}{9 \cdot 10} \cdot 10 \cdot 11 + \tfrac{2}{9 \cdot 10 \cdot 11} \cdot 11 \cdot 12))$$
$$= 10^{(9)}(10 - \tfrac{2 \cdot 11}{9} + \tfrac{2 \cdot 12}{9 \cdot 10})$$
$$= 28385280$$

21. Let $\Delta F(x) = f(x)$. Then
$$\Delta \sum_{i=x}^{c} f(i) = \Delta(F(c+1) - F(x))$$
$$= \Delta(F(c+1) - \Delta F(x))$$
$$= 0 - f(x)$$
$$= -f(x)$$

Chapter 12—Section 12.1

1. $k^n = 3^7 = 2187$
3. $C(n+k-1, n) = C(7+3-1, 7) = C(9, 7) = 36$
5. $S_1^{(7)} + S_2^{(7)} + S_3^{(7)} = 1 + 63 + 301 = 365$
7. $k^n = 4^{10} = 1048576$
9. $C(n+k-1, n) = C(10+4-1, 10) = C(13, 10) = 286$
11. $S_1^{(10)} + S_2^{(10)} + S_3^{(10)} + S_4^{(10)} = 1 + 511 + 9330 + 34105 = 43947$
13.

n\k	0	1	2	3	4	5	6
11	0	1	1023	28501	145750	246730	179487
12	0	1	2047	86526	611501	1379400	1323652

n\k	7	8	9	10	11	12
11	63987	11880	1155	55	1	
12	627396	159027	22275	1705	66	1

15. $k^n = 4^{12} = 16777216$
17. $C(n+k-1, n) = C(15, 12) = 455$
19. $S_1^{(12)} + S_2^{(12)} + S_3^{(12)} + S_4^{(12)} = 1 + 2047 + 86526 + 611501 = 70075$
21. $k^n = 6^{14}$
23. $C(n+k-1, n) = C(19, 14) = 11628$
25. $S_1^{(14)} + S_2^{(14)} + S_3^{(14)} + S_4^{(14)} + S_5^{(14)} + S_6^{(14)}$
27. $S(n, 0) = \frac{1}{0!}(-1)^0 \binom{0}{0}(0-0)^n = 0$

Chapter 12—Section 12.2

1. $C_5 = \frac{1}{6}\binom{10}{5} = \frac{252}{6} = 42$
3. $C_7 = \frac{1}{8}\binom{14}{7} = \frac{3432}{8} = 429$
5. $C_{15} = \frac{1}{16}\binom{30}{15} = 9694845$
7. C_n

Chapter 12—Section 12.3

1. $0 = (1-1)^n = 1 - \binom{n}{1} + \binom{n}{2} - \binom{n}{3} + \cdots + (-1)^n \binom{n}{n}$
3. $D_{15} = 15!(1 - \frac{1}{1!} + \frac{1}{2!} - \frac{1}{3!} + \frac{1}{4!} - \frac{1}{5!} + \frac{1}{6!} - \frac{1}{7!} + \frac{1}{8!} - \frac{1}{9!} + \cdots + \frac{1}{14!} - \frac{1}{15!})$
   $\approx 4.8 \times 10^{11}$
5. $7!(1 - 1 + \frac{1}{2} - \frac{1}{6} + \frac{1}{24} - \frac{1}{120} + \frac{1}{720} - \frac{1}{5040}) = 1854$
7. $7!(1 - 1 + \frac{1}{2} - \frac{1}{6} + \frac{1}{24} - \frac{1}{120} + \frac{1}{720}) = 1855$
9. $D_{12} = 12!(1 - \frac{1}{1!} + \frac{1}{2!} - \frac{1}{3!} + \frac{1}{4!} - \frac{1}{5!} + \frac{1}{6!} - \frac{1}{7!} + \frac{1}{8!} - \frac{1}{9!} + \cdots - \frac{1}{11!} + \frac{1}{12!})$
   $= 176214841$
11. $7!(1 - 1 + \frac{1}{2} - \frac{1}{6} + \frac{1}{24} - \frac{1}{120} + \frac{1}{720}) = 1855$
13. $7! - 1855 - 1854 = 1331$
15. $8! - 4 \cdot 7! + \binom{4}{2} \cdot 6! - \binom{4}{3} 5! + 4! = 24024$
17. $\binom{8}{4} D_4 = \binom{8}{4} 4!(1 - 1 + \frac{1}{2} - \frac{1}{6} + \frac{1}{24}) = 630$
19. Number divisible by 7, 2, or 5 is
    $1150 + 460 + 328 - 230 - 164 - 65 + 32 = 1511$
    Number relatively prime to 700 is
    $2300 - 1511 = 789$

Chapter 12—Section 12.4

1. $R(x, C) = 1 + 6x + 6x^2$
3. $R(x, C) = (1 + 4x + 2x^2)^2$
5. $R(x, C) = (1 + 3x + x^2)(1 + 2x) + x(1 + 3x + 2x^2)$
7. $R(x, C) = (1 + 3x + x^2)^2(1 + x)$
9. $R(x, C) =$ rook polynomial of [board] $+ x \cdot$ rook polynomial of [board]
   $=$ rook polynomial of [board] $+ x \cdot$ rook polynomial of [board]
   $+ x \cdot$ rook polynomial of [board]
   $= 1 + 6x + 6x^2 + x(1 + 4x + 2x^2) + x(1 + 5x + 4x^2)$
   $= 1 + 8x + 15x^2 + 6x^3$
11. $R(x, C) =$ rook polynomial of [board]
    $+ x \cdot$ rook polynomial of [board]
    $= (1+x)(1+2x)(1+3x+x^2) + x(1+x)(1+2x)(1+x)$
    $= (1+2x)(1+x)(1+4x+2x^2)$

13. $R(x, C) = (1 + 2x)^2(1 + 10x + 24x^2 + 12x^3)$
    $= 1 + 14x + 68x^2 + 148x^3 + 144x^4 + 48x^5$
    number of arrangements is
    $8! - 14 \cdot 7! + 68 \cdot 6! - 148 \cdot 5!$
    $+ 144 \cdot 4! - 48 \cdot 3! = 4128$

15. $R(x, C) = (1 + 3x)(1 + 2x)(1 + x)$
    $= 1 + 6x + 11x^2 + 6x^3$
    number of arrangements is
    $5! - 6 \cdot 4! + 11 \cdot 3! - 6 \cdot 2! = 30$

17. $R(x, C) = (1 + 4x + 2x^2)^3$
    $= 1 + 12x + 54x^2 + 112x^3 + 108x^4$
    $+ 48x^5 + 8x^6$
    number of arrangements is
    $6! - 112 \cdot 5! + 54 \cdot 4! - 112 \cdot 3! + 108 \cdot 2!$
    $- 48 \cdot 1! + 8 \cdot 0! = 80$

19.
```
       X       X
        X X
        X X
       X       X
```
    $R(x, C) = 1 + 8x + 20x^2 + 16x^3 + 4x^4$
    number of arrangements is
    $4! - 8 \cdot 3! + 20 \cdot 2! - 16 \cdot 1! + 4 \cdot 0! = 4$

21.
```
              X   X
              X X X
              X
         X X
         X X
         X X X
```
    $R(x, C) = (1 + 5x + 4x^2)(1 + 8x + 16x^2 + 8x^3)$
    $= 1 + 13x + 60x^2 + 120x^3 + 104x^4 + 32x^5$
    number of arrangements is
    $6! - 13 \cdot 5! + 60 \cdot 4! - 120 \cdot 3! + 104 \cdot 2! - 32 = 56$

23.
```
         X X
          X X
          X X
         X X
```
    $R(x, C) = (1 + 4x + 2x^2)^2$
    $= 1 + 8x + 20x^2 + 16x^3 + 4x^4$
    number of paths is
    $4! - 8 \cdot 3! + 20 \cdot 2! - 16 \cdot 1! + 4 = 4$

25.
```
         X X
         X   X
          X X
           X X
           X X
```
    $R(x, C) = (1 + 6x + 9x^2 + 2x^3)(1 + 4x + 2x^2)$
    $= 1 + 10x + 35x^2 + 50x^3 + 26x^4 + 4x^5$
    number of arrangements is
    $5! - 10 \cdot 4! + 35 \cdot 3! - 50 \cdot 2! + 26 \cdot 1! - 4 = 12$

27.
```
              X             X
              X X
               X X
                X X
                 X X
```
    $R(x, C) = $ rook poly of
```
              X
              X X
               X X
                X X
                 X X
```
    $+ x \cdot$ rook poly of
```
              X X
              X X
               X X
                X
```
    $= $ rook poly of
```
              X
              X X
              X
               X X
                X X
```
    $+ x \cdot$ rook poly of
```
              X
              X X
               X
                X X
```
    $+ x \cdot$ rook poly of
```
              X X
              X
               X X
                X
```
    $+ x^2 \cdot$ rook poly of
```
              X X
               X
               X
```
    $= (1 + 4x + 3x^2)^2 + x(1 + 3x + x^2)^2 + x(1 + 3x + x^2)^2 + x^2(1 + 2x)$
    $= 1 + 10x + 35x^2 + 50x^3 + 25x^4 + 2x^5$

    number of arrangements is
    $5! - 10 \cdot 4! + 35 \cdot 3! - 50 \cdot 2! + 25 \cdot 1! - 2 = 13$

Chapter 13—Section 13.2

1. $\dfrac{1}{(x+3)(x+4)} = \dfrac{A}{x+4} + \dfrac{B}{x+3}$

Multiplying both sides by $(x+3)(x+4)$, $1 = A(x+3) + B(x+4)$

Letting $x = -3$, $B = 1$.

Letting $x = -4$, $A = -1$.

Therefore, $\dfrac{1}{(x+3)(x+4)} = \dfrac{-1}{x+4} + \dfrac{1}{x+3}$.

3. $\dfrac{x^2 + 2x + 3}{(x-1)^2(x-2)} = \dfrac{A}{x-1} + \dfrac{B}{(x-1)^2} + \dfrac{C}{x-2}$

Multiplying both sides by $(x-1)^2(x-2)$,

$x^2 + 2x + 3 = A(x-1)(x-2) + B(x-2) + C(x-1)^2$

Letting $x = 1$, $B = -6$.

Letting $x = 2$, $C = 11$.

Letting $x = 0$, $3 = A(-1)(-2) - 6(-2) + 11$ and $A = -10$.

Therefore,

$\dfrac{x^2 + 2x + 3}{(x-1)^2(x-2)} = \dfrac{-10}{x-1} - \dfrac{6}{(x-1)^2} + \dfrac{11}{x-2}$

5. $\dfrac{3x^2 + 1}{(x-1)^2(x-2)^2} = \dfrac{A}{x-1} + \dfrac{B}{(x-1)^2} + \dfrac{C}{x-2} + \dfrac{D}{(x-2)^2}$

Multiplying both sides by $(x-1)^2(x-2)^2$,

$3x^2 + 1 = A(x-1)(x-2)^2 + B(x-2)^2 + C(x-1)^2(x-2) + D(x-1)^2$

Letting $x = 1$, $B = 4$.

Letting $x = 2$, $D = 13$.

Letting $x = 0$, $2A + C = 14$.

Letting $x = 3$, $A + 2C = -14$.

Solving the last two equations simultaneously, $A = 14$ and $C = -14$.

Therefore,

$\dfrac{3x^2 + 1}{(x-1)^2(x-2)^2}$
$= \dfrac{14}{x-1} + \dfrac{4}{(x-1)^2} - \dfrac{14}{x-2} + \dfrac{13}{(x-2)^2}$

7. $\dfrac{2x - 3}{(x+4)(x-2)} = \dfrac{A}{x-2} + \dfrac{B}{x+4}$

Multiplying both sides by $(x-2)(x+4)$,

$2x - 3 = A(x+4) + B(x-2)$

Letting $x = -4$, $B = \dfrac{11}{6}$.

Letting $x = 2$, $A = \dfrac{1}{6}$.

Therefore,

$\dfrac{2x - 3}{(x+4)(x-2)} = \dfrac{\tfrac{1}{6}}{x-2} + \dfrac{\tfrac{11}{6}}{x+4}$

9. $\dfrac{x^2}{(x-1)(x+3)(x-5)} = \dfrac{A}{x-1} + \dfrac{B}{x+3} + \dfrac{C}{x-5}$

Multiplying both sides by $(x-1)(x+3)(x-5)$,

$x^2 = A(x+3)(x-5) + B(x-1)(x-5) + C(x-1)(x+3)$

Letting $x = 1$, $A = -\dfrac{1}{16}$.

Letting $x = -3$, $B = \dfrac{9}{32}$.

Letting $x = 5$, $C = \dfrac{25}{32}$.

Therefore,

$\dfrac{x^2}{(x-1)(x+3)(x-5)} = \dfrac{-\tfrac{1}{16}}{x-1} + \dfrac{\tfrac{9}{32}}{x+3} + \dfrac{\tfrac{25}{32}}{x-5}$

11. $\dfrac{1}{x+4}$
$= \dfrac{\tfrac{1}{4}}{\tfrac{x}{4} + 1}$
$= \dfrac{1}{4} \dfrac{1}{1 - (-\tfrac{x}{4})}$
$= \tfrac{1}{4}(1 - \tfrac{1}{4}x + \tfrac{1}{16}x^2 - \tfrac{1}{64}x^3 + \cdots + (-\tfrac{1}{4})^n x^n + \cdots)$

13. $\dfrac{1}{(x-1)^3} = -\dfrac{1}{(1-x)^3}$
$= -(1 + 3x + 6x^2 + 10x^3 + \cdots + \dfrac{n(n+1)}{2}x^n - \cdots$
$= -1 - 3x - 6x^2 - 10x^3 - \cdots - \dfrac{n(n+1)}{2}x^n - \cdots$

15. $\dfrac{3x^2 + 1}{(x-1)^2(x-2)} = \dfrac{A}{(1-x)} + \dfrac{B}{(1-x)^2} + \dfrac{C}{(2-x)}$

Multiplying by $(x-1)^2(x-2)$,

$3x^2 + 1 = A(2-x)(1-x) + B(2-x) + C(1-x)^2$

Letting $x = 1$, $B = -4$.

Letting $x = 2$, $C = -13$.

Letting $x = 0$, $A = 10$.

$$\frac{3x^2+1}{(x-1)^2(x-2)}$$
$$= \frac{10}{(1-x)} - 4 \cdot \frac{1}{(1-x)^2} - 13 \frac{1}{(2-x)}$$
$$= \frac{10}{(1-x)} - 4 \cdot \frac{1}{(1-x)^2} - \frac{13}{2} \frac{1}{\left(1-\frac{x}{2}\right)}$$
$$= 10 \cdot (1 + x + x^2 + x^3 + \cdots + x^n + \cdots)$$
$$- 4 \cdot (1 + 2x + 3x^2 + 4x^3 + \cdots + (n+1)x^n + \cdots)$$
$$- \frac{13}{2}\left(1 + \frac{1}{2}x + \frac{1}{4}x^2 + \cdots + \frac{1}{2^n}x^n + \cdots\right)$$
$$= -\tfrac{1}{2} - \tfrac{5}{4}x - \tfrac{29}{8}x^2 - \cdots - \left(4n - 6 + \frac{13}{2^{n+1}}\right)x^n - \cdots$$

17.
$$\frac{x}{(x-2)^2} = x \cdot \frac{1}{(2-x)^2}$$
$$= \frac{1}{4} \cdot x \cdot \frac{1}{\left(1-\frac{x}{2}\right)^2}$$
$$= \frac{1}{4} \cdot x \cdot \left(1 + 2 \cdot \frac{1}{2}x + 3 \cdot \frac{1}{2^2}x^2 + 4 \cdot \frac{1}{2^3}x^3 \right.$$
$$\left. + \cdots + (n+1)\frac{1}{2^n}x^n + \cdots\right)$$
$$= \frac{1}{4}\left(x + x^2 + \frac{3}{2^2}x^3 + \frac{4}{2^3}x^4 \right.$$
$$\left. + \cdots + \frac{n+1}{2^n}x^{n+1} + \cdots\right)$$

19.
$$\frac{x}{(x+1)(x-2)(x+3)} = \frac{A}{1+x} + \frac{B}{2-x} + \frac{C}{3+x}$$
Multiplying by $(1+x)(2-x)(3+x)$,
$$-x = A(2-x)(3+x) + B(1+x)(3+x)$$
$$+ C(1+x)(2-x)$$
Letting $x = 2$, $B = -\frac{2}{15}$.

Letting $x = -3$, $C = -\frac{3}{10}$.

Letting $x = -1$, $A = \frac{1}{6}$.

Therefore,
$$\frac{x}{(x-1)(x-2)(x+3)}$$
$$= \frac{1}{6} \cdot \frac{1}{1+x} - \frac{2}{15} \cdot \frac{1}{2-x} - \frac{3}{10} \cdot \frac{1}{3+x}$$
$$= \frac{1}{6} \cdot \frac{1}{1-(-x)} - \frac{1}{15} \cdot \frac{1}{\left(1-\frac{x}{2}\right)} - \frac{1}{10} \cdot \frac{1}{1-\left(-\frac{x}{3}\right)}$$
$$= \frac{1}{6}(1 - x + x^2 - x^3 + \cdots + (-1)^n x^n + \cdots)$$
$$- \frac{1}{15}\cdot\left(1 + \tfrac{1}{2}x + \tfrac{1}{4}x^2 + \cdots + \tfrac{1}{2^n}x^n + \cdots\right)$$

$$- \frac{1}{10} \cdot \left(1 - \frac{1}{3}x + \frac{1}{9}x^2 + \cdots + \left(\frac{-1}{3}\right)^n x^n + \cdots\right)$$
$$= -\frac{1}{6}x + \frac{25}{90}x^2 + \cdots$$
$$+ \left(\frac{(-1)^n}{6} - \frac{1}{15 \cdot 2^n} - \frac{1}{10}\cdot\left(\frac{-1}{3}\right)^n\right)x^n + \cdots$$

21. $f(x) = a_0 + a_1x + a_2x^2 + \cdots + a_nx^n + \cdots$
$2xf(x) = 2a_0x + 2a_1x^2 + 2a_2x^3 + \cdots + 2a_nx^{n+1} + \cdots$
$$\frac{1}{1-3x} = 1 + 3x + 9x^2 + \cdots + 3^nx^n + \cdots$$
$$f(x) - 2xf(x) - \frac{1}{1-3x} = a_0 - 1 = 0$$
Therefore,
$$(1-2x)f(x) = \frac{1}{1-3x}$$
and
$$f(x) = \frac{1}{(1-3x)(1-2x)} = \frac{A}{(1-3x)} + \frac{B}{(1-2x)}$$
Multiplying by $(1-3x)(1-2x)$, we have
$$1 = A(1-2x) + B(1-3x)$$
Letting $x = \frac{1}{2}$, we get $B = -2$.
Letting $x = \frac{1}{3}$, we get $A = 3$.
Therefore,
$$f(x) = \frac{3}{(1-3x)} + \frac{-2}{(1-2x)}$$
$$= 3(1 + 3x + 9x^2 + \cdots + 3^nx^n + \cdots)$$
$$- 2(1 + 2x + 4x^2 + \cdots + 2^nx^n + \cdots)$$
so
$$a_n = 3 \cdot 3^n - 2 \cdot 2^n = 3^{n+1} - 2^{n+1}$$

23. $f(x) = a_0 + a_1x + a_2x^2 + \cdots + a_nx^n + \cdots$
$2xf(x) = 2a_0x + 2a_1x^2 + 2a_2x^3 + \cdots + 2a_nx^{n+1} + \cdots$
$3x^2f(x) = 3a_0x^2 + 3a_1x^3 + 3a_2x^4 + \cdots + 3a_nx^{n+2} + \cdots$
$f(x) - 2xf(x) - 3x^2f(x) = a_0 + a_1x - 2a_0x = 8$
so
$$(1 - 2x - 3x^2)f(x) = 8$$
and
$$f(x) = \frac{8}{(1-3x)(1+x)} = \frac{A}{(1-3x)} + \frac{B}{1+x}$$
Multiplying by $(1-3x)(1+x)$, we have
$$8 = A(1+x) + B(1-3x)$$
Letting $x = -1$, we get $B = 2$.

Letting $x = \frac{1}{3}$, we get $A = 6$.

Therefore,

$$f(x) = \frac{8}{(1-3x)(1+x)}$$
$$= \frac{6}{(1-3x)} + \frac{2}{1+x}$$
$$= 6(1 + 3x + 9x^2 + \cdots + 3^n x^n + \cdots)$$
$$\quad + 2(1 - x + x^2 + \cdots + (-1)^n x^n + \cdots)$$

and $a_n = 6 \cdot 3^n + 2(-1)^n$.

25. $f(x) = a_0 + a_1 x + a_2 x^2 + \cdots + a_n x^n + \cdots$

$2xf(x) = 2a_0 x + 2a_1 x^2 + 2a_2 x^3 + \cdots + 2a_n x^{n+1} + \cdots$

$\dfrac{3}{1-x} = 3 + 3x + 3x^2 + \cdots + 3x^n + \cdots$

so

$f(x) - 2xf(x) + \dfrac{3}{1-x} = a_0 + 3 = 7$

$(1 - 2x)f(x) = 7 - \dfrac{3}{1-x}$

$= \dfrac{7 - 7x - 3}{1 - x}$

$= \dfrac{4 - 7x}{1 - x}$

and

$f(x) = \dfrac{4 - 7x}{(1-x)(1-2x)} = \dfrac{A}{1-2x} + \dfrac{B}{1-x}$

Multiplying by $(1-x)(1-2x)$, we have

$4 - 7x = A(1 - x) + B(1 - 2x)$

Letting $x = 1$, we get $B = 3$.

Letting $x = 0$, we get $A = 1$.

$f(x) = \dfrac{1}{1-2x} + \dfrac{3}{(1-2x)^2}$
$= (1 + 2x + 4x^2 + \cdots + 2^n x^n + \cdots)$
$\quad + 3(1 + 2 \cdot 2x + 3 \cdot 2^2 x^2$
$\quad + \cdots + (n+1)2^n x^n + \cdots)$

and $a_n = 2^n + 3(n+1)2^n$.

27. $f(x) = a_0 + a_1 x + a_2 x^2 + \cdots + a_n x^n + \cdots$

$4xf(x) = 4a_0 x + 4a_1 x^2 + 4a_2 x^3$
$\quad + \cdots + 4a_n x^{n+1} + \cdots$

$\dfrac{x}{(1-x)^2} = x + 2x^2 + \cdots + nx^n + \cdots$

so

$f(x) - 4xf(x) - \dfrac{2x}{(1-x)^2} = a_0 = 1$

and

$(1 - 4x)f(x) = 1 + \dfrac{2x}{(1-x)^2}$

$= \dfrac{x^2 - 2x + 1 + 2x}{(1-x)^2}$

$= \dfrac{x^2 + 1}{(1-x)^2}$

Thus,

$f(x) = \dfrac{x^2 + 1}{(1-4x)(1-x)^2}$

$= \dfrac{A}{1-x} + \dfrac{B}{(1-x)^2} + \dfrac{C}{(1-4x)}$

and multiplying by $(1-x)^2(1-4x)$, we have

$x^2 + 1 = A(1-x)(1-4x) + B(1-4x) + C(1-x)^2$

Letting $x = 1$, we get $B = \dfrac{-2}{3}$.

Letting $x = \dfrac{1}{4}$, we get $C = \dfrac{17}{19}$.

Letting $x = 0$, we get $A + B + C = 1$ so $A = \dfrac{-2}{9}$.

$f(x) = \dfrac{-2}{9} \dfrac{1}{1-x} - \dfrac{2}{3} \dfrac{1}{(1-x)^2} + \dfrac{17}{19} \dfrac{1}{(1-4x)}$

$= \dfrac{-2}{9}(1 + x + x^2 + \cdots + x^n + \cdots)$

$\quad - \dfrac{2}{3}(1 + 2x + 3x^2 + \cdots + (n+1)x^n + \cdots)$

$\quad + \dfrac{17}{19}(1 + 4x + 16x^2 + \cdots + 4^n x^n + \cdots)$

Therefore,

$a_n = \dfrac{-2}{9} - \dfrac{2}{3}(n+1) + \dfrac{17}{19} \cdot 4^n = -\dfrac{2}{3}n - \dfrac{8}{9} + \dfrac{17}{19} \cdot 4^n$

29. $f(x) = a_0 + a_1 x + a_2 x^2 + \cdots + a_n x^n + \cdots$

$7xf(x) = 7a_0 x + 7a_1 x^2 + 7a_2 x^3 = \cdots + 7a_n x^{n+1} + \cdots$

$10x^2 f(x) = 10a_0 x^2 + 10a_1 x^3 + 10a_2 x^4 + \cdots$
$\quad + 10a_n x^{n+2} + \cdots$

so

$f(x) - 7xf(x) + 10x^2 f(x) = a_0 + a_1 x - 7a_0 x$
$= 1 - 4x$

and

$(1 - 7x + 10x^2)f(x) = 1 - 4x$

and

$f(x) = \dfrac{1 - 4x}{(1-2x)(1-5x)} = \dfrac{A}{1-2x} + \dfrac{B}{1-5x}$

Multiplying by $(1-2x)(1-5x)$, we have
$$1 - 4x = A(1 - 5x) + B(1 - 2x)$$
Letting $x = 0$, we get $A + B = 1$.
Letting $x = 1$, we get $-4 \cdot A - B = -3$.
Solving, we have $A = \frac{2}{3}$, $B = \frac{1}{3}$.
$$f(x) = \frac{2}{3} \cdot \frac{1}{1-2x} + \frac{1}{3} \cdot \frac{1}{(1-5x)}$$
$$= \frac{2}{3} \cdot (1 + 2x + 4x^2 + \cdots + 2^n x^n + \cdots)$$
$$+ \frac{1}{3} \cdot (1 + 5x + 25x^2 + \cdots + 5^n x^n + \cdots)$$
and $a_n = \frac{2}{3} \cdot 2^n + \frac{1}{3} \cdot 5^n$

31.
$$f(x) = a_0 + a_1 x + a_2 x^2 + \cdots + a_n x^n + \cdots$$
$$xf(x) = a_0 x + a_1 x^2 + a_2 x^3 + \cdots + a_n x^{n+1} + \cdots$$
$$\frac{x}{(1-x)^2} = x + 2x^2 + \cdots + nx^n + \cdots$$
so
$$f(x) - xf(x) - \frac{x}{(1-x)^2} = a_0 = 1$$
and
$$(1-x)f(x) = \frac{x}{(1-x)^2} + 1$$
$$= \frac{x^2 - 2x + 1 + x}{(1-x)^2}$$
$$= \frac{x^2 - x + 1}{(1-x)^2}$$
Thus,
$$f(x) = \frac{x^2 - x + 1}{(1-x)^3} = \frac{A}{1-x} + \frac{B}{(1-x)^2} + \frac{C}{(1-x)^3}$$
and multiplying by $(1-x)^3$, we have
$$x^2 - x + 1 = A(1-x)^2 + B(1-x) + C$$
Letting $x = 1$, we get $C = 1$.
Letting $x = 0$, we get $A + B + C = 1$.
Letting $x = 2$, we get $A - B + C = 3$ so $A = 1$ and $B = -1$.
$$f(x) = \frac{1}{1-x} - \frac{1}{(1-x)^2} + \frac{1}{(1-x)^3}$$
$$= 1(1 + x + x^2 + \cdots + x^n + \cdots)$$
$$-(1 + 2x + 3x^2 + \cdots + (n+1)x^n + \cdots)$$
$$+(1 + 3x + \cdots + \frac{(n+1)(n+2)}{2} x^n + \cdots)$$

Therefore,
$$a_n = 1 - (n+1) + \frac{(n+1)(n+2)}{2}$$
$$= -n + \frac{n^2 + 3n + 2}{2}$$
$$= \frac{n^2 + n + 2}{2}$$

33.
$$f(x) = a_0 + a_1 x + a_2 x^2 + \cdots + a_n x^n + \cdots$$
$$2xf(x) = 2a_0 x + 2a_1 x^2 + 2a_2 x^3$$
$$+ \cdots + 2a_n x^{n+1} + \cdots$$
$$\frac{1}{1-2x} = 1 + 2x + 4x^2 + \cdots + 2^n x^n + \cdots$$
$$\frac{x}{(1-x)^2} = x + 2x^2 + 3x^2 + \cdots + nx^n + \cdots$$
so
$$f(x) - 2xf(x) - \frac{x}{(1-x)^2} - \frac{1}{1-2x} = a_0 - 1 = 0$$
and
$$(1 - 2x)f(x) = \frac{x}{(1-x)^2} + \frac{1}{1-2x}$$
$$= \frac{x(1 - 2x) + (1 - x)^2}{(1-x)^2(1-2x)}$$
$$= \frac{1 - x - x^2}{(1-x)^2(1-2x)}$$
Thus,
$$f(x) = \frac{1 - x - x^2}{(1-x)^2(1-2x)^2} = \frac{A}{(1-2x)} + \frac{B}{(1-2x)^2}$$
$$+ \frac{C}{(1-x)} + \frac{D}{(1-x)^2}$$
and multiplying by $(1-x)^2(1-2x)^2$, we have
$$1 - x - x^2 = A(1-x)^2(1-2x) + B(1-x)^2$$
$$+ C(1-x)(1-2x)^2 + D(1-2x)^2$$
Letting $x = 1$, we get $D = -1$.
Letting $x = \frac{1}{2}$, we get $B = 1$.
Letting $x = 0$, we get $A + B + C + D = 1$.
Letting $x = -1$, we get $12A + 4B + 18C + 9D = 1$.
Solving, we get $A = 2$, $C = -2$.
$$f(x) = 2\frac{1}{1-2x} + \frac{1}{(1-2x)^2} - \frac{2}{1-x} - \frac{1}{(1-x)^2}$$
$$= 2(1 + 2x + 4x^2 + \cdots + 2^n x^n + \cdots)$$
$$+ (1 + 2 \cdot 2x + \cdots + (n+1)2^n x^n + \cdots)$$
$$-2(1 + x + x^2 + \cdots + x^n + \cdots)$$
$$-(1 + 2x + 3x^2 + \cdots + (n+1)x^n + \cdots)$$

Therefore,
$$a_n = 2 \cdot 2^n + (n+1) \cdot 2^n - 2 - (n+1)$$
$$= (n+3)2^n - n - 3$$

Chapter 13—Section 13.3

1. $(1 + x + x^2 + \cdots)^2(1 + x + x^2)(1 + x + x^2 + x^3 + x^4)$

3. $(x + x^3 + x^5 + x^7 + \cdots)(1 + x^2 + x^4 + x^6 + \cdots)(1 + x + x^2 + \cdots)(x^4 + x^5 + x^6 + x^7 + \cdots)$

5. $(1 + x + x^2 + \cdots)(1 + x^2 + x^4 + \cdots)(1 + x^3 + x^6 + \cdots)(1 + x^4 + x^8 + \cdots)$

7. $(1 + x + x^2 + x^3)^3$

9. $(x^3 + x^4 + x^5 + \cdots)^4$

11. $(1 + x + x^2 + x^3 + x^4)(1 + x + x^2 + x^3)(1 + x + x^2 + \cdots + x^6)(1 + x + x^2)$

13. $(1 + x + x^2 + \cdots + x^6)(x^4 + x^5 + \cdots + x^{11})(1 + x + x^2 + \cdots + x^7)(1 + x^2 + x^4 + \cdots x^{10})$

15. $(1 + x + x^2 + \cdots + x^6)(1 + x + x^2 + \cdots + x^5)(1 + x + x^2 + x^3 + x^4)(1 + x + x^2 + x^3)$

17. $(x^3 + x^4 + x^5)(x + x^2 + x^3 + x^4)(1 + x^2 + x^4 + x^6)(x^2 + x^3 + x^4 + x^5 + x^6)$

19. $(1 + x + x^2 + \cdots)(1 + x^5 + x^{10} + \cdots)(1 + x^{10} + x^{20} + \cdots)(1 + x^{25} + x^{50} + \cdots)$

21. $(1 + x^3 + x^6 + \cdots)(1 + x^5 + x^{10} + \cdots)$ disregard first three terms.

23. $(x + x^2 + x^3 + x^4 + x^5 + x^6)^k$

25.
$$f(x) = \frac{x^4}{(1-x)^5} = x^4\left(1 + \binom{5}{1}x + \binom{6}{2}x^2 + \cdots + \binom{5+n-1}{n}x^n + \cdots\right).$$

We want coefficient of x^6 in $(1 + \binom{5}{1}x + \binom{6}{2}x^2 + \cdots + \binom{5+n-1}{n}x^n + \cdots)$, which is $\binom{10}{6}$.

27. From the equation in the previous exercise, the coefficient of $x^{19} = 1$.

29.
$$(x^2 + x^3 + x^4 + \cdots)^4$$
$$= x^8(1 + x + x^2 + x^3 \cdots)^4$$
$$= x^8 \frac{1}{(1-x)^4}$$
$$= x^8\left(1 + 4x + \binom{5}{2}x^2 + \cdots + \binom{4+n-1}{n}x^n + \cdots\right)$$

The coefficient of x^4 in $1 + 4x + \binom{5}{2}x^2 + \cdots + \binom{4+n-1}{n}x^n + \cdots$ is $\binom{7}{4}$.

31.
$$\frac{x+3}{1 - 2x + x^2} = \frac{x+3}{(1-x)^2}$$
$$= (x+3)(1 + 2x + 3x^2 + \cdots + (n+1)x^n + \cdots).$$

The coefficient of x^{12} is $12 + 3 \cdot 13 = 51$.

33.
$$\frac{1}{(1-3x)^4} = \left(1 + 4 \cdot 3x + \binom{5}{2}3^2x^2 + \cdots + \binom{4+n-1}{n}3^n x^n\right) + \cdots.$$

The coefficient of x^{12} is
$$\binom{4+12-1}{12}3^{12} = \binom{15}{12} \cdot 3^{12}$$

35. The generating function is
$$f(x) = (x + x^2 + x^3 + \cdots)^3$$
$$= x^3(1 + x + x^2 + x^3 + \cdots)^3$$
$$= x^3 \frac{1}{(1-x)^3}$$
$$= x^3\left(1 + 3x + 6x^2 + \cdots + \frac{(n+1)(n+2)}{2}x^n + \cdots\right)$$

The coefficient of x^{10} is $\frac{(8)(9)}{2} = 36$.

37. The generating function is
$$f(x) = (1 + x + x^2 + x^3 + x^4)^5$$
$$= \left(\frac{1-x^5}{1-x}\right)^5$$
$$= (1 - x^5)^5(1 + 5x + \cdots + \binom{n+4}{n}x^n + \cdots)$$
$$= (1 - 5x^5 + 10x^{10} - 10x^{15} \cdots)$$
$$\times (1 + 5x + \cdots + \binom{n+4}{n}x^n + \cdots)$$

The coefficient of x^{15} is $\binom{15+4}{15} - 5\binom{10+4}{10} + 10\binom{5+4}{5} - 10 = 121$.

39. The generating function is
$$f(x) = (x^2 + x^3 + x^4 + x^5 + x^6)^5$$
$$= x^{10}(1 + x + x^2 + x^3 + x^4)^5$$
$$= x^{10}\left(\frac{1-x^5}{1-x}\right)^5$$
$$x^{10}(1-x^5)^5 \frac{1}{(1-x)^5}$$
$$= x^{10}(1 - 5x^5 + 10x^{10} - 10x^{15} \cdots)$$
$$\times (1 + 5x + \cdots + \binom{n+4}{n}x^n + \cdots)$$

The coefficient of x^{16} is $\binom{6+4}{6} - 5 \cdot 5 = \binom{10}{6} - 25 = 185$.

Chapter 13—Section 13.4

1. The generating function is
$$f(x) = (1 + x + x^2 + x^3 + x^4 + x^5 + x^6 + x^7)$$
$$(1 + x^2 + x^4 + x^6)(1 + x^3 + x^6)$$
$$(1 + x^4)(1 + x^5)(1 + x^6)(1 + x^7)$$

The coefficient of x^7 is 15.

3. The generating function is $\dfrac{1}{1-x} \cdot \dfrac{1}{1-x^2} \cdot \dfrac{1}{1-x^3} \cdot \dfrac{1}{1-x^4} \cdot \dfrac{1}{1-x^5} \cdot \dfrac{1}{1-x^6}$.

5. The generating function is $\dfrac{1}{1-x^2} \cdot \dfrac{1}{1-x^4} \cdot \dfrac{1}{1-x^6} \cdot \dfrac{1}{1-x^8} \cdot \dfrac{1}{1-x^{10}} \cdots$.

7. In the rows, beginning at the top, represent each part of the partition into even sizes. Now looking at the columns we see the parts represented occur in pairs of decreasing size. Since the increase in a row involves an increase of an even number, this produces a pair or pairs of the same size for the columns.

9. Let the columns represent the partition of $2n$ objects into n parts.

Since none of these parts are empty, the bottom row contains n elements. Ignoring this row, we see that the other rows form a partition of n elements.

11. $\dfrac{1}{1-x} \cdot \dfrac{1}{1-x^2} \cdot \dfrac{1}{1-x^3} \cdot \dfrac{1}{1-x^4} \cdot \dfrac{1}{1-x^5} \cdot \dfrac{1}{1-x^6} \cdot \dfrac{1}{1-x^7}$.

Chapter 13—Section 13.5

1.
$$\left(\frac{x^2}{2!} + \frac{x^3}{3!} + \frac{x^4}{4!} + \cdots\right)$$
$$\times \left(1 + \frac{x}{1!} + \frac{x^2}{2!} + \frac{x^3}{3!} + \frac{x^4}{4!} + \cdots\right)^2$$

3. $\left(1 + \dfrac{x^2}{2!} + \cdots\right)\left(x + \dfrac{x^3}{3!} + \dfrac{x^5}{5!} + \dfrac{x^7}{7!} + \dfrac{x^9}{9!} + \cdots\right)^2$

5. $\left(1 + \dfrac{x}{1!} + \dfrac{x^2}{2!} + \dfrac{x^3}{3!} + \dfrac{x^4}{4!}\right)\left(1 + \dfrac{x}{1!} + \dfrac{x^2}{2!} + \dfrac{x^3}{3!}\right)$
$\times \left(1 + \dfrac{x}{1!} + \cdots + \dfrac{x^7}{7!}\right)\left(1 + \dfrac{x}{1!} + \cdots + \dfrac{x^6}{6!}\right)$

7. e^x

9.
$$f(x) = \left(\frac{x}{1!} + \frac{x^2}{2!} + \frac{x^3}{3!} + \frac{x^4}{4!} + \cdots\right)^2$$
$$\times \left(1 + \frac{x}{1!} + \frac{x^2}{2!} + \frac{x^3}{3!} + \frac{x^4}{4!} + \cdots\right)^3$$
$$= (e^x - 1)^2 e^{3x}$$
$$= e^{5x} - 2e^{4x} + e^{3x}$$

Therefore, the number of ways is $5^n - 2 \cdot 4^n + 3^n$.

11.
$$f(x) = (e^x - 1)^3$$
$$= e^{3x} - 3 \cdot e^{2x} + 3 \cdot e^x + 1$$

and the number of ways is $3^{30} - 3 \cdot 2^{30} + 3$.

13.
$$f(x) = \left(\frac{e^x + e^{-x}}{2}\right)\left(\frac{e^x - e^{-x}}{2}\right)(e^x - 1)^2$$
$$= \tfrac{1}{4}(e^{2x} - e^{-2x})(e^{2x} - 2e^x + 1)$$
$$= \tfrac{1}{4}(e^{4x} - 2e^{3x} + e^{2x} - 1 + 2e^{-x} - e^{-2x})$$

Number for 25 objects is $\tfrac{1}{4}(4^{25} - 2 \cdot 3^{25} + 2^{25} - 2 + 2^{25})$.

15. Generating function $\left(1 + x + \dfrac{x^2}{2!} + \cdots\right)$
$\left(1 + x^5 + \dfrac{x^{10}}{2!} + \cdots\right)\left(1 + x^{10} + \dfrac{x^{20}}{2!} + \cdots\right)$
$\left(1 + x^{25} + \dfrac{x^{50}}{2!} + \cdots\right) a_n$ is the coefficient of
$\sum \dfrac{x^{n_1 + 5n_2 + 10n_3 + 25n_4}}{(n_1 + n_2 + n_3 + n_4)!}$ such that $n_1 + 5n_2 + 10n_3 + 25n_4 = n$.

17. $(e^x - 1)(e^{2x} - 1)(e^{3x} - 1) \cdots (e^{kx} - 1)$

19. (a) $(e^x - 1)$
$= \left(1 + \dfrac{x}{1!} + \dfrac{x^2}{2!} + \cdots\right) - 1 = \dfrac{x}{1!} + \dfrac{x^2}{2!} + \cdots$

(b)
$$\frac{e^x + e^{-x}}{2} = \tfrac{1}{2}\left(1 + \frac{x}{1!} + \frac{x^2}{2!} + \cdots\right)$$
$$+ \tfrac{1}{2}\left(1 - \frac{x}{1!} + \frac{x^2}{2!} - \frac{x^3}{3!} \cdots\right)$$
$$= \tfrac{1}{2}\left(2 + 2\frac{x^2}{2!} + 2\frac{x^4}{4!} \cdots\right)$$
$$= \left(1 + \frac{x^2}{2!} + \frac{x^4}{4!} + \cdots\right)$$

(c)
$$\frac{e^x - e^{-x}}{2} = \tfrac{1}{2}\left(1 + \frac{x}{1!} + \frac{x^2}{2!} + \cdots\right)$$
$$- \tfrac{1}{2}\left(1 - \frac{x}{1!} + \frac{x^2}{2!} - \frac{x^3}{3!} \cdots\right)$$
$$= \frac{1}{2}(2x + 2\frac{x^3}{3!} + 2\frac{x^5}{5!} \cdots)$$
$$= \left(x + \frac{x^3}{3!} + \frac{x^5}{5!} + \cdots\right)$$

Chapter 14—Section 14.1

1. $f(v_1) = f(v_2) = a$
$f(v_3) = f(v_4) = b$
$f(v_5) = c$

3. $f(v_3) = c$
$f(v_2) = b$
$f(v_1) = f(v_4) = a$

5. $f(v_1) = f(v_5) = c$
$f(v_3) = f(v_6) = b$
$f(v_2) = f(v_4) = a$

7. No, in second figure b has degree 4.

9. No, first figure has two vertices of degree 3, while second figure has 4.

11. $f(v_1) = a$
$f(v_2) = f$
$f(v_3) = c$
$f(v_4) = h$
$f(v_5) = d$
$f(v_6) = g$
$f(v_7) = b$
$f(v_8) = e$

13. No, different number of vertices in second figure.

15. $f(v_6) = a$
$f(v_1) = d$
$f(v_3) = e$
$f(v_4) = b$
$f(v_2) = c$
$f(v_5) = f$

17. No, d has degree 4, no vertex of the first graph has degree 4.

19. $f(v_1) = d$
$f(v_2) = e$
$f(v_3) = f$
$f(v_4) = b$
$f(v_5) = a$
$f(v_6) = c$

21. No, second figure has cycles of length 4.

23. $f(v_1) = a$
$f(v_2) = c$
$f(v_3) = e$
$f(v_4) = b$
$f(v_5) = d$

25. yes.

27. yes

29. no

31. no

33. yes

35. yes

37. yes

39. No, Different number of odd degree vertices.

41. No, Different number of odd degree vertices.

43. No, Different number of odd degree vertices.

45. union

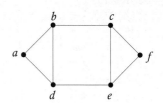

intersection

47. union

intersection

49. union

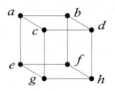

intersection

51.

53.

55.

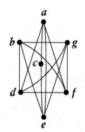

57.

59.

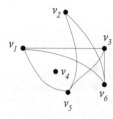

61.

63. yes

65. yes

67. yes

69. No, two vertices missing.

71. If $f(v_i) = f(v_j)$ for $i \neq j$, then there is no edge between $f(v_i)$ and $f(v_j)$ so f is not a homomorphism. Therefore, $f(v_i) \neq f(v_j)$, and f is an isomorphism.

73. Assume G is connected and let u,v belong to $f(G)$. Let $u = f(a)$ and $v = f(b)$. Since G is connected, there is a path $av_1v_2 \cdots v_kb$ from a to b. Therefore, $uf(v_1)f(v_2) \cdots f(v_k)v$ is a path from u to v, and $f(G)$ is connected.

75. Assume G_1 and G_2 are not pairwise disjoint. Let a be a vertex in both G_1 and G_2. There is a path from every vertex in G_1 to a and from every vertex in G_2 to a. Therefore, there is a path from every vertex of G_1 to every vertex in G_2, and $G_1 \cup G_2$ is connected. But this contradicts the assumption that G_1 is a component and, hence, a maximal connected graph.

77. If every edge in the cycle $v_0v_1v_2v_3v_4 \cdots v_0$ were in E', then T would contain a cycle. But a tree cannot contain a cycle. Hence, at least one of the edges must be in $E - E'$.

79. Let G be the cyclic graph $abcda$ and G' be the path xyz. Define $f: G \to G'$ by $f(a) = f(c) = y$, $f(b) = x$, $f(d) = z$. Then $f(b)$ has degree 1 in $f(G)$, while b has degree 2 in G. Degree $f(b)$ in $f(G)$ is less than or equal to degree b in G. The degree of $f(b)$ in G' may be either larger, equal to, or less than the degree of b in G.

81. If $f: G(E, V) \to G'(E', V')$ is an isomorphism and G is connected, then by Problem 16, G' is connected. Similarly, $f^{-1}: G'(E', V') \to G(E, V)$ is an isomorphism so if G' is connected, then G is connected.

83. (v, a) is an edge of G iff $(f(v), f(a))$ is an edge of G'. Therefore, $\deg(a) \leq \deg(f(a))$. Similarly, $\deg(f(a)) \leq \deg(a)$.

85. Let $v_1v_2v_3 \cdots v_k$ be a proper Euler path in G. Then certainly $f(v_1)f(v_2)f(v_3) \cdots f(v_k)$ is a path in G'. Further, since every edge of G is in the Euler path, every edge of G' must be in the path $f(v_1)f(v_2)f(v_3) \cdots f(v_k)$ since every edge in G' is the image of a path in G. If $(f(v_i), f(v_j))$ and $(f(v_k), f(v_l))$ are the same edge in G', then (v_i, v_j) and (v_k, v_l) are the same edge in G. Hence, each edge in the path $f(v_1)f(v_2)f(v_3) \cdots f(v_k)$ occurs only once. Finally, since $v_1 \neq v_k$, $f(v_1) \neq f(v_k)$, so $f(v_1)f(v_2)f(v_3) \cdots f(v_k)$ is a proper path.

87. From the previous problem, we know that G' is bipartite. Let $V = A \cup B$ where there is no edge between a vertex in A and a vertex in B. Then $V' = f(A) \cup f(B)$. Let $a' \in f(A)$ and $b' \in f(B)$. Then $a' = a$ for some $a \in A$ and $b' = b$ for some $b \in B$. Since G is complete, there exists an edge (a, b) in G. Therefore, there is an edge $(f(A), f(B))$ in G', and G' is complete.

89. Let $a, b \in V$. If there is no edge (a, b) in G, then by definition of complement, there must be an edge (a, b) in the complement G^c of G. Therefore, the edge (a, b) is in $G \cup G^c$, and $G \cup G^c$ is complete.

91.

93. Since G_1^c is isomorphic to G^c, there is a one-to-one correspondence between the vertices of G_1^c and those of G^c. Hence, there is a one-to-one correspondence between the vertices of G_1 and those of G. Let $f: G_1^c \to G^c$. Define $g: G_1 \to G$ by $g(v) = f(v)$ for all $v \in V$. $(a, b) \notin G_1^c$, iff $(f(a), f(b)) \notin G^c$. Hence, $(a, b) \in G_1$ iff $(f(a), f(b)) \in G^c$. Define $g((a, b)) = (f(a), f(b))$ for all $(a, b) \in G_1$. Thus, $g: G_1 \to G$ is the desired isomorphism.

95. The vertex d is an articulation point. There are no cut edges. Cut sets include $\{\{a, d\}, \{d, f\}\}$, $\{\{b, d\}, \{d, g\}\}$, $\{\{a, d\}, \{d, b\}\}$.

97. $\{\{c,d\}, \{a,h\}, \{g,h\}\}$, $\{\{i,h\}, \{i,g\}\}$, and $\{\{c, d\}, \{a, h\}\}$. There are no cut edges. Articulation points are e, f, b, i.

99. If C is a cut set in G, then with the removal of C, there are vertices a and b in G that have no path between them. Hence, every path between a and b must contain an edge from C. Since every path in T is a path in G, every path between a and b in T must contain an edge of C. Therefore, since T is connected and there must be a path between a and b in T, there must be an edge of C in the edges of T. Therefore, $C \cap E' \neq \emptyset$.

101. Using the definition of intersection in Definition 14.15, trivially any graph of $\bigcap_{i=1}^{n} G_i$ is a subgraph of G_j. Also by this definition, if every element of G_j'' is contained in every graph G_j, then every element of G_j'' is contained in the intersection $\bigcap_{i=1}^{n} G_i$, and $G_j'' \preceq \bigcap_{i=1}^{n} G_i$. The two definitions are equivalent if there is only one graph satisfying the definition of graph given in the problem. But if there are two graphs G' and G'' satisfying the definition, then $G' \preceq G''$ and $G'' \preceq G'$ so $G'' = G'$.

Chapter 14—Section 14.2

1. Yes, it is already drawn so that no lines cross.

3. No, it has K_5 as a subgraph.

5. Yes, it can be drawn as follows:

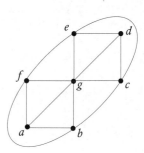

7. Yes, it can be drawn as follows:

9. Yes, it can be drawn as follows:

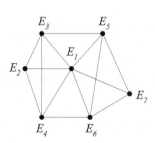

11. Yes, it can be drawn as shown in Problem 7 above.

13. If there are 12 vertices of degree 3, then the sum of the degrees of the vertices is 36, so there are 18 edges. Since $v - e + f = 2$. $12 - 18 + f = 2$ and $f = 8$.

15. (a) 4, (b) 6, (c) $2k$.

17. Since a graph with four vertices cannot be homeomorphic to a subgraph homeomorphic to K_5 or $K_{3,3}$ it must be planar. Alternatively, it is easy to show that K_4 is planar and, hence, since any graph with four vertices is a subgraph of K_4, it is planar.

19. Use induction of the number of edges in the tree. Certainly, any tree with one edge is planar. Assume any tree with k edges is planar. Let T be a tree with $k + 1$ edges. Every tree contains a vertex with degree 1. Pick such a vertex and the edge incident to it and remove them. The remaining tree is planar by the induction hypothesis. The removed edge and vertex can again be added so the edge crosses no other edges. Hence, T is planar.

Chapter 14—Section 14.3

1. $E_1 \bullet \longrightarrow \bullet E_2$

3.

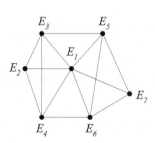

5.

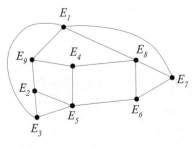

7.

9.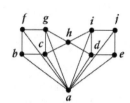

11. 2

13. 1

15. 4

17. 2

19. 4

21. $K_{m,n}$ contains a set of m vertices and a set of n vertices. The only edges are between the two sets. Therefore, paint one set a given color and paint the other set a second color.

23. The chromatic polynomial for

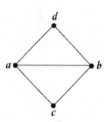

is equal to the chromatic polynomial for

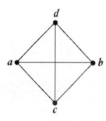

+ the chromatic polynomial for

Therefore,

$$C_G(\lambda) = \lambda(\lambda - 1)(\lambda - 2)(\lambda - 3) \\ + \lambda(\lambda - 1)(\lambda - 2) \\ = \lambda(\lambda - 1)(\lambda - 2)^2$$

A-74 Answers

25. The chromatic polynomial for

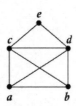

is equal to the chromatic polynomial for

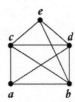

+ the chromatic polynomial for

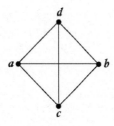

equals the chromatic polynomial for

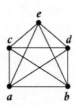

+ twice the chromatic polynomial for

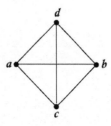

Therefore,
$$C_G(\lambda) = \lambda(\lambda - 1)(\lambda - 2)(\lambda - 3)(\lambda - 4)$$
$$+ 2\lambda(\lambda - 1)(\lambda - 2)(\lambda - 3)$$
$$= \lambda(\lambda - 1)(\lambda - 2)^2(\lambda - 3)$$

27. The chromatic polynomial for

is equal to the chromatic polynomial for

+ the chromatic polynomial for

Therefore,
$$C_G(\lambda) = \lambda(\lambda - 1)(\lambda - 2)(\lambda - 3) + \lambda(\lambda - 1)(\lambda - 2)$$
$$= \lambda(\lambda - 1)(\lambda - 2)^2$$

29. The chromatic polynomial for

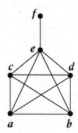

is equal to the chromatic polynomial for

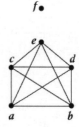

− the chromatic polynomial for

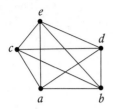

Therefore,
$$C_G(\lambda) = \lambda^2(\lambda - 1)(\lambda - 2)(\lambda - 3)(\lambda - 4)$$
$$- \lambda(\lambda - 1)(\lambda - 2)(\lambda - 3)(\lambda - 4)$$
$$= \lambda(\lambda - 1)^2(\lambda - 2)((\lambda - 3)(\lambda - 4))$$

31. The chromatic polynomial for

is equal to the chromatic polynomial for

− the chromatic polynomial for

But chromatic polynomial for

is equal to the chromatic polynomial for

− the chromatic polynomial for

The chromatic polynomial for

is
$$\lambda(\lambda - 1)^2(\lambda - 2)$$

Therefore, the chromatic polynomial for

is
$$\lambda^2(\lambda - 1)^2(\lambda - 2)$$

Therefore, the chromatic polynomial for

is
$$\lambda^2(\lambda - 1)^2(\lambda - 2) - .\lambda(\lambda - 1)^2(\lambda - 2)$$
$$= .\lambda(\lambda - 1)^3(\lambda - 2)$$

Therefore,
$$C_G(\lambda) = .\lambda(\lambda - 1)^3(\lambda - 2) - \lambda(\lambda - 1)^2(\lambda - 2)$$
$$= \lambda(\lambda - 1)^2(\lambda - 2)((\lambda - 1) - 1)$$
$$= \lambda(\lambda - 1)^2(\lambda - 2)^2$$

33. If a planar graph is isomorphic to its dual, then the number of faces equals the number of vertices. Since $v - e + f = 2$, $n - e + n = 2$ and $e = 2n - 2$.

Chapter 14—Section 14.4

1. $acgfedba$

3. $abihfdcega$

5. Does not exist.

7. Does not exist.

9. Does not exist.

11. $abcde$

13. $fdcegabih$

15. $adbcefgh$

17. $acjhebdgi$

19. Does not exist.

21. If a graph G has a Hamiltonian cycle, then any edge of the cycle cannot be a cut edge. On the other hand, let e be an edge in G that is not in the Hamiltonian cycle. Then e cannot be a cut edge since the removal of e still leaves G connected via the Hamiltonian cycle. Let $\{a, b\}$ be the cut edge and components C_1 and C_2 be components with Hamiltonian cycles, say, $av_1v_2v_3\cdots v_{k-1}a$ and $bv'_1v'_2v'_3\cdots v'_{m-1}b$. Then $v_1v_2v_3\cdots v_{k-1}abv'_1v'_2v'_3\cdots v'_{m-1}$ is a Hamiltonian path.

23.

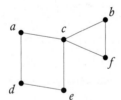

Chapter 14—Section 14.5

1. Dijkstra's algorithm produces the following graphs:

   Steps 1 and 2

   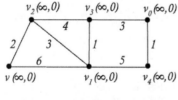

   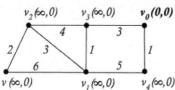

   Steps 3 and 4

   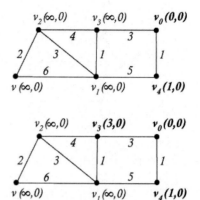

   Steps 5 and 6

   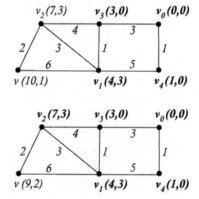

   giving the answer

   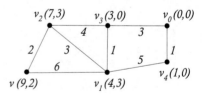

3. Steps 1 and 2

   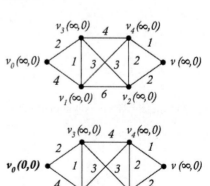

Steps 3 and 4

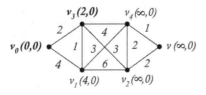

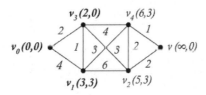

Steps 5 and 6

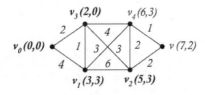

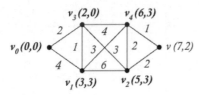

giving the answer

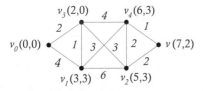

5. Steps 1 and 2

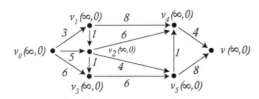

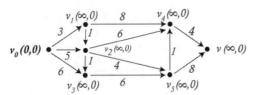

Steps 3 and 4

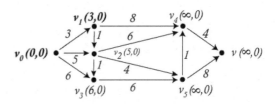

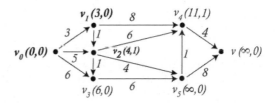

Steps 5 and 6

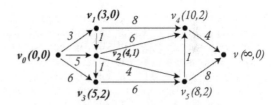

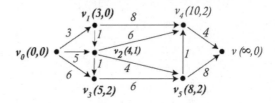

Steps 7

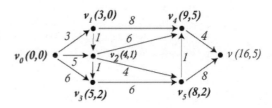

giving the answer

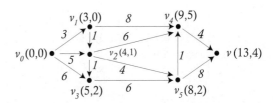

7.

$$A = \begin{bmatrix} 0 & 4 & 2 & \infty & \infty & 6 & \infty & \infty & \infty \\ 4 & 0 & 1 & \infty & 4 & \infty & \infty & \infty & \infty \\ 2 & 1 & 0 & 2 & \infty & \infty & \infty & \infty & \infty \\ \infty & \infty & 2 & 0 & 4 & 1 & \infty & 2 & \infty \\ \infty & 4 & \infty & 4 & 0 & \infty & \infty & \infty & 4 \\ 6 & \infty & \infty & 1 & \infty & 0 & 1 & \infty & \infty \\ \infty & \infty & \infty & \infty & \infty & 1 & 0 & 2 & \infty \\ \infty & \infty & \infty & 2 & \infty & \infty & 2 & 0 & 3 \\ \infty & \infty & \infty & \infty & 4 & \infty & \infty & 3 & 0 \end{bmatrix}$$

$$A^{(1)} = \begin{bmatrix} 0 & 4 & 2 & \infty & \infty & 6 & \infty & \infty & \infty \\ 4 & 0 & 1 & \infty & 4 & 10 & \infty & \infty & \infty \\ 2 & 1 & 0 & 2 & \infty & 8 & \infty & \infty & \infty \\ \infty & \infty & 2 & 0 & 4 & 1 & \infty & 2 & \infty \\ \infty & 4 & \infty & 4 & 0 & \infty & \infty & \infty & 4 \\ 6 & 10 & 8 & 1 & \infty & 0 & 1 & \infty & \infty \\ \infty & \infty & \infty & \infty & \infty & 1 & 0 & 2 & \infty \\ \infty & \infty & \infty & 2 & \infty & \infty & 2 & 0 & 3 \\ \infty & \infty & \infty & \infty & 4 & \infty & \infty & 3 & 0 \end{bmatrix}$$

$$A^{(2)} = \begin{bmatrix} 0 & 4 & 2 & \infty & 8 & 6 & \infty & \infty & \infty \\ 4 & 0 & 1 & \infty & 4 & 10 & \infty & \infty & \infty \\ 2 & 1 & 0 & 2 & 5 & 11 & \infty & \infty & \infty \\ \infty & \infty & 2 & 0 & 4 & 1 & \infty & 2 & \infty \\ 8 & 4 & 5 & 4 & 0 & 14 & \infty & \infty & 4 \\ 6 & 10 & 11 & 1 & 14 & 0 & 1 & \infty & \infty \\ \infty & \infty & \infty & \infty & \infty & 1 & 0 & 2 & \infty \\ \infty & \infty & \infty & 2 & \infty & \infty & 2 & 0 & 3 \\ \infty & \infty & \infty & \infty & 4 & \infty & \infty & 3 & 0 \end{bmatrix}$$

$$A^{(3)} = \begin{bmatrix} 0 & 3 & 2 & 4 & 7 & 6 & \infty & \infty & \infty \\ 3 & 0 & 1 & 3 & 4 & 10 & \infty & \infty & \infty \\ 2 & 1 & 0 & 2 & 5 & 11 & \infty & \infty & \infty \\ 4 & 3 & 2 & 0 & 4 & 1 & \infty & 2 & \infty \\ 7 & 4 & 5 & 4 & 0 & 14 & \infty & \infty & 4 \\ 6 & 10 & 11 & 1 & 14 & 0 & 1 & \infty & \infty \\ \infty & \infty & \infty & \infty & \infty & 1 & 0 & 2 & \infty \\ \infty & \infty & \infty & 2 & \infty & \infty & 2 & 0 & 3 \\ \infty & \infty & \infty & \infty & 4 & \infty & \infty & 3 & 0 \end{bmatrix}$$

$$A^{(4)} = \begin{bmatrix} 0 & 3 & 2 & 4 & 7 & 5 & \infty & 6 & \infty \\ 3 & 0 & 1 & 3 & 4 & 4 & \infty & 5 & \infty \\ 2 & 1 & 0 & 2 & 5 & 3 & \infty & 4 & \infty \\ 4 & 3 & 2 & 0 & 4 & 1 & \infty & 2 & \infty \\ 7 & 4 & 5 & 4 & 0 & 5 & \infty & 6 & 4 \\ 5 & 4 & 3 & 1 & 5 & 0 & 1 & 3 & \infty \\ \infty & \infty & \infty & \infty & \infty & 1 & 0 & 2 & \infty \\ 6 & 5 & 4 & 2 & 6 & 3 & 2 & 0 & 3 \\ \infty & \infty & \infty & \infty & 4 & \infty & \infty & 3 & 0 \end{bmatrix}$$

$$A^{(5)} = \begin{bmatrix} 0 & 3 & 2 & 4 & 7 & 5 & \infty & 6 & 11 \\ 3 & 0 & 1 & 3 & 4 & 4 & \infty & 5 & 8 \\ 2 & 1 & 0 & 2 & 5 & 3 & \infty & 4 & 9 \\ 4 & 3 & 2 & 0 & 4 & 1 & \infty & 2 & 8 \\ 7 & 4 & 5 & 4 & 0 & 5 & \infty & 6 & 4 \\ 5 & 4 & 3 & 1 & 5 & 0 & 1 & 3 & 9 \\ \infty & \infty & \infty & \infty & \infty & 1 & 0 & 2 & \infty \\ 6 & 5 & 4 & 2 & 6 & 3 & 2 & 0 & 3 \\ 11 & 8 & 9 & 8 & 4 & 9 & \infty & 3 & 0 \end{bmatrix}$$

$$A^{(6)} = \begin{bmatrix} 0 & 3 & 2 & 4 & 7 & 5 & 6 & 6 & 11 \\ 3 & 0 & 1 & 3 & 4 & 4 & 5 & 5 & 8 \\ 2 & 1 & 0 & 2 & 5 & 3 & 6 & 4 & 9 \\ 4 & 3 & 2 & 0 & 4 & 1 & 2 & 2 & 8 \\ 7 & 4 & 5 & 4 & 0 & 5 & 6 & 6 & 4 \\ 5 & 4 & 3 & 1 & 5 & 0 & 1 & 3 & 9 \\ 6 & 5 & 6 & 2 & 6 & 1 & 0 & 2 & 10 \\ 6 & 5 & 4 & 2 & 6 & 3 & 2 & 0 & 3 \\ 11 & 8 & 9 & 8 & 4 & 9 & 10 & 3 & 0 \end{bmatrix}$$

$$A^{(7)} = \begin{bmatrix} 0 & 3 & 2 & 4 & 7 & 5 & 6 & 6 & 9 \\ 3 & 0 & 1 & 3 & 4 & 4 & 5 & 5 & 8 \\ 2 & 1 & 0 & 2 & 5 & 3 & 6 & 4 & 9 \\ 4 & 3 & 2 & 0 & 4 & 1 & 2 & 2 & 8 \\ 7 & 4 & 5 & 4 & 0 & 5 & 6 & 6 & 4 \\ 5 & 4 & 3 & 1 & 5 & 0 & 1 & 3 & 9 \\ 6 & 5 & 6 & 2 & 6 & 1 & 0 & 2 & 10 \\ 6 & 5 & 4 & 2 & 6 & 3 & 2 & 0 & 3 \\ 9 & 8 & 9 & 8 & 4 & 9 & 10 & 3 & 0 \end{bmatrix}$$

$$A^{(8)} = \begin{bmatrix} 0 & 3 & 2 & 4 & 7 & 5 & 6 & 6 & 9 \\ 3 & 0 & 1 & 3 & 4 & 4 & 5 & 5 & 8 \\ 2 & 1 & 0 & 2 & 5 & 3 & 6 & 4 & 7 \\ 4 & 3 & 2 & 0 & 4 & 1 & 2 & 2 & 5 \\ 7 & 4 & 5 & 4 & 0 & 5 & 6 & 6 & 4 \\ 5 & 4 & 3 & 1 & 5 & 0 & 1 & 3 & 6 \\ 6 & 5 & 6 & 2 & 6 & 1 & 0 & 2 & 5 \\ 6 & 5 & 4 & 2 & 6 & 3 & 2 & 0 & 3 \\ 9 & 8 & 7 & 8 & 4 & 6 & 5 & 3 & 0 \end{bmatrix}$$

$$A^{(9)} = \begin{bmatrix} 0 & 3 & 2 & 4 & 7 & 5 & 6 & 6 & 9 \\ 3 & 0 & 1 & 3 & 4 & 4 & 5 & 5 & 8 \\ 2 & 1 & 0 & 2 & 5 & 3 & 6 & 4 & 7 \\ 4 & 3 & 2 & 0 & 4 & 1 & 2 & 2 & 5 \\ 7 & 4 & 5 & 4 & 0 & 5 & 6 & 6 & 4 \\ 5 & 4 & 3 & 1 & 5 & 0 & 1 & 3 & 6 \\ 6 & 5 & 6 & 2 & 6 & 1 & 0 & 2 & 5 \\ 6 & 5 & 4 & 2 & 6 & 3 & 2 & 0 & 3 \\ 9 & 8 & 7 & 8 & 4 & 6 & 5 & 3 & 0 \end{bmatrix}$$

9.

$$A = \begin{bmatrix} 0 & 2 & \infty & 1 & \infty & \infty \\ 2 & 0 & 2 & 3 & 2 & 4 \\ \infty & 2 & 0 & \infty & \infty & 1 \\ 1 & 3 & \infty & 0 & 4 & \infty \\ \infty & 2 & \infty & 4 & 0 & 3 \\ \infty & 4 & 1 & \infty & 3 & 0 \end{bmatrix}$$

$$A^{(1)} = \begin{bmatrix} 0 & 2 & \infty & 1 & \infty & \infty \\ 2 & 0 & 2 & 3 & 2 & 4 \\ \infty & 2 & 0 & \infty & \infty & 1 \\ 1 & 3 & \infty & 0 & 4 & \infty \\ \infty & 2 & \infty & 4 & 0 & 3 \\ \infty & 4 & 1 & \infty & 3 & 0 \end{bmatrix}$$

$$A^{(2)} = \begin{bmatrix} 0 & 2 & 4 & 1 & 4 & 6 \\ 2 & 0 & 2 & 3 & 2 & 3 \\ 4 & 2 & 0 & 5 & 4 & 1 \\ 1 & 3 & 5 & 0 & 4 & 6 \\ 4 & 2 & 4 & 4 & 0 & 3 \\ 6 & 4 & 1 & 7 & 3 & 0 \end{bmatrix}$$

$$A^{(3)} = \begin{bmatrix} 0 & 2 & 4 & 1 & 4 & 5 \\ 2 & 0 & 2 & 3 & 2 & 3 \\ 4 & 2 & 0 & 5 & 4 & 1 \\ 1 & 3 & 5 & 0 & 4 & 6 \\ 4 & 2 & 4 & 4 & 0 & 3 \\ 5 & 3 & 1 & 6 & 3 & 0 \end{bmatrix}$$

Answers **A-79**

$$A^{(4)} = \begin{bmatrix} 0 & 2 & 4 & 1 & 4 & 5 \\ 2 & 0 & 2 & 3 & 2 & 3 \\ 4 & 2 & 0 & 5 & 4 & 1 \\ 1 & 3 & 5 & 0 & 4 & 6 \\ 4 & 2 & 4 & 4 & 0 & 3 \\ 5 & 3 & 1 & 6 & 3 & 0 \end{bmatrix}$$

$$A^{(5)} = \begin{bmatrix} 0 & 2 & 4 & 1 & 4 & 5 \\ 2 & 0 & 2 & 3 & 2 & 3 \\ 4 & 2 & 0 & 5 & 4 & 1 \\ 1 & 3 & 5 & 0 & 4 & 6 \\ 4 & 2 & 4 & 4 & 0 & 3 \\ 5 & 3 & 1 & 6 & 3 & 0 \end{bmatrix}$$

$$A^{(6)} = \begin{bmatrix} 0 & 2 & 4 & 1 & 4 & 5 \\ 2 & 0 & 2 & 3 & 2 & 3 \\ 4 & 2 & 0 & 5 & 4 & 1 \\ 1 & 3 & 5 & 0 & 4 & 6 \\ 4 & 2 & 4 & 4 & 0 & 3 \\ 5 & 3 & 1 & 6 & 3 & 0 \end{bmatrix}$$

11. Steps 1 and 2

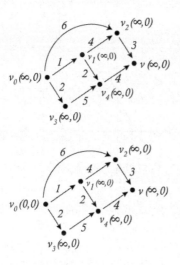

Steps 5 and 6

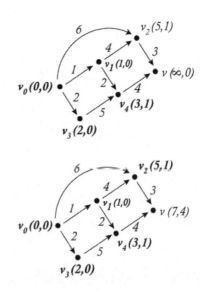

giving the answer

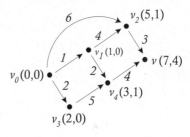

13. Steps 1 and 2

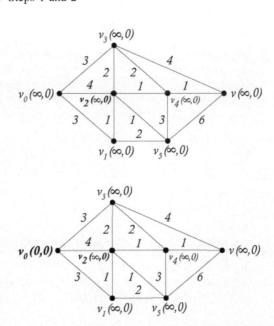

Steps 3 and 4

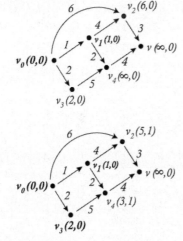

A-80 Answers

Steps 3 and 4

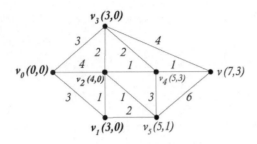

Steps 5 and 6

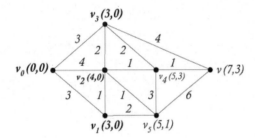

Step 7

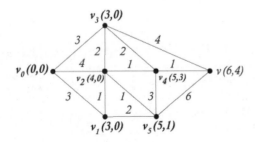

giving the answer

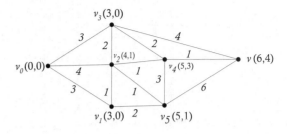

14. Step 1

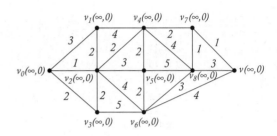

Step 2

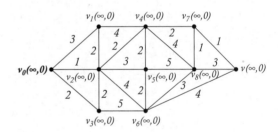

Step 3

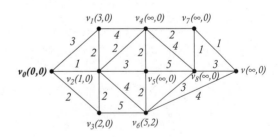

Step 4

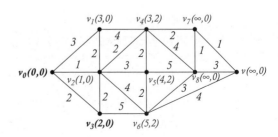

Step 5

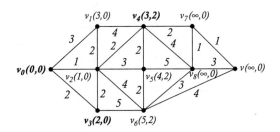

Step 6

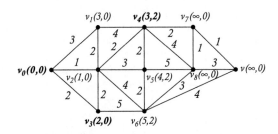

Step 7

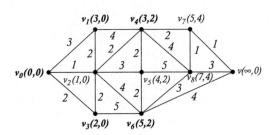

Step 8

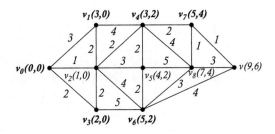

giving the answer

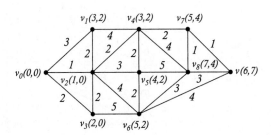

15. Steps 1 and 2

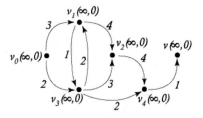

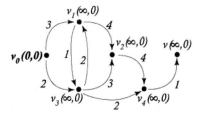

Steps 3 and 4

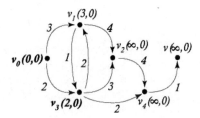

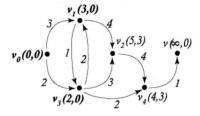

Steps 5 and 6

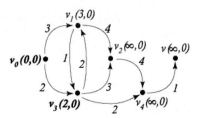

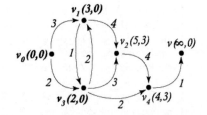

Step 7

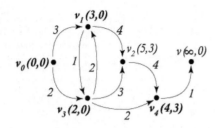

$$A^{(3)} = \begin{bmatrix} 0 & 6 & 7 & 10 & 11 & 8 & 1 & \infty & \infty & \infty \\ 6 & 0 & 1 & 4 & 5 & 2 & 7 & \infty & \infty & \infty \\ 7 & 1 & 0 & 3 & 4 & 3 & 8 & \infty & \infty & \infty \\ 10 & 4 & 3 & 0 & 7 & 6 & 11 & \infty & \infty & 4 \\ 11 & 5 & 4 & 7 & 0 & 1 & 12 & 2 & \infty & \infty \\ 8 & 2 & 3 & 6 & 1 & 0 & 2 & \infty & 1 & \infty \\ 1 & 7 & 8 & 11 & 12 & 2 & 0 & \infty & 1 & 3 \\ \infty & \infty & \infty & \infty & 2 & \infty & \infty & 0 & 2 & \infty & 2 \\ \infty & \infty & \infty & \infty & \infty & 1 & 1 & 2 & 0 & 2 & 3 \\ \infty & \infty & \infty & \infty & \infty & \infty & 3 & \infty & 2 & 0 & 4 \\ \infty & \infty & \infty & 4 & \infty & \infty & \infty & 2 & 3 & 4 & 0 \end{bmatrix}$$

giving the answer

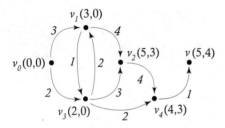

$$A^{(4)} = \begin{bmatrix} 0 & 6 & 7 & 10 & 11 & 8 & 1 & \infty & \infty & 14 \\ 6 & 0 & 1 & 4 & 5 & 2 & 7 & \infty & \infty & 8 \\ 7 & 1 & 0 & 3 & 4 & 3 & 8 & \infty & \infty & 7 \\ 10 & 4 & 3 & 0 & 7 & 6 & 11 & \infty & \infty & 4 \\ 11 & 5 & 4 & 7 & 0 & 1 & 12 & 2 & \infty & 11 \\ 8 & 2 & 3 & 6 & 1 & 0 & 2 & \infty & 1 & 10 \\ 1 & 7 & 8 & 11 & 12 & 2 & 0 & \infty & 1 & 3 & 15 \\ \infty & \infty & \infty & \infty & 2 & \infty & \infty & 0 & 2 & \infty & 2 \\ \infty & \infty & \infty & \infty & \infty & 1 & 1 & 2 & 0 & 2 & 3 \\ \infty & \infty & \infty & \infty & \infty & \infty & 3 & \infty & 2 & 0 & 4 \\ 14 & 8 & 7 & 4 & 11 & 10 & 15 & 2 & 3 & 4 & 0 \end{bmatrix}$$

17.

$$A = \begin{bmatrix} 0 & 6 & \infty & \infty & \infty & \infty & 1 & \infty & \infty & \infty \\ 6 & 0 & 1 & \infty & \infty & 2 & \infty & \infty & \infty & \infty \\ \infty & 1 & 0 & 3 & 4 & \infty & \infty & \infty & \infty & \infty \\ \infty & \infty & 3 & 0 & \infty & \infty & \infty & \infty & \infty & 4 \\ \infty & \infty & 4 & \infty & 0 & 1 & \infty & 2 & \infty & \infty \\ \infty & 2 & \infty & \infty & 1 & 0 & 2 & \infty & 1 & \infty \\ 1 & \infty & \infty & \infty & \infty & 2 & 0 & \infty & 1 & 3 & \infty \\ \infty & \infty & \infty & \infty & 2 & \infty & \infty & 0 & 2 & \infty & 2 \\ \infty & \infty & \infty & \infty & \infty & 1 & 1 & 2 & 0 & 2 & 3 \\ \infty & \infty & \infty & \infty & \infty & \infty & 3 & \infty & 2 & 0 & 4 \\ \infty & \infty & \infty & 4 & \infty & \infty & \infty & 2 & 3 & 4 & 0 \end{bmatrix}$$

$$A^{(1)} = \begin{bmatrix} 0 & 6 & \infty & \infty & \infty & \infty & 1 & \infty & \infty & \infty \\ 6 & 0 & 1 & \infty & \infty & 2 & 7 & \infty & \infty & \infty \\ \infty & 1 & 0 & 3 & 4 & \infty & \infty & \infty & \infty & \infty \\ \infty & \infty & 3 & 0 & \infty & \infty & \infty & \infty & \infty & 4 \\ \infty & \infty & 4 & \infty & 0 & 1 & \infty & 2 & \infty & \infty \\ \infty & 2 & \infty & \infty & 1 & 0 & 2 & \infty & 1 & \infty \\ 1 & 7 & \infty & \infty & \infty & 2 & 0 & \infty & 1 & 3 & \infty \\ \infty & \infty & \infty & \infty & 2 & \infty & \infty & 0 & 2 & \infty & 2 \\ \infty & \infty & \infty & \infty & \infty & 1 & 1 & 2 & 0 & 2 & 3 \\ \infty & \infty & \infty & \infty & \infty & \infty & 3 & \infty & 2 & 0 & 4 \\ \infty & \infty & \infty & 4 & \infty & \infty & \infty & 2 & 3 & 4 & 0 \end{bmatrix}$$

$$A^{(2)} = \begin{bmatrix} 0 & 6 & 7 & \infty & \infty & 8 & 1 & \infty & \infty & \infty \\ 6 & 0 & 1 & \infty & \infty & 2 & 7 & \infty & \infty & \infty \\ 7 & 1 & 0 & 3 & 4 & 3 & 8 & \infty & \infty & \infty \\ \infty & \infty & 3 & 0 & \infty & \infty & \infty & \infty & \infty & 4 \\ \infty & \infty & 4 & \infty & 0 & 1 & \infty & 2 & \infty & \infty \\ 8 & 2 & 3 & \infty & 1 & 0 & 2 & \infty & 1 & \infty \\ 1 & 7 & 8 & \infty & \infty & 2 & 0 & \infty & 1 & 3 & \infty \\ \infty & \infty & \infty & \infty & 2 & \infty & \infty & 0 & 2 & \infty & 2 \\ \infty & \infty & \infty & \infty & \infty & 1 & 1 & 2 & 0 & 2 & 3 \\ \infty & \infty & \infty & \infty & \infty & \infty & 3 & \infty & 2 & 0 & 4 \\ \infty & \infty & \infty & 4 & \infty & \infty & \infty & 2 & 3 & 4 & 0 \end{bmatrix}$$

$$A^{(5)} = \begin{bmatrix} 0 & 6 & 7 & 10 & 11 & 8 & 1 & 13 & \infty & \infty & 14 \\ 6 & 0 & 1 & 4 & 5 & 2 & 7 & 7 & \infty & \infty & 8 \\ 7 & 1 & 0 & 3 & 4 & 3 & 8 & 6 & \infty & \infty & 7 \\ 10 & 4 & 3 & 0 & 7 & 6 & 11 & 9 & \infty & \infty & 4 \\ 11 & 5 & 4 & 7 & 0 & 1 & 12 & 2 & \infty & \infty & 11 \\ 8 & 2 & 3 & 6 & 1 & 0 & 2 & 3 & 1 & \infty & 10 \\ 1 & 7 & 8 & 11 & 12 & 2 & 0 & 14 & 1 & 3 & 15 \\ 13 & 7 & 6 & 9 & 2 & 3 & 14 & 0 & 2 & \infty & 2 \\ \infty & \infty & \infty & \infty & \infty & 1 & 1 & 2 & 0 & 2 & 3 \\ \infty & \infty & \infty & \infty & \infty & \infty & 3 & \infty & 2 & 0 & 4 \\ 14 & 8 & 7 & 4 & 11 & 10 & 15 & 2 & 3 & 4 & 0 \end{bmatrix}$$

$$A^{(6)} = \begin{bmatrix} 0 & 6 & 7 & 10 & 9 & 8 & 1 & 11 & 9 & \infty & 14 \\ 6 & 0 & 1 & 4 & 3 & 2 & 4 & 5 & 3 & \infty & 8 \\ 7 & 1 & 0 & 3 & 4 & 3 & 5 & 6 & 4 & \infty & 7 \\ 10 & 4 & 3 & 0 & 7 & 6 & 8 & 9 & 7 & \infty & 4 \\ 9 & 3 & 4 & 7 & 0 & 1 & 3 & 4 & 2 & \infty & 11 \\ 8 & 2 & 3 & 6 & 1 & 0 & 2 & 3 & 1 & \infty & 10 \\ 1 & 4 & 5 & 8 & 3 & 2 & 0 & 5 & 1 & 3 & 12 \\ 11 & 5 & 6 & 9 & 4 & 3 & 5 & 0 & 2 & \infty & 2 \\ 9 & 3 & 4 & 7 & 2 & 1 & 1 & 2 & 0 & 2 & 3 \\ \infty & \infty & \infty & \infty & \infty & \infty & 3 & \infty & 2 & 0 & 4 \\ 14 & 8 & 7 & 4 & 11 & 10 & 12 & 2 & 3 & 4 & 0 \end{bmatrix}$$

$$A^{(7)} = \begin{bmatrix} 0 & 5 & 6 & 9 & 4 & 3 & 1 & 6 & 2 & 4 & 13 \\ 5 & 0 & 1 & 4 & 3 & 2 & 4 & 5 & 3 & 7 & 8 \\ 6 & 1 & 0 & 3 & 4 & 3 & 5 & 6 & 4 & 8 & 7 \\ 9 & 4 & 3 & 0 & 7 & 6 & 8 & 9 & 7 & 11 & 4 \\ 4 & 3 & 4 & 7 & 0 & 1 & 3 & 4 & 2 & 6 & 11 \\ 3 & 2 & 3 & 6 & 1 & 0 & 2 & 3 & 1 & 5 & 10 \\ 1 & 4 & 5 & 8 & 3 & 2 & 0 & 5 & 1 & 3 & 12 \\ 6 & 5 & 6 & 9 & 4 & 3 & 5 & 0 & 2 & 8 & 2 \\ 2 & 3 & 4 & 7 & 2 & 1 & 1 & 2 & 0 & 2 & 3 \\ 4 & 7 & 8 & 11 & 6 & 5 & 3 & 8 & 2 & 0 & 4 \\ 13 & 8 & 7 & 4 & 11 & 10 & 12 & 2 & 3 & 4 & 0 \end{bmatrix}$$

$$A^{(8)} = \begin{bmatrix} 0 & 5 & 6 & 9 & 4 & 3 & 1 & 6 & 2 & 4 & 8 \\ 5 & 0 & 1 & 4 & 3 & 2 & 4 & 5 & 3 & 7 & 7 \\ 6 & 1 & 0 & 3 & 4 & 3 & 5 & 6 & 4 & 8 & 7 \\ 9 & 4 & 3 & 0 & 7 & 6 & 8 & 9 & 7 & 11 & 4 \\ 4 & 3 & 4 & 7 & 0 & 1 & 3 & 4 & 2 & 6 & 6 \\ 3 & 2 & 3 & 6 & 1 & 0 & 2 & 3 & 1 & 5 & 5 \\ 1 & 4 & 5 & 8 & 3 & 2 & 0 & 5 & 1 & 3 & 7 \\ 6 & 5 & 6 & 9 & 4 & 3 & 5 & 0 & 2 & 8 & 2 \\ 2 & 3 & 4 & 7 & 2 & 1 & 1 & 2 & 0 & 2 & 3 \\ 4 & 7 & 8 & 11 & 6 & 5 & 3 & 8 & 2 & 0 & 4 \\ 8 & 7 & 7 & 4 & 6 & 5 & 7 & 2 & 3 & 4 & 0 \end{bmatrix}$$

$$A^{(1)} = \begin{bmatrix} 0 & 3 & 1 & 2 & \infty & \infty & \infty & \infty & \infty & \infty \\ 3 & 0 & 2 & 5 & 4 & \infty & \infty & \infty & \infty & \infty \\ 1 & 2 & 0 & 2 & 2 & 3 & 4 & \infty & \infty & \infty \\ 2 & 5 & 2 & 0 & \infty & \infty & 5 & \infty & \infty & \infty \\ \infty & 4 & 2 & \infty & 0 & 2 & \infty & 2 & 4 & \infty \\ \infty & \infty & 3 & \infty & 2 & 0 & 2 & \infty & 5 & \infty \\ \infty & \infty & 4 & 5 & \infty & 2 & 0 & \infty & 3 & 4 \\ \infty & \infty & \infty & \infty & 2 & \infty & \infty & 0 & 1 & 1 \\ \infty & \infty & \infty & \infty & 4 & 5 & 3 & 1 & 0 & 3 \\ \infty & \infty & \infty & \infty & \infty & \infty & 4 & 1 & 3 & 0 \end{bmatrix}$$

$$A^{(9)} = \begin{bmatrix} 0 & 5 & 6 & 9 & 4 & 3 & 1 & 4 & 2 & 4 & 5 \\ 5 & 0 & 1 & 4 & 3 & 2 & 4 & 5 & 3 & 5 & 6 \\ 6 & 1 & 0 & 3 & 4 & 3 & 5 & 6 & 4 & 6 & 7 \\ 9 & 4 & 3 & 0 & 7 & 6 & 8 & 9 & 7 & 9 & 4 \\ 4 & 3 & 4 & 7 & 0 & 1 & 3 & 4 & 2 & 4 & 5 \\ 3 & 2 & 3 & 6 & 1 & 0 & 2 & 3 & 1 & 3 & 4 \\ 1 & 4 & 5 & 8 & 3 & 2 & 0 & 3 & 1 & 3 & 4 \\ 4 & 5 & 6 & 9 & 4 & 3 & 3 & 0 & 2 & 4 & 2 \\ 2 & 3 & 4 & 7 & 2 & 1 & 1 & 2 & 0 & 2 & 3 \\ 4 & 5 & 6 & 9 & 4 & 3 & 3 & 4 & 2 & 0 & 4 \\ 5 & 6 & 7 & 4 & 5 & 4 & 4 & 2 & 3 & 4 & 0 \end{bmatrix}$$

$$A^{(2)} = \begin{bmatrix} 0 & 3 & 1 & 2 & 7 & \infty & \infty & \infty & \infty & \infty \\ 3 & 0 & 2 & 5 & 4 & \infty & \infty & \infty & \infty & \infty \\ 1 & 2 & 0 & 2 & 2 & 3 & 4 & \infty & \infty & \infty \\ 2 & 5 & 2 & 0 & 9 & \infty & 5 & \infty & \infty & \infty \\ 7 & 4 & 2 & 9 & 0 & 2 & \infty & 2 & 4 & \infty \\ \infty & \infty & 3 & \infty & 2 & 0 & 2 & \infty & 5 & \infty \\ \infty & \infty & 4 & 5 & \infty & 2 & 0 & \infty & 3 & 4 \\ \infty & \infty & \infty & \infty & 2 & \infty & \infty & 0 & 1 & 1 \\ \infty & \infty & \infty & \infty & 4 & 5 & 3 & 1 & 0 & 3 \\ \infty & \infty & \infty & \infty & \infty & \infty & 4 & 1 & 3 & 0 \end{bmatrix}$$

$$A^{(10)} = \begin{bmatrix} 0 & 5 & 6 & 9 & 4 & 3 & 1 & 4 & 2 & 4 & 5 \\ 5 & 0 & 1 & 4 & 3 & 2 & 4 & 5 & 3 & 5 & 6 \\ 6 & 1 & 0 & 3 & 4 & 3 & 5 & 6 & 4 & 6 & 7 \\ 9 & 4 & 3 & 0 & 7 & 6 & 8 & 9 & 7 & 9 & 4 \\ 4 & 3 & 4 & 7 & 0 & 1 & 3 & 4 & 2 & 4 & 5 \\ 3 & 2 & 3 & 6 & 1 & 0 & 2 & 3 & 1 & 3 & 4 \\ 1 & 4 & 5 & 8 & 3 & 2 & 0 & 3 & 1 & 3 & 4 \\ 4 & 5 & 6 & 9 & 4 & 3 & 3 & 0 & 2 & 4 & 2 \\ 2 & 3 & 4 & 7 & 2 & 1 & 1 & 2 & 0 & 2 & 3 \\ 4 & 5 & 6 & 9 & 4 & 3 & 3 & 4 & 2 & 0 & 4 \\ 5 & 6 & 7 & 4 & 5 & 4 & 4 & 2 & 3 & 4 & 0 \end{bmatrix}$$

$$A^{(3)} = \begin{bmatrix} 0 & 3 & 1 & 2 & 3 & 4 & 5 & \infty & \infty & \infty \\ 3 & 0 & 2 & 4 & 4 & 5 & 6 & \infty & \infty & \infty \\ 1 & 2 & 0 & 2 & 2 & 3 & 4 & \infty & \infty & \infty \\ 2 & 4 & 2 & 0 & 4 & 5 & 5 & \infty & \infty & \infty \\ 3 & 4 & 2 & 4 & 0 & 2 & 6 & 2 & 4 & \infty \\ 4 & 5 & 3 & 5 & 2 & 0 & 2 & \infty & 5 & \infty \\ 5 & 6 & 4 & 5 & 6 & 2 & 0 & \infty & 3 & 4 \\ \infty & \infty & \infty & \infty & 2 & \infty & \infty & 0 & 1 & 1 \\ \infty & \infty & \infty & \infty & 4 & 5 & 3 & 1 & 0 & 3 \\ \infty & \infty & \infty & \infty & \infty & \infty & 4 & 1 & 3 & 0 \end{bmatrix}$$

$$A^{(11)} = \begin{bmatrix} 0 & 5 & 6 & 9 & 4 & 3 & 1 & 4 & 2 & 4 & 5 \\ 5 & 0 & 1 & 4 & 3 & 2 & 4 & 5 & 3 & 5 & 6 \\ 6 & 1 & 0 & 3 & 4 & 3 & 5 & 6 & 4 & 6 & 7 \\ 9 & 4 & 3 & 0 & 7 & 6 & 8 & 6 & 7 & 8 & 4 \\ 4 & 3 & 4 & 7 & 0 & 1 & 3 & 4 & 2 & 4 & 5 \\ 3 & 2 & 3 & 6 & 1 & 0 & 2 & 3 & 1 & 3 & 4 \\ 1 & 4 & 5 & 8 & 3 & 2 & 0 & 3 & 1 & 3 & 4 \\ 4 & 5 & 6 & 6 & 4 & 3 & 3 & 0 & 2 & 4 & 2 \\ 2 & 3 & 4 & 7 & 2 & 1 & 1 & 2 & 0 & 2 & 3 \\ 4 & 5 & 6 & 8 & 4 & 3 & 3 & 4 & 2 & 0 & 4 \\ *5 & 6 & 7 & 4 & 5 & 4 & 4 & 2 & 3 & 4 & 0 \end{bmatrix}$$

$$A^{(4)} = \begin{bmatrix} 0 & 3 & 1 & 2 & 3 & 4 & 5 & \infty & \infty & \infty \\ 3 & 0 & 2 & 4 & 4 & 5 & 6 & \infty & \infty & \infty \\ 1 & 2 & 0 & 2 & 2 & 3 & 4 & \infty & \infty & \infty \\ 2 & 4 & 2 & 0 & 4 & 5 & 5 & \infty & \infty & \infty \\ 3 & 4 & 2 & 4 & 0 & 2 & 6 & 2 & 4 & \infty \\ 4 & 5 & 3 & 5 & 2 & 0 & 2 & \infty & 5 & \infty \\ 5 & 6 & 4 & 5 & 6 & 2 & 0 & \infty & 3 & 4 \\ \infty & \infty & \infty & \infty & 2 & \infty & \infty & 0 & 1 & 1 \\ \infty & \infty & \infty & \infty & 4 & 5 & 3 & 1 & 0 & 3 \\ \infty & \infty & \infty & \infty & \infty & \infty & 4 & 1 & 3 & 0 \end{bmatrix}$$

19.

$$A = \begin{bmatrix} 0 & 3 & 1 & 2 & \infty & \infty & \infty & \infty & \infty & \infty \\ 3 & 0 & 2 & \infty & 4 & \infty & \infty & \infty & \infty & \infty \\ 1 & 2 & 0 & 2 & 2 & 3 & 4 & \infty & \infty & \infty \\ 2 & \infty & 2 & 0 & \infty & \infty & 5 & \infty & \infty & \infty \\ \infty & 4 & 2 & \infty & 0 & 2 & \infty & 2 & 4 & \infty \\ \infty & \infty & 3 & \infty & 2 & 0 & 2 & \infty & 5 & \infty \\ \infty & \infty & 4 & 5 & \infty & 2 & 0 & \infty & 3 & 4 \\ \infty & \infty & \infty & \infty & 2 & \infty & \infty & 0 & 1 & 1 \\ \infty & \infty & \infty & \infty & 4 & 5 & 3 & 1 & 0 & 3 \\ \infty & \infty & \infty & \infty & \infty & \infty & 4 & 1 & 3 & 0 \end{bmatrix}$$

$$A^{(5)} = \begin{bmatrix} 0 & 3 & 1 & 2 & 3 & 4 & 5 & 5 & 7 & \infty \\ 3 & 0 & 2 & 4 & 4 & 5 & 6 & 6 & 8 & \infty \\ 1 & 2 & 0 & 2 & 2 & 3 & 4 & 4 & 6 & \infty \\ 2 & 4 & 2 & 0 & 4 & 5 & 5 & 6 & 8 & \infty \\ 3 & 4 & 2 & 4 & 0 & 2 & 6 & 2 & 4 & \infty \\ 4 & 5 & 3 & 5 & 2 & 0 & 2 & 4 & 5 & \infty \\ 5 & 6 & 4 & 5 & 6 & 2 & 0 & 8 & 3 & 4 \\ 5 & 6 & 4 & 6 & 2 & 4 & 8 & 0 & 1 & 1 \\ 7 & 8 & 6 & 8 & 4 & 5 & 3 & 1 & 0 & 3 \\ \infty & \infty & \infty & \infty & \infty & \infty & 4 & 1 & 3 & 0 \end{bmatrix}$$

$$A^{(6)} = \begin{bmatrix} 0 & 3 & 1 & 2 & 3 & 4 & 5 & 5 & 7 & \infty \\ 3 & 0 & 2 & 4 & 4 & 5 & 6 & 6 & 8 & \infty \\ 1 & 2 & 0 & 2 & 2 & 3 & 4 & 4 & 6 & \infty \\ 2 & 4 & 2 & 0 & 4 & 5 & 5 & 6 & 8 & \infty \\ 3 & 4 & 2 & 4 & 0 & 2 & 4 & 2 & 4 & \infty \\ 4 & 5 & 3 & 5 & 2 & 0 & 2 & 4 & 5 & \infty \\ 5 & 6 & 4 & 5 & 4 & 2 & 0 & 6 & 3 & 4 \\ 5 & 6 & 4 & 6 & 2 & 4 & 6 & 0 & 1 & 1 \\ 7 & 8 & 6 & 8 & 4 & 5 & 3 & 1 & 0 & 3 \\ \infty & \infty & \infty & \infty & \infty & \infty & 4 & 1 & 3 & 0 \end{bmatrix}$$

$$A^{(7)} = \begin{bmatrix} 0 & 3 & 1 & 2 & 3 & 4 & 5 & 5 & 7 & 9 \\ 3 & 0 & 2 & 4 & 4 & 5 & 6 & 6 & 8 & 10 \\ 1 & 2 & 0 & 2 & 2 & 3 & 4 & 4 & 6 & 8 \\ 2 & 4 & 2 & 0 & 4 & 5 & 5 & 6 & 8 & 9 \\ 3 & 4 & 2 & 4 & 0 & 2 & 4 & 2 & 4 & 8 \\ 4 & 5 & 3 & 5 & 2 & 0 & 2 & 4 & 5 & 6 \\ 5 & 6 & 4 & 5 & 4 & 2 & 0 & 6 & 3 & 4 \\ 5 & 6 & 4 & 6 & 2 & 4 & 6 & 0 & 1 & 1 \\ 7 & 8 & 6 & 8 & 4 & 5 & 3 & 1 & 0 & 3 \\ 9 & 10 & 8 & 9 & 8 & 6 & 4 & 1 & 3 & 0 \end{bmatrix}$$

$$A^{(8)} = \begin{bmatrix} 0 & 3 & 1 & 2 & 3 & 4 & 5 & 5 & 6 & 6 \\ 3 & 0 & 2 & 4 & 4 & 5 & 6 & 6 & 7 & 7 \\ 1 & 2 & 0 & 2 & 2 & 3 & 4 & 4 & 5 & 5 \\ 2 & 4 & 2 & 0 & 4 & 5 & 5 & 6 & 7 & 7 \\ 3 & 4 & 2 & 4 & 0 & 2 & 4 & 2 & 3 & 3 \\ 4 & 5 & 3 & 5 & 2 & 0 & 2 & 4 & 5 & 5 \\ 5 & 6 & 4 & 5 & 4 & 2 & 0 & 6 & 3 & 4 \\ 5 & 6 & 4 & 6 & 2 & 4 & 6 & 0 & 1 & 1 \\ 6 & 7 & 5 & 7 & 3 & 5 & 3 & 1 & 0 & 2 \\ 6 & 7 & 5 & 7 & 3 & 5 & 4 & 1 & 2 & 0 \end{bmatrix}$$

$$A^{(9)} = \begin{bmatrix} 0 & 3 & 1 & 2 & 3 & 4 & 5 & 5 & 6 & 6 \\ 3 & 0 & 2 & 4 & 4 & 5 & 6 & 6 & 7 & 7 \\ 1 & 2 & 0 & 2 & 2 & 3 & 4 & 4 & 5 & 5 \\ 2 & 4 & 2 & 0 & 4 & 5 & 5 & 6 & 7 & 7 \\ 3 & 4 & 2 & 4 & 0 & 2 & 4 & 2 & 3 & 3 \\ 4 & 5 & 3 & 5 & 2 & 0 & 2 & 4 & 5 & 5 \\ 5 & 6 & 4 & 5 & 4 & 2 & 0 & 4 & 3 & 4 \\ 5 & 6 & 4 & 6 & 2 & 4 & 4 & 0 & 1 & 1 \\ 6 & 7 & 5 & 7 & 3 & 5 & 3 & 1 & 0 & 2 \\ 6 & 7 & 5 & 7 & 3 & 5 & 4 & 1 & 2 & 0 \end{bmatrix}$$

$$A^{(10)} = \begin{bmatrix} 0 & 3 & 1 & 2 & 3 & 4 & 5 & 5 & 6 & 6 \\ 3 & 0 & 2 & 4 & 4 & 5 & 6 & 6 & 7 & 7 \\ 1 & 2 & 0 & 2 & 2 & 3 & 4 & 4 & 5 & 5 \\ 2 & 4 & 2 & 0 & 4 & 5 & 5 & 6 & 7 & 7 \\ 3 & 4 & 2 & 4 & 0 & 2 & 4 & 2 & 3 & 3 \\ 4 & 5 & 3 & 5 & 2 & 0 & 2 & 4 & 5 & 5 \\ 5 & 6 & 4 & 5 & 4 & 2 & 0 & 4 & 3 & 4 \\ 5 & 6 & 4 & 6 & 2 & 4 & 4 & 0 & 1 & 1 \\ 6 & 7 & 5 & 7 & 3 & 5 & 3 & 1 & 0 & 2 \\ 6 & 7 & 5 & 7 & 3 & 5 & 4 & 1 & 2 & 0 \end{bmatrix}$$

Chapter 15—Section 15.1

1.

3.

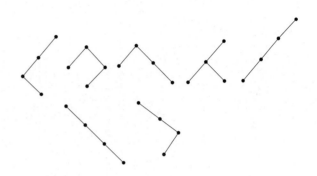

5. $C_6 = \dfrac{1}{7} \cdot \dfrac{12!}{6!6!} = 132$

7. $C_{10} = \dfrac{1}{11} \cdot \dfrac{20!}{10!10!} = 16796$

9. 63 vertices, 32 leaves, and 31 internal vertices.

11. 511 vertices, 256 leaves, and 255 internal vertices.

13. 10 vertices, 1 leaf, and 9 internal vertices.

15.

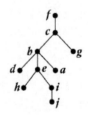

(i) 5, (ii) 3, (iii) 2,
(iv) 3, (v) e, (vi) d, a, e

17.

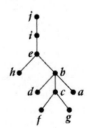

(i) 5, (ii) 2, (iii) 5,
(iv) 4, (v) j, (vi) d, a, c

19.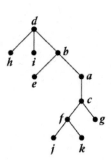

(i) 5, (ii) 2, (iii) 4,
(iv) 2, (v) d, (vi) e, a.

21.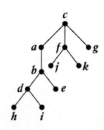

(i) 4, (ii) 3, (iii) 1,
(iv) 1, (v) d, (vi) e, d.

23.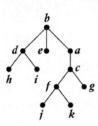

(i) 4, (ii) 1, (iii) 3,
(iv) 1, (v) d, (vi) a, e, d.

25. balanced

27. balanced, full

29. balanced

31. balanced

33. neither

35. $v = 2^6 - 1 = 63$

37. $v = 2^8 - 1 = 255$

39. Since the number of leaves is less than or equal to 2^h and only equals 2^h when the tree is full,

$$2^h \geq l$$
$$\therefore \log_2 2^h \geq \log_2 l \quad \text{since } l \text{ is an increasing function}$$
$$\text{and } h \geq \log_2 l$$

41.
$$v \leq 2^{h+1} - 1$$
since a full tree has the maximum number of vertices
$$\therefore v + 1 \leq 2^{h+1}$$
since l is an increasing function
$$\therefore \log_2(v+1) \leq h + 1$$
since l is an increasing function
and $h \geq \log_2(v+1) - 1$

43. Assume that G is a tree. Let the edge $\{a, b\}$ be added to the tree. Since the tree was connected, there was already a path $av_1v_2v_3\cdots v_{k-1}b$ from a to b. Thus, $av_1v_2v_3\cdots v_{k-1}ba$ is a cycle.

Assume adding $\{a, b\}$ produces two cycles, $av_1v_2v_3\cdots v_{k-1}b$ and $av'_1v'_2v'_3\cdots v'_{m-1}b$. Then $av_1v_2v_3\cdots v_{k-1}bbv'_{m-1}\cdots v'_3v'_2v'_1a$ is already a cycle, contradicting the assumption that G is a tree.

Assume that G is not a tree. Then it already contains a cycle. Assume edge $\{a, b\}$ is added to G. There was already a path from a to b, since G is connected. There are now two paths from a to b so $\{a, b\}$ is part of a cycle. It was not part of the original cycle so there are now at least two cycles in G with the added edge.

45. False. On a tree, a leaf is not a cut vertex.

47. G has an Euler cycle.

49. We must show that $\hat{a}(vv') = \hat{a}(v)\hat{a}(v')$.

$$\hat{a}(vv') = a(vv')$$
$$= (aa)(vv')$$
since elements of a semilattice are idempotents
$$= avav'$$
since a semilattice is associative and commutative
$$= \hat{a}(v)\hat{a}(v').$$

Chapter 15—Section 15.2

1.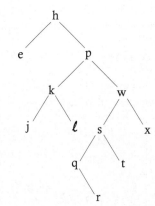

3.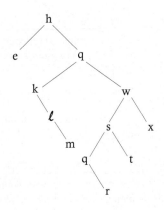

5. 4

7.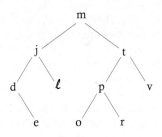

9.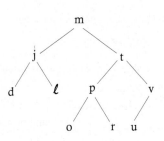

11. 2

13. chain

15. See Figure below.

17.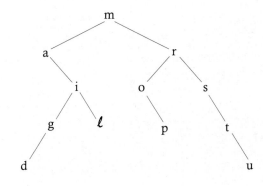

19.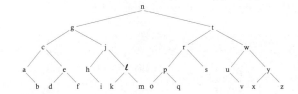

21.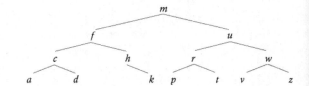

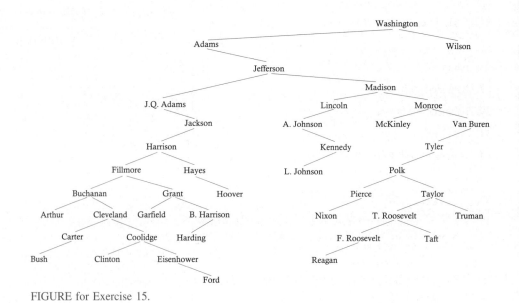

FIGURE for Exercise 15.

23.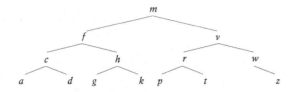

25.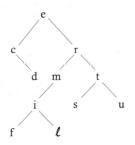

27.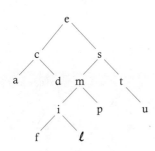

29.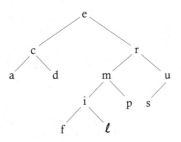

31. Procedure Search(a, r)
   If $a = r$
      Found = yes
   endif
   If $a < r$
     If r has left son u
       call Search(a, u)
     endif
     Found = no
   endif
   If $a > r$
     If r has right son w
       call Search(a, w)
     endif
     Found = no
   endif

Chapter 15—Section 15.3

1. **(a)** 101110001001011110100100010

   (b) 001011010110101

   (c) decided

   (d) issues

3. Weight of code 1 equals $7 \cdot 3 + 8 \cdot 3 + 10 \cdot 2 + 2 \cdot 5 + 9 \cdot 4 + 4 \cdot 4 = 127$.

    Weight of code 2 equals $7 \cdot 3 + 8 \cdot 2 + 10 \cdot 3 + 2 \cdot 3 + 9 \cdot 3 + 4 \cdot 3 = 112$. Weight of code 3 equals $7 \cdot 3 + 8 \cdot 3 + 10 \cdot 3 + 2 \cdot 5 + 9 \cdot 3 + 4 \cdot 6 = 136$.
Select code 2.

5. **(a)** Step 1

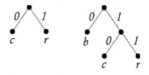

Step 2

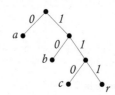

   (b) $a = 0$, $b = 10$, $c = 110$, $r = 111$

   (c) Weight equals $1 \cdot 8 + 2 \cdot 6 + 3 \cdot 3 + 3 \cdot 4 = 41$.

   (d) 110111010

   (e) 100111

   (f) car

   (g) barb

7. **(a)** Steps 1 and 2

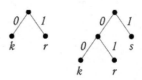

Step 3

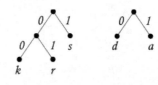

Step 4

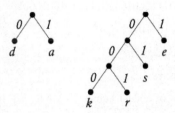

Step 5

giving the answer

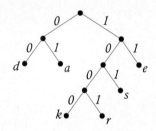

(b) $a = 01$, $d = 00$, $e = 11$, $k = 1000$, $r = 1001$, $s = 101$

(c) Weight equals $2 \cdot 8 + 2 \cdot 7 + 2 \cdot 12 + 4 \cdot 1 + 4 \cdot 4 + 3 \cdot 6 = 92$.

(d) 11100101101111001

(e) 000110011000111001

(f) raked

(g) dearer

9. (a) Steps 1 and 2

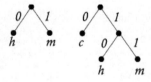

Step 3

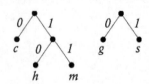

Step 4

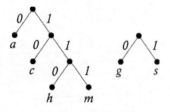

Step 5

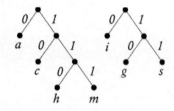

Step 6

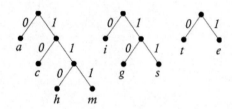

Step 7

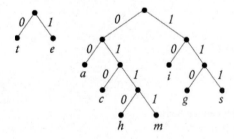

Answer

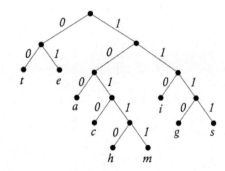

(b) $a = 100$, $c = 1010$, $e = 01$, $i = 110$, $g = 1110$, $h = 10110$, $m = 10111$, $s = 1111$, $t = 00$.

(c) Weight equals 118.

11. (a) Steps 1 and 2

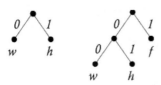

Step 3

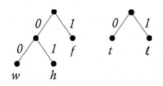

Step 4

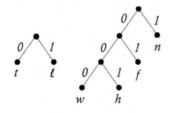

Step 5

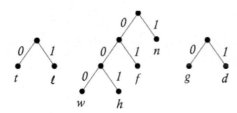

Step 6

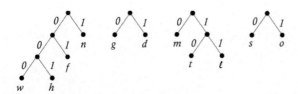

Step 7

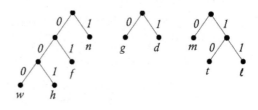

Step 8

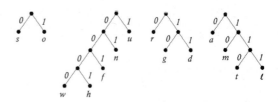

Step 9

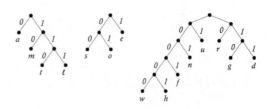

Step 10

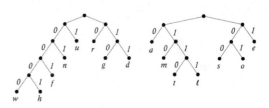

giving the answer

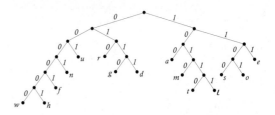

(b) $a = 100$, $d = 0111$, $e = 111$, $f = 00001$, $g = 0110$, $h = 000001$, $l = 10111$, $m = 1010$, $n = 0001$, $o = 1101$, $r = 010$, $s = 1100$, $t = 10110$, $u = 001$, $w = 000000$.

(c) Weight equals 547.

(d) 111100110010110110100001101100000011111100001000110000010111

00000011111001011011010000110110000001111101101110100001

(e) 11000000001111001011011010

13. Consider two elements to be encoded. They agree only as long as they continue the same path. If their paths differ at the nth edge in the path, then they differ in the nth binary bit. Their paths must differ before one encoding reaches a leaf. Otherwise they would both reach the leaf and stop. Hence, neither encoding is an initial string of the other and we have a prefix code.

Chapter 15—Section 15.4

1. (a) $abdce$
   (b) $bacde$
   (c) $adecb$
3. (a) $dacbfegijhkt$
   (b) $gfdcabehijkl$
   (c) $abcdefjilkhg$
5. (a) $deacbgfhjlkmi$
   (b) $hgedcabfijklm$
   (c) $dabcefglmkjih$
7. (a)
   (b) $\times + abc$
   (c) $ab + c\times$
9. (a)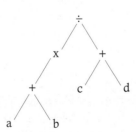
   (b) $\div \times + abc + cd$
   (c) $ab + c \times cd + \div$

11. (a)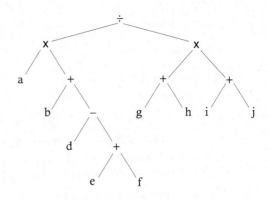
    (b) $\div \times a + b - d + ef \times +gh + ij$
    (c) $abcdef + - + \times gh + ij + \times \div$

13. (a)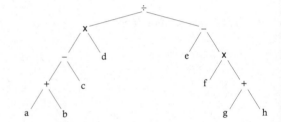
    (b) $\div \times - + abcd - e \times f + gh$
    (c) $ab + c - d \times efgh + \times - \div$

15.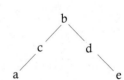

17.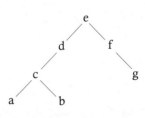

19.

21.

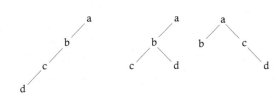

(b)

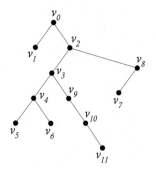

23. (d)

25. Every arithmetic operation must be preceded by two more numbers than arithmetic operations, and the last symbol must be an arithmetic operation.

27. After b has been processed in both trees, Step 4 calls for IBT to process the left child of b on both trees. This is c on both trees and in both trees, the vertex c has no children. The IBT returns to b in both trees and in Step 5 calls IBT to process the right child of b in both trees. When in Step 2, $Iso = F$, the trees are shown not to be isomorphic.

(c)

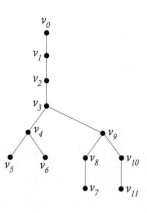

Chapter 15—Section 15.5

1. (a)

(b)

5. (a)

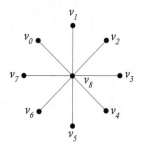

(c)

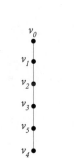

(b)

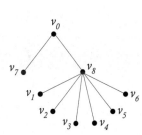

(c)

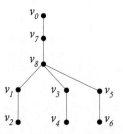

3. (a)

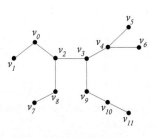

7. (a)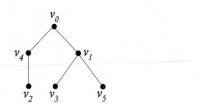

(b)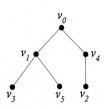

(c)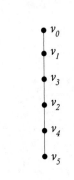

9. (a)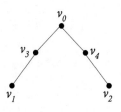

(b)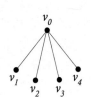

(c)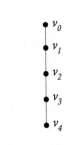

11. (a)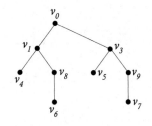

(b)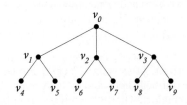

(c)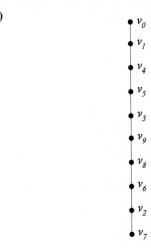

13.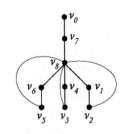

v	$CV(v)$	$BE(v)$
v_0	1	1
v_7	2	1
v_8	3	1
v_6	4	3
v_5	5	3
v_4	6	3
v_3	7	3
v_1	8	3
v_2	9	3

articulation point v_8; bicomponents $v_8 v_4 v_3$, $v_8 v_1 v_2$, $v_0 v_7 v_8$, $v_8 v_6 v_5$

15.

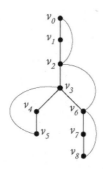

v	$CV(v)$	$BE(v)$
v_0	1	1
v_1	2	1
v_2	3	1
v_3	4	3
v_4	5	4
v_5	6	4
v_6	7	3
v_7	8	7
v_8	9	7

articulation point v_2, v_3, v_6 bicomponents $v_0v_1v_2$, $v_3v_4v_5$, $v_6v_7v_8$

17. If $v_0v_1v_2\cdots v_{k-1}$ is the cycle, then if we begin with v_0, the breadth-first tree consists of two simple paths, $v_0v_1v_2\cdots$ and $v_0v_{k-1}v_{k-2}v_{k-3}\cdots$ so it has the form $\cdots v_2v_1v_0v_{k-1}v_{k-2}v_{k-3}\cdots$. The depth-first tree has the form $v_0v_1v_2\cdots v_{k-1}$. There are k trees, each with one of the k edges removed.

19. Let $\{a,b\}$ be an edge on the graph. If it is not a cut edge, then if $\{a,b\}$ is removed, then the graph is still connected. Hence, there is a spanning tree for the graph not containing $\{a,b\}$. If $\{a,b\}$ is a cut edge, then there exists vertices v_0 and v_1 such that every path from v_0 to v_1 contains the edge $\{a,b\}$. Since v_0 and v_1 belong to any spanning tree and the tree is connected, there must be a path on the tree from v_0 to v_1, and this path must contain the edge $\{a,b\}$. Therefore, $\{a,b\}$ is an edge on every spanning tree.

21. The depth-first tree of K_n is a simple path of the form $v_1v_2v_3\cdots v_k$ or any other order. Since we can get from any vertex to any other vertex, any vertex may be selected at any time in the depth-first search. If a vertex v_i is selected as the beginning vertex in the breadth-first search, then the other vertices are all children of this vertex, since all are adjacent to this vertex.

23. Let E' be a cut set of the graph G. If E' is removed from the graph, then the graph is no longer connected and there exist vertices a and b so that there is no path from a to b. Hence, in the graph G, every path from a to b must contain edges of E'. In any spanning tree of G, a and b belong to the tree. Since the spanning tree is connected, there must be a path from a to b. But this path must contain edges of E'.

25. Kirchoff's matrix $= \begin{bmatrix} n & 0 & -1 & -1 & -1 & \cdots & -1 \\ 0 & n & -1 & -1 & -1 & \cdots & -1 \\ -1 & -1 & n & 0 & 0 & \cdots & 0 \\ -1 & -1 & 0 & n & 0 & \cdots & 0 \\ -1 & -1 & 0 & 0 & n & \cdots & 0 \\ \vdots & \vdots & \vdots & \vdots & & \ddots & 0 \\ -1 & -1 & 0 & 0 & & \cdots & n \end{bmatrix}$

$\begin{vmatrix} n & -1 & -1 & \cdots & -1 \\ -1 & 2 & 0 & \cdots & 0 \\ -1 & 0 & 2 & \cdots & 0 \\ \vdots & \vdots & \vdots & \ddots & 0 \\ -1 & 0 & 0 & \cdots & 2 \end{vmatrix}$

$= \dfrac{1}{2}\begin{vmatrix} 2n & -2 & -2 & -2 & \cdots & -2 \\ -1 & 2 & 0 & 0 & \cdots & 0 \\ -1 & 0 & 2 & 0 & \cdots & 0 \\ -1 & 0 & 0 & 2 & \cdots & 0 \\ \vdots & \vdots & \vdots & \vdots & \ddots & 0 \\ -1 & 0 & 0 & 0 & \cdots & 2 \end{vmatrix}$

$= \dfrac{1}{2}\begin{vmatrix} n & 0 & 0 & 0 & \cdots & 0 \\ -1 & 2 & 0 & 0 & \cdots & 0 \\ -1 & 0 & 2 & 0 & \cdots & 0 \\ -1 & 0 & 0 & 2 & \cdots & 0 \\ \vdots & \vdots & \vdots & \vdots & \ddots & 0 \\ -1 & 0 & 0 & 0 & \cdots & 2 \end{vmatrix}$

$= \dfrac{1}{2}n \times 2^n$

$= n \times 2^{n-1}$

27. Assume G has $n+1$ vertices. Therefore, T and T' both have n edges. Assume that they have k edges in common, so that each tree has $n-k$ edges not in the other. Let G' be the graph with n edges of T' together with the $n-k-1$ edges of T_e not in T'. G' has $n+n-k-1 = 2n-n-1$ edges. To form a spanning tree, $n-k-1$ edges can be removed from circuits. No edge of T_e need be removed since there is always an edge in each circuit not in T_e. This leaves a spanning tree with all of the edges of T_e together with one edge from T'.

29. Remove the edge $\{v_1,v_3\}$ so the first integer in the sequence is 1, and the new tree is the second below. Next remove $\{v_2,v_4\}$ so the sequence is 1, 2, and the new tree is the third below.

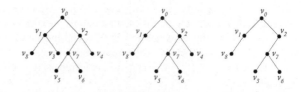

Now remove edge $\{v_7, v_5\}$, so the sequence is 1, 2, 7 and the new tree is the first below. Next remove $\{v_6, v_7\}$, so the sequence is 1, 2, 7, 7 and the new tree is the second below. Then remove $\{v_7, v_2\}$, so the sequence is 1, 2, 7, 7, 2 and the new tree is the third below.

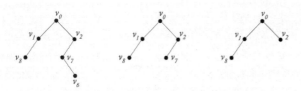

Now remove edge $\{v_0, v_2\}$, so the sequence is 1, 2, 7, 7, 2, 0 and the new tree is given below.

Finally, remove $\{v_0, v_1\}$, so the sequence is 1, 2, 7, 7, 2, 0, 1 and the remaining edge is $\{v_1, v_8\}$. Thus, the answer is 1, 2, 7, 7, 2, 0, 1.

31. Remove the edge $\{v_0, v_2\}$ so the first integer in the sequence is 2, and the new tree is the second below. Next remove $\{v_1, v_2\}$ so the sequence is 2, 2, and the new tree is the third below.

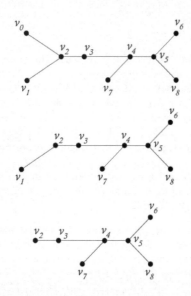

Now remove edge $\{v_2, v_3\}$, so the sequence is 2, 2, 3 and the new tree is the first below. Next remove $\{v_3, v_4\}$, so the sequence is 2, 2, 3, 4 and the new tree is the second below. Then remove $\{v_5, v_6\}$, so the sequence is 2, 2, 3, 4, 5 and the new tree is the third below.

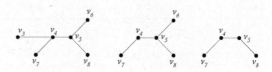

Now remove edge $\{v_4, v_7\}$, so the sequence is 2, 2, 3, 4, 5, 4 and the new tree is given below.

Finally, remove $\{v_4, v_5\}$, so the sequence is 2, 2, 3, 4, 5, 4, 5 and the remaining edge is $\{v_5, v_8\}$. Thus, the answer is 2, 2, 3, 4, 5, 4, 5.

33. Remove the edge $\{v_0, v_2\}$ so the first integer in the sequence is 2, and the new tree is the second below. Next remove $\{v_1, v_2\}$ so the sequence is 22, and the new tree is the third below.

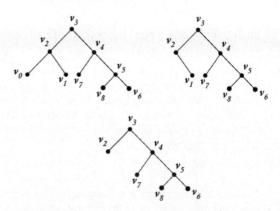

Now remove edge $\{v_2, v_3\}$, so the sequence is 2, 2, 3 and the new tree is the first below. Next remove $\{v_3, v_4\}$, so the sequence is 2, 2, 3, 4 and the new tree is the second below. Then remove $\{v_5, v_6\}$, so the sequence is 2, 2, 3, 4, 5 and the new tree is the third below.

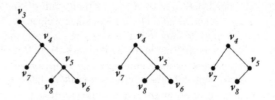

Now remove edge $\{v_7, v_4\}$, so the sequence is 2, 2, 3, 4, 5, 4 and the new tree is given below.

Finally, remove $\{v_4, v_5\}$, so the sequence is 2, 2, 3, 4, 5, 4, 5 and the remaining edge is $\{v_5, v_8\}$. Thus, the answer is 2, 2, 3, 4, 5, 4, 5.

35. Remove the edge $\{v_0, v_2\}$ so the first integer in the sequence is 2, and the new tree is the second below.

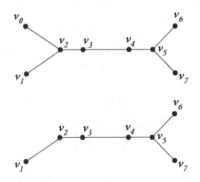

Now remove edge $\{v_1, v_2\}$, so the sequence is 2, 2 and the new tree is the first below. Next remove $\{v_2, v_3\}$, so the sequence is 2, 2, 3 and the new tree is the second below.

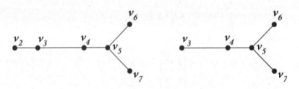

Now remove edge $\{v_3, v_4\}$, so the sequence is 2, 2, 3, 4 and the new tree is the first below. Next remove $\{v_4, v_5\}$, so the sequence is 2, 2, 3, 4, 5 and the new tree is the second below. Finally, remove $\{v_6, v_5\}$, so the sequence is 2, 2, 3, 4, 5, 5 and the new tree is the third below.

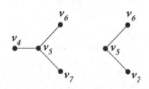

37. Given the sequence 1, 3, 3, 5, 5, 4, 4, we have $\deg(1) = 2$, $\deg(3) = \deg(4) = \deg(5) = 3$, and $\deg(2) = \deg(6) = \deg(7) = \deg(8) = \deg(9) = 1$. Thus, we have the 9-tuple (2, 1, 3, 3, 3, 1, 1, 1, 1). First form edge $\{a_1, a_2\}$, reducing the 9-tuple to (1, 0, 3, 3, 3, 1, 1, 1, 1). Next form the edge $\{a_1, a_3\}$, reducing the 9-tuple to (0, 0, 2, 3, 3, 1, 1, 1, 1). Then form the edge $\{a_3, a_6\}$, reducing the 9-tuple to (0, 0, 1, 3, 3, 0, 1, 1, 1). We now have figure

Then form the edge $\{a_3, a_5\}$, reducing the 9-tuple to (0, 0, 0, 3, 2, 0, 1, 1, 1). Next form the edge $\{a_5, a_7\}$, reducing the 9-tuple to (0, 0, 0, 3, 1, 0, 0, 1, 1). Now form the edge $\{a_4, a_8\}$, reducing the 9-tuple to (0, 0, 0, 2, 1, 0, 0, 0, 1). We now have figure

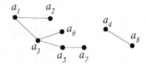

Now form edge $\{a_4, a_5\}$, reducing the 9-tuple to (0, 0, 0, 1, 0, 0, 0, 0, 1). Finally, form the edge $\{a_4, a_9\}$ to finish the tree.

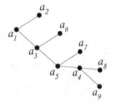

39. Given the sequence 1, 4, 2, 3, 2, 3, we have $\deg(1) = \deg(4) = 2$, $\deg(2) = \deg(3) = 3$ and $\deg(5) = \deg(6) = \deg(7) = \deg(8) = 1$. Thus, we have the 8-tuple (2, 3, 3, 2, 1, 1, 1, 1). First form edge $\{a_1, a_5\}$, reducing the 8-tuple to (1, 3, 3, 2, 0, 1, 1, 1). Then form edge $\{a_1, a_4\}$, reducing the 8-tuple to (0, 3, 3, 1, 0, 1, 1, 1). Next form the edge $\{a_2, a_4\}$, reducing the 8-tuple to (0, 2, 3, 0, 0, 1, 1, 1). We now have figure

Now form the edge $\{a_3, a_6\}$, reducing the 8-tuple to (0, 2, 2, 0, 0, 0, 1, 1). Next form the edge $\{a_2, a_7\}$, reducing the 8-tuple to (0, 1, 2, 0, 0, 0, 0, 1). We now have figure

Now form the edge $\{a_2, a_3\}$, reducing the 8-tuple to (0, 0, 1, 0, 0, 0, 0, 1). Finally, form the edge $\{a_3, a_8\}$ to finish the tree.

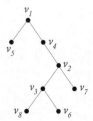

41. Given the sequence $1, 5, 2, 5, 3, 6, 5, 5$, we have $\deg(1) = \deg(2) = \deg(3) = \deg(6) = 2$, $\deg(5) = 5$, and $\deg(4) = \deg(7) = \deg(8) = \deg(9) = \deg(10) = 1$. Thus, we have the 10-tuple $(2, 2, 2, 1, 5, 2, 1, 1, 1, 1)$. First form edge $\{a_1, a_4\}$, reducing the 10-tuple to $(1, 2, 2, 0, 5, 2, 1, 1, 1, 1)$. Now form edge $\{a_1, a_5\}$, reducing the 10-tuple to $(0, 2, 2, 0, 4, 2, 1, 1, 1, 1)$. Next form edge $\{a_2, a_7\}$, reducing the 10-tuple to $(0, 1, 2, 0, 4, 2, 0, 1, 1, 1)$. Then form edge $\{a_2, a_5\}$, reducing the 10-tuple to $(0, 0, 2, 0, 3, 2, 0, 1, 1, 1)$. We now have figure

Now form edge $\{a_3, a_8\}$, reducing the 10-tuple to $(0, 0, 1, 0, 3, 2, 0, 0, 1, 1)$. Next form edge $\{a_3, a_6\}$, reducing the 10-tuple to $(0, 0, 0, 0, 3, 1, 0, 0, 1, 1)$. Then form edge $\{a_5, a_6\}$, reducing the 10-tuple to $(0, 0, 0, 0, 2, 0, 0, 0, 1, 1)$. We now have

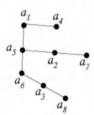

Next form edge $\{a_5, a_9\}$, reducing the 10-tuple to $(0, 0, 0, 0, 1, 0, 0, 0, 1, 0)$. Finally form edge $\{a_5, a_{10}\}$ to form the tree in the following figure.

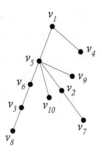

43.
$$K = \begin{bmatrix} 2 & -1 & -1 & 0 & 0 \\ -1 & 2 & -1 & 0 & 0 \\ -1 & -1 & 4 & -1 & -1 \\ 0 & 0 & -1 & 1 & 0 \\ 0 & 0 & -1 & 0 & 1 \end{bmatrix}$$

The cofactor
$$\begin{vmatrix} 2 & -1 & 0 & 0 \\ -1 & 4 & -1 & -1 \\ 0 & -1 & 1 & 0 \\ 0 & -1 & 0 & 1 \end{vmatrix} = 2 \cdot \begin{vmatrix} 4 & -1 & -1 \\ -1 & 1 & 0 \\ -1 & 0 & 1 \end{vmatrix}$$
$$+ 1 \begin{vmatrix} -1 & 0 & 0 \\ -1 & 1 & 0 \\ -1 & 0 & 1 \end{vmatrix}$$
$$= 2 \cdot 2 + (-1)$$
$$= 3$$

Therefore, there are three spanning trees.

45.
$$K = \begin{bmatrix} 2 & -1 & -1 & 0 \\ -1 & 2 & 0 & -1 \\ -1 & 0 & 2 & -1 \\ 0 & -1 & -1 & 2 \end{bmatrix}$$

The cofactor
$$\begin{vmatrix} 2 & 0 & -1 \\ 0 & 2 & -1 \\ -1 & -1 & 2 \end{vmatrix} = 2 \cdot 3 - 1(2) = 4$$

47.
$$K = \begin{bmatrix} 3 & -1 & -1 & -1 & 0 \\ -1 & 3 & -1 & 0 & -1 \\ -1 & -1 & 3 & 0 & -1 \\ -1 & 0 & 0 & 2 & -1 \\ 0 & -1 & -1 & -1 & 3 \end{bmatrix}$$

The cofactor
$$\begin{vmatrix} 3 & -1 & 0 & -1 \\ -1 & 3 & 0 & -1 \\ 0 & 0 & 2 & -1 \\ -1 & -1 & -1 & 3 \end{vmatrix} = 2 \begin{vmatrix} 3 & -1 & -1 \\ -1 & 3 & -1 \\ -1 & -1 & 3 \end{vmatrix}$$
$$+ \begin{vmatrix} 3 & -1 & 0 \\ -1 & 3 & 0 \\ -1 & -1 & -1 \end{vmatrix}$$
$$= 2 \cdot 16 - 8$$
$$= 24$$

There are 24 spanning trees.

Chapter 15—Section 15.6

1. First figure. In sequence, select edges $\{d, c\}$, $\{b, d\}$, and $\{g, h\}$ to give figure

Then select in order $\{e, h\}, \{e, d\}$, and $\{c, f\}$ to give figure

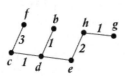

Finally, select edge $\{a,b\}$ to give figure

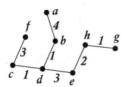

Second figure. In sequence, select edges $\{d, i\}, \{e, f\}$, and $\{e, g\}$ to give figure

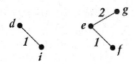

Then select in order $\{a, d\}, \{i, f\}$, and $\{c, d\}$ to give figure

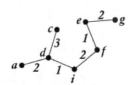

Finally, select edges $\{g, h\}$ and $\{b, d\}$ to give figure

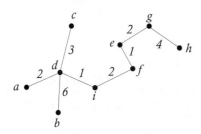

3. First figure. The matrices occur in the following order:

$$\begin{bmatrix} 0 & 4 & 6 & 0 & 0 & 0 & 0 & 0 \\ 4 & 0 & 0 & 1 & 3 & 0 & 0 & 0 \\ 6 & 0 & 0 & 1 & 0 & 3 & 0 & 0 \\ 0 & 1 & 1 & 0 & 3 & 4 & 0 & 0 \\ 0 & 3 & 0 & 3 & 0 & 0 & 5 & 2 \\ 0 & 0 & 3 & 4 & 0 & 0 & 0 & 5 \\ 0 & 0 & 0 & 0 & 5 & 0 & 0 & 1 \\ 0 & 0 & 0 & 0 & 2 & 5 & 1 & 0 \end{bmatrix}$$

$$\Longrightarrow \begin{bmatrix} 0 & 4 & 6 & 0 & 0 & 0 & 0 & * \\ 0 & 0 & 0 & 1 & 3 & 0 & 0 & 0 \\ 0 & 0 & 0 & 1 & 0 & 3 & 0 & 0 \\ 0 & 1 & 1 & 0 & 3 & 4 & 0 & 0 \\ 0 & 3 & 0 & 3 & 0 & 0 & 5 & 2 \\ 0 & 0 & 3 & 4 & 0 & 0 & 0 & 5 \\ 0 & 0 & 0 & 0 & 5 & 0 & 0 & 1 \\ 0 & 0 & 0 & 0 & 2 & 5 & 1 & 0 \\ U & & & & & & & \end{bmatrix}$$

$$\Longrightarrow \begin{bmatrix} 0 & 4 & 6 & 0 & 0 & 0 & 0 & * \\ 0 & 0 & 0 & 1 & 3 & 0 & 0 & * \\ 0 & 0 & 0 & 1 & 0 & 3 & 0 & 0 \\ 0 & 0 & 1 & 0 & 3 & 4 & 0 & 0 \\ 0 & 0 & 0 & 3 & 0 & 0 & 5 & 2 \\ 0 & 0 & 3 & 4 & 0 & 0 & 0 & 5 \\ 0 & 0 & 0 & 0 & 5 & 0 & 0 & 1 \\ 0 & 0 & 0 & 0 & 2 & 5 & 1 & 0 \\ U & U & & & & & & \end{bmatrix}$$

$$\Longrightarrow \begin{bmatrix} 0 & 4 & 6 & 0 & 0 & 0 & 0 & * \\ 0 & 0 & 0 & 1 & 3 & 0 & 0 & * \\ 0 & 0 & 0 & 0 & 0 & 3 & 0 & 0 \\ 0 & 0 & 1 & 0 & 3 & 4 & 0 & 0 & * \\ 0 & 0 & 0 & 0 & 0 & 0 & 5 & 2 \\ 0 & 0 & 3 & 0 & 0 & 0 & 0 & 5 \\ 0 & 0 & 0 & 0 & 5 & 0 & 0 & 1 \\ 0 & 0 & 0 & 0 & 2 & 5 & 1 & 0 \\ U & U & & U & & & & \end{bmatrix}$$

$$\Longrightarrow \begin{bmatrix} 0 & 4 & 0 & 0 & 0 & 0 & 0 & * \\ 0 & 0 & 0 & 1 & 3 & 0 & 0 & * \\ 0 & 0 & 0 & 0 & 0 & 3 & 0 & * \\ 0 & 0 & 1 & 0 & 3 & 4 & 0 & 0 & * \\ 0 & 0 & 0 & 0 & 0 & 0 & 5 & 2 \\ 0 & 0 & 0 & 0 & 0 & 0 & 0 & 5 \\ 0 & 0 & 0 & 0 & 5 & 0 & 0 & 1 \\ 0 & 0 & 0 & 0 & 2 & 5 & 1 & 0 \\ U & U & U & U & & & & \end{bmatrix}$$

$$\Longrightarrow \begin{bmatrix} 0 & 4 & 0 & 0 & 0 & 0 & 0 & * \\ 0 & 0 & 0 & 1 & 3 & 0 & 0 & * \\ 0 & 0 & 0 & 0 & 0 & 3 & 0 & * \\ 0 & 0 & 1 & 0 & 0 & 4 & 0 & 0 & * \\ 0 & 0 & 0 & 0 & 0 & 0 & 5 & 2 & * \\ 0 & 0 & 0 & 0 & 0 & 0 & 0 & 5 \\ 0 & 0 & 0 & 0 & 0 & 0 & 0 & 1 \\ 0 & 0 & 0 & 0 & 0 & 5 & 1 & 0 \\ U & U & U & U & U & & & \end{bmatrix}$$

$$\Rightarrow \begin{bmatrix} 0 & 4 & 0 & 0 & 0 & 0 & 0 & 0 & * \\ 0 & 0 & 0 & 1 & 3 & 0 & 0 & 0 & * \\ 0 & 0 & 0 & 0 & 0 & 3 & 0 & 0 & * \\ 0 & 0 & 1 & 0 & 0 & 4 & 0 & 0 & * \\ 0 & 0 & 0 & 0 & 0 & 0 & 5 & 2 & * \\ 0 & 0 & 0 & 0 & 0 & 0 & 0 & 0 & \\ 0 & 0 & 0 & 0 & 0 & 0 & 0 & 0 & \\ 0 & 0 & 0 & 0 & 0 & 5 & 1 & 0 & * \\ U & U & U & U & U & & & U & \end{bmatrix}$$

$$\Rightarrow \begin{bmatrix} 0 & 4 & 0 & 0 & 0 & 0 & 0 & 0 & * \\ 0 & 0 & 0 & 1 & 3 & 0 & 0 & 0 & * \\ 0 & 0 & 0 & 0 & 0 & 3 & 0 & 0 & * \\ 0 & 0 & 1 & 0 & 0 & 4 & 0 & 0 & * \\ 0 & 0 & 0 & 0 & 0 & 0 & 0 & 2 & * \\ 0 & 0 & 0 & 0 & 0 & 0 & 0 & 0 & \\ 0 & 0 & 0 & 0 & 0 & 0 & 0 & 0 & * \\ 0 & 0 & 0 & 0 & 0 & 5 & 1 & 0 & * \\ U & U & U & U & U & & U & U & \end{bmatrix}$$

$$\Rightarrow \begin{bmatrix} 0 & 4 & 0 & 0 & 0 & 0 & 0 & 0 & * \\ 0 & 0 & 0 & 1 & 3 & 0 & 0 & 0 & * \\ 0 & 0 & 0 & 0 & 0 & 3 & 0 & 0 & * \\ 0 & 0 & 1 & 0 & 0 & 0 & 0 & 0 & * \\ 0 & 0 & 0 & 0 & 0 & 0 & 0 & 2 & * \\ 0 & 0 & 0 & 0 & 0 & 0 & 0 & 0 & * \\ 0 & 0 & 0 & 0 & 0 & 0 & 0 & 0 & * \\ 0 & 0 & 0 & 0 & 0 & 0 & 1 & 0 & * \\ U & U & U & U & U & U & U & U & \end{bmatrix}$$

giving the figure

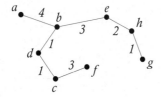

Second figure. The matrices occur in the following order:

$$\begin{bmatrix} 0 & 8 & 4 & 2 & 0 & 0 & 0 & 0 & * \\ 0 & 0 & 0 & 6 & 0 & 0 & 0 & 0 & \\ 0 & 0 & 0 & 3 & 0 & 0 & 0 & 0 & \\ 0 & 6 & 3 & 0 & 4 & 3 & 0 & 0 & 1 \\ 0 & 0 & 0 & 4 & 0 & 1 & 2 & 5 & 0 \\ 0 & 0 & 0 & 3 & 1 & 0 & 6 & 8 & 2 \\ 0 & 0 & 0 & 0 & 2 & 6 & 0 & 4 & 0 \\ 0 & 0 & 0 & 0 & 5 & 8 & 4 & 0 & 0 \\ 0 & 0 & 0 & 1 & 0 & 2 & 0 & 0 & 0 \\ U & & & & & & & & \end{bmatrix}$$

$$\Rightarrow \begin{bmatrix} 0 & 8 & 4 & 2 & 0 & 0 & 0 & 0 & * \\ 0 & 0 & 0 & 0 & 0 & 0 & 0 & 0 & \\ 0 & 0 & 0 & 0 & 0 & 0 & 0 & 0 & \\ 0 & 6 & 3 & 0 & 4 & 3 & 0 & 0 & 1 & * \\ 0 & 0 & 0 & 0 & 0 & 1 & 2 & 5 & 0 \\ 0 & 0 & 0 & 0 & 1 & 0 & 6 & 8 & 2 \\ 0 & 0 & 0 & 0 & 2 & 6 & 0 & 4 & 0 \\ 0 & 0 & 0 & 0 & 5 & 8 & 4 & 0 & 0 \\ 0 & 0 & 0 & 0 & 0 & 2 & 0 & 0 & 0 \\ U & & & U & & & & & \end{bmatrix}$$

$$\Rightarrow \begin{bmatrix} 0 & 8 & 4 & 2 & 0 & 0 & 0 & 0 & * \\ 0 & 0 & 0 & 0 & 0 & 0 & 0 & 0 & \\ 0 & 0 & 0 & 0 & 0 & 0 & 0 & 0 & \\ 0 & 6 & 3 & 0 & 4 & 3 & 0 & 0 & 1 & * \\ 0 & 0 & 0 & 0 & 0 & 1 & 2 & 5 & 0 \\ 0 & 0 & 0 & 0 & 1 & 0 & 6 & 8 & 0 \\ 0 & 0 & 0 & 0 & 2 & 6 & 0 & 4 & 0 \\ 0 & 0 & 0 & 0 & 5 & 8 & 4 & 0 & 0 \\ 0 & 0 & 0 & 0 & 0 & 2 & 0 & 0 & 0 & * \\ U & & & U & & & & & U & \end{bmatrix}$$

$$\Rightarrow \begin{bmatrix} 0 & 8 & 4 & 2 & 0 & 0 & 0 & 0 & * \\ 0 & 0 & 0 & 0 & 0 & 0 & 0 & 0 & \\ 0 & 0 & 0 & 0 & 0 & 0 & 0 & 0 & \\ 0 & 6 & 3 & 0 & 4 & 0 & 0 & 0 & 1 & * \\ 0 & 0 & 0 & 0 & 0 & 0 & 2 & 5 & 0 \\ 0 & 0 & 0 & 0 & 1 & 0 & 6 & 8 & 0 & * \\ 0 & 0 & 0 & 0 & 2 & 0 & 0 & 4 & 0 \\ 0 & 0 & 0 & 0 & 5 & 0 & 4 & 0 & 0 \\ 0 & 0 & 0 & 0 & 0 & 2 & 0 & 0 & 0 & * \\ U & & & U & & U & & & U & \end{bmatrix}$$

$$\Rightarrow \begin{bmatrix} 0 & 8 & 4 & 2 & 0 & 0 & 0 & 0 & * \\ 0 & 0 & 0 & 0 & 0 & 0 & 0 & 0 & \\ 0 & 0 & 0 & 0 & 0 & 0 & 0 & 0 & \\ 0 & 6 & 3 & 0 & 0 & 0 & 0 & 0 & 1 & * \\ 0 & 0 & 0 & 0 & 0 & 0 & 2 & 5 & 0 & * \\ 0 & 0 & 0 & 0 & 1 & 0 & 6 & 8 & 0 & * \\ 0 & 0 & 0 & 0 & 0 & 0 & 0 & 4 & 0 \\ 0 & 0 & 0 & 0 & 0 & 0 & 4 & 0 & 0 \\ 0 & 0 & 0 & 0 & 0 & 2 & 0 & 0 & 0 & * \\ U & & & U & U & U & & & U & \end{bmatrix}$$

$$\Rightarrow \begin{bmatrix} 0 & 8 & 4 & 2 & 0 & 0 & 0 & 0 & * \\ 0 & 0 & 0 & 0 & 0 & 0 & 0 & 0 & \\ 0 & 0 & 0 & 0 & 0 & 0 & 0 & 0 & \\ 0 & 6 & 3 & 0 & 0 & 0 & 0 & 0 & 1 & * \\ 0 & 0 & 0 & 0 & 0 & 0 & 2 & 5 & 0 & * \\ 0 & 0 & 0 & 0 & 1 & 0 & 0 & 8 & 0 & * \\ 0 & 0 & 0 & 0 & 0 & 0 & 0 & 4 & 0 & * \\ 0 & 0 & 0 & 0 & 0 & 0 & 0 & 0 & 0 \\ 0 & 0 & 0 & 0 & 0 & 2 & 0 & 0 & 0 & * \\ U & & & U & U & U & & & U & \end{bmatrix}$$

$$\Rightarrow \begin{bmatrix} 0 & 8 & 0 & 2 & 0 & 0 & 0 & 0 & * \\ 0 & 0 & 0 & 0 & 0 & 0 & 0 & 0 & \\ 0 & 0 & 0 & 0 & 0 & 0 & 0 & 0 & * \\ 0 & 6 & 3 & 0 & 0 & 0 & 0 & 0 & 1 & * \\ 0 & 0 & 0 & 0 & 0 & 0 & 2 & 5 & 0 & * \\ 0 & 0 & 0 & 0 & 1 & 0 & 0 & 8 & 0 & * \\ 0 & 0 & 0 & 0 & 0 & 0 & 0 & 4 & 0 & * \\ 0 & 0 & 0 & 0 & 0 & 0 & 0 & 0 & 0 & \\ 0 & 0 & 0 & 0 & 0 & 2 & 0 & 0 & 0 & * \\ U & U & U & U & U & U & & U & \end{bmatrix}$$

$$\Rightarrow \begin{bmatrix} 0 & 8 & 0 & 2 & 0 & 0 & 0 & 0 & * \\ 0 & 0 & 0 & 0 & 0 & 0 & 0 & 0 & \\ 0 & 0 & 0 & 0 & 0 & 0 & 0 & 0 & * \\ 0 & 6 & 3 & 0 & 0 & 0 & 0 & 0 & 1 & * \\ 0 & 0 & 0 & 0 & 0 & 0 & 2 & 0 & 0 & * \\ 0 & 0 & 0 & 0 & 1 & 0 & 0 & 0 & 0 & * \\ 0 & 0 & 0 & 0 & 0 & 0 & 0 & 4 & 0 & * \\ 0 & 0 & 0 & 0 & 0 & 0 & 0 & 0 & 0 & * \\ 0 & 0 & 0 & 0 & 0 & 2 & 0 & 0 & 0 & * \\ U & U & U & U & U & U & U & U & \end{bmatrix}$$

$$\Rightarrow \begin{bmatrix} 0 & 0 & 0 & 2 & 0 & 0 & 0 & 0 & * \\ 0 & 0 & 0 & 0 & 0 & 0 & 0 & 0 & * \\ 0 & 0 & 0 & 0 & 0 & 0 & 0 & 0 & * \\ 0 & 6 & 3 & 0 & 0 & 0 & 0 & 0 & 1 & * \\ 0 & 0 & 0 & 0 & 0 & 0 & 2 & 0 & 0 & * \\ 0 & 0 & 0 & 0 & 1 & 0 & 0 & 0 & 0 & * \\ 0 & 0 & 0 & 0 & 0 & 0 & 0 & 4 & 0 & * \\ 0 & 0 & 0 & 0 & 0 & 0 & 0 & 0 & 0 & * \\ 0 & 0 & 0 & 0 & 0 & 2 & 0 & 0 & 0 & * \\ U & U & U & U & U & U & U & U & \end{bmatrix}$$

which gives the same tree.

5. First figure. In sequence, select edges $\{f, h\}$, $\{g, h\}$, and $\{i, f\}$ to give figure

Then select in order $\{e, h\}$, $\{e, d\}$, and $\{a, d\}$ to give figure

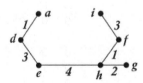

Finally, select edges $\{a, b\}$ and $\{b, c\}$ to give the same figure as before.

Second figure. In sequence, select edges $\{a, b\}$, $\{a, c\}$, and $\{c, d\}$ to give figure

Then select in order $\{d, e\}$, $\{e, f\}$, and $\{d, g\}$ to give figure

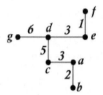

Finally, select edges $\{a, b\}$ and $\{b, c\}$ to give the same figure as before.

7. First figure. In sequence, select edges $\{c, e\}$, $\{f, h\}$, and $\{b, c\}$ to give figure

Then select in order $\{e, h\}$, $\{d, f\}$, and $\{a, d\}$ to give figure

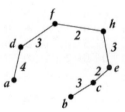

Finally, select edge $\{f, g\}$ to give figure

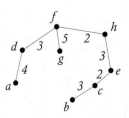

A-100 Answers

Second figure. In sequence, select edges $\{a, d\}, \{d, f\}$, and $\{f, h\}$. Then select in order $\{e, h\}, \{c, e\}$, and $\{b, c\}$ to give figure

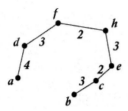

Finally, select edge $\{f, g\}$ to give the same figure as before.

9. First figure. The matrices occur in the following order.

$$\begin{bmatrix} 0 & 0 & 6 & 4 & 0 & 0 & 0 & 0 \\ 0 & 0 & 3 & 0 & 8 & 0 & 0 & 0 \\ 6 & 3 & 0 & 7 & 2 & 0 & 0 & 0 \\ 4 & 0 & 7 & 2 & 0 & 3 & 12 & 0 \\ 0 & 8 & 2 & 0 & 0 & 4 & 0 & 3 \\ 0 & 0 & 0 & 3 & 4 & 0 & 5 & 2 \\ 0 & 0 & 0 & 12 & 0 & 5 & 0 & 6 \\ 0 & 0 & 0 & 0 & 3 & 2 & 6 & 0 \end{bmatrix}$$

$$\Longrightarrow \begin{bmatrix} 0 & 0 & 6 & 4 & 0 & 0 & 0 & 0 & * \\ 0 & 0 & 3 & 0 & 8 & 0 & 0 & 0 \\ 0 & 3 & 0 & 7 & 2 & 0 & 0 & 0 \\ 0 & 0 & 7 & 2 & 0 & 3 & 12 & 0 \\ 0 & 8 & 2 & 0 & 0 & 4 & 0 & 3 \\ 0 & 0 & 0 & 3 & 4 & 0 & 5 & 2 \\ 0 & 0 & 0 & 12 & 0 & 5 & 0 & 6 \\ 0 & 0 & 0 & 0 & 3 & 2 & 6 & 0 \\ U & & & & & & & \end{bmatrix}$$

$$\Longrightarrow \begin{bmatrix} 0 & 0 & 6 & 4 & 0 & 0 & 0 & 0 & * \\ 0 & 0 & 3 & 0 & 8 & 0 & 0 & 0 \\ 0 & 3 & 0 & 0 & 2 & 0 & 0 & 0 \\ 0 & 0 & 7 & 0 & 0 & 3 & 12 & 0 & * \\ 0 & 8 & 2 & 0 & 0 & 4 & 0 & 3 \\ 0 & 0 & 0 & 0 & 4 & 0 & 5 & 2 \\ 0 & 0 & 0 & 0 & 0 & 5 & 0 & 6 \\ 0 & 0 & 0 & 0 & 3 & 2 & 6 & 0 \\ U & & U & & & & & \end{bmatrix}$$

$$\Longrightarrow \begin{bmatrix} 0 & 0 & 6 & 4 & 0 & 0 & 0 & 0 & * \\ 0 & 0 & 3 & 0 & 8 & 0 & 0 & 0 \\ 0 & 3 & 0 & 0 & 2 & 0 & 0 & 0 \\ 0 & 0 & 7 & 0 & 0 & 3 & 12 & 0 & * \\ 0 & 8 & 2 & 0 & 0 & 0 & 0 & 3 \\ 0 & 0 & 0 & 0 & 4 & 0 & 5 & 2 & * \\ 0 & 0 & 0 & 0 & 0 & 0 & 0 & 6 \\ 0 & 0 & 0 & 0 & 3 & 0 & 6 & 0 \\ U & & U & U & & & & \end{bmatrix}$$

$$\Longrightarrow \begin{bmatrix} 0 & 0 & 6 & 4 & 0 & 0 & 0 & 0 & * \\ 0 & 0 & 3 & 0 & 8 & 0 & 0 & 0 \\ 0 & 3 & 0 & 0 & 2 & 0 & 0 & 0 \\ 0 & 0 & 7 & 0 & 0 & 3 & 12 & 0 & * \\ 0 & 8 & 2 & 0 & 0 & 0 & 0 & 0 \\ 0 & 0 & 0 & 0 & 4 & 0 & 5 & 2 & * \\ 0 & 0 & 0 & 0 & 0 & 0 & 0 & 0 \\ 0 & 0 & 0 & 0 & 3 & 0 & 6 & 0 & * \\ U & & U & U & & & & \end{bmatrix}$$

$$\Longrightarrow \begin{bmatrix} 0 & 0 & 6 & 4 & 0 & 0 & 0 & 0 & * \\ 0 & 0 & 3 & 0 & 8 & 0 & 0 & 0 \\ 0 & 3 & 0 & 0 & 0 & 0 & 0 & 0 \\ 0 & 0 & 7 & 0 & 0 & 3 & 12 & 0 & * \\ 0 & 8 & 2 & 0 & 0 & 0 & 0 & 0 & * \\ 0 & 0 & 0 & 0 & 0 & 0 & 5 & 2 & * \\ 0 & 0 & 0 & 0 & 0 & 0 & 0 & 0 \\ 0 & 0 & 0 & 0 & 3 & 0 & 6 & 0 & * \\ U & & U & U & & & U & \end{bmatrix}$$

$$\Longrightarrow \begin{bmatrix} 0 & 0 & 0 & 4 & 0 & 0 & 0 & 0 & * \\ 0 & 0 & 0 & 0 & 0 & 0 & 0 & 0 & * \\ 0 & 3 & 0 & 0 & 0 & 0 & 0 & 0 & * \\ 0 & 0 & 0 & 0 & 0 & 3 & 12 & 0 & * \\ 0 & 8 & 2 & 0 & 0 & 0 & 0 & 0 & * \\ 0 & 0 & 0 & 0 & 0 & 0 & 5 & 2 & * \\ 0 & 0 & 0 & 0 & 0 & 0 & 0 & 0 \\ 0 & 0 & 0 & 0 & 3 & 0 & 6 & 0 & * \\ U & & U & U & U & & & \end{bmatrix}$$

$$\Longrightarrow \begin{bmatrix} 0 & 0 & 0 & 4 & 0 & 0 & 0 & 0 & * \\ 0 & 0 & 0 & 0 & 0 & 0 & 0 & 0 \\ 0 & 3 & 0 & 0 & 0 & 0 & 0 & 0 & * \\ 0 & 0 & 0 & 0 & 0 & 3 & 12 & 0 & * \\ 0 & 0 & 2 & 0 & 0 & 0 & 0 & 0 & * \\ 0 & 0 & 0 & 0 & 0 & 0 & 5 & 2 & * \\ 0 & 0 & 0 & 0 & 0 & 0 & 0 & 0 \\ 0 & 0 & 0 & 0 & 3 & 0 & 6 & 0 & * \\ U & U & U & U & U & U & & U \end{bmatrix}$$

$$\Longrightarrow \begin{bmatrix} 0 & 0 & 0 & 4 & 0 & 0 & 0 & 0 & * \\ 0 & 0 & 0 & 0 & 0 & 0 & 0 & 0 & * \\ 0 & 3 & 0 & 0 & 0 & 0 & 0 & 0 & * \\ 0 & 0 & 0 & 0 & 0 & 3 & 0 & 0 & * \\ 0 & 0 & 2 & 0 & 0 & 0 & 0 & 0 & * \\ 0 & 0 & 0 & 0 & 0 & 0 & 5 & 2 & * \\ 0 & 0 & 0 & 0 & 0 & 0 & 0 & 0 \\ 0 & 0 & 0 & 0 & 3 & 0 & 0 & 0 & * \\ U & U & U & U & U & U & U & U \end{bmatrix}$$

giving the same tree.

Second figure. The matrices occur in the following order:

$$\begin{bmatrix} 0 & 0 & 0 & 2 & 1 & 3 \\ 0 & 0 & 0 & 1 & 3 & 4 \\ 0 & 0 & 0 & 6 & 7 & 4 \\ 2 & 1 & 6 & 0 & 0 & 0 \\ 1 & 3 & 7 & 0 & 0 & 0 \\ 3 & 4 & 4 & 0 & 0 & 0 \end{bmatrix}$$

$$\Rightarrow \begin{bmatrix} 0 & 0 & 0 & 2 & 1 & 3 & * \\ 0 & 0 & 0 & 1 & 3 & 4 & \\ 0 & 0 & 0 & 6 & 7 & 4 & \\ 0 & 1 & 6 & 0 & 0 & 0 & \\ 0 & 3 & 7 & 0 & 0 & 0 & \\ 0 & 4 & 4 & 0 & 0 & 0 & \\ U & & & & & & \end{bmatrix}$$

$$\Rightarrow \begin{bmatrix} 0 & 0 & 0 & 2 & 1 & 3 & * \\ 0 & 0 & 0 & 1 & 0 & 4 & \\ 0 & 0 & 0 & 6 & 0 & 4 & \\ 0 & 1 & 6 & 0 & 0 & 0 & \\ 0 & 3 & 7 & 0 & 0 & 0 & * \\ 0 & 4 & 4 & 0 & 0 & 0 & \\ U & & & U & & & \end{bmatrix}$$

$$\Rightarrow \begin{bmatrix} 0 & 0 & 0 & 2 & 1 & 3 & * \\ 0 & 0 & 0 & 0 & 0 & 4 & \\ 0 & 0 & 0 & 0 & 0 & 4 & \\ 0 & 1 & 6 & 0 & 0 & 0 & * \\ 0 & 3 & 7 & 0 & 0 & 0 & * \\ 0 & 4 & 4 & 0 & 0 & 0 & \\ U & & & U & U & & \end{bmatrix}$$

$$\Rightarrow \begin{bmatrix} 0 & 0 & 0 & 2 & 1 & 3 & * \\ 0 & 0 & 0 & 0 & 0 & 4 & * \\ 0 & 0 & 0 & 0 & 0 & 4 & \\ 0 & 1 & 6 & 0 & 0 & 0 & * \\ 0 & 0 & 7 & 0 & 0 & 0 & * \\ 0 & 0 & 4 & 0 & 0 & 0 & \\ U & U & & U & U & & \end{bmatrix}$$

$$\Rightarrow \begin{bmatrix} 0 & 0 & 0 & 2 & 1 & 3 & * \\ 0 & 0 & 0 & 0 & 0 & 0 & * \\ 0 & 0 & 0 & 0 & 0 & 0 & \\ 0 & 1 & 6 & 0 & 0 & 0 & * \\ 0 & 0 & 7 & 0 & 0 & 0 & * \\ 0 & 0 & 4 & 0 & 0 & 0 & * \\ U & U & & U & U & U & \end{bmatrix}$$

$$\Rightarrow \begin{bmatrix} 0 & 0 & 0 & 2 & 1 & 3 & * \\ 0 & 0 & 0 & 0 & 0 & 0 & * \\ 0 & 0 & 0 & 0 & 0 & 0 & * \\ 0 & 1 & 0 & 0 & 0 & 0 & * \\ 0 & 0 & 0 & 0 & 0 & 0 & * \\ 0 & 0 & 4 & 0 & 0 & 0 & * \\ U & U & U & U & U & U & \end{bmatrix}$$

giving the same tree.

11. First figure. In sequence, select edges $\{a, c\}, \{c, i\}, \{b, i\}$, $\{g, i\}$, and $\{g, j\}$ to give figure

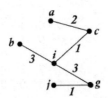

Then select in order $\{g, e\}, \{j, h\}$, and $\{f, j\}$ to give figure

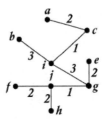

Finally, select edges $\{a, k\}$ and $\{d, f\}$ to give the same figure as before.

Second figure. In sequence, select edges $\{a, d\}, \{d, b\}$, and $\{b, c\}$ to give figure

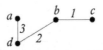

Then select in order $\{c, f\}, \{e, f\}$, and $\{e, g\}$ to give figure

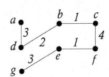

Finally, select edges $\{g, h\}$ and $\{h, i\}$ to give the same figure as before.

Chapter 16—Section 16.1

1.

vertex	flow in	edge	flow out	edge
b	2	(a, b)	3	(b, c)
	1	(e, b)		
c	3	(b, c)	2	(c, z)
			1	(c, d)
d	3	(a, d)	4	(d, e)
	1	(c, d)		

3.

edge	capacity
(d, e)	5
(c, z)	4
total	9

5.

edge	capacity
(b, c)	3
(d, e)	5
total	8

7. 8

9.
edge	capacity
(b, c)	10
(e, z)	8
total	18

11.

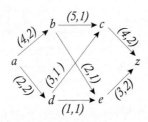

13. Begin with figure

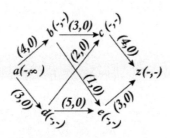

using path *abcz* we have figure

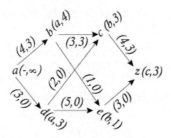

using path *adez* we have figure

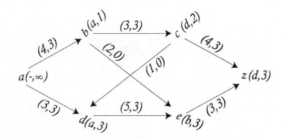

using path *abedcz* we have figure

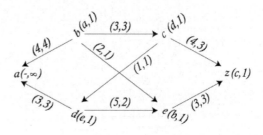

so we have figure

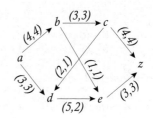

15. Begin with figure

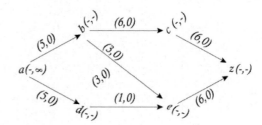

using path *abcz* we have figure

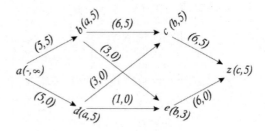

using path *adez* we have figure

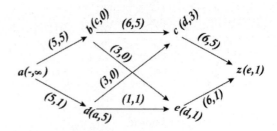

using path *adcbez* we have figure

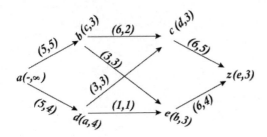

which gives the correct figure.

17. $\{a\}$

19. $\{a, d\}$

Chapter 16—Section 16.2

1. Assume no edges are matching. In the following figure

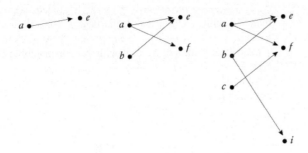

Since there is no match for a, match a to e. Since there is no match for b, match b to e, return to a, and match a to f. Thus, we have $b \to e \leftarrow a \to f$. Using this matching, there is no match for c. Match c with f, return to a, match a with e, and return to b. Then match b with i. Thus, we have $c \to f \leftarrow a \to e \leftarrow b \to i$. Finally there is no match for d, so map d to i and return along the previous route, then map c to h, giving $d \to i \leftarrow b \to e \leftarrow a \to f \leftarrow c \to h$.

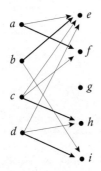

The matching is complete.

3. Assume we have reached the matching in figure

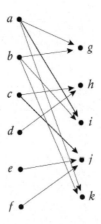

Since e is not matched, match e with j. Return to c and match c with h. This produces figure

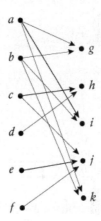

Since b is not matched, match b with i. Return to a and match a with g.

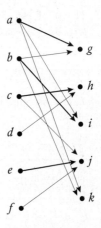

The matching is not complete.

5. Suppose we have arrived at matching $c \to i$, $b \to h$, $d \to k$, $a \to f$, described by matrix

	f	g	h	i	j	k	
a	[1]	0	0	1	0	0	
b	0	1	[1]	0	1	0	
c	1	0	0	[1]	1	0	
d	0	0	0	1	0	[1]	
e	0	0	1	0	0	1	#

After the first pass, we have

	f	g	h	i	j	k	
a	[1]	0	0	1	0	0	
b	0	1	[1]	0	1	0	h
c	1	0	0	[1]	1	0	
d	0	0	0	1	0	[1]	k
e	0	0	1	0	0	1	#
			e		e		

After the second pass, we have

	f	g	h	i	j	k	
a	[1]	0	0	1	0	0	
b	0	1	[1]	0	1	0	h
c	1	0	0	[1]	1	0	
d	0	0	0	1	0	[1]	k
e	0	0	1	0	0	1	#
		b	e	d	b	e	

There is no [1] in column g so g has no match. i has a match c so we have

	f	g	h	i	j	k	
a	[1]	0	0	1	0	0	
b	0	1	[1]	0	1	0	h
c	1	0	0	[1]	1	0	i
d	0	0	0	1	0	[1]	k
e	0	0	1	0	0	1	#
	c	b	e	d	b	e	

Thus, $g \leftarrow b \to h \leftarrow e$ becomes $g \to b \leftarrow h \to e$ and our matrix changes to

	f	g	h	i	j	k	
a	[1]	0	0	1	0	0	
b	0	[1]	1	0	1	0	h
c	1	0	0	[1]	1	0	i
d	0	0	0	1	0	[1]	k
e	0	0	[1]	0	0	1	#
	c	b	e	d	b	e	

and we have matching. It is complete.

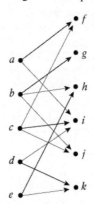

7. Assume we have the matching $c \to k$, $d \to n$, $g \to o$, $h \to p$, $i \to q$. Since a is not matched, we match it with k. We then return to c, and match it with m. Our matching is now $d \to n$, $g \to o$, $h \to p$, $i \to q$, $a \to k, c \to m$. Since e is not matched, we match it with o. We then return to g, and match it with r.

Our matching is now $e \to o, d \to n, h \to p, i \to q, a \to k, c \to m, g \to r$. f is not matched, so we match it with l. Our matching is now $e \to o, d \to n, h \to p, i \to q, a \to k, c \to m, g \to r, f \to l$. Since j is not matched, we match it with p. Our matching is now $e \to o, d \to n, , h \to p, i \to q, a \to k, c \to m, g \to r, f \to l, j \to p$. This matching is maximal. It is not complete.

9. yes

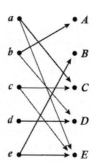

11. yes

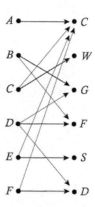

13. Algorithm matrixmatch, (M, m, n)

    M is an $m \times n$ adjacency matrix with brackets around 1's where the corresponding edge is a matching edge. Assume column 0 contains list of vertices to be matched and row 0 contains the vertices to which they are to be matched. Also assume there are row $m + 1$ and column $n + 1$ containing 0's.

    1. Select a vertex v_j from column 0, which is not a matching edge. Place a # in $M(j, n + 1)$. Set $l_j = j$.

    2. For all k such that $M(l_j, k) = 1$, if $M(m + 1, k) = 0$, set $M(m + 1, k) = l_j$ and $r_k = k$.

        For all j, if $M(j, r_k) = [1]$, and $M(j, n + 1) = 0$, set $M(j, n + 1) = l_j$. Set $l_j = j$ and repeat step (2).

    3. If for all j, $M(j, r_k) = 1$,

        (a) set $(l_j, r_k) = [1]$, and $r_k = (l_j, n + 1)$

        (b) set $(l_j, r_k) = 1$, and $l_j = (m + 1, r_k)$

        repeat steps (a) and (b)

(v)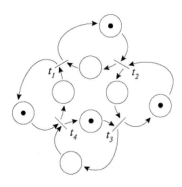

(vi)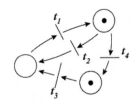

(vii)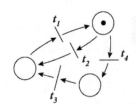

Chapter 16—Section 16.3

1. (i), (iii), (iv), (v), (vii)

3. (i), (ii), (iii), (iv), (v), (vi), (vii)

5. (v), (vi), (vii)

7. (iii), (iv), (v), (vi), (vii)

9. (i) cannot occur

   (ii) cannot occur

   (iii)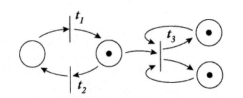

   (iv)

11. (i) original state

    (ii) figure

    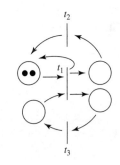

    (iii)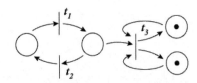

    (iv)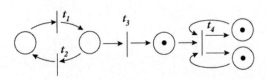

(v) none

(vi) none

(vii) none

13.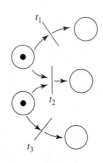

15.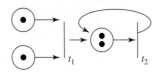

17.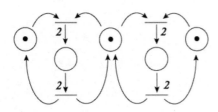

19. yes

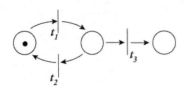

21.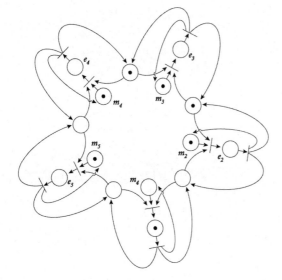

Chapter 17—Section 17.1

1. (a) $\{aba, aca, ada\}$

   (b) $\{a\}^* \circ \{b\}^* \circ \{c\}$, $\{c, abc, aabc, abbc, aabbc, ac, bc, aac, bbc, aaabbbc\}$

   (c) $\{ac, ad, bc, bd\}$

   (d) $(\{a\} \circ \{b\}^* \circ \{\lambda\}) \cup \{cd\}^*$, $\{a, ab, abb, abbb, \lambda, cd, cdcd, cdcdcd, cdcdcdcd, abbbb\}$

   (e) $\{a\} \circ \{bc\}^* \circ \{d\}$, $\{ad, abcd, abcbcd, abcbcbcd, abcbcbcbcd, abcbcbcbcbcd, abcbcbcbcbcbcd\}$

3. (a) $a(b \vee c \vee d)$ (b) $(a \vee b)(b \vee c)$

   (c) ab^* (d) $ab(cd)^*cd$

   (e) $a(\lambda \vee b \vee a \vee ab)b$

5. (a) $(a \vee c)^*b(a \vee c)^*b(a \vee c)^*$

   (b) $a^*ba^*ba^*ca^*ca^* \vee a^*ba^*ca^*ba^*ca^* \vee a^*ca^*ba^*ba^*ca^* \vee a^*ca^*ca^*ba^*ba^* \vee a^*ca^*ba^*ca^*ba^* \vee a^*ba^*ca^*ca^*ba^*$

   (c) $(a \vee b \vee c)^*b(a \vee b \vee c)^*b(a \vee b \vee c)^*$

   (d) $a(a \vee b \vee c)^*b(a \vee b \vee c)^*c(a \vee b \vee c)^* \vee a(a \vee b \vee c)^*c(a \vee b \vee c)^*a)b(a \vee b \vee c)^*$

   (e) $aa^*bb^*cc^*$

7. (b), (c), (e)

9. Uniquely decipherable (a), (b), (c), (d), (e)

   Suffix codes (a), (c), (e)

11. (a) Not uniquely decipherable; 11 followed by 1011 same as 1110 followed by 11

    (b) Uniquely decipherable, prefix code

    (c) Not uniquely decipherable; 10101 followed by 010 same as 1010 followed by 1010

    (d) Uniquely decipherable; suppose there are two different ways of combining code to form the same string. The final code word of one string must be a terminal substring of the final code word of the other. But then the code word preceding the shortest word must end with 0, which is impossible.

    (e) uniquely decipherable, prefix code

13. (a), (b)

15. (d)

17. No, 110011, 1001 is a code that is suffix and prefix, but not infix.

19. (a), (c), (e)

21. (a), (b)

23. No. Suppose one-letter string is 1. Possible three-letter strings are 011, 010, 000, and 001. Suppose 011, 010, and 000 are chosen. Then possible four-letter strings must begin with 001 and are 0010 or 0011. If 0010 is chosen, then five-letter strings must begin with 0011, which is impossible.

25. Let w and w' be elements of a key code. There exists an element of the alphabet that is in w and not in w'. Also exists an element of the alphabet that is in w' and not in w. Hence, neither string can be the initial or terminal string of the other.

Chapter 17—Section 17.2

1. (b), (c), (d)

3. $a^*b(aa^*b)^*$

5. $(a \vee ba)^*$

7.

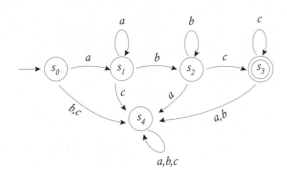

9.

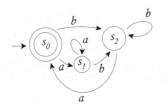

11.

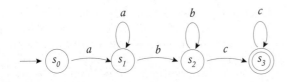

13.

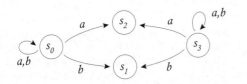

15.

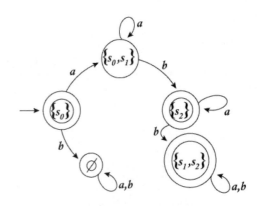

which may be simplified to

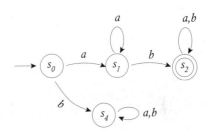

17.

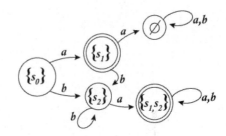

Chapter 17—Section 17.3

1. $A = \{a, b\}$, $\Sigma = \{0, 1\}$, $S = \{s_0, s_1, s_2, s_3\}$. F is given

by the table

F	s_0	s_1	s_2	s_3
a	s_0	s_2	s_3	s_3
b	s_1	s_1	s_0	s_3

. ϕ is given by

the table

s	$\phi(s)$
s_0	1
s_1	0
s_2	1
s_3	1

.

3. 11001110

5. 1000111

7. 01011001

9. 01000110

11.

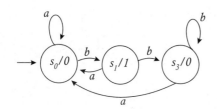

13.

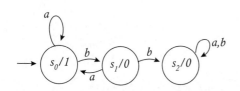

15. $A = \{a, b\}$, $\Sigma = \{0, 1\}$, $S = \{s_0, s_1, s_2\}$. F is given by

the table
F	s_0	s_1	s_2
a	s_1	s_0	s_0
b	s_2	s_2	s_1

. ϕ is given by the

table
ϕ	s_0	s_1	s_2
a	1	0	1
b	1	0	1

.

17. 10110101

19. 1011011

21. 11111000

23. 1111100

25.

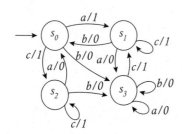

27.

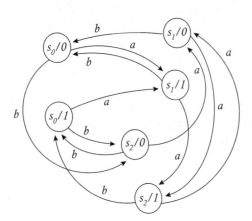

29. To subtract binary string s from binary string t, first use the one's complement Mealy machine to find the 1's complement of s (Example 25) to get string s'. Then use the add one Mealy machine to add one s' to get the 2's complement (Example 26). Call this string s''. Then use the Mealy machine M_+ (Example 27) to add s'' to t.

Chapter 17—Section 17.4

1.

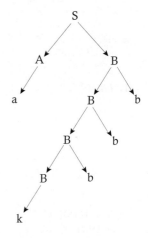

3.

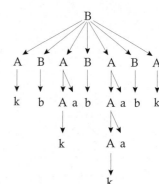

5. **ab(ab)***

7. **b*a b*a (b*ab*ab*)***

Chapter 17—Section 17.5

1. Assume the alphabet is $A = \{a, b, c\}$, $S = \{s_0, s_1, h, s_a, s_b, s_c\}$, c is in the nth square and the string has length m. We then use **go-right**(n) to get to c. Assume we are in state s_0 at this point. Use rule $(s_0, c, s_1, \#, R)$ to replace c with $\#$. Use rules $(s_1, c, s_c, \#, L)$, $(s_1, a, s_a, \#, L)$, $(s_1, b, s_b, \#, L)$ to erase the next letter after c and move back to previous block. Use $(s_c, \#, s_1, c, R)$, $(s_a, \#, s_1, a, R)$, $(s_b, \#, s_1, b, R)$ to place this letter in place of the blank in the previous block and go right to the new blank that has been created. Use $(s_2, \#, s_1, \#, R)$

to get to the next letter. Continue this process until the end of the string, then use $(s_1, \#, h, \#, \#)$ to halt the process.

3. Assume $A = \{a, b\}$, $S = \{s_1, s_2, \ldots, s_{n+1}\}$. The rules for **go-left**(n) are $(s_{n+1}, a, s_n, a, L), \ldots, (s_3, a, s_2, a, L), (s_2, a, s_1, a, L), \ldots, (s_{n+1}, b, s_n, b, L), \ldots, (s_3, b, s_2, b, L), (s_2, b, s_1, b, L)$.

5.

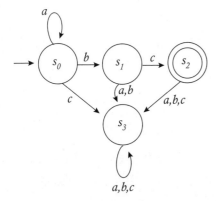

The rules for the Turing machine are $(s_0, a, s_0, \#, R)$, $(s_0, b, s_1, \#, R)$, $(s_0, c, s_4, \#, R)$, $(s_1, c, s_2, \#, R)$, $(s_1, b, s_4, \#, R)$, $(s_1, a, s_4, \#, R)$, $(s_2, a, s_4, \#, R)$, $(s_2, b, s_4, \#, R)$, $(s_2, c, s_b, \#, R)$, $(s_3, a, s_3, \#, R)$, $(s_3, b, s_3, \#, R)(s_3, c, s_3, \#, R)$.

7.

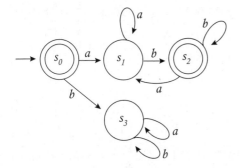

The rules for the Turing machine are $(s_0, a, s_1, \#, R)$, $(s_0, b, s_3, \#, R)$, $(s_1, a, s_1, \#, R)$, $(s_1, b, s_2, \#, R)$, $((s_2, a, s_1, \#, R)$, $(s_2, b, s_2, \#, R)$, $(s_3, a, s_3, \#, R)$, $(s_3, b, s_3, \#, R),(s_2, \#, h, \#, \#), (s_0, \#, h, \#, \#)$.

Chapter 18—Section 18.2

1. (c)

3. (a)
$$G^\perp = \begin{bmatrix} 1 & 1 & 1 & 0 & 1 & 0 & 0 \\ 1 & 1 & 0 & 1 & 0 & 1 & 0 \\ 1 & 0 & 0 & 1 & 0 & 0 & 1 \end{bmatrix}$$

(b)
$$(1111) \begin{bmatrix} 1 & 0 & 0 & 0 & 1 & 1 & 1 \\ 0 & 1 & 0 & 0 & 1 & 1 & 0 \\ 0 & 0 & 1 & 0 & 1 & 0 & 0 \\ 0 & 0 & 0 & 1 & 0 & 1 & 1 \end{bmatrix} = (1111110)$$

$$(0101) \begin{bmatrix} 1 & 0 & 0 & 0 & 1 & 1 & 1 \\ 0 & 1 & 0 & 0 & 1 & 1 & 0 \\ 0 & 0 & 1 & 0 & 1 & 0 & 0 \\ 0 & 0 & 0 & 1 & 0 & 1 & 1 \end{bmatrix} = (0101101)$$

$$(1001) \begin{bmatrix} 1 & 0 & 0 & 0 & 1 & 1 & 1 \\ 0 & 1 & 0 & 0 & 1 & 1 & 0 \\ 0 & 0 & 1 & 0 & 1 & 0 & 0 \\ 0 & 0 & 0 & 1 & 0 & 1 & 1 \end{bmatrix} = (1001100)$$

$$(1010) \begin{bmatrix} 1 & 0 & 0 & 0 & 1 & 1 & 1 \\ 0 & 1 & 0 & 0 & 1 & 1 & 0 \\ 0 & 0 & 1 & 0 & 1 & 0 & 0 \\ 0 & 0 & 0 & 1 & 0 & 1 & 1 \end{bmatrix} = (1010011)$$

(c) 1111, 0111, 0111, 1011

(d)
$$\begin{bmatrix} 1 & 1 & 1 & 0 & 1 & 0 & 0 \\ 1 & 1 & 0 & 1 & 0 & 1 & 0 \\ 1 & 0 & 0 & 1 & 0 & 0 & 1 \end{bmatrix} \begin{bmatrix} 1 \\ 1 \\ 1 \\ 1 \\ 1 \\ 1 \\ 0 \end{bmatrix} = \begin{bmatrix} 0 \\ 0 \\ 0 \end{bmatrix} \text{ yes}$$

$$\begin{bmatrix} 1 & 1 & 1 & 0 & 1 & 0 & 0 \\ 1 & 1 & 0 & 1 & 0 & 1 & 0 \\ 1 & 0 & 0 & 1 & 0 & 0 & 1 \end{bmatrix} \begin{bmatrix} 0 \\ 1 \\ 1 \\ 1 \\ 0 \\ 1 \\ 0 \end{bmatrix} = \begin{bmatrix} 0 \\ 1 \\ 0 \end{bmatrix} \text{ no}$$

$$\begin{bmatrix} 1 & 1 & 1 & 0 & 1 & 0 & 0 \\ 1 & 1 & 0 & 1 & 0 & 1 & 0 \\ 1 & 0 & 0 & 1 & 0 & 0 & 1 \end{bmatrix} \begin{bmatrix} 0 \\ 1 \\ 1 \\ 1 \\ 1 \\ 0 \\ 1 \end{bmatrix} = \begin{bmatrix} 1 \\ 0 \\ 0 \end{bmatrix} \text{ no}$$

$$\begin{bmatrix} 1 & 1 & 1 & 0 & 1 & 0 & 0 \\ 1 & 1 & 0 & 1 & 0 & 1 & 0 \\ 1 & 0 & 0 & 1 & 0 & 0 & 1 \end{bmatrix} \begin{bmatrix} 1 \\ 0 \\ 1 \\ 1 \\ 1 \\ 1 \\ 0 \end{bmatrix} = \begin{bmatrix} 1 \\ 1 \\ 0 \end{bmatrix} \text{ no}$$

5. (a) $G^\perp = \begin{bmatrix} 0 & 1 & 1 & 1 & 0 & 0 \\ 1 & 0 & 1 & 0 & 1 & 0 \\ 1 & 1 & 0 & 0 & 0 & 1 \end{bmatrix}$

(b)

000000	100011	010101	001110	110110	011011	101101	111000
100000	000011	110101	101110	010110	111011	001101	011000
010000	110011	000101	011110	100110	001011	111101	101000
001000	101011	011101	000110	111110	010011	100101	110000
000100	100111	010001	001010	110010	011111	101001	111100
000010	100001	010111	001100	110100	011001	101111	111010
000001	100010	010100	001111	110111	011010	101100	111001
100100	000111	110001	101010	010010	111111	001001	011100

(c) 011001 should be 011011. 111000. 110111 should be 110110. 101100 should be 101101.

7. (a) $G^\perp = \begin{bmatrix} 1 & 0 & 1 & 1 & 0 & 0 \\ 0 & 1 & 1 & 0 & 1 & 0 \\ 1 & 1 & 1 & 0 & 0 & 1 \end{bmatrix}$

(b) See Table A.

(c) 111101 should be 111001. 111001 correct. 110010 should be 110110.

9. 101101, 011011, 110110 $G^\perp = \begin{bmatrix} 1 & 0 & 1 & 1 & 0 & 1 \\ 0 & 1 & 1 & 0 & 1 & 1 \\ 1 & 1 & 0 & 1 & 1 & 0 \end{bmatrix}$

11. Associativity. Let $a = (a_1, a_2, \cdots, a_n)$, $b = (b_1, b_2, \cdots, b_n)$, and $c = (c_1, c_2, \cdots, c_n)$ be elements of B_n.

$(a + b) + c$
$= ((a_1, a_2, \cdots, a_n) + (b_1, b_2, \cdots, b_n))$
$\quad + (c_1, c_2, \cdots, c_n)$
$= (a_1 + b_1, a_2 + b_2, \cdots, a_n + b_n) + (c_1, c_2, \cdots, c_n)$
$= ((a_1 + b_1) + c_1, (a_2 + b_2) + c_2, \cdots, (a_n + b_n) + c_n)$
$= (a_1 + (b_1 + c_1), a_2 + (b_2 + c_2), \cdots, a_n(+b_n + c_n))$
$= (a_1, a_2, \cdots, a_n) + (b_1 + c_1, b_2 + c_2, \cdots, b_n + c_n)$
$= (a_1, a_2, \cdots, a_n) + ((b_1, b_2, \cdots, b_n)$
$\quad + (c_1, c_2, \cdots, c_n))$
$= a + (b + c)$

The identity is $(0, 0, \cdots, 0)$. Each element is its own inverse. Closure follows from the definition of addition. Therefore, B_n is a group.

13. Since $C^\perp$ is a subset of B_n, it is associative. Each element is its own inverse. Therefore, all we need to show is closure. Let $u, v \in C^\perp$. Then $u \circ c = v \circ c = \mathbf{0}$ for all $c \in C$. Therefore, $(u + v) \circ c = u \circ c + v \circ c = \mathbf{0} + \mathbf{0} = \mathbf{0}$ for all $c \in C$, and $u + v \in C^\perp$. Hence, $C^\perp$ is a group.

Chapter 18—Section 18.3

1. $G_H = \begin{bmatrix} 1 & 0 & 0 & 0 & 1 & 1 & 1 \\ 0 & 1 & 0 & 0 & 1 & 1 & 1 \\ 0 & 0 & 1 & 0 & 0 & 1 & 0 \\ 0 & 0 & 0 & 1 & 1 & 0 & 1 \end{bmatrix}$

000000	000	100101	010011	001111	110110	011100	101010	111001
100000	101	000101	110011	101111	010110	111100	001010	011001
010000	011	110101	000011	011111	100110	001100	111010	101001
001000	111	101101	011011	000111	111110	010100	100010	110001
000100	100	100001	010111	001011	110010	011000	101110	111101
000010	010	100111	010001	001101	110100	011110	101010	111001
000001	001	100100	010010	001110	110111	011101	101011	111000
110000	110	010101	100011	111111	000110	101100	011010	001001

TABLE A.

3. $\delta(110010101, 010101111) = 5$

5. 001100110, 101100111, 100000111

7. (a)
$$G_H^\perp = \begin{bmatrix} 1 & 0 & 1 & 1 & 1 & 0 & 0 \\ 1 & 1 & 1 & 0 & 0 & 1 & 0 \\ 1 & 1 & 0 & 1 & 0 & 0 & 1 \end{bmatrix}$$

(b)
$$\begin{bmatrix} 1 & 0 & 1 & 1 & 1 & 0 & 0 \\ 1 & 1 & 1 & 0 & 0 & 1 & 0 \\ 1 & 1 & 0 & 1 & 0 & 0 & 1 \end{bmatrix} \begin{bmatrix} 1 \\ 1 \\ 1 \\ 0 \\ 1 \\ 1 \\ 0 \end{bmatrix} = \begin{bmatrix} 1 \\ 0 \\ 0 \end{bmatrix}$$

1110110 should be 1110010

$$\begin{bmatrix} 1 & 0 & 1 & 1 & 1 & 0 & 0 \\ 1 & 1 & 1 & 0 & 0 & 1 & 0 \\ 1 & 1 & 0 & 1 & 0 & 0 & 1 \end{bmatrix} \begin{bmatrix} 1 \\ 0 \\ 1 \\ 1 \\ 0 \\ 0 \\ 1 \end{bmatrix} = \begin{bmatrix} 1 \\ 0 \\ 1 \end{bmatrix}$$

1011001 should be 1010001

$$\begin{bmatrix} 1 & 0 & 1 & 1 & 1 & 0 & 0 \\ 1 & 1 & 1 & 0 & 0 & 1 & 0 \\ 1 & 1 & 0 & 1 & 0 & 0 & 1 \end{bmatrix} \begin{bmatrix} 1 \\ 1 \\ 0 \\ 1 \\ 0 \\ 1 \\ 0 \end{bmatrix} = \begin{bmatrix} 0 \\ 1 \\ 1 \end{bmatrix}$$

1101010 should be 1001010

$$\begin{bmatrix} 1 & 0 & 1 & 1 & 1 & 0 & 0 \\ 1 & 1 & 1 & 0 & 0 & 1 & 0 \\ 1 & 1 & 0 & 1 & 0 & 0 & 1 \end{bmatrix} \begin{bmatrix} 1 \\ 1 \\ 1 \\ 1 \\ 1 \\ 1 \\ 0 \end{bmatrix} = \begin{bmatrix} 0 \\ 0 \\ 1 \end{bmatrix}$$

1111110 should be 1111111

9. The generating function in Exercise 7 changes 1001001 to 1101001.

The generating function in Exercise 8 changes 1001001 to 1011001. Both generating functions "correct" the code to produce incorrect code.

11. (a) If $c = c'$, then there is no place where the corresponding bits are different. Hence, there are no places where one bit is 1 and the other 0. Therefore, $\delta(c, c') = 0$. Conversely, if $\delta(c, c') = 0$, there is no place where one bit is 1 and the corresponding bit is 0. Therefore, all of the corresponding bits must be the same and $c = c'$.

(b) Let $\delta(c, c') = n$. There are then n places where the corresponding bits of c and c' are different. But then there are n places where the corresponding bits of c' and c are different, and $\delta(c', c) = n$.

13.
$$H_2 = \begin{bmatrix} 1 & 1 \\ 1 & -1 \end{bmatrix} \quad H_4 = \begin{bmatrix} 1 & 1 & 1 & 1 \\ 1 & -1 & 1 & -1 \\ 1 & 1 & -1 & -1 \\ 1 & -1 & -1 & 1 \end{bmatrix}$$

$$H_8 = \begin{bmatrix} 1 & 1 & 1 & 1 & 1 & 1 & 1 & 1 \\ 1 & -1 & 1 & -1 & 1 & -1 & 1 & -1 \\ 1 & 1 & -1 & -1 & 1 & 1 & -1 & -1 \\ 1 & -1 & -1 & 1 & 1 & -1 & -1 & 1 \\ 1 & 1 & 1 & 1 & -1 & -1 & -1 & -1 \\ 1 & -1 & 1 & -1 & -1 & 1 & -1 & 1 \\ 1 & 1 & -1 & -1 & -1 & -1 & 1 & 1 \\ 1 & -1 & -1 & 1 & -1 & 1 & 1 & -1 \end{bmatrix}$$

For H_{16}, see below.

$$H_{16} = \begin{bmatrix} 1 & 1 & 1 & 1 & 1 & 1 & 1 & 1 & 1 & 1 & 1 & 1 & 1 & 1 & 1 & 1 \\ 1 & -1 & 1 & -1 & 1 & -1 & 1 & -1 & 1 & -1 & 1 & -1 & 1 & -1 & 1 & -1 \\ 1 & 1 & -1 & -1 & 1 & 1 & -1 & -1 & 1 & 1 & -1 & -1 & 1 & 1 & -1 & -1 \\ 1 & -1 & -1 & 1 & 1 & -1 & -1 & 1 & 1 & -1 & -1 & 1 & 1 & -1 & -1 & 1 \\ 1 & 1 & 1 & 1 & -1 & -1 & -1 & -1 & 1 & 1 & 1 & 1 & -1 & -1 & -1 & -1 \\ 1 & -1 & 1 & -1 & -1 & 1 & -1 & 1 & 1 & -1 & 1 & -1 & -1 & 1 & -1 & 1 \\ 1 & 1 & -1 & -1 & -1 & -1 & 1 & 1 & 1 & 1 & -1 & -1 & -1 & -1 & 1 & 1 \\ 1 & -1 & -1 & 1 & -1 & 1 & 1 & -1 & 1 & -1 & -1 & 1 & -1 & 1 & 1 & -1 \\ 1 & 1 & 1 & 1 & 1 & 1 & 1 & 1 & -1 & -1 & -1 & -1 & -1 & -1 & -1 & -1 \\ 1 & -1 & 1 & -1 & 1 & -1 & 1 & -1 & -1 & 1 & -1 & 1 & -1 & 1 & -1 & 1 \\ 1 & 1 & -1 & -1 & 1 & 1 & -1 & -1 & -1 & -1 & 1 & 1 & -1 & -1 & 1 & 1 \\ 1 & -1 & -1 & 1 & 1 & -1 & -1 & 1 & -1 & 1 & 1 & -1 & -1 & 1 & 1 & -1 \\ 1 & 1 & 1 & 1 & -1 & -1 & -1 & -1 & -1 & -1 & -1 & -1 & 1 & 1 & 1 & 1 \\ 1 & -1 & 1 & -1 & -1 & 1 & -1 & 1 & -1 & 1 & -1 & 1 & 1 & -1 & 1 & -1 \\ 1 & 1 & -1 & -1 & -1 & -1 & 1 & 1 & -1 & -1 & 1 & 1 & 1 & 1 & -1 & -1 \\ 1 & -1 & -1 & 1 & -1 & 1 & 1 & -1 & -1 & 1 & 1 & -1 & 1 & -1 & -1 & 1 \end{bmatrix}$$

Matrix H_{16} for problem 13, Section 18.3.

Chapter 19—Section 19.1

1. Let $I = \begin{pmatrix} 1 & 2 & 3 \\ 1 & 2 & 3 \end{pmatrix}$, $\rho_1 = \begin{pmatrix} 1 & 2 & 3 \\ 2 & 3 & 1 \end{pmatrix}$,

$\rho_2 = \begin{pmatrix} 1 & 2 & 3 \\ 3 & 1 & 2 \end{pmatrix}$, $\sigma_1 = \begin{pmatrix} 1 & 2 & 3 \\ 1 & 3 & 2 \end{pmatrix}$,

$\sigma_2 = \begin{pmatrix} 1 & 2 & 3 \\ 3 & 2 & 1 \end{pmatrix}$, $\sigma_3 = \begin{pmatrix} 1 & 2 & 3 \\ 2 & 1 & 3 \end{pmatrix}$.

Then we have the following multiplication table.

	I	ρ_1	ρ_2	σ_1	σ_2	σ_3
I	I	ρ_1	ρ_2	σ_1	σ_2	σ_3
ρ_1	ρ_1	ρ_2	I	σ_3	σ_1	σ_2
ρ_2	ρ_2	I	ρ_1	σ_2	σ_3	σ_1
σ_1	σ_1	σ_2	σ_3	I	ρ_1	ρ_2
σ_2	σ_2	σ_3	σ_1	ρ_2	I	ρ_1
σ_3	σ_3	σ_1	σ_2	ρ_1	ρ_2	I

3. $I = \begin{pmatrix} 1 & 2 & 3 & 4 & 5 & 6 & 7 & 8 \\ 1 & 2 & 3 & 4 & 5 & 6 & 7 & 8 \end{pmatrix}$,

$\rho_1 = \begin{pmatrix} 1 & 2 & 3 & 4 & 5 & 6 & 7 & 8 \\ 2 & 3 & 4 & 5 & 6 & 7 & 8 & 1 \end{pmatrix}$,

$\rho_2 = \begin{pmatrix} 1 & 2 & 3 & 4 & 5 & 6 & 7 & 8 \\ 3 & 4 & 5 & 6 & 7 & 8 & 1 & 2 \end{pmatrix}$,

$\rho_3 = \begin{pmatrix} 1 & 2 & 3 & 4 & 5 & 6 & 7 & 8 \\ 4 & 5 & 6 & 7 & 8 & 1 & 2 & 3 \end{pmatrix}$,

$\rho_4 = \begin{pmatrix} 1 & 2 & 3 & 4 & 5 & 6 & 7 & 8 \\ 5 & 6 & 7 & 8 & 1 & 2 & 3 & 4 \end{pmatrix}$,

$\rho_5 = \begin{pmatrix} 1 & 2 & 3 & 4 & 5 & 6 & 7 & 8 \\ 6 & 7 & 8 & 1 & 2 & 3 & 4 & 5 \end{pmatrix}$,

$\rho_6 = \begin{pmatrix} 1 & 2 & 3 & 4 & 5 & 6 & 7 & 8 \\ 7 & 8 & 1 & 2 & 3 & 4 & 5 & 6 \end{pmatrix}$,

$\rho_7 = \begin{pmatrix} 1 & 2 & 3 & 4 & 5 & 6 & 7 & 8 \\ 8 & 1 & 2 & 3 & 4 & 5 & 6 & 7 \end{pmatrix}$,

$\delta_1 = \begin{pmatrix} 1 & 2 & 3 & 4 & 5 & 6 & 7 & 8 \\ 2 & 1 & 8 & 7 & 6 & 5 & 4 & 3 \end{pmatrix}$,

$\delta_2 = \begin{pmatrix} 1 & 2 & 3 & 4 & 5 & 6 & 7 & 8 \\ 4 & 3 & 2 & 1 & 8 & 7 & 6 & 5 \end{pmatrix}$,

$\delta_3 = \begin{pmatrix} 1 & 2 & 3 & 4 & 5 & 6 & 7 & 8 \\ 6 & 5 & 4 & 3 & 2 & 1 & 8 & 7 \end{pmatrix}$,

$\delta_4 = \begin{pmatrix} 1 & 2 & 3 & 4 & 5 & 6 & 7 & 8 \\ 8 & 7 & 6 & 5 & 4 & 3 & 2 & 1 \end{pmatrix}$,

$\eta_1 = \begin{pmatrix} 1 & 2 & 3 & 4 & 5 & 6 & 7 & 8 \\ 1 & 8 & 7 & 6 & 5 & 4 & 3 & 2 \end{pmatrix}$,

$\eta_2 = \begin{pmatrix} 1 & 2 & 3 & 4 & 5 & 6 & 7 & 8 \\ 3 & 2 & 1 & 8 & 7 & 6 & 5 & 4 \end{pmatrix}$,

$\eta_3 = \begin{pmatrix} 1 & 2 & 3 & 4 & 5 & 6 & 7 & 8 \\ 5 & 4 & 3 & 2 & 1 & 8 & 7 & 6 \end{pmatrix}$,

$\eta_4 = \begin{pmatrix} 1 & 2 & 3 & 4 & 5 & 6 & 7 & 8 \\ 7 & 6 & 5 & 4 & 3 & 2 & 1 & 8 \end{pmatrix}$

5. 32 colorings are left fixed by the identity matrix. For the ones left fixed by the reflection permutation, the center band can be either color, the two next to the center can be either color and the two outer bands can be any color. Thus, there are 8 colorings left fixed by the reflection permutation. There are, thus, $\frac{32+8}{2} = 20$ different colorings.

7. Permutations are

$$I = \begin{pmatrix} 1 & 2 & 3 & 4 \\ 1 & 2 & 3 & 4 \end{pmatrix},$$

$$\sigma_1 = \begin{pmatrix} 1 & 2 & 3 & 4 \\ 3 & 4 & 1 & 2 \end{pmatrix},$$

$$\sigma_2 = \begin{pmatrix} 1 & 2 & 3 & 4 \\ 2 & 1 & 4 & 3 \end{pmatrix}$$

and

$$\sigma_3 = \begin{pmatrix} 1 & 2 & 3 & 4 \\ 4 & 3 & 2 & 1 \end{pmatrix}.$$

I leaves 16 colorings fixed, σ_1 leaves 4 colorings fixed, σ_2 leaves 4 colorings fixed, and σ_3 leaves 4 colorings fixed. Therefore, the number of colorings is $\frac{16+4+4+4}{4} = 7$.

Chapter 19—Section 19.2

1.

Permutation	Cycle structure	Number of 2-colorings	Number of 3-colorings
$I = (1)(2)(3)(4)(5)$	c_1^5	32	243
$\rho_1 = (12345)$	c_5	2	3
$\rho_2 = (13524)$	c_5	2	3
$\rho_3 = (14253)$	c_5	2	3
$\rho_4 = (15432)$	c_5	2	3
$\delta_1 = (1)(25)(34)$	$c_1 c_2^2$	8	27
$\delta_2 = (13)(2)(45)$	$c_1 c_2^2$	8	27
$\delta_3 = (15)(3)(24)$	$c_1 c_2^2$	8	27
$\delta_4 = (12)(35)(4)$	$c_1 c_2^2$	8	27
$\delta_5 = (14)(23)(5)$	$c_1 c_2^2$	8	27
Total		80	390

There are $\frac{80}{10} = 8$ two-colorings. There are $\frac{390}{10} = 39$ three-colorings.

3.

Permutation	Structure	Number of 3-colorings
$I = (1)(2)(3)$	c_1^3	27
$\rho_1 = (123)$	c_3	3
$\rho_2 = (132)$	c_3	3
$\sigma_1 = (1)(23)$	$c_1 c_2$	9
$\sigma_2 = (2)(13)$	$c_1 c_2$	9
$\sigma_3 = (12)(3)$	$c_1 c_2$	9
Total		60

There are $\dfrac{60}{6} = 10$ different colorings.

5.

Permutation	Structure	Number of 2-colorings
$I = (1)(2)(3)(4)(5)$	c_1^5	32
$\rho_1 = (12345)$	c_5	2
$\rho_2 = (13524)$	c_5	2
$\rho_3 = (14253)$	c_5	2
$\rho_4 = (15432)$	c_5	2
$\sigma_1 = (1)(25)(34)$	$c_1 c_2^2$	8
$\sigma_2 = (2)(13)(45)$	$c_1 c_2^2$	8
$\sigma_3 = (3)(24)(15)$	$c_1 c_2^2$	8
$\sigma_4 = (4)(35)(12)$	$c_1 c_2^2$	8
$\sigma_5 = (3)(14)(23)$	$c_1 c_2^2$	8
Total		80

Total number of colorings is equal to $\dfrac{80}{10} = 8$.

7.

Permutation	Structure	Number of 2-colorings
$I = (1)(2)(3)(4)(5)(6)(7)(8)$	c_1^8	256
$\rho_1 = (12345678)$	c_8	2
$\rho_2 = (1357)(2468)$	c_4^2	4
$\rho_3 = (14725836)$	c_8	2
$\rho_4 = (15)(26)(37)(48)$	c_2^4	16
$\rho_5 = (16385274)$	c_8	2
$\rho_6 = (1753)(2864)$	c_4^2	4
$\rho_7 = (18765432)$	c_8	2
$\delta_1 = (12)(38)(47)(56)$	c_2^4	16
$\delta_2 = (14)(23)(58)(67)$	c_2^4	16
$\delta_3 = (16)(25)(34)(78)$	c_2^4	16
$\delta_4 = (18)(27)(36)(45)$	c_2^4	16

Permutation	Structure	Number of 2-colorings
$\eta_1 = (1)(28)(37)(46)(5)$	$c_1^2 c_2^3$	32
$\eta_2 = (13)(2)(48)(57)(6)$	$c_1^2 c_2^3$	32
$\eta_3 = (15)(24)(3)(68)(7)$	$c_1^2 c_2^3$	32
$\eta_4 = (17)(26)(53)(4)(8)$	$c_1^2 c_2^3$	32
Total		480

Total number of colorings is equal to $\dfrac{480}{16} = 30$.

9.

Permutation	Cycle structure	Inventory of fixed colorings
$I = (1)(2)(3)(4)$	c_1^4	$(r+b)^4$
$\sigma_1 = (12)(34)$	c_2^2	$(r^2+b^2)^2$
$\sigma_2 = (13)(24)$	c_2^2	$(r^2+b^2)^2$
$\rho = (14)(23)$	c_2^2	$(r^2+b^2)^2$
Total	$c_1^4 + 4c_2^2$	$4r^4 + 4r^3b + 12r^2b^2$ $+4rb^3 + 4b^4$

The inventory of nonequivalent patterns is $r^4 + r^3 b + 3r^2 b^2 + rb^3 + b^4$.

11.

Permutation	Cycle structure	Inventory of fixed colorings
$I = (1)(2)(3)(4)(5)$	c_1^5	$(r+b)^5$
$\rho_1 = (12345)$	c_5	$r^5 + b^5$
$\rho_2 = (13524)$	c_5	$r^5 + b^5$
$\rho_3 = (14253)$	c_5	$r^5 + b^5$
$\rho_4 = (15432)$	c_5	$r^5 + b^5$
$\sigma_1 = (1)(25)(34)$	$c_1 c_2^2$	$(r+b)(r^2+b^2)^2$
$\sigma_2 = (2)(13)(45)$	$c_1 c_2^2$	$(r+b)(r^2+b^2)^2$
$\sigma_3 = (3)(24)(15)$	$c_1 c_2^2$	$(r+b)(r^2+b^2)^2$
$\sigma_4 = (4)(35)(12)$	$c_1 c_2^2$	$(r+b)(r^2+b^2)^2$
$\sigma_5 = (3)(14)(23)$	$c_1 c_2^2$	$(r+b)(r^2+b^2)^2$
Total	$c_1^5 + 4c_5$ $+5c_1 c_2^2$	$10r^5 + 10r^4 b$ $+20r^3 b^2 + 20r^2 b^3$ $+10rb^4 + 10b^5$

The inventory of nonequivalent patterns is $r^5 + r^4 b + 2r^3 b^2 + 2r^2 b^3 + rb^4 + b^5$.

13.

Permutation	Cycle structure	Inventory of fixed colorings
$I = (1)(2)(3)(4)(5)(6)(7)(8)(9)$	c_1^9	$(r+b)^9$
$\rho_1 = (1397)(2684)(5)$	$c_4^2 c_1$	$(r^4+b^4)^2(r+b)$
$\rho_2 = (19)(28)(37)(46)(5)$	$c_2^4 c_1$	$(r^2+b^2)^4(r+b)$
$\rho_3 = (1793)(2486)(5)$	$c_4^2 c_1$	$(r^4+b^4)^2(r+b)$
Total	$c_1^9 + 2c_4^2 c_1 + c_2^4 c_1$	

Total inventory of fix colorings $4r^9 + 12r^8 b + 40r^7 b^2 + 88r^6 b^3 + 136r^5 b^4 + 136r^4 b^5 + 88r^3 b^6 + 40r^2 b^7 + 12r b^8 + 4b^9$.

The inventory of nonequivalent patterns is $r^9 + 3r^8 b + 10r^7 b^2 + 22r^6 b^3 + 34r^5 b^4 + 34r^4 b^5 + 22r^3 b^6 + 10r^2 b^7 + 3r b^8 + b^9$.

15. When only rotations are allowed, we have

Permutation	Cycle structure	Inventory of fixed colorings
$I = (1)(2)(3)(4)(5)(6)(7)(8)(9)(10)$	c_1^{10}	$(r+b)^{10}$
$\rho_1 = (1\ 7\ 10)(286)(349)(5)$	$c_3^3 c_1$	$(r+b)(r^3+b^3)^3$
$\rho_2 = (1\ 10\ 7)(268)(394)(5)$	$c_3^3 c_1$	$(r+b)(r^3+b^3)^3$
Total	$2c_3^3 c_1$	$(r+b)^{10} + 2(r^3+b^3)^3$

The inventory of nonequivalent patterns is $r^{10} + 4r^9 b + 15r^8 b^2 + 42r^7 b^3 + 72r^6 b^4 + 84r^5 b^5 + 72r^4 b^6 + 42r^3 b^7 + 15r^2 b^8 + 4r b^9 + b^{10}$.

Chapter 20—Section 20.1

1. If $ax = b$, then $a^{-1}ax = a^{-1}b$, and $x = a^{-1}b$. If $am = b$ and $an = b$, then $am = an$ so $a^{-1}am = a^{-1}an$ and $m = n$.

3. Assume ab is a zero divisor. Then $(ab)c = 0$ for some c. Hence, $a(bc) = 0$. If $bc = 0$, then b is a zero divisor. If $bc \neq 0$, then a is a zero divisor. Hence, if a and b are not zero divisors, their product is not a zero divisor.

5. 0

7. Assume $ab = ac$ implies that $b = c$ for all nonzero a in R. If $ad = 0 = a0$, and $a \neq 0$, then $d = 0$, and R is an integral domain. Conversely, if R is an integral domain, and $ab = ac$ for $a \neq 0$, then $a(b-c) = 0$. Hence, $b - c = 0$, and $b = c$.

9. Let $i + j$ and $i' + j'$ belong to $I + J$. Then $i + j + i' + j' = i + i' + j + j'$. Since $i + i' \in I$ and $j + j' \in J$, then $i + j + i' + j' \in I + J$. Also $a(i+j) = ai + aj$. Since $ai \in I$ and $aj \in J$, $a(i+j) \in I + J$.

11. $\langle 3 \rangle$

13. $f(0) = 0$. Therefore, $0 \in f(R)$. Let $a, b \in f(R)$. Then $a = f(c)$ and $b = f(d)$ for some $c, d \in R$. Therefore, $c - d \in R$ and $f(c-d) = f(c) - f(d)$. Hence, $f(c) - f(d) \in f(R)$. Similarly, if $a, b \in f(R)$, $a = f(c)$ and $b = f(d)$ for some $c, d \in R$, then $cd \in R$ and $f(cd) = f(c)f(d)$. Therefore, $ab \in f(R)$. Hence, $f(R)$ is a ring.

15. No, consider $f : Z \to Z_{10}$.

17. False. [2] and [3] are zero divisors of Z_{12}. However, $[2] + [3] = [5]$ is not a zero divisor of Z_{12}.

19. Let R be an integral domain. If $a^2 = a$, then $a \cdot a = a \cdot 1$, and by a previous problem, if $a \neq 0$, we can cancel and $a = 1$.

21. For all $a, b \in R$,

$$(-a)(b) + ab = (-a + a)b$$
$$= 0b$$
$$= 0$$

Therefore, $(-a)(b) = -ab$. Similarly, $a(-b) = -ab$. For all $a, b \in R$,

$$(-a)(-b) + (-a)(b) = (-a)(-b + b)$$
$$= (-a)0$$
$$= 0$$

Therefore, $(-a)(-b) + (-a)(b) = ab + (-a)(b)$ and $(-a)(-b) = ab$.

Chapter 20—Section 20.2

1. We first show that the intersection of subrings with unity is a ring with unity. Since the multiplicative identity is in all of the subrings, it is in the intersection. Let a, b belong to the intersection of the subrings, then a, b belong to each of the subrings. Therefore, $a - b$ is in all of the subrings and, hence, in the intersection of the subrings. Similarly, ab is in the intersection of all of the subrings. Therefore, the intersection of subrings with unity is a ring with unity. We next show that the intersection of integral domains is an integral domain. Assume a and b are in the intersection of all of the subrings, $ab = 0$ and $a \neq 0$. But a and b are in each of the integral domains. Therefore, $b = 0$.

3. Either $a > 0$ or $-a > 0$. If $a > 0$, then by definition of the ordering, $a^2 > 0$. If $-a > 0$, then $(-a)^2 > 0$. But $(-a)^2 = a^2$, so $a^2 > 0$. Since $1^2 = 1$, $1 > 0$.

Chapter 20—Section 20.3

1. Let $f = (a_i)^*$, $g = (b_i)^*$, and $h = (c_i)^*$.

$$f + (g + h) = (a_i)^* + (b_i + c_i)^*$$
$$= (a_i + (b_i + c_i))^*$$
$$= ((a_i + b_i) + c_i)^*$$
$$= (a_i + b_i)^* + (c_i)^*$$
$$= (f + g) + h$$

so addition is associative. The additive identity is $(d_i)^*$ where $d_i = 0$ for all i. The additive inverse of $(a_i)^*$ is $(-a_i)^*$.

$$f + g = (a_i)^* + (b_i)^*$$
$$= (a_i + b_i)^*$$
$$= (b_i + a_i)^*$$
$$= (b_i)^* + (a_i)^*$$
$$= g + f$$

and addition is commutative. Therefore, the set of polynomials is a commutative group under addition.

$$f(gh) = (a_i)^*((b_i)^*(c_i)^*)$$
$$= (a_i)^* \Big(\sum_{i+j=k} b_i c_j \Big)^*$$
$$= \Big(\sum_{m+k=n} a_m \sum_{i+j=k} b_i c_j \Big)^*$$
$$= \Big(\sum_{m+i+j=n} a_m b_i c_j \Big)^*$$
$$= \Big(\sum_{s+j=n} \Big(\sum_{m+i=s} a_m b_i \Big) c_j \Big)^*$$
$$= \Big(\sum_{m+i=s} a_m b_i \Big)^* (c_j)^*$$
$$((a_i)^*(b_i)^*)(c_i)^*$$
$$= (fg)h$$

and multiplication is associative. Finally

$$f \cdot (g + h) = (a_i)^*(b_i + c_i)^*$$
$$= \Big(\sum_{i+j=k} a_i(b_j + c_j) \Big)^*$$
$$= \Big(\sum_{i+j=k} (a_i b_j + a_i c_j) \Big)^*$$
$$= \Big(\sum_{i+j=k} a_i b_j \Big)^* + \Big(\sum_{i+j=k} a_i c_j \Big)^*$$
$$= fg + fh$$

3. False. Let $f = (1, 1, 1, 1, 0, 0, \cdots)$ and $g = (1, 1, -1, -1, 0, 0, \cdots)$. Then $f + g = (1, 1, 0, 0, 0, 0, \cdots)$. $\deg(f) = \deg(g) = 3$ while $\deg(f + g) = 1$.

5. In Z_5, $f(x) = x^4 - 1$ is the zero function while f has degree 4.

Chapter 20—Section 20.4

1. Let $r = a + bi$ and $s = c + di$ be Gaussian integers. Then $r - s = (a - c) + (b - d)i$ is also a Gaussian integer. Also $rs = (a + bi)(c + di) = (ac - bd) + (ad + bc)i$ is also a Gaussian integer. Therefore, the set of Gaussian integers is a subring.

3. $\sin(n\pi) = 0$ for every integer n. Therefore, it has an infinite number of zeroes and so $\sin(x)$ cannot be a polynomial function.

5. Let $A = Z$, and $f, g \in A[x]$ be defined by $f(x) = x^2 + 1$ and $g(x) = x^2 - 3$, respectively, then f and g are prime, but $f + g$ is not prime since $f + g(x) = 2x^2 - 2$.

7. Let $c = a + bi$ be a Gaussian number. Then

$$f(x) = (x - (a + bi))(x - (a - bi))$$
$$= (x - a) + bi)(x - a) - bi)$$
$$= (x - a)^2 + b^2$$

so $f \in Q[x]$ and $f(c) = 0$.

9. The subring consisting of all numbers of the form $a + b\sqrt{5}$ for rational numbers a, b.

11. The subring consisting of all numbers of the form $a + b\sqrt{3}i$ for rational numbers a, b.

Chapter 21—Section 21.2

1. $1 = \chi(gg^{-1}) = \chi(g)\chi(g^{-1}) = (a + bi)\chi(g^{-1})$. Therefore,

$$\chi(g^{-1}) = \frac{1}{a + bi} = \frac{a - bi}{a^2 + b^2}$$

But since $a + bi$ is on the unit circle, $a^2 + b^2 = 1$, so $\chi(g^{-1}) = a - bi$.

3. If $g \neq g'$, then $g_1^{j(1)} g_2^{j(2)} g_3^{j(3)} \cdots g_m^{j(m)}$ and $g' = g_1^{j'(1)} g_2^{j'(2)} g_3^{j'(3)} \cdots g_m^{j'(m)}$, where for some i, $g_i^{j(i)} \neq g_i^{j'(i)}$. Then $\chi_g(g_i) = a_i^{j(i)}$ and $\chi_{g'}(g_i) = a_i^{j'(i)}$. Since $a_i^{j(i)} \neq a_i^{j'(i)}$, $\chi_g(g_i) \neq \chi_{g'}(g_i)$ and $\chi_g \neq \chi_{g'}$.

Chapter 21—Section 21.3

1. Let $i \in I$ and $s \in S$. $\chi(i \cdot s) = \chi(i) \cdot \chi(s) = 0 \cdot \chi(s) = 0$, and $i \cdot s \in I$. Therefore, I is an ideal.

3. Let $C = [C_{ij}] = \sigma\sigma'$. Then

$$C_{ij} = \sum_{k=1}^{n} \sigma_{ik}\sigma'$$

$$= \sum_{k=1}^{n} \sigma_{ik}\text{conj}(\sigma_{jk})$$

$$= \sum_{k=1}^{n} \chi_i(g_k)\text{conj}(\chi_j(g_k))$$

$$= \sum_{k=1}^{n} \chi_i(g_k)(\chi_j(g_k))^{-1}$$

$$= n\delta_{ij}$$

by Theorem 21.19.

Chapter 22—Section 22.1

1. Let $N(W) = x_1 2^{n-1} + x_2 2^{n-2} + \cdots + x_{n-1} 2^1 + x_n$
$N(V) = y_1 2^{n-1} + y_2 2^{n-2} + \cdots + y_{n-1} 2^1 + y_n$.
Then
$$|N(W) - N(V)| \leq |x_1 - y_1| 2^{n-1} + |x_2 - y_2| 2^{n-2}$$
$$+ \cdots + |x_{n-1} - y_{n-1}| 2^1$$
$$+ |(x_n - y_n)|$$
$$= \sum_{i=1}^{n} |x_i - y_i| 2^{n-i}$$
$$\leq \sum_{i=1}^{n} 2^{n-i}$$
$$= 2^n - 1$$

and $2^n - 1$ is less than 2^n.

3. $w = 7, 8, 11, 12, 13, 14, 15, 16$

5. k words for storage of k primes.

k words for storage of $N_p(X)$ for k primes p.

k words for storage of $N_p(Y(i))$ for k primes p.

k words for storage of $[[2^{n-1}]]_p$

Small amount for loops, etc.

Therefore, approximately $4k$ words.

Chapter 22—Section 22.2

1. If $d|a$ and $d|b$, then $d|a - b$. Therefore, $d|\gcd(a - b, a)$. Therefore, $\gcd(a, b) | \gcd(a - b, a)$. Conversely, if $d|a$ and $d|a - b$, then $d|b$, and, therefore, $d|\gcd(a, b)$. Thus, $\gcd(a - b, a) | \gcd(a, b)$. Therefore, $\gcd(a - b, a) = \gcd(a, b)$.

3. $D = \text{lcm}(5, 10, 15, 20) = 60$.

5.
$$\frac{(i-1)m_i}{n} \leq \left\lceil \frac{(i-1)m_i}{n} \right\rceil$$
$$\leq \frac{(i-1)m_i}{n} + 1 < \frac{(i-1)m_i}{n} + \frac{m_i}{n}$$

since $m_i > n$. Therefore,

$$(i-1)m_i \leq n\left\lceil \frac{(i-1)m_i}{n} \right\rceil < \frac{im_i}{n}$$

If $kn < n\left\lceil \frac{(i-1)m_i}{n} \right\rceil$ then $k < \left\lceil \frac{(i-1)m_i}{n} \right\rceil$. But since k is an integer, $k < \frac{(i-1)m_i}{n}$ and $kn < (i-1)m_i$.

7.

i	k_i	m_i	M_i	b_i	N_i	$M_i b_i N_i$
1	12	145	26551252	532	0	0
2	14	169	22780660	5	43	4897841900
3	15	181	21270340	711	91	1376212268340
4	18	217	17741620	81	163	234242608860

$D = 12$, $C = 2231403840$,

$$f(12) = \left[\left[\left\lfloor \frac{2231403840}{12 \times 12 + 1} \right\rfloor\right]\right] = 0,$$

$$f(14) = \left[\left[\left\lfloor \frac{2231403840}{12 \times 14 + 1} \right\rfloor\right]\right] = 1,$$

$$f(15) = \left[\left[\left\lfloor \frac{2231403840}{12 \times 15 + 1} \right\rfloor\right]\right] = 2,$$

$$f(18) = \left[\left[\left\lfloor \frac{2231403840}{12 \times 18 + 1} \right\rfloor\right]\right] = 3$$

Chapter 22—Section 22.3

1.
$$[[1683^{217}]]_{2993} = 77$$
$$[[79^{217}]]_{2993} = 79$$
$$[[2560^{217}]]_{2993} = 78$$
$$[[872^{217}]]_{2993} = 69$$
$$[[2571^{217}]]_{2993} = 89$$

3. To encrypt one must have values of e and n. If one knows $\phi(n)$, then using the formulas in Theorem 22.15, one can find d and then be able to decrypt.

INDEX

∨, 151
∧, 151
$\Gamma(L)$, 673
δ, 701
↓, 29, 40
≡, 129
∞, 551
λ, 648
|, 29
ϕ function, 404
ϕ, 401
$\pi(w)$, 764
σ^2, 343
$\sqrt{2}$, 27
$\triangle$, 434
$\triangle(x^{-n})$, 443
$\triangle^n f$, 434
$\Omega(g(n))$, 182
$\Theta(g(n))$, 182
↔, 29
⊙, 132
⊕, 132
£(a), 649
≺, 161
≺, 313
⪯, 161
~, 723
$\triangle^2 f(x)$, 434
$\triangle^n(f(x))$, 435
$\triangle f(x)$, 434
∨, 29
1's complement, 666
2's complement, 666

A^*, 648
$[A]_R$, 84
a-arrow, 653
$A \times B$, 58, 72
$A \wedge B$, 551
$a > b$, 108
$a \geq b$, 108
$\langle a, b \rangle$, 725
Abelian, 360
Abelian group, 752
about the set N, 107
Absolute value, 751
$A \odot C$, 152
acceptance state, 684
Ack, 174
Ackerman's function, 174
acyclic, 564

Addition counting principle, 292
additive identity, 105
additive inverse, 105
adj(A), 216
Adjacency matrix, 250, 529, 540
adjacent, 233
adjective, 676
adjoint, 216
Adleman, L. M., 773
a is related to b, 72
Alfred Bay Kempe, 539
Algebraic integer, 736, 742
Algebraic number, 736, 742
Algebraic over, 742
algebraic structures, 724
Algorithm Isomorphic Binary Tree, 586
Algorithm, 168, 172, 271
Algorithms
    Breadth-first search, 589
    Bubble Sort, 188
    Depth-first search, 591
    Dijkstra's, 554
    Euclidean, 274
    Floyd-Warshall, 556
    Horner's, 169
    Huffman's, 576
    Kirchoff, 602
    maximum flow, 626
    Merge Sort, 191
    Topological Sort, 169
    Tree to sequence, 598
    Warshall's, 253
Alphabet, 361, 648
$a \mid m$, 276
Amicable numbers, 123
annuity, 175
antidifference operator, 447
approximation of area, 434
Argument, 19
    addition, 23
    case elimination, 24
    cases, 24
Argument
    conjunction, 24
    contrapositive, 107
    hypothesis, 19
    law of detachment, 23
    modus ponens, 23
    modus tollens, 23
    non sequitur, 25

    premise, 19
    proof, 25
    reductio ad absurdum, 24, 229
    specialization, 24
    syllogism, 23
    valid, 19
Arguments, 19
arithmetic sequence, 144
article, 676
Articulation point, 521, 594, 596
ASCII code, 692
associative, 105, 372
Associativity, 367
asymmetric, 237
atom, 82, 368
augmenting the flow, 625
Augustus De Morgan, 539
automata, 640, 652
automation, 234
Automaton, 652
    accepted by, 653
    deterministic, 655
    language accepted by, 652
    nondeterministic, 655
    sink state, 654
    state diagram, 653
    transition function, 652
automaton, 652, 662, 683
automorphism, 382
Axiom system
    equality, 105
    Gödel-Hilbert-Bernays, 51
    integers, 105
    Russell-Whitehead, 51
    Zermelo-Fraenkel-von Neumann, 51
Axiomatic Systems, 19, 51
axioms, 1, 19, 107

back edge, 591
Bayes' Theorem, 346
Bernoulli trials, 339, 341
Bernoulli, Jean, 404
bicomponents, 605
big-Oh, 182
big-Omega, 182
big-Theta, 182
bijection, 89
Binary, 67, 199
binary connectives, 4
Binary operation, 142, 360, 361

I-1

binary search tree, 569
Binary trees, 239, 566, 569
    binary search tree, 569
    isomorphic binary tree, 567
    traversing, 582
Binomial distribution, 339
Binomial theorem, 307, 691
bipartite graph, 529, 542, 543, 632, 634
birthday problem, 320
Bit String, 292
Block code, 689
Boolean algebra, 38, 67, 367
    absorption laws, 68
    associative properties, 67
    commutative properties, 67
    complement of identities laws, 69
    Complement properties, 68
    De Morgan's laws, 69
    distributive properties, 68
    Identity properties, 68
    idempotent laws, 68
    involution laws, 69
    null laws, 68
Boolean algebra notation, 358
Boolean Algebras, 67
Boolean expression, 45
Boolean operation, 151
bounded distributive lattice, 367
bounded homomorphisms, 366
Boyer, R. S., 766
breadth-first search method, 589
Breadth-first tree search, 589
Burali-Forti, 51
Burnside counting lemma, 712
Burnside's equation, 712

$c + A$, 551
$C^{\perp}$, 695
$C_G(\lambda)$, 536
C_n, 567
$C(n, r)$, 305
C. C. Chang, 768
C. E. Shannon, 38
calculus, 404, 434
Cantor, Georg, 51
capacity, 620, 632
Cardinality, 156
    countable, 157, 159
    uncountable, 159
cardinality, 161
cardinality of a finite set, A, 53
cardinality of sets, 158
Cartesian coordinate system, 751
Cat(0), 173
Cat($n + 1$), 173
Catalan number, 173, 424, 459, 464, 567

Cayley's Theorem, 383
Cayley's tree formula, 597
ceiling function, 769
chain, 625
chains of matrices, 154
Chang, C. C., 769
Character
    Dirichlet, 756
    group, 751
    principal, 752
    semigroup, 755
characteristic equation, 419
characteristic polynomial, 418
Chebyshev's Inequality, 344, 345
children, 566
Chinese Remainder Theorem, 396, 397, 768
Chromatic number, 536
Chromatic polynomial, 536, 542
chromatic rectangle, 717
Church's thesis, 683
Ciphertext, 772
Circuit diagram, 38
    and gate, 39
    full-adder, 42
    half-adder, 40
    nand gate, 40
    negation gate, 39
    nor gate, 40
    or gate, 39
    switching circuit, 38
Circuit diagram, 29, 38, 45
class of regular expressions, 649
cl(G), 547
closed, 105
closed form, 416
closure, 372, 547
coconuts, 397
Code, 575, 650
    ASCII code, 692
    block code, 689
    comma code, 690
    dual code, 695
    error correcting codes, 690
    error detecting codes, 690
    Golay code, 703
    Gray code, 691
    group code, 692
    Hadamard code, 704
    Hamming code, 700
    Huffman code, 690
    infix code, 652
    instantaneous code, 690
    key code, 650
    linear code, 692
    Morse code, 690
    prefix code, 575, 650, 690

    Reed-Muller code, 704
    suffix code, 650
    uniquely decipherable, 575, 650
    weight, 693
    weight of, 576
code, 689
code theory, 689
codomain, 87
coefficient, 308
cofactor, 214
cofactor matrix, 216
cofinite, 71
collision, 134
colored edges, 332
Coloring Graphs, 534
Colorings, 536, 708
    inventory, 715
column matrix, 147
Combinations, 305, 326
    generalized, 322
    generating, 314
    with repetition, 327
common divisor, 121
common multiple, 122
common simple cycle, 522
commutative, 105
commutative monoid, 361
commutative ring with unity, 731
commutative semigroup, 360
Commutativity, 367
commutators, 382
compiler, 767
Complements, 68, 317, 334, 367, 368
complement graph, 529
complete, 109, 516, 566
complete bipartite graph, 295, 529, 531
complete graphs, 332
Complete matching, 632, 634
complete set of reduced residues modulo n, 411
Complex number, 750
    absolute value, 751
    conjugate, 751
    imaginary part, 750
    length, 751
    real part, 750
complex numbers, 421, 722, 736, 750, 753
Complex plane, 751
Complexity, 181
    big-oh, 182
Complexity of Algorithms, 181
component, 147
Composite, 125, 406
composition, 380
compound statement, 2

Concatenation, 361, 648
conclusion, 20
concurrent processes, 639
conditional, 13
Conditional statements, 9, 18
Congruence, 129, 393
Congruence equations, 393
Congruence Relations, 129, 362, 363
conj(c), 751
connected, 75, 227, 516
connected graph, 546, 564
connected planar graph, 530
connected subgraph, 231
Connective, 1
    and, 1
    biconditional, 10, 17
    conditional, 9
    conjunction, 2
    disjunction, 2
    exclusive or, 4
    if ... then, 1
    if and only if, 1
    nand, 30
    negation, 2, 4
    nor, 30
    not, 1
    only if, 1
    or, 1
connectives, 1
consecutive and alternate convergents, 287
conservation of flow, 625
consistent, 109
Constant, 731
Constant complexity, 185
context-free grammars, 680
Continued fraction
    finite, 281
    simple, 281
Continued Fractions, 278, 282
continued fraction representation, 281, 283
Contradiction, 15
Contrapositive, 13m 15, 18, 26
Convergents, 281, 283
converse, 13
Converse statement, 18
convex polygon, 463
Coset, 375
    left, 739
coset, 695
coset leaders, 694
countable, 158
countable set, 158
countably infinite, 157, 158
countably infinite set, 163
counterexample, 111

Counting, 290, 316, 455, 485, 496
counting techniques, 308
Counting tree, 290
Cramer's rule, 213
Cryptanalysis, 772
Cryptography, 772, 689
Cut, 623
    capacity of, 623
    minimal, 624
cut (S, T), 630
cut edge, 545
cut vertex, 521
cut vertices, 569
$Cy(\sigma)$, 713
Cycle, 519
    Euler, 244
    Hamiltonian, 543
Cycle index, 714
Cycle structure, 713
cyclic group, 374, 404, 411
cyclic semigroup, 362
cyclic subgroups, 752

De Moivre, 421
De Morgan's law for set theory, 57
deadlocked, 642
decipherment, 772
decompose a continued fraction, 281
Decryption, 772, 773
Deducible, 19
definite integrals, 449
definite sums, 449
$\deg(f)$, 731
degree, 518
Degree matrix, 602
denial of a universally quantified predicate, 99
depth-first search method, 589
depth-first spanning tree, 594
Depth-first tree search, 591
Derangements, 464, 466
Derivable, 19
derivation of combinatorial functions, 434
derivation tree, 679
derived, 673
DES
    U. S. Data Encryption Standard, 776
descendant, 594
$\det(A)$, 214
determinant, 213, 364, 377
deterministic algorithms, 185
Deterministic automata, 655
DFT articulation point search algorithm, 605
difference relation, 447

Diffie, W., 775
Digital Signatures, 773
digraphs, 156
Dijkstra's Algorithm, 551, 554
Dijkstra's second algorithm, 553
dimension, 147
dining philosophers' problem, 642, 644
Diophantine Equations, 391
direct proof, 26
Direct sum, 752
directed arrow, 653
directed cycle, 236
directed edge, 76, 233, 653
Directed graph, 76, 233
    connected, 235
    directed cycle, 246
    directed edge, 76, 233
    directed graph, 234
    indegree of vertex, 233
    initial vertex, 76, 233
    labeled, 234
    outdegree of vertex, 233
    sink, 233
    source, 233
    strongly connected, 235
    subgraph, 234
    terminal vertex, 76, 233
    underlying graph, 235
directed graph, 233, 564, 620, 653
directed multigraph, 234
Directed tree, 564
    balanced, 566
    full m-ary, 566
    height, 565
    homomorphism, 567
    isomorphism, 567
    level, 565
    m-ary, 566
    rooted directed tree, 564
directed tree, 239
Dirichlet characters, 758
Dirichlet drawer principle, 330
Disjoint, 85, 382
disjoint events, 347
Disjunctive normal form, 31, 32, 34
distance, 700
distance function, 701
distribution counting, 195
distributive, 105
Distributive Laws, 367
Divide and conquer, 192
Divisibility, 119
Division
    quotient, 120
    remainder, 120

Division Algorithm, 120, 273, 278, 396, 725
dodecahedron, 543
domain, 87
Domain of discourse, 98
dominance, 183
dominates, 182
Dot product, 148
Douglas Shier, 294
Dual code, 695

E, 434
$E(R)$, 340
E. F. Moore, 662
edges, 76, 233, 518, 620
edge set, 227
efficient code, 690
electrical circuits, 38
element, 147
empty set, 156
encipherment, 772
Encryption, 772, 773
endpoints, 227
enigma code, 689
entry, 147
Epimorphism
    group, 377
    ring, 721
epimorphism, 379
equal roots, 420
equal term by term, 281
Equivalence
    class, 84
    relation, 83
equivalence classes, 131, 363, 375, 521
equivalence classes modulo m, 394
equivalence classes modulo n, 131
Equivalence Relations, 83, 521, 710, 738
equivalent colorings, 710
Equivalent Statements, 13
Eratosthenes
    sieve of, 271
Erdös and Szekeres, 330, 331, 333
Error correcting code, 690
Error detecting codes, 690
error detection, 434
Euclid
    theorem on primes, 125
Euclid's algorithm, 277
Euclidean Algorithm, 274, 278, 391
Euclidean domain, 737
    norm, 737
Euclidean geometry, 19
Euler, 228
Euler cycle, 244, 246, 543

Euler cycle (proper path), 518
Euler diagram, 102
Euler path, 244, 245
Euler's ϕ function, 401
Euler's formula, 530
Euler's Theorem, 406
Euler's totient function, 401
Euler, Léonard, 404
Euler-phi, 411
even-numbered convergents, 285
event, 316
Existential
    generalization, 101
    instantiation, 101
expansion of $\frac{f(x)}{g(x)}$, 487
Exponential complexity, 185
Exponential generating function, 486, 507
Expression, 649
    regular, 649
extension field, 742
extrapolation, 434

Factorial complexity, 185
factorial function, 172
factorial polynomials, 437
Factorial Polynomial Algorithm, 439
factoring primes, 272
false match, 764
family, 237
Fermat number, 413
Fermat's, 404
Fermat's Factorization Method, 272
Fermat's Little Theorem, 406, 412
Ferrer's Diagram, 504
Fibonacci numbers, 312, 425, 501
Fibonacci sequence, 173, 277, 417, 420
Field, 722
    extension, 736
    fraction, 724
    of rational functions, 742
    subfield, 736, 750
Fingerprint, 762
finite automaton, 664
finite Boolean algebra, 368
finite code, 650
finite commutative semigroup, 755
Finite Differences, 434
    factor polynomials, 438
    product rule, 436
    quotient rule, 436
    summation, 447
    summation by parts, 449
finite graph, 75
finite group, 372
finite sequences, 284

finite set of cardinality n, 156
finite set of cardinality 0, 156
Finite sets, 51
finite state diagram, 665
finite state machines, 640
Finite State Machines with Output, 662
first difference, 434
first isomorphism theorem for groups, 379
First principle of induction, 729
fitting polynomials to data, 434
fixed colorings, 716
floor function, 145
Flow, 621
    conservation, 621
    maximal, 623
    maximum flow algorithm, 626
    value of, 623
flow, 625
Floyd-Warshall Algorithm, 550, 555, 556, 558
Forbidden positions, 470
Ford-Fulkerson algorithm, 626
four-color problem, 534, 539
Francis Gutherie, 539
fully parenthesized, 197
Function
    binary operation, 142, 360
    ceiling, 142
    Complexity, 181
    factorial, 142
    floor, 142
    hashing, 767
Functions, 87
Fundamental Theorem
    of Arithmetic, 126
Fundamental Theorem of Algebra, 751

$\gcd(a, b)$, 274
G. Brassard, 154
G. D. Birkhoff, 535
Gödel Incompleteness Theorem, 109, 161
gate, 39
Gauss, 401, 740, 751
Gaussian integers, 736
Generalized set operation
    indexing set, 60
    intersection, 60
    union, 60
generated, 650, 693
Generating combinations, 314
Generating Functions, 485–487, 496
    counting, 496
    exponential, 507

partitions, 501
    recurrence relation, 486, 487
Generating permutations, 314
Generator matrix, 693
generator of I, 738
generators, 752
geometric sequence, 144
George H. Mealy, 665
go-end, 682
go-left, 682
go-right, 682
Golay code, 703
Grammar, 672
    context-sensitive, 680
    corresponding tree, 679
    formal, 673
    language generated by, 673
    nonterminal symbols, 673
    phrase structure, 673
    regular, 681
    start symbol, 673
    terminal symbols, 673
Graph, 75, 227, 516
    adjacency matrix, 250
    adjacent edges, 227
    adjacent vertices, 227
    articulation point, 521
    bicomponent, 521
    biconnected, 521
    bipartite, 231
    chromatic number, 536
    chromatic polynomial, 536
    coloring, 536
    complement, 519
    complete, 231
    complete bipartite, 231
    component, 231
    connected, 230
    cut edge, 520
    cut set, 520
    cycle, 231
    degree of vertex, 229
    derived, 517
    diameter, 258
    distance, 258
    dual graph, 535
    edge, 75, 227
    Euler path, 244
    extension, 517
    faces, 530
    grid, 262
    homeomorphic, 517
    homomorphism, 516
    incidence matrix, 250
    intersection, 518
    isomorphism, 517
    multigraph, 228
    pairwise disjoint, 519
    path, 230
    planar graph, 530
    simple cycle, 231
    simple path, 230
    spanning graph, 519
    spanning tree, 519
    union, 518
    vertex, 75, 227
    Warshall's Algorithm, 253
    weighted graph, 550
Graph theory, 228
graph with loops, 228
Graphs, 227
Gray code, 261, 691
Greatest common divisor, 121, 274, 741
greatest element, 357
greatest integer function, 280, 769
greedy algorithm, 540
Group, 372
    Abelian, 372
    character, 752
    commutative, 372
    coset, 375
    cyclic group, 374
    direct sum, 752
    epimorphism, 377
    factor group, 379
    finite cyclic group, 374
    homomorphism, 377
    identity, 372
    inverse, 372
    isomorphism, 377
    kernel, 378
    left coset, 375
    monomorphism, 377
    normal subgroup, 378
    order, 372
    order of an element, 373
    subgroup, 373
    symmetric group, 381
    symmetries of the square, 709
group, 721
Group characters, 751, 753
Group code, 372, 692
group of permutations on four elements, 709
Groups, 371

Hadamard code, 704
Hadamard matrix, 704
Hall's Theorem, 634
halt state, 682
halting problem, 110
Hamiltonian Cycle, 185, 543
Hamiltonian Paths, 543

Hamming code, 700
Hamming distance, 700, 704
Hamming matrix, 700
hash function, 134
Hashing function, 767
    minimum perfect, 767
Hellman, M. E., 775, 777
Hexadecimal, 199, 201
$H \circ K$, 378
Homeomorphic, 517, 532
Homogeneous Linear Recurrence Relations, 416
Homogeneous recurrence relation, 416, 429
Homomorphism
    graph, 516
    group, 377
    ring, 721
    semigroup, 362
homomorphism, 366, 379
horizontal vectors, 383, 384
Horner's algorithm, 169
Horner's polynomial evaluation algorithm, 185
Horner's Rule, 410
Huffman code, 690
Huffman's algorithm, 576, 578
Huffman's code, 576
Huffman's tree, 579
Hypercube, 259, 542
Hypothesis, 19, 20

Ideal, 724, 755
    maximal, 738
    prime, 726
Idempotent, 363
Identities, 368
identity, 360, 373
identity function, 90
identity matrix, 219
identity permutation, 381
identity transformation, 384
image, 87
incidence matrix, 250, 253
incident, 227, 233
Inclusion-Exclusion, 296, 464, 465
indeg(a), 565
independent trials, 339
Indeterminate, 734
Induction
    First principle of, 729
    Induction Principle for Integers, 117
    Principle of, 111
    Second Principle of, 116, 729
    Well-ordering principle, 729
Infinite descent, 119

infinite integral domain, 735
infinite loop, 109
infinite sequence, 143
Infix code, 652
Infix form, 583, 584
Infix notation, 196
initial vertex, 621
injective, 89
inner automorphisms, 382
Inner product, 148, 693, 697
inorder transversal, 583
Instant insanity, 242
integer overflow, 477
Integers
    composite, 125
    congruent, 129
    divisor, 119
    factor, 119
    Gaussian, 743
    greatest common divisor, 121
    least, 116
    least common multiple, 122, 276
    multiple, 119
    order of, 406
    prime, 125, 728
    relatively prime, 122
    set of integers modulo n, 129
Integral domain, 721
    embedded, 724
    irreducible element, 736
    minimal domain, 730
    reducible element, 736
    well-ordered, 728
integral domain, 737
integral equations, 274
integral solution, 391
interpolation, 434
intersection, 529
Inventory of colorings, 715
inverse, 13
inverse fallacy, 25
inverse of a matrix, 213, 219
inverse relation, 89
Inverse statement, 18
Irreducible, 726
irreducible element, 739
isolated, 229
isomorphic, 517, 567
Isomorphism, 366
    group, 377
    ring, 722

J. C. Shieh, 768
Jaeschke, G., 767
joined, 75, 227

Königsberg bridge problem, 228, 244

Kahn, David, 777
Karnaugh map, 31, 33, 45, 260
Karp and Rabin algorithm, 763
Karp, R. M., 760
k cycles, 458
k-cube, 261
Key
    cryptographic, 772
    deciphering, 775
    enciphering, 775
Kirchoff's matrix tree formula, 602, 603, 607
k-list, 261
Kleene star, 649
Kleene's Theorem, 658
Kneiphof Island, 228
Knuth, D. E., 766, 767
k-path, 252
Kronecker delta, 733
Kruskal's algorithm, 608
Kuratowski's Theorem, 532
Kurt Gödel, 109

Léonard Euler, 404
labeled digraphs, 653
labeled directed graph, 234
labeled graph, 228, 234
labeled vertices, 597
Lagrange, 376
Language, 648
language accepted by the automaton, 654
latin square, 480
Lattice, 365
    bounded, 366
    distributive, 367
    join, 365
    meet, 365
    sublattice, 366
Lattices, 365
lcm(a, b), 276
leaf path code, 575
Least common multiple, 122, 276
left child, 239
length of a directed path, 234
lexicographical ordering, 82
lexicographical ordering $\preceq$, 313
LIFO, 197
Linear Algebra, 383
Linear code, 692
Linear Equations, 391
    integral solutions, 391
linear equations, 219
linear order, 569
linear property of matrice, 155
Linear recurrence relation, 416
    constant coefficients, 417

    homogeneous, 416
    nonhomogeneous, 425
linear recurrence relations, 423
linear recursive functions, 416
linear recursive relation, 417
linear transformation, 384
linearly ordered, 313
linearly ordered set, 313
live, 642
Logarithmic complexity, 185
Logical argument, 19
Logical equivalence
    associative properties, 14
    commutative properties, 14
    contrapositive, 15
    de Morgan's Laws, 14
    distributive properties, 15
    double negation, 14
    idempotent law, 14
Logically equivalent, 13
Logically false, 15
Logically true, 15
loop, 76, 228
loop-free, 564
lower bound, 357
lower semilattice, 363
Lucas sequence, 425
Lucas' Primality Test, 409

M. Plotkin, 704
mapping, 87
Markov Chains, 348
    states, 348
    transition matrix, 349
matched, 632
Matching, 632
    complete, 632
    maximal, 632
Matching edges, 632, 633
Matching network, 632
Mathematical Induction, 111
mathematical system, 19
Matrices, 146, 213
Matrix, 146, 169
    adjacency, 250
    Boolean, 152
    degree matrix, 602
    determinant, 214
    generator matrix, 693
    Hadamard Matrix, 704
    Hamming, 700
    identity, 215
    incidence, 250
    inverse, 216
    parity check, 696
    permutation, 153
    product, 148, 149, 170

representation, 151
   sum, 148, 170
   symmetric, 151
   transpose, 151
matrix, 146, 377, 529
matrix multiplication, 364
matrix of the coefficients, 219
Matyas, S. M., 776
maximal connected subgraph, 231
maximal element, 366
Maximal ideal, 738, 739
Maximal matching, 632
maximal simple path, 546
maxterms, 32
m-cycle, 458
Mealy machine, 662, 665
mean, 342
Merge Sort, 191
Mersenne number, 409
Meyer, C. H., 776
minimal distance, 701
minimal element, 169, 366
minimal generating set, 374
minimal set, 693
Minimal spanning tree, 608
   Kruskal's algorithm, 608
   Prim's algorithm, 609, 610
minimal weight, 609
minimum perfect hashing function, 771
minimum weight, 695
Miniterm, 31, 32
minor, 214
minterm, 33
$M(L)$, 652
modular exponentiation, 777
Modulo, 129
Monoid, 360, 365, 371
Monomorphism
   group, 377
   ring, 721
monomorphism, 732
Moore automaton, 662
Moore machine, 662
Moore, J. S., 766
Morris, J. H., 766
Morse code, 690
multigraph, 234, 244, 653
Multiplication counting principal, 291, 455, 535
multiplicative cancellation property, 105
multiplicative homomorphism, 743
multiplicative identity, 105
multiplicative inverse, 216
multiplicity, 711

mutual exclusion problem, 642
Mutually exclusive sets, 85

n pairs of parentheses, 174
n-cycle, 231
n-place predicate, 97
natural logarithm function, 182
ncr-board, 717
nearly full, 566
necessary, 16
negation of an existentially quantified predicate, 99
negative weights, 550
Network, 620
   cut, 623
   flow, 621
   matching, 632
network, 620, 630
$n \log(n)$ complexity, 185
noise, 690
nondeterministic algorithm, 185
Nondeterministic automata, 655
Nonhomogeneous recurrence relation, 425
nonisomorphic rooted binary trees, 567
nonmatching edges, 633
nonplanar, 530
nonsingular matrices, 378
nonzero reduced residue classes, 403
normal subgroup, 380
not connected, 230
not p, 39
noun phrase, 676
NP, 185
NP complete, 185
nth difference of f, 434
null set, 385
number of spanning trees, 597
Number Theory, 271, 372
numerical analysis, 434

Occupancy, 455
octic group, 381, 709
odd degree, 518
odd-numbered convergents, 285
$O(g(n))$, 182
one-place predicate, 97
one-to-one, 89, 366, 721
one-to-one correspondence, 89
one-to-one functions, 291
onto, 89, 366, 721
onto functions, 457
operator, 434
$\text{ord}_n a$, 412, 407
order R on a directed tree, 242
Order of an Integer, 405, 406
Ordered pair, 58

orthogonal, 695
orthogonality properties, 757
output alphabet, 682
overflow, 211, 311

P, 185
$P(A)$, 317
P. Bratley, 154
P. J. Heawood, 539
pairwise disjoint events, 335
pairwise relatively prime integers, 769
parity bit, 691
Parity check matrix, 696
Partial fractions, 489
Partial ordering, 80, 169, 357, 569
partial ordering of the integers, 108
partial organizational, 237
Partial quotient, 279, 281
Partially ordered set, 80, 357
   chain, 80
   comparable, 80, 357
   greatest lower bound, 357
   Hasse diagram, 81
   least element, 357
   least upper bound, 357
   maximal element, 357
   minimal element, 357
   topological sorting, 169
   total ordering, 80, 357
Partially Ordered Sets, 80
partition, 347, 375
partitioned into n disjoint sets, 132
partitioning, 456
Partitions, 85, 501
Pascal triangle, 308, 310, 456
Pascal's identity, 488
Path
   Euler, 244
   Hamiltonian, 543
path, 230
Pattern matching, 760
Pattern Search Algorithm, 763
Peirce's Arrow, 29, 40
perfect number, 123
permanent vertex, 553
permissible arrangements, 475
permutation, 89, 466
permutation matrix, 221, 469, 475
Permutations, 141, 302, 303, 313, 372, 380, 535
   cycle, 381
   forbidden positions, 470
   generalized, 322
   generating, 314
   orbit, 381
   with repetition, 326

Petersen graph, 533, 544
Petri net, 639
    bounded, 643
    conservative, 643
    deadlock, 642
    enabled, 640
    firing, 639
    immediately reachable, 641
    marking, 639
    place, 639
    safe, 642
    token, 639
    transition, 639
Pigeonhole principle, 330
    strong form, 331
Placing of n objects in k boxes, 457
Plaintext, 772
Planar graph, 530, 537
Polish form, 584
Polish notation, 196
Polya's counting theorem, 716
Polya's theorem, 712
Polynomial, 731
    coefficient, 731
    degree, 731
    division, 732
    equality, 732
    factors of, 732
    form, 731, 734
    function, 734
    irreducible, 736
    leading coefficient, 731
    monic, 731
    prime, 737
    primitive, 732
    product, 731
    reducible, 736
    sum, 731
polynomial, 730
Polynomial complexity, 185
Polynomial equation
    solution, 734
polynomial reducibility, 185
poset, 80, 357, 363, 365
postfix, 675
Postfix form, 585
postfix notation, 196
postorder transversal, 585
postulates, 19
power set, 160, 162, 365, 371
Pratt, V. R., 766
predecessor, 626
$pred(i)$, 553
Predicate, 97
    to satisfy, 97
Predicate Calculus, 97
Prefix code, 575, 690

Prefix form, 584
Prefix notation, 196
Pregel River, 228
preimage, 87
Premise, 19
preorder transversal, 584
preposition, 676
Prim's algorithm, 609, 610
primary reduced residue system, 407
primary residues, 764
prime, 402, 721
prime factorization, 403
prime factorization of integers, 125
Prime Factorization Theorem, 126
prime ideal, 739
prime integer, 124, 728
Primes
    integer, 125
    pseudo-prime, 406
    relatively prime, 122
Primitive n-th root, 751
primitive polynomial, 740
primitive root, 409, 411
principal character, 754
Principal ideal domain, 726, 738
principle of induction, 171
prints, 663
probabilistic algorithm, 763
probabilistic methodology, 763
probabilistic methods, 760
Probability, 317, 334, 764
    binomial distribution, 339
    conditional, 335
    disjoint, 317
    expected value, 340
    independent events, 338
    mathematical expectation, 340
    mean, 343
    sample space, 316
probability, 316
probability function, 334
probability of B given A, 335
productions, 673
Proof
    case proof, 109
    reductio as absurdum, 108
proof by contradiction, 24
proof by induction, 111
Proofs, 19
proper descendant, 595
proper encodings, 699
proper Euler path, 246, 529
proper ideal, 724
proper sentences, 676
Proposition, 1
proxy integers, 762
"proxy" number, 762

pseudograph, 228, 244
Public Key Cryptosystems, 773, 775
Purdum, Jack, 766

Q, 724
quadratic characteristic polynomial, 431
Quantifier
    existential, 98
    universal, 98
quotient group, 379
quotient semigroup, 363

Rabin, M. O., 760
Ramsey number, 332
Ramsey property, 332
Ramsey's Theorem, 332, 333
random primes, 760
random sampling, 342
Random variable, 339
range, 87
range of A, 634
Rational numbers, 721, 724
reachable, 641
reads, 663
$real(a + bi)$, 750
real numbers, 721
reciprocal hashing function, 772
recognizes, 653
Recurrence relation
    generating function, 486, 487
    linear, 416
recurrence relation, 447
Recursion, 171, 416, 311
Recursive, 172
    definition, 172
    relation, 172
    solving, 177
recursive function, 416
recursive relation, 416
recursively enumerable, 684
reduced residue group, 758
reduced residue set, 411
reduced residue system modulo n, 134
reduced residues modulo n, 401
reductio ad absurdum method, 159
Reed-Muller code, 704
Reed-Muller matrix, 704
reflections, 708
reflexive, 83
reflexive relation, 232
Regular expression, 649
Regular grammar, 681
Regular Languages, 648
regular set, 650
Relation, 72
    antireflexive, 74
    antisymmetric, 74

composition, 73
congruence, 739
domain, 72
equivalence, 83
inverse, 73
partial ordering, 80, 357
range, 72
reflexive, 74
reflexive closure, 75
symmetric, 74
symmetric closure, 75
transitive, 74
transitive closure, 75
relation R between A and B, 72
relation on A, 72
Relations, 72
relatively prime, 129, 397, 402, 721, 768
repetition of independent trials, 338
representation matrices, 152
Residue, 132
    complete residue system, 132
    primary residue system, 132
    reduced residue system, 132
Retract, 364, 650
Retraction map, 364, 650
Reverend Thomas Bayes, 346
Reverse Polish form, 584, 585
Reverse Polish notation, 196
reverse synthetic division, 442
right child, 239
Ring, 720
    commutative, 720
    group of units, 726
    homomorphism, 721
    ideal, 724
    irreducible, 726
    ordered, 728
    polynomial, 731
    principal ideal, 724
    ring with unity, 720
    subring, 722
    unit, 726
Rivest, R. L., 773
Rook polynomial, 470
rooks, 469
root, 742
root vertex, 571
rooted binary tree, 569
Rooted directed tree, 564
rooted tree, 569
Rosser, J. B., 764
rotations, 708
row matrices, 551
Roy-Warshall's algorithm, 253
RSA cryptosystem, 773
rules of inference, 19

Russell's paradox, 161
Russell, Bertrand, 51

Scalar, 146, 170
scalar multiplications, 154
Scalar product, 148
scalars, 383
Schoenfeld, L., 764
search algorithm, 760
second difference of f, 434
Second Principle of induction, 729
self-complementary, 529
Semigroup, 360
    closure, 360
    free semigroup, 362
    homomorphism, 362
    ideal, 755
    minimal generating set, 362
    monoid, 360
    prime ideal, 755
semigroup, 363, 721
semigroup homomorphism, 569
semigroup with identity, 360
Semigroups, 360
Semilattice, 363
    lower semilattice, 358
    upper semilattice, 358
Semilattice, 360, 365
    geometric, 175
sequence, 143
Sequence to tree algorithm, 599
Set, 51
    belongs to, 52
    Cartesian product, 58
    complement, 56
    element of, 52
    empty set, 52
    indexed, 60
    indexing, 60
    power set, 57
    universal, 52
Set builder notation, 51
Set operation
    difference, 56
    generalized intersection, 60
    generalized union, 60
    intersection, 54
    symmetric difference, 56
    union, 55
Set properties
    associative properties, 65
    commutative properties, 65
    Complement properties, 65
    de Morgan's Laws, 65
    distributive properties, 65
    Double complement, 65

Idempotent properties, 65
Identity properties, 65
Set relation
    disjoint, 85
    equality, 52
    mutually exclusive, 85
    proper subset, 52
    subset, 52
Shamir, A., 773
Sheffer Stroke, 29, 40
Shieh, J. C., 769
Shier collection, 495
Sieve of Eratosthenes, 271
signatures, 776
Signed Numbers, 210
simple connected directed graph, 620
simple cycle, 521
simple directed cycle, 236
simple directed path, 236
simple graphs, 228
simple statement, 2
simultaneous equations, 213, 218
sink, 620
slack, 626
Sociable groups of order n, 123
solving recurrence relations, 485
Sorting Algorithms, 187
    bubble sort, 188
    insertion sort, 192
    merge sort, 190
    quick sort, 188
    selection sort, 187
Sorting arrays, 193
source, 620
spanning forest, 594
Spanning tree, 589, 519, 589
    minimal, 608
spanning tree, 295, 495
square matrix, 147, 154, 252
Stabilizer, 710
Stack, 197
    pop, 197
    push, 197
stacks, 172
start state, 682
state, 640
Steven Cole Kleene, 649
Stirling numbers
    first kind, 444, 445, 458
    second kind, 444, 445, 456, 511
String, 361, 648
    concatenation, 648
string-matching problem, 766
strongly connected, 248
Subgraph, 229, 516, 518
subspace, 383
sufficient, 16

sum of ideals, 727
summands of atoms, 369
summation notation, 144, 168
summation of functions, 434
Sunday, D. M., 766
surjective, 89
switching circuit, 45
symmetric, 83
symmetric group, 380
symmetric matrix, 151
Symmetries of the square, 709
Syndromes, 697
synthetic division, 440
system of congruences, 399

Tautology, 15
Taylor expansion, 468
terminal vertex, 621
terms of the continued fraction, 281
the atoms of a Boolean algebra, 368
the induction hypothesis, 115
Theorem, 19
three house–three utilities problem, 229
three-way switch, 40
Tom Head, 650
topology, 404
total ordering, 169
Tower of Hanoi, 175, 181
transition function, 662
transitive, 83
transitive closure, 252
transpose, 217, 696
Traveling Salesman Problem, 185
Tree, 237, 564
    ancestor, 239
    binary search, 569
    breadth-first search, 589
    child, 239
    counting, 290
    depth-first search, 591

descendent, 239
directed, 237
directed tree, 564
forest, 237
height, 239
internal vertex, 237
leaves, 237
level, 239
m-ary, 239
parent, 239
parse, 679
root, 239
rooted directed tree, 239
rooted tree, 239, 240
sibling, 239
spanning tree, 240, 589
weight of, 576
tree, 237, 294, 564
Tree to sequence algorithm, 598
Triangle for Stirling numbers of the first kind, 458
Triangle for Stirling numbers of the second kind, 456
triangulated matrix, 604
triangulation, 463
Trichotomy axiom, 108
Truth table, 1, 2, 38
truth value, 1
Turing machine, 682
Two's complement, 210
two-place predicate, 97

unary, 67
unary connective, 4
undirected graph, 540, 637
(undirected) trees, 237
union, 529
Unique factorization domain, 727, 740
unique path, 564
Uniquely decipherable code, 650, 689
Uniqueness of Complement Law, 69
Unit, 726

Unit circle, 751
unit delay machine, 665
unity, 68
Universal generalization, 101
Universal instantiation, 101
Universe, 98
Update formula, 762
upper bound, 357
upper semilattice, 363, 569

$val(f)$, 630
Vandermonde's theorem, 309
variance, 343
vector space, 383, 385
Venn diagram, 61, 298
verb phrase, 676
vertex, 518
vertical vectors, 383
vertices, 620

Warshall's algorithm, 253, 555
weight function $c : E \to N$, 620
weight of a spanning tree, 608
weight of the leaf, 576
weight of the tree, 578
Weighted graph, 550, 608
weighted tree, 576, 575, 690
weights of strings, 700
well defined, 723
Well ordered, 728
well-ordered integral domain, 728
Well-ordering principle, 116, 120, 729
William Rowan Hamilton, 543
Wilson's Theorem, 403
Word, 361, 648
word processing, 760

Z_n, 129
zero, 68, 105
zero divisor, 727
zero matrix, 155

85.	AB	matrix product	page 149
86.	A^t	transpose of matrix A	page 151
87.	$A \odot B$	Boolean product of matrices	page 152
88.	$\overline{A_{ij}}$	matrix obtained by removing ith row and jth column from A	page 213
89.	$|A| = \det(A)$	determinant of the matrix A	page 214
90.	I_n	$n \times n$ identity matrix	page 215
91.	A^{-1}	multiplicative inverse of the matrix A	page 216
92.	$Adj(A)$	adjoint matrix of A	page 216

■ RECURSION AND COUNTING

93.	$n!$	n factorial	page 142
94.	$\sum_{i=1}^{n} a_i = a_1 + a_2 + \cdots + a_n$	summation notation	page 144
95.	$P(n, r) = \frac{n!}{(n-r)!}$	number of r permutations on n objects	page 303
96.	$C(n, r) = \binom{n}{r} = \frac{n!}{r!(n-r)!}$	number of r combinations on n objects	page 305
97.	$P(n : n_1, n_2, \ldots, n_k) = \frac{n!}{n_1!n_2!\ldots n_k!}$	generalized permutation	page 336
98.	$C(n : n_1, n_2, \ldots, n_k) = \frac{n!}{n_1!n_2!\ldots n_k!}$	generalized combination	page 324
99.	$P(A)$	probability of A	page 317
100.	$P(B \mid A)$	probability of B given A	page 335
101.	$E(R)$	expected value of random variable R	page 340
102.	μ	mean of random variable R	page 343
103.	σ^2	variance of random variable R	page 343
104.	$R(p, q)$	Ramsey number	page 332
105.	$\text{Fib}(n)$	nth Fibonacci number	page 173
106.	$\text{Cat}(n) = C_n$	nth Catalan number	page 173
107.	$G(C)$	the stabilizer of coloring C	page 710
108.	$\Delta^k f(x)$	kth difference of $f(x)$	page 434
109.	E	operator defined by $E(f(x)) = f(x+1)$	page 434
110.	$x^{(n)}$	$x(x-1)(x-2)\cdots(x-n+1)$	page 437
111.	$a_n x^{(n)} + a_{n-1} x^{(n-1)} + \cdots + a_2 x^{(2)} + a_1 x + a_0$ factorial polynomial		page 438
112.	$x^{(-m)}$	$\frac{1}{(x+m)^m}$	page 443
113.	$\binom{x}{n}$	$\frac{x^{(n)}}{n!}$	page 444
114.	$s_k^{(n)}$	Stirling number of the first kind	page 444
115.	$S_k^{(n)}$	Stirling number of the second kind	page 444
116.	$\sum$	summation operator	page 447
117.	L_n	nth lucas number	page 425
118.	D_n	the number of derangements of n distinct ordered symbols	page 466
119.	$r_k(c)$	the number of ways of placing k rooks on board C in nonattacking position	page 470
120.	$R(x, C)$	rook polynomial on board C	page 470
121.	$[t_0 : t_1, t_2, \ldots, t_n]$	continued fraction	page 281
122.	$T_k^{(n)}$	the number of ways of placing n distinguishable objects in k distinguishable boxes with no box empty	page 510

■ ALGEBRA

123.	$(A, \vee)$ or $(A, +)$	upper semilattice A	page 358
124.	$(A, \wedge)$ or $(A, \cdot)$	lower semilattice A	page 358
125.	$(S, \vee, \wedge)$	lattice S	page 365
126.	$a \circ H$	left coset	page 375
127.	$G_1 \oplus G_2$	direct sum of groups G_1 and G_2	page 752